calculus

an introduction to applied mathematics

second edition

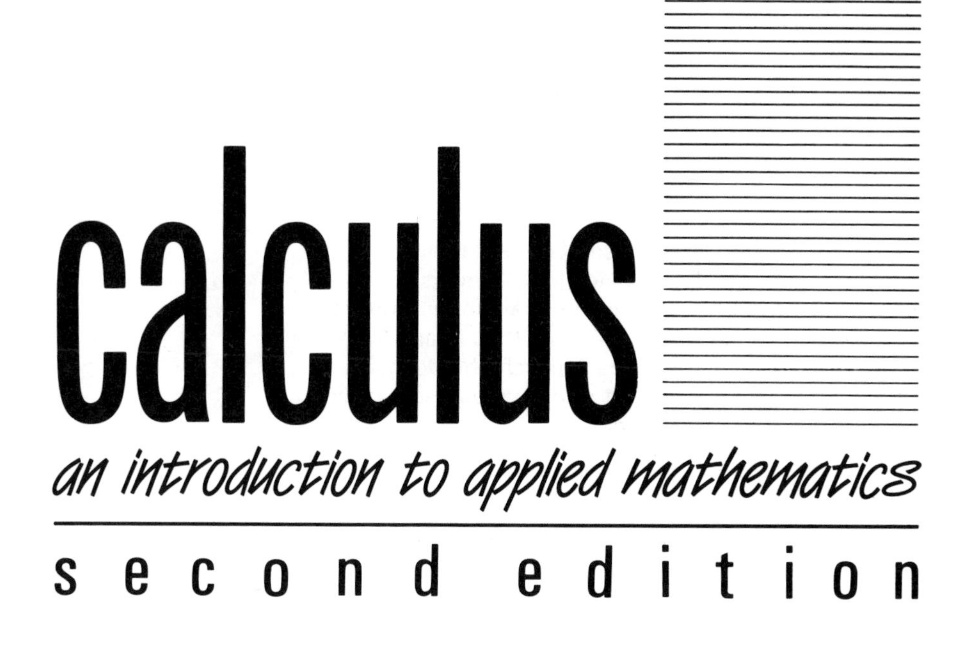

Harvey P. Greenspan
David J. Benney
Professors of Applied Mathematics
Massachusetts Institute of Technology

Revised by:
James E. Turner
Professor of Mathematics
McGill University

McGraw-Hill Ryerson Limited

Toronto Montreal New York Auckland Bogotá
Cairo Guatemala Hamburg Johannesburg Lisbon
London Madrid Mexico New Delhi Panama Paris
San Juan São Paulo Singapore Sydney Tokyo

CALCULUS: AN INTRODUCTION TO APPLIED MATHEMATICS, SECOND EDITION

ISBN 0-07-548926-0

1 2 3 4 5 6 7 8 9 0 AP 5 4 3 2 1 0 9 8 7 6

Printed and bound in Canada

Canadian Cataloguing in Publication Data

Greenspan, Harvey P. (Harvey Philip).
Calculus, an introduction to applied mathematics

Includes index.
ISBN 0-07-548926-0

1. Calculus. I. Benney, David J. II. Turner, James E.
III. Title.

QA303.G75 1986 515 C86-093380-6

Contents

Preface

In presenting calculus as an introduction to applied mathematics, we begin an educational curriculum whose basic objective is the twofold capability of formulating a problem mathematically and extracting, by any means, useful and desirable information. The selection of subject matter, the emphasis placed, and the priorities assigned are all predicated to this end, and, in a more personal sense, they reflect our experience and judgment as scientists. Accordingly, our values are quite different from those most commonly expressed in mathematics texts. But in a larger sense we really return to the traditional values that marked the vital development and vigorous application of calculus. The magnitude and difficulty of the task—dealing with nature and the social sciences—requires its own approach, an attitude stressing intuition, versatility, a willingness to explore and to test, and a dedication which will not be subverted by comparatively minor issues. It is a no-holds-barred contest where the ends often justify the means, where the sources of motivation and intuition are usually experiment and observation, and where the final vindication of procedure is comparison with reality. In order to convey the process of discovery realistically, this book is written in the discursive style of scientific literature. This form provides the vast majority of the calculus audience with a typical and most useful account of how mathematics is actually created and employed in practice.

The intent here is a sound presentation that is sometimes intuitive, often physically inspired, but always relevant and utilitarian. Sound mathematics is basically correct mathematics, developed to the point of optimum return in a manner most easily understood, assimilated, and applied. We could have been precise without fault, had we so desired, but we deliberately chose not to spend precious time on that which is obvious, minor, strictly pedantic, or without interest or purpose. Complete rigor is often unnecessary from the scientific viewpoint and may even be counter-productive. The conversion of this discursive exposition into the deductive and conservative mathematical format of axioms, theorems, and proofs is an appropriate enterprise at a more specialized and advanced level of study.

In this spirit, the concept of a function is treated simply, befitting a first course, and the historic and extraordinarily useful notation $y = f(x)$ is adopted rather than the "modern" nomenclature. The treatment of increments as infinitesimal elements, and the full use of illustrations in their explanation, is entirely consistent with our stated aims. The understanding and appreciation of how such approximations are and have been used is of paramount importance in learning calculus. A discussion of the subtleties involved is pedagogically sound after basic ideas are assimilated and practical methods are mastered.

Some candid and critical questions must be raised concerning the present college mathematics curricula. Is calculus presented as an impressive and meaningful intellectual achievement? Can it be said that the subject matter selected meets the real needs of the engineering and scientific community? Or has calculus become almost exclusively an introduction to analysis for that small minority destined for pure mathematics? Our answers culminated in this text.

Let us turn to more practical matters concerning the organization and use of the material in this text. Chapter 1 is by design a rapid survey of calculus pivoted about the central concept of a limit, which is illustrated in a variety of contexts. It also provides time to establish a common foundation for a class despite the different backgrounds involved. All objectives are met if a good working knowledge of limit operations is acquired, as well as a review of certain prerequisites and an exposure to the principal themes of calculus. An overview of the subject gives an immediate appreciation of its content, structure, and purpose; goals can be identified and kept in sight from then on. But it is not necessary to achieve a deep understanding of any concept introduced here, other than that of a limit, because each is considered anew in subsequent chapters. Even a modest familiarity with fundamentals and basic attitudes is an effective aid and stimulant for further study. Of course, this survey can be modified, sections omitted, or deferred and integrated at a later time with related material in the general development.

Reasonable facility is assumed with algebra and trigonometry, but very few basic facts about analytic geometry (given in Appendix 5) are actually used. It is not essential to have any prior knowledge of calculus.

Approximate techniques are emphasized, and the basis is set for computer use. No specific programming language is advocated or adopted, nor have we included many associated topics which are best left as a separate option.

There are three kinds of exercises: problems to develop facility with techniques already illustrated; elaborations on material in the text; applications. These are easily recognized, and, roughly speaking, they are listed consecutively. Several assigned problems per lecture, properly and systematically done, are an essential facet of the learning process. To learn calculus "in principle" only but to be unable to use it is to have learned nothing.

The material content is adequate for a two, three, and even four term sequence of courses from introductory to advanced calculus. Sections are sufficiently independent to accommodate many arrangements, and reference to the numbers in the margin allows subdivisions into even smaller natural units. For example, the theoretical ε, δ description of continuity can be deleted from Sec. 1.5 by simply omitting subsections 2 and 4, and this leaves a purely verbal and descriptive discussion. Most sections can be manipulated this way to satisfy particular requirements. Obviously, a very large number of schedules are possible.

In preparing the second edition, we have had the benefit of our experience in using the text at MIT and McGill University. There has been some reordering of the material in the early chapters with simpler motivating examples where needed. The greatest changes will be found in the exercise sets. We have attempted to put these exercises in the most natural order for learning. For the more difficult problems, a greater use of hints has been made. More detailed answers to the odd-numbered exercises have been provided.

Harvey P. Greenspan
David J. Benney
James E. Turner

1
Limits

1.1 Introduction

Science seeks to understand the world we live in, and mathematics is a primary means to this end. By an often beautiful and intrinsically mysterious process, the laws of nature can be expressed as mathematical equations and manipulated to yield new and even unsuspected knowledge. Mathematics becomes the "automatic machine" that frees the mind from the burden of overwhelming detail and allows thereby the perception of what is really fundamental. Calculus is in this respect, or by any other standard, a major intellectual triumph of civilization, which has been a mainstay of scientific and technological progress since its creation in the seventeenth century by Isaac Newton (1642-1727) and Gottfried Leibniz (1646-1716).

This study of the subject will focus on the areas of greatest application, which deal with the mathematical formulation of a problem and its solution by approximate and exact methods. The objective is the twofold ability to translate a problem into a correct mathematical statement and to effect a useful solution by approximate methods, mainly those which can by systematic improvement provide the exact answer in some limiting sense.

The concept of a *limit* is so fundamental that calculus may be justly defined as this extraordinary idea and its consequences. On the other hand, the basic features of the limit process are easily revealed and do not require any "deep" mathematical techniques (which can and do come later). In fact, there is much to be said for a first exposure on such a qualitative and intuitive level because the related mathematics will then appear as natural developments.

Consider then, as a first example, the practical problem of calculating the area A of a geometrical figure such as that shown in Fig. 1.1. (This could be the floor plan of some exotic room which is to be resurfaced.)

The knowledge that the area of a rectangle is its length multiplied by its width is sufficient information to obtain an approximate value of A, which most importantly can be improved to any desired accuracy. A first estimate is made by placing squares of unit area *within* A (think of them as tiles, but none are cut to fit because finding the area of a partial tile is equivalent to the problem under discussion). According to Fig. 1.1, the area A will accommodate 26 unit squares (each, say, is 1 ft.^{2}), but considerable space is left uncovered. Clearly, A is larger than the area of the inscribed figure: $A > 26$ ft.^{2}; however inaccurate this estimate is, it at least provides a *lower bound*. Moreover, this first approximation can be improved greatly by placing smaller tiles in the remaining area. If the area of each tile used in this second state is $\frac{1}{4}$ ft.^{2}, then 26 of them can be fitted in the residual space within A as shown. This adds a correction of 6.5 ft.^{2} to the first estimate, so that now

$$A > 32.5 \text{ ft.}^2 = 26 \text{ ft.}^2 + 6.5 \text{ ft.}^2.$$

A more refined bound can be obtained by repeating the procedure again with still smaller tiles, say 16 to the square foot. A careful count shows that 64 of these can be added to the previous inscribed figure to boost the estimate of the area by another 4 ft.^{2};

$$A > 36.5 \text{ ft.}^2.$$

Obviously, it is possible to continue this process by using ever smaller tiles, and in just a few more steps the actual task would certainly require a microscope and other delicate instruments. The area of each inscribed figure is less than A but larger than any previous approximation, so that the estimates become increasingly accurate.

The method can also be viewed as a sequence of coverings with uniform-sized tiles; each step would then require a *larger number* of *smaller* and *smaller* squares. In this light, 26 large unit tiles are placed first; 130 of the $\frac{1}{4}$ ft.^{2} tiles are accommodated at the second stage; and 584 tiles, each with area of $\frac{1}{16}$ ft.^{2} are used in the third covering. Still larger numbers of individually smaller tiles would be placed in succeeding steps.

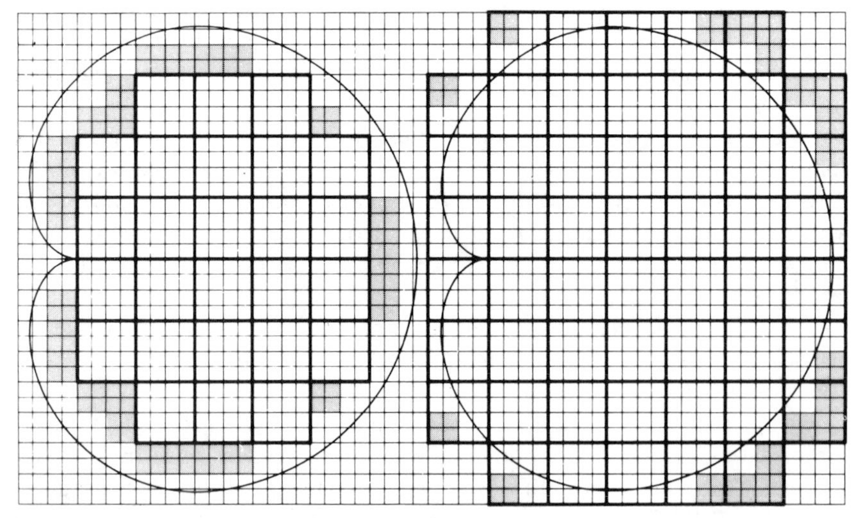

FIGURE 1.1 FIGURE 1.2

To obtain a different estimate, and a crude assessment of accuracy, the area A is also approximated from the outside by enclosing or circumscribing it within a pattern of squares shown in Fig. 1.2. This provides an *upper bound* for A, and it is found that with the largest tiles $A < 52$ ft.^{2}. This estimate is quickly improved in the second stage to $A < 47$ ft.^{2}; with the smallest tiles ($^1/_{16}$ ft.^{2}) 3 ft.^{2} can be eliminated, to make the bound $A < 44$ ft.^{2}. The numerical results of the first three stages of approximation are summarized as follows:

Stage 1 (large squares):	26 ft.$^2 < A < 52$ ft.^{2}
Stage 2 (medium squares):	32.5 ft.$^2 < A < 47$ ft.^{2}
Stage 3 (small squares):	36.5 ft.$^2 < A < 44$ ft.^{2}

The difference between the upper and lower bounds is systematically reduced, which means that the method delivers improved estimates of the area. (The differences are, respectively, 26, 14.5, and 7.5 ft.^{2}.) By continuing the process, the estimates of A, as drawn, can be made as accurate as physical skill will permit. This is a natural restriction, but it is *conceptually* possible to continue beyond this point, in which case the procedure can be carried to its ultimate limit. As the number of tiles used becomes ever larger and their individual size diminishes to zero, the upper and lower bounds focus upon a single number, called the *limit* (here 39.58 ... ft.^{2}), which is the exact area of A. In other words, the *limit* is attained only in a conceptual or abstract manner—by conceiving the region to be covered with an *infinite* number of *infinitesimally* small squares. Since the *exact* geometrical figure is itself an abstraction, it is not surprising that the precise area can be determined only by thought alone and not by real measurements.

Of course, many other approximation schemes are possible. In Fig. 1.3 or 1.4

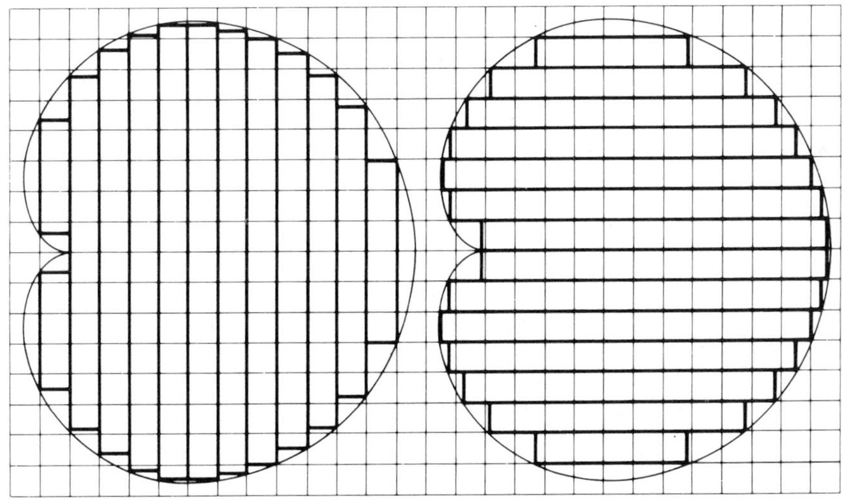

FIGURE 1.3　　FIGURE 1.4

rectangular slats (a wooden floor this time) are placed within A to form an inscribed area. By calculating the surface area of each slat and adding all together an estimate of A is obtained. As the width of the slats is decreased, the accuracy improves. The ultimate approximation will again involve the *limit* of an *infinite* number of *infinitesimally* thin slats which together give area A exactly. A mathematical expression for the inscribed area at any stage of approximation can be determined rather easily (see Exercise 2).

It will take considerable time and effort to formalize this procedure in mathematical terms, and the task will not be completed until Chap. 7. However, the fundamental idea of slicing an area into small calculable segments is quite simple and has been used since its discovery by Archimedes (287-212 B.C.).

In the remainder of this chapter we will examine the notion of a limit as it occurs throughout the calculus. This also provides an overview for the reader, a brief explanatory and perhaps motivational study of the entire subject.

Material is reviewed, concepts introduced, directions suggested, and a philosophy, or at least an attitude, expounded. Since basic ideas are reconsidered in subsequent chapters, the knowledgeable reader can proceed to Chap. 2 at any time.

Exercises 1.1

1. Average the upper and lower bounds obtained for the area A in Figs. 1.1 and 1.2 and compare with the exact value given. In this case why are the averages more accurate than the bounds?

2. Suppose that a total of n numbered slats are used in the covering of A shown in Figs. 1.3 and 1.4. Let the length of slat numbered j be l_j and its width w_j. Show that the area of all n slats, denoted by A_n, is given as

$$A_n = \sum_{j=1}^{n} l_j w_j = l_1 w_1 + l_2 w_2 + \cdots + l_n w_n .$$

Simplify this expression for the particular case in which all the slats are the same width w. Calculate the areas shown and compare them with previous results.

3. Construct an upper bound for the area A using an arrangement of slats similar to that in Fig. 1.3. How could this estimate be improved? Can you detail more systematic instructions for this purpose?

4. The area of A can also be calculated as the sum of approximating triangles which form an inscribed polygon (Fig. 1.5). Select an interior point O and divide the interior angle of 360° into equal parts; the figure shows 12 angles each of 30°. Form triangles whose vertices are O and the points of intersection of the rays with the curve, as shown. Use the fact that the area of a triangle is half its base multiplied by its height and, with aid of a ruler, calculate the area of the approximating polygon. Compare this lower bound with previous results. Does this procedure always yield a lower bound?

5. Can you give instructions for improving the approximation shown in Fig. 1.5? How would you construct an upper bound this way?

6. The perimeter of a circle can be calculated approximately from the perimeter of an inscribed polygon (Fig. 1.6). As the number of sides of the inscribed polygon increases, its perimeter

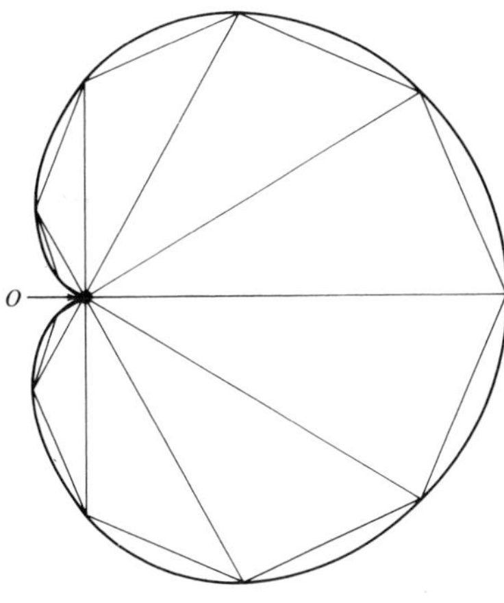

FIGURE 1.5

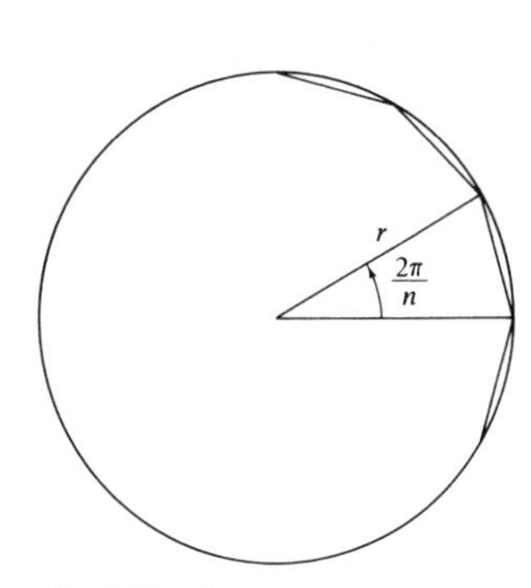

FIGURE 1.6

becomes a closer approximation to that of the circle. With a ruler, compute the first two approximations and compare them with the exact answer $2\pi r$ (r is the radius).

7. The approximate values calculated in Exercise 6 are lower bounds for the perimeter of a circle. Devise a procedure which yields upper bounds only.

8. The areas of the inscribed polygons (Fig. 1.6) are also approximations for the area of the circle. As in Exercise 4, calculate the two approximations indicated. Compare these with the exact answer, πr^2.

9. Figure 1.7 motivates a false "proof" that the length of the hypotenuse of a right triangle *equals* the sum of the lengths of each side. The "argument" is as follows. Bisect the two

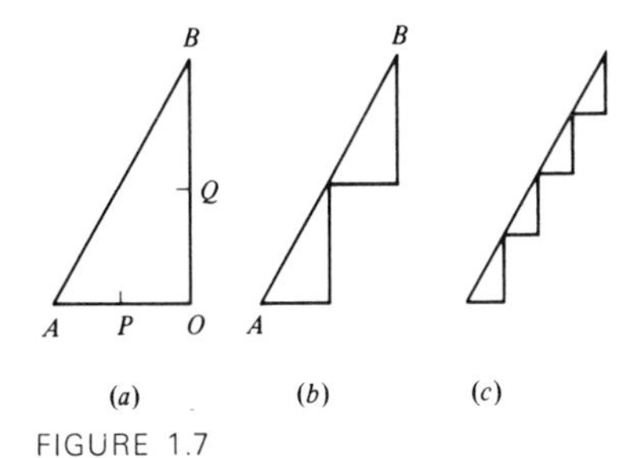

FIGURE 1.7

legs of the triangle and rearrange them as shown in Fig. 1.7*b*. The total length $AP + PO + OQ + QB$ is unchanged and equal to $AO + OB$. Repeat the process (Fig. 1.7*c*) and continue in this fashion. The bisected curves become progressively closer to the hypotenuse, and eventually they will be indistinguishable to the naked eye. Since the curves coalesce, their lengths must be the same, which proves the theorem. What is wrong? In Fig. 1.6, both areas and perimeters tend to correct values for the circle. Can you guess why?

1.2 Variables and Functions

1

A *variable* in science is a quantity that can be measured according to some well-defined procedure. A *function* is a definite procedure or rule to determine one number or variable (the *dependent variable*) when another or several others (the *independent variables*) are given. Functions are used to describe in quantitative mathematical terms the natural laws which relate different bodies of fact, and such "intuitive" concepts as that of a limit.

The simplest functions to study are functions of one independent variable. The area A of a circle is determined if the radius r is known. Symbolically, the rule for determining the area from the radius can be written

$$A = f(r) \tag{1}$$

where f indicates the function. In this case, the procedure or rule to calculate the area is well-known, simply square the radius and multiply by π. The function f is given by the rule $f(r) = \pi r^2$.

To study functions of one variable, the following common notation is used,

$$y = f(x). \tag{2}$$

The function f denotes the rule by which the number y is calculated from the number x. The particular value of y corresponding to $x = x_0$ is written

$$y_0 = f(x_0). \tag{3}$$

The values that x can take on form the *domain* of f; the corresponding values that y takes on form the *range* of f. Since the variable x can be assigned freely, it is called the *independent variable*; y is the *dependent variable* whose value follows by calculation. On occasion, we will write $y = y(x)$, the meaning of which is equivalent to $y = f(x)$.

2

Although an explicit formula is usually the best description of a function, it is also advantageous to have a picture of $f(x)$ in order to transfer the maximum information

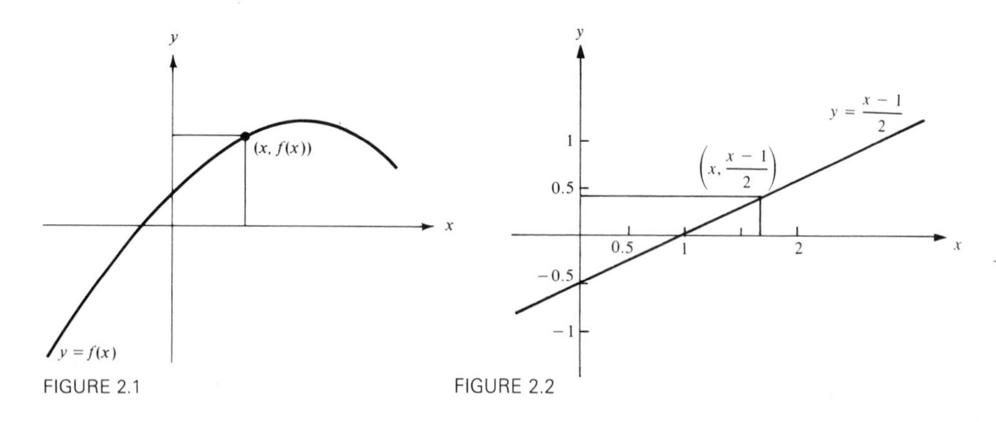

FIGURE 2.1

FIGURE 2.2

in the shortest time. A convenient and informative way of displaying a function is by means of a *graph* in the *cartesian plane*. The independent variable x is measured along the horizontal axis while the dependent variable y is measured along the vertical scale. The correspondence of y to x is then represented by the point $(x, y) = (x, f(x))$, as shown in Fig. 2.1. All such points describe the function completely, even when the analytical expression is not known, as is the case in many experimental situations. As long as one variable can be determined from another, the two are functionally related since all the criteria for this relationship are met.

Example 1 Graph the function $y = (x - 1)/2$.

This very simple function associates to any number x the number $(x - 1)/2$. When $x = 0$, $y = -\frac{1}{2}$; when $x = 1$, $y = 0$; and so on. The graph of the function (Fig. 2.2) is a straight line that intersects the y axis at $(0, -\frac{1}{2})$ (the *y intercept*) and the x axis at $(1,0)$ (the *x intercept*). The domain is the set of all real numbers, $-\infty < x < \infty$, and the range is also the set of all reals, $-\infty < y < \infty$.

Example 2 Graph the function

$$y = 1 + x + x^2 \qquad \text{for} \qquad 0 \le x \le 1.$$

Eleven corresponding values of y and x are tabulated. The first point

TABLE 2.1

x	0	0.1	0.2	0.3	0.4	0.5	0.6	0.7	0.8	0.9	1
y	1	1.11	1.24	1.39	1.56	1.75	1.96	2.19	2.44	2.71	3

$x = 0$, $y = 1$, that is, $(0, 1)$, is located on the y axis 1 unit above the origin. The second, $(0.1, 1.11)$ is located 0.1 unit to the right on the x axis and 1.11 units vertically above that. The rest of the points are located similarly, as shown in Fig. 2.3. Additional points between these can be calculated if necessary, but a smooth curve drawn through the known positions provides an excellent approximation of the function for all intermediate values. The approximate value of y at $x = 0.85$ can be read from the curve as $y = 2.57$. The exact value calculated from the formula is 2.5725. (A method for finding in-between values of a function from given data is called *interpolation*.)

Example 3 Graph the function $y = \sqrt{x - 1}$.

The function $\sqrt{x - 1}$ is the positive square root of $x - 1$ and is defined for $x \ge 1$ (the domain of the function). The range is the set of all non-negative numbers, $y \ge 0$. Using a calculator, many points on the graph can be determined quickly. The graph (Fig. 2.4) is confined to the first quadrant in the cartesian plane.

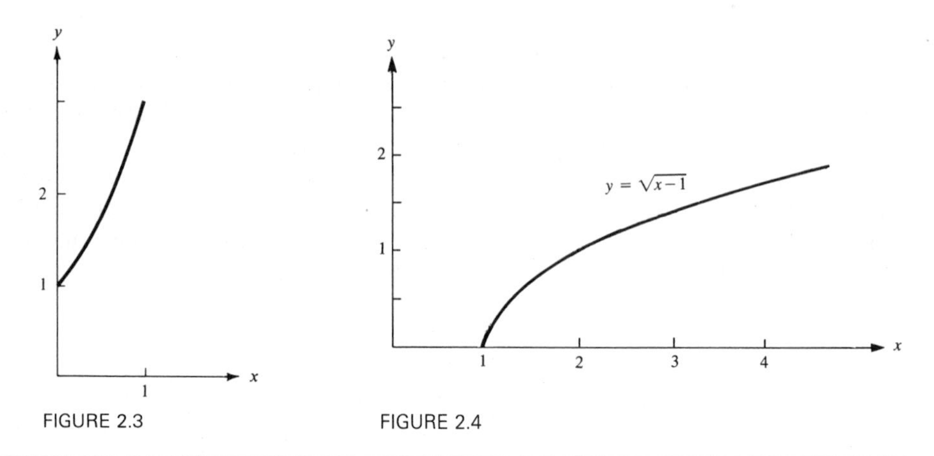

FIGURE 2.3

FIGURE 2.4

3

Since the number of conceivable functions is infinite, some limitations must be imposed in order to make the concept viable. A natural restriction is to discuss only those which are important by virtue of their frequent occurrence and widespread application.

A *simple function* is a familiar one about which much is known (which is the reason it is also termed a *known function*). The information required to make any specific function available for general use as part of a mathematics education is a complete tabulation of values (or a method for this) and the description of its special properties. As the level of sophistication and mathematical maturity increases, more and more functions will enter this privileged category, although the total number so designated must remain small to be truly useful. The aim of much analytical work then is to express and comprehend any "unknown" function that is encountered in terms of the information or language at hand, i.e., the known functions. "Simple" is, of course, a relative description that depends on an individual's knowledge.

The *polynomials of degree n* are important examples of simple functions. They can be written

$$y = P_n(x) = a_0 + a_1x + a_2x^2 + \ldots + a_nx^n \tag{4}$$

This function is formed from the independent variable x and the given numbers $a_0, a_1, \ldots, a_n$ by the processes of addition, subtraction, and multiplication. The $n+1$ numbers $a_0, a_1, \ldots, a_n$ are called the *coefficients* of the polynomial; a_k is the coefficient of x^k $(k = 0, 1, 2, \ldots, n)$. The polynomial is completely determined when these coefficients are known.

The *roots* of the polynomial are the values of x that satisfy $P_n(x) = 0$. Quoting a result from algebra, a polynomial of degree n can have at most n real roots. For example, the polynomial $P_2(x) = 6 - 5x + x^2$ has exactly two roots, $x = 2$ and $x = 3$. This is easily seen by writing $P_2(x) = 6 - 5x + x^2 = (x-2)\ (x-3)$. The polynomial $P_4(x) = 1 + x^2 + x^4$ clearly has no real roots since $1 + x^2 + x^4 \geq 1$ for all real x. If a polynomial of degree n does have exactly n real roots $x_1, x_2, \ldots, x_n$, then it can be written as a product of n factors

$$P_n(x) = a_n\,(x - x_1)\,(x - x_2)\,\ldots\,(x - x_n), \tag{5}$$

where x_i is the ith of n roots, i.e., a solution of

$$P_n(x_i) = a_0 + a_1 x_i + \ldots + a_n x_i^{\,n} = 0. \tag{6}$$

The graphs of the simplest polynomials are familiar geometrical curves. For example, the polynomial of first degree,

$$y = a_0 + a_1 x, \tag{7}$$

represents a *straight line* in the x, y plane which passes through the point $(0, a_0)$ and $(-a_0/a_1, 0)$ (Fig. 2.5). The coefficient a_1 determines the *slope* of the line and is so named for this reason.

The graph of the general polynomial of degree 2,

$$y = a_0 + a_1 x + a_2 x^2, \tag{8}$$

is a *parabola*, a fact made more transparent by rearranging the expression somewhat. Equation (8) is equivalent to

$$y = a_2\left(x + \frac{a_1}{2a_2}\right)^2 + \left(a_0 - \frac{a_1^{\,2}}{4a_2}\right). \tag{9}$$

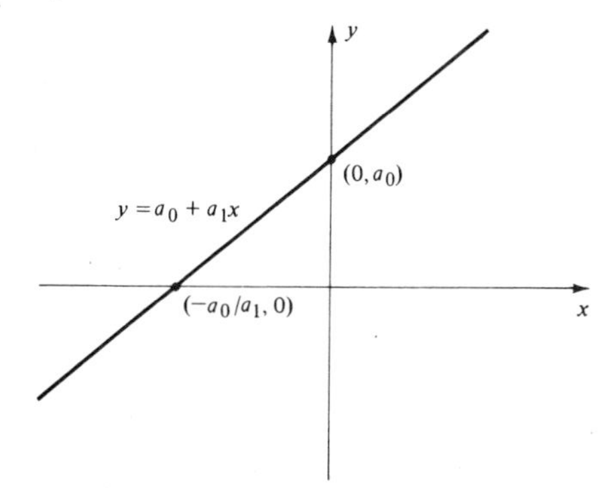

FIGURE 2.5

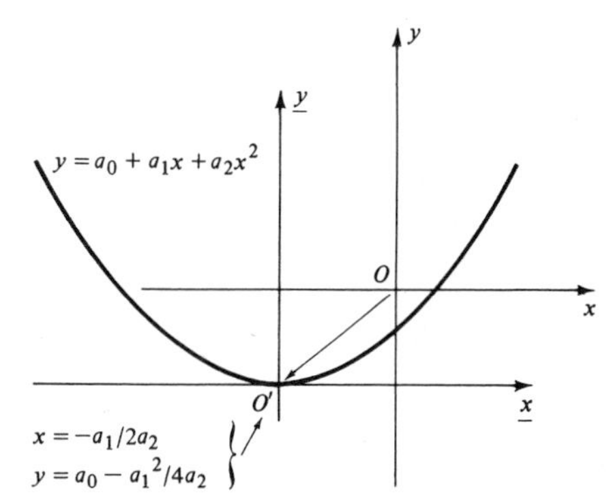

FIGURE 2.6

If now we write

$$\underline{x} = x + \frac{a_1}{2a_2} \quad \text{and} \quad \underline{y} = y + \frac{a_1^{\,2}}{4a_2} - a_0, \tag{10}$$

then, upon substitution for x and y, (9) becomes

$$\underline{y} = a_2\,\underline{x}^2,$$

which is the standard form for the parabola. A formula that seems complicated with certain variables can simplify drastically with others, a lesson that should be carefully noted. The relationship of variables $\underline{x}$ and $\underline{y}$ to x and y expressed by (10) can be visualized as describing a horizontal and vertical translation of the coordinate axes. Thus if the origin O in Fig. 2.6 is moved $a_1/2a_2$ units horizontally to the left and $a_1^2/4a_2 - a_0$ units vertically down, it will coincide with O'. In general, every *change of variables*

$$x = \underline{x} + a, \qquad y = \underline{y} + b$$

represents a *translation* of the x, y coordinates to new axes whose origin lies at the point (a, b). The proper choice of a coordinate origin, which is an open prerogative, can make life easier.

Example 4 Sketch the graph of the function $y = 1 + x + x^2$.

In terms of the variables

$$\underline{x} = x + {}^1\!/_2,\ \underline{y} = y - {}^3\!/_4,$$

the expression which can be written as $y = (x + {}^1\!/_2)^2 + {}^3\!/_4$ reduces to

$$\underline{y} = \underline{x}^2.$$

The point $(-{}^1\!/_2, {}^3\!/_4)$ is the center of the new $\underline{x}$, $\underline{y}$ coordinate system in the x, y plane (Fig. 2.7). In the reference frame the parabola is *symmetric* about the vertical axis, i.e., if $(\underline{x},\underline{y})$ is on the parabola, then so is $(-\underline{x},\underline{y})$. This symmetry property makes the graph relatively easy to draw.

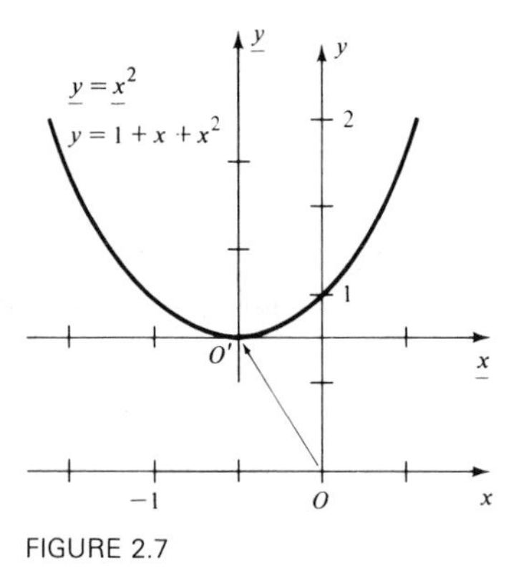

FIGURE 2.7

---

4 The *trigonometric functions* sine and cosine and their ratios (tangent, cotangent, secant and cosecant) play a fundamental role in mathematics. These relatively simple functions defined originally to study problems in geometry arise naturally throughout science.

To define the sine and cosine functions, consider any point P on the circle with radius 1 and center the origin in the cartesian plane (Fig. 2.8). The location of the point is completely determined if the angle θ measured counterclockwise from the x

axis is given. In calculus, angles are measured in *radians*; one degree is $\pi/180$ radians. The angle θ radians corresponds to the arc of length θ on the circle from the point $(1,0)$ to the point P. Each point P corresponds to a unique value of θ in the interval $0 \le \theta < 2\pi$.

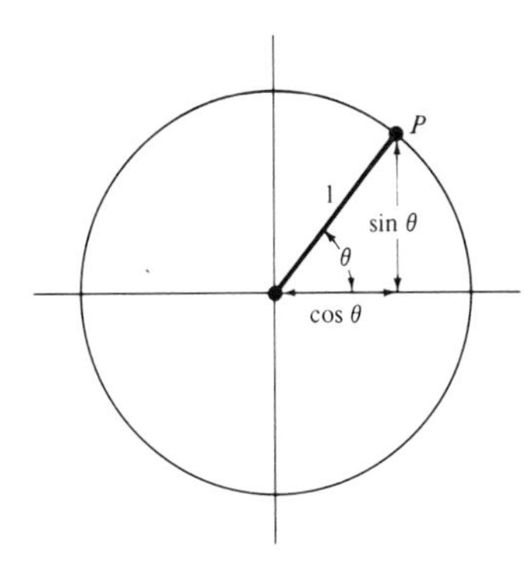

FIGURE 2.8

By definition, $\sin\theta$ is the y coordinate of P and $\cos\theta$ is the x coordinate, $P=(x,y)=(\cos\theta, \sin\theta)$. In principle, we could find the values of $\sin\theta$ and $\cos\theta$ by drawing an accurate diagram and measuring. As a check on the accuracy of these estimates, they should satisfy the equation

$$(\sin\theta)^2 + (\cos\theta)^2 = \sin^2\theta + \cos^2\theta = 1.$$

This identity follows from the theorem of Pythagoras applied to the right-angled triangle with sides $\sin\theta$, $\cos\theta$ and hypotenuse 1.

If the point P moves counterclockwise along the circle from the initial point $(1,0)$, the angle θ increases from the value 0. When $\theta = 2\pi$, the point P has returned to the initial position and, as θ increases beyond 2π, the point P in Fig. 2.8 begins retracing its path and the functional values are repeated. In mathematical terms, this means that the addition of 2π radians to the angle θ leaves the values of $\sin\theta$ and $\cos\theta$ unchanged:

$$\sin\theta = \sin(\theta + 2\pi), \qquad \cos\theta = \cos(\theta + 2\pi). \tag{11}$$

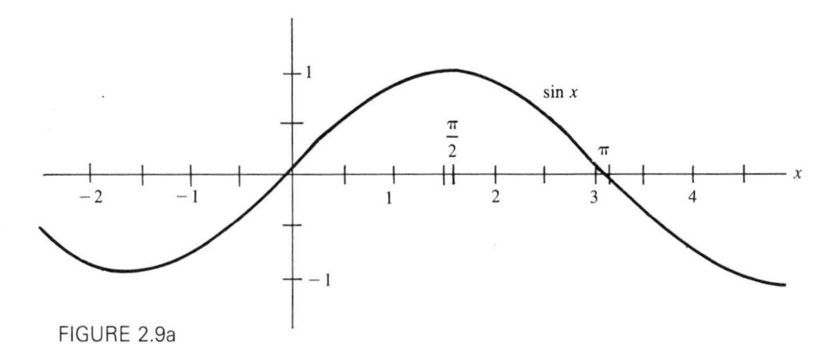

FIGURE 2.9a

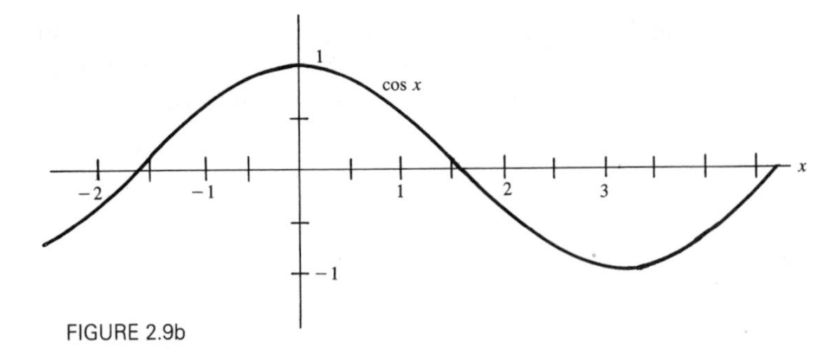

FIGURE 2.9b

A function with this property is said to be *periodic* with period 2π. Equation (11) is used to define and evaluate the functions for any value of θ,

$$-\infty < \theta < \infty,$$

and the graphs are shown in Fig. 2.9. Although θ has a particular geometrical connotation, any other notation can be used now that the functions are completely specified. In other words, the functions $\sin x$, $\cos x$ "exist" quite independently of the manner (or diagrams) by which they were first introduced.

Many important properties of the sine and cosine functions are summarized in Table 2.2, which should serve as a short review in conjunction with the exercises.

TABLE 2.2 *Some trigonometric formulas*

1. $\sin^2 x + \cos^2 x = 1$
2. $\sin(x_1 + x_2) = \sin x_1 \cos x_2 + \cos x_1 \sin x_2$
3. $\cos(x_1 + x_2) = \cos x_1 \cos x_2 - \sin x_1 \sin x_2$
4. $\sin(-x) = -\sin x$
5. $\cos(-x) = \cos x$
6. $\sin x = 0$ at $x = n\pi$, n an integer
7. $\cos x = 0$ at $x = (n + 1/2)\pi$
8. $\sin 2x = 2 \sin x \cos x$
9. $\sin 3x = 3 \sin x - 4 \sin^3 x$
10. $\sin 4x = 8 \cos^3 x \sin x - 4 \cos x \sin x$
11. $\cos 2x = 2 \cos^2 x - 1 = \cos^2 x - \sin^2 x$
12. $\cos 3x = 4 \cos^3 x - 3 \cos x$
13. $\cos 4x = 8 \cos^4 x - 8 \cos^2 x + 1$
14. $\sin x_1 \pm \sin x_2 = 2 \sin 1/2(x_1 \pm x_2) \cos 1/2(x_1 \pm x_2)$
15. $\cos x_1 + \cos x_2 = 2 \cos 1/2(x_1 + x_2) \cos 1/2(x_1 - x_2)$
16. $\cos x_1 - \cos x_2 = -2 \sin 1/2(x_1 + x_2) \sin 1/2(x_1 - x_2)$

Example 5 Find the values of the independent variable for which the sine function is zero.

The "zeros" or roots of an arbitrary function $f(x)$ are the particular values of x for which

$$f(x) = 0.$$

It follows from the basic definition that the equation

$$\sin x = 0$$

is satisfied at $x = 0,\ \pi,\ 2\pi$. (The sine is the height of point P above the horizontal axis in Fig. 2.8.) Equation (11) indicates that these zeros are repeated in every 2π interval, so that the formula

$$x = n\pi, \qquad n = 0,\ \pm 1,\ \pm 2,\ \ldots,$$

locates *all* the roots.

Example 6 Prove that $\sin x$ is not equal to any polynomial.

Suppose that $\sin x$ is a polynomial of degree n,

$$\sin\ x = P_n(x) = a_0 + a_1 x + a_2 x^2 + \ldots + a_n x^n$$

Then $\sin x$ has at most n real roots. But this is not correct since $\sin x = 0$ when $x = 0,\ \pm\pi,\ \pm 2\pi,\ \ldots$. Since there are an infinite number of roots, we conclude that $\sin x$ cannot be equal to any polynomial.

5

In the cartesian plane, a point P corresponds uniquely to a pair of numbers (x, y), its cartesian coordinates. Once the coordinate axes have been chosen, the two numbers x and y determine the location of the point. Another pair of numbers, the *polar coordinates* r and θ, may also be used to locate the point P. In Fig. 2.10, r is the

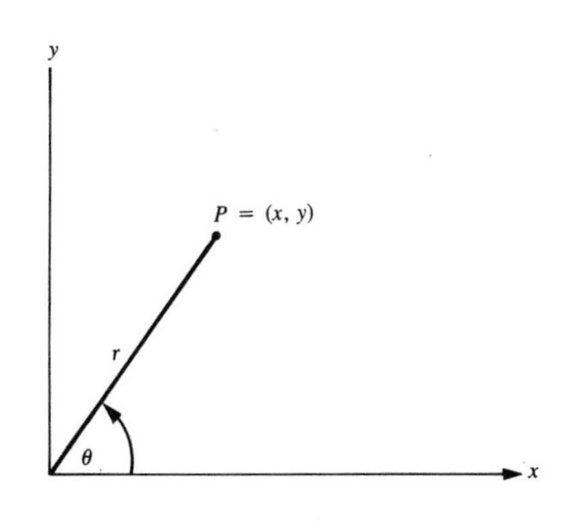

FIGURE 2.10

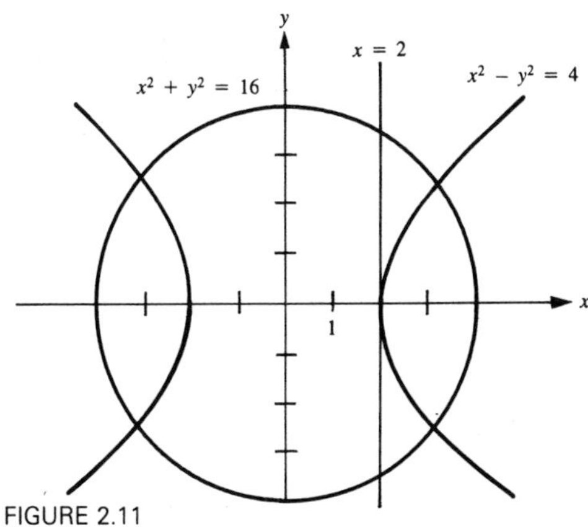

FIGURE 2.11

distance OP and θ is the angle between the x axis and the line segment OP measured counterclockwise. The two coordinate systems are related by

$$x = r\cos\theta,\ y = r\sin\theta,\ x^2 + y^2 = r^2.$$

For example, the point (1,1) has polar coordinates $r = \sqrt{2}$ and $\theta = \pi/4$; the point $(0, -2)$ has polar coordinates $r = 2$, $\theta = 3\pi/2$. In some problems, it will be more convenient to work in polar coordinates than in cartesian coordinates.

Example 7 Write the following equations in polar coordinates.

(i) $x = 2$ **(ii)** $x^2 + y^2 = 16$ **(iii)** $x^2 - y^2 = 4$

These three equations represent a straight line, a circle and a hyperbola in the cartesian plane (Fig. 2.11). Substituting $x = r\cos\theta$, $y = r\sin\theta$, equation (i) becomes $r\cos\theta = 2$. Equation (ii) is $r^2\cos^2\theta + r^2\sin^2\theta = 16$ or $r = 4$ in polar coordinates. Equation (iii) is $r^2(\cos^2\theta - \sin^2\theta) = 4$ or $r^2\cos 2\theta = 4$.

Exercises 1.2

1. Express the following as mathematical equations.
(i) the temperature in degrees Celsius as a function of the Fahrenheit temperature
(ii) the hypotenuse of a 90° triangle as a function of the other sides
(iii) the volume of a rectangular box as a function of length, width and height
(iv) the surface area of the rectangular box.

2. Graph the following functions in the cartesian plane
(i) $y = 1 - x$ **(ii)** $y = 1 - x^2$ **(iii)** $y = \sqrt{1 - x}$

3. Determine the domains and ranges of the functions in the previous problem.

4. For the linear function $y = mx + b$, show that, when the independent variable x changes by a given amount h, the dependent variable y changes by an amount mh. Illustrate with several examples.

5. If $f(x) = mx + b$ and it is known that $f(1) = 2$ and $f(2) = 3$, find the constants m and b.

6. A box is x meters long, $x + 1$ wide and y high. If the volume is 2 m^3, find y as a function of x. By solving a quadratic equation, find x as a function of y. Draw a graph of $y = y(x)$.

7. A rectangle of dimensions $2x$ by $2y$ is inscribed in the circle of unit radius $x^2 + y^2 = 1$. (The corners of the rectangle are the points $(\pm x, \pm y)$ on the circle.) Are x and y independent variables or is one a function of the other? Draw a graph of the area as a function of x. What value of x produces the largest area?

8. For the function $y = f(x) = x^3 - x + 1$, evaluate $f(-1)$, $f(0)$, $f(1)$ and $f(2)$. Determine $f(1 + h)$ and $f(2 + h)$ as functions of the parameter h.

9. If $y = f(x)$ is a polynomial of degree 1 whose graph in the cartesian plane passes through the points (1,2) and (2,3), determine the polynomial and draw its graph.

10. If $y = f(x)$ is a polynomial of degree 2 whose graph passes through (0,3), (1,2) and (2,3), determine the polynomial and draw its graph.

11. If $y = f(x)$ is a polynomial of degree n and c is any constant, prove that $y = f(cx)$ is also a polynomial of degree n except when $c = 0$. What happens when $c = 0$?

12. Draw graphs of the parabola $y = x^2$ and the lines $y = 2x - a$ for $a = 0$, ½, 1. Determine the points of intersection of the lines with the parabola. What is special about the case $a = 1$?
13. Show that the line $y = mx + b$ is tangent to the parabola $y = x^2$ if $m^2 + 4b = 0$.
14. Show that the line $y = mx + b$ is tangent to the circle $x^2 + y^2 = 1$ if $b^2 = m^2 + 1$.
15. Determine two tangent lines to the parabola $y = x^2 + 1$ that pass through the origin.
16. Graph the following functions for the given domains. Determine the ranges.

(i) $y = \dfrac{1}{1 + x^2} \qquad -\infty < x < \infty$ **(ii)** $y = \begin{cases} x & 0 \le x \le 1 \\ 2 - x & 1 \le x \le 2 \end{cases}$

(iii) $y = (x - 1)^{1/2} \qquad x \ge 1$ **(iv)** $y = (x^2 - 1)^{1/2} \qquad x^2 \ge 1$.

17. Given the graph of a function $y = f(x)$, show how to obtain the graphs of the function $y = -f(x)$ and the function $y = f(-x)$. Illustrate with the example $y = x^2 + x$.
18. Show that the graph of the function $y = f(ax)$ where a is a given positive constant can be obtained from the graph of $y = f(x)$ by a change of scale on the x axis. Illustrate with the examples $y = x$ and $y = x^2$ and the values $a =$ ½, 1, 2.
19. If $0 < a < 1$, show that $y = f(x/a)$, $y = af(x)$ and $y = f(x)/a$ can be interpreted as a stretching or magnification of the scale of the x axis, a contraction of the vertical scale, and a magnification of the vertical scale, respectively. Illustrate with the examples $y = x^2$ and $y = \sin x$.
20. Suppose that (x_0, y_0) and (x_1, y_1) are two points on the graph of $y = f(x)$. Show that the straight line joining these points is given by

$$y - y_0 = \frac{y_1 - y_0}{x_1 - x_0}(x - x_0)$$

This formula can be used to find the approximate value of a function $y = f(x)$ when $x_0 < x < x_1$, a procedure called *linear interpolation*.
21. The points (0.9, 0.81) and (1.1, 1.21) lie on the parabola $y = x^2$. Determine the straight line through these points. Use linear interpolation to estimate the value of y at $x = 1$ on the parabola. What is the error? How much improvement is there in the estimate if the closer points (0.95, 0.9025) and (1.05, 1.1025) are used? Draw graphs to illustrate this problem.
22. Show that the distance in the cartesian plane between the points (x_0, y_0) and (x_1, y_1) is $[(x_1 - x_0)^2 + (y_1 - y_0)^2]^{1/2}$. Show that the equation $(x - x_0)^2 + (y - y_0)^2 = r^2$ describes the circle of radius r with center at (x_0, y_0).
23. Determine the centers and radii of the following circles

(i) $x^2 + y^2 + 2x - 2y = 3$ **(ii)** $x^2 + y^2 - x + 3y = 5$
(iii) $x^2 + y^2 + 4x + 4y = 12$ **(iv)** $x^2 + y^2 - 2x - 4y = 5$

24. Show that any point on the parabola $x^2 = 4ay$ is at equal distances from the point $(0, a)$ and the straight line $y = -a$.
25. Show that the sum of the distances from any point on the ellipse $x^2/a^2 + y^2/b^2 = 1$ to the foci $(\pm(a^2 - b^2)^{1/2}, 0)$ is $2a$.
26. The *sign function*, sgn(x), is defined to be $+1$ if x is positive, -1 if x is negative, and 0 if $x = 0$. Draw a graph of sgn(x). Show that $\text{sgn}(ax) = \text{sgn}(a)\,\text{sgn}(x)$. What is the function $\text{sgn}(x - 5)$? the function $\text{sgn}(x^2 - x)$?
27. The *unit step function* (or Heaviside function) H(x) is 0 for $x < 0$ and 1 for $x \ge 0$. Draw a graph of H(x). Describe the functions $y = \text{H}(x - 5)$ and $y = \text{H}(x^2 - x)$. Prove that $\text{sgn}(x) = \text{H}(x) - \text{H}(-x)$.
28. The *absolute value function* $|x|$ is the numerical magnitude of x; if $x \ge 0$, $|x| = x$ and, if $x < 0$, $|x| = -x$. Draw a graph of $|x|$ and show that it satisfies the following properties.

(i) $|x| = x\,\text{sgn}(x)$ **(ii)** $|x^2| = |x|^2 = x^2$
(iii) $|x| \le L$ is equivalent to $-L \le x \le L$.

29. If a and b are any two real numbers, prove the following.
(i) $|ab| = |a|\ |b|$ **(ii)** $ab \leq |a|\ |b|$ **(iii)** $|a+b|^2 = (a+b)^2$
(iv) $|a+b| \leq |a| + |b|$ **(v)** $|a| - |b| \leq |a-b|$
[*Hint*: Use **(iii)** to prove **(iv)**. In **(v)**, note $a = (a-b) + b$ and use **(iv)**.]

30. Show that the equation $|x| + |y| = 1$ represents the sides of a square with center at (0, 0) and corners on the coordinate axes. What does the equation $|x - x_0| + |y - y_0| = r$ describe?

31. If $x_1, x_2, \ldots, x_n$ are real numbers, show that

$$|x_1 + x_2 + \ldots + x_n| = |\sum_{i=1}^{n} x_i| \leq \sum_{i=1}^{n} |x_i| = |x_1| + |x_2| + \ldots + |x_n|$$

When does equality hold?

32. Graph the following functions.
(i) $|x^2 + x|$ **(ii)** sgn $(\sin x)$ **(iii)** $|\sin x|$
(iv) $\mathrm{H}(\sin x)$ **(v)** $\sin x + \cos x$ **(vi)** $\sin 2x$

33. From the identity $\sin(x_1 + x_2) = \sin x_1 \cos x_2 + \cos x_1 \sin x_2$, prove the following.
(i) $\sin 2x = 2 \sin x \cos x$ **(ii)** $\sin 3x = 3 \sin x - 4 \sin^3 x$
(iii) $\sin x_1 + \sin x_2 = 2 \sin (x_1 + x_2)/2 \cos (x_1 - x_2)/2$.

34. If $u = \tan \theta$, prove the following
(i) $\sin 2\theta = \dfrac{2u}{1+u^2}$ **(ii)** $\cos 2\theta = \dfrac{1-u^2}{1+u^2}$ **(iii)** $\tan 2\theta = \dfrac{2u}{1-u^2}$

35. Graph the following curves given in polar coordinates.
(i) $r = 2\theta$ (spiral) **(ii)** $r = 3 \cos \theta$ (circle) **(iii)** $r = \dfrac{2}{2 - \cos \theta}$ (ellipse)

36. Show that every function of the form $y = a \cos x + b \sin x$ where a and b are given constants can be written $y = A \sin (x + \epsilon)$. Determine A and ϵ in terms of a and b. Use this result to evaluate $2 \cos \dfrac{\pi}{9} + 3 \sin \dfrac{\pi}{9}$ and $4 \sin \dfrac{\pi}{10} - \cos \dfrac{\pi}{10}$.

37. Verify the following identities
(i) $\sec^2 x = 1 + \tan^2 x$ **(ii)** $\csc^2 x = 1 + \cot^2 x$
(iii) $\tan (x_1 + x_2) = \dfrac{\tan x_1 + \tan x_2}{1 - \tan x_1 \tan x_2}$ **(iv)** $\tan \dfrac{x}{2} = \dfrac{\sin x}{1 + \cos x}$
(v) $\sin 2x = \dfrac{2 \tan x}{1 + \tan^2 x}$ **(vi)** $\cos 2x = \dfrac{1 - \tan^2 x}{1 + \tan^2 x}$

38. The two functions $y = f(u)$ and $u = g(x)$ define y as a function of x, $y = F(x) = f(g(x))$. The function $F = f \circ g$ is called a *composite function*. The substitution $u = g(x)$ is called a *change of variable*. In the following examples, determine y as a function of x.
(i) $y = 1 + u^2$, $u = 1 + x$ **(ii)** $y = (1 + u)^3$, $u = x^2 - 1$
(iii) $y = u^{1/2}$, $u = 1/(4 + x^2)$ **(iv)** $y = \tan u$, $u = 3x^2$
(v) $y = \sin u$, $u = (1 - x^2)^{1/2}$ **(vi)** $y = \sin u + \cos u$, $u = x + \pi/2$.

39. In the following examples, express $y = f(x)$ as a composite function by defining $u = G(x)$ and $y = F(u)$. Show that G and F are not uniquely determined by f.
(i) $y = (x^2 + 2x)^3$ **(ii)** $y = x^6 + x^3 + 1$ **(iii)** $y = (1 + x^2)^{-2}$
(iv) $y = \sin (1 - x^2)$ **(v)** $y = \cos^2(1 + x)$ **(vi)** $y = (1 + \tan^2 x)^{-1}$

1.3 Limits and functions

Even the evaluation of a function can require a limit procedure. For example, $\sin x$ and x are each well defined, and their quotient,

$$y = \frac{\sin x}{x}, \tag{1}$$

can be calculated easily at every point except $x = 0$. There both functions are zero, and the indeterminate ratio $0/0$ presents a certain measure of difficulty. However, the use of a calculator for small values of x yields Table 3.1. The graph in Fig. 3.1

TABLE 3.1

x	$\dfrac{\sin x}{x}$
1.0	0.84147
0.9	0.87036
0.8	0.89670
0.7	0.92031
0.6	0.94107
0.5	0.95885
0.4	0.97355
0.3	0.98507
0.2	0.99335
0.1	0.99833

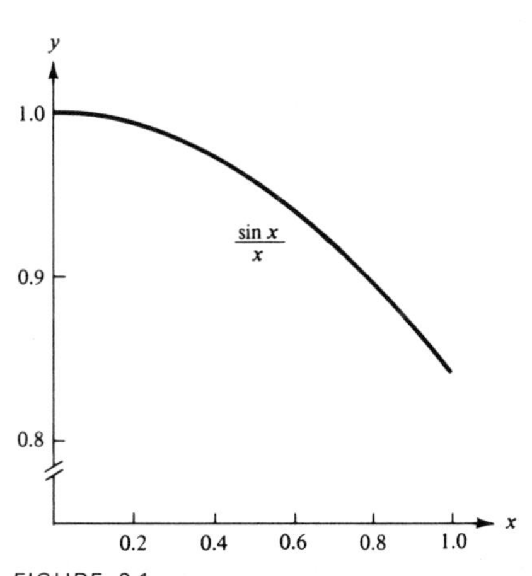

FIGURE 3.1

indicates that the function approaches the value of $+1$ as x tends to zero. Although this is excellent *evidence*, a proof of the result should be given. This is provided by finding the limit of the actual computational procedure as the value of x is made smaller and smaller. In formal language, we evaluate the function in the *limit as x approaches zero*, a process symbolically indicated by

$$\lim_{x \to 0} \frac{\sin x}{x}.$$

The analysis relies upon the definition of the sine function in terms of the unit circle. To relate more easily with the diagram given here, the ratio $(\sin \theta)/\theta$ as θ becomes zero is considered instead, i.e.,

$$\lim_{\theta \to 0} \frac{\sin \theta}{\theta}.$$

The area of a sector of a *unit* circle (Fig. 3.2) which has central angle θ is the fraction $\theta/2\pi$ of the total area π. Thus:

$$\text{Area of circular sector } OBC = \frac{\theta}{2\pi}\pi = \frac{\theta}{2}, \tag{2}$$

$$\text{Area of interior triangle } OAC = \tfrac{1}{2}\cos\theta\sin\theta,$$

$$\text{Area of exterior triangle } OBD = \frac{1}{2}\frac{\sin\theta}{\cos\theta}. \tag{3}$$

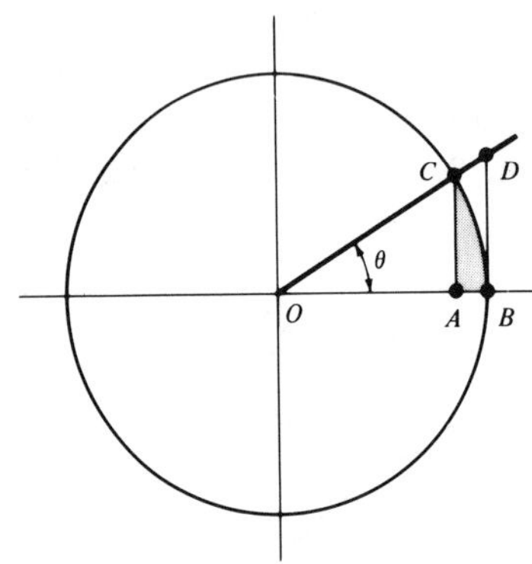

FIGURE 3.2

However, the area of the sector is, by definition, larger than the first of these triangles and smaller than the second, a fact expressed by the inequality

$$\frac{1}{2}\cos\theta\sin\theta \le \frac{\theta}{2} \le \frac{1}{2}\frac{\sin\theta}{\cos\theta}. \tag{4}$$

For θ positive, each term above can be divided by $\frac{1}{2}\sin\theta$ without disturbing the inequality, to obtain

$$\cos\theta \le \frac{\theta}{\sin\theta} \le \frac{1}{\cos\theta}. \tag{5}$$

At $\theta = 0$, $\cos\theta$ has a definite and unambiguous value, namely, $\cos 0 = 1$, which means that both left- and right-hand terms of (5) approach 1 as θ becomes zero. Thus the value of ratio $\theta/(\sin\theta)$ is constrained to lie within an interval which, as θ decreases, contracts to the single point $+1$. The meaning of $\theta/(\sin\theta)$ as θ becomes zero *must* then be 1; there can be no other interpretation. But if this is true, then the inverse ratio *must* also become 1 in this limit:

$$\lim_{\theta\to 0}\frac{\sin\theta}{\theta} = \lim_{x\to 0}\frac{\sin x}{x} = 1, \tag{6}$$

and this completes the analysis.

2 The limit process is very straightforward in many cases, especially when the *limit of the function at a point equals its assigned value there*:

$$\lim_{x\to a} f(x) = f(a).$$

For example,

$$\lim_{x\to 3} x = 3, \qquad \lim_{x\to 1} (1 + x + 2x^3) = 1 + 1 + 2 = 4,$$

$$\lim_{x\to \pi/2} \sin 2x = \sin \pi = 0, \qquad \lim_{x\to 0} \frac{(2 + x^4)\cos x}{1 + 2x + 3x^2} = \frac{2 \cos 0}{1} = 2.$$

The graphs of these functions would show smooth and continuous curves with no "breaks" or "holes."

On the other hand, the limit can fail to exist. Since $|x| = x$ for $x>0$ and $|x| = -x$ for $x<0$, the value of the *sign function*

$$\text{sgn}(x) = \frac{x}{|x|} \tag{7}$$

is $+1$ or -1, depending on whether x is positive or negative. This steplike function is shown in Fig. 3.3. The value of the function sgn(x) at the origin is the indeterminate ratio 0/0, but in this case the limit procedure does not resolve the difficulty. If the

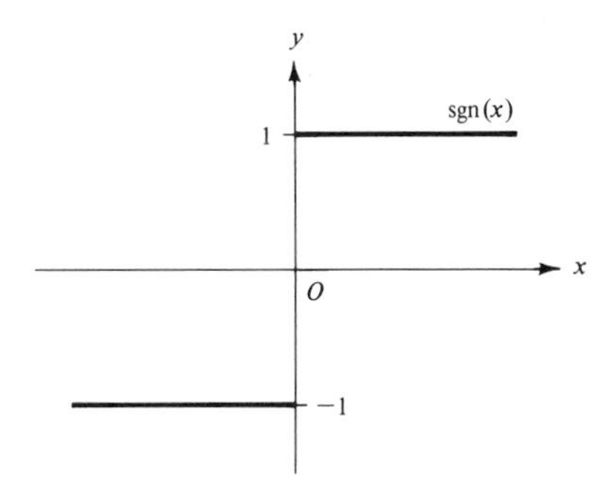

FIGURE 3.3

variable x is allowed to approach zero through positive values only, the function is and remains 1 identically, which implies that the limit must be unity. But if x tends to zero through negative values, the conclusion is that the limit is also equal to -1. Obviously, there is no single answer; a unique limit does not exist. Of course, the notion of a limit can be generalized to include a definite approach to zero from either the right or the left, from positive or negative values, and this is denoted by

$$\lim_{x\to 0+} \qquad \text{or} \qquad \lim_{x\to 0-} .$$

Under this restricted interpretation, the jump in the step function is described by

$$\lim_{x\to 0+} \frac{x}{|x|} = 1, \qquad \lim_{x\to 0-} \frac{x}{|x|} = -1;$$

the value of the function at $x = 0$ makes little difference and can be arbitrarily prescribed. Similarly, right and left limits can be defined for any point a; for example,

$$\lim_{x \to a+} \frac{x - a}{|x - a|} = 1, \qquad \lim_{x \to a-} \frac{x - a}{|x - a|} = -1.$$

But if a limit of the function does exist, then by definition all approaches to a yield the same result.

---

3 Another manifestation of abnormal behavior is illustrated by the function

$$y = \frac{1}{|x|}$$

as x approaches zero, i.e.,

$$\lim_{x \to 0} \frac{1}{|x|}.$$

Here, $1/|x|$ can be made arbitrarily large, as shown in Fig. 3.4, by selecting x small enough. The "value" of the function at $x = 0$ is then larger than *any* positive number, which means that it is infinite, a condition written

$$\lim_{x \to 0} \frac{1}{|x|} = \infty.$$

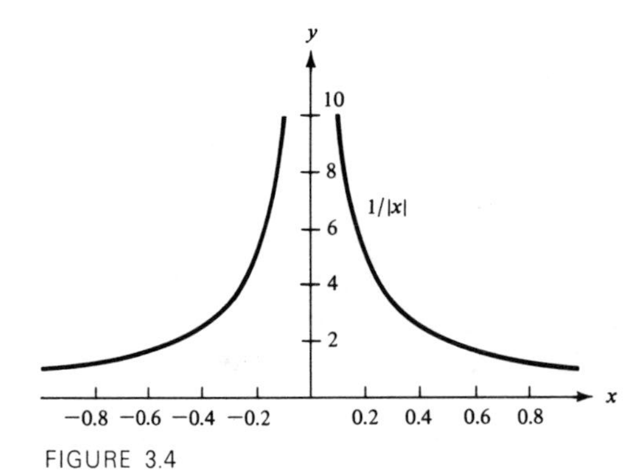

FIGURE 3.4

For this reason, the origin, $x = 0$, is called a *singularity* of the function $1/|x|$. Likewise $1/|x - a|$, is singular at $x = a$.

The function

$$y = \frac{1}{x}$$

has all the aforementioned idiosyncrasies. The limit exists at any point $x \neq 0$ and is easily determined. The origin is a singular point, as seen in Fig. 3.5, because the

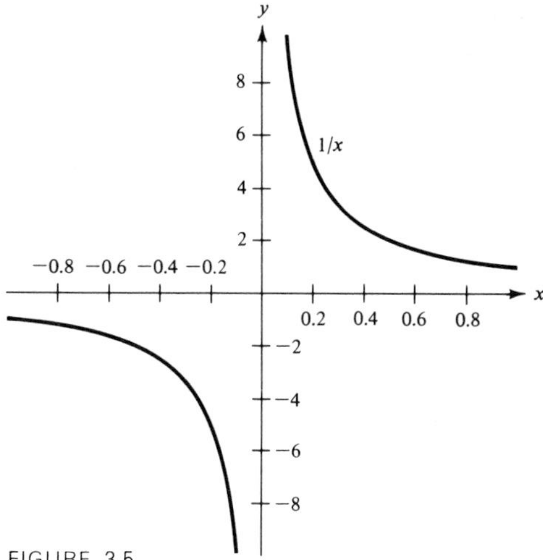

FIGURE 3.5

function is unbounded. However, the behavior depends on whether x is positive or negative, a state best described by right- and left-hand limits:

$$\lim_{x \to 0+} \frac{1}{x} = +\infty, \qquad \lim_{x \to 0-} \frac{1}{x} = -\infty.$$

4

A *rational function*, which is the ratio of two polynomials

$$R(x) = \frac{a_0 + a_1 x + \cdots + a_n x^n}{b_0 + b_1 x + \cdots + b_m x^m},$$

is singular wherever the denominator *is* zero but the numerator *is not*. For example,

$$R(x) = \frac{x}{2 + 3x + x^2} = \frac{x}{(x + 2)(x + 1)}$$

is infinite at $x = -2$ and $x = -1$, which are the roots of the denominator polynomial, but the function

$$R(x) = \frac{x + x^2}{2 + 3x + x^2} = \frac{x(x + 1)}{(x + 1)(x + 2)} = \frac{x}{x + 2}$$

is singular only at $x = -2$. In the last example, the factor $(x + 1)$, which is common to both numerator and denominator, can and should be cancelled since

$$\lim_{x \to -1} \frac{x + 1}{x + 1} = 1.$$

The *tangent function*, defined by

$$\tan x = \frac{\sin x}{\cos x},$$

is readily calculated from the known values of sin x and cos x (Table 3.2); its graph is shown in Fig. 3.6. The tangent exhibits singular behavior whenever cos $x = 0$, that is,

$$x = (n + \tfrac{1}{2})\pi.$$

The behavior in the neighborhood of these zeros is entirely similar to that of $1/x$ near the origin, a comparison more deeply probed in the exercises.

TABLE 3.2

x, deg	*x, rad*	*tan x*
0	0	0
5	.08727	.08749
10	.17453	.17633
15	.26180	.26795
20	.34807	.36397
25	.43633	.46631
30	.52360	.57735
35	.61087	.70021
40	.69813	.83910
45	.78540	1.00000
50	.87266	1.19175
55	.25323	1.42815
60	1.04720	1.73205
65	1.13446	2.14451
70	1.22173	2.74748
75	1.30800	3.73205
80	1.38626	5.67128
85	1.48353	11.43005

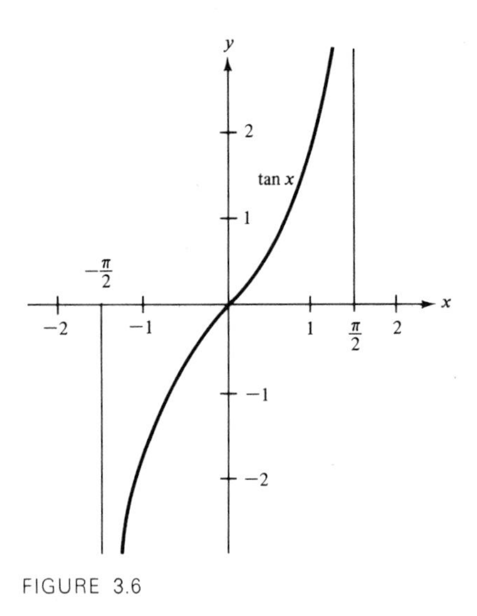

FIGURE 3.6

Exercises 1.3

1. Using a calculator, study the following limits.

(i) $\lim_{x\to 1} \dfrac{1-x^2}{1-x}$ **(ii)** $\lim_{x\to 2} \dfrac{x^3-8}{x^2-4}$ **(iii)** $\lim_{x\to 0} \dfrac{\sin x^2}{x^2}$

2. Using a calculator, estimate the following limits.

(i) $\lim_{x\to 0} \dfrac{\tan 2x}{x}$ **(ii)** $\lim_{x\to\infty} \dfrac{x^2}{1+x^2}$ **(iii)** $\lim_{x\to\pi/4} \dfrac{\tan x - 1}{x - \pi/4}$

3. Provide the details of the proof that

$$\frac{\sin x}{x} \to 1 \qquad \text{as } x \to 0$$

when x is always negative.
[*Hint:* Give a geometrical proof or use the fact that $\sin x = -\sin(-x)$.]

4. Verify the following limits.

(i) $\lim_{x\to 0} \frac{\sin 2x}{x} = 2$ **(ii)** $\lim_{x\to 0} \frac{1-\cos^2 x}{x^2} = 1$ **(iii)** $\lim_{x\to 0} \frac{\tan x}{x} = 1$

(iv) $\lim_{x\to 0} \frac{1-\cos x}{x} = 0$ **(v)** $\lim_{x\to 0} \frac{\sin 3x}{\sin 2x} = \frac{3}{2}$ **(vi)** $\lim_{x\to 0} \frac{\sin x - x}{x} = 0$

5. Use a calculator to construct a table of values of the function $\frac{\sin(2\sin x)}{x}$ near $x = 0$. What do you think the limit is as $x\to 0$? Can you prove it?

6. Introduce new variables to verify the following limits.

(i) $\lim_{\theta\to 0} \frac{\sin p\theta}{q\theta} = \frac{p}{q}$ **(ii)** $\lim_{x\to 0} \frac{\sin(\sin x)}{\sin x} = 1$ **(iii)** $\lim_{x\to 0} \frac{\tan ax}{\tan bx} = \frac{a}{b}$

7. Evaluate the following limits.

(i) $\lim_{h\to 0} (1 + 4h + h^3 \sin h)$ **(ii)** $\lim_{\theta\to 0} \frac{\sin\theta + \cos\theta}{\sin\theta - \cos\theta}$ **(iii)** $\lim_{x\to 1} \frac{3x^2+5}{x-1}$

(iv) $\lim_{s\to 2^+} \frac{(s^2-4)^{1/2}}{s^2+1}$ **(v)** $\lim_{t\to 0} \frac{t^3-1}{t-1}$ **(vi)** $\lim_{t\to 4} \sin 4\pi t$

8. A function $f(x)$ is *periodic* if there exists a number Ω, called the *period*, so that, for any x,

$$f(x+\Omega) = f(x).$$

Show that $\sin x$, $\cos x$, $\tan x$, and $\sin[\sin(x/2)]$ are periodic functions and find Ω in each case.

9. A function $f(x)$ is *bounded* in the interval $a \le x \le b$ if there is a constant M such that $|f(x)| < M$ for all x in the interval. Which of the following functions are bounded for $|x| < \infty$?

(i) $\sin x$ **(ii)** $\cos \frac{1}{2}x$ **(iii)** $x/(1+x^2)$

(iv) $1/(1+\tan x)$ **(v)** $\sin(\tan x)$ **(vi)** $\tan(\sin x)$

10. Prove that the product of two bounded functions is bounded. What can be said about their quotient?

11. Identify the singularities and draw a graph of the rational function

$$R(x) = \frac{x^2-1}{2x^3+3x^2-2x}.$$

Evaluate the following limits.

(i) $\lim_{x\to 0} xR(x)$ **(ii)** $\lim_{x\to 1/2} (x-\frac{1}{2})R(x)$ **(iii)** $\lim_{x\to -2} (x+2)R(x)$

12. If $\lim_{x\to a} [f(x)/g(x)] = 1$, $f(x)$ is said to be approximately equal to $g(x)$ when x is near a. This is written $f(x) \approx g(x)$ for x near a. Prove the following.

(i) $\sin x \approx x$ for x near zero **(ii)** $\tan x \approx x$ for x near zero

(iii) $\frac{x^2-1}{x^2+4} \approx \frac{2}{5}(x-1)$ for x near 1 **(iv)** $\frac{x^2-1}{x-1} \approx 2$ for x near 1.

13. The meaning of $f(x) \approx \frac{A}{x-a}$ for x near a is $\lim_{x\to a} (x-a)f(x) = A$. Show that

$$R(x) = \frac{x^2+2x+5}{x(x-1)(x+2)} \approx \begin{cases} -\frac{5}{2x} & \text{for } x \text{ near } 0 \\ \frac{8}{3(x-1)} & \text{for } x \text{ near } 1 \\ \frac{5}{6(x+2)} & \text{for } x \text{ near } -2. \end{cases}$$

Draw a graph of $y = R(x)$.

14. Find the zeros and singular points of the rational function

$$R(x) = \frac{x^3-x^2+x-1}{x^2-2x-3}.$$

Write an approximate formula for $R(x)$ in the neighborhood of each singularity. Draw a graph of $y = R(x)$.

15. If $|x|$ is very large, show that the approximate value of

$$R(x) = \frac{a_0 + a_1 x + \dots + a_n x^n}{b_0 + b_1 x + \dots + b_m x^m} = \frac{P(x)}{Q(x)}$$

is

$$R(x) \approx \frac{a_n}{b_m} x^{n-m}.$$

Estimate the relative error made in this approximation and show that it approaches zero as the magnitude of $|x|$ increases beyond bound.

16. If $z = 1/x$, show that the rational function in Exercise 15 can be written

$$R = \frac{a_0 z^n + a_1 z^{n-1} + \dots + a_n}{z^{n-m}(b_0 z^m + b_1 z^{m-1} + \dots + b_m)}$$

Relate the behavior of R for large $|x|$ to its behavior in the neighborhood of $z = 0$. The "point at infinity," $|x| = \infty$, corresponds to the point $z = 0$. Under what conditions is the point at infinity a singular point of the function?

17. Is $x = \infty$ a singular point of the rational function $R(x) = (x^3 - 1)/(x - 2)$? Illustrate with a graph.

18. If $R(x) = P(x)/Q(x)$ and $Q(x) = (x - a)^2 S(x)$, with $S(a) \neq 0, P(a) \neq 0$, show that for $|x - a|$ small

$$R(x) \approx \frac{P(a)}{S(a)} \frac{1}{(x-a)^2}.$$

19. Graph the functions $f(x) = \sin(1/x)$ and $g(x) = x \sin(1/x)$ and examine their behaviors near $x = 0$.

1.4 Rate of change

1

The motion of an automobile on a highway is essentially determined by specifying its position s as a function of time t:

$$s = S(t) \tag{1}$$

(In this idealization, position refers to any definite point marked on the car. The graph of the function $S(t)$ as a curve in the t, s plane is called the *trajectory* of the vehicle. For example, Fig. 4.1 shows the position of a car which approaches and passes a constriction in the highway. The trajectory could be constructed from multiple photographs taken from above in which the time of each exposure is recorded, say, in seconds and the scale of distance is determined by road side markers which might be utility poles spaced 50 m apart. The corresponding values of s and t can be calculated. The choice of coordinate origin is settled by making $s = 0$ at $t = 0$. A very rapid sequence of photos would yield enough points for the trajectory to be drawn as the smooth curve shown. The data is recorded as in Table 4.1.

An important and quite fundamental problem of calculus is involved in the determination of the velocity v of the automobile as a function of time from the knowledge

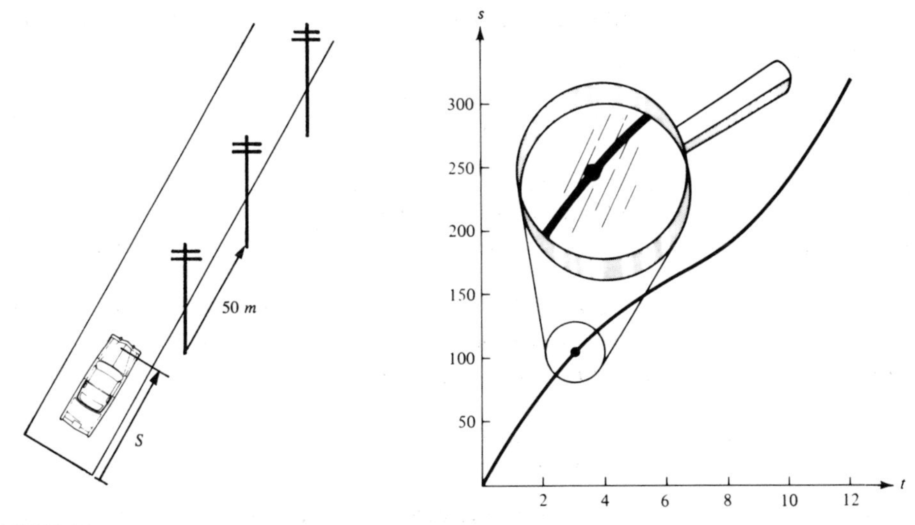

FIGURE 4.1

of the trajectory $S(t)$. If the velocity were constant, it would be calculated as the distance travelled divided by the elapsed time.

$$v = \frac{s}{t}. \tag{2}$$

The trajectory for such motion would then be the straight line

$$s = vt. \tag{3}$$

Equivalently, if any part of the trajectory is straight, the speed in that time interval is constant. In the particular case illustrated, the observed velocity is clearly not constant *anywhere*, for any time span, and this is the exact point of difficulty. How is a continuously changing velocity determined? What approximate method can be devised for this purpose which in a limiting sense provides the exact velocity?

The answer is to think small: over a sufficiently small interval of time, the velocity does not vary much from a constant value. In other words, very small segments of the trajectory do indeed appear to be almost straight when magnified (see Fig. 4.1). In order to compute the speed at time t and position s, the velocity is taken to be constant over some small interval of time designated by Δt (read "delta t") during which the distance traversed is Δs. The position of the car is then $s + \Delta s$, and the distance travelled is obtained from (1). Since

$$s = S(t)$$

and

$$s + \Delta s = S(t + \Delta t), \tag{4}$$

the difference yields

$$\Delta s = S(t + \Delta t) - S(t).$$

The constant velocity consistent with this rate of change is obtained from the total distance moved in the short period of time as

$$v = \frac{s + \Delta s - s}{t + \Delta t - t} = \frac{\Delta s}{\Delta t} = \frac{S(t + \Delta t) - S(t)}{\Delta t}. \tag{5}$$

For a small value of Δt, the last formula yields an adequate approximation that will be used soon as a means of numerical computation. A radar determination of velocity utilizes the same formula, with space and time increments that are truly minute.

As Δt, and hence Δs, is made smaller and smaller, (5) becomes an increasingly accurate approximation. The instantaneous *velocity* v at time t is defined as the limiting value of the ratio $\Delta s/\Delta t$ as Δt approaches zero. The formal statement is

$$v(t) = \lim_{\Delta t \to 0} \frac{\Delta s}{\Delta t} = \lim_{\Delta t \to 0} \frac{S(t + \Delta t) - S(t)}{\Delta t}. \tag{6}$$

(Read "v is the limit, as delta t approaches zero, of the ratio delta s divided by delta t.")

2

Equations (5) and (6) have an important and informative graphical interpretation. Figure 4.2 is an enlargement of the trajectory curve in the neighborhood of the arbitrary point (t, s) labelled A. The neighboring point $(t + \Delta t, s + \Delta s)$ used in the computation of the velocity is shown as B. Since Δs and Δt are, respectively, the vertical

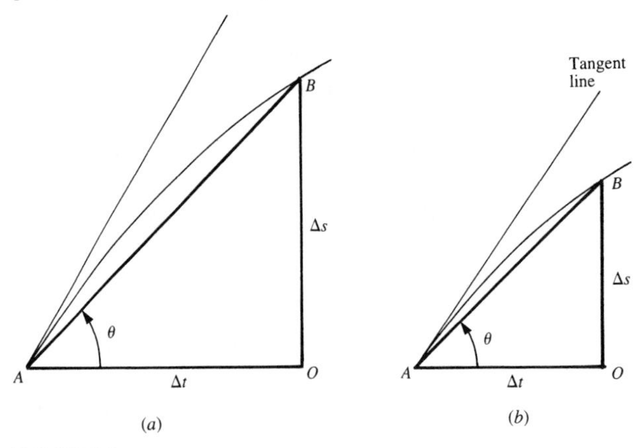

FIGURE 4.2

and horizontal legs of triangle AOB, the velocity v, according to (5), is simply the ratio of these lengths, which is the tangent of the interior angle of the triangle. In other words, the approximation actually yields the tangent of the angle made by the chord AB with the horizontal. The value of v calculated in this way can be systematically improved by making the time increment Δt shorter and shorter, i.e., by choosing point B nearer to A, as shown in Fig. 4.2b. As the limit or end result of this procedure, the chord angle θ becomes the angle of the line that is *tangent to the trajectory curve* at A. Therefore, the *exact velocity* at any point, as given by (6), is the slope of the tangent line to the trajectory curve.

3

This limit procedure for calculating the velocity is called *differentiation* and its fundamental nature justifies a special symbolism (due to Leibniz):

$$v(t) = \lim_{\Delta t \to 0} \frac{\Delta s}{\Delta t} = \frac{ds}{dt}. \tag{7}$$

(It pays to think of the quantities ds and dt as the infinitesimal forms of the small but finite increments Δs and Δt.) The velocity v is then said to be the *derivative* of the distance travelled s with respect to elapsed time. *The derivative expresses the instantaneous rate of change of the dependent variable with respect to the independent variable.*

The velocity can now be calculated from the graph with the aid of a ruler or analytically if the function $S(t)$ is given explicitly. For example, the velocity at $t = 3$ sec is the slope of the tangent line there, which is found to be 27 m/sec.

The accuracy of the approximate formula (5) can be gauged by actually carrying out a few computations indicated by the limit process. To find v at $t = 3$ sec, let $\Delta t = 1$ sec, in which case the corresponding distances (given in Table 4.1) are $s = 105.46$ m and $\Delta s = 23.26$ m [note that $S(t + \Delta t) = S(4 \text{ sec}) = 128.72$ m]. Substitution in (5) yields the estimate

$$v = \frac{\Delta s}{\Delta t} = \frac{23.26 \text{ m}}{1 \text{ sec}} = 23.26 \text{ m/sec}.$$

Suppose now that the time increment is reduced to $\Delta t = 0.5$ sec, which corresponds to the new distance increment $\Delta s = 12.47$ m. The revised estimate for the velocity is then

$$v = \frac{12.47 \text{ m}}{0.5 \text{ sec}} = 24.94 \text{ m/sec}.$$

which is a somewhat improved value. The results of these and additional computations for smaller increments are tabulated.

TABLE 4.1 *Calculation for* $t = 3$ sec, $s = S(3) = 105.46$ m

t, sec	$s = S(t)$, m	Δt, sec	$S(t + \Delta t)$, m	$\Delta s = S(t + \Delta t) - S(t)$, m	$v = \frac{\Delta s}{\Delta t}$, m/sec
3	105.46	1.0	128.72	23.26	23.26
		0.5	117.93	12.47	24.94
		0.2	110.66	5.20	26.00
		0.1	108.10	2.64	26.40
		−0.1	102.76	−2.70	27.00
		−0.2	99.99	−5.47	27.35
		−0.5	91.26	−14.20	28.40
		−1.0	75.39	−30.07	30.07

Note that both positive and negative increments can be used in the calculation. The limit exists no matter how $\Delta t \to 0$, and there is no restriction on the sign.

Since the velocity can be determined at any time, v and t are functionally related:

$$v = V(t), \tag{8}$$

and a graph can be constructed. An explicit formula will be obtained when similar information about $S(t)$ is available.

4

The original data are usually not given as a complete trajectory but as a discrete number of points, the positions at certain definite times. For example, the information obtained from the "photograph", Fig. 4.1*b*, yields only 12 data points. The number could be a thousand or a million, but it would still be finite. Since the real trajectory passes through all these known positions, it is reasonably approximated by any smooth curve that does the same if the number of points is large. Curve fitting and interpolation are important problems which we shall return to later. However, the original data can be analyzed directly without inquiring about the entire trajectory, but of course only a restricted amount of information is obtainable. To illustrate this process, the data from a sequence of photographs is first assembled in a more systematic form.

The time of each exposure is denoted by t_i, where index i runs from 0 to 12. The position of the vehicle at time t_i is recorded as

$$s_i = S(t_i), \tag{9}$$

and the corresponding values are given in Table 4.2.

TABLE 4.2

i	t_i, sec	s_i, m	v_i, m/sec
0	0	0	39.40
1	1	39.40	35.99
2	2	75.39	30.07
3	3	105.46	23.26
4	4	128.72	17.35
5	5	146.07	13.93
6	6	160.00	13.93
7	7	173.93	17.35
8	8	191.28	23.26
9	9	214.54	30.07
10	10	244.61	35.99
11	11	280.60	39.40
12	12	320.00	

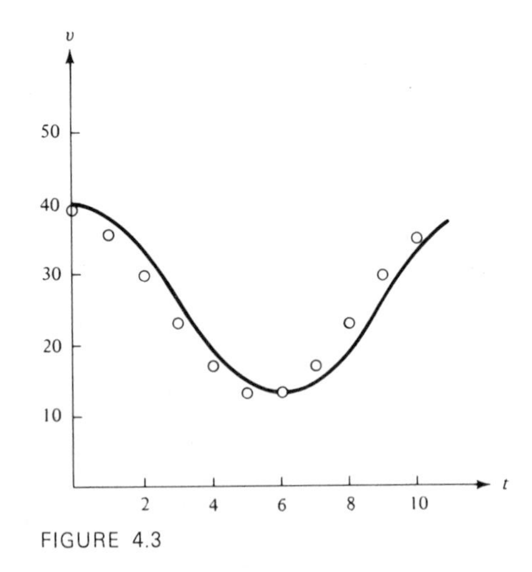

FIGURE 4.3

In order to find the velocity at time t_i, that is,

$$v_i = V(t_i), \tag{10}$$

(5) is converted into a formula for each numerical calculation. If the distance and time increments at each step are identified as

$$\Delta s = s_{i+1} - s_i \qquad \text{and} \qquad \Delta t = t_{i+1} - t_i, \tag{11}$$

then a simple approximation for the velocity is

$$v_i = \frac{\Delta s}{\Delta t} = \frac{s_{i+1} - s_i}{t_{i+1} - t_i} \tag{12}$$

This equation, which uses information at times later than t_i, is called a *forward difference* in the parlance of numerical analysts. The error involved depends on the size of the time step; for a given increment this can be minimized by using a more elaborate difference formula in place of (12).

The magnitude of v_i is easily calculated. For example, at $i = 3$

$$v_3 = \frac{s_4 - s_3}{t_4 - t_3} = \frac{128.72 \text{ m} - 105.46 \text{ m}}{1 \text{ sec}} = 23.26 \text{ m/sec};$$

the remaining values are similarly determined and recorded in the table. The points (t_i, v_i) are circled in Fig. 4.3 so that the errors incurred can be easily assayed from inspection with the exact curve, which will be given later. Approximate values of v at intermediate times, say 3.5 sec, can be obtained from the numerical data by simply drawing a smooth curve through the calculated points, or we may assume v to be a constant in an entire time increment so that $v = v_i$ for $t_i \le t \le t_{i+1}$. More accurate methods can and will be devised.

---

5 Now that the velocity at time t has been defined, a similar argument leads to a definition of *instantaneous acceleration*. The ordinary concept of the average acceleration is the rate of change of velocity with time. The formula for the average acceleration from time t_1 to time t_2 is the ratio

$$\frac{v(t_2) - v(t_1)}{t_2 - t_1} \tag{13}$$

In Fig. 4.4, a graph of the velocity function $v = v(t)$ is drawn. Geometrically, the average acceleration is the slope of the chord joining the points $(t_1, v(t_1))$ and $(t_2, v(t_2))$. The instantaneous acceleration is defined as a limit,

$$a(t) = \lim_{\Delta t \to 0} \frac{\Delta v}{\Delta t} = \lim_{\Delta t \to 0} \frac{v(t + \Delta t) - v(t)}{\Delta t}$$

In the diagram, this is interpreted as the slope of the tangent line at $(t, v(t))$.

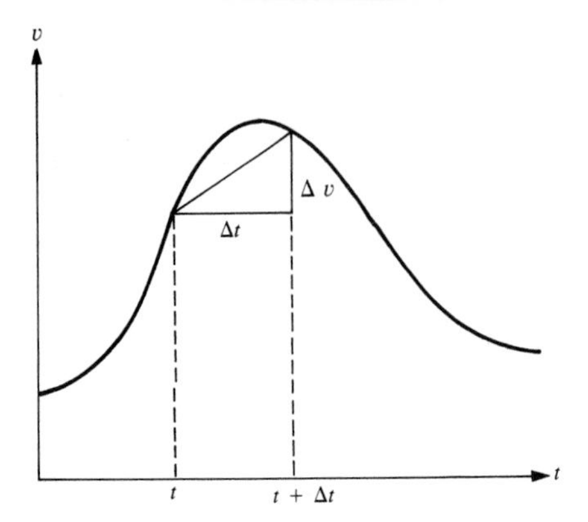

FIGURE 4.4

Example 1 If the average acceleration is constant, determine $v = v(t)$ and $s = s(t)$.

Suppose that a is the constant average acceleration, the same in each time interval. Choosing $t_1 = 0$ and $t_2 = t$, we have

$$a = \frac{v(t) - v(0)}{t} \text{ and } v(t) = v(0) + at \tag{14}$$

Over the time interval from 0 to t, the velocity increases linearly from $v(0)$ to $v(0) + at$. Since the increase is uniform over the time interval, the average velocity is the velocity at time $t/2$ (the midpoint of the time interval). This implies that

$$\frac{s(t) - s(0)}{t} = v(0) + \frac{at}{2} \tag{15}$$

Solving for $s(t)$, we have

$$s(t) = s(0) + v(0)t + \tfrac{1}{2}at^2 \tag{16}$$

The trajectory of a particle moving with constant acceleration is a parabola.

Example 2 Determine formulas for the velocity (m $\sec^{-1}$) and acceleration (m $\sec^{-2}$) of a particle whose position $s(t)$ (m) on a line at time t (sec) is given by

$$s(t) = \begin{cases} t^2 & 0 \leq t \leq 10 \\ 100 + 20(t - 10) & t > 10 \end{cases}$$

The trajectory of the particle $s = s(t)$ is parabolic for $0 \leq t \leq 10$ and linear for $t > 10$ (Fig. 4.5). Since the velocity is constant after the first 10 seconds, its magnitude is simply the slope of the straight line portion of the trajectory. Therefore, $v = 20$ for $t \geq 10$. To calculate the velocity for $0 \leq t \leq 10$, the limit definition is used,

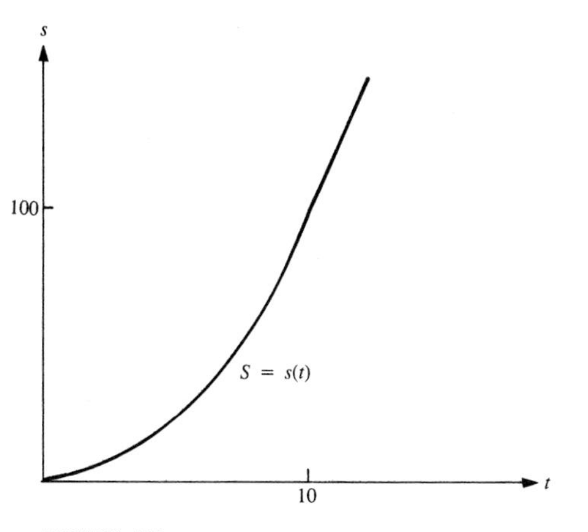

FIGURE 4.5

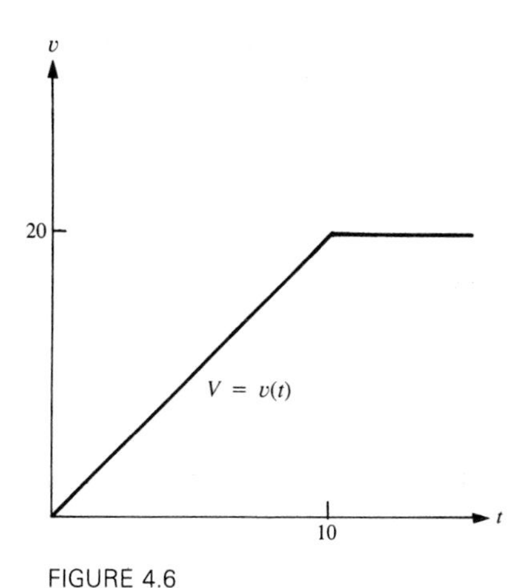

FIGURE 4.6

$$v(t) = \frac{ds}{dt} = \lim_{\Delta t \to 0} \frac{\Delta s}{\Delta t} = \lim_{\Delta t \to 0} \frac{s(t + \Delta t) - s(t)}{\Delta t}$$

In this example,

$$\begin{aligned} v(t) &= \lim_{\Delta t \to 0} \frac{(t + \Delta t)^2 - t^2}{\Delta t} \\ &= \lim_{\Delta t \to 0} \frac{2t\,\Delta t + (\Delta t)^2}{\Delta t} \\ &= \lim_{\Delta t \to 0} (2t + \Delta t) = 2t \end{aligned}$$

The complete velocity function for $t \geq 0$ (Fig. 4.6) is then

$$v(t) = \begin{cases} 2t & 0 \leq t \leq 10 \\ 20 & t > 10 \end{cases}$$

To calculate the acceleration $a(t)$, we note that the velocity is constant for $t \geq 10$ and therefore the acceleration is zero. For $0 \leq t \leq 10$,

$$\begin{aligned} a(t) &= \lim_{\Delta t \to 0} \frac{v(t + \Delta t) - v(t)}{\Delta t} \\ &= \lim_{\Delta t \to 0} \frac{2(t + \Delta t) - 2t}{\Delta t} = 2 \end{aligned}$$

The acceleration for $t \geq 0$ is given by

$$a(t) = \begin{cases} 2 & 0 \leq t \leq 10 \\ 0 & t > 10 \end{cases}$$

The particle accelerates uniformly for the first ten seconds and then continues to move with constant velocity.

Example 3 An athlete runs a 100-meter race in 10.0 seconds. He accelerates at a constant rate for the first five seconds and then runs at constant velocity. Determine the acceleration, velocity and distance as functions of time.

As in Example 2, the acceleration is a step function, $a(t) = a$ for $0 \le t \le 5$ and $a(t) = 0$ for $5 < t \le 10$ where a is a constant to be determined. For the first five seconds, the velocity increases linearly ($v(t) = at$ for $0 \le t \le 5$) and then is constant ($v(t) = 5a$ for $5 < t \le 10$). Using equation (16), the distance covered is

$$s(t) = \begin{cases} \frac{1}{2} at^2 & 0 \le t \le 5 \\ 25a/2 + 5a\,(t-5) & 5 < t \le 10. \end{cases}$$

Therefore, $s(10) = 25a/2 + 25a = 75a/2 = 100$ and $a = 200/75 = 8/3$. The acceleration in the first five seconds is $8/3$ m/sec^2. The maximum speed is $v(5) = 5a = 5(8/3) = 40/3$ m/sec, reached after 5 seconds. When the runner reaches maximum speed, he has covered $s(5) = \frac{1}{2}(8/3)5^2 = 33.\dot{3}$ m or one third of the total distance. The acceleration, velocity and distance functions are graphed in Fig. 4.7.

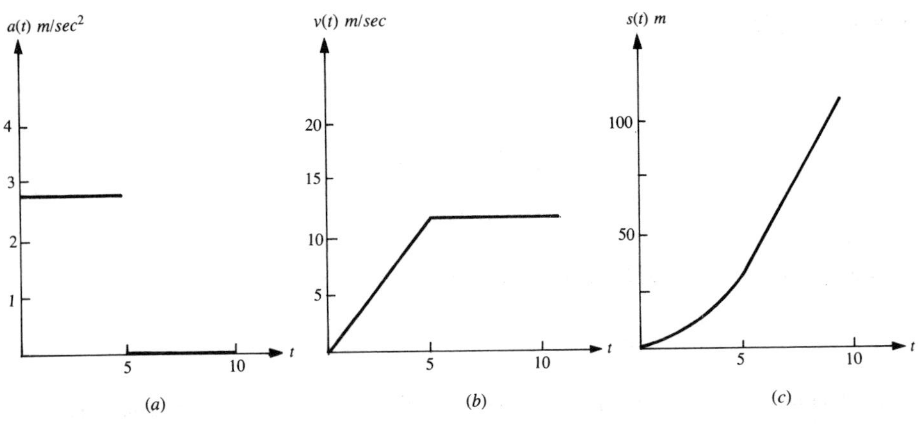

FIGURE 4.7

Exercises 1.4

1. Graph the following particle trajectories $s = s(t)$. Determine the average velocity over the time interval from t to $t + \Delta t$. Evaluate the (instantaneous) velocity at time t.

(i) $s(t) = 3 + 2t$ **(ii)** $s(t) = 3t - t^2$ **(iii)** $s(t) = 5 - t + t^2$

2. In the previous problem, evaluate the average acceleration over the time interval from t to $t + \Delta t$. Evaluate the (instantaneous) acceleration at time t. Illustrate the solutions with graphs.

3. A bacteria population at time t is given by the function $n = n(t)$. Define the average growth rate over the time interval t to $t + \Delta t$ and the (instantaneous) growth rate at time t. Evaluate both the average and instantaneous growth rates for the following populations.

(i) $n(t) = 10^4 + 10^3 t$ **(ii)** $n(t) = 10^4\,(5t - t^2)$ **(iii)** $n(t) = 10^6 - 10^5 t + 10^4 t^2$

4. The position $s(t)$ at time t of a particle moving with constant acceleration a along a straight line is given by $s(t) = s(0) + ut + \frac{1}{2}at^2$ where $s(0)$ and u are the position and velocity at time 0. If $v(t) = ds/dt$ is the velocity at time t, verify the following.

(i) $v(t) = u + at$ **(ii)** $s(t) - s(0) = \dfrac{u + v(t)}{2}t$ **(iii)** $(v(t))^2 = u^2 + 2as(t)$

5. The height $s(t)$ at time t of an object falling from a height h above the surface of the earth is given by $s(t) = h - \frac{1}{2}gt^2$ where g is the constant acceleration due to gravity. Draw graphs of the particle trajectory, the velocity and the acceleration. When does the object hit the ground?

6. An object falling from a height h is one-third of the way there after three seconds. When does the object hit the ground and what was the height h? (Use $g = 9.8$ m sec^{-2}).

7. For a particle falling from a height h, determine the time $t(x)$ to fall a distance x, $0 \leq x \leq h$. Draw graphs of $t = t(x)$ and $x = x(t)$.

8. A projectile is launched vertically upward from the surface of the earth with velocity $v(0)$ at time $t = 0$. The height $s(t)$ of the projectile at time t is given by $s(t) = v(0)t - \frac{1}{2}gt^2$. Draw a graph of the projectile trajectory. Determine the velocity $v(t)$ and the acceleration $a(t)$. Determine the maximum height reached. When does the projectile return to the surface of the earth? What is its velocity at that moment?

9. Suppose that the projectile in the previous problem has mass m. Verify that the sum of the potential energy mgs and the kinetic energy $\frac{1}{2}mv^2$ is a constant.

10. A baseball is caught by an outfielder six seconds after it is hit. Estimate the maximum height reached by the ball. (Explain any assumptions made to solve this problem.)

11. Suppose that the runner in Example 3 reaches maximum speed after 3.0 seconds. Determine the runner's acceleration, velocity and distance as functions of time. How far has the runner travelled after 3 seconds? after 5 seconds? When is the instantaneous velocity equal to the average velocity?

12. Draw graphs of the following trajectories $s = s(t)$ for $t > 0$. Determine the average velocity over the time interval from t to $t + \Delta t$ and the instantaneous velocity $v(t)$. When is the particle moving to the right ($v(t) > 0$) and when it is moving to the left?

(i) $s(t) = 3 + 2t - t^2$ **(ii)** $s(t) = 1 - t + t^2 - t^3$

(iii) $s(t) = 4t - t^2$ **(iv)** $s(t) = 5 - |t - 5|$

13. A car is moving along a straight highway. In a one minute time interval, it is observed to travel exactly one kilometer. Is it intuitively obvious that, at least once during the time interval, the instantaneous velocity of the car was exactly one kilometer per minute? Draw graphs of possible trajectories to show why this is correct.

14. A car accelerates uniformly from 0 to 30 m sec^{-1} in exactly 10 seconds. How far does it travel in this time interval? Graph the trajectory $s = s(t)$ for $0 \leq t \leq 10$.

15. An aircraft lands at speed 50 m sec^{-1} and decelerates at a constant rate to come to a stop 1000 m down the runaway. What is the deceleration rate and how long does the plane roll?

16. Calculate the velocity v_i at time t_i for the data in Table 4.1 using the backward difference

$$v_i = \frac{s_i - s_{i-1}}{t_i - t_{i-1}}$$

and the centered difference

$$v_i = \frac{s_{i+1} - s_{i-1}}{t_{i+1} - t_{i-1}}.$$

Compare these with the exact curve shown in Fig. 4.3.

17. Calculate from Fig. 4.3 and Table 4.2 the acceleration of the vehicle using the numerical formula

$$a_i = \frac{v_{i+1} - v_i}{t_{i+1} - t_i},$$

and graphically by finding the slope of the tangent line to the velocity-time curve. Find the acceleration from the original data on distance in Table 4.2 using the numerical formula

$$a_i = 4\,\frac{s_{i+1} - 2s_i + s_{i-1}}{(t_{i+1} - t_{i-1})^2}.$$

1.5 *Limits and Continuity*

1

So far, we have discussed a few central problems and ideas from a fairly general viewpoint, without being able to carry out the details of calculation in any but the simplest cases. The time has come to narrow this gap somewhat, and we begin by converting our intuitive notion of the limit process into a precise mathematical statement. A secure foundation allows a step-by-step advance through the subject that utilizes all the information and methods made available. Of course, a completely new situation dictates a return to fundamental concepts.

The interpretation of

$$\lim_{x \to a} f(x) = A \tag{1}$$

has been that as x approaches the value a, $f(x)$ will approach A. This is an entirely correct *qualitative* description, but a *quantitative* criterion is required for analytical work. Since the only relevant quantities that can be measured are the differences $f(x) - A$ and $x - a$, the limit must be described in terms of these numbers.

2

The language of a precise statement, one without flaws or loopholes, seems strange initially but some thought and a little practice will make it familiar and useful. Frequent reference will be made to "arbitrarily small numbers," and it is convenient to have special names for this purpose. Following established usage, let ε (epsilon) denote a small number, which we are free to select at will, and δ (delta) another which is then determined. In other words, ε is *given* and δ is *found.* (For example, if $\varepsilon = 0.0001$, it might turn out that $\delta = 0.003$.) So armed, we can give the precise meaning of (1).

The function $f(x)$ is said to have the limit A as x approaches a if, for any given positive number ε, another positive number δ can be found such that

$$|f(x) - A| < \varepsilon, \tag{2}$$

for values of x in the intervals

$$0 < |x - a| < \delta. \tag{3}$$

In other words, no matter how small ε is chosen, the quantity $|f(x) - A|$ can be made less than this, *provided* x is sufficiently near to a, that is, $|x - a| < \delta$. The number ε is a measure of how close we wish $f(x)$ to be to A. The number δ is then a measure of an interval about a in which any x will satisfy this requirement, as illustrated in Fig. 5.1.

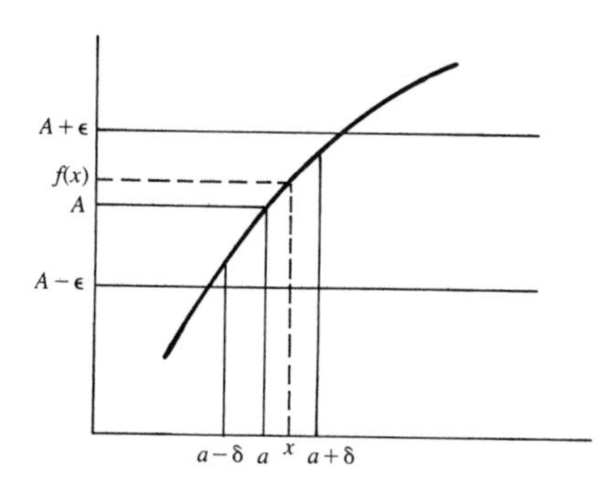

FIGURE 5.1

The procedure for calculating a limit, (1), begins with the quantity $|f(x) - A|$. By manipulating this, the magnitude of δ is then determined which guarantees that the difference (2) is less than the preassigned arbitrarily small ε. The existence of a limit requires only that a number δ be found; it does not have to be an optimum value, such as the *largest* δ corresponding to a given ε.

Example 1 Prove that $\lim_{x \to 2} x^2 = 4$.

Form the difference

$$|x^2 - 4| = |(x - 2)(x + 2)| = |x - 2|\,|x + 2|. \tag{4}$$

Since the limit calculation concerns values of x near 2, we may restrict the discussion to any interval about this point, say

$$0 < x < 4.$$

This is a useful restriction because it implies

$$|x + 2| < 6, \tag{5}$$

and allows (4) to be rewritten as the inequality

$$|x^2 - 4| < 6|x - 2|. \tag{6}$$

This expression is very nearly the form required by the limit criterion. To complete the argument, let

$$|x - 2| < \delta, \tag{7}$$

in which case (6) becomes

$$|x^2 - 4| < 6|x - 2| < 6\delta. \tag{8}$$

Moreover, a particular choice of δ for which $6\delta < \varepsilon$ now guarantees the result

$$|x^2 - 4| < \varepsilon, \tag{9}$$

and this completes the proof. In other words, the difference (4) can be made less than any preassigned number ε for $\delta < \varepsilon/6$; that is, when x is in the interval $2 - \varepsilon/6 < x < 2 + \varepsilon/6$.

---

3 Several of the most useful properties of limits are now set forth. For this purpose, consider two functions each having a limit as $x \to a$:

$$\lim_{x \to a} f(x) = A, \qquad \lim_{x \to a} g(x) = B. \tag{10}$$

It follows that

(i) $\lim\limits_{x \to a}[f(x) + g(x)] = \lim\limits_{x \to a} f(x) + \lim\limits_{x \to a} g(x) = A + B$

(ii) $\lim\limits_{x \to a}[f(x) - g(x)] = \lim\limits_{x \to a} f(x) - \lim\limits_{x \to a} g(x) = A - B$

(iii) $\lim\limits_{x \to a}[f(x) \cdot g(x)] = \lim\limits_{x \to a} f(x) \cdot \lim\limits_{x \to a} g(x) = AB$

(iv) $\lim\limits_{x \to a} \dfrac{f(x)}{g(x)} = \dfrac{\lim\limits_{x \to a} f(x)}{\lim\limits_{x \to a} g(x)} = \dfrac{A}{B}$ provided that $B \neq 0$.

---

4 Formula (iii) is proved here to illustrate the general approach, but the others are left as exercises.

Simple manipulations and use of the triangle inequality, $|x_0 + x_1| \leq |x_0| + |x_1|$, allow us to write

$$\begin{aligned} |f(x)g(x) - AB| &= |[f(x) - A]g(x) + A[g(x) - B]| \\ &\leq |[f(x) - A]g(x)| + |A[g(x) - B]| \\ &\leq |f(x) - A|\,|g(x)| + |A|\,|g(x) - B|. \end{aligned} \tag{11}$$

Since $f(x)$ and $g(x)$ both possess limits, a *single* number δ can be found for which

$$|f(x) - A| \leq \varepsilon, \qquad |g(x) - B| \leq \varepsilon$$

when

$$|x - a| < \delta.$$

(Of the two possible numbers for δ, choose the smaller.) Furthermore, within this interval about a, $|g(x)|$ and $|A|$ are bounded; i.e., both are less than *some* positive constant M:

$$|g(x)| < M, \qquad |A| < M.$$

These results are used to strengthen inequality (11) so that it reads

$$|f(x)g(x) - AB| \leq 2M\varepsilon \qquad \text{for} \quad |x - a| < \delta. \tag{12}$$

Formula (iii) follows from the observation that when ε is arbitrarily small, so too is

$$2M\varepsilon = \varepsilon_*. \tag{13}$$

Therefore the difference (12) can be made less than any arbitrarily small number ε_* by choosing x appropriately, and the limit is established.

---

5 Formulas (i) to (iv) are used effectively to determine new limits from those already established, and the basic ε, δ method is reserved for the special situations that require a return to fundamentals. For most purposes, then, the ε, δ approach is relegated to a distinctly secondary position, although as the foundation it supports everything else.

Example 2 Show that $\lim\limits_{x \to 1} \dfrac{2x^2 + 3x - 4}{5x^3 - 1} = \dfrac{1}{4}$.

Use formula (iv), followed by applications of (i), (ii), and (iii). The limit of the quotient is the quotient of the limits:

$$\lim_{x \to 1} \frac{2x^2 + 3x - 4}{5x^3 - 1} = \frac{\lim\limits_{x \to 1}(2x^2 + 3x - 4)}{\lim\limits_{x \to 1}(5x^3 - 1)}.$$

Since

$$\lim_{x \to 1} (2x^2 + 3x - 4) = 2 \lim_{x \to 1} x^2 + 3 \lim_{x \to 1} x - 4 = 2 + 3 - 4 = 1,$$

and

$$\lim_{x \to 1} (5x^3 - 1) = 4,$$

the limit in question is $\frac{1}{4}$.

Example 3 Evaluate $\lim\limits_{\theta \to 1} \dfrac{\theta^3 - 1}{\theta - 1}$.

Note the factorization

$$\theta^3 - 1 = (\theta - 1)(\theta^2 + \theta + 1),$$

so that

$$\frac{\theta^3 - 1}{\theta - 1} = \theta^2 + \theta + 1.$$

Therefore

$$\lim_{\theta \to 1} \frac{\theta^3 - 1}{\theta - 1} = \lim_{\theta \to 1} (\theta^2 + \theta + 1) = 3.$$

The main point here is to recognize that

$$\lim_{\theta \to 1} \frac{\theta - 1}{\theta - 1} = 1,$$

since the ratio is obviously 1 for all $\theta \neq 1$. Algebraic cancellation should precede the limit taking, in order to avoid (or to interpret properly) a 0/0 ratio.

Example 4 Evaluate $\lim_{x \to 0} \frac{\sin^2 kx}{x}$.

Formula (iii) is applied to yield

$$\lim_{x \to 0} \frac{\sin^2 kx}{x} = \left(\lim_{x \to 0} \sin kx\right)\left(\lim_{x \to 0} \frac{\sin kx}{x}\right).$$

Since

$$\lim_{x \to 0} \frac{\sin kx}{x} = k \lim_{x \to 0} \frac{\sin kx}{kx} = k \lim_{z \to 0} \frac{\sin z}{z} = k \cdot 1 = k,$$

and

$$\lim_{x \to 0} \sin kx = 0,$$

the result is

$$\lim_{x \to 0} \frac{\sin^2 kx}{x} = 0 \cdot k = 0. \tag{14}$$

Note the use of the variable $z = kx$ to reduce this problem to one already solved.

Example 5 Evaluate $\lim_{x \to 0} \frac{\sin^2 kx}{x^2}$.

In this case, write

$$\lim_{x \to 0} \frac{\sin^2 kx}{x^2} = k^2\left(\lim_{x \to 0} \frac{\sin kx}{kx}\right)\left(\lim_{x \to 0} \frac{\sin kx}{kx}\right)$$

$$= k^2\left(\lim_{z \to 0} \frac{\sin z}{z}\right)^2;$$

the limit on the right-hand side is 1, so that

$$\lim_{x \to 0} \frac{\sin^2 kx}{x^2} = k^2. \tag{15}$$

Example 6 Evaluate $\lim_{\theta \to 0} \frac{1 - \cos \theta}{\theta}$.

The identity

$$1 - \cos \theta = 2 \sin^2 \tfrac{1}{2}\theta$$

is used to express the limit as

$$\lim_{\theta \to 0} \frac{1 - \cos \theta}{\theta} = \lim_{\theta \to 0} \frac{\sin^2 \frac{1}{2}\theta}{\frac{1}{2}\theta}.$$

But the right-hand limit is zero according to the result of Example 4 (set $x = \frac{1}{2}\theta$), so that

$$\lim_{\theta \to 0} \frac{1 - \cos \theta}{\theta} = 0. \tag{16}$$

Example 7 Examine the behavior of

$$f(x) = \frac{1}{1 + x^2}$$

for very large x.

As x increases in magnitude, $f(x)$ decreases toward zero (Fig. 5.2). In fact, zero is the *limit* of the function as x approaches infinity, a result described by

$$\lim_{x \to \infty} \frac{1}{1 + x^2} = 0.$$

This can be demonstrated easily by introducing a new variable

$$z = \frac{1}{x}, \tag{17}$$

so that *large* x will correspond to *small* z. In this case

$$\frac{1}{1 + x^2} = \frac{z^2}{1 + z^2},$$

and by definition (Fig. 5.3)

$$\lim_{x \to \infty} \frac{1}{1 + x^2} = \lim_{z \to 0} \frac{z^2}{1 + z^2} = 0.$$

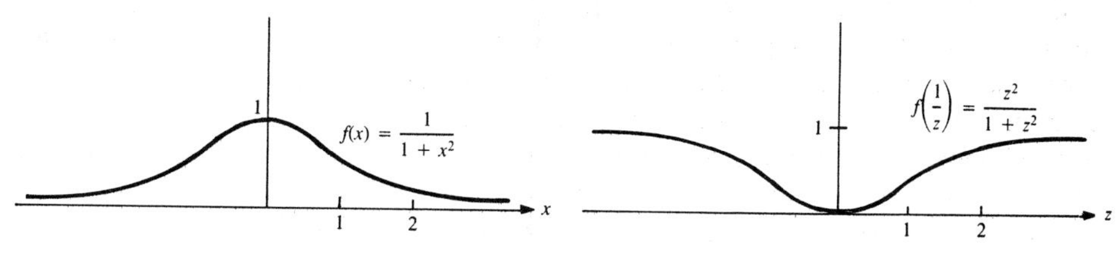

FIGURE 5.2

FIGURE 5.3

Example 8 Show that $$\lim_{x\to\infty} \frac{x^3}{1 + x - x^2 + 2x^3} = \frac{1}{2}.$$

Use the substitution $z = \frac{1}{x}$ to write

$$\lim_{x\to\infty} \frac{x^3}{1 + x - x^2 + 2x^3} = \lim_{z\to 0} \frac{1}{2 - z + z^2 + z^3} = \frac{1}{2}.$$

Example 9 If $f(x) \le h(x) \le g(x)$ and $A = \lim_{x\to a} f(x) = \lim_{x\to a} g(x)$, show that $\lim_{x\to a} h(x) = A$.

Write

$$\begin{aligned} |A - h(x)| &= |A - g(x) + g(x) - h(x)| \\ &\le |A - g(x)| + |g(x) - h(x)| \\ &\le |A - g(x)| + |g(x) - f(x)|. \end{aligned} \tag{18}$$

Since by assumption $\lim_{x\to a} [g(x) - f(x)] = 0$ and $\lim_{x\to a} g(x) = A$, both terms on the right-hand side of (18) can be made less than an arbitrarily small number $\varepsilon/2$ for some δ, $|x - a| < \delta$. Replacement of these terms by $\varepsilon/2$ means that the last inequality can be made still stronger:

$$|A - h(x)| \le \frac{\varepsilon}{2} + \frac{\varepsilon}{2} = \varepsilon,$$

and this completes the proof since the criteria for a limit are met. This technique of bounding the limit between converging extremes was used in Sec. 1.1 to estimate the area of a plane figure and in Sec. 1.3 to evaluate the limit of $\sin x/x$ as x approaches zero.

6

Everyone has an intuitive understanding of what is meant by a continuous curve, and a verbal description would probably use such words as "connected," "without breaks," "no holes," etc. A precise analytical definition of a continuous function will be given now in terms of limits. The motivation for such a mathematical statement is drawn from nature, where very slight changes usually produce correspondingly small effects. By analogy, a continuous function $f(x)$ must be defined as one that experiences a small variation for a slight change in the independent variable x. The continuity of $f(x)$ at $x = x_0$ therefore depends on its local behavior in some small interval about this point (see Fig. 5.4). Specifically, if x is any point near x_0 so that $|x - x_0|$ is very small, then the corresponding values of the function, $f(x_0)$ and $f(x)$, should be almost equal. In other words, continuity requires the difference $|f(x) - f(x_0)|$ to be small when $|x - x_0|$ is also small, which in effect means that

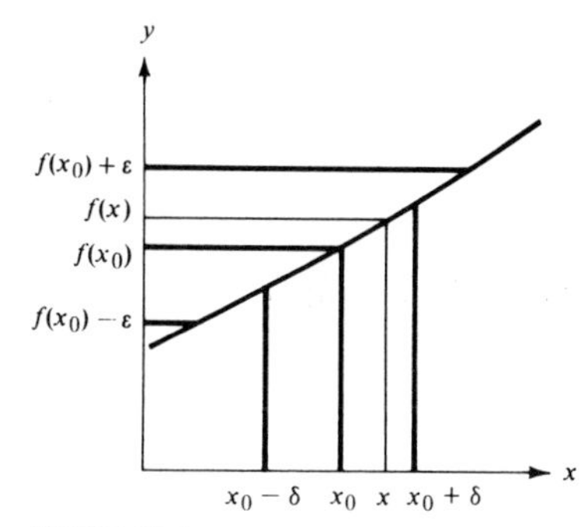

FIGURE 5.4

$$\lim_{x \to x_0} f(x) = f(x_0). \tag{19}$$

Therefore, a function $f(x)$ is continuous at a point x_0 if:

1. It has a limit as x approaches x_0.
2. The limit is the value of the function $f(x_0)$.

The essential feature is that the assigned value of $f(x_0)$ must be consistent with the value obtained from the limit, and this will assure a smooth variation.

7

The equivalent statement of continuity in the ε, δ terminology is as follows: *a function $f(x)$ is continuous at $x = x_0$ if, for any arbitrarily small positive number ε, another positive number δ can be found such that*

$$|f(x) - f(x_0)| < \varepsilon \tag{20}$$

whenever

$$0 < |x - x_0| < \delta. \tag{21}$$

The general procedure to prove that a function is continuous at a point is the same as that used to show the existence of a limit. Start with $|f(x) - f(x_0)|$ and manipulate this quantity to determine the magnitude of δ which guarantees that the functional difference is less than the preassigned, arbitrarily small ε. Most of the examples of limits in this section also demonstrate that the functions involved are continuous.

Example 10 Prove that $f(x) = \sin x$ is a continuous function.

We examine continuity at an arbitrary point x and let the neighboring point be $x + \Delta x$, where the increment $|\Delta x|$ is small. In this notation, the criterion for continuity takes the form

$$|\sin(x + \Delta x) - \sin x| < \varepsilon, \tag{22}$$

for

$$|x + \Delta x - x| = |\Delta x| < \delta. \tag{23}$$

The following familiar facts will be used:

(i) $\sin(x + h) - \sin x = 2 \cos(x + h/2) \sin h/2;$

(ii) $|\cos x| \leq 1;$

(iii) $|\sin h| \leq |h|.$

So armed, we can easily manipulate the functional difference (20) to prove continuity in the sequence of simple steps shown next:

$$|\sin(x + \Delta x) - \sin x| = 2\left|\sin \frac{\Delta x}{2} \cos\left(x + \frac{\Delta x}{2}\right)\right|, \qquad \text{[fact (i)]}$$

$$\leq 2\left|\sin \frac{\Delta x}{2}\right|, \qquad \text{[fact (ii)]}$$

$$< |\Delta x|. \qquad \text{[fact (iii)]}$$

If $|\Delta x| < \delta$ and $\delta < \varepsilon$, it follows that

$$|\sin(x + \Delta x) - \sin x| < \varepsilon, \tag{24}$$

which completes the proof.

Example 11 Show that

$$H(x) = \begin{cases} 1, & x \geq 0; \\ 0, & x < 0; \end{cases}$$

is not a continuous function at $x = 0$ but that $xH(x)$ is.

The step function does not have a unique limit at $x = 0$, and this proves the first part of the statement. However,

$$\lim_{x \to 0} xH(x) = 0,$$

and zero is the value of $xH(x)$ no matter what finite value is assigned arbitrarily to $H(0)$. Since the limit agrees with the value at $x = 0$, the product function $xH(x)$ is continuous.

Except for a finite jump at $x = 0$, the step function is continuous at all other points. A function which is continuous except for a finite number of finite jumps, as shown in Fig. 5.5, said to be piecewise continuous.

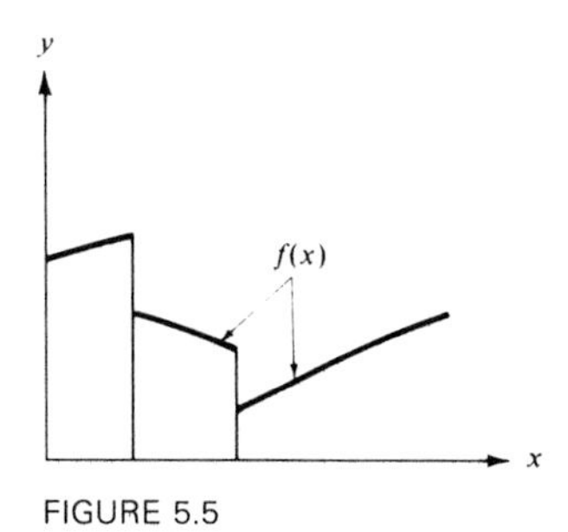

FIGURE 5.5

Exercises 1.5

1. Evaluate $\lim_{x\to 2}(3x+1)$ using ε, δ notation, i.e., given $\varepsilon>0$, find δ such that $|(3x+1)-7|<\varepsilon$ when $0<|x-2|<\delta$. Illustrate with a graph.

2. Generalize the previous example to find $\lim_{x\to a}(mx+b)$ using ε, δ notation.

3. If $\lim_{x\to a} f(x)=A$ and $\lim_{x\to a} g(x)=B$, verify the following results:

(i) $\lim_{x\to a}[f(x)+g(x)]=A+B$ **(ii)** $\lim_{x\to a}[f(x)-g(x)]=A-B$ **(iii)** $\lim_{x\to a}\dfrac{f(x)}{g(x)}=\dfrac{A}{B}$.

4. Evaluate $\lim_{x\to 2}(x^2-2x+1)$ using ε, δ notation. (Given $\varepsilon>0$, find δ such that $|(x^2-2x+1)-1|<\varepsilon$ when $0<|x-2|<\delta$.) Illustrate with a graph and the special case $\varepsilon=0.1$.

5. Generalize the previous example to find $\lim_{x\to x_o}(ax^2+bx+c)$ using ε, δ notation.

6. What is the limit of the function

$$f(t)=\begin{cases} t & 0\le t<1 \\ 2-t & 1<t\le 2\end{cases}$$

as $t\to 1$? Illustrate with a graph. If $f(t)$ represents the trajectory of a particle, show that the corresponding velocity function does not have a well-defined limit as $t\to 1$.

7. Evaluate the following limits:

(i) $\lim_{x\to 0}\dfrac{x(x+3)}{\sin x}$ **(ii)** $\lim_{\theta\to 0}\dfrac{\theta+2\theta^2}{\sin\theta}$ **(iii)** $\lim_{h\to 0}\dfrac{\cos(x+h)-\cos x}{h}$

(iv) $\lim_{t\to\infty}\dfrac{t^3+4t-5}{t^4+2}$ **(v)** $\lim_{s\to a}\dfrac{1/s-1/a}{s-a}$ **(vi)** $\lim_{t\to\infty}\dfrac{\sin t}{t}$

(vii) $\lim_{s\to a}\dfrac{1/s^2-1/a^2}{s-a}$ **(viii)** $\lim_{x\to 0}x\sin\left(\dfrac{1}{x}\right)$ **(ix)** $\lim_{y\to 1}\dfrac{|y^2-1|^{1/2}}{(y^2+1)^2}$

8. If the curve $y=f(x)$ approaches the straight line $y=mx+b$ as either x or y becomes infinite, the line is said to be an *asymptote* of the curve. Show that

$$y=\frac{2x^3+3x^2+5}{x^2+1}$$

has the line $y=2x+3$ as an asymptote. Illustrate with a graph.
[*Hint:* Note that $y=2x+3-10(x+1)/(x^2+1)$.]

9. Graph the following functions $y=f(x)$ and determine all asymptotes.

(i) $(x^2+1)/(x^2-1)$ **(ii)** $(2x+3)/(x-1)$ **(iii)** $(x^4+2x^3+3)/(x^3+2x^2)$

10. Find the coordinates of the point of intersection of the straight lines $2x + y = 1$, and $(1 + c)x + c^2y = 1$ and determine its limiting position as $c \to 1$.
11. Prove that if $f(x) > g(x)$ for all x and $\lim_{x \to a} g(x) = M$, then $\lim_{x \to a} f(x) \geq M$, if this limit exists. Construct an example for which the equality sign holds in this equation.
12. Verify that $f(x) = x^2$ is a continuous function using ε, δ notation.
13. Verify that the following functions are continuous at any point $x > 0$.
 (i) $1/x^2$ **(ii)** $x^{1/2}$ **(iii)** $|x|$
14. Prove that $\cos x$, $\cos kx$ and $\sin kx$ are continuous functions where k is an arbitrary constant.
15. If $f(x)$ is a continuous function in some interval of the x axis containing x_0 and if $f(x_0) > 0$, show that the function is positive in some small interval about x_0. (*Hint:* Assume no such interval exists and deduce a contradiction.)
16. Suppose that, in an interval $a \leq x \leq b$, a continuous function $f(x)$ assumes both positive and negative values. Prove that there is at least one point c in $a < c < b$ for which $f(c) = 0$.
17. Define $f(x) = x$ when x is rational and $f(x) = 1$ when x is irrational. Is there any point x at which $f(x)$ is continuous?
18. If $f(x)$ and $g(x)$ are both continuous functions, show that $f(x) + g(x)$ and $f(x)g(x)$ are continuous functions. Is $f(x)/g(x)$ continuous? Is the composite function $f(g(x))$ continuous? Illustrate with examples.

1.6 *Tangents and derivatives*

1

The problem of calculating a velocity was shown in Sec. 1.4 to be the same as finding the tangent to a curve. By divorcing the notation from a particular physical context, the latter problem can be treated in a more general form.

Let

$$y = f(x), \tag{1}$$

be a general function whose graph is a smooth curve like that shown in Fig. 6.1. If P denotes point (x, y) and Q is $(x + \Delta x, y + \Delta y)$, where

$$y + \Delta y = f(x + \Delta x),$$

then the slope of the line tangent to $f(x)$ at P is determined from the limit, as $\Delta x \to 0$, of the ratio

$$\frac{\Delta y}{\Delta x} = \frac{f(x + \Delta x) - f(x)}{\Delta x}. \tag{2}$$

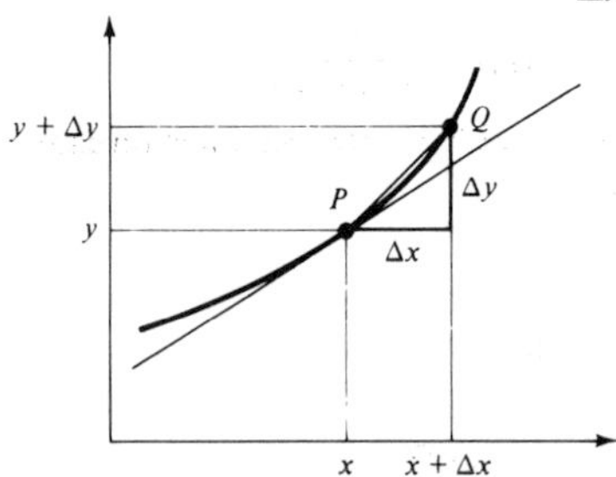

FIGURE 6.1

This limit is by definition the *derivative* of the function and is denoted by

$$\frac{dy}{dx} = \lim_{\Delta x \to 0} \frac{f(x + \Delta x) - f(x)}{\Delta x}. \tag{3}$$

The slope of the tangent line and the derivative of the function are one and the same at the point in question. The calculation of a derivative is called *differentiation.*

Example 1 Find the derivative of the function

$$y = mx + b. \tag{4}$$

Since

$$y + \Delta y = f(x + \Delta x) = mx + m\,\Delta x + b,$$

it follows that

$$\Delta y = f(x + \Delta x) - f(x) = m\,\Delta x,$$

and

$$\frac{\Delta y}{\Delta x} = m. \tag{5}$$

The ratio of increments is a constant and remains so in the limit as $\Delta x \to 0$; therefore,

$$\frac{dy}{dx} = \frac{d}{dx}(mx + b) = m. \tag{6}$$

Since (4) represents a straight line, the derivative is the constant slope of this line. This is clearly consistent with the interpretation of the derivative as the slope of the tangent line. Two elementary results inherent in (6) are that the derivative of a constant is zero,

$$\frac{d}{dx}K = 0,$$

and the derivative of a constant multiplied by x is the constant,

$$\frac{d}{dx}Kx = K\frac{d}{dx}x = K.$$

Example 2 Differentiate the function

$$y = x^2. \tag{7}$$

The result of forming the ratio (2) in this case is

$$\frac{\Delta y}{\Delta x} = \frac{(x + \Delta x)^2 - x^2}{\Delta x} = 2x + \Delta x. \tag{8}$$

The derivative is the limit of this expression as $\Delta x \to 0$, and the result is

$$\frac{dy}{dx} = \frac{d}{dx} x^2 = 2x. \tag{9}$$

This formula also supplies the slope of the tangent line to the parabola $y = x^2$ at any point (x, y).

Example 3 Determine the *equation* of the straight line tangent to the parabola $y = x^2$ at the point $x = 2$, $y = 4$.

First the straight line in question must pass through the point (2, 4), which implies that the equation sought is of the general form

$$y - 4 = m(x - 2).$$

Of all the possible lines with this property shown in Fig. 6.2, the tangent line is the one whose slope m equals the derivative of x^2 evaluated at the point of contact, $x = 2$. According to (9), this means that $m = 4$, and the equation of the tangent line is consequently

$$y - 4 = 4(x - 2). \tag{10}$$

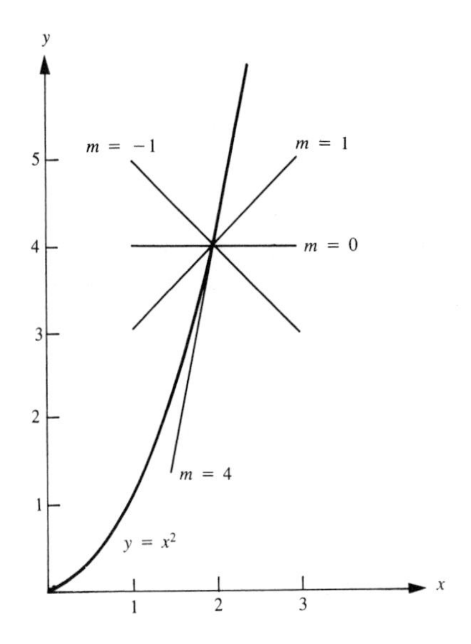

FIGURE 6.2

2 **Example 4** Show that

$$\frac{d}{dx} \sin kx = k \cos kx, \tag{11}$$

where k is a constant.

By definition
$$\frac{d \sin kx}{dx} = \lim_{\Delta x \to 0} \frac{\sin k(x + \Delta x) - \sin kx}{\Delta x}. \tag{12}$$

The limit calculation is greatly facilitated by writing

$$\sin k(x + \Delta x) = \sin kx \cos(k\,\Delta x) + \cos kx \sin(k\,\Delta x); \tag{13}$$

the result of substituting this in (12) is

$$\frac{d}{dx} \sin kx = \sin kx \lim_{\Delta x \to 0} \frac{\cos(k\,\Delta x) - 1}{\Delta x} + \cos kx \lim_{\Delta x \to 0} \frac{\sin(k\,\Delta x)}{\Delta x}. \tag{14}$$

But it has already been established that

$$\lim_{h \to 0} \frac{\cos h - 1}{h} = 0 \qquad \text{and} \qquad \lim_{h \to 0} \frac{\sin h}{h} = 1$$

and (14) therefore reduces to

$$\frac{d}{dx} \sin kx = k \cos kx.$$

In particular, the derivative of $\sin 2x$ is $2 \cos 2x$. The presence of the parameter k makes this a very general result.

Example 5 Show that

$$\frac{d}{dx}[f(x) + g(x)] = \frac{d}{dx} f(x) + \frac{d}{dx} g(x).$$

This identity follows from the basic definition:

$$\begin{aligned}
\frac{d}{dx}[f(x) + g(x)] &= \lim_{\Delta x \to 0} \frac{f(x + \Delta x) + g(x + \Delta x) - f(x) - g(x)}{\Delta x} \\
&= \lim_{\Delta x \to 0} \left[\frac{f(x + \Delta x) - f(x)}{\Delta x} + \frac{g(x + \Delta x) - g(x)}{\Delta x} \right] \\
&= \lim_{\Delta x \to 0} \frac{f(x + \Delta x) - f(x)}{\Delta x} + \lim_{\Delta x \to 0} \frac{g(x + \Delta x) - g(x)}{\Delta x} \\
&= \frac{d}{dx} f(x) + \frac{d}{dx} g(x).
\end{aligned}$$

Applying this identity to an example, the derivative of $x^2 + \sin x$ is $2x + \cos x$.

Exercises 1.6

1. Using the limit definition, evaluate the derivatives of the following functions:

(i) $2x+1$ **(ii)** $3x^2-4$ **(iii)** x^2-x+7

(iv) x^5 **(v)** $2x^2+3x+4$ **(vi)** $\frac{1}{6}x^2+5$

(vii) $ax+b$ **(viii)** ax^2+bx+c **(ix)** ax^3+bx^2+cx+d

2. Graph the function $y=\sqrt{x}$ defined for $x \geq 0$. Determine the intersection points of this curve with the line $y=mx+b$. For what values of m and b is this line a tangent line to the curve?

3. From first principles, find the derivative of $y=\sqrt{x}$ defined for $x \geq 0$. What happens at $x=0$?

4. Graph the function $y=|x|$ defined for all x. From the graph, evaluate dy/dx. What happens at $x=0$?

5. Graph the function $y=x^3+9x^2+24x+1$. Draw the tangent lines at $x=-1, 0, 1$.

6. Verify the following derivatives:

(i) $\dfrac{d}{dx}\dfrac{1}{x}=-\dfrac{1}{x^2}$ **(ii)** $\dfrac{d}{dx}\dfrac{1}{x^2}=-\dfrac{2}{x^3}$ **(iii)** $\dfrac{d}{dx}\dfrac{1}{x^{1/2}}=-\dfrac{1}{2x^{3/2}}$

7. Verify the following derivatives:

(i) $\dfrac{d}{dx}\cos kx=-k\sin kx$ **(ii)** $\dfrac{d}{dx}\sin(x+a)=\cos(x+a)$ **(iii)** $\dfrac{d}{dx}\cos(x+a)=-\sin(x+a)$.

8. Evaluate the derivatives of $\sin kx$ and $\cos kx$ using the following identities:

(i) $\sin\theta-\sin\phi=2\sin\dfrac{\theta-\phi}{2}\cos\dfrac{\theta+\phi}{2}$ **(ii)** $\cos\theta-\cos\phi=-2\sin\dfrac{\theta-\phi}{2}\sin\dfrac{\theta+\phi}{2}$

9. Using the identities $\cos 2\theta=2\cos^2\theta-1=1-2\sin^2\theta$, verify the following:

(i) $\dfrac{d}{d\theta}\cos^2\theta=-2\sin\theta\cos\theta$ **(ii)** $\dfrac{d}{d\theta}\sin^2\theta=2\sin\theta\cos\theta$.

10. Using the limit definition, find the derivative of $y=\tan x=(\sin x)/(\cos x)$. Determine the tangent lines at $x=0, \pm\pi/3, \pm\pi/4$, and $\pm\pi/6$ and illustrate with a graph.

11. Find the constants a, b, c so that the parabola $y=ax^2+bx+c$ passes through the points (1,0) and (2,0) and its tangent at (2,0) has slope 1. Graph the curve and the tangent lines at these two points.

12. Evaluate the derivatives of the following functions:

(i) $(x+1)^2$ **(ii)** $(x-1)^3$ **(iii)** $(x+1)^2+(x-1)^2$

(iv) $\left(x+\dfrac{1}{x}\right)^2$ **(v)** $(x^{1/2}+x^{-1/2})^2$ **(vi)** $(mx+b)^2$

13. For what values of x do the functions $y=2x^3+15x-1$ and $y=-3/x$ have the same derivative? Illustrate with graphs. Determine the tangent lines for these values of x.

14. The derivative of a function $y=f(x)$ is a new function written $dy/dx=f'(x)$. From the definition of a derivative, verify that the derivative of $y=f(x+a)$ is $dy/dx=f'(x+a)$. (For example, $f(x)=x^2$ has derivative $f'(x)=2x$ and $f(x+a)=(x+a)^2$ has derivative $f'(x+a)=2(x+a)$.)

15. Evaluate the derivatives of the following functions:

(i) $(x+a)^2$ **(ii)** $(x+a)^{1/2}$ **(iii)** $(x+a)^{-1}$

(iv) $(x+a)^{-2}$ **(v)** $\sin(x+\pi/2)$ **(vi)** $\cos(x+\pi/2)$

16. From the definition of a derivative, verify that the derivative of $y=f(ax)$ is $dy/dx=af'(ax)$. (For example, $f(x)=x^2$ has derivative $f'(x)=2x$ and $f(ax)=(ax)^2$ has derivative $af'(ax)=a\,(2ax)=2a^2x$.)

17. Evaluate the derivatives of the following functions:

(i) $(ax)^3$ **(ii)** $(ax)^{-1}$ **(iii)** $(ax)^{1/2}$

(iv) $\sin 3x$ **(v)** $\cos 5x$ **(vi)** $\tan (ax)$

18. Show that $\dfrac{d}{dx}[f(x)+g(x)+h(x)] = \dfrac{d}{dx}f(x) + \dfrac{d}{dx}g(x) + \dfrac{d}{dx}h(x)$. What is the general result?

1.7 Sequences

1

The calculation of velocity requires the evaluation of the limit of the ratio $\Delta s/\Delta t$ as Δt approaches zero. To make this limiting procedure definite, the ratio could be evaluated successively for $\Delta t = 1, 1/2, 1/3, \ldots, 1/n, \ldots$. The calculated ratios then form a sequence of numbers that approach the instantaneous velocity.

An *infinite sequence* is an ordered succession of numbers

$$a_1, a_2, \ldots, a_n, \ldots \tag{1}$$

where a_n is said to be the *general element* or *nth term* of the sequence. The notation $\{a_n\}$ is used to describe the sequence. For example, if $a_n = 1/n$, the corresponding sequence is $\{a_n\} = \{1/n\} = 1, 1/2, 1/3, 1/4, \ldots$. With $a_n = n/(1+n)$, the corresponding sequence is $\{n/(1+n)\} = 1/2, 2/3, 3/4, 4/5, \ldots$.

Other examples of sequences are the following:

(i) $\{n\} = 1, 2, 3, \ldots, n, \ldots$ **(ii)** $\{(-1)^n\} = -1, 1, -1, \ldots, (-1)^n, \ldots$

(iii) $\{1-(0.1)^n\} = 0.9, 0.99, 0.999, \ldots$ **(iv)** $\{1 + 1/n^2\} = 2, 5/4, 10/9, 17/16, \ldots$ (2)

(v) $\{(1 + 1/n)^n\} = 2, 9/4, 64/27, \ldots$ **(vi)** $\left\{\dfrac{n^2(n+1)}{n^3+4}\right\} = 2/5, 12/12, 36/31, 80/68, \ldots$

2

The elements of any sequence $\{a_n\}$ can be represented as points on the real line. In Fig. 7.1, the elements of the sequence $\{1/n\}$ are shown and it is apparent that these points accumulate at the origin as n increases. This is equivalent to saying that the limit of this sequence is zero.

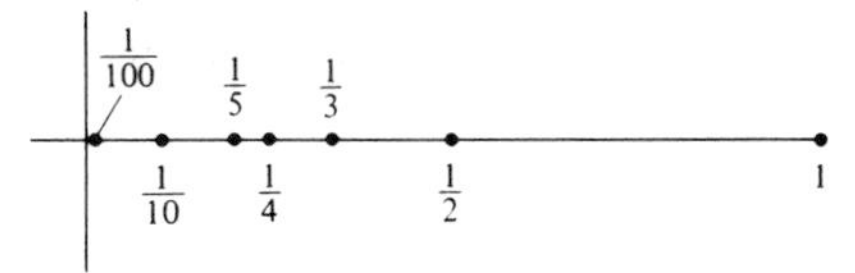

FIGURE 7.1

This observation can be translated into a precise definition of the limit of a sequence. The sequence $\{a_n\}$ has the *limit a* if, given any $\epsilon>0$, there is an integer N such that the inequality

$$|a - a_n| < \epsilon \tag{3}$$

is true for all $n>N$. An equivalent statement is

$$\lim_{n\to\infty} a_n = a.$$

The sequence is also said to *converge* to a. A sequence that does not converge is said to *diverge* or to be *divergent*.

Example 1 Show that 1 is the limit of the sequence $\{n/(1+n)\}$.

This will be demonstrated first using the formal definition of a limit. The particular form of (3) can be written as

$$|a - a_n| = \left|1 - \frac{n}{1+n}\right| = \left|\frac{1}{1+n}\right| < \frac{1}{n}$$

from which it follows that, for any $\epsilon>0$,

$$\left|1 - \frac{n}{1+n}\right| < \epsilon \qquad \text{whenever } n > \frac{1}{\epsilon}.$$

For example, if $\epsilon = 10^{-3}$, choose $N = 1000 = 1/\epsilon$. Then, whenever $n>N=1000$,

$$\left|1 - \frac{n}{1+n}\right| < \frac{1}{1000}.$$

In general, if N is any integer greater than $1/\epsilon$, the inequality holds for all $n>N$. Therefore, 1 is the limit of the sequence.

This conclusion can also be established by the following argument using properties of limits discussed in Sec. 1.5.

$$\lim_{n\to\infty} \frac{n}{1+n} = \lim_{n\to\infty} \frac{1}{1+1/n} = \frac{1}{\lim\limits_{n\to\infty}(1+1/n)} = \frac{1}{1+\lim\limits_{n\to\infty}(1/n)} = \frac{1}{1+0} = 1.$$

Example 2 Examine the sequences in (2) for convergence or divergence.

(i) The sequence $\{a_n\} = \{n\}$ is clearly divergent since its successive terms increase without any bound. The sequence is said to be unbounded or, symbolically,

$$\lim_{n\to\infty} n = \infty\,.$$

(ii) The numbers $(-1)^n$ are either $+1$ or -1 and no unique limit exists. This sequence diverges even though every term is bounded above and below, i.e., $-1 \le a_n \le 1$ for all n.

(iii) The sequence $\{1-(0.1)^n\}$ is convergent to the limit 1.

$$|1-(1-(0.1)^n| = |(0.1)^n| = (0.1)^n.$$

By taking n large enough, the difference $(0.1)^n$ can be made smaller than any pre-assigned positive number ϵ.

(iv) The sequence $\{1 + {}^1\!/_{n^2}\}$ is convergent to the limit 1 since

$$\left|1 - (1 + {}^1\!/_{n^2})\right| = \left|-\frac{1}{n^2}\right| = \frac{1}{n^2}.$$

Given any $\epsilon > 0$, then ${}^1\!/_{n^2} < \epsilon$ whenever $n > 1/\epsilon^{1/2}$.

(v) The sequence $\{(1 + {}^1\!/_n)^n\}$ has as its limit the number $e = 2.71828\ldots$. This calculation is more difficult and will be postponed until more powerful methods are available. For the moment, the existence of this limit can be made plausible by using a calculator to find values of $(1 + {}^1\!/_n)^n$ for large integers n.

(vi) Divide numerator and denominator by n^3 to write the general term of the sequence as

$$\frac{n^2(n+1)}{n^3+4} = \frac{1(1 + {}^1\!/_n)}{1 + {}^4\!/_{n^3}}$$

and then take the limit as follows:

$$\lim_{n\to\infty} \frac{1 + {}^1\!/_n}{1 + {}^4\!/_{n^3}} = \frac{\lim\limits_{n\to\infty} (1 + {}^1\!/_n)}{\lim\limits_{n\to\infty} (1 + {}^4\!/_{n^3})} = \frac{1}{1} = 1.$$

Example 3 Using a calculator, discuss the convergence of the sequence $\{a_n\}$ defined by $a_1 = 1$ and $a_n = \sin a_{n-1}$ for $n = 2, 3, 4, \ldots$.

The equation $a_n = \sin a_{n-1}$ is called a recurrence relation. Since $a_1 = 1$, $a_2 = \sin a_1 = \sin 1 = 0.84147\ldots$, $a_3 = \sin a_2 = 0.745624\ldots$, $a_4 = \sin a_3 = 0.678431\ldots$, etc. By pressing the sin key repeatedly, further terms of the sequence can be determined. It is evident that successive terms decrease in value since $a_n = \sin a_{n-1} < a_{n-1}$ but the rate of decrease slows down and it is not clear what limit this sequence approaches. A simple geometrical argument shows that the limit is zero. If a limit exists, it must satisfy $a = \sin a$. The only solution of this equation is $a = 0$.

Exercises 1.7

1. Using ϵ, N notation, verify the following limits:

(i) $\lim\limits_{n\to\infty}\left(3 - \frac{1}{n}\right) = 3$ **(ii)** $\lim\limits_{n\to\infty} \frac{n}{n+2} = 1$ **(iii)** $\lim\limits_{n\to\infty} \frac{1}{\sqrt{n^2+1}} = 0$

(*Hint:* In **(iii)**, use $n^2 < n^2 + 1$.)

2. Choosing $\epsilon = 10^{-3}$, find a and an integer $N \geq 1$ such that $|a - a_n| < 10^{-3}$ for $n > N$ in each of the following sequences.

(i) $\left\{\frac{1}{5n}\right\}$ **(ii)** $\left\{\frac{n}{2n+1}\right\}$ **(iii)** $\left\{\frac{1}{2^n}\right\}$

3. Prove that each of the sequences in the previous problem is convergent.

4. Which of the following sequences are convergent and which are divergent?

(i) $\{(-1)^{2n}\}$ (ii) $\left\{\left(-\frac{1}{2}\right)^{2n}\right\}$ (iii) $\left\{\frac{1}{n}\sin\frac{n\pi}{2}\right\}$

(iv) $\left\{\cos\frac{n\pi}{2}\right\}$ (v) $\left\{\sin\frac{2n\pi}{3}\right\}$ (vi) $\{1+(-1)^{n+1}\}$

5. Determine the convergent sequences and their limits among the following.

(i) $\left\{\frac{n+2}{2n+1}\right\}$ (ii) $\left\{\frac{n^2-1}{3n^2+n}\right\}$ (iii) $\left\{n\sin\frac{1}{n}\right\}$

(iv) $\left\{\sqrt{n+1}-\sqrt{n}\right\}$ (v) $\left\{\frac{1}{\sqrt{n}}-\frac{1}{\sqrt{n+1}}\right\}$ (vi) $\left\{\frac{2^n}{n}\right\}$

(vii) $\left\{\frac{n}{2^n}\right\}$ (viii) $\left\{\frac{(-n+2)^3}{3n^3}\right\}$ (ix) $\{n^2(1+n^2)^{1/2}-n^3\}$

6. For the following sequences $\{a_n\}$, find the limits as $n\to\infty$ or show divergence.

(i) $a_n=\frac{n^4+3n^2}{2n^2+1}$ (ii) $a_n=\frac{(n+1/n)^4}{(2n^2-1/n)^2}$ (iii) $a_n=\frac{n-\sqrt{n}}{n+\sqrt{n}}$

(iv) $a_n=\frac{(-1)^n}{n+1}$ (v) $a_n=\frac{3n^2}{n^2+(-1)^n}$ (vi) $a_n=\left(1+\frac{1}{n}\right)^3-1$

(vii) $a_n=(-1)^n\cos n\pi$ (viii) $a_n=n\sin n\pi$ (ix) $a_n=\sin(2n+1)\pi$

7. If the sequence $\{a_n\}$ converges to the limit a, show that the sequence $\{a_n^2\}$ converges to the limit a^2. (*Hint:* $a^2-a_n^2=(a-a_n)(a+a_n)$.)

8. If the sequences $\{a_n\}$ and $\{b_n\}$ converge to the limits a and b respectively, show that $\{a_n+b_n\}$ converges to $a+b$ and $\{a_nb_n\}$ converges to ab.
(*Hint*: $ab-a_nb_n=a(b-b_n)+b_n(a-a_n)$.)

9. Show that, if $\lim_{n\to\infty} a_n=a\neq 0$, then $\lim_{n\to\infty}\frac{a_{n+1}}{a_n}=1$.

10. Show that, if $\lim_{n\to\infty} a_n=a$, then $\lim_{n\to\infty}\frac{a_1+a_2+\ldots+a_n}{n}=a$.

11. Consider the sequence $\{a_n\}$ defined by the recurrence relation $a_{n+1}=\sqrt{1+a_n}$, $a_1=1$. Use a calulator to find the first ten terms of the sequence. Show that $\lim_{n\to\infty} a_n=\frac{1+\sqrt{5}}{2}$.

12. Determine the first ten terms of the sequence $\{a_n\}$ defined by the recurrence relation $a_{n+1}=\cos a_n$, $a_1=1$. Evaluate the limit a as $n\to\infty$ and interpret this limit geometrically.

13. Determine the first ten terms of the sequence $\{a_n\}$ defined by the recurrence relation $a_{n+1}=(3a_n)^{1/2}$, $a_1=1$. Show that the sequence converges to $a=3$.

14. Use a calculator to find the terms $a_n=(1+1/n)^n$ and $b_n=(1-1/n)^n$ for $n=1$, 10, 100, 1000 and 10,000. For these values of n, find a_nb_n.

1.8 Series

1 An interesting and very common method of forming one sequence $\{s_n\}$ from another $\{a_n\}$ is by the consecutive addition of terms:

$$s_1 = a_1,$$
$$s_2 = a_1 + a_2 = s_1 + a_2,$$
$$s_3 = a_1 + a_2 + a_3 = s_2 + a_3,$$
$$s_n = a_1 + a_2 + a_3 + \cdots + a_n = \sum_{i=1}^{n} a_i = s_{n-1} + a_n. \tag{1}$$

The sum s_n, of the first n terms, is called a *partial sum* or *finite series*. The convergence of the sequence of partial sums $\{s_n\}$ to a definite limit gives meaning to the concept of an *infinite series*

$$s = \sum_{i=1}^{\infty} a_i.$$

An infinite series is said to *converge* if the sequence of its partial sums has a limit. *The sum s of the series is the limit of the sequence* $\{s_n\}$:

$$s = \lim_{n\to\infty} s_n = \sum_{i=1}^{\infty} a_i. \tag{2}$$

Example 1 Examine the sequence of partial sums formed from $\{1/2^{n-1}\}$.
The first few terms are

$$s_1 = 1,$$
$$s_2 = 1 + \frac{1}{2},$$
$$s_3 = 1 + \frac{1}{2} + \frac{1}{2^2}.$$

The general term of this finite geometric series,

$$s_n = 1 + \frac{1}{2} + \frac{1}{2^2} + \cdots + \frac{1}{2^{n-1}},$$

is actually summable to

$$s_n = 2\left(1 - \frac{1}{2^n}\right). \tag{3}$$

This follows from the algebraic identity (obtained by long division)

$$\frac{1}{1-r} = 1 + r + r^2 + \cdots + r^n + \frac{r^{n+1}}{1-r},$$

by setting $r = \frac{1}{2}$. Since the sequence $\{s_n\}$ has a limit,

$$s = \lim_{n\to\infty} s_n = 2 - \lim_{n\to\infty} \frac{1}{2^{n-1}} = 2,$$

the infinite geometric series converges:

$$1 + \frac{1}{2} + \frac{1}{2^2} + \cdots = \sum_{i=0}^{\infty} \frac{1}{2^i} = 2. \tag{4}$$

Of course, it is not usually possible to write so simple a formula for the partial sum s_n, and the proof of convergence will in general require more sophisticated tests. However, these advanced methods, which form the content of Chap. 4, all derive from the basic concept of convergence set forth here.

Example 2 Examine the sequence of partial sums formed from $\{a_n\} = \{n\}$.

The first few terms are $s_1 = 1$, $s_2 = 1 + 2 = 3$, $s_3 = 1 + 2 + 3 = 6$. The general term, which is the sum of the first n integers,

$$s_n = 1 + 2 + 3 + \cdots + n, \tag{5}$$

can be summed concisely by an ingenious maneuver. If the formula is reversed,

$$s_n = n + (n - 1) + \cdots + 1,$$

and then added to the first representation, (5), the result is remarkably simple in appearance:

$$2s_n = \underbrace{(n + 1) + (n + 1) + \cdots + (n + 1)}_{n \text{ terms}} = n(n + 1).$$

Thus

$$s_n = \frac{n(n + 1)}{2}. \tag{6}$$

Since $\lim_{n \to \infty} s_n = \infty$, this proves the obvious: $\sum_{i=1}^{\infty} i$ does not converge to a finite value. A series that does not converge is said to *diverge* or to be *divergent*.

Although this technique is a complete success here, the chances of solving other problems of this type in this way are extremely remote. It is not a generally applicable method since it is not generally possible to write concise formulas for the partial sums. Methods to prove convergence of infinite series are discussed in Chap. 4.

---

2 Even the simplest calculation of an area may involve a difficult summation, e.g., the sum of the squares of the first n integers:

$$s_n = 1 + 4 + 9 + \cdots + n^2 = \sum_{i=1}^{n} i^2. \tag{7}$$

A concise formula for this is

$$s_n = \frac{n(n + 1)(2n + 1)}{6}; \tag{8}$$

a constructive method of obtaining this result is sketched in the exercises. Here we digress briefly in order to prove (8) by the sophisticated but simple method of *mathematical induction.*

Suppose that after a good deal of experimentation, i.e., the computation and inspection of the partial sums given by (7),

$$s_1 = 1, \qquad s_2 = 5, \qquad s_3 = 14, \qquad s_4 = 30, \qquad s_5 = 55, \qquad \ldots,$$

we are able to make the remarkable conjecture expressed in equation (8). This guess can be checked easily enough if n is a relatively small integer, and indeed it seems true:

$$s_1 = \frac{1\cdot(2)\cdot(3)}{6} = 1, \qquad s_2 = \frac{2\cdot(3)\cdot(5)}{6} = 5, \qquad \ldots.$$

But it is possible to verify the formula by direct calculation only in a relatively small number of cases, and we can never be certain in this way that a failure does not occur for some untested value of n. How can a general confirmation be made? This question can be settled conclusively if the truth of the formula for s_{n+1} follows from the established rule for s_n. This would provide a logical step-by-step procedure of proving the proposition in each and every case from the results and conclusions already established. Thus the truth of the proposition for $n = 2$ would be a consequence of the statement for $n = 1$, the formula which is certified valid by direct calculation; the formula for s_3 would follow from that for s_2, just as $s_{1,000}$ would be a consequence of s_{999}. No case could escape from this methodical advance through the integers, and the proposition would be proved correct. (Climbing a ladder is a good analogy because one rung is used to step to another.) In the case at hand, we attempt to prove that the formula for the sum of squares of the first $n + 1$ integers,

$$s_{n+1} = \frac{(n+1)(n+2)(2n+3)}{6}, \tag{9}$$

follows from the assumption that (8) itself is true. The proof will then be complete since s_1 has already been verified. The reasoning may be sophisticated, but the execution is almost trivial since (9) does follow almost directly from the definition:

$$s_{n+1} = 1 + 2^2 + \cdots + n^2 + (n+1)^2 = s_n + (n+1)^2. \tag{10}$$

If (8), which is assumed valid, is used to replace s_n here, the result is

$$s_{n+1} = \frac{n(n+1)(2n+1)}{6} + (n+1)^2, \tag{11}$$

and the two terms of this expression can be combined and simplified to yield the desired conclusion, Eq. (9).

The general features of a proof by mathematical induction are that (i) the proposition to be proved, call it P_n, is verified directly in one case, P_1; (ii) the proposition

P_n is assumed true and the validity of P_{n+1} is established from this. It follows that P_n is true for all n.

Mathematical induction is certainly a brilliant and widely applicable method, but it is of no help whatsoever in discovering the formula to be proved. That is where most of the real creative work lies.

Example 3 Use mathematical induction to prove that

$$s_n = 1 + 2 + \cdots + n = \frac{n(n+1)}{2}. \tag{12}$$

First, the formula is checked and found to be true for $n = 1$:

$$s_1 = 1 = \frac{1 \cdot 2}{2}.$$

Second, if the formula is assumed true for s_n, then it also holds for the next case, s_{n+1}, since

$$\begin{aligned} s_{n+1} &= 1 + 2 + \cdots + n + (n+1) = s_n + (n+1) \\ &= \frac{n(n+1)}{2} + (n+1) \\ &= \frac{(n+1)(n+2)}{2}. \end{aligned} \tag{13}$$

Equation (13) is in the form required because it is obtained from (12) by replacing n everywhere with $n + 1$. This completes the proof.

Exercises 1.8

1. Write out the first few terms of the sequence $\{a_n\} = \{a + (n-1)d\}$ where a and d are given constants. By two methods, verify that the sequence of partial sums is
$\{s_n\} = \{na + \frac{n(n-1)}{2} d\}$.

2. Prove that the sequence of partial sums formed from the sequence $\{a_n\} = \{r^{n-1}\}$ is given by $s_n = (1 - r^n)/(1 - r)$. If $0 \le |r| < 1$, verify that the sequence of partial sums converges to $s = (1 - r)^{-1}$.

3. Write out the first few terms of the sequence $\{a_n\} = \{nr^{n-1}\}$ where r is a given constant. By two methods, verify that the sum of the first n terms is
$s_n = [1 - (n+1)r^n + nr^{n+1}]/(1 - r)^2$. If $0 \le |r| < 1$, verify that the sequence of partial sums converges to $s = (1 - r)^{-2}$.

4. Evaluate the following sums:

(i) $1+4+7+\ldots+31$ **(ii)** $-2+3+8+\ldots+98$

(iii) $1+2+2^2+\ldots+2^9$ **(iv)** $1.01010101\ldots$

(v) $1+1/3+1/3^2+\ldots+1/3^{12}$ **(vi)** $1+3/4+(3/4)^2+\ldots$

(vii) $1\cdot1+2\cdot3+3\cdot3^2+\ldots+10\cdot3^9$ **(viii)** $1\cdot1+2(1/2)+3(1/2)^2+\ldots$

5. Prove that the product $(1+x)^n$ can be written as a polynomial of degree n in the variable x for every positive integer n.

6. By induction, prove that $(1+x)^n \geq 1+nx$ where $x>-1$ and n is any positive integer greater than 1 with equality holding only for $x=0$.

7. Evaluate the partial sums s_n defined by the sequence $\{a_n\}=\{1/n(n+1)\}$. Verify the result by mathematical induction. (*Hint:* $\dfrac{1}{n(n+1)}=\dfrac{1}{n}-\dfrac{1}{n+1}$.)

8. Evaluate the partial sums s_n defined by the sequence $\{a_n\}=\{n(n+1)\}$. Verify by mathematical induction. (*Hint*: $n(n+1)=n^2+n$.)

9. The sum of the first n integers is a quadratic function of n with quadratic term $n^2/2$ and constant term zero. The sum of the squares is a third degree polynomial with cubic term $n^3/3$ and constant zero. Make a reasonable conjecture for the sum of the cubes of the first n integers. Use this conjecture to obtain an exact formula and verify by induction.

10. Show that $(x+h)^n-x^n$ is divisible by h for every positive integer n. Deduce that $(x+h)^n-(x-h)^n$ is divisible by $2h$ and b^n-a^n is divisible by $b-a$.

11. If $f(x)$ is any polynomial, show that $f(x+h)-f(x)$ is divisible by h.

12. Show that $(x+h)^n-2x^n+(x-h)^n$ is divisible by h^2 for any positive integer n. What is the general result for polynomials?

13. By induction, show that $\cos nx$ can be written as a polynomial of degree n in the variable $\cos x$ for every positive integer n. Determine the polynomials for $\cos 2x$, $\cos 3x$ and $\cos 4x$. (*Hint:* $\cos(n+1)x+\cos(n-1)x=2\cos x\cos nx$.)

14. Extend the result of the previous problem to prove the following:

(i) The polynomial for $\cos 2nx$ contains even powers of $\cos x$ only.

(ii) The polynomial for $\cos(2n+1)x$ contains odd powers of $\cos x$ only.

(iii) The function $\sin(2n+1)x$ can be written as a polynomial of degree $2n+1$ in the variable $\sin x$ containing only odd powers of $\sin x$.
{*Hint:* Expand $\cos[(2n+1)(x+\pi/2)]$ in powers of $\cos(x+\pi/2)=-\sin x$.}

15. By induction, verify the binomial expansion:

$$(1+x)^n=1+nx+\frac{n(n-1)}{2!}x^2+\frac{n(n-1)(n-2)}{3!}x^3+\ldots+x^n$$

16. Write out the binomial expansions of the following products:

(i) $(1+x)^5$ **(ii)** $(a+b)^4$ **(iii)** $(1-x)^3$

(iv) $(a^2+x^2)^3$ **(v)** $(2+3x)^4$ **(vi)** $(1+ax)^6$

17. Use the binomial expansion to show that

$$2<\left(1+\frac{1}{n}\right)^n<1+1+\frac{1}{2!}+\frac{1}{3!}+\ldots+\frac{1}{n!}.$$

Since $1/n!\leq 1/2^{n-1}$ for $n=2,3,4,\ldots$, show next that

$$\left(1+\frac{1}{n}\right)^n<1+1+\frac{1}{2}+\frac{1}{2^2}+\ldots+\frac{1}{2^{n-1}}<3.$$

18. Two opposing armies move toward each other at the relative rate of 1.5 km/h. When they are 20 km apart, a messenger whose running speed is 5 km/h is sent from one to the other. He continues to run back and forth at this speed with no rest until the armies meet. By two methods, determine how far he runs.

19. The elastic properties of a rubber ball are such that it returns to a height rh $(0<r<1)$ if it is dropped from a height h. Define a sequence $\{a_n\}$ where a_n is the distance travelled on the nth bounce (fall and return). Determine the total distance s_n travelled on the first n bounces. What is the total distance travelled before the ball comes to rest?

20. Assuming that the rebound is instantaneous in the previous problem, determine the length of time before the ball comes to rest. Examine the special case $r=0.8$, $h=5$ m. [*Hint:* The time to fall from height h is $(2h/g)^{1/2}$.]

1.9 Area

This chapter began with the problem of calculating a specific area. The basic method was introduced, and very loose bounds for the exact answer were derived, but the techniques did not then exist to carry the calculation to a conclusion. Sufficient information has been accumulated for this purpose, at least for simple cases, and two such problems are now solved in complete detail.

The first area A to be determined by a systematic limit procedure is the shaded region shown in Fig. 9.1(a) bounded by the line $y=x$, the vertical line $x=X$, and the x axis. (Of course, this region is a triangle whose area $X^2/2$ can be found by elementary methods.) An estimate for A is obtained by calculating the total area $\underline{A}_n$ of the shaded rectangles below the line $y=x$ in Fig. 9.1(b). From the diagram, it is clear that the estimate is a *lower bound* for A, i.e. an approximation that is less than or equal to the accurate value.

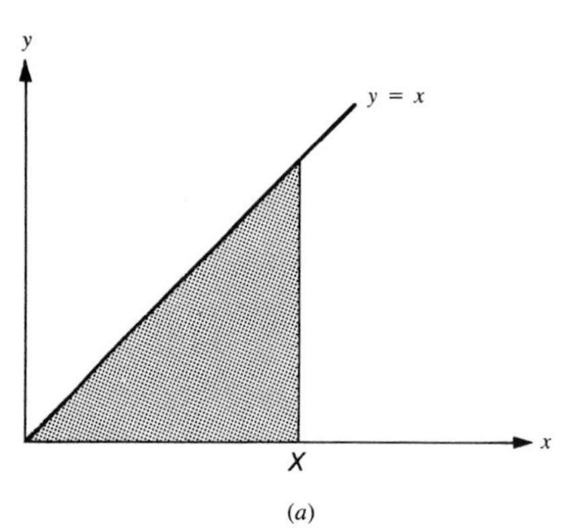

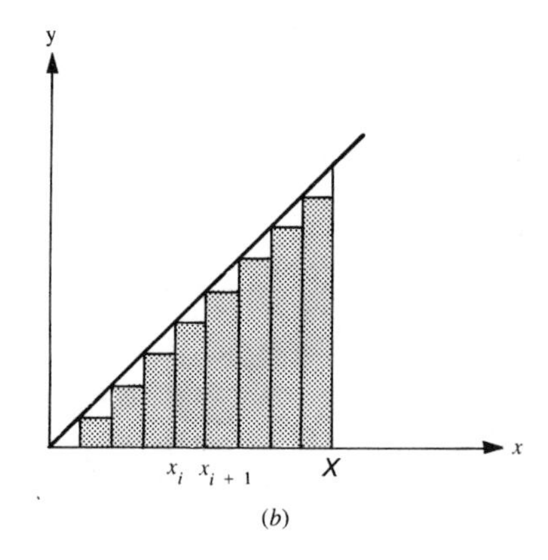

FIGURE 9.1

The base length X is subdivided into n equal intervals with points of subdivision

$$x_0=0,\ x_1=\frac{X}{n},\ \ldots,\ x_i=i\,\frac{X}{n},\ \ldots,\ x_n=n\frac{X}{n}=X \tag{1}$$

The interval size or increment $\Delta x = X/n$ is the width of each rectangle. The particular rectangle with base from x_i to x_{i+1} has height x_i and area $x_i \Delta x_i = x_i(x_{i+1} - x_i)$. The total area of the n rectangles is

$$\underline{A}_n = \sum_{i=0}^{n-1} x_i \Delta x = \sum_{i=0}^{n-1} i \frac{X}{n}\frac{X}{n} = \frac{X^2}{n^2} \sum_{i=0}^{n-1} i \tag{2}$$

Recalling that the sum of the first $n - 1$ integers is $n(n - 1)/2$, we have

$$\underline{A}_n = \frac{X^2}{n^2}\frac{n(n-1)}{2} = \frac{X^2}{2}\left(1 - \frac{1}{n}\right).$$

The approximation to the exact area improves as n increases. The sequence $\{\underline{A}_n\}$ converges to the limit $A = X^2/2$ which is the correct result.

The second area A to be determined is that shown in Fig. 9.2(a) as the shaded region. Its boundary lines are the parabola $y = x^2$, the vertical line $x = X$, and the horizontal axis $y = 0$. An estimate and genuine lower bound for A is obtained by calculating the area of the figure of interior rectangles illustrated in Fig. 9.2(b). The base length X is divided into n equal intervals with the points of subdivision given in (1).

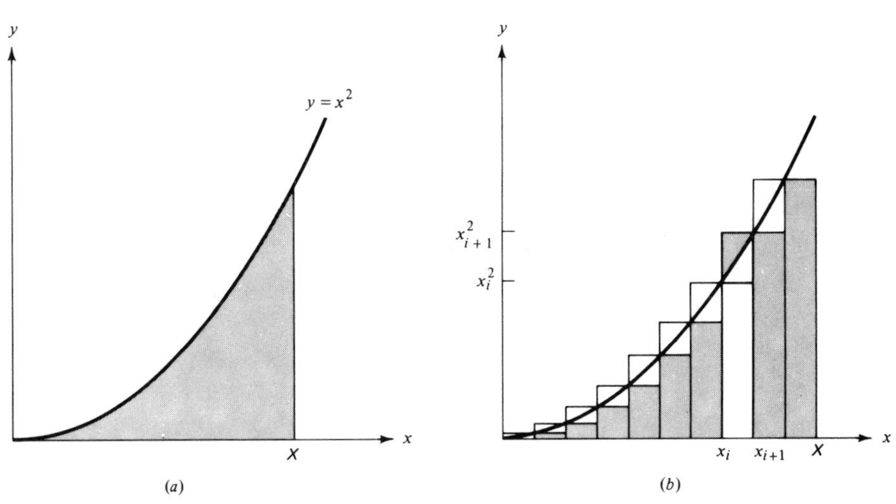

FIGURE 9.2

The height of rectangle $i + 1$, whose base length is the interval $x_{i+1} - x_i = \Delta x$, is chosen as $y = x_i^2$, the vertical distance to the parabola at $x = x_i$. This is the smallest height possible in the entire interval, and, accordingly, the corresponding rectangle has the smallest area. Therefore the area of the interior rectangle $i + 1$ is

$$x_i^2 \, \Delta x = \left(\frac{iX}{n}\right)^2 \frac{X}{n} = \frac{i^2}{n^3} X^3, \tag{3}$$

and the area of all n interior rectangles which constitute the approximating figure is

$$\underline{A}_n = \sum_{i=0}^{n-1} \frac{i^2 X^3}{n^3} = \frac{X^3}{n^3} \sum_{i=0}^{n-1} i^2. \tag{4}$$

The notation signifies as before a *lower bound*. The sum involved was evaluated in the last section, where it was found that

$$\sum_{i=0}^{n-1} i^2 = 1 + 4 + \cdots + (n-1)^2 = \frac{n(n-1)(2n-1)}{6}. \tag{5}$$

Upon substituting this in (4), a concise and explicit *lower bound* is obtained,

$$A > \underline{A}_n = \frac{X^3}{n^3} \frac{n(n-1)(2n-1)}{6}, \tag{6}$$

which holds for *all n*. The best estimate is attained in the limit as $n \to \infty$, which corresponds to a covering by an infinite number of rectangles each, according to (3), with an infinitesimal area. In fact, intuition suggests that the limit should result in the exact value of A, and this will now be proved. Therefore

$$A \geq \lim_{n\to\infty} \underline{A}_n = \lim_{n\to\infty} \frac{X^3}{6} \frac{n(n-1)(2n-1)}{n^3} = \frac{X^3}{3}, \tag{7}$$

so that

$$A \geq \frac{X^3}{3}. \tag{8}$$

An *upper bound* is obtained by selecting the height of rectangle $i + 1$ to be

$$y = x_{i+1}^2 = (i+1)^2 \frac{X^2}{n^2}, \tag{9}$$

so that its area is

$$y \, \Delta x = (i+1)^2 \frac{X^3}{n^3}. \tag{10}$$

The area of all n rectangles constituting the exterior figure is

$$\bar{A}_n = \sum_{i=0}^{n-1} \frac{(i+1)^2 X^3}{n^3} = \frac{X^3}{n^3} \sum_{i=0}^{n-1} (i+1)^2, \tag{11}$$

where, for *all n*,

$$\bar{A}_n > A. \tag{12}$$

Since

$$\sum_{i=0}^{n-1} (i+1)^2 = 1 + 4 + \cdots + n^2 = \frac{n(n+1)(2n+1)}{6},$$

it follows that

$$\bar{A}_n = \frac{X^3}{6} \frac{n(n+1)(2n+1)}{n^3} > A, \tag{13}$$

a bound that must also apply in the limit as $n \to \infty$:

$$\lim_{n \to \infty} \bar{A}_n = \frac{X^3}{3} \geq A. \tag{14}$$

According to (8) and (14), the upper and lower bounds for A are *identical*, which proves that the exact area is

$$A = \frac{X^3}{3}. \tag{15}$$

Several observations are noteworthy:

1. Both upper and lower bounds converge to the same, unique answer. Either limit procedure could have been used alone to calculate the area.
2. Let A_n be an area approximation consisting of n rectangles, with the height of the ith chosen as *any* value of $y = x^2$ for $x_i < x < x_i + 1$. As long as the number of rectangles increases to infinity and the area of each diminishes to zero, the estimate A_n will, in the limit, yield the exact area:

$$\lim_{n \to \infty} A_n = A. \tag{16}$$

This conclusion follows from the inequality

$$\underline{A}_n < A_n < \bar{A}_n,$$

which is valid for all n and in particular as $n \to \infty$ (see Example 9, page 40).
3. The approximations form a sequence $\{A_n\}$, which has A as a limit.
4. Our simple intuitive approach to the calculation of an area has been fully justified in this one case. Since there was little real doubt on this score, we will dispense with a general proof concerning the validity or convergence of the basic procedure.

2 The area under many curves can be determined in a similar manner. If the area is bounded by

$$y = f(x) \geq 0, \tag{17}$$

and the lines $y = 0$, $x = a$, and $x = b$, then a sequence of approximating figures of rectangles is constructed and the limit assessed. The interval from a to b is divided into n equal steps of length

$$\Delta x = \frac{b - a}{n};$$

the points

$$x_0 = a, \qquad x_1 = a + \left(\frac{b - a}{n}\right), \qquad \ldots, \qquad x_i = a + i\left(\frac{b - a}{n}\right), \qquad \ldots,$$

$$x_n = a + n\left(\frac{b - a}{n}\right) = b,$$

mark the subdivision (see Fig. 9.3). The area of the $(i + 1)$st rectangle is $y\ \Delta x$. Its height can be chosen as any value of $f(x)$ in the interval $x_i \leq x \leq x_{i+1}$, but the simplest choice, $y = f(x_i)$, suffices since all differences introduced this way eventually disappear as the limit is taken.

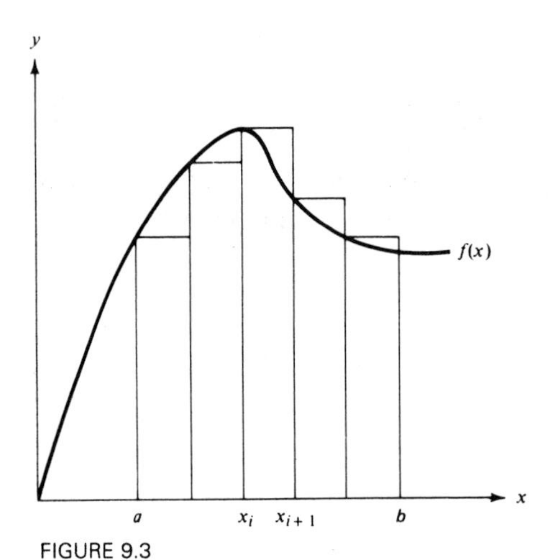

FIGURE 9.3

The total area of the n rectangles constituting the approximation A_n is

$$A_n = \sum_{i=0}^{n-1} f(x_i)(x_{i+1} - x_i) = \sum_{i=0}^{n-1} f(x_i)\ \Delta x. \tag{18}$$

As n approaches infinity, Δx approaches zero (but $x_0 = a$ and $x_n = b$ for all n) and the limit yields the exact area

$$A = \lim_{n \to \infty} \sum_{i=0}^{n-1} f(x_i)(x_{i+1} - x_i) = \lim_{n \to \infty} \sum_{i=0}^{n-1} f(x_i)\ \Delta x \qquad \text{for } \Delta x \to 0. \tag{19}$$

This limit procedure occurs frequently and is of fundamental importance in the calculus. For this reason a special notation (due to Leibniz) is introduced:

$$A = \int_a^b f(x)\,dx, \tag{20}$$

which is read "A is the *integral* of $f(x)$ from $x = a$ to $x = b$." The integral sign, $\int$, represents an elongated S, the conceptual extension of a sum; a and b are the *limits of integration* and indicate where to start and stop the process; $f(x)$ is called the *integrand*. The integral represents the infinite sum of infinitesimals, and the notation reflects this fact: the finite increment Δx is transformed into the differential element dx, and x itself plays the role corresponding to the index of summation.

The exact area under the parabola (Fig. 9.2) in the interval $0 \leq x \leq X$ is represented by

$$A = \int_0^X x^2\,dx, \tag{21}$$

where 0 and X are the lower and upper limits of integration and x^2 is the integrand. The specific evaluation of this integral in (15) showed that

$$\int_0^X x^2\,dx = \tfrac{1}{3}X^3. \tag{22}$$

Since the integral clearly depends on the value of X, it is a function of this variable, identified in (22) as a cubic polynomial.

3

Although the area under a curve is exactly expressed as an integral, this may be an empty victory indeed if the integration is too difficult to be expressed explicitly in terms of simple known functions. This is, as a matter of fact, the usual circumstance, and the only recourse then is to determine an accurate approximation by numerical methods. This, in effect, means that the sequence of approximate values $\{A_n\}$ is calculated until one is obtained whose accuracy is deemed satisfactory.

Example 1 Use a numerical method to find the area bounded by

$$y = \sin \frac{\pi}{2} x, \tag{23}$$

and the lines $x = 0$, $x = 1$, $y = 0$. The area is shown in Fig. 9.4. The numerical formula given in (18) will be used, and in this case it reads

$$A_n = \sum_{i=0}^{n-1} \sin \frac{\pi}{2} x_i\,\Delta x.$$

With $\Delta x = 1/n$ and $x_i = i/n$, this becomes

$$A_n = \frac{1}{n} \sum_{i=0}^{n-1} \sin \frac{\pi i}{2n}. \tag{24}$$

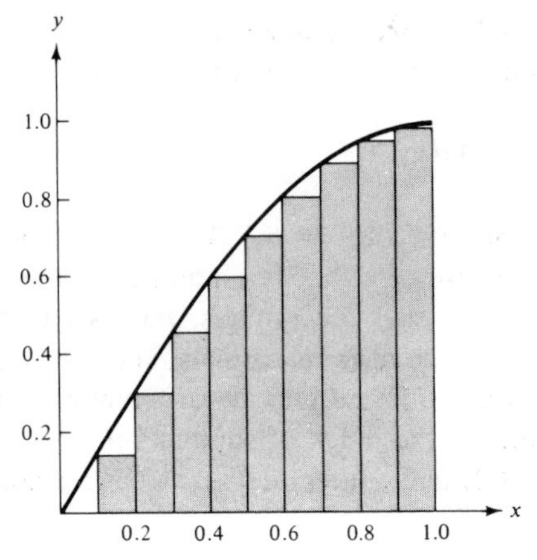

FIGURE 9.4

Values of the sequence $\{A_n\}$ computed this way are presented in Table 9.1 and indicate a fairly slow convergence to the actual limit $2/\pi$:

$$\lim_{n\to\infty} A_n = A_\infty = \frac{2}{\pi} = 0.636619. \tag{25}$$

TABLE 9.1

n	A_n	n	A_n	n	A_n
10	0.585311	100	0.631607	2,000	0.636368
20	0.611293	500	0.635619	5,000	0.636514
50	0.626568	1,000	0.636119		

Exercises 1.9

1. Using interior rectangles, show that $\int_0^X 1\,dx = X$ and $\int_a^b 1\,dx = b - a$.

2. Prove that the area of a right triangle of base X and height H is $1/2\ XH$ by using the approximation illustrated in Fig. 9.5 and carrying the calculation to a limit. Show that this result implies $\int_0^X x\,dx = 1/2\,X^2$.

3. Draw each of the following curves in the interval stated. In each case, divide the interval into four equal subintervals and calculate upper and lower bounds for the area under the curve by using upper and lower rectangles.

(i) $y = 2x + 1,\ a = 0,\ b = 1$ **(ii)** $y = \sin x,\ a = 0,\ b = \pi$

(iii) $y = 1/x,\ a = 1,\ b = 2$ **(iv)** $y = \sqrt{x},\ a = 0,\ b = 4$

4. Show that $\int_0^X (mx + b)\,dx = 1/2\,mX^2 + bX$.

5. Show that $\int_0^X (ax^2 + bx + c)\, dx = 1/3\, aX^3 + 1/2\, bX^2 + cX$.

6. Find the area bounded by the coordinate axes and the line $x + y = 1$.

7. Show that $\int_0^X x^3 dx = 1/4\, X^4$. (*Hint:* The result $\sum_{i=1}^{n} i^3 = 1/4\, n^2(n+1)^2$ will be needed.)

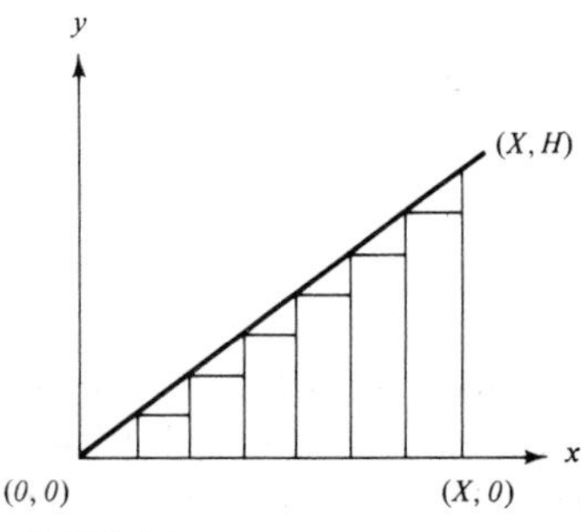

FIGURE 9.5

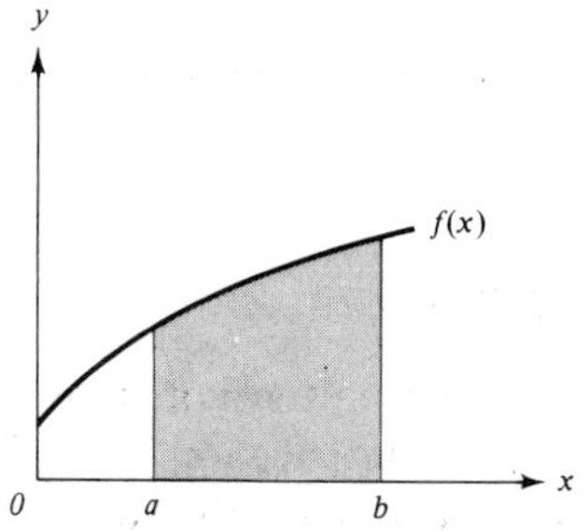

FIGURE 9.6

8. From a consideration of the areas under the curve $y = f(x)$ in Fig. 9.6, show that

$$\int_a^b f(x)\, dx = \int_0^b f(x)\, dx - \int_0^a f(x)\, dx.$$

Use this result to show the following:

(i) $\int_a^b 1\, dx = b - a$ **(ii)** $\int_a^b x\, dx = 1/2(b^2 - a^2)$ **(iii)** $\int_a^b x^2 dx = 1/3(b^3 - a^3)$.

9. Evaluate the following integrals:

(i) $\int_1^2 x\, dx$ **(ii)** $\int_2^4 x^2\, dx$ **(iii)** $\int_1^5 (1 + x + x^2/3)\, dx$.

10. Calculate the area between the curve $y = 4 - x^2$ and the x-axis from $x = -2$ to $x = 2$.

11. Find the area between the curves $y = x$ and $y = x^2/2$ from $x = 0$ to $x = 2$.

12. Show that the area between two curves $y = f(x)$ and $y = g(x)$ from $x = a$ to $x = b$ where $f(x) \geq g(x)$ (Fig. 9.7) is given by

$$A = \int_a^b [f(x) - g(x)]\, dx = \int_a^b f(x)\, dx - \int_a^b g(x)\, dx.$$

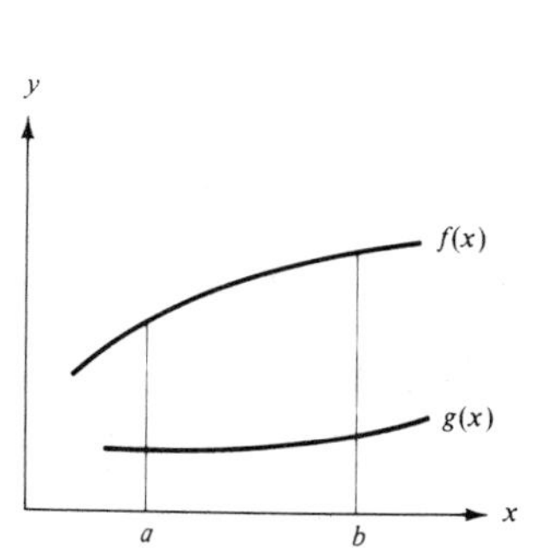

FIGURE 9.7

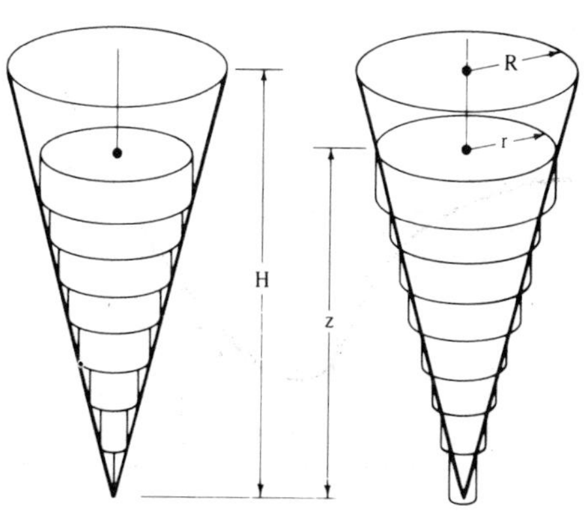

FIGURE 9.8

13. Calculate the volume of a circular cone of height H and radius R by approximating the volume with both stacks of thin cylindrical disks shown in Fig. 9.8. Obtain the exact answer in the limit as the number of disks used becomes infinite and their individual thickness approaches zero. (*Hint:* The volume of a disk of thickness Δz at height z is

$$\Delta V = \pi r^2(z)\,\Delta z = \frac{\pi R^2}{H^2} z^2\,\Delta z.$$

The limiting sum for the volume is an integral that has already been evaluated!)

1.10 Integration

Integration is a limiting form of summation which can be used to calculate areas, as demonstrated in the last section. This is only one application of the method, which, like differentiation, has a truly fundamental position in mathematics and science. In order to illustrate its role in another context, a particle problem is considered.

Suppose that the velocity of a particle (or an automobile as in Sec. 1.4) is measured as a function of time. The problem is then to determine from this information the position S of the particle at a definite time T, that is, $S(T)$. The data can be in the form of discrete measurements or preferably a complete graph of the velocity function, as shown in Fig. 10.1. The method of solution is to devise an accurate approximation scheme which in a limiting sense will lead to the exact answer.

Divide the time interval $0 \le t \le T$ into n equal parts so that

$$t_0 = 0, \qquad t_1 = \frac{T}{n}, \qquad t_2 = \frac{2T}{n}, \qquad \ldots, \qquad t_i = \frac{iT}{n}, \qquad \ldots, \qquad t_n = T. \quad (1)$$

In each small time span, $\Delta t = t_{i+1} - t_i = T/n$, the velocity is approximated by a constant v_i, which can be the function $v(t)$ evaluated at any point in the interval. Physically, this means that one measurement of the velocity must be made in each time interval. Although the efficacy of numerical routines will depend on how v_i is

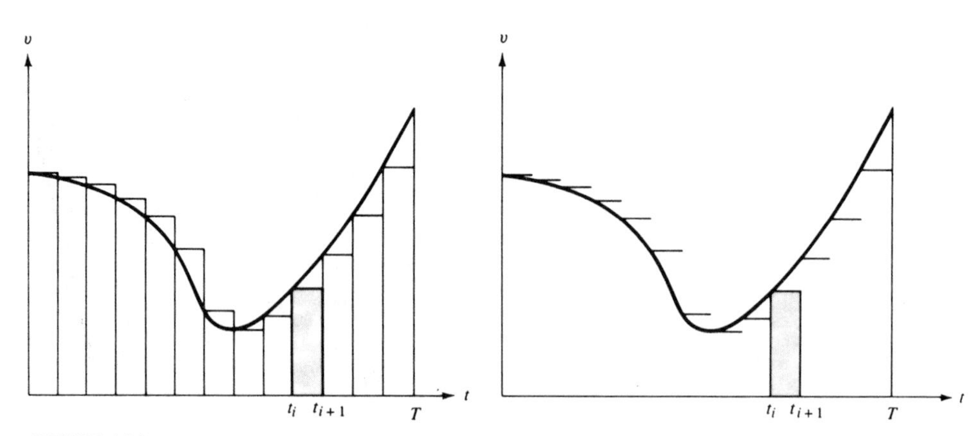

FIGURE 10.1

chosen, the errors introduced this way will eventually diminish to zero as the limit $n \to \infty$, $\Delta t \to 0$ is taken. For the sake of simplicity then, we evaluate the velocity at the first point of each interval

$$v_i = v(t_i), \tag{2}$$

which has the effect of replacing the continuous curve by a broken sequence of horizontal lines. Obviously, the true curve could be approximated more closely than this at any stage, but this would not affect the final answer.

The approximate distance travelled in the first small time interval $s_1 - s_0$ is simply the velocity, $v_0 = v(t_0)$, multiplied by the elapsed time, $t_1 - t_0$:

$$s_1 - s_0 = v_0(t_1 - t_0). \tag{3}$$

[The starting position $s_0 = S(0)$ could be made zero by referring all distances to this origin, just as time is measured from this instant. However, the initial distance s_0 will be carried along as a constant in this analysis so that the relative travel distance is $S - s_0$.] In the second time increment the distance moved is

$$s_2 - s_1 = v_1(t_2 - t_1),$$

whereas in the $(i + 1)$st interval the increment of distance is

$$s_{i+1} - s_i = v_i(t_{i+1} - t_i), \tag{4}$$

and in the last it is

$$s_n - s_{n-1} = v_{n-1}(t_n - t_{n-1}).$$

Note that (4) is exactly the same expression used earlier to find the velocity,

$$v = \frac{\Delta s}{\Delta t},$$

converted here into a formula for the unknown distance increment

$$\Delta s = v\, \Delta t. \tag{5}$$

By adding all the distances traversed in each of the small increments of time an approximation is computed for the total distance moved, $s_n - s_0$, in the elapsed time $T = t_n$. The result is

$$\begin{aligned} s_n - s_0 &= (s_1 - s_0) + (s_2 - s_1) + (s_3 - s_2) + \cdots + (s_{n-1} - s_{n-2}) + (s_n - s_{n-1}), \\ &= v_0(t_1 - t_0) + v_2(t_2 - t_1) + \cdots + v_{n-1}(T - t_{n-1}), \end{aligned} \tag{6}$$

or

$$s_n - s_0 = \sum_{i=0}^{n-1} v_i(t_{i+1} - t_i). \tag{7}$$

The last expression is especially useful for numerical calculation based on n time increments. A smaller choice of $\Delta t = T/n$, corresponding to a larger value of n, would give a different and improved result. The approximations $s_n - s_0$ form a sequence $\{s_n - s_0\}$ whose limit is the exact travel distance:

$$S - s_0 = \lim_{n\to\infty} (s_n - s_0) = \lim_{n\to\infty} \sum_{i=0}^{n-1} v_i(t_{i+1} - t_i). \tag{8}$$

But the limit of the sum is by definition the integral

$$\int_0^T v(t)\, dt,$$

where the velocity is the integrand. The position of the car at time T is precisely given as

$$S = \int_0^T v(t)\, dt + s_0, \tag{9}$$

which asserts that the position at time T is the initial position s_0 plus the distance travelled.

This result can also be interpreted as the *area* in the t, v plane lying under the velocity function and bounded by the lines $t = 0$, $t = T$, as shown in Fig. 10.1. Note that the distance increment given by (4) or (5) is equivalent to the area of the ith rectangle in this figure and the sum of all these areas, which is $s_n - s_0$, is just an approximation to the *total* area under the curve for $0 \le t \le T$. The distance travelled in a given time is the area under the velocity function.

2

The role of t in the above integral is similar to the index of summation i in (7). These variables have no special significance, and just as j can replace i in a summation without affecting the sum, so may t_* or any other symbol replace t in (9). The advantage is that t is freed to be used again without confusion as the real time variable, in which case (9) can be written as

$$s = s(t) = \int_0^t v(t_*)\, dt_* + s_0. \tag{10}$$

Here s is the position and t is the time; there is really no longer any need for S, T, and a return to conventional notation is warranted.

Equation (10) expresses distance in terms of time, and the integral provides the functional relationship. Since distance s can be computed at any time t, the upper limit in the integral (10) is not specific and for this reason it is called an *indefinite integral*. However, the calculation of s corresponding to a definite choice of t, say $t =$ 100, means that a *definite* integral must be evaluated, in this case

$$s(100) - s(0) = \int_0^{100} v(t_*)\, dt_*.$$

3

In an earlier section, the velocity was determined from the distance $s(t)$ by differentiation:

$$v = \frac{ds}{dt}. \tag{11}$$

Now we have shown that distance is determined from a known velocity $v(t)$ by integration

$$s = \int_0^t v(t_*)\, dt_* + s_0,$$

where $v(t)$ is the integrand and s_0 is the initial position at $t = 0$. We call (10) the integral of $v(t)$. Apparently differentiation and integration are inverse operations to each other, and the latter is often referred to as anti-differentiation. Anti-integration would also be appropriate but is not used. The important point is that the integration of (11) gives (10) and the differentiation of (10) must yield (11). In other words, $s(t)$ is a function whose derivative is $v(t)$, and $v(t)$ is a function whose integral is $s(t)$.

The time rate of change of the integral (10) is its integrand. This intimate relationship of integration to differentiation will be formalized later as the *fundamental theorem of calculus.*

Example 1 Show that the derivative of $s(t)$ in (10) is the velocity $v(t)$, that is, the integrand of the integral. (Let $s_0 = 0$.)

This is the first time that an integral is to be differentiated and since we have no precedent, a return to first principles is in order. A word of caution is appropriate: the arguments given next are correct, but a more rigorous justification is postponed until a later section.

Since

$$s(t + \Delta t) = \int_0^{t+\Delta t} v(t_*)\, dt_*,$$

and

$$s(t) = \int_0^t v(t_*)\, dt_*,$$

it follows that

$$\frac{\Delta s}{\Delta t} = \frac{s(t + \Delta t) - s(t)}{\Delta t} = \frac{1}{\Delta t}\left[\int_0^{t+\Delta t} v(t_*)\, dt_* - \int_0^t v(t_*)\, dt_*\right]. \tag{12}$$

The first integral in the brackets above represents the *area* under the velocity curve in the time interval 0 to $t + \Delta t$, and the second is the area in the slightly shorter interval 0 to t. The difference is then the area under the curve between time t and $t + \Delta t$, which is very nearly $v(t)\,\Delta t$, as shown in Fig. 10.2. As Δt diminishes to zero, the left-hand side of (12) becomes the derivative and the right-hand side approaches $[v(t)\,\Delta t]/\Delta t$:

$$\frac{ds}{dt} = \lim_{\Delta t \to 0} \frac{\Delta s}{\Delta t} = \lim_{\Delta t \to 0} \frac{1}{\Delta t}[v(t)\,\Delta t] = v(t),$$

which is the desired result.

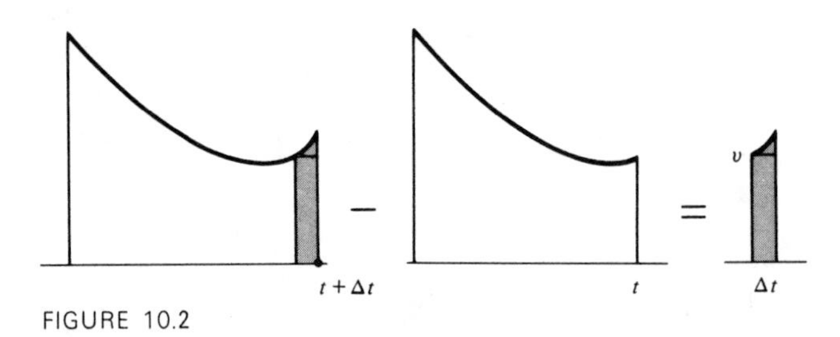

FIGURE 10.2

4 **Example 2** If the velocity is given as

$$v = \begin{cases} 10t & \text{for} \quad 0 \le t \le 10; \\ 100 & \text{for} \quad t > 10; \end{cases}$$

find the relative distance travelled as a function of time.

For convenience let the starting position be the origin, that is, $s_0 = 0$. The distance travelled in time t is then the area under the corresponding part of the velocity curve shown in Fig. 10.3. Although this area in general must be computed by integration, the situation here is so simple that the calculation is almost

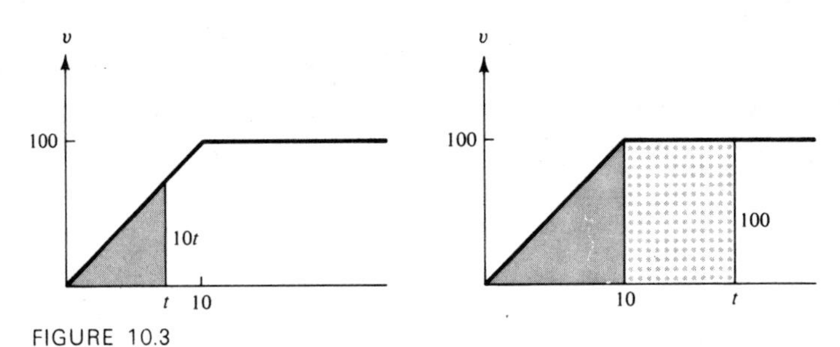

FIGURE 10.3

trivial. The area under the curve at any time $t < 10$ is that of a triangle of base t and height $10t$. In other words

$$s = \tfrac{1}{2}t(10t) = 5t^2 \qquad \text{for} \quad t \le 10.$$

After this period of acceleration, the area is composed of the complete triangle plus a rectangle of base $t - 10$, altitude 100 and area $100(t - 10)$, so that

$$s = 500 + 100(t - 10) \qquad \text{for} \quad t \ge 10.$$

Example 3 If $v = t^2$, find the distance travelled in time t.

The travel distance is given by the integral

$$s - s_0 = \int_0^t t_*^2 \, dt_*,$$

which was evaluated in the last section. With the notation changed for current usage, the result is

$$s - s_0 = \tfrac{1}{3}t^3.$$

In other words, $\frac{1}{3}t^3$ is a function whose time derivative is t^2.

Exercises 1.10

1. Graph the following velocity functions $v = v(t)$ for $t \geq 0$. Assuming $s(0) = 0$, determine the position $s(t)$ of the particle at time t. Draw graphs of $s = s(t)$ for $t \geq 0$.

(i) $v(t) = 2 + 3t$ **(ii)** $v(t) = 1 + t + t^2$ **(iii)** $v(t) = (2 + t)^2$

2. The velocity of a particle (in meters per second) is given by

$$v = \begin{cases} t(10 - t), & t \leq 10; \\ 0, & t > 10. \end{cases}$$

Compute the distance X travelled in time T as follows: subdivide the interval $0 - T$ into n equal parts (Fig. 10.4) and determine the area under this section of the curve in the limit $n \to \infty$. Take $t_i = iT/n$, $\Delta t = T/n$ and the area of the ith rectangle as

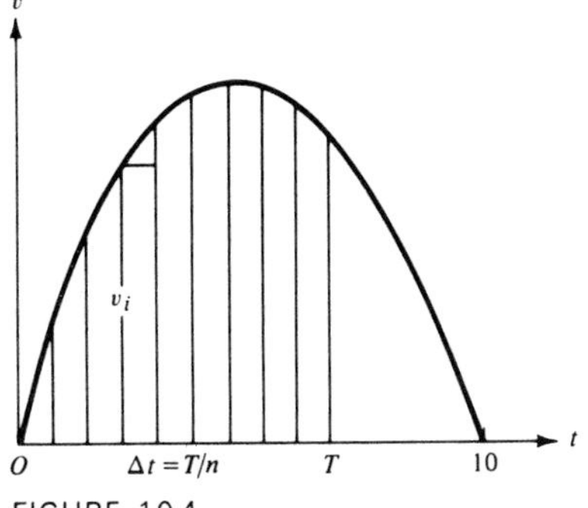

FIGURE 10.4

$$\frac{T}{n} v_i = \frac{T}{n}\frac{iT}{n}\left(10 - \frac{iT}{n}\right) = \frac{T^2}{n^2}\left(10i - \frac{i^2T}{n}\right).$$

The area of all n rectangles is then

$$A_n = \frac{T^2}{n^2}\sum_{i=0}^{n-1}\left(10i - \frac{i^2T}{n}\right) = \frac{T^2}{n^2}\left(10\sum_{i=0}^{n-1} i - \frac{T}{n}\sum_{i=0}^{n-1} i^2\right).$$

3. The distance travelled in time t by a particle moving in a straight line is given by

$$s(t) = \int_0^t (1 + \tfrac{1}{2}t_* + t_*^2)\, dt_*.$$

Use the known values of the integrals $\int_0^t t_* \, dt_*$ and $\int_0^t t_*^2 \, dt_*$ to calculate $s(2)$ and the distance travelled between $t = 2$ and $t = 5$. Find the velocity $v = ds/dt$ at these times. Draw graphs of $s = s(t)$ and $v = v(t)$.

4. If the velocity of a particle is given by

$$v = \begin{cases} U(t), & t \leq t_1; \\ W(t), & t_1 < t; \end{cases}$$

show that the distance

$$s = \int_0^t v(t_*) \, dt_*,$$

travelled from the initial instant is

$$s = \begin{cases} \int_0^t U(t_*) \, dt_* & \text{for } t \leq t_1; \\ \int_0^{t_1} U(t_*) \, dt_* + \int_{t_1}^t W(t_*) \, dt_* & \text{for } t_1 < t. \end{cases}$$

Interpret this solution in terms of areas under the curve $v = v(t)$.

5. Use the results of Exercise 4 to find the distance travelled by a vehicle whose velocity history is

$$v = \begin{cases} 4t^2 & \text{for } 0 \leq t \leq 5; \\ 100 - 15(t - 5) & \text{for } 5 < t. \end{cases}$$

Draw graphs of $v = v(t)$ and $s = s(t)$.

6. A cylindrical drum of cross-sectional area A is filled with a fluid at a rate of $Q(t)$ m^3/sec. If h is the instantaneous height of the fluid within the drum, show that

$$\frac{dh}{dt} = \frac{Q(t)}{A}.$$

Why is the problem of finding $h(t)$ essentially equivalent to that of determining distance from velocity? Express h as an integral which corresponds to the area under the curve $Q(t)$. If the initial height of fluid is h_0, find the h at a later time.

1.11 Continuity and approximate methods

The work of this chapter has, for the most part, involved functions of only one independent variable:

$$y = f(x). \tag{1}$$

However, the explicit functional form frequently contained an arbitrary constant, called a *parameter*, such as m, a, and k in the following:

$$\textbf{(i)}\quad y = mx; \qquad \textbf{(ii)}\quad y = \frac{x^2 - a^2}{x^2 + a^2}; \qquad \textbf{(iii)}\quad y = \cos(kx + k^2).$$

In order to calculate and graph a function of this kind it is first necessary to choose a definite numerical value for the parameter. For each such choice, there will be a different curve, and these can be graphed separately or all together if the particular parametric values are indicated on the composite. The totality of curves that can be obtained this way is known as a *family* of curves. A function with a parameter represents not one but an entire family of curves, as illustrated in Fig. 11.1.

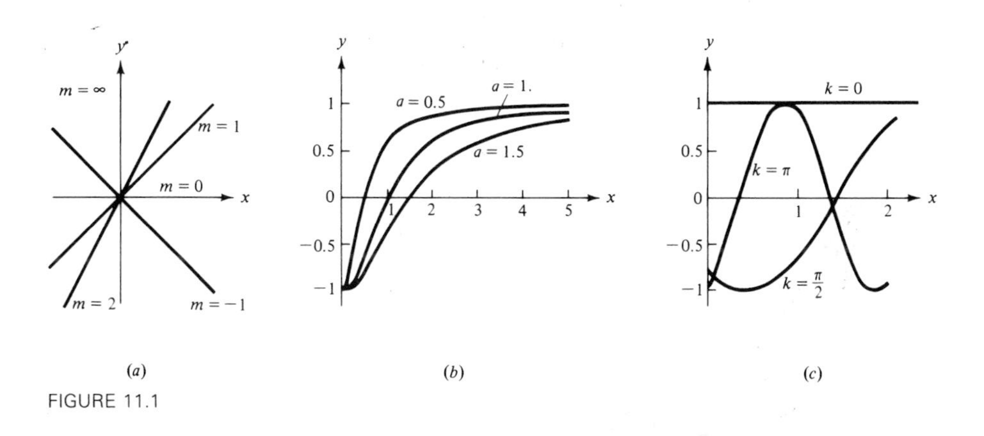

FIGURE 11.1

The fact that y is a function of x with a as a parameter is indicated clearly by modifying (1) to read

$$y = f(x, a). \tag{2}$$

This makes the crucial role of a apparent and, in effect, states that a and the variable x must both be specified in order to compute the corresponding numerical value of y. Viewed in this manner, the parameter can be interpreted as just another independent variable because it is assigned in (2) as freely as x is. This equation is then a relationship between one dependent variable and *two* independent variables; $f(x, a)$ is a function of two variables.

Alternatively, we can at any time regard x as the parameter and a as the independent variable if there is any particular advantage in this tactic. The function y can then be graphed versus a for particular values of x and its continuity and limiting behavior with respect to a examined.

Example 1 Discuss the function

$$y = \frac{x^2 - a^2}{x^2 + a^2},$$

treating a as the variable and x as the parameter.

The family of curves is shown in Fig. 11.2 for different values of x. This rational function is continuous for all a, and certain informative limits are

$$\lim_{a \to 1} \frac{x^2 - a^2}{x^2 + a^2} = \frac{x^2 - 1}{x^2 + 1},$$

$$\lim_{a \to 0} \frac{x^2 - a^2}{x^2 + a^2} = 1,$$

$$\lim_{a \to \infty} \frac{x^2 - a^2}{x^2 + a^2} = \lim_{a \to \infty} \frac{x^2/a^2 - 1}{x^2/a^2 + 1} = -1.$$

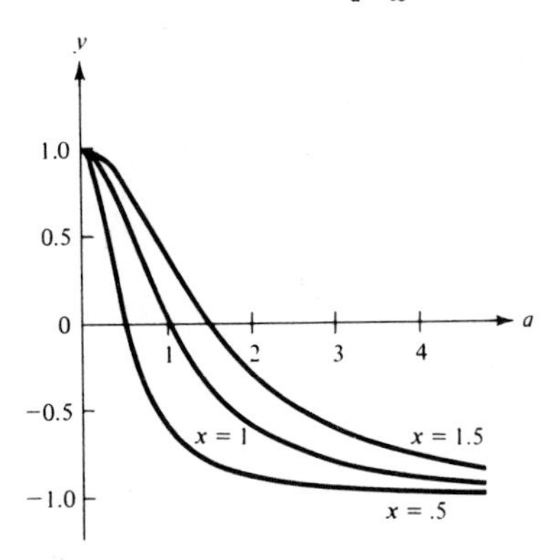

FIGURE 11.2

Example 2 Discuss the power function

$$y = a^x. \tag{3}$$

Most readers have probably encountered this function before, but since its meaning may still be obscure, a short discussion is warranted. The power function arises as the generalization of the identity for integral exponents

$$a \cdot a \cdots a = a^n.$$

With a restricted to positive values only, the definition is extended to cover all fractional powers of the form

$$y = a^{n/m},$$

by interpreting y to be the real number which raised to the mth power is equal to a^n:

$$y^m = a^n.$$

Further generalization is based on the fact that *any* real number x is the limit of some sequence of fractions $\{r_i\}$

$$\lim_{i \to \infty} r_i = x.$$

Since the power function is well defined for each fractional exponent, i.e.,

$$y_i = a^{r_i},$$

the limit of the sequence $\{y_i\}$ defines the value of a^x:

$$y = \lim_{i \to \infty} y_i = \lim_{i \to \infty} a^{r_i} = a^x.$$

To illustrate, the value of 2^π is determined, accurate to two decimal places. The fractions $r_0 = 3$, $r_1 = 3.1 = 3 + \frac{1}{10}$, $r_2 = 3.14 = 3 + \frac{1}{10} + \frac{4}{100}$, etc., formed from the decimal expansion of π, obviously have the correct limit, and the sequence $\{2^{r_i}\}$ defines the value of 2^π. The calculations that result are

$$2^{r_0} = 8, \qquad 2^{r_1} = 8.57, \qquad 2^{r_2} = 8.81, \qquad \ldots, \qquad 2^\pi = 8.82 \ldots.$$

The important algebraic rules for power functions are

$$a^{x_1} a^{x_2} = a^{x_1 + x_2}, \qquad a^{x_1}/a^{x_2} = a^{x_1 - x_2}, \qquad (a^{x_1})^{x_2} = a^{x_1 x_2}, \qquad (a^{x_1})^{1/x_2} = a^{x_1/x_2}. \tag{4}$$

The proofs are constructed as limiting forms of the analogous theorems for fractional exponents and are developed in the exercises.

The power function for various values of a is shown in Fig. 11.3.

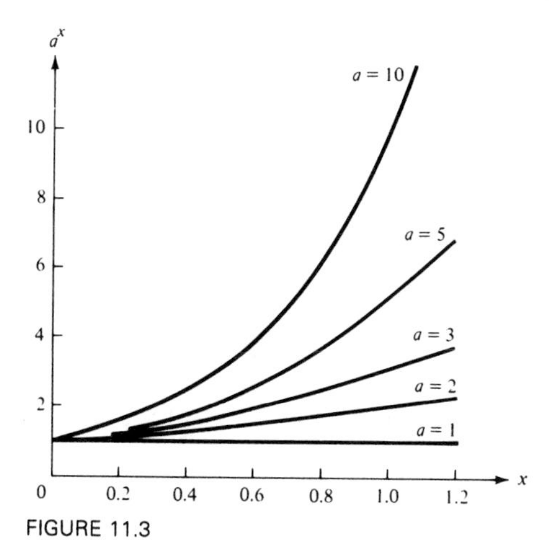

FIGURE 11.3

Physical problems are usually characterized by parameters whose magnitudes may provide the motivation for good approximate methods of solution. Suppose, for example, that the solution y of a particular problem is known to depend on the variable x and the parameter ε, which is a very small number:

$$y = f(x, \varepsilon). \tag{5}$$

There are usually good reasons to assume that the function is continuous in its dependence on ε, in which case the solution for small ε resembles the member of the same family of curves, $\bar{y}$, which corresponds to $\varepsilon = 0$, that is,

$$\bar{y} = f(x, 0), \tag{6}$$

and

$$\bar{y}(x) = \lim_{\varepsilon \to 0} f(x, \varepsilon). \tag{7}$$

After all, not many physical processes can detect sensible differences between the values of say $\varepsilon = 0$ and $\varepsilon = 10^{-15}$, and the corresponding functions of this family should be almost indistinguishable. Small variations of a parameter are expected, as a matter of experience, to produce equally small changes in the behavior of the function. This expresses a deep conviction that functions representing natural phenomena are, in the main, continuous in that all variations are smooth. Since physical measurements always involve some error, their value would certainly be nil if imperceptible parameter changes implied drastic differences. In fact, very little could ever be learned if all experiments were incredibly sensitive to environmental conditions, which in truth can never be held fixed or repeated exactly. This faith is not always justified, and "irregularities" do occur which are immediately referred to as "interesting." After all, a pathological case requires some new thought, whereas the common situation reduces to an almost standard routine called *perturbation theory*. A few simple examples will suffice for now to illustrate how this assumption of continuous behavior is actually applied, but we shall return to this subject again.

Example 3 Find the positive root x_* of the polynomial

$$y(x) = x^2 + \varepsilon x - 4. \tag{8}$$

The root x_* is a solution of

$$y(x_*) = 0;$$

its value obviously depends on the magnitude of ε, which implies a functional relationship:

$$x_* = F(\varepsilon).$$

If the parameter is very small, then x_* should be very nearly the same as the root obtained from (8) by setting $\varepsilon = 0$, that is,

$$x_* \approx F(0) = 2.$$

But how accurate is this approximation, and how can *systematic* improvements be made? Accuracy can be checked by computing the actual value of $y(x)$

at $x = 2$ to see by what amount it deviates from zero. This calculation yields $y(2) = 2\varepsilon$, which is a small but nonzero error. In order to reduce the discrepancy, we try to locate the root more accurately by setting

$$x_* = 2 + a_1\varepsilon. \tag{9}$$

The constant a_1 is then chosen to make $y(x_*)$ as close to zero as possible. The result of substituting this in (8) and rearranging terms is

$$\begin{aligned} y(x_*) &= 4a_1\varepsilon + a_1{}^2\varepsilon^2 + 2\varepsilon + a_1\varepsilon^2, \\ &= (4a_1 + 2)\varepsilon + (a_1{}^2 + a_1)\varepsilon^2. \end{aligned}$$

The largest contribution to the total error in this expression comes from the term $(4a_1 + 2)\varepsilon$, which can be eliminated by selecting $a_1 = -\frac{1}{2}$. The corrected root position is then $x_* = 2 - \varepsilon/2$ with an incurred error

$$y(x_*) = -\tfrac{1}{4}\varepsilon^2,$$

whereas formerly the discrepancy was larger, that is, 2ε. For example, if $\varepsilon = 0.1$, the revised root,

$$x_* = 2 - \tfrac{1}{2}(0.1) = 1.95,$$

is more accurate than the first estimate, $x_* = 2$.

The process can be continued by setting

$$x_* = 2 + a_1\varepsilon + a_2\varepsilon^2, \tag{10}$$

where both a_1 and a_2 are chosen to make $y(x_*)$ as close to zero as possible. The result of inserting (10) into (8) is

$$y(x_*) = (4a_1 + 2)\varepsilon + (4a_2 + a_1 + a_1{}^2)\varepsilon^2 + (a_2 + 2a_1a_2)\varepsilon^3 + a_2{}^2\varepsilon^4. \tag{11}$$

The largest terms are those multiplied by ε and ε^2, and these can be eliminated by choosing $a_1 = -\frac{1}{2}$, $a_2 = \frac{1}{16}$, in which case

$$y(x_*) = \tfrac{1}{256}\varepsilon^4. \tag{12}$$

The incurred error is minimized this way, and the corrected root position is now

$$x_* = 2 - \tfrac{1}{2}\varepsilon + \tfrac{1}{16}\varepsilon^2. \tag{13}$$

In particular, $x_* = 1.950625$ for $\varepsilon = 0.1$.

To summarize the analysis thus far, the three successive approximations are

$$\begin{aligned} &x_* = 2 && \text{with } y(x_*) = 2\varepsilon; \\ &x_* = 2 - \tfrac{1}{2}\varepsilon && \text{with } y(x_*) = \tfrac{1}{4}\varepsilon^2; \\ &x_* = 2 - \tfrac{1}{2}\varepsilon + \tfrac{1}{16}\varepsilon^2 && \text{with } y(x_*) = \tfrac{1}{256}\varepsilon^4. \end{aligned}$$

A general recipe for further approximation is to write the root as

$$x_* = 2 + \sum_{i=1}^{n} a_i \varepsilon^i = 2 + a_1 \varepsilon + a_2 \varepsilon^2 + \cdots + a_n \varepsilon^n, \tag{14}$$

and to select the numbers a_i so that the smallest value of $|y(x_*)|$ is obtained. Eventually, the number of terms used could become infinite, in which case (14) is a *perturbation series* in powers of the parameter ε. If the successive approximations to the root given by (14) are labelled x_n, they would form a sequence $\{x_n\}$ whose limit is the *exact* root x_* of the original quadratic. In this simple case,

$$x_* = \frac{-\varepsilon}{2} + 2\left(1 + \frac{\varepsilon^2}{16}\right)^{1/2}, \tag{15}$$

and in particular $x_* = 1.9506248$ for $\varepsilon = 0.1$.

The representation of a slight deviation of a function $f(x, \varepsilon)$ as a series in powers of the small parameter ε is the basis of *perturbation theory*. Aspects of this important technique will be explored and developed later.

Example 4 Find an approximate solution of the equation

$$y + \varepsilon \sin 2y = x, \tag{16}$$

where ε is very small.

By solution we mean the conversion of this *implicit* equation into an explicit functional statement of the form

$$y = f(x, \varepsilon). \tag{17}$$

In this case, $f(x, \varepsilon)$ is not a simple known function (nor does it occur frequently enough to make it one). However, for small ε, it should closely resemble

$$y = x.$$

In other words, Eq. (16) represents a small but complicated distortion of a straight line. It does not take much of a deviation to turn a simple function into something quite difficult to describe analytically.

For given ε, the tabulation of functional values is easily completed by reversing the roles of the variables, i.e., choose y and compute x. The graph of $f(x, \varepsilon)$ for different values of the parameter ε is shown in Fig. 11.4. The complete tabulation of a function with two independent variables would require a large book, each page corresponding to a value of ε. An analytical formula that elegantly or succinctly summarizes this information is very desirable. One means of achieving this, for small ε at least, is by the method of *iteration*, a procedure by which an initial guess (good or bad) is corrected and improved systematically until a desired accuracy is attained. To illustrate, Eq. (16) is first rearranged as

$$y = x - \varepsilon \sin 2y, \tag{18}$$

so that the small disturbing quantity which causes all the difficulty, $\varepsilon \sin 2y$, is

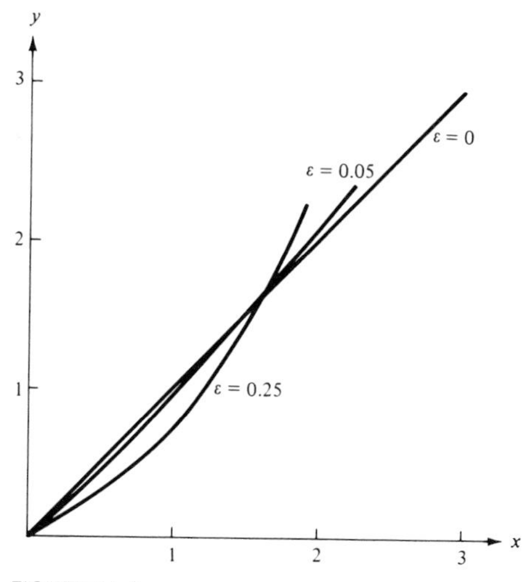

FIGURE 11.4

positioned as shown. If a guess is substituted for y on the right-hand side, the equation itself (the left-hand side) gives a revised estimate of the function. This estimate is then used as the next guess, and the *iteration* continues.

Let the initial guess be denoted $y_{(0)}$; the first correction to this, $y_{(1)}$, is obtained from the formula as

$$y_{(1)} = x - \varepsilon \sin 2y_{(0)}. \tag{19}$$

If $y_{(n-1)}$ is iteration $n - 1$, the next approximation in the procedure is obtained from

$$y_{(n)} = x - \varepsilon \sin 2y_{(n-1)}. \tag{20}$$

An initial guess which is not terribly imaginative or accurate is

$$y_{(0)} = 0. \tag{21}$$

The first iteration based on this is

$$y_{(1)} = x;$$

the second is

$$y_{(2)} = x - \varepsilon \sin 2x; \tag{22}$$

and the third is

$$y_{(3)} = x - \varepsilon \sin 2(x - \varepsilon \sin 2x). \tag{23}$$

There is of course no assurance as yet that this procedure actually works, or if it does how fast or accurate the calculations are. Table 11.1 shows the values calculated for $\varepsilon = 0.05$ corresponding to each of the first five iterations. In this case, the last set of entries is indeed accurate to 0.0001, which is evidence at least that the scheme is actually operating successfully. To have full confidence in an iteration procedure, a *convergence proof* is required. This would show conclusively that as $n \to \infty$, $y_{(n)} \to y_{(\infty)} = f(x, \varepsilon)$, and it might also indicate the range of values of ε for which the method is appropriate.

TABLE 11.1 *Calculations for* $\varepsilon = 0.05$.

$x = y_{(1)}$	$y_{(2)}$	$y_{(3)}$	$y_{(4)}$	$y_{(5)}$
0.1000	0.0901	0.0910	0.0909	0.0909
0.2000	0.1805	0.1823	0.1822	0.1822
0.3000	0.2718	0.2741	0.2739	0.2740
0.4000	0.3641	0.3667	0.3665	0.3665
0.5000	0.4579	0.4603	0.4602	0.4602
0.6000	0.5534	0.5553	0.5552	0.5552
0.7000	0.6507	0.6518	0.6518	0.6518
0.8000	0.7500	0.7501	0.7501	0.7501
0.9000	0.8513	0.8504	0.8504	0.8504
1.0000	0.9545	0.9528	0.9528	0.9528
1.1000	1.0596	1.0573	1.0572	1.0572
1.2000	1.1662	1.1638	1.1636	1.1636
1.3000	1.2742	1.2721	1.2719	1.2719
1.4000	1.3833	1.3817	1.3815	1.3815
1.5000	1.4929	1.4922	1.4922	1.4922
1.6000	1.6029	1.6032	1.6032	1.6032
1.7000	1.7128	1.7140	1.7141	1.7141
1.8000	1.8221	1.8241	1.8243	1.8243
1.9000	1.9036	1.9330	1.9331	1.9331

---

5 Examples of continuous functions that arise in applications range from simple ones like the dependence of the distance, velocity, or acceleration of a vehicle on time to more sophisticated relationships like the variation of a magnetic or gravitational field with distance from the earth. On the other hand there are also many real situations that *must* be described by functions with discontinuities. Piecewise continuous functions arise in counting processes where changes are only in integer units. Examples include the time variation of the world population, the number of stocks exchanged, or a bank-account balance. Finally, there are many processes, though basically of one kind, say discrete, which are in fact more efficiently described by continuous functions. The reverse is also true, and the motivation behind such important approximations will be introduced now and discussed frequently in the chapters to follow.

Many natural systems (media such as water or air, social groupings, etc.) are organizations of a very large but a finite integer number of individual units (in particular, molecules and people). Although these smaller entities react, for the most part, with their immediate neighbors, the net effect of the aggregate is usually a distinctive new phenomenon that completely transcends individual behavior. Waves and currents observed in the ocean have little direct connection with microscopic molecular processes. Are the characteristics of society—inflation, recession, war, or peace—any more sensitive to individual action or desires? In both illustrations, certain phenomena describe the behavior of the total population—these are called *macroscopic* properties —just as there are local or *microscopic* effects. Although macroscopic behavior is surely a manifestation of microscopic processes, the connection is by no means clear or direct. Fortunately, a knowledge of the position and velocity of every molecule of water in the ocean is not required in order to study ocean waves. Even if it were possible to provide such exact data (which it is not), it would be a useless, blinding, and overwhelming mess. It pays then to disregard the molecular viewpoint in a study of the macroscopic features of fluid motion, and this approach is generally valid.

The fluid medium is best described as continuous, just as it looks and feels to us. This is an approximation which is appropriate (accurate) as long as our interests do not extend to events taking place in the molecular range. A "small" volume within this approximation must still be large enough to contain many molecules and form a recognizable sample of water.

Macroscopic phenomena are best described in terms of macroscopic variables. The gross properties of a material or a fluid at a point in space are defined in terms of a small incremental volume ΔV centered at this position. For example, an extremely useful concept is that of *density*, or the mass of fluid in ΔV divided by the volume. In practice, the density can be calculated in just this way; theoretically, its molecular significance or connection is as the average

$$\frac{\sum m_i}{\Delta V},$$

where m_i denotes the mass of the ith molecule in ΔV and $\sum m_i$ is the total mass contained therein. The density represents a continuous approximation to a discrete mass distribution and in effect smears out the mass of each molecule over the whole incremental volume. The fluid velocity and pressure at a point are also averages over the same incremental volume. The macroscopic variables of pressure, density, and velocity defined this way are continuous functions of position and provide the means of portraying large-scale phenomena.

Similar macroscopic variables can be introduced as easily to describe society in the large, although in this case the fundamental units are not nearly as numerous as molecules and the size of an increment, which is still recognizable as a sample of the whole, may be larger than desirable. However, many of the ideas and approximations drawn from fluid dynamics still apply: phenomena can be analyzed without consideration of individual behavior. Although we all value our individuality, it is sometimes depressing to see just how many of our actions and total behavior patterns are actually part of the movements of society as a whole. The point is that many features of human society are probably unavoidable consequences, i.e., "natural phenomena" of the system, just as waves appear in bodies of water.

It should also be noted that continuous processes are often approximated to great advantage by piecewise continuous functions. Thus, the continuous time history of an automobile's velocity or a rocket's altitude is approximated by a sequence of short constant-velocity segments in order to compute the distance travelled (see page 66). Most digital numerical procedures discretize continuous functions in much the same way as area and position were computed earlier in this chapter.

Natural phenomena are often the result of an interplay between several physical processes, each of which has its own particular characteristics. This is usually reflected in the mathematical formulation by functions whose variations occur over widely different length scales. For example, the function pictured in Fig. 11.5(a) shows two sinusoidal waves connected by a sharp transition zone. This might represent the wave height of a river bore or the pressure distribution across a shock wave caused by supersonic flight. (The narrow transition zones in these examples are regions of internal dissipation.) The function shown is certainly continuous, but if the "long" waves are the object of study, it makes sense to approximate the transition zone by a discontinuous step, Fig. 11.5(b). On the other hand, a proper analysis of the "shock" requires that this region be magnified by changing the length scale, in which case the function is again continuous but the wave motion is now hard to discern, Fig. 11.5(c).

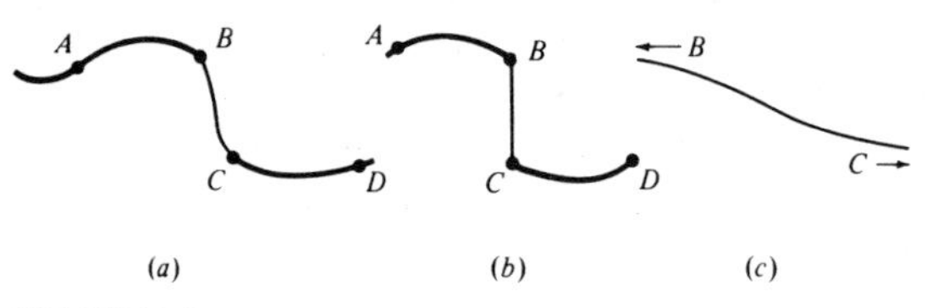

FIGURE 11.5

There are phenomena whose time scales are seconds, or years, and still others which involve all of these. Many examples with multiple scales will be presented in the course of this work. A certain flexibility of attitude is required because situations will be encountered in which a continuous function is best represented by one with jumps or vice versa. Remember that mathematics is used to learn something about nature and in this endeavor no holds are barred. The ability to make the right approximation simplifies the task enormously and is one of the most important skills to be acquired by a scientist.

Exercises 1.11

1. The equation $y = ax + a$ describes a one parameter family of straight lines. Graph the lines corresponding to $a = 0, \pm 1/2, \pm 1, \pm 2$. Verify that all lines of the family pass through the point $(-1, 0)$.

2. The equation $y = ax^2$ describes a one parameter family of parabolas. Graph the members of this family corresponding to $a = \pm 1/2, \pm 1, \pm 2$. Graph the three limiting cases $a = 0, \pm\infty$.

3. The equation $y = 2ax - a^2$ describes a family of straight lines for different values of the parameter a. Graph the lines corresponding to $a = 0, \pm 1/2, \pm 1$. On the same diagram, plot the parabola $y = x^2$. What relation does the parabola have to the straight lines? Prove your statements.

4. Graph the family of straight lines $(\cos\alpha)x + (\sin\alpha)y = 1$ for the parameter values $\alpha = 0, \pm\pi/4, \pm\pi/2, \pm\pi$. Verify that the line corresponding to the parameter value α is tangent to the circle $x^2 + y^2 = 1$ at the point $(\cos\alpha, \sin\alpha)$. (*Hint*: Where does the line intersect the circle?)

5. Determine the family of straight lines that are tangent to the hyperbola $xy = 1$. (*Hint*: When does the line $y = mx + b$ intersect the hyperbola at one point?)

6. Show that the one parameter family of curves $y = \dfrac{x^2 - a^2}{x^2 + a^2}$ is related to the single curve $y = \dfrac{x^2 - 1}{x^2 + 1}$ by a change in the length scale (except for the special case $a = 0$). Illustrate with graphs for several values of a. What happens to the interval $-a \leq x \leq a$ after the change in the length scale?

7. For the arbitrary scale change $x = lz$, show that the function in the previous problem becomes $y = \dfrac{z^2 - (a/l)^2}{z^2 + (a/l)^2}$. If l is much larger than a $(l \gg a)$, define a parameter $\epsilon = a/l$ and determine a useful approximation for $y(z)$ based on the smallness of ϵ. For what values of z is the approximation reasonable? For l much smaller than a $(l \ll a)$, define another small parameter $\mu = l/a$ and repeat the analysis. Draw graphs for the special cases $a = 1, l = 20$ and $a = 3, l = 1/30$.

8. Draw the curves $y = \dfrac{x}{a + |x|}$ corresponding to the parameter values $a = 0.1$, 1 and 10. Change the horizontal length scale by defining $x = lz$ and show that $y = \dfrac{z}{a/l + |z|}$. For what values of the ratio $\lambda = a/l$ is the function y reasonably approximated by a step function? For what values of λ and z is the function approximately linear?

9. Prove the algebraic rules for exponents.

(i) $a^{x_1}a^{x_2} = a^{x_1 + x_2}$ **(ii)** $\dfrac{a^{x_1}}{a^{x_2}} = a^{x_1 - x_2}$ **(iii)** $(a^{x_1})^{x_2} = a^{x_1 x_2}$

10. Graph the function $y = (x + \epsilon)^2$ where ϵ is a small parameter. By comparing graphs and using algebra, analyze the accuracy of the approximation $\bar{y} = x^2$. Examine the abolute error $|y - \bar{y}|$ and the relative error $(y - \bar{y})/\bar{y}$.

11. Figure 11.6 is a weather map showing air pressure on the surface of the eastern United States. What are the variables in this plot? What is the parameter?

12. If ϵ is a small parameter, find an approximation to the root of the cubic polynomial $x^3 - 1 + \epsilon x = 0$ which has the form $x_* = 1 + a_1\epsilon + a_2\epsilon^2$. Study this problem geometrically by drawing graphs of $y = x^3$ and the family of straight lines $y = 1 - \epsilon x$.

13. Find an approximation to the root of the fifth degree polynomial $x^5 + \epsilon x^4 - 1 = 0$ which lies closest to $x = 1$. If $\epsilon = 0.01$, compute this root accurate to three decimal places. Study this problem geometrically by drawing graphs of $y = x^5$ and $y = 1 - \epsilon x^4$.

14. Some of the difficulties of perturbation theory become apparent in the attempt to find *both* roots of the quadratic $\epsilon x^2 + x - 4 = 0$. Show that one root is very nearly $x_* = 4$ but that the other is large and quite sensitive to the value of ϵ. Relate this difficulty to the fact that for $\epsilon = 0$ the quadratic reduces to a linear equation. (*Hint*: Solve the quadratic exactly.)

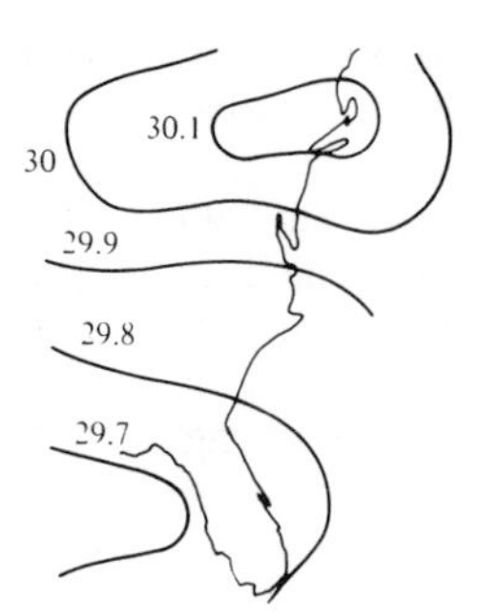

FIGURE 11.6

15. Find the roots of the quadratic equation $x^2 + \epsilon x - 4 = 0$ by the following iteration scheme. Since $x^2 - 4 = (x+2)(x-2) = -\epsilon x$, the equation for the positive root $x = 2 - \dfrac{\epsilon x}{x+2}$ suggests the iteration formula $x_{n+1} = 2 - \dfrac{\epsilon x_n}{x_n + 2}$. Choose $x_0 = 0$ and calculate x_1, x_2 and x_3. Show that $x_3 = 2 - \epsilon/2 + \epsilon^2/(16 - 2\epsilon)$ and compare this with the result in the text.

16. Analyze the problem of determining the depth of a well by dropping a stone and measuring the time interval until a splash is heard. Assume first that the velocity of sound (330 m $\sec^{-1}$) is large and then determine correction terms.

1.12 Units and Dimensions

The equations encountered in science and engineering describe relationships between physical variables such as force and time and therefore they must be dimensionally consistent. It is impossible to equate a distance to an area or a mass to a force and have a correct description valid in all systems of units. Dimensional arguments provide a valuable tool for verifying whether a particular equation can be a valid description of a physical problem.

The basic dimensions of length, mass and time are designated by capital letters L, M and T. Once the choice of units of measurement has been made (for example, meters, kilograms and seconds), then the units of measurement of all physical quantities are known. An area A has dimension L^2, measured in square meters or m^2. The dimension of volume is L^3, measured in cubic meters or m^3. An angle θ, measured in radians, is considered to be dimensionless since θ is defined as the ratio of distances, the length of an arc on the circumference divided by the radius of the circle.

Velocity and acceleration have dimensions LT^{-1} and LT^{-2} measured in m $\sec^{-1}$ and m $\sec^{-2}$. The equation discussed in Sec. 1.4

$$s(t) = s(0) + v(0)t + \tfrac{1}{2}at^2$$

describing linear motion with constant acceleration a is dimensionally correct since each term has units of length. This is obvious for the terms $s(t)$ and $s(0)$. The term $v(0)t$ has dimensions $LT^{-1}T = L$ and the term $\frac{1}{2}at^2$ has dimensions $LT^{-2}T^2 = L$.

The most complicated physical quantities have dimensions that can be expressed in the form $L^{\alpha}T^{\beta}M^{\gamma}$ where α, β and γ are constants. Force, defined as mass times acceleration, has dimensions MLT^{-2}. In this example, $\alpha = 1$, $\beta = -2$ and $\gamma = 1$. Pressure is defined as force per unit area with dimensions $MLT^{-2}L^{-2} = L^{-1}T^{-2}M^1$ ($\alpha = -1$, $\beta = -2$, $\gamma = 1$). Force is measured in units of kg m sec^{-2} (or newtons) and pressure is measured in units of kg m^{-1} sec^{-2} (newtons per square meter or pascals). Many other examples are given in the exercises.

An equation $y = f(x)$ in a physical problem gives a relationship between a measurement x and a measurement y. If, for example, x is a time interval measured in seconds and y is a distance measured in meters, the formula must be dimensionally correct converting units properly from seconds to meters. An easy way to ensure this is to measure all variables and constants in a fixed system of units. In effect, this restores every variable to the status of a pure number, because each is interpreted as a multiple of the relevant measure adopted ($t = 6$ might, by agreement, mean $t = 6$ years). A change to another set of units is effected by means of conversion factors, relating corresponding scales of measurement. As an illustration, if x and x_* are measurements of the same distance but in different units, then

$$x = ax_*$$

where a is the conversion factor of the scale change. For example, if x is a length in centimeters and x_* is the same length in meters, then $x = 100x_*$.

Example 1 In a constant gravitational field with acceleration $g = 9.806$ m/sec^2, the distance d travelled in time t by an object falling from rest is given by $d = \frac{1}{2}gt^2$. Find the distance d at $t = 3$ sec and $t = 1.5$ min.

Since g is given in units of meters and seconds, it is convenient to express all quantities this way. Therefore, at $t = 3$ sec,

$$d = \tfrac{1}{2}(9.806 \text{ m sec}^{-2})(3 \text{ sec})^2 = \tfrac{1}{2}(9.806)(9) \text{ m} = 44.127 \text{ m}$$

Note that the units of time cancel, leaving the answer expressed appropriately in meters.

The next case requires that the time $t = 1.5$ minutes be converted into seconds in order to use the formula. Alternatively, the gravitational acceleration g could be calculated in units of meters and minutes. Since $t = 90$ seconds, it follows that

$$d = \tfrac{1}{2}(9.806)(90)^2 \text{ m} = 39{,}714.3 \text{ m}$$

Since 1 minute = 60 seconds, the gravitational constant is

$$g = 9.806 \text{ m sec}^{-2} = 9.806 \text{ m } (\tfrac{1}{60}\text{ min})^{-2} = 35{,}301.6 \text{ m min}^{-2}.$$

In these units, the distance corresponding to $t = 3/2$ minutes is

$$d = \tfrac{1}{2}(35{,}301.6 \text{ m min}^{-2})(\tfrac{3}{2} \text{ min})^2 = 39{,}714.3 \text{ m}.$$

Dimensional arguments can play a valuable role in calculus problems. The derivative of a function $y = y(x)$ expresses the rate of change of the dependent variable y with respect to the independent variable x,

$$\frac{dy}{dx} = \lim_{h \to 0} \frac{y(x+h) - y(x)}{h}.$$

From the limit definition, it is clear that the dimensions of dy/dx are the dimensions of y divided by the dimensions of x. This observation can be used to give a simple check on calculations of derivatives. For example, the function $y = x^2$ has derivative $dy/dx = 2x$. This is dimensionally correct since $2x$ does have dimensions equal to the dimensions of y divided by the dimensions of x.

As a second example, the function $y = \sin ax$ is considered to be dimensionless since the sine function is defined as the ratio of quantities of the same dimensions. The argument ax of the sine function is also considered dimensionless since it represents an angle. If the variable x has dimensions of length, then the constant a must have dimensions $(\text{length})^{-1}$ in order that ax be dimensionless. Therefore, the derivative of $y = \sin ax$ must have dimensions L^{-1} (dimensions of y divided by dimensions of x). Of course, $dy/dx = a \cos ax$ which does have dimensions L^{-1} as required.

In later sections, when derivatives and integrals of complicated functions are calculated, verifying that the resulting formulas are dimensionally correct often provides a useful check on the calculations. Even in problems that do not involve physical dimensions, it may be helpful to assign dimensions arbitrarily to the variables. Dimensional checks can then be applied to verify the consistency of all calculations.

Exercises 1.12

1. In the following formulas, x and r are distance variables, A is area, V is volume and t, v, a, m and θ represent time, velocity, acceleration, mass and angle (radians). Determine the dimensions of the following physical quantities in the form $L^\alpha T^\beta M^\gamma$.

(i) Density $\rho = m/V$
(ii) Momentum $\rho = mv$
(iii) Force $F = ma$
(iv) Pressure $P = F/A$
(v) Work $W = Fx$
(vi) Angular velocity $\omega = d\theta/dt$
(vii) Kinetic energy $E = \tfrac{1}{2} mv^2$
(viii) Moment of inertia $I = mr^2$
(ix) Angular momentum $h = mvr$
(x) Rotational kinetic energy $E = \tfrac{1}{2} mr^2\omega^2$

2. A train travels at 60 mph. Find its speed in km/h, ft $\sec^{-1}$ and m $\sec^{-1}$.

3. Determine the dimensions of the gravitational constant G in Newton's law of gravitation $F = Gm_1m_2/r^2$.

4. Suppose that the formula $s = 4t^2 - t + 1$ describes the motion of a particle along a straight line where the distance s is measured in meters and time t is measured in seconds. Since the equation is not dimensionally consistent (correct in all systems of units), define s_0 and t_0 to be the basic units of length and time, $s_0 = 1$ m, $t_0 = 1$ sec. Show that the equation

$$\frac{s}{s_0} = 4\left(\frac{t}{t_0}\right)^2 - \left(\frac{t}{t_0}\right) + 1$$

is dimensionally consistent. Calculate the velocity and acceleration and show that their formulas are dimensionally consistent.

5. The formula $V(t) = 40 - 3t - t^2$ gives the volume of water in liters remaining in a bathtub t seconds after the plug is pulled. Defining V_0 and t_0 to be the basic units of volume and time, write the formula for $V(t)$ in dimensionally consistent form. When is the bathtub empty?

6. In the previous problem, solve the dimensionally consistent equation for $V(t)$ to determine $t = t(V)$, the time when the volume left is V. Show that the equation for t is dimensionally consistent.

7. Convert $d = \frac{1}{2}\, gt^2$ where $g = 32.1725$ ft sec^{-2} to an expression appropriate for the measurement of distance in meters. More generally, define $a = 1$ foot and $b = 1$ second to be the basic units of length and time. Define d_* and t_*, dimensionless variables of length and time by $d_* = d/a$, $t_* = t/b$. Show that the original equation becomes $d_* = gb^2 t_*^{\,2}/2a$. Examine the units of gb^2/a and show it is a dimensionless constant with the same value in any system of units.

8. The distance travelled by a particle moving along a straight line with constant acceleration a is given by $s(t) = s(0) + v(0)t + \frac{1}{2}\, at^2$. By solving a quadratic equation, find the inverse function $t = t(s)$. Verify that this function has correct dimensions and interpret its meaning.

9. In a study of the motion of a particle, the velocity v is found to be a quadratic function of the distance s along the path of motion, $v = as^2 + bs + c$. Determine the dimensions of the constants a, b and c. Verify that the roots of $v = 0$ have dimensions of length.

10. The position of a weight suspended at the end of a spring is given by $h(t) = 20 + 5 \sin \omega t$ at time t. If $h(t)$ is measured in centimeters and t in seconds, what are the dimensions of the parameter ω? Determine the period of oscillation. Calculate the velocity dh/dt and the acceleration d^2h/dt^2 and verify that the formulas are dimensionally correct.

11. Given a function $y = f(x)$, draw and compare the graphs of $y = f(x)$, $y = f(x - 1)$ and $y = f(x - 2)$. Show that the graph of the function $y = f(x - t)$ where t is a parameter is the graph of $y = f(x)$ shifted t units to the right. Identifying the parameter t as a time variable, interpret the function $y = f(x - t)$ as a wave which moves to the right with velocity one as t increases. Show that $y = f(x + 2t)$ is a wave that moves to the left with velocity 2.

12. If x and t are variables of distance and time, show that $y = f(x - ct)$ is dimensionally consistent when c has dimensions of velocity. Interpret this function as a wave moving to the right with velocity c. Interpret the function $y = g(x + ct)$. Illustrate with specific examples.

13. In the function $y = A \sin(kx - \omega t)$, the variables x and y represent distances, t represents time, A, k and ω are constants. Determine the dimensions of the constants. By writing the function in the form $y = f(x - ct)$, identify the wave speed c in terms of k and ω.

14. Verify the dimensional consistency of the following derivatives. (Assign dimensions to the variables and constants and check that the formulas for the derivatives are dimensionally consistent.)

(i) $\dfrac{d}{dx}(mx + b) = m$ **(ii)** $\dfrac{d}{dx}(ax^2 + bx + c) = 2ax + b$

(iii) $\dfrac{d}{dx}(ax^3 + bx^2 + cx + d) = 3ax^2 + 2bx + c$ **(iv)** $\dfrac{d}{dx}(ax + b)^{1/2} = \dfrac{a}{2(ax + b)^{1/2}}$

15. Verify the dimensional consistency of the following integrals.

(i) $\int_a^b x\,dx = ^1\!/_2\,(b^2 - a^2)$ **(ii)** $\int_a^b x^2 dx = ^1\!/_3\,(b^3 - a^3)$

(iii) $\int_0^X (mx + b)\,dx = ^1\!/_2\,mX^2 + bX$ **(iv)** $\int_0^X (ax^2 + bx + c)\,dx = ^1\!/_3\,aX^3 + ^1\!/_2\,bX^2 + cX$

2
Differentiation

2.1 The derivative

1 The derivative of a function $u(x)$ at point x is defined as the limit

$$\frac{du}{dx} = \lim_{\Delta x \to 0} \frac{u(x + \Delta x) - u(x)}{\Delta x} = \lim_{\Delta x \to 0} \frac{\Delta u}{\Delta x}. \tag{1}$$

Other common notations are

$$\frac{du(x)}{dx} = u'(x) = Du(x), \tag{2}$$

and the adoption of one of these over another depends on the particular usage, a special interpretation, or visual (typographic) clarity. For example, the first (due to Leibnitz) is especially relevant to the geometrical interpretation of the derivative as the slope of the tangent line to a curve (Fig. 1.1). In this case, the separate quantities du and dx are understood to be the limiting forms of increments Δu and Δx; du/dx is then interpreted as the *ratio* of infinitesimal elements.

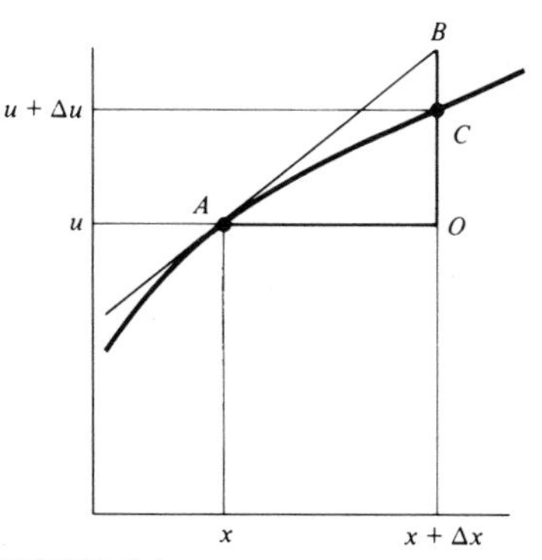

FIGURE 1.1

The second form of the derivative in (2) minimizes the symbolism required and allows cumbersome formulas to be written in an abbreviated form. The last expression will facilitate manipulation in a later section when the derivative is viewed as an operation on a function which possesses certain algebraic properties.

Several derivatives were determined in the first chapter, among them

$$\frac{d}{dx}(mx+b) = m, \qquad \frac{d}{dx}\sin x = \cos x, \qquad \frac{d}{dx}x^2 = 2x. \tag{3}$$

To review the basic procedure, a few new results are added to the foregoing.

Example 1 Find the derivative of $u = x^{1/2}$.

By definition

$$\frac{du}{dx} = \frac{d}{dx}x^{1/2} = \lim_{\Delta x \to 0} \frac{(x+\Delta x)^{1/2} - x^{1/2}}{\Delta x}.$$

The limit can be evaluated upon multiplying numerator and denominator by $(x+\Delta x)^{1/2} + x^{1/2}$, in which case

$$\frac{(x+\Delta x)^{1/2} - x^{1/2}}{\Delta x} = \frac{x + \Delta x - x}{\Delta x[(x+\Delta x)^{1/2} + x^{1/2}]} = \frac{1}{(x+\Delta x)^{1/2} + x^{1/2}}.$$

Therefore

$$\lim_{\Delta x \to 0} \frac{(x+\Delta x)^{1/2} - x^{1/2}}{\Delta x} = \lim_{\Delta x \to 0} \frac{1}{(x+\Delta x)^{1/2} + x^{1/2}} = \frac{1}{2x^{1/2}},$$

and

$$\frac{d}{dx}x^{1/2} = \frac{1}{2x^{1/2}}. \tag{4}$$

Example 2 Show that

$$\frac{d}{dx}\frac{1}{x} = -\frac{1}{x^2}. \tag{5}$$

Since

$$\frac{1}{x+\Delta x} - \frac{1}{x} = -\frac{\Delta x}{(x+\Delta x)x},$$

it follows that

$$\frac{d}{dx}\frac{1}{x} = \lim_{\Delta x \to 0} \frac{1}{\Delta x}\left(\frac{1}{x+\Delta x} - \frac{1}{x}\right) = \lim_{\Delta x \to 0} \frac{-1}{(x+\Delta x)x} = -\frac{1}{x^2}.$$

Example 3 If n is a positive integer, show that

$$\frac{d}{dx}x^n = nx^{n-1}. \tag{6}$$

The derivative is given by

$$\frac{d}{dx}x^n = \lim_{h\to 0}\frac{(x+h)^n - x^n}{h},$$

where for convenience h replaces Δx. The binomial expansion

$$(x+h)^n = x^n + nhx^{n-1} + \frac{n(n-1)}{2!}h^2x^{n-2} + \cdots + nh^{n-1}x + h^n,$$

can be used to write

$$\frac{(x+h)^n - x^n}{h} = nx^{n-1} + h\left\{\frac{n(n-1)}{2!}x^{n-2} + \cdots + nh^{n-3}x + h^{n-2}\right\}.$$

In this form the limit $h \to 0$ is straightforward, and the result is

$$\frac{d}{dx}x^n = \lim_{h\to 0}\frac{(x+h)^n - x^n}{h} = nx^{n-1} + \lim_{h\to 0} h\{\ \ \} = nx^{n-1}.$$

Example 4 Show that

$$\frac{d}{d\theta}\cos k\theta = -k\sin k\theta. \tag{7}$$

Once again the basic definition is applied. The reduction, which makes use of the expansion formula for $\cos(\theta_1 + \theta_2)$, is as follows:

$$\begin{aligned}\frac{d}{d\theta}\cos k\theta &= \lim_{\Delta\theta\to 0}\frac{\cos k(\theta+\Delta\theta) - \cos k\theta}{\Delta\theta}\\ &= \lim_{\Delta\theta\to 0}\frac{1}{\Delta\theta}[\cos k\theta\cos(k\,\Delta\theta) - \sin k\theta\sin(k\,\Delta\theta) - \cos k\theta]\\ &= \cos k\theta\lim_{\Delta\theta\to 0}\frac{\cos(k\,\Delta\theta) - 1}{\Delta\theta} - \sin k\theta\lim_{\Delta\theta\to 0}\frac{\sin(k\,\Delta\theta)}{\Delta\theta}.\end{aligned}$$

Both limits involved were evaluated in Chap. 1, where it was found that

$$\lim_{\theta\to 0}\frac{\cos k\theta - 1}{\theta} = 0, \qquad \lim_{\theta\to 0}\frac{\sin k\theta}{\theta} = k.$$

Therefore

$$\frac{d}{d\theta}\cos k\theta = -k\sin k\theta.$$

2

The derivative of a function expresses its rate of change with respect to the independent variable. In essence, the definition, (1), expresses exactly this, for it employs the increment of change in the function Δu that occurs in a distance Δx. On the other hand, if the derivative is known, Eq. (1) can be turned around and used to calculate the approximate value of $u(x)$ at a neighboring point. A simple formula for $u(x + \Delta x)$, based on a knowledge of $u(x)$ and $u'(x)$, is obtained by removing the limit sign and assuming "near equality" for small Δx:

$$u(x + \Delta x) \approx u(x) + u'(x)\,\Delta x, \qquad \text{or} \qquad \Delta u \approx u'(x)\,\Delta x. \tag{8}$$

(The symbol $\approx$ is read "is approximately equal to.") As Δx, and hence Δu, diminishes in magnitude, the accuracy of this formula improves, but the relationship is *exact* only for infinitesimally small increments, called *differentials*, which are denoted as dx and du:

$$du = u'(x)\,dx. \tag{9}$$

For the practical purpose of calculation, this *differential statement* relating infinitesimal quantities is equivalent to (8). However, it is extremely useful to treat du and dx as definite entities subject to the same algebraic rules that can be applied to the small finite increments Δu and Δx.

Example 5 Find the value of $\sqrt{100.2}$.

The key to problems of this type is to use a much simpler calculation at a nearby point. Since $\sqrt{100}$ is trivial to evaluate, this readily accessible information is exploited, using (9), to compute $\sqrt{100 + 0.2}$. First, the function involved is identified:

$$u(x) = \sqrt{x},$$

and differentiated [using (5)]:

$$u'(x) = \frac{1}{2\sqrt{x}}.$$

Next we choose $x = 100$ (because $\sqrt{100}$ is so easy), and this means that the increment in the problem at hand is

$$\Delta x = 0.2.$$

Application of (9), the practical version of (10), yields the desired approximation:

$$u(100.2) = u(100 + 0.2) \approx u(100) + u'(100)(0.2).$$

Upon substituting the values $u(100) = \sqrt{100} = 10$, $u'(100) = 1/(2\sqrt{100}) = 1/20$, we obtain

$$\sqrt{100.2} \approx 10 + 1/20(0.2) = 10.01.$$

Since $(10.01)^2 = 100.2001$, the result is an exceedingly accurate approximation for small increments.

Example 6 Estimate the value of sin 46°.

The appropriate comparison point is 45° or $\pi/4$ radians since $\sin 45° = 1/\sqrt{2}$.

Define $y(x) = \sin x$ with x measured in radians. Since the derivative of $\sin x$ is $\cos x$, we have

$$\sin(x + \Delta x) \approx \sin x + (\cos x)\,\Delta x$$

Substituting $x = \pi/4$ and $\Delta x = \pi/180$,

$$\sin 46° \approx \sin \pi/4 + \cos \pi/4(\pi/180) = (1/\sqrt{2})\,[1 + \pi/180] = 0.719448\ldots$$

The accurate value is 0.7193398... .

3

The problem of finding the error in a numerical formula (or, at the very least, a bound for it) and methods for systematically improving accuracy will be considered later. For now, we must content ourselves with some rather obvious statements about the error involved in using (9). Note first that in the vicinity of a fixed point x the expression

$$u(x + \Delta x) - u(x) - u'(x)\,\Delta x$$

is a function only of the increment Δx, which we will write $R(\Delta x)\,\Delta x$. Therefore

$$u(x + \Delta x) = u(x) + u'(x)\,\Delta x + R(\Delta x)\,\Delta x, \tag{10}$$

and since this is the exact analogy of (9), the quantity $R(\Delta x)\,\Delta x$ can be identified as the error in that approximation. Upon rearrangement, (10) can be written

$$R(\Delta x) = \frac{u(x + \Delta x) - u(x)}{\Delta x} - u'(x),$$

which must also hold in the limit,

$$\lim_{\Delta x \to 0} R(\Delta x) = \lim_{\Delta x \to 0} \frac{u(x + \Delta x) - u(x)}{\Delta x} - u'(x) = 0. \tag{11}$$

Thus the error is seen to vanish in the limit, which is not terribly surprising since that was the basis of constructing (8). In fact, until $R(\Delta x)$ is determined so that the magnitude of the error can be estimated quantitatively, we are not much better off than before, although (10) has a theoretical application in the next section.

The approximation given by (8) is the result of a local replacement of the function $u(x)$ by its tangent line at the point x, as shown in Fig. 1.1. The extent to which the real curve deviates from this line over a distance Δx is then the exact error incurred, i.e., the term $R(\Delta x)\,\Delta x$ in (10) or segment BC in the figure. The use of Eq. (8) is called *linear interpolation*, an approximate method the accuracy of which improves as Δx diminishes.

Example 7 Show that the area of a circle of radius r is πr^2.

The constant π is defined as the ratio of the circumference of the circle to the diameter $C = \pi d = 2\pi r$ (Fig. 1.2). The area of the circle must be proportional to r^2, $A(r) = ar^2$ for some constant a. A simple calculus argument determines this constant. Suppose that the radius is increased by a small amount Δr. The change in area is

$$A(r + \Delta r) - A(r) = a(r + \Delta r)^2 - ar^2 = 2ar\Delta r + a(\Delta r)^2.$$

If Δr is small (relative to r), then the change in area is approximately $2ar\Delta r$. But, from the diagram, the change in area is approximately $2\pi r\Delta r$, the circumference multiplied by Δr. Comparing the two approximations, we must have $a = \pi$. This proves the familiar result $A(r) = \pi r^2$.

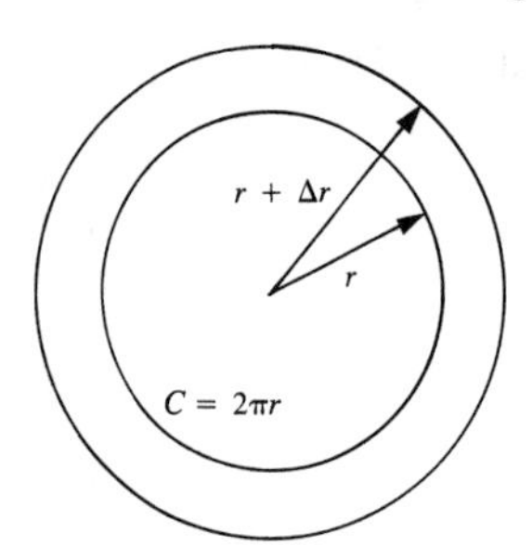

FIGURE 1.2

Exercises 2.1

1. Use the limit definition to find the derivatives of x^3, x^4 and x^5.

2. Verify the algebraic identity

$$\frac{(x+h)^{n+1} - x^{n+1}}{h} = (x+h)\left[\frac{(x+h)^n - x^n}{h}\right] + x^n$$

Use this identity and mathematical induction to give an alternative proof that the derivative of x^n is nx^{n-1} for every positive integer n.

3. Given that the derivative of $x^{1/2}$ is $x^{-1/2}/2$, use the identity of the previous problem to find the derivatives of $x^{3/2}$, $x^{5/2}$ and $x^{7/2}$.

4. Verify the algebraic identity

$$\frac{(x+h)^n - x^n}{h} = x^{n-1}\left[1 + \left(1 + \frac{h}{x}\right) + \left(1 + \frac{h}{x}\right)^2 + \ldots + \left(1 + \frac{h}{x}\right)^{n-1}\right]$$

(valid for n a positive integer) by comparing to the sum of the finite geometric series

$$1 + r + r^2 + \ldots + r^{n-1} = (1 - r^n)/(1 - r).$$

Use this result to evaluate the derivative of x^n for every positive integer n.

5. Use the definition $\lim_{\Delta x \to 0} \dfrac{f(x + \Delta x) - f(x)}{\Delta x} = f'(x)$ to deduce the following.

(i) $\lim_{\Delta x \to 0} \dfrac{f(x) - f(x - \Delta x)}{\Delta x} = f'(x)$ **(ii)** $\lim_{\Delta x \to 0} \dfrac{f(x + \Delta x) - f(x - \Delta x)}{2\Delta x} = f'(x)$

Interpret these limits graphically.

6. If n is a positive integer and c is any constant, use the limit definition of the derivative to show that the derivative of cx^n is cnx^{n-1}. Is this a general property of derivatives?
7. If n is a positive integer, use the basic definition of the derivative to show that

$$\frac{d}{dx}\left(\frac{1}{x^n}\right) = \frac{-n}{x^{n+1}}.$$

State the general result for the derivative of x^n where n is any integer.
8. Use the limit definition to determine the derivatives of the following functions.

(i) $(1+x)^2$ **(ii)** $(2-x)^3$ **(iii)** $(1+x)^{1/2}$
(iv) $x^{3/2}$ **(v)** $x^{1/3}$ **(vi)** $(1+x)^{-1}$

9. From the definition of a derivative, verify the following identities.

(i) $\frac{d}{dx}[x+f(x)] = 1+f'(x)$ **(ii)** $\frac{d}{dx}[xf(x)] = xf'(x)+f(x)$

(iii) $\frac{d}{dx}(f(x))^2 = 2f(x)f'(x)$ **(iv)** $\frac{d}{dx}\left(\frac{1}{f(x)}\right) = \frac{-f'(x)}{(f(x))^2}$

(v) $\frac{d}{dx}(f(x)+g(x)) = f'(x)+g'(x)$ **(vi)** $\frac{d}{dx}(f(x)-g(x)) = f'(x)-g'(x)$

10. Find, wherever possible, the derivatives of the following functions. Illustrate with graphs.

(i) $f(x) = |x|$ **(ii)** $f(x) = |x|^n$ (n a positive integer)

(iii) $f(x) = |x| + |x-1|$ **(iv)** $f(x) = \cos|x|$

(v) $f(x) = \begin{cases} 2x & x \leq 1 \\ 2x^2 & x > 1 \end{cases}$ **(vi)** $f(x) = \frac{1}{2x+3}$

11. For what values of x do the functions $y = 2x^3 + 10x - 1$ and $y = -4/x$ have the same derivative? Draw graphs of these functions and draw the tangent lines at these points.
12. Find the point where the tangent line to the curve $y = x^3$ at $x = 2$ crosses the line $x = -2$. Illustrate with a graph.
13. For what values of the constants m and b is the line $y = mx + b$ tangent to the parabola $y = x^2 + 1$? (Solve by an algebraic method and by a calculus method.) Find the two tangent lines that pass through the origin. Illustrate with a graph.
14. The path of an electron in a magnetic field is given by the curve $y = 2 - 3x^2$. When the particle is at $x = 2$, it is ejected from the field and proceeds along the tanget line. Find the equation of the tangent line and graph the trajectory of the electron. Where does the electron cross the line $x = 8$?
15. Evaluate the derivatives of the following functions.

(i) $\cos 2\theta$ **(ii)** $\sin^2\theta$ **(iii)** $\cos\theta - \sin\theta$
(iv) $x + 1/x$ **(v)** $(a^2+x^2)^2$ **(vi)** $\cos^2\theta - \sin^2\theta$

16. Using differentials, compute approximate values of the following numbers.

(i) $\sqrt{24}$ **(ii)** $\sqrt{165}$ **(iii)** $(1.01)^5$
(iv) $(3.99)^3$ **(v)** $(1.02)^{-3}$ **(vi)** $(0.98)^6$

With a calculator, determine the accurate values.
17. Estimate the values of sin 31° and cos 31° using differentials. Determine the absolute and the relative errors in these estimates.
18. Find an approximation to $(49+\epsilon)^{1/2}$ valid when ϵ is small. Use this approximation to estimate $\sqrt{50}$ and $\sqrt{48}$. Using a calculator, determine the absolute and relative errors in these estimates.

19. Find a linear approximation to the function $y = x^2$ valid near $x = 1$. What is the error in the approximation when $x = 1.05$? when $x = 1.1$? What is the error as a function of x? Indicate this error on a graph of $y = x^2$.
20. For the cubic polynomial $y = x^3 - 3x + 1$, use differentials to approximate its values at $x = 1.03$, 2.01 and 3.02. Compare the estimates to the accurate values.
21. Use differentials to approximate the following functions at $x = 0.98$ and $x = 1.03$. Compare the approximations to the accurate values.

(i) $x^2 + 2x + 1$ **(ii)** $x^3 + 3x^2 + 3x + 1$ **(iii)** $x^{1/2}$
(iv) $\sin^2 x$ **(v)** $\cos^2 x$ **(vi)** $x + x^{-1}$

22. The volume of a sphere of radius r is $V = \frac{4}{3}\pi r^2$. Show that $dV = 4\pi r^2 dr$. Interpret this geometrically by drawing two concentric spheres of radii r and $r + dr$.
23. One half of the total mass of the atmosphere lies below an altitude of 5500 meters. Use differentials to estimate the volume of this part of the atmosphere. (The radius of the earth is 6370 km.)

2.2 *Formulas for differentiation*

1

This section is concerned almost exclusively with the development of technique and analytical skill. Applications demand technical proficiency, and it is almost meaningless in this context to speak of a solution "in principle" but not in fact. The object is to gain knowledge—to learn—and the extraction of useful information from a complicated analysis is often the major task confronting the applied mathematician.

The differentiation of complicated functions is greatly simplified by using general formulas, five of which are established next. If $u(x)$ and $v(x)$ are arbitrary differentiable functions and c is any constant, then:

(i) $$\frac{d}{dx}(u + v) = \frac{du}{dx} + \frac{dv}{dx}$$ Addition Rule

(ii) $$\frac{d}{dx}(cu) = c\frac{du}{dx}$$ Constant Rule

(iii) $$\frac{d}{dx}(uv) = u\frac{dv}{dx} + \frac{du}{dx}v$$ Product Rule

(iv) $$\frac{d}{dx}\left(\frac{u}{v}\right) = \frac{1}{v^2}\left(v\frac{du}{dx} - u\frac{dv}{dx}\right)$$ Quotient Rule

(v) $$\frac{d}{dx}u(v(x)) = \frac{du(v)}{dv}\frac{dv(x)}{dx}$$ Chain Rule

2 The first two results are simple exercises in manipulating limits

$$\begin{aligned}\frac{d}{dx}(u(x)+v(x)) &= \lim_{h\to 0}\frac{u(x+h)+v(x+h)-u(x)-v(x)}{h}\\ &= \lim_{h\to 0}\frac{u(x+h)-u(x)}{h}+\lim_{h\to 0}\frac{v(x+h)-v(x)}{h}\\ &= \frac{du}{dx}+\frac{dv}{dx}.\\ \frac{d}{dx}(cu(x)) &= \lim_{h\to 0}\frac{cu(x+h)-cu(x)}{h}\\ &= c\lim_{h\to 0}\frac{u(x+h)-u(x)}{h}\\ &= c\frac{du}{dx}.\end{aligned}$$

To prove the *product rule* requires some algebraic rearrangement:

$$\begin{aligned}\frac{d}{dx}(u(x)\,v(x)) &= \lim_{h\to 0}\frac{u(x+h)\,v(x+h)-u(x)\,v(x)}{h}\\ &= \lim_{h\to 0}\left\{u(x+h)\left[\frac{v(x+h)-v(x)}{h}\right]+v(x)\left[\frac{u(x+h)-u(x)}{h}\right]\right\}\\ &= u(x)\frac{dv}{dx}+v(x)\frac{du}{dx}.\end{aligned}$$

The *quotient rule* can be proved by evaluating a limit or, more simply, by a clever application of the product rule. Since the function u is the product of the functions v and u/v, we have

$$\frac{du}{dx}=\frac{d}{dx}\left(v\frac{u}{v}\right)=v\frac{d}{dx}\left(\frac{u}{v}\right)+\frac{u}{v}\frac{dv}{dx}.$$

Solving for the derivative of the quotient,

$$\frac{d}{dx}\left(\frac{u}{v}\right)=\frac{1}{v^2}\left(v\frac{du}{dx}-u\frac{dv}{dx}\right).$$

This formula is valid if $v(x) \neq 0$.

The *chain rule* expresses the derivative of a composite function of x. In the function $u(v(x))$, u is a function of v which is a function of x. We know that

$$\frac{du}{dv}=\lim_{\Delta v\to 0}\frac{u(v+\Delta v)-u(v)}{\Delta v} \quad\text{and}\quad \frac{dv}{dx}=\lim_{h\to 0}\frac{v(x+h)-v(x)}{h}.$$

From the limit definition, the derivative of $u(v(x))$ is

$$\begin{aligned}\frac{d}{dx}u(v(x)) &= \lim_{h\to 0}\frac{u(v(x+h)) - u(v(x))}{h}\\ &= \lim_{h\to 0}\frac{u(v(x+h) - u(v(x))}{v(x+h) - v(x)}\,\frac{v(x+h) - v(x)}{h}\\ &= \lim_{\Delta v\to 0}\frac{u(v+\Delta v) - u(v)}{\Delta v}\lim_{h\to 0}\frac{v(x+h) - v(x)}{h}\\ &= \frac{du}{dv}\frac{dv}{dx}.\end{aligned}$$

We have used the identity $\Delta v = v(x+h) - v(x)$ and the fact that, as h approaches 0, Δv also approaches zero.

---

3 These five general formulas are now used to differentiate a number of functions in order to illustrate the techniques and methods involved. With some imagination, these rules allow the derivatives of many functions to be calculated with very little effort.

Example 1 Differentiate the polynomial

$$u = 1 - 2x + 2x^2 - 2x^3 + x^4.$$

It follows rather simply from rule (i) that the derivative of a finite sum of functions is the sum of the individual derivatives (see Exercise 8). Therefore

$$\frac{du}{dx} = \frac{d}{dx}1 - \frac{d}{dx}2x + \frac{d}{dx}2x^2 - \frac{d}{dx}2x^3 + \frac{d}{dx}x^4.$$

The calculation is completed using rule (ii) and the general formula

$$\frac{d}{dx}x^n = nx^{n-1};$$

the final result is

$$\frac{du}{dx} = 0 - 2 + 4x - 6x^2 + 4x^3,$$

or

$$\frac{d}{dx}(1 - 2x + 2x^2 - 2x^3 + x^4) = -2(1 - 2x + 3x^2 - 2x^3).$$

Example 2 Show that

$$\frac{d}{dx}\sin kx = k\cos kx. \tag{5}$$

This becomes a simple illustration of the chain rule if we identify

$$u = \sin v \qquad \text{and} \qquad v = kx.$$

Since

$$\frac{du}{dv} = \cos v = \cos kx \qquad \text{and} \qquad \frac{dv}{dx} = k,$$

it follows that

$$\frac{d}{dx} \sin kx = \frac{du}{dv}\frac{dv}{dx} = k \cos kx.$$

Example 3 Show that

$$\frac{d}{dx} \tan ax = a \sec^2 ax. \tag{6}$$

In this case write

$$\tan ax = \frac{\sin ax}{\cos ax},$$

and apply formula (iv) with $u = \sin ax$ and $v = \cos ax$. Since

$$\frac{d}{dx} \sin ax = a \cos ax \qquad \text{and} \qquad \frac{d}{dx} \cos ax = -a \sin ax,$$

it follows that

$$\frac{d}{dx} \tan ax = \frac{d}{dx}\frac{\sin ax}{\cos ax} = \frac{a}{\cos^2 ax}(\cos^2 ax + \sin^2 ax)$$

$$= \frac{a}{\cos^2 ax} = a \sec^2 ax.$$

Example 4 Differentiate the functions

$$(1 - 2x + x^2)(1 + x^2) \qquad \text{and} \qquad \frac{1 - 2x + x^2}{1 + x^2}.$$

In both these cases, set

$$u(x) = 1 - 2x + x^2 = (1 - x)^2,$$

and

$$v(x) = 1 + x^2,$$

where, according to rules (i) and (ii),

$$\frac{du}{dx} = -2 + 2x = -2(1 - x), \qquad \frac{dv}{dx} = 2x. \tag{7}$$

The differentiation of the product $u(x) \cdot v(x)$ is accomplished using rule (iii) and Eq. (7):

$$\frac{d}{dx} uv = \frac{d}{dx}[(1 - 2x + x^2)(1 + x^2)] = u\frac{dv}{dx} + v\frac{du}{dx}$$
$$= (1 - x)^2 2x - 2(1 + x^2)(1 - x).$$

The final result is

$$\frac{d}{dx}(1 - 2x + x^2)(1 + x^2) = -2(1 - x)(1 - x + 2x^2), \tag{8}$$

which agrees with the answer from Example 1, where the derivative was calculated by first completing the multiplication of u and v.

The quotient, $u(x)/v(x)$, is differentiated using rule (iv) and (7):

$$\frac{d}{dx}\frac{u}{v} = \frac{1}{v^2}\left(v\frac{du}{dx} - u\frac{dv}{dx}\right).$$

Therefore

$$\frac{d}{dx}\frac{(1 - x)^2}{1 + x^2} = \frac{1}{(1 + x^2)^2}\{(1 + x^2)[-2(1 - x)] - 2(1 - x)^2 x\}$$
$$= \frac{-2(1 - x^2)}{(1 + x^2)^2}. \tag{9}$$

Example 5 Differentiate

$$u(x) = (1 + x^2)^{10}. \tag{10}$$

In this case, make the identifications

$$v(x) = 1 + x^2 \qquad \text{and} \qquad u(v) = v^{10}.$$

To apply rule (v) we must determine the individual components of the formula

$$\frac{du}{dx} = \frac{du}{dv}\frac{dv}{dx}.$$

Since

$$\frac{du}{dv} = 10v^9 \qquad \text{and} \qquad \frac{dv}{dx} = 2x,$$

the result is

$$\frac{d}{dx}(1 + x^2)^{10} = (10v^9)(2x) = 20x(1 + x^2)^9.$$

Example 6 Differentiate $\sin^n[z + (1 + z)^{1/2}]$.

Let $v(z) = z + (1 + z)^{1/2}$ and $u = \sin v$; then the derivative to be calculated is

$$\frac{d}{dz}\sin^n[z + (1+z)^{1/2}] = \frac{d}{dz}[u(v)]^n.$$

Successive application of the chain rule yields

$$\frac{du^n}{dz} = \frac{du^n}{du}\cdot\frac{du}{dz} = \frac{du^n}{du}\cdot\frac{du}{dv}\cdot\frac{dv}{dz}.$$

Since

$$\frac{du^n}{du} = nu^{n-1}, \qquad \frac{du}{dv} = \cos v, \qquad \text{and} \qquad \frac{dv}{dz} = 1 + \frac{1}{2(1+z)^{1/2}},$$

and the final result is

$$\frac{d}{dz}\sin^n[z + (1+z)^{1/2}]$$

$$= n\{\sin^{n-1}[z + (1+z)^{1/2}]\}\{\cos[z + (1+z)^{1/2}]\}\left[1 + \frac{1}{2(1+z)^{1/2}}\right].$$

Example 7 Derive a general formula for the derivative of a triple product

$$y = u(x)\cdot v(x)\cdot w(x).$$

If $v(x)\cdot w(x)$ is first treated as a single function, rule (iii) can be applied to the product

$$y = u(x)\cdot[v(x)\cdot w(x)],$$

with the result

$$\frac{dy}{dx} = \frac{du}{dx}[v(x)\cdot w(x)] + u(x)\frac{d}{dx}[v(x)\cdot w(x)].$$

But the same rule implies that

$$\frac{d}{dx}[v(x)\cdot w(x)] = \frac{dv}{dx}w + v\frac{dw}{dx},$$

and the substitution of this in the preceding yields the desired formula:

$$y'(x) = u'(x)v(x)w(x) + u(x)v'(x)w(x) + u(x)v(x)w'(x). \tag{11}$$

Exercises 2.2

1. Calculate the derivatives of the following functions:

(i) $1 + 2x + x^2$ **(ii)** $1 + 3x + 3x^2 + x^3$ **(iii)** $1 - 4x^2 + 4x^4$

(iv) $x^7 - x^5 + x^3 - x$ **(v)** $(x^2 + 1)^4$ **(vi)** $(3x + 3x^2)^5$

(vii) $4x^2 + \dfrac{1}{4x^2}$ **(viii)** $\dfrac{x^2 - 1}{x^2 + 1}$ **(ix)** $(x^2 + 1)^{-4}$

2. Verify the following:

(i) $\dfrac{d}{dx}\sec x = \sec x \tan x$ **(ii)** $\dfrac{d}{dx}\csc x = -\csc x \cot x$

(iii) $\frac{d}{dx}\cot x = -\csc^2 x$ **(iv)** $\frac{d}{dx}\sin^2 x = \sin 2x$

3. Determine the derivatives of sec ax, csc ax and cot ax where a is any constant.

4. Differentiate $x^{1/3}$ by writing $(x^{1/3})^3 = x$ and using the chain rule or the product rule.

5. Differentiate $x^{1/m}$ where m is any positive integer by writing $(x^{1/m})^m = x$.

6. Differentiate $x^{n/m}$ (n and m positive integers) by writing $(x^{1/m})^n = x^{n/m}$.

7. Differentiate the following functions:

(i) $x^{1/3}$ **(ii)** $3x^{1/4}$ **(iii)** $x^{1/2} + x^{-1/2}$
(iv) $x^{3/5} + x + x^{7/5}$ **(v)** $2x^{1/2} + 4x^{1/4} + 6x^{1/6}$ **(vi)** $(x^{1/2} + x^{-1/2})^2$

8. Prove the general addition rule.

$$\frac{d}{dx}[u_1(x) + u_2(x) + \ldots + u_n(x)] = \frac{du_1}{dx} + \frac{du_2}{dx} + \ldots + \frac{du_n}{dx}.$$

9. Differentiate the following functions:

(i) $(x^2 - 1)(x^2 + 1)$ **(ii)** $(2x + x^2)^{1/2}$ **(iii)** $x^3/(1 + x^3)$
(iv) $(3 - 2x)^{1/2}$ **(v)** $(1 + x^2)^{5/2}$ **(vi)** $(1 + x^2)^{-10}$
(vii) $x^3/(1 + x^2)^3$ **(viii)** $(1 + 2x + x^2)^{1/2}$ **(ix)** $(1 - 3x + 3x^2 - x^3)^{1/3}$

10. Differentiate the following functions:

(i) $\cos x + \sec x$ **(ii)** $\tan 2x$ **(iii)** $x \sin x + \cos x$
(iv) $\tan x^3$ **(v)** $\sin^2 x + \cos^2 x$ **(vi)** $\tan(\cos x)$
(vii) $\frac{1 + \sin x}{1 - \sin x}$ **(viii)** $\frac{1 + \tan x}{1 - \tan x}$ **(ix)** $\frac{\sin 2x}{2 \sin x}$

11. If x is measured in degrees, determine the derivatives of the six trigonometric functions $\sin x$, $\cos x$, $\tan x$, $\csc x$, $\sec x$ and $\cot x$.

12. Extend the chain rule to apply to composite functions of the form $u(v(w(x)))$. Use this result to differentiate the following functions. Identify u, v and w.

(i) $(\sqrt{2x + x^2})^3$ **(ii)** $[\sin(1 + x^2)]^{-1}$ **(iii)** $(1 + x^2)^5$
(iv) $(\sqrt{x} - 1)^{3/2}$ **(v)** $[\tan(x + x^2)]^2$ **(vi)** $\sin(\cos(1 + x^2))$

13. Differentiate the following functions:

(i) $y = \frac{1}{x + x^2}$ **(ii)** $u = \frac{1}{\sqrt{1 - s^2}}$ **(iii)** $u = \sin\frac{t}{1 - t}$
(iv) $v = (\sin t^{1/2} + 1)^{1/2}$ **(v)** $w = \sin 3x - \sin 4x$ **(vi)** $y = x(1 - x^2)^{-1/2}$
(vii) $y = \sqrt{1 + \sin x}$ **(viii)** $y = \frac{2 \tan x}{1 - \tan^2 x}$ **(ix)** $y = \left(\frac{1 + x}{1 - x}\right)^{1/2}$

14. Two functions $f(x)$ and $g(x)$ are said to be *inverse functions* if $f(g(x)) = g(f(x)) = x$. For example, $f(x) = x^3$ and $g(x) = x^{1/3}$ are inverse functions. Use the chain rule to show that $g'(x) = 1/df/dg = dg/df = 1/f'(x)$.

15. Differentiate both sides of the following binomial expansion and verify the result.

$$(1 + x)^n = 1 + nx + \frac{n(n-1)}{2!}x^2 + \ldots + nx^{n-1} + x^n$$

16. Differentiate the following identities and verify the results.

(i) $x^{3/2} = x^2/x^{1/2}$ **(ii)** $x^3 = (x^{3/4})^4$ **(iii)** $\sin^2 x + \cos^2 x = 1$
(iv) $\sin x = \sqrt{1 - \cos^2 x}$ **(v)** $\sin x = \cos x \tan x$ **(vi)** $\sec^2 x = 1 + \tan^2 x$

17. Differentiate the function $f(x)/f(x) = f(x)(1/f(x))$ by the quotient rule and the product rule and verify that the derivative is zero.

2.3 *Higher derivatives: stationary points*

The derivative of a function is itself a legitimate function which can also be differentiated. The second derivative, or the derivative of a derivative, can then be obtained and denoted by any of the following:

$$\frac{d}{dx}\left[\frac{d}{dx}u(x)\right] = \frac{d^2}{dx^2}u(x) = u''(x) = D^2u(x).$$

Moreover, the numerical value, say at $x = a$, is written

$$\left.\frac{d^2u}{dx^2}\right]_{x=a}, \qquad u''(a), \qquad \text{or} \qquad D^2u(a).$$

Obviously, the definition can be generalized to even higher orders, and the kth derivative of $u(x)$ is

$$\underbrace{\frac{d}{dx}\frac{d}{dx}\cdots\frac{d}{dx}}_{k\text{ times}}u(x) = \frac{d^k}{dx^k}u(x) = u^{(k)}(x) = D^ku(x). \tag{1}$$

Example 1 Find the kth derivative of the function $u(x) = x^n$, where n is a positive integer.

The answer is obtained by repeatedly applying the formula for the derivative of an integral power of x. The first derivative is

$$\frac{du}{dx} = nx^{n-1},$$

and the second is

$$\frac{d^2u}{dx^2} = \frac{d}{dx}\left(\frac{du}{dx}\right) = \frac{d}{dx}nx^{n-1} = n\frac{d}{dx}x^{n-1} = n(n-1)x^{n-2}.$$

Continuing in this fashion, we find that the kth derivative is

$$\frac{d^ku}{dx^k} = n(n-1)\cdots(n-k+1)x^{n-k} = \frac{n!}{(n-k)!}x^{n-k}. \tag{2}$$

In particular, for $k = n$,

$$\frac{d^n}{dx^n}x^n = n!, \tag{3}$$

and all higher derivatives beyond this are identically zero:

$$\frac{d^k}{dx^k}x^n = 0 \qquad \text{for} \qquad k > n. \tag{4}$$

The rules established in the last section apply at each stage of higher-order differentiation, and it is rather easy to prove, for example, that the kth derivative of a sum of functions is just the sum of the kth derivatives:

$$\frac{d^k}{dx^k}[u_1(x) + \cdots + u_m(x)] = \frac{d^k}{dx^k}u_1(x) + \cdots + \frac{d^k}{dx^k}u_m(x),$$

or, written more succinctly using the summation notation,

$$\frac{d^k}{dx^k}\sum_{i=1}^{m} u_i(x) = \sum_{i=1}^{m} \frac{d^k}{dx^k} u_i(x).$$

Formulas for product differentiation are also simply derived but cumbersome as a typical calculation illustrates:

$$\begin{aligned}\frac{d^2}{dx^2}u(x)\cdot v(x) &= \frac{d}{dx}\left(\frac{d}{dx}u\cdot v\right) = \frac{d}{dx}\left(u\frac{dv}{dx} + v\frac{du}{dx}\right)\\ &= \frac{d}{dx}\left(u\frac{dv}{dx}\right) + \frac{d}{dx}\left(v\frac{du}{dx}\right)\\ &= u\frac{d^2v}{dx^2} + 2\frac{du}{dx}\frac{dv}{dx} + v\frac{d^2u}{dx^2}.\end{aligned} \tag{5}$$

The third derivative of a product has a similar formula

$$\frac{d^3}{dx^3}(uv) = u\frac{d^3v}{dx^3} + 3\frac{du}{dx}\frac{d^2v}{dx^2} + 3\frac{d^2u}{dx^2}\frac{dv}{dx} + \frac{d^3u}{dx^3}v,$$

and the general result, known as Leibniz's rule, is

$$\frac{d^n}{dx^n}(uv) = \sum_{k=0}^{n}\binom{n}{k}\frac{d^k}{dx^k}u\cdot\frac{d^{n-k}v}{dx^{n-k}} \qquad \text{where } \binom{n}{k} = \frac{n!}{k!(n-k)!}$$

It is not necessary to commit such formulas to memory; it is important instead to understand the basic procedure well enough so that results like (5) can be derived whenever needed.

Example 2 Find all the derivatives of the polynomial $p(x) = x^4 - x^3$.
Repeated differentiation yields

$$\begin{aligned}p'(x) &= 4x^3 - 3x^2,\\ p''(x) &= 12x^2 - 6x,\\ p'''(x) &= 24x - 6,\\ p^{(4)}(x) &= 24,\end{aligned}$$

and all higher derivatives are zero. Of course, any one of these formulas could be obtained directly from (2) and (4). To illustrate, consider the calculation of the third derivative:

$$\frac{d^3}{dx^3}\,p(x) = \frac{d^3x^4}{dx^3} - \frac{d^3x^3}{dx^3}.$$

The two terms on the right-hand side are both special cases of (2) which correspond, respectively, to $k = 3$, $n = 4$ and $k = 3$, $n = 3$. A minor calculation reproduces the expression obtained above.

Example 3 Find the first, second and third derivatives of $x^2 \sin x$.

Defining $u(x) = x^2$, $v(x) = \sin x$, the Leibniz rule could be used to find the derivatives. It is simpler to differentiate directly using the product rule.

$$\begin{aligned}
\frac{d}{dx}(x^2 \sin x) &= 2x \sin x + x^2 \cos x \\
\frac{d^2}{dx^2}(x^2 \sin x) &= \frac{d}{dx}(2x \sin x + x^2 \cos x) \\
&= 2 \sin x + 4x \cos x - x^2 \sin x \\
\frac{d^3}{dx^3}(x^2 \sin x) &= \frac{d}{dx}(2 \sin x + 4x \cos x - x^2 \sin x) \\
&= 6 \cos x - 6x \sin x - x^2 \cos x.
\end{aligned}$$

2

An alternative statement of Eq. (4) is that the kth derivative of *any* polynomial of degree $k - 1$ or lower is identically zero. It is also useful to express this property in reverse: a function which satisfies the condition that its kth derivative is everywhere zero,

$$\frac{d^k}{dx^k}\,u(x) = 0, \tag{6}$$

is the general polynomial

$$p_{k-1}(x) = a_{k-1}x^{k-1} + a_{k-2}\,x^{k-2} + \cdots + a_0. \tag{7}$$

The information contained in (6) specifies only the kth derivative of an unknown function, $u(x)$, and for this reason it is called a *differential equation*. The polynomial is a *solution* of this differential equation. In fact, it is *the* general solution: every function which satisfies (6) is a polynomial corresponding to particular choices for the coefficients $a_0, a_1, \ldots, a_{k-1}$. (In particular, the solution of $u' = 0$ is $u =$ constant.)

Example 4 Find the second derivatives of the trigonometric functions $u(x) = \sin \omega x$, $v(x) = \cos \omega x$.

Since it has already been established that

$$\frac{du}{dx} = \omega \cos \omega x = \omega v(x), \qquad \frac{dv}{dx} = -\omega \sin \omega x = -\omega u(x),$$

the second derivatives are now easily calculated:

$$\frac{d^2u}{dx^2} = \omega \frac{dv}{dx} = -\omega^2 \sin \omega x = -\omega^2 u(x),$$

$$\frac{d^2v}{dx^2} = -\omega \frac{du}{dx} = -\omega^2 \cos \omega x = -\omega^2 v(x).$$

Both trigonometric functions satisfy the same differential equation

$$\frac{d^2y}{dx^2} + \omega^2 y = 0, \tag{8}$$

and the linear combination (with A and B arbitrary constants)

$$y = A \sin \omega x + B \cos \omega x$$

turns out to be the most general solution.

Example 5 A new function $E(x)$ is specified by the property that it is equal to its derivative at every point. Find the higher derivatives of this function and evaluate them at $x = 0$, assuming further that $E(0) = 1$.

By definition

$$\frac{d}{dx} E(x) = E(x). \tag{9}$$

The differentiation of this equation yields

$$\frac{d^2}{dx^2} E(x) = \frac{d}{dx} E(x),$$

or, using (9),

$$\frac{d^2}{dx^2} E(x) = E(x).$$

Evidently, the second derivative is also equal to the original function, and this is generally true for all the higher derivatives:

$$\frac{d^k}{dx^k} E(x) = E(x). \tag{10}$$

Since $E(0) = 1$, it follows that the value of every higher derivative at the origin is unity:

$$\frac{d^k}{dx^k} E(x)\bigg]_{x=0} = E^{(k)}(0) = 1.$$

The function $E(x)$ satisfies the simplest of differential equations, and for this and other reasons it is studied in detail in Sec. 2.8, named the *exponential function*, and incorporated into the select group of *known* functions.

---

3 The rate at which a function $y = f(x)$ changes with respect to the independent variable x is expressed by the first derivative, $f'(x)$. When $f'(x) > 0$, the value of the function increases with increasing x; likewise $f(x)$ decreases as x increases wherever $f'(x) < 0$.

As a specific illustration of these general remarks, consider the function

$$F(x) = x^3 + 3x^2 - 1, \tag{11}$$

with first and second derivatives

$$F'(x) = 3(x^2 + 2x) \qquad \text{and} \qquad F''(x) = 6(x + 1).$$

These functions are shown in Fig. 3.1, from which it is apparent that $F'(x)$ is positive on segments AB and DE but negative on BD. This corresponds to the observation that $F(x)$ increases with x on AB and DE but decreases on BD.

At the points where its first derivative is exactly zero, a function exhibits almost no local change in value. A point where $f'(x) = 0$ *may* be a relative maximum (point B or $x = -2$ in Fig. 3.1) or a relative minimum (point D or $x = 0$). Both types are jointly referred to as *relative extrema* or *extremal points*. The adjective relative refers to a comparison with nearby values of the function. (Sometimes a more subtle variation is encountered where $f'(x) = 0$, and this will be described shortly.)

Obviously, the nature of an extremal point is dictated by the local behavior of the function. A point $x = x_*$, is a *relative maximum* if $f(x_*)$ is larger than $f(x)$ for all x in some small neighborhood about this position. More precisely, a relative maximum exists at x_* if there is a number δ such that

$$f(x_*) > f(x) \qquad \text{for} \qquad 0 < |x - x_*| < \delta.$$

A *relative minimum* occurs at x_* if

$$f(x_*) < f(x) \qquad \text{for} \qquad 0 < |x - x_*| < \delta.$$

If the function $f(x)$ has a continuous derivative in the neighborhood of $x = x_*$, then either of the foregoing conditions for relative extrema implies that $f'(x_*) = 0$, a cri-

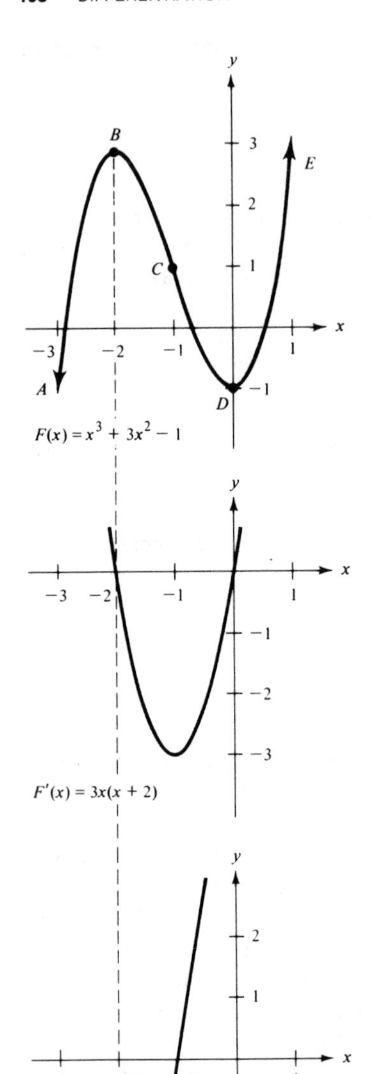

FIGURE 3.1

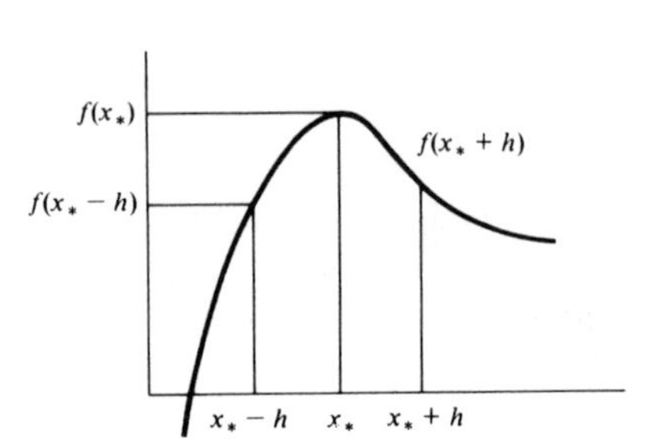

FIGURE 3.2

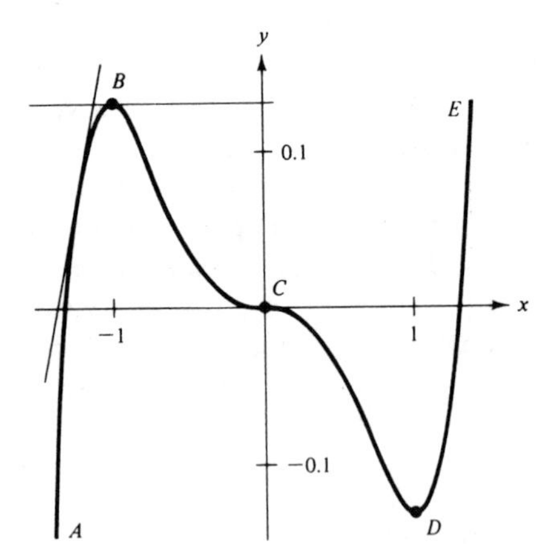

FIGURE 3.3

terion that is quite obvious from a geometrical viewpoint. The proof is as follows. Consider the relative maximum at x_* (Fig. 3.2) and two neighboring points, $x_* - h$ and $x_* + h$; then, by definition,

$$f(x_*) - f(x_* - h) > 0 \qquad \text{and} \qquad f(x_* + h) - f(x_*) < 0.$$

The derivative at x_* can be calculated either as a limit from the left, in which case

$$f'(x_*) = \lim_{h \to 0} \frac{f(x_*) - f(x_* - h)}{h} \geq 0,$$

or from the right,

$$f'(x_*) = \lim_{h \to 0} \frac{f(x_* + h) - f(x_*)}{h} \leq 0.$$

Since $f'(x)$ is assumed to be continuous, the only consistent possibility is

$$f'(x_*) = 0. \tag{12}$$

If the local difference $f(x) - f(x_*)$ is either positive or negative, depending on whether x is on one or the other side of x_*, then x_* cannot be a relative extremum *even when* $f'(x_*) = 0$. Equation (12) is a necessary but not a sufficient condition.

Values of x which satisfy $f'(x) = 0$ are collectively called *stationary points*. We have now demonstrated that all relative extrema are stationary points, but the reverse is not necessarily true.

An examination of the local behavior of a function is *the* foolproof method of showing that a point is (or is not) a relative extremum. However, a simpler test is desirable, even if it is not all-inclusive. A very effective test for extremal points is based on the sign of the second derivative $f''(x)$.

The sign of $f''(x)$ is related to the *concavity* of the curve $y = f(x)$. In a region where $f''(x) > 0$, $f'(x)$ must increase as x increases, and this means the curve is *concave upward*. This is the case in the neighborhood of the relative minimum point D in Fig. 3.1. An upward concavity typifies the relative minimum. On the other hand, wherever $f''(x) < 0$, $f'(x)$ is a decreasing function of x and the curve is *concave downward*. As shown by point B in the illustration, a relative maximum is a position of downward concavity. We conclude then that extremal points are characterized by $f'(x) = 0$ and the local concavity of the curve. These conditions are summarized as follows:

If $f'(x) = 0$ and $f''(x) < 0$, x is a relative maximum.

If $f'(x) = 0$ and $f''(x) > 0$, x is a relative minimum.

Relative extrema of $F(x) = x^3 + 3x^2 - 1$ are located among the stationary values given by

$$F'(x) = 3(x^2 + 2x) = 0,$$

that is, $x = -2$, $x = 0$. Since $F''(-2) = -6$ and $F''(0) = 6$, the point $x = -2$ is a relative maximum while $x = 0$ is a relative minimum, both rather obvious features in the graph.

This test for extrema is not informative when the second derivative is zero at a stationary point, that is, $f''(x) = 0$ *and* $f'(x) = 0$. A return to more fundamental ideas is then in order.

Since the sign of $f''(x)$ determines the concavity of the curve, points which satisfy $f''(x) = 0$ *may* be positions where the concavity changes. In that event, the location is called a *point of inflection*, a position where the curve crosses its tangent line. By definition, x_* is a point of inflection if $f''(x_*) = 0$ *and* the second derivative changes sign there.

For the illustrative example, $F''(x) = 6(x + 1)$, which implies that $x = -1$ is the only possible inflection point. The concavity does change there, and this certifies point C as an inflection point.

A stationary point x_* may, if $f''(x_*) = 0$, be an inflection point or an extremum. Only closer inspection of the functional variation about this point can resolve the question.

Of course, this discussion presumes that the function involved has continuous derivatives at least to second order. However, it is possible for the extremal points of a function to be located either at positions where the higher derivatives are, in fact, discontinuous or, more likely, at the end point of its domain of definition. For example, $y = 1 - |x|$, a continuous function, has an absolute maximum at $x = 0$, where the first derivative is discontinuous. In the interval $1 < x < 3$, say, the absolute maximum is at $x = 1$, where $y = 0$ and $y'(1) = -1$, and the absolute minimum is at the other end, where $x = 3$ and $y = -2$, $y'(3) = -1$.

Example 6 Graph the function $f(x) = x^5/5 - x^3/3$.

The object here is not to transfer a complete tabulation of functional values to graph paper but to gain a good qualitative and quantitative understanding with as little effort as possible. To this end, we note first that the function is *odd*,

$$f(x) = -f(-x),$$

which makes it sufficient to restrict the analysis to $x > 0$. Indeed, the curve in the left half-plane is the negative reflection of that in the right half-plane. Our task has been halved by this simple observation.

First the roots or zeros of the function are located by solving

$$f(x) = \frac{x^3}{5}(x^2 - 5/3) = 0,$$

which yields $x = 0$, $x = \pm(5/3)^{1/2}$.

The stationary points, determined from the solutions of

$$f'(x) = x^4 - x^2 = 0,$$

are $x = 0$ and $x = \pm 1$; their type is inferred from the corresponding signs of the second derivative

$$f''(x) = 4x^3 - 2x,$$

that is,

$$f''(0) = 0, \qquad f''(1) = 2, \qquad f''(-1) = -2.$$

According to the derivative test, $x = 1$ is a relative minimum with $f(1) = -2/15$, and $x = -1$ is a relative maximum with $f(-1) = +2/15$. Upon closer inspection, we ascertain that $x = 0$ is actually an inflection point because $f''(x)$ changes sign at the origin. Other inflection points, which are located where the concavity changes, are $x = \pm 2^{-1/2}$.

For large values of x, the function is approximately given by

$$f(x) \approx x^5/5$$

since the term $x^3/3$ is then negligible compared to the term $x^5/5$.

The precise version of this statement is the following limit,

$$\lim_{|x| \to \infty} \frac{f(x)}{x^5} = \frac{1}{5}.$$

If now all the information is accumulated concerning oddness, zeros, extrema, and approximate values for large $|x|$, a good sketch of the curve can be made (Fig. 3.3). Additional data, such as the values of $f(x)$ and $f'(x)$ at a few easy to calculate points, improve the figure greatly.

4

The principal reason for curve plotting is to gain a rapid impression or understanding of a function's behavior by identifying its singularities and other gross features. This is attained by providing the following information:

1. Check for certain basic properties such as oddness, evenness, boundedness, periodicity, etc.
2. Locate the zeros, $f(x) = 0$.
3. Locate all singularities, $f(x) = \pm \infty$.
4. Locate all stationary points, $f'(x) = 0$, and determine relative maxima and minima by evaluating $f''(x)$. Locate all inflection points among the values of x which satisfy $f''(x) = 0$.
5. Determine limiting properties and approximations valid for large or small $|x|$.
6. Calculate the values of the function, its derivative, and concavity wherever this is especially easy.

Example 7 Graph the function $y = F(x) = \dfrac{x^2}{x^2 - 4}$.

As set forth above, the following information is obtained:

1. The function is even, $F(-x) = F(x)$, which means that it is symmetric about the y axis.

2. The point $x = 0$ is the only zero of the function.

3. The function becomes infinite at $x = \pm 2$, and

$$\lim_{x \to 2+} F(x) = +\infty, \qquad \lim_{x \to 2-} F(x) = -\infty.$$

4. The higher derivatives are

$$F'(x) = \frac{-8x}{(x^2 - 4)^2}, \qquad F''(x) = \frac{8(3x^2 + 4)}{(x^2 - 4)^3}.$$

The only stationary point is $x = 0$, which must be a relative maximum since $F''(0) = -\frac{1}{2} < 0$. There are no inflection points; the lines $x = \pm 2$ are asymptotes of the curve.

5. By rewriting the function as

$$F(x) = \frac{1}{1 - 4/x^2},$$

we see that for large $|x|$ the limiting value is 1. Indeed $\lim_{|x| \to \infty} F(x) = 1$, and the curve in question also has the horizontal line, $y = 1$, as an asymptote.

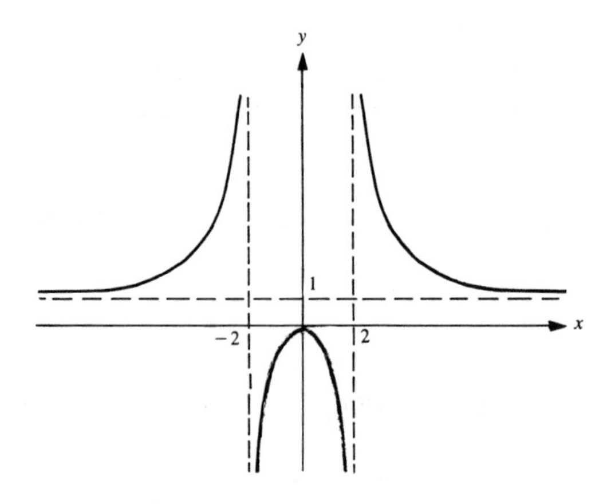

FIGURE 3.4

6. It is not difficult to compute the values

$$F(1) = -\tfrac{1}{3}, \qquad F'(1) = -\tfrac{8}{9};$$

$$F(3) = \tfrac{9}{5}, \qquad F'(3) = -\tfrac{24}{25}.$$

A reasonably good plot is obtained by transferring all this to graph paper, as shown in Fig. 3.4.

Exercises 2.3

1. Find the first and second derivatives of the following functions:

(i) $\sin x - \cos x$ **(ii)** $\sin x^2$ **(iii)** $\cos^2 5x$

(iv) $\cos(4 - x^2)$ **(v)** $\tan x$ **(vi)** $\cot x$

(vii) $\sin^2 2x$ **(viii)** $\tan x + \cot x$ **(ix)** $x^2 \sin x$

2. Graph the cubic polynomial $y = x^3 - 2x^2 - x$. Determine the stationary points and the inflection point. Determine the intervals of the x axis where y is increasing or decreasing. For what values of x are the first and second derivatives equal?

3. Find the stationary points and inflection points of the following functions. Illustrate with graphs.

(i) $\sin 3x$ **(ii)** $\cos 2x$ **(iii)** $x(x^2 - 4)$

(iv) $\dfrac{1}{x^2 + 1}$ **(v)** $\dfrac{x}{x^2 + 1}$ **(vi)** $\dfrac{x^2}{x^2 + 1}$

4. Determine the even and odd functions among the following. Evaluate their first and second derivatives at $x = 0$.

(i) $\sin x$ **(ii)** $\cos x$ **(iii)** x^3

(iv) $x^2 + x^4$ **(v)** $x \sin x$ **(vi)** $x + 2x^2 + x^3$

(vii) $x^3 + x \sin x^3$ **(viii)** $(x + x^5)\cos^2 x$ **(ix)** $|x| + x^2$

5. Show that the derivative of an odd function is an even function and conversely. What is the result for the second derivative? Illustrate with graphs.

6. Describe the stationary points and inflection points of the following functions. Determine the intervals of the x axis where the functions are increasing or decreasing.

(i) $3x^2 - 6x^3$ **(ii)** $x - \sqrt{x}$ **(iii)** $2x^3 - 9x^2 + 12$

(iv) $x^2 + a^3/4x$ **(v)** $2x^3 - 15x^2 + 24x + 6$ **(vi)** $2 + x - x^3$

(vii) $x^3 (x + 3)^2$ **(viii)** $(x + 1)/(x - 1)$ **(ix)** $(x + 1)/(x^2 + 2)$

7. Determine which of the following statements are correct. Explain with examples.

(i) The relative maxima of $y = f(x)$ are the relative minima of $y = 1/f(x)$.

(ii) The points where the graph of $y = f(x)$ has slope 1 are the stationary points of $y = x - f(x)$.

(iii) If $y = f(x)$ has exactly two inflection points, it has exactly one relative extremum.

8. Graph the following cubic polynomials. Determine the concavity (up or down) of the following curves at $x = 0$ and $x = 3$. Plot the tangent lines at these points.

(i) $y = 4(x - 3)^2$ **(ii)** $y = x^3 + 3x + 1$ **(iii)** $y = x^3 - 6x^2 + 3x$

9. Graph the following functions. Find all stationary points and inflection points. Determine the intervals of the x axis where the functions are increasing or decreasing.

(i) $\dfrac{x^2+1}{x^2-1}$ **(ii)** $\dfrac{x(x-1)}{x^2-4}$ **(iii)** $\dfrac{(x+1)^2}{(2x+1)^2(x-1)}$

(iv) $\dfrac{x^3+8}{x^2}$ **(v)** $\dfrac{x^3-2}{x^2-2}$ **(vi)** $2\sin x + \cos 2x$

(vii) $5\cos^3 x - 3\cos x$ **(viii)** $\dfrac{x(x-2)(x-4)}{(x^2-1)(x-3)}$ **(ix)** $x\tan x$

10. Sketch the curve $y = f(x) = \cos x - 1 + \frac{1}{2}x^2$ and calculate $f(0)$, $f'(0)$, $f''(0)$ and $f'''(0)$. Determine $f^{(n)}(0)$ for $n > 3$.

11. Show that the relative rate of change of a product $y(x) = y_1(x)\, y_2(x) \ldots y_n(x)$ is the sum of the relative rates of change of the factors,

$$\frac{y'}{y} = \frac{y_1'}{y_1} + \frac{y_2'}{y_2} + \cdots + \frac{y_n'}{y_n}.$$

12. Determine the relative rates of change y'/y of the following functions:

(i) $(x-1)^2(x+1)^2$ **(ii)** $(x^2-1)(x+2)^2\,(x+3)^2$ **(iii)** $(x^2+1)\cos^2 x$

(iv) $(1+x)^{1/2}\,(1+x^2)^{3/2}$ **(v)** $x^3 \sin x \tan x$ **(vi)** $(a^2+x^2)^m\,(b^2+x^2)^n$

13. Show that the nth derivative of $1/x$ is $(-1)^n\, n!/x^{n+1}$. Interpret with graphs the assertion that the singularity of the function $1/x$ gets worse with each higher differentiation.

14. Find expressions for the nth derivatives of the following functions:

(i) $\sin ax$ **(ii)** $\cos ax$ **(iii)** $\sin^2 x$

(iv) $\cos^2 ax$ **(v)** $(1+x)^{-1}$ **(vi)** $(1+2x)^{1/2}$

(vii) $x\sin x$ **(viii)** $(x^2-1)^{-1}$ **(ix)** $(x^2-3x+2)^{-1}$

15. Verify the Leibniz rule for $n = 3$,

$$\frac{d^3}{dx^3}(uv) = \frac{d^3u}{dx^3}v + 3\frac{d^2u}{dx^2}\frac{dv}{dx} + 3\frac{du}{dx}\frac{d^2v}{dx^2} + u\frac{d^3v}{dx^3}.$$

16. Calculate the first, second and third derivatives of the following products.

(i) $(x^2-1)(x^2+1)$ **(ii)** $\sin x \cos x$ **(iii)** $x\sin x$

17. Show that the graph of a cubic polynomial $y = ax^3 + bx^2 + cx + d$ has exactly one inflection point. Show also that, if there are two relative extrema, the inflection point is halfway between the extrema on the x axis.

18. Show that the graph of a cubic polynomial $y = ax^3 + bx^2 + cx + d$ has a relative extremum if and only if $b^2 > 3ac$.

19. Plot the following curves. In each case, determine all lines $y = mx + b$ which are asymptotes to the curve.

(i) $x^2 - y^2 = 4$ **(ii)** $y(x^2-1) = 1$ **(iii)** $x^2y + xy^2 = 16$

(iv) $y^2 = x(x-2)$ **(v)** $y^3 = x(x^2-4)$ **(vi)** $y^3 + yx^2 - x^2 = 0$

20. Show by direct substitution that $y = A\sin\omega x + B\cos\omega x$ is a solution of $y'' + \omega^2 y = 0$. Find the particular solution for which $y(0) = 1$, $y'(0) = 0$.

21. Explain why the function $E(x)$ defined by $E'(x) = E(x)$, $E(0) = 1$ cannot be a polynomial.

2.4 Applications: rates and differentials

The first derivative of a function measures its rate of change with respect to the independent variable; the second derivative provides the rate of change of the first, and higher derivatives are similarly interpreted. Relationships among derivatives of different orders account for most formulations of applied problems in mathematical terms. The object is to find the functions which satisfy such an equation (or several such equations). However, in some situations the derivatives are themselves of direct interest, and the most commonplace of these concern the deviations from, or control of, some basic state. For example, a business may replenish its inventory in accordance with the *rate of sales*; a public utility may expand its facilities to serve a *growing* population; altitude and *rate of descent* are vital information to the pilot; the pollster is certainly interested in opinion *trends*; an equilibrium chemical process requires the removal of heat at the same *rate* as it is produced; the economist examines changes and *rates* of change in the economy, and even politicians seem acutely sensitive to the higher derivatives of selected economic indicators—especially at election time. Several idealized examples on rates are presented next in order to gain some experience with problems of this sort.

Example 1 A pebble dropped into a still pool of water sends out concentric ripples. If the outermost ripple spreads at a rate of 4 m/sec, how rapidly is the area of disturbance increasing after t seconds?

The area of disturbance at time t is the area of a circle of radius $4t$, $A(t) = \pi(4t)^2 = 16\pi t^2$ (m^2). The rate of change is $A'(t) = 32\pi t$ (m^2 sec^{-1}). After 3 seconds, for example, the area of disturbance is 144π m^2, increasing at the rate 96π m^2 sec^{-1}.

Example 2 An aircraft lands at a speed of 60 m/sec, decelerates at a constant rate and comes to a stop 1200 m down the runway. What is the deceleration rate and how long does the plane roll before it stops?

Define $s(t)$ to be the distance that the plane has rolled along the runway t seconds after landing. Since the deceleration is constant, $d^2s/dt^2 = -a$ and, as can be checked by differentiating twice, $s(t)$ is a quadratic function of t,

$$s(t) = s(0) + ut - (a/2)\, t^2.$$

In this example, $s(0) = 0$ and $u = 60$ m sec^{-1}. Therefore,

$$s(t) = 60t - (a/2)\, t^2.$$

The velocity at time t is given by

$$v(t) = ds/dt = 60 - at.$$

The plane comes to a stop when $t = {}^{60}\!/_{a}$ and $s(t) = 1200$. This implies

$$s({}^{60}\!/_{a}) = 60({}^{60}\!/_{a}) - ({}^{a}\!/_{2})({}^{60}\!/_{a})^2 = 1800/a = 1200.$$

The deceleration rate is $a = 1800/1200 = 1.5$ m/sec^2. The time required to come to a stop is $t = {}^{60}\!/_{a} = 40$ seconds.

Example 3 A spherical balloon is blown up at the constant rate a cm^3/sec. Determine the rates of change of the radius and surface area.

The volume of the balloon at time t is $V(t) = at$. If $R(t)$ and $S(t)$ are the radius and surface area at time t, then

$$V(t) = at = {}^{4}\!/_{3}\,\pi(R(t))^3 \quad \text{and} \quad S(t) = 4\pi(R(t))^2.$$

Solving for the radius and surface area as functions of time,

$$R(t) = \left(\frac{3at}{4\pi}\right)^{1/3} \quad \text{and} \quad S(t) = 4\pi\left(\frac{3at}{4\pi}\right)^{2/3} = (4\pi)^{1/3}\,(3at)^{2/3}.$$

Differentiating, the rates of change of the radius and surface area are

$$R'(t) = \left(\frac{3a}{4\pi}\right)^{1/3}\frac{1}{3}t^{-2/3} \quad \text{and} \quad S'(t) = (4\pi)^{1/3}\,(3a)^{2/3}\,\frac{2}{3}t^{-1/3}.$$

The volume increases at a constant rate but the radius and surface area increase at slowing rates. This is a familiar observation to anyone who has blown up a balloon.

Example 4 An oil spill from a stricken tanker results in the release of V m^3 of oil in a very short time. At what rate does the oil spread on the sea surface?

We are not yet able to calculate the transient process of oil contamination, and the best that can be done now is to make some simplifying assumptions: the sea is calm, and the oil spreads in a circular pattern of uniform thickness, which at time t is $h(t)$. The radius of the circle $R(t)$ is determined from the fact that the volume of oil on the sea surface is always equal to the original volume of the spill:

$$\pi R^2 h = V. \tag{1}$$

Since V is constant, the rate of change of the *volume* of oil is zero,

$$\frac{dV}{dt} = \frac{d}{dt}\pi R^2(t)h(t) = 0,$$

so that

$$2R(t)h(t)R'(t) + R^2(t)h'(t) = 0,$$

or

$$\frac{2R'(t)}{R(t)} = -\frac{h'(t)}{h(t)}. \tag{2}$$

The rate of radial spread can be determined if, somehow, the thickness of the layer of oil is made known. Laboratory experiments indicate that in the important phase of oil spread when viscous and buoyancy forces predominate

$$h(t) = \frac{k}{t^{1/2}}, \qquad t > 0,$$

so that

$$R = \left(\frac{Vt^{1/2}}{\pi k}\right)^{1/2}.$$

The calculation yields

$$R'(t) = \frac{1}{4}\left(\frac{V}{\pi k}\right)^{1/2} t^{-3/4}.$$

---

2 **Example 5** The design of a ground-controlled automatic landing system calls for a landing approach to a runway similar to that in Fig. 4.1. Additional conditions

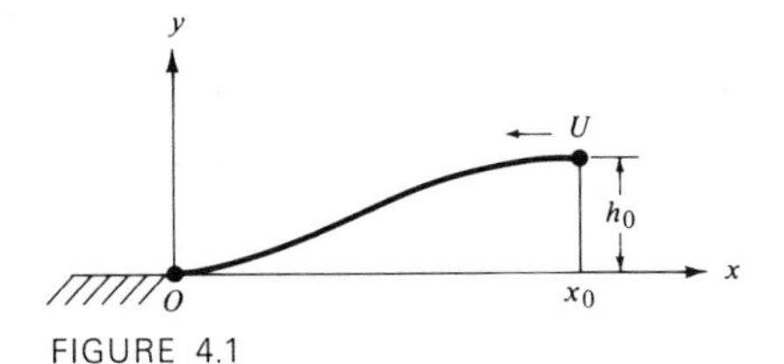

FIGURE 4.1

imposed are that the altitude is h_0 when descent commences; $x = 0$ is the point of touchdown; the maximum absolute vertical acceleration allowed is $g/10$; a constant horizontal airspeed $-U$ is maintained (the plane approaches from the right). Find a cubic polynomial, $y = f(x)$, which gives an acceptable trajectory.

A constant horizontal velocity $dx/dt = -U$ means that the horizontal position of the aircraft at time t, $t > 0$, is

$$x = x_0 - Ut. \tag{3}$$

Touchdown at $x = 0$ occurs then at $t = x_0/U$.

Assume that the trajectory is given by a cubic polynomial

$$y = ax^3 + bx^2 + cx + d,$$

so that

$$y'(x) = 3ax^2 + 2bx + c.$$

The four constants a, b, c and d are evaluated from the four imposed conditions, which require that the approach pattern be tangent to the original flight path at one end, $x = x_0$, and to the runway at the other, $x = 0$. The conditions are then $y(0) = 0 = y'(0)$, and $y(x_0) = h_0$, $y'(x_0) = 0$. From the first two we find that $d = 0$ and $c = 0$, whereas the last two imply $a = -2h_0/x_0^3$, $b = 3h_0/x_0^2$. Therefore, the trajectory is

$$y = \frac{h_0}{x_0^2} x^2 \left(3 - \frac{2x}{x_0}\right). \tag{4}$$

The final constraint is on maximum acceleration, and this determines the minimum distance x_0 at which descent from altitude h_0 is permissible. Since y is a function of x and x is a function of t, the vertical velocity dy/dt is calculated using the chain rule for differentiation:

$$\frac{d}{dt} y(x(t)) = \frac{dy}{dx}\frac{dx}{dt}.$$

Upon substituting from (3) and (4), this becomes

$$\frac{dy}{dt} = -U\frac{dy}{dx} = \frac{-6Uh_0}{x_0^2} x\left(1 - \frac{x}{x_0}\right).$$

[The vertical velocity is zero at the end points of the descent trajectory, and this is, of course, equivalent to the previous conditions on $y'(x)$.] Finally, the vertical acceleration is given by

$$\frac{d^2y}{dt^2} = \frac{d}{dt}\left(-U\frac{dy}{dx}\right) = -U\frac{d}{dx}\left(\frac{dy}{dx}\right)\frac{dx}{dt} = U^2\frac{d^2y}{dx^2},$$

or

$$\frac{d^2y}{dt^2} = \frac{6U^2h_0}{x_0^2}\left(1 - \frac{2x}{x_0}\right).$$

The maximum vertical acceleration is seen to occur at the end points $x = 0$, $x = x_0$ (why?), where its value is

$$\left|\frac{d^2y}{dt^2}\right| = \frac{6U^2h_0}{x_0^2}.$$

In order not to exceed the allowable acceleration $g/10$, descent can commence only when

$$\left|\frac{d^2y}{dt^2}\right| = \frac{6U^2h_0}{x_0^2} \le \frac{g}{10} \qquad \text{or} \qquad x_0 \ge U\left(\frac{60h_0}{g}\right)^{1/2}.$$

For $U = 200$ km/h, $h_0 = 1$ km, $g = 9.8$ m/sec^2, we find that $x_0 = 4.35$ km is the shortest distance from the runway for a permissible descent.

Example 6 The spring runoff from the melting of snow in a region is accumulated in a reservoir whose shape is approximately a trough l m long with apex angle 2α (Fig. 4.2). If the daily inflow of water is $Q(t)$ m^3/day how fast will the water level in the dam rise?

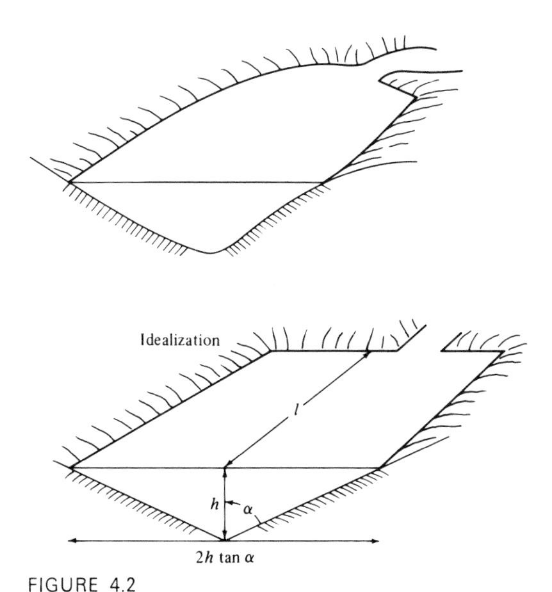

FIGURE 4.2

If h is the instantaneous depth of the water, the volume of water in the trough (reservoir) is

$$V = lh^2 \tan \alpha.$$

The rate of change of this volume with time must account for the inflow Q, and since l and α are constants,

$$\frac{dV}{dt} = 2lh \tan \alpha \frac{dh}{dt} = Q(t).$$

[Note that $2hl \tan \alpha \, \Delta h$ is the incremental volume increase that occurs in time Δt (Fig. 4.3).] The rate at which depth increases, in meters per day, is then

$$\frac{dh}{dt} = \frac{Q}{2lh \tan \alpha}.$$

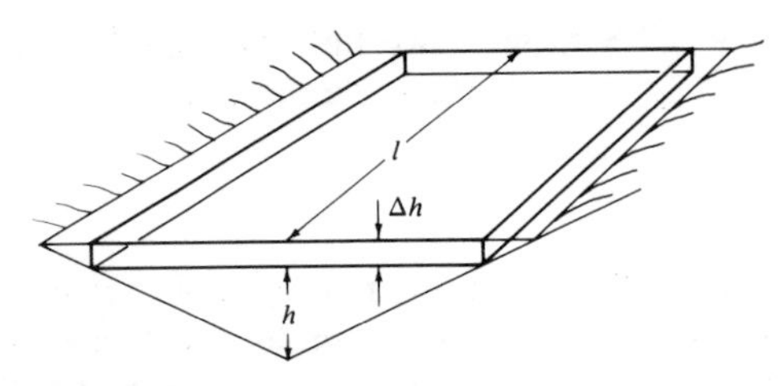

FIGURE 4.3

Suppose Q is a constant and the initial depth of water is h_0; then the depth after t days is determined from the total increase in volume,

$$lh^2 \tan \alpha - lh_0{}^2 \tan \alpha = Qt,$$

so that

$$h = \left(h_0{}^2 + \frac{Qt}{l \tan \alpha}\right)^{1/2}.$$

In this case we find that

$$\frac{dh}{dt} = \frac{Q}{2l \tan \alpha}\left(h_0{}^2 + \frac{Qt}{l \tan \alpha}\right)^{-1/2}.$$

---

3 **Example 7** The function $E(x)$ is defined by the equation

$$\frac{dE}{dx} = E, \tag{5}$$

with $E(0) = 1$. Determine the value $E(1) = e$.

In the last section, this function was shown to have the property

$$\frac{d^n}{dx^n} E(x) = E(x),$$

so that in particular $E^{(n)}(0) = 1$. Actually, a great deal of information is known about $E(x)$ but only at one single point $x = 0$, and the problem is to use this wealth of data to calculate the value $E(1)$. One way of doing this is to approximate $E(x)$ by a polynomial,

$$P_n(x) = a_0 + a_1 x + \cdots + a_n x^n,$$

and to determine the coefficients a_k so that the first n derivatives of both $P_n(x)$ and $E(x)$ agree at $x = 0$. This requirement is met by choosing

$$P_n(0) = 1, \qquad P_n{}'(0) = 1, \qquad \cdots, \qquad P_n{}^{(n)}(0) = 1.$$

Since the derivatives of the polynomial are

$$P_n'(x) = a_1 + 2a_2 x + \cdots + na_n x^{n-1},$$
$$P_n''(x) = 2a_2 + 3 \cdot 2a_3 x + \cdots + n(n-1)a_n x^{n-2},$$
$$\cdots\cdots\cdots\cdots\cdots\cdots\cdots\cdots\cdots\cdots,$$
$$P_n^{(n)}(x) = n!a_n,$$

the conditions $P_n^{(k)}(0) = 1$ for $k \leq n$, yield

$$a_0 = 1, \qquad a_1 = 1, \qquad a_2 = \tfrac{1}{2}, \qquad \ldots, \qquad a_k = \frac{1}{k!}, \qquad \ldots, \qquad a_n = \frac{1}{n!}.$$

Thus, the polynomial

$$P_n(x) = 1 + x + \frac{x^2}{2!} + \cdots + \frac{x^n}{n!}$$

and the function $E(x)$ at the origin have identical higher derivatives to order n. The polynomial is our approximation of the function $E(x)$, and by writing

$$E(x) = P_n(x) + R_n(x)$$

we would eventually hope to estimate the error involved, as expressed by the *remainder* $R_n(x)$, proving it to be a negligible correction for n sufficiently large. [We could replace the equality in this equation by an approximate equality, $E(x) \approx P_n(x)$, by neglecting $R_n(x)$ entirely.] No restrictions have been placed on the degree of the approximating polynomial, and we can therefore consider the limiting case $n \to \infty$. If, indeed, it turns out that $\lim_{n \to \infty} R_n(x) = 0$, then the function $E(x)$ can be represented as an *infinite series* whose nth partial sum is just the polynomial $P_n(x)$:

$$E(x) = \lim_{n \to \infty} P_n(x),$$

or

$$E(x) = 1 + x + \frac{x^2}{2!} + \cdots + \frac{x^n}{n!} + \cdots = 1 + \sum_{n=1}^{\infty} \frac{x^n}{n!}. \tag{6}$$

The value at $x = 1$ is obtained from (6) as

$$e = E(1) = 1 + 1 + \tfrac{1}{2} + \tfrac{1}{6} + \tfrac{1}{24} + \cdots. \tag{7}$$

The addition process seems rapidly convergent upon execution, and this is strong evidence for the convergence of the series; the result is

$$e = 2.7182818284\ldots.$$

All of this is, of course, motivational in spirit but highly constructive nonetheless. A short digression would be sufficient to prove the results rigorously true, but this is not the intention at this time. However, it is of interest to conclude in the same fashion and show, by direct substitution, that (6) is actually a solution of the differential equation $E'(x) = E(x)$. This is accomplished by differentiating the series for E *term by term*, as follows:

$$\begin{aligned}\frac{d}{dx}E(x) &= \frac{d}{dx}\left(1 + x + \frac{x^2}{2!} + \cdots + \frac{x^n}{n!} + \cdots\right) \\ &= \frac{d}{dx}1 + \frac{d}{dx}x + \frac{d}{dx}\frac{x^2}{2!} + \cdots + \frac{d}{dx}\frac{x^n}{n!} + \cdots \\ &= 1 + x + \cdots + \frac{x^{n-1}}{(n-1)!} + \cdots.\end{aligned}$$

Examination of the right-hand side shows that the derivative of the series is just the function $E(x)$, which is what we set out to demonstrate.

A great many subtle questions are involved in the preceding analysis, such as the convergence and differentiation of infinite series and the existence and uniqueness of solutions of differential equations. These issues are examined more closely in chapters to follow, but the results obtained here will indeed be found valid. The lack of a rigorous proof should never be an obstacle to constructive analysis. In fact, without such productive forays as these, there would be very little grist for the mill of rigor. As long as the results and predictions of theory and analysis make sense and can be tested, ad hoc methods can and should be used freely.

---

4 The five basic rules for differentiation discussed in Sec. 2.2 can be interpreted in terms of differentials which relate the various infinitesimal increments of change. If $u(x)$ and $v(x)$ are differentiable functions, then

$$\begin{aligned}d(u + v) &= du + dv \\ d(cu) &= cdu \\ d(uv) &= vdu + udv \\ d\left(\frac{u}{v}\right) &= \frac{1}{v^2}(vdu - udv) \\ du(v) &= \left(\frac{du}{dv}\right)dv.\end{aligned}$$

In each case, the differential dx appearing in the five integration rules is simply cancelled throughout like a common algebraic factor. The next three examples illustrate the application of differentials to problems involving estimating errors or the changes produced by small changes in the independent variable.

Example 8 The radius of a sphere is measured with an accuracy of 0.1%. Determine the possible errors in the calculated values of the surface area and volume.

If r is the measured value of the radius, then the true radius lies between $(0.99)\,r$ and $(1.01)\,r$. The estimates for surface area $S = 4\pi r^2$ and volume $V = 4/3\,\pi r^3$ are also subject to error. Suppose that Δr is the difference between the true radius and the measured value and ΔS and ΔV are the corresponding differences for surface area and volume.

$$\Delta S = 4\pi(r+\Delta r)^2 - 4\pi r^2 = 8\pi r\Delta r + 4\pi(\Delta r)^2$$

$$\Delta V = 4/3\pi(r+\Delta r)^3 - 4/3\pi r^3 = 4\pi r^2\Delta r + 4\pi r(\Delta r)^2 + 4/3\pi(\Delta r)^3$$

Since Δr is small compared to r, terms involving $(\Delta r)^2$ and $(\Delta r)^3$ are relatively small and are neglected to produce simple approximations,

$$\Delta S \approx 8\pi r\,\Delta r \quad \text{and} \quad \Delta V \approx 4\pi r^2\,\Delta r.$$

Alternatively, the corresponding differential formulas could have been used

$$dS = 8\pi r\,dr \quad \text{and} \quad dV = 4\pi r^2\,dr.$$

Since $|\Delta r| \le (0.01)r$, we have

$$|\Delta S| \le (0.08)\pi r^2 = (0.02)\,4\pi r^2 = (0.02)\,S$$

$$|\Delta V| \le (0.04)\pi r^3 = (0.03)\,4/3\,\pi r^3 = (0.03)\,V.$$

These formulas state that a 1% error in the radius produces a 2% error in the surface area and a 3% error in the volume.

Example 9 The gravitational force per unit mass exerted on a body by the earth is described by the inverse square law $f = \mu/r^2$. Here μ is the gravitational constant and r is the distance of the body from the center of the earth. Estimate the relative change in weight of a person in an airplane that climbs from sea level to an altitude of 7 km. (Weight is mass M multiplied by gravitational force.)

Define $w(r)$ to be the weight of the person r km from the center of the earth. Assume that the radius of the earth is 6370 km. The change in weight produced by a small change in r is $\Delta w = w(r+\Delta r) - w(r)$ or, approximating by differentials,

$$dw = \frac{dw}{dr}\,dr = \frac{-2M\mu}{r^3}\,dr.$$

The relative change in weight is

$$\frac{dw}{w} = -\frac{2\mu}{r^3}\,dr\,\frac{r^2}{\mu} = -2\frac{dr}{r} = -2\cdot\frac{7}{6370} = -0.0022\ldots$$

The weight of the person decreases by 0.22%.

Example 10 The area of an ellipse of semimajor axis a and semiminor axis b is $A = \pi ab$. If the errors made in measuring a and b are Δa and Δb, respectively, find the approximate error in the area.

For the measured values of the axes, $a + \Delta a$, $b + \Delta b$, the formula for the area of the ellipse yields a value which is also in error

$$A + \Delta A = \pi(a + \Delta a)(b + \Delta b) = \pi ab + \pi(a\,\Delta b + b\,\Delta a + \Delta a\,\Delta b).$$

The magnitude of the area error ΔA, produced by deviations Δa, Δb, is then

$$\Delta A = \pi(a\,\Delta b + b\,\Delta a + \Delta a\,\Delta b).$$

If the increments Δa and Δb are both small, $\Delta a\,\Delta b$ is negligible compared to either of them and a reasonable approximation is

$$\Delta A \approx \pi(a\,\Delta b + b\,\Delta a).$$

In fact, for infinitesimal changes, we have the exact differential statement

$$\pi d(ab) = dA = \pi(a\,db + b\,da),$$

and the incremental approximation could have been developed directly from this formula.

Since Δa and Δb can be positive or negative, it is also useful to bound the absolute error:

$$|\Delta A| \leq \pi b|\Delta a| + \pi a|\Delta b|.$$

For example, let $a = 4$ cm, $b = 2$ cm, $\Delta a = -0.15$ cm, and $\Delta b = 0.1$ cm; then

$$A = 8\pi \text{ cm}^2, \qquad \text{and} \qquad \Delta A = 0.1\pi \text{ cm}^2.$$

If, however, it is known only that $|\Delta a| < 0.15$ cm and $|\Delta b| < 0.1$ cm, a comparable bound for the error in area is

$$|\Delta A| < 0.7\pi \text{ cm}^2.$$

Exercises 2.4

1. A car travelling at 30 m/sec brakes suddenly and comes to a stop over a distance of 100 m. Assuming constant deceleration, how long does the car take to come to a stop?
2. A spherical balloon is blown up at the variable rate $a/(1 + t)$ cm^3/sec where t is measured in seconds. Determine the rates of change of the radius and surface area.
3. The height $h(t)$ in meters of a ball t seconds after it is thrown vertically upward is given by $h(t) = 21t - 4.9t^2$. Draw a graph of $h = h(t)$ for $t \geq 0$. Find the velocity and acceleration at $t = 0$, 1 and 2. Find the greatest height attained. What is the velocity at this point?
4. In the previous problem, solve for $t = t(h)$ and draw a graph of this function for $h \geq 0$. Explain why there are two values of t for each value of h. Calculate dt/dh and interpret the meaning of this derivative.
5. Verify the following identities.

(i) $\dfrac{d}{dt} y(x(t)) = \dfrac{dy}{dx}\dfrac{dx}{dt}$ **(ii)** $\dfrac{d^2}{dt^2} y(x(t)) = \dfrac{d^2y}{dx^2}\left(\dfrac{dx}{dt}\right)^2 + \dfrac{dy}{dx}\dfrac{d^2x}{dt^2}.$

6. Find dy/dt and d^2y/dt^2 by two methods.
(i) $x = 2t$, $y = x^2 + x^3$ **(ii)** $x = \sin t$, $y = x^2 - x$ **(iii)** $x = \cos t$, $y = (1 - x^2)^{1/2}$

7. A particle moves along the curve $y = \sqrt{x}$ for $x \geq 0$. At time t, the particle is at the point with coordinates $(x(t), y(t))$. Find the point on the curve where both coordinates are changing at the same rate. Determine the tangent line at this point and illustrate with a graph.

8. A baseball player 15 ft off first base tries to steal second. His running speed is 22 ft/sec and the bases are 90 ft apart. After a 1.5 second delay, the catcher throws the ball at 100 ft/sec to the second baseman. At what rate does the ball approach the runner? Is the steal successful?

9. Use a calculator to find the value of $e = E(1)$ to four decimal places from the series

$$e = 1 + 1 + \frac{1}{2!} + \frac{1}{3!} + \cdots + \frac{1}{n!} + \cdots .$$

10. From the infinite series for the function $E(x)$, find the values of $E(-1)$, $E(-1/2)$, $E(0)$ and $E(1/2)$ correct to four decimal places. Verify that the estimates suggest that
$E(-1)\,E(1) = E(-1/2)\,E(1/2) = E(0) = 1$.

11. A function $F(x)$ is defined by $dF/dx = -F(x)$ with $F(0) = 1$. Determine the polynomials $P_n(x)$ of degree n that have the same first n derivatives as $F(x)$ at $x = 0$. Evaluate $F(1)$ to four decimal places.

12. Find the quadratic polynomials $p(x) = ax^2 + bx + c$ that satisfy the following conditions and draw their graphs.
(i) $p(-1) = 0$, $p(0) = 1$, $p(1) = 0$ **(ii)** $p'(-1) = 2$, $p(0) = 0$, $p'(1) = -2$
(iii) $p(1) = 0$, $p(2) = 1$, $p(4) = 0$ **(iv)** $p'(1) = 1$, $p(3) = 4$, $p'(5) = -1$

13. Find the cubic polynomials $p(x) = ax^3 + bx^2 + cx + d$ that satisfy the following conditions and draw their graphs.
(i) $p(0) = p(1) = 0$, $p'(0) = 1$, $p'(1) = -2$ **(ii)** $p(-1) = p(1) = p(3) = 0$, $p'(0) = -1$
(iii) $p(0) = p'(0) = 0$, $p(1) = -1$, $p(2) = 0$ **(iv)** $p(0) = p(2) = 0$, $p'(1) = -1$, $p'(2) = 8$.

14. Find the polynomial of smallest degree which satisfies all the constraints on the landing trajectory in Example 5 and which also corresponds to zero vertical acceleration at the points where descent commences and touchdown occurs.

15. A spherical balloon has been blown up to an initial volume $V(0)$ m^3. Air is released beginning at $t = 0$ at the rate $1/(1 + t^2)$ m^3 sec^{-1}. What is the rate of change of the radius of the balloon?

16. A chemical firm can sell x kg of sulfur per month at a price of z cents per kilogram to produce total monthly revenue $P(x)$ where

$$z = 15 - \frac{x}{10{,}000} \quad \text{and} \quad P(x) = zx = 15x - \frac{x^2}{10{,}000}.$$

Calculate the *marginal revenue* dP/dx for production levels $x = 5000$, 10,000 and 20,000 kg of sulfur per month. What production level maximizes the total monthly revenue? Illustrate with graphs of $P(x)$ and dP/dx.

17. Two planes are flying on parallel courses 10 km apart in opposite directions. One flies at 150 km/h and the other at 200 km/h. The planes are 20 km apart at time $t = 0$. Determine a formula for the separation distance $r(t)$ at time t (hours) for $t > 0$. When are the planes at minimum separation?

18. In the previous problem, determine the relative speed dr/dt and the relative acceleration d^2r/dt^2. Show that $\lim_{t\to\infty} \frac{dr}{dt} = 350$ and $\lim_{t\to\infty} \frac{d^2r}{dt^2} = 0$.

19. The motion of a particle on the x axis given by $x(t) = A \sin \omega t$ is called *simple harmonic motion* with *constant amplitude A* and *constant angular frequency* ω. Show that $x(t)$ is periodic with period $T = 2\pi/\omega$. Calculate the velocity and acceleration and illustrate with graphs. Show that $x(t) = A \cos(\omega t + \theta_0)$ also represents simple harmonic motion.

20. Write the polynomial

$$y = 1 + x + 3x^2 + 5x^3$$

as

$$y = c_0 + c_1(x-1) + c_2(x-1)^2 + c_3(x-1)^3$$

and determine the coefficients by equating both expressions and their higher derivatives at $x = 1$.

21. Write the polynomial

$$P_n(x) = a_0 + a_1 x + \ldots + a_n x^n$$

as

$$P_n(x) = c_0 + c_1(x-a) + \ldots + c_n(x-a)^n$$

and identify the coefficients $c_0, \ldots, c_n$ by equating the derivatives of both forms at $x = a$.

22. Approximate $\cos x$ in the range $-\pi \le x \le \pi$ by the polynomial $a_0 + a_1 x + a_2 x^2$, where the three coefficients are determined from the conditions

$$\left.\frac{d^n}{dx^n}\cos x\right]_{x=0} = \left.\frac{d^n}{dx^n}(a_0 + a_1 x + a_2 x^2)\right]_{x=0}$$

for $n = 0, 1, 2$. How would you generalize this procedure? What infinite series is implied? Calculate cos 15° this way and compare your answer with the accurate value.

23. Approximate $\sin x$ in the vicinity of 30° ($= 0.5236 = \pi/6$ rad), by a cubic polynomial in powers of $x - \pi/6$. (*Hint:* Equate derivatives of both functions.) Find the approximate value of sin 40° in this way.

24. Show that the value of the function $f(x)$ and the polynomial of nth degree

$$P_n(x) = f(a) + f'(a)(x-a) + \frac{1}{2!}f''(a)(x-a)^2 + \cdots + \frac{1}{n!}f^{(n)}(a)(x-a)^n$$

and their respective derivatives to order n are equal to each other at the point $x = a$. Find the approximations for $\sin x$ and $\cos x$ of the form

$$f(x) \approx P_n(x) = f(a) + f'(a)(x-a) + \cdots + \frac{1}{n!}f^{(n)}(a)(x-a)^n.$$

If $a = \pi/4$, how many terms are necessary to find $\sin \pi/6$ and $\cos \pi/6$ accurate to 10^{-3}?

25. If $u = u(x)$ and $v = v(x)$, determine the following differentials in terms of du and dv.

(i) $d(u^2 + v^2)$ **(ii)** $d(u^2v^2)$ **(iii)** $d(\sqrt{u^2 + v^2})$

(iv) $d\left(\dfrac{u}{u+v}\right)$ **(v)** $d(\sin uv)$ **(vi)** $d\left(\tan \dfrac{u}{v}\right)$

26. Estimate the following numbers using differentials. Compare to the accurate values.

(i) sin 32° **(ii)** sin 28° **(iii)** cos 59°

(iv) cos 61° **(v)** tan 43° **(vi)** tan 47°

27. Using differentials, find approximate values of the following functions at the points indicated. Compare to the accurate values.

(i) $x^2 + x$, $x = 1.04$ **(ii)** $\cos x$, $x = 0.02$ **(iii)** $(1-x)^3$, $x = 2.001$

(iv) $\sin^2 x$, $x = 31°$ **(v)** $\tan^2 x$, $x = 46°$ **(vi)** $\sin x + \cos x$, $x = 46°$.

28. A right circular cone of radius r and height h has volume $V = \frac{1}{3}\pi r^2 h$. If r and h are measured with slight inaccuracies Δr and Δh, estimate the absolute and relative errors in the volume.

29. The total surface area of a right circular cylinder of radius r and height h is given by $A = 2\pi r^2 + 2\pi rh$. If r and h are measured with slight inaccuracies Δr and Δh, estimate the absolute and relative errors in the area.

30. The time T for a complete swing of a pendulum is proportional to the square root of its length l. Show that a small change Δl in the length produces a small change in the period ΔT satisfying

$$\frac{\Delta T}{T} \approx \frac{1}{2}\frac{\Delta l}{l}.$$

31. Suppose that a pendulum clock loses 15 seconds per hour. How should the length of the pendulum be adjusted?

32. The heat output (watts) of an electric heater is given by $H = kV^2/R$ where V is the voltage (volts), R is the resistance (ohms) and k is a dimensionless constant measuring the efficiency of the heater. Determine the approximate change ΔH in heat output produced by small fluctuations ΔV and ΔR in the voltage and resistance. If $V = 220$, $R = 12$, $\Delta V = 10$ and $\Delta R = 0.1$ with $k = 0.9$, determine ΔH and compare to the accurate value.

33. Points P and Q in Fig. 4.4 are connected by a rigid rod of length l; point P moves around the circle of radius r, but Q is constrained to move along the horizontal line through O. If angle θ is changed slightly by an amount $\Delta\theta$, find the corresponding change in the distance OQ.

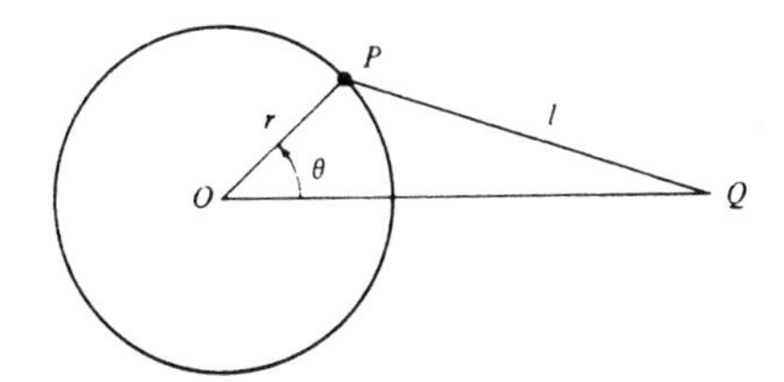

FIGURE 4.4

34. Under quiescent atmospheric conditions, an aircraft crosses the Atlantic Ocean (3,000 miles) at an airspeed of 550 mph. Estimate the time saved with a 35 mph tail wind. If the cost of the trip per passenger as a function of the groundspeed u of the plane is

$$100 + \frac{u}{10} + \frac{36{,}000}{u} \quad \text{dollars},$$

find the approximate savings attributable to the tail wind. (How much extra per passenger does it cost the company if there is 50-mph head wind?)

2.5 Applications: maxima and minima

A selection of problems is presented now which deal with the practical applications of maxima and minima. In each case a problem is formulated mathematically, and the extrema of some relevant function are found and categorized. The first example is purely geometrical in character.

Example 1 Find the largest rectangle which can be inscribed in the area bounded by $y = 0$ and the semi-ellipse

$$y = b\left[1 - \left(\frac{x}{a}\right)^2\right]^{1/2}.$$

Let the base of an inscribed rectangle be $2x$, as shown in Fig. 5.1; its height is then $b[1 - (x/a)^2]^{1/2}$, which makes the area

$$A(x) = 2bx\left[1 - \left(\frac{x}{a}\right)^2\right]^{1/2}.$$

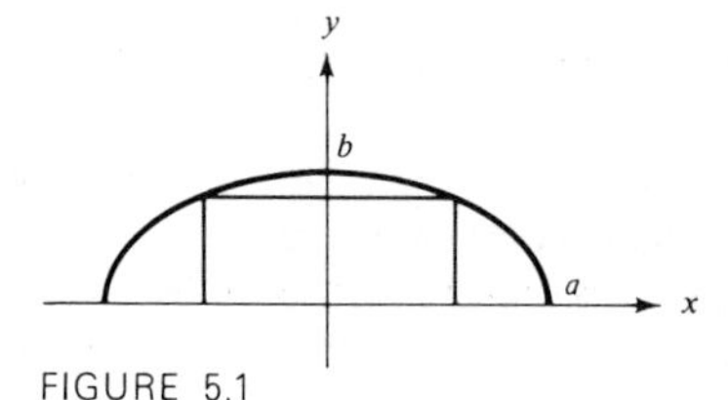

FIGURE 5.1

Extremal points are determined from the zeros of the derivative of $A(x)$, which is calculated as follows:

$$\frac{dA}{dx} = \left[1 - \left(\frac{x}{a}\right)^2\right]^{1/2} \frac{d}{dx}(2bx) + 2bx\frac{d}{dx}\left[1 - \left(\frac{x}{a}\right)^2\right]^{1/2}$$

$$= 2b\left[1 - \left(\frac{x}{a}\right)^2\right]^{1/2} - \frac{2b}{a^2}\frac{x^2}{[1 - (x/a)^2]^{1/2}} = \frac{2b[1 - 2(x/a)^2]}{[1 - (x/a)^2]^{1/2}}.$$

The only stationary point, for which $A'(x) = 0$, is

$$x = 2^{-1/2}a,$$

and this is a relative maximum. [Since there is an inscribed rectangle of zero area, the stationary point clearly must correspond to a maximum, and this commonsense conclusion is readily proved by computing the second derivative.] The dimensions of the largest inscribed rectangle are

$$\text{base} = 2x = \sqrt{2}a, \qquad \text{height} = y = \frac{b}{\sqrt{2}},$$

and its area is

$$A = 2xy = ab.$$

Example 2 For migrating fish swimming at a constant speed v relative to the water about it, the energy expenditure per unit time is proportional to v^3. Determine the optimum swimming speed against a steady current of speed u.

This problem is not completely specified since, for example, if the fish is being pursued by a predator, the optimum speed is the physiological maximum. The usual problem of migrating fish however is to minimize the total energy required to swim a fixed distance. With swimming speed v, the resulting speed against the current is $v - u$ and the time required to swim a fixed distance l upstream is $l/(v - u)$. The total energy expenditure to swim this distance is

$$E(v) = av^3 \frac{l}{v - u}$$

where a is a proportionality constant. The value of v that minimizes $E(v)$ is determined by differentiating,

$$E'(v) = al \frac{(v - u)\, 3v^2 - v^3}{(v - u)^2} = al \frac{v^2\,(2v - 3u)}{(v - u)^2}.$$

The stationary point of interest is $v = 3/2u$. Since $E'(v)$ is an increasing function of v at $v = 3/2u$, $E''(v)$ must be positive there so that this speed produces the minimum energy expenditure. Migrating fish seem to be able to solve this calculus problem; they swim against a current with a speed 50% greater than the current speed.

Example 3 A publishing firm will price a new novel to maximize its net profit. The cost of production and marketing x copies of this book is known to be

$$C = 25{,}000 + 5x \qquad \text{dollars.}$$

Since it costs at least \$5 to produce one book, the sales price P must substantially exceed this figure or a loss will be incurred. A good empirical formula for the number of books that can be sold at a price $P \geq 5$ is

$$\frac{x}{1{,}000} = 6\left(1 - \frac{P}{30}\right).$$

What should the price of the book be?

According to the formula, the price per copy in order to sell x books, must be

$$P = 30\left(1 - \frac{x}{6{,}000}\right),$$

and this sale would entail a gross return

$$xP = 30x\left(1 - \frac{x}{6{,}000}\right).$$

The gross less the total cost C is the net profit T;

$$T(x) = xP - C = 25x - \frac{x^2}{200} - 25{,}000.$$

The greatest net profit is achieved when $T'(x) = 0$, that is,

$$T'(x) = 25 - \frac{x}{100} = 0.$$

Therefore, $x = 2{,}500$ is the optimum sales figure. [This stationary point is a maximum because $T''(2{,}500) < 0$.] A total sale of 2,500 copies, for which

$P = \$17.50$, realizes the largest net profit of $T = \$6{,}250$. The corresponding gross is \$43,750 and $C = \$37{,}500$.

Example 4 In a particular manufacturing process, it is known that the number of rejects y depends on the total daily output x, that is, $y = y(x)$. The firm makes a profit A for each item sold, but loses $A/3$ for each reject. What should the daily output be in order to maximize profits?

Of the x units produced daily, $x - y(x)$ are first-quality merchandise and $y(x)$ are rejects, which makes the total profit realizable

$$T(x) = A(x - y) - \tfrac{1}{3}Ay.$$

This is a maximum at an acceptable value of x for which

$$T'(x) = A[1 - \tfrac{4}{3}y'(x)] = 0$$

and

$$T''(x) = -\tfrac{4}{3}Ay''(x) < 0.$$

These conditions are equivalent to

$$y'(x) = \tfrac{3}{4}, \qquad y''(x) > 0. \tag{1}$$

For definiteness suppose that

$$y = \begin{cases} \dfrac{x}{101 - x} & \text{for} \quad x \leq 100; \\ x & \text{for} \quad x > 100. \end{cases}$$

Obviously, the company loses money if $x > 100$ since each unit manufactured must then be rejected. The maximum profit is obtained for $x \leq 100$, at the value for which

$$y'(x) = \frac{101}{(101 - x)^2} = \frac{3}{4},$$

that is,

$$x = 89.4.$$

But the production rate must be an integer quantity, which means that the number of units manufactured should be either 89 or 90, whichever yields the larger profit. Since $T(89) = 79.11A$ and $T(90) = 79.09A$, a production rate of 89 units per day will yield the maximum profit.

2

Example 5 An underground pipeline is to be constructed between two cities labelled A and B in Fig. 5.2. Analysis of the subsurface rock strata indicates that the construction cost per mile will be c_1 in region I, $y>0$, and c_2 in region II, $y \leq 0$. What should be the route of the pipeline to minimize the total cost of construction?

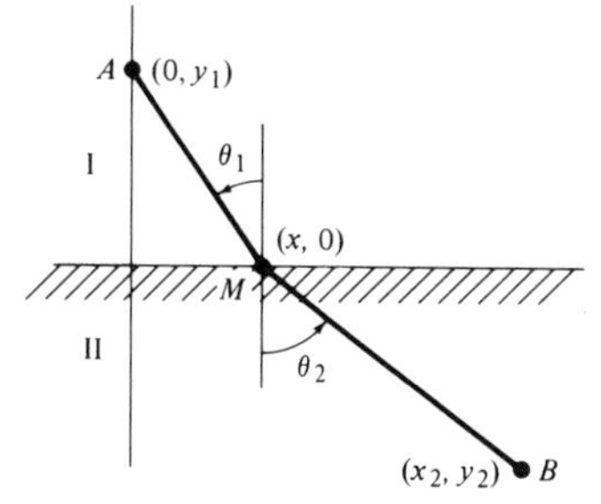

FIGURE 5.2

The costs in each region are directly proportional to the respective lengths of the pipeline, and the routes should therefore be as short as possible, i.e., straight-line segments. The route from A to B will then consist at most of two straight lines, which join at some point M on $y = 0$. If, as shown in Fig. 5.2, the coordinates are chosen so that $A = (0, y_1)$, $B = (x_2, y_2)$, $M = (x, 0)$, the relevant distances are

$$AM = (x^2 + y_1{}^2)^{1/2}, \qquad BM = [(x_2 - x)^2 + y_2{}^2]^{1/2}.$$

The total cost of the pipeline from A to B,

$$C(x) = c_1(x^2 + y_1{}^2)^{1/2} + c_2[(x_2 - x)^2 + y_2{}^2]^{1/2}, \tag{2}$$

is a minimum when

$$\frac{d}{dx}C(x) = \frac{c_1 x}{(x^2 + y_1{}^2)^{1/2}} - \frac{c_2(x_2 - x)}{[(x_2 - x)^2 + y_2{}^2]^{1/2}} = 0.$$

The last equation must be solved for x, a task not quite as formidable as it appears. In fact, a rather simple geometrical interpretation is possible if the individual functions are identified from the figure as

$$\sin\theta_1 = \frac{x}{(x^2 + y_1{}^2)^{1/2}}, \qquad \sin\theta_2 = \frac{x_2 - x}{[(x_2 - x)^2 + y_2{}^2]^{1/2}},$$

for then the condition for the minimum becomes

$$c_1 \sin\theta_1 = c_2 \sin\theta_2. \tag{3}$$

Consider the special case in which city B is located on the fault line, $y = 0$. If the costs are such that $c_2 > c_1$, it is obviously cheaper to build a pipeline in a straight line directly from A to B and to keep entirely in the less expensive terrain. On the other hand, for $c_1 > c_2$ it may be advantageous to have a route consisting of two segments, as shown in Fig. 5.3. In this situation, $\theta_2 = \pi/2$, and it follows from (3) that

$$\sin\theta_1 = \frac{c_2}{c_1},$$

or equivalently

$$x = y_1\left[\left(\frac{c_1}{c_2}\right)^2 - 1\right]^{-1/2}. \tag{4}$$

Of course, we require that $0 \le x \le x_2$; a value $x > x_2$ is not acceptable because the direct route AB would once again be cheaper.

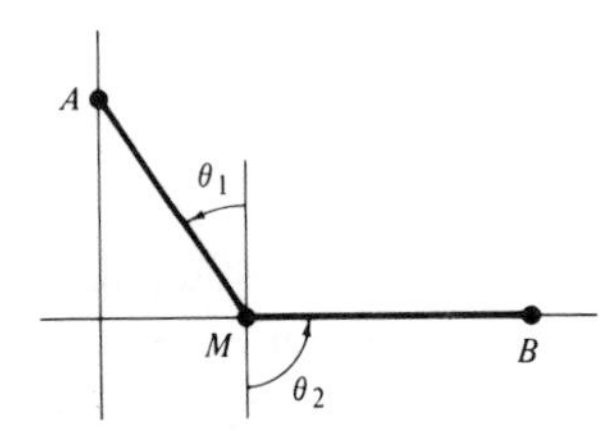

FIGURE 5.3

Example 6 The geological situation is the same as in Example 5, but now a pipeline is required between the cities A and P (Fig. 5.4). Find the path of minimum cost if $c_2 < c_1$.

We assume that the cities are sufficiently far apart so that the path $AMNP$, which takes advantage of lower construction cost on MN, may indeed be cheaper than a direct connection, line segment AP. For the longer route to be economically feasible, the intercity distance must greatly exceed the lengths of the spurs AM and PN if the additional construction costs are to be recouped.

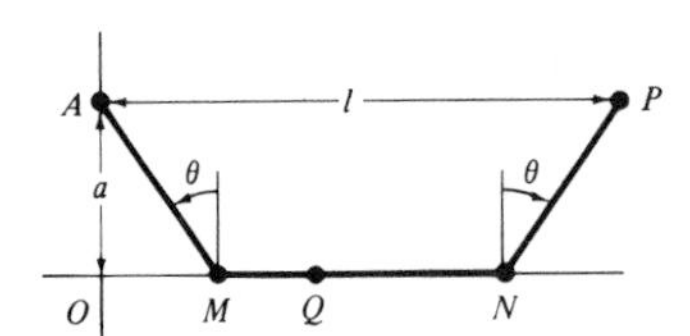

FIGURE 5.4

The approach then is to assume that the segmented route is feasible and, in the course of the solution, to delineate the restrictions that must hold for this to be true. This will quantify the meaning of *sufficiently far apart.*

In Example 5 it was shown that the path of minimum cost from A to point Q on the horizontal, for which

$$OQ > OM = x = a\left[\left(\frac{c_1}{c_2}\right)^2 - 1\right]^{-1/2},$$

consists of line segments AM, MQ, where $\theta = \theta_1$ and $\sin \theta_1 = c_2/c_1$. Likewise, the minimum-cost route between points Q and P follows the horizontal from Q to N and then proceeds along the segment NP, which also makes an angle θ_1 with the vertical. Note that

$$MN = l - 2a\left[\left(\frac{c_1}{c_2}\right)^2 - 1\right]^{-1/2}.$$

This analysis shows that the minimum-cost path from A to P which uses a segment of the horizontal axis is completely symmetric about the centerline $x = l/2$. (After all, there is really nothing to distinguish city A from city P, and the solution that takes this into account has mirror symmetry.) The construction cost for route $AMNP$ is then a minimum and is given by

$$C = \frac{2c_1 a}{\cos \theta_1} + c_2(l - 2a \tan \theta_1). \tag{5}$$

Thus far, our arguments have been based entirely on previous results. The conclusion that (5) is the least cost can be established directly by finding the minima of the cost function associated with the arbitrary *symmetric* path shown in Fig. 5.4:

$$C = \frac{2c_1 a}{\cos \theta} + c_2(l - 2a \tan \theta),$$

where

$$\sin \theta = \frac{x}{(a^2 + x^2)^{1/2}}.$$

In this case, the derivative is

$$\frac{dC}{dx} = \left(\frac{2c_1 a \sin \theta}{\cos^2 \theta} - \frac{2ac_2}{\cos^2 \theta}\right)\frac{d\theta}{dx}.$$

The only stationary point is the minimum located at

$$\sin \theta = \frac{c_2}{c_1},$$

which proves that $\theta = \theta_1$ corresponds to the optimal path.

The condition that route $AMNP$ costs less than the straight path between cities is

$$\frac{2c_1 a}{\cos \theta_1} + c_2(l - 2a \tan \theta_1) < c_1 l.$$

This inequality can be manipulated to read

$$\frac{l}{a} > \left(1 - \frac{c_2}{c_1}\right)^{-1}\left(\frac{2}{\cos\theta_1} - \frac{2c_2}{c_1}\tan\theta_1\right) = 2\left(\frac{c_1 + c_2}{c_1 - c_2}\right)^{1/2}. \tag{6}$$

The original specification that the cities be sufficiently far apart is then taken to mean that the values of l and a satisfy (6); the multisegment path is then actually cheapest. To illustrate, let $c_1 = 2c_2$; then (6) becomes

$$\frac{l}{a} > 2\sqrt{3} = 3.464.$$

The direct intercity distance must, in this case, be at least 3.464 times the shortest distance to the region of low-construction cost for the segmented path to be economically advantageous.

---

3 The examples discussed so far have been devised to yield functions that are readily differentiated, with stationary points that are also easily found. Maxima and minima problems encountered in real situations are far more complicated to treat, although the underlying theory is the same. The practical circumstances demand greater facility with intricate differentiations and may require the aid of a computer. Numerical methods are certainly indispensable in the task of locating the roots of an arbitrary function. In the next few sections we continue to develop the techniques of differentiation; certain relevant numerical techniques are introduced in Sec. 2.10.

It is important to realize that many optimization problems are not solvable just by differentiating some function to find isolated points of relative extrema. In one type of problem the optimum value may be an *absolute* maximum or minimum, and at that point the derivative is not necessarily zero (if it exists at all). In another situation, a *solution* may mean an entire curve and not just a single functional value at one point. Both these exceptions are easy to illustrate, and we consider the latter first.

Examples 5 and 6 dealt with minimum-cost routes where the cost of laying pipeline was the constant c_1 or c_2 in one or the other region. The solutions required recognition that the optimal path consists of *straight-line* segments, and with this established, the points of connection were determined from the criteria of relative minima. Suppose now, however, that the terrain varies continuously, making the cost per mile actually a function of position $c = c(y)$ or, more generally, $c = c(x, y)$. In this case, the minimum-cost route is no longer a succession of straight-line segments—indeed, the solution really requires an explicit formula for the optimal path, such as

$$y = f(x).$$

The question posed asks which curve, of the infinite number that connect two points, provides the optimum value, in this case the minimum cost. This is a typical problem of the *calculus of variations*. Far more is required here than in the simple versions analyzed earlier, where it was implicit from the outset that at a constant cost the straight line is the cheapest path to connect two points. This is the assumption that obviated the need for a more sophisticated theory.

4

The following elementary problem from the subject of *linear programming* will illustrate the occurrence of absolute extrema.

Example 7 An automobile manufacturer has 4,500 specially equipped cars ready for export; 3,500 are in Detroit, and 1,000 are in Wichita. Of these, 2,000 are to be shipped overseas from New York and the remainder from San Francisco. The transportation charges for each car from the points of origin to the ports of embarkation are as follows: Detroit to New York, \$70; Detroit to San Francisco, \$95; Wichita to New York, \$65; Wichita to San Francisco, \$85. How should the shipment be scheduled to minimize the total transportation cost?

This problem is simple enough to solve without great fanfare, but a more elaborate format is used, which is analogous to that required in a more general case (where there are, say, 20 storage locations and 10 ports of embarkation).

Let x and y be the number of cars sent from Detroit and Wichita to New York; $3{,}500 - x$ and $1{,}000 - y$ are the numbers transported from these cities to San Francisco. All the data are arranged in easy visual form in Table 5.1.

TABLE 5.1

To / *From*	*New York*	*San Francisco*	*Total*
Detroit	70 x	95 $3{,}500 - x$	3,500
Wichita	65 y	85 $1{,}000 - y$	1,000
Total	2,000	2,500	

The number of cars sent from one city to another is written in the large boxes, the corresponding transportation charge appears in the small triangles, and the total number of cars at each point of origin or final destination is placed in the same column or row as the name of the city. The transportation cost for all cars from Detroit is then

$$70x + 95(3{,}500 - x),$$

while from Wichita it is

$$65y + 85(1{,}000 - y),$$

which makes the total cost

$$C = 70x + 95(3{,}500 - x) + 65y + 85(1{,}000 - y). \tag{7}$$

Variables x and y are related because the number of cars to be exported from New York is 2,000:

$$x + y = 2{,}000. \tag{8}$$

There are other conditions as well, since both x and y are non-negative and bounded by the number of autos available at each point of origin:

$$0 \leq x \leq 3{,}500, \qquad 0 \leq y \leq 1{,}000.$$

These restrictions combined with (8) are equivalent to

$$1{,}000 \leq x \leq 2{,}000, \qquad y = 2{,}000 - x, \tag{9}$$

and the minimum cost must be found for x in this range. Upon substituting for y, the total cost becomes

$$C = \$377{,}500 - 5x,$$

and this is minimized when x is chosen as large as possible. According to (9), this means $x = 2{,}000$. [Note that $C(x)$ has no relative minimum.] The optimal solution is then $x = 2{,}000$, $y = 0$, $C = \$367{,}500$. In words, the transportation schedule calls for Detroit to supply New York exclusively, while everything else goes to San Francisco. (The maximum cost is incurred when x assumes its lower bound, $x = 1{,}000$, in which case $C = \$372{,}500$. The total savings involved is only \$5,000.)

Other problems of this type are considered later in Sec. 6.9.

Exercises 2.5

1. Determine the quadratic polynomial $y = ax^2 + bx + c$ that passes through the points $(0,1)$ and $(1,0)$ and has a maximum at $x = -1$.

2. Determine the cubic polynomial $y = x^3 + ax^2 + bx + c$ which has a relative maximum at $x = -1$, a minimum at $x = 1$ and passes through the point $(0, 1)$. Draw a graph of this curve and determine the inflection point. From the graph, estimate the three roots of the cubic polynomial.

3. By completing the square and by calculus methods, show that $y = 10x^2 + 14x + 5$ is positive for all x. What is its minimum value?

4. Graph the function $y = x + 1/x$. Show that $|x + 1/x| \geq 2$ by completing the square and by calculus methods.

5. Graph the cubic polynomial $y = x^3 - 3x^2 - 9x + 7$. Determine the stationary points and inflection point. From the graph, estimate the three roots.

6. Verify the following inequalities valid for $x > 0$. Illustrate with graphs.

(i) $1 - x < \dfrac{1}{1+x} < 1 - x + x^2$ **(ii)** $(1+x)^{1/2} < 1 + \dfrac{x}{2}$

(iii) $1-\frac{x^2}{2}<\cos x$ **(iv)** $x-\frac{x^3}{6}<\sin x<x$.

7. Determine the relative extrema of the following functions by completing the square and by calculus methods.

(i) $2-x-x^2$ **(ii)** x^2+5x-1 **(iii)** $2x^2-3x+1$

8. Find the relative extrema of the following functions. Illustrate with graphs.

(i) $\frac{x}{x^2+1}$ **(ii)** $\frac{(2x-1)(x-8)}{(x-1)(x-4)}$ **(iii)** $\sin x+\cos x$

9. The cubic polynomial $f(x)=x^3+px+q$ must have at least one real root. Show that it will have three real distinct roots provided that $4p^3+27q^2<0$. (*Hint:* Consider the possible stationary points.)

10. Find the relative extrema and inflection points of the fifth degree polynomial $y=x^5-5x^3+10x-3$. Use these points to draw an accurate graph. From the graph, estimate the roots of the polynomial.

11. Find the relative extrema and inflection points of the function

$$f(x)=\begin{cases}2x(x-1) & x<1\\ (x-1)(x-2)(x-3) & x\geq 1.\end{cases}$$

Draw an accurate graph. Does this function have a well-defined derivative at $x=1$?

12. If $f(x)$ and $g(x)$ both have relative maxima (or relative minima) at the same point $x=a$, is this also true of the product $h(x)=f(x)g(x)$? If they both have stationary points at $x=a$, is this true of the product? Illustrate your answer with specific examples.

13. Find the area of the largest rectangle with sides parallel to the coordinate axes and with vertices lying on the curves $y(x^2+1)=1$, $y(x^2+1)=-1$. Determine the inscribed circle, the circle with center at the origin that touches the two curves at four points.

14. Graph the closed curve $x^{2/3}+y^{2/3}=2$. Find the area of the largest rectangle with sides parallel to the coordinate axes and with vertices on the given curve. Find the radius of the inscribed circle.

15. A length of wire is cut into two pieces, one of which is bent to form a circle, the other to form a square. How should the wire be cut if the area enclosed by the two curves is to be a minimum? a maximum?

16. The power output of a battery is given in terms of its voltage V and resistance R by $P=VI-RI^2$, where I is the current. If V and R are constants, find the current corresponding to maximum power output. Determine the maximum power output. If the current producing maximum power is subject to a small fluctuation ΔI, estimate the change ΔP in power output.

17. Design a circular cone of a given slant edge l for which the volume is a maximum. Design the cone to make the surface area a maximum.
(*Hint:* $V=\frac{1}{3}\pi r^2h$, $S=\pi r^2+\pi rl$, $l=(r^2+h^2)^{1/2}$.)

18. Four equal squares are cut from the corners of a square piece of cardboard 48 cm on a side. An open box is to be made by folding the flaps as shown in Fig. 5.5 on page 138. What size squares should be cut to obtain the largest volume possible?

19. After its printing is completed the publishing firm in Example 3 learns that a tax of 50 cents will be imposed on each book sold. How much of this tax should be passed on to the purchaser and how much should be absorbed? Why? How is the firm's profit affected?

20. The cost of operating a diesel truck is $21+v/5$ cents per kilometer when it travels at v km/h. If the driver earns $12.75 per hour, what is the most economical speed at which to operate during a 1,200 kilometer trip?

21. A man who can run 3 times as fast as he can swim rushes to the aid of a swimmer in a

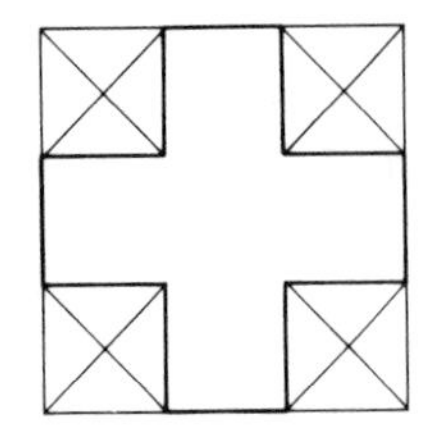

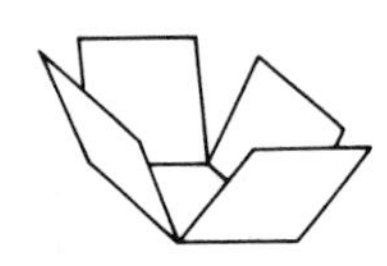

FIGURE 5.5

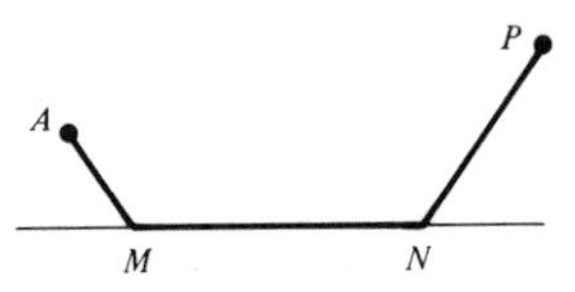

FIGURE 5.6

large circular pool of radius R. In terms of a coordinate system whose origin is at the center of the pool, the initial position of the swimmer is $(-aR, 0)$, where $0<a<1$, while that of his rescuer is $(R, 0)$. How would you direct the rescue in the shortest possible time?

22. Two corridors of widths 2 and 3 m meet at right angles. What is the longest pole that can be carried around the corner horizontally?

23. According to an empirical study, the separation distance l m between automobiles travelling in a single-lane tunnel is related to their velocity v m/sec by $l=6+3v+v^2/3$. (All cars move at the same speed v.) How many cars pass a given point in an hour when the traffic speed is v? For what traffic speed is the traffic volume a maximum?

24. The location of two cities on a map scaled in kilometers are $A=(2,0)$, $B=(1,-5)$. Find the cheapest route for a pipeline between these cities if the construction cost for $y \geq 0$ is \$$10^5$ per kilometer whereas the cost if $y<0$ is \$$3 \times 10^5$ per kilometer.

25. Light travels between two points along the path that requires the least time. Let the velocity of light in two adjoining media such as air and water be v_1 and v_2. Calculate the path of a light ray which travels from point P in one medium to point Q in the other. Discuss the analogy of this problem to the pipe-laying problem in the text.

26. As in Example 6, a pipeline is to be constructed between two cities A and P, shown in Fig. 5.6. Assume that the cities are sufficiently far apart for it to pay to use a route like $AMNP$ and find, by any method, the path of minimum cost. What does sufficiently far apart mean quantitatively?

27. Find the least expensive route for a pipeline between cities located at points $A=(0,5)$, $B=(25,5)$ if the construction cost in $y>0$ is $c_1=\$10^5$ per kilometer whereas that in $y \leq 0$ is $c_2=\$5 \times 10^4$ per kilometer. Find the smallest value of c_2 for which the straight route is cheapest.

28. A ship travelling at a constant speed v expends energy at a rate proportional to v^2. Determine the optimum speed against a steady current of speed u. (Minimize the energy required to travel a fixed distance.)

29. An oil company has 200,000 barrels of oil stored in Kuwait (on the Persian Gulf), 150,000 barrels stored in Galveston, Texas, and 100,000 barrels stored in Caracas, Venezuela. A customer in New York would like 250,000 barrels, and a customer in London would like the remaining 200,000 barrels. The shipping costs in cents per barrel are shown below. Find the minimum-cost shipment schedule.

	Kuwait	*Galveston*	*Caracas*
New York	36¢	10¢	18¢
London	34	22	25

2.6 *Implicit and inverse functions*

A function has been defined as a rule by which one number, say y, is obtained from another, x, and the relationship has usually been given in the explicit form

$$y = f(x).$$

However, functional dependence can also be described without necessarily solving for one variable in terms of the other by writing

$$F(x, y) = 0. \tag{1}$$

A typical example of such a relationship is

$$x^2 + y^2 - 1 = 0. \tag{2}$$

This defines y implicitly as a function of x (or vice versa) but not uniquely. There are two continuous functions, $y = \sqrt{1 - x^2}$ and $y = -\sqrt{1 - x^2}$, defined by this implicit equation when $-1<x<1$.

The derivative of an implicit function is calculated in a completely straightforward fashion using the general rules for differentiation, especially the chain rule. To find $y'(x)$ in (2), for example, each side of the equation is differentiated term by term with respect to x. The differentiation of the right-hand side obviously gives zero, and since

$$\frac{d}{dx}(x^2 + y^2 - 1) = \frac{d}{dx}x^2 + \frac{d}{dx}y^2 - \frac{d}{dx}1 = 2x + 2y\frac{dy}{dx},$$

it follows that

$$x + y\frac{dy}{dx} = 0,$$

or

$$\frac{dy}{dx} = -\frac{x}{y}. \tag{3}$$

The chain rule is used in the reduction, $\dfrac{dy^2}{dx} = \dfrac{dy^2}{dy}\dfrac{dy}{dx} = 2y\dfrac{dy}{dx}$.

This calculation of dy/dx applies to both continuous solutions but it is only meaningful for x between -1 and $+1$ (and for $x^2 + y^2 = 1$). The second derivative, $\dfrac{d^2y}{dx^2}$, can be obtained in a similar manner from (3) or its predecessor. For example,

$$\frac{d}{dx}\left(x + y\frac{dy}{dx}\right) = 1 + \left(\frac{dy}{dx}\right)^2 + y\frac{d^2y}{dx^2} = 0,$$

and it follows that

$$\frac{d^2y}{dx^2} = -\frac{1 + \left(\dfrac{dy}{dx}\right)^2}{y} = -\frac{x^2 + y^2}{y^3} = -\frac{1}{y^3}.$$

If x is regarded as the dependent variable instead of y, then the derivative of x with respect to y can be computed in the same manner and the result is

$$\frac{dx}{dy} = -\frac{y}{x}. \tag{4}$$

From (3) and (4), we note the identity

$$\frac{dy}{dx} = \frac{1}{dx/dy}, \tag{5}$$

a relationship that further enhances the interpretation of the differentials dx, dy as real infinitesimal elements which are manipulated like ordinary numbers. From this viewpoint, Eq. (5) represents a generally valid algebraic manipulation in which numerator and denominator are divided by dy,

$$\frac{dy}{dx} = \frac{\dfrac{dy}{dy}}{\dfrac{dx}{dy}} = \frac{1}{\dfrac{dx}{dy}},$$

in exact analogy with the fraction p/q written as $1/(q/p)$.

Example 1 If $y^2 + xy + x^2 = 3$, find dy/dx and dx/dy.

By solving a quadratic equation, y can be determined as an explicit function of x,

$$y = y(x) = \frac{-x \pm \sqrt{12 - 3x^2}}{2}$$

A symmetric calculation yields x as a function of y,

$$x = x(y) = \frac{-y \pm \sqrt{12 - 3y^2}}{2}$$

Using these formulas the derivatives dy/dx and dx/dy can be evaluated. A simpler method is to differentiate the original equation with respect to x and y.

$$\frac{d}{dx}(x^2 + xy + y^2) = 2x + y + x\frac{dy}{dx} + 2y\frac{dy}{dx} = 0$$

$$\frac{d}{dy}(x^2 + xy + y^2) = 2x\frac{dx}{dy} + y\frac{dx}{dy} + x + 2y = 0$$

Solving, $$\frac{dy}{dx} = -\frac{2x+y}{x+2y}, \quad \frac{dx}{dy} = -\frac{x+2y}{2x+y}.$$

One advantage of the explicit formulas is that they show that the function $y = y(x)$ is defined for $|x| \leq 2$ and $x = x(y)$ is defined for $|y| \leq 2$.

Example 2 Find dy/dx and dx/dy from the implicit relationship

$$y^2 + \frac{1}{x+y} - \frac{1}{x^2} + 3 = 0.$$

In this case, differentiating the equation yields

$$2y\frac{dy}{dx} - \frac{1}{(x+y)^2}\left(1 + \frac{dy}{dx}\right) + \frac{2}{x^3} = 0,$$

which can be solved to yield

$$\frac{dy}{dx} = \left[\frac{1}{(x+y)^2} - \frac{2}{x^3}\right] \Big/ \left[2y - \frac{1}{(x+y)^2}\right].$$

The expression for dx/dy is just the reciprocal of this formula.

Example 3 Differentiate the function $y = x^{m/n}$, where m and n are integers. The equivalent implicit form of this function is

$$y^n = x^m,$$

and the result of differentiating both sides of this equation is

$$ny^{n-1}\frac{dy}{dx} = mx^{m-1}.$$

By substituting for y, this becomes

$$\frac{dy}{dx} = \frac{m}{n}\frac{x^{m-1}}{x^{(m/n)(n-1)}} = \frac{m}{n}x^{m/n-1},$$

or

$$\frac{d}{dx}x^{m/n} = \frac{m}{n}x^{m/n-1}. \tag{6}$$

If $\{r_n\}$ is a sequence of rational numbers (fractions) having the real number a as a limit, then $\lim_{n\to\infty} x^{r_n}$ defines the power function x^a. Since

$$\frac{d}{dx}x^{r_n} = r_n x^{r_n - 1}$$

holds for every fraction r_n, this formula is also valid in the limit, so that

$$\frac{d}{dx}x^a = ax^{a-1}. \tag{7}$$

The implicit form of a functional relationship shows clearly that the designation of one variable x as "independent" is really an arbitrary decision. Indeed it is possible, desirable, and often necessary to reverse the roles assigned to the variables, i.e., to determine x from a knowledge of y. This procedure is equivalent to expressing x as a function of y. If, for example,

$$y = f(x), \tag{8}$$

then x can also be written as a function of y:

$$x = g(y), \tag{9}$$

and, in particular

$$x = g(f(x)) \qquad \text{and} \qquad y = f(g(y)).$$

We merely locate y in the table of values and note the corresponding value of x, assuming that the correspondence is one to one. For example, if $y = f(x) = x^3$, then $x = g(y) = y^{1/3}$. Clearly, $x = (x^3)^{1/3}$ and $y = (y^{1/3})^3$. The function $f(x)$ can be viewed as transforming x into y, in which case $g(y)$ changes y back into x; g is called the *inverse* function of f (and vice versa). (One is the poison, the other the antidote.) A notation which emphasizes the neutralizing property of the inverse function is

$$g(y) = f^{-1}(y), \tag{10}$$

for then we can write

$$f^{-1}(f(x)) = f(f^{-1}(x)) = x. \tag{11}$$

The combined operations $f^{-1}f$ are self-cancelling, in analogy with the law of exponents. Note carefully that

$$f^{-1}(y) \neq 1/f(y).$$

In order to study an inverse function in detail, it is convenient to restore traditional roles to x and y. This means that the variables in (9) are interchanged, so that the inverse function is written as

$$y = g(x). \tag{12}$$

Indeed, this interchange of variables can be made a step earlier in which case the inverse function of

$$y = f(x)$$

is obtained by solving

$$x = f(y) \tag{13}$$

for y to obtain (12) directly. Moreover, if (a, b) is a point on the graph of $f(x)$, then the interchanged coordinates (b, a) give a point on the curve of the inverse function $y = g(x)$. This implies that the curves of a function and its inverse are just mirror images in the 45° line of slope 1 that passes through the origin (Fig. 6.1). To prove this, we must show that the line through the two points is perpendicular to the "mirror" and that the distances of the images from the reflecting line are the same.

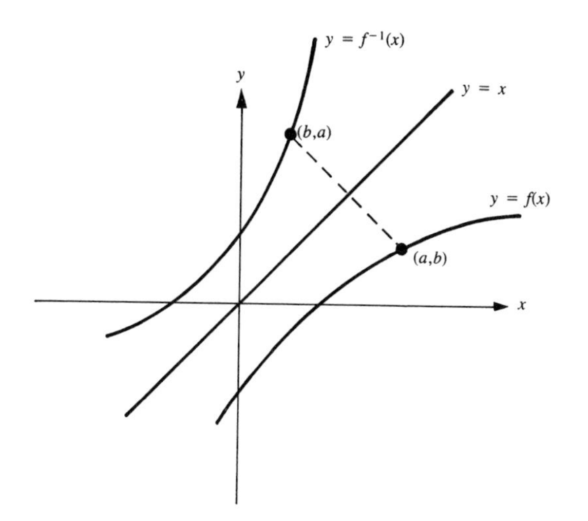

FIGURE 6.1

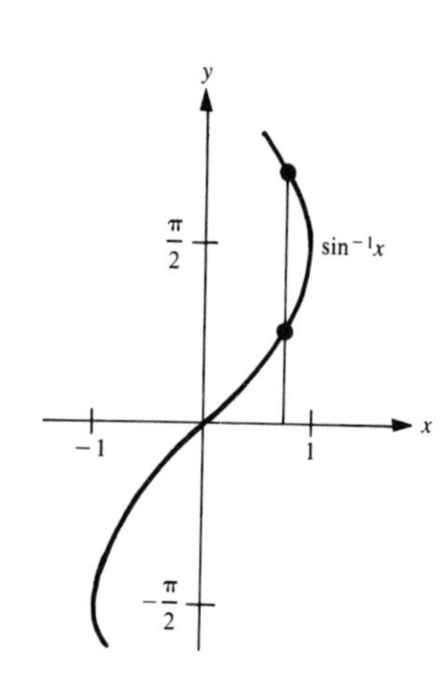

FIGURE 6.2

The first part is established by noting that the line connecting (a,b) and (b,a) has the slope $(b-a)/(a-b) = -1$ and therefore makes an angle of 135° with the horizontal. The two lines intersect at $x = y = \dfrac{a+b}{2}$. The distance, l, from this point $\left(\dfrac{a+b}{2}, \dfrac{a+b}{2}\right)$ to either of the image points (a,b) or (b,a) is obtained from the Pythagorean theorem: $l^2 = \left(a - \dfrac{a+b}{2}\right)^2 + \left(b - \dfrac{a+b}{2}\right)^2 = \dfrac{1}{2}\left(a-b\right)^2.$

The inverse trigonometric functions occur frequently enough to warrant special attention.

The *inverse sine function*, also called the *arc sine*, is written

$$y = \sin^{-1} x. \tag{14}$$

Here, the value of y is sought, the trigonometric sine of which is equal to another number x, that is,

$$\sin y = x. \tag{15}$$

Clearly, the *sine* of the *inverse sine* of a number is the number itself:

$$\sin(\sin^{-1} x) = x.$$

The graph of the arc sine (Fig. 6.2) obtained by reflecting the sine curve in the 45° line indicates that one value of x in the interval $(-1, 1)$ corresponds to an infinite number of values of y; the arc sine is said to be a *multivalued function*. On the other hand, there is no real value of y which corresponds to $|x| > 1$.

Restrictions on the inherent multivaluedness can be set by arbitrarily limiting the discussion of $\sin^{-1} x$ to the range of values within an interval of the y axis of length

π. We choose to define the function as

$$y = \sin^{-1} x \qquad \text{for} \qquad -\frac{\pi}{2} \leq y \leq \frac{\pi}{2}, \tag{16}$$

and this is called the *principal value* to distinguish it from all other possibilities.

Equation (15) is an implicit relationship, but the derivative of the arc sine can be calculated rather easily as follows

$$\frac{d}{dx} \sin y = \frac{d}{dx} x = 1,$$

which yields

$$\cos y \frac{dy}{dx} = 1 \qquad \text{or} \qquad \frac{dy}{dx} = \frac{1}{\cos y}.$$

Since

$$\cos y = (1 - \sin^2 y)^{1/2} = (1 - x^2)^{1/2}$$

in the interval $|y| \leq \pi/2$, the final result is

$$\frac{d}{dx} \sin^{-1} x = \frac{1}{(1 - x^2)^{1/2}}. \tag{17}$$

The *inverse cosine function* is *defined* by the relation

$$\cos^{-1} x = \frac{\pi}{2} - \sin^{-1} x, \tag{18}$$

which implies that $0 \leq \cos^{-1} x \leq \pi$. The *inverse tangent*,

$$y = \tan^{-1} x, \qquad |y| \leq \frac{\pi}{2}, \tag{19}$$

is the equivalent of

$$\tan y = x. \tag{20}$$

The remaining inverse trigonometric functions are defined in the exercises.

The derivative of the arc cosine is obtained from its definition and Eq. (18):

$$\frac{d}{dx} \cos^{-1} x = \frac{d}{dx}\left(\frac{\pi}{2} - \sin^{-1} x\right) = -\frac{d}{dx} \sin^{-1} x = \frac{-1}{(1 - x^2)^{1/2}}.$$

The derivative of the arc tangent function is calculated from (20), by implicit differentiation, as follows:

$$\frac{d}{dx} \tan y = 1 \qquad \text{or} \qquad \sec^2 y \frac{dy}{dx} = 1.$$

In terms of x alone, the last equation may be written

$$\frac{d}{dx} \tan^{-1} x = \frac{1}{1 + x^2}.$$

In summary,

$$\left.\begin{aligned}\frac{d}{dx}\sin^{-1}x &= (1-x^2)^{-1/2},\\ \frac{d}{dx}\cos^{-1}x &= -(1-x^2)^{-1/2},\\ \frac{d}{dx}\tan^{-1}x &= (1+x^2)^{-1}.\end{aligned}\right\} \tag{21}$$

Example 4 Find the values of the functions $\sin^{-1}x$, $\cos^{-1}x$, $\tan^{-1}x$, and their derivatives at $x = \frac{1}{2}$.

From the table of trigonometric functions,

$$\sin^{-1}\tfrac{1}{2} = \frac{\pi}{6},$$

$$\cos^{-1}\tfrac{1}{2} = \frac{\pi}{2} - \sin^{-1}\tfrac{1}{2} = \frac{\pi}{3},$$

$$\tan^{-1}\tfrac{1}{2} = 0.4636.$$

The values of the derivatives are obtained from (21) by setting $x = \frac{1}{2}$ in each formula:

$$\frac{d}{dx}\sin^{-1}x = \frac{2}{\sqrt{3}}, \qquad \frac{d}{dx}\cos^{-1}x = -\frac{2}{\sqrt{3}}, \qquad \frac{d}{dx}\tan^{-1}x = \frac{4}{5}.$$

---

4 Consider the function $E(x)$, which was defined as the particular solution of the differential equation

$$\frac{dy}{dx} = y \qquad \text{with} \qquad y(0) = 1. \tag{22}$$

Since the inverse function of $y = E(x)$ is obtained by solving for x as a function of y, $x = L(y)$, it must also be the solution of the differential equation

$$\frac{dx}{dy} = \frac{1}{y}, \tag{23}$$

which is just (22) rewritten so that y appears as the independent variable.

If, as discussed earlier, the roles of the variables are interchanged to conform with common usage, the inverse function, written now as $y = L(x)$, satisfies the differential equation

$$\frac{dy}{dx} = \frac{d}{dx}L(x) = \frac{1}{x}. \tag{24}$$

By definition

$$L(E(x)) = x, \tag{25}$$

and since $E(0) = 1$, it follows that

$$L(E(0)) = L(1) = 0. \tag{26}$$

The function $L(x)$ can be characterized as the particular solution of (24) with the value given in (26). Its name is the *natural logarithm*, and it will be studied more closely later, where it is shown that $E(L(x)) = x$ is strictly valid for $x \geq 0$.

---

5 The functional relationship between two variables x and y is often conveniently described in terms of a parameter or an auxiliary variable, as for example in the *parametric equations*

$$x = x(t), \qquad y = y(t). \tag{27}$$

Here the corresponding values of x and y are calculated directly from t, a procedure that may be much simpler than having an explicit statement

$$y = f(x).$$

This explicit form can be obtained by solving the first equation in (27) for t as a function of x, $t = t(x)$, and then substituting this in the second:

$$y = y(t(x)) = f(x).$$

The derivative dy/dx is determined by implicit differentiation and use of the chain rule, with x assigned the role of the independent variable and y and t the dependent variables. If both expressions in (27) are differentiated with respect to x, the results are

$$1 = \frac{dx}{dt}\frac{dt}{dx},$$

$$\frac{dy}{dx} = \frac{dy}{dt}\frac{dt}{dx},$$

and by eliminating dt/dx between these we obtain

$$\frac{dy}{dx} = \frac{dy}{dt}\bigg/\frac{dx}{dt}. \tag{28}$$

The derivative $y'(x)$ is given as the ratio of derivatives with respect to the parametric variable.

Note that (28) reduces to an identity by manipulating the differentials algebraically to eliminate dt:

$$\frac{dy}{dx} = \frac{dy}{dt}\bigg/\frac{dx}{dt} = \frac{dy}{dt}\frac{dt}{dx} = \frac{dy}{dx}.$$

Another way of obtaining (28) is to divide the incremental formulas

$$dx = x'(t)\,dt, \qquad dy = y'(t)\,dt,$$

thereby eliminating dt. Once again we see that it is perfectly valid to handle differentials like real numbers.

Example 5 Find dy/dx if $x = t + 1$, $y = t^2 - t$.

Since $t = x - 1$, $y = (x-1)^2 - (x-1) = x^2 - 3x + 2$ and $dy/dx = 2x - 3$. The same result is obtained by calculating $dx/dt = 1$ and $dy/dt = 2t - 1$. Then $dy/dx = (dy/dt)/(dx/dt) = (2t-1)/1 = 2t - 1 = 2x + 3$, because $x = t + 1$.

Example 6 Find dy/dx, if

$$x = t - \sin t \qquad \text{and} \qquad y = 1 - \cos t.$$

Compare the ease of calculating x and y using these parametric equations (Table 6.1) with the difficulty inherent in an explicit functional relationship. (The tabulated values of t are integral multiples of $\pi/10$.)

TABLE 6.1

t	x	y	t	x	y	t	x	y
0.	0.	0.	2.19911	1.39010	1.58779	4.39823	5.34929	1.30902
0.31416	0.00514	0.04894	2.51327	1.92549	1.80902	4.71239	5.71239	1.00000
0.62832	0.04053	0.19098	2.82743	2.51842	1.95106	5.02655	5.97760	0.69098
0.94248	0.13346	0.41221	3.14159	3.14159	2.00000	5.34071	6.14972	0.41222
1.25664	0.30558	0.69098	3.45575	3.76477	1.95106	5.65487	6.24265	0.19098
1.57080	0.57080	1.00000	3.76991	4.35770	1.80902	5.96903	6.27804	0.04894
1.88496	0.93390	1.30902	4.08407	4.89309	1.58779	6.28318	6.28319	0.00000

In this case, the function $y = f(x)$ is not very easy to determine, and the inverse,

$$x = \cos^{-1}(1-y) - \sin[\cos^{-1}(1-y)],$$

is quite complicated because of its multivaluedness.

Since

$$\frac{dx}{dt} = 1 - \cos t \qquad \text{and} \qquad \frac{dy}{dt} = \sin t,$$

dividing the second by the first yields

$$\frac{dy}{dx} = \frac{\sin t}{1 - \cos t}.$$

Further development of technique is left to the exercises.

Exercises 2.6

1. Find the inverse functions for the following functions:

(i) $y = x^2$ **(ii)** $y = x^2 + x$ **(iii)** $y = x^{1/3}$

Illustrate by graphs the reasons why the inverses of **(i)** and **(ii)** are not unique.

2. Show how the graph of the function inverse to $y = f(x)$ can be obtained from the graph of $y = f(x)$ by reflection in the line $y = x$. Illustrate with the example $y = x^3$. What function is equal to its own inverse?

3. Differentiate each of the following equations to find dy/dx and dx/dy.

(i) $y + (xy)^{1/2} + x^2 = 0$ **(ii)** $\dfrac{x^2}{a^2} - \dfrac{y^2}{b^2} = 1$ **(iii)** $y^3 = 8(x^2 + y^2)$

(iv) $x^2y + xy^2 = 2a^3$ **(v)** $x^{1/2} + y^{1/2} = a^{1/2}$ **(vi)** $\tan^{-1}y = x^2 + y^2$

(vii) $y = x(x^2 + y^2)$ **(viii)** $\sin(xy) = (xy)^3$ **(ix)** $\cos(x + y) = 1 - xy$

4. Show that the equation of the line tangent to the parabola $y^2 = 4px$ at (x_0, y_0) is $y_0y = 2p(x + x_0)$. Illustrate your answer with a graph drawing the tangent lines at $(0,0)$, $(p, 2p)$, $(p, -2p)$, $(4p, 4p)$ and $(4p, -4p)$.

5. Find the slope of the curve $x^{2/3} + y^{2/3} = a^{2/3}$ at (x_0, y_0) and determine the tangent line. At which points is the slope 0, 1 or -1?

6. Find the slope of the curve $(x + y)^{2/3} + (x - y)^{2/3} = 2b^{2/3}$ at the points for which $x = 0$, $x = y$ and $x = -y$. Determine the tangent lines at these points.

7. The total surface area of a right circular cylinder of radius r and height h is $A = 2\pi r^2 + 2\pi rh$. If A remains constant while r and h vary, determine dr/dh and dh/dr by implicit differentiation. Interpret the meaning of these derivatives.

8. The total surface area of a right circular cone of radius r and height h is $A = \pi r^2 + \pi r(r^2 + h^2)^{1/2}$. Repeat the calculations of the previous problem for this surface.

9. Graph the function $y = x + \sqrt{x^2 + 1}$ and determine the inverse function. Evaluate dy/dx and dx/dy. Repeat with the function $y = x - \sqrt{x^2 + 1}$.

10. The derivatives dy/dx and dx/dy are reciprocals of each other. What is the relation between d^2y/dx^2 and d^2x/dy^2?

11. If $x = f(t)$ and $y = g(t)$, find dy/dx, dx/dy, d^2y/dx^2 and d^2x/dy^2 in terms of $f'(t)$, $g'(t)$, $f''(t)$ and $g''(t)$. Illustrate with the special case $f(t) = t^2 + 1$, $g(t) = t^2 + t$.

12. The circle $x^2 + y^2 = 4$ can be represented parametrically by $x = 2\cos t$, $y = 2\sin t$. Find dy/dx and dx/dy by differentiation with respect to the parameter. Determine d^2y/dx^2 and d^2x/dy^2.

13. Sketch the curve given parametrically by $x = s/(1 + s)$, $y = s^2/(1 + s)$. Find dy/dx and d^2y/dx^2 by differentiating with respect to the parameter. Solve for $y = f(x)$ explicitly and calculate dy/dx and d^2y/dx^2 by explicit differentiation.

14. Find dy/dx, dx/dy, d^2y/dx^2 and d^2x/dy^2 in each of the following:

(i) $x = t$, $y = 4t^2 + 1$ **(ii)** $x = (1 + s^2)^{1/2}$, $y = (1 - s^2)^{1/2}$

(iii) $x = a\cos^3\theta$, $y = a\sin^3\theta$ **(iv)** $x = \theta - \sin\theta$, $y = 1 - \cos\theta$

(v) $x = \cos(1 + t)$, $y = \sin(1 - t)$ **(vi)** $x = \tan(t^2)$, $y = \cot(t^2)$

15. Show that the definitions of $\cos^{-1}x$ and $\sin^{-1}x$ are consistent with the fact that the interior acute angles of a right triangle must add to $\pi/2$.

16. Find the values of the following:

(i) $\sin^{-1}(0.5)$ **(ii)** $\tan^{-1} 2$ **(iii)** $\tan^{-1}(\tan \pi/3)$

(iv) $\sin^{-1}(\cos \pi/4)$ **(v)** $\cos^{-1}(\sqrt{3}/2)$ **(vi)** $\sin^{-1}(\sin \pi/4)$

17. Define the principal value of $\cot^{-1}x$ and draw its graph. Show that its derivative is $-1/(1+x^2)$ and explain why $\tan^{-1}x + \cot^{-1}x = \pi/2$ if $0<x<\pi/2$.

18. Evaluate the derivatives of $\sin^{-1}x$, $\cos^{-1}x$ and $\tan^{-1}x$ by using the chain rule to differentiate $\sin(\sin^{-1}x)$, $\cos(\cos^{-1}x)$ and $\tan(\tan^{-1}x)$.

19. Verify the following identities.

(i) $\tan^{-1}\dfrac{x+a}{1-ax} = \tan^{-1}x + \tan^{-1}a$ **(ii)** $\tan^{-1}x = \sin^{-1}\dfrac{x}{\sqrt{1+x^2}} = \cos^{-1}\dfrac{1}{\sqrt{1+x^2}}$

(iii) $\tan(\sin^{-1}x) = \dfrac{x}{\sqrt{1-x^2}}$ **(iv)** $\tan(\cos^{-1}x) = \dfrac{\sqrt{1-x^2}}{x}$

20. Define $\sec^{-1}x = \cos^{-1}(1/x)$ and $\csc^{-1}x = \sin^{-1}(1/x)$ and show the following.

(i) $\dfrac{d}{dx}(\sec^{-1}x) = \dfrac{1}{|x|\,(x^2-1)^{1/2}}$ **(ii)** $\dfrac{d}{dx}(\csc^{-1}x) = \dfrac{-1}{|x|\,(x^2-1)^{1/2}}$

(iii) $\sec^{-1}x = \csc^{-1}\dfrac{x}{(x^2-1)^{1/2}}$ **(iv)** $\sec^{-1}x + \csc^{-1}x = \dfrac{\pi}{2}$.

21. Using principal values, find the following numbers.

(i) $\cos^{-1}(-1)$ **(ii)** $\cos^{-1}(1/2)$ **(iii)** $\cos^{-1}1 - \sin^{-1}1$

(iv) $\sin^{-1}\left(\dfrac{1}{\sqrt{2}}\right)$ **(v)** $\lim\limits_{x\to\pi+}\tan^{-1}\left(2\tan\dfrac{x}{2}\right)$ **(vi)** $\lim\limits_{x\to\pi-}\tan^{-1}\left(2\tan\dfrac{x}{2}\right)$

(vii) $\csc^{-1}(\sqrt{2})$ **(viii)** $\cot^{-1}2$ **(ix)** $\sec^{-1}1$

22. Differentiate the following functions.

(i) $\sec^{-1}(x+2)^{1/2}$ **(ii)** $\tan^{-1}(2\sin x)$ **(iii)** $\sin^{-1}(\cos x)$

(iv) $\cos^{-1}(\sin x)$ **(v)** $\sin^{-1}(\tan x)$ **(vi)** $\sec^{-1}(\cos x)$

23. Graph the following curves given in polar coordinates for the special cases $a=1$. Determine their equations in cartesian coordinates $x = r\cos\theta$, $y = r\sin\theta$.

(i) $r = 2a\cos\theta$ (circle) **(ii)** $r = a\sin 3\theta$ (rose)

(iii) $r = a(1-\sin\theta)$ (cardioid) **(iv)** $r = a\theta$ (spiral)

(v) $r = a/\theta$ (spiral) **(vi)** $r = a/(2-\cos\theta)$ (ellipse)

24. Graph the cardioid given in polar coordinates by $r = a(1-\cos\theta)$. Show that the curve is also described parametrically by $x = r\cos\theta = a(\cos\theta - \cos^2\theta)$ and $y = r\sin\theta = a(\sin\theta - \sin\theta\cos\theta)$. Find the values of $dx/d\theta$, $dy/d\theta$ and dy/dx at $\theta = \pi/2, \pi/3, \pi/4, \pi/6$. Determine the tangent lines at these points.

25. The boom of a derrick is 30 m long. A cable attached to the end of the boom ranges over a pulley at the top of a 35-m pole as shown (Fig. 6.3). If the cable is drawn at the rate of 2 m/sec to lift the weight, how fast is the angle between the boom and the pole changing when $\theta = 12°$? How fast is the weight being lifted at this instant?

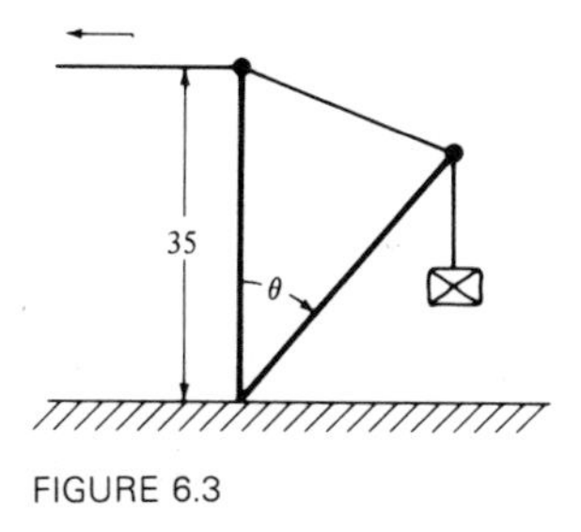

FIGURE 6.3

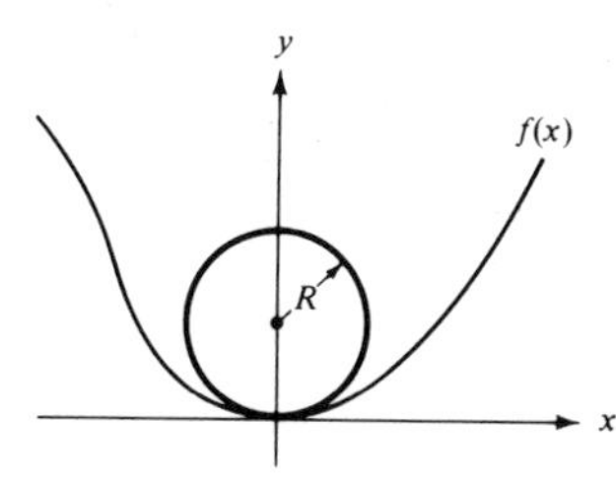

FIGURE 6.4

26. Find the radius of the circle $x^2 + (y-R)^2 = R^2$ which (as shown in Fig. 6.4) fits the curve

$y = f(x)$ smoothly at the origin where $f(0) = 0$ and $f'(0) = 0$. (Hint: Make the first and second derivatives of both curves identical at the origin.) The value of R is called the *radius of curvature* of the curve $y = f(x)$ at the origin.

27. Determine the radius of curvature at the origin for the following functions. Illustrate with graphs.

(i) $y = x^2$ **(ii)** $y = \sin^2 x$ **(iii)** $y = x \sin x$.

28. A ladder 10 m long leans against the side of a house. The lower end slips away from the house at a constant rate of 1.5 m/sec. When the lower end is 5 m from the wall, find the rate at which the other end slides downward.

29. A revolving light in a lighthouse 2 km off shore makes 5 revolutions per minute. If the shoreline is straight, find the velocity of the beam as it sweeps along the coast. In particular, find the velocity when the angle of the beam to the coastline is 60°.

30. The volume of a sphere of radius r is $V = 4/3\, \pi r^3$. What is the approximate volume of a thin spherical shell of thickness $\Delta r \ll r$? Silver weighs 10.5 gm/cm^3 and gold weighs 19.3 gm/cm^3. What are the radii of the smallest spherical shells 1 cm thick that can float in water (density 1 gm/cm^3)?

2.7 *The mean-value theorem*

The mean-value theorem is a valuable theoretical tool with many important and constructive applications, as succeeding sections will demonstrate. The theorem states that if $f(x)$ is continuous and has a derivative at every point in some interval $a \leq x \leq b$, then there is at least one number ξ, $a < \xi < b$, such that

$$f(b) - f(a) = f'(\xi)(b - a). \tag{1}$$

Since

$$\frac{f(b) - f(a)}{b - a}$$

is actually the slope of the chord AB, shown in Fig. 7.1, the graphical interpretation of the theorem is that there is at least one point C where the tangent line to $f(x)$ has this very same slope. (In fact, there are two such points in Fig. 7.1.) Note carefully

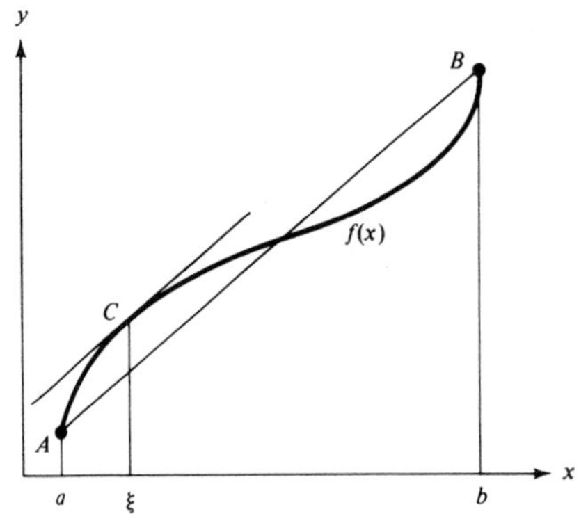

FIGURE 7.1

that although (1) is an *exact* equation, *nothing* is said about how to find the point ξ in question; the theorem guarantees only its existence.

First we consider the special case for which

$$f(a) = f(b) = 0$$

and prove that

$$f'(\xi) = 0, \tag{2}$$

for some value $a<\xi<b$ (see Fig. 7.2). This result (known as Rolle's theorem) is certainly simple and geometrically obvious. There are only two possibilities. If the function $f(x)$ is identically zero, then $f'(x) = 0$ for all x and (2) is certainly true. Alternatively, if $f(x)$ is nonzero somewhere in the interval, the function has a relative maximum or minimum. The derivative $f'(x)$ is then zero at the extremal point, which we identify as ξ, and the validity of (2) is established.

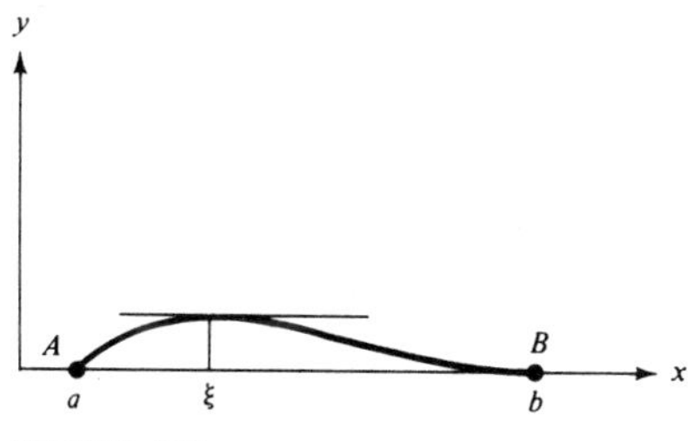

FIGURE 7.2

The foundation is now set for an attack on the more general theorem, where $f(a)$ and $f(b)$ can be nonzero. The trick is to reduce this set of circumstances to the simpler case covered in (2), a feat not difficult to accomplish.

The straight line joining the points $(a, f(a))$ and $(b, f(b))$ has the equation

$$y = f(a) + \frac{f(b) - f(a)}{b - a}(x - a)$$

This line and the curve $y = f(x)$ both pass through points A and B, Fig. 7.1, and the difference between them,

$$\theta(x) = f(x) - \left[f(a) + \frac{f(b) - f(a)}{b - a}(x - a)\right], \tag{3}$$

constitutes a function which is, in fact, zero at these end positions; $\theta(a) = \theta(b) = 0$. As such, $\theta(x)$ satisfies all the criteria underlying Eq. (2), and consequently there must be an intermediate value, $x = \xi$, for which

$$\theta'(\xi) = 0. \tag{4}$$

According to (3), however, this implies

$$\theta'(\xi) = f'(\xi) - \frac{f(b) - f(a)}{b - a} = 0,$$

which on rearrangement is just Eq. (1). The proof of the mean-value theorem is complete.

Another interpretation of (4) is that $\theta(x)$, which represents the vertical distance between the curve $f(x)$ and the chord, has a relative extremum at $x = \xi$.

Example 1 The function $f(x) = x^2 - x$ satisfies $f(0) = f(1) = 0$. Find ξ in $0 < \xi < 1$ such that $f'(\xi) = 0$.

Since $f'(x) = 2x - 1$, the required number ξ is determined by $2\xi - 1 = 0$ or $\xi = 1/2$.

Example 2 If $f'(x) = 0$ for all x in some interval, show that $f(x)$ is a constant.

This conclusion is rather obvious and has been accepted as such till now. The proof is a trivial consequence of (1).

By assumption, $f'(x) = 0$ for $a \le x \le b$, so that, in particular, $f'(\xi) = 0$, in which case (1) becomes

$$f(b) - f(a) = 0 \qquad \text{or} \qquad f(b) = f(a).$$

Since b and a are *any* two points, it follows that $f(x)$ is a constant.

Example 3 If $u'(x) = v'(x)$, show that $u(x) = v(x) + \text{constant}$.

In this case, write

$$\frac{d}{dx}[u(x) - v(x)] = 0$$

and apply the conclusion in Example 2, identifying $f(x) = u(x) - v(x)$.

Example 4 If $f(x) = \sin x$ and $0 \le x \le \pi/2$, find the value of ξ for which the mean-value theorem holds.

Here, $a = 0$, $b = \pi/2$, $f(0) = 0$, $f(\pi/2) = 1$, and $f'(x) = \cos x$, so that (1) becomes

$$\frac{1 - 0}{\pi/2 - 0} = \cos \xi = \frac{2}{\pi}.$$

It follows that $\xi = 0.8805$ (50°27′).

---

2 Equation (1) can also be written

$$f(x) = f(a) + f'(\xi)(x - a), \tag{5}$$

where $a < \xi < x$. The fact that ξ depends on x, which is one of the end points, makes (5) unsuitable for practical numerical work; its exactness is a small virtue indeed. An approximate but useful version of this formula is the familiar relationship

$$f(x) \approx f(a) + f'(a)(x - a), \tag{6}$$

and the error involved in this is now seen to be associated with the relative magnitudes of $f'(\xi)$ and $f'(a)$. Certainly for $|x - a|$ small enough the error is negligible.

A more accurate approximation than (6) and an exact description of the error are the natural consequences of an extended form of the mean-value theorem. To this end, we begin by finding a second-degree polynomial in the interval $a \leq x \leq b$ which passes through the same end points as the curve $y = f(x)$ and the derivative of which equals $f'(a)$ at $x = a$. The polynomial with these properties is

$$P(x) = f(a) + f'(a)(x - a) + \kappa(x - a)^2, \tag{7}$$

where

$$(b - a)^2\kappa = f(b) - f(a) - f'(a)(b - a). \tag{8}$$

As a check, note that $P(a) = f(a)$, $P(b) = f(b)$, and $P'(a) = f'(a)$. The function

$$\Phi(x) = f(x) - P(x)$$

is then zero at $x = a$ and $x = b$, and from (2) its derivative must vanish at some intermediate point, $a < \xi < b$;

$$\Phi'(\xi) = f'(\xi) - P'(\xi) = 0.$$

By construction, $\Phi'(a) = 0$, and since it has been demonstrated that $\Phi'(\xi) = 0$, the mean-value theorem can be applied once more to the function $\Phi'(x)$. Hence, there is yet another point ξ_* in the subinterval $a < \xi_* < \xi$ at which

$$\Phi''(\xi_*) = f''(\xi_*) - P''(\xi_*) = 0.$$

But

$$P''(\xi_*) = 2\kappa,$$

and it follows that

$$\kappa = \tfrac{1}{2}f''(\xi_*),$$

in which case (8) becomes

$$f(b) = f(a) + f'(a)(b - a) + \tfrac{1}{2}f''(\xi_*)(b - a)^2. \tag{9}$$

Point b is arbitrary and can be renamed x if we like, to make the last equation read

$$f(x) = f(a) + f'(a)(x - a) + \tfrac{1}{2}f''(\xi)(x - a)^2 \qquad \text{for} \qquad a < \xi < x. \tag{10}$$

An analogous procedure can be used to establish the most general statement of the mean-value theorem, a polynomial approximation of $f(x)$ about $x = a$ known as *Taylor's formula*

$$f(x) = f(a) + f'(a)(x - a) + \frac{1}{2!}f''(a)(x - a)^2 + \cdots + \frac{1}{n!}f^{(n)}(a)(x - a)^n$$

$$+ \frac{1}{(n + 1)!}f^{(n+1)}(\xi)(x - a)^{n+1} = P_n(x) + R_n(x), \tag{11}$$

where $a < \xi < x$. In other words, the error made in approximating $f(x)$ (about $x = a$) by the polynomial $P_n(x)$ is the remainder

$$R_n(x) = \frac{1}{(n+1)!} f^{(n+1)}(\xi)(x-a)^{n+1} \tag{12}$$

If this remainder can be shown to approach zero as n becomes infinite but x is held fixed, $\lim_{n\to\infty} R_n(x) = 0$, then the function $P_n(x)$ actually represents a partial sum, one of a sequence of polynomials which in the limit defines the convergent infinite series

$$f(x) = \sum_{n=0}^{\infty} \frac{f^{(n)}(a)}{n!}(x-a)^n. \tag{13}$$

These series, known as Taylor series, will occupy our attention later in Chap. 4.

Example 4 Find the value of sin 37°.

The calculation is based on the information available at 30°, that is, $a = 0.523599$ in radian measure. Since

$$\left.\frac{d}{dx}\sin x\right]_a = \cos a, \qquad \left.\frac{d^2}{dx^2}\sin x\right]_a = -\sin a,$$

$$\left.\frac{d^3}{dx^3}\sin x\right]_a = -\cos a, \qquad \left.\frac{d^4}{dx^4}\sin x\right]_a = \sin a,$$

Taylor's two-, three-, and four-term formulas (without remainders) are

$\sin x \approx \sin a + \cos a\,(x-a)$, two-term;
$\sin x \approx \sin a + \cos a\,(x-a) - \frac{1}{2}\sin a\,(x-a)^2$, three-term;
$\sin x \approx \sin a + \cos a\,(x-a) - \frac{1}{2}\sin a\,(x-a)^2 - \frac{1}{6}\cos a\,(x-a)^3$, four-term.

In particular $\sin 30° = 0.5$, $\cos 30° = 0.866025$; we let $x = 0.645772$ (37°), so that $x - a = 0.122173$. Upon substituting these numbers in the preceding formulas, we obtain

$$\sin 37° \approx \begin{cases} 0.605805, & \text{two-term;} \\ 0.602009, & \text{three-term;} \\ 0.601746, & \text{four-term.} \end{cases}$$

It would have been just as easy to use an expansion centered at $a = 0.785398$ (45°), in which case $\sin a = \cos a = 0.7071068$, and the above formulas give

$$\sin 37° \approx \begin{cases} 0.608376, & \text{two-term;} \\ 0.601483, & \text{three-term;} \\ 0.601804, & \text{four-term.} \end{cases}$$

3

An important application of the mean-value theorem is in the evaluation of limits, especially indeterminate ratios such as $\lim_{x \to a} [f(x)/g(x)]$, where $f(a) = 0$, $g(a) = 0$. If both functions are continuous and have first derivatives, each can be written in the form (5):

$$f(x) = f(a) + f'(\xi)(x - a) = f'(\xi)(x - a), \tag{14}$$

$$g(x) = g(a) + g'(\eta)(x - a) = g'(\eta)(x - a). \tag{15}$$

Although the points ξ and η are generally different, both are in the same interval

$$a < \xi < x, \qquad a < \eta < x; \tag{16}$$

and it follows that

$$\frac{f(x)}{g(x)} = \frac{f'(\xi)}{g'(\eta)}. \tag{17}$$

According to (16), $x \to a$ implies $\xi \to a$ and $\eta \to a$, and the limiting value of the ratio is

$$\lim_{x \to a} \frac{f(x)}{g(x)} = \frac{f'(a)}{g'(a)}. \tag{18}$$

This formula is known as *l'Hôpital's rule*, and if the right-hand side is not also indeterminate, it provides an efficient method of computing indeterminate forms.

Example 5 Use l'Hôpital's rule to evaluate $\lim_{x \to 1} \dfrac{x^3 - 1}{x^2 - 1}$.

The functions $f(x) = x^3 - 1$ and $g(x) = x^2 - 1$ are both zero at $x = 1$. By l'Hôpital's rule, since $f'(x) = 3x^2$ and $g'(x) = 2x$,

$$\lim_{x \to 1} \frac{x^3 - 1}{x^2 - 1} = \frac{3}{2}.$$

In this case, the result can be verified by factoring. If $x \neq 1$,

$$\frac{x^3 - 1}{x^2 - 1} = \frac{(x - 1)(x^2 + x + 1)}{(x - 1)(x + 1)} = \frac{x^2 + x + 1}{x + 1}$$

and

$$\lim_{x \to 1} \frac{x^3 - 1}{x^2 - 1} = \lim_{x \to 1} \frac{x^2 + x + 1}{x + 1} = \frac{3}{2}.$$

4

When the ratio of derivatives happens to be another indeterminate form, we can prove that the correct generalization of (18) is

$$\lim_{x \to a} \frac{f(x)}{g(x)} = \lim_{x \to a} \frac{f'(x)}{g'(x)}.$$

To this end, consider the function

$$\theta(x) = g(x)f(b) - f(x)g(b),$$

which, since $f(a) = g(a) = 0$, has the values

$$\theta(a) = \theta(b) = 0.$$

It follows from the mean-value theorem that at $x = \xi$

$$\theta'(\xi) = 0 = g'(\xi)f(b) - f'(\xi)g(b),$$

or

$$\frac{f(b)}{g(b)} = \frac{f'(\xi)}{g'(\xi)}.$$

Since point b is arbitrary, the preceding equation is, with a slight change of notation, equivalent to

$$\frac{f(x)}{g(x)} = \frac{f'(\xi)}{g'(\xi)} \qquad \text{for some } \xi \text{ in } a < \xi < x. \tag{19}$$

The virtue of this relationship over (17) is that it involves only a single intermediate value ξ (and not the two, ξ and η, required above). Therefore the limit $x \to a$ can now be taken, with the result

$$\lim_{x \to a} \frac{f(x)}{g(x)} = \lim_{\xi \to a} \frac{f'(\xi)}{g'(\xi)} = \lim_{x \to a} \frac{f'(x)}{g'(x)}. \tag{20}$$

If the ratio of derivatives is still indeterminate, obviously the rule can be applied once more:

$$\lim_{x \to a} \frac{f'(x)}{g'(x)} = \lim_{x \to a} \frac{f''(x)}{g''(x)}.$$

In fact, the procedure can be continued until a ratio is obtained that is not indeterminate. Thus, if $f(x)$ and $g(x)$ have the properties $f(a) = g(a) = 0, f^{(k)}(a) = g^{(k)}(a) = 0$, for $k = 1, \ldots, n$ [but $f^{(n+1)}(a)$ or $g^{(n+1)}(a)$ is nonzero], then

$$\lim_{x \to a} \frac{f(x)}{g(x)} = \lim_{x \to a} \frac{f^{(n+1)}(x)}{g^{(n+1)}(x)}.$$

Example 6 Illustrate l'Hôpital's rule by evaluating $\lim_{x \to 0} \dfrac{\sin x}{x}$.

In this case, $f(x) = \sin x$, $g(x) = x$, and $a = 0$, so that by (18) or (20)

$$\lim_{x \to 0} \frac{\sin x}{x} = \lim_{x \to 0} \frac{\frac{d}{dx} \sin x}{\frac{d}{dx} x} = \lim_{x \to 0} \frac{\cos x}{1} = 1.$$

This is the result proved by a geometrical argument in Chapter 1.

Example 7 Evaluate $\displaystyle\lim_{\theta\to\pi/2}\frac{(4+\cos\theta)^{1/2}-2}{\theta-\pi/2}$.

Here

$$f(\theta)=(4+\cos\theta)^{1/2}-2, \qquad g(\theta)=\theta-\frac{\pi}{2},$$

so that

$$\lim_{\theta\to\pi/2}\frac{f(\theta)}{g(\theta)}=\lim_{\theta\to\pi/2}\frac{f'(\theta)}{g'(\theta)}=\lim_{\theta\to\pi/2}\frac{-\sin\theta}{2(4+\cos\theta)^{1/2}}=-\frac{1}{4}.$$

Example 8 Evaluate $\displaystyle\lim_{x\to 0}\frac{x^3}{1-\cos x^2}$.

Identify $f(x)=x^3$, $g(x)=1-\cos x^2$ and note that both functions are zero at the origin, as are their first and second derivatives:

$$\begin{aligned} f'(x)&=3x^2, \qquad & g'(x)&=2x\sin x^2,\\ f''(x)&=6x, & g''(x)&=2\sin x^2+4x^2\cos x^2,\\ f'''(x)&=6, & g'''(x)&=12x\cos x^2-8x^3\sin x^2. \end{aligned}$$

Therefore

$$\lim_{x\to 0}\frac{f(x)}{g(x)}=\lim_{x\to 0}\frac{f'''(x)}{g'''(x)}=\infty.$$

5

Another very common type of indeterminacy is ∞/∞, the usual form being $\lim_{x\to a+}[f(x)/g(x)]$, where $\lim_{x\to a+} f(x)=\infty$ and $\lim_{x\to a+} g(x)=\infty$. Sometimes the minor transformation

$$\frac{f(x)}{g(x)}=\frac{1/g(x)}{1/f(x)},$$

will convert the original ratio to the 0/0 indeterminacy just studied. However, this technique does not always work, and a proof is now *sketched* that l'Hôpital's rule applies directly to the new form:

$$\lim_{x\to a+}\frac{f(x)}{g(x)}=\lim_{x\to a+}\frac{f'(x)}{g'(x)}. \tag{21}$$

Let x and z be any two points such that $a<x<z$; then the mean-value theorem as expressed in (19) implies more generally that

$$\frac{f(x) - f(z)}{g(x) - g(z)} = \frac{f'(\xi)}{g'(\xi)}, \qquad \text{where } x < \xi < z$$

(see Exercise 18). This can be manipulated to form the expression

$$\frac{f(x)}{g(x)} = \frac{1 - g(z)/g(x)}{1 - f(z)/f(x)} \frac{f'(\xi)}{g'(\xi)}.$$

A novel feature is introduced now. Variables x and z are both allowed to approach $a+$, but x is made to do this much more rapidly than z, so that

$$\lim_{x \to a+} [g(z)/g(x)] = \lim_{x \to a+} [f(z)/f(x)] = 0.$$

This is possible because the two functions f and g become infinite in the limit. Since in the same process $\xi \to a+$, the result is

$$\lim_{x \to a+} \frac{f(x)}{g(x)} = \lim_{x \to a+} \frac{1 - g(z)/g(x)}{1 - f(z)/f(x)} \cdot \lim_{x \to a+} \frac{f'(\xi)}{g'(\xi)} = \lim_{x \to a+} \frac{f'(x)}{g'(x)},$$

which is (21).

Many different indeterminate limits can be manipulated into either the forms 0/0 or ∞/∞, for which l'Hôpital's rule is appropriate. For example, suppose $F(a) = 0$ and $G(a) = \infty$, then $\lim_{x \to a} F(x) \cdot G(x)$, an indeterminacy of the form $0 \cdot \infty$, is analyzed by expressing the product as either of the quotients

$$\frac{F(x)}{1/G(x)} \qquad \text{or} \qquad \frac{G(x)}{1/F(x)}.$$

(The former is a 0/0 form, whereas the latter is ∞/∞.) Similarly, if $U(a) = 0$, $V(a) = 0$, the difficulty with the $\infty - \infty$ indeterminacy,

$$\frac{1}{U(x)} - \frac{1}{V(x)},$$

may yield in the form

$$\frac{V(x) - U(x)}{U(x)V(x)}.$$

Example 9 Evaluate $\lim_{x \to \infty} x \sin(1/x)$.

This limit has an $\infty \cdot 0$ indeterminacy which can be resolved by writing $x \sin(1/x) = \sin(1/x)/(1/x)$. Then,

$$\lim_{x \to \infty} x \sin(1/x) = \lim_{x \to \infty} \frac{\sin(1/x)}{1/x} = \lim_{x \to \infty} \frac{(-1/x^2)\cos(1/x)}{(-1/x^2)} = 1.$$

(An alternative method is to introduce a change of variable $z = 1/x$.)

Example 10 Evaluate $\lim_{s\to\pi/4} (s - \pi/4)\tan 2s$.

The indeterminacy involved is of the form $0 \cdot \infty$, and to evaluate the limit rewrite it as

$$\lim_{s\to\pi/4} \frac{s-\pi/4}{\cot 2s},$$

so that (20) can be applied. Therefore

$$\lim_{s\to\pi/4} \frac{s-\pi/4}{\cot 2s} = \lim_{s\to\pi/4} \frac{\dfrac{d}{ds}(s-\pi/4)}{\dfrac{d}{ds}\cot 2s} = \lim_{s\to\pi/4} \frac{1}{-2\csc^2 2s} = -\frac{1}{2}.$$

Other indeterminate forms will be examined when our listing of known functions has been suitably expanded.

Exercises 2.7

1. The function $f(x) = x^4 - x^2$ is equal to zero at $x = -1, 0, 1$. Determine the values of ξ such that $f'(\xi) = 0$. Illustrate with a graph.

2. The function $y = x^{2/3}$ takes on the value 1 at $a = -1$ and $b = 1$ but its derivative is never zero at any point. Does this example contradict the mean-value theorem?

3. Apply the mean-value theorem to each of the following functions in the given intervals. Determine the values of ξ and illustrate with graphs.

(i) $f(x) = x^3$ $\quad a = 1,\ b = 4$ $\qquad$ **(ii)** $f(x) = \sin 2x$ $\quad a = 0,\ b = \pi/2$

(iii) $f(x) = \sin^{-1}x$ $\quad a = 0,\ b = 1$ $\qquad$ **(iv)** $f(x) = \tan x$ $\quad a = -\pi/4,\ b = \pi/4$

(v) $f(x) = x^2(x^2 - 2)$ $\quad a = -1,\ b = 1$ $\qquad$ **(vi)** $f(x) = x^3 + 2x$ $\quad a = 0,\ b = 1$

4. By means of a graph, show that the mean-value theorem cannot be applied to the function $f(x) = |x|$ in the interval $-1 \le x \le 1$.

5. Use the mean-value theorem to prove that a function $y = f(x)$ that satisfies $f'(x) > 0$ must be an increasing function as x increases. (*Hint:* Assume $f(x)$ decreases in some interval and deduce a contradiction.)

6. If $m < |f'(x)| < M$ in the interval (a, b), prove that $m(b-a) < |f(b) - f(a)| < M|b - a|$. Use this result to show the following:

(i) $5.009 < (25.1)^{1/2} < 5.01$ $\qquad$ **(ii)** $1 + \dfrac{h}{2\sqrt{1+h}} < \sqrt{1+h} < 1 + \dfrac{h}{2},\ h > 0$

(iii) $|\sin b - \sin a| < |b - a|$ $\qquad$ **(iv)** $\dfrac{b-a}{1+b^2} < \tan^{-1}b - \tan^{-1}a < \dfrac{b-a}{1+a^2}.$

7. Use l'Hôpital's rule to calculate the following limits:

(i) $\displaystyle\lim_{x\to\pi/2} \frac{\sin x - 1}{x - \pi/2}$ $\qquad$ **(ii)** $\displaystyle\lim_{x\to 0} \frac{x(x-1)}{\sin x}$ $\qquad$ **(iii)** $\displaystyle\lim_{x\to 1} \frac{x^2 - 3x + 2}{x - 1}$

(iv) $\lim_{x\to\pi/4} \dfrac{1-\tan x}{x-\pi/4}$ (v) $\lim_{x\to 1} \dfrac{\sin \pi x}{x-1}$ (vi) $\lim_{t\to 2} \dfrac{t^5-32}{t-2}$

(vii) $\lim_{x\to 0} \dfrac{\tan^{-1}x - x}{\sin^{-1}x - x}$ (viii) $\lim_{x\to 0} \dfrac{\cos x - 1}{x^2}$ (ix) $\lim_{x\to 0} \dfrac{\tan ax}{x}$

(x) $\lim_{x\to 0} \dfrac{\sin^2 ax}{\cos ax - 1}$ (xi) $\lim_{x\to 0} \dfrac{\sin ax}{\tan bx}$ (xii) $\lim_{x\to 0} \dfrac{E(x)-1}{x}$

8. Determine the following limits:

(i) $\lim_{x\to 0} \dfrac{\cos x - \cos 3x}{\sin x^2}$ (ii) $\lim_{x\to 0} \dfrac{\sin^{-1}x}{x}$ (iii) $\lim_{x\to\pi/2} \dfrac{\cos x}{x-\pi/2}$

(iv) $\lim_{x\to 0} \dfrac{6x - 2\sin 3x}{x^3}$ (v) $\lim_{x\to 0} \dfrac{\cos(\sin x) - 1}{3x^2}$ (vi) $\lim_{x\to 0} \dfrac{\sqrt{a^2+x^2}-a}{x}$

(vii) $\lim_{x\to 0} \dfrac{\sqrt{1+x}-1}{x}$ (viii) $\lim_{h\to 0} \dfrac{\sin^2(x+h) - \sin^2 x}{h}$ (ix) $\lim_{x\to a} \dfrac{x^2-a^2}{x^3-a^3}$

(x) $\lim_{x\to 1} \dfrac{x^{m/n}-1}{x^{p/q}-1}$ (xi) $\lim_{x\to\pi/4} \dfrac{\sin x - \cos x}{(x-\pi/4)^2}$ (xii) $\lim_{x\to\pi} \dfrac{\sin x}{\pi^2 - x^2}$

9. Suppose that $f(x)$ is a polynomial of degree n. Prove the following statements.

(i) If $f(x)$ has only real roots, the derivatives $f'(x), f''(x), \ldots, f^{(n-1)}(x)$ have only real roots.

(ii) If all roots of $f(x)$ are positive, all roots of the derivatives are positive.

(iii) If $f(x)$ has n distinct real roots, the kth derivative has $(n-k)$ distinct real roots for $k = 1, 2, 3, \ldots, n-1$.

10. For any polynomial $f(x)$ of degree n, show that the Taylor polynomial of degree n for $f(x)$ expanded about $x = a$ is exact. Expand the following polynomials in powers of $x - a$.

(i) $x^2 + 4x - 6$ $a = 1$ (ii) $x^3 - x^2 + x - 1$ $a = 2$

(iii) $x^3 + 3x^2 + 3x + 1$ $a = -1$ (iv) $x^4 - 4x^3 + x^2$ $a = 4$

11. Using Taylor's formula, expand $f(x) = (x^2 - x + 1)^2$ in a series of powers of x. Verify the answer by multiplication.

12. Determine the Taylor polynomials of the following functions including exactly three non-zero terms.

(i) $\sin x$, $a = \pi/3$ (ii) $\cos x$, $a = \pi/6$ (iii) x^3, $a = 3$

Use these polynomials to approximate sin 59°, cos 31° and $(3.01)^3$. Compare to the accurate values.

13. Determine the following limits:

(i) $\lim_{x\to 0} x\cot x$ (ii) $\lim_{x\to 0} \dfrac{\cot x}{\cot 2x}$ (iii) $\lim_{x\to 0} \left(\dfrac{1}{x} - \dfrac{1}{\sin x}\right)$

(iv) $\lim_{x\to 0} \left(\dfrac{1}{\sin x} - \dfrac{1}{\tan x}\right)$ (v) $\lim_{x\to\pi/2} (x-\pi/2)\tan x$ (vi) $\lim_{x\to 0} \dfrac{\cos^{-1}x - \pi/2}{x}$

(vii) $\lim_{x\to 0} \dfrac{\tan x - x}{x - \sin x}$ (viii) $\lim_{x\to 0} \dfrac{x^2 \cos x}{1 - \sec x}$ (ix) $\lim_{x\to 0} \dfrac{\sin x^2 - \sin^2 x}{x^4}$

(x) $\lim_{x\to\infty} \dfrac{\tan^{-1}x}{\cot^{-1}x}$ (xi) $\lim_{x\to\infty} x(\tan^{-1}x - \pi/2)$ (xii) $\lim_{x\to\infty} (\sqrt{1+x} - \sqrt{x-1})$

14. From Taylor's formula (about $x = 0$) find a polynomial approximation of $1/(1+x)$. What infinite series is implied? What can you infer from the behavior of the function as $x \to -1$?

15. Find explicit expressions for the remainders in the Taylor polynomial expansions about $x = 0$ of the following functions. Find useful bounds for the remainders.

(i) $f(x) = \sin x$ (ii) $f(x) = \cos x$ (iii) $f(x) = \dfrac{1}{1+x}$

16. From the polynomial expansion of $\sin x$ about $x=0$ find, by a change of variable, the corresponding expansions for $\sin x^2$ and $\sin(1/x)$. Where is the latter expansion most useful?
17. Use Taylor's formula to obtain an approximation for $(a^2+x)^{1/n}$ when x is small (a and n are given constants). Determine approximate values for the following numbers correct to three decimal places. Compare to the accurate values.
(i) $(0.98)^{1/3}$ **(ii)** $(1.45)^{1/2}$ **(iii)** $(16.1)^{1/4}$ **(iv)** $(1.22)^{1/3}$
18. If $F(x)=f(x)-f(a)$, $G(x)=g(x)-g(a)$ and $\theta(x)=F(x)G(b)-G(x)F(b)$, use the mean value theorem to show that

$$\frac{f(x)-f(a)}{g(x)-g(a)}=\frac{f'(\xi)}{g'(\xi)} \quad \text{for } a<\xi<x.$$

19. Determine the following limits.
(i) $\lim_{x\to\infty}(\sqrt{x^2+1}-\sqrt{x^2-1})$ **(ii)** $\lim_{x\to\infty}[(x^3+x)^{1/3}-x]$ **(iii)** $\lim_{x\to\infty}[(x^9-x^6)^{1/3}-x^3]$
(iv) $\lim_{x\to\infty}x^2[(x^4+1)^{1/2}-x^2]$ **(v)** $\lim_{x\to\infty}\left(\frac{x^4}{x^2+1}-\frac{x^3}{x+1}\right)$ **(vi)** $\lim_{x\to\infty}\left(\frac{x^5}{x^2+a^2}-x^3\right)$
20. Prove Taylor's form of the mean-value theorem

$$f(b)=f(a)+f'(a)(b-a)+\cdots+\frac{1}{n!}f^{(n)}(a)(b-a)^n+\frac{1}{(n+1)!}f^{(n+1)}(\xi)(b-a)^{n+1}$$

by considering the function

$$\theta(x)=f(x)-f(a)-f'(a)(x-a)-\cdots-\frac{1}{n!}f^{(n)}(a)(x-a)^n-K(x-a)^{n+1}$$

and choosing K so that $\theta(b)=0$. Apply the mean-value theorem to $\theta(x)$ to show that there is a number ξ_1 such that $\theta'(\xi_1)=0$. Since $\theta'(a)=0$, apply the mean-value theorem again to $\theta'(x)$ and infer the existence of another number ξ_2 for which $\theta''(\xi_2)=0$. Repeat the procedure consecutively and deduce that

$$a<\xi_{n+1}<\xi_n<\ldots<\xi_2<\xi_1<b, \quad \text{where } \theta^{(k)}(\xi_k)=0,\ k=1,\ \ldots\ n$$

and

$$K=\frac{1}{(n+1)!}f^{(n+1)}(\xi_{n+1}).$$

By a change of notation write

$$f(x)=f(a)+f'(a)(x-a)+\ldots+\frac{1}{n!}f^{(n)}(a)(x-a)^n+\frac{1}{(n+1)!}f^{(n+1)}(\xi)(x-a)^{n+1}$$

with

$$a<\xi=\xi_{n+1}<x.$$

2.8 *The exponential function*

The exponential function, $E(x)$, has already been introduced as the solution of the differential equation

$$\frac{d}{dx}y(x)=y(x), \tag{1}$$

which has the particular value $y(0) = 1$. The properties of this function are delineated here, and most of the claims made earlier are substantiated.

As a preliminary, two important characteristics of the differential equation are noted. Let $y(x)$ be *any* solution of (1); then the equation satisfied by $y(z(x))$ is obtained by direct substitution in (1) using chain-rule differentiation:

$$\frac{d}{dx}\,y(z(x)) = \frac{dy(z)}{dz}\,\frac{dz(x)}{dx} = y(z)\,\frac{dz(x)}{dx},$$

or

$$\frac{dy(z)}{dx} = y(z)\,\frac{dz}{dx}.$$

In particular, let $z = x + a$; then

$$\frac{d}{dx}\,y(x + a) = y(x + a), \tag{2}$$

and the equation is seen to be unaffected by a shift or translation in x. Second, let $z = kx$, which represents a scale change; it follows that

$$\frac{d}{dx}\,y(kx) = ky(kx). \tag{3}$$

With these preliminaries completed, the first order of business is to validate the assertion that $E(x)$ is *the* solution which assumes the value unity at the origin. In other words, we must prove (i) that there is a solution and (ii) that it is unique.

The question of existence is disposed of by actually constructing the solution. Uniqueness, on the other hand, is established by considering the possibility that there are two *different* solutions and showing this to be an untenable assumption. To this end, consider the combination

$$w(x) = \psi(-x)\theta(x + a), \tag{4}$$

where $\psi(x)$, $\theta(x)$ are, supposedly, any two different solutions of (1), with $\psi(0) = \theta(0) = 1$. The result of differentiating this product is

$$\frac{dw(x)}{dx} = \theta(x + a)\,\frac{d}{dx}\,\psi(-x) + \psi(-x)\,\frac{d}{dx}\,\theta(x + a).$$

Furthermore, since both functions are solutions of (1), it follows from (2) and (3) (set $k = -1$) that

$$\frac{dw(x)}{dx} = -\psi(-x)\theta(x + a) + \psi(-x)\theta(x + a) = 0.$$

According to the mean-value theorem, the function $w(x)$ is a constant (see Example 2 in Sec. 2.7), and from the known values at $x = 0$ Eq. (4) becomes

$$\psi(-x)\theta(x + a) = \theta(a). \tag{5}$$

But *exactly* the same analysis applies to the combination $\theta(-x)\theta(x + a)$; and the special case of (5), wherein $\theta(-x)$ replaces $\psi(-x)$, is

$$\theta(-x)\theta(x + a) = \theta(a). \tag{6}$$

However, the identical forms of (5) and (6) imply that

$$\psi(-x) = \theta(-x)$$

or

$$\psi(x) = \theta(x).$$

Thus, the hypothesis that $\psi(x)$ and $\theta(x)$ are different solutions leads to a contradiction, and it turns out that they must indeed be one and the same. The solution of the problem is then a unique function, we choose to call the *exponential function*, $E(x)$.

Equation (6), now expressed as

$$E(-x)E(x + a) = E(a), \tag{7}$$

has several important consequences. First set $a = 0$ to obtain

$$E(-x)E(x) = 1.$$

The substitution of this for $E(-x)$ in (7) yields the important addition formula

$$E(x + a) = E(a)E(x), \tag{8}$$

from which almost everything else follows. For example,

$$E(2a) = E(a) \cdot E(a) = [E(a)]^2,$$

which is indicative of the more general equation, valid for any integer n,

$$E(na) = [E(a)]^n. \tag{9}$$

As a matter of fact, this *addition formula* is true for any real power r:

$$E(rx) = [E(x)]^r. \tag{10}$$

One way to show this is to generalize (9) so that it is valid for fractional powers and from there to advance to real numbers by a limit process. However, it is simpler and more instructive to return to the original differential equation. In this approach, (1) is first multiplied by E^{r-1}, to obtain

$$E^{r-1}\frac{dE}{dx} = E^r. \tag{11}$$

But according to the chain rule for differentiation,

$$\frac{d}{dx}E^r = rE^{r-1}\frac{dE}{dx},$$

so that (11) may be written

$$\frac{1}{r}\frac{d}{dx}E^r = E^r \qquad \text{or} \qquad \frac{d}{dx}E^r = rE^r.$$

Comparison of the last formula with (3) shows that the function $[E(x)]^r$ is just another form of the exponential function with a scaled argument, i.e.,

$$[E(x)]^r = E(rx),$$

and the proof is complete.

The name exponential function can now be justified. Set $x = 1$ in (10) so that

$$E(r) = [E(1)]^r = e^r,$$

where the value $e = E(1) = 2.71828\ldots$ was calculated earlier. By making a notational exchange of x for r this becomes

$$E(x) = e^x; \tag{12}$$

the exponential function $E(x)$ is then just the real number e raised to the x power.

In terms of the numbers $x = x_1$, $a = x_2$, the addition property (8) is

$$e^{x_1 + x_2} = e^{x_1}e^{x_2},$$

while (10) reads

$$(e^{x_1})^{x_2} = e^{x_1 \cdot x_2};$$

these are the usual rules for power multiplication and exponentiation. Two important properties, left as exercises, are

$$e^x > 0 \qquad \text{for} \qquad \text{all } x; \tag{13}$$

$$e^{x_1} > e^{x_2} \qquad \text{for} \qquad x_1 > x_2. \tag{14}$$

---

2 An efficient means of computing the exponential function is to use Taylor's formula, given in Eq. (11) Sec. 2.7. The expansion about $x = 0$ is

$$E(x) = E(0) + E'(0)x + \frac{1}{2!}E''(0)x^2 + \cdots + \frac{1}{n!}E^{(n)}(0)x^n + \frac{1}{(n+1)!}E^{(n+1)}(\xi)x^{n+1}.$$

Since

$$E^{(n)}(0) = 1 \qquad \text{for all } n$$

[recall that $E^{(n)}(x) = E(x)$], the preceding equation becomes

$$e^x = 1 + x + \frac{x^2}{2!} + \cdots + \frac{x^n}{n!} + R_n(x). \tag{15}$$

The values of e^x are calculated from this by simply ignoring the remainder, R_n, having chosen n sufficiently large to obtain whatever numerical accuracy is desired. The computational evidence is overwhelming in support of this procedure, and later it will be proved that $\lim_{n \to \infty} R_n(x) = 0$, which is equivalent to establishing the convergence of the infinite series

$$e^x = \sum_{n=0}^{\infty} \frac{x^n}{n!}. \tag{16}$$

Figure 8.1 and Table 8.1 on page 166 are produced in this manner.

---

3 **Example 1** Differentiate the functions x^2e^x, $xe^{-\sin x}$, and $e^{(e^x)}$.

Each of these is of the form $v(x)e^{u(x)}$, the derivative of which is

$$\frac{d}{dx} ve^u = \frac{dv}{dx} e^u + v \frac{d}{dx} e^u$$

$$= v'e^u + v\left(\frac{d}{du} e^u\right) \frac{du}{dx}$$

$$= (v' + vu')e^u.$$

Therefore,

$$\frac{d}{dx} x^2e^x = (2x + x^2)e^x, \qquad v = x^2, \qquad u = x;$$

$$\frac{d}{dx} xe^{-\sin x} = (1 - x\cos x)e^{-\sin x}, \qquad v = x, \qquad u = -\sin x;$$

$$\frac{d}{dx} e^{(e^x)} = e^x e^{(e^x)}, \qquad v = 1, \qquad u = e^x.$$

(The reader should supply all the details in each case.)

TABLE 8.1

x	e^x	e^{-x}
0	1.000000	1.000000
0.100000	1.105171	0.904837
0.200000	1.221403	0.818731
0.300000	1.349859	0.740818
0.400000	1.491825	0.670320
0.500000	1.648721	0.606531
0.600000	1.822119	0.548812
0.700000	2.013753	0.496585
0.800000	2.225541	0.449329
0.900000	2.459603	0.406570
1.000000	2.718282	0.367879
1.100000	3.004166	0.332871
1.200000	3.320117	0.301194
1.300000	3.669297	0.272532
1.400000	4.055200	0.246597
1.500000	4.481689	0.223130
1.600000	4.953032	0.201897
1.700000	5.473947	0.182684
1.800000	6.049647	0.165299
1.900000	6.685894	0.149569
2.000000	7.389056	0.135335
2.100000	8.166170	0.122456
2.200000	9.025013	0.110803
2.300000	9.974182	0.100259
2.400000	11.023176	0.090718
2.500000	12.182494	0.082085
2.600000	13.463738	0.074274
2.700000	14.879732	0.067206
2.800000	16.444647	0.060810
2.900000	18.174145	0.055023
3.000000	20.085537	0.049787

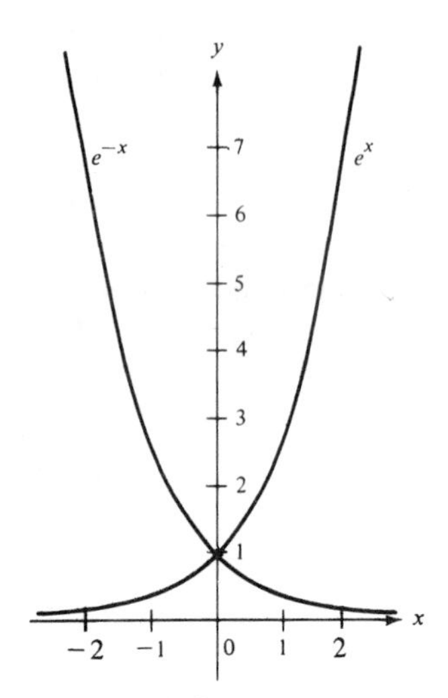

FIGURE 8.1

Example 2 Find the extremal points of $y = e^{-x} - e^{-2x}$.
The derivative of this function is

$$\frac{dy}{dx} = -e^{-x} + 2e^{-2x},$$

and the stationary points are the zeros of the right-hand side;

$$e^{-x} = 2e^{-2x} \qquad \text{or} \qquad e^x = 2.$$

Only one such point exists, $x \approx 0.693$, and this is a relative maximum because the second derivative is negative for this value:

$$\left.\frac{d^2y}{dx^2}\right]_{e^x=2} = (e^{-x} - 4e^{-2x})\Big]_{e^x=2} = \tfrac{1}{2} - 4(\tfrac{1}{2})^2 = -\tfrac{1}{2}.$$

Example 3 Find the solution of $dy/dx = y$ with $y(a) = A$.

The function Ce^x satisfies this differential equation:

$$\frac{d}{dx}Ce^x = C\frac{d}{dx}e^x = Ce^x,$$

and the constant factor C can be chosen to satisfy the condition at $x = a$,

$$Ce^a = A.$$

Upon replacing C, the solution is

$$y = Ae^{x-a}.$$

Example 4 Find the solution of $dy/dx = ky$ with $y(a) = A$.

According to (3), e^{kx} is the solution of the differential equation, as is any constant multiple of this. Therefore, in general,

$$y = Ce^{kx},$$

and the value of C is determined from the condition at $x = a$:

$$Ce^{ka} = A.$$

The solution is then

$$y = Ae^{k(x-a)}.$$

Example 5 The rate of decay of a radioactive material is proportional to the amount present. If the initial quantity is M_0 gm, how many grams will remain after t days?

The decay law must be formulated in mathematical terms. Let $M(t)$ be the number of grams at time t (in days); then its time rate of change is dM/dt. The fact that this is proportional to $M(t)$ is expressed by

$$\frac{dM}{dt} = -qM,$$

where q is the decay rate (units are days^{-1}) and the negative sign accounts for a *decay*. Since $M(0) = M_0$, the explicit solution of this equation is, according to the results of the last problem,

$$M(t) = M_0 e^{-qt}.$$

The *half-life*, T, of a radioactive substance is the time for $M(t)$ to disintegrate to half its initial value, i.e.,

$$\tfrac{1}{2}M_0 = M_0 e^{-qT} \qquad \text{or} \qquad e^{-qT} = \tfrac{1}{2}.$$

This implies $T = 0.6931/q$.

To illustrate this law quantitatively, consider the decay of radium 224, which has a half-life of $T = 3.6$ days. In this case, $q = 0.1925$ day^{-1} and the amount of radium t days later is

$$M(t) = M_0 e^{-(0.1925)t}.$$

Example 6 The atmospheric pressure p at any altitude is related to the density ρ there by the perfect-gas law

$$p = \rho RT, \tag{17}$$

where T is the absolute temperature and R is a constant. Find the pressure and density as functions of altitude z.

The pressure at altitude z, denoted by $p(z)$, measures the force per unit area exerted by the weight of gas above this height. Suppose that the mass of gas above z in a cylindrical column of cross-sectional area A (Fig. 8.2) is M; then its weight is gM, and the pressure on the baseplate of the cylinder (at altitude z) is

$$p(z) = \frac{gM}{A}.$$

Of course, we do not know M as yet, but it will not actually be needed because the variation of pressure can be determined as follows. Let the pressure at a slightly greater altitude $z + \Delta z$ be $p(z + \Delta z)$. The reduction in pressure over the distance Δz is the incremental difference in weight divided by the area A.

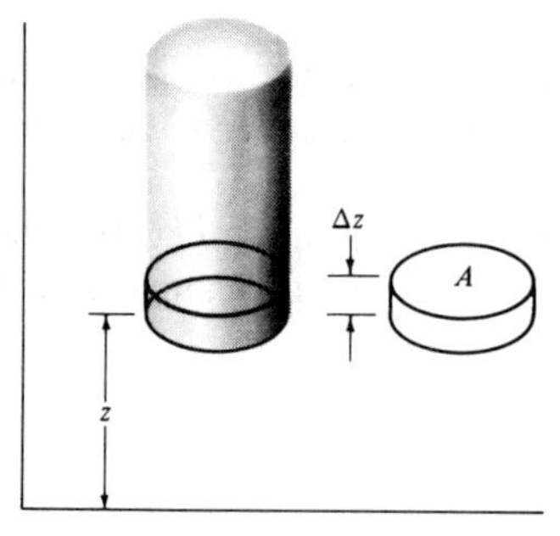

FIGURE 8.2

Since the volume involved is $A\,\Delta z$, the weight of gas that occupies this space, $\rho g A\,\Delta z$, equals the difference in the weights of the two columns, $Ap(z) - Ap(z + \Delta z)$. Therefore

$$A[p(z) - p(z + \Delta z)] = \rho g\, A\Delta z,$$

or

$$\frac{p(z + \Delta z) - p(z)}{\Delta z} = -\rho g,$$

which in the limit $\Delta z \to 0$ becomes

$$\frac{dp}{dz} = -\rho(z)g. \tag{18}$$

This formula expresses the *rate* of change of pressure with height. If Eq. (17) is now used to replace the density in the preceding, the result is

$$\frac{dp}{dz} = -\frac{g}{RT}p,$$

which is a special case of (3). For T constant, it follows that

$$p = p_0 e^{-(g/RT)z},$$

where p_0 is the surface air pressure. The density is then

$$\rho = \rho_0 e^{-(g/RT)z}.$$

---

4 Certain combinations of exponential functions occur frequently enough to merit special attention, albeit brief. These are known as the *hyperbolic functions*, the principal ones being the *hyperbolic sine*

$$\sinh x = \tfrac{1}{2}(e^x - e^{-x}), \tag{19}$$

and the *hyperbolic cosine*

$$\cosh x = \tfrac{1}{2}(e^x + e^{-x}). \tag{20}$$

The others are defined in terms of these; e.g., the *hyperbolic tangent* is

$$\tanh x = \frac{\sinh x}{\cosh x}. \tag{21}$$

The hyperbolic sine is an odd function, as shown by the reduction

$$\sinh(-x) = \tfrac{1}{2}(e^{-x} - e^{+x}) = -\tfrac{1}{2}(e^x - e^{-x}) = -\sinh x.$$

A similar manipulation establishes that $\cosh x$ is even. The derivatives are readily obtained:

$$\begin{aligned} \frac{d}{dx}\sinh x &= \tfrac{1}{2}\left(\frac{d}{dx}e^x - \frac{d}{dx}e^{-x}\right) = \tfrac{1}{2}(e^x + e^{-x}) = \cosh x, \\ \frac{d}{dx}\cosh x &= \tfrac{1}{2}\left(\frac{d}{dx}e^x + \frac{d}{dx}e^{-x}\right) = \tfrac{1}{2}(e^x - e^{-x}) = \sinh x. \end{aligned} \tag{22}$$

Most of the properties of the ordinary trigonometric functions have their counterparts for hyperbolic functions, and the more important ones are

$$\cosh^2 x - \sinh^2 x = 1, \tag{23}$$

$$\sinh(x_1 + x_2) = \sinh x_1 \cosh x_2 + \cosh x_1 \sinh x_2, \tag{24}$$

$$\cosh(x_1 + x_2) = \cosh x_1 \cosh x_2 + \sinh x_1 \sinh x_2. \tag{25}$$

These and others are developed in the exercises, but evidence of their utility must await practical experience.

Exercises 2.8

1. Estimate the values of e^x for $x = 0.1$, 0.2, 0.5 and 1.0 by calculating the values of the Taylor polynomial of degree 3. Compare to the accurate values obtained using a calculator.
2. Show that $y = 1 + ce^{-x}$ is a solution of the differential equation $y' = 1 - y$ for all values of the constant c. Graph this one parameter family of solutions. Show that all solutions are asymptotic to the line $y = 1$ as $x \to \infty$.
3. Differentiate the following functions:
(i) e^{-3x} **(ii)** $e^{x \sin x}$ **(iii)** $e^{-(x+1)^2}$
(iv) $e^{(f(x))^2}$ **(v)** $e^{\cos f(x)}$ **(vi)** $e^{\tan f(x)}$
4. Show that $y = Ae^x - 1 - x$ is a solution of the differential equation $y' = x + y$ for any value of the constant A. Graph this one parameter family of functions. Show that all solutions are asymptotic to the line $x + y + 1 = 0$.
5. Prove that $e^x > 0$ for every x and $e^{x_1} > e^{x_2}$ if $x_1 > x_2$. (Prove that e^x is an increasing function that is never equal to zero.)
6. Prove that the solution of $y'(x) = y(x)$ with $y(a) = A$ is unique. Verify that the solution is $y(x) = Ae^{x-a}$. (Follow the development presented in the text.)
7. Graph the following functions. Determine all stationary points and points of inflection.
(i) $y = e^{-x^2}$ **(ii)** $y = xe^{-x^2}$ **(iii)** $y = x^2e^{-x^2}$
8. Solve the following differential equations and graph the solutions.
(i) $y' = 2y$ $\quad y(0) = 1$ **(ii)** $y' = 4y$ $\quad y(2) = 3$
(iii) $y' = -3y$ $\quad y'(0) = 3$ **(iv)** $y' = -y$ $\quad y(3) = 2$
9. Sketch the graphs of $\sinh x$, $\cosh x$ and $\tanh x$ and their reciprocals,

$$\operatorname{csch} x = \frac{1}{\sinh x}, \quad \operatorname{sech} x = \frac{1}{\cosh x}, \quad \coth x = \frac{1}{\tanh x}.$$

10. Determine the derivatives of the six hyperbolic functions. Compare these derivatives to the corresponding derivatives of the trigonometric functions.
11. Determine the Taylor polynomial expansions about $x = 0$ for $y = \cosh x$ and $y = \sinh x$ of degrees $2n$ and $2n + 1$ respectively.
12. For $n = 2$ in the previous problem, estimate $\cosh x$ and $\sinh x$ for $x = 0.1$, 0.2 and 0.5 using the Taylor polynomials. What are the errors in these estimates?
13. Verify the following properties of the hyperbolic functions.
(i) $\cosh x + \sinh x = e^x$ **(ii)** $\cosh x - \sinh x = e^{-x}$ **(iii)** $\cosh^2 x - \sinh^2 x = 1$
(iv) $1 - \tanh^2 x = \operatorname{sech}^2 x$ **(v)** $\coth^2 x - 1 = \operatorname{csch}^2 x$ **(vi)** $|\tanh x| < 1$.

14. Verify the addition formulas for the hyperbolic functions.

(i) $\sinh(x_1 + x_2) = \sinh x_1 \cosh x_2 + \cosh x_1 \sinh x_2$

(ii) $\cosh(x_1 + x_2) = \cosh x_1 \cosh x_2 + \sinh x_1 \sinh x_2$

(iii) $\sinh x_1 + \sinh x_2 = 2\sinh\dfrac{x_1 + x_2}{2}\cosh\dfrac{x_1 - x_2}{2}$

(iv) $\cosh x_1 + \cosh x_2 = 2\cosh\dfrac{x_1 + x_2}{2}\cosh\dfrac{x_1 - x_2}{2}$

(v) $\cosh x_1 - \cosh x_2 = 2\sinh\dfrac{x_1 + x_2}{2}\sinh\dfrac{x_1 - x_2}{2}$

15. Evaluate the derivatives of the following functions.

(i) $y = e^{3x}$ **(ii)** $y = x^3 e^{\cos x}$ **(iii)** $y = \sinh^2 x$

(iv) $y = \cosh(1 + x^2)$ **(v)** $y = (e^x + 1)^{3/2}$ **(vi)** $y = \tanh^2 x$

(vii) $y = \sin(e^{-x})$ **(viii)** $y = (\sinh x)^{1/2}$ **(ix)** $y = e^{\cosh x}$

16. Use implicit differentiation to determine the derivatives of the six inverse hyperbolic functions. Compare the derivatives to the corresponding formulas for the inverse trigonometric functions.

17. Show that $y = \cosh x$ satisfies $y'' = [1 + (y')^2]^{1/2}$. What similar equation does $y = a\cosh(x/a)$ satisfy? Plot these curves for $a = 1/2$, 1 and 2. Explain how the curves are related by changes of scale on the coordinate axes.

18. Evaluate the derivatives dy/dx using the chain rule.

(i) $y = f(e^x)$ **(ii)** $y = \sinh u(x)$ **(iii)** $y = \cosh u(x)$

(iv) $y = e^{\cosh u(x)}$ **(v)** $y = e^{-\tanh x}$ **(vi)** $y = \sin(\tanh x)$

(vii) $y = (1 + \sinh^2 x)^{1/2}$ **(viii)** $y = u(e^{-x^2})$ **(ix)** $y = u(v(\cosh x))$

19. Evaluate the derivatives dy/dx using implicit differentiation:

(i) $\cosh x + \sinh y = e^x$ **(ii)** $\tanh(x + y) = e^{-x}$ **(iii)** $x = \cosh(y^2)$

(iv) $e^{2y} = \sin(x + 2y)$ **(v)** $\sin^{-1} y = e^{x+y}$ **(vi)** $e^{x^2 - y^2} = \cosh(x + y)$

20. Use l'Hôpital's rule to evaluate the following limits:

(i) $\lim\limits_{x\to 0}\dfrac{1 - e^{-x}}{x}$ **(ii)** $\lim\limits_{x\to 0}\dfrac{e^x - 1 - x}{x^2}$ **(iii)** $\lim\limits_{x\to 0}\dfrac{e^x - e^{2x}}{x}$

(iv) $\lim\limits_{x\to a}\dfrac{e^x - e^a}{x - a}$ **(v)** $\lim\limits_{x\to 0}\dfrac{\tanh x - x}{\sin x - x}$ **(vi)** $\lim\limits_{x\to 0}\dfrac{e^{-x^2} - 1 + x^2}{x^4}$

(vii) $\lim\limits_{x\to 0}\dfrac{e^{\sin x} - e^x}{x}$ **(viii)** $\lim\limits_{x\to a}\dfrac{\sinh x - \sinh a}{x - a}$ **(ix)** $\lim\limits_{x\to 0}\dfrac{1 - \cosh x}{1 - \cos x}$

21. Evaluate the following limits:

(i) $\lim\limits_{x\to\infty} x e^{-x}$ **(ii)** $\lim\limits_{x\to\infty} x^4 e^{-x}$ **(iii)** $\lim\limits_{x\to\infty}\dfrac{e^{2x}}{x^2}$

(iv) $\lim\limits_{x\to\infty} e^{-x}\tanh x$ **(v)** $\lim\limits_{x\to\infty} e^{e^{-x}}$ **(vi)** $\lim\limits_{x\to\infty} e^{1/x}$

(vii) $\lim\limits_{x\to\infty}(\cosh x - \sinh x)$ **(viii)** $\lim\limits_{x\to\infty}\dfrac{\tan^{-1} x - \pi/2}{x - \pi/2}$ **(ix)** $\lim\limits_{x\to\infty}\left(\dfrac{1}{\sinh x} - \dfrac{1}{\cosh x}\right)$

22. Evaluate the following sums:

(i) $\cosh x + \cosh 2x + \ldots + \cosh nx$ **(ii)** $\sinh x + \sinh 2x + \ldots + \sinh nx$

(iii) $\cosh x + 2\cosh 2x + \ldots + n\cosh nx$ **(iv)** $\sinh x + 2\sinh 2x + \ldots + n\sinh nx$

23. From the Taylor's formula for e^x expanded about $x = 0$, find corresponding polynomial expansions of e^{-x}, e^{-x^2} and $e^{1/x}$ using appropriate changes of variables.

24. Use the mean-value theorem to verify the following inequalities.
(i) If $a<b$, $e^a(b-a)<e^b-e^a<e^b(b-a)$.
(ii) If $0<x<1$, $1+x<e^x<1/(1-x)$.
Use the second result to obtain bounds for $e^{0.1}$ and $e^{0.01}$. What are the accurate values?
25. A radioactive tracer drug used in medical diagnosis loses 40% of its radioactivity in 5 hours. Assuming an exponential decay of the radioactivity, determine the proportions of radioactivity lost after 10 hours and after 20 hours. What is the half-life of the radioactivity?
26. Carbon 14 is a radioactive isotope of carbon 12 produced at a constant rate by the action of cosmic rays in the atmosphere. In living organic matter, it occurs in the proportion of one part in 10^{12}. After the death of the organic matter, this level is reduced by radioactive decay with a half-life of 5570 years. Suppose that the carbon 14 level in a fossil specimen is found to be 5 parts in 10^{14}. Estimate the age of the fossil.
27. The atmospheric pressure $P(z)$ decreases exponentially with height z above the surface of the earth. At 5500 meters, it is reduced to one-half the surface pressure. Draw a graph of $P(z)$. At what height is the pressure 60% of its surface magnitude?
28. A radioactive tracer drug with half-life λ_1 is introduced into an organ whose cells have a biological half-life λ_2 (half the cells are replaced in time λ_2). Show that the *effective half-life* λ of the radioactivity is $\lambda_1\lambda_2/(\lambda_1+\lambda_2)$.
29. In the previous problem, suppose that the biological half-life is much longer than the radioactive half-life. Show that $\lambda\approx\lambda_1$ and determine the first order correction.
30. Newton's law of cooling asserts that the rate at which a body loses heat through radiation is proportional to the difference between its temperature and that of the surrounding medium. Formulate this mathematically and find the law explicitly.
31. As water fills a hopper, the inflow valve is slowly pressed shut by a float-lever arrangement so that when a desired water level is achieved, the valve closes completely. The rate at which the water level h increases with time is proportional to the volume influx Q; Q is itself proportional to h_0-h, where h_0 is the level at which the flow is to cease ($Q=0$). Formulate a differential equation for h, and solve it to find the water height as an explicit function of time.
32. After a small puncture in a tire, the interior air pressure decreases at a rate proportional to the difference between the tire pressure and the atmospheric pressure. Formulate this mathematically and find the law of decrease of the tire pressure. If the pressure difference is reduced 50% of its initial value ten minutes after the puncture, when will it be reduced 90%?

2.9 Logarithms

The exponential function e^x is a positive increasing function with a unique inverse function. The inverse of e^x is called the *natural logarithm* and is denoted by

$$y=\log x, \tag{1}$$

defined for $x>0$. By the definition of an inverse function, the following properties hold

$$\log(e^x)=x, \qquad e^{\log x}=x. \tag{2}$$

In effect, Eq. (1), rephrased, says: find a number y such that e raised to this power will equal a prescribed value x, that is,

$$e^y=x. \tag{3}$$

Equations (1) and (3) are equivalent, and the process by which the latter is obtained from the former is called *exponentiation*. In this way, the exponential function provides the graph of $\log x$ and also its complete tabulation. The log function is clearly defined only for positive arguments, and the most noteworthy values are $\log 1 = 0$ and $\log e = 1$. Figure 9.1 shows that the origin is a singularity and $\lim_{x \to 0} \log x = -\infty$.

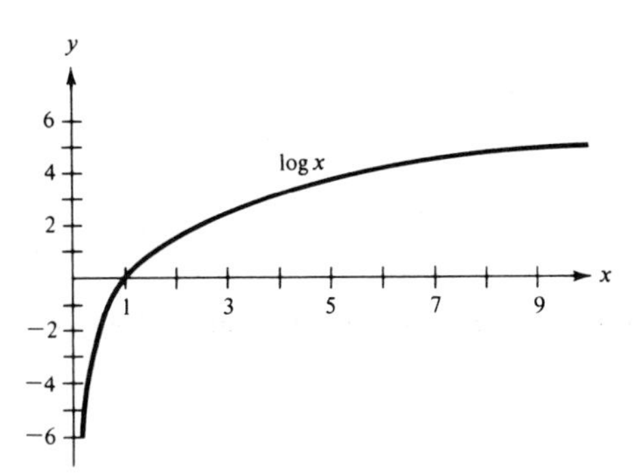

FIGURE 9.1

The derivative of $\log x$ is obtained from (3) as follows:

$$\frac{d}{dx} e^y = \frac{d}{dx} x = 1,$$

or

$$\frac{d}{dx} e^y = e^y \frac{dy}{dx} = x \frac{dy}{dx} = 1.$$

In other words,

$$\frac{d}{dx} \log x = \frac{1}{x}. \tag{4}$$

Indeed, the log function was previously defined as the particular solution of this differential equation which had the value zero at $x = 1$.

The familiar properties of the logarithm are obtained from the corresponding formulas for the exponential function. For example, let x_1 and x_2 be any two positive numbers and y_1 and y_2 be the values for which

$$e^{y_1} = x_1, \qquad e^{y_2} = x_2,$$

so that, by definition, $\log x_1 = y_1$ and $\log x_2 = y_2$. Consider now the product

$$x_1 x_2 = e^{y_1} e^{y_2} = e^{y_1 + y_2},$$

which by (3) implies that

$$\log x_1 x_2 = y_1 + y_2 = \log x_1 + \log x_2. \tag{5}$$

Likewise we can show that

$$\log \frac{x_1}{x_2} = \log x_1 - \log x_2. \tag{6}$$

The formula

$$\log x_1^{x_2} = x_2 \log x_1$$

is easily established by writing

$$x_1^{x_2} = (e^{y_1})^{x_2} = e^{y_1 x_2},$$

so that,

$$\log x_1^{x_2} = x_2 y_1 \doteq x_2 \log x_1.$$

2

The function

$$y = a^x, \tag{7}$$

can be interpreted in terms of the natural log and exponential functions. Since for *any* positive number

$$a = e^{\log a},$$

Eq. (7) is equivalent to

$$y = a^x = e^{x \log a}, \tag{8}$$

and this can be taken as an alternative definition of the function. The derivative of (7) is then calculated in a straightforward way:

$$\frac{d}{dx} a^x = \frac{d}{dx} e^{x \log a} = (\log a) e^{x \log a},$$

or more succinctly

$$\frac{d}{dx} a^x = a^x \log a. \tag{9}$$

The function inverse to a^x is *the logarithm to base* a

$$y = \log_a x, \tag{10}$$

which means nothing more than

$$a^y = x. \tag{11}$$

In this nomenclature, the *natural logarithm* should be written with the base e attached, $\log_e x$, but this distinction will not usually be adopted. Of course, the properties of $\log_a x$ are analogous to those of the natural logarithm, e.g.,

$$\log_a x_1 x_2 = \log_a x_1 + \log_a x_2 .$$

Example 1 Find the derivative of $y = \log_a x$.
The derivative is determined from (11):

$$\frac{d}{dx} x = 1 = \frac{d}{dx} a^y = \frac{d}{dx} e^{y \log_e a} = (\log_e a) e^{y \log_e a} \frac{dy}{dx},$$

so that

$$1 = x \log_e a \frac{dy}{dx},$$

or

$$\frac{dy}{dx} = \frac{d}{dx} \log_a x = \frac{1}{x \log_e a}. \tag{12}$$

Example 2 Given $\log 3 = 1.0986123$, estimate $\log 2.99$ and $\log 3.02$.

Using differentials, $\log(x + dx) \approx \log x + \dfrac{d}{dx}(\log x)dx = \log x + \dfrac{dx}{x}$.

With $x = 3$ and $dx = -0.01$ or 0.02,

$$\log 2.99 \approx \log 3 - \frac{0.01}{3} = 1.095279\ldots$$

$$\log 3.02 \approx \log 3 + \frac{0.02}{3} = 1.105279\ldots$$

The accurate values are 1.0952734 and 1.1052568.

The generalization of the mean-value theorem expressed in Taylor's formula relates a function at x to its value and the values of its derivatives at position x_0 via

$$f(x) = f(x_0) + f'(x_0)(x - x_0) + \cdots + \frac{f^{(n)}(x_0)}{n!}(x - x_0)^n + \frac{f^{(n+1)}(\xi)}{(n+1)!}(x - x_0)^{n+1}, \tag{13}$$

with $x_0 < \xi < x$. This formula can be used to develop an approximation for $\log x$, and a convenient point of departure is $x_0 = 1$, where the function and all its derivatives are easily calculated. Since

$$\frac{d}{dx} \log x = \frac{1}{x}, \qquad \frac{d^2}{dx^2} \log x = -\frac{1}{x^2}, \qquad \frac{d^3}{dx^3} \log x = \frac{2}{x^3},$$

and in general

$$\frac{d^k}{dx^k} \log x = (-1)^{k-1} \frac{(k-1)!}{x^k}.$$

The specific values of these derivatives at $x_0 = 1$ are

$$\log 1 = 0, \qquad \left.\frac{d^k}{dx^k} \log x\right]_{x=1} = (-1)^{k-1}(k-1)!.$$

The substitution of these for the coefficients in (13) yields

$$\log x = (x - 1) - \tfrac{1}{2}(x - 1)^2 + \tfrac{1}{3}(x - 1)^3 + \cdots + \frac{(-1)^{n-1}}{n}(x - 1)^n + R_n(\xi). \tag{14}$$

An approximate formula for computational purposes is obtained by merely neglecting the remainder. (The log function is singular at $x = 0$, and we can anticipate trouble there on this account.) A more commonly quoted version of this formula utilizes the substitution $x = 1 + h$, in which case

$$\log(1 + h) = h - \frac{h^2}{2} + \frac{h^3}{3} + \cdots + \frac{(-1)^{n-1}}{n} h^n + R_n. \tag{15}$$

It remains to be shown that $\lim_{n\to\infty} R_n(\xi) = 0$, which would justify the infinite-series expansion

$$\log(1 + x) = x - \frac{x^2}{2} + \cdots + \frac{(-1)^{n-1}}{n} x^n + \cdots = \sum_{n=1}^{\infty} \frac{(-1)^{n-1} x^n}{n}, \tag{16}$$

for $|x| < 1$, but, once again, computational evidence is strong on this point as long as the restrictions imposed on x are observed. Moreover, we can anticipate from the form of (16) that the singularity of $\log(1 + x)$ at $x = -1$ implies the divergence of the *harmonic series*

$$1 + \tfrac{1}{2} + \tfrac{1}{3} + \cdots = \sum_{n=1}^{\infty} \frac{1}{n}.$$

A proof of this is given in Chap. 4.

---

4 The derivative of $\log x$ when calculated from first principles (see Exercise 15) leads to a new description of the exponential function as the limit

$$e^x = \lim_{n\to\infty} \left(1 + \frac{x}{n}\right)^n. \tag{17}$$

Here, we choose to establish this formula as an illustration of l'Hôpital's rule.

Since

$$\left(1 + \frac{x}{n}\right)^n = \exp\left[n \log\left(1 + \frac{x}{n}\right)\right],^\dagger$$

the limit in (17) can be expressed as

$$\lim_{n\to\infty} \left(1 + \frac{x}{n}\right)^n = \exp\left[\lim_{n\to\infty} n \log\left(1 + \frac{x}{n}\right)\right]. \tag{18}$$

The exponent involves an $\infty \cdot 0$ indeterminacy, and to apply l'Hôpital's rule we manipulate this expression into a proper form:

$$\lim_{n\to\infty} n \log\left(1 + \frac{x}{n}\right) = \lim_{n\to\infty} \frac{\log(1 + x/n)}{1/n} = \lim_{s\to 0} \frac{\log(1 + sx)}{s},$$

† For convenience we often write e^u as exp (u).

where the approach of s to zero is regarded as a smooth process. (It is, of course, not really necessary to replace integer n by the continuous variable s; note too that x is held fixed in this limit.) The derivatives relevant to the application of the rule are

$$\frac{d}{ds}\log(1 + sx) = \frac{x}{1 + sx}; \qquad \frac{d}{ds}s = 1,$$

and it follows that

$$\lim_{s\to 0}\frac{\log(1 + sx)}{s} = \lim_{s\to 0}\frac{x}{1 + sx} = x.$$

Substitution of this result in (18) reproduces Eq. (17).

Example 3 Evaluate (i) $\lim_{x\to 0} x \log x$ and (ii) $\lim_{x\to 0} x^x$.

In case (i) the expression is first rewritten as

$$\lim_{x\to 0}\frac{\log x}{1/x},$$

which, as an ∞/∞ indeterminacy, can be calculated by l'Hôpital's rule. Therefore

$$\lim_{x\to 0}\frac{\log x}{1/x} = \lim_{x\to 0}\frac{1/x}{-1/x^2} = \lim_{x\to 0} - x = 0,$$

and the answer is

$$\lim_{x\to 0} x \log x = 0. \tag{19}$$

In case (ii) write $x^x = e^{x \log x}$, and, as a consequence of the result just established in (19),

$$\lim_{x\to 0} x^x = \exp\left(\lim_{x\to 0} x \log x\right) = e^0 = 1.$$

Example 4 A bank pays a yearly interest of r percent, compounded and paid n times. What is the value at the end of 1 year of an initial deposit of A dollars?

At the time of the first interest payment, an amount $(r/n)A$ is added to the initial deposit A to make the account worth $(1 + r/n)A$. After the second payment, this becomes

$$\left(1 + \frac{r}{n}\right)\left[\left(1 + \frac{r}{n}\right)A\right] = \left(1 + \frac{r}{n}\right)^2 A,$$

and at the year's end, following n compoundings, the value of the account is

$$P = \left(1 + \frac{r}{n}\right)^n A.$$

If the bank were to compound interest *continuously*, the amount after 1 year would be

$$P = \lim_{n \to \infty} \left(1 + \frac{r}{n}\right)^n A = e^r A.$$

To illustrate, let $r = 0.06$; then the value of $(1 + r/n)^n$ for $n = 1, 2, 4, 12$, and ∞ is, respectively, 1.06, 1.0609, 1.0614, 1.0617, and 1.0618. Unless the deposit is substantial, continuous versus monthly or even quarterly compounding is mainly an advertising gimmick.

---

5 The properties of the logarithm make it a natural tool in calculating the derivatives of certain types of functions, especially those which are products or powers of others. The technique, called *logarithmic differentiation*, consists of transforming

$$y = f(x) \qquad \text{to} \qquad \log y = \log f(x),$$

which is then differentiated implicitly to obtain

$$\frac{1}{y} y'(x) = \frac{1}{f} f'(x).$$

Suppose that

$$y(x) = \frac{u^a v^b}{w^c},$$

where u, v, w are functions of x; then

$$\log y = a \log u + b \log v - c \log w,$$

and

$$\frac{1}{y} y' = a \frac{u'}{u} + b \frac{v'}{v} - c \frac{w'}{w}.$$

The calculation is completed by multiplying through by $y(x)$.

Example 5 Find the derivative of $y = (x^2 + 4)(x^3 + 8)(x^4 + 16)$.

This derivative can be evaluated by using the product rule or by first calculating the logarithm of y and then differentiating.

$$\log y = \log(x^2 + 4) + \log(x^3 + 8) + \log(x^4 + 16)$$

$$\frac{1}{y}\frac{dy}{dx} = \frac{2x}{x^2 + 4} + \frac{3x^2}{x^3 + 8} + \frac{4x^3}{x^4 + 16}$$

Multiplying by y, the derivative is

$$\frac{dy}{dx} = 2x(x^3 + 8)(x^4 + 16) + 3x^2(x^2 + 4)(x^4 + 16) + 4x^3(x^2 + 4)(x^3 + 8).$$

Example 6 Find the derivative of

$$y = \left[\frac{(x^2+4)(x^3+1)}{2x+1}\right]^{1/2}$$

The logarithm of the function is

$$\log y = \tfrac{1}{2}\log(x^2+4) + \tfrac{1}{2}\log(x^3+1) - \tfrac{1}{2}\log(2x+1),$$

and implicit differentiation yields

$$\frac{1}{y}y' = \frac{x}{x^2+4} + \frac{3x^2}{2(x^3+1)} - \frac{1}{2x+1}.$$

Multiplication of each side by $y(x)$ gives dy/dx explicitly.

Exercises 2.9

1. Given $\log 7 = 1.94590$, estimate $\log 6.97$ and $\log 7.05$. Compare the estimates to the accurate values obtained using a calculator.

2. Estimate the values of $\log(1+x)$ for $x = -0.1, 0.1, 0.2$ and 0.5 by calculating the values of the Taylor polynomial of degree 3. Compare to the accurate values.

3. Differentiate the following functions:

(i) $\log x^2$ **(ii)** $\log(x \cos x)$ **(iii)** $\log(1+e^x)$

(iv) $\log(f(x))^3$ **(v)** $\log(\cos f(x))$ **(vi)** $\log(f(\cos x))$.

4. Graph the function $y = x \log x$ defined for $x \geq 0$. On the same graph, plot the first and second derivatives. What is the slope at $x = 0$?

5. Graph the function $y = x/\log x$ defined for $x \geq 0$ and plot the first derivative on the same graph. What is the slope at $x = 0$?

6. Apply the mean value theorem to $f(x) = \log(1+x)$ on the interval $0 \leq x \leq h$ to prove

$$\frac{h}{1+h} < \log(1+h) < h.$$

Use this result to estimate $\log 1.05$ and $\log 1.005$. Compare to the accurate values.

7. State the mean value theorem $f(b) - f(a) = f'(\xi)(b-a)$ for the following functions and find the values of ξ. Illustrate with graphs.

(i) $f(x) = xe^{-x}$ $a = 0, b = 1$ **(ii)** $f(x) = \log x$ $a = 1, b = 2$

(iii) $f(x) = x \log x$ $a = 1, b = 1.5$ **(iv)** $f(x) = e^x/(1+e^x)$ $a = 0, b = 10$

8. Differentiate Taylor's formula for $\log x$ expanded about $x = 1$ to find the corresponding formulas for $1/x$ and $1/x^2$. Estimate $(1.01)^{-1}$ and $(1.01)^{-2}$ by using four terms and compare to the accurate values.

9. Prove the following inequalities, valid for $x > 0$ and any positive integer n.

(i) $1 - x + x^2 - \cdots - x^{2n-1} < \dfrac{1}{1+x} < 1 - x + x^2 - \cdots + x^{2n}$

(ii) $x-\frac{x^2}{2}+\frac{x^3}{3}-\cdots-\frac{x^{2n}}{n}<\log(1+x)<x-\frac{x^2}{2}+\frac{x^3}{3}-\cdots+\frac{x^{2n+1}}{2n+1}$.

Deduce that the error in the Taylor polynomial for $(1+x)^{-1}$ or $\log(1+x)$ is less than the first term omitted.

10. Prove the following properties of the logarithm, valid for $x_1>0$ and $x_2>0$.

(i) $\log_a x_1 x_2=\log_a x_1+\log_a x_2$ **(ii)** $x_2 \log_a x_1=\log_a x_1{}^{x_2}$

(iii) $\log_a \frac{x_1}{x_2}=\log_a x_1-\log_a x_2$ **(iv)** $\frac{1}{x_2}\log_a x_1=\log_a x_1{}^{1/x_2}$

(v) $\log_a \frac{1}{x}=-\log_a x$ **(vi)** $\log_a x \log_x a=1$.

11. Evaluate the derivatives of the following functions:

(i) $x^3\log x$ **(ii)** $\cos(\log x)$ **(iii)** 10^x

(iv) $(\log x)^n$ **(v)** $\log(1-x^2)$ **(vi)** $\log[x+(x^2+1)^{1/2}]$

(vii) $\cosh(\log x)$ **(viii)** $\tanh(\log x)$ **(ix)** 2^{3^x}

12. Determine explicit formulas for the inverse hyperbolic functions, $\sinh^{-1}x$ and $\cosh^{-1}x$. (*Hint:* $y=\sinh x=(e^x-e^{-x})/2$ implies $e^{2x}-2e^xy-1=0$, a quadratic equation for e^x.) Evaluate the derivatives of the inverse hyperbolic functions using the explicit formulas and by implicit differentiation.

13. Approximate the following numbers and compare to the accurate values.

(i) $(.99)^{100}$ **(ii)** $(1.01)^{200}$ **(iii)** $(1.001)^{500}$

(iv) $(1.002)^{3000}$ **(v)** $1000\log(1.002)$ **(vi)** $1000\log(0.999)$

14. Find the derivatives of the following functions by logarithmic differentiation.

(i) $(x^3+4)^{1/3}$ **(ii)** x^x **(iii)** $(x-1)^5(x-2)^6(x-3)^7$

(iv) 2^{4x} **(v)** 10^{-4x} **(vi)** $(e^x)^x$

(vii) e^{x^x} **(viii)** $e^{\cosh^{-1}x}$ **(ix)** $e^{\tanh(\log x)}$

15. Calculate the derivative of $\log x$ from first principles and establish the formula $e^x=\lim_{h\to 0}(1+hx)^{1/h}$. (*Hint:* Verify that $x=\lim_{h\to 0}\frac{1}{h}\log(1+hx)$.)

16. Find the following limits:

(i) $\lim_{x\to 0^+}\frac{x\log x}{\log(1+2x)}$ **(ii)** $\lim_{x\to 0} x\,|\log x|^a$ **(iii)** $\lim_{x\to 0}\frac{\log(1-x)}{x}$

(iv) $\lim_{x\to 0}\frac{\log(1+x)}{e^x-1}$ **(v)** $\lim_{x\to 0^+}\frac{\log(\sin x)-\log x}{x^2}$ **(vi)** $\lim_{x\to 0^+}\frac{\log\sin 2x}{\log\sin x}$

(vii) $\lim_{x\to 0^+} x^x$ **(viii)** $\lim_{x\to 0^+}(\sin x)^{\tan x}$ **(ix)** $\lim_{x\to 0}\left(\frac{\tan x}{x}\right)^{1/x^2}$

17. Find the following limits:

(i) $\lim_{x\to 1/3}\frac{\log 3x}{3x-1}$ **(ii)** $\lim_{z\to 1/2^+}\frac{\tan z\pi}{\log(2z-1)}$ **(iii)** $\lim_{x\to\pi/2^+}|\tan x|^{\tan x}$

(iv) $\lim_{x\to 1^+}(\log x)^{\sin(x-1)}$ **(v)** $\lim_{x\to\infty} x^{-1/2}\log|\log x|$ **(vi)** $\lim_{x\to\infty} xe^{-\beta x}$

(vii) $\lim_{x\to\infty}\frac{\log(1+1/x)}{\sin(1/x)}$ **(viii)** $\lim_{x\to 0}\frac{a^x-b^x}{x}$ **(ix)** $\lim_{x\to 0}\frac{a^x-b^x}{c^x-d^x}$

18. Graph the functions $y=x^x$ and $y=x^{1/x}$ defined for $x>0$. Determine their relative extrema and inflection points.

19. A bank pays interest at the rate of r percent per year compounded n times. How much will an initial deposit of P dollars be worth N years later? Calculate the limiting value of this amount as n increases.

20. Bond dealers estimate the doubling time of money by the formula $70/r$ where r is the annual interest rate in percent per year. Explain the basis of this rule and discuss accuracy.
21. Three banks offer annual interest rates of 5.25% simple interest, 5% compounded quarterly and 4.8% compounded monthly. Which would you choose to deposit your money for one year?
22. An empirical formula for the surface area $S(m^2)$ in terms of weight $W(kg)$ and height $H(cm)$ of adult humans is given by $S = (0.072)\ W^{0.425}\ H^{0.725}$. Calculate your surface area. If your weight increases by 2% and your height by 1%, what is the percent increase in surface area?
23. The Weber-Fechner law in psychology describes many stimulus-response experiments. It states that the change in perceived response (or sensation) is proportional to the relative change in a stimulus (for example heat, light or sound intensity). If x and y represent the stimulus and sensation, show that the mathematical formulation of this law is $dy/dx = a/x$ where a is a constant. (It is assumed that the stimulus and response are continuous variables.) Show that $y = a \log x + b$. How are the constants a and b determined?
24. The intensity of light passing through a semi-transparent medium decreases exponentially with distance of penetration into the medium. In certain smog conditions, the light intensity decreases by 2% per meter. At what distance is the intensity of a light source decreased by 50%? by 95%?

2.10 Numerical methods

Consider the problem of finding the derivative of a function $y = f(x)$ when only a few of its values are known,

$$y_i = f(x_i) \qquad \text{for} \qquad i = 0, 1, \ldots, n. \tag{1}$$

Let us assume that the information at hand is sufficient to permit drawing a smooth curve, which passes through the points (x_i, y_i). The graph is, in effect, an approximation of the function at every point in a certain region of the x axis. Of course, it is assumed that the function varies smoothly and continuously, as in Fig. 10.1*a*, and does *not* exhibit the erratic behavior between points illustrated in Fig. 10.1*b*. Whenever there is no evidence to the contrary, our first assumption will always be the simplest.

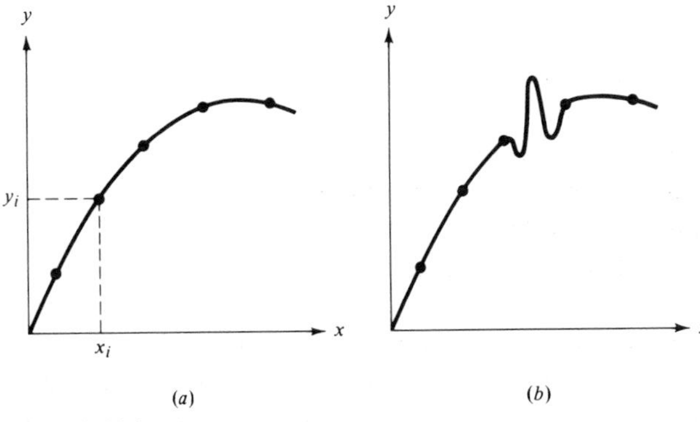

FIGURE 10.1

The derivative at a point can be calculated from the graph as the slope of the tangent line to the curve, and this is the method employed in the earliest encounter with differentiation in Chap. 1. The theoretical analog of this procedure is to construct the approximation as a simple or known function which has the same values given in (1) and then to differentiate this function analytically (the additional effort required should pay off with increased accuracy). The simplest approximation is formed from the *polynomial* of smallest degree, which assumes the $n + 1$ known values of the original function. This will generally be a polynomial of degree n, call it

$$\mathscr{L}_n(x) = a_0 + a_1 x + \cdots + a_n x^n, \tag{2}$$

for which

$$\mathscr{L}_n(x_i) = y_i, \qquad i = 0, 1, \ldots, n. \tag{3}$$

This set of conditions results in $n + 1$ linear equations to determine the coefficients:

$$\begin{aligned} a_0 + a_1 x_0 + a_2 x_0{}^2 + \cdots + a_n x_0{}^n &= y_0, \\ a_0 + a_1 x_1 + a_2 x_1{}^2 + \cdots + a_n x_1{}^n &= y_1, \\ \cdots\cdots\cdots\cdots\cdots\cdots\cdots\cdots\cdots\cdots \\ a_0 + a_1 x_n + a_2 x_n{}^2 + \cdots + a_n x_n{}^n &= y_n. \end{aligned} \tag{4}$$

The solution of this system is laborious and difficult to handle; the final result is more simply constructed by inspection and a little thought. The polynomial

$$\begin{aligned} y = \mathscr{L}_n(x) = {} & \frac{(x - x_1)(x - x_2) \cdots (x - x_n)}{(x_0 - x_1)(x_0 - x_2) \cdots (x_0 - x_n)} y_0 \\ & + \cdots + \frac{(x - x_0) \cdots (x - x_{i-1})(x - x_{i+1}) \cdots (x - x_n)}{(x_i - x_0) \cdots (x_i - x_{i-1})(x_i - x_{i+1}) \cdots (x_i - x_n)} y_i \\ & + \cdots + \frac{(x - x_0) \cdots (x - x_{n-1})}{(x_n - x_0) \cdots (x_n - x_{n-1})} y_n, \end{aligned} \tag{5}$$

satisfies all the required conditions and is the one sought. To verify this formula, simply set x equal to the $n + 1$ given values. For example, if $x = x_0$, the formula reduces to $y = y_0$.

This equation, known as *Lagrange's interpolation formula*, makes it possible to calculate a function at an arbitrary point from the very limited information available. The derivative $f'(x)$ is then approximated by $\mathscr{L}_n'(x)$, but a rather lengthy calculation is involved, which is greatly facilitated by some additional notation to be set forth in the exercises.

Lagrange's formula is most useful for interpolation between precisely known values of a function. As such, it is of little benefit in fitting a curve to experimental data, where each point may be specified only to within some inherent, tolerated error.

Clearly, there is no advantage to finding a particular polynomial curve that passes through definite points which are themselves inaccurately specified (the same experiment repeated will produce slightly different data). What is really needed in this situation is an interpolation formula that minimizes experimental errors (even if the resultant curve does not pass through a single data point). This is called a *least squares fit* and will be examined later in Sec. 6.9.

Example 1 Find the polynomial of smallest degree which passes through (0, 0), (1, 1), (2, 1).

According to Eq. (5), the polynomial of smallest degree is the quadratic

$$y = \frac{(x-1)(x-2)}{(0-1)(0-2)} \cdot 0 + \frac{x(x-2)}{1(1-2)} \cdot 1 + \frac{x(x-1)}{2(2-1)} \cdot 1 = \tfrac{1}{2}x(3-x).$$

Example 2 Find the value of $y = \sin x$ and its derivative at $x = 0.6458$ (37°) from the information $\sin 0.5236 = 0.5000$ ($x = 30°$); $\sin 0.6981 = 0.6428$ ($x = 40°$); $\sin 0.8727 = 0.7660$ ($x = 50°$).

Although it is simpler to use an interpolation formula in which angles are measured in degrees, we will confine ourselves to radian measure to avoid any confusion about units that might arise in the process of differentiation.

The most common and familiar procedure to find the value sin 0.6458 is by linear interpolation between the two surrounding points (at 30° and 40°). This is, of course, equivalent to Lagrange's two-point formula

$$y = \frac{x - x_1}{x_0 - x_1} y_0 + \frac{x - x_0}{x_1 - x_0} y_1 = y_0 + \frac{y_1 - y_0}{x_1 - x_0}(x - x_0).$$

With $(x_0, y_0) = (0.5236, 0.5000)$ and $(x_1, y_1) = (0.6981, 0.6428)$, the preceding formula becomes

$$y = 0.5000 + 0.8183(x - 0.5236),$$

in which case we find that at $x = 0.6458$

$$y = \sin 0.6458 = 0.6000, \qquad y' = 0.8183.$$

The use of the three data points allows the determination of a quadratic interpolation formula,

$$y = \frac{(x-x_1)(x-x_2)}{(x_0-x_1)(x_0-x_2)} y_0 + \frac{(x-x_0)(x-x_2)}{(x_1-x_0)(x_1-x_2)} y_1 + \frac{(x-x_0)(x-x_1)}{(x_2-x_0)(x_2-x_1)} y_2 .$$

With $(x_2, y_2) = (0.8727, 0.7660)$ and the other points as stated above, this becomes

$$y = -0.3218x^2 + 1.213x - 0.0466,$$

so that at $x = 0.6458$,

$$y = 0.6024 \qquad \text{and} \qquad y' = 0.7994.$$

The exact answers are $y = 0.6018$, $y' = 0.7986$; the sine curve and both linear and quadratic approximations are shown in Fig. 10.2.

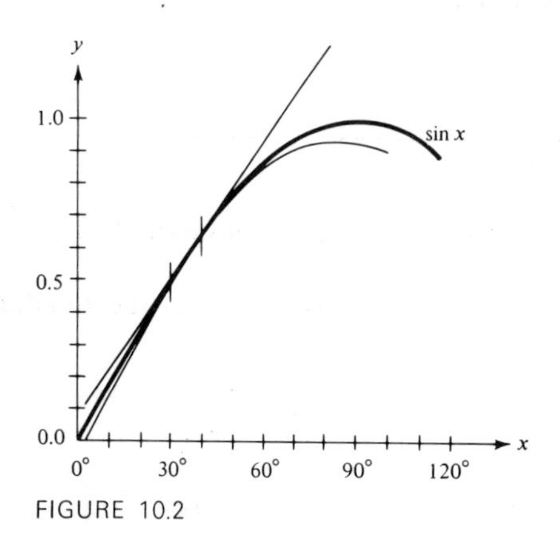

FIGURE 10.2

Taylor's formula for a two- and three-term series was used earlier (page 154) to evaluate $\sin 37°$ from the known values at 30°. It is of interest to compare the results of that analysis with those obtained here (Table 10.1).

TABLE 10.1

Values of sin 37°

Lagrange		*Taylor*		
Two-point	*Three-point*	*Two-term*	*Three-term*	*Exact*
0.6000	0.6024	0.6058	0.6020	0.6018

Example 3 The times of sunrise and sunset at 30-day intervals are as shown in Table 10.2. Find the "longest day" of the year.

TABLE 10.2

	May 1	*May* 31	*June* 30
Sunrise	4 : 51	4 : 17	4 : 16
Sunset	19 : 04	19 : 38	19 : 50

It is convenient to establish a day count x, starting with May 1 as day zero ($x = 0$) and, furthermore, to measure the length of each day (hours of daylight) as 14 hr 13 min + T. The data can then be represented by points (x, T), the day number and its length beyond a certain value. In particular, the three dates recorded correspond to the points (0, 0), (30, 68 min) and (60, 81 min). (Note that measurements relative to May 1 amount to choosing the first point as the coordinate origin.) Lagrange's formula for three points is then

$$T = \frac{(x-30)(x-60)}{(-30)\cdot(-60)}\cdot 0 + \frac{x(x-60)}{30\cdot(-30)}\cdot 68 + \frac{x(x-30)}{60\cdot 30}\cdot 81,$$

or

$$T = \frac{x(-55x + 5{,}730)}{1{,}800}.$$

The maximum value of T is located by setting $T'(x) = 0$, which yields

$$x = \frac{5{,}730}{110} = 52.09.$$

The longest day is then 52 days past May 1, or June 22 to be precise, when $T = 83$ min. The length of this day or the number of hours between sunrise and sunset is 15 hr 36 min.

---

2 Numerical differentiation is a tricky process, highly susceptible to error. Discrete approximations of the basic limits such as

$$\frac{f(x+h) - f(x)}{h}$$

may involve the subtraction of large numbers divided by a small one. It is desirable to avoid such a situation whenever possible, and, in any event, all calculations should be executed with a maximum regard for accuracy. In this connection, we examine the errors involved in some of the common finite-difference formulas used to simulate derivatives.

Consider the three neighboring points $x = a - h$, a, and $a + h$, for which the corresponding functional values are $f(a-h)$, $f(a)$, and $f(a+h)$, as shown in Fig. 10.3.

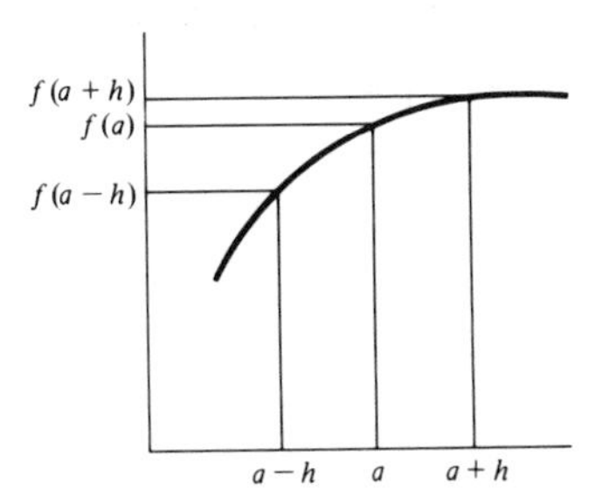

FIGURE 10.3

For h small, these values are used to form the *forward* and *backward differences*

$$\frac{f(a+h)-f(a)}{h} \quad \text{and} \quad \frac{f(a)-f(a-h)}{h},$$

which are perhaps the simplest approximations of the derivative $f'(a)$. Each of these formulas introduces an error whose magnitude is proportional to the size of the increment h, and this is established using the mean-value theorem. Since the analyses are similar, only the forward difference is examined, and to this end we write

$$f(a+h) = f(a) + f'(a)h + f''(\xi)\frac{h^2}{2}, \tag{6}$$

where $a < \xi < a + h$. A little manipulation brings this to the form

$$\frac{1}{h}[f(a+h) - f(a)] - f'(a) = \tfrac{1}{2}f''(\xi)h,$$

where the left-hand side represents the discrepancy between the derivative and its numerical analog. If M is a bound for the second derivative in some larger interval about a, then

$$|f''(\xi)| < M.$$

This allows (6) to be expressed as an inequality:

$$\left|\frac{1}{h}[f(a+h) - f(a)] - f'(a)\right| \leq \tfrac{1}{2}Mh, \tag{7}$$

which is a bound on the error and, in fact, the conclusion stated above.

The error incurred in approximating the derivative by finite differences can be substantially reduced by using a symmetrical numerical formula that gives equal importance to *both sides* of a. One such approximation, called the *centered difference*, is obtained by averaging the forward and backward-difference formulas above:

$$\frac{1}{2}\left\{\frac{1}{h}[f(a+h) - f(a)] + \frac{1}{h}[f(a) - f(a-h)]\right\} = \frac{f(a+h) - f(a-h)}{2h}. \tag{8}$$

The bound for the error, in this case, is determined by the mean-value theorem:

$$\begin{aligned} f(a+h) &= f(a) + f'(a)h + \tfrac{1}{2}f''(a)h^2 + \tfrac{1}{6}f'''(\xi)h^3, \\ f(a-h) &= f(a) - f'(a)h + \tfrac{1}{2}f''(a)h^2 - \tfrac{1}{6}f'''(\eta)h^3, \end{aligned} \tag{9}$$

where

$$a < \xi < a + h \quad \text{and} \quad a - h < \eta < a.$$

The result of subtracting these expressions, when arranged as

$$\frac{1}{2h}[f(a+h) - f(a-h)] - f'(a) = \tfrac{1}{12}[f'''(\xi) + f'''(\eta)]h^2,$$

yields the bound

$$\left| \frac{1}{2h} [f(a+h) - f(a-h)] - f'(a) \right| \leq \frac{\overline{M}h^2}{6}, \tag{10}$$

where

$$|f'''(\eta)| \leq \overline{M} \qquad \text{and} \qquad |f'''(\xi)| \leq \overline{M}.$$

Therefore (10) shows that the error introduced by this symmetric difference formula is less than a constant multiple of h^2; the error is said to be the order of h^2 [written $O(h^2)$]. In (7) the error was only the order of h, $O(h)$, and a substantial improvement of accuracy has been obtained for sufficiently small h.

A finite-difference approximation to the second derivative $f''(a)$ can be constructed from the forward and backward-difference approximations to $f'(a)$ as follows:

$$\frac{1}{h}\left[\frac{f(a+h) - f(a)}{h} - \frac{f(a) - f(a-h)}{h}\right] = \frac{1}{h^2}[f(a+h) - 2f(a) + f(a-h)]. \tag{11}$$

This expression can be reworked using the appropriate form of the four-term Taylor formula:

$$f(a+h) = f(a) + f'(a)h + \tfrac{1}{2}f''(a)h^2 + \tfrac{1}{6}f'''(a)h^3 + \tfrac{1}{24}f^{(4)}(\xi)h^4,$$

to show that

$$\left| \frac{1}{h^2}[f(a+h) - 2f(a) + f(a-h)] - f''(a) \right| \leq \tfrac{1}{12}h^2 M^*, \tag{12}$$

where M^* is a bound for the fourth derivative in the interval $a - h \leq x \leq a + h$. In other words, the error made in replacing the second derivative by (11) is $O(h^2)$ (less than a multiple of h^2).

3

We turn now to the problem of finding the roots of a function, which is of such crucial importance in locating singularities and stationary points.

A purely algebraic procedure, lengthy but surefire, is to find an interval in which the sign of $f(x)$ changes and then systematically to reduce the size of the interval until it focuses on the precise root. For example, let $f(l_0) < 0$ but $f(l_1) > 0$, in which case the root c lies somewhere in the interval (l_0, l_1), $l_0 < c < l_1$. The value of the function at the midpoint, $f(\tfrac{1}{2}(l_0 + l_1))$, is then either zero, positive, or negative. In the first circumstance the root is found, and the procedure ends; in either of the latter two, the method is repeated once more. Suppose that $f(\tfrac{1}{2}(l_0 + l_1)) > 0$; then

$$l_0 < c < \tfrac{1}{2}(l_0 + l_1),$$

and the bounding interval has been reduced in length from $l_1 - l_0$ to $\tfrac{1}{2}(l_1 - l_0)$. To continue, the value of the function at the midpoint is calculated anew, and stricter upper and lower bounds are determined. A sequence of nested intervals, each contained within the preceding, is constructed in this manner, which, in the limit of an infinite number of repetitions, picks out the exact zero.

Once a sign change has been found, it is much faster to locate the root by the *Newton-Raphson* method. The essence of this technique is to use the function and its derivative to refine some estimated value of the root as the next step of a convergent iterative sequence. Let c be the precise root, $f(c) = 0$, and suppose that x_0 is our first guess, calculated by a quick and dirty search procedure based on the sign changes of the function. A revised estimate of the root is given by the point where the tangent line to the curve at x_0, $y_0 = f(x_0)$, crosses the x axis (see Fig. 10.4). This line,

$$y - y_0 = f'(x_0)(x - x_0),$$

intersects the horizontal axis $y = 0$ at

$$x_1 = x_0 - \frac{f(x_0)}{f'(x_0)}.$$

The values of $f(x_1)$ and $f'(x_1)$ now provide the data necessary to repeat the iteration, and, in general, the $(n + 1)$st iterative correction is obtained from the previous one by the recursion formula

$$x_{n+1} = x_n - \frac{f(x_n)}{f'(x_n)}. \tag{13}$$

This process is indicated symbolically in Fig. 10.4 by arrows which designate the sequence of steps. For example, guessing x_0 means ($\rightarrow$) start at point A_0; calculate

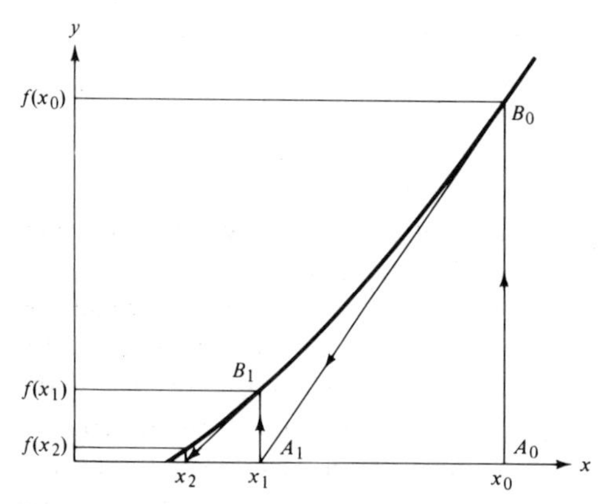

FIGURE 10.4

$y_0 \rightarrow$ go to B_0; iterate to find $x_1 \rightarrow$ slide down the tangent to A_1; calculate $y_1 \rightarrow$ go to B_1; etc.

Example 4 Use the Newton-Raphson method to locate the root of

$$f(x) = 2x^4 - 3x^3 - 9x^2 + 10x$$

which lies between 2 and 3.

Note that $f(2) = -8$ and $f(3) = 30$, so that there is indeed a root in the stated interval (the precise location is 2.5). Let $x_0 = 3$ be the first guess at its value; then (13), which is easily programmed for automatic computation, provides the revised estimates in Table 10.3.

TABLE 10.3

x_1	x_2	x_3	x_4
2.67033	2.52865	2.50101	2.50000

The Newton-Raphson method provides speed at the expense of absolute reliability: it does not always work. If the derivative is small, which can happen in the neighborhood of a stationary point, or if the existence of a nearby root is not confirmed by an independent check of sign changes, numerical instability or divergence can result. The most dramatic illustration of this occurs for a function which has no zero, as in Fig. 10.5. Obviously, the initial guess x_0 must be bad. The scheme produces a wild oscillation about the stationary point, and the first three iterations illustrated show no tendency to converge, as well they should not.

Instability can be more subtle in that convergence or divergence of the iteration procedure depends on the magnitude of the ratio

$$\beta = \frac{f''(c)}{f'(c)},$$

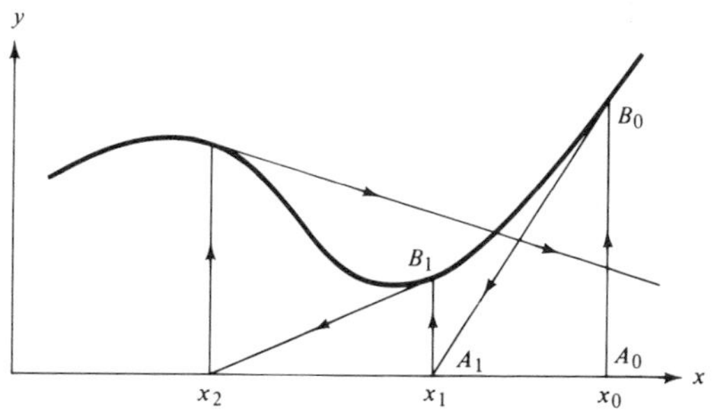

FIGURE 10.5

which in a sense describes the local features of the curve near the zero. This is established from an analysis of convergence based on the mean-value theorem, as follows. By introducing the error between the nth iterative approximation and the exact root,

$$\delta_n = x_n - c,$$

Eq. (13) can be written as

$$\delta_{n+1} = \delta_n - \frac{f(x_n)}{f'(x_n)}.$$

Taylor's three-term formula for $f(x)$ about $x = c$ is used to write

$$\delta_{n+1} = \delta_n - \frac{f(c) + f'(c)(x_n - c) + \frac{1}{2}f''(c)(x_n - c)^2 + \text{remainder}}{f'(c) + f''(c)(x_n - c) + \text{remainder}},$$

or since c is a root,

$$\delta_{n+1} = \delta_n - \frac{f'(c)\,\delta_n + \frac{1}{2}f''(c)\,\delta_n^{\,2} + \text{remainder}}{f'(c) + f''(c)\,\delta_n + \text{remainder}}.$$

If δ_n is very small, the preceding is approximately the same as

$$\frac{\delta_{n+1}}{\delta_n} \approx 1 - \frac{1 + \frac{1}{2}\,\delta_n f''(c)/f'(c)}{1 + \delta_n f''(c)/f'(c)}, \tag{14}$$

where the remainder terms have been neglected. To this degree of accuracy, (14) is equivalent to

$$\frac{\delta_{n+1}}{\delta_n} = 1 - \left[1 + \tfrac{1}{2}\,\delta_n \frac{f''(c)}{f'(c)}\right]\left[1 - \delta_n \frac{f''(c)}{f'(c)}\right],$$

or

$$\frac{\delta_{n+1}}{\delta_n} = \tfrac{1}{2}\,\delta_n \frac{f''(c)}{f'(c)} = \tfrac{1}{2}\beta\delta_n,$$

so that

$$\delta_{n+1} = \tfrac{1}{2}\,\beta\delta_n^{\,2}.$$

Let the initial error be δ_0; then $\delta_1 = \frac{1}{2}\beta\,\delta_0^{\,2}$ and $\delta_2 = \frac{1}{2}\beta\,\delta_1^{\,2} = \frac{1}{8}\beta^3\,\delta_0^{\,4}$, and in general

$$\delta_n = \frac{2}{\beta}\left(\tfrac{1}{2}\,\delta_0\,\beta\right)^{2^n}.$$

The last formula implies that the sequence $\{\delta_n\}$ converges to zero if $\frac{1}{2}\,\delta_0\,\beta < 1$. Therefore, convergence of the Newton-Raphson method depends in this way on the shape of the curve (as expressed by β) and the magnitude of the initial error δ_0.

Because of its basic importance and broad application in applied mathematics, the concept of stability warrants a brief introduction. A commonsense definition would be that a system, whatever it is, is *unstable* if small errors or deviations lead to large or violent changes. Otherwise, we can call it stable. For example, an iterative procedure is unstable when a small initial error grows larger and larger in succeeding stages. (Does or should stability also imply convergence?) A pencil standing on end is unstable since the slightest perturbation will topple it. An airplane wing is a stable structure capable of responding to slight pressure fluctuations without breaking off. An example pertinent to root-finding problems illustrates that the positions of the zeros of a polynomial may be quite sensitive (unstable) to slight changes in the polynomial's coefficients. Consider the polynomial

$$(x + 1)(x + 2) \cdots (x + 20) = x^{20} + 210x^{19} + \cdots + 20!.$$

It can be shown that a terribly small perturbation in the single coefficient of x^{19}, from 210 to $210 + 2^{-23}$, leads to drastic changes in 11 of the 20 roots (specifically those at $-9, -10, \ldots, -20$, ten of which become complex).

If a function is represented as

$$f(x) = x - g(x)$$

its roots can be interpreted as the points of intersection of two curves $y = x$, $y = g(x)$ (Fig. 10.6). An iterative procedure for locating the zero might then be

$$x_{n+1} = g(x_n),$$

and the first few sequential steps are indicated in the diagram. The separation indicated above can be accomplished in an infinite number of ways, some of which are stable and converge, while others, like that shown, are unstable and diverge. This will be pursued further in the exercises.

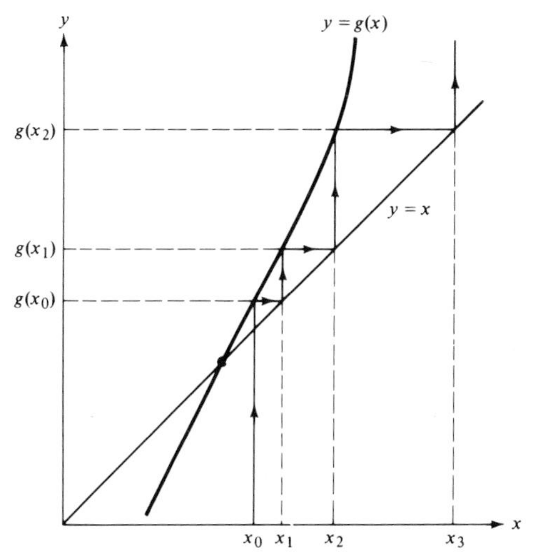

FIGURE 10.6

Exercises 2.10

1. Find a bound for the error made by approximating $f'(x)$ by

$$\frac{f(x) - f(x-h)}{h} \qquad \text{for small } h.$$

2. From the graph of $f(x)$ and the points x, $x+h$, and $x-h$, interpret the meaning of the finite-difference approximation to $f''(x)$ given by

$$\frac{1}{h}\left[\frac{f(x+h)-f(x)}{h} - \frac{f(x)-f(x-h)}{h}\right] = \frac{f(x+h) - 2f(x) + f(x-h)}{h^2}.$$

3. Solve for c_0, c_1, c_2, c_3, and c_4 so that

$$c_0 f(a+2h) + c_1 f(a+h) + c_2 f(a) + c_3 f(a-h) + c_4 f(a-2h)$$

is a finite-difference approximation to $f'''(a)$ whose error is of the order of h^2.

4. Find the polynomial of lowest degree which passes through the three points $(0,1)$, $(2,-1)$, and $(4,5)$.

5. Some of the values of a function are given by

x	0.35	0.40	0.45	0.50	0.55	0.60	0.65
y	1.5215	1.5059	1.4880	1.4675	1.4442	1.4181	1.3887

Find the derivative at $x = 0.5$, using a finite-difference formula. Repeat the calculation with a three-point Lagrange interpolation polynomial. Find the second derivative $y''(0.5)$.

6. From the information in Example 3, find the days which have the earliest sunrise and the latest sunset. (*Hint:* Use the method of Example 3.)

7. Derive an *inverse* interpolation formula that expresses x as a polynomial in y. For the three points $(0,0)$, $(0.5,0.125)$, and $(1.5,3.375)$ use a three-point interpolation formula to express y as a polynomial in x and vice versa. Find the approximate value of y when $x = 1$ and of x when $y = 1$. Compare these values with the exact formula $y = x^3$ ($x = y^{1/3}$) and discuss the implications.

8. Draw a graph to locate approximately the roots of the equation $\cos x = x^2$. Use the Newton-Raphson method or another iteration method to determine the roots accurately.

9. Find the smallest positive roots of the following equations:

(i) $\sin x = x/2$ **(ii)** $\tan x + \sin x = 0.5$ **(iii)** $\tan x = \cos^2 x$.

10. Determine the point on the curve $y = \log x$ that is closest to the origin.

11. Show that the equation $125x^3 + 5x = 1$ has a unique real root. Determine this root by the Newton-Raphson method or another iteration method. (*Hint:* First rescale the variable to be $z = 5x$.)

12. Determine the solutions of the following equations. Illustrate with graphs.

(i) $x = e^{-x}$ **(ii)** $4x = e^x$ **(iii)** $2x = \sinh x$

13. Use the Newton-Raphson method to locate the root of $f(x) = x^3 - x - 1$ that lies between 1.0 and 1.5.

14. Examine the following iterative schemes $x_{n+1} = g(x_n)$ for determining the root in the previous problem.

(i) $x_{n+1} = x_n^3 - 1$ **(ii)** $x_{n+1} = 1/(x_n^2 - 1)$ **(iii)** $x_{n+1} = (x_n + 1)^{1/3}$

In each case, graph the functions $y = x$, $y = g(x)$ and find the point of intersection. Map the successive steps of the iteration on the diagram. Which of the three examples provides a stable, convergent iteration formula?

15. Show graphically that the iteration formula $x_{n+1} = g(x_n)$ will converge to a root of the equation $x = g(x)$ if the derivative satisfies $g'(x) < 1$ in a neighborhood of the root. Explain why this condition on the slope is required. (*Hint:* Draw a diagram similar to Fig. 10.6. Note that, by the mean value theorem, $x_{n+1} - x_n = g(x_n) - g(x_{n-1}) = g'(\xi)(x_n - x_{n-1})$ for some ξ in $x_{n-1} < \xi < x_n$.)

16. Find the roots of $\cos x = x^2$ by using the iteration formula $x_{n+1} = (\cos x_n)^{1/2}$, $x_1 = 1$. Why does the iteration $x_{n+1} = \cos^{-1}(x_n^2)$, $x_1 = 1$ not work?

17. Find the smallest non-zero root of the equation $\sin x = \epsilon x$ where $|\epsilon|$ is a small parameter ($|\epsilon| \ll 1$). Illustrate with a graph. (*Hint:* Write $x = \pi + a_1\epsilon + a_2\epsilon^2 + \cdots$ and determine the first few coefficients.)

18. Let h be a small number and for n an integer define $x_n = nh$, $y_n = f(x_n)$. Show that in this terminology the forward difference approximation for $y'(x)$ is

$$\frac{y(x+h) - y(x)}{h} = \frac{y_{n+1} - y_n}{h}.$$

Approximate the differential equation $y' = f(x)$ by the difference equation

$$y_{n+1} - y_n = hf_n, \qquad f_n = f(x_n)$$

and use this formula to find the solution of $dy/dx = 1/x$ with $y(1) = 0$. Take $h = 0.1$, $x_0 = 1$, $y_0 = 0$, and find the value of y at $x = 2$ (which corresponds to $n = 10$). Repeat this calculation with $h = 0.05$ and compare the results.

19. Develop a simple numerical method to solve $y'(x) = F(x, y)$ with $y(a) = b$. Test it on $F(x, y) = x + y$ with $y(0) = 0$, for which the exact answer is $e^x - 1 - x$.

20. A mechanism for opening a window (Fig. 10.7) is made of rigid thin rods. If $AO = 3$ units and $AS = 4$ units, show that the height x units of the slider above the fixed point O is given by

$$x^2 - 6x\cos\theta - 7 = 0.$$

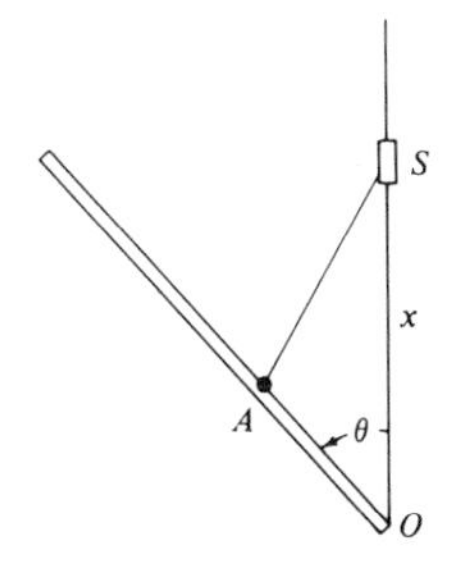

FIGURE 10.7

Find the relationship between the vertical velocity of the slider dx/dt and the rate of change of angle θ, that is, the angular velocity $d\theta/dt$. Calculate $d\theta/dt$ when $dx/dt = 1$ unit/sec. Find an approximate linear equation that relates θ to x when x is nearly 5 units; use this to calculate the change in the angle θ in order to raise the slider S 0.1 units.

21. Obtain the Lagrange polynomials determined by the following points. Draw their graphs.

(i) $(0,1)$, $(3,7)$

(ii) $(0,1)$, $(1,2)$, $(3,7)$

(iii) $(0,1)$, $(1,2)$, $(3,7)$, $(6,21)$

(iv) $(-2,-3)$, $(0,1)$, $(1,2)$, $(3,7)$, $(6,21)$

22. Let

$$F(x) = (x - x_1) \cdots (x - x_n) = \prod_{i=1}^{n} (x - x_i)$$

and

$$F_k(x) = (x - x_1) \cdots (x - x_{k-1})(x - x_{k+1}) \cdots (x - x_n),$$

so that

$$F_k(x) = \frac{1}{x - x_k} F(x).$$

Show that Lagrange's interpolation formula for n points can be written

$$\mathcal{L}_n(x) = \sum_{k=1}^{n} \frac{F_k(x)}{F_k(x_k)} y_k.$$

23. From the definitions in Exercise 22 show by logarithmic differentiation (or otherwise) that

$$F'(x_k) = F_k(x_k),$$

so that

$$\mathcal{L}_n(x) = \sum_{k=1}^{n} \frac{F(x)}{(x - x_k)F'(x_k)} y_k.$$

24. Obtain the result

$$\frac{d}{dx} \mathcal{L}_n(x) = \sum_{k=1}^{n} \frac{(x - x_k)F'(x) - F(x)}{(x - x_k)^2 F'(x_k)} y_k.$$

2.11 Integration and other matters

From our present vantage point, we could proceed logically in any one of several different directions. For this reason, and in order to survey the general terrain ahead, this section will be brief and somewhat exploratory in nature.

The mathematical formulation of a scientific problem often leads to a relationship among the various derivatives of some function $y(x)$. The simplest prototype is

$$\frac{dy}{dx} = f(x), \tag{1}$$

the solution of which requires finding the function $y(x)$ whose derivative is the known function $f(x)$. [Since the operation inverse to differentiation is involved here, we note in passing that this is equivalent to finding the integral of $f(x)$.]

One method of examining the solution in a gross qualitative way, but without much effort, is a graphical procedure called the *direction field.* This technique makes use of the little sections of the lines tangent to the curves which satisfy (1). For example, at a point $P = (x,y)$, the slope of the curve $y(x)$ is $f(x)$, and in Fig. 11.1 this is indicated by a small line segment having this direction at P. According to (1), the slope at a fixed value of x would necessarily be the same for all y, as shown in Fig. 11.1*b*. If now the slopes at other positions are also entered upon the graph (Fig. 11.1*c*), the emergent pattern is sufficient to draw, freehand, the particular curve which not only satisfies (1) but also passes through a prescribed point. It seems clear that there is a unique curve with this property. This curve is called the *solution curve*.

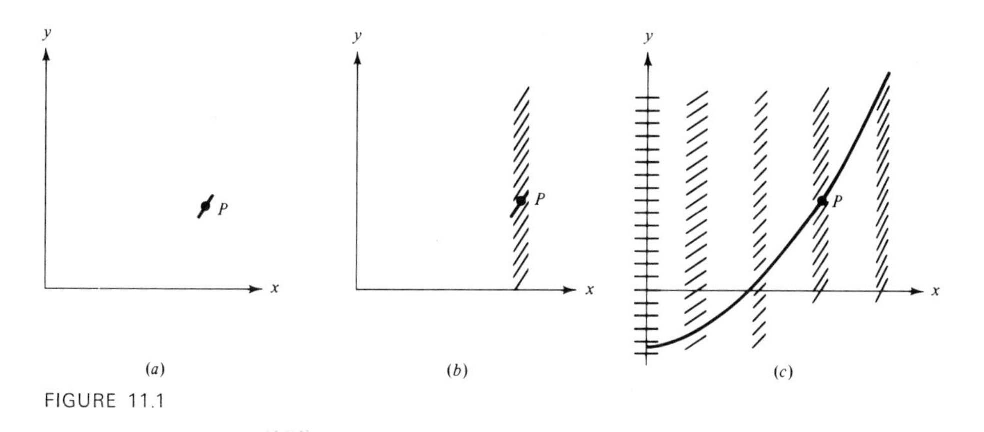

FIGURE 11.1

Example 1 Construct the direction field for $dy/dx = 1/x$.

The value of $1/x$ gives the slope at all points (x, y), and the field directed line segment is as shown in Fig. 11.2. In particular, the curve through $(1, 0)$ is the natural logarithm $y = \log x$. Note that all solutions in the half-plane $x > 0$ are singular at the origin. All solution curves are of the form $y = \log|x| + a$ where a is a constant.

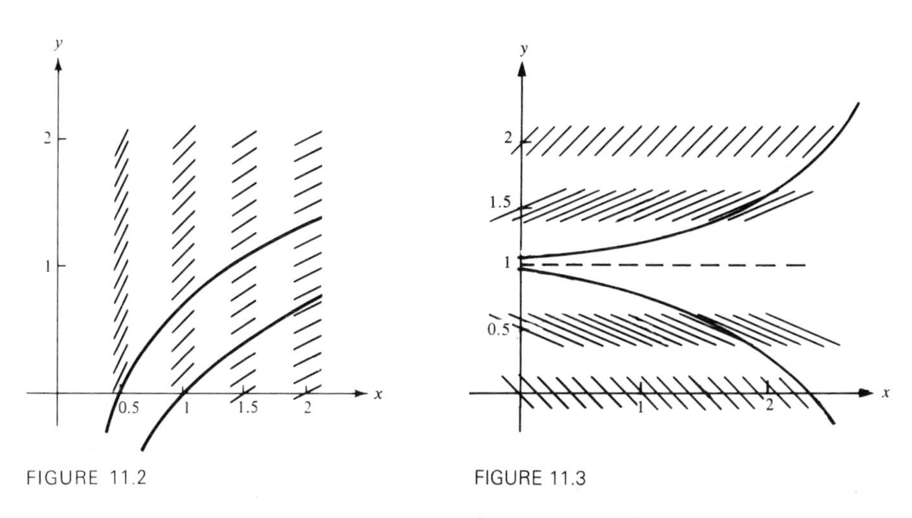

FIGURE 11.2

FIGURE 11.3

Example 2 Construct the direction field for $dy/dx = y - 1$.

The direction field is drawn in Fig. 11.3. The special case $y = 1$ is the only constant solution. If y is greater than 1 at some point, then y is an increasing function of x. If y is less than 1 at any point, y is a decreasing function. All solution curves are given by $y = 1 + ae^x$ where a is a constant.

The solutions of much more difficult problems can be sketched by this method, which is especially suited for equations of the form $y' = F(x,y)$. Here the slope of the solution curve varies with position in the xy plane according to the given function of two variables $F(x,y)$. The procedure, however, is the same, in that the slopes at a sufficient number of points are placed on the graph paper and an appropriate curve is drawn consistent with the total pattern. The curves $F(x,y) = c$ where c is an arbitrary constant are called the *isoclines*. The direction field has slope c at every point on the isocline $F(x,y) = c$.

Example 3 Sketch the solution of $y' = x + y$ which passes through (0,0).

The direction field and approximate solution are shown in Fig. 11.4. By first plotting the isoclines $x + y = c$, the direction field can be drawn very quickly.

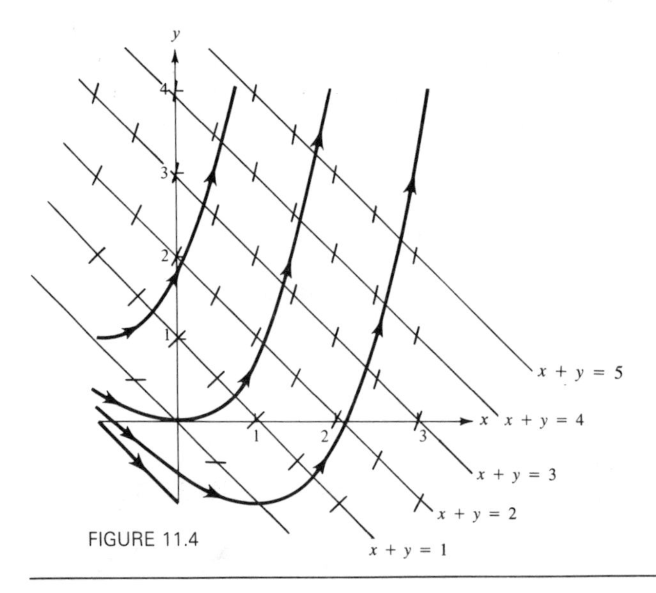

FIGURE 11.4

It is easy enough to translate this graphical procedure into a quantitative numerical technique for the solution of differential equations, but this will be pursued later.

---

2 Differentiation of a function of more than one variable, such as

$$z = f(x, y),$$

can be defined with respect to either of the independent variables. Thus, the ratio

$$\frac{f(x + \Delta x, y) - f(x, y)}{\Delta x}$$

for fixed y becomes, in the limit $\Delta x \to 0$, the derivative of $f(x, y)$ with respect to x. In order to indicate that derivatives with respect to other variables are possible, this limit is called a *partial derivative* and is denoted by

$$\frac{\partial f}{\partial x} = \lim_{\Delta x \to 0} \frac{f(x + \Delta x, y) - f(x, y)}{\Delta x}.$$

(Read $\partial f/\partial x$ as "the partial derivative of f with respect to x.") Similarly,

$$\frac{\partial f}{\partial y} = \lim_{\Delta y \to 0} \frac{f(x, y + \Delta y) - f(x, y)}{\Delta y}.$$

The definition is easily extended to higher partial derivatives as well, and

$$\frac{\partial^2 f}{\partial x\, \partial y}$$

means that $f(x, y)$ is differentiated twice, once with respect to x and once with respect to y; the order in which this is executed is not important, but this fact will require proof, i.e.,

$$\frac{\partial}{\partial y}\left(\frac{\partial f}{\partial x}\right) = \frac{\partial}{\partial x}\left(\frac{\partial f}{\partial y}\right).$$

Example 4 Find

$$\frac{\partial f}{\partial x}, \frac{\partial f}{\partial y}, \frac{\partial^2 f}{\partial x\, \partial y} \qquad \text{for} \qquad f = 2 + x^2 + 3xy^2 + y^3 + 2x^3y^4.$$

The partial derivative with respect to x is calculated as if x were the sole independent variable (and y is some constant). Therefore

$$\frac{\partial f}{\partial x} = 2x + 3y^2 + 6x^2y^4.$$

Likewise, the partial derivative with respect to y (holding x fixed) is

$$\frac{\partial f}{\partial y} = 6xy + 3y^2 + 8x^3y^3.$$

If now the process is repeated, we find that

$$\frac{\partial}{\partial y}\left(\frac{\partial f}{\partial x}\right) = \frac{\partial}{\partial y}(2x + 3y^2 + 6x^2y^4) = 6y + 24x^2y^3$$

whereas

$$\frac{\partial}{\partial x}\left(\frac{\partial f}{\partial y}\right) = \frac{\partial}{\partial x}(6xy + 6y^2 + 8x^3y^3) = 6y + 24x^2y^3.$$

The result is seen not to depend on the ordering of derivatives, and $\partial^2 f/\partial x\, \partial y$ is defined uniquely.

Example 5 Calculate

$$\frac{\partial z}{\partial x}, \frac{\partial z}{\partial y}, \qquad \text{where} \qquad x^2z + 2y^2z^2 + y = 0.$$

Here the function $z = f(x, y)$ is defined implicitly, and the partial derivatives will be obtained by implicit differentiation. Holding y fixed, we find that

$$2xz + x^2\frac{\partial z}{\partial x} + 4y^2z\frac{\partial z}{\partial x} = 0$$

so that

$$\frac{\partial z}{\partial x} = \frac{-2xz}{x^2 + 4y^2z}.$$

A similar calculation yields

$$\frac{\partial z}{\partial y} = -\frac{1 + 4yz^2}{x^2 + 4y^2z}.$$

All the topics discussed in connection with ordinary differentiation have analogs and generalizations for partial derivatives; these will be taken up in Chap. 6.

Exercises 2.11

1. Sketch the direction fields for the following differential equations. Compare the approximate graphical solutions with the known exact solutions.

(i) $\frac{dy}{dx} = y$ **(ii)** $\frac{dy}{dx} = x$ **(iii)** $\frac{dy}{dx} = \cos x$

(iv) $\frac{dy}{dx} = \frac{1}{x^2}$ **(v)** $\frac{dy}{dx} = e^{-x}$ **(vi)** $\frac{dy}{dx} = \sinh x$

2. Construct the direction fields for the following equations and plot the solution that passes through the point $(0, 1)$. Where do the solutions have maxima and minima?

(i) $\frac{dy}{dx} = 1 - x - y$ **(ii)** $\frac{dy}{dx} = x^2 + y^2$ **(iii)** $\frac{dy}{dx} = \tanh x$

3. Show that the solutions of $dy/dx = f(y/x)$ have the same slope on any line passing through the origin. Sketch the direction fields of the following equations and plot the solution that passes through the point $(1,0)$.

(i) $\frac{dy}{dx} = \frac{y}{x}$ **(ii)** $\frac{dy}{dx} = \frac{-x}{y}$ **(iii)** $\frac{dy}{dx} = \frac{y}{x} + \left(\frac{y}{x}\right)^2$

4. Construct the direction field for the differential equation $dy/dx = y^{1/2}$ defined for $y \geq 0$. Show that there are two solutions that pass through $(0, 0)$.

5. From the direction field for the equation $y' = x + y$ (Example 3), show that all solutions are asymptotic to the line $x + y = -1$.

6. How does the direction field method relate to the finite difference approximation of $y' = F(x,y)$ given by $y_{n+1} - y_n = hF_n = hF(x_n, y_n)$ where $x_n = nh$? Illustrate your answer with the example $y' = y$ on $0 \le x \le 1$.
7. Construct the direction field for the equation $y' = y + 1 - x$ and draw the approximate solution curve that passes through the origin. Compare the approximate solution to the exact solution $y = x$.
8. Construct the direction field for the equation $y' = y^2$ and draw the approximate solution curve through the point $(0, 1)$. Compare the approximation to the graph of the exact solution $y = 1/(1 - x)$.
9. Sketch the direction field for the equation $y' = y(1 - y)$. Show that no solution curve can cross either the line $y = 0$ or the line $y = 1$. Show that solutions in the upper half plane approach the line $y = 1$ as x increases and solutions in the lower half plane approach the line $y = 0$ as x decreases.
10. Find the partial derivatives $\partial z/\partial x$ and $\partial z/\partial y$ for the following functions:

(i) $z = x^2 + y^2$ **(ii)** $z = x + y + xy$ **(iii)** $z = (x^2 + y^2)^2$
(iv) $z = \sin x \cos y$ **(v)** $z = ye^x$ **(vi)** $z = \log xy$
(vii) $z = \sinh(x + y)$ **(viii)** $z = e^x \cosh y$ **(ix)** $z = \tanh(x^2 - y^2)$

11. For the functions $z = z(x,y)$ in the previous exercise, calculate the second derivatives $\dfrac{\partial^2 z}{\partial x^2}, \dfrac{\partial^2 z}{\partial y^2}, \dfrac{\partial}{\partial y}\left(\dfrac{\partial z}{\partial x}\right)$ and $\dfrac{\partial}{\partial x}\left(\dfrac{\partial z}{\partial y}\right)$. Verify that the last two derivatives are equal.
12. If $z = x^2 + y^2 + \sin xy$, find the values at the origin of the following derivatives:

(i) $\dfrac{\partial z}{\partial x}$ **(ii)** $\dfrac{\partial z}{\partial y}$ **(iii)** $\dfrac{\partial}{\partial x}\left(\dfrac{\partial z}{\partial x}\right) = \dfrac{\partial^2 z}{\partial x^2}$

(iv) $\dfrac{\partial}{\partial y}\left(\dfrac{\partial z}{\partial y}\right) = \dfrac{\partial^2 z}{\partial y^2}$ **(v)** $\dfrac{\partial}{\partial y}\left(\dfrac{\partial z}{\partial x}\right) = \dfrac{\partial^2 z}{\partial x \partial y}$ **(vi)** $\dfrac{\partial}{\partial x}\left(\dfrac{\partial z}{\partial y}\right) = \dfrac{\partial^2 z}{\partial y \partial x}$.

13. Find $\partial w/\partial x$ and $\partial w/\partial y$ for each of the following functions:

(i) $w = (x^3 - y)^2 + xy$ **(ii)** $w^2 = x^2 + y^2$ **(iii)** $e^w = x^2 + y^2$
(iv) $\log w = \sinh xy$ **(v)** $xyw = 1$ **(vi)** $e^w + (x^2 + y^2)w + 6 = 0$

14. If $\sin xz = y$, find the values of the following derivatives at $x = 1$, $y = 1/2$, $z = \pi/6$. Evaluate these derivatives by implicit differentiation and by solving for $z = z(x,y)$.

(i) $\dfrac{\partial z}{\partial x}$ **(ii)** $\dfrac{\partial z}{\partial y}$ **(iii)** $\dfrac{\partial^2 z}{\partial x^2}$

(iv) $\dfrac{\partial^2 z}{\partial y^2}$ **(v)** $\dfrac{\partial}{\partial y}\left(\dfrac{\partial z}{\partial x}\right)$ **(vi)** $\dfrac{\partial}{\partial x}\left(\dfrac{\partial z}{\partial y}\right)$

15. If $u = u(x,t) = \sin(x - ct)$ where c is a constant, show that $\partial u/\partial t + c\partial u/\partial x = 0$. Show that this result holds for $u(x,t) = f(x - ct)$ where f is any differentiable function.
16. If $u(x,t) = f(x - ct) + g(x + ct)$ where c is a constant and f and g are differentiable functions, show that $\partial^2 u/\partial t^2 = c^2 \partial^2 u/\partial x^2$.
17. In Example 5, $z = z(x,y)$ is defined implicitly by the equation $x^2 z + 2y^2 z^2 + y = 0$. Solve for z as an explicit function of x and y and verify that $z = z(x,y)$ is defined only if $x^4 \ge 8y^3$. Draw a graph to indicate this domain of definition.
18. Given a function $z = z(x,y)$, suppose that $x = x(t)$ and $y = y(t)$ are functions of a variable t. This defines z as a function of t, $z(t) = z(x(t), y(t))$. Derive, from first principles, the *chain rule* for partial differentiation:

$$\frac{dz}{dt} = \frac{\partial z}{\partial x}\frac{dx}{dt} + \frac{\partial z}{\partial y}\frac{dy}{dt}.$$

19. Calculate dz/dt for the following functions $z = z(x,y)$, $x = x(t)$, $y = y(t)$:

(i) $z = x^2 + y^2$, $x = e^t$, $y = e^{-t}$ **(ii)** $z = \log(x + y)$, $x = t$, $y = t^{-1}$

(iii) $z = \cosh xy$, $x = t^{1/2}$, $y = t^{1/2}$ **(iv)** $z = e^{x^2+y^2}$, $x = \cos t$, $y = \sin t$

20. The chain rule gives a formula for the differential dz of $z = z(x,y)$ in terms of dx and dy; $dz = (\partial z/\partial x)dx + (\partial z/\partial y)dy$. Evaluate the differentials dz for the functions defined in the previous problem.

3 Integration

3.1 The integral

1

A method was developed in Chap. 1 for calculating an arbitrary area as the limit of a sequence of simple approximating figures. The concept of an integral was introduced in this way as a generalized form of summation which involved "an infinite number of infinitesimally small" areas. Although the discussion there was brief and exploratory in nature, it did touch upon several important topics, such as existence and uniqueness of integrals, some general properties of definite and indefinite integrals, and the relationship of integration to differentiation. Moreover, the integrals of several elementary functions were calculated from first principles and their use illustrated in specific applications dealing with areas, volumes, distances and rates of change. All of that can be regarded as motivation for this chapter, with the added advantage that the results already obtained can be used now to facilitate further development of the subject. However, we essentially begin again because the concepts are so important.

Consider anew the calculation of the area A under the curve $y = f(x)$ bounded by the lines $x = a$, $x = b$, and $y = 0$, as shown in Fig. 1.1. The interval $a \leq x \leq b$ is divided into n sections or partitioned into n subintervals. The points of subdivision are denoted by

$$a = x_0 < x_1 < \cdots < x_{i=1} < x_i < \cdots < x_{n-1} < x_n = b. \tag{1}$$

The simplest choice is to make the subintervals

$$\Delta x_i = x_i - x_{i-1} \qquad \text{for} \qquad i = 1, 2, \ldots, n$$

of equal length. In this case, we set

$$x_i = a + \frac{i}{n}(b - a), \tag{2}$$

so that

$$\Delta x_i = \Delta x = \frac{b - a}{n}.$$

Furthermore, let ξ_i be any point in the subinterval $x_{i-1} \le \xi_i \le x_i$; then $f(\xi_i)(x_i - x_{i-1})$ is the area of the ith approximating rectangle shown in the figure. The sum of the areas of all n such rectangles,

$$A_n = \sum_{i=1}^{n} f(\xi_i)(x_i - x_{i-1}), \tag{3}$$

constitutes an approximation which in the limit $n \to \infty$ and $|x_i - x_{i-1}| \to 0$ (for each i) yields the exact area in question. This limit, which is the area A, defines more generally the *definite integral* of $f(x)$ between a and b, that is,

$$A = \int_a^b f(x)\, dx. \tag{4}$$

Here, $f(x)$ is called the *integrand* and a, b are the *lower and upper limits* of integration. In the particular sequence of approximations for which all distance increments are the same length, the definition becomes

$$A = \lim_{n\to\infty} \sum_{i=1}^{n} f(\xi_i) \frac{(b-a)}{n} = \lim_{n\to\infty} \sum_{i=1}^{n} f(\xi_i)\, \Delta x = \int_a^b f(x)\, dx. \tag{5}$$

The integral notation reflects the deployment of symbols in the finite summation, and the position of b, above, makes it the upper limit of integration while a becomes the lower limit of integration (and Δx is transformed to dx). In order to determine the correct limits of integration, it may be helpful to imagine that the area is an assemblage of adjacent vertical lines or strips of different lengths added one by one until they exactly cover the figure (and nothing else). Each strip in Fig. 1.2 is then the graphical representation of the infinitesimal area $f(x)\, dx$. The placement positions of the first line, or strip, on the extreme left and the last line on the extreme right of the area correspond respectively to the lower and upper limits of integration.

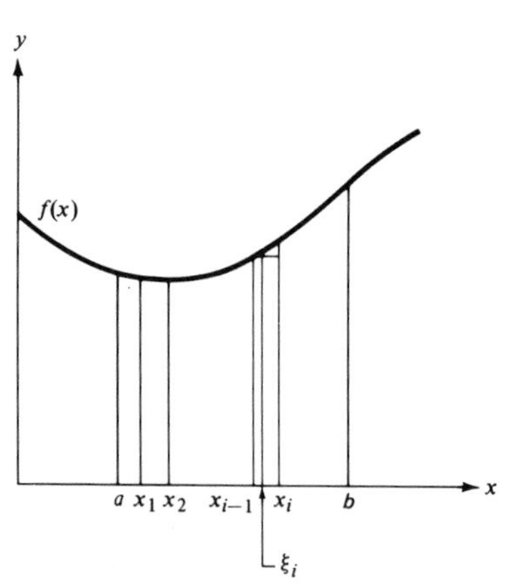

FIGURE 1.1

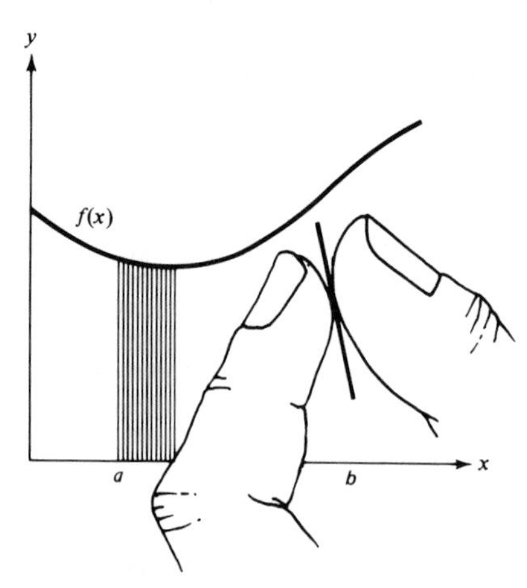

FIGURE 1.2

In Section 1.9, the integrals of $y = x$ and $y = x^2$ were found using the limit definition. Formulas for the integrals of more complicated functions can sometimes be found by this method although the effort may be very considerable as illustrated in the following examples.

Example 1 Evaluate $A = \int_a^b e^x dx$.

This area (Fig. 1.3) is evaluated by dividing the interval $[a,b]$ into n subintervals of equal length $h = (b-a)/n$ with endpoints $x_i = a + ih$ for $i = 0,1,2,\ldots,n$. Choosing $\xi_i = x_i$ in each subinterval, the integral is approximated by the sum of the areas of n rectangles, each with base h and height $e^{\xi_i} = e^{x_i}$. The area A_n of these rectangles is

$$A_n = \sum_{i=1}^{n} f(\xi_i)(x_i - x_{i-1}) = \sum_{i=1}^{n} e^{a+ih}h = e^a h \sum_{i=1}^{n} e^{ih}.$$

The last sum is a finite geometic series and, therefore,

$$A_n = e^a he^h \frac{e^{nh} - 1}{e^h - 1} = e^a(e^{b-a} - 1)\frac{he^h}{e^h - 1}.$$

This is an exact formula for A_n. As n increases, $h = (b-a)/n$ approaches zero. Applying l'Hôpital's rule,

$$\begin{aligned}\lim_{n\to\infty} A_n &= (e^b - e^a)\lim_{h\to 0}\frac{he^h}{e^h - 1}\\ &= (e^b - e^a)\lim_{h\to 0}\frac{e^h + he^h}{e^h} = e^b - e^a.\end{aligned}$$

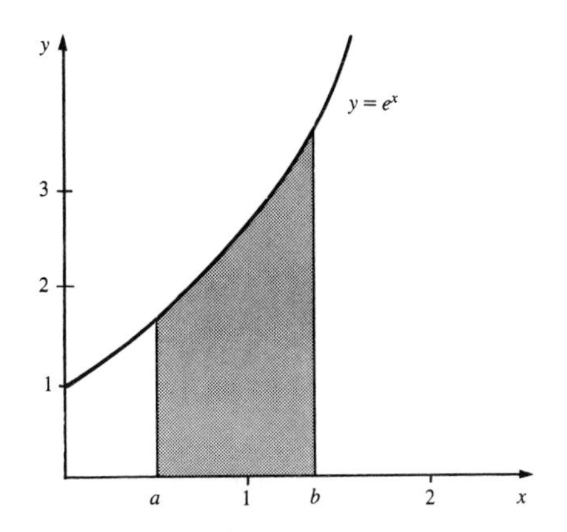

FIGURE 1.3

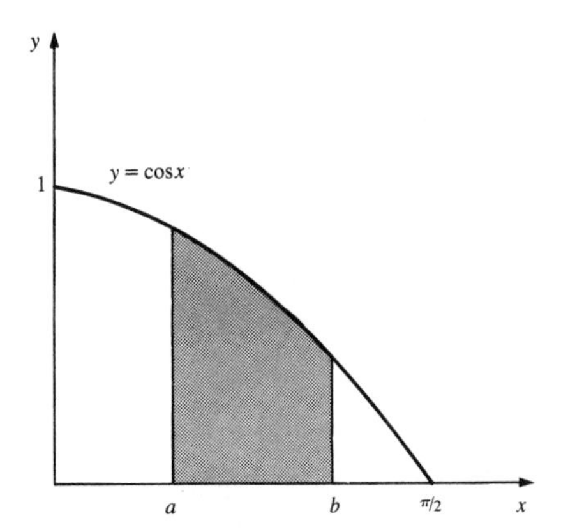

FIGURE 1.4

The final result has a very simple formula,

$$A = \lim_{n\to\infty} A_n = \int_a^b e^x dx = e^b - e^a.$$

Example 2 Evaluate $A = \int_a^b \cos x\, dx$.

Using the same partition of the interval $[a, b]$, (Fig. 1.4)

$$A_n = \sum_{i=1}^{n} f(\xi_i)(x_i - x_{i-1}) = h\sum_{i=1}^{n} \cos(a + ih).$$

To evaluate this sum exactly requires some ingenuity. Recalling the identity

$$2\cos\left(\frac{\theta+\phi}{2}\right)\sin\left(\frac{\theta-\phi}{2}\right) = \sin\theta - \sin\phi,$$

we write

$$2\cos(a + ih)\sin\left(\frac{h}{2}\right) = \sin\left(a + \frac{(2i+1)h}{2}\right) - \sin\left(a + \frac{(2i-1)h}{2}\right).$$

Using this result, the formula for A_n simplifies considerably.

$$\begin{aligned} A_n &= \frac{h}{2\sin(h/2)} \sum_{i=1}^{n} \left[\sin\left(a + \frac{(2i+1)h}{2}\right) - \sin\left(a + \frac{(2i-1)h}{2}\right)\right] \\ &= \frac{h}{2\sin(h/2)} \left[\sin\left(b + \frac{h}{2}\right) - \sin\left(a + \frac{h}{2}\right)\right]. \end{aligned}$$

An easy application of l'Hôpital's rule gives

$$\lim_{h\to 0} \frac{h}{2\sin(h/2)} = \lim_{h\to 0} \frac{1}{\cos(h/2)} = 1.$$

Therefore, in the limit $n\to\infty$,

$$A = \lim_{n\to\infty} A_n = \int_a^b \cos x\, dx = \sin b - \sin a.$$

Once again, the final result is given by a simple formula.

In the previous chapter, a few general rules of differentiation made it possible to calculate many derivatives with very little effort. Using the limit definition to calculate the derivative of each new function would have been much more time consuming. This is also true of the problem of evaluating integrals which have a much more complicated limit definition. The purpose of this chapter is to develop efficient methods of calculating integrals by establishing the general properties of the integration process.

2

That the limit posed in (3) or (5) is uniquely defined no matter how the numbers ξ_i and Δx_i are chosen really requires proof, although the interpretation of an integral as an area is very convincing in this regard. As a matter of fact, consideration of the areas represented by integrals suggests many of their important properties. For example, the total area below the curve $y = f(x)$ shown in Fig. 1.5 can be viewed as the sum of two (or more) smaller areas corresponding to $a \le x \le c$ and $c \le x \le b$. This implies the relationship

$$\int_a^b f(x)\,dx = \int_a^c f(x)\,dx + \int_c^b f(x)\,dx.$$

Moreover, if $m \le f(x) \le M$ for $a \le x \le b$ (m and M are lower and upper bounds for the function) then from Fig. 1.6 it is clear that the area under the curve is larger than that of the interior rectangle but smaller than the area of the exterior rectangle:

$$m(b - a) \le \int_a^b f(x)\,dx \le M(b - a). \tag{6}$$

The mathematical statement

$$\int_c^c f(x)\,dx = 0$$

is equivalent to the observation that the area under a single point (say P in Fig. 1.5) is obviously zero.

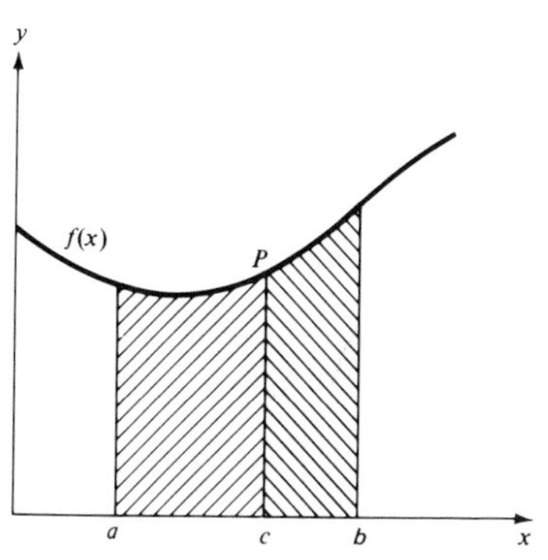

FIGURE 1.5

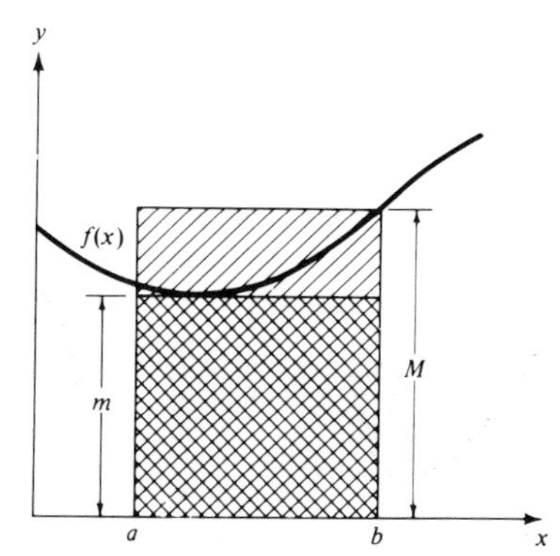

FIGURE 1.6

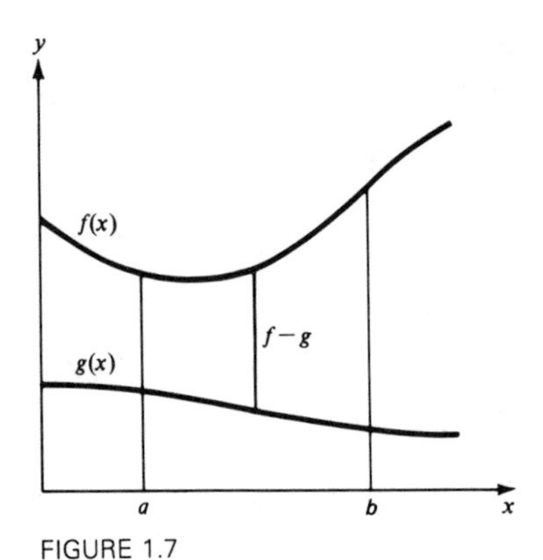

FIGURE 1.7

The area between the curves $y = f(x)$ and $y = g(x)$ and $x = a, x = b$ (Fig. 1.7) can be calculated as the area under the curve $f(x)$ (above $y = 0$) less that under the curve $g(x)$. (Imagine vertical strips of length $f - g$ placed side by side from $x = a$ to $x = b$ until they completely cover the area in question.) Formally

$$\int_a^b [f(x) - g(x)]\,dx = \int_a^b f(x)\,dx - \int_a^b g(x)\,dx.$$

A slight qualification is required in as much as $f(x)$ can be negative but area is always a positive quantity. If $f(x) < 0$ in the interval $\alpha \leq x \leq \beta$, the integral over this range, $\int_\alpha^\beta f(x)\,dx$, represents the *negative* of the area bounded by the curves $y = f(x)$, $y = 0$, $x = \alpha$, $x = \beta$. The integral of a function which changes sign in an interval (Fig. 1.8*a*) is equal to the area lying above the interval $[a, c]$ minus that area lying below the section $[c, b]$, and as such the final value can be positive, negative, or zero. In this case, the total area bounded by $y = f(x)$, $y = 0$ and $x = a$, $x = b$ is determined from the integral of the positive function $|f(x)|$ (see Fig. 1.8*b*) as

$$\int_a^b |f(x)|\,dx.$$

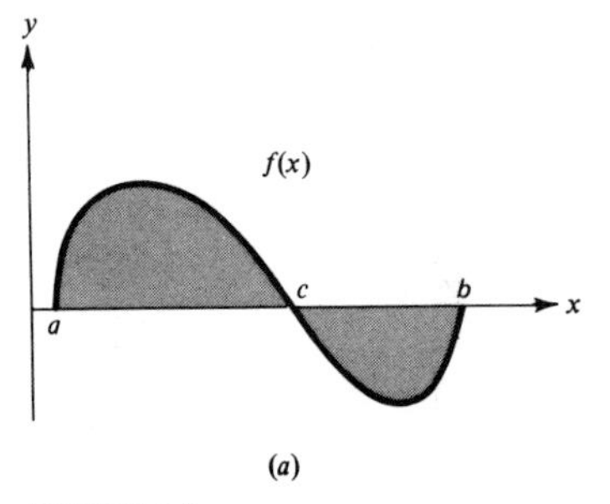

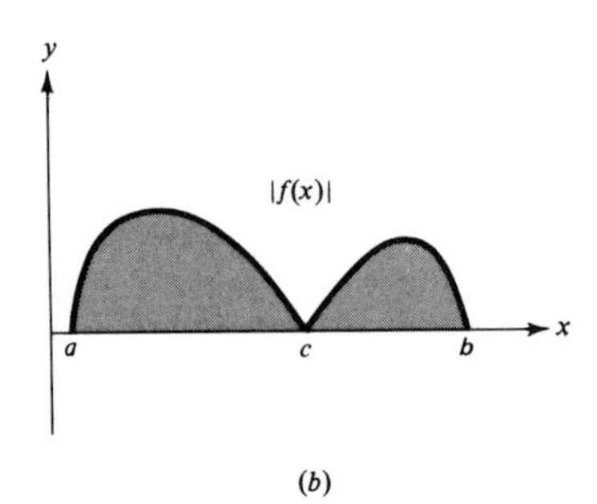

FIGURE 1.8

Clearly then

$$\left|\int_a^b f(x)\,dx\right| \le \int_a^b |f(x)|\,dx,$$

since the first integral here involves a possible cancellation of areas and the latter does not.

Table 1.1 lists simple properties of integrals inferred this way.

TABLE 1.1

1. $\displaystyle\int_a^b Cf(x)\,dx = C\int_a^b f(x)\,dx$
2. $\displaystyle\int_a^b [f(x) \pm g(x)]\,dx = \int_a^b f(x)\,dx \pm \int_a^b g(x)\,dx$
3. $\displaystyle\int_a^b f(x)\,dx = \int_a^c f(x)\,dx + \int_c^b f(x)\,dx$
4. $\displaystyle\int_a^b f(x)\,dx = -\int_b^a f(x)\,dx$
5. $\displaystyle\left|\int_a^b f(x)\,dx\right| \le \int_a^b |f(x)|\,dx$ for $b > a$
6. $\displaystyle\int_a^b f(x)\,dx \le \int_a^b g(x)\,dx$ if $f(x) \le g(x)$ in $a \le x \le b$
7. $\displaystyle\int_a^a f(x)\,dx = 0$

3

The formal proofs of these are rather easy and follow from the analogous properties for finite sums. For simplicity, we assume a partition of interval $[a, b]$ in increments of uniform length Δx. To illustrate, consider formula **1** in Table 1.1:

$$\int_a^b Cf(x)\,dx = \lim_{n\to\infty}\sum_{i=1}^n Cf(\xi_i)\,\Delta x = \lim_{n\to\infty} C\sum_{i=1}^n f(\xi_i)\,\Delta x,$$

$$= C\lim_{n\to\infty}\sum_{i=1}^n f(\xi_i)\,\Delta x = C\int_a^b f(x)\,dx.$$

Formula **4** follows by rewriting the fundamental definition as

$$\int_a^b f(x)\,dx = \lim_{n\to\infty}\sum_{i=1}^n f(\xi_i)\frac{b-a}{n} = \lim_{n\to\infty} -\sum_{i=1}^n f(\xi_i)\frac{a-b}{n}$$

$$= -\left[\lim_{n\to\infty}\sum_{i=1}^n f(\xi_i)\frac{a-b}{n}\right]. \tag{7}$$

The finite sum in (7) is an addition over *all n* subsegments which make up the (negative) length from b to a, and the points ξ_i are distributed, by definition, one to each of the subintervals. The placement of a and b indicates that the limit involved is, by (5), the integral

$$\int_b^a f(x)\,dx,$$

so that the preceding equation then reads

$$\int_a^b f(x)\,dx = -\int_b^a f(x)\,dx.$$

[The right-hand integral here can be interpreted as having a negative differential dx, which is the limit of $\Delta x = (a - b)/n$. In this view, the integral is equivalent to a *reverse calculation* of area starting at b and proceeding to a.]

Formula **5** follows from the generalized triangle inequality;

$$\left|\sum_{i=1}^n f(\xi_i)(x_i - x_{i-1})\right| \le \sum_{i=1}^n |f(\xi_i)|\,|x_i - x_{i-1}|,$$

which is also valid in the limit as $n \to \infty$:

$$\left|\int_a^b f(x)\,dx\right| = \left|\lim_{n\to\infty}\sum_{i=1}^n f(\xi_i)\,\Delta x\right| = \lim_{n\to\infty}\left|\sum_{i=1}^n f(\xi_i)\,\Delta x\right|$$

$$\le \lim_{n\to\infty}\sum_{i=1}^n |f(\xi_i)|\,|\Delta x| = \int_a^b |f(\xi_i)|\,dx.$$

(The assumption $b > a$ implies $|\Delta x| = \Delta x$ and $|dx| = dx$.)

If $f(x) \le g(x)$ for all x in the integration range, it follows that

$$\sum_{i=1}^n f(\xi_i)(x_i - x_{i-1}) \le \sum_{i=1}^n g(\xi_i)(x_i - x_{i-1}),$$

and this yields formula **6** in the limit as $n \to \infty$.

---

4 The *indefinite integral*

$$F(x) = \int_a^x f(\mathbf{x})\,d\mathbf{x}, \tag{8}$$

has a variable upper limit of integration x and is therefore a function of x. It represents the area under the curve between two vertical lines, one of which is prescribed while the other may vary. The integration variable has been changed to distinguish it from the arbitrary limit of integration. This frees x for use again as the important independent variable. Although *any* symbol could be employed instead of $\mathbf{x}$ for the *dummy variable* (read $\mathbf{x}$ as x-bold) this choice is made for obvious reasons. It pays to think of the indefinite integral as an area in the xy plane that is bounded by $y = f(x)$, $y = 0$ and the vertical lines $x = a$, $x = x$ (Fig. 1.9). In this view, a particular value of x is just a point on the x axis.

The indefinite integral is a legitimate function $F(x)$ of the independent variable x, which simply means that for any particular value, say $x = b$, $F(b)$ can be calculated. The definite integral is then just an indefinite integral that is evaluated at a particular point. The properties established for definite integrals are valid for indefinite integrals as well, and only a notational exchange of x for b is required, e.g.,

$$\int_a^x f(x)\,dx = -\int_x^a f(x)\,dx.$$

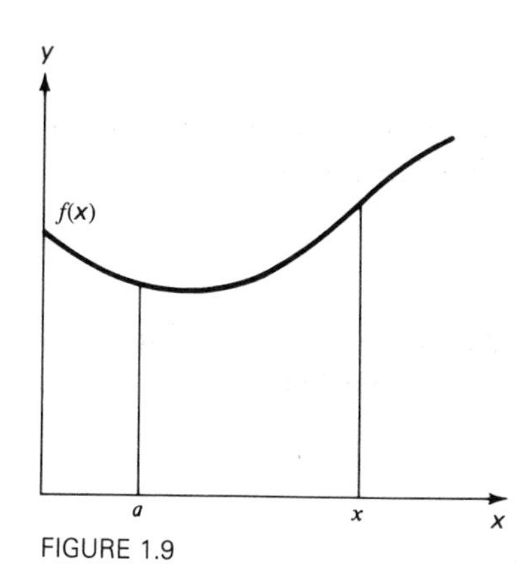

FIGURE 1.9

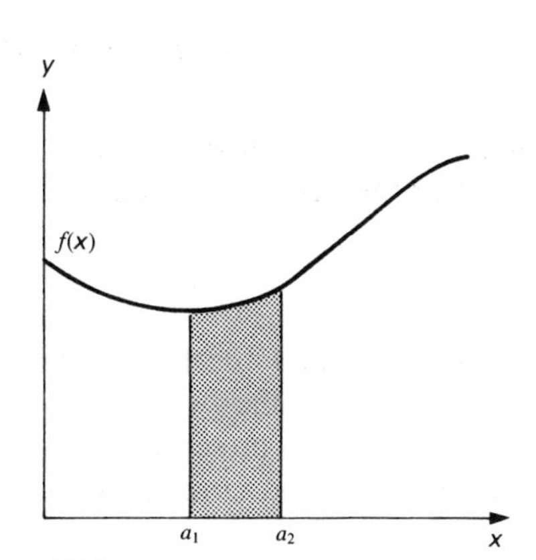

FIGURE 1.10

The indefinite integral also depends on its lower limit of integration a, which plays the role of a parameter. A family of indefinite integrals is obtained by varying a. However, any two indefinite integrals of the same function differ by only a constant, so that the family relationship is fairly simple. This result is easily established if we set

$$F_1(x) = \int_{a_1}^x f(x)\,dx, \qquad F_2(x) = \int_{a_2}^x f(x)\,dx,$$

and use formulas **3** and **4** (Table 1.1) to write

$$\begin{aligned} F_1(x) - F_2(x) &= \int_{a_1}^x f(x)\,dx - \int_{a_2}^x f(x)\,dx = \int_{a_1}^x f(x)\,dx + \int_x^{a_2} f(x)\,dx \\ &= \int_{a_1}^{a_2} f(x)\,dx = C. \end{aligned} \tag{9}$$

Equation (9) is the desired conclusion. The constant C is the shaded area in Fig. 1.10, the area between $x = a_1$ and $x = a_2$. To evaluate the constant C, simply set x equal to any value, say $x = \beta$. Then

$$F_1(\beta) - F_2(\beta) = C,$$

and the substitution of this in (9) yields a symmetric formula

$$F_1(x) - F_1(\beta) = F_2(x) - F_2(\beta). \tag{10}$$

The significance of all this is that a definite integral, $\int_r^s f(\mathsf{x})\, d\mathsf{x}$, can be evaluated in terms of *any indefinite integral* like (8), and the choice of a lower limit of integration has absolutely no effect on this particular calculation. To demonstrate this, formulas **3** and **4** are used once again in the manipulation

$$\int_r^s f(\mathsf{x})\, d\mathsf{x} = \int_a^s f(\mathsf{x})\, d\mathsf{x} - \int_a^r f(\mathsf{x})\, d\mathsf{x},$$

which by (8) implies

$$\int_r^s f(\mathsf{x})\, d\mathsf{x} = F(s) - F(r).$$

But according to (10), the right-hand side is a constant independent of a, and the proof is complete. For this reason, it is useful to write the indefinite integral as

$$\int^x f(\mathsf{x})\, d\mathsf{x} = F(x) + C, \tag{11}$$

because this separates the important functional dependence on x from the arbitrary additive constant [the specification of which designates one indefinite integral of the family and is equivalent to prescribing a in (8)].

Another convenient shorthand notation is

$$F(\mathsf{x})\Big]_a^b = F(b) - F(a),$$

and using this, we can write

$$\int_a^b f(\mathsf{x})\, d\mathsf{x} = F(\mathsf{x})\Big]_a^b = F(b) - F(a), \tag{12}$$

where $F(\mathsf{x})$ is *any* indefinite integral of the function $f(\mathsf{x})$. In particular, note that $F(\mathsf{x})]_a^b = F(x)]_a^b$, because the name of the dummy variable employed in this computation is irrelevant. Indeed the regular variable x can, and should replace its bold counterpart x in (12)—and everywhere else for that matter—because two similar type styles are troublesome to print and to write. However, this notational distinction is adopted *temporarily* to avoid possible confusion about the integration variable and the variable upper limit of an indefinite integral. Both symbols are used in this section, often interchangeably, but x is dropped shortly as soon as these new concepts are more familiar. Thereafter, the indefinite integral is written simply as $\int f(x)\, dx$.

Example 3 From the results calculated in Sec. 1.9, namely,

(i) $\int_0^x d\mathsf{x} = x$, (ii) $\int_0^x \mathsf{x}\, d\mathsf{x} = \tfrac{1}{2}x^2$, (iii) $\int_0^x \mathsf{x}^2\, d\mathsf{x} = \tfrac{1}{3}x^3$, find the values of

$$\int_a^x dx, \qquad \int_a^x x\,dx, \qquad \int_a^x x^2\,dx.$$

Equation (12) is applied in each case. In (i), we identify $F(x) = x$, so that $F(a) = a$ and

$$\int_a^x dx = F(x) - F(a) = x - a. \tag{13}$$

In (ii), let $F(x) = \frac{1}{2}x^2$; since $F(a) = \frac{1}{2}a^2$, we obtain

$$\int_a^x x\,dx = \tfrac{1}{2}(x^2 - a^2). \tag{14}$$

In (iii), let $F(x) = \frac{1}{3}x^3$, so that $F(a) = \frac{1}{3}a^3$ and, consequently,

$$\int_a^x x^2\,dx = \tfrac{1}{3}(x^3 - a^3). \tag{15}$$

All these results can be expressed in the form of Eq. (11):

$$\int^x dx = x + C, \qquad \int^x x\,dx = \tfrac{1}{2}x^2 + C, \qquad \int^x x^2\,dx = \tfrac{1}{3}x^3 + C, \tag{16}$$

where C in each case is an arbitrary constant.

Example 4 Evaluate

$$\textbf{(i)}\ \int_{-1}^1 dx, \qquad \textbf{(ii)}\ \int_{-1}^1 x\,dx, \qquad \textbf{(iii)}\ \int_{-1}^1 x^2\,dx.$$

Let $x = 1$, $a = -1$ in (13) to (15) to obtain:

$$\int_{-1}^1 dx = x\Big]_{-1}^1 = 1 - (-1) = 2,$$

$$\int_{-1}^1 x\,dx = \tfrac{1}{2}x^2\Big]_{-1}^1 = \tfrac{1}{2}[1 - (-1)^2] = 0,$$

$$\int_{-1}^1 x^2\,dx = \tfrac{1}{3}x^3\Big]_{-1}^1 = \tfrac{1}{3}[1 - (-1)^3] = \tfrac{2}{3}.$$

Example 5 If $f(x) = 3 + 2x - x^2$, find the indefinite integral $\int^x f(x)\,dx$ and evaluate the definite integral $\int_1^4 f(x)\,dx$.

Formula **2** (Table 1.1) and the results of Example 3 are used in writing

$$\int^x (3 + 2x - x^2)\,dx = 3\int^x dx + 2\int^x x\,dx - \int^x x^2\,dx$$

$$= 3x + x^2 - \tfrac{1}{3}x^3 + C$$

$$= F(x) + C.$$

It follows that

$$\int_1^4 (3 + 2x - x^2)\, dx = F(x)\Big]_1^4 = F(4) - F(1) = 3.$$

Example 6 Find the *area* bounded by $y = 3 + 2x - x^2$, $y = 0$, $x = 1$, and $x = 4$.

Note that the function $f(x) = 3 + 2x - x^2$ is negative in the range $x > 3$, so that the integral $\int_1^4 f(x)\, dx$, represents the *difference* of the areas labelled I and II in Fig. 1.11 and *not* their sum, which is the answer to this problem.

The total shaded area is given by

$$A = \int_1^4 |f(x)|\, dx,$$

and since

$$|f(x)| = \begin{cases} f(x) & \text{for} \quad 1 \le x \le 3, \\ -f(x) & \text{for} \quad 3 \le x \le 4, \end{cases}$$

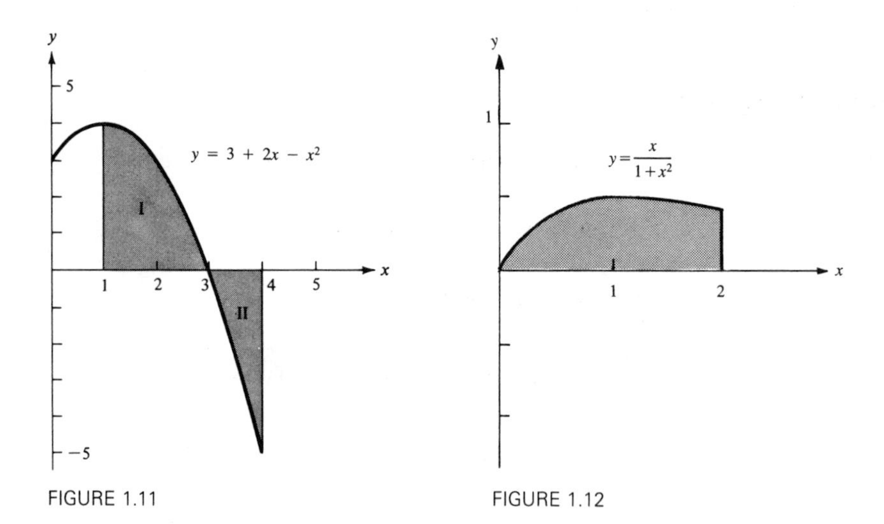

FIGURE 1.11

FIGURE 1.12

this can be expressed as

$$A = \int_1^3 |f(x)|\, dx + \int_3^4 |f(x)|\, dx$$

or

$$A = \int_1^3 f(x)\, dx - \int_3^4 f(x)\, dx.$$

The two terms on the right-hand side represent the areas of I and II, respectively, and their sum correctly accounts for the sign change of the function. These definite integrals are evaluated as in the preceding example:

$$\int_1^3 (3 + 2x - x^2)\, dx = (3x + x^2 - \tfrac{1}{3}x^3)\Big]_1^3 = \tfrac{16}{3},$$

$$\int_3^4 (3 + 2x - x^2)\, dx = (3x + x^2 - \tfrac{1}{3}x^3)\Big]_3^4 = -\tfrac{7}{3},$$

and the area is

$$A = \tfrac{16}{3} - (-\tfrac{7}{3}) = \tfrac{23}{3}.$$

Example 7 Find a bound for $I = \int_0^2 \frac{x}{1 + x^2}\, dx$.

The integrand (Fig. 1.12) is positive in the range of integration, and we can therefore dispense with the absolute-value signs required by formula **5**. The maximum value of $f(x) = x/(1 + x^2)$ occurs at $x = +1$ [where $f'(+1) = 0$] and is $f(1) = \frac{1}{2}$. Therefore, for $x > 0$, $x/(1 + x^2) \le \frac{1}{2}$ and by formula **6**

$$\int_0^2 \frac{x}{1 + x^2}\, dx \le \int_0^2 \tfrac{1}{2}\, dx = \tfrac{1}{2}(2) = 1,$$

or

$$I \le 1.$$

(The exact value is $I = \frac{1}{2} \log 5 = 0.80472\ldots$.)

Exercises 3.1

1. If p and q are constants, show that

$$\int_a^x [pf(x) + qg(x)]\, dx = p\int_a^x f(x)\, dx + q\int_a^x g(x)dx.$$

(Use the results in Table 1.1.)

2. Verify the results in Table 1.1 for $f(x) = x$, $g(x) = 1 + x^2$, $a = 1$, $b = 2$. Illustrate your answer with appropriate graphs.

3. Write as definite integrals the following limits.

(i) $\lim_{n\to\infty} \sum_{i=1}^{n} G(\xi_i)\frac{b - a}{n}$

(ii) $\lim_{n\to\infty} \left[\frac{1}{n+1} + \frac{1}{n+2} + \frac{1}{n+3} + \cdots + \frac{1}{2n}\right]$

(iii) $\lim_{n\to\infty}\left[\frac{n}{n^2+1^2}+\frac{n}{n^2+2^2}+\cdots+\frac{n}{2n^2}\right]$

4. Evaluate the following limits:

(i) $\lim_{n\to\infty}\left[\frac{n+1}{n^2}+\frac{n+2}{n^2}+\cdots+\frac{2n}{n^2}\right]$ **(ii)** $\lim_{n\to\infty}\frac{1^4+2^4+3^4+\cdots+n^4}{n^5}$

5. Are the following true or false? Explain your answers.

(i) $\int_a^b f(x)\,dx=0$ implies $f(x)=0$ in $a<x<b$.

(ii) $\int_a^b (f(x))^2\,dx=0$ implies $f(x)=0$ in $a<x<b$.

(iii) $\int_0^2 x\,dx<\int_0^2 xe^x\,dx$ **(iv)** $\int_0^{\pi/2}\sin x\,dx=\frac{1}{2}\int_0^{\pi}\sin x\,dx$.

6. Simplify the following expressions:

(i) $\int_{-3}^{-2} f(x)\,dx+\int_{-2}^{1} f(x)\,dx$ **(ii)** $\int_2^{-1} f(x)\,dx+\int_{-1}^{3} f(x)\,dx$

(iii) $\int_a^{x+\Delta x} f(z)\,dz-\int_a^{x} f(z)\,dz$ **(iv)** $\int_a^b f(x)\,dx-\int_b^a f(x)\,dx$

7. Draw a diagram of $y=mx+p$ where $ma+p$ and $mb+p$ are both positive. Use this to prove that

$$\int_a^b (mx+p)\,dx=(\tfrac{1}{2}mb^2+pb)-(\tfrac{1}{2}ma^2+pa).$$

If $ma+p<0$, what modifications are required?

8. The function $f(x)$ is periodic with period Ω if $f(x+\Omega)=f(x)$ for all x. Show that, if n is any positive integer,

$$\int_0^{n\Omega} f(x)\,dx=n\int_0^{\Omega} f(x)\,dx.$$

Draw a graph to illustrate this result.

9. If $f(x)$ is periodic with period Ω and $f(x-\Omega/2)=-f(\Omega/2-x)$, draw a graph of $f(x)$ and demonstrate that $\int_0^{\Omega} f(x)\,dx=0$. Use the example $f(x)=\sin x$ to illustrate your answer.

10. Evaluate the following expressions:

(i) $\left.\pi x^4\right]_a^b$ **(ii)** $\left.\frac{1}{x}\right]_a^b$ **(iii)** $\left[x^{1/2}-3x^2+1\right]_0^2$

(iv) $\left.f(r)\right]_{r_1}^{r_2}$ **(v)** $\left.e^{\theta}(1+\sin\theta)\right]_0^{2\pi}$ **(vi)** $\left.\log(\cos z)\right]_0^{\pi/4}$

11. Draw graphs to illustrate the areas represented by the following definite integrals. Evaluate the integrals.

(i) $\int_1^3 (x+3)\,dx$ **(ii)** $\int_1^2 |3x-4|\,dx$ **(iii)** $\int_{-1}^2 x^2\,dx$

12. The Fresnel sine integral is the function $I(x)=\int_0^x \sin t^2\,dt$. Evaluate the following in terms of $I(x)$.

(i) $\int_{\pi}^{2\pi}\sin t^2\,dt$ **(ii)** $\int_0^{1+\epsilon}\sin t^2\,dt$ **(iii)** $\frac{1}{\Delta x}\int_x^{x+\Delta x}\sin t^2\,dt$

13. Find the closed areas formed by the x axis and the curve $y=(x+3)(x-2)(x-5)$.

14. Evaluate the following definite integrals.

(i) $\int_2^3 (1-x)^2\,dx$ **(ii)** $\int_{-4}^{0} (6-x-\frac{1}{2}x^2)\,dx$ **(iii)** $\int_{\pi/4}^{\pi/2} \cos x\,dx$

15. From the areas associated with the following integrals, verify the following:

(i) $\int_0^{2\pi} \sin x\,dx = 0$ **(ii)** $\int_{-\pi/3}^{\pi/3} \sin x\,dx = 0$ **(iii)** $\int_0^{\pi/2} (\sin x - \cos x)\,dx = 0$

16. Draw a graph of the area bounded by the curve $y=x(x-1)^2$ and the lines $y=0$, $x=-\frac{1}{2}$, $x=1$. What does the definite integral $\int_{-1/2}^{2} x(x-1)^2\,dx$ represent?

17. Find bounds for the following integrals:

(i) $\int_0^3 \frac{x^2}{1+x^2}\,dx$ **(ii)** $\int_1^2 \frac{dx}{1+x^4}$ **(iii)** $\int_0^{\pi/4} \cos^2 x\,dx$

(iv) $\int_2^5 \frac{e^{-x^2}}{x^2-1}\,dx$ **(v)** $\int_0^{2\pi} e^{-x}\sin x\,dx$ **(vi)** $\int_0^{\pi/2} \frac{\cos x}{1+x^2}\,dx$

18. Express the areas bounded by the following curves as definite integrals and evaluate.

(i) $y=2+x^2$, $y=0$, $x=1$, $x=4$ **(ii)** $y=x-4x^2$, $y=0$, $x=1$, $x=3$

19. Since $2x/\pi < \sin x < x$ for $0 < x < \pi/2$, show that

$$1 < \int_0^{\pi/2} \frac{\sin x}{x}\,dx < \pi/2.$$

20. By adapting the methods of Examples 1 and 2, prove the following:

(i) $\int_a^b e^{-x}\,dx = e^{-a}-e^{-b}$ **(ii)** $\int_a^b e^{kx}\,dx = (e^{kb}-e^{ka})/k$ ($k=$ constant)

(iii) $\int_a^b 2^x\,dx = (2^b-2^a)/\log 2$ **(iv)** $\int_a^b e^{x+k}\,dx = e^{b+k}-e^{a+k}$

(v) $\int_a^b \cos(x+k)\,dx = \sin(b+k)-\sin(a+k)$ **(vi)** $\int_a^b \cos kx\,dx = (\sin kb - \sin ka)/k$

(vii) $\int_a^b \sin x\,dx = \cos a - \cos b$

(viii) $\int_a^b \cos^2 x\,dx = \frac{b-a}{2} + \frac{\sin 2b - \sin 2a}{2}$ (*Hint:* $2\cos^2 x = \cos 2x + 1$.)

21. In the limit definition of an integral, choose ξ_i to satisfy the mean-value theorem

$$F(x_i) - F(x_{i-1}) = F'(\xi_i)(x_i - x_{i-1})$$

and then prove that the integral of $F'(x)$ satisfies

$$\int_a^b F'(x)\,dx = F(b) - F(a).$$

(By using these ξ_i's, the approximation of the integral of $F'(x)$ by the sum of n rectangles becomes exact.)

22. Illustrate the result of the previous problem with the following functions:

(i) $F(x)=x^2+2x+5$ $a=0$, $b=1$

(ii) $F(x)=e^x$ $a=1$, $b=3$

(iii) $F(x)=\sin x$ $a=\pi/6$, $b=\pi/3$

23. If $g(x)=f(x)$ for $a \leq x \leq b$ except at one point $x=c$ where $g(c) \neq f(c)$ and $a \leq c \leq b$, show that

$$\int_a^b g(x)\,dx = \int_a^b f(x)\,dx.$$

Illustrate graphically why changing the value of a function at a few points does not change the value of the integral.

24. Prove the Cauchy-Schwarz inequality for integrals

$$\left(\int_a^b f(x)g(x)\,dx\right)^2 \leq \int_a^b f^2(x)\,dx \int_a^b g^2(x)\,dx.$$

(*Hint:* Write $F(t) = \int_a^b [tf(x) + g(x)]^2\,dx$ as a quadratic function of t. It is clear that $F(t) \geq 0$. What condition guarantees that a quadratic function is never negative?)

25. Prove, using the result of the previous problem,

$$\left(\int_a^b f(x)\,dx\right)^2 \leq (b-a)\int_a^b f^2(x)\,dx.$$

For what functions $f(x)$ is this an equality? Illustrate with the examples $f(x)=x$ and $f(x)=\sin x$ on $0 \leq x \leq \pi$.

26. Integrate the following inequalities (valid for $x>0$) from 0 to x to derive new inequalities.

(i) $\cos x \leq 1$ **(ii)** $\sin x < x$ **(iii)** $1 < (1+x)^2$

(iv) $0 < \sinh x$ **(v)** $e^{-x} < 1$ **(vi)** $1 < e^x$

27. Verify the following integrals of the hyperbolic functions.

(i) $\int_a^b \cosh x\,dx = \sinh b - \sinh a$ **(ii)** $\int_a^b \sinh x\,dx = \cosh b - \cosh a.$

3.2 Fundamental theorem

1 Consider the continuous function $f(x)$ in the interval $a \leq x \leq b$. Let its smallest and largest values in this range be m and M (Fig. 2.1), so that

$$m \leq f(x) \leq M, \tag{1}$$

(each equality sign holds at, at least, one point). Area considerations in the previous section [see Eq. (6) there] imply that

$$m(b-a) \leq \int_a^b f(x)\,dx \leq M(b-a), \tag{2}$$

which is equivalent to

$$m \leq \frac{1}{b-a}\int_a^b f(x)\,dx \leq M. \tag{3}$$

Since $y = f(x)$ is a continuous function, it follows that for every number y, $m \leq y \leq M$, there must be at least one corresponding value of x in the interval $a \leq x \leq b$. In particular then, (3) shows that the specific number

$$y = \eta = \frac{1}{b-a}\int_a^b f(x)\,dx \tag{4}$$

lies between the extreme values of the function $f(x)$. We conclude that there is a point in the interval, $x = \xi$, at which

$$\eta = f(\xi)$$

or

$$(b-a)f(\xi) = \int_a^b f(x)\,dx. \tag{5}$$

This conclusion is called the *mean-value theorem for integrals*. Its geometrical interpretation is evident from Fig. 2.2; we have found the dimensions of a rectangle [length, $b - a$; height $\eta = f(\xi)$] which has the *same* area as that under the curve $f(x)$ in the interval $a \le x \le b$.

The number η is called the arithmetic mean of the values of $f(x)$ in the prescribed interval, since it is the obvious generalization of the corresponding definition for finite sums (see Exercise 1). It is the average height of the function $f(x)$ in the interval.

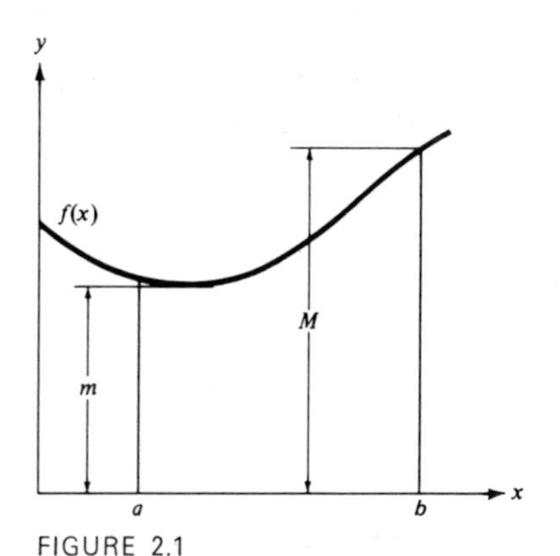

FIGURE 2.1

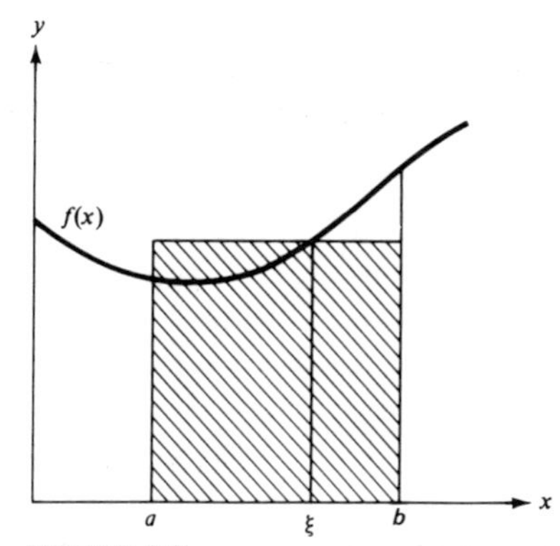

FIGURE 2.2

Example 1 Find the arithmetic mean η and the value of ξ when $f(x) = x^2$ and $a = 0$, $b = 3$.

Since

$$\int_0^3 x^2\,dx = \tfrac{1}{3}x^3\Big]_0^3 = 9,$$

it follows from (4) that

$$\eta = \frac{1}{3-0}\int_0^3 x^2\,dx = \tfrac{1}{3}(9) = 3.$$

Moreover, $f(\xi) = \xi^2 = 3$ implies $\xi = \sqrt{3}$. Therefore, the rectangle of height 3 and length 3 has the same area as that bounded by $y = x^2$, $y = 0$, $x = 0$, and $x = 3$ (see Fig. 2.3).

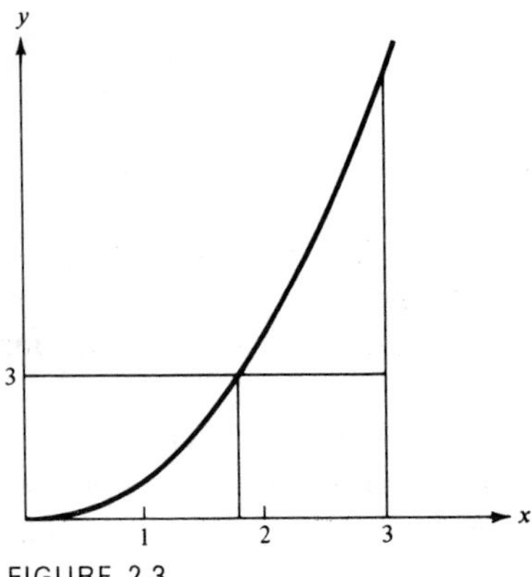

FIGURE 2.3

2

The mean-value theorem can be used at once to establish the relationship between indefinite integrals and differentiation, known as the *fundamental theorem of calculus*.

Let

$$F(x) = \int_a^x f(x)\,dx, \tag{6}$$

so that

$$F(x + \Delta x) = \int_a^{x+\Delta x} f(x)\,dx.$$

In order to calculate the derivative of the integral, we return to basic concepts and consider the quotient

$$\frac{F(x + \Delta x) - F(x)}{\Delta x} = \frac{1}{\Delta x}\left[\int_a^{x+\Delta x} f(x)\,dx - \int_a^x f(x)\,dx\right],$$

which can be contracted to read

$$\frac{F(x + \Delta x) - F(x)}{\Delta x} = \frac{1}{\Delta x}\int_x^{x+\Delta x} f(x)\,dx. \tag{7}$$

[This manipulation is based on formula **3** of Table 1.1. Note, too, that the lower limit of integration in (6) does not appear in (7), which is valid for any indefinite integral of the same family.]

The mean-value theorem applies to the integral in (7) and with $a = x$, $b = x + \Delta x$, the result is

$$\int_x^{x+\Delta x} f(x)\,dx = f(\xi)(x + \Delta x - x) = f(\xi)\,\Delta x, \tag{8}$$

where $x \le \xi \le x + \Delta x$. For convenience, we adopt the representation $\xi = x + \theta\,\Delta x$ where $0 \le \theta \le 1$; for example, $\theta = 0$ if $\xi = x$, and $\theta = 1$ if $\xi = x + \Delta x$. The substitution of (8) in (7) yields

$$\frac{F(x + \Delta x) - F(x)}{\Delta x} = f(\xi) = f(x + \theta\,\Delta x),$$

which in the limit $\Delta x \to 0$ becomes

$$F'(x) = \lim_{\Delta x \to 0} \frac{F(x + \Delta x) - F(x)}{\Delta x} = \lim_{\Delta x \to 0} f(x + \theta\,\Delta x) = f(x)$$

or

$$F'(x) = \frac{d}{dx}\int_a^x f(x)\,dx = f(x). \tag{9}$$

Thus, *the derivative of an indefinite integral, like* (6), *is its integrand*, and this is *the fundamental theorem.*

The converse statement is equally important: the function whose derivative is $f(x)$, that is,

$$\frac{dy}{dx} = f(x), \tag{10}$$

is any one of the family of indefinite integrals represented by

$$y(x) = \int^x f(x)\,dx = F(x) + C, \tag{11}$$

(where the additive constant replaces the arbitrary lower limit of integration). This follows by simply differentiating $y(x)$ and using (9) to arrive at (10). Alternatively, we can say that the *solution of the differential equation* (10) *is the indefinite integral* in (11). An especially useful version of the last formula is the following.

$$\int^x F'(x)\,dx = F(x) + C. \tag{12}$$

All the information accumulated in Chap. 2 on the differentiation of specific functions can be translated via (12) into a compilation of indefinite integrals.

Clearly such a list obviates the laborious approach to integration that begins each time with the basic definition of an integral. For example, the formula

$$\cos x = \frac{d}{dx} \sin x$$

converts to

$$\int^x \cos x\, dx = \sin x + C,$$

when the identifications $F(x) = \sin x$, $F'(x) = \cos x$, are made. Similarly

$$\sin x = -\frac{d}{dx} \cos x$$

implies

$$\int^x \sin x\, dx = -\cos x + C.$$

Further results from Chap. 2 are given in Table 2.1 where, for brevity, the constant of integration is omitted but its presence is understood. All of these results should be verified by differentiation. For example, the fifth formula follows from

$$\frac{d}{dx} \int^x x^n\, dx = x^n = \frac{d}{dx} \frac{x^{n+1}}{n+1}.$$

It is important to recognize that the information obtainable from earlier work on differentiation is limited in extent. The development of constructive methods for the direct calculation of indefinite integrals is imperative.

TABLE 2.1

1. $\int^x e^x\, dx = e^x$	**6.** $\int^x \sinh kx\, dx = \frac{1}{k} \cosh kx$
2. $\int^x e^{kx}\, dx = \frac{1}{k} e^{kx}$	**7.** $\int^x \cosh kx\, dx = \frac{1}{k} \sinh kx$
3. $\int^x \sin kx\, dx = \frac{-1}{k} \cos kx$	**8.** $\int^x \frac{1}{x}\, dx = \log\lvert x\rvert$
4. $\int^x \cos kx\, dx = \frac{1}{k} \sin kx$	**9.** $\int^x (1 - x^2)^{-1/2}\, dx = \sin^{-1} x$
5. $\int^x x^n\, dx = \frac{1}{n+1} x^{n+1}, \quad n \neq -1$	**10.** $\int^x \frac{1}{1 + x^2}\, dx = \tan^{-1} x$

Example 2 Evaluate $\int_0^1 \sin \frac{\pi}{2} x \, dx$.

The indefinite integral is, according to formula **3** (Table 2.1),

$$\int^x \sin \frac{\pi}{2} x \, dx = -\frac{2}{\pi} \cos \frac{\pi}{2} x + C;$$

we now use the general formula

$$\int_a^b f(x) \, dx = F(b) - F(a) = \int_a^b F'(x) \, d(x), \tag{13}$$

[the last equality follows from (12)] and identify

$$f(x) = \sin \frac{\pi}{2} x, \qquad F(x) = -\frac{2}{\pi} \cos \frac{\pi}{2} x, \qquad a = 0, \, b = 1,$$

so that

$$\int_0^1 \sin \frac{\pi}{2} x \, dx = -\frac{2}{\pi} \cos \frac{\pi}{2} x \Big]_0^1 = \frac{2}{\pi}.$$

This definite integral was calculated numerically in Chap. 1.

Example 3 Evaluate $\int_0^{\pi/2} \cos x \sin^3 x \, dx$.

The corresponding indefinite integral is *not* listed above, and the only recourse, at present, is either to search through all of Chap. 2 to find the function whose derivative is the integrand or by experimentation (and luck?) deduce the answer. By such means, we might come upon the example of the chain rule for differentiation which contains the gem

$$\frac{d}{dx} \sin^n x = n \cos x \sin^{n-1} x.$$

For $n = 4$ this implies

$$\frac{d}{dx} \sin^4 x = 4 \cos x \sin^3 x.$$

It follows that

$$\int^x \cos x \sin^3 x \, dx = \tfrac{1}{4} \sin^4 x + C$$

and

$$\int_0^{\pi/2} \cos x \sin^3 x\, dx = \tfrac{1}{4} \sin^4 x \Big]_0^{\pi/2} = \tfrac{1}{4}.$$

Example 4 Find the area bounded by the curves $y = x/(1 + x^2)$, $y = 0$, $x = 0$ and $x = 2$.

The area is given by

$$A = \int_0^2 \frac{x}{1 + x^2}\, dx;$$

a bound for this integral was determined in Example 7 of Sec. 3.1. Having no direct method of calculation, we must once again search for the indefinite integral or resort to ingenuity or guesswork. With some luck, we find the formula

$$\frac{1}{2}\frac{d}{dx} \log(1 + x^2) = \frac{x}{1 + x^2},$$

so that by (13)

$$A = \int_0^2 \frac{x}{1 + x^2}\, dx = \tfrac{1}{2} \log(1 + x^2)\Big]_0^2 = \tfrac{1}{2} \log 5.$$

---

3 If $f(x)$ is positive, the indefinite integral

$$A = \int_a^x f(x)\, dx$$

represents the area (in the xy plane) under the curve $y = f(x)$ between vertical lines $x = a$ and $x = x$, as shown in Fig. 2.4. The fundamental theorem asserts that

$$\frac{dA}{dx} = f(x),$$

and the rate of change of area with respect to x is proportional to the height of the curve above the x axis. From the equivalent differential form,

$$dA = f(x)\, dx,$$

we see that the increment of area is just that of the infinitesimal "rectangle" at x, of height $f(x)$ and width dx. This was the "loose" argument invoked in Chap. 1, when the indefinite integral was first differentiated, and that analysis is now fully vindicated. It frequently takes a great deal of work to substantiate the obvious, which is why this task is often left to others.

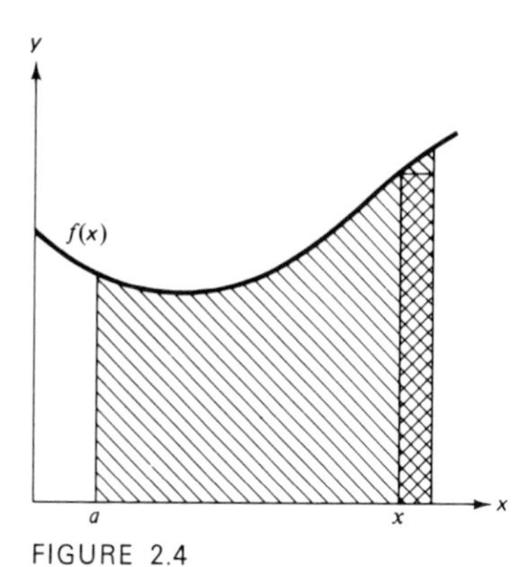

FIGURE 2.4

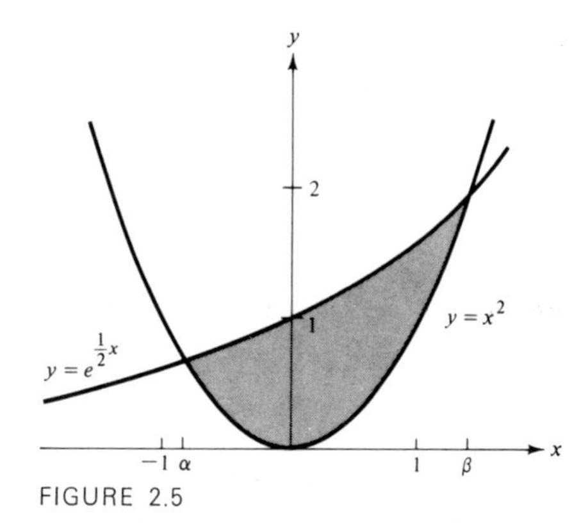

FIGURE 2.5

Example 5 Find the area bounded by the curves $y = e^{x/2}$ and $y = x^2$.

The area A, shown in Fig. 2.5, is given by the integral

$$A = \int_{\alpha}^{\beta} (e^{x/2} - x^2)\, dx,$$

where α and β are the horizontal coordinates of the left- and right-hand points of intersection of the two curves. (Imagine a complete covering of this figure formed by placing vertical lines side by side.) The graph yields the reasonably accurate values for the integration limits

$$\alpha \approx -0.82, \qquad \beta \approx 1.42,$$

and these estimates can be improved by using the Newton-Raphson method. With

$$g(x) = e^{x/2} - x^2$$

the iterative sequence that locates the exact roots of this function is

$$x_{n+1} = x_n - \frac{g(x_n)}{g'(x_n)} = x_n - \frac{e^{x_n/2} - x_n^{\,2}}{\frac{1}{2}e^{x_n/2} - 2x_n}.$$

The values obtained graphically provide the initial guesses, and after only three iterations at each position the roots are determined accurate to six decimal places:

$$\alpha = -0.815553, \qquad \beta = 1.429612.$$

Now that the "exact" limits of integration are given, the definite integral can be evaluated with equal precision. It follows from the list of indefinite integrals above (and it can be checked by differentiation) that

$$\int^{x} (e^{x/2} - x^2)\, dx = 2e^{x/2} - \frac{x^3}{3} + C,$$

so that

$$A = \int_{\alpha}^{\beta} (e^{x/2} - x^2)\, dx = \left[2e^{x/2} - \frac{x^3}{3}\right]_{-0.815553}^{1.429612} = 1.602567.$$

Example 6 Find the solution of the differential equation

$$\frac{dy}{dx} = \frac{2}{x} + \frac{1}{2} e^{-3x}$$

which passes through the point $x = 1$, $y = 1$.

According to the fundamental theorem, the indefinite integral

$$y = \int^{x} \left(\frac{2}{x} + \frac{1}{2} e^{-3x}\right) dx$$

is the solution of this differential equation. From formulas **2** and **8** in Table 2.1 we find that

$$y = 2\int^{x} \frac{1}{x}\, dx + \frac{1}{2}\int^{x} e^{-3x}\, dx = 2 \log x - \tfrac{1}{6} e^{-3x} + C.$$

The constant C is determined from the condition that the curve passes through $x = 1$, $y = 1$; substitution of these values in the preceding expression yields

$$1 = 2 \log 1 - \tfrac{1}{6} e^{-3} + C$$

or

$$C = 1 + \tfrac{1}{6} e^{-3}.$$

Therefore the solution is

$$y = 2 \log x - \tfrac{1}{6}(e^{-3x} - e^{-3}) + 1.$$

The more general problem of this type is to solve

$$\frac{dy}{dx} = f(x), \tag{14}$$

subject to the condition $y = y_0$ at $x = x_0$. In this case, the solution is the integral

$$y = \int^{x} f(x)\, dx = F(x) + C.$$

It follows upon substituting x_0 and y_0 for x and y that

$$C = y_0 - F(x_0).$$

Therefore

$$y = y_0 + F(x) - F(x_0),$$

or

$$y - y_0 = \int_{x_0}^{x} f(x)\,dx. \tag{15}$$

An equivalent procedure is to express (14) as $dy = f(x)\,dx$ and then to integrate both sides of this equation, the result of which is (15) in the form

$$\int_{y_0}^{y} dy = \int_{x_0}^{x} f(x)\,dx. \tag{16}$$

Exercises 3.2

1. The *arithmetic mean* (or average) of n numbers $y_1, y_2, \ldots, y_n$ is defined as

$$\eta = \frac{1}{n}(y_1 + y_2 + \cdots + y_n) = \frac{1}{n}\sum_{i=1}^{n} y_i.$$

If $y_i = f(x_i)$ and $n = (b-a)/\Delta x$, show that in the limit as $n \to \infty$ or $\Delta x \to 0$

$$\eta = \lim_{\Delta x \to 0} \frac{1}{b-a}\sum_{i=1}^{n} y_i\,\Delta x = \frac{1}{b-a}\int_a^b f(x)\,dx.$$

(The number η is the *average value* of $f(x)$ in the interval $[a,b]$.)

2. Determine the average values of the following functions over the given intervals.

(i) $f(x) = 1 + x^2$ $\quad a = 1,\ b = 3$ **(ii)** $\sin x$ $\quad a = 0,\ b = \pi/2$

(iii) e^x $\quad a = 2,\ b = 5$ **(iv)** $x/(1+x^2)$ $\quad a = 0,\ b = 2$

3. Employ the definition of an integral to prove the following:

(i) $\displaystyle\lim_{n\to\infty}\left[\frac{1}{n+1} + \frac{1}{n+2} + \cdots + \frac{1}{2n}\right] = \log 2$

(ii) $\displaystyle\lim_{n\to\infty}\frac{1}{n}\left[2^{1/n} + 2^{2/n} + \cdots + 2^{n/n}\right] = \frac{1}{\log 2}$

4. Find the values of ξ and $\eta = f(\xi)$ in the mean-value theorem for each of the following. Illustrate with appropriate graphs.

(i) $f(x) = e^{2x}$ in the interval $[0,1]$ **(ii)** $f(x) = x^3$ in $[0,2]$

(iii) $f(x) = x^{1/2}$ in $[1,2]$ **(iv)** $f(x) = \sin \pi x/2$ in $[0,1]$.

5. If $f(x)$ is a continuous function and $\displaystyle\int_a^b f(x)\,dx = 0$ for all values of a and b, show that $f(x) = 0$. (*Hint*: Suppose that $f(x)$ is not zero at some point.)

6. Evaluate the following integrals:

(i) $\displaystyle\int_0^1 (x^2 - x)\,dx$ **(ii)** $\displaystyle\int_0^{\pi} \sin x\,dx$ **(iii)** $\displaystyle\int_0^1 e^x\,dx$

(iv) $\displaystyle\int_1^2 \frac{dx}{x}$ **(v)** $\displaystyle\int_{-2}^{-1}\left(x + \frac{1}{x}\right)dx$ **(vi)** $\displaystyle\int_{-\pi/2}^{\pi/2} (x + \cos x)\,dx$

7. Show, by differentiating, the following results:

(i) $\displaystyle\int^x xe^x\,dx = xe^x - e^x + C$

(ii) $\int^x \sin 2x\, dx = \sin^2 x + C$ **(iii)** $\int^x x \log x\, dx = \frac{x^2}{2} \log x - \frac{x^2}{4} + C.$

8. Use the results of the previous problem to evaluate the following integrals.

(i) $\int_0^1 xe^x\, dx$ **(ii)** $\int_0^{\pi/4} \sin 2x\, dx$ **(iii)** $\int_1^e x \log x\, dx$

9. By differentiating, verify the following results:

(i) $\int^x \frac{dx}{x \log x} = \log(\log x) + C$

(ii) $\int^x \log(a^2 + x^2)\, dx = x \log(a^2 + x^2) - 2x + 2a \tan^{-1} x/a + C$

(iii) $\int^x \frac{\tan x\, dx}{(1 + 2\tan^2 x)^{1/2}} = \cos^{-1}\left(\frac{1}{\sqrt{2}} \cos x\right) + C.$

12. Find the areas bounded by the given curves:

(i) $y = \sin 2x,\ y = 0,\ x = \frac{\pi}{2},\ x = \pi$ **(ii)** $y = \frac{1}{x^2 + 1},\ y = (x - 2)^2$

(iii) $y = \frac{2}{x},\ x + y = 3$ **(iv)** $y = \frac{x}{x^2 + 1},\ x = 0,\ y = 1/2.$

13. Find all solutions of the following differential equations:

(i) $\frac{dy}{dx} = \frac{2}{x} + x^2 + \sin x$ **(ii)** $\frac{dy}{dx} = \frac{1}{(1 - x^2)^{1/2}}$

(iii) $\frac{dy}{dx} = x + x^2 + 2x^3$ **(iv)** $\frac{dy}{dx} = xe^x + e^x$

14. Graph the region bounded by the curves $y = x$ and $y = e^x/4$. Evaluate this area as a definite integral. (Use numerical methods to determine the limits of integration.)

15. Use the fundamental theorem to evaluate the following derivatives.

(i) $\frac{d}{dt}\int_0^t e^{-x^2}\, dx$ **(ii)** $\frac{d}{dt}\int_t^{t^2} \frac{dx}{1 + x}$ **(iii)** $\frac{d}{dt}\int_t^{2t} f(z)\, dz$

16. Use the chain rule for differentiation to show that

$$\frac{d}{dx}\int_a^{g(x)} f(z)\, dz = f(g(x))g'(x).$$

Evaluate the following derivatives.

(i) $\frac{d}{dt}\int_0^{\log t} e^{-x}\, dx$ **(ii)** $\frac{d}{dt}\int_0^{\sin^{-1} t} (1 + x)\, dx$ **(iii)** $\frac{d}{dt}\int_0^{\sin t^2} (1 - z^4)^{1/2}\, dz$

17. A particle moving along a straight line has velocity $v(t)$ at time t given by $v(t) = a\omega \cos \omega t$. Find the distance $s(t) = \int_0^t v(t)dt$ travelled from time 0 to t. Check that the answer is dimensionally consistent and show that the distance travelled is a periodic function of t. (In this problem, a and ω are given constants.)

18. The acceleration of a particle as a function of time *is* given by

$$a(t) = \frac{t}{(1 + t^2)^2} \text{ for } t \geq 0.$$

Suppose that the initial velocity is zero. Find the velocity $v(t)$ and the distance travelled $s(t)$ at time t. Determine the average acceleration and the average velocity over the time interval $[0, t]$. Draw graphs of the three functions.

19. The velocity of a particle at time t is given by

$$v(t) = \frac{30\, e^{2t}}{1 + e^{2t}} \qquad \text{(m/sec) for } t \geq 0.$$

Write, as an integral, the distance travelled between $t = 1.5$ and $t = 6.2$ (sec). What is the average velocity in this time interval? Draw graphs of the acceleration, velocity and distance functions.

20. If a body moves along a straight line with constant acceleration a, derive formulas for its velocity and displacement at time t using integration methods. Show that the average velocity in a time interval is the average of the initial and final velocities.

21. A ball is thrown vertically upwards with an initial velocity of 20 m/sec. Since $dv/dt = -g = -9.8$ m/sec^2 and $v = ds/dt$, determine $s(t)$, the height of the ball at time t. Find the maximum height of the ball. When does it hit the ground?

3.3 *Techniques of integration: I*

Many scientific problems involve integrals of the form $\int_a^x f(x)\, dx$, and the extent to which these functions can be simply described in quantitative or qualitative terms is often synonymous with the value and meaning of a "solution". The extraction of information from such formulas is a primary goal; ideally, we would like to find an exact or closed-form expression for every integral that arises, as a combination of the simple functions known to us. Failing this, which is most likely in view of the range of possibilities, we might attempt to develop an accurate approximation of the same form. If this is also impractical, the integral can be evaluated numerically. This is a reasonable approach but hardly a rigid prescription. The methods adopted must always be geared to the particular problem, and we must constantly keep in mind the objectives of the analysis. A numerical calculation is appropriate in one case, a formula in another. Although certain techniques will become favorites by virtue of their simplicity, utility, or elegance, none should ever be scorned. Aside from textbook problems, integration is a difficult operation requiring considerable skill and ingenuity; there is no panacea, no automatic crank-grinding procedure that, with more labor than brains, always yields the solution. Every method has its time and place when it becomes the most efficient means available, and for these reasons many techniques will be studied. The experience so acquired is the greatest resource for dealing with practical problems.

Table 3.1 is a list of integrals culled from the corresponding results on differentiation. It will be exploited fully since many integrals can be reduced to combinations of these. Here and in the future the notation for an indefinite integral is simplified to read

$$\int f(x)\,dx = F(x) + C.$$

The difference between a variable limit of integration and the integration variable should be clear by now, and unless special emphasis is required, no further distinction will be made.

We now begin a discussion of integration techniques which will continue through the better part of the next three sections.

TABLE 3.1

1. $\int x^n\,dx = \frac{x^{n+1}}{n+1} + C, \qquad n \neq -1$	**7.** $\int \sin ax\,dx = -\frac{1}{a}\cos ax + C$
2. $\int \frac{dx}{x} = \log\|x\| + C$	**8.** $\int \cos ax\,dx = \frac{1}{a}\sin ax + C$
3. $\int \frac{dx}{x-a} = \log\|x-a\| + C$	**9.** $\int \sec^2 x\,dx = \tan x + C$
4. $\int \frac{dx}{x^2+a^2} = \frac{1}{a}\tan^{-1}\frac{x}{a} + C$	**10.** $\int e^{ax}\,dx = \frac{e^{ax}}{a} + C$
5. $\int \frac{x\,dx}{x^2+a^2} = \frac{1}{2}\log(x^2+a^2) + C$	**11.** $\int \sinh ax\,dx = \frac{1}{a}\cosh ax + C$
6. $\int \frac{dx}{(a^2-x^2)^{1/2}} = \sin^{-1}\frac{x}{a} + C$	**12.** $\int \cosh ax\,dx = \frac{1}{a}\sinh ax + C$

2 ALGEBRAIC SIMPLIFICATION

Certain seemingly complicated integrals can be made known by the simple expedient of judicious manipulations, in which the integrand is reworked algebraically. This tactic requires the recognition and full use of established identities, as the following examples illustrate.

Example 1 Evaluate the indefinite integrals

$$\int \sin^2 x\,dx, \qquad \int \cos^2 x\,dx, \qquad \int \sin^3 x\,dx.$$

The identity

$$\sin^2 x = \tfrac{1}{2}(1 - \cos 2x)$$

is used to write

$$\int \sin^2 x\,dx = \frac{1}{2}\int dx - \frac{1}{2}\int \cos 2x\,dx,$$

and from formulas **1** and **8** of Table 3.1 we obtain

$$\int \sin^2 x\,dx = \frac{x}{2} - \frac{1}{4}\sin 2x + C.$$

Since

$$\cos^2 x = 1 - \sin^2 x,$$

it follows that

$$\int \cos^2 x\,dx = \int dx - \int \sin^2 x\,dx = \frac{x}{2} + \frac{1}{4}\sin 2x + C.$$

The identity

$$\sin^3 x = \tfrac{3}{4}\sin x - \tfrac{1}{4}\sin 3x$$

allows the third integral to be expressed as

$$\int \sin^3 x\,dx = \tfrac{3}{4}\int \sin x\,dx - \tfrac{1}{4}\int \sin 3x\,dx,$$

and it follows that

$$\int \sin^3 x\,dx = -\tfrac{3}{4}\cos x + \tfrac{1}{12}\cos 3x + C.$$

All of these results may be verified by differentiating.

Example 2 Find $\int \frac{x}{x+1}\,dx$.

Write

$$\frac{x}{x+1} = \frac{x+1-1}{x+1} = 1 - \frac{1}{x+1},$$

so that

$$\int \frac{x}{x+1}\,dx = \int dx - \int \frac{dx}{x+1} = x - \log|x+1| + C.$$

Example 3 Evaluate $\int_0^1 \frac{1}{x^2+3x+2}\,dx$.

The integrand written as

$$\frac{1}{x^2+3x+2} = \frac{1}{(x+1)(x+2)} = \frac{1}{x+1} - \frac{1}{x+2}$$

permits speedy calculation of the indefinite integral:

$$\int \frac{dx}{x^2 + 3x + 2} = \int \frac{dx}{x+1} - \int \frac{dx}{x+2} = \log|x+1| - \log|x+2| + C,$$

$$= \log\left|\frac{x+1}{x+2}\right| + C.$$

The definite integral is then

$$\int_0^1 \frac{dx}{x^2 + 3x + 2} = \log\left|\frac{x+1}{x+2}\right|\Bigg]_0^1 = \log \tfrac{2}{3} - \log \tfrac{1}{2} = \log \tfrac{4}{3}.$$

---

3 CHANGE OF VARIABLE

The basis of Table 3.1 is the observation that every differentiation formula has its analog as an indefinite integral. The relationship

$$f(x) = \frac{d}{dx} F(x), \tag{1}$$

corresponds to the indefinite integral

$$\int f(x)\, dx = F(x) + C. \tag{2}$$

A more widely applicable set of integration formulas is easily obtained by considering the implications of chain-rule differentiation. The formula

$$\frac{d}{dx} F(u(x)) = \frac{dF(u)}{du}\frac{du}{dx} = f(u)u'(x), \tag{3}$$

implies that

$$\int f(u(x))u'(x)\, dx = F(u) + C, \tag{4}$$

just as (1) gives (2). Therefore, the tabulated integral, represented by (2), enables us to find the more general integral in (4). However, it would be redundant to list all the possible variations because Eqs. (4) and (2) are really the same statements, described in different variables (with u replacing x). Another way of seeing this is to write

$$du = u'(x)\, dx,$$

in which case

$$f(u(x))u'(x)\, dx = f(u)\, du,$$

and

$$\int f(u(x))u'(x)\,dx = \int f(u)\,du.$$

The indefinite integral on the right-hand side is that given in (2) (with u instead of x); upon replacing it there, we recover (4). In effect, the integral in Eq. (4) is handled by first changing to the variable u, performing the integration, and then reverting to the original variable.

Example 4 Find $\int \cos x \sin^n x\,dx$, $n \neq -1$.

Identify $u(x) = \sin x$ so that $du = \cos x\,dx$. The result of a changeover to the variable u and the use of formula **1** of Table 3.1 is

$$\int \cos x \sin^n x\,dx = \int u^n\,du = \frac{u^{n+1}}{n+1} + C = \frac{\sin^{n+1} x}{n+1} + C.$$

Example 5 Integrate $\int \frac{e^x}{1+e^{2x}}\,dx$.

In this case, let $u(x) = e^x$, so that $du = e^x\,dx$. Therefore

$$\int \frac{e^x}{1+e^{2x}}\,dx = \int \frac{du}{1+u^2} = \tan^{-1} u + C = \tan^{-1} e^x + C.$$

A change to a new integration variable which simplifies the appearance of an integrand can be a decisive move (and is always worth trying). The substitution may be explicit, $x = x(u)$, or implicit, $u = u(x)$, depending on which is the more natural and easier to employ. In the former case, $dx = x'(u)\,du$, and the effect of this variable change on the integral in (2) is

$$\int f(x)\,dx = \int f(x(u))x'(u)\,du.$$

Similarly, the implicit relationship yields

$$\int f(x)\,dx = \int f(x(u))\frac{du}{u'(x)}.$$

A few examples will illustrate the procedure.

Example 6 Find $\int \frac{x}{x^2 + a^2}\,dx$.

Let $u = x^2 + a^2$, in which case $du = 2x\,dx$ (or $dx = du/2x = du/u'(x)$). Upon transforming the integral to the new variable, we obtain

$$\int \frac{x}{x^2 + a^2}\,dx = \int \frac{du}{2u} = \tfrac{1}{2}\log u + C = \tfrac{1}{2}\log(x^2 + a^2) + C.$$

Example 7 Find $\int \frac{x\,dx}{x^2 + x + 1}$.

First the denominator is rewritten by completing the square:

$$x^2 + x + 1 = x^2 + x + \tfrac{1}{4} + (1 - \tfrac{1}{4}) = (x + \tfrac{1}{2})^2 + \tfrac{3}{4}.$$

In the next step we introduce the simple variable change

$$u = x + \tfrac{1}{2}, \qquad \text{for which} \qquad du = dx.$$

With this transformation the integral becomes

$$\int \frac{x\,dx}{x^2 + x + 1} = \int \frac{x\,dx}{(x + \frac{1}{2})^2 + \frac{3}{4}} = \int \frac{(u - \frac{1}{2})\,du}{u^2 + \frac{3}{4}} = \int \frac{u}{u^2 + \frac{3}{4}}\,du - \frac{1}{2}\int \frac{du}{u^2 + \frac{3}{4}}.$$

The first integral on the extreme right-hand side was determined in Exercise 6. (If this were not the case, we would have to repeat that analysis at this point.) The result is

$$\int \frac{u\,du}{u^2 + \frac{3}{4}} = \tfrac{1}{2}\log(u^2 + \tfrac{3}{4}).$$

The remaining integral is tabulated (see formula **4** in Table 3.1);

$$\int \frac{du}{u^2 + \frac{3}{4}} = \frac{2}{\sqrt{3}}\tan^{-1}\frac{2u}{\sqrt{3}}.$$

Upon collecting results and reverting to the original variable x, we obtain

$$\int \frac{x\,dx}{x^2 + x + 1} = \tfrac{1}{2}\log(x^2 + x + 1) - \frac{1}{\sqrt{3}}\tan^{-1}\left[\frac{2}{\sqrt{3}}(x + \tfrac{1}{2})\right] + C.$$

Example 8 Find $\int \tan x\,dx$.

The integral is much more suggestive when it is written

$$\int \frac{\sin x}{\cos x}\,dx.$$

In this form, even the novice might observe that the numerator of the integrand,

$\sin x$, is just the derivative of the denominator (with the appropriate sign), i.e.,

$$d(\cos x) = -\sin x\, dx.$$

This motivates the substitution

$$u = \cos x \qquad (\text{with } du = -\sin x\, dx),$$

and this change of variable transforms the integral to a standard form:

$$\int \tan x\, dx = -\int \frac{\sin x}{\cos x}\frac{du}{\sin x} = -\int \frac{du}{u} = -\log|u| + C = -\log|\cos x| + C.$$

(Here and in Example 6 the use of the explicit relationship, $x(u)$, only complicates the algebra.) This transformation, which works beautifully, is born of experience, the experience gained by going down so many blind alleys that we learn which to try and which to avoid. Sometimes alternatives exist, perhaps not equally simple or elegant, but viable nonetheless. The reader who would not deprive himself of pain should consider the substitution $v = \sin x$, which will also work. More effort is required, and eventually he will be led right back to the first substitution.

Example 9 Find $\int \sec x\, dx$.

After some experimentation with different changes of variable, the substitution

$$u = \sin x$$

is found to be promising; i.e., it does not lead to immediate rejection. Since $du = \cos x\, dx$, the integration is performed as follows:

$$\int \sec x\, dx = \int \frac{dx}{\cos x} = \int \frac{du}{1-u^2},$$

$$= \frac{1}{2}\int \left(\frac{1}{1+u} + \frac{1}{1-u}\right) du = \frac{1}{2}\int \frac{du}{1+u} + \frac{1}{2}\int \frac{du}{1-u},$$

$$= \tfrac{1}{2}(\log|1+u| - \log|1-u|) + C = \tfrac{1}{2}\log\left|\frac{1+u}{1-u}\right| + C,$$

$$= \log\left|\frac{1+\sin x}{1-\sin x}\right|^{1/2} + C.$$

The trigonometric identity

$$\left|\frac{1+\sin x}{1-\sin x}\right|^{1/2} = \left|\frac{(1+\sin x)^2}{1-\sin^2 x}\right|^{1/2} = \left|\frac{1+\sin x}{\cos x}\right| = |\sec x + \tan x|$$

allows the last formula to be expressed succinctly as

$$\int \sec x\, dx = \log|\sec x + \tan x| + C.$$

This integral is often presented as the consequence of a simple trick in which we write

$$\int \sec x \, dx = \int \frac{\sec x (\sec x + \tan x)}{\sec x + \tan x} dx = \int \frac{\sec^2 x + \sec x \tan x}{\sec x + \tan x} dx,$$

$$= \int \frac{d(\sec x + \tan x)}{\sec x + \tan x} = \int \frac{dv}{v} = \log|v| + C,$$

$$= \log|\sec x + \tan x| + C.$$

However, this maneuver is undoubtedly hindsight in action; i.e., the answer as evaluated in a direct manner is used to devise a new tactic. We could just as well have guessed the final formula (an obviously brilliant stratagem) and then checked it by differentiation.

Exercises 3.3

1. Simplify and integrate. Verify the answers by differentiating.

(i) $\int x(x+2)\, dx$ **(ii)** $\int (x-1)^2(x+1)^2\, dx$ **(iii)** $\int x^{1/2}(1-x^{1/2})\, dx$

2. Simplify and integrate.

(i) $\int \frac{dx}{x^2+4}$ **(ii)** $\int \frac{dx}{x^2+2x+2}$ **(iii)** $\int \frac{dx}{x^2+4x+3}$

(iv) $\int \frac{x^2+2}{1+x^2}\, dx$ **(v)** $\int \frac{3x+1}{(x+2)^2}\, dx$ **(vi)** $\int \frac{x+10}{x^2-x-12}\, dx$

3. Simplify and integrate.

(i) $\int \sin 3x\, dx$ **(ii)** $\int \cos x \sin^7 x\, dx$ **(iii)** $\int \sin^3 x\, dx$

(iv) $\int \sin^4 x\, dx$ **(v)** $\int \cos^3 x\, dx$ **(vi)** $\int \cos^4 x\, dx$

(*Hint:* $\sin^3 x = \sin x(1-\cos^2 x)$ or $\sin^3 x = \frac{3}{4}\sin x - \frac{1}{4}\sin 3x$.)

4. Evaluate the following definite integrals.

(i) $\int_0^2 x^3\, dx$ **(ii)** $\int_0^{\pi/4} \sin x\, dx$ **(iii)** $\int_1^2 x^{-3}\, dx$

(iv) $\int_{-1}^1 (2-x^3)\, dx$ **(v)** $\int_{-\pi/2}^{\pi/2} (\cos x - \sin x)\, dx$ **(vi)** $\int_0^n x^n\, dx,\ n > -1$

(vii) $\int_0^{\pi/2} \cos^3 x\, dx$ **(viii)** $\int_{-4}^{-5} (x^2+2)\, dx$ **(ix)** $\int_0^{2\pi} \cos(x/8)\, dx.$

5. Simplify and integrate.

(i) $\int \frac{dx}{x+3}$ **(ii)** $\int \frac{dx}{3x+1}$ **(iii)** $\int \frac{dx}{ax+b}$

(iv) $\int \frac{dx}{(x-1)(x-2)}$ **(v)** $\int \frac{dx}{x(1-x^2)}$ **(vi)** $\int \frac{x \log(1-x^2)}{1-x^2}\, dx$

6. Evaluate the following derivatives and limits.

(i) $\frac{d}{dx}\int_0^x e^{t^2} \sin t\, dt$ **(ii)** $\frac{d}{dx}\int_2^x z(a^2-z^2)^{1/2}\, dz$ **(iii)** $\frac{d}{dx}\int_x^5 \sin(\cos\theta)\, d\theta$

(iv) $\lim_{x\to 0} \frac{\int_0^x \sin t\, dt}{\sin x^2}$ **(v)** $\lim_{x\to 0} \frac{\int_0^x \sin t^2\, dt}{x^2}$ **(vi)** $\lim_{x\to 0} \frac{\int_0^x f(t)\, dt}{x}$

7. Evaluate the following integrals.

(i) $\int \sqrt{2x+3}\, dx$ **(ii)** $\int (ax+b)^{1/2}\, dx$ **(iii)** $\int (ax+b)^n\, dx$

(iv) $\int \tan(2x+1)\, dx$ **(v)** $\int x^2 e^{3x^3}\, dx$ **(vi)** $\int e^{\sin x} \cos x\, dx$

(vii) $\int \sin^3 x \cos x\, dx$ **(viii)** $\int \tan^2 x \sec^2 x\, dx$ **(ix)** $\int \tan^2 x\, dx$

8. Use the substitution indicated to evaluate the following integrals.

(i) $\int \frac{\sin x}{2+\cos x}\, dx, \quad u = \cos x$ **(ii)** $\int x^{1/3}(x^{4/3}+2)^{1/2}\, dx, \quad u = x^{4/3}+2$

(iii) $\int \frac{1}{(1-x^2)^{1/2}}\, dx, \quad t = \sin^{-1} x$ **(iv)** $\int 5^x\, dx, \quad u = 5^x$

9. Make a change of variables and integrate.

(i) $\int s e^{-s^2}\, ds$ **(ii)** $\int \frac{\tan^{-1} x}{1+x^2}\, dx$ **(iii)** $\int \tan^3 x \sec^2 x\, dx$

(iv) $\int \frac{\sin^{-1} x}{(1-x^2)^{1/2}}\, dx$ **(v)** $\int (\cos x)\, e^{\sin x}\, dx$ **(vi)** $\int \frac{x}{(x^2+4)^{1/2}} \cos\sqrt{x^2+4}\, dx$

10. Evaluate the following definite integrals.

(i) $\int_0^{\pi/4} \sec x\, dx$ **(ii)** $\int_{\pi/6}^{\pi/3} \tan x\, dx$ **(iii)** $\int_1^5 \frac{x\, dx}{x^2+x+1}$

(iv) $\int_0^3 \frac{dx}{x^2+9}$ **(v)** $\int_2^3 \frac{x-2}{(x-1)(x-4)}\, dx$ **(vi)** $\int_{-1}^1 \frac{e^x}{e^x+1}\, dx$

11. Using the identities

$$\cos\theta + \cos\phi = 2\cos\frac{\theta+\phi}{2}\cos\frac{\theta-\phi}{2}, \quad \cos\theta - \cos\phi = -2\sin\frac{\theta+\phi}{2}\sin\frac{\theta-\phi}{2},$$

evaluate the following integrals.

(i) $\int \cos(x+a)\cos(x-a)\, dx$ **(ii)** $\int \sin(x+a)\sin(x-a)\, dx$

12. In the previous problem, show that the second integral can be evaluated by setting $a = \alpha + \pi/2$ in the first integral.

13. Using the formulas

$$\begin{aligned} 2\sin\theta\sin\phi &= \cos(\theta-\phi) - \cos(\theta+\phi) \\ 2\sin\theta\cos\phi &= \sin(\theta+\phi) + \sin(\theta-\phi) \\ 2\cos\theta\cos\phi &= \cos(\theta-\phi) + \cos(\theta+\phi), \end{aligned}$$

evaluate the following integrals:

(i) $\int \sin ax \cos bx\, dx$ **(ii)** $\int \sin ax \sin bx\, dx$ **(iii)** $\int \cos ax \cos bx\, dx$.

14. Evaluate the following integrals:

(i) $\int \sin 2x \cos 3x\, dx$ **(ii)** $\int \sin x \sin 4x\, dx$ **(iii)** $\int \cos x \cos 2x\, dx$

15. Complete the square and integrate.

(i) $\int \frac{x\, dx}{x^2+4x+1}$ **(ii)** $\int \frac{dx}{x^2+x-6}$ **(iii)** $\int \frac{e^x\, dx}{e^{2x}+2e^x-3}$

(iv) $\int \frac{x\, dx}{x^4+4x^2+3}$ **(v)** $\int \frac{dz}{(z-2)^2+2z}$ **(vi)** $\int \frac{\cos\theta\, d\theta}{2+2\sin\theta-\cos^2\theta}$

16. The resale value R (dollars) of a certain machine decreases with time (weeks) according to

$$R(t) = \frac{3A}{4} e^{-t/96}$$

where A is the initial cost. The profit rate (profit per unit time) realized from use of the machine at any time t is given by

$$p(t) = \frac{A}{4} e^{-t/48}.$$

When should the machine be sold in order to maximize the *total* profit obtainable from its use? What is the profit? How much is the machine sold for?

3.4 Techniques of integration: II

We continue discussion of the integration technique based on a judicious change of variable by applying this method to the evaluation of definite integrals. This is already within our capability, but a new interpretation adds insight and reduces the effort involved. As a review, the orthodox approach is given first.

Example 1 Evaluate $I = \int_0^1 x(1+x)^{1/2}\,dx$.

The numerical value will be determined from the explicit expression for the corresponding indefinite integral. An appropriate substitution for this purpose is

$$u = 1 + x, \tag{1}$$

or $x = u - 1$, so that $dx = du$. It follows that

$$\int x(1+x)^{1/2}\,dx = \int (u-1)u^{1/2}\,du = \int u^{3/2}\,du - \int u^{1/2}\,du, \tag{2}$$

$$= \tfrac{2}{5}u^{5/2} - \tfrac{2}{3}u^{3/2} + C, \tag{3}$$

$$= \tfrac{2}{5}(1+x)^{5/2} - \tfrac{2}{3}(1+x)^{3/2} + C. \tag{4}$$

Therefore

$$\int_0^1 x(1+x)^{1/2}\,dx = \left[\tfrac{2}{5}(1+x)^{5/2} - \tfrac{2}{3}(1+x)^{3/2}\right]_0^1, \tag{5}$$

$$= \tfrac{4}{15}(\sqrt{2} + 1).$$

The procedure illustrated here calls for a reversion to the variable x before final evaluation. This is unnecessary because the indefinite integral can be left as the function of u in (3) and evaluated directly at the two points, $u = 2$, $u = 1$, which correspond to the original limits of integration, $x = 1$, $x = 0$. In effect, then,

$$\int_0^1 x(1+x)^{1/2}\,dx = \left[\tfrac{2}{5}u - \tfrac{2}{3}u^{3/2}\right]_1^2; \tag{6}$$

but according to (2), the right-hand side of this equation can be identified as the *definite* integral

$$\int_1^2 (u-1)u^{1/2}\,du.$$

Therefore, the transformation $u = x + 1$ is complete when the integrand, differential, and limits of integration are properly converted, in which case, the original definite integral changes into an equivalent one in the new variable:

$$\int_0^1 x(1 + x)^{1/2}\,dx = \int_1^2 (u - 1)u^{1/2}\,du. \tag{7}$$

In the more general situation, the transformation $x = x(u)$ is applied to the integral $\int_a^b f(x)\,dx$, and everything in sight must be properly converted. Let $u = \alpha, \beta$ correspond to the values $x = a, b$; that is, $a = x(\alpha)$, $b = x(\beta)$. Then with

$$f(x) = f(x(u)) \qquad \text{and} \qquad dx = x'(u)\,du$$

it follows that

$$\int_a^b f(x)\,dx = \int_\alpha^\beta f(x(u))x'(u)\,du = \int_\alpha^\beta g(u)\,du, \tag{8}$$

where

$$g(u) = f(x(u))x'(u). \tag{9}$$

The graphical interpretation of this transformation is that the area under the curve $f(x)$ (Fig. 4.1) between $x = a$ and $x = b$ is *the same* as the area under the

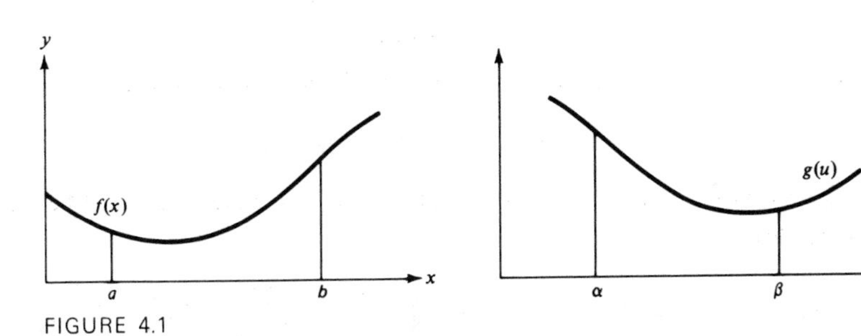

FIGURE 4.1

curve $g(u)$ bounded by $u = \alpha$ and $u = \beta$. We can, if we wish, now discard the original integral, change notation, and consider *anew* the evaluation of

$$\int_\alpha^\beta g(x)\,dx.$$

Example 2 Evaluate $\displaystyle\int_0^{1/4} \left(\frac{x}{1 - x}\right)^{1/2} dx.$

A mathematical analysis is usually tidied up and made "elegant" before its final presentation, which may in no way reflect any of the intuition or motivation that led to the solution. This is understandable but unfortunate, for these factors are the essence of creative work. In order to display a truer picture of the actual procedure, we will do this integration in as realistic a manner as possible.

Method 1 The ratio of two functions looks nasty, and some modification is attempted with a change of variable. A hint of what this might be is obtained from the separation

$$\left(\frac{x}{1-x}\right)^{1/2} dx = x^{1/2}\frac{dx}{(1-x)^{1/2}} \sim x^{1/2}\, d(1-x)^{1/2},$$

where $\sim$ means "proportional to." On this basis, we *try* (for there are no guarantees) the substitution

$$u = (1-x)^{1/2} \qquad \text{or} \qquad x = 1 - u^2.$$

Since $du = -dx/[2(1-x)^{1/2}]$ (or $dx = -2u\, du$) and $x = 0$, $x = 1/4$ correspond to $u = 1$, $u = \sqrt{3}/2$, the integral transforms as follows:

$$\int_0^{1/4}\left(\frac{x}{1-x}\right)^{1/2} dx = -2\int_1^{\sqrt{3}/2}(1-u^2)^{1/2}\, du = 2\int_{\sqrt{3}/2}^{1}(1-u^2)^{1/2}\, du. \quad (10)$$

It may seem that we have not gained much, and this would not be surprising. However, the integrand on the extreme right side can be dealt with effectively by another transformation based on the familiar trigonometric identity

$$1 - \sin^2 z = \cos^2 z.$$

Therefore, let $u = \sin z$. The essential details of the substitution are

$$(1-u^2)^{1/2} = \cos z, \qquad du = \cos z\, dz;$$

$u = \sqrt{3}/2$ becomes $z = \sin^{-1}\sqrt{3}/2 = \pi/3$; likewise, $u = 1$ is $z = \pi/2$. The transformation yields

$$\int_{\sqrt{3}/2}^{1}(1-u^2)^{1/2}\, du = \int_{\pi/3}^{\pi/2}\cos^2 z\, dz = \left[\frac{z}{2} + \frac{1}{4}\sin 2z\right]_{\pi/3}^{\pi/2} = \frac{1}{4}\left(\frac{\pi}{3} - \frac{\sqrt{3}}{2}\right),$$

where the last integral has already been calculated. It follows, upon replacing this expression in (10), that

$$\int_0^{1/4}\left(\frac{x}{1-x}\right)^{1/2} dx = \frac{1}{2}\left(\frac{\pi}{3} - \frac{\sqrt{3}}{2}\right).$$

Method 2 If we had been cleverer (and we will be in the future), the two substitutions made, $x = 1 - u^2$, and $u = \sin z$, could have been contracted into one at the very outset:

$$x = 1 - \sin^2 z = \cos^2 z.$$

In this case,

$$dx = -2\cos z \sin z\, dz, \qquad \left(\frac{x}{1-x}\right)^{1/2} = \frac{\cos z}{\sin z},$$

and the limits $x = 0$ and $x = \frac{1}{4}$ become $z = \pi/2$ and $\pi/3$, respectively. The transformation is then

$$\int_0^{1/4} \left(\frac{x}{1-x}\right)^{1/2} dx = \int_{\pi/2}^{\pi/3} \frac{\cos z}{\sin z}(-2\cos z \sin z)\, dz = 2\int_{\pi/3}^{\pi/2} \cos^2 z\, dz,$$

which is of course the same result as before. A general procedure for integrals of this type is presented in Sec. 3.5.

Example 3 Show that

$$\int_{-a}^{a} f(x)\, dx = \begin{cases} 0 & \text{if } f(x) \text{ is an odd function;} \\ 2\int_0^a f(x)\, dx & \text{if } f(x) \text{ is even.} \end{cases}$$

Examination of the areas involved (Fig. 4.2) makes the statements rather obvious: the areas on either side of $x = 0$ cancel for an odd function and add for an even one. The formal derivation is based on the decomposition

$$\int_{-a}^{a} f(x)\, dx = \underbrace{\int_{-a}^{0} f(x)\, dx}_{\text{I}} + \underbrace{\int_0^a f(x)\, dx}_{\text{II}}.$$

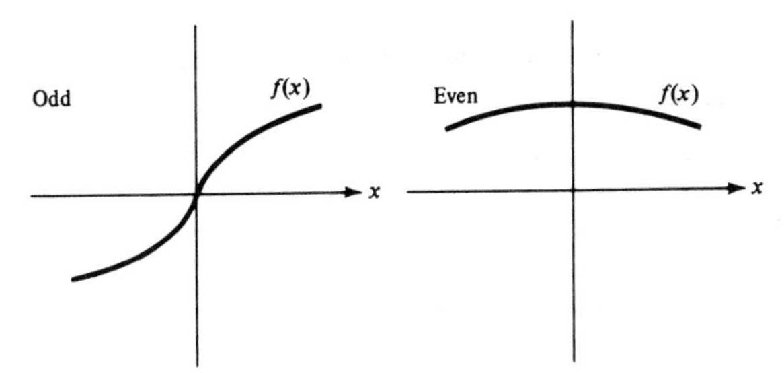

FIGURE 4.2

Apply the substitution $z = -x$ in the integral labelled I, so that

$$\int_{-a}^{0} f(x)\, dx = \int_a^0 f(-z)(-dz) = -\int_a^0 f(-z)\, dz,$$

$$= \int_0^a f(-z)\, dz = \int_0^a f(-x)\, dx.$$

Here, the exchange of notation for the dummy integration variable facilitates comparison of the two integrals. Therefore,

$$\int_{-a}^{a} f(x)\, dx = \int_0^a f(-x)\, dx + \int_0^a f(x)\, dx,$$

or

$$\int_{-a}^{a} f(x)\,dx = \int_{0}^{a} [f(-x) + f(x)]\,dx. \tag{11}$$

Now, for an odd function

$$f(-x) = -f(x) \qquad \text{or} \qquad f(-x) + f(x) = 0,$$

in which case the preceding integrand is zero. Therefore the integral is zero, and the first assertion is verified. For an even function

$$f(x) = f(-x) \qquad \text{or} \qquad f(x) + f(-x) = 2f(x),$$

which implies

$$\int_{0}^{a} [f(x) + f(-x)]\,dx = 2\int_{0}^{a} f(x)\,dx,$$

and this establishes the second result.

Example 4 Evaluate $J = \int_{-1}^{1} \sin(x \sin x^2) \cos^2 x(1 + x^2)^{-1/2} e^{x^2}\,dx$.

Fortunately, the integrand is an odd function, so that $J = 0$ (see Example 3).

---

2 INTEGRATION BY PARTS

This method, which is a natural and essential concomitant to the change-of-variable technique, is based on the differential formula

$$d(uv) = u\,dv + v\,du. \tag{12}$$

The corresponding integration formula,

$$uv = \int u\,dv + \int v\,du,$$

is applied in the form

$$\int u\,dv = uv - \int v\,du. \tag{13}$$

The procedure involves, then, the identification of u and dv and their manipulation into the form (13). Of course, the integral on the right-hand side must be more amenable than the one we start with or no benefit accrues. Note, too, that the function v must be easy to determine in any event. If a definite integral is involved

$$\int_{a}^{b} u\frac{dv}{dx}\,dx = uv\Big]_{a}^{b} - \int_{a}^{b} v\frac{du}{dx}\,dx. \tag{14}$$

Example 5 Find $\int \log x\,dx$.

Here, identify

$$u = \log x, \qquad dv = dx,$$

so that

$$du = \frac{1}{x}\,dx, \qquad v = x.$$

The substitution of these in (13) yields

$$\int \log x\,dx = x \log x - \int x \cdot \frac{1}{x}\,dx,$$
$$= x \log x - x + C.$$

Example 6 Find $\int x \sin ax\,dx$.

Let

$$u = x, \qquad dv = \sin ax\,dx,$$

so that

$$du = dx, \qquad v = -\frac{1}{a}\cos ax.$$

Integration by parts, with this decomposition, yields

$$\int x \sin ax\,dx = -\frac{x}{a}\cos ax - \int \left(-\frac{1}{a}\cos ax\right) dx,$$
$$= -\frac{x}{a}\cos ax + \frac{1}{a}\int \cos ax\,dx = -\frac{x}{a}\cos ax + \frac{1}{a^2}\sin ax + C.$$

Example 7 Find the area, shown in Fig. 4.3, bounded by the curve $y = \sin^{-1}x$ and the lines $x = 1$ and $y = 0$.

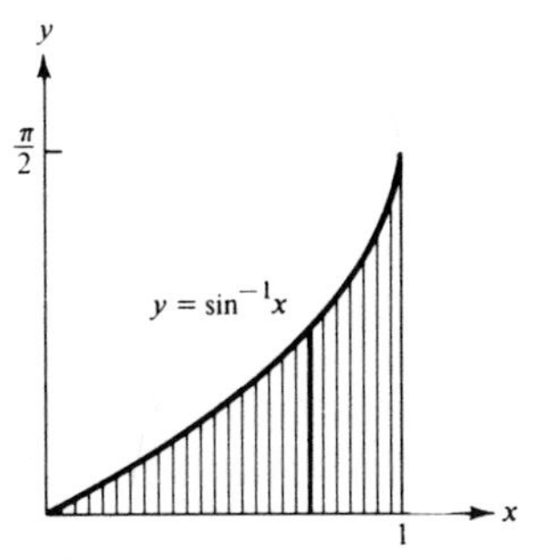

FIGURE 4.3

The area is given by the integral

$$A = \int_0^1 \sin^{-1} x \, dx$$

(where we imagine vertical strips placed side by side from $x = 0$ to $x = 1$). To find A, we integrate by parts using

$$u = \sin^{-1} x, \qquad dv = dx,$$

$$du = \frac{1}{(1 - x^2)^{1/2}} dx, \qquad v = x,$$

and, simultaneously, evaluate the functions that arise (instead of waiting to do this at the very end). Therefore by (14)

$$\begin{aligned} A = \int_0^1 \sin^{-1} x \, dx &= x \sin^{-1} x \Big]_0^1 - \int_0^1 \frac{x}{(1 - x^2)^{1/2}} dx, \\ &= \frac{\pi}{2} - \int_0^1 \frac{x}{(1 - x^2)^{1/2}} dx = \frac{\pi}{2} + (1 - x^2)^{1/2} \Big]_0^1, \\ &= \frac{\pi}{2} - 1. \end{aligned}$$

This area could have been computed more easily by imagining it to consist of a stack of horizontal strips, as in Fig. 4.4. In this view, the element of rectangular area is $dA = (1 - x)\,dy$, where $x = \sin y$ and the integration variable y runs from zero to $\pi/2$. Therefore

$$\begin{aligned} A &= \int_0^{\pi/2} (1 - \sin y)\,dy, \\ &= [y + \cos y]_0^{\pi/2} = \frac{\pi}{2} - 1. \end{aligned}$$

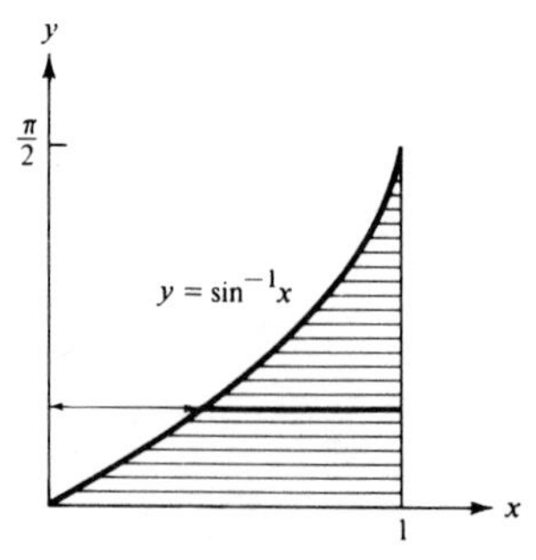

FIGURE 4.4

Example 8 Find $\int x^{1/2}e^{x^{1/2}}\,dx$.

A switch to the variable $z = x^{1/2}$ simplifies the integrand and facilitates further manipulation. Since $dz = dx/(2x^{1/2})$, we find that

$$\int x^{1/2}e^{x^{1/2}}\,dx = 2\int z^2e^z\,dz.$$

Integration by parts is now in order to eliminate z^2 in the last integral, and to this end we identify

$$u = z^2, \qquad dv = e^z\,dz,$$

so that

$$du = 2z\,dz, \qquad v = e^z.$$

The result is

$$\int z^2e^z\,dz = z^2e^z - 2\int ze^z\,dz.$$

Another integration by parts is required, this time with

$$u = z, \qquad dv = e^z\,dz,$$

and

$$du = dz, \qquad v = e^z,$$

in which case

$$\int ze^z\,dz = ze^z - \int e^z\,dz = ze^z - e^z + C.$$

If all the results are collected together, we obtain

$$\begin{aligned}\int x^{1/2}e^{x^{1/2}}\,dx &= 2[z^2e^z - 2(ze^z - e^z)] + C^* \\ &= 2e^{x^{1/2}}(x - 2x^{1/2} + 2) + C^*.\end{aligned}$$

Example 9 Derive the *reduction formula*

$$\int x^ne^{ax}\,dx = \frac{1}{a}x^ne^{ax} - \frac{n}{a}\int x^{n-1}e^{ax}\,dx.$$

Write

$$u = x^n, \qquad dv = e^{ax}\,dx,$$

so that

$$du = nx^{n-1}\,dx, \qquad v = \frac{1}{a}e^{ax};$$

the reduction formula is just an integration by parts for this particular separation. Note that the procedure reduces by 1 the power of x in the integrand. If, in particular, n is a positive integer, then after n repetitions, the simple integral $\int e^{ax}\,dx$ will emerge and we will have an explicit formula for the original integral.

Example 10 Find $\int e^{-x^2}\,dx$.

This integral *cannot* be expressed in terms of the elementary functions available at present. It will be studied in detail later.

Example 11 Find $\int e^x \sin x\,dx$.

An integration by parts with $u = \sin x$, $v = e^x$ yields

$$\int e^x \sin x\,dx = e^x \sin x - \int e^x \cos x\,dx,$$

which does not seem to be much of an improvement. However, a second integration by parts with $u = \cos x$, $v = e^x$ leads to the reappearance of the integral in question:

$$\int e^x \sin x\,dx = e^x \sin x - e^x \cos x - \int e^x \sin x\,dx.$$

This equation can now be solved for the unknown integral:

$$\int e^x \sin x\,dx = \tfrac{1}{2}\, e^x(\sin x - \cos x).$$

This formula, like all indefinite integrals, is readily checked by differentiating the answer to recover the integrand.

Exercises 3.4

1. Write the following definite integrals in terms of the new variable u and evaluate.

(i) $\displaystyle\int_0^4 \frac{x\,dx}{(5x+1)^{1/2}}$ $\quad u = 5x + 1$ **(ii)** $\displaystyle\int_1^3 \frac{x^2}{(2x-1)^{1/3}}\,dx$ $\quad u = 2x - 1$

(iii) $\displaystyle\int_0^2 x^2(9 - x^3)^{1/2}\,dx$ $\quad u = 9 - x^3$ **(iv)** $\displaystyle\int_0^1 \frac{x^{1/2}}{1 + x^{1/4}}\,dx$ $\quad u = 1 + x^{1/4}$

(v) $\displaystyle\int_0^{\pi/2} \frac{4\sin\theta}{2 + 9\cos\theta}\,d\theta$ $\quad u = \cos\theta$ **(vi)** $\displaystyle\int_2^3 \frac{1}{x\log x}\,dx$ $\quad u = \log x$

2. For the following integrals, graph the area $\int_a^b f(x)\,dx$ in the xy plane and the corresponding area $\int_{u(a)}^{u(b)} g(u)\,du$ produced by the transformation $u = u(x)$.

(i) $\displaystyle\int_0^5 \frac{x}{(x^2+1)^{1/2}}\,dx$ $\quad u = x^2 + 1$ **(ii)** $\displaystyle\int_0^2 xe^{-x^2}\,dx$ $\quad u = x^2$

(iii) $\int_0^1 \frac{e^x}{e^x+1}\,dx$ $\quad u = e^x + 1$ $\quad$ **(iv)** $\int_0^1 \tan x\,dx$ $\quad u = \cos x$

3. Make a change of variables and evaluate the following integrals.

(i) $\int_{1/2}^1 (1+x)^{1/3}(2-x)\,dx$ $\quad$ **(ii)** $\int_2^3 \frac{x}{(x-1)^{1/2}}\,dx$ $\quad$ **(iii)** $\int_0^2 \frac{x\,dx}{(x^2+1)^{2/3}}$

(iv) $\int_0^2 x(1+x)^{1/3}\,dx$ $\quad$ **(v)** $\int_1^2 \frac{x}{(1+x)^{1/2}}\,dx$ $\quad$ **(vi)** $\int_{-1}^1 x^3 e^{-x^4}\,dx$

4. Integrate by parts to evaluate the following integrals:

(i) $\int x^2 \cos ax\,dx$ $\quad$ **(ii)** $\int x^3 \sin ax\,dx$ $\quad$ **(iii)** $\int x^4 \cos ax\,dx$

5. Show that $\int_0^a f(x)\,dx = \int_0^a f(a-x)\,dx$. Explain by means of a graph why this result is obvious.

6. Show that $\int_0^\pi xf(\sin x)\,dx = \pi \int_0^{\pi/2} f(\sin x)\,dx$ and use this result to evaluate the following definite integrals.

(i) $\int_0^\pi x \sin x\,dx$ $\quad$ **(ii)** $\int_0^\pi x \sin^2 x\,dx$ $\quad$ **(iii)** $\int_0^\pi \frac{x \sin x}{1+\cos^2 x}\,dx$

7. Using the change of variable $x = t + 1/t$, show that

$$\int_{1/2}^2 \left(1 - \frac{1}{t^2}\right) f\left(t + \frac{1}{t}\right) dt = \int_{5/2}^2 f(x)\,dx + \int_2^{5/2} f(x)\,dx = 0.$$

Evaluate the indefinite integral

$$\int \frac{t^2-1}{t^4+1}\,dt$$

and verify the result for the integral from $t = 1/2$ to $t = 2$.

8. Show that $\int \csc x\,dx = \log|\csc x - \cot x| + C$.

9. Evaluate the following integrals.

(i) $\int x \tan^{-1} x\,dx$ $\quad$ **(ii)** $\int \tan^{-1} x\,dx$ $\quad$ **(iii)** $\int \sin^{-1} x\,dx$

(iv) $\int x \sec^2 x\,dx$ $\quad$ **(v)** $\int x^2 e^{ax}\,dx$ $\quad$ **(vi)** $\int (\log x)^2\,dx$

(vii) $\int x^{1/2} \log x\,dx$ $\quad$ **(viii)** $\int \sin(\log x)\,dx$ $\quad$ **(ix)** $\int x^{1/2} \cos x^{1/2}\,dx$

10. Define the integrals $I_1 = \int e^{ax} \sin bx\,dx$ and $I_2 = \int e^{ax} \cos bx\,dx$. By integration by parts, verify that $aI_1 + bI_2 = e^{ax} \sin bx + c_1$ and $bI_1 - aI_2 = -e^{ax} \cos bx + c_2$ where c_1 and c_2 are constants. Use these results to evaluate I_1 and I_2.

11. Evaluate the following integrals:

(i) $\int e^{2x} \cos 3x\,dx$ $\quad$ **(ii)** $\int e^{-x} \sin x\,dx$ $\quad$ **(iii)** $\int e^{3x} \sin 5x\,dx$

12. Express $\int \frac{\cos x}{x^2}\,dx$ in terms of $\int \frac{\sin x}{x}\,dx$ (not a simple function). Similarly, express $\int \frac{\sin x}{x^2}\,dx$ in terms of $\int \frac{\cos x}{x}\,dx$.

13. Using the inequalities $16 < (4 + \sin x)^2 < (4 + x)^2$ valid for $0 < x < \pi/2$, find bounds for the integral $\int_0^{\pi/2} \frac{dx}{(4 + \sin x)^2}$.

14. Verify the bounds for the following integrals.

(i) $0.5 < \int_0^{1/2} (1 - x^4)^{-1/2}\,dx < 0.524$ $\quad$ **(ii)** $\pi/3 < \int_0^{\pi/2} (\sin x)^{1/2}\,dx < 2/3\,(\pi/2)^{3/2}$

15. Integrate by parts to establish the reduction formula ($n>1$)

$$\int \frac{dx}{(x^2+a^2)^n} = \frac{x}{(2n-2)a^2(x^2+a^2)^{n-1}} + \frac{2n-3}{(2n-2)a^2}\int \frac{dx}{(x^2+a^2)^{n-1}}.$$

16. Use the result of the last problem to evaluate the following integrals.

(i) $\int \frac{dx}{(1+x^2)^2}$ **(ii)** $\int \frac{dx}{(4x^2+25)^2}$ **(iii)** $\int_0^2 \frac{dx}{(x^2+x+1)^2}$

17. Integrate by parts to derive the following reduction formulas.

(i) $\int x^m(\log x)^n\,dx = \frac{x^{m+1}(\log x)^n}{m+1} - \frac{n}{m+1}\int x^m(\log x)^{n-1}\,dx$

(ii) $\int \sin^n x\,dx = \frac{-1}{n}\sin^{n-1}x\cos x + \frac{n-1}{n}\int \sin^{n-2}x\,dx$

(iii) $\int \cos^n x\,dx = \frac{1}{n}\cos^{n-1}x\sin x + \frac{n-1}{n}\int \cos^{n-2}x\,dx$

18. From the result of the last problem, establish the formula

$$\int_0^{\pi/2} \sin^{2n}x\,dx = \int_0^{\pi/2} \cos^{2n}x\,dx = \frac{(2n-1)(2n-3)\ldots 3\cdot 1}{(2n)(2n-2)\ldots 4\cdot 2}\,\frac{\pi}{2}.$$

19. Find the value of $\int_0^{\pi/2} \cos^{2n+1}x\,dx = \int_0^{\pi/2} \sin^{2n+1}x\,dx$.

20. Evaluate $\int \sin(\log ax)\,dx$ and $\int \cos(\log ax)\,dx$.

21. Integrate by parts to establish reduction formulas for $\int x^n \cos ax\,dx$ and $\int x^n \sin ax\,dx$. Evaluate $\int x^2 \cos 3x\,dx$ and $\int x^2 \sin 3x\,dx$.

3.5 *Techniques of integration: III*

1 The entries in a table of integrals† are usually arranged according to the functional form of the integrand. As in any dictionary, this arrangement is for ease in locating a particular listing. But the ordering also correlates surprisingly well with a tabulation of different methods, because a special technique of integration is usually associated with each category. In this section, many of the methods introduced earlier in isolated examples are discussed again from a more general vantage point. The objective is to learn how to select a promising technique based on the general structure of an integral.

Certain types of integrals are so firmly connected with a definite approach that it has become possible in recent years to have a computing machine perform the integration symbolically. In other words, a procedure may be so automatic and the directions made so explicit that the task is reduced to one of labor alone (and a great deal of that). Needless to say, most integrals are not of this type.

COMPLETING THE SQUARE

2 An integral which involves the expression $ax^2 + bx + c$ can be simplified by completing the square and making an appropriate change of variable. The simplification results from the identity

$$ax^2 + bx + c = a\left[\left(x + \frac{b}{2a}\right)^2 + \frac{4ac - b^2}{4a^2}\right], \tag{1}$$

† Our table, which appears as Table 5.1 in Sec. A.5 of the Appendix, is only a partial compilation.

and the further substitution $u = x + b/2a$. The sign of $q = (4ac - b^2)/4a^2$ plays a crucial role in the form of the solution, as shown next.

Example 1 Find $\int \frac{dx}{ax^2 + bx + c}$.

According to (1), this can be written

$$\int \frac{dx}{ax^2 + bx + c} = \frac{1}{a} \int \frac{du}{u^2 + q}.$$

For q a positive number, set $q = \omega^2$, so that with $v = u/\omega$

$$\int \frac{du}{u^2 + \omega^2} = \frac{1}{\omega} \int \frac{dv}{v^2 + 1} = \frac{1}{\omega} \tan^{-1} v + C = \frac{1}{\omega} \tan^{-1} \frac{u}{\omega} + C,$$

or

$$\int \frac{dx}{ax^2 + bx + c} = \frac{1}{a\sqrt{q}} \tan^{-1} \frac{x + b/2a}{\sqrt{q}} + C, \qquad q > 0.$$

On the other hand, for q negative, let $-q = \omega^2$, so that

$$\int \frac{du}{u^2 - \omega^2} = \frac{1}{2\omega} \left(\int \frac{du}{u - \omega} - \int \frac{du}{u + \omega} \right) = \frac{1}{2\omega} \log \left| \frac{u - \omega}{u + \omega} \right| + C,$$

and finally

$$\int \frac{dx}{ax^2 + bx + c} = \frac{1}{2a\sqrt{-q}} \log \left| \frac{x + b/2a - \sqrt{-q}}{x + b/2a + \sqrt{-q}} \right| + C, \qquad q < 0.$$

Example 2 Find $I_1 = \int \frac{dx}{x^2 + 2x + 2}$ and $I_2 = \int \frac{dx}{4x^2 + 4x - 3}$.

These integrals are special cases of the previous example. Completing the squares, we have $x^2 + 2x + 2 = (x + 1)^2 + 1$ and $4x^2 + 4x - 3 = 4\left[\left(x + \frac{1}{2}\right)^2 - 1\right]$. In the first integral, define $u = x + 1$ and, in the second, define $u = x + \frac{1}{2}$. Therefore,

$$\begin{aligned} I_1 &= \int \frac{dx}{(x + 1)^2 + 1} = \int \frac{du}{u^2 + 1} \\ &= \tan^{-1} u + C = \tan^{-1}(x + 1) + C, \end{aligned}$$

and

$$\begin{aligned} I_2 &= \int \frac{dx}{4[(x + 1/2)^2 - 1]} = \frac{1}{4} \int \frac{du}{u^2 - 1} \\ &= \frac{1}{8} \log \left| \frac{u - 1}{u + 1} \right| + C = \frac{1}{8} \log \left| \frac{x - 1/2}{x + 3/2} \right| + C. \end{aligned}$$

In each case, C is an arbitrary constant. These results are verified easily by differentiating.

---

3 PARTIAL FRACTIONS

A polynomial of degree n

$$P_n(x) = a_0 + a_1 x + \cdots + a_n x^n, \tag{2}$$

can be written as a product of factors

$$P_n(x) = a_n(x - x_1)(x - x_2) \ldots (x - x_n),$$

where x_i is the ith root: $P_n(x_i) = 0.$

The roots which are complex occur in conjugate pairs, and in order to avoid any imaginary values in this analysis, these factors are combined into real quadratic expressions of the form $x^2 + \alpha x + \beta$. The decomposition of a polynomial using only real numbers involves, then, a product of linear terms, $x - x_i$, where x_i is a real root, multiplied by a succession of quadratic expressions which represent the coalescence of the complex factors. If there are r real roots, then $n - r = 2k$ (an even number) are complex and

$$P_n(x) = a_n(x - x_1)(x - x_2) \cdots (x - x_r)(x^2 + \alpha_1 x + \beta_1) \cdots (x^2 + \alpha_k x + \beta_k). \tag{3}$$

The integration of rational functions is based on this representation.

By the process of long division, any rational function can be written as a polynomial plus a remainder which is called a *proper rational function* if the degree of its numerator polynomial is less than that of its denominator. Since the integration of a polynomial is, by now, trivial, we concentrate on the latter. Let

$$R(x) = \frac{b_0 + b_1 x + \cdots + b_m x^m}{a_0 + a_1 x + \cdots + a_n x^n}, \tag{4}$$

where $m < n$, and assume for the moment that the denominator has no multiple roots. It follows, then, using the factorization in (3), that this proper rational function can be written as the *partial-fraction expansion*

$$R(x) = \frac{A_1}{x - x_1} + \cdots + \frac{A_r}{x - x_r} + \frac{B_1 x + C_1}{x^2 + \alpha_1 x + \beta_1} + \cdots + \frac{B_k x + C_k}{x^2 + \alpha_k x + \beta_k}. \tag{5}$$

The coefficients $A_1, \ldots, A_r; B_1, \ldots, B_k; C_1, \ldots, C_k$ can be determined by collapsing (5) back into (4) (with the aid of a common denominator) and by then identifying like powers of x in each numerator. This is, in general, a lengthy task; many short cuts have been devised, but these important algebraic techniques will not concern us here, although some are discussed in the exercises.

If the denominator in (4) has a multiple root, $(x - x_1)^l$, then the partial-fraction representation (5) must include the terms

$$\frac{A_{11}}{x - x_1} + \frac{A_{12}}{(x - x_1)^2} + \cdots + \frac{A_{1l}}{(x - x_1)^l};$$

likewise the occurrence of the factor $(x^2 + \alpha_1 x + \beta_1)^l$ implies that the decomposition includes the terms

$$\frac{B_{11}x + C_{11}}{x^2 + \alpha_1 x + \beta_1} + \cdots + \frac{B_{1l}x + C_{1l}}{(x^2 + \alpha_1 x + \beta_1)^l}.$$

The integral of a rational function can be reduced to a series of simpler integrations upon replacing the integrand by its partial-fraction expansion. The general types of integrals which arise in this process are

$$\int \frac{dx}{(x - \alpha)^l} \quad \text{and} \quad \int \frac{Ax + B}{(x^2 + \alpha x + \beta)^l}\,dx.$$

As a consequence, we will show that the indefinite integral of a rational function can be completely described by rational functions, logarithms, and arc tangents. It is for these reasons that an algorithm for symbolic integration of any rational function can be programmed for high-speed computers.

---

4 **Example 3** Integrate $\displaystyle\int \frac{x}{(x - 1)(x^2 + 4)}\,dx.$

The integrand is a proper rational function and can be written

$$\frac{x}{(x - 1)(x^2 + 4)} = \frac{A}{x - 1} + \frac{Bx + C}{x^2 + 4}.$$

The coefficients A, B, C are determined by collecting terms,

$$\frac{A}{x - 1} + \frac{Bx + C}{x^2 + 4} = \frac{(A + B)x^2 + (-B + C)x + (4A - C)}{(x - 1)(x^2 + 4)},$$

and equating the polynomial in the numerator with that of the original function. This implies

$$x = (A + B)x^2 + (-B + C)x + 4A - C;$$

and since like powers of x must be the same, we see that

$$A + B = 0, \qquad -B + C = 1, \qquad 4A - C = 0$$

or

$$A = 1/5, \qquad B = -1/5, \qquad C = 4/5.$$

Therefore

$$\frac{x}{(x-1)(x^2+4)} = \frac{1}{5}\frac{1}{(x-1)} - \frac{1}{5}\frac{x-4}{x^2+4},$$

and consequently

$$\int \frac{x\,dx}{(x-1)(x^2+4)} = \frac{1}{5}\int \frac{dx}{x-1} - \frac{1}{5}\int \frac{x\,dx}{x^2+4} + \frac{4}{5}\int \frac{dx}{x^2+4},$$

$$= \tfrac{1}{5}\log|x-1| - \tfrac{1}{10}\log|x^2+4| + \tfrac{2}{5}\tan^{-1}\frac{x}{2} + C.$$

Example 4 Integrate $\int \frac{x^3}{x^3-3x+2}\,dx.$

The integrand is first made a proper rational function by long division:

$$\frac{x^3}{x^3-3x+2} = 1 + \frac{3x-2}{x^3-3x+2}.$$

Second, the denominator is factored

$$x^3 - 3x + 2 = (x-1)^2(x+2),$$

and a double root is noted. It follows that

$$\frac{3x-2}{x^3-3x+2} = \frac{8}{9}\frac{1}{x-1} + \frac{1}{3}\frac{1}{(x-1)^2} - \frac{8}{9}\frac{1}{x+2}.$$

(This calculation by the method of the last example involves a system of three linear equations and is already sufficiently tedious to motivate a search for a more efficient technique.) Consequently,

$$\int \frac{x^3}{x^3-3x+2}\,dx = \int dx + \frac{8}{9}\int \frac{dx}{x-1} + \frac{1}{3}\int \frac{dx}{(x-1)^2} - \frac{8}{9}\int \frac{dx}{x+2},$$

$$= x + \frac{8}{9}\log|x-1| - \frac{1}{3(x-1)} - \frac{8}{9}\log|x+2| + C.$$

---

5 **Example 5** Find explicit formulas, in terms of known functions, for the general integrals which underlie the method of partial fractions, namely,

$$\int \frac{dx}{(x-a)^l}; \qquad \int \frac{Ax+B}{(x^2+\alpha x+\beta)^l}\,dx.$$

The first of these is handled with the substitution $u = x - a$:

$$\int \frac{dx}{(x-a)^l} = \begin{cases} \int \frac{du}{u^l} = \frac{-1}{(l-1)u^{l-1}} + C = \frac{-1}{(l-1)(x-a)^{l-1}} + C & \text{for } l > 1; \\ \log|x-a| + C & \text{for } l = 1. \end{cases}$$

In the second integral the roots of the quadratic are assumed to be complex, which is equivalent to the assertion

$$\beta - \frac{\alpha^2}{4} > 0.$$

Set $\omega^2 = \beta - \alpha^2/4$ in order to complete the square of the quadratic:

$$x^2 + \alpha x + \beta = \left(x + \frac{\alpha}{2}\right)^2 + \omega^2.$$

Furthermore, the substitution $u = x + \alpha/2$ transforms the integral to

$$\int \frac{Ax+B}{(x^2+\alpha x+\beta)^l}\,dx = A\int \frac{u\,du}{(u^2+\omega^2)^l} + \left(B - \frac{A\alpha}{2}\right)\int \frac{du}{(u^2+\omega^2)^l}.$$

The first integral on the right-hand side is simply executed:

$$\int \frac{u}{(u^2+\omega^2)^l}\,du = \begin{cases} \frac{-1}{2(l-1)(u^2+\omega^2)^{l-1}} + C & \text{for } l > 1; \\ \tfrac{1}{2}\log(u^2+\omega^2) + C & \text{for } l = 1. \end{cases} \qquad (6)$$

A general reduction formula for the last integral is obtained by integration by parts, but this is left to the exercises. [Note, however, that for $l = 1$ the integral is $\omega^{-1}\tan^{-1}(u/\omega)$.]

TRIGONOMETRIC SUBSTITUTIONS

6 Integrals involving the expressions $(a^2 - x^2)^{1/2}$ and $(a^2 + x^2)^{1/2}$ may be susceptible to a trigonometric substitution which converts the radical to an ordinary function of some angle variable θ. In doing this, it is convenient to represent the variables as the appropriate sides of a right triangle, for then the relationship to x of any trigonometric function that may arise can be deduced easily. For example, let $x = a\sin\theta$, so that

$$(a^2 - x^2)^{1/2} = a(1 - \sin^2\theta)^{1/2} = a\cos\theta, \qquad dx = a\cos\theta\,d\theta$$

[and according to Fig. 5.1, $\tan\theta = x/(a^2 - x^2)^{1/2}$, etc.].

The term $(a^2 + x^2)^{1/2}$ is transformed using $x = a\tan\theta$ (see Fig. 5.2), in which case

$$(a^2 + x^2)^{1/2} = a\sec\theta, \qquad dx = a\sec^2\theta\,d\theta.$$

Lastly, the term $(x^2 - a^2)^{1/2}$, for $x > a$, yields to the substitution $x = a\sec\theta$ (see Fig. 5.3), so that

$$(x^2 - a^2)^{1/2} = a\tan\theta, \qquad dx = a\sec\theta\tan\theta\,d\theta.$$

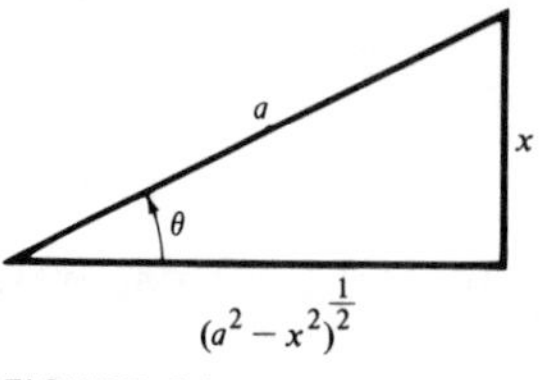

FIGURE 5.1

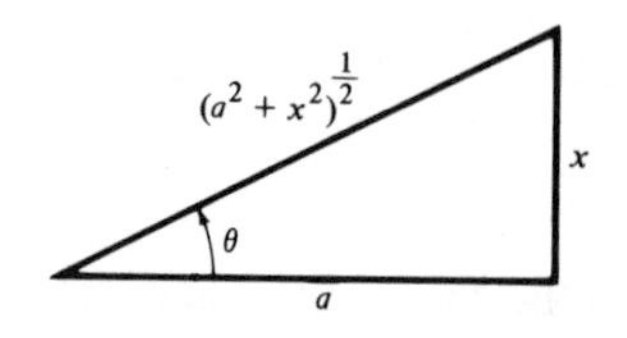

FIGURE 5.2

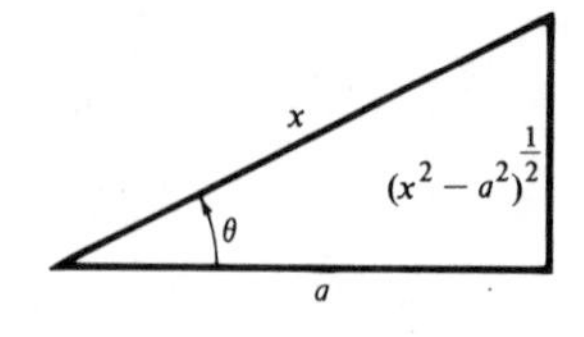

FIGURE 5.3

Example 6 Integrate $I_1 = \int (a^2 - x^2)^{1/2}\, dx$ and $I_2 = \int (a^2 + x^2)^{-1/2}\, dx$.

In the first integral, define $x = a \sin\theta$, in which case $(a^2 - x^2)^{1/2} = a \cos\theta$ and $dx = a \cos\theta\, d\theta$. With this substitution, the integral becomes

$$\begin{aligned}
I_1 &= \int a^2 \cos^2\theta\, d\theta = a^2 \int \frac{\cos 2\theta + 1}{2}\, d\theta \\
&= \frac{a^2}{4} \sin 2\theta + \frac{a^2}{2}\theta + C = \frac{a^2}{2} \sin\theta \cos\theta + \frac{a^2}{2}\theta + C \\
&= \frac{a^2}{2}\frac{x}{a}\left(1 - \frac{x^2}{a^2}\right)^{1/2} + \frac{a^2}{2}\sin^{-1}\frac{x}{a} + C \\
&= \frac{x}{2}(a^2 - x^2)^{1/2} + \frac{a^2}{2}\sin^{-1}\frac{x}{a} + C.
\end{aligned}$$

For the second integral, define $x = a \tan\theta$. Then $a^2 + x^2 = a^2 \sec^2\theta$ and $dx = a \sec^2\theta\, d\theta$. The integral is

$$\begin{aligned}
I_2 &= \int (a^2 \sec^2\theta)^{-1/2} a \sec^2\theta\, d\theta = \int \sec\theta\, d\theta \\
&= \log|\sec\theta + \tan\theta| + C = \log\left|\frac{x}{a} + \sqrt{1 + \frac{x^2}{a^2}}\right| + C \\
&= \log\left|\frac{x + \sqrt{x^2 + a^2}}{a}\right| + C.
\end{aligned}$$

In both calculations, C is the arbitrary constant of integration.

Example 7 Integrate $\int x^3(4 - x^2)^{1/2}\, dx$.

As shown in Fig. 5.1, we introduce the change of variable $x = 2 \sin\theta$. Since $dx = 2 \cos\theta\, d\theta$, $x^3 = 8 \sin^3\theta$, and $(4 - x^2)^{1/2} = 2 \cos\theta$, it follows that

$$\begin{aligned}
\int x^3(4 - x^2)^{1/2}\, dx &= \int 8 \sin^3\theta\, (2 \cos\theta)(2 \cos\theta)\, d\theta, \\
&= 32 \int \sin^3\theta \cos^2\theta\, d\theta.
\end{aligned}$$

The further change of variable, $u = \cos\theta$, $du = -\sin\theta\, d\theta$, enables us to write

$$\int \sin^3\theta\cos^2\theta\, d\theta = -\int (1-u^2)u^2\, du = -\frac{u^3}{3} + \frac{u^5}{5} + C,$$

$$= -\tfrac{1}{3}\cos^3\theta + \tfrac{1}{5}\cos^5\theta + C.$$

Upon collecting these results and converting back to x, it follows that

$$\int x^3(4-x^2)^{1/2}\, dx = 32\left[\frac{1}{5}\left(1-\frac{x^2}{4}\right)^{5/2} - \frac{1}{3}\left(1-\frac{x^2}{4}\right)^{3/2}\right] + C^*.$$

Example 8 Find the area of the ellipse $x^2/a^2 + y^2/b^2 = 1$, shown in Fig. 5.4.

The total area A is 4 times that in the first quadrant. By adding vertical strips of incremental area $y\, dx$, from $x = 0$ to $x = a$ we find that

$$A = 4\int_0^a y\, dx = 4b\int_0^a \left(1-\frac{x^2}{a^2}\right)^{1/2} dx.$$

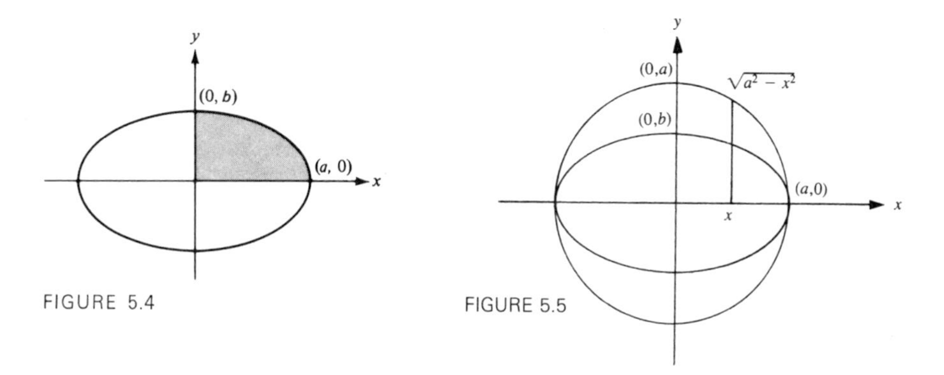

FIGURE 5.4

FIGURE 5.5

Let $x = a\sin\theta$, so that $dx = a\cos\theta\, d\theta$; $x = 0$ and $x = a$ correspond then to $\theta = 0$ and $\pi/2$. This transformation yields

$$A = 4b\int_0^a \left(1-\frac{x^2}{a^2}\right)^{1/2} dx = 4ab\int_0^{\pi/2} \cos^2\theta\, d\theta$$

and

$$A = 4ab\left[\frac{\theta}{2} + \frac{1}{4}\sin 2\theta\right]_0^{\pi/2} = \pi ab.$$

Archimedes gave a simple proof of this result by comparing heights of the ellipse and the circle $x^2 + y^2 = a^2$ (Fig. 5.5). The height on the ellipse corresponding to x is $(b/a)\sqrt{a^2 - x^2}$. But the y coordinate on the circle is

$\sqrt{a^2 - x^2}$. Therefore, the area of the ellipse is the area of the circle multiplied by b/a, $A = (\pi a^2)(b/a) = \pi ab$.

Example 9 Integrate $I(x) = \int (1 + x^2)^{1/2}\, dx$.

Integration by parts, with $du = dx$, and $v = (1 + x^2)^{1/2}$ yields

$$I(x) = \int (1 + x^2)^{1/2}\, dx = x(1 + x^2)^{1/2} - \int \frac{x^2}{(1 + x^2)^{1/2}}\, dx,$$

which can be expressed as

$$\begin{aligned} I(x) &= x(1 + x^2)^{1/2} - \int \frac{1 + x^2 - 1}{(1 + x^2)^{1/2}}\, dx, \\ &= x(1 + x^2)^{1/2} - \int (1 + x^2)^{1/2}\, dx + \int \frac{dx}{(1 + x^2)^{1/2}}, \\ &= x(1 + x^2)^{1/2} - I(x) + \int \frac{dx}{(1 + x^2)^{1/2}}. \end{aligned}$$

Upon rearrangement, we obtain

$$I(x) = \frac{1}{2}\left[x(1 + x^2)^{1/2} + \int \frac{dx}{(1 + x^2)^{1/2}}\right].$$

The integral still on the right-hand side of this equation succumbs to the trigonometric substitution

$$x = \tan\theta,\ dx = \sec^2\theta\, d\theta$$

in which case

$$\begin{aligned} \int \frac{dx}{(1 + x^2)^{1/2}} &= \int \frac{\sec^2\theta\, d\theta}{\sec\theta} = \int \sec\theta\, d\theta \\ &= \log(\tan\theta + \sec\theta) \\ &= \log(x + (1 + x^2)^{1/2}). \end{aligned}$$

Therefore,

$$I(x) = \frac{1}{2}\left[x(1 + x^2)^{1/2} + \log(x + (1 + x^2)^{1/2})\right].$$

---

7 The integral of any rational function of $\sin x$ and $\cos x$, a simple one being

$$\int \frac{dx}{a + b\cos x},$$

can always be transformed into a rational function of a new variable z, defined by

$$z = \tan x/2. \tag{7}$$

Since

$$\sin x = 2 \sin\frac{x}{2} \cos\frac{x}{2} = 2\left(\sin\frac{x}{2} \bigg/ \cos\frac{x}{2}\right)\cos^2\frac{x}{2}$$

and

$$\cos x = \cos^2\frac{x}{2} - \sin^2\frac{x}{2} = \cos^2\frac{x}{2}\left(1 - \sin^2\frac{x}{2} \bigg/ \cos^2\frac{x}{2}\right)$$

we conclude that

$$\sin x = \frac{2\tan x/2}{1 + \tan^2 x/2} = \frac{2z}{1+z^2} \qquad \text{and} \qquad \cos x = \frac{1 - \tan^2 x/2}{1 + \tan^2 x/2} = \frac{1 - z^2}{1 + z^2}.$$

These identities are the foundation of this reduction, the details of which are illustrated by example.

Example 10 Integrate $\displaystyle\int \frac{dx}{2 + 3\cos x}$.

The transformation $z = \tan x/2$, implies that

$$dz = \tfrac{1}{2}\sec^2 x/2\, dx \qquad \text{or} \qquad dx = \frac{2dz}{1+z^2}$$

and

$$\frac{1}{2 + 3\cos x} = \frac{1}{2 + 3(1 - z^2)/(1 + z^2)} = \frac{1 + z^2}{5 - z^2}.$$

Therefore

$$\begin{aligned}\int \frac{dx}{2 + 3\cos x} &= \int \frac{2dz}{5 - z^2} = \frac{1}{\sqrt{5}}\left(\int \frac{dz}{\sqrt{5} - z} + \int \frac{dz}{\sqrt{5} + z}\right),\\ &= \frac{1}{\sqrt{5}}\log\left|\frac{\sqrt{5} + z}{\sqrt{5} - z}\right| + C,\\ &= \frac{1}{\sqrt{5}}\log\left|\frac{\sqrt{5} + \tan x/2}{\sqrt{5} - \tan x/2}\right| + C.\end{aligned}$$

Further elucidation of these and other techniques is left to the exercises and the applications in sections to follow. A compilation of the most common integrals is given in Appendix Table 5.1.

Exercises 3.5

1. Find the indefinite integral by completing the square or by partial fractions.

(i) $\int \frac{dx}{x^2+2x+5}$ **(ii)** $\int \frac{dx}{x^2-4x+4}$ **(iii)** $\int \frac{dx}{2x^2+2x+5}$

(iv) $\int \frac{dx}{3x^2-2x}$ **(v)** $\int \frac{dx}{x^2-x+1}$ **(vi)** $\int \frac{dx}{5x^2+9x-2}$

2. Show that

$$\frac{3x-2}{x^3-3x+2} = \frac{8}{9}\frac{1}{x-1} + \frac{1}{3}\frac{1}{(x-1)^2} - \frac{8}{9}\frac{1}{x+2}.$$

Evaluate $\int \frac{3x-2}{x^3-3x+2}\,dx$.

3. Integrate by the method of partial fractions.

(i) $\int \frac{2x-7}{x^2-3x+2}\,dx$ **(ii)** $\int \frac{dx}{x^2(x+1)}$ **(iii)** $\int \frac{dx}{(x-2)(x^2+1)}$

(iv) $\int \frac{dx}{4x^2-25}$ **(v)** $\int \frac{9x}{(x-1)^2(x+2)}\,dx$ **(vi)** $\int \frac{x+2}{(x+1)(x^2+x+1)}\,dx$

(vii) $\int \frac{dx}{x^2(x^2+1)}$ **(viii)** $\int \frac{dx}{9-x^2}$ **(ix)** $\int \frac{e^x\,dx}{e^{2x}-2e^x+1}$

4. Evaluate the following definite integrals.

(i) $\int_0^1 \frac{dx}{x^2+2x+2}$ **(ii)** $\int_{-1}^1 \frac{1+x^3}{1+x^2}\,dx$ **(iii)** $\int_1^{3/2} \frac{x+1}{x(x^2-4)}\,dx$

(iv) $\int_1^2 \frac{dx}{4x^2-25}$ **(v)** $\int_{-1}^1 \frac{x+x^3}{1+x^2}\,dx$ **(vi)** $\int_2^4 \frac{1}{x^2(x^2+1)}\,dx$

5. Factor the denominators to evaluate the following integrals.

(i) $\int \frac{dx}{1+x^3}$ **(ii)** $\int \frac{dx}{1+x^4}$ **(iii)** $\int \frac{x^4\,dx}{1+x^4}$

6. Integrate using the transformation indicated.

(i) $\int \frac{e^x\,dx}{(1-e^{2x})^{1/2}}$, $u=e^x$ **(ii)** $\int \frac{x\,dx}{x^2-1}$, $u=x^2-1$

(iii) $\int (\sin^{-1}x)^2\,dx$, $u=\sin^{-1}x$ **(iv)** $\int \frac{dx}{(4+x^2)^2}$, $u=\tan^{-1}\frac{x}{2}$

(v) $\int \frac{dx}{e^x+e^{-x}}$, $u=e^x$ **(vi)** $\int \frac{\tan^{-1}x}{1+x^2}$, $u=\tan^{-1}x$.

7. Integrate using a trigonometric substitution.

(i) $\int \frac{dx}{(4-x^2)^{1/2}}$ **(ii)** $\int \frac{dx}{9+x^2}$ **(iii)** $\int (1-x^2)^{1/2}\,dx$

(iv) $\int (25-x^2)^{3/2}\,dx$ **(v)** $\int \frac{x\,dx}{(1-x^2)^{3/2}}$ **(vi)** $\int \frac{dx}{(a^2+x^2)^{3/2}}$

(vii) $\int \frac{dx}{(16+x^2)^2}$ **(viii)** $\int \frac{dx}{x(5+x^2)^{1/2}}$ **(ix)** $\int \frac{x}{(9+x^2)^{1/3}}\,dx$

(x) $\int \frac{x^2}{(16-x^2)^{1/2}}\,dx$ **(xi)** $\int \frac{dx}{(9-4x^2)^{1/2}}$ **(xii)** $\int \frac{dx}{x(9-4x^2)^{1/2}}$

8. Using the substitution $z = \tan x/2$, evaluate the following integrals:

(i) $\int \sec x \, dx$ (ii) $\int \csc x \, dx$ (iii) $\int \frac{dx}{2 + \sin x}$

(iv) $\int \frac{dx}{2 - \cos x}$ (v) $\int \frac{\cos x}{2 - \cos x} dx$ (vi) $\int \frac{dx}{2 + \sin x + \cos x}$

9. Point out the mistake in the following argument:

$$\int_0^{\pi} \frac{dx}{\cos^2 x + \sin^2 x} = \int_0^{\pi} \frac{\sec^2 x \, dx}{1 + \tan^2 x} = \tan^{-1}(\tan x) \Big|_0^{\pi} = 0.$$

What is the correct value of the integral?

10. If $x>0$, show that the sum $\int_0^x \frac{dt}{1+t^2} + \int_0^{1/x} \frac{dt}{1+t^2}$ is independent of x. What is its value?

11. Evaluate by any method the following integrals:

(i) $\int \frac{x \, dx}{4 - x^2}$ (ii) $\int x(x^2 + 1)^3 \, dx$ (iii) $\int x^2(1 - x)^{1/2} \, dx$

(iv) $\int x^3(x^2 + 1)^{1/2} \, dx$ (v) $\int \frac{x \, dx}{(1 + x)^{1/5}}$ (vi) $\int \frac{dx}{(3 + 4x - 4x^2)^{1/2}}$

(vii) $\int x e^{-x^{1/3}} \, dx$ (viii) $\int \log(1 + x^2) \, dx$ (ix) $\int x \log(1 + x^2) \, dx$

(x) $\int \frac{dx}{x(x + 1)(x + 2)}$ (xi) $\int \frac{x^3 + 2x^2}{x^2 + x + 1} dx$ (xii) $\int \log[x + (x^2 - 1)^{1/2}] \, dx$

12. Evaluate the following integrals:

(i) $\int \tan^{-1} x \, dx$ (ii) $\int x \tan^2 x \, dx$ (iii) $\int \frac{\tan^{-1} x}{x^2} dx$

(iv) $\int \frac{1}{1 + \cos^2 x} dx$ (v) $\int x \sin 2x \cos 2x \, dx$ (vi) $\int x \sin^{-1} x \, dx$

(vii) $\int \frac{1 + \sin x}{x - \cos x} dx$ (viii) $\int \frac{\cos^2 x}{1 + \cos^2 x} dx$ (ix) $\int (\sin^{-1} x)^3 \, dx$

(x) $\int \sin \sqrt{x + 2} \, dx$ (xi) $\int \frac{d\theta}{2 + \sin\theta - \cos\theta}$ (xii) $\int \frac{3 + \sin\theta}{4 + \cos\theta} d\theta$

13. Evaluate the following definite integrals:

(i) $\int_5^{11} (11 + 10x - x^2)^{1/2} \, dx$ (ii) $\int_{-1}^{7} (7 + 6x - x^2)^{1/2} \, dx$ (iii) $\int_0^4 (4x - x^2)^{1/2} \, dx$

14. Evaluate the definite integrals in the previous problem by plotting graphs of the integrands.

15. Show that the integral of any rational function of $\sinh x$ and $\cosh x$ can always be transformed into the integral of a rational function of a new variable $z = \tanh(x/2)$. With this change of variable, show that $\sinh x = 2z/(1 - z^2)$, $\cosh x = (1 + z^2)/(1 - z^2)$ and $dx = 2 \, dz/(1 - z^2)$.

16. Using the substitution $z = \tanh(x/2)$, evaluate the following integrals:

(i) $\int \operatorname{sech} x \, dx$ (ii) $\int \operatorname{csch} x \, dx$ (iii) $\int \frac{dx}{1 + \cosh x}$

(iv) $\int \frac{dx}{\sinh x + \cosh x}$ (v) $\int \frac{dx}{(\sinh x + \cosh x)^2}$ (vi) $\int (\cosh x - 1)^{1/2} \, dx$

3.6 *Area and arc length*

The representation of an area as an integral (and vice versa) has been the focus of attention from the very outset. The technique of integration is now sufficiently far advanced for the areas of many plane figures to be calculated exactly, and in this connection it seems fitting to include some general advice. An area can be formulated as an integral in a number of different ways, and it pays to give careful thought beforehand to which might be the simplest or require the least effort. The integral should then be meticulously set up and the limits of integration double-checked so that their correctness is absolutely assured. The integration is performed using *any* of the techniques available or by resorting to a good standard table; every device, every symmetry should be exploited to reduce the actual computation required. Finally, it is a good idea to examine the answer, and the very least that can be done is to make sure that its *sign* is correct. A very common error, camouflaged by the intricacies of the analysis, is to wind up with a negative sign for a definitely positive quantity such as area, or length, etc. A wrong sign is the first and most apparent symptom of disorder; another is a dimensional inconsistency. A few advanced problems will illustrate the procedure and the flexibility required.

Example 1 Find the area A bounded by the parabolas

$$x = 5y^2 \qquad \text{and} \qquad x = 1 + y^2.$$

Consideration of the symmetry involved implies that the total area is twice that in the first quadrant shown in Fig. 6.1. It is simplest to calculate this area by visualizing *horizontal* strips stacked between the horizontal axis and the point of intersection of the two curves. Since the incremental area of the typical strip is

$$dA = [(1 + y^2) - 5y^2]\,dy = (1 - 4y^2)\,dy,$$

the total area in question is

$$A = 2\int_0^{1/2} (1 - 4y^2)\,dy = 2(y - \tfrac{4}{3}y^3)\Big]_0^{1/2} = \tfrac{2}{3}.$$

It is a little more difficult to compute the area using vertical strips, as illustrated in Fig. 6.2. Although symmetry arguments still apply, the formula for the

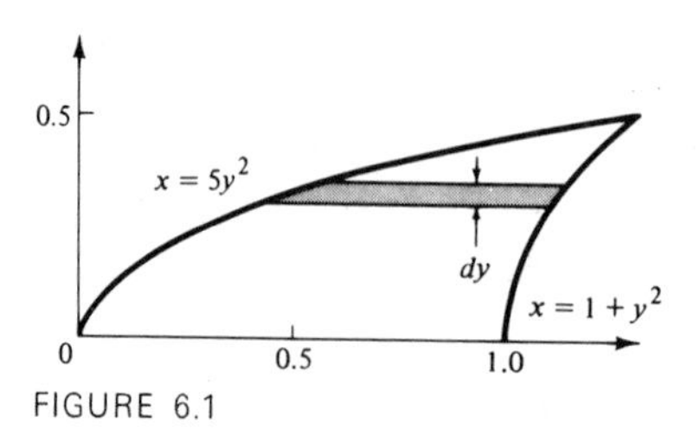

FIGURE 6.1

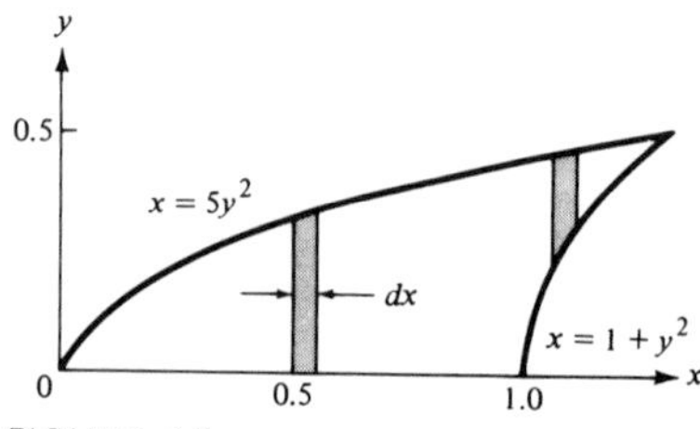

FIGURE 6.2

vertical height of a rectangular strip changes abruptly across the line $x = 1$. For $0 \leq x \leq 1$, the incremental area is $(x/5)^{1/2}\,dx$, whereas in $1 \leq x \leq \frac{5}{4}$ it is $[(x/5)^{1/2} - (x-1)^{1/2}]\,dx$. Therefore, the total area, which is twice that in the first quadrant, is given by

$$A = 2\int_0^1 \left(\frac{x}{5}\right)^{1/2} dx + 2\int_1^{5/4} \left[\left(\frac{x}{5}\right)^{1/2} - (x-1)^{1/2}\right] dx$$

$$= \frac{2}{5^{1/2}}\int_0^{5/4} x^{1/2}\,dx - 2\int_1^{5/4} (x-1)^{1/2}\,dx$$

$$= \frac{4}{3\cdot 5^{1/2}}\,x^{3/2}\Big]_0^{5/4} - \frac{4}{3}(x-1)^{3/2}\Big]_1^{5/4} = \frac{2}{3}.$$

---

2 When the boundary of an area is most easily described in polar coordinates, like the curve $r = r(\theta)$, it makes good sense to calculate the area in a manner that exploits this description without, at any stage, reverting to cartesian coordinates. This can be done by approximating the area with an array of small adjoining triangles, as shown in Fig. 6.3. Here, the polar angle is divided into equal increments of magnitude $\Delta\theta$, so that the elemental triangle at angle θ has a base length $r(\theta)$, height $r\tan\Delta\theta \approx r\,\Delta\theta$, and area

$$\Delta A \approx \tfrac{1}{2}r^2(\theta)\,\Delta\theta. \tag{1}$$

In the limit as $\Delta\theta \to 0$ this area increment becomes the differential

$$dA = \tfrac{1}{2}r^2(\theta)\,d\theta, \tag{2}$$

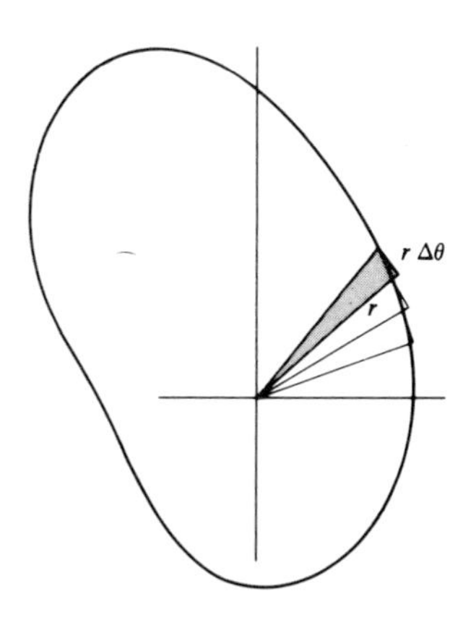

FIGURE 6.3

and the total area is then given by

$$A = \int_{\alpha}^{\beta} \tfrac{1}{2} r^2 \, d\theta, \tag{3}$$

where the integration range for θ is taken to be bounded by the two rays $\theta = \alpha$, and $\theta = \beta$.

From now on, an integral will usually be written as soon as its differential is identified, for there is no longer any need to justify the limit from a theoretical standpoint. Numerical calculation may still follow the conventional route, in which an approximation is determined by actually adding together all the small triangles of a subdivision. In this case, the central angle in Fig. 6.3 would be divided into equal increments, $\Delta\theta = 2\pi/n$, by the rays $\theta_i = (2\pi/n)i$, for $i = 0, \ldots, n$; and with $r_i = r(\theta_i)$, the area approximation is, explicitly,

$$A_n = \sum_{i=1}^{n} \tfrac{1}{2} r_i^2 \, \Delta\theta.$$

Moreover $A = \lim_{n \to \infty} A_n$.

The appropriate limits of integration in Fig. 6.3 are $\alpha = 0$, $\beta = 2\pi$, which means that the angle θ (or the ray segment from point 0 to the curve) makes one complete sweep about the origin. This is certainly not the case in every example, and the bounding rays which are the limits of integration may require careful calculation. The procedure is illustrated by several examples.

Example 2 Find the area A in the upper half-plane bounded by the cardioid

$$r = a(1 + \cos\theta).$$

Before we proceed, note that the description of this curve in cartesian coordinates, as $y = f(x)$, is extremely complicated. Consequently, a calculation of the area by the conventional approach, using vertical or horizontal strips, must lead to a difficult integration. Any doubt about this can be dispelled by actually trying this route.

The closed curve is shown in Fig. 6.4; the differential element of "triangular" area is

$$dA = \tfrac{1}{2} r^2 \, d\theta = \tfrac{1}{2} a^2 (1 + \cos\theta)^2 \, d\theta.$$

The area is covered by ray strips like that from the origin to a point P on the curve; the placement starts at the initial line $\theta = 0$ and ends at the ray $\theta = \pi$. The area A is then given by the sum of all differential elements as

$$A = \int_0^{\pi} \tfrac{1}{2} a^2 (1 + \cos\theta)^2 \, d\theta = \tfrac{1}{2} a^2 \int_0^{\pi} (1 + \cos\theta)^2 \, d\theta.$$

The integral can be evaluated by simply rewriting the integrand:

$$\int_0^{\pi} (1 + \cos\theta)^2 \, d\theta = \int_0^{\pi} (3/2 + 2\cos\theta + 1/2 \cos 2\theta) \, d\theta,$$

$$= [3/2\theta + 2\sin\theta + 1/4 \sin 2\theta]_0^{\pi} = \frac{3\pi}{2},$$

and it follows that

$$A = \frac{3\pi a^2}{4}.$$

[The total area within the cardioid is, by symmetry, twice that in the upper half-plane, or $(3\pi/2)a^2$ to be precise. The particular calculation carried out in Sec. 1.1 (see Figs. 1.1 to 1.3) corresponds to the cardioid for which $a = 2.9$, the area of which is 39.584 ... square units.]

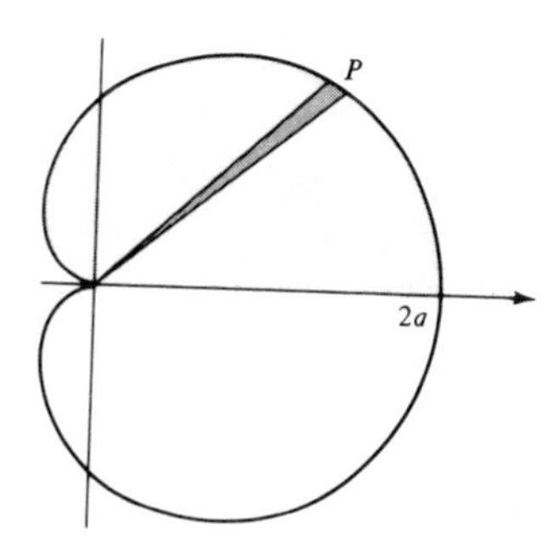

FIGURE 6.4

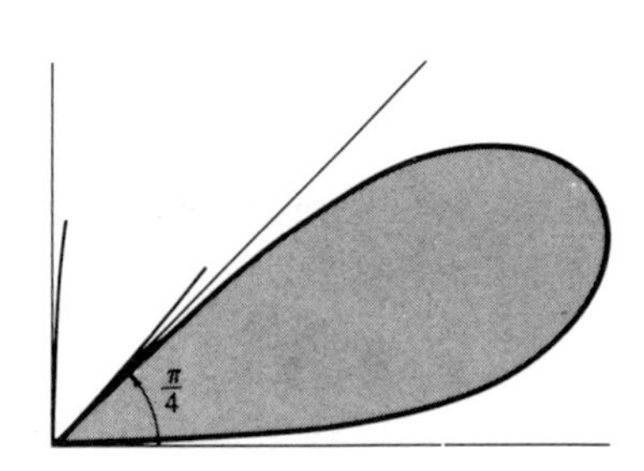

FIGURE 6.5

Example 3 Find the area enclosed by one loop of the curve $r = \sin 4\theta$.

The curve consists of four petals of a "rose." One lobe, shown in Fig. 6.5, is traced out as θ varies from 0 to $\pi/4$ (which establishes the proper range of integration), and since

$$dA = 1/2 r^2 \, d\theta = 1/2 \sin^2 4\theta \, d\theta,$$

its area is

$$A = \int_0^{\pi/4} \frac{1}{2} \sin^2 4\theta \, d\theta = \int_0^{\pi/4} \frac{1}{2}\left(\frac{1}{2} - \frac{1}{2}\cos 8\theta\right) d\theta = \frac{1}{2}\left(\frac{\theta}{2} - \frac{1}{16}\sin 8\theta\right)\Bigg]_0^{\pi/4} = \frac{\pi}{16}.$$

Example 4 Find the area lying in the first quadrant inside the circle $r = 2\cos\theta$ but outside the cardioid $r = 2(1 - \cos\theta)$, as shown in Fig. 6.6.

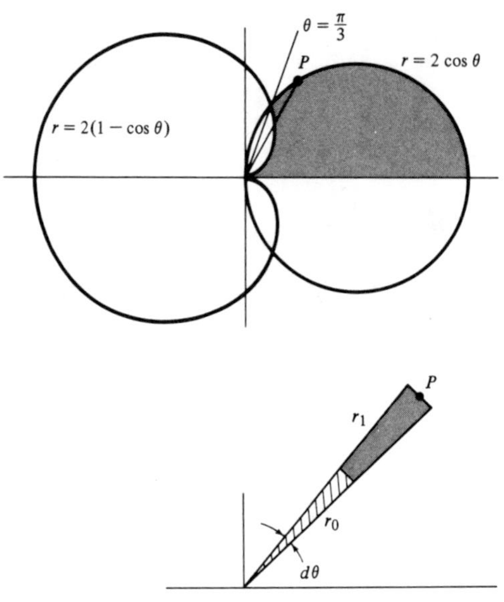

FIGURE 6.6

The element of area is seen to be formed as the difference of the respective incremental triangular strips to each curve:

$$dA = dA_1 - dA_0,$$

where

$$dA_1 = \tfrac{1}{2}r_1^2\, d\theta = \tfrac{1}{2}(2\cos\theta)^2\, d\theta,$$

$$dA_0 = \tfrac{1}{2}r_0^2\, d\theta = \tfrac{1}{2}[2(1 - \cos\theta)]^2\, d\theta.$$

Therefore

$$dA = 2[\cos^2\theta - (1 - \cos\theta)^2]\, d\theta = 2(2\cos\theta - 1)\, d\theta.$$

The range of integration in the first quadrant extends from the horizontal axis, $\theta = 0$, to the ray $\theta = \pi/3$, which passes through the point of intersection of the two curves. The total area is then

$$A = \int_0^{\pi/3} 2(2\cos\theta - 1)\, d\theta = 2(2\sin\theta - \theta)\Big]_0^{\pi/3} = 2\left(\sqrt{3} - \frac{\pi}{3}\right).$$

Of course, certain problems are more appropriately handled in cartesian coordinates no matter how they are specified, and to maintain polar representations would only complicate matters. The point is that the approach should be carefully selected; if the reader has any doubt on this score, the next example should be convincing.

Example 5 Find the area bounded by $r(1 + \cos\theta) = 3$ and the line $r\cos\theta = 1$. In cartesian coordinates the two curves are the parabola $x = 3/2 - 1/6\, y^2$ and the line $x = 1$. The area formed by these boundaries is shown in Fig. 6.7.

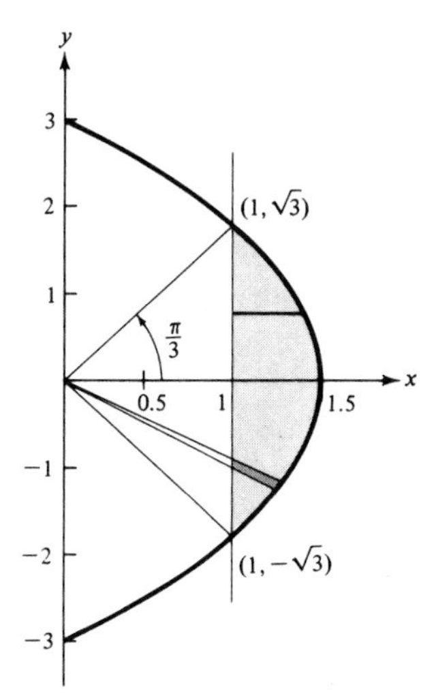

FIGURE 6.7

The simplest method of calculating the enclosed area is to employ horizontal strips, in which case

$$A = \int_{-\sqrt{3}}^{\sqrt{3}} [(3/2 - 1/6\,y^2) - 1]\,dy = \int_{-\sqrt{3}}^{\sqrt{3}} (1/2 - 1/6\,y^2)\,dy.$$

Since this integrand is even, i.e., the area is symmetrical about $y = 0$, it follows that

$$A = 2\int_0^{\sqrt{3}} (1/2 - 1/6\,y^2)\,dy = \left[y - \frac{y^3}{9}\right]_0^{\sqrt{3}} = \frac{2}{\sqrt{3}}. \tag{4}$$

Consider next the same calculation as performed in polar coordinates in which case the element of area is

$$dA = \frac{1}{2}\left[\frac{9}{(1 + \cos\theta)^2} - \frac{1}{\cos^2\theta}\right] d\theta.$$

Since the two curves intersect at points $r = 2$, $\theta = \pm\pi/3$, the rays $\theta = \pm\pi/3$ are the appropriate limits of integration. The total area is then

$$A = \frac{1}{2}\int_{-\pi/3}^{\pi/3} \left[\frac{9}{(1 + \cos\theta)^2} - \frac{1}{\cos^2\theta}\right] d\theta$$

or

$$A = \int_0^{\pi/3} \left[\frac{9}{(1 + \cos\theta)^2} - \frac{1}{\cos^2\theta}\right] d\theta.$$

The techniques of the last section are adequate to evaluate this integral, but the operation is certainly not as simple as that executed above.

4

The length of an arc of a plane curve $y = f(x)$ is calculated as the limit of a sequence of polygonal approximations (Fig. 6.8). As illustrated, the length of a small section of the curve is approximately given by the Pythagorean formula

$$\Delta s = [(\Delta x)^2 + (\Delta y)^2]^{1/2},$$

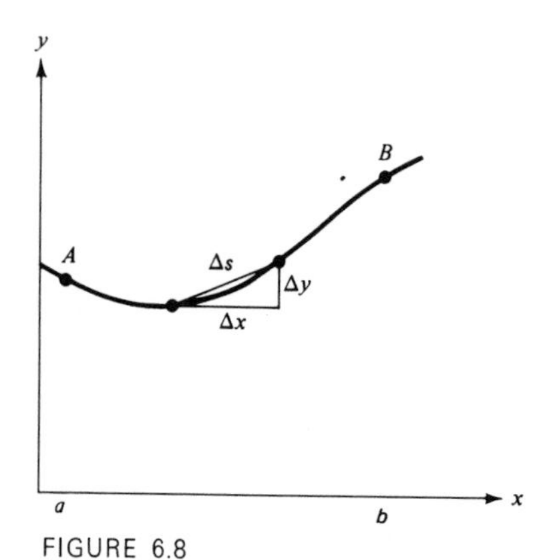

FIGURE 6.8

which in the limit becomes the differential expression

$$ds = (dx^2 + dy^2)^{1/2} \qquad \text{or} \qquad ds^2 = dx^2 + dy^2. \tag{5}$$

It is also convenient to write

$$ds = \left[1 + \left(\frac{dy}{dx}\right)^2\right]^{1/2} dx = \left[1 + \left(\frac{dx}{dy}\right)^2\right]^{1/2} dy, \tag{6}$$

and

$$1 = \left(\frac{dx}{ds}\right)^2 + \left(\frac{dy}{ds}\right)^2. \tag{7}$$

The sum of all the infinitesimals ds from, say, $x = a$ to $x = b$ is the arc length of the curve in this range:

$$s = \int_a^b \left[1 + \left(\frac{dy}{dx}\right)^2\right]^{1/2} dx. \tag{8}$$

[In order to calculate s numerically, we could revert to the definition of an integral and establish the particular sequence of approximations $\{s_n\}$, where $s = \lim_{n\to\infty} s_n$,

$$s_n = \sum_{i=1}^{n} \{1 + [y'(x_i)]^2\}^{1/2}\, \Delta x, \tag{9}$$

and $\Delta x = (b - a)/n$, $x_i = i(b - a)/n$.]

If the inverse function of $f(x)$ is single-valued, we could just as easily write

$$s = \int_{f(a)}^{f(b)} \left[1 + \left(\frac{dx}{dy}\right)^2\right]^{1/2} dy;$$

in the case of a multivalued function, the lengths of different sections would have to be computed separately. These difficulties are often avoided by specifying the plane curve parametrically, as

$$x = x(t), \qquad y = y(t). \tag{10}$$

In this situation, the formula for the differential ds follows by substituting

$$dx = x'(t)\,dt, \qquad dy = y'(t)\,dt$$

in (5):

$$ds^2 = dx^2 + dy^2 = \{[x'(t)]^2 + [y'(t)]^2\}\,dt^2,$$

or

$$ds = (x'^2 + y'^2)^{1/2}\,dt. \tag{11}$$

If $t = t_0$ and $t = t_1$ are the parametric values corresponding to the initial and final points of the arc (A, B in Fig. 6.8), then

$$s = \int_{t_0}^{t_1} \left[\left(\frac{dx}{dt}\right)^2 + \left(\frac{dy}{dt}\right)^2\right]^{1/2} dt. \tag{12}$$

The variable s may itself be the parameter which specifies the location of a point on the curve; e.g., take $t_0 = 0$, and $t_1 = s$. In this instance (12) reduces by (7) to the obvious identity

$$s = \int_0^s \left[\left(\frac{dx}{ds}\right)^2 + \left(\frac{dy}{ds}\right)^2\right]^{1/2} ds = \int_0^s ds.$$

---

5 It might be advantageous on occasion to compute arc length using polar coordinates. The change of variable

$$x = r\cos\theta, \qquad y = r\sin\theta$$

implies the following relationships among increments

$$dx = \cos\theta\,dr - r\sin\theta\,d\theta,$$
$$dy = \sin\theta\,dr + r\cos\theta\,d\theta.$$

Upon substituting for dx and dy in (5), we obtain

$$ds^2 = (\cos\theta\, dr - r\sin\theta\, d\theta)^2 + (\sin\theta\, dr + r\cos\theta\, d\theta)^2$$

or

$$ds^2 = dr^2 + r^2\, d\theta^2. \tag{13}$$

The cross terms $dr\, d\theta$ have cancelled. Therefore, the arc length of the curve $r = r(\theta)$ from $\theta = \alpha$ to $\theta = \beta$ is

$$s = \int ds = \int_\alpha^\beta \left[r^2 + \left(\frac{dr}{d\theta}\right)^2\right]^{1/2} d\theta. \tag{14}$$

The integration can also be performed with respect to r:

$$s = \int_{r_0}^{r_1} \left[r^2\left(\frac{d\theta}{dr}\right)^2 + 1\right]^{1/2} dr, \tag{15}$$

where $r_0 = r(\alpha)$ and $r_1 = r(\beta)$.

6

Example 6 Calculate the arc length of the parabola $y = x^2$ from the origin to $(\frac{1}{2}, \frac{1}{4})$.

Since $dy/dx = 2x$, the length of this arc is given by

$$s = \int_0^{1/2} \left[1 + \left(\frac{dy}{dx}\right)^2\right]^{1/2} dx = \int_0^{1/2} (1 + 4x^2)^{1/2}\, dx.$$

To integrate, we set $u = 2x$, which reduces the integral to a tabulated form (see also Ex. 9, Sec. 3.5),

$$s = \int_0^1 (1 + u^2)^{1/2} \frac{du}{2} = \frac{1}{4}\left[u(1 + u^2)^{1/2} + \log(u + (1 + u^2)^{1/2})\right]_0^1,$$

$$s = \frac{1}{4}[\sqrt{2} + \log(1 + \sqrt{2})].$$

Example 7 Find the arc length of the curve $x = e^t \cos t$, $y = e^t \sin t$ from $t = 1$ to $t = 3$.

Since

$$dx = e^t(\cos t - \sin t)\, dt \qquad \text{and} \qquad dy = e^t(\cos t + \sin t)\, dt,$$

the differential of arc length is

$$ds = (dx^2 + dy^2)^{1/2} = \sqrt{2}e^t\, dt.$$

The arc length is then

$$s = \int_1^3 \sqrt{2}e^t\, dt = \sqrt{2}e^t]_1^3 = \sqrt{2}e(e^2 - 1).$$

7

Example 8 Find the total arc length of the cardioid

$$r = a(1 - \cos\theta).$$

In this case

$$dr = a \sin\theta \, d\theta,$$

and according to (13), the increment of arc length is

$$ds = \sqrt{2}a(1 - \cos\theta)^{1/2} \, d\theta.$$

The total arc length is calculated by integrating from $\theta = 0$ to $\theta = 2\pi$,

$$s = \sqrt{2}a \int_0^{2\pi} (1 - \cos\theta)^{1/2} \, d\theta,$$

and by invoking symmetry arguments, or more formally by transforming the integral, we see that

$$s = 2\sqrt{2}a \int_0^{\pi} (1 - \cos\theta)^{1/2} \, d\theta.$$

The trigonometric identity

$$(1 - \cos\theta)^{1/2} = \sqrt{2} \sin \tfrac{1}{2}\theta$$

makes rapid integration possible:

$$s = 4a \int_0^{\pi} \sin\frac{\theta}{2} \, d\theta = -8a \cos\frac{\theta}{2}\Big]_0^{\pi} = 8a.$$

Example 9 Find the arc length of the curve $r = \theta^2$ from $\theta = 0$ to $\theta = \pi/2$.

In this case, the problem will be done in two ways, once by integration with respect to θ and then again as an integration over r.

Since $dr = 2\theta \, d\theta$ and $d\theta = dr/(2\sqrt{r})$, the differential formula

$$ds^2 = dr^2 + r^2 \, d\theta^2$$

can be written as either

$$ds^2 = (\theta^4 + 4\theta^2) \, d\theta^2 \qquad \text{or} \qquad ds^2 = \left(1 + \frac{r}{4}\right) dr^2.$$

As point P in Fig. 6.9 moves along the particular section of curve from beginning to end, θ varies from 0 to $\pi/2$ and r varies from 0 to $\pi^2/4$. These values are the

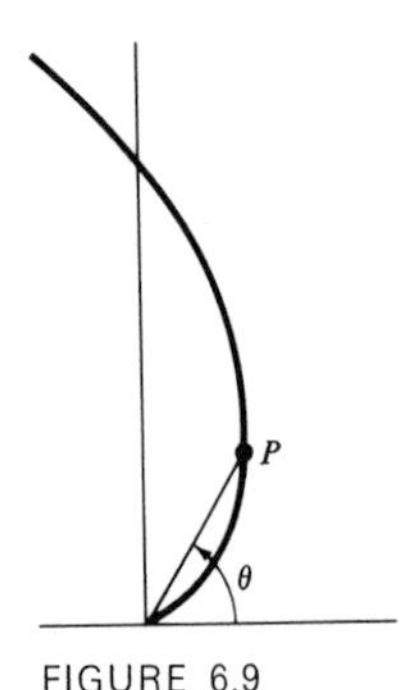

FIGURE 6.9

respective limits of integration for each variable. Therefore, we can now write the arc length as

$$s = \int_0^{\pi/2} \theta(4+\theta^2)^{1/2}\, d\theta = \frac{1}{3}(4+\theta^2)^{3/2}\Big]_0^{\pi/2}$$

or

$$s = \int_0^{\pi^2/4} \left(1+\frac{r}{4}\right)^{1/2} dr = \frac{8}{3}\left(1+\frac{r}{4}\right)^{3/2}\Big]_0^{\pi^2/4};$$

and from either formula it follows that

$$s = \frac{8}{3}\left[\left(1+\frac{\pi^2}{16}\right)^{3/2} - 1\right].$$

Exercises 3.6

1. Find the area bounded by the curves $x = 5y^2$ and $x = 1 + 2y^2$ using two different integration methods.

2. Find the area bounded by the curves $y = ax^2$ and $y = c + bx^2$ $(0<b<a,\ c>0)$. If x and y have dimensions of length, verify that the answer has the correct dimensions.

3. Find the area between the curve $y = x^2e^{-x}$ and the x axis from $x = 0$ to the x coordinate of the maximum of the curve. Determine the area under the curve from $x = 0$ to ∞.

4. Find the area bounded by the hyperbola $xy = 1$ and the line $2x + 2y = 5$. Evaluate this area by integrating the x variable and by integrating the y variable.

5. Find the area enclosed by the curves $y^2 = \pi x$ and $x^2 + y^2 = 2\pi^2$.

6. Find the area bounded by the curves $y = \tan 2x$, $y = \log x$, $y = 1/2$ and $y = 0$. (*Hint:* Integrate the y variable between the curves $x = e^y$ and $x = \tan^{-1}y/2$.)

7. Find the areas bounded by the following curves given in polar coordinates. Verify that the answers have the correct dimensions.

(i) $r = 2a\cos\theta$ **(ii)** $r = a\sin 3\theta$ **(iii)** $r = a(1 - \sin\theta)$

(iv) $r = a + b\sin 2\theta$ **(v)** $r = 2a/(2 - \cos\theta)$ **(vi)** $r = a\theta$ $\quad 0 \le \theta \le \pi/2$

8. Find the area common to the circles $r = 2\cos\theta$ and $r = 3\sin\theta$.

9. Find the area common to the cardioids $r = 1 - \cos\theta$ and $r = 1 + \sin\theta$.

10. Using polar coordinates, find the area bounded by $r(1 + \cos\theta) = 3$ and $r\cos\theta = 1$. Repeat the calculation in cartesian coordinates.

11. Find the arc lengths of the following curves over the given intervals.

(i) $y = x^{3/2}$, $0 \le x \le 2$ **(ii)** $y = \log x$, $\sqrt{3} \le x \le \sqrt{8}$

(iii) $y = x^2$, $0 \le x \le 1$ **(iv)** $y = 2|x|$, $-1 \le x \le 1$

(v) $y = \sin^{-1} 2e^x$, $\pi/6 \le y \le \pi/3$ **(vi)** $x = y^{2/3}$, $0 \le y \le 8$

12. The *center of mass* of the curve $y = y(x)$ from $x = a$ to $x = b$ is the point $(\bar{x}, \bar{y})$ where

$$\bar{x} = \frac{\int_a^b x[1 + (y')^2]^{1/2}\, dx}{\int_a^b [1 + (y')^2]^{1/2}\, dx}, \qquad \bar{y} = \frac{\int_a^b y[1 + (y')^2]^{1/2}\, dx}{\int_a^b [1 + (y')^2]^{1/2}\, dx}.$$

Determine the centers of mass of the circle $x^2 + y^2 = r^2$ and of the semi-circle $x^2 + y^2 = r^2$, $y \ge 0$.

13. Find the arc lengths of the following curves over the given intervals. Illustrate with graphs.

(i) $x = t^3$, $y = 3t^2$, $0 \le t \le 2$ **(ii)** $x = 1 + 4t$, $y = t^2 + 2$, $-1 \le t \le 1$

(iii) $x = e^t \cos t$, $y = e^t \sin t$, $0 \le t \le 1$ **(iv)** $x = t^2 - 1$, $y = t^2 + 1$, $0 \le t \le 1$

14. Determine the lengths of the following curves given in polar coordinates. Illustrate with graphs. Verify that the answers have the correct dimensions.

(i) $r = a\theta$, $0 \le \theta \le \pi$ **(ii)** $x = a \cos^3\theta$, $y = a \sin^3\theta$, $0 \le \theta \le 2\pi$

(iii) $r = ae^{b\theta}$, $0 \le \theta \le 2\pi$ **(iv)** $r = a \sec b\theta$, $0 \le \theta \le \pi/4$

3.7 Volumes and surfaces of revolution

The problem of finding an arbitrary volume (say that of a potato) is genuinely a three-dimensional calculation involving variables of depth, width, and height. However, so far this development of the calculus has been based mainly on the relationships among only two variables, one independent and the other dependent. As a consequence, the volumes that can be calculated by means of the existing theory are rather special configurations, mainly those that possess a high degree of symmetry. This category contains every body of revolution which is generated by revolving or turning a plane area about one of the coordinate axes (see Fig. 7.1). The volume cut out of space by this circular motion is completely symmetrical with respect to the rotation axis. If the body is sliced perpendicular to this symmetry axis, the cross section is a circle; likewise, all longitudinal slices along the axis look like the original area.

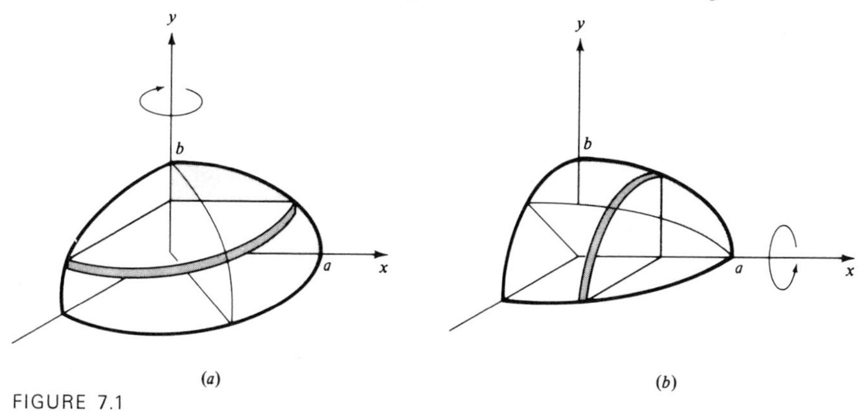

(a) (b)

FIGURE 7.1

The volume of a body of revolution can be calculated by the slicing method of Archimedes, discussed earlier in Chap. 1. In this approach, we imagine that the body is sliced into thin sections and compute the incremental volume ΔV of each wafer as approximately that of a cylinder of the same radius x and height Δy (see Fig. 7.2). In this way we obtain

$$\Delta V = \pi x^2 \, \Delta y.$$

Equivalently, the volume can be approximated by a stack of cylindrical wafers, each of an appropriate radius to fit the body. The limit, in either case, is the exact volume.

For definiteness, consider the volume of revolution about the y axis, produced by the rotation of the area in the first quadrant [bounded by $y = f(x)$ or $x = g(y)$] illustrated in Fig. 7.2. The differential element of volume is

$$dV = \pi x^2 \, dy, \tag{1}$$

the total volume being

$$V = \int dV = \int_0^b \pi x^2 \, dy, \tag{2}$$

where y ranges from the base plate, $y = 0$, to the pinnacle, $y = b$. Similarly, the volume of revolution about the x axis (shown in Fig. 7.1b) is given by

$$V = \int dV = \int_0^a \pi y^2 \, dx. \tag{3}$$

Example 1 Find the volumes of revolution formed from the rotation about the coordinate axes of the area in the first quadrant bounded by

$$\frac{x^2}{a^2} + \frac{y^2}{b^2} = 1, \qquad x = 0, \qquad y = 0.$$

The volume generated about the y axis is determined from (2):

$$V = \int_0^b \pi a^2 \left(1 - \frac{y^2}{b^2}\right) dy = \pi a^2 \left(y - \frac{y^3}{3b^2}\right)\Bigg]_0^b = \tfrac{2}{3}\pi a^2 b.$$

The volume formed by rotating the same area about the x axis is

$$V = \int_0^a \pi b^2 \left(1 - \frac{x^2}{a^2}\right) dx = \tfrac{2}{3}\pi a b^2.$$

(Note that V has the dimensions L^3 since a and b are both lengths.) In the special case, $a = b$, we find that the volume of a hemisphere is $\tfrac{2}{3}\pi a^3$; the volume of a full sphere is then twice this, or $\tfrac{4}{3}\pi a^3$.

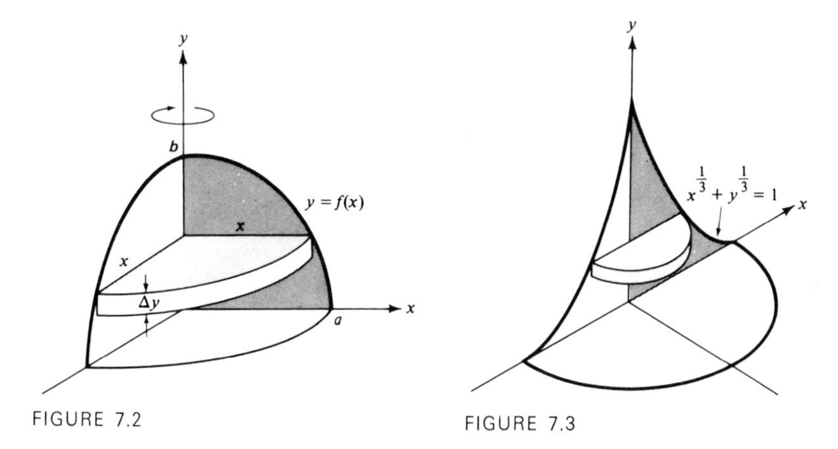

FIGURE 7.2

FIGURE 7.3

Example 2 Find the volume of revolution (half of which is shown in Fig. 7.3) with bounding curve $x^{1/3} + y^{1/3} = 1$.

In this case $x = (1 - y^{1/3})^3$, and, by (2),

$$V = \pi \int_0^1 (1 - y^{1/3})^6 \, dy.$$

With $u = 1 - y^{1/3}$ this becomes

$$V = 3\pi \int_0^1 u^6(1-u)^2 \, du = 3\pi \int_0^1 (u^6 - 2u^7 + u^8) \, du,$$

$$= 3\pi(1/7 - 1/4 + 1/9) = \pi/84.$$

Example 3 Find the volume of revolution about the y axis formed from the area bounded by the parabola $y^2 = x - 1$ and the line $y - 1 = 2(x - 2)$.

The area is shown in Fig. 7.4, which also gives the coordinates of the points of intersection of the two curves. The differential element of volume is now that of a washer with an outer radius $x_o = 1/2(y + 3)$ and inner radius $x_i = y^2 + 1$; that is,

$$dV = \pi(x_o^2 - x_i^2) \, dy = \pi[1/4(y+3)^2 - (y^2+1)^2] \, dy.$$

The volume is therefore

$$V = \pi \int_{-1/2}^1 [1/4(y+3)^2 - (y^2+1)^2] \, dy,$$

and the integration yields

$$V = 63\pi/40.$$

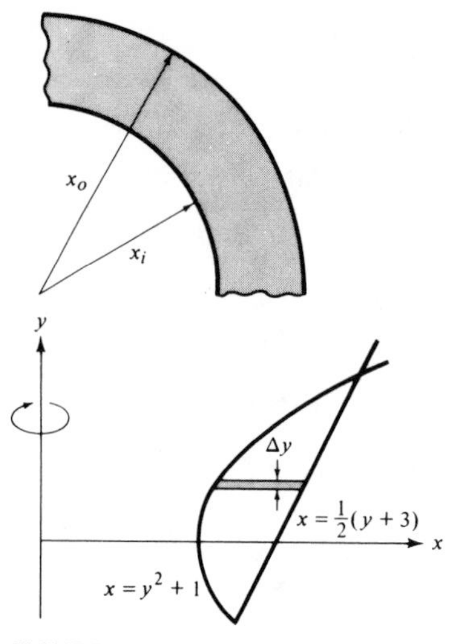

FIGURE 7.4

---

2 The volume of revolution can also be calculated as the limit of a sum of thin cylindrical shells, shown in Fig. 7.5*a*, whose incremental volume is the product of the height y, circumference $2\pi x$, and thickness Δx:

$$\Delta V = 2\pi xy\, \Delta x.$$

The differential element of volume is then

$$dV = 2\pi xy\, dx,$$

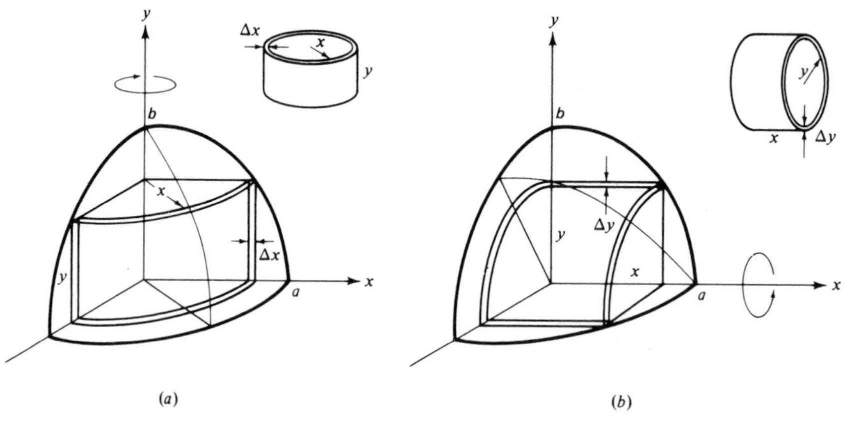

FIGURE 7.5

and from Fig. 7.5 it follows that

$$V = \int dV = \int_0^a 2\pi xy\, dx.$$

The volume of revolution about the other axis (Fig. 7.5*b*) is determined by a similar argument to be

$$V = \int_0^b 2\pi xy\, dy.$$

Example 4 Use the cylindrical-shell method to calculate the volume of revolution about the y axis of the area bounded by the ellipse

$$\frac{x^2}{a^2} + \frac{y^2}{b^2} = 1, \qquad x = 0, \qquad y = 0.$$

The volume, shown in Fig. 7.5*a*, is given by

$$V = \int_0^a 2\pi x b\left(1 - \frac{x^2}{a^2}\right)^{1/2} dx$$

$$= -\tfrac{2}{3}\pi a^2 b\left(1 - \frac{x^2}{a^2}\right)^{3/2}\Bigg]_0^a = \tfrac{2}{3}\pi a^2 b.$$

Available techniques are adequate to calculate a volume by the slicing method provided its cross-sectional area can be described as a function of *one* variable, $A = A(x)$ (see Fig. 7.6). In this case, the differential element of volume is clearly

$$dV = A(x)\, dx,$$

and the total volume is

$$V = \int_a^b A(x)\, dx.$$

A typical situation is for all the slices of a body to be similar in shape to each other but of different size.

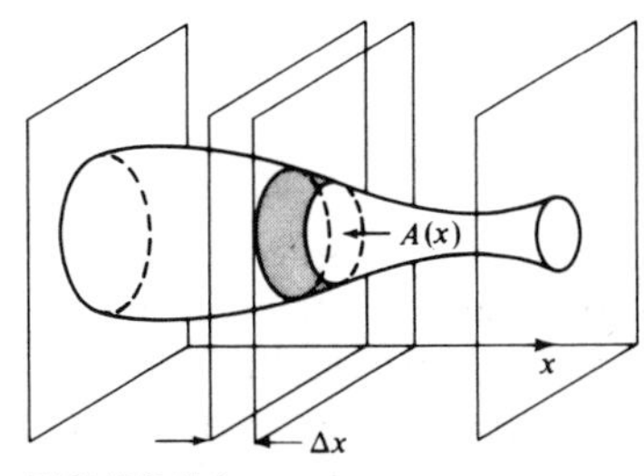

FIGURE 7.6

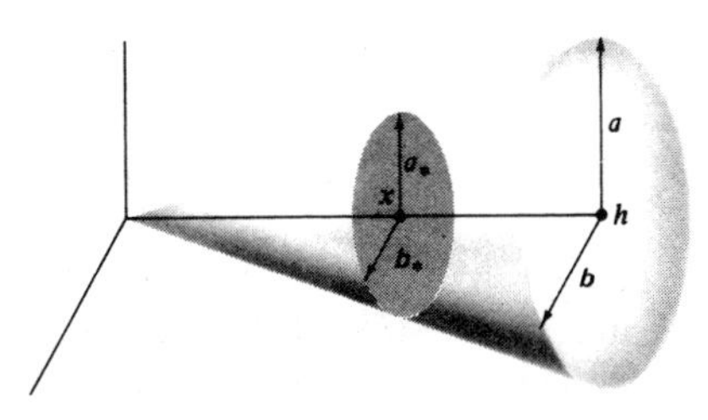

FIGURE 7.7

Example 5 Calculate the volume of the elliptical cone in Fig. 7.7.

Since the cross section at an arbitrary point x is an ellipse with major and minor semi-axes

$$a_* = \frac{ax}{h} \qquad \text{and} \qquad b_* = \frac{bx}{h},$$

the area of this cut is

$$A(x) = \pi a_* b_* = \frac{\pi abx^2}{h^2}.$$

The differential element of volume is then

$$dV = \frac{\pi abx^2}{h^2}\, dx,$$

making the total volume between the apex $x = 0$ and the plate $x = h$

$$V = \int_0^h \frac{\pi abx^2}{h^2}\, dx = \frac{\pi ab}{3h^2} x^3 \Big]_0^h = \frac{\pi abh}{3}.$$

As a check on this calculation, note that V reduces correctly to $\pi r^2 h/3$ if $a = b = r$. This is the familiar result for a right circular cone.

4 The conceptual process of slicing a volume of revolution into very thin wafers also enables us to calculate the surface area of the body. We imagine that the outer surface of each wafer is peeled off (like the skin of an onion) and laid out straight so that its area can be measured (see Fig. 7.8). (This process is also equivalent to wrapping the surface with very thin strips of tape.) Since the length of the strip is approximately equal to the perimeter of a circle of radius x and its width is the increment of arc length Δs, the differential element of surface area is

$$dS = 2\pi x\, ds = 2\pi x(dx^2 + dy^2)^{1/2}.$$

The lateral surface area (excluding the area of the baseplate) is then

$$S = 2\pi \int x\, ds = \int_0^b 2\pi x \left[1 + \left(\frac{dx}{dy}\right)^2\right]^{1/2} dy.$$

Similarly the surface of revolution about the x axis is

$$S = 2\pi \int y\, ds = \int_0^a 2\pi y \left[1 + \left(\frac{dy}{dx}\right)^2\right]^{1/2} dx.$$

Note carefully that we have *not* approximated the surface area as the sum of the lateral areas of a stack of cylindrical shells (which were used to find the volume). In order to obtain the correct value, it is *essential* that the strips actually be in contact with the surface and tangent to it. (The strip element of area for a cylindrical

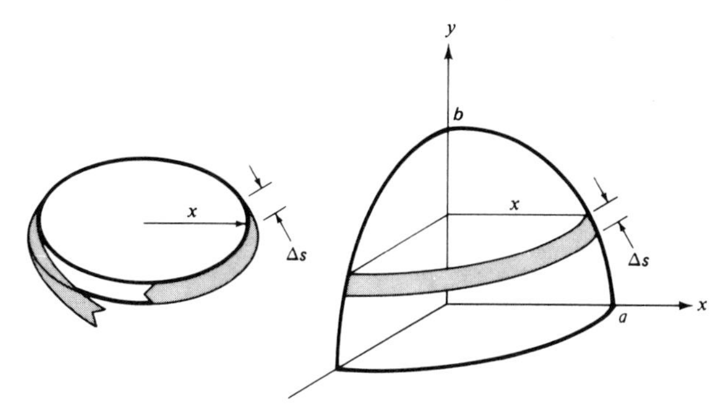

FIGURE 7.8

slab is $2\pi x\, dy$, whereas that on the surface is $2\pi x\, ds$; the difficulty is similar to that discussed in Exercise 9 of Sec. 1.1.)

Example 6 Find the surface area formed when the curve $y = \sin x$ from $x = 0$ to $x = \pi$ is revolved about the x axis.

The area (Fig. 7.9) is given by

$$S = 2\pi \int_0^\pi y\left[1 + \left(\frac{dy}{dx}\right)^2\right]^{1/2} dx,$$

$$= 2\pi \int_0^\pi (1 + \cos^2 x)^{1/2} \sin x\, dx.$$

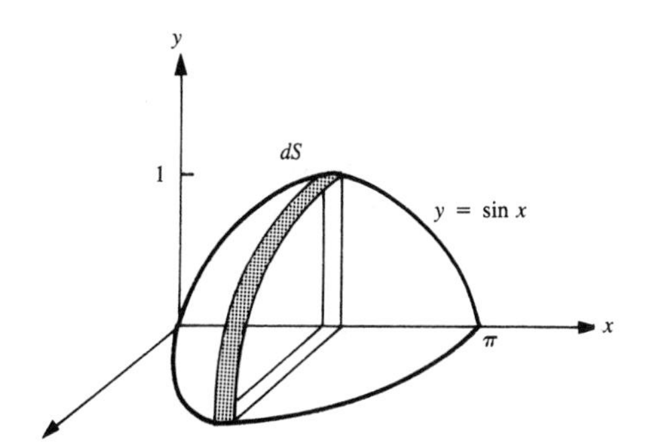

FIGURE 7.9

The integral is reduced to a standard form with the substitution $u = \cos x$, so that

$$S = 2\pi \int_{-1}^{1} (1 + u^2)^{1/2}\, du = 4\pi \int_0^1 (1 + u^2)^{1/2}\, du.$$

According to formula **15** of Appendix Table 5.1, the value is

$$S = 2\pi[2^{1/2} + \log(1 + 2^{1/2})].$$

Example 7 Find the area of the surface of revolution formed by rotating the ellipse $x = a \cos t$, $y = b \sin t$, $b < a$, about the x axis (Fig. 7.10).

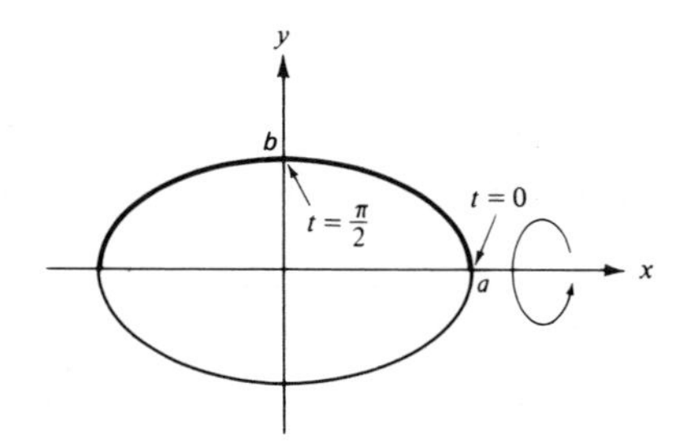

FIGURE 7.10

The formula for surface area is first converted in order that we may take full advantage of the parametric description of the curve. From $ds = (dx^2 + dy^2)^{1/2}$ and

$$dx = -a \sin t \, dt, \qquad dy = b \cos t \, dt,$$

it follows that

$$ds = (a^2 \sin^2 t + b^2 \cos^2 t)^{1/2} \, dt.$$

Since this is a calculation of surface area *and not of arc length*, only the segment of the ellipse in the top half-plane is required to generate the entire surface. The range of integration for the variable t is then $0 \leq t \leq \pi$, and it follows that

$$S = \int 2\pi y \, ds = 2\pi b \int_0^{\pi} \sin t \, (a^2 \sin^2 t + b^2 \cos^2 t)^{1/2} \, dt.$$

The transformation $u = \cos t$ allows this to be expressed as

$$S = 2\pi b \int_{-1}^{1} [a^2 - (a^2 - b^2)u^2]^{1/2} \, du,$$

$$= 4\pi ab \int_0^1 (1 - k^2u^2)^{1/2} \, du,$$

where $k = [(a^2 - b^2)/a^2]^{1/2}$ is called the *eccentricity* of the ellipse. With $z = ku$, this integral reduces to a standard form:

$$S = \frac{4\pi ba}{k}\int_0^k (1 - z^2)^{1/2}\,dz,$$

$$= \frac{2\pi ba}{k}\,[z(1 - z^2)^{1/2} + \sin^{-1} z]_0^k,$$

$$= 2\pi ba\left[(1 - k^2)^{1/2} + \frac{1}{k}\sin^{-1} k\right] = 2\pi b^2 + \frac{2\pi ba}{k}\sin^{-1} k.$$

Use l'Hôpital's rule to verify the special case $k = 0$ where $S = 4\pi a^2$.

Exercises 3.7

1. Show that the volume of revolution about the x axis generated by $y = f(x)$ for $a_1 \leq x \leq a_2$ is given by $V = \int_{a_1}^{a_2} \pi y^2\,dx$. Calculate the volumes corresponding to the following curves. Illustrate with diagrams.
(i) $y = mx + b$, $0 \leq x \leq 1$ **(ii)** $y = ce^{kx}$, $1 \leq x \leq 2$ **(iii)** $y = c\sin kx$, $0 \leq x \leq \pi k/2$

2. Use the cylindrical-shell method to calculate the volume of a right circular cone with radius r and height h.

3. Find the volume of a sphere by the shell method.

4. The area bounded by the parabola $y = ax^2$ and the line $y = h$ is revolved around the y axis. Find the volume of the paraboloid generated and verify dimensions. (Assume $a > 0$ and $h > 0$.)

5. The area bounded by the hyperbola $xy = 2$ and the lines $x = 1$, $x = 2$, $y = 0$ is revolved around the y axis. Find the volume generated. Solve the more general problem for the area bounded by $xy = a^2$, $x = a_1$, $x = a_2$ where $0 < a_1 < a_2$.

6. Find the volume generated by rotating the area bounded by the parabolas $y = 2(x - 1)^2$, $y = 4(x - 2)^2$ and the line $y = 0$ about the x axis. State and solve a generalization of this problem.

7. Find the following volumes of revolution. Illustrate with diagrams.
(i) The area between $xy = 1$ and $x + 2y = 4$ rotated about the x axis.
(ii) The same area rotated about the y axis.
(iii) The area bounded by the curve $x = 2(\theta - \sin\theta)$, $y = 2(1 - \cos\theta)$ for $0 \leq \theta \leq 2\pi$ rotated about the x axis.

8. Calculate the volume of a pyramid with all cross-sections equilateral triangles. The height of the pyramid is equal to the length of a side of the base. Repeat the calculation for square cross-sections.

9. The area in the first quadrant bounded by the curve $y = \log x$ and the lines $y = 0$, $x = e$ is revolved around the y axis. Find the volume generated.

10. A hole of radius b is drilled through the center of a sphere of radius $a > b$. Find the volume that remains. Verify your answer by checking that the dimensions and the special cases $b = 0$, $b = a$ are correct.

11. Find the surface area generated by revolving a circular arc of inner angle Φ about the y axis, as shown in Fig. 7.11. If the Arctic Ocean is considered to be circular in shape centered at the North Pole with $\Phi = 15°$ and $R = 6360$ km, what is its approximate area?

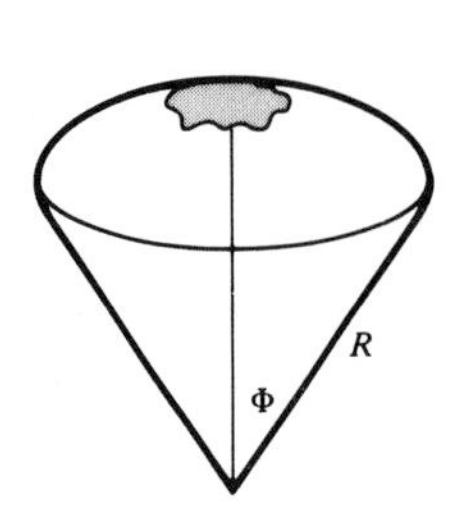

FIGURE 7.11

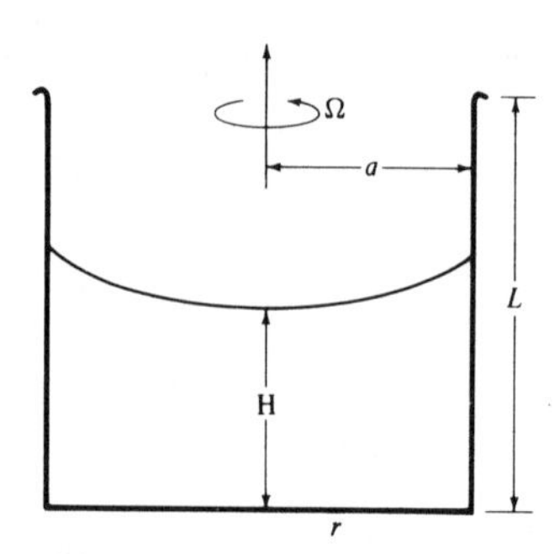

FIGURE 7.12

12. Suppose that two parallel planes a distance d apart both intersect a sphere of radius r. Show that the surface area on the sphere between the two planes is $2\pi rd$. (*Hint:* Use the previous problem. This result is due to Archimedes.)

13. Find the surface areas generated by rotating the following curves about the given axes.

(i) $y = \cos x$, $0 \le x \le \pi$, about the x axis

(ii) $y = \sin 2x$, $0 \le x \le \pi/4$, about the x axis

(iii) $y = x^3$, $0 \le x \le 1$, about the y axis

(iv) $y = e^x$, $0 \le x \le 1$, about the x axis

14. Find the surface areas generated by revolving the ellipse $x = a \cos t$, $y = b \sin t$ about each coordinate axis. Verify dimensions and the special case $a = b$. How can one of the areas be found from the formula for the other area?

15. Determine the eccentricities of the following ellipses. Illustrate with graphs.

(i) $r = a$ **(ii)** $r = 2a \cos\theta$ **(iii)** $r \cos\theta = a$

(iv) $4x^2 + y^2 = 16$ **(v)** $r = 4/(2 - \cos\theta)$ **(vi)** $r = 8/(2 + \sin\theta)$

16. Find the surface area of a torus generated by revolving the circle $x^2 + (y - b)^2 = r^2$, $b > r$, about the x axis. Verify the dimensions of the answer.

17. The horizontal cross-section of a pyramid z cm above its base is a square of side $(100 - z)$ cm. Calculate the volume of the pyramid. Calculate the volume between heights z_1 and z_2 where $0 < z_1 < z_2 < 100$.

18. Water partially fills a cylindrical pail of radius a and height L. When the pail is rotated about its symmetry axis with angular speed Ω, the surface of the water assumes a parabolic shape the profile of which (Fig. 7.12) is

$$z = H + \frac{\Omega^2}{2g} r^2,$$

where $g = 9.8$ m/sec^2. For a fixed volume of water V how does H depend on Ω? At what value of Ω does the water surface touch bottom? When does water spill over the top?

19. Find the volume of a figure whose base is a circle of unit radius if every plane cross-section perpendicular to a fixed diameter of the circle is an isosceles triangle with height twice its base length.

20. A barrel of length L has the shape of an ellipsoid of revolution the ends of which are chopped off and replaced by disks of radius r. Find the volume of the barrel if the radius at the center is R. Verify dimensions and the special case $r = 0$. (Many problems of this type were solved by Kepler (1571-1630) in his studies of the volumes of wine barrels.)

3.8 Physical applications

1

This discussion will be limited to a few physical applications that are closest to the average reader's experience. The scope will be broadened considerably in subsequent chapters.

The *density* of a material ρ is defined as the mass of a small but recognizable sample divided by the volume it occupies. Inherent in this definition is a restriction on how really small the sample can be, but as long as our interest is not with molecular processes, we can consider ρ to be a continuous macroscopic function of position. For a homogeneous medium, ρ is constant, and the mass of any volume V is just ρV. However, media of variable density can be formed as mixtures or compositions of homogeneous materials. For example, the rates of which two constant density fluids flow into a tank can be adjusted with time to create a continuously stratified liquid whose density varies with height.

If the density of a medium is not constant, the total mass in a volume is calculated by summing the incremental masses of every differential element of volume dV. Since

$$dm = \rho\, dV,$$

the total mass is given by the integral

$$m = \int \rho\, dV.$$

The average density is

$$\bar{\rho} = m/V = \int \rho dV / \int dV.$$

At present, it is possible to perform these calculations when the density distribution is simple and the body is symmetrical.

Example 1 The density of the fluid filling a conical glass (Fig. 8.1) varies with height according to the formula

$$\rho = \rho_0\left(1 - \frac{1}{2}\frac{z}{H}\right).$$

Find the mass of the contained fluid. What is the average density?

The radius of the glass at height z is $r = (R/H)z$, so that the increment of volume there in the thin cylindrical wafer is

$$\Delta V = \pi r^2\, \Delta z = \frac{\pi R^2}{H^2}\, z^2\, \Delta z.$$

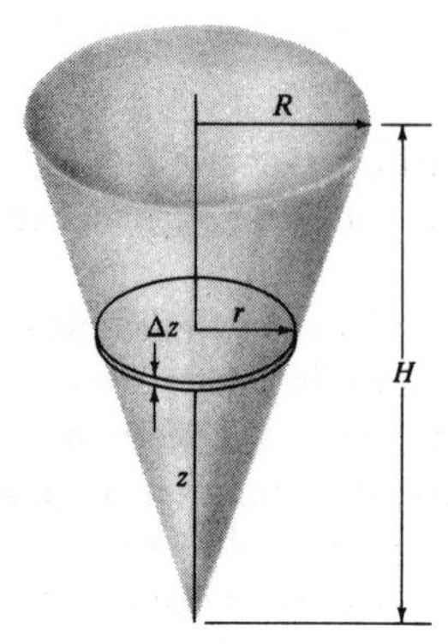

FIGURE 8.1

The differential element of mass is then

$$dm = \rho\, dV = \frac{\pi R^2}{H^2}\rho_0\left(1 - \frac{1}{2}\frac{z}{H}\right)z^2\, dz$$

so that the total mass of the fluid in the glass is

$$m = \int dm = \frac{\pi R^2}{H^2}\rho_0 \int_0^H \left(1 - \frac{1}{2}\frac{z}{H}\right)z^2\, dz = \frac{5}{24}\pi R^2 H \rho_0 .$$

Since the volume of the glass is $\frac{1}{3}\pi R^2 H$, the average density is

$$\bar{\rho} = m/V = \frac{5}{8}\rho_0 .$$

---

2 Balance on a seesaw (Fig. 8.2) is obtained by distributing two weights in such a way that any tendency of the system to rotate is nullified. The disposition to turn under a load is measured by the torque, which is the product of weight and distance, wl. Equilibrium requires that the torques produced by each weight balance identically:

$$w_1 l_1 = w_2 l_2; \tag{1}$$

this principle, discovered by Archimedes, is known as the *law of the lever.*

As the figure shows, weight w_1 induces a clockwise rotation, which is counteracted by the counterclockwise rotation caused by w_2. If we rather arbitrarily agree to measure clockwise rotations as positive, so that counterclockwise motions are negative, and if further we establish a coordinate origin at the fulcrum O, then (1) can be expressed as

$$\sum_{i=1}^{2} w_i x_i = 0,$$

where, in particular, $x_1 = l_1$, $x_2 = -l_2$. More generally, if n weights are placed on the lever, a balance is obtained only when

$$\sum_{i=1}^{n} w_i x_i = 0. \tag{2}$$

Suppose now that the situation is reversed and we seek the point of balance for an arbitrary distribution of n weights on a weightless, i.e., very light, rod (Fig. 8.3). With respect to the origin at O, let $\bar{x}$ be the distance to this balance point P. (The

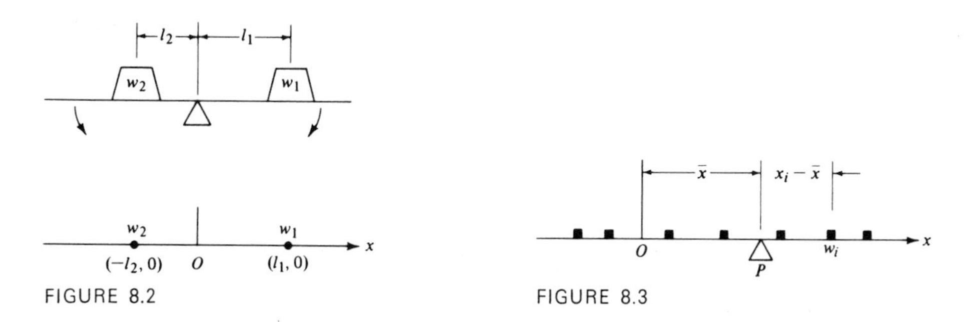

FIGURE 8.2

FIGURE 8.3

rod would remain horizontal if suspended from a string or supported by a fulcrum at this position.) Equilibrium requires that the total torque about P be zero:

$$\sum_{i=1}^{n} w_i(x_i - \bar{x}) = 0, \tag{3}$$

and this equation can be solved to find

$$\bar{x} = \frac{\sum_{i=1}^{n} w_i x_i}{\sum_{i=1}^{n} w_i}. \tag{4}$$

Point P is known as the *center of gravity*; for constant gravitational acceleration g it is also called the *center of mass* because weight and mass are proportional, $w_i = gm_i$. (The center of gravity refers to a force field, but the center of mass is a geometrical property of the configuration of matter. The two are the same for constant g, and we will not draw any further distinction.)

Consider next the problem of finding the center of gravity (or mass) $\bar{x}$, of a straight rod of uniform cross-sectional area A if its density varies continuously with length. Take the origin at one end (Fig. 8.4) and let the x axis pass through the center line of the rod. The differential element of mass at position x is

$$dm = \rho(x)\, dV = \rho(x) A\, dx,$$

and this produces a torque about $\bar{x}$ given by

$$g(x - \bar{x})\, dm = g(x - \bar{x})\rho(x) A\, dx.$$

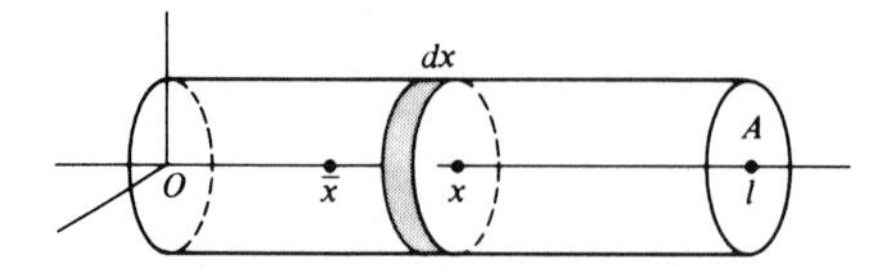

FIGURE 8.4

The condition for equilibrium is that the total moment about $\bar{x}$ be zero, which implies that

$$\int_0^l g\rho(x)(x - \bar{x})A\,dx = 0.$$

Therefore

$$\int_0^l (x - \bar{x})\rho(x)\,dx = \int_0^l x\rho(x)\,dx - \bar{x}\int_0^l \rho(x)\,dx = 0;$$

the result of solving this equation is

$$\bar{x} = \frac{\int_0^l x\rho(x)\,dx}{\int_0^l \rho(x)\,dx}, \tag{5}$$

which is the "continuous" analog of (4). Since the denominator of this expression is the total mass M of the rod, this can also be written

$$\bar{x} = \frac{1}{M}\int_0^l x\rho(x)\,dx.$$

For constant density, $\rho = \rho_0$, (5) reduces to

$$\bar{x} = \frac{1}{2}\,l,$$

which is as it should be, as a little thought makes evident.

Consider next the center of mass, or *centroid*, of a plane sheet of *homogeneous* material, Fig. 8.5. A sheet is, simply speaking, a thin slab of material of very small but uniform thickness h. The idealization obtained as $h \to 0$ is called a *lamina*.

We take the top and bottom boundaries of the sheet, referred to the origin at O, to be $y_T = y_T(x)$ and $y_B = y_B(x)$. Next, the lamina is divided into thin strips, each of incremental mass

$$dm = \rho\,dV.$$

Since

$$dV = h\,dA = h[y_T(x) - y_B(x)]\,dx = hY(x)\,dx,$$

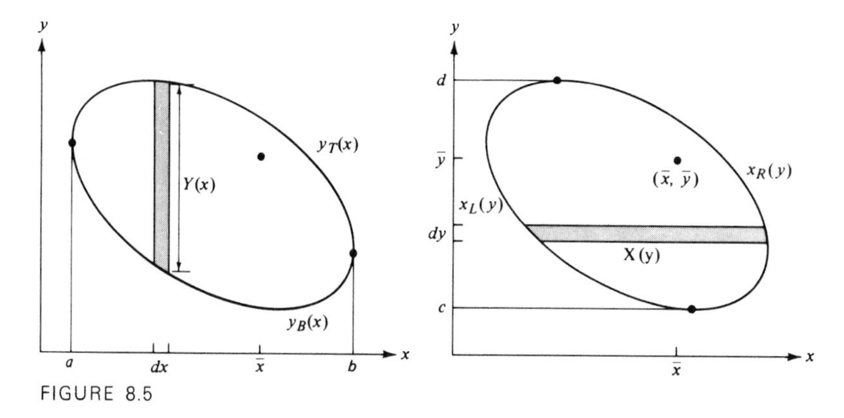

FIGURE 8.5

the differential element of mass is $dm = \rho h Y(x)\, dx$. If the mass of each vertical strip were concentrated on the x axis, the center of mass of such a distribution (on a weightless rod) would be

$$\bar{x} = \frac{\int_a^b x[y_T(x) - y_B(x)]\, dx}{\int_a^b [y_T(x) - y_B(x)]\, dx} = \frac{\int_a^b x\, Y(x)\, dx}{A}, \tag{6}$$

where A is the actual area of the figure. The value of $\bar{x}$ defines the horizontal coordinate of the centroid of the lamina. The vertical coordinate of the mass center $\bar{y}$ is calculated in a similar fashion by imagining the sheet divided into horizontal strips and concentrating the increments of mass on the y axis. With notation as defined in Fig. 8.5, we obtain

$$\bar{y} = \frac{\int_c^d y[x_R(y) - x_L(y)]\, dy}{\int_c^d [x_R(y) - x_L(y)]\, dy} = \frac{\int_c^d yX(y)\, dy}{A}. \tag{7}$$

The coordinates of the centroid or center of mass are $(\bar{x}, \bar{y})$. Neither the thickness of the plate nor its density (both constants) enters the final formulas for the point $(\bar{x}, \bar{y})$ and this information need not be given. With the assumed conditions, the centroid is a purely geometrical property of the configuration.

Example 2 Find the center of mass of a rod of length l if its density is given by $\rho = \rho_0[1 + \sin(\pi x/2l)]$.

The center is given by

$$\bar{x} = \frac{\int_0^l x\rho(x)\, dx}{\int_0^l \rho(x)\, dx} = \frac{\int_0^l x[1 + \sin(\pi x/2l)]\, dx}{\int_0^l [1 + \sin(\pi x/2l)]\, dx}.$$

The most difficult integral here is evaluated by an integration by parts;

$$\int_0^l x \sin \frac{\pi x}{2l}\, dx = -\frac{2lx}{\pi} \cos \frac{\pi x}{2l}\Big]_0^l + \frac{2l}{\pi}\int_0^l \cos \frac{\pi x}{2l}\, dx,$$

$$= \frac{4l^2}{\pi^2} \sin \frac{\pi x}{2l}\Big]_0^l = \frac{4l^2}{\pi^2}.$$

All other integrals are straightforward, and it follows that

$$\bar{x} = \frac{l^2/2 + 4l^2/\pi^2}{l + 2l/\pi} = \frac{l}{2\pi}\,\frac{\pi^2 + 8}{\pi + 2} = \frac{l}{2}\,(1.1063\ldots).$$

Example 3 Find the centroid of the area in the first quadrant bounded by the coordinate axes and by the circles $x^2 + y^2 = r^2$ and $x^2 + y^2 = 1$, $r < 1$; see Fig. 8.6.

The vertical length between the boundaries is

$$Y(x) = y_T(x) - y_B(x) = \begin{cases} (1 - x^2)^{1/2} - (r^2 - x^2)^{1/2} & \text{for } 0 < x < r; \\ (1 - x^2)^{1/2} & \text{for } r < x < 1, \end{cases}$$

and the horizontal coordinate of the center of mass is then

$$\bar{x} = \frac{\int_0^1 x\,Y(x)\,dx}{\int_0^1 Y(x)\,dx}.$$

The integral in the numerator is

$$\int_0^r [(1 - x^2)^{1/2} - (r^2 - x^2)^{1/2}]x\,dx + \int_r^1 (1 - x^2)^{1/2} x\,dx$$

$$= \int_0^1 (1 - x^2)^{1/2} x\,dx - \int_0^r (r^2 - x^2)^{1/2} x\,dx,$$

$$= -\tfrac{1}{3}(1 - x^2)^{3/2}]_0^1 + \tfrac{1}{3}(r^2 - x^2)^{3/2}]_0^r = \tfrac{1}{3}(1 - r^3).$$

The integral in the denominator can be calculated directly, but since the area represented is one-quarter of the total area between two circles of radii 1 and r, we see immediately that

$$\int_0^1 Y(x)\,dx = A = \frac{1}{4}(\pi - \pi r^2) = \frac{\pi}{4}(1 - r^2).$$

Therefore

$$\bar{x} = \frac{4}{3\pi}\,\frac{1 - r^3}{1 - r^2} = \frac{4}{3\pi}\,\frac{1 + r + r^2}{1 + r}.$$

The vertical coordinate of the center of mass can be determined in a similar manner, using Eq. (7), but it is far simpler to invoke a symmetry argument. The configuration is completely symmetric with respect to the ray $\theta = \pi/4$,

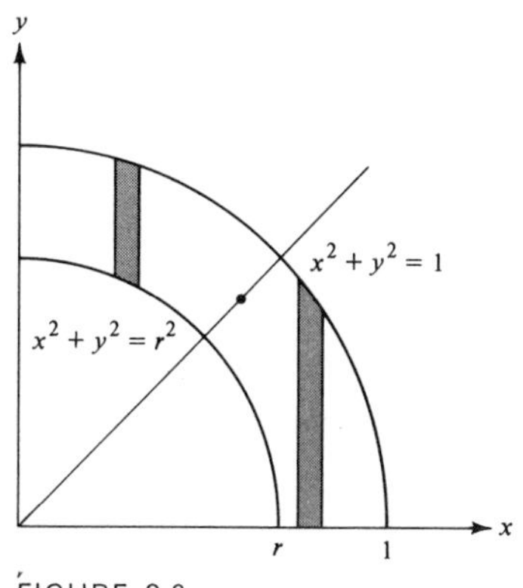

FIGURE 8.6

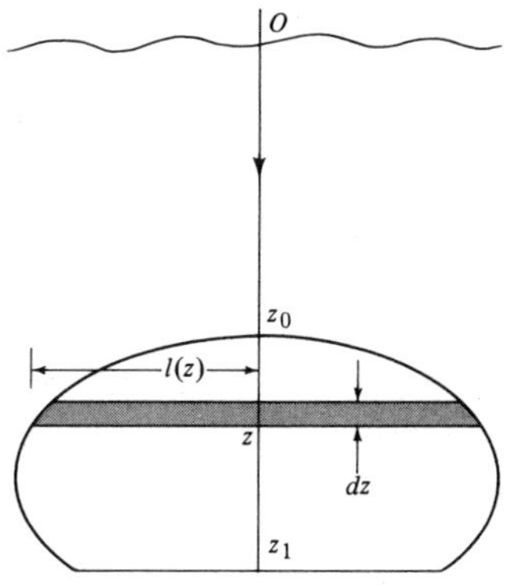

FIGURE 8.7

and this implies that the centroid lies on this line. Hence

$$\bar{y} = \bar{x} = \frac{4}{3\pi} \frac{1 + r + r^2}{1 + r}.$$

Example 4 Find the total force exerted on an observation "window" of a submarine used to explore the bottom of the ocean.

The statement of the problem is not terribly specific because the design of the window may be influenced by the analysis of the forces involved. In any event, we assume that the window is vertical and its shape symmetrical (Fig. 8.7); the fluid pressure (in excess of the sea-level atmospheric pressure, which will be neglected) is given by the hydrostatic law,

$$p = \rho g z,$$

where z is the depth below the surface of the sea. (The density ρ, and g are taken as constants.) The element of force on the incremental rectangle at depth z, with area $dA = 2l(z)\, dz$, is then

$$dF = p\, dA = 2\rho g z l(z)\, dz.$$

Therefore, the total force on the surface is

$$F = \int dF = 2\rho g \int_{z_0}^{z_1} z l(z)\, dz.$$

This formula can be rewritten in terms of the centroid and the area of the window. Since

$$A = 2 \int_{z_0}^{z_1} l(z)\, dz$$

and, by definition,

$$\bar{z} = \frac{2}{A}\int_{z_0}^{z_1} zl(z)\,dz,$$

it follows that

$$F = \rho g \bar{z} A.$$

But $\rho g \bar{z}$ is just the hydrostatic pressure at depth $\bar{z}$, and so the total force on the surface is the product of the exposed area of the window and the pressure at its centroid.

As a concrete illustration, let the window be a circular porthole (the most likely figure) of radius 7.5 cm and take $\rho g = 1200$ kg/m² sec², $\bar{z} = 2000$ m, in which case

$$F = (1200 \text{ kg/m}^2 \text{ sec}^2)(\pi(.075)^2 \text{ m}^2)(2000 \text{ m}) = 42{,}400 \text{ kg m/sec}^2.$$

Example 5 The kinetic energy of a mass m that moves with velocity v is $E = \frac{1}{2}mv^2$. Use this formula to find the kinetic energy of a thin disk of radius R, thickness H, and density ρ which rotates about its symmetry axis at a constant rate Ω.

This problem could refer to a record turntable revolving at $33\frac{1}{3}$ rpm. An arbitrary point P on such a disk moves about the origin in a circular path of constant radius r. As such, the distance P travels from its initial position P' (see Fig. 8.8) is the arc length s, which is given in terms of the central angle θ by

$$s = r\theta.$$

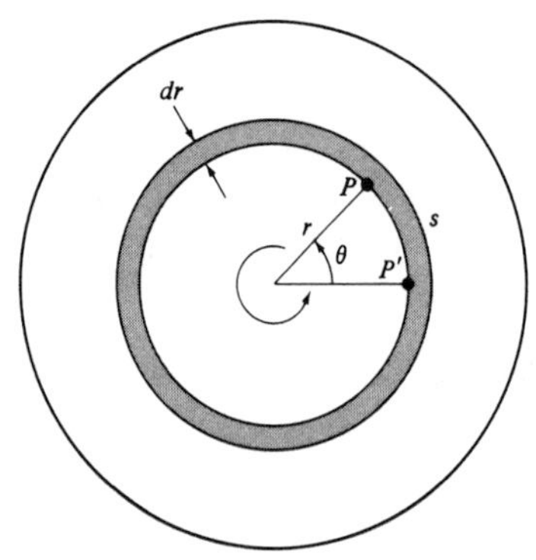

FIGURE 8.8

The circular speed at which P moves is computed as the rate of change of the distance travelled with respect to time, i.e.,

$$v = \frac{ds}{dt}.$$

Since r is fixed and $d\theta/dt$ is also assumed constant, it follows that

$$v = \frac{d}{dt}(r\theta) = r\frac{d\theta}{dt} = r\Omega. \tag{8}$$

Therefore point P travels at a constant speed; moreover, every point on the circle radius r travels at the very same speed. As a consequence, the increment of mass dm contained in the rim or annular region between r and $r + dr$ moves with constant velocity $v = r\Omega$ and has the kinetic energy

$$dE = \tfrac{1}{2}v^2\,dm. \tag{9}$$

Since the element of mass can be written

$$dm = \rho\,dV = \rho H\,dA = \rho H 2\pi r\,dr,$$

it follows that

$$dE = (\pi\rho H\Omega^2)r^3\,dr.$$

The total kinetic energy of the disk is then

$$E = \int dE = \pi\rho\Omega^2 H \int_0^R r^3\,dr = \tfrac{1}{4}\pi\rho\Omega^2 HR^4.$$

In terms of the total mass of the disk,

$$M = \int dm = 2\pi\rho H \int_0^R r\,dr = \pi\rho HR^2,$$

the kinetic energy of rotation can be expressed as

$$E = \tfrac{1}{4}MR^2\Omega^2.$$

Exercises 3.8

1. Find the mass of fluid contained in a vertical right circular cylinder of height H and radius R if the fluid density varies with height z $(0 \le z \le H)$ according to $\rho(z) = \rho_0(1 + z/H)^{-1}$. What is the average density of the fluid? Determine the center of mass.

2. Determine the center of mass of the fluid filling the conical glass in Example 1.

3. Find the mass of fluid contained in a spherical centrifuge (a sphere of radius R spinning

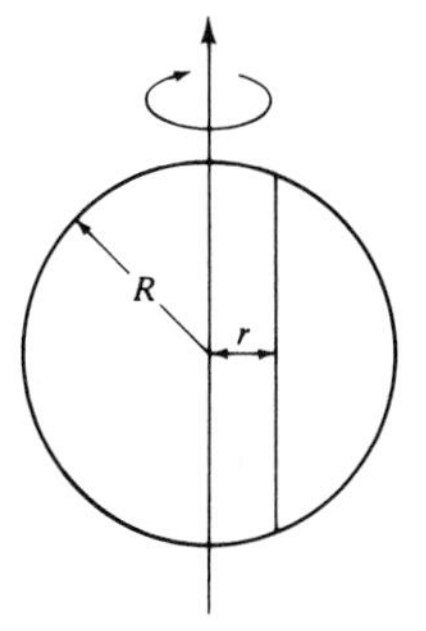

FIGURE 8.9

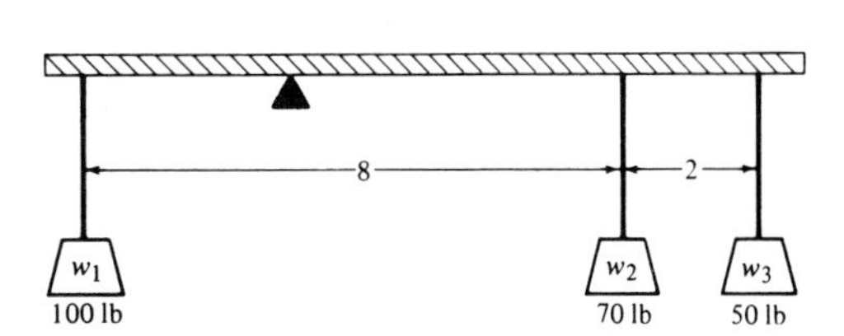

FIGURE 8.10

rapidly about its axis) if the fluid density is given by $\rho = \beta \exp\{-[1-(r/R)^2]^{1/2}\}$, where r is the distance from the rotation axis (Fig. 8.9).

4. Where should the fulcrum be placed in order to have the seesaw in Fig. 8.10 in balance? If the weight w of the seesaw cannot be neglected, where should the fulcrum be placed? (Assume w_1 and w_3 are suspended from the ends.)

5. The density of a thin rod of constant cross-sectional area and length l varies with distance x from one end according to $\rho = \rho_0[1+(x/l)^3]$. Find the mass of the rod and its center of gravity.

6. Consider the annulus $a^2 \leq x^2 + y^2 \leq b^2$. Find the centroids (or centers of mass) of the portions of the annulus in the upper half plane and in the first quadrant. Explain the positions of these centroids in the limit $b \to a$.

7. Find the centroid of the larger area bounded by $x^2+y^2=a^2$ and $x+y=a$.

8. Find the center of mass of a uniform triangular lamina with vertices at $(0,0)$, $(a,0)$ and (b,c) where a, b, c are positive numbers.

9. Find the volume of water in a rectangular swimming pool 50 m long and 20 m wide if the shallow end is 1 m below the surface and the deep end is 3 m below the surface. Assume the bottom slope is constant.

10. In the previous problem, what is the minimum energy required to empty the pool?

11. A horizontal cylindrical tank is filled with water to a height h above its lowest point. What force is exerted in the end plate of the tank, which is a circle of radius a?

12. The outlet gate of a reservoir is a (vertical) circular hole of radius 1 m; its center is 15 m beneath the surface of the water. Find the total fluid force exerted on the gate when it is closed.

13. A circular disk of radius R and thickness H rotates with constant angular velocity Ω. The density ρ of the disk material varies with the radial distance r as $\rho = \rho_0(1+r/R)^{-1}$. What is the average density? Find the kinetic energy of the disk. Check your answer for dimensional consistency.

14. If the density of seawater, $\rho = \rho(z)$, varies with depth below the sea surface, show that the hydrostatic pressure p is governed by $dp/dz = \rho(z)g$. Let the pressure and density at $z=0$ be p_0 and ρ_0; write the pressure at depth z as an integral. If $\rho = \rho_0\, e^{z/H}$, find as an integral the total force exerted on a vertical circular porthole of radius a if its center is a distance $L > a$ below the surface.

15. A solid cone of uniform density ρ, height H and base radius R rotates at constant angular velocity Ω about its axis. Find the kinetic energy of the cone. (*Hint:* Use cylindrical shells.)

16. Derive a formula in terms of integrals for the kinetic energy of a flywheel which is a volume of revolution rotating about its symmetry axis with a constant angular velocity. Assume first that the density of the flywheel is constant and then consider the general problem. Examine some special cases.

3.9 *Improper integrals*

An integral $\int_a^b f(x)\,dx$ is called *improper* when either limit of integration is infinite or when the integrand $f(x)$ is singular somewhere in the range of integration $a \leq x \leq b$. The integrals

$$\int_0^\infty e^{-x}\,dx \qquad \text{and} \qquad \int_0^1 x^{-3/2}\,dx$$

are prototypes of each impropriety, and, as Fig. 9.1 indicates, an improper integral represents an unbounded area. However, it is not at all obvious from these diagrams whether improper integrals (or, equivalently, the areas beneath unbounded curves) have finite or infinite values. The only way to resolve this question in general is to consider every improper integral to be the relevant limit of a proper integral, as exemplified by

$$\int_0^\infty e^{-x}\,dx = \lim_{b\to\infty}\int_0^b e^{-x}\,dx, \qquad \int_0^1 x^{-3/2}\,dx = \lim_{a\to 0}\int_a^1 x^{-3/2}\,dx.$$

If the limit is a finite number, the improper integral is said to be *convergent*; conversely, the integral is *divergent* if the limit is infinite or does not exist.

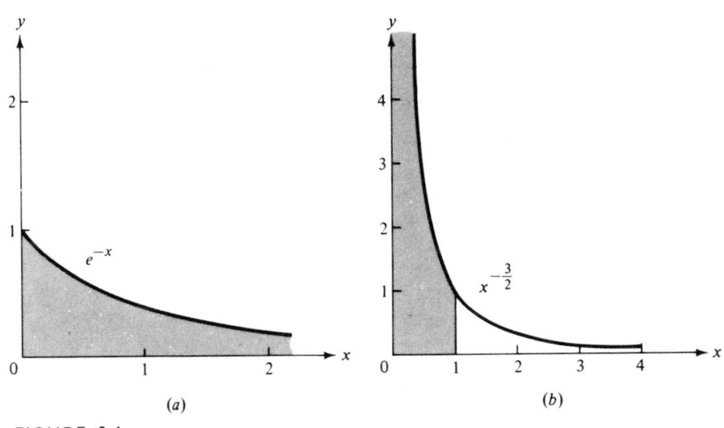

FIGURE 9.1

The values of the two integrals cited above as illustrations are easily determined. Since

$$\int_0^b e^{-x}\,dx = -e^{-x}\Big]_0^b = 1 - e^{-b},$$

it follows that

$$\int_0^\infty e^{-x}\,dx = \lim_{b\to\infty}(1 - e^{-b}) = 1$$

(see Fig. 9.2). Furthermore,

$$\int_0^1 x^{-3/2}\,dx = \lim_{a\to 0}\int_a^1 x^{-3/2}\,dx = \lim_{a\to 0}(-2 + 2a^{-1/2}) = \infty.$$

In the former case, the unbounded area has a finite value, and the corresponding integral is convergent; in the latter, the integral is divergent because the limiting area is infinite.

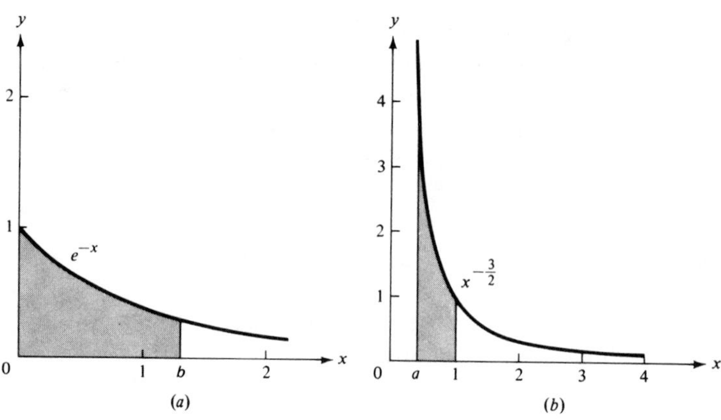

FIGURE 9.2

Example 1 Evaluate $I = \int_0^1 x^{-\alpha}\,dx$ for $\alpha > 0$.

Let

$$I = \lim_{\varepsilon \to 0} \int_\varepsilon^1 x^{-\alpha}\,dx.$$

Since

$$\int_\varepsilon^1 x^{-\alpha}\,dx = \begin{cases} \left.\dfrac{x^{1-\alpha}}{1-\alpha}\right]_\varepsilon^1 = \dfrac{1-\varepsilon^{1-\alpha}}{1-\alpha}, & \alpha \neq 1; \\ \log x\Big]_\varepsilon^1 = -\log \varepsilon, & \alpha = 1, \end{cases}$$

the limit is finite, i.e., the integral is convergent, only when $\alpha < 1$, and in this case

$$I = \lim_{\varepsilon \to 0} \frac{1-\varepsilon^{1-\alpha}}{1-\alpha} = \frac{1}{1-\alpha}.$$

A singular point x_* of a function $f(x)$ is said to be *integrable* if the improper integral $\int_{x_*}^b f(x)\,dx$ is convergent. The conclusion drawn from the last example is that $x = 0$ is an integrable singularity of $x^{-\alpha}$ for $0 < \alpha < 1$.

Example 2 Evaluate $\displaystyle\int_2^\infty \frac{dx}{x \log x}$.

Let

$$I = \lim_{s \to \infty} \int_2^s \frac{dx}{x \log x};$$

Figure 9.3 shows the area represented by this proper integral. The integration can be performed by introducing the variable $u = \log x$, in which case

$$\int_2^s \frac{dx}{x \log x} = \int_{\log 2}^{\log s} \frac{du}{u} = \log \frac{\log s}{\log 2}.$$

Therefore

$$I = \lim_{s \to \infty} \log \frac{\log s}{\log 2} = \infty,$$

and the integral is divergent; i.e., an infinite area lies under the curve for $x \geq 2$.

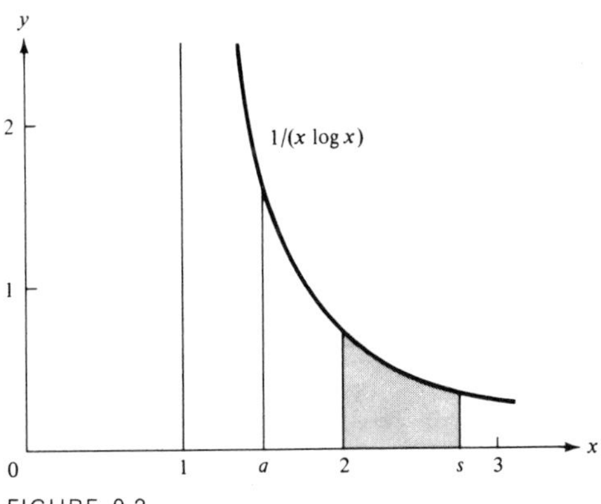

FIGURE 9.3

Example 3 Find $J = \int_1^2 \frac{dx}{x \log x}$.

In this case, we consider

$$\int_a^2 \frac{dx}{x \log x} = \log(\log 2) - \log(\log a),$$

so that

$$J = \lim_{a \to 1} [\log(\log 2) - \log(\log a)] = \infty.$$

The integral is divergent.

Example 4 Show that $\int_0^\infty \sin x\, dx$ is a divergent integral.

By definition

$$\int_0^\infty \sin x\, dx = \lim_{b\to\infty} \int_0^b \sin x\, dx = \lim_{b\to\infty} (1 - \cos b).$$

Since the last limit on the right does not exist, the integral is divergent.

Example 5 Is $\int_{-1}^{1} \frac{1 - \cos x}{x^2}\, dx$ an improper integral?

The integrand, $f(x) = (1 - \cos x)/x^2$, is actually not singular anywhere in the range of integration. Its value at the origin, the only questionable point, is determined by using l'Hopital's rule:

$$f(0) = \lim_{x\to 0} \frac{1 - \cos x}{x^2} = \lim_{x\to 0} \frac{\sin x}{2x} = \frac{1}{2}.$$

Therefore the integral, despite appearances, is proper. (The separation

$$\int_{-1}^{1} \frac{1 - \cos x}{x^2}\, dx = \int_{-1}^{1} \frac{dx}{x^2} - \int_{-1}^{1} \frac{\cos x}{x^2}\, dx$$

shows that this proper integral can be written as the sum of two *divergent integrals*; such a decomposition should obviously be avoided.)

---

2 We should be able to determine whether an improper integral is convergent or not without having to integrate it explicitly in terms of known functions. After all, most integrals cannot be so simply expressed. One method of doing this is to compare the integral with another having properties easier to calculate or already established. For example, it may be possible to analyze the integral $\int_a^\infty f(x)\, dx$, with $f(x) \geq 0$, by comparing it to a simpler one, denoted by $\int_a^\infty g(x)\, dx$, and in this way obtain bounds which automatically infer convergence or divergence. To illustrate, suppose that $\int_a^\infty g(x)\, dx$ is known to be convergent; then $\int_a^\infty f(x)\, dx$ is also convergent *if*

$$0 \leq f(x) \leq g(x) \qquad \text{for} \qquad a \leq x \leq \infty, \tag{1}$$

because this condition immediately implies

$$\int_a^\infty f(x)\, dx \leq \int_a^\infty g(x)\, dx < \infty. \tag{2}$$

Conversely, if $f(x) \geq r(x) \geq 0$ and $\int_a^\infty r(x)\, dx$ is a divergent integral, then $\int_a^\infty f(x)\, dx$ *must* also be divergent since

$$\int_a^\infty f(x)\,dx \geq \int_a^\infty r(x)\,dx. \tag{3}$$

The decomposition

$$\int_a^\infty f(x)\,dx = \int_a^b f(x)\,dx + \int_b^\infty f(x)\,dx, \tag{4}$$

implies that the question of convergence or divergence arising from the infinite range of integration depends only on the last integral on the right-hand side [and the behavior of $f(x)$ for large x]. Thus, if the integrand is known to be positive (or negative) only, say, for $x > b$, the comparison method can be applied to the integral over the range $[b, \infty]$.

Corresponding results can be derived to test an improper integral whose integrand is singular at a finite point. Once again the conditions $0 \leq f(x) \leq g(x)$ for $a \leq x \leq b$ and $\int_a^b g(x)\,dx < \infty$ imply that $\int_a^b f(x)\,dx$ is convergent, whereas the conditions $0 \leq r(x) \leq f(x)$, $|\int_a^b r(x)\,dx| = \infty$ mean that the integral is divergent.

Example 6 Show that $\int_0^\infty e^{-x^2}\,dx$ is a convergent integral.

Write

$$\int_0^\infty e^{-x^2}\,dx = \int_0^1 e^{-x^2}\,dx + \int_1^\infty e^{-x^2}\,dx,$$

where the integral between 0 and 1 is quite proper and has some constant value M. Since $e^{-x^2} < e^{-x}$ for $x > 1$, it follows that

$$\int_1^\infty e^{-x^2}\,dx \leq \int_1^\infty e^{-x}\,dx = \frac{1}{e}.$$

Therefore

$$\left|\int_0^\infty e^{-x^2}\,dx\right| \leq \left|\int_0^1 e^{-x^2}\,dx\right| + \left|\int_1^\infty e^{-x^2}\,dx\right| \leq M + \frac{1}{e},$$

which establishes the convergence of the integral. A rough numerical bound is obtained using the inequality $e^{-x^2} < 1$, in which case

$$M = \int_0^1 e^{-x^2}\,dx < \int_0^1 dx = 1.$$

The result of substituting this bound for M is

$$\int_0^\infty e^{-x^2}\,dx \leq 1 + \frac{1}{e} = 1.36.$$

(Later, it will be shown that the exact value of the integral is $\pi^{1/2}/2 = 0.88623\ldots$.)

Example 7 Show that $\int_0^\infty \frac{dx}{(x+x^3)^{1/2}}$ is convergent by finding a bound for it.

The integral is improper for two reasons: the integrand is singular at the origin and the range of integration is infinite. For these reasons, it will be broken in two, each piece to be analyzed separately:

$$\int_0^\infty \frac{dx}{(x+x^3)^{1/2}} = \int_0^1 \frac{dx}{(x+x^3)^{1/2}} + \int_1^\infty \frac{dx}{(x+x^3)^{1/2}}. \tag{5}$$

There should not be much doubt that the integral is convergent, since at the origin the integrand is singular like $x^{-1/2}$ whereas at infinity it is comparable to $x^{-3/2}$. These observations are the motivation for the approach adopted.

To bound the first integral we use the fact $|1+x^2| \geq 1$, in which case

$$\int_0^1 \frac{dx}{(x+x^3)^{1/2}} = \int_0^1 \frac{dx}{x^{1/2}(1+x^2)^{1/2}} \leq \int_0^1 \frac{dx}{x^{1/2}} = 2.$$

The second integral is treated similarly by writing

$$(x+x^3)^{1/2} = x^{3/2}\left(1+\frac{1}{x^2}\right)^{1/2} > x^{3/2}$$

or

$$\frac{1}{(x+x^3)^{1/2}} < \frac{1}{x^{3/2}},$$

so that

$$\int_1^\infty \frac{dx}{(x+x^3)^{1/2}} < \int_1^\infty \frac{dx}{x^{3/2}} = \left.\frac{-2}{x^{1/2}}\right]_1^\infty = 2.$$

Finally, these results can be substituted in (5) to obtain the bound

$$\int_0^\infty \frac{dx}{(x+x^3)^{1/2}} < 2+2=4,$$

which proves convergence.

---

3 Improper integrals may arise from many different sources. They can be a direct consequence of some change of variable or an intrinsic feature in the formulation of a physical problem. A simple illustration of the first possibility is afforded by the transformation $x = e^{-y}$, which converts $\int_0^a f(x)\,dx$ to $\int_{-\log a}^\infty f(e^{-y})e^{-y}\,dy$. Examples of the latter possibility occur in almost all transient problems in which there is an evolution in time from an initial condition at $t = 0$ to a final state at $t = \infty$.

The definition of a physical quantity may itself be expressed as an improper integral, a good example being the formula for the *period of oscillation*. In simple harmonic motion an object moves to and fro between two extreme positions. The case of a weight attached to a spring is illustrated in Fig. 9.4. The period T is the time required

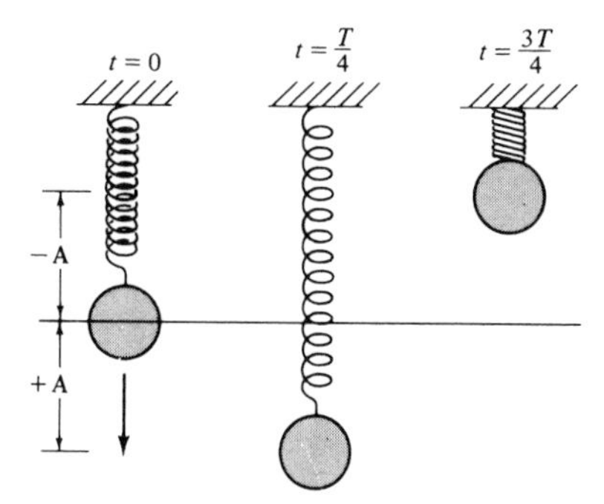

FIGURE 9.4

for the object to return to its original state, and the point of maximum displacement from equilibrium is a convenient place to make such measurements. The period is then the time for the object to travel from, say, $-A$ to $+A$ *and back again*, and this is obviously twice the time of a one-way trip. If v represents velocity, then dx/v is the differential increment of time required to traverse a distance dx and the period is therefore

$$T = 2\int_{-A}^{A} \frac{dx}{v}.$$

However, the velocity at the end points is zero, $v(\pm A) = 0$, which means that the preceding integral is improper—but necessarily convergent if the motion is truly periodic.

Improper integrals are also the consequences of the approximations and idealizations used to model physical phenomena. A small sphere regarded as a point, a large volume taken as infinite in extent, certain effects and interactions assumed negligible—all these can produce improper integrals. If, however, the essence of the phenomena under study is still adequately described by the mathematical model, the theory will do the best it can, and this often means integrable singularities. Nature is terribly complex, and approximations and models are not only inevitable but an indispensible part of science; ergo, the common appearance of improper integrals.

Exercises 3.9

1. Discuss the following improper integrals and when possible find their values.

(i) $\displaystyle\int_0^2 \frac{1}{(2-x)^{1/2}}\,dx$ **(ii)** $\displaystyle\int_{-1}^{1} \frac{dx}{x^2}$ **(iii)** $\displaystyle\int_1^{\infty} \frac{dx}{x^{2/3}}$

(iv) $\displaystyle\int_0^{\infty} e^{-x}\,dx$ **(v)** $\displaystyle\int_{-\infty}^{0} xe^x\,dx$ **(vi)** $\displaystyle\int_1^{\infty} \frac{dx}{x^2-1}$

(vii) $\displaystyle\int_0^{\infty} \frac{dx}{(x+1)(x+2)}$ **(viii)** $\displaystyle\int_{-1}^{2} \frac{dx}{x^3}$ **(ix)** $\displaystyle\int_1^{\infty} \frac{dx}{x(x-1)}$

(x) $\displaystyle\int_1^2 \frac{dx}{[(2-x)(x-1)]^{1/2}}$ (xi) $\displaystyle\int_0^\infty xe^{-x^2}\,dx$ (xii) $\displaystyle\int_{-\infty}^\infty \frac{dx}{1+x^2}$

2. Which of the following are convergent? Evaluate when possible.

(i) $\displaystyle\int_0^\infty \frac{dx}{(1+x)^4}$ (ii) $\displaystyle\int_0^\infty e^{-x}\cos x\,dx$ (iii) $\displaystyle\int_1^2 \frac{dx}{x(\log x)^2}$

(iv) $\displaystyle\int_0^\infty \frac{1-\cos x}{x^2}\,dx$ (v) $\displaystyle\int_0^\infty \frac{\sin z}{z}\,dz$ (vi) $\displaystyle\int_0^1 x(\log x)^2\,dx$

(vii) $\displaystyle\int_0^\infty \frac{dx}{e^x+1}$ (vii) $\displaystyle\int_0^\infty \frac{dx}{x^2+4x+5}$ (ix) $\displaystyle\int_0^{1/2} \frac{dx}{x(\log x)^2}$

(x) $\displaystyle\int_0^\infty \sin t^4\,dt$ (xi) $\displaystyle\int_0^2 \frac{e^{-x}}{x}\,dx$ (xii) $\displaystyle\int_0^1 \frac{\log x}{x}\,dx$

3. Evaluate or find bounds for the convergent improper integrals among the following:

(i) $\displaystyle\int_1^2 \frac{dx}{x(\log x)^{1/2}}$ (ii) $\displaystyle\int_0^\infty \frac{dx}{(1+x)^4}$ (iii) $\displaystyle\int_0^\infty \frac{\sin x\,dx}{x^2}$

(iv) $\displaystyle\int_3^\infty \frac{dx}{x(x-1)^{1/2}}$ (v) $\displaystyle\int_0^1 x\csc x\,dx$ (vi) $\displaystyle\int_1^\infty \frac{dx}{x(1+\log x)^2}$

(vii) $\displaystyle\int_0^1 \log x\,dx$ (viii) $\displaystyle\int_{-1}^1 \left(\frac{1-x}{1+x}\right)^{1/2} dx$ (ix) $\displaystyle\int_0^\infty \frac{dx}{(2x+x^6)^{1/3}}$

(x) $\displaystyle\int_0^{\pi/2} \frac{\cot x}{\log(\sin x)}\,dx$ (xi) $\displaystyle\int_0^\infty \frac{\sin x}{(x+1)(x+2)}\,dx$ (xii) $\displaystyle\int_0^{\pi/2} \frac{\cos^3 x}{\sqrt{\sin x}}\,dx$

4. Are the integrals $\int_0^\infty \sin x^2\,dx$ and $\int_0^\infty \cos x^2\,dx$ convergent?

5. Show that $\displaystyle\int_0^\infty \frac{1}{1+x^4}\,dx = \int_0^\infty \frac{x^2}{1+x^4}\,dx$ by making a change of variable. Prove that the common value is $\pi\sqrt{2}/4$. [*Hint:* $x^4+1=(x^2+\sqrt{2}\,x+1)(x^2-\sqrt{2}\,x+1)$.]

6. Show that $\displaystyle\int_0^\infty \frac{x\log x}{(1+x^2)^2}\,dx = 0$. (*Hint:* Integrate from 0 to 1 and from 1 to ∞ separately.)

7. Show that $\displaystyle\int_0^{\pi/2} \log(\sin x)\,dx = \int_0^{\pi/2} \log(\cos x)\,dx = -(\pi\log 2)/2.$
(*Hint:* $\sin 2x = 2\sin x\cos x$). Evaluate $\displaystyle\int_0^{\pi/2} \log(a\sin x)\,dx$ and $\displaystyle\int_0^{\pi/2} \log(a\tan x)\,dx$.

8. For what values of the parameter a is $\int_0^\infty t^a e^{-t}\,dt$ convergent? If n is a non-negative integer, verify that $\int_0^\infty x^n e^{-x}\,dx = n!$.

9. Show that $\int_{-\infty}^\infty f(t)g(a-t)\,dt = \int_{-\infty}^\infty f(a-t)g(t)\,dt$, assuming that both integrals are convergent. Illustrate this identity with the special case $f(t)=e^{-t^2}$, $g(t)=t$.

10. Test for convergence or divergence.

(i) $\displaystyle\int_1^3 \frac{dx}{x^2-1}$ (ii) $\displaystyle\int_0^{\pi/2} \tan x\,dx$ (iii) $\displaystyle\int_{-1}^1 \frac{\sin x}{x^2}\,dx$

(iv) $\displaystyle\int_0^\infty \frac{x^3}{4x^4+1}\,dx$ (v) $\displaystyle\int_0^\infty \tan^{-1}x\,dx$ (vi) $\displaystyle\int_0^\infty x^{1/2}e^{-2x}\,dx$

11. Show that the integral $\displaystyle\int_0^\infty \frac{\sin ax}{x}\,dx$ is convergent and independent of $|a|$. Interpret this result graphically by plotting $(\sin ax)/x$ for several values of a.

12. What is wrong with the following calculation?

$$\int_{-1}^{1} \frac{dx}{x^2} = -\int_{-1}^{1} u^2 \left(-\frac{1}{u^2}\right) du = -\int_{-1}^{1} du = -2 \qquad \text{where } u = 1/x.$$

13. Evaluate $\int_{-1}^{1} \frac{1}{x}\, dx$ by the limit

$$\int_{-1}^{1} \frac{1}{x}\, dx = \lim_{\epsilon \to 0+} \left(\int_{-1}^{-\epsilon} \frac{1}{x}\, dx + \int_{\epsilon}^{1} \frac{1}{x}\, dx \right).$$

Does this make any sense? Try an alternative approach given by

$$\int_{-1}^{1} \frac{1}{x}\, dx = \lim_{a \to 0+} \int_{-1}^{-a} \frac{1}{x}\, dx + \lim_{b \to 0+} \int_{b}^{1} \frac{1}{x}\, dx.$$

Why is an infinite value obtained this way?

14. The density of the atmosphere at altitude z km above a particular point is given by $\rho = \rho_0 e^{-z/l}$ where ρ_0 is the surface density. What is the atmospheric pressure on the surface of the planet?

15. The curve $y = e^{-x}$ from $x = 0$ to $x = \infty$ is rotated about the x axis. Find the volume and surface area generated. Solve the more general problem with the curve $y = ae^{-bx}$ $(b > 0)$ and verify the dimensional consistency of the answers.

16. Find the period of motion of a particle whose position on the x axis at time t is given by $x(t) = 4 + 2\cos^2 2t$. Do this first by calculating the velocity and using the integral formula. Repeat the calculation by an elementary method.

17. The formula relating kinetic energy to work (derived from $F = ma = mv\,dv/ds$) is

$$\int_{s_0}^{s} F(s)ds = \tfrac{1}{2} mv^2 \Big|_{s_0}^{s} = \tfrac{1}{2} m(v(s))^2 - \tfrac{1}{2} m(v(s_0))^2.$$

Find the minimum initial kinetic energy that must be given to a body of mass m starting at s_0 with speed $v_0 = v(s_0)$ for it to reach infinity $(s = \infty)$ against a retarding force $F = F(s)$.

18. If a force $F(x) = m/(1 + x^2)$ is acting on a body of mass m initially at rest at $x = 0$, find the speed of the object when it passes $x = a$. What is the limiting velocity as $x \to \infty$?

19. The gravitational force on a body of mass m at a distance r from the center of the earth is $F(r) = mg(r_0/r)^2$ where $g = 9.806$ m/sec^2 and $r_0 = 6370$ km. What is the minimal work $W(r_0)$ that must be done against the force of gravity for the body to escape the earth's gravitational field? If all this energy is supplied as initial kinetic energy, find the initial velocity $v_0(r_0)$ that the body must have to escape the earth. (Assume the body moves along a straight line from the center of the earth. Initially, the body is on the earth's surface.)

3.10 Approximate methods

The numerical value of the definite integral $\int_a^b f(x)\, dx$ can be calculated approximately in many ways, all of which involve, in some sense, the replacement of the integrand $f(x)$ by a simpler function.

Consider first a sequence of numerical methods, each closely related to Lagrangian interpolation. (The order of presentation will be seen to correspond with the degree of an interpolating polynomial.) The interval $[a, b]$ is subdivided into n increments each of length $\Delta x = h = (b - a)/n$, the end points of which are given by

$$x_i = a + ih, \qquad i = 0, 1, \ldots, n,$$

where $x_0 = a$, $x_n = b$. Furthermore, define

$$f_i = f(x_i).$$

The $n + 1$ points (x_i, f_i) are used to construct approximations for the integrand in the entire range $[a, b]$ which form the basis of the numerical calculation of the integral. The simplest approximation for the area is the familiar one consisting of incremental rectangles, each of width h and height $f(x_i)$, as shown in Fig. 10.1a. This yields the fundamental estimate

$$\int_a^b f(x)\,dx \approx \sum_{i=1}^{n} f(x_{i-1})h = h(f_0 + \cdots + f_{n-1}). \tag{1}$$

This formula can also be interpreted as the result of replacing the function $f(x)$ in each subinterval $x_{i-1} \le x \le x_i$ by the *constant* $f(x_{i-1}) = f_{i-1}$ and, as such, corresponds to "interpolation" with a polynomial of degree zero. Of course, as h tends to zero and n approaches infinity, the estimate given in (1) becomes exact. For simple problems, this formula is already adequate, and it certainly does not pay to invest much effort developing a more sophisticated routine that will only save a few seconds of machine time—unless your own labor comes cheap. However, the interesting problem, as opposed to a routine one, is almost by definition that which taxes or challenges everyone's and everything's ability to cope with it. To this end, we would like to have as efficient an approximation scheme as possible, in terms of speed and accuracy, for this enlarges the range of possible applications that can be handled by available means.

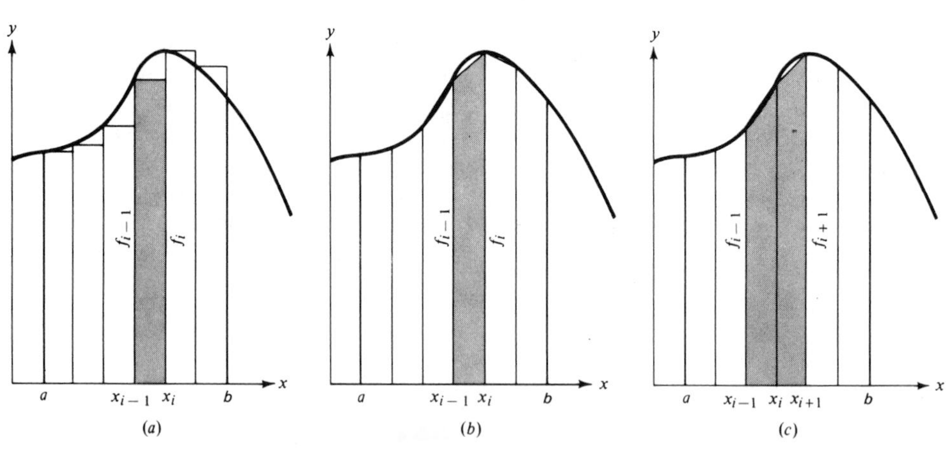

FIGURE 10.1

Using the same information, a more accurate approximation to the integral is obtained by replacing the rectangle in subinterval $[x_{i-1}, x_i]$ by a trapezoid, as shown in

Fig. 10.1b. The area of this trapezoid is $f_{i-1}h + \frac{1}{2}(f_i - f_{i-1})h = \frac{1}{2}(f_i + f_{i-1})h$, and the approximation formed by adding *all* these increments together is

$$\int_a^b f(x)\,dx \approx [\tfrac{1}{2}(f_1 + f_0) + \tfrac{1}{2}(f_2 + f_1) + \cdots + \tfrac{1}{2}(f_n + f_{n-1})]h$$

or

$$\int_a^b f(x)\,dx \approx \frac{h}{2}\left[f_0 + 2(f_1 + f_2 + \cdots + f_{n-1}) + f_n\right]. \tag{2}$$

This trapezoid formula can also be interpreted as a consequence of replacing the function $f(x)$ in each interval $x_{i-1} \leq x \leq x_i$ by the straight line

$$y = \frac{f_i - f_{i-1}}{x_i - x_{i-1}}(x - x_{i-1}) + f_{i-1}. \tag{3}$$

[The total effect is to approximate $f(x)$ by a continuous polygonal curve of straight-line segments, or chords, as shown in Fig. 10.1b.] Equation (3) is the Lagrangian interpolation polynomial of the first degree between the two points (x_{i-1}, f_{i-1}) and (x_i, f_i).

2

In the next stage of approximation, the curve $y = f(x)$ is replaced by parabolic arcs, and the corresponding areas are calculated from the information at our disposal (Fig. 10.1c). Since three points determine a parabola uniquely, we can compute one such arc using x_0, x_1, x_2 and another from x_2, x_3, x_4, etc., which shows that an *odd* number of points is required for a perfect subdivision of interval $[a, b]$ or an *end correction* is necessary. Surely this would be no tragedy, but for the sake of clarity, we now take the points of the subdivision to run from x_0 to x_{2n} so that there are $2n + 1$ points and exactly $2n$ intervals. The basic increment, or step-size, is then

$$h = \frac{b - a}{2n}.$$

If we focus our attention on the three points (x_0, f_0), (x_1, f_1), and (x_2, f_2), where $x_1 = x_0 + h$ and $x_2 = x_0 + 2h$, the first item of business is to find the parabola which passes through these positions. This can be done by returning to fundamentals, which would require the identification of the coefficients of a quadratic $y = \alpha_0 + \alpha_1 x + \alpha_2 x^2$; more simply, we can use Lagrange's interpolation formula for three points, derived in general on page 182. In either way, the parabola is found to be

$$\mathscr{L}(x) = \frac{1}{2h^2}[f_0(x - x_1)(x - x_2) - 2f_1(x - x_0)(x - x_2) + f_2(x - x_0)(x - x_1)]. \tag{4}$$

The area under the parabolic arc from x_0 to $x_2 = x_0 + 2h$ is an approximation to the integral of $f(x)$ over the same range. This integral can be calculated efficiently using the variable $z = x - x_1 = x - x_0 - h$, which represents a translation of the coordinate origin to the centerline of the interval $x = x_1$. Therefore

$$\int_{x_0}^{x_2} \mathscr{L}(x)\,dx = \frac{1}{2h^2}\int_{-h}^{h} \mathscr{L}(z + x_0 + h)\,dz,$$

and it follows that

$$\int_{x_0}^{x_2} \mathscr{L}(x)\,dx = \frac{1}{2h^2}\int_{-h}^{h} [(f_0 - 2f_1 + f_2)z^2 + (f_2 - f_0)zh + 2f_1h^2]\,dz$$

or

$$\int_{x_0}^{x_2} \mathscr{L}(x)\,dx = \tfrac{1}{3}h(f_0 + 4f_1 + f_2). \tag{5}$$

The last formula (and the area it represents) is the desired approximation to $\int_{x_0}^{x_2} f(x)\,dx$. As a consequence, the sum of all such approximations for every three-point interval yields *Simpson's rule*:

$$\int_a^b f(x)\,dx = \int_{x_0}^{x_{2n}} f(x)\,dx \approx \tfrac{1}{3}h[(f_0 + 4f_1 + f_2) + (f_2 + 4f_3 + f_4) + \cdots + (f_{2n-2} + 4f_{2n-1} + f_{2n})],$$

or

$$\int_a^b f(x)\,dx \approx \tfrac{1}{3}h[f_0 + 4(f_1 + f_3 + \cdots + f_{2n-1}) + 2(f_2 + f_4 + \cdots + f_{2n-2}) + f_{2n}]. \tag{6}$$

A useful mnemonic is

$$\int_a^b f(x)\,dx \approx \tfrac{1}{3}h(\text{"ends"} + 4\ \text{"odds"} + 2\ \text{"evens"}),$$

but no one should be required to memorize this.

The procedure can be continued using higher-order interpolation polynomials, but this is as far as we will take the sequence of formulas.

3

In terms of a subdivision of the interval $[a, b]$ into $2n$ increments each of length h (the step size), the three formulas for numerical integration derived thus far are:

Rectangular rule:
$$\int_a^b f(x)\,dx \approx h(f_0 + \cdots + f_{2n-1}); \tag{7}$$

Trapezoidal rule:
$$\int_a^b f(x)\,dx \approx \tfrac{1}{2}h[f_0 + 2(f_1 + \cdots + f_{2n-1}) + f_{2n}]; \tag{8}$$

Simpson's or parabolic rule:

$$\int_a^b f(x)\,dx \approx \tfrac{1}{3}h[f_0 + 4(f_1 + \cdots + f_{2n-1}) + 2(f_2 + \cdots + f_{2n-2}) + f_{2n}]. \quad (9)$$

Example 1 Apply each of the formulas above to evaluate

$$I = \int_0^1 \sin\frac{\pi}{2}x\,dx \qquad \left(= \frac{2}{\pi} = 0.6366197\ldots\right).$$

A routine problem is chosen in order to check and compare various procedures with a known result. Although for small values of integer n the computations can be completed by hand with a desk calculator, a fast digital computer is necessary in most applications. The formulas are easily programmed in any of the machine languages available, and for $n = 10$, 100, 1,000, the results are presented in Table 10.1.

TABLE 10.1

Formula	$n = 10$	$n = 100$	$n = 1{,}000$
(7)	0.585310	0.631607	0.636118
(8)	0.635310	0.636607	0.636618
(9)	0.636622	0.636620	0.636620

Example 2 Use Simpson's rule to evaluate $\dfrac{2}{\pi^{1/2}}\displaystyle\int_0^3 e^{-x^2}\,dx$.

Formula (9) is applied using the subdivisions $n = 30$, 300, and 3,000, which correspond to increment sizes $2h = 0.1$, 0.01, and 0.001. The arithmetic is performed by a computer, and the results are

$$\frac{2}{\pi^{1/2}}\int_0^3 e^{-x^2}\,dx = \begin{cases} 0.99998, & 2h = 0.1; \\ 0.99998, & 2h = 0.01; \\ 0.99998, & 2h = 0.001. \end{cases}$$

All entries are exact to five decimal places. (The purpose of the multiplicative factor $2/\pi^{1/2}$ should be clear now from the value of the integral.)

---

4 The indefinite integral

$$Y(x) = \int_a^x f(x)\,dx, \tag{10}$$

can be computed, once x is assigned a value, by using any one of the formulas (7) to (9). However, since the tabulation of an entire function is required, we would not want to repeat this cumbersome procedure anew at each position. For this reason, the analysis is modified slightly in order to develop a *marching method* of computation, in which the complete calculation advances one step at a time. This is accomplished quite easily by subtracting the values of $Y(x)$ at two neighboring points. Let h be the step size of the advance, and set

$$x_i = a + ih, \qquad Y_i = Y(x_i),$$

and so forth [where, if needed, $x_{i+1/2} = a + (i + \tfrac{1}{2})h$]. Therefore

$$Y_{i+1} = Y(x_{i+1}) = \int_a^{x_{i+1}} f(x)\,dx,$$

$$Y_i = Y(x_i) = \int_a^{x_i} f(x)\,dx,$$

and it follows that

$$Y_{i+1} = Y_i + \int_{x_i}^{x_{i+1}} f(x)\,dx. \tag{11}$$

The integral over the small interval of incremental length $x_{i+1} - x_i = h$ can be approximated by any of the methods introduced thus far. In particular, the rectangular and trapezoidal rules,

$$\int_{x_i}^{x_{i+1}} f(x)\,dx \approx \begin{cases} hf_i; \\ \tfrac{1}{2}h(f_i + f_{i+1}), \end{cases}$$

yield the approximations

$$Y_{i+1} = Y_i + hf_i, \tag{12}$$

and

$$Y_{i+1} = Y_i + \tfrac{1}{2}h(f_i + f_{i+1}). \tag{13}$$

Moreover, the adoption of Eq. (5) for present use (which involves only a rescaling of increment size, as shown in Exercise 1) leads to the finite-difference formula

$$Y_{i+1} = Y_i + \tfrac{1}{6}h(f_i + 4f_{i+1/2} + f_{i+1}), \tag{14}$$

where

$$f_{i+1/2} = f(a + (i + \tfrac{1}{2})h).$$

The initial value, $Y_0 = Y(a) = 0$, starts the process. Upon substituting the numerical values of $f(x)$ required in any of the three finite-difference equations, the value of the indefinite integral is determined at $x = h$, that is, $Y_1 = Y(h)$. The process is repeated, using this number, to find $Y_2 = Y(2h)$, and the advance continues, restricted only by accumulated errors and inaccuracies (or failing interest).

The indefinite integral (10) also represents the solution of the *differential equation*

$$\frac{d}{dx} Y = f(x), \tag{15}$$

which satisfies the initial condition $Y = 0$ at $x = a$. Therefore, formulas (12) to (14) provide the numerical solution of this differential equation as well, since the problem is entirely equivalent to the calculation already undertaken. Moreover, the definite integral $\int_a^b f(x)\,dx$ can be determined from the indefinite integral by continuing the marching process until we reach $x_n = a + nh = b$, whereupon the value of $Y(x_n) = Y(b)$ is automatically generated. [Clearly, we can also view every definite integral $Y(b)$ as the solution of the associated differential equation evaluated at the particular point $x_n = b$.]

Example 3 Tabulate the function defined by $\dfrac{2}{\pi^{1/2}}\displaystyle\int_0^x e^{-z^2}\,dz$.

This function occurs frequently enough in science to be made "known" once and for all. It is called the *error function*, a name which stems from statistical applications, and it will be written as

$$\operatorname{erf}(x) = \frac{2}{\pi^{1/2}}\int_0^x e^{-z^2}\,dz. \tag{16}$$

This integral has been examined previously where we have shown that $\lim_{x\to\infty} \operatorname{erf}(x) < \infty$ and $\operatorname{erf}(3) = 0.99998$. The complete tabulation is obtained using Simpson's rule, as expressed in (14), for the solution of the associated differential equation

$$\frac{d}{dx}\operatorname{erf}(x) = \frac{2}{\pi^{1/2}}\,e^{-x^2} \qquad \text{with } \operatorname{erf}(0) = 0.$$

The specific numerical formula is

$$\operatorname{erf}(x_{i+1}) = \operatorname{erf}(x_i) + \frac{2}{6\pi^{1/2}}\,h\left(e^{-x_i^2} + 4e^{-(x_{i+1/2})^2} + e^{-(x_{i+1})^2}\right),$$

where $x_i = ih$. Since the error function is odd, $\operatorname{erf}(-x) = -\operatorname{erf}(x)$, the calculation can be restricted to positive values of x only. Moreover, the computation need not extend beyond $x = 3$, because the functional value there is already very nearly equal to the limit, $\operatorname{erf}(\infty) = 1$ (see Example 2). This assertion about the limit requires proof, which will be given later, but present evidence is very strong. Moreover, strict bounds on $\operatorname{erf}(\infty)$ can be derived now which almost end possible doubt about this prediction. The formula

$$\operatorname{erf}(\infty) - \operatorname{erf}(3) = \frac{2}{\pi^{1/2}}\left(\int_0^\infty e^{-z^2}\,dz - \int_0^3 e^{-z^2}\,dz\right) = \frac{2}{\pi^{1/2}}\int_3^\infty e^{-z^2}\,dz$$

implies that $\operatorname{erf}(\infty) > \operatorname{erf}(3) > 0$, and since

$$e^{-z^2} < e^{-3z} \qquad \text{for} \qquad z > 3,$$

it follows that

$$\operatorname{erf}(\infty) - \operatorname{erf}(3) \leq \frac{2}{\pi^{1/2}} \int_3^\infty e^{-3z}\, dz = \frac{2e^{-9}}{3\pi^{1/2}} = 4.64 \times 10^{-5}.$$

Therefore

$$\operatorname{erf}(3) \leq \operatorname{erf}(\infty) \leq \operatorname{erf}(3) + 4.64 \times 10^{-5},$$

or, in particular,

$$0.99998 \leq \operatorname{erf}(\infty) \leq 1.00002.$$

Henceforth, the result

$$\operatorname{erf}(\infty) = \frac{2}{\pi^{1/2}} \int_0^\infty e^{-z^2}\, dz = 1, \tag{17}$$

will be used without hesitation, subject to absolute confirmation later in Sec. 7.2. Table 10.2 gives values of erf(x) for $0 \leq x \leq 3$ (see Fig. 10.2).

TABLE 10.2

x	erf (x)
0.0	0.
0.1	0.11246
0.2	0.22270
0.3	0.32863
0.4	0.42839
0.6	0.60386
0.8	0.74210
1.0	0.84270
1.5	0.96611
2.0	0.99532
3.0	0.99998

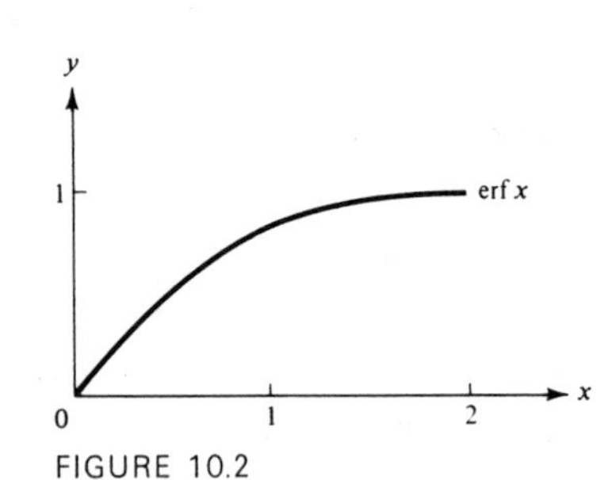

FIGURE 10.2

5

Another method of evaluating

$$I = \int_a^b f(x)\, dx, \tag{18}$$

is to replace the integrand (or any part of it) by a polynomial approximation based on Taylor's formula and the extended mean-value theorem. In this technique, the integrand is expressed as a polynomial expansion about some convenient point, $x = c$, as

$$f(x) = f(c) + f'(c)(x - c) + \frac{1}{2}f''(c)(x - c)^2 + \cdots + \frac{1}{n!}f^{(n)}(c)(x - c)^n + R_n. \quad (19)$$

The remainder term is carried along if we wish to analyze the error involved in this type of approximation. Otherwise, we can simply dispense with it and live dangerously. (The error incurred in the finite-difference methods can also be estimated by the mean-value theorem, but this is not our intention.) The substitution of (19) in the integral (18) and consequent integration, which is a relatively easy task, lead to an analytical approximation. In the limit as the polynomial becomes an infinite series, this approximation becomes an exact representation; this will be studied at length in the next chapter. For the present, the technique is illustrated by example.

Example 4 Evaluate $I = \int_0^1 \sin \frac{\pi}{2} t \, dt$.

With the minor substitution $x = (\pi/2)t$, this is brought into a form that is a little simpler to handle algebraically:

$$I = \frac{2}{\pi}\int_0^{\pi/2} \sin x \, dx.$$

Taylor's formula for $\sin x$ (about $x = 0$) is

$$\sin x = x - \frac{x^3}{3!} + \frac{x^5}{5!} + \cdots + \frac{(-1)^{n+1}x^{2n-1}}{(2n-1)!} + R_{2n-1}, \quad (20)$$

and this expression is substituted in I. The remainder is dropped, and this yields the approximation

$$I \approx \frac{2}{\pi}\int_0^{\pi/2} \left[x - \frac{x^3}{3!} + \cdots + \frac{(-1)^{n+1}x^{2n-1}}{(2n-1)!}\right] dx,$$

which upon integrating term by term becomes

$$I \approx \frac{2}{\pi}\left[\frac{x^2}{2!} - \frac{x^4}{4!} + \cdots + \frac{(-1)^{n+1}x^{2n}}{(2n)!}\right]_0^{\pi/2}$$

or

$$I \approx \frac{2}{\pi}\left[\frac{1}{2!}\left(\frac{\pi}{2}\right)^2 - \frac{1}{4!}\left(\frac{\pi}{2}\right)^4 + \cdots + \frac{(-1)^{n+1}}{(2n)!}\left(\frac{\pi}{2}\right)^{2n}\right].$$

The results to five decimal places for increasing values of n are as follows:

$$I \approx \begin{cases} 0.78540 & \text{with one term of series } (n = 1), \\ 0.62391 & \text{with two terms of series } (n = 2), \\ 0.63719 & \text{for } n = 3, \\ 0.63660 & \text{for } n = 4, \\ 0.63662 & \text{for } n = 5. \end{cases}$$

As in Example 1, the exact value is $2/\pi = 0.63662\ldots$. This method would have been even more effective had the upper limit of integration been closer to zero (or at least less than 1), for then consecutive terms of the series would diminish rapidly in magnitude.

Example 5 Find the value of $\int_0^{1/2} \frac{\sin x}{x}\, dx$.

From (20) it follows that

$$\frac{\sin x}{x} \approx 1 - \frac{x^2}{3!} + \frac{x^4}{5!} + \cdots + \frac{(-1)^{n-1}x^{2n-2}}{(2n-1)!},$$

and the substitution of this in the integral yields

$$\int_0^{1/2} \frac{\sin x}{x}\, dx \approx \left[x - \frac{x^3}{3 \cdot 3!} + \frac{x^5}{5 \cdot 5!} + \cdots + \frac{(-1)^{n-1}x^{2n-1}}{(2n-1)(2n-1)!}\right]_0^{1/2},$$

$$\approx \frac{1}{2} - \frac{1}{3 \cdot 3!}\left(\frac{1}{2}\right)^3 + \cdots + \frac{(-1)^{n-1}}{(2n-1)(2n-1)!}\left(\frac{1}{2}\right)^{2n-1}.$$

The exact value of the integral to five decimal places is obtained by summing five terms of this series. Convergence of the procedure is indicated by the results obtained using one, two, and three terms (Table 10.3).

TABLE 10.3

n	1	2	3
$\int_0^{1/2} \frac{\sin x}{x}\, dx$	0.5000	0.49306	0.49311

Example 6 Apply the series method to develop an analytical approximation to erf(x) valid for small x.

By definition

$$\operatorname{erf}(x) = \frac{2}{\pi^{1/2}} \int_0^x e^{-z^2}\, dz.$$

To apply the series method, we replace the integrand with the polynomial approximation obtained from Taylor's formula about $z = 0$, that is,

$$e^{-z^2} \approx 1 - z^2 + \frac{z^4}{2!} - \frac{z^6}{3!} + \cdots + \frac{(-1)^n z^{2n}}{n!}.$$

Therefore

$$\frac{\pi^{1/2}}{2}\operatorname{erf}(x) = \int_0^x e^{-z^2}\,dz \approx \int_0^x \left[1 - z^2 + \cdots + \frac{(-1)^n(z^2)^n}{n!}\right] dz,$$

$$\approx x - \frac{x^3}{3} + \frac{x^5}{5\cdot 2!} + \cdots + \frac{(-1)^n x^{2n+1}}{(2n+1)(n)!}.$$

Exactly the same formula results from differentiating erf(x) directly, as required in the extended mean-value theorem. For example,

$$\frac{\pi^{1/2}}{2}\frac{d}{dx}\operatorname{erf}(x) = e^{-x^2}, \qquad \text{and} \qquad \frac{\pi^{1/2}}{2}\frac{d}{dx}\operatorname{erf}(x)\bigg]_{x=0} = 1;$$

$$\frac{\pi^{1/2}}{2}\frac{d^2}{dx^2}\operatorname{erf}(x) = -2xe^{-x^2}, \qquad \text{and} \qquad \frac{\pi^{1/2}}{2}\frac{d^2}{dx^2}\operatorname{erf}(x)\bigg]_{x=0} = 0;$$

$$\frac{\pi^{1/2}}{2}\frac{d^3}{dx^3}\operatorname{erf}(x) = (-2+4x^2)e^{-x^2}, \qquad \text{and} \qquad \frac{\pi^{1/2}}{2}\frac{d^3}{dx^3}\operatorname{erf}(x)\bigg]_{x=0} = -2;$$

higher derivatives are calculated in the same fashion. Taylor's formula (about $x = 0$) is then

$$\operatorname{erf}(x) = \operatorname{erf}(x)\bigg]_{x=0} + \frac{d}{dx}\operatorname{erf}(x)\bigg]_{x=0} x + \cdots,$$

and

$$\frac{\pi^{1/2}}{2}\operatorname{erf}(x) = x - \tfrac{1}{3}x^3 + \cdots.$$

It is evident that Taylor's formula for the error function is more easily derived via integration than by direct differentiation, and this is often the case.

There are other types of possible approximations for the integrand, generally in the form of a sum of simple functions

$$f(x) \approx u_1(x) + u_2(x) + \cdots.$$

This concept underlies the entire subject of Fourier series; the reader should be aware that the analysis of this section is only a gentle probe into a vast area.

Exercises 3.10

1. Verify that the parabola passing through points (x_0, f_0), (x_1, f_1), and (x_2, f_2) is that given in Eq. (4). Show that the area below this arc (above the x axis) is $\frac{1}{3}h(f_0 + 4f_1 + f_2)$, [see (5)]. Derive Eq. (14).

2. Evaluate $\int_{0.5}^{1} \tan x^{1/2}\, dx$ to two decimal places using the rectangular and trapezoidal rules.

3. Divide the integration range into 10 intervals and evaluate the following integrals using Simpson's rule:

(i) $\int_0^{1.5} x^4\,dx$; **(ii)** $\int_0^{\pi/2} e^{-\sin x}\,dx$;

(iii) $\int_{-2}^{-1} \frac{dx}{x^2+1}$; **(iv)** $\int_0^1 \left(\sin\frac{\pi}{2}x\right)^{1/3} dx$.

Repeat the calculations using the rectangular and trapezoidal rules and compare the three answers obtained.

4. Evaluate the following integrals to four decimal places using Simpson's rule:

(i) $\int_{0.5}^{3} \frac{e^{-x}}{x}\,dx$; **(ii)** $\int_{-\pi/2}^{\pi/2} (1-\tfrac{1}{4}\sin^2\theta)^{1/2}\,d\theta$;

(iii) $\int_{1.5}^{2} (1+x^2)^{1/2}\sin(\log x)\,dx$.

5. In Exercise 4, formulate each problem as a differential equation. Use Simpson's rule and the marching method to obtain the solutions required.

6. Solve the following differential equations numerically:

(i) $\frac{dy}{dx} = 1 - xe^{-x^2}$ with $y(1) = 1$; find $y(2)$;

(ii) $\frac{dy}{dx} = \frac{\sin(\cos x)}{(1+x^2)^{1/2}}$ with $y(0) = 0$; find $y(1)$.

If you can, find an explicit solution and check your numerical calculation.

7. Find the relative error in using Simpson's rule with only two intervals to obtain a numerical estimate for $\int_0^b x^{1/2}\,dx$. Why is the error so great?

8. Show that the error involved in using the trapezoidal rule to evaluate $\int_0^b x^2\,dx$ is $b^3/6n^2$, where n is the number of trapezoids used.

9. Show that the error function is an odd function. Evaluate $\lim_{x\to 0} \frac{\text{erf}(x)}{x}$.

10. Find the stationary points of $\frac{\text{erf}(x)}{x}$ and $\text{erf}(x) + \frac{2}{\sqrt{\pi}}e^{-x^2}$.

11. Find the value of $\int_0^{3/4} \frac{\sin x}{x}\,dx$ by using Taylor's formula for the integrand.

12. Find the value of $\int_0^{1/2} \frac{1-\cos x}{x^2}\,dx$ by the series method, i.e., using Taylor's formula to replace the integrand.

3.11 Special techniques

An infinite range of integration or an integrand that is singular wreaks havoc with a finite-difference calculation of an improper integral. These difficulties are often quickly remedied by removing the singularity, by changing variables, or by expressing the integral as a sum of two or more integrals each of which can be handled differently. As a matter of fact, the same techniques also apply to proper integrals and can improve the speed, accuracy, and general efficiency of numerical calculations. It will better serve our purpose to illustrate these techniques by example than to indulge in a lengthy exposition.

Example 1 Evaluate $\int_0^1 \frac{\cos x}{x^{1/2}}\,dx$.

Simpson's rule could be applied to $\int_\varepsilon^1 \frac{\cos x}{x^{1/2}}\,dx$, where ε and the step size h are necessarily very small numbers, but this would be an inefficient procedure. In searching for a better method, we note that the singularity of the integrand at the origin is like $1/x^{1/2}$, a function that can be integrated easily. With this motivation, we simply subtract this term from the integrand and write

$$\frac{\cos x}{x^{1/2}} = \frac{\cos x - 1}{x^{1/2}} + \frac{1}{x^{1/2}},$$

so that

$$\int_0^1 \frac{\cos x}{x^{1/2}}\,dx = \int_0^1 \frac{\cos x - 1}{x^{1/2}}\,dx + \int_0^1 \frac{dx}{x^{1/2}}.$$

The first integral on the right-hand side is quite proper, and there is no special difficulty in using any numerical method for its evaluation. The original singularity has been separated out and appears isolated in the second integral, which, *by design*, can be integrated in a straightforward manner;

$$\int_0^1 \frac{dx}{x^{1/2}} = 2x^{1/2}\Big]_0^1 = 2.$$

Therefore

$$\int_0^1 \frac{\cos x}{x^{1/2}}\,dx = \int_0^1 \frac{\cos x - 1}{x^{1/2}}\,dx + 2,$$

and this is as far as we will carry the analysis, since the integral has been reduced to a routine chore. (In this simple example, the singularity can be eliminated by the substitution $z = x^{1/2}$. Another technique would be to approximate $\cos x$ by Taylor's formula; see Exercises 24 and 25.)

Any integral of the form $\int_0^b x^{-1/2} f(x)\,dx$ can be handled this way, and the removal of the singularity leads to the formula

$$\begin{aligned}\int_0^b x^{-1/2} f(x)\,dx &= \int_0^b x^{-1/2}[f(x) - f(0)]\,dx + f(0)\int_0^b x^{-1/2}\,dx,\\ &= \int_0^b x^{-1/2}[f(x) - f(0)]\,dx + 2f(0)b^{1/2}. \qquad (1)\end{aligned}$$

Clearly the technique is not restricted to square-root singularities, and it is readily generalized.

Example 2 Convert $I = \int_0^1 \frac{dx}{[x(1-x^2)]^{1/2}}$ to a form more amenable for numerical integration.

The integral is improper and has integrable singularities at $x = 0$ and $x = 1$. One method of proceeding would be to follow the analysis in the last example and remove the singularities. For small x, the integrand is approximately

$$[x(1-x^2)]^{-1/2} \approx x^{-1/2},$$

whereas for x nearly unity

$$[x(1-x^2)]^{-1/2} \approx 2^{-1/2}(1-x)^{-1/2}.$$

Therefore, we can write

$$I = \int_0^1 \left\{ \frac{1}{[x(1-x^2)]^{1/2}} - \frac{1}{x^{1/2}} - \frac{1}{2^{1/2}(1-x)^{1/2}} \right\} dx + \int_0^1 \frac{dx}{x^{1/2}} + \frac{1}{2^{1/2}} \int_0^1 \frac{dx}{(1-x)^{1/2}}$$

and proceed as before, employing Simpson's rule on the first proper integral and performing the last two analytically. A more efficient procedure, which eliminates differences of large numbers, is developed in Exercise 23.

Another tactic is to eliminate the singularities completely by a change of variable or a partial integration. Thus, the transformation $x = \sin^2 \theta$ $(dx = 2 \sin \theta \cos \theta \, d\theta)$ converts the integral to

$$I = \int_0^{\pi/2} \frac{2 \cos \theta \, d\theta}{(1 - \sin^4 \theta)^{1/2}} = 2 \int_0^{\pi/2} \frac{d\theta}{(1 + \sin^2 \theta)^{1/2}},$$

which is no longer singular. (There is, of course, more than one way of accomplishing this.) In this form, numerical integration is straightforward, and the result using Simpson's rule with a step size $h = \pi/200$ is $I = 2.622057$, whereas for $h = \pi/2{,}000$, $I = 2.622054$.

The integral, in either its original or final form, can also be evaluated using Taylor's formula to replace a part of the integrand by a polynomial approximation. This approach will be examined in the exercises.

Example 3 Arrange $J = \int_0^\infty \frac{dx}{(2 + x^2 + x^4)^{1/2}}$ in a form for efficient numerical calculation.

In this case, the integral is written

$$J = \int_0^1 \frac{dx}{(2 + x^2 + x^4)^{1/2}} + \int_1^\infty \frac{dx}{(2 + x^2 + x^4)^{1/2}},$$

where only the second term on the right is difficult because of the infinite range of integration. This can be remedied via the change of variable $z = 1/x$, in which case

$$\int_1^\infty \frac{dx}{(2 + x^2 + x^4)^{1/2}} = \int_1^0 \frac{-dz}{z^2(2 + z^{-2} + z^{-4})^{1/2}} = \int_0^1 \frac{dz}{(1 + z^2 + 2z^4)^{1/2}}.$$

Therefore, the original integral can be expressed as two proper integrals

$$J = \int_0^1 \frac{dx}{(2 + x^2 + x^4)^{1/2}} + \int_0^1 \frac{dx}{(1 + x^2 + 2x^4)^{1/2}},$$

where we have used the same dummy integration variable for each because both integrals can then be combined neatly into one:

$$J = \int_0^1 \left[\frac{1}{(2 + x^2 + x^4)^{1/2}} + \frac{1}{(1 + x^2 + 2x^4)^{1/2}}\right] dx.$$

In this form, J is readily evaluated by Simpson's rule, and for the step size $h = 0.01$ we obtain $J = 1.452821$; there are no complications.

---

2 An indefinite integral or a definite integral whose integrand involves a parameter defines a function of one variable, of which typical examples are

$$F(x) = \int_a^x f(t)\, dt; \qquad I(\varepsilon) = \int_a^b g(t, \varepsilon)\, dt,$$

A complete table of values of F versus x or I versus ε, can be calculated by the numerical methods available. These data, arranged as a graph perhaps, are a rather complete description of the function in question. Moreover, this format is especially appropriate when the integral is a final result, i.e., the solution of a problem. However, it is just as likely that such a function will be used in further theoretical work, and for this purpose we would want an analytical description or approximation in simpler terms. (Indeed, a representation of this kind is frequently the very means of numerical calculation.) This can be a difficult task, and the attack usually begins with an analysis of the function for small or large values of the independent variable. The assumption that some variable is either small or large is of extreme importance in science and applied mathematics because it underlies much and sometimes all of the progress that can be made on the difficult problems of the real world. The functional descriptions obtained in this manner apply to restricted ranges of the independent variable, but these special formulas may frequently be combined into a single overall approximation that is uniformly valid. The methods are illustrated next by example.

Example 4 Find analytical approximations for

$$\operatorname{erf}(x) = \frac{2}{\pi^{1/2}} \int_0^x e^{-z^2}\, dz, \tag{2}$$

which are valid for both small and large x.

In the last section, for small x, a formula for the error function was developed by replacing the exponential with its Taylor expansion about the origin. The substitution of

$$e^{-z^2} = 1 - z^2 + \frac{(z^2)^2}{2!} - \frac{(z^2)^3}{3!} + \cdots + \frac{(-1)^n}{n!}(z^2)^n + R_n(z)$$

and subsequent integration, term by term, yields

$$\operatorname{erf}(x) = \frac{2}{\pi^{1/2}} \left[x - \frac{x^3}{3 \cdot 1!} + \frac{x^5}{5 \cdot 2!} - \frac{x^7}{7 \cdot 3!} + \cdots + \frac{(-1)^n x^{2n+1}}{(2n+1)n!} \right] + \frac{2}{\pi^{1/2}} \int_0^x R_n(z)\, dz.$$

The term with the remainder R_n will be analyzed in the next chapter to establish the validity of the infinite series obtained as $n \to \infty$,

$$\operatorname{erf}(x) = \frac{2}{\pi^{1/2}} \sum_{n=0}^{\infty} \frac{(-1)^n x^{2n+1}}{(2n+1)n!}. \tag{3}$$

For the present, we will simply neglect this correction and use the polynomial truncation as the desired approximation for small x. Figure 11.1*a* shows the curves of the approximating polynomials corresponding to one and three terms compared to the exact function. Gross inaccuracies are evident for $x > 0.9$.

An analytical approximation for large values of x is obtained by first writing

$$\begin{aligned} \operatorname{erf}(x) &= \frac{2}{\pi^{1/2}} \int_0^\infty e^{-z^2}\, dz - \frac{2}{\pi^{1/2}} \int_x^\infty e^{-z^2}\, dz, \\ &= 1 - \frac{2}{\pi^{1/2}} \int_x^\infty e^{-z^2}\, dz, \end{aligned} \tag{4}$$

and by then manipulating the last integral (which is truly small when x is large). The manipulation is a clever use of partial integration applied to

$$\int_x^\infty e^{-z^2}\, dz = \int_x^\infty \frac{1}{z}(z e^{-z^2}\, dz)$$

with the choice $u = 1/z$, $dv = z e^{-z^2}\, dz$ so that $du = -dz/z^2$ and $v = -\tfrac{1}{2} e^{-z^2}$. Integration by parts yields

$$\int_x^\infty e^{-z^2}\, dz = -\frac{1}{2z} e^{-z^2} \Big]_x^\infty - \int_x^\infty \frac{e^{-z^2}}{2z^2}\, dz,$$

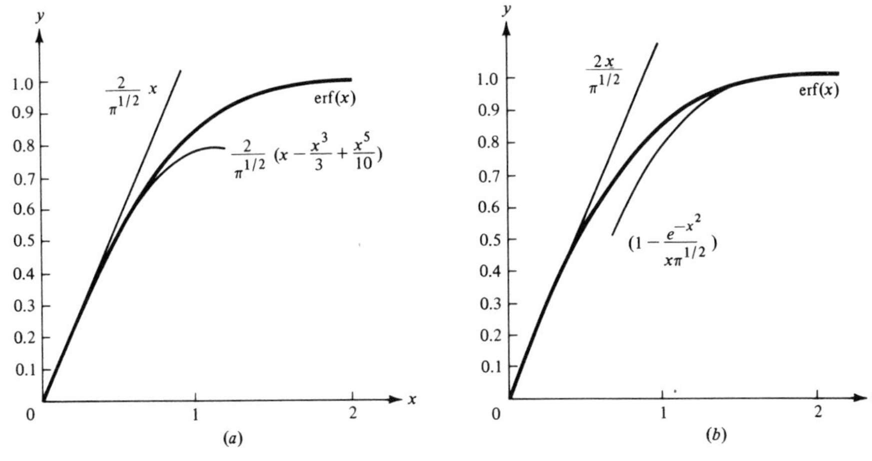

FIGURE 11.1

so that

$$\int_x^\infty e^{-z^2}\,dz = \frac{e^{-x^2}}{2x} - \frac{1}{2}\int_x^\infty \frac{e^{-z^2}}{z^2}\,dz. \tag{5}$$

The net effect of this reduction is to convert the small integral we started with in (4) into a simple known function plus *an even smaller* integral. This assertion is substantiated easily because for $z > x$ the following bound is valid:

$$\int_x^\infty \frac{e^{-z^2}}{z^2}\,dz < \frac{1}{x^2}\int_x^\infty e^{-z^2}\,dz.$$

This means that the integral on the right-hand side of (5) is less than that on the left by at least the small multiplicative factor $1/x^2$. The result of substituting (5) in (4) is the formula

$$\operatorname{erf}(x) = 1 - \frac{e^{-x^2}}{x\pi^{1/2}} + \frac{1}{\pi^{1/2}}\int_x^\infty \frac{e^{-z^2}}{z^2}. \tag{6}$$

Equation (4) can be interpreted as the explicit approximation of the error function for large x given by $\operatorname{erf}(x) \approx 1$, with a small correction or error represented by the integral $\frac{-2}{\pi^{1/2}}\int_x^\infty e^{-z^2}\,dz$. The accuracy of this approximation can be assessed by simple bounds

$$|\operatorname{erf}(x) - 1| = \frac{2}{\pi^{1/2}}\int_x^\infty e^{-z^2}\,dz = \frac{2}{\pi^{1/2}}\int_x^\infty \frac{ze^{-z^2}}{z}\,dz \le \frac{2}{\pi^{1/2}x}\int_x^\infty ze^{-z^2}\,dz$$

or upon completing the last integration

$$|\operatorname{erf}(x) - 1| \le \frac{e^{-x^2}}{x\pi^{1/2}}.$$

Thus for $x = 3$ the error incurred is less than $e^{-9}/3\pi^{1/2}$, which is a slightly better bound than that derived in Sec. 3.10. In this view, (6) can also be interpreted as the approximation

$$\operatorname{erf}(x) \approx 1 - \frac{e^{-x^2}}{x\pi^{1/2}}, \tag{7}$$

with the small correction given by the integral that appears there. The error in this formula is then

$$\left|\operatorname{erf}(x) - \left(1 - \frac{e^{-x^2}}{x\pi^{1/2}}\right)\right| = \frac{1}{\pi^{1/2}} \int_x^\infty \frac{e^{-z^2}}{z^2}\, dz = \frac{1}{\pi^{1/2}} \int_x^\infty \frac{ze^{-z^2}}{z^3}\, dz$$

or

$$\left|\operatorname{erf}(x) - \left(1 - \frac{e^{-x^2}}{x\pi^{1/2}}\right)\right| < \frac{e^{-x^2}}{2\pi^{1/2}x^3},$$

which shows (7) to be a better approximation than its predecessor. Just how good this approximation is can be seen from Fig. 11.1*b*, where the function $1 - e^{-x^2}/(x\pi^{1/2})$ is graphed and compared to the error function. Note that the two curves are substantially the same at $x = 1.5$, which is not such a terribly large number after all. Figure 11.1*b* shows how well the approximations for small and large x cover the entire range. The coverage is improved by taking more terms of each approximation.

The success of this tactic encourages us to apply it repeatedly, and the result after three such partial integrations is

$$\operatorname{erf}(x) = 1 - \frac{e^{-x^2}}{x\pi^{1/2}}\left[1 - \frac{1}{2x^2} + \frac{1 \cdot 3}{(2x^2)^2}\right] + \frac{5 \cdot 3}{2^2\pi^{1/2}} \int_x^\infty \frac{e^{-z^2}\, dz}{z^6}.$$

The general formula has some very interesting features which will be studied in greater detail in the next chapter.

Example 5 Derive an analytical formula for

$$I(\varepsilon) = \int_0^{\pi/2} \cos x \sin(\varepsilon \cos x)\, dx,$$

where ε is a small number.

Taylor's formula for the expansion of $\sin x$ about the origin is

$$\sin x = x - \frac{x^3}{3!} + \frac{x^5}{5!} + \cdots + \frac{(-1)^{n-1}x^{2n-1}}{(2n-1)!} + R_n.$$

By substituting $\varepsilon \cos x$ for x and dropping the remainder we obtain the approximation

$$\sin(\varepsilon \cos x) \approx \varepsilon \cos x - \frac{\varepsilon^3 \cos^3 x}{3!} + \frac{\varepsilon^5 \cos^5 x}{5!} + \cdots + \frac{(-1)^{n-1}\varepsilon^{2n-1} \cos^{2n-1} x}{(2n-1)!}.$$

Therefore

$$I(\varepsilon) \approx \int_0^{\pi/2} \cos x\left(\varepsilon \cos x - \frac{\varepsilon^3 \cos^3 x}{3!} + \cdots\right) dx,$$

$$\approx \varepsilon \int_0^{\pi/2} \cos^2 x\, dx - \frac{\varepsilon^3}{3!}\int_0^{\pi/2} \cos^4 x\, dx + \cdots + \frac{(-1)^{n-1}\varepsilon^{2n-1}}{(2n-1)!}\int_0^{\pi/2} \cos^{2n} x\, dx.$$

The general integral involved was calculated previously (Exercise 18 of Sec. 3.4) and shown to be

$$\int_0^{\pi/2} \cos^{2n} x\, dx = \frac{1 \cdot 3 \cdot 5 \cdots (2n-1)}{2 \cdot 4 \cdot 6 \cdots 2n}\frac{\pi}{2};$$

it follows that

$$I(\varepsilon) \approx \frac{\pi}{2}\left(\frac{\varepsilon}{2} - \frac{1 \cdot 3}{2 \cdot 4}\frac{\varepsilon^3}{3!} + \cdots\right),$$

where as many terms can be incorporated as desired.

---

3 So far the techniques employed have been fairly clean—fit, so to speak, for respectable use. However, we consider next a few nasty integrals to illustrate methods equal to the task—rough, dirty, fast, and efficient. Once the information desired is extracted from the integral in this way, the analysis can be cleansed and presented to anyone's satisfaction. The novel features introduced can be explored, pathologies examined, and plausible arguments made rigorous (if this is necessary!). But more often than not in practice, the vindication of a basically sound procedure is a sensible physical answer.

Example 6 Find an analytical expression for integrals of the type

$$J(\lambda) = \int_0^{\infty} f(x)e^{-\lambda x^2}\, dx, \tag{8}$$

where λ is a very large number.

Let $z = \lambda^{1/2}x$ so that

$$J = \frac{1}{\lambda^{1/2}}\int_0^{\infty} f\left(\frac{z}{\lambda^{1/2}}\right)e^{-z^2}\, dz.$$

It is assumed that the behavior of the integrand for large z is dominated by the exponential. When z is moderate, $z/\lambda^{1/2}$ is small and $f(z/\lambda^{1/2})$ can be replaced by its Taylor formula about zero:

$$f\left(\frac{z}{\lambda^{1/2}}\right) = f(0) + f'(0)\frac{z}{\lambda^{1/2}} + \cdots.$$

When $z/\lambda^{1/2}$ is not small, the integrand is almost zero anyway because of the presence of the exponential term. Therefore, we anticipate that a reasonably good approximation for the integral is obtained by substituting the preceding equation for $f(x)$ in the entire range of integration:

$$J(\lambda) \approx \frac{1}{\lambda^{1/2}}\int_0^\infty \left[f(0) + f'(0)\frac{z}{\lambda^{1/2}} + \cdots\right]e^{-z^2}\,dz,$$

$$\approx \frac{f(0)}{\lambda^{1/2}}\int_0^\infty e^{-z^2}\,dz + \frac{f'(0)}{\lambda}\int_0^\infty ze^{-z^2}\,dz + \cdots,$$

or, using the result of Eq. (17) in the preceding section,

$$J(\lambda) \approx \frac{\pi^{1/2}f(0)}{2\lambda^{1/2}} + \frac{f'(0)}{2\lambda} + \cdots. \tag{9}$$

Example 7 Evaluate $K(\varepsilon) = \int_0^\infty \frac{dx}{x^2 + \varepsilon\cos x}$ for small ε.

The difficulties here are twofold: the integral is improper, and if we blithely set $\varepsilon = 0$, it becomes divergent instead of remaining convergent. The ordinary concept of a perturbation has to be modified somewhat, and to this end a change of variable is introduced which shifts the location of the parameter. Let $x = \varepsilon^{1/2}t$, in which case

$$K(\varepsilon) = \int_0^\infty \frac{\varepsilon^{1/2}\,dt}{\varepsilon t^2 + \varepsilon\cos\varepsilon^{1/2}t},$$

$$= \frac{1}{\varepsilon^{1/2}}\int_0^\infty \frac{dt}{t^2 + \cos\varepsilon^{1/2}t}.$$

For ε small and t moderate, the integrand can be approximated by

$$\frac{1}{t^2 + \cos\varepsilon^{1/2}t} \approx \frac{1}{t^2 + 1},$$

since the cosine of a small number is very nearly 1. This approximation should have very little effect when t is large because the integrand is then small anyway, as is the contribution to the total integral from this range of integration. (Admittedly, this is not as convincing a case as that in the preceding example.) Therefore we anticipate that an accurate estimate of $K(\varepsilon)$ for small ε is

$$K(\varepsilon) \approx \frac{1}{\varepsilon^{1/2}}\int_0^\infty \frac{dt}{t^2 + 1} = \frac{\pi}{2\varepsilon^{1/2}}.$$

Accuracy in this context means that the relative error is small.

As a final illustration of the many techniques that remain unmentioned, a definite integral will be evaluated by throwing darts, a probabilistic procedure called the *Monte Carlo method*. For definiteness, consider the evaluation of

$$A = \int_0^1 \sin \frac{\pi}{2} x \, dx,$$

which represents the area shown in Fig. 11.2. If indeed we actually do throw a large number of darts at the figure bounded by the bold lines, in a random sort of way (without skill), then the number which land in the shaded region will depend on the

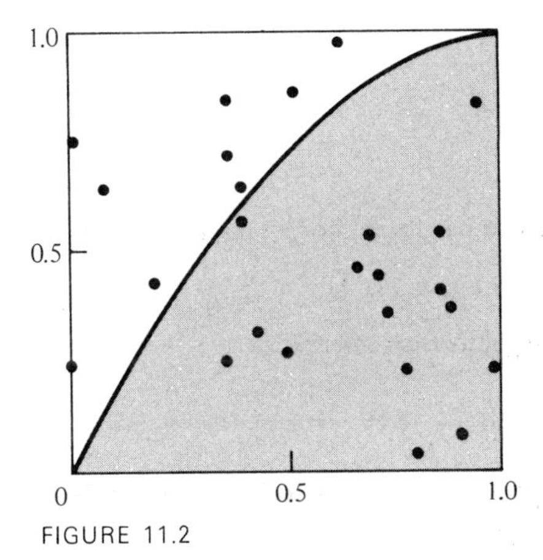

FIGURE 11.2

size of this area compared to the total area of the target. Suppose that N darts are tossed and that n land in the area to be calculated; then since the target area is 1 square unit by construction, it follows that

$$\frac{A}{1} \approx \frac{n}{N},$$

or more precisely $A = \lim_{N \to \infty} (n/N)$. (The limit is in fact the probability that a dart will land in the area A.) The actual process of dart tossing is replaced by a table of random numbers (Table 11.1), where each entry provides the two coordinates (x, y) of a

TABLE 11.1

3885	8004	5997	7336	5287	4767	4102	8229	2643	8737
4066	4332	8737	8641	9584	2559	5413	9418	4230	0736
4058	9008	3772	0866	3725	2031	5331	8098	3290	3209
7823	8655	5027	2043	0024	0230	7102	4993	2324	0086
9824	6747	7145	6954	0176	0332	6701	9254	9797	5272

dart's puncture point on the target. This is done by identifying the first two digits of each random number as 10^2x and the last two as 10^2y. For example, the first entry 3885 is converted into the point $x = 0.38$, $y = 0.85$. Figure 11.2 shows the target as pierced by 25 points; since 16 have landed within A, the probabilistic estimate of the area is

$$A \approx \frac{n}{N} = 16/25 = 0.64.$$

If 25 extra points, i.e., darts, are added, it turns out that once again 16 of these fall within A and so the area estimate of A remains unchanged. The exact answer is, of course $A = 2/\pi = 0.635\ldots$, and the approximation is fairly good.

Exercises 3.11

1. If $a < c < b$ and $f(x)$ is a continuous function, generalize the method in the text to remove the singularity from the integral $I = \int_a^b \frac{f(x)}{|x-c|^\alpha}\,dx$, with $0 < \alpha < 1$.

2. Convert the following to forms more amenable to numerical integration:

(i) $\int_0^1 \frac{\sin x}{|x - 0.25|^{1/3}}\,dx$; **(ii)** $\int_0^2 \frac{e^{-x}}{x^{1/2}}\,dx$; **(iii)** $\int_1^\infty \frac{dx}{(1 + x + x^5)^{1/2}}$.

3. Show how to improve the efficiency of numerical evaluations of the following integrals by removing (or treating) the singularity of the integrand:

(i) $\int_{0.001}^{1.0} \frac{1}{e^x - 1}\,dx$; **(ii)** $\int_0^2 \frac{e^{-x}}{(2-x)^{1/2}}\,dx$; **(iii)** $\int_0^{\pi/2.1} (\tan x)^{1/3}\,dx$.

4. Evaluate $\int_0^{\pi/2} \frac{d\theta}{(1 + \sin^2\theta)^{1/2}}$ by using Taylor's formula for $(1 + h)^{-1/2}$ to replace $(1 + \sin^2\theta)^{-1/2}$. Integrate the polynomial approximation term by term to obtain an estimate for the integral. Redo the problem making use of the identity $1 + \sin^2\theta = \frac{1}{2}(3 - \cos 2\theta)$ and compare this for efficiency with the foregoing.

5. Find an analytical approximation for $Ei(x) = \int_x^\infty \frac{e^{-z}}{z}\,dz$ valid for x large. Does $Ei(0)$ exist? What is the value of $Ei(1)$?

6. Use Taylor's formula to expand the function $Ei(x)$ about $x = 1$.

7. How would you compute $Ei(x)$ for small x? Consider the expression

$$Ei(x) = \int_x^1 \frac{e^{-z}}{z}\,dz + \int_1^\infty \frac{e^{-z}}{z}\,dz.$$

Try to evaluate the first integral by replacing e^{-z} by its Taylor formula about zero.

8. By repeated partial integrations show that

$$\operatorname{erf}(x) = 1 - \frac{e^{-x^2}}{x\pi^{1/2}}\left[1 + \sum_{i=1}^{n}(-1)^i\frac{1\cdot 3\cdots(2i-1)}{(2x^2)^i}\right] + (-1)^n\frac{1\cdot 3\cdots(2n+1)}{2^n\pi^{1/2}}\int_x^\infty \frac{e^{-z^2}}{z^{2n+2}}\,dz.$$

9. Use perturbation theory to evaluate

$$\int_0^{\pi/2} \sin(x + \varepsilon \cos x)\, dx \qquad \text{and} \qquad \int_0^{\pi} \sin(x + \varepsilon\, e^{-x})\, dx.$$

10. Evaluate $\int_0^{\pi/2} \sin x \sin(1/10 \cos x)\, dx$ exactly and by a perturbation series.
11. Evaluate approximately using Eq. (9):

(i) $\displaystyle\int_0^{\infty} \frac{x^2}{1+x^5} e^{-100x^2}\, dx$; **(ii)** $\displaystyle\int_0^{5} e^{-|\sin x|} e^{-25x^2}\, dx$;

(iii) $\displaystyle\int_0^{\infty} x e^{-10x^2}\, dx$ (compare with the exact answer).

12. Determine an approximate formula for $\int_a^\infty f(x)e^{-\lambda x^2}\, dx$, where λ is large and positive. Test this by examples that can be evaluated exactly.
13. Suppose that the integral $I(\lambda) = \int_a^b f(x)g(x)e^{-\lambda x^2}\, dx$, $0 < a < b$, cannot be expressed in simple terms but that $S(\lambda) = \int_a^b f(x)e^{-\lambda x^2}\, dx$ is a known function. Make a plausible argument for the approximation

$$\int_a^b f(x)g(x)e^{-\lambda x^2}\, dx \approx g(a)S(\lambda) \qquad \text{for} \qquad \lambda \gg 1.$$

14. With the same conditions as in Exercise 13, discuss the approximation

$$\int_a^b h(x)e^{-\lambda x^2}\, dx \approx \frac{h(a)}{f(a)} \int_a^b f(x)e^{-\lambda x^2}\, dx = \frac{h(a)}{f(a)} S(\lambda).$$

15. Show that $\displaystyle\int_1^\infty \frac{dx}{1 + x^2 + e^{-x}}$ is a convergent integral and find an estimate for the "tail," that is, $\displaystyle\int_R^\infty \frac{dx}{1 + x^2 + e^{-x}}$ for large R.

16. Evaluate $\displaystyle\int_0^1 \frac{dx}{(x+\varepsilon)^{1/2}}$ in closed form by first changing the scale of the integration variable. Can this integral be evaluated as a routine perturbation? Explain.

17. Evaluate $\displaystyle I(a) = \int_{-\infty}^{\infty} \frac{a}{\pi(a^2 + x^2)}\, dx$ for $a > 0$ and show that

$$I(a) = \int_{-\infty}^{\infty} \frac{a\, dx}{\pi[a^2 + (x - x_*)^2]}$$

for any value of x_*. Discuss the limiting situations,

$$\lim_{a \to 0} \frac{a}{\pi(a^2 + x^2)} \qquad \text{and} \qquad \lim_{a \to 0} I(a).$$

18. Discuss the approximations,

$$\int_{-\infty}^{\infty} \frac{a}{\pi(a^2 + x^2)} f(x)\, dx \approx f(0), \qquad \text{for } a \text{ very small},$$

or

$$\int_{-\infty}^{\infty} \frac{af(x)}{\pi[a^2 + (x - x_*)^2]}\, dx \approx f(x_*).$$

If we write

$$\delta(x - x_*) = \lim_{a \to 0} \frac{a}{\pi[a^2 + (x - x_*)^2]},$$

then

$$\int_{-\infty}^{\infty} \delta(x - x_*) f(x)\, dx = f(x_*).$$

Is $\delta(x - x_*)$ (called a delta function) a legitimate function? Does its use make sense?

19. Evaluate $\int_0^1 \left(\sin \frac{\pi x}{2}\right)^{1/2} dx$ by the Monte Carlo method. Does the estimate improve as more points are included?

20. Evaluate $\int_0^{\pi/2} (1 - \frac{1}{4} \sin^2\theta)^{1/2}\, d\theta$ by the Monte Carlo method. Compare your answer to the computed value 1.4675. (First rescale the integration variable so that the range of integration is the interval [0, 1].)

21. Evaluate $\int_0^1 t^2 \log t\, dt$ by differentiating the associated integral

$$I(a) = \int_0^1 t^2 t^a\, dt = \int_0^1 t^2 e^{a \log t}\, dt$$

with respect to a. [Note that $I(0) = \frac{1}{3}$.]

22. Show that $I(a) = \int_0^\infty \frac{e^{-t^2}}{t} \sin at\, dt$ can be expressed as an error function. Do this by differentiating with respect to a and then by integrating. [*Hint:* $I''(a) = -(a/2)I'(a)$, and $I(0) = 0$, $I'(0) = \pi^{1/2}/2$.]

23. Write the integral of Example 2 as

$$I = \int_0^{1/2} \frac{dx}{[x(1 - x^2)]^{1/2}} + \int_{1/2}^{1} \frac{dx}{[x(1 - x^2)]^{1/2}}.$$

Let $x = z^2$ in the first integral to convert it to a proper integral. Make a similar change of variable in the second integral and in this way find the value of I accurate to three decimal places.

24. Show that the change of variables $z = x^{1/2}$ converts $\int_0^1 \frac{\cos x}{x^{1/2}}\, dx$ from an improper to a proper integral.

25. Perform the integration in the last problem by replacing $\cos x$ by its Taylor approximation, i.e.,

$$\cos x = 1 - \frac{x^2}{2!} + \cdots + \frac{(-1)^n x^{2n}}{(2n)!} + R_n.$$

26. Develop an approximate formula for $\int_0^\infty \frac{dx}{x^2 + \cos \varepsilon x}$ valid when ε is small.

4
Series

4.1 Fundamentals

1

From the very beginning, series of numbers or functions have arisen quite naturally in this presentation of calculus. The calculation of an area—indeed the basic definition of an integral—numerical computations, polynomial approximations, interpolation and perturbation procedures—all these involve incipient infinite series, truncated for the most part after a finite number of terms. The concept of an infinite series, its convergence or divergence, has been broached, explored, and even applied, and with this background, we launch a more definitive and detailed study in the next three sections. The remainder of the chapter will then deal with the use of infinite series in a variety of contexts, including symbolic methods and differential equations.

An infinite series

$$a_1 + a_2 + \cdots + a_n + \cdots, \tag{1}$$

is convergent if the sequence of partial sums

$$\begin{aligned} s_1 &= a_1, \\ s_2 &= a_1 + a_2, \\ &\cdots\cdots\cdots\cdots\cdots\cdots \\ s_n &= a_1 + a_2 + \cdots + a_n = \sum_{k=1}^{n} a_k, \end{aligned}$$

has a finite limit as $n \to \infty$. In this way, the limit of the sequence $\{s_n\}$, denoted by

$$s = \lim_{n\to\infty} s_n = \lim_{n\to\infty} \sum_{k=1}^{n} a_k, \tag{2}$$

gives a precise meaning to the infinite series in (1), and we write

$$s = \sum_{k=1}^{\infty} a_k. \tag{3}$$

2 The formal definition of convergence for a series is that the sequence of partial sums $\{s_n\}$ has a limit s if for any arbitrarily small positive number ε an integer N can be found so that for all $n>N$

$$|s - s_n| < \varepsilon.$$

This is equivalent to

$$\left| \sum_{k=n+1}^{\infty} a_k \right| < \varepsilon, \tag{4}$$

which, simply put, asserts that the infinite tail of a convergent series can be made arbitrarily small. A series that does not converge is said to diverge or to be divergent.

The major difficulty in applying this formal definition in practice is that a knowledge of the limit s is required, and this is not only unknown but a primary objective as well. In order to eliminate this dependence on the exact limit, tests for convergence will be devised that employ only the information at hand.

---

3 In order for a series (1) to converge, it is necessary that the general term a_n approach zero as n becomes infinite:

$$\lim_{n \to \infty} a_n = 0. \tag{5}$$

This basic fact follows directly from (2), since by definition

$$a_n = s_n - s_{n-1}$$

and

$$\lim_{n \to \infty} a_n = \lim_{n \to \infty} s_n - \lim_{n \to \infty} s_{n-1} = s - s = 0.$$

Alternatively, if $\lim_{n \to \infty} a_n \neq 0$, the series diverges. [A series can diverge even though (5) is true; this equation is a necessary but not a sufficient condition for convergence.]

Note carefully that a finite number of terms can always be added to or eliminated from an infinite series without affecting its convergence or divergence. These are intrinsic properties of the infinite number of terms involved. Clearly, the definition of convergence, (4), pays no attention to the first n terms, and n is an arbitrary integer.

Example 1 Show that the geometric series $\sum_{n=1}^{\infty} r^n$ converges for $|r|<1$.

The formula (obtained by long division)

$$\frac{1}{1-r} = 1 + r + r^2 + \cdots + r^n + \frac{r^{n+1}}{1-r}$$

allows us to find an explicit expression for the nth partial sum $(n=0,1,2,\ldots)$,

$$s_n = \frac{1}{1-r} - \frac{r^{n+1}}{1-r} = \frac{1-r^{n+1}}{1-r}.$$

For $|r| < 1$, the limit as n becomes infinite is readily calculated:

$$s = \lim_{n \to \infty} s_n = \frac{1 - \lim_{n \to \infty} r^{n+1}}{1 - r} = \frac{1}{1 - r},$$

and this establishes convergence. The series diverges for $|r| \geq 1$.

Example 2 Show that the harmonic series $\sum_{n=1}^{\infty} \frac{1}{n}$ diverges.

The nth term $a_n = 1/n$ clearly approaches zero as n increases. If the series did converge, many more than 1000 terms would be required to have an accuracy of at least three decimal places. In fact, the series diverges *very* slowly to infinity. This is seen most easily by grouping terms.

$$\sum_{n=1}^{\infty} \frac{1}{n} = 1 + \left(\frac{1}{2}\right) + \left(\frac{1}{3} + \frac{1}{4}\right) + \left(\frac{1}{5} + \frac{1}{6} + \frac{1}{7} + \frac{1}{8}\right) + \cdots$$

In each bracket, the sum is at least $1/2$. Therefore,

$$\sum_{n=1}^{\infty} \frac{1}{n} > 1 + \frac{1}{2} + \frac{1}{2} + \frac{1}{2} + \cdots = \infty,$$

and the series must diverge.

Example 3 Show that $\sum_{n=1}^{\infty} \left(\frac{1}{n}\right)^{1/n}$ diverges.

The general term is

$$a_n = \left(\frac{1}{n}\right)^{1/n},$$

and divergence is established by showing that $\lim_{n \to \infty} a_n \neq 0$. This is done by introducing $x = 1/n$ and considering instead

$$\lim_{n \to \infty} a_n = \lim_{x \to 0} x^x = \lim_{x \to 0} \exp(x \log x),$$

which is evaluated using l'Hôpital's rule. It follows that

$$\lim_{x \to 0} x \log x = 0,$$

which implies

$$\lim_{n \to \infty} a_n = 1,$$

proving that the series is divergent.

4

A series $\sum_{k=1}^{\infty} a_k$ is called *absolutely convergent* when

$$S = \lim_{n\to\infty} S_n = \lim_{n\to\infty} \sum_{k=1}^{n} |a_k| = \sum_{k=1}^{\infty} |a_k| \tag{6}$$

converges; i.e., the corresponding series of only positive terms converges.

Absolute convergence is a very strong condition. A series that is absolutely convergent automatically converges in the ordinary sense of Eq. (2). To prove this using only methods presently available, we proceed with some adroit definitions. Let

$$p_k = \begin{cases} a_k & \text{for} \quad a_k > 0; \\ 0 & \text{for } a_k \leq 0 \end{cases} \qquad \text{and} \qquad q_k = \begin{cases} -a_k & \text{for } a_k < 0; \\ 0 & \text{for } a_k \geq 0. \end{cases}$$

Clearly, p_k and q_k are always positive, and $a_k = p_k - q_k$ for all k. Moreover

$$0 \leq p_k \leq |a_k|, \qquad \text{and} \qquad 0 \leq q_k \leq |a_k|.$$

In terms of the partial sums $P_n = \sum_{k=1}^{n} p_k$ and $Q_n = \sum_{k=1}^{n} q_k$, it follows that

$$s_n = \sum_{k=1}^{n} a_k = P_n - Q_n. \tag{7}$$

But if the series (3) is absolutely convergent,

$$S_n = \sum_{k=1}^{n} |a_k| \leq \sum_{k=1}^{\infty} |a_k| = S$$

and $P_n \leq S_n \leq S$, $Q_n \leq S_n \leq S$. The sums P_n and Q_n are then nondecreasing and bounded as n increases, and this implies convergence of both sequences $\{P_n\}$, $\{Q_n\}$, that is,

$$\lim_{n\to\infty} P_n = P; \qquad \lim_{n\to\infty} Q_n = Q.$$

Therefore, according to (7),

$$\lim_{n\to\infty} s_n = P - Q.$$

A convergent series, $s = \sum_{n=1}^{\infty} a_n$, is called *conditionally convergent*, if it does not converge absolutely; that is, $S = \sum_{n=1}^{\infty} |a_n| = \infty$.

A series of positive terms only,

$$\sum_{n=1}^{\infty} c_n \qquad \text{with} \qquad c_n > 0,$$

is, of course, equivalent in form to (6), and its convergence is synonymous with absolute convergence.

An *alternating series* consists of positive terms but has a changing or alternating sign:

$$c_1 - c_2 + c_3 - c_4 + c_5 - \cdots = \sum_{n=1}^{\infty} (-1)^{n+1} c_n,$$

with

$$c_n > 0. \tag{8}$$

These occur often enough in practice to merit special attention. Convergence of an alternating series is readily established under the following set of conditions:

$$c_n > c_{n+1}, \quad \text{or} \quad c_1 > c_2 > c_3 > \cdots; \tag{9}$$

$$\lim_{n\to\infty} c_n = 0. \tag{10}$$

Here, (10) is just a repetition of (5), and (9) is a new constraint which states that every term is smaller than its immediate precedessor. In order to demonstrate convergence, we consider separately the partial sums of even and odd indices. Since, in general,

$$s_{2n} = (c_1 - c_2) + (c_3 - c_4) + \cdots + (c_{2n-1} - c_{2n})$$

and every term in parentheses is positive, it is evident that

$$s_2 < s_4 < s_6 < \cdots < s_{2n} < \cdots. \tag{11}$$

The sequence of even partial sums, $\{s_{2n}\}$, is *strictly increasing.*

Similarly

$$s_{2n+1} = c_1 - (c_2 - c_3) - (c_4 - c_5) - \cdots - (c_{2n} - c_{2n+1}),$$

and once again every parenthesis is a positive term, which shows that

$$s_1 > s_3 > s_5 > \cdots > s_{2n+1} > \cdots. \tag{12}$$

The odd partial sums form a *strictly decreasing* sequence of numbers, $\{s_{2n-1}\}$. Finally, the relationship

$$s_{2n+1} = s_{2n} + c_{2n+1}, \tag{13}$$

implies that

$$s_{2n} < s_{2n+1}. \tag{14}$$

Indeed a much stronger statement can be made now (see Exercise 13): *every* even partial sum is less than *every* odd partial sum. That is to say, $s_{2n} < s_{2m+1}$ for any two integers n, m.

The inequality, $s_2 < s_{2n} < s_{2n+1} < s_1$, which is obtained from (11), (12), and (14), shows in particular that the terms of the sequence $\{s_{2n}\}$ are strictly increasing but have the upper bound s_1. Therefore this sequence must have a limit; call it $\bar{s}$ (see Fig. 1.1):

$$\lim_{n\to\infty} s_{2n} = \bar{s}.$$

Similarly the sequence $\{s_{2n+1}\}$ is strictly decreasing, and since it has the lower bound s_2, a limit exists

$$\lim_{n\to\infty} s_{2n+1} = \underline{s}.$$

But the limit of equation (13) [using (10)] shows that $\underline{s}$ and $\bar{s}$ are one and the same:

$$\lim_{n\to\infty} s_{2n+1} = \lim_{n\to\infty} s_{2n} + \lim_{n\to\infty} c_{2n+1},$$

or

$$\bar{s} = \underline{s} = s.$$

Therefore an alternating series which satisfies conditions (9) and (10) is convergent.

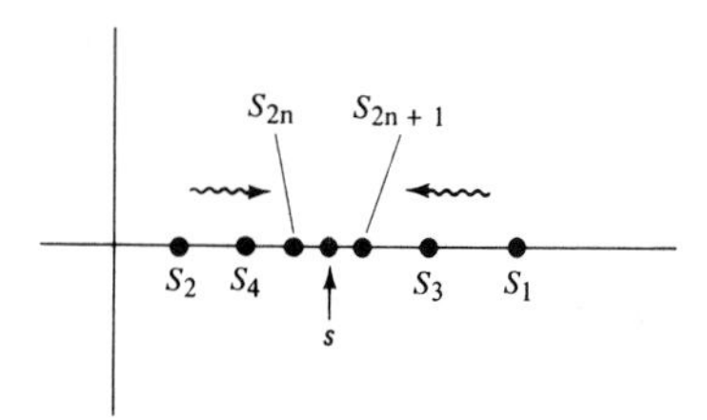

FIGURE 1.1

The sum of an alternating series can be written

$$s = \sum_{k=1}^{n} (-1)^{k+1} c_k + \sum_{k=n+1}^{\infty} (-1)^{k+1} c_k$$

or

$$s = s_n + R_n, \tag{15}$$

in which case a bound for the remainder after n terms is readily determined. The number s lies between consecutive partial sums; that is, $s_n < s < s_{n+1}$ if n is even, or $s_{n+1} < s < s_n$ if n is odd. In either case, the consequence is

$$|R_n| = |s - s_n| \leq |s_{n+1} - s_n|.$$

Since

$$|s_{n+1} - s_n| = c_{n+1},$$

it follows that

$$|R_n| \leq c_{n+1}. \tag{16}$$

In other words, *the magnitude of the remainder after n terms is less than (or equal to) the first term not taken, c_{n+1}.*

The arguments invoked above depend crucially on the assertion that a bounded sequence of increasing positive numbers has a finite limit point. The mathematical statement is that if the elements of the sequence $\{s_n\}$ satisfy the conditions $0 < s_n < M$ and $s_n < s_{n+1}$ for all n, then as a consequence, a limit exists,

$$s = \lim_{n\to\infty} s_n \leq M.$$

This is certainly geometrically and perhaps intuitively obvious, and it will be accepted as a property of real numbers. It can be proved rigorously, but this requires an assumption about the validity of another obvious property of real numbers and we will not belabor this point.

Example 4 Show that $\sum_{n=1}^{\infty} \frac{(-1)^{n+1}}{n} = 1 - 1/2 + 1/3 - 1/4 + \cdots$ converges.

This is an alternating series whose general element is $c_n = 1/n$. Since $c_n > c_{n+1}$ and $\lim_{n\to\infty} c_n = 0$, the three conditions (8), (9), and (10) are certainly satisfied and the series converges. (We have seen in Example 2 that $\sum_{n=1}^{\infty} \frac{1}{n}$ is actually divergent. The alternating series above is therefore conditionally convergent.) The first few partial sums and the bounds for the corresponding remainders, using (16), are

$$s_1 = 1, \qquad |R_1| < \tfrac{1}{2};$$
$$s_2 = 1 - \tfrac{1}{2} = 0.5, \qquad |R_2| \leq \tfrac{1}{3};$$
$$s_3 = 1 - \tfrac{1}{2} + \tfrac{1}{3} = 0.8333, \qquad |R_3| \leq \tfrac{1}{4}.$$

In order to attain an accuracy to three decimal places, that is, $|R_n| < 0.001$, we must sum more than 1,000 terms.

Example 5 Show that $I = \sum_{n=1}^{\infty} (-1)^n \frac{\log n}{n^{1/2}}$ converges.

This is an alternating series with $c_n = n^{-1/3} \log n$. To show that $c_n > c_{n+1}$, we consider the function $f(x) = x^{-1/3} \log x$, which assumes the values c_n at $x = n$. Since the derivative

$$f'(x) = x^{-4/3}(1 - \tfrac{1}{3} \log x)$$

is *negative* for $x > e^3$, the function decreases continually beyond this value, as must the consecutive terms of the series:

$$c_n > c_{n+1} \qquad \text{for} \qquad n > 21 > e^3.$$

Moreover,

$$\lim_{n\to\infty} c_n = \lim_{x\to\infty} x^{-1/3} \log x = 0,$$

where l'Hôpital's rule is employed in the last limit calculation. Therefore, the particular alternating series $J = \sum_{n=21}^{\infty} (-1)^n n^{-1/3} \log n$ satisfies all the criteria for convergence and this implies convergence of the series in question, which has only a few additional terms: $I = -c_1 + c_2 - c_3 + \cdots + c_{20} + J$.

6

In this section we have established a few general conclusions which are most useful in developing more decisive tests for convergence. In summary, the principal points are:

1. A necessary condition for convergence of $\sum_{n=1}^{\infty} a_n$ is $\lim_{n\to\infty} a_n = 0$.
2. $\sum_{n=0}^{\infty} r^n$ is convergent for $|r| < 1$.
3. Absolute convergence implies convergence.
4. An alternating series $\sum_{n=1}^{\infty} (-1)^{n+1} c_n$ (with $c_n > 0$) converges if $c_{n+1} < c_n$ and $\lim_{n\to\infty} c_n = 0$.

Exercises 4.1

1. Write out the following sums and determine their values.

(i) $\sum_{n=1}^{6} (n^2 + 2n)$ **(ii)** $\sum_{j=0}^{4} \frac{1}{j+1}$ **(iii)** $\sum_{i=1}^{3} i^i$

(iv) $\sum_{n=1}^{\infty} \frac{1}{2^n} - \sum_{k=3}^{\infty} \frac{1}{2^k}$ **(v)** $\sum_{k=0}^{3} \frac{1}{(k+2)^2 + 3}$ **(vi)** $\sum_{n=0}^{\infty} \frac{1}{2^{2n}}$

2. Express the following geometric series in the summation notation and evaluate.

(i) $1 + 99/100 + (99/100)^2 + \cdots$ **(ii)** $1 + 4/3 + (4/3)^2 + \cdots$

(iii) $0.474747\cdots$ **(iv)** $5.635635635\cdots$

(v) $r^2 + r^5 + r^8 + \cdots$ **(vi)** $10 + 1 + (0.1) + \cdots$

(vii) $1 - \frac{1}{2^4} + \frac{1}{2^8} - \cdots$ **(viii)** $\frac{r^2}{1+r^2} + \frac{r^2}{(1+r^2)^2} + \frac{r^2}{(1+r^2)^3} + \cdots$

3. Write the general term of each of the following series and express the sum in the summation notation. Using a calculator, evaluate each series correct to two decimal places.

(i) $1 - \frac{1}{2!} + \frac{1}{4!} - \frac{1}{6!} + \cdots$ **(ii)** $\frac{1}{2^4} - \frac{1}{3^4} + \frac{1}{4^4} - \cdots$

(iii) $\frac{1}{2^4} - \frac{1}{3^6} + \frac{1}{4^8} - \cdots$ **(iv)** $\frac{1}{2} + \frac{2!}{4!} + \frac{3!}{6!} + \cdots$

(v) $1 + \frac{(0.1)}{1!} + \frac{(0.1)^2}{2!} + \cdots$ **(vi)** $1 - \frac{1}{2^2} + \frac{1}{2^4} - \cdots$

4. Show that a series $\sum_{n=1}^{\infty} a_n$ diverges if $\lim_{n\to\infty} a_n \neq 0$. Give examples. Estimate the sums of the first one million terms of the following series.

(i) $\sum_{n=1}^{\infty} \frac{n}{2n+1}$ **(ii)** $\sum_{n=1}^{\infty} n^2 \sin^2 \frac{\pi}{n}$ **(iii)** $\sum_{n=1}^{\infty} \left(1 + \frac{1}{n}\right)^{-n}$

5. The terms of a sequence $\{a_n\}$ are 1, 1/2, 1/2, 1/3, 1/3, 1/3, 1/4, 1/4, 1/4, 1/4, 1/5, $\cdots$ where $a_n = 1/k$ if $k(k-1)/2 < n \leq k(k+1)/2$. Is the series formed from these terms convergent or divergent? Determine the exact sums of the first thousand and the first million terms.

6. Given a sequence $\{a_n\}$, define the *sequence of differences* $\{d_n\}$ by $d_n = a_n - a_{n+1}$.

(i) Verify that $\sum_{k=1}^{n} d_k = a_1 - a_{n+1}$.

(ii) When does the series $\sum_{k=1}^{\infty} d_k$ converge? What is its sum?

7. Using the previous problem, evaluate the following series:

(i) $\sum_{n=1}^{\infty} \frac{1}{n(n+1)}$ **(ii)** $\sum_{n=1}^{\infty} \frac{1}{(n+4)(n+5)}$ **(iii)** $\sum_{n=1}^{\infty} \frac{1}{(2n-1)(2n+1)}$

(iv) $\sum_{n=1}^{\infty} (1-r)r^{n-1}$ **(v)** $\sum_{n=1}^{\infty} \frac{2n+1}{n^2(n+1)^2}$ **(vi)** $\sum_{n=1}^{\infty} \frac{1}{n(n+1)(n+2)}$

8. Show that $\sum_{n=2}^{\infty} \log\left(1 - \frac{1}{n}\right)$ is divergent but $\sum_{n=2}^{\infty} \log\left(1 - \frac{1}{n^2}\right) = -\log 2$. (*Hint:* Identify each series as a series of differences.)

9. **(i)** If $\sum_{n=1}^{\infty} a_n = s$, prove that $\sum_{n=1}^{\infty} (a_n + a_{n+1}) = 2s - a_1$.

(ii) If $\sum_{n=1}^{\infty} a_n = s$ and $\sum_{n=1}^{\infty} b_n = t$, prove that $\sum_{n=1}^{\infty} (a_n + b_n) = s + t$.

(*Hint:* Consider partial sums.)

10. Consider two sequences $\{a_n\}$ and $\{b_n\}$ where $\sum_{n=1}^{\infty} a_n$ is convergent.

(i) If $\lim_{n\to\infty} (a_n - b_n) = 0$, does it follow that $\sum_{n=1}^{\infty} b_n$ is convergent?

(ii) If $\lim_{n\to\infty} \frac{a_n}{b_n} = 1$, is $\sum_{n=1}^{\infty} b_n$ convergent?

11. For the following c_n, test the alternating series $\sum_{n=1}^{\infty} (-1)^n c_n$ for convergence.

(i) $\frac{1}{2n-1}$ **(ii)** $\frac{n^2+n}{n^3+3}$ **(iii)** $\frac{n}{e^{n^2}}$

(iv) $\log\frac{n}{n+1}$ **(v)** $\frac{n^n}{(n+1)^n}$ **(vi)** $\frac{\log n}{n^{1/2}}$

(vii) $\frac{1}{n(\log\log(n+1))^3}$ **(viii)** $\sin\frac{1}{n}$ **(ix)** $e^{-n}\cosh n$

12. For the convergent series in the previous example, determine upper bounds for the magnitudes of the errors made by truncating the series after four terms.

13. Prove the inequalities for the partial sums of alternating series $\sum_{n=1}^{\infty} (-1)^{n+1} c_n$, $c_n > 0$.

(i) $s_{2m+1} > s_{2m} > s_{2n}$ if $m > n$
(ii) $s_{2n} < s_{2n+1} < s_{2m+1}$ if $m > n$
(iii) $s_{2n} < s_{2m+1}$ for any integers m, n.

14. Show the following:

(i) $\sum_{k=m}^{n} a_k = \sum_{k=m+p}^{n+p} a_{k-p}$ $m < n$, any p **(ii)** $\sum_{k=m+1}^{n} (a_k - a_{k-1}) = a_n - a_m$

(iii) $\left|\sum_{k=m}^{n} a_k\right| \le \sum_{k=m}^{n} |a_k|$ **(iv)** $\sum_{k=1}^{\infty} a_k = \sum_{k=1}^{n} a_k + \sum_{k=1}^{\infty} a_{n+k}$

15. By relabelling the summation indexes, verify that

$$\sum_{k=0}^{\infty} a_{k+2} + 2\sum_{k=1}^{\infty} (-1)^k a_k = \sum_{n=1}^{\infty} [a_{n+1} + 2(-1)^n a_n].$$

Collect each of the following under one summation sign.

(i) $\sum_{n=1}^{\infty} a_n + \sum_{n=3}^{\infty} a_{n-2}$

(ii) $\sum_{k=1}^{\infty} (-1)^{k+1} c_k + \sum_{k=2}^{\infty} (-1)^{k+1} c_{k-1}$

(iii) $\sum_{k=2}^{\infty} (2r)^k + \sum_{k=3}^{\infty} (3r)^k$

(iv) $\sum_{n=2}^{\infty} \frac{1}{n^2} - \sum_{n=3}^{\infty} \frac{1}{n(n-1)}$

4.2 *Convergence tests*

1 We develop now a few tests for absolute convergence of a series (or equivalently for the convergence of a series with positive terms only). The fundamental idea is that of comparing the unknown series with another whose convergence (or divergence) is already established. Consider first a direct *comparison test*.

Let $\sum_{k=1}^{\infty} c_k$ be a series with positive terms *which is known to converge*. The series $\sum_{k=1}^{\infty} a_k$ will then converge absolutely if

$$|a_n| \le c_n \qquad \text{for} \qquad n = 1, 2, \ldots \tag{1}$$

(or for all integers n beyond a particular value, $n > N$). This conclusion is established by calculating a bound for $|\sum_{k=n+1}^{\infty} a_k|$ as follows:

$$\left|\sum_{k=n+1}^{n+m} a_k\right| \le \sum_{k=n+1}^{n+m} |a_k| \le \sum_{k=n+1}^{n+m} c_k \le \sum_{k=n+1}^{\infty} c_k < \varepsilon. \tag{2}$$

This reduction is based solely on (1) and the known convergence of the comparison series, a fact used in the last step of (2). The preceding equation holds for all integers m, and in the limit $m \to \infty$ we obtain

$$\left|\sum_{k=n+1}^{\infty} a_k\right| \le \sum_{k=n+1}^{\infty} |a_k| < \varepsilon,$$

which establishes absolute convergence. (To justify this operation, refer to the analysis in Sec. 4.1.) Note, once again, that a finite number of terms added to or discarded from the original series has no effect on its convergence. [For example, if Eq. (1) were to apply only for $n > 10^6$, the series would still converge.]

By similar arguments, we can prove that if $|a_n| \ge c_n > 0$ and $\sum_{n=1}^{\infty} c_n$ is *known to diverge*, then the series $\sum_{n=1}^{\infty} |a_n|$ is divergent too. (But $\sum_{n=1}^{\infty} a_n$ may be conditionally convergent.)

Example 1 Show that $\sum_{n=1}^{\infty} \frac{\log n}{n^{1/2} 2^n}$ converges.

The series consists only of positive terms, so that absolute convergence is synonomous with ordinary convergence. First we observe that for $n > 1$, $n^{-1/2} \log n < 1$. [To see this, consider the function $f(x) = x^{-1/2} \log x$ and show that the maximum value of $f(x)$ is $f(e^2) = 2/e < 1$.] Therefore

$$\frac{\log n}{n^{1/2} 2^n} \leq \frac{1}{2^n},$$

which is the equivalent of (1). With $c_n = 1/2^n$, we note that the comparison series is a *convergent* geometric series and in fact

$$\sum_{n=1}^{\infty} \frac{1}{2^n} = 1,$$

which proves absolute convergence of the series in question.

---

2 The *ratio test* also involves comparison with a geometric series. Suppose that upon examining a series we establish that

$$\lim_{n \to \infty} \left| \frac{a_{n+1}}{a_n} \right| = r < 1. \tag{3}$$

It follows from the definition of a limit that for sufficiently large n ($n > N$)

$$\left| \frac{a_{n+1}}{a_n} \right| < t < 1, \qquad r < t, \tag{4}$$

which in effect means

$$|a_{n+1}| < t|a_n| \qquad \text{for all } n > N.$$

Therefore

$$|a_{N+1}| < t|a_N|, \qquad |a_{N+2}| < t|a_{N+1}| < t^2|a_N|,$$

and in general

$$|a_{N+m}| < t^m|a_N|. \tag{5}$$

Consider next the infinite series

$$\sum_{k=N}^{\infty} |a_k| = \sum_{m=0}^{\infty} |a_{N+m}|;$$

it follows from (5) that

$$\sum_{m=0}^{\infty} |a_{N+m}| \leq |a_N| \sum_{m=0}^{\infty} t^m = \frac{|a_N|}{1-t}. \tag{6}$$

This bound shows that $\sum_{k=1}^{\infty} a_k$ is absolutely convergent (because the terms $n \leq N$ are quite unimportant in this regard).

On the other hand, if $\lim_{n\to\infty} |a_{n+1}/a_n| = r > 1$, we can write

$$\left|\frac{a_{n+1}}{a_n}\right| > t > 1 \qquad \text{for} \qquad n > N.$$

As in Eq. (5), it follows that

$$|a_{N+m}| > t^m |a_N|$$

for all m. Consequently,

$$\lim_{n\to\infty} |a_n| = \lim_{m\to\infty} |a_{N+m}| = \infty,$$

which shows that the series $\sum_{n=0}^{\infty} a_n$ is divergent in this case (because the basic condition for convergence, $\lim_{n\to\infty} a_n = 0$, is violated).

If these results are collected, a summary statement can be made. The ratio test of the series $\sum_{n=1}^{\infty} a_n$ consists of evaluating the limit of the ratio of consecutive terms:

$$\lim_{n\to\infty} \left|\frac{a_{n+1}}{a_n}\right| = r. \tag{7}$$

If $r < 1$, the series converges absolutely. If $r > 1$, the series diverges. If $r = 1$, the test gives no information whatsoever.

Example 2 Apply the ratio test to the series $\sum_{n=1}^{\infty} \frac{n}{10^n}$.

Since

$$a_n = \frac{n}{10^n} \qquad \text{and} \qquad a_{n+1} = \frac{n+1}{10^{n+1}},$$

Eq. (7) becomes

$$r = \lim_{n\to\infty} \left|\frac{a_{n+1}}{a_n}\right| = \lim_{n\to\infty} \left|\frac{n+1}{n} \frac{10^n}{10^{n+1}}\right| = \frac{1}{10} \lim_{n\to\infty} \left|1 + \frac{1}{n}\right| = \frac{1}{10}.$$

Therefore, the ratio of consecutive elements is less than 1 in the limit, and the series converges.

Example 3 Apply the ratio test to the series $\sum_{n=1}^{\infty} \frac{n^n}{3^n n!}$.

In this case,

$$a_n = \frac{n^n}{3^n n!}, \qquad \text{and} \qquad a_{n+1} = \frac{(n+1)^{n+1}}{3^{n+1}(n+1)!} = \frac{(n+1)^n}{3^{n+1} n!},$$

so that

$$\left|\frac{a_{n+1}}{a_n}\right| = \frac{1}{3}\left(\frac{n+1}{n}\right)^n = \frac{1}{3}\left(1+\frac{1}{n}\right)^n.$$

Therefore

$$\lim_{n\to\infty}\left|\frac{a_{n+1}}{a_n}\right| = \lim_{n\to\infty}\frac{1}{3}\left(1+\frac{1}{n}\right)^n = \frac{e}{3} < 1$$

(where the limit was evaluated by l'Hôpital's rule in Sec. 2.9). The series is convergent. Note that since

$$\lim_{n\to\infty} a_n = 0$$

is a necessary condition for convergence, we have also shown that $n!$ increases faster than $(n/3)^n$; that is,

$$\lim_{n\to\infty}\frac{1}{n!}\left(\frac{n}{3}\right)^n = 0. \tag{8}$$

3

In the *integral test*, the series $\sum_{n=1}^{\infty}|a_n|$ is interpreted as an area and compared with another area represented as an integral. The test can be applied whenever there is a positive, continuous function $f(x)$ which *decreases* as x increases and which has the particular values

$$f(n) = |a_n|.$$

Implicit in this assumption about the behavior of $f(x)$ is the fact that

$$f(n) = |a_n| > |a_{n+1}| = f(n+1).$$

As an illustration, the harmonic series $\sum_{n=1}^{\infty}\frac{1}{n}$ may be associated with the function $f(x) = 1/x$, which for successive integer values of x generates all the terms of the series, i.e.,

$$f(n) = \frac{1}{n} = a_n.$$

Moreover, the function is positive, continuous, and decreasing, and certainly $a_n > a_{n+1}$.

The assumptions made about $f(x)$ imply that its graph is, in general, as shown in Fig. 2.1. It is evident that in the interval of unit length,

$$n \le x \le n+1, \tag{9}$$

the function satisfies

$$|a_{n+1}| \le f(x) \le |a_n|. \tag{10}$$

This inequality can be converted into a relationship among areas by integrating over the range $[n, n + 1]$. Since all quantities are positive,

$$\int_n^{n+1} |a_{n+1}|\,dx \le \int_n^{n+1} f(x)\,dx \le \int_n^{n+1} |a_n|\,dx,$$

so that

$$|a_{n+1}| \le \int_n^{n+1} f(x)\,dx \le |a_n|. \tag{11}$$

This equation asserts, as is made clear by Fig. 2.2, that the area under the curve $f(x)$

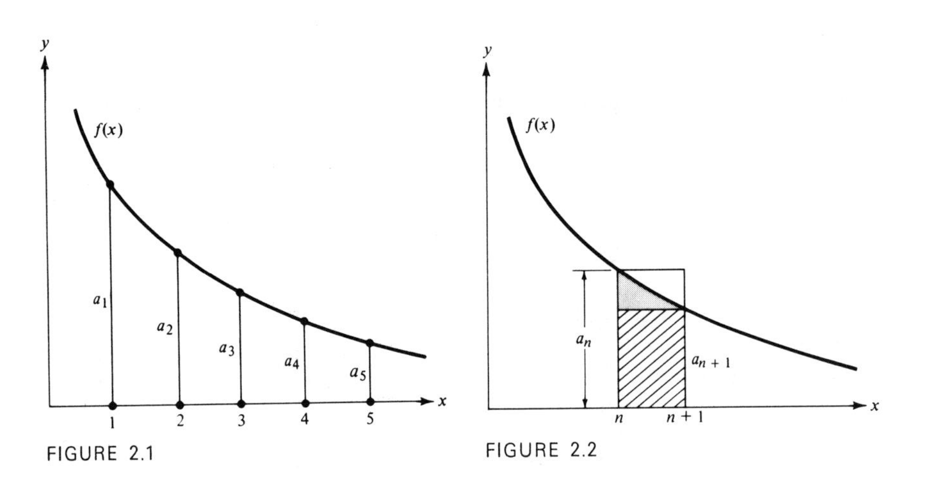

FIGURE 2.1

FIGURE 2.2

above the unit interval $[n, n + 1]$ is *larger* than the rectangle of height $|a_{n+1}|$ and unit width but *smaller* than the rectangle of height $|a_n|$ and unit width. Furthermore, (11) holds in every such unit interval:

$$\begin{aligned} |a_2| &\le \int_1^2 f(x)\,dx \le |a_1|, \\ |a_3| &\le \int_2^3 f(x)\,dx \le |a_2|, \\ &\cdots\cdots\cdots\cdots\cdots\cdots \\ |a_{n+1}| &\le \int_n^{n+1} f(x)\,dx \le |a_n|, \end{aligned} \tag{12}$$

and by adding all these together, we obtain bounds for the nth partial sum of the series

$$S_{n+1} - |a_1| = |a_2| + \cdots + |a_{n+1}| \le \int_1^{n+1} f(x)\,dx \le |a_1| + \cdots + |a_n| = S_n.$$

This can also be expressed as

$$S_n \le |a_1| + \int_1^n f(x)\,dx \qquad \text{and} \qquad S_n \ge \int_1^{n+1} f(x)\,dx; \tag{13}$$

and these formulas form the basis of the integral test. If the improper integral $\int_1^\infty f(x)\,dx$ converges, the first of these relationships shows that the series also converges (and provides a bound as well):

$$S = \lim_{n\to\infty} S_n \le |a_1| + \int_1^\infty f(x)\,dx.$$

On the other hand, if the integral diverges, the second relationship implies that the series also diverges since

$$S = \lim_{n\to\infty} S_n \ge \int_1^\infty f(x)\,dx.$$

Therefore the series of positive terms $\sum_{n=1}^\infty |a_n|$ (under the assumed restrictions) *converges* or *diverges* as the integral $\int_1^\infty f(x)\,dx$ converges or diverges. This is the integral test.

Example 4 Use the integral test to show that the harmonic series $\sum_{k=1}^\infty \frac{1}{k}$ diverges.

The integral test is appropriate in this case. According to (13),

$$S_n = \sum_{k=1}^n \frac{1}{k} \ge \int_1^{n+1} \frac{dx}{x} = \log(n+1),$$

so that

$$S = \lim_{n\to\infty} S_n \ge \lim_{n\to\infty} \log(n+1) = \infty.$$

The series diverges, as does the improper integral $\int_1^\infty \frac{dx}{x}$ (but note that $\lim_{n\to\infty} a_n = 0$). [Furthermore, we have now proved that $\sum_{k=1}^\infty \frac{(-1)^k}{k}$ is conditionally convergent.]

Example 5 For what values of α does the series $\sum_{n=1}^\infty \frac{1}{n^\alpha}$ converge.

For $\alpha \le 0$ the series obviously diverges. For $\alpha > 0$ the function $f(x) = 1/x^\alpha$ satisfies all the criteria for the integral test. Since

$$\int_1^\infty \frac{1}{x^\alpha}\,dx = \begin{cases} \infty & \text{for } \alpha \le 1; \\ (\alpha-1)^{-1} & \text{for } \alpha > 1, \end{cases}$$

the series $\sum_{n=1}^\infty \frac{1}{n^\alpha}$ also converges for $\alpha > 1$ and diverges for $\alpha \le 1$.

Example 6 Use (13) to estimate the sum $\sum_{n=1}^{\infty} \frac{1}{n^2}$.

This equation can be written

$$\int_1^{n+1} f(x)\,dx \le S_n \le |a_1| + \int_1^n f(x)\,dx,$$

which in the limit $n \to \infty$ provides bounds for the series given by

$$\int_1^{\infty} f(x)\,dx \le S \le |a_1| + \int_1^{\infty} f(x)\,dx.$$

Since $\int_1^{\infty} \frac{dx}{x^2} = 1$, these bounds become

$$1 \le S \le 2,$$

which are not particularly sharp. The estimates can be improved greatly by simply summing the first few terms and using the integral bounds for the remaining series. For example, we write

$$\sum_{n=1}^{\infty} \frac{1}{n^2} = 1 + \frac{1}{4} + \frac{1}{9} + \sum_{n=4}^{\infty} \frac{1}{n^2}.$$

Since, by (13),

$$\int_4^{\infty} \frac{dx}{x^2} \le \sum_{n=4}^{\infty} \frac{1}{n^2} \le \frac{1}{4^2} + \int_4^{\infty} \frac{dx}{x^2}$$

or

$$\frac{1}{4} \le \sum_{n=4}^{\infty} \frac{1}{n^2} \le \frac{1}{4^2} + \frac{1}{4},$$

we find upon replacing the series above that

$$1 + \frac{1}{4} + \frac{1}{9} + \left(\frac{1}{4}\right) \le \sum_{n=1}^{\infty} \frac{1}{n^2} \le 1 + \frac{1}{4} + \frac{1}{9} + \left(\frac{1}{4^2} + \frac{1}{4}\right).$$

Therefore

$$1.6111 \ldots \le \sum_{n=1}^{\infty} \frac{1}{n^2} \le 1.6736 \ldots.$$

The exact answer is $\pi^2/6 = 1.6449 \ldots$.

---

4 The assortment of problems presented next will illustrate the methods introduced thus far and conclude this section.

Example 7 Determine by any method whether the following series are convergent or divergent:

(i) $\sum_{n=1}^{\infty} \frac{2^n}{3^{n+1}}$; **(ii)** $\sum_{n=1}^{\infty} \frac{1}{(2n-1)n}$; **(iii)** $\sum_{n=1}^{\infty} \frac{(-1)^n x^n}{n}$;

(iv) $\sum_{n=2}^{\infty} \frac{(-1)^n}{n \log n}$; **(v)** $\sum_{n=1}^{\infty} n \sin \frac{\pi}{n}$; **(vi)** $\sum_{n=1}^{\infty} \frac{1}{n^3} \sin n$.

In case (i) a direct comparison test with the geometric series works well. Since

$$\frac{2^n}{3^{n+1}} = \tfrac{1}{3}(\tfrac{2}{3})^n < (\tfrac{2}{3})^n$$

and $\sum_{n=1}^{\infty} (\frac{2}{3})^n$ converges, the series (i) is also convergent.

For case (ii) the ratio test will fail to give an answer, but the integral test is successful. Let

$$f(x) = \frac{1}{(2x-1)x} = \frac{1}{2(x-\frac{1}{2})x},$$

which is positive, continuous, and decreasing and as such satisfies all the requisite criteria. Since

$$\int_1^{\infty} f(x)\,dx = \lim_{t\to\infty} \frac{1}{2}\int_1^t \frac{dx}{(x-\frac{1}{2})x} = \lim_{t\to\infty}\left(\log\frac{t-\frac{1}{2}}{t} + \log 2\right) = \log 2,$$

the series is convergent. $\left(\text{An alternative approach is to compare the series with } \sum_1^{\infty} \frac{1}{n^2} \text{ since } \frac{1}{(2n-1)n} \le \frac{1}{n^2}.\right)$

In case (iii) the variable x is a parameter in the series $\sum_{n=1}^{\infty} \frac{(-1)^n x^n}{n}$, and its value is a crucial factor to the question of divergence or convergence. The ratio test is applied, and with

$$a_n = \frac{(-1)^n x^n}{n}, \qquad a_{n+1} = \frac{(-1)^{n+1} x^{n+1}}{n+1},$$

it follows that

$$\lim_{n\to\infty}\left|\frac{a_{n+1}}{a_n}\right| = \lim_{n\to\infty}\left|\frac{xn}{n+1}\right| = |x|.$$

Therefore, for $|x| < 1$, the series converges absolutely, but for $|x| > 1$ the series diverges. The ratio test fails when $|x| = 1$. However for $x = 1$ the resultant series is known to be conditionally convergent, and for $x = -1$ the series diverges, as shown in Example 4.

In case (iv) the alternating series is certainly convergent because

$$\lim_{n\to\infty} \frac{1}{n \log n} = 0$$

and

$$\frac{1}{n \log n} > \frac{1}{(n+1)\log(n+1)}.$$

That the series does not converge absolutely is shown easily by the integral test, using $f(x) = 1/(x \log x)$ for $x > 2$. Since the associated improper integral diverges,

$$\int_2^\infty \frac{1}{x \log x}\, dx = \int_{\log 2}^\infty \frac{du}{u} = \log u \Big]_{\log 2}^\infty = \infty,$$

so must the series of absolute values.

A necessary condition for convergence is that the nth term, a_n, approach zero as n becomes infinite. However, in case (v)

$$\lim_{n\to\infty} n \sin\frac{\pi}{n} = \lim_{n\to\infty} \frac{\sin(\pi/n)}{1/n} = \pi,$$

and this series diverges as a consequence.

In case (vi) since $\sum_{n=1}^{\infty} \frac{1}{n^3}$ is known to be convergent (see Example 5), the series in question may be compared to this to prove absolute convergence; i.e.,

$$\left|\frac{\sin n}{n^3}\right| \le \frac{1}{n^3}.$$

Exercises 4.2

1. Use the direct comparison test to determine whether the following series converge.

(i) $\sum_{n=1}^{\infty} \frac{1}{n2^n}$ **(ii)** $\sum_{n=1}^{\infty} \frac{1}{n^2+1}$ **(iii)** $\sum_{n=1}^{\infty} \frac{1}{(n+1)^{1/2}}$

(iv) $\sum_{n=1}^{\infty} \frac{n+1}{n^2+2}$ **(v)** $\sum_{n=1}^{\infty} \frac{n}{(2n+1)^{3/2}}$ **(vi)** $\sum_{n=0}^{\infty} \frac{1}{(n+4)^{4/3}}$

(vii) $\sum_{n=0}^{\infty} \frac{n^2+4}{n^4-6}$ **(viii)** $\sum_{n=1}^{\infty} \frac{e^n}{(n+2)^n}$ **(ix)** $\sum_{n=0}^{\infty} \frac{1}{(n+2)(n+3)}$

2. Use the integral test to determine whether the following series converge.

(i) $\sum_{n=1}^{\infty} \frac{1}{n^4+1}$ **(ii)** $\sum_{n=1}^{\infty} \frac{n^3}{n^4+1}$ **(iii)** $\sum_{n=2}^{\infty} \frac{\log n}{n}$

(iv) $\sum_{n=2}^{\infty} \frac{\log n}{n^2}$ **(v)** $\sum_{n=2}^{\infty} \frac{1}{n^2-1}$ **(vi)** $\sum_{n=1}^{\infty} \frac{n^2}{3^n}$

(vii) $\sum_{n=1}^{\infty} \frac{1}{4+n^2}$ **(viii)** $\sum_{n=1}^{\infty} \frac{1}{n^2-2n+2}$ **(ix)** $\sum_{n=1}^{\infty} \frac{1}{n(n^2+1)^{1/2}}$

3. Write the following series in the summation notation. Determine which series are convergent.

(i) $1 + \frac{e}{\pi} + \left(\frac{e}{\pi}\right)^2 + \cdots$ **(ii)** $\frac{1}{\sqrt{1}} - \frac{1}{\sqrt{2}} + \frac{1}{\sqrt{3}} - \cdots$ **(iii)** $\frac{\log 2}{\sqrt{2}} + \frac{\log 3}{\sqrt{3}} + \frac{\log 4}{\sqrt{4}} + \cdots$

(iv) $1/2 + 2/3 + 3/4 + \cdots$ **(v)** $1 - 1/3 + 1/5 - \cdots$ **(vi)** $\sin 1 - \sin 1/2 + \sin 1/3 - \cdots$

4. Write the following using factorials. (For example, $2n(2n-2)(2n-4)\cdots 2 = 2^n n!$)

(i) $(2n+1)(2n-1)(2n-3)\cdots 1$ **(ii)** $2n(2n-1)\cdots(n+2)(n+1)$

(iii) $(4n+1)(4n-3)(4n-7)\cdots 1$

(iv) $(3n+1)(3n+2)(3n+4)(3n+5)\cdots(6n-2)(6n-1)$

5. If $0<|r|<1$, show the following:

(i) $\sum_{n=1}^{\infty} nr^n = \frac{r}{(1-r)^2}$ **(ii)** $\sum_{n=1}^{\infty} n^2 r^n = \frac{r(1+r)}{(1-r)^3}$ **(iii)** $\sum_{n=1}^{\infty} n^3 r^n = \frac{r(1+4r+r^2)}{(1-r)^4}$

(*Hint:* Differentiate the geometric series or note $(1-r)\sum_{n=1}^{\infty} nr^n = \sum_{n=1}^{\infty} nr^n - \sum_{n=1}^{\infty} nr^{n+1} = \sum_{n=1}^{\infty} r^n$ and $(1-r)\sum_{n=1}^{\infty} n^2 r^n = \sum_{n=1}^{\infty} (2n-1)r^n$.)

6. Evaluate the following series:

(i) $\sum_{n=1}^{\infty} \frac{n}{10^n}$ **(ii)** $\sum_{n=1}^{\infty} n(.99)^n$ **(iii)** $\sum_{n=1}^{\infty} \frac{n+1}{2^n}$

(iv) $\sum_{n=1}^{\infty} \frac{n^2}{10^n}$ **(v)** $\sum_{n=1}^{\infty} n^2(.99)^n$ **(vi)** $\sum_{n=1}^{\infty} \frac{(n+1)^2}{2^n}$

7. Test for absolute or conditional convergence by any method.

(i) $\sum_{n=1}^{\infty} (-1)^n \frac{\log n}{n}$ **(ii)** $\sum_{n=1}^{\infty} (-1)^n \frac{n}{n+1}$ **(iii)** $\sum_{n=1}^{\infty} \frac{2^n}{n!}$

(iv) $\sum_{n=1}^{\infty} \frac{1}{n^2}\left(1+\frac{1}{n}\right)^n$ **(v)** $\sum_{n=1}^{\infty} (-1)^n \frac{(2n)!}{2^{2n}(n!)^2}$ **(vi)** $\sum_{n=1}^{\infty} \frac{3\cdot 5 \cdots (2n+1)}{2\cdot 4\cdot 6\cdots 2n}$

(vii) $\sum_{n=1}^{\infty} \frac{\cos(n\pi)}{n\pi}$ **(viii)** $\sum_{n=1}^{\infty} \frac{2^n}{1+2^{2n}}$ **(ix)** $\sum_{n=1}^{\infty} (-1)^n \frac{\tanh n}{n}$

8. Use the integral test to determine whether the following series converge.

(i) $\sum_{n=1}^{\infty} \frac{n}{3^n}$ **(ii)** $\sum_{n=1}^{\infty} n^{10} e^{-n/2}$ **(iii)** $\sum_{n=1}^{\infty} \frac{1}{n}\log\left(1+\frac{1}{n}\right)$

(iv) $\sum_{n=1}^{\infty} \frac{n}{n^3+1}$ **(v)** $\sum_{n=1}^{\infty} \frac{\log(n+1)}{n^2+n}$ **(vi)** $\sum_{n=2}^{\infty} \frac{1}{(\log n)^2}$

9. Use the ratio test to determine whether the following series converge.

(i) $\sum_{n=1}^{\infty} \frac{2^n n!}{(2n)!}$ **(ii)** $\sum_{n=1}^{\infty} (-1)^{n+1} \frac{n!}{10^n}$ **(iii)** $\sum_{n=1}^{\infty} \frac{n^3}{n!}$

(iv) $\sum_{n=1}^{\infty} \frac{(n+1)!}{n!\,5^n}$ **(v)** $\sum_{n=1}^{\infty} \sin\left(\frac{\pi}{2^n}\right)$ **(vi)** $\sum_{n=1}^{\infty} \frac{\cos n\pi}{\cosh n\pi}$

(vii) $\sum_{n=1}^{\infty} \frac{\tan(\pi/2 - 1/n)}{2^n}$ **(viii)** $\sum_{n=1}^{\infty} \frac{n!}{n^n}$ **(ix)** $\sum_{n=1}^{\infty} \left(\sin\frac{1}{n}\right)^n$

10. For what values of x do the following series converge?

(i) $\sum_{n=0}^{\infty} \left(\frac{x-2}{3}\right)^n$ **(ii)** $\sum_{n=1}^{\infty} n(x-5)^n$ **(iii)** $\sum_{n=0}^{\infty} (-1)^n \frac{x^n}{n!}$

(iv) $\sum_{n=1}^{\infty} (-1)^n \frac{(x-3)^n}{n}$ **(v)** $\sum_{n=0}^{\infty} \frac{x^{2n}}{(2n)!}$ **(vi)** $\sum_{n=0}^{\infty} \frac{x^n}{(n+1)(n+2)}$

(vii) $\sum_{n=0}^{\infty} \left(\frac{x}{1+x}\right)^n$ **(viii)** $\sum_{n=0}^{\infty} \frac{x^{3n+1}}{2^n}$ **(ix)** $\sum_{n=1}^{\infty} \log\left(1+\frac{x^2}{n^2}\right)$

11. Prove that $\sum_{n=1}^{\infty} a_n$ converges absolutely if $\lim_{n\to\infty} |a_n|^{1/n} = r < 1$ and diverges if $r > 1$. Test the following series for convergence.

(i) $\sum_{n=1}^{\infty} \frac{n}{2^n}$ **(ii)** $\sum_{n=2}^{\infty} \frac{1}{(\log n)^n}$ **(iii)** $\sum_{n=1}^{\infty} \left(\frac{2n}{1+3n}\right)^n$

12. Prove that $\sum_{n=1}^{\infty} a_n$ converges absolutely if $\lim_{n\to\infty} n^\alpha a_n = A$ for some $\alpha > 1$. Test the following series for convergence.

(i) $\sum_{n=1}^{\infty} \frac{(-1)^{n+1}}{n^2}$ **(ii)** $\sum_{n=1}^{\infty} \sin\left(\frac{\pi}{n^2}\right)$ **(iii)** $\sum_{n=1}^{\infty} n^{-1/2} \log\left(1+\frac{1}{n}\right)$

13. By comparison with the harmonic series, prove that $\sum_{n=1}^{\infty} a_n$ diverges if $\lim_{n\to\infty} n a_n = A \neq 0$. Use this test to prove that the following series diverge.

(i) $\sum_{n=1}^{\infty} \sin\frac{\pi}{n}$ **(ii)** $\sum_{n=1}^{\infty} \log\left(1+\frac{2}{n}\right)$ **(iii)** $\sum_{n=1}^{\infty} \frac{n \log n}{1+n^2 \log n^2}$

14. The ratio test for convergence gives no information when $\lim_{n\to\infty} |a_{n+1}/a_n| = 1$. Derive the following test to deal in part with this situation: if $a_n > 0$ and $\lim_{n\to\infty} n\left(\frac{a_{n+1}}{a_n} - 1\right) = C$, then $\sum_{n=1}^{\infty} a_n$ converges if $C < -1$ and diverges if $C > -1$. (When $C = -1$, a more refined test is needed.) Use this test on the following series.

(i) $\sum_{n=1}^{\infty} \frac{1}{n^2+1}$ **(ii)** $\sum_{n=1}^{\infty} \frac{\log n}{n^2}$ **(iii)** $\sum_{n=1}^{\infty} \frac{n^3}{n^4+1}$

15. Using the integral test, discuss the rate of divergence of the series $\sum_{n=1}^{\infty} \frac{1}{n}$ and $\sum_{n=2}^{\infty} \frac{1}{n \log n}$. How many terms are necessary to produce sums close to 100?

4.3 *Taylor series*

The generalization of the mean-value theorem led to the representation of a smooth function as a polynomial plus a remainder term, known as *Taylor's formula*. The formula for $f(x)$ expanded about the point $x = a$ is

$$f(x) = f(a) + f'(a)(x-a) + \frac{f''(a)}{2!}(x-a)^2 + \cdots + \frac{f^{(n)}(a)}{n!}(x-a)^n + R_n, \quad (1)$$

where the remainder is given as a derivative,

$$R_n = \frac{1}{(n+1)!} f^{(n+1)}(\xi)(x-a)^{n+1} \qquad \text{for} \qquad a < \xi < x. \quad (2)$$

The polynomial part of (1) was used to approximate the function $f(x)$ on many occasions, for various purposes, and the natural extension of this procedure is to consider the limit, $n \to \infty$. If the remainder can be shown to approach zero, i.e.,

$$\lim_{n \to \infty} R_n = 0, \tag{3}$$

then the infinite series that evolves is identical to the function and in fact just another form of it, known as its *Taylor series*:

$$f(x) = f(a) + f'(a)(x-a) + \cdots + \frac{f^{(n)}(a)(x-a)^n}{n!} + \cdots = \sum_{n=0}^{\infty} \frac{f^{(n)}(a)(x-a)^n}{n!} \tag{4}$$

Since increasing powers of $x - a$ are involved, the Taylor series is also called a *power-series expansion*. (The special case $a = 0$ is sometimes named a *Maclaurin series*.)

The validity of this series representation depends critically on the limit calculation in (3), and for this purpose it is sometimes useful to have an alternative form of the remainder, expressed as an integral. To this end, Taylor's formula is now rederived by a very clever application of successive integration by parts starting with the identity

$$f(x) - f(a) = \int_a^x f'(z)\,dz. \tag{5}$$

Upon setting $u = f'(z)$, $dv = dz$, so that $du = f''(z)\,dz$ and $v = z - x$ (x is a constant in this procedure), the partial integration yields

$$\int_a^x f'(z)\,dz = f'(z)(z-x)\Big]_a^x - \int_a^x (z-x)f''(z)\,dz.$$

Equation (5) can then be written

$$f(x) = f(a) + f'(a)(x-a) + \int_a^x (x-z)f''(z)\,dz.$$

Once again, the integral involved here is manipulated by partial integration, and with $u = f''(z)$, $dv = (x-z)\,dz$, $du = f'''(z)\,dz$, $v = -\tfrac{1}{2}(z-x)^2$, the result is

$$f(x) = f(a) + f'(a)(x-a) + \frac{1}{2} f''(a)(x-a)^2 + \frac{1}{2}\int_a^x f'''(z)(x-z)^2\,dz.$$

Clearly, the process can be continued indefinitely, and after n repetitions the resultant formula is

$$f(x) = f(a) + f'(a)(x-a) + \cdots + \frac{1}{n!} f^{(n)}(a)(x-a)^n + \frac{1}{n!}\int_a^x f^{(n+1)}(z)(x-z)^n\,dz. \tag{6}$$

This is identical in form with (1), but the remainder is expressed now as the integral

$$R_n = \frac{1}{n!}\int_a^x f^{(n+1)}(z)(x-z)^n\,dz. \tag{7}$$

[Equation (7) implies (2), as shown in the exercises.]

The limit involved in (3) will be calculated in specific cases, but the general issue of the convergence of Taylor series is left open. (Either form of the remainder may be used in the limit process, whichever is more advantageous.)

2

Example 1 Find the Taylor series of e^x about $x = 0$.

The derivatives of higher order, evaluated at $x = 0$, are $(d^n/dx^n)\, e^x]_{x=0} = 1$. The Taylor series is then

$$e^x = 1 + x + \frac{x^2}{2!} + \cdots + \frac{x^n}{n!} + \cdots, \tag{8}$$

and one way of justifying this expansion (and a difficult way too) is to show that the remainder after n terms approaches zero as n becomes infinite. According to (7),

$$R_n = \frac{1}{n!}\int_0^x e^z(x-z)^n\, dz,$$

and since $e^z < e^x$ for $z < x$, it follows that for $x > 0$

$$|R_n| \le \frac{e^x}{n!}\int_0^x (x-z)^n\, dz = \frac{e^x x^{n+1}}{(n+1)!}.$$

On the other hand, when the upper limit x is negative, $e^z < 1$ in the range of integration. Consequently,

$$|R_n| \le \frac{|x|^{n+1}}{(n+1)!}.$$

In either case, we may write

$$|R_n| \le \frac{M|x|^{n+1}}{(n+1)!}, \tag{9}$$

where M is taken as the larger of the two numbers e^x and 1.

To complete the proof that $\lim_{n\to\infty} |R_n| = 0$, we call upon an earlier result obtained in Sec. 4.2:

$$\lim_{n\to\infty} \frac{1}{n!}\left(\frac{n}{3}\right)^n = 0.$$

Since $|x|$ is a fixed number, it follows that

$$|x|^n \le \left(\frac{n}{3}\right)^n$$

for n sufficiently large; that is, $n > 3x$. Therefore

$$0 \le \frac{|x|^n}{n!} \le \left(\frac{n}{3}\right)^n \frac{1}{n!} \qquad \text{for } n > N > 3x,$$

and
$$0 \leq \lim_{n\to\infty} \frac{|x|^n}{n!} \leq \lim_{n\to\infty} \left(\frac{n}{3}\right)^n \frac{1}{n!} = 0,$$

or
$$\lim_{n\to\infty} \frac{|x|^n}{n!} = 0 \qquad \text{for } all\ x.$$

This and the limit of (9) imply that

$$\lim_{n\to\infty} |R_n| = 0, \tag{10}$$

which establishes the validity of the Taylor series expansion (8) *for all* x. It is very much simpler to prove convergence of this Taylor series by applying the ratio test directly, and we do this shortly. The analysis here, then, is mainly in illustration of the basic principle involved and not the most practical means of demonstrating convergence.

Example 2 Find the Taylor series for $\sin x$ about $x = \pi/2$.

In this case

$$\frac{d}{dx}\sin x = \cos x, \qquad \frac{d^2}{dx^2}\sin x = -\sin x,$$

and in general

$$\frac{d^{2n}}{dx^{2n}}\sin x = (-1)^n \sin x, \qquad \frac{d^{2n+1}}{dx^{2n+1}}\sin x = (-1)^n \cos x.$$

The specific values

$$\left.\frac{d^{2n}}{dx^{2n}}\sin x\right]_{x=\pi/2} = (-1)^n, \qquad \left.\frac{d^{2n+1}}{dx^{2n+1}}\sin x\right]_{x=\pi/2} = 0$$

are required to determine the Taylor series, which is, accordingly,

$$\sin x = 1 - \frac{1}{2!}\left(x - \frac{\pi}{2}\right)^2 + \frac{1}{4!}\left(x - \frac{\pi}{2}\right)^4 + \cdots + \frac{(-1)^n}{(2n)!}\left(x - \frac{\pi}{2}\right)^{2n} + \cdots. \tag{11}$$

Since $|d^n \sin x/dx^n| \leq 1$ for all x, the remainder as given in (2) can be bounded simply:

$$|R_n| \leq \frac{|x - \pi/2|^{n+1}}{(n+1)!}.$$

But this is similar to the bound that arose in the discussion of the last problem [in particular see Eqs. (9) and (10)]. Therefore it follows that

$$\lim_{n\to\infty} R_n = 0,$$

and (11) is justified for *all* x.

By setting $x = \pi/2 + h$, so that $\sin x = \sin(\pi/2 + h) = \cos h$, (11) reduces to the corresponding Taylor series of $\cos h$ about $h = 0$:

$$\cos h = 1 - \frac{h^2}{2!} + \frac{h^4}{4!} + \cdots. \tag{12}$$

---

3 Sometimes a power-series representation of a function can be determined without resorting to the formal recipe, which can involve a tedious evaluation of higher derivatives. This can be achieved by algebraic manipulation in which series are added, subtracted, multiplied, and divided; by variable substitution; or by integration and differentiation. The main point is that no matter how a power series for a function is arrived at, it is *the* Taylor's series expansion. Power series are a unique form of representation, and this fundamental result is established first.

We suppose that a function can be expressed as two different series expansions about a point a and show that this leads to a contradiction. For convenience a is taken as zero, and we assume then that

$$f(x) = a_0 + a_1 x + a_2 x^2 + \cdots = b_0 + b_1 x + b_2 x^2 + \cdots. \tag{13}$$

This equation *must* hold for *all* x in some interval about the origin where the series converge. If, in particular, the series are evaluated at $x = 0$, we see at once that

$$a_0 = b_0. \tag{14}$$

The first coefficients of both series are therefore the same and may be cancelled in (13), leaving the identity

$$a_1 x + a_2 x^2 + \cdots = b_1 x + b_2 x^2 + \cdots.$$

The result of dividing this equation by x is

$$a_1 + a_2 x + \cdots = b_1 + b_2 x + \cdots,$$

and repetition of the argument leading to (14), i.e., evaluation at $x = 0$, yields

$$a_1 = b_1.$$

Clearly, the procedure can be continued indefinitely to show that

$$a_n = b_n,$$

which, in effect, means that the series in (13) are identical to each other and cannot be different as assumed. The power series of a function is therefore unique:

$$f(x) = a_0 + a_1 x + a_2 x^2 + \cdots + a_n x^n + \cdots. \tag{15}$$

Moreover, by repeatedly differentiating this series and upon evaluating the resultant equations successively at $x = 0$ we find (see Exercise 15) that

$$f^{(n)}(0) = n! a_n.$$

Thus, (15) is actually *the* Taylor series of $f(x)$ about $x = 0$.

4

The convergence of a power series $\sum_{n=0}^{\infty} a_n(x-a)^n$ can be determined by any of the comparison tests of the last section. Of course, convergence or divergence will now depend on the value of $x - a$. Generally speaking, convergence will occur in some interval given by

$$|x - a| < l,$$

called the region or *interval of convergence*.

Example 3 Find the Taylor series for e^{-x^2} about $x = 0$.

Since

$$e^h = 1 + h + \frac{h^2}{2!} + \cdots + \frac{h^n}{n!} + \cdots, \tag{16}$$

we substitute $-x^2$ for h to obtain

$$e^{-x^2} = 1 - x^2 + \frac{x^4}{2!} + \cdots + \frac{(-1)^n x^{2n}}{n!} + \cdots,$$

which is the Taylor series of the function. Convergence of this series is checked directly by the ratio test. The limit of the ratio of consecutive terms is

$$\lim_{n\to\infty}\left|\frac{(-1)^{n+1}x^{2(n+1)}}{(n+1)!}\bigg/\frac{(-1)^n x^{2n}}{n!}\right| = \lim_{n\to\infty}\frac{x^2}{n+1} = 0,$$

and this shows that the series converges for all x. Note how much simpler this is than the analysis of the remainder term in Example 1.

Example 4 Find the Taylor series for $\sinh kx$ about $x = 0$.

By definition

$$\sinh kx = \tfrac{1}{2}(e^{kx} - e^{-kx}).$$

The result of replacing each exponential by its Taylor series [set $h = kx$ and $h = -kx$ in (16)] is

$$\sinh kx = \tfrac{1}{2}\left[\left(1 + kx + \frac{k^2x^2}{2!} + \cdots\right) - \left(1 - kx + \frac{k^2x^2}{2!} + \cdots\right)\right].$$

These series are added together, term by term, to yield the Taylor series

$$\sinh kx = kx + \frac{k^3x^3}{3!} + \cdots + \frac{k^{2n+1}x^{2n+1}}{(2n+1)!} + \cdots.$$

Example 5 Find the Taylor series for $\log x$ about $x = 1$.

The derivatives of the log function evaluated at $x = 1$ are

$$\frac{d^n}{dx^n} \log x \Big]_{x=1} = (-1)^{n+1}(n-1)!,$$

so that the Taylor series is

$$\log x = (x-1) - \frac{1}{2}(x-1)^2 + \cdots + \frac{(-1)^{n+1}}{n}(x-1)^n + \cdots,$$

$$= \sum_{n=1}^{\infty} \frac{(-1)^{n+1}}{n}(x-1)^n. \tag{17}$$

Convergence of this series can be analyzed by the ratio test. The ratio of consecutive terms is given by

$$\left| \frac{(-1)^{n+2}(x-1)^{n+1}}{n+1} \Big/ \frac{(-1)^{n+1}(x-1)^n}{n} \right| = \left| \frac{n(x-1)}{n+1} \right|.$$

Since

$$\lim_{n\to\infty} \left| \frac{n(x-1)}{n+1} \right| = |x-1|,$$

the series converges for $|x-1|<1$ or $0<x<2$, which is its interval of convergence. For $x=2$, the series converges but not for $x=0$.

In terms of the variable $h = x - 1$, (17) becomes

$$\log(1+h) = h - \frac{h^2}{2} + \cdots + \frac{(-1)^{n+1}h^n}{n} + \cdots, \tag{18}$$

which is actually a more commonly quoted formula.

Example 6 Find the Taylor series of $1/(1-x)$ about $x = 0$.

The simplest approach is long division

$$\begin{array}{r|l} & 1 + x + x^2 + \cdots \\ \hline 1 - x & 1 \\ & \underline{1 - x} \\ & \quad x \\ & \quad \underline{x - x^2} \\ & \qquad x^2 \\ & \qquad \cdots \end{array}$$

which yields

$$\frac{1}{1-x} = 1 + x + x^2 + \cdots. \tag{19}$$

Another tactic is to write (18) in the form

$$-\log(1-x) = x + \frac{1}{2}x^2 + \frac{1}{3}x^3 + \cdots + \frac{1}{n}x^n + \cdots$$

and to differentiate both sides of the equation, producing

$$\frac{1}{1-x} = 1 + x + x^2 + \cdots.$$

The series in (19) can be shown to be convergent for $|x| < 1$. A series valid for $|x| > 1$ is obtained by rearranging the function to read

$$\frac{1}{1-x} = \frac{-1}{x(1-1/x)}.$$

The term on the extreme right is then developed as a legitimate series in powers of $1/x$, and since

$$\frac{1}{1-1/x} = 1 + \frac{1}{x} + \frac{1}{x^2} + \cdots,$$

it follows that

$$\frac{1}{1-x} = -\frac{1}{x} - \frac{1}{x^2} - \frac{1}{x^3} - \cdots,$$

which clearly converges for $|x| > 1$.

Example 7 Expand $1/(1+x)$ about $x = 1$.

The function is rewritten as

$$\frac{1}{1+x} = \frac{1}{2+(x-1)} = \frac{1}{2}\,\frac{1}{1+(x-1)/2},$$

and (19) is then used to expand $1/(1-h)$, where

$$h = \frac{1-x}{2}.$$

Therefore

$$\begin{aligned}\frac{1}{1+x} &= \frac{1}{2}\left[1 - \frac{x-1}{2} + \left(\frac{x-1}{2}\right)^2 + \cdots\right],\\ &= \frac{1}{2} - \frac{x-1}{2^2} + \cdots + \frac{(-1)^n(x-1)^n}{2^{n+1}} + \cdots.\end{aligned}$$

Convergence is guaranteed by the ratio test for $\left|\dfrac{x-1}{2}\right| < 1$ or $-1 < x < 3$.

Example 8 Find the Taylor series of $\tan^{-1} x$ about $x = 0$.

The series can be obtained from the formula

$$\frac{d}{dx}\tan^{-1} x = \frac{1}{1 + x^2}$$

by substituting

$$\frac{1}{1 + x^2} = 1 - x^2 + x^4 + \cdots + (-1)^n x^{2n} + \cdots$$

and integrating term by term. The details are left to the exercises, but the result is

$$\tan^{-1} x = x - \frac{x^3}{3} + \frac{x^5}{5} + \cdots + \frac{(-1)^n}{2n + 1} x^{2n+1} + \cdots + C.$$

The arbitrary constant of integration is evaluated from the condition that $\tan^{-1} 0 = 0$, so that $C = 0$ and

$$\tan^{-1} x = x - \frac{x^3}{3} + \frac{x^5}{5} + \cdots + \frac{(-1)^n x^{2n+1}}{2n + 1} + \cdots. \tag{20}$$

This formula can be used to evaluate π by simply setting $x = 1$, in which case

$$\tan^{-1} 1 = \frac{\pi}{4} = 1 - \frac{1}{3} + \cdots + \frac{(-1)^n}{2n + 1} + \cdots.$$

The rate of convergence is improved greatly by using instead the identity

$$\tan^{-1} 1 = \tan^{-1}\left(\frac{1/2 + 1/3}{1 - 1/2\ 1/3}\right) = \tan^{-1} 1/2 + \tan^{-1} 1/3 = \frac{\pi}{4},$$

and Eq. (20) to evaluate each of the trigonometric functions involved.

5

Example 9 Evaluate $\lim_{x \to 0} \dfrac{2 \log(1 + x) - \sin 2x}{x^2}$.

L'Hôpital's rule can be applied directly, or, as an alternative, successive differentiations can be avoided by replacing each function with its power-series expansion about the origin. Since

$$2 \log(1 + x) - \sin 2x = 2\left(x - \frac{x^2}{2} + \frac{x^3}{3} - \cdots\right) - \left(2x - \frac{(2x)^3}{6} + \cdots\right),$$

it follows that

$$\lim_{x \to 0} \frac{2 \log(1 + x) - \sin 2x}{x^2} = \lim_{x \to 0} \frac{-x^2 + 2x^3 + \cdots}{x^2} = -1 + \lim_{x \to 0}(2x + \cdots) = -1.$$

Only as many terms as needed from each series are written explicitly, and this is frequently a much easier approach in actual calculations.

Example 10 Find the first few terms of the Taylor series of $e^{x^2}\tan^{-1}x$ about $x = 0$.
Since

$$e^{x^2} = 1 + x^2 + \frac{x^4}{2!} + \frac{x^6}{3!} + \cdots$$

and, by (20),

$$\tan^{-1}x = x - \frac{x^3}{3} + \frac{x^5}{5} + \cdots,$$

the two series can be multiplied together term by term and the product rearranged in consecutive powers of x. The calculation involved is

$$\begin{array}{l} 1 + x^2 + \dfrac{x^4}{2} + \cdots \\ \times\, x - \dfrac{x^3}{3} + \dfrac{x^5}{5} + \cdots \\ \hline x + x^3 + \dfrac{x^5}{2} + \cdots \\ \quad - \dfrac{x^3}{3} - \dfrac{x^5}{3} - \dfrac{x^7}{6} + \cdots \\ \qquad + \dfrac{x^5}{5} + \dfrac{x^7}{5} + \cdots \\ \hline x + \tfrac{2}{3}x^3 + \tfrac{11}{30}x^5 + \cdots. \end{array}$$

Consequently,

$$e^{x^2}\tan^{-1}x = x + \tfrac{2}{3}x^3 + \tfrac{11}{30}x^5 + \cdots.$$

It is possible, but not very easy, to find a concise general formula for the nth term of this series. It is shown in the exercises that when two series are multiplied to form a third, for example,

$$\left(\sum_{i=0}^{\infty} a_i x^i\right)\left(\sum_{j=0}^{\infty} b_j x^j\right) = \sum_{k=0}^{\infty} c_k x^k,$$

the general coefficient of the resultant series is given by

$$c_k = \sum_{i=0}^{k} a_i b_{k-i}. \tag{21}$$

This is not an especially convenient form to test convergence, and the difficulties inherent in such cumbersome formulas constitute a major block to completely rigorous presentations of applied problems.

Example 11 Find the power-series expansion about $x = 0$ of $f(x) = (1 + x)^\alpha$ for α an arbitrary number.

Clearly $f(0) = 1$. Repeated differentiations yield

$$f'(x) = \alpha(1 + x)^{\alpha - 1}, \qquad f'(0) = \alpha;$$

$$f''(x) = \alpha(\alpha - 1)(1 + x)^{\alpha - 2} \qquad f''(0) = \alpha(\alpha - 1);$$

$$\cdots\cdots\cdots\cdots\cdots\cdots\cdots\cdots\cdots\cdots\cdots\cdots$$

$$f^{(n)}(x) = \alpha(\alpha - 1)\cdots[\alpha - (n-1)](1 + x)^{\alpha - n}, \qquad f^{(n)}(0) = \alpha(\alpha - 1)\cdots[\alpha - (n-1)].$$

From the definition of Taylor series we find that

$$(1 + x)^\alpha = 1 + \alpha x + \frac{\alpha(\alpha - 1)}{2!}x^2 + \cdots + \frac{\alpha(\alpha - 1)\cdots(\alpha - n + 1)x^n}{n!} + \cdots,$$

or

$$(1 + x)^\alpha = 1 + \sum_{n=1}^{\infty} \frac{\alpha(\alpha - 1)\cdots(\alpha - n + 1)}{n!} x^n, \tag{22}$$

and this result is known as the *binomial theorem*.

---

6 For $\alpha = -\frac{1}{2}$, the series in (22) is

$$(1 + x)^{-1/2} = 1 - \tfrac{1}{2}x + \frac{1}{2}\cdot\frac{3}{2}\frac{x^2}{2!} + \cdots,$$

$$= 1 + \sum_{n=1}^{\infty}(-1)^n \frac{(2n)!}{2^{2n}(n!)^2} x^n. \tag{23}$$

The ratio test readily shows that this is convergent for

$$-1 < x < 1,$$

but no information is given at the end points, which must be considered separately. For example, at $x = 1$, (23) becomes the alternating series

$$1 + \sum_{n=1}^{\infty} \frac{(-1)^n(2n)!}{2^{2n}(n!)^2} = 1 + \sum_{n=1}^{\infty}(-1)^n c_n,$$

and convergence can be established by showing that $c_n > c_{n+1}$, $\lim_{n\to\infty} c_n = 0$. Since

$$c_{n+1} = \frac{(2n+2)!}{2^{2n+2}[(n+1)!]^2} = \frac{(2n+2)(2n+1)(2n)!}{4\cdot 2^{2n}(n+1)^2(n!)^2} = \frac{n + \frac{1}{2}}{n + 1} c_n < c_n,$$

the first requisite condition is found to hold. The second condition follows by writing

$$c_n = \frac{(2n)!}{2^{2n}(n!)^2} = \frac{1\cdot 3\cdots(2n-1)}{2\cdot 4\cdots 2n} = \frac{1}{2}\cdot\frac{3}{4}\cdots\frac{2n-1}{2n}.$$

Each term of the product on the right is replaced by a larger number chosen in accordance with the inequality $(2n - 1)/2n < 2n/(2n + 1)$. For example, $\frac{1}{2} < \frac{2}{3}$, $\frac{3}{4} < \frac{4}{5}$, ..., and if the positive fractions $\frac{1}{2}$, $\frac{3}{4}$, etc., are replaced by $\frac{2}{3}$, $\frac{4}{5}$, etc., it follows that

$$c_n = \frac{1}{2}\cdot\frac{3}{4}\cdots\frac{2n-1}{2n} < \frac{2}{3}\cdot\frac{4}{5}\cdots\frac{2n}{2n+1} = \frac{1}{(2n+1)c_n}.$$

Therefore

$$c_n{}^2 < \frac{1}{2n+1}, \qquad \text{or} \qquad c_n < \frac{1}{(2n+1)^{1/2}},$$

and this clearly implies $\lim_{n\to\infty} c_n = 0$. The series is convergent at $x = 1$, but analysis shows that the series diverges at $x = -1$.

Operations with infinite series require justification, which will not be presented. Some caution is advisable, as evidenced by the fact that while integration of a series often improves its rate of convergence, differentiation has the opposite effect and can produce divergence. A general rule of thumb is that an operation is valid if the resultant series converges, and there the matter will rest for the time being.

Exercises 4.3

1. Write the following power series in the summation notation. Determine their sums and their intervals of convergence.

(i) $1 + \frac{x}{4} + \frac{x^2}{4^2} + \cdots + \frac{x^n}{4^n} + \cdots$ **(ii)** $1 - 5x + 5^2x^2 - \cdots + (-1)^n 5^n x^n + \cdots$

(iii) $1 + e^x + e^{2x} + \cdots + e^{nx} + \cdots$ **(iv)** $1 - \frac{1}{2!}\left(\frac{x}{2}\right)^2 + \frac{1}{4!}\left(\frac{x}{2}\right)^4 - \frac{1}{6!}\left(\frac{x}{2}\right)^6 + \cdots$

(v) $1 + 2x + 3x^2 + \cdots + nx^{n-1} + \cdots$ **(vi)** $1 + 2^2x + 3^2x^2 + \cdots + n^2x^{n-1} + \cdots$

(vii) $1 + \frac{x}{1} + \frac{x^2}{2} + \cdots + \frac{x^n}{n} + \cdots$ **(viii)** $\tan x - \frac{(\tan x)^3}{3} + \frac{(\tan x)^5}{5} - \frac{(\tan x)^7}{7} + \cdots$

2. Use the mean value theorem applied to the function $f^{(n+1)}(x)$ on the interval $[a,x]$ to prove the equivalence of the three forms of the remainder term $R_n(x,a)$.

3. Find the Taylor series about $x = 0$ for each of the following functions:

(i) $\sin^{-1}x$ **(ii)** $\cos^{-1}x$ **(iii)** $\cosh x$

(iv) $\sinh x$ **(v)** $\tan x$ **(vi)** $\tanh x$

4. By differentiating repeatedly and setting $x = 0$, show that $f(x) = \sum_{n=0}^{\infty} a_n x^n$ implies $a_n = f^{(n)}(0)/n!$. (This means that the Taylor series is the unique power series representation of a function.)

5. Evaluate $f^{(4)}(0)$ and $f^{(5)}(0)$ for the following functions:

(i) $\sin 2x$ **(ii)** $\cosh 3x$ **(iii)** $1/(1+x)$

(iv) $\log(1+x)$ **(v)** $(1-x)^{1/2}$ **(vi)** $1 + 2x + 3x^2 + 4x^3 + \cdots$

6. Find the intervals of convergence of the following series. Discuss possible convergence at the endpoints.

(i) $\sum_{n=0}^{\infty} \frac{x^n}{n^2+1}$ **(ii)** $\sum_{n=0}^{\infty} \frac{(x+1)^n}{n!}$ **(iii)** $\sum_{n=1}^{\infty} \frac{n(x+1)^n}{3^n}$

(iv) $\sum_{n=0}^{\infty} \frac{(-1)^n(x-2)^n}{2n+1}$ **(v)** $\sum_{n=0}^{\infty} \frac{(-x)^n}{n2^n}$ **(vi)** $\sum_{n=0}^{\infty} \frac{2^n x^{2n}}{n!}$

7. Differentiate or integrate the Taylor series of $1/(1-x)$ about $x=0$ to obtain the following results:

(i) $\dfrac{1}{(1-x)^2} = 1 + 2x + 3x^2 + \cdots$ (ii) $\log(1-x) = -x - \dfrac{x^2}{2} - \dfrac{x^3}{3} - \cdots$

Verify that the three series have the same intervals of convergence.

8. Show that $\dfrac{1}{2}\log\dfrac{1+x}{1-x} = x + \dfrac{x^3}{3} + \dfrac{x^5}{5} + \cdots + \dfrac{x^{2n+1}}{2n+1} + \cdots$.

Use this result to evaluate log 2, log 3 and log 5 correct to four decimal places.

9. Show that $\sum_{k=0}^{\infty} a_k x^k \sum_{j=0}^{\infty} b_j x^j = \sum_{n=0}^{\infty} c_n x^n$ where $c_n = \sum_{k=0}^{n} a_k b_{n-k}$. Find the first few terms of the Taylor series about $x=0$ of $\log(1-x)\log(1+x)$.

10. By multiplying series together, verify the following identities.

(i) $e^x e^x = e^{2x}$ (ii) $\sin x \cos x = \dfrac{1}{2}\sin 2x$ (iii) $\cos^2 x = \dfrac{1}{2}(\cos 2x + 1)$

11. Find the Taylor series expansions about $x=0$ for the following functions from their definitions as integrals. Determine the intervals of convergence.

(i) $\operatorname{erf}(x) = \dfrac{2}{\pi^{1/2}}\displaystyle\int_0^x e^{-t^2}\,dt$ (ii) $\sin^{-1}x = \displaystyle\int_0^x \dfrac{dt}{\sqrt{1-t^2}}$

(iii) $\sinh^{-1}x = \displaystyle\int_0^x \dfrac{dt}{\sqrt{1+t^2}}$ (iv) $\tanh^{-1}x = \displaystyle\int_0^x \dfrac{dt}{1-t^2}$

12. Evaluate the limits as $x \to 0$ of the following functions by considering the first few terms of Taylor's series expansions about $x=0$.

(i) $\dfrac{\log(1+x)}{e^x - 1}$ (ii) $\dfrac{x(\sqrt{1+x^2}-1)}{\sin^3 x}$ (iii) $\dfrac{\cos x - \sqrt{1-x^2}}{x^4}$

(iv) $\dfrac{1-\cosh x}{1-\cos x}$ (v) $\dfrac{\log[x + (1+x^2)^{1/2}]}{\log(1+x)}$ (vi) $\dfrac{x^2 + 2\log\cos x}{x^2 + 6\log([\sin x]/x)}$

13. Use partial fractions to find the Taylor series about $x=0$ of the following functions. Find the values of x for which the Taylor series converge.

(i) $\dfrac{x}{x^2-3x+2}$ (ii) $\dfrac{2x+1}{x^2+4x+3}$ (iii) $\dfrac{1}{x^2+x-2}$

14. Find the Taylor series expansions about $x=0$ and $x=2$ of $f(x) = 1/(x-1)(x-4)$. Is there a valid expansion of this function in powers of $x-1$ or $x-4$?

15. Use the binomial theorem to find the Taylor series expansions of $(1+x^2)^{1/2}$ and $(1+x^2)^{-1/2}$ about $x=0$. Show that the second series can be obtained by differentiating the first series.

16. If the power series of a function $y = f(x) = x + a_2x^2 + a_3x^3 + \cdots$ is given, show how to find the power series expansion of the inverse function $x = f^{-1}(y) = y + b_2y^2 + b_3y^3 + \cdots$. (This process is known as *inversion* of series.)

17. By the method of inversion, determine power series expansions of the inverses of the following functions:

(i) $\sin x$ (ii) $\tan^{-1}x$ (iii) $\log(1+x)$

18. Find the Taylor series expansions about $x=0$ of the functions $\sin(x+\pi/6)$ and $\cos(x+\pi/4)$. Evaluate sin 31° and cos 46° correct to four decimal places.

19. Find the first three terms of the Taylor series expansion about $x=0$ of the function $y = (1+x)^{1/x}$. (Note: Use $\log y = [\log(1+x)]/x$ to evaluate derivatives of y.)

20. Use Taylor series expansions to evaluate the following definite integrals correct to three decimal places. Compare with the exact values where possible.

(i) $\int_0^{1/2} e^{-x^2}\,dx$ **(ii)** $\int_0^1 xe^{-x^2}\,dx$ **(iii)** $\int_0^{1/4} \log(1+x)\,dx$

(iv) $\int_0^{1/2} \frac{dx}{(1-x^2)^{1/2}}$ **(v)** $\int_0^{1/2} \frac{\log(1+x)}{x}\,dx$ **(vi)** $\int_0^1 \frac{dx}{(4+x^2)^{1/2}}$

21. Find the first three terms of the Taylor series about $x=0$ of the following functions:

(i) $\log\cos x$ **(ii)** $\log\frac{\sin x}{x}$ **(iii)** $\log\frac{\tan x}{x}$

How are these three series related?

22. Verify that the series $\sum_{n=1}^{\infty}\frac{\sin nx}{n}$ converges for all x but its derivative does not converge for any value of x. $\left[\textit{Hint: } 2\sin nx \sin\frac{x}{2} = \cos\left(\frac{2n-1}{2}x\right) - \cos\left(\frac{2n+1}{2}x\right).\right]$

23. Integrate the inequality $\cos x \le 1$ from 0 to x (any positive number) to obtain the inequality $\sin x < x$. Integrating again, show that $1 - x^2/2 < \cos x \le 1$. By repeated integration from 0 to x, verify the following:

(i) $x - \frac{x^3}{3!} + \frac{x^5}{5!} - \cdots - \frac{x^{4n-1}}{(4n-1)!} < \sin x < x - \frac{x^3}{3!} + \frac{x^5}{5!} - \cdots + \frac{x^{4n+1}}{(4n+1)!}$

(ii) $1 - \frac{x^2}{2!} + \frac{x^4}{4!} - \cdots - \frac{x^{4n-2}}{(4n-2)!} < \cos x < 1 - \frac{x^2}{2!} + \frac{x^4}{4!} - \cdots + \frac{x^{4n}}{(4n)!}$

24. Beginning with the inequality $e^{-x} < 1$ for $x > 0$, derive the Taylor series expansion of e^{-x} by repeated integration. Show that the truncation error is less than the first term omitted.

25. Derive the binomial expansion of $(1+x)^{\alpha}$ for $x>0$ and $\alpha<0$ by the following method. Integrate the inequality $(1+x)^{\alpha-1} < 1$ from 0 to x to obtain the result $1+\alpha x < (1+x)^{\alpha}$. Since this result holds for any $\alpha < 0$, it is valid with α replaced by $\alpha - 1$, $1 + (\alpha-1)x < (1+x)^{\alpha-1}$. Integrate to find $(1+x)^{\alpha} < 1 + \alpha x + \frac{\alpha(\alpha-1)}{2}x^2$. Repeat this argument to obtain the binomial series.

4.4 Applications: calculations

In this section we illustrate the use of power series in a few common situations. For the most part, the problems examined are purged of any physical content so that attention can be directed to the analytical techniques involved.

Consider first the use of infinite series to find the solution of the simple differential equation

$$y'' - y = 0. \tag{1}$$

In order to solve this, we assume that the unknown function, $y(x)$, can be represented as an infinite series

$$y(x) = a_0 + a_1 x + a_2 x^2 + \cdots = \sum_{n=0}^{\infty} a_n x^n, \tag{2}$$

and use (1) to determine the coefficients. Since

$$y''(x) = 2a_2 + 3\cdot 2a_3 x + 4\cdot 3a_4 x^2 + \cdots = \sum_{n=2}^{\infty} a_n n(n-1)x^{n-2},$$

the result of substituting the series for y and y'' in (1) is

$$(2a_2 + 3 \cdot 2a_3 x + 4 \cdot 3a_4 x^2 + \cdots) - (a_0 + a_1 x + a_2 x^2 + \cdots) = 0.$$

Upon rearranging this in powers of x we obtain

$$(2a_2 - a_0) + (3 \cdot 2a_3 - a_1)x + (4 \cdot 3a_4 - a_2)x^2 + \cdots \\ + [n(n-1)a_n - a_{n-2}]x^{n-2} + \cdots = 0. \quad (3)$$

According to the identity theorem for series, the coefficient of each power of x must be zero. (The "series" on the right is $0 + 0 \cdot x + 0 \cdot x^2 + \cdots$.) Therefore

$$2 \cdot 1a_2 = a_0, \qquad 3 \cdot 2a_3 = a_1,$$
$$4 \cdot 3a_4 = a_2, \qquad 5 \cdot 4a_5 = a_3,$$

and in general $n(n-1)a_n = a_{n-2}$. The coefficients with even indices can be given entirely in terms of a_0, and we find that

$$a_2 = \frac{1}{2}a_0, \qquad a_4 = \frac{1}{4 \cdot 3}a_2 = \frac{a_0}{4!}, \qquad \ldots, \qquad a_{2n} = \frac{a_0}{(2n)!}.$$

Likewise, the coefficients with odd indices are given by

$$a_{2n+1} = \frac{a_1}{(2n+1)!}.$$

The solution of the basic equation is then

$$y = a_0\left(1 + \frac{x^2}{2!} + \frac{x^4}{4!} + \frac{x^6}{6!} + \cdots\right) + a_1\left(x + \frac{x^3}{3!} + \frac{x^5}{5!} + \cdots\right), \quad (4)$$

where the two coefficients a_0, a_1 are completely arbitrary. If we write

$$a_0 = A + B, \qquad a_1 = A - B,$$

(4) can be expressed as

$$y = A\left(1 + x + \frac{x^2}{2!} + \frac{x^3}{3!} + \cdots\right) + B\left(1 - x + \frac{x^2}{2!} - \frac{x^3}{3!} + \cdots\right). \quad (5)$$

Each infinite series is then a recognizable function, and the general solution is seen to be a sum of two exponentials

$$y(x) = Ae^x + Be^{-x}. \quad (6)$$

Of course, this can now be checked by direct substitution. Note that two conditions must be given in order to determine both arbitrary constants; for example, $y(0) = 1$, $y'(0) = 0$ implies $A + B = 1$, $A - B = 0$ or $A = B = 1/2$. (The solution then is $y(x) = \cosh x$.)

The general method, as illustrated in this simple example, is a little longer than strictly necessary, but it is a procedure applicable to very complicated problems. Moreover, the expansion can be centered about any point, $x = a$, in order to develop representations valid in different ranges of the independent variable. (There are, however, points about which the function cannot be expressed as a power series; these are called, appropriately enough, *singular points.*)

2

Example 1 Find the general solution of $y'' - xy' - y = 0$.

To solve this differential equation, we write

$$y = a_0 + a_1 x + a_2 x^2 + \cdots + a_n x^n + \cdots,$$

so that

$$y' = a_1 + 2a_2 x + \cdots + na_n x^{n-1} + \cdots,$$
$$y'' = 2a_2 + 3 \cdot 2a_3 x + \cdots + n(n-1)a_n x^{n-2} + \cdots.$$

The substitution of these series in the differential equation yields

$$2a_2 + 3 \cdot 2a_3 x + \cdots - x(a_1 + 2a_2 x + \cdots) - (a_0 + a_1 x + a_2 x^2 + \cdots) = 0,$$

which can be arranged in ascending powers of x to obtain

$$(2a_2 - a_0) + (3 \cdot 2a_3 - 2a_1)x + \cdots + [n(n-1)a_n - (n-1)a_{n-2}]x^{n-2} + \cdots = 0.$$

Therefore

$$a_2 = \frac{a_0}{2}, \qquad a_3 = \frac{a_1}{3}, \qquad a_4 = \frac{a_2}{4}, \qquad \ldots, \qquad a_n = \frac{1}{n} a_{n-2}.$$

The coefficients with even indices are

$$a_2 = \frac{a_0}{2}, \qquad a_4 = \frac{a_2}{4} = \frac{a_0}{4 \cdot 2}, \qquad \ldots, \qquad a_{2n} = \frac{a_0}{2^n n!}, \qquad \ldots;$$

similarly, we find that

$$a_3 = \frac{a_1}{3}, \qquad a_5 = \frac{a_3}{5} = \frac{2^2 2}{5!} a_1, \qquad \ldots, \qquad a_{2n+1} = \frac{2^n n!}{(2n+1)!} a_1.$$

Consequently, the general solution can be written

$$y(x) = a_0\left[1 + \frac{x^2}{2} + \frac{x^4}{2!2^2} + \cdots + \frac{1}{n!}\left(\frac{x^2}{2}\right)^n + \cdots\right]$$
$$+ a_1\left[x + \frac{x^3}{3} + \frac{x^5}{5 \cdot 3} + \cdots + \frac{2^n n!}{(2n+1)!} x^{2n+1} + \cdots\right].$$

The series in the first brackets is readily identified as the function $e^{x^2/2}$, which can be shown to be a solution of the differential equation by direct substitution. The second series represents a new function, call it $Y(x)$, which is not expressible in simple terms. The ratio test shows that the series for $Y(x)$ converges for all x.

The arbitrary constants a_0, a_1 can be determined by setting two conditions, for example, $y(0) = 1$, $y'(0) = 0$ or perhaps $y(0) = 1$, $y(1) = 0$, or some other combination.

---

3 A solution of a problem that contains a small number ε is often developed as a series in powers of the parameter, called a *perturbation series*. In this circumstance, the coefficient of the general term ε^n is really a function of all other variables in the problem. As a simple illustration, we return to a problem considered in Chap. 1, page 76, that of locating the roots of the quadratic

$$x^2 + \varepsilon x - 4 = 0. \tag{7}$$

For ε equal to zero, the roots are clearly ± 2, and for ε small, these roots should change only slightly, as demonstrated earlier. The formal approach now is to set

$$x = a_0 + a_1\varepsilon + a_2\varepsilon^2 + \cdots + a_n\varepsilon^n + \cdots$$

and to determine the coefficients directly from the quadratic, (7). Since

$$\begin{aligned} x^2 &= (a_0 + a_1\varepsilon + \cdots)(a_0 + a_1\varepsilon + \cdots), \\ &= a_0^2 + 2a_1a_0\varepsilon + (2a_2a_0 + a_1^2)\varepsilon^2 + \cdots, \end{aligned}$$

and

$$\varepsilon x = a_0\varepsilon + a_1\varepsilon^2 + a_2\varepsilon^3 + \cdots,$$

it follows that

$$x^2 + \varepsilon x - 4 = (a_0^2 - 4) + (2a_1a_0 + a_0)\varepsilon + (2a_2a_0 + a_1^2 + a_1)\varepsilon^2 + \cdots.$$

In order for this series to equal zero *identically* for all values of ε, the coefficient of each power ε^n must itself be zero. Therefore $a_0^2 - 4 = 0$, which implies that $a_0 = \pm 2$; and

$$\begin{aligned} 2a_1a_0 + a_0 &= 0 \qquad \text{implies} \qquad a_1 = -\tfrac{1}{2}; \\ 2a_2a_0 + a_1^2 + a_1 &= 0 \qquad \text{implies} \qquad a_2 = \pm\tfrac{1}{16}; \\ &\cdots\cdots\cdots\cdots\cdots\cdots \end{aligned}$$

The two roots are then given by

$$x = \pm 2 - \tfrac{1}{2}\varepsilon \pm \tfrac{1}{16}\varepsilon^2 \pm \tfrac{1}{1{,}024}\varepsilon^4 + \cdots. \tag{8}$$

We already know how to solve quadratic equations, and so it is easy to identify this series as the function

$$x = -\frac{\varepsilon}{2} \pm 2\left(1 + \frac{\varepsilon^2}{16}\right)^{1/2}. \tag{9}$$

Otherwise, it would take a great deal of effort to make this identification by finding the general term of the expansion (8) (which would be necessary anyway in order to prove convergence).

4 Series expansions provide an important means of evaluating integrals. One procedure is to replace the integrand (or a part of it) by the associated Taylor series expan-

sion about a convenient point. To illustrate this technique, we consider the evaluation of

$$I(x) = \int_0^x \sin t^2 \, dt, \tag{10}$$

which cannot be expressed in closed form, i.e., as a finite composition of known functions.

Certainly for sufficiently small x it pays to develop the integral as a series expansion about zero. Since

$$\sin t^2 = t^2 - \frac{(t^2)^3}{3!} + \cdots + \frac{(-1)^{n-1}(t^2)^{2n-1}}{(2n-1)!} + \cdots, \tag{11}$$

the substitution of this series in (10) followed by a term-by-term integration yields

$$I(x) = \int_0^x t^2 \, dt - \frac{1}{3!}\int_0^x t^6 \, dt + \cdots + \frac{(-1)^{n-1}}{(2n-1)!}\int_0^x t^{4n-2} \, dt + \cdots$$

or

$$I(x) = \frac{x^3}{3} - \frac{x^7}{7 \cdot 3!} + \cdots + \frac{(-1)^{n-1}x^{4n-1}}{(4n-1)(2n-1)!} + \cdots. \tag{12}$$

Application of the ratio test and evaluation of the relevant limit,

$$\lim_{n\to\infty}\left| \frac{(-1)^n x^{4n+3}}{(4n+3)(2n+1)!} \bigg/ \frac{(-1)^{n-1}x^{4n-1}}{(4n-1)(2n-1)!} \right| = \lim_{n\to\infty}\left| \frac{4n-1}{4n+3} \frac{x^4}{(2n+1)(2n)} \right| = 0,$$

show that this alternating series converges absolutely for all x. Moreover, convergence is fairly rapid, as shown by the bound for the remainder after n terms (which is less than the first term not taken):

$$|R_n| < \frac{|x|^{4n+3}}{(4n+3)(2n+1)!}.$$

However, for *very large* values of $|x|$, n also must be relatively large in order to obtain sufficient accuracy. For $x = 1$, two terms of the series yield a sum that is accurate to 0.001, or, to be precise,

$$|R_2| < \frac{1}{11 \cdot 5!} = 0.000757575\cdots.$$

In order to achieve the same accuracy when $x = 3$, approximately 14 terms of the series must be summed. This is not difficult for a computing machine, but it requires some effort by hand calculation. The graph of $I(x)$ is shown in Fig. 4.1. As x increases, the computations yield a limiting value

$$\lim_{x\to\infty} I(x) = 0.6266\cdots.$$

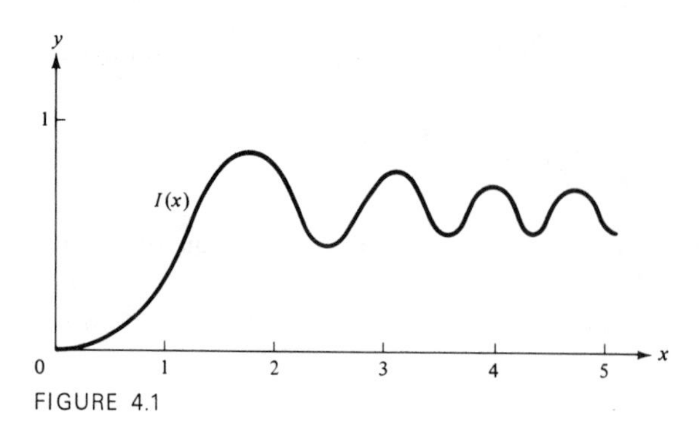

FIGURE 4.1

It can be shown that $\int_0^\infty \sin t^2\, dt = \tfrac{1}{2}(\pi/2)^{1/2}$. The fact that this improper integral is actually convergent can be established by first writing

$$I(\infty) = \int_0^\infty \sin t^2\, dt = \int_0^1 \sin t^2\, dt + \int_1^\infty \sin t^2\, dt$$

and then performing a partial integration in the last integral, with $u = 1/t$, $dv = t \sin t^2\, dt$, to obtain

$$I(\infty) = \int_0^1 \sin t^2\, dt - \frac{\cos t^2}{2t}\bigg]_1^\infty - \frac{1}{2}\int_1^\infty \frac{\cos t^2}{t^2}\, dt.$$

The improper integral appearing on the extreme right in this expression is easily bounded:

$$\left|\int_1^\infty \frac{\cos t^2}{t^2}\, dt\right| \le \int_1^\infty \frac{dt}{t^2} = 1,$$

and this completes the demonstration of convergence.

The integral can also be calculated numerically from the differential equation

$$\frac{dI}{dx} = \sin x^2, \qquad I(0) = 0,$$

using Simpson's rule.

Example 2 Find the arc length of the ellipse given by $x = a \cos t$, $y = b \sin t$ with $b < a$ (Fig. 4.2).

The element of arc length is given by

$$ds^2 = dx^2 + dy^2 = (a^2 \sin^2 t + b^2 \cos^2 t)\, dt^2$$

or

$$ds = (a^2 \sin^2 t + b^2 \cos^2 t)^{1/2}\, dt.$$

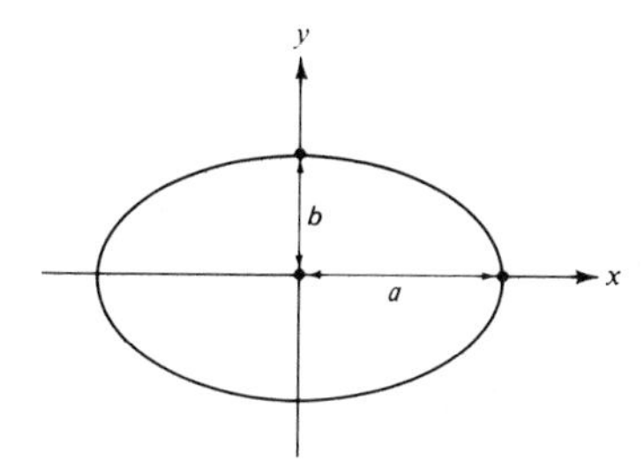

FIGURE 4.2

The arc length of the curve is 4 times that in the first quadrant, so that

$$s = 4\int_0^{\pi/2} (a^2 \sin^2 t + b^2 \cos^2 t)^{1/2}\, dt$$

or

$$s = 4a \int_0^{\pi/2} \left[1 - \left(1 - \frac{b^2}{a^2}\right)\cos^2 t\right]^{1/2} dt. \tag{13}$$

The integral that appears here is a function of the eccentricity k defined by

$$k^2 = 1 - \frac{b^2}{a^2}, \tag{14}$$

and may be written in the equivalent form

$$\mathscr{E}(k) = \int_0^{\pi/2} (1 - k^2 \cos^2 t)^{1/2}\, dt = \int_0^{\pi/2} (1 - k^2 \sin^2 t)^{1/2}\, dt. \tag{15}$$

The function $\mathscr{E}(k)$, called a *complete elliptic integral*, cannot be expressed in simple terms. In this terminology,

$$s = 4a\mathscr{E}(k).$$

In order to evaluate $\mathscr{E}(k)$, two modes of analysis are possible: numerical integration for a range of values

$$0 \le k \le 1$$

or utilization of an infinite series, basically a perturbation series that is valid for small k (when $a \approx b$ and the ellipse is almost circular).

Consider the latter approach first and let k be a small positive number. The integrand of either integral in (15) can be expanded in powers of k^2 using the binomial series:

$$(1 - h)^{1/2} = 1 - \frac{h}{2} - \sum_{n=2}^{\infty} \frac{(2n)!}{(2^n n!)^2(2n - 1)} h^n.$$

If, in particular, we set $h = k^2 \sin^2 t$ (with $h < 1$) and replace the integrand of (15) by its series expansion, the result is

$$\begin{aligned}
\mathscr{E}(k) &= \int_0^{\pi/2} (1 - k^2 \sin^2 t)^{1/2}\, dt, \\
&= \int_0^{\pi/2} \left[1 - \frac{k^2}{2}\sin^2 t - \sum_{n=2}^{\infty} \frac{(2n)!}{(2^n n!)^2 (2n-1)} k^{2n} \sin^{2n} t\right] dt, \\
&= \frac{\pi}{2} - \frac{k^2}{2}\int_0^{\pi/2} \sin^2 t\, dt - \sum_{n=2}^{\infty} \frac{(2n)!\, k^{2n}}{(2^n n!)^2 (2n-1)} \int_0^{\pi/2} \sin^{2n} t\, dt.
\end{aligned}$$

Since

$$\int_0^{\pi/2} \sin^{2n} t\, dt = \frac{1 \cdot 3 \cdots (2n-1)}{2 \cdot 4 \cdot 6 \cdots (2n)} \frac{\pi}{2} = \frac{(2n)!}{(2^n n!)^2} \frac{\pi}{2},$$

the final formula for the elliptic integral is

$$\mathscr{E}(k) = \frac{\pi}{2}\left\{1 - \frac{k^2}{4} - \sum_{n=2}^{\infty} \left[\frac{(2n)!}{(2^n n!)^2}\right]^2 \frac{k^{2n}}{2n-1}\right\}. \tag{16}$$

Another limiting case occurs when k is nearly 1 (or b is very small) and the ellipse is almost flat, as shown in Fig. 4.3. The illustration indicates clearly that the arc length should be nearly double the distance from a to $-a$, that is, $s = 4a$. This result is indeed obtained at once by setting $k = 1$ in (13) and performing the integration. But *it is not possible* to find a series representation for $\mathscr{E}(k)$ in powers of $1 - k$. The perturbation takes another quite complicated form. However, for definite values of k, in this vicinity and elsewhere, the integral can be computed numerically, using Simpson's rule (Sec. 3.10). The two methods are combined in practice to give an accurate description of $\mathscr{E}(k)$ over the entire range $|k| \le 1$, and the elliptic integral $\mathscr{E}(k)$ is graphed in Fig. 4.4.

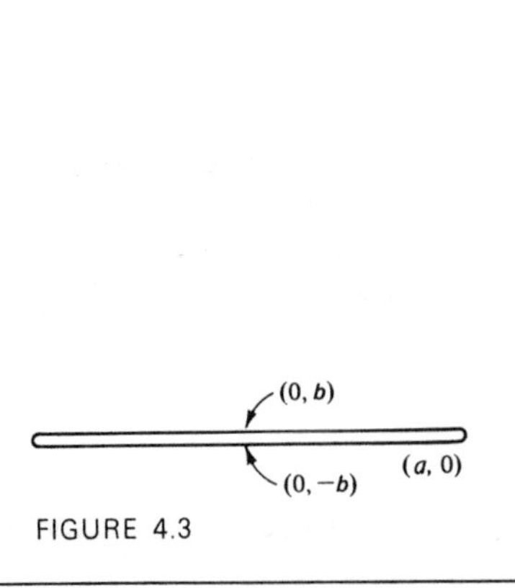

FIGURE 4.3

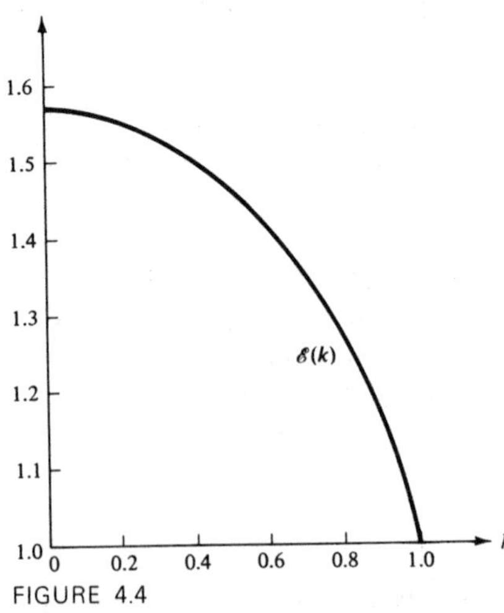

FIGURE 4.4

6

We conclude this discussion by illustrating a few techniques for actually summing series. Closed-form expressions are sometimes possible, but a more realistic enterprise is to work to improve the rate of convergence so that any series can be evaluated accurately with *as few terms as possible.* The advantages in doing this are rather obvious.

Certain simple series can be evaluated by rearranging terms. For example

$$S = \sum_{n=1}^{\infty} \frac{1}{n(n+1)}$$

can be written

$$S = \sum_{n=1}^{\infty} \left(\frac{1}{n} - \frac{1}{n+1} \right).$$

In this form, we observe that consecutive terms cancel automatically:

$$\begin{aligned} S &= (1 - \tfrac{1}{2}) + (\tfrac{1}{2} - \tfrac{1}{3}) + (\tfrac{1}{3} - \tfrac{1}{4}) + \cdots \\ &= 1 + (\tfrac{1}{2} - \tfrac{1}{2}) + (\tfrac{1}{3} - \tfrac{1}{3}) + (\tfrac{1}{4} - \tfrac{1}{4}) + \cdots = 1. \end{aligned}$$

Therefore

$$\sum_{n=1}^{\infty} \frac{1}{n(n+1)} = 1. \tag{17}$$

In a similar fashion, we can prove (see Exercise 18) that

$$\sum_{n=1}^{\infty} \frac{1}{n(n+1)(n+2)} = \frac{1}{4}, \tag{18}$$

or

$$\sum_{n=1}^{\infty} \frac{1}{n(n+1)(n+2)(n+3)} = \frac{1}{18}. \tag{19}$$

More generally,

$$\sum_{n=1}^{\infty} \frac{1}{(a+n-1)(a+n)} = \frac{1}{a} \qquad \text{for } a \neq 0, -1, -2, \ldots,$$

and with the same restrictions on a,

$$\sum_{n=1}^{\infty} \frac{1}{(a+n-1)(a+n)(a+n+1)} = \frac{1}{2a(a+1)}.$$

7

The evaluation of many series can be determined from the known Taylor series expansions of simple functions, which, like integrals, are tabulated and made accessible. Thus, the expansion

$$\log(1 + x) = x - \frac{x^2}{2} + \frac{x^3}{3} + \cdots$$

yields the specific formula

$$\log 2 = 1 - \tfrac{1}{2} + \tfrac{1}{3} - \tfrac{1}{4} + \cdots = \sum_{n=1}^{\infty} \frac{(-1)^{n+1}}{n}. \tag{20}$$

Integrating or differentiating a power series will often lead to an explicit formula for a particular series. For example, the value of

$$S = 1 - \tfrac{1}{4} + \tfrac{1}{7} + \cdots + \frac{(-1)^n}{3n + 1} + \cdots$$

is determined from

$$J(x) = x - \frac{x^4}{4} + \cdots = \sum_{n=0}^{\infty} \frac{(-1)^n x^{3n+1}}{3n + 1}, \tag{21}$$

by setting $x = 1$, that is,

$$S = \lim_{x \to 1} J(x) = J(1).$$

However, $J(x)$ can be manipulated and actually summed, and this in turn yields a simple formula for S. The series obtained by differentiating $J(x)$ is readily identified as a known function:

$$J'(x) = \sum_{n=0}^{\infty} (-1)^n x^{3n} = 1 - x^3 + x^6 + \cdots = \frac{1}{1 + x^3}.$$

Upon reversing the procedure and integrating, we obtain a representation of the series (21) as the integral

$$J(x) = \int_0^x \frac{dx}{1 + x^3},$$

where the particular value $J(0) = 0$ has been used to set the lower integration limit. Therefore, the sum of the original series is given by the *definite* integral,

$$S = \int_0^1 \frac{dx}{1 + x^3},$$

which can be evaluated by a partial-fraction expansion. It follows from the identity

$$\frac{1}{x^3 + 1} = \frac{1}{(x + 1)(x^2 - x + 1)} = \frac{1}{3(x + 1)} - \frac{1}{3}\frac{x - 2}{x^2 - x + 1}$$

that

$$S = \frac{1}{3}\int_0^1 \frac{dx}{x+1} - \frac{1}{3}\int_0^1 \frac{x-2}{x^2-x+1}\,dx.$$

The calculation yields

$$S = \sum_{n=0}^{\infty} \frac{(-1)^n}{3n+1} = \frac{1}{3}\log 2 + \frac{\pi}{3\sqrt{3}}. \tag{22}$$

---

8 Once a series is summed, such as any of those above, it can be used to improve the rate of convergence of other closely related series. Consider the evaluation of $I = \sum_{n=1}^{\infty} u_n$ and let $S = \sum_{n=1}^{\infty} a_n$ be a known series related to this in the sense that

$$\lim_{n\to\infty} \frac{u_n}{a_n} = \beta \neq 0.$$

If we then write

$$\sum_{n=1}^{\infty} u_n = \sum_{n=1}^{\infty} (u_n - \beta a_n) + \beta \sum_{n=1}^{\infty} a_n,$$

$$= \sum_{n=1}^{\infty} \left(1 - \frac{\beta a_n}{u_n}\right) u_n + \beta S, \tag{23}$$

the new series on the right will converge *faster* than the original one because a small coefficient of u_n, that is, $1 - \beta a_n/u_n$, has been introduced which approaches zero as n becomes infinite.

Example 3 Evaluate $I = \sum_{n=1}^{\infty} \frac{1}{n^2}$.

In this case convergence is improved using the series in (17)

$$S = \sum_{1}^{\infty} \frac{1}{n(n+1)} = 1.$$

In the terminology of (23), we identify $a_n = 1/n(n+1)$ and $u_n = 1/n^2$ so that

$$\lim_{n\to\infty} \frac{u_n}{a_n} = \beta = 1.$$

Therefore, the series I, like that in (23), is rewritten as

$$I = \sum_{n=1}^{\infty} \frac{1}{n^2} = \sum_{n=1}^{\infty} \left[\frac{1}{n^2} - \frac{1}{n(n+1)}\right] + 1,$$

$$= \sum_{n=1}^{\infty} \frac{1}{n^2(n+1)} + 1 = 1 + \tfrac{1}{2} + \tfrac{1}{12} + \tfrac{1}{36} + \cdots.$$

The general term of the new infinite series decays to zero like $1/n^3$ for n large, which is a much faster rate than what we started with, namely, $1/n^2$. Equation (18) can now be used to improve the convergence still further, but this is left to the exercises.

In the course of present and future work, many series will be evaluated explicitly. For example, certain advanced techniques yield the exact result $\sum_{n=1}^{\infty} \frac{1}{n^2} = \frac{\pi^2}{6}$, as well as a host of others, all of which can be used in this type of analysis. A short list of known series is given in Table 4.1. We make no attempt here to derive these formulas, but they are useful nonetheless.

TABLE 4.1

Sums of series $\sum_{n=1}^{\infty} f(n)$

$f(n)$	Sum	$f(n)$	Sum
$\frac{1}{n^2}$	$\frac{\pi^2}{6}$	$\frac{1}{n^2+a^2}$	$\frac{\pi}{2a}\left(\coth \pi a - \frac{1}{\pi a}\right)$
$\frac{1}{(2n-1)^2}$	$\frac{\pi^2}{8}$	$\frac{1}{n^2-a^2}$	$\frac{\pi}{2a}\left(\frac{1}{\pi a} - \cot \pi a\right)$
$\frac{1}{(2n)^2}$	$\frac{\pi^2}{24}$	$\frac{(-1)^{n-1}}{n}$	$\log 2$
$\frac{1}{n^4}$	$\frac{\pi^4}{90}$	$\frac{(-1)^{n-1}}{n^2}$	$\frac{\pi^2}{12}$
$\frac{1}{(2n-1)^4}$	$\frac{\pi^4}{96}$	$\frac{(-1)^{n-1}}{n^4}$	$\frac{7}{720}\pi^4$
$\frac{1}{(2n)^4}$	$\frac{\pi^4}{1{,}440}$		

Exercises 4.4

1. Show by direct substitution that $y = e^{x^2/2}$ is a solution of $y'' - xy' - y = 0$. From the differential equation and the initial conditions $y(0) = 1$, $y'(0) = 0$, derive the Taylor series expansion of $e^{x^2/2}$ about $x = 0$.

2. Show that if $n(n-1)a_n = a_{n-2}$ for $n \geq 2$, then

$$a_{2n} = \frac{a_0}{2n!} \text{ and } a_{2n+1} = \frac{a_1}{(2n+1)!}.$$

Determine the functions $\sum_{n=0}^{\infty} a_{2n}x^{2n}$ and $\sum_{n=0}^{\infty} a_{2n+1}x^{2n+1}$ in closed form.

3. Find the interval of convergence of the series

$$Y(x) = \sum_{n=0}^{\infty} \frac{2^n n!}{(2n+1)!} x^{2n+1}.$$

Show that $Y'(x) = xY(x) + 1$.

4. Use a power-series expansion to solve $y'(x) - 2y(x) = x$ with $y(0) = 2$. What is the interval of convergence? Write the equation as

$$\frac{d}{dx}[e^{-2x}y(x)] = e^{-2x}x$$

and integrate to find the solution explicitly. Check your first answer.

5. Use power series to find the general solution of $u''(x) - (1+x^2)u(x) = 0$. Determine the intervals of convergence of the two power series solutions.

6. Solve, using a power-series expansion, $y' + 2y + xy^2 = 0$ with $y(0) = 1/2$. In this case, show also that the substitution $u = 1/y$, reduces this problem to Exercise 4.

7. Use a series expansion to evaluate the Fresnal cosine integral $J(x) = \int_0^x \cos t^2 \, dt$. Find $J(\pi/4)$ to three decimal places.

8. Calculate the elliptic integral $\mathscr{E}(k) = \int_0^{\pi/2} (1 - k^2\cos^2 t)^{-1/2} \, dt$ using a series expansion of the integrand in powers of $h = k^2\cos^2 t$. Determine the interval of convergence of this series.

9. Calculate the arc length of the parabola $y = px^2$, $p > 0$, which lies below the line $y = 1$. Evaluate the integral exactly and also approximately for p very small or very large. Discuss the accuracy and utility of the approximate procedures. Illustrate with appropriate graphs.

10. Use series expansions to evaluate

(i) $\int_0^{0.5} \frac{x^{1/2}\,dx}{1+e^x}$ **(ii)** $\int_0^{0.5} \frac{e^{-x}}{1+x^2}\,dx$ **(iii)** $\int_0^1 \frac{x}{8+x^3}\,dx$

11. Use a perturbation series $x = \sum_{n=0}^{\infty} a_n\epsilon^n$ to find the root of $x^5 + \epsilon x^4 - 1 = 0$ which lies closest to $x = 1$. Is this series a Taylor series?

12. Use a perturbation series $y(x) = \sum_{n=0}^{\infty} \epsilon^n y_n(x)$ to solve the implicit equation $y + \epsilon \sin 2y = x$. Check your solution against that obtained by iteration in Sec. 1.11.

13. Sum the series

$$S = \frac{1}{2} + \frac{1}{3}\left(\frac{1}{2}\right)^3 + \frac{1}{5}\left(\frac{1}{2}\right)^5 + \cdots + \frac{1}{(2n-1)}\left(\frac{1}{2}\right)^{2n-1} + \cdots$$

by manipulating the power series

$$f(x) = x + \frac{x^3}{3} + \cdots + \frac{x^{2n-1}}{2n-1} + \cdots.$$

14. Evaluate $S = 1 - 1/5 + 1/9 - 1/13 + \cdots$ by considering the series

$$J(x) = x - \frac{x^5}{5} + \frac{x^9}{9} + \cdots = \sum_{n=0}^{\infty} \frac{(-1)^n x^{4n+1}}{4n+1}.$$

15. Use the power series for $\log(1+x)$ to calculate $\log 2$ by setting $x = 1$ and $x = -\tfrac{1}{2}$. Which is the more efficient series? Why?

16. Use the expansion $\log\dfrac{1+x}{1-x} = 2\left[x + \dfrac{x^3}{3} + \dfrac{x^5}{5} + \cdots\right]$ to evaluate $\log(\tfrac{4}{3})$ and $\log(\tfrac{9}{8})$ to seven decimal places. Use these numbers to evaluate $\log 2$ and $\log 3$ to at least six decimal places.

17. Find the sum of

$$\frac{1}{1\cdot 4\cdot 7} + \frac{1}{4\cdot 7\cdot 10} + \cdots + \frac{1}{(3n-2)(3n+1)(3n+4)}$$

by writing

$$\frac{1}{(3n-2)(3n+1)} - \frac{1}{(3n+1)(3n+4)} = \frac{6}{(3n-2)(3n+1)(3n+4)}.$$

18. Show that:

(i) $\displaystyle\sum_{n=1}^{\infty} \frac{1}{n(n+1)(n+2)} = \frac{1}{4}$ (ii) $\displaystyle\sum_{n=1}^{\infty} \frac{1}{n(n+1)(n+2)(n+3)} = \frac{1}{18}$

(iii) $\displaystyle\sum_{n=1}^{\infty} \frac{1}{(a+n-1)(a+n)} = \frac{1}{a}, \qquad a \neq 0, -1, \ldots$

(iv) $\displaystyle\sum_{n=1}^{\infty} \frac{1}{(a+n-1)(a+n)(a+n+1)} = \frac{1}{2a(a+1)}, \qquad a \neq 0, -1, \ldots$

19. The series $\displaystyle\sum_{n=1}^{\infty} \frac{1}{n(n+1)}$ was used to improve the convergence of $\displaystyle\sum_{n=1}^{\infty} \frac{1}{n^2}$. Use the series $\displaystyle\sum_{n=1}^{\infty} \frac{1}{n(n+1)(n+2)}$ to improve convergence still further. Evaluate the original sum correct to four decimal places.

20. Show that the series $\displaystyle\sum_{n=1}^{\infty} \frac{1}{(n-\tfrac{1}{2})(n+\tfrac{1}{2})}$ is a better choice to improve the convergence of $\displaystyle\sum_{n=1}^{\infty} \frac{1}{n^2}$. Evaluate the original sum correct to four decimal places.

21. Improve the convergence of the following series making use of series that have already been summed:

(i) $\displaystyle\sum_{n=1}^{\infty} (-1)^n \frac{n+1}{n^2+2}$ (ii) $\displaystyle\sum_{n=1}^{\infty} \frac{n-1}{n^3+n^2}$ (iii) $\displaystyle\sum_{n=1}^{\infty} \frac{(-1)^n}{n^2+2}$

4.5 *Asymptotic series*

1 In our earlier study of the error function (Sec. 3.10),

$$\operatorname{erf}(x) = \frac{2}{\pi^{1/2}} \int_0^x e^{-z^2}\, dz = 1 - \frac{2}{\pi^{1/2}} \int_x^{\infty} e^{-z^2}\, dz, \tag{1}$$

a formula was derived permitting fast and efficient calculation when x is a large number. That analysis was based on repeated integration by parts, utilizing the formula

$$\int_x^{\infty} \frac{e^{-z^2}}{z^n}\, dz = \frac{e^{-x^2}}{2x^{n+1}} - \frac{n+1}{2} \int_x^{\infty} \frac{e^{-z^2}}{z^{n+2}}\, dz. \tag{2}$$

The result of using this in (1) is

$$\operatorname{erf}(x) = 1 - \frac{e^{-x^2}}{x\pi^{1/2}}\left[1 - \frac{1}{2x^2} + \frac{1\cdot 3}{(2x^2)^2} + \cdots + \frac{(-1)^n 1\cdot 3\cdot 5\cdots(2n-1)}{(2x^2)^n}\right] + R_n(x), \quad (3)$$

where the remainder is given by

$$R_n(x) = (-1)^{n+1}\frac{1\cdot 3\cdot 5\cdots(2n+1)}{2^n\pi^{1/2}}\int_x^\infty \frac{e^{-z^2}}{z^{2n+2}}\,dz. \quad (4)$$

Equation (3), without the remainder, was shown to provide an extremely accurate approximation for large x, but it was noted earlier, without explanation, that certain peculiarities were apparent. These can now be discussed, and to begin we examine the convergence of the infinite series that results in the formal limit as n approaches infinity. Naturally, we expect a valid representation of the error function to emerge:

$$\operatorname{erf}(x) = 1 - \frac{e^{-x^2}}{x\pi^{1/2}}\left[1 - \frac{1}{2x^2} + \cdots + (-1)^n\frac{1\cdot 3\cdot 5\cdots(2n-1)}{(2x^2)^n} + \cdots\right]. \quad (5)$$

However, the infinite series in the brackets can be shown to diverge by the ratio test, and to facilitate this analysis the general term is written concisely as

$$(-1)^n\frac{1\cdot 3\cdot 5\cdots(2n-1)}{(2x^2)^n} = (-1)^n\frac{(2n)!}{n!(2x)^{2n}}.$$

Therefore, the limit of the ratio of consecutive terms is

$$\lim_{n\to\infty}\left|\frac{(2n+2)!}{(n+1)!(2x)^{2n+2}}\Big/\frac{(2n)!}{n!(2x)^{2n}}\right| = \lim_{n\to\infty}\left|\frac{(2n+2)(2n+1)}{(n+1)(2x)^2}\right|,$$
$$= \frac{1}{x^2}\lim_{n\to\infty}\left(n + \frac{1}{2}\right) = \infty,$$

and as a consequence the series (5) is seen to be *divergent* for *all* x.

We are now faced with a real dilemma—a divergent series that seems to be extremely efficient in calculating the values of a function. The source of this difficulty and its resolution lie in the nature of the remainder given in Eq. (4), which must be examined more closely. The integral that appears there is easily bounded

$$\int_x^\infty \frac{e^{-z^2}}{z^{2n+2}}\,dz \le \frac{1}{x^{2n+3}}\int_x^\infty ze^{-z^2}\,dz = \frac{e^{-x^2}}{2x^{2n+3}},$$

so that

$$|R_n(x)| \le \frac{(2n+1)!}{2^{2n+1}n!\pi^{1/2}}\frac{e^{-x^2}}{|x|^{2n+3}}. \quad (6)$$

For a *fixed integer value of* n the remainder clearly approaches zero as $x \to \infty$;

$$\lim_{x\to\infty} R_n(x) = 0.$$

This is the reason that the *finite* sum in (3) is an accurate approximation of the function for large x.

On the other hand, for a *fixed value* of x, the remainder eventually increases beyond bound as n *becomes infinite* because of the dominance of the factor $(2n+1)!$ in the numerator of (4). This is described by

$$\lim_{n\to\infty} R_n(x) = \infty$$

and is equivalent to the assertion that the infinite series diverges. The peculiar behavior of the series can be ascribed to the fact that

$$\lim_{n\to\infty}\lim_{x\to\infty} R_n(x) \neq \lim_{x\to\infty}\lim_{n\to\infty} R_n(x). \tag{7}$$

Thus the two limits are *not* interchangeable (one is infinite; the other is zero), and for this reason an excellent approximation is obtained for large x by retaining only a finite number of terms whereas the infinite series diverges. In fact, for a definite value of x there is always an optimum number of terms of the series which yields the most accurate estimate. As the value of n increases beyond this optimum, the series becomes inaccurate and begins to diverge. However, note very carefully that the estimate obtained with only a few terms of the series may already be entirely satisfactory in regard to accuracy.

---

2 Although the series (5) does not converge, the equation still makes sense because the difference between the function and any partial sum can be made arbitrarily small when x is sufficiently large. This is the defining property of an *asymptotic series*, which is sometimes also called semiconvergent. Indeed, a series

$$f(x) = a_0 + \frac{a_1}{x} + \frac{a_2}{x^2} + \cdots, \tag{8}$$

is said to be asymptotic in x if

$$\lim_{x\to\infty} x^n[f(x) - s_n] = 0, \tag{9}$$

where s_n is the partial sum

$$s_n = a_0 + \frac{a_1}{x} + \cdots + \frac{a_n}{x^n}. \tag{10}$$

An equivalent statement to (9) is then

$$\lim_{x\to\infty} x^n R_n(x) = 0. \tag{11}$$

Asymptotic series may be added term by term, multiplied, and integrated, but differentiation can present problems. Although it is common to denote an asymptotic series by

$$f(x) \sim a_0 + \frac{a_1}{x} + \cdots,$$

we will retain the equality sign.

As (5) indicates, the asymptotic expansion of a function sometimes takes the form of a product of simple functions and an asymptotic series

$$f(x) = \Phi(x)\left(a_0 + \frac{a_1}{x} + \cdots\right).$$

In this case, the definition (9) is applied to the quotient $f(x)/\Phi(x)$.

The series for the error function in (5) is genuinely an asymptotic series in inverse powers of x^2. According to (6),

$$|(x^2)^n e^{x^2} R_n(x)| \leq \frac{(2n+1)!}{2^{2n+1} n! \pi^{1/2} |x^3|},$$

and it follows that

$$\lim_{x \to \infty} (x^2)^n e^{x^2} R_n(x) = 0,$$

which means that the definition applies. For $x = \sqrt{10}$ the estimates obtained by summing $n = 1, \ldots, 28$ terms of the series are presented in Table 5.1. Compared to the exact value of $xe^{x^2}[1 - \text{erf}(x)]$ at $x = \sqrt{10}$, which is 0.5394141, we see that $n = 9$ gives the best asymptotic estimate accurate to four decimal places.

TABLE 5.1

Asymptotic evaluation of $xe^{x^2}[1 - \text{erf}(x)]$ *for* $x = \sqrt{10}$ *using* (3)

n	*Asymptotic value*	n	*Asymptotic value*
0	0.5641895750	15	0.5393502710
1	0.5359800970	16	0.5395154720
2	0.5402115130	17	0.5392428860
3	0.5391536570	18	0.5397199010
4	0.5395238990	19	0.5388374100
5	0.5393572890	20	0.5405582710
6	0.5394489160	21	0.5370305030
7	0.5393893560	22	0.5446151940
8	0.5394340230	23	0.5275496310
9	0.5393960550	24	0.5676536850
10	0.5394321230	25	0.4693987510
11	0.5393942440	26	0.7199488430
12	0.5394378000	27	0.0559911099
13	0.5393833510	28	1.8818748290
14	0.5394568590		

3

An expansion in terms of a parameter, which may be small or large, is often asymptotic in nature. For example, an approximation for the Laplace integral

$$I(\lambda) = \int_0^\infty e^{-\lambda x} f(x)\, dx, \tag{12}$$

can be obtained easily on the assumption that λ is large and positive. The variable change $z = \lambda x$ yields

$$I(\lambda) = \frac{1}{\lambda} \int_0^\infty e^{-z} f\left(\frac{z}{\lambda}\right) dz, \tag{13}$$

and this motivates the next step. The argument z/λ is small in the important range of integration where e^{-z} is moderately valued, and if the Taylor series

$$f(h) = f(0) + hf'(0) + \cdots + \frac{h^n}{n!} f^{(n)}(0) + \cdots$$

replaces this function in the integrand, term-by-term integration leads to the formula

$$I(\lambda) = \frac{f(0)}{\lambda} + \sum_{n=1}^{\infty} \frac{f^{(n)}(0)}{n!\,\lambda^{n+1}} \int_0^\infty e^{-z} z^n\, dz$$

or

$$I(\lambda) = \sum_{n=0}^{\infty} \frac{f^{(n)}(0)}{\lambda^{n+1}}. \tag{14}$$

This is an asymptotic series which may or may not converge. Note that if $f(x)$ is a polynomial the asymptotic series $I(\lambda)$ has a finite number of terms and is convergent.

Example 1 If λ is large and positive, determine the asymptotic series for

$$J(\lambda) = \int_0^\infty \frac{e^{-\lambda x}}{1 + x^2}\, dx. \tag{15}$$

It is not necessary for the first step of the analysis to be a variable change, and we can simply substitute the series

$$\frac{1}{1 + x^2} = 1 - x^2 + x^4 - x^6 + \cdots$$

directly into the integral. (Note that this series converges for $|x| < 1$ only, whereas the range of integration is infinite.) Therefore

$$J(\lambda) = \int_0^\infty e^{-\lambda x} \sum_{n=0}^{\infty} (-1)^n x^{2n}\, dx,$$

$$= \sum_{n=0}^{\infty} (-1)^n \int_0^\infty e^{-\lambda x} x^{2n}\, dx,$$

or

$$J(\lambda) = \sum_{n=0}^{\infty} (-1)^n \frac{(2n)!}{\lambda^{2n+1}},$$

which is the asymptotic series desired. This typifies the formal procedure found in practice. However, to be absolutely sure about the classification as asymptotic, definition (9) or (11) must be applied. For this purpose we use the identity

$$\frac{1}{1+x^2} = 1 - x^2 + x^4 - \cdots + (-1)^n x^{2n} + \frac{(-1)^{n+1} x^{2n+2}}{1+x^2}$$

instead of the infinite series in (15). This leads to the exact formula

$$J(\lambda) = \frac{1}{\lambda} - \frac{2}{\lambda^3} + \cdots + \frac{(-1)^n (2n)!}{\lambda^{2n+1}} + (-1)^{n+1} \int_0^\infty \frac{x^{2n+2}}{1+x^2} e^{-\lambda x}\, dx.$$

A bound for the remainder is given by

$$\left| \int_0^\infty \frac{x^{2n+2}}{1+x^2} e^{-\lambda x}\, dx \right| \leq \int_0^\infty x^{2n+2} e^{-\lambda x}\, dx = \frac{(2n+2)!}{\lambda^{2n+3}},$$

which clearly satisfies all the criteria of an asymptotic series.

The practical advantages of asymptotic series make them a highly desirable and common form of solution to problems. Analyses centered on the condition that a parameter or variable be small (or large) naturally lead to series of this type, and since so much of science proceeds from such assumptions, the importance of asymptotic series cannot be overestimated. It is paradoxical, to say the least, that so much effort is expended in studying convergent series when in practice convergence is rarely proved, and divergence, in the form of asymptotic series, may actually be preferred.

Exercises 4.5

1. Consider the integral $I(x) = \int_0^x \frac{dt}{1+t^2}$ for $x>0$. Obtain approximate values of this integral valid when x is small and when x is large. Compare these results with the exact values obtained by integrating. Use $x = 0.1$ and $x = 100$ as numerical examples.

2. Consider the integral $I(x) = \int_0^x \frac{t}{1+t^3}\, dt$ for $x>0$. Obtain approximate values valid when x is small and when x is large. Compare with the exact values obtained by integration using partial fractions. Estimate $I(0.1)$ and $I(100)$ and compare to the accurate values.

3. The exponential integral is defined by

$$Ei(x) = \int_x^\infty \frac{e^{-z}}{z}\, dz.$$

Show that $Ei(0)$ is a divergent improper integral. By repeated partial integrations calculate a formula for $Ei(x)$ in inverse powers of x valid for x large. Does this series converge? Show that this series is asymptotic by examining the remainder after n terms.

4. Use the asymptotic series for $Ei(x)$ to calculate the functional value at $x = 4$.

5. Develop a numerical procedure to calculate $Ei(x)$, for $x > 3$, accurate to four decimal places.

6. Find a formula for $Ei(x)$ for small x by writing

$$Ei(x) = \int_1^\infty \frac{e^{-t}}{t}\,dt + \int_x^1 \frac{e^{-t}}{t}\,dt.$$

7. Derive the formula

$$\operatorname{erf}(x) = 1 - \frac{e^{-x^2}}{x\pi^{1/2}}\left[1 - \frac{1}{2x^2} + \cdots + (-1)^n \frac{1\cdot 3\cdot 5\cdots(2n-1)}{(2x^2)^n}\right] + (-1)^{n+1}\frac{1\cdot 3\cdot 5\cdots(2n+1)}{2^n\pi^{1/2}}\int_x^\infty \frac{e^{-z^2}}{z^{2n+2}}\,dz.$$

8. Since $\dfrac{2}{\pi^{1/2}}\displaystyle\int_0^\infty e^{-t^2}\,dt = 1$, show that for $\alpha > 0$, $\lambda > 0$:

(i) $\displaystyle\int_0^\infty t^2 e^{-\alpha t^2}\,dt = \frac{\pi^{1/2}}{4}\alpha^{-3/2}$ **(ii)** $\displaystyle\int_0^\infty t[\operatorname{erf}(\alpha t)]e^{-\lambda t^2}\,dt = \frac{\alpha}{2\lambda}(\alpha^2+\lambda)^{-1/2}$

9. Develop an approximate formula for $I(x) = \displaystyle\int_x^\infty e^{-t^3}\,dt$ valid for $x \gg 0$. Verify that the formula is an asymptotic expansion. Estimate $I(10)$.

10. Develop an approximate formula for $I(\lambda) = \displaystyle\int_0^\infty \frac{te^{-\lambda t}}{1+t^2}\,dt$ valid for $\lambda \gg 0$. Is this an asymptotic series? Estimate $I(10)$.

11. Determine asymptotic series for the following integrals valid for $\lambda \gg 0$.

(i) $\displaystyle\int_0^\infty \frac{e^{-\lambda x}}{1+x^4}\,dx$ **(ii)** $\displaystyle\int_1^\infty \frac{e^{-\lambda x}}{x^2}\,dx$ **(iii)** $\displaystyle\int_0^\infty \frac{x^2 e^{-\lambda x}}{1+x^4}\,dx$

4.6 *Symbolic methods*

This section and the next will provide a cursory introduction to the *calculus of operators*. The approach will be pretty much the way the subject was originally developed—formal, utilitarian, and nonrigorous. The methods presented constitute an extremely powerful means of finding and devising approximate (and exact) formulas for numerical calculations, series summation, and other tasks as well.

A rule by which a function is changed into another function is called an *operation*. An *operator* is a symbol that designates this transformation. For example, the operation defined by

$$Du(x) = u'(x), \tag{1}$$

produces the derivative of the function. Other basic operations are

$$Eu(x) = u(x+h); \tag{2}$$

$$\Delta u(x) = u(x+h) - u(x); \tag{3}$$

$$cu(x) = cu(x). \tag{4}$$

E is called a *shift operator* because the value of the independent variable is displaced by an increment h; Δ transforms a function into a forward difference and c is just multiplication by a constant.

Example 1 If $u(x) = x^3$ and $h = 0.5$, find $\Delta u(3)$, $Eu(2)$, $Du(1)$.

In the first case

$$\Delta u(3) = u(3 + h) - u(3) = u(3.5) - u(3),$$
$$= (3.5)^3 - 3^3 = 15.875.$$

Furthermore,

$$Eu(2) = u(2 + h) = u(2.5) = 15.625,$$

and

$$Du(1) = \frac{d}{dx}x^3\Big]_1 = 3x^2\Big]_1 = 3.$$

Operators can be applied sequentially and, for example, $DE\,\Delta u(x)$ means that $\Delta u(x)$ is computed first, followed by a shift, $E[\Delta u(x)]$ and finally the result is differentiated: $D\{E[\Delta u(x)]\}$. Specifically, the details are

$$E\,\Delta u(x) = E[u(x + h) - u(x)] = u(x + 2h) - u(x + h)$$

and

$$DE\,\Delta u(x) = Du(x + 2h) - Du(x + h) = u'(x + 2h) - u'(x + h).$$

Similarly, an operation can be repeated several times, e.g.,

$$E[Eu(x)] = Eu(x + h) = u(x + 2h)$$

or

$$\Delta[\Delta u(x)] = \Delta[u(x + h) - u(x)] = u(x + 2h) - 2u(x + h) + u(x).$$

This invites the use of exponents to write $EE = E^2$ so that

$$E^2u(x) = u(x + 2h),$$

or more generally, for any positive integer n,

$$E^nu(x) = u(x + nh).$$

The algebraic rules for combining integer exponents are found to be valid:

$$E^mE^nu(x) = E^mu(x + nh) = u(x + nh + mh),$$
$$= u(x + (m + n)h) = E^{m+n}u(x),$$

that is,

$$E^m E^n = E^{m+n}.$$

The extension of this to cover fractional powers is straightforward, and we set

$$E^{1/2}u(x) = u(x + \tfrac{1}{2}h)$$

and

$$E^{p/q}u(x) = u\left(x + \frac{p}{q}h\right).$$

where p and q are integers. In this way, the operator E^r where r is a real number, may be defined, by analogy with a power function x^r, as

$$E^r u(x) = u(x + rh).$$

It follows that, for any real numbers r_1 and r_2,

$$E^{r_1}E^{r_2} = E^{r_1+r_2}. \tag{5}$$

2

By such simple arguments, we can verify that operators conform to the basic rules of algebra. If A, B, C, are three operators as defined above, then

$$A + B = B + A, \qquad \text{that is} \qquad (A + B)u = Au + Bu;$$

$$A + (B + C) = (A + B) + C;$$

$$AB = BA;$$

$$(AB)C = A(BC);$$

$$A(B + C) = AB + AC.$$

Various operations can now be combined according to these rules of algebra, and typical examples are

$$E^2 + E - 2 = (E - 1)(E + 2)$$

and

$$(E^{1/2}D - \Delta)(E^{1/2}D + \Delta) = ED^2 - \Delta^2.$$

Example 2 Show that $E^r\,\Delta u(x) = \Delta E^r u(x)$.

Since $\Delta u = u(x + h) - u(x)$ it follows that

$$\begin{aligned} E^r(\Delta u) &= E^r u(x + h) - E^r u(x), \\ &= u(x + (r + 1)h) - u(x + rh). \end{aligned}$$

But

$$\Delta E^r u(x) = \Delta u(x + rh) = u(x + (r + 1)h) - u(x + rh).$$

Example 3 Show that $E = 1 + \Delta$.

The difference $\Delta u(x) = u(x + h) - u(x)$ can obviously be written as the sum of operators

$$(\Delta + 1)u(x) = u(x + h) = Eu(x);$$

therefore

$$\Delta + 1 = E. \tag{6}$$

3

When the values of a function $u(x)$ are given at $x = a, a + h, \ldots, a + mh$, a table of differences can be arranged.

TABLE 6.1

x	$u(x)$	$\Delta u(x)$	$\Delta^2 u(x)$	$\Delta^m u(x)$
a	$u(a)$	$\Delta u(a)$	$\Delta^2 u(a)$	$\Delta^m u(a)$
$a + h$	$u(a + h)$	$\Delta u(a + h)$	$\Delta^2 u(a + h)$	
$a + 2h$	$u(a + 2h)$			
$\cdots$	$\cdots$	$\cdots$	$\cdots$	
$a + (m - 1)h$	$u(a + (m - 1)h)$	$\Delta u(a + (m - 1)h)$		
$a + mh$	$u(a + mh)$			

The $m + 1$ values of the function provide m values for Δu, $m - 1$ values for $\Delta^2 u$, and, ultimately, only one value for $\Delta^m u$. The difference operation is applied to each column successively until the entries are all completed.

Example 4 Arrange a difference table for $u = x^3$ with $a = 5$, $h = 2$, $m = 5$.

The information given requires a tabulation of x^3 at $x = 5, 7, \ldots, 15 = a + mh$. This is given in Table 6.2. The first difference Δx^3 is formed by subtracting consecutive numbers in the column under x^3, according to the definition

$$\Delta u(a) = u(a + h) - u(a).$$

Higher differences are formed by repeating the procedure, but the process stops at Δ^4 because all differences beyond this are zero.

TABLE 6.2

x	x^3	Δx^3	$\Delta^2 x^3$	$\Delta^3 x^3$	$\Delta^4 x^3$
5	125	218	168	48	0
7	343	386	216	48	0
9	729	602	264	48	
11	1,331	866	312		
13	2,197	1,178			
15	3,375				

Example 5 Show that the kth difference of a polynomial of degree n is a polynomial of degree $n - k$.

Let

$$P(x) = a_n x^n + a_{n-1}x^{n-1} + \cdots + a_0,$$

so that

$$\begin{aligned} P(x+h) &= a_n(x+h)^n + a_{n-1}(x+h)^{n-1} + \cdots + a_0, \\ &= a_n(x^n + nhx^{n-1} + \cdots) + a_{n-1}[x^{n-1} + (n-1)hx^{n-2} + \cdots] + \cdots + a_0, \\ &= a_n x^n + (nha_n + a_{n-1})x^{n-1} + \cdots. \end{aligned}$$

Therefore

$$\Delta P(x) = P(x+h) - P(x) = nha_n x^{n-1} + \text{terms of lower degree.}$$

The first difference reduces the degree of the polynomial by 1. It follows that after k differences the degree will be reduced to $n - k$; to a constant after n differences, and to zero afterward.

---

4 A fundamental relationship exists among the operators defined in (1) to (4), and it is derived by expressing the Taylor series representation of a function in the new symbolism. In the expansion

$$f(a+h) = f(a) + f'(a)h + \cdots + \frac{f^{(n)}(a)}{n!}h^n + \cdots, \tag{7}$$

the left-hand side is identified as $Ef(a)$ [see (2)], and every derivative on the right can be expressed as an appropriate power of D. Therefore

$$\begin{aligned} Ef(a) &= f(a) + hDf(a) + \cdots + \frac{h^n}{n!}D^n f(a) + \cdots, \\ &= \left(1 + hD + \frac{1}{2!}h^2D^2 + \cdots + \frac{1}{n!}h^nD^n + \cdots\right)f(a). \end{aligned} \tag{8}$$

The operators left and right produce the same function and must be identical, so that

$$E = 1 + hD + \frac{1}{2!}h^2D^2 + \cdots + \frac{1}{n!}h^nD^n + \cdots. \tag{9}$$

This can be expressed succinctly as

$$E = e^{hD}, \tag{10}$$

where the definition of the exponential is equivalent to the infinite series. In doing this, the term hD is simply treated as a number, for which the series expansion is certainly legitimate. The justification for this procedure is to be found in the validity of the result, Eq. (8). As a matter of fact, *the foundation and utility of the operational calculus lies in handling operators just as we would real numbers.* Upon combining (10) with (6), we obtain the fundamental identities

$$E = 1 + \Delta = e^{hD}. \tag{11}$$

---

5 The few notions introduced so far are already adequate to treat the problem of interpolation. Specifically, we wish to calculate the value of a function $f(x)$ at a point

$$x = a + rh, \tag{12}$$

where r is arbitrary, given the values at $m + 1$ points, $f(a), f(a + h), \ldots, f(a + mh)$. The value in question may be expressed

$$f(a + rh) = E^r f(a) = (1 + \Delta)^r f(a), \tag{13}$$

and, as above, *we manipulate this equation by treating the operator* Δ *as if it were a real number.* This, as we mentioned, is the practical basis of symbolic methods. According to the binomial theorem,

$$(1 + \Delta)^r = 1 + r\Delta + \frac{r(r-1)\Delta^2}{2!} + \frac{r(r-1)(r-2)\Delta^3}{3!} + \cdots,$$

and the substitution of this in (13) yields

$$f(a + rh) = f(a) + r\,\Delta f(a) + \frac{r(r-1)}{2!}\Delta^2 f(a) + \cdots. \tag{14}$$

But when $m + 1$ values of the function are given, it is possible to calculate the higher differences only to order m, that is, $\Delta^m f(a)$. Since differences to all orders are required in (14), a useful *approximate interpolation formula* is obtained by simply discarding terms involving differences $\Delta^n f(a)$ for $n \geq m + 1$. The *approximation* to (14) is then

$$f(a + rh) \approx f(a) + r\,\Delta f(a) + \frac{r(r-1)}{2!}\Delta^2 f(a)$$
$$+ \cdots + \frac{r(r-1)(r-2)\cdots(r-m+1)}{m!}\Delta^m f(a). \tag{15}$$

This is an interesting formula because it reproduces the exact, prescribed values of the function at the points corresponding to $r = 1, 2, \ldots, m$. The reason for this is made evident upon replacing r everywhere by x, that is,

$$r = \frac{x - a}{h},$$

for then (15) becomes a *polynomial in x* of degree m, which has the same values as the function $f(x)$ at $m + 1$ given points. Since this is the polynomial of lowest degree with this property, it is indeed none other than *Lagrange's interpolation polynomial* (see page 182). Thus the approximation given by Eq. (15) is, in reality, equivalent to earlier interpolation formulas, but it is very much easier to use. [Equation (15) is also known as *Newton's interpolation formula*.]

Example 6 Find the value of sin 37° from the information sin 25° = 0.4226, sin 30° = 0.5000, sin 35° = 0.5736, sin 40° = 0.6428, sin 45° = 0.7071.

TABLE 6.3

x	$\sin x$	Δ	Δ^2	Δ^3	Δ^4
25°	0.4226	0.0774	−0.0038	−0.0006	+0.0001
30°	0.5000	0.0736	−0.0044	−0.0005	
35°	0.5736	0.0692	−0.0049		
40°	0.6428	0.0643			
45°	0.7071				

The data are arranged in difference Table 6.3, where the basic increment is $h = 5°$; and $a = 25°$. It follows that $r = 2.4$. With these values and the differences to fourth order $[\Delta^4 f(a)]$ we can apply (15) in the form

$$\sin 37° = \sin 25° + (2.4)(\Delta \sin 25°) + \frac{(2.4)(1.4)}{2}\Delta^2 \sin 25° + \frac{(2.4)(1.4)(0.4)}{3 \cdot 2 \cdot 1}\Delta^3 \sin 25° + \frac{(2.4)(1.4)(0.4)(-0.6)}{4 \cdot 3 \cdot 2 \cdot 1}\Delta^4 \sin 25°.$$

Inserting the relevant numbers from the table, we obtain

$$\sin 37° = 0.4226 + (2.4)(0.774) - \frac{(2.4)(1.4)}{2}(0.0038) - \frac{(2.4)(1.4)(0.4)}{6}(0.0006) - \frac{(2.4)(1.4)(0.4)(0.6)}{24}(0.0001)$$

or $\sin 37° = 0.60186$, rounded to five decimal places.

This calculation based on differences is equivalent to Lagrangian interpolation (using five points) and makes short work of it. The accurate value is $0.601815 \cdots$.

---

6 Operator methods can frequently be used to find a simple formula for the sum of a finite number of terms such as

$$S_n = \sum_{k=0}^{n-1} f(a + kh) = f(a) + f(a + h) + \cdots + f(a + (n - 1)h).$$

This is done by substituting

$$f(a + kh) = E^k f(a),$$

in which case

$$S_n = \sum_{k=0}^{n-1} E^k f(a) = \left(\sum_{k=0}^{n-1} E^k\right) f(a).$$

The operator can be manipulated further, and use of the identity

$$1 + E + E^2 + \cdots + E^{n-1} = \frac{E^n - 1}{E - 1} = \frac{(1 + \Delta)^n - 1}{\Delta}$$

yields the formula

$$S_n = \frac{E^n - 1}{\Delta} f(a) = \frac{(1 + \Delta)^n - 1}{\Delta} f(a). \tag{16}$$

Another form of this is obtained by expanding $(1 + \Delta)^n$:

$$S_n = \sum_{k=0}^{n-1} f(a + kh) = \left[n + \frac{n(n-1)\Delta}{2!} + \frac{n(n-1)(n-2)}{3!}\Delta^2 + \cdots + \Delta^{n-1}\right] f(a), \tag{17}$$

and this is especially convenient when the function $f(x)$ is a polynomial of degree m, for then differences of order greater than m are identically zero.

Example 7 Find the sum of the squares of the first n integers.

With $f(x) = x^2$ and $h = 1$, $a = 1$, a difference table can be constucted.

TABLE 6.4

x	x^2	Δ	Δ^2	Δ^3
1	1	3	2	0
2	4	5	2	0
3	9	7	2	
4	16	9		
5	25			

Inserting the appropriate values in (17), we obtain

$$1 + 4 + 9 + 16 + \cdots + n^2 = n + \frac{n(n-1)}{2}(3) + \frac{n(n-1)(n-2)}{6}(2),$$

$$= \frac{n(n+1)(2n+1)}{6},$$

a result that required some ingenuity earlier in Sec. 1.8.

Exercises 4.6

1. If $u(x) = x^2 + 2x + 5$ and $h = 1/2$, evaluate:
(i) Eu **(ii)** Δu **(iii)** $\Delta^2 u$ **(iv)** $D\,\Delta u$ **(v)** $E^{5/3}u$

2. Show that:
(i) $E\,\Delta(Du) = ED(\Delta u) = D\,\Delta(Eu)$ **(ii)** $E + (\Delta + D) = (E + \Delta) + D$
(iii) $(\Delta + E)D = D(\Delta + E)$ **(iv)** $E^{r_1}E^{r_2} = E^{r_1+r_2}$

3. Write out the functional form of:

(i) $(E^2 + E - 2)f(a)$ **(ii)** $(E^{1/2}\Delta - 1)(E^{1/2} - D)f(a)$

4. Write out the functional form of:

(i) $e^{hD^2}f(x)$ **(ii)** $(1 + \Delta)^3 f(x)$ **(iii)** $\log(2 + \Delta)u(x)$

5. Construct a difference table for x^4 with $h = 1$ starting with $x = 1$ and up to $x = 10$.
6. Construct a difference table for $\cos x$ from the values at 20, 25, 30, 35, and 40°.
7. Construct a difference table for e^x from the values at $x = 0$, 0.5, 1, 1.5, and 2.
8. Write in operator form $f'(x + h) - 2f(x - 2h) + f(x)$.
9. The values of a function at several points are:

x	0.35	0.40	0.45	0.50	0.55
y	1.5215	1.5059	1.4880	1.4675	1.4442

Find the approximate value of y at $x = 0.48$.
10. Find explicit formulas for:

(i) $\sum_{k=1}^{n} k^3$; **(ii)** $\sum_{k=1}^{n} k(k + 4)$; **(iii)** $\sum_{k=1}^{n} (k - 1)^2(k - 2)^2$.

11. Sum to n terms: $1(5^2) + 5(9^2) + 9(13^2) + \cdots$
12. Let $m = 2$ in Eq. (15) and write

$$f(a + rh) \approx f(a) + r\,\Delta f(a) + \frac{r(r - 1)}{2!}\,\Delta^2 f(a)$$

in terms of x, using $r = (x - a)/h$. Show that this quadratic passes through the points $(a, f(a))$, $(a + h, f(a + h))$, $(a + 2h, f(a + 2h))$.
13. From the data

θ, deg	30	33	36	39	42
$\tan\theta$	0.5774	0.6494	0.7265	0.8098	0.9004

find the approximate value of tan 34°. Compare to the accurate value.
14. Obtain a formula for $f'(x)$ at $x = a + rh$ by replacing r in Eq. (14) and differentiating this with respect to x.

15. Define the backward-difference operation

$$\nabla u = u(x) - u(x - h).$$

Find formulas relating ∇ to E and Δ. Write ∇u as a series in powers of Δ.

4.7 *Numerical formulas*

The fundamental relationship

$$E = 1 + \Delta = e^{hD}, \tag{1}$$

allows the operators E and D to be expressed in terms of Δ alone

$$E = 1 + \Delta \qquad \text{and} \qquad D = \frac{1}{h}\log(1 + \Delta). \tag{2}$$

Once again, the operator Δ is treated just like a number, in which case we may use the Taylor series for the logarithm to write

$$D = \frac{1}{h}\log(1 + \Delta) = \frac{1}{h}\left[\Delta - \frac{\Delta^2}{2} + \frac{\Delta^3}{3} + \cdots \frac{(-1)^{n+1}\Delta^n}{n} + \cdots\right]. \tag{3}$$

The derivative of $f(x)$ at a point $x = a$ can then be evaluated strictly in terms of the *forward differences* as

$$Df(a) = f'(a) = \frac{1}{h}\left[\Delta - \frac{\Delta^2}{2} + \cdots + \frac{(-1)^{n+1}\Delta^n}{n} + \cdots\right]f(a). \tag{4}$$

If only $m + 1$ values of the function are given, an approximate numerical formula for the derivative at $x = a$ is obtained by simply discarding differences of order higher than m, that is, Δ^{m+1}, Δ^{m+2}, etc.

To find the derivative of the function at a point $x = a + rh$ we write

$$f'(x) = f'(a + rh) = Df(a + rh) = DE^r f(a)$$

or, using (2),

$$f'(x) = (1 + \Delta)^r\left[\frac{1}{h}\log(1 + \Delta)\right]f(a).$$

The order of operations is immaterial and is taken for convenience of calculation only. Substitution of (3) in the foregoing and use of the series expansion for $(1 + \Delta)^r$ yields

$$f'(a + rh) = \left[1 + r\Delta + \frac{r(r-1)\Delta^2}{2} + \cdots\right]\frac{1}{h}\left(\Delta - \frac{\Delta^2}{2} + \frac{\Delta^3}{3} + \cdots\right)f(a).$$

Upon completing the multiplication of these series, we obtain

$$f'(a + rh) = \frac{1}{h}\left\{\Delta + \left(r - \frac{1}{2}\right)\Delta^2 + \left[\frac{r}{2}(r - 2) + \frac{1}{3}\right]\Delta^2 + \cdots\right\} f(a), \tag{5}$$

which is as far as this process will be taken.

Example 1 Find the derivative of $f(x) = \sin x$ at $x = 37°$ using the data in Table 7.1. Neglect differences of order 4 and higher.

TABLE 7.1

x	$\sin x$	Δ	Δ^2	Δ^3
25°	0.4226	0.0774	−0.0038	−0.0006
30°	0.5000	0.0736	−0.0044	−0.0005
35°	0.5736	0.0692	−0.0049	
40°	0.6428	0.0643		
45°	0.7071			

Since $a = 25°$, $h = 5° = 0.087266$ rad, and $a + rh = 37°$, it follows that $r = 2.4$. The result of substituting this information in (5) is

$$f'(37°) = \cos 37° = \frac{1}{0.087266}[0.0774 + (1.9)(-0.0038) + (0.81333)(-0.0006)],$$

$$= 0.7986,$$

which is the exact value to four decimal places.

2

The inverse operators D^{-1} and Δ^{-1} are, by definition, operators which undo the changes produced by D and Δ, that is,

$$\begin{aligned} DD^{-1}u(x) &= u(x) = D^{-1}Du(x), \\ \Delta\Delta^{-1}u(x) &= u(x) = \Delta^{-1}\Delta u(x). \end{aligned} \tag{6}$$

Clearly D^{-1} may be identified with the indefinite integral:

$$D^{-1}u(x) = \int u(x)\, dx, \tag{7}$$

and to take full advantage of the algebraic properties of operators we also write

$$D^{-1}u = \frac{1}{D}u = \int u\, dx. \tag{8}$$

The operation $\Delta^{-1}u$ yields a function $U(x)$ such that

$$\Delta U(x) = u(x)$$

or

$$U(x+h) - U(x) = u(x).$$

Alternate forms of these inverse operators are

$$D^{-1} = \frac{1}{D} = \frac{h}{\log(1+\Delta)}, \qquad \Delta^{-1} = \frac{1}{\Delta} = \frac{1}{E-1},$$

both of which may be subjected to the same type of manipulation we indulged in previously.

---

3 Consider the evaluation of the definite integral

$$I = \int_{a+kh}^{a+nh} f(x)\,dx, \tag{9}$$

and set $x = a + rh$, so that with this change of variable

$$I = h\int_k^n f(a+rh)\,dr. \tag{10}$$

The integrand can now be expressed in operator form as

$$f(a+rh) = E^r f(a) = e^{r\log E} f(a),$$

and upon substitution we write

$$I = h\left(\int_k^n e^{r\log E}\,dr\right) f(a).$$

The original integral has become an operator integral on $f(a)$ which can be evaluated explicitly. (Note that E is treated like a real number at all times.) Upon integrating the result is

$$I = h\,\frac{e^{n\log E} - e^{k\log E}}{\log E}\,f(a), \tag{11}$$

or

$$I = \frac{E^n - E^k}{D}\,f(a).$$

If E and D are expressed in terms of Δ alone, then

$$I = \int_{a+kh}^{a+nh} f(x)\,dx = h\,\frac{(1+\Delta)^n - (1+\Delta)^k}{\log(1+\Delta)}\,f(a). \tag{12}$$

Various approximate formulas for integrals are obtained from this equation by expanding the operators involved in powers of Δ and by truncating the series at some appropriate point. This cutoff corresponds to the substitution of a polynomial approximation for the integrand $f(x)$. To illustrate, Simpson's rule (page 300) is quickly rederived. Let $n = 2$ and $k = 0$, in which case (12) becomes

$$\int_a^{a+2h} f(x)\,dx = h\,\frac{(1+\Delta)^2 - 1}{\log(1+\Delta)}\,f(a), \tag{13}$$

$$= h\,\frac{2\Delta + \Delta^2}{\Delta - \Delta^2/2 + \Delta^3/3 + \cdots}\,f(a) = h\,\frac{2 + \Delta}{1 - \Delta/2 + \Delta^2/3 + \cdots}\,f(a),$$

whence by division

$$\int_a^{a+2h} f(x)\,dx = h\left(2 + 2\Delta + \frac{\Delta^2}{3} - \frac{\Delta^4}{90} + \cdots\right)f(a). \tag{14}$$

If differences of order 4 and higher are dropped, we obtain a formula valid for a polynomial approximation of the integrand, $f(x)$, which is of degree less than 4. (Recall that differences of a polynomial of higher order than its degree are identically zero.) The truncated series is

$$\int_a^{a+2h} f(x)\,dx \approx h(2 + 2\,\Delta + \tfrac{1}{3}\,\Delta^2)f(a) = h[2 + 2(E-1) + \tfrac{1}{3}(E-1)^2]f(a)$$

or

$$\int_a^{a+2h} f(x)\,dx \approx \frac{h}{3}(1 + 4E + E^2)f(a). \tag{15}$$

In other words,

$$\int_a^{a+2h} f(x)\,dx \approx \frac{h}{3}\,[f(a) + 4f(a+h) + f(a+2h)], \tag{16}$$

which is Simpson's rule.

4

By changing tactics slightly, a formula can be determined for the asymptotic evaluation of sums and integrals. So far, the operators D and E have been expanded in powers of Δ, but now we reverse procedure and express Δ in terms of D. It is convenient to start with

$$\frac{1}{\Delta} = \frac{1}{e^{hD} - 1} = \frac{1}{hD + h^2D^2/2! + h^3D^3/3! + \cdots}, \tag{17}$$

and by long division we obtain

$$\frac{1}{\Delta} = \frac{1}{hD} - \tfrac{1}{2} + \tfrac{1}{12}hD - \tfrac{1}{720}h^3D^3 + \cdots. \tag{18}$$

At this point, we note two formulas already established:

$$\frac{E^n - 1}{D}\,f(a) = \int_a^{a+nh} f(x)\,dx$$

and $$\frac{E^n - 1}{\Delta} f(a) = \sum_{k=0}^{n-1} f(a + kh)$$

[see Eq. (16) in Sec. 4.6]. The multiplication of (18) by $(E^n - 1)$ yields the *Euler-Maclaurin* formula

$$\frac{E^n - 1}{\Delta} = \frac{E^n - 1}{hD} + (E^n - 1)(-\tfrac{1}{2} + \tfrac{1}{12}hD - \tfrac{1}{720}h^3 D^3 + \cdots),$$

or, equivalently, in functional form

$$\sum_{k=0}^{n-1} f(a + kh) = \frac{1}{h}\int_a^{a+nh} f(x)\,dx - \frac{1}{2}[f(a + nh) - f(a)]$$
$$+ \frac{h}{12}[f'(a + nh) - f'(a)] - \frac{h^3}{720}[f'''(a + nh) - f'''(a)] + \cdots. \quad (19)$$

This formula is useful to express a finite sum as an integral and vice versa, and in either case the power series in h is asymptotic in nature.

One application of the Euler-Maclaurin series is to estimate the value of $n!$ for large n. This can be done by recognizing that

$$\log n! = \log n + \log(n - 1) + \cdots + \log 2 + \log 1,$$

which, with $f(x) = \log x$, can be identified as the left-hand side of (19). In particular, let $a = 1$, $h = 1$, so that the Euler-Maclaurin series becomes

$$\sum_{k=0}^{n-1} \log(1 + k) = \int_1^{1+n} \log x\,dx - \tfrac{1}{2}[\log(n + 1) - \log 1]$$
$$+ \frac{1}{12}\left(\frac{1}{n + 1} - 1\right) - \frac{1}{720}\left[\frac{2}{(n + 1)^3} - 2\right] + \cdots. \quad (20)$$

Therefore upon completing the integration and collecting terms

$$\log n! = (n + \tfrac{1}{2})\log(n + 1) - n + \frac{1}{12}\left(\frac{1}{n + 1} - 1\right) - \frac{1}{360}\left[\frac{1}{(n + 1)^3} - 1\right] + \cdots. \quad (21)$$

We are primarily interested in an approximate formula for n very large, and we intend to discard all terms like $1/n$, $1/(n + 1)$ or smaller. Since

$$(n + \tfrac{1}{2})\log(n + 1) = (n + \tfrac{1}{2})\log n + (n + \tfrac{1}{2})\log\left(1 + \frac{1}{n}\right)$$
$$= (n + \tfrac{1}{2})\log n + 1 + \tfrac{1}{12}n^2 + \cdots,$$

(21) can also be written

$$\log n! = (n + \tfrac{1}{2})\log n - n + 1 - \tfrac{1}{12} + \cdots$$

or $$\log n! = (n + \tfrac{1}{2})\log n - n + \log A,$$

where $\log A$ represents the constant part of this asymptotic expansion in inverse powers of n. Therefore the final results, valid for large n, are $A \approx 2.507$ and

$$n! \approx A n^{n+1/2} e^{-n}. \tag{22}$$

The *exact* value of A is difficult to determine this way, for not everything in life comes easy. Having used operator techniques to discover the Euler-Maclaurin formula, it would now pay to rederive this equation by another method, one that would allow an estimate of the error involved. Although the capability exists, we choose not to enter such a protracted discussion at this time. However, it turns out that

$$A = (2\pi)^{1/2},$$

and substitution of this in (22) yields Stirling's formula (circa 1730):

$$n! \approx (2\pi)^{1/2} n^{n+1/2} e^{-n}. \tag{23}$$

A more precise statement is

$$\lim_{n\to\infty} \frac{n!}{n^{n+1/2}e^{-n}} = (2\pi)^{1/2}.$$

Table 7.2 compares values of $n!$ with the asymptotic approximation. Evidently, the asymptotic formula is quite accurate even at $n = 1$.

TABLE 7.2

n	$n!$	$\dfrac{n!}{(2\pi)^{1/2}n^{n+1/2}e^{-n}}$
1	1	1.04220712
2	2	1.02806452
3	6	1.02100830
4	24	1.01678399
5	120	1.01397285
6	720	1.01196776
7	5,040	1.01046565
8	40,320	1.00929843
9	362,880	1.00836536
10	3,628,800	1.00760243

This is only a brief introduction to operational methods, but it should demonstrate their tremendous power in applications.

Exercises 4.7

1. Use operator methods to derive the three-eighths rule for numerical integration

$$\int_a^{a+3h} f(x)\,dx = \frac{3h}{8}\,[f(a) + 3f(a+h) + 3f(a+2h) + f(a+3h)].$$

[*Hint:* Use (12) and drop fourth and higher differences.]

2. Evaluate $\int_1^2 \frac{dx}{x}$ by taking $h = 0.1$ and using the procedures of this section. Compare to the accurate value $\log 2$.

3. Arrange a difference table from the data

θ, deg	10	20	30	40	50
$K(\theta)$	1.5828	1.6200	1.6858	1.7868	1.9356

Find approximate values for $K(18°)$, $\int_0^{2\pi/9} K(\theta)\, d\theta$.

4. Values of the Bessel function $J_0(x)$ are as follows:

x	0.0	0.5	1.0	1.5	2.0	2.5	3.0
$J_0(x)$	1.00000	0.93847	0.76520	0.51183	0.22389	−0.04838	−0.26005

Find approximate values for $J_0(1.3)$, $\int_0^3 J_0(x)\, dx$, and the root location, $J_0(x) = 0$.

5. Using the difference table for $K(\theta)$ in Exercise 3, find the sum $K(25°) + K(35°) + K(45°)$.

6. Express D in terms of ∇, where $\nabla u = u(x) - u(x - h)$.

7. Use the Euler-Maclaurin formula to show that

$$\sum_{n=1}^{\infty} \frac{1}{n^4} = 1 + \frac{1}{2^4} + \frac{1}{3^4} + \frac{1}{4^4} + \int_5^{\infty} \frac{dx}{x^4} + \frac{1}{2 \cdot 5^4} + \frac{1}{3 \cdot 5^5} + \cdots = 1.082323\ldots, \ (= \pi^4/90).$$

8. Show that

$$S = u_0 - u_1 + u_2 - u_3 + \cdots + (-1)^n u_n + \cdots = \sum_{n=0}^{\infty} (-1)^n u_n$$

can also be written

$$S = u_0 - u_1 + \cdots + (-1)^{n-1} u_{n-1} + (-1)^n \sum_{k=0}^{\infty} (-1)^k u_{n+k}.$$

Note that, despite appearances, S is not necessarily an alternating series because u_n is not definitely positive.

9. If $u(x)$ is a function such that

$$u(a) = u_0,$$
$$u(a + nh) = u(x_n) = u_n,$$

show that

$$u_{n+k} = u(x_n + kh) \qquad \text{and} \qquad u_{n+k} = E^k u_n.$$

Show next that the sum in Exercise 8 can be written in operator form as

$$\sum_{k=0}^{\infty} (-1)^k u_{n+k} = \frac{1}{1+E} u_n = \frac{1}{2}\left(1 + \frac{\Delta}{2}\right)^{-1} u_n.$$

10. Since S in Exercise 8 can be written

$$S = u_0 - u_1 + \cdots + (-1)^{n-1}u_{n-1} + \frac{1}{2}\left(1 + \frac{\Delta}{2}\right)^{-1} u_n,$$

show upon expanding $(1 + \Delta/2)^{-1}$ that

$$S = u_0 - u_1 + \cdots + (-1)^{n-1}u_{n-1} + \tfrac{1}{2}(-1)^n(u_n - \tfrac{1}{2}\Delta u_n + \tfrac{1}{4}\Delta^2 u_n + \cdots).$$

This is known as *Euler's transformation* and is most useful in improving the convergence of slowly converging series.

11. Use Euler's transformation to speed up the convergence of $\sum_{n=1}^{\infty} \frac{(-1)^{n+1}}{n}$. In particular, show that this method yields the formula

$$\sum_{n=1}^{\infty} \frac{(-1)^{n+1}}{n} = \sum_{n=1}^{\infty} \frac{1}{n2^n}.$$

12. Evaluate

$$S = 1 - \frac{1}{2^2} + \frac{1}{3^2} - \frac{1}{4^2} + \cdots = \sum_{n=0}^{\infty} \frac{(-1)^n}{(n+1)^2}$$

by writing

$$S = 1 - \frac{1}{2^2} + \cdots - \frac{1}{8^2} + \sum_{k=0}^{\infty} \frac{(-1)^k}{(9+k)^2}$$

and applying Euler's transformation to the last sum. Construct a difference table and use it to evaluate S to four decimal places.

13. Use the asymptotic formula for $n!$ to determine the interval of convergence of the following series (including end points):

(i) $\sum_{n=1}^{\infty} \frac{1}{n!}\left(\frac{nx}{e}\right)^n$ **(ii)** $\sum_{n=1}^{\infty} \frac{(n+1)x^n}{(n!)^{1/n}}$.

4.8 Differential equations

The mathematical formulation of a scientific problem often leads to a *differential equation*, a relationship among an unknown function $y(x)$, some of its higher derivatives $y'(x)$, $y''(x)$, ..., and the independent variable x. Typical examples of these are

(i) $y' = \cos x$; **(ii)** $xy' + 2y = e^{-x}$; **(iii)** $y' = \dfrac{f(x)}{g(y)}$;

(iv) $y' = 1 - \varepsilon y^2$; **(v)** $y'' - y = 0$; **(vi)** $y''' = \sin y$.

The objective in each case is to determine the function $y(x)$ explicitly, i.e., in an analytical or tabulated form. When this is done, the equation is said to be solved. Usually the solution must also satisfy certain extra conditions included in the original statement of the scientific problem. In the simplest case this might require the solution, which represents a curve in the xy plane, to pass through a definite point.

The resolution of all the inherent issues—regarding the existence, uniqueness, and continuity of a solution, the nature and number of subsidiary conditions in a well-posed problem, and constructive methods of solution—provides the subject matter for a complete course in differential equations (which commonly follows calculus). However, the objectives of this section are not that grand. In keeping with the frequent and rather natural appearance of differential equations throughout this text, we will continue to illustrate the methods and techniques of the calculus as they apply to such problems. The information derived from several specific examples reveals, with inspired interpretation, many of the basic theoretical results alluded to above. At the very least, the motivation will be given for a more comprehensive study later.

The *order* of a differential equation is given by the highest derivative that appears in the equation. For example (i) to (iv) are first-order differential equations, (v) is a second-order equation, and (vi) is a third-order equation. It seems reasonable to conjecture that as the order increases, the task of finding a solution will become more difficult; for this reason initial explorations are limited to first-order equations.

The most general form of a first-order differential equation is

$$F(x, y, y') = 0, \tag{1}$$

or, alternatively,

$$y' = G(x, y). \tag{2}$$

Not much can be said about the solution of so general a problem without first describing and otherwise severely restricting the forms of the functions F and G. Even then the task is difficult, and so we start with a problem that can be done and introduce complications gradually.

2

Equation (i) is a simple prototype of an entire class of problems given by

$$\frac{dy}{dx} = f(x), \tag{3}$$

where $f(x)$ is a known function. Of course, the solution in this case is the indefinite integral

$$y = \int f(x)\, dx + C, \tag{4}$$

and the methods by which such integrations are executed formed the content of Chap. 3. The single constant C involved in this solution allows the specification of one extra condition. If this is taken to be $y = y_0$ at $x = x_0$ [so that the curve $y(x)$ passes through point (x_0, y_0)], then the constant can be determined and (4) becomes

$$y - y_0 = \int_{x_0}^{x} f(x)\, dx. \tag{5}$$

For example, the complete solution of (i), that is, $y' = \cos x$, is

$$y = \sin x + C.$$

This is a one parameter family of solutions. If in addition we require that $y = 0$ at $x = \pi/2$, then $C = -1$ and

$$y = \sin x - 1$$

is the appropriate solution curve of the family which passes through the given point.
The differential equation (ii),

$$xy' + 2y = e^{-x},$$

presents a greater challenge, but it too is readily solved by a broadly applicable technique. If (ii) is multiplied by x, the resultant expression,

$$x^2y' + 2xy = xe^{-x},$$

can be written as the derivative of a product:

$$\frac{d}{dx}(x^2y) = xe^{-x}. \tag{6}$$

This is now analogous to (3) (with x^2y replacing y), and it follows that

$$x^2y = \int xe^{-x}\,dx + C$$

or

$$y = \frac{1}{x^2}\int xe^{-x}\,dx + \frac{C}{x^2}, \tag{7}$$

and finally

$$y = -\frac{x+1}{x^2}e^{-x} + \frac{C}{x^2}. \tag{8}$$

Again a single arbitrary constant appears in the complete solution which consists of two separate and distinct terms,

$$y = y_P(x) + Cy_H(x). \tag{9}$$

The function

$$y_P(x) = -\frac{x+1}{x^2}e^{-x}$$

is itself a *particular solution* of the differential equation (ii) or (6), as can be checked by direct calculation. The other function, $y_H(x) = 1/x^2$, satisfies

$$xy'_H + 2y_H = 0, \tag{10}$$

an equation which differs from (ii) only in that the term e^{-x} on the right-hand side is replaced by zero. This extraneous term, which alone does not multiply y or y', is called an *inhomogeneity* of the equation, and for this reason (10) is referred to as the *homogeneous* form of (ii). In this language, (9) asserts that the complete solution is the sum of any particular solution and an arbitrary multiple of the homogeneous solution. The only justification for such a formal pronouncement is that it happens to be true (and useful) in a more general situation.

---

3 Equation (ii) is a special case of the more general differential equation

$$y'(x) + p(x)y(x) = g(x), \tag{11}$$

where $p(x)$ and $g(x)$ are given functions. [In particular, (ii) is reproduced by choosing $p(x) = 2/x$, $g(x) = e^{-x}/x$.] The function $g(x)$ is the inhomogeneous term, and the associated homogeneous equation is

$$y'(x) + p(x)y(x) = 0. \tag{12}$$

Equation (11) can be solved as easily as (ii) if we recognize that the crucial step in the solution of that special case was to express it as the derivative of a product, (6), which could then be integrated directly. This maneuver for (ii) was accomplished by multiplying it through by x, and although the resultant contraction may have appeared fortuitous, a systematic procedure can in fact be devised for this purpose.

The aim is then to manipulate the left-hand side of (11) into the form

$$\frac{d}{dx}[a(x)y(x)],$$

where $a(x)$ is a special function that depends on $p(x)$. Since, by the product rule,

$$\frac{d}{dx}[a(x)y(x)] = ay' + a'y = a\left(y' + \frac{a'}{a}y\right),$$

the comparison of the last expression in parentheses with $y'(x) + p(x)y(x)$ shows that the relationship

$$\frac{a'(x)}{a(x)} = p(x), \tag{13}$$

will permit the desired regrouping of terms.

The solution of (11) can now be determined by substituting (13) in the differential equation as follows:

$$y'(x) + \frac{a'(x)}{a(x)}y(x) = g(x),$$

so that

$$a(x)y'(x) + a'(x)y(x) = a(x)g(x),$$

and finally

$$\frac{d}{dx}[a(x)y(x)] = a(x)g(x). \tag{14}$$

The last expression can be immediately integrated, and the result is

$$a(x)y(x) = \int a(x)g(x)\,dx + C$$

or

$$y(x) = \frac{1}{a(x)} \int a(x)g(x)\,dx + \frac{C}{a(x)}. \tag{15}$$

Since the integration of (13) yields

$$\log a(x) = \int p(x)\,dx \qquad \text{or} \qquad a(x) = \exp\left[\int p(x)\,dx\right], \tag{16}$$

(15) is expressed in terms of the original function $p(x)$ as

$$y(x) = \exp\left[-\int^x p(x)\,dx\right] \int^x \exp\left[\int^x p(\xi)\,d\xi\right] g(x)\,dx + C \exp\left[-\int^x p(x)\,dx\right]. \tag{17}$$

(Here some extra notation is introduced in the indefinite integrals to avoid any possible confusion concerning the roles of the variables.) This complete solution is again the sum of a particular solution of (11) and an arbitrary multiple of the solution of the homogeneous equation (12) (which are respectively the first and the second terms on the right-hand side). The verification of these statements is left as an exercise, as are the discussions of alternative approaches for the derivation of (17). The student should realize that understanding the basic idea will make it forever unnecessary to commit this formula to memory.

Before passing on to definite examples, a few remarks and observations should be made about this analysis:

1. The appearance of just one arbitrary constant in the solution of the first-order differential equation (11) means that a single extra condition can be imposed. For example, the curve $y(x)$ can be made to pass through any one point in the plane.
2. Although (17) is the formal solution, the integrations involved are generally quite difficult. Some ingenuity and hard work are usually required to make the formula truly useful and informative.
3. The function $a(x)$ is called an *integrating factor*.
4. One of the principal reasons that (11) can be solved so readily is that it is a rather special type called a *linear* first-order differential equation. This means that the function

$$F(x, y, y') = y' + p(x)y - g(x)$$

is a polynomial of the *first* degree in y and y', that is, a linear combination of these dependent variables. An equation that is not linear is called *nonlinear*, appropriately enough. For example, (i) and (ii) are linear first-order equations, (iii) is generally nonlinear unless $g(y) = \alpha/y$, and (iv) is nonlinear because the term y^2 appears. [Equation (v) is linear in the variables y'' and y while (vi) is a nonlinear third-order equation.] Linear equations of any order are, relatively speaking, the simplest kind, and their theory is the most highly developed and well understood. Nonlinear equations are almost another subject, still under intense study.

Example 1 Find the solution of

$$xy'(x) + \alpha y(x) = 1 + x^2 \qquad \text{with} \qquad y(1) = 1.$$

This is a first-order differential equation; its general solution, for any α, is expressed in (17).

For the purposes of illustration we will parallel the steps of the general analysis. The equation is rewritten as

$$y'(x) + \frac{\alpha}{x} y(x) = \frac{1 + x^2}{x}, \tag{18}$$

which should be compared to (11). The function $p(x) = \alpha/x$ is expressed in terms of the integrating factor as

$$\frac{a'(x)}{a(x)} = \frac{\alpha}{x} = p(x),$$

which implies that

$$\log a(x) = \int \frac{\alpha}{x}\, dx = \alpha \log x$$

and

$$a(x) = e^{\alpha \log x} = x^{\alpha}.$$

Therefore, the result of multiplying (18) by $a(x)$ is

$$x^{\alpha} y'(x) + \alpha x^{\alpha - 1} y(x) = (1 + x^2) x^{\alpha - 1},$$

which condenses to

$$\frac{d}{dx} [x^{\alpha} y(x)] = (1 + x^2) x^{\alpha - 1}.$$

Integration yields

$$x^{\alpha} y(x) = \int (1 + x^2) x^{\alpha - 1}\, dx + C$$

or

$$y(x) = \frac{1}{x^{\alpha}} \int (1 + x^2) x^{\alpha - 1}\, dx + \frac{C}{x^{\alpha}},$$

and the explicit solution is

$$y(x) = \frac{1}{\alpha} + \frac{x^2}{\alpha + 2} + \frac{C}{x^{\alpha}}.$$

In order that $y(1) = 1$, as prescribed, the constant must be

$$C = 1 - \frac{1}{\alpha} - \frac{1}{\alpha + 2},$$

and the required solution is then

$$y(x) = \frac{1}{\alpha} + \frac{x^2}{\alpha + 2} + \left(1 - \frac{1}{\alpha} - \frac{1}{\alpha + 2}\right)\frac{1}{x^\alpha}. \tag{19}$$

4

Some difficulty seems to arise when $\alpha = 0$ or $\alpha = -2$, for then individual terms in the preceding formula are infinite. On closer inspection, we see that a delicate limit process is involved wherein large terms cancel. For α small, the solution is more conveniently written

$$y(x) = \frac{x^\alpha - 1}{\alpha x^\alpha} + \frac{x^2}{\alpha + 2} + \left(1 - \frac{1}{\alpha + 2}\right)\frac{1}{x^\alpha},$$

and here the potential difficulty at $\alpha = 0$ is concentrated in the first term on the right. But, as $\alpha \to 0$, the limit can be found using l'Hôpital's rule. Since

$$\lim_{\alpha \to 0} \frac{x^\alpha - 1}{\alpha} = \lim_{\alpha \to 0} \frac{e^{\alpha \log x} - 1}{\alpha} = \lim_{\alpha \to 0} \frac{\dfrac{d}{d\alpha}(e^{\alpha \log x} - 1)}{\dfrac{d}{d\alpha}\alpha} = \log x$$

and all other limits are regular, the solution for $\alpha = 0$ is

$$y = \lim_{\alpha \to 0} \left[\frac{x^\alpha - 1}{\alpha x^\alpha} + \frac{x^2}{\alpha + 2} + \left(1 - \frac{1}{\alpha + 2}\right)\frac{1}{x^\alpha}\right],$$

$$y = \log x + \tfrac{1}{2}x^2 + \tfrac{1}{2}.$$

Of course, the same answer is as readily found by first setting $\alpha = 0$ in (18) and then integrating the remaining terms:

$$y' = \frac{1 + x^2}{x}.$$

The limiting form of solution in the other exceptional case, $\alpha = -2$, is handled similarly. The equation is rearranged so that the singular terms are grouped together:

$$y(x) = \frac{1}{\alpha + 2}\left(x^2 - \frac{1}{x^\alpha}\right) + \frac{1}{\alpha} + \left(1 - \frac{1}{\alpha}\right)\frac{1}{x^\alpha}$$

The limit $\alpha \to -2$ can then be taken using l'Hôpital's rule (see Exercise 5), and the result is

$$y = x^2 \log x - \tfrac{1}{2} + \tfrac{3}{2}x^2.$$

This can be confirmed by showing that this function satisfies (18) with $\alpha = -2$.

Equation (iii), a separable, nonlinear differential equation of the first order, illustrates the fact that the designations of independent and dependent variables are, and should be, self-serving. The equation can be rearranged in differential form as

$$g(y)\,dy = f(x)\,dx, \tag{20}$$

and it is then said to be *separated*. Note that y and x have comparable roles in this format. Each side of this separated equation can be integrated, and with the initial condition $y = y_0$ when $x = x_0$ it follows that

$$\int_{y_0}^{y} g(y)\,dy = \int_{x_0}^{x} f(x)\,dx$$

or

$$\int^{y} g(y)\,dy = \int^{x} f(x)\,dx + C.$$

The solution is an implicit equation for y as a function of x

$$G(y) = F(x),$$

where G and F are the appropriate indefinite integrals.

Example 2 Solve $y' = 3e^{y^2}x^2$ with $y = 0$ at $x = 0$.

This equation separates into

$$e^{-y^2}\,dy = 3x^2\,dx,$$

and integration yields

$$\frac{\pi^{1/2}}{2}\operatorname{erf}(y) = \int_0^y e^{-y^2}\,dy = x^3 + C.$$

Moreover, the imposed condition implies $C = 0$. In this case, it is simpler to describe x as a function of y, rather than vice versa, and the solution is

$$x = [\tfrac{1}{2}\pi^{1/2}\operatorname{erf}(y)]^{1/3}.$$

Example 3 Natural factors, such as space and available food supply, inhibit the growth of any population. Let P represent the population of a species and M the maximum size of a stable population (for which the death rate is just equal to the birth rate). Discuss the model of population growth given by

$$\frac{dP}{dt} = \alpha P(M - P)$$

and find $P(t)$ if $P = P_0 < M$ at $t = 0$.

For small P the model states that growth is proportional to the size of the population: $dP/dt \approx \alpha MP$. Once P exceeds M, the growth rate turns negative and the population decreases. The population remains constant when $M = P$, that is, $dP/dt = 0$. The differential equation can be separated

$$\frac{dP}{P(M-P)} = \alpha\, dt$$

and integrated to yield

$$\int \frac{dP}{P(M-P)} = \alpha t + C.$$

Since

$$\frac{1}{P(M-P)} = \frac{1}{M}\left(\frac{1}{P} + \frac{1}{M-P}\right)$$

it follows that

$$\int \frac{dP}{P(M-P)} = \frac{1}{M} \log \frac{P}{M-P}.$$

Therefore

$$\frac{1}{M} \log \frac{P}{M-P} = \alpha t + C,$$

or

$$\frac{P}{M-P} = C^* e^{\alpha M t}, \qquad \text{where} \quad C^* = e^{CM}.$$

Since $P = P_0$ at $t = 0$, the constant is

$$C^* = \frac{P_0}{M - P_0},$$

and the explicit solution for P is

$$P = \frac{M}{1 + (M/P_0 - 1)e^{-\alpha M t}}.$$

Note that for $\alpha > 0$, $P \to M$ as $t \to \infty$.

Equation (iv) is exactly solved, for any ε, by making x the dependent variable and y the independent variable. Since

$$\frac{dy}{dx} = \frac{1}{dx/dy},$$

the equation can be written

$$(1 - \varepsilon y^2)\frac{dx}{dy} = 1.$$

Note that a nonlinear equation for y becomes a linear differential equation for x. Separation and integration yield

$$x = \int \frac{dy}{1 - \varepsilon y^2} + C,$$

$$x = \frac{1}{\sqrt{\varepsilon}} \tanh^{-1} \sqrt{\varepsilon} y + C.$$

The last expression can be inverted, to obtain

$$y = \frac{1}{\sqrt{\varepsilon}} \tanh \sqrt{\varepsilon}(x - C),$$

which is the complete solution of the original nonlinear equation.

Suppose next that problem (iv) is modified slightly in a way that makes the particular method just employed useless. It does not take much of a change to accomplish this, and an equation of the form

$$y' = f(x) - \varepsilon y^2 \tag{21}$$

will do nicely. The question now is to develop a technique—any technique—of finding a solution or a good approximation to one. One method, suggested by the notation, is to try a perturbation series based on the assumption that ε is small, because this will pivot the general problem about the simpler and feasible calculation $\varepsilon = 0$. In effect, the solution $y(x, \varepsilon)$ is expressed as a series in powers of the now small perturbation parameter ε:

$$y(x, \varepsilon) = y_0(x) + \varepsilon y_1(x) + \varepsilon^2 y_2(x) + \cdots. \tag{22}$$

The substitution of this in (21) yields

$$\begin{aligned} y_0'(x) + \varepsilon y_1'(x) + \varepsilon y_2'(x) + \cdots &= f(x) - \varepsilon\{y_0{}^2(x) + 2\varepsilon y_0(x) y_1(x) + \cdots\}, \\ &= f(x) - \varepsilon y_0{}^2(x) - 2\varepsilon^2 y_0(x) y_1(x) + \cdots. \end{aligned}$$

Upon equating the coefficients of like powers of ε in the preceding expression an infinite number of linear problems is obtained:

$$\begin{aligned} y_0'(x) &= f(x), \\ y_1'(x) &= -\, y_0{}^2(x), \\ y_2'(x) &= -2 y_0(x) y_1(x), \\ &\cdots\cdots\cdots\cdots \end{aligned} \tag{23}$$

These are solved sequentially, and each successive problem depends on its predecessors. For example,

$$y_0(x) = \int f(x)\, dx, \qquad y_1(x) = -\int y_0{}^2(x)\, dx, \qquad \text{etc.}$$

A condition of the type $y = \beta$ when $x = \alpha$, that is,

$$y(\alpha, \varepsilon) = y_0(\alpha) + \varepsilon y_2(\alpha) + \varepsilon^2 y_2(\alpha) + \cdots = \beta,$$

implies that

$$y_0(\alpha) = \beta, \qquad y_1(\alpha) = 0, \qquad y_2(\alpha) = 0, \qquad \text{etc.}$$

In this situation every function $y_n(x)$ must satisfy a definite constraint, and the particular solution becomes

$$y_0(x) = \beta + \int_\alpha^x f(x)\,dx, \qquad y_1(x) = -\int_\alpha^x y_0^{\,2}(x)\,dx, \ldots. \tag{24}$$

To illustrate, (iv) is solved anew by a perturbation series, and for the sake of simplicity, the condition $y(0) = 0$ is imposed. The identification $f(x) = 1$ and $\alpha = 0$, $\beta = 0$ in (21) and (24) and substitution in the preceding equations lead to the results

$$y_0(x) = \beta + \int_0^x dx = x, \qquad y_1(x) = -\frac{x^3}{3}, \qquad \ldots.$$

Therefore the perturbation solution is

$$y = x - \frac{\varepsilon x^3}{3} + \cdots,$$

which is the correct power-series expansion of the exact solution

$$y = \frac{1}{\sqrt{\varepsilon}} \tanh \sqrt{\varepsilon} x.$$

The perturbation series replaces a single difficult problem with an infinite number of simpler calculations. In analogy with the mechanical advantage derived from certain devices, less effort is applied over a more extended period of time.

7

Finally, attempts can be made to solve the first-order equation (2) numerically (Sec. 6.10), by iteration (Exercise 16), or by series expansion, as illustrated next.

Example 4 Solve

$$y' = f(x) - \varepsilon y^2 \qquad \text{with} \qquad y(0) = 0$$

by series expansion about $x = 0$.

The function $f(x)$ is replaced by its Taylor series expansion

$$f(x) = f(0) + f'(0)x + \frac{f''(0)}{2}x^2 + \cdots,$$

and a solution is assumed of the form

$$y = a_1 x + a_2 x^2 + a_3 x^3 + \cdots = \sum_{n=1}^{\infty} a_n x^n,$$

which automatically satisfies the imposed condition. Since

$$y'(x) = a_1 + 2a_2 x + 3a_3 x^2 + \cdots$$

and

$$[y(x)]^2 = a_1{}^2 x^2 + 2a_1 a_2 x^3 + \cdots,$$

the replacement of all these series in the differential equation yields

$$a_1 + 2a_2 x + 3a_3 x^2 + \cdots = f(0) + f'(0)x + \frac{f''(0)}{2} x^2 + \cdots - \varepsilon(a_1{}^2 x^2 + \cdots).$$

Upon equating the coefficients of like powers of x, we obtain

$$\begin{aligned} a_1 &= f(0), \\ 2a_2 &= f'(0), \\ 3a_3 &= \frac{f''(0)}{2} - \varepsilon a_1{}^2 = \frac{f''(0)}{2} - \varepsilon[f(0)]^2, \\ &\cdots\cdots\cdots\cdots\cdots\cdots \end{aligned}$$

The coefficients a_n can then be determined, and the power series solution is

$$y = f(0)x + \frac{f'(0)}{2} x^2 + \frac{1}{3}\left[\frac{f''(0)}{2} - \varepsilon f^2(0)\right] x^3 + \cdots.$$

If, in particular, $f(x) = 1$, it follows that

$$y = x - \frac{\varepsilon}{3} x^3 + \cdots,$$

which is in agreement with earlier results, $y = (1/\sqrt{\varepsilon}) \tanh \sqrt{\varepsilon}x$.

3

Higher-order equations like (v) and (vi) are equivalent to *systems of first-order equations.* For example,

$$y'' - y = 0$$

is obtained from the first-order equations

$$y' - v = 0 \qquad \text{and} \qquad v' - y = 0$$

by eliminating v. It seems reasonable to conjecture that the solution of a system of first-order equations will contain one arbitrary constant for each equation. Therefore, two constants should then be associated with the second-order equation; the general solution of this particular equation determined on page 354,

$$y = Ae^x + Be^{-x},$$

shows that this is indeed the case.

Linear equations of any order such as (v) (inhomogeneous in form as well) can often be efficiently solved by the operational methods of Secs. 4.6 and 4.7. The relevant technique is developed in any good course in differential equations.

The third-order equation (vi) is equivalent to three first-order differential equations. For example (vi) is obtained from the system

$$y' = v,$$
$$v' = u,$$
$$u' = \sin y$$

by eliminating u and v. The general solution should contain three arbitrary constants, and we might guess now that the number of such constants will always be the same as the order of the equation. The solution of an nth order differential equation

$$F(x, y, y', y'', \ldots, y^{(n)}) = 0$$

will involve n arbitrary constants, which can be determined by setting n subsidiary conditions on the function $y(x)$.

Exercises 4.8

1. Determine the orders of the following differential equations. Which are linear?

(i) $y'' - 8y = 4x^2 + 1$ **(ii)** $y'y''' - (y'')^2 = 1$ **(iii)** $yy' = \cos^2 x$

(iv) $x^2y' - x^3y = e^x$ **(v)** $y' = \sin y''$ **(vi)** $y''' - 2y' = x \sin x$

(vii) $y^{(v)} + e^y = y''' + 3y^2$ **(viii)** $y^{(vi)} = 4x^3 + 3x + 2$ **(ix)** $xy' - y = y^2$

2. Determine first order differential equations satisfied by the following one-parameter families of functions.

(i) $x^2 + y^2 = a^2$ **(ii)** $y = ax + a$ **(iii)** $y = 1/(x + a)$

3. Construct the direction fields for the following equations.

(i) $y' = \cos x$ **(ii)** $xy' + 2y = e^{-x}$ **(iii)** $y' = 1 - \frac{1}{10}y^2$

4. Solve the differential equations in the previous example and determine the solutions that satisfy $y(1) = \frac{1}{2}$.

5. Solve $xy' + \alpha y = 1 + x^2$ with $y(1) = 1$ in the particular case $\alpha = -2$. Compare the answer with that obtained by the limit process in Example 1.

6. Solve the equation $(y')^3 + xy' - y = 0$. (*Hint:* First differentiate the equation.) Show that a family of straight lines forms solutions. Are there any other solutions?

7. Solve by separation of variables.

(i) $\dfrac{dy}{dx} = \dfrac{-x}{y}$ **(ii)** $\dfrac{dy}{dx} = \dfrac{y}{x}$ **(iii)** $\dfrac{dy}{dx} = e^x y$

(iv) $\dfrac{dy}{dx} = y \cos x$ **(v)** $e^{y+x^2} \dfrac{dy}{dx} = \dfrac{-x}{y}$ **(vi)** $\dfrac{d\theta}{dt} = \dfrac{\sin t}{\cos^2 \theta}$

(vii) $y' + xy = y$ **(viii)** $y' = x^2(1 + y^2)$ **(ix)** $\dfrac{dr}{dt} = ar \sin \omega t$

8. Use integrating factors to solve the following equations. In each case, identify a particular solution and the general solution of the associated homogeneous equation.

(i) $y' + y = 3$ **(ii)** $xy' + y = x$ **(iii)** $y' - 3y = 2x + 5$
(iv) $y' + 2xy = e^{-x^2}$ **(v)** $x^2y' + 3xy = \log x$ **(vi)** $(\sin x)y' + (2\cos x)y = \cot x$

9. Show that an equation of the form $y' = f(y/x)$ is converted to a separable equation by the substitution $y = vx$. Use this method to solve the following equations.

(i) $\dfrac{dy}{dx} = \dfrac{y}{x+y}$ **(ii)** $\dfrac{dy}{dx} = \dfrac{x-y}{x+y}$ **(iii)** $\dfrac{dy}{dx} = \dfrac{y}{x} + \cos\left(\dfrac{y}{x}\right)$

(iv) $\dfrac{dy}{dx} = \dfrac{x^2+y^2}{xy}$ **(v)** $\dfrac{dy}{dx} = \dfrac{y}{x}\log\left(\dfrac{y}{x}\right)$ **(vi)** $x^2\dfrac{dy}{dx} = y^2 - 6x^2$

10. Show that an equation of the form $y' + p(x)y = g(x)y^m$ is reduced to a linear equation by the substitution $u = y^{1-m}$. Use this method to solve the following equations.

(i) $y' - y = e^x y^2$ **(ii)** $y' - 2y = xy^3$ **(iii)** $y' - 3y = e^{3x}y^4$
(iv) $y' - xy = e^{-x^2}y^3$ **(v)** $xy' - y = x^3y^2$ **(vi)** $y' - x^2y = x^2y^4$

11. Set up a numerical scheme to solve $y'(x) + p(x)y(x) = g(x)$ with $y(x_0) = y_0$ by introducing a finite difference approximation for the first derivative. Examine $y' + y = e^{-x}$, $y(0) = 1$ as a special case.

12. If $y(x, \epsilon) = \sin(x + \epsilon)$ is expanded in powers of the small parameter ϵ, $y(x, \epsilon) = y_0(x) + \epsilon y_1(x) + \epsilon^2 y_2(x) + \cdots$, determine a formula for the functions $y_n(x)$.

13. In the perturbation expansion $y(x, \epsilon) = y_0(x) + \epsilon y_1(x) + \epsilon^2 y_2(x) + \cdots$ where ϵ is a small parameter and $y(\alpha, \epsilon) = \beta$, show that $y_0(\alpha) = \beta$, $y_1(\alpha) = 0$, $y_2(\alpha) = 0$, $\cdots$. By comparison with a Taylor series, derive a formula for $y_n(x)$.

14. Construct an iteration procedure to solve the initial value problem

$$y' = f(x) - \epsilon y^2 \qquad y(\alpha) = \beta,\ 0 < \epsilon \ll 1$$

(Write $y'_{n+1} = f(x) - \epsilon y_n^2$ with $y_0(x) = \beta$.) Test this procedure in the special case $f(x) = 1$, $\alpha = \beta = 0$ by comparing the sequence of iterative solutions to the exact solution.

15. Find the solution of $\epsilon y'(x) + y(x) = 1$ which satisfies $y(0) = 0$ where $0 < \epsilon \ll 1$. (Do this using an integrating factor or by integrating $y'/(1-y) = 1/\epsilon$.) Attempt a perturbation expansion and discuss the inherent difficulties. Draw graphs of the solution for several values of ϵ.

16. Find solutions of the following initial value problems expressed as power series in x including terms up to x^5.

(i) $\dfrac{dy}{dx} = x - y$ $y(0) = 1$ **(ii)** $\dfrac{dy}{dx} = x^2 - y^2$ $y(0) = 1$

(iii) $\dfrac{dy}{dx} = 2y + xy^2$ $y(0) = 1$ **(iv)** $\dfrac{dy}{dx} = 2x + \sin y$ $y(0) = \dfrac{\pi}{2}$

17. Find solutions of the form $y = e^{mx}$ for the following differential equations.

(i) $y'' + 2y' - 3y = 0$ **(ii)** $y'' - 4y = 0$ **(iii)** $y'' - 8y' + 12y = 0$
(iv) $y''' + 9y' = 0$ **(v)** $y''' - 3y'' + 2y' = 0$ **(vi)** $y''' - 3y'' + 3y' - y = 0$

18. Find solutions of the form $y = x^m$ for the following differential equations.

(i) $xy' + 2y = 0$ **(ii)** $x^2y'' + 4xy' - 10y = 0$ **(iii)** $x^2y'' - 3xy' + 4y = 0$
(iv) $x^2y'' - 2y = 0$ **(v)** $x^2y'' - xy' + y = 0$ **(vi)** $x^3y''' - xy' = 0$

19. The volume $V(t)$ of the lungs at time t during maximal expiration satisfies the first order equation $aV'(t) = -k(V - r)$. The constants a, k and r measure the airway resistance, the elastic driving force, and the residual lung volume. If $k/a = 2(\sec^{-1})$, $r = 1.2$ (liters) and $V(0) = 3.0$ (liters), solve for $V(t)$. When are the lungs half empty? Draw a graph of $V(t)$.

20. A chemical dissolves into a solution at a rate that is proportional to the amount of undissolved chemical present and to the difference between the concentration of a saturated solution and that of the solution. Express this information in a first order differential equation and solve the equation.

21. In a first order chemical reaction $A + B \to X$, the chemical B is abundant relative to A. If $A(t)$ represents the concentration of reagent A at time t, then $A'(t) = -kA(t)$ where k is the reaction constant. Solve for $A(t)$ in terms of the initial concentration $a = A(0)$. If $x(t)$ is the concentration of the product X at time t, then $x'(t) = k(a - x)$. Assuming $x(0) = 0$, solve for $x(t)$. Draw graphs of $A(t)$ and $x(t)$.

22. In the previous problem, if the reaction is 90% completed after 20 minutes, when is it 99% completed? Determine the reaction constant.

23. In a second order chemical reaction $A + B \to X$, neither reagent is abundant relative to the other. If $A(t)$ and $B(t)$ are the concentrations of A and B at time t, then $A'(t) = B'(t) = -kAB$ where k is a reaction constant. If $x(t)$ is the product concentration at time t, then $x'(t) = kAB = k(a - x)(b - x)$ where $a = A(0)$ and $b = B(0)$. Assuming $x(0) = 0$, solve for $x(t)$. Verify that $\lim_{t\to\infty} x(t)$ is the smaller of a and b. Draw a graph of $x(t)$.

24. Water runs out of a container through a small opening on the side of the container near its bottom. The rate of outflow is proportional to the square root of the height of the water. (Potential energy proportional to height is being converted to kinetic energy proportional to velocity squared.) If the container is a right circular cylinder with axis vertical, determine an equation for $V(t)$, the volume of water at time t (min). If $V(0) = 100$ (liters) and $V(20) = 50$, when is $V(t) = 25$? When will the container be empty? (*Hint:* Find an equation for $h(t)$, the height of water at time t.)

25. The general solution of $y''(x) - y(x) = 0$ is $y(x) = Ae^x + Be^{-x}$. Solve $\epsilon y''(x) - y(x) = 0$, $\epsilon > 0$, by rescaling the independent variable, defining ξ by $x = \sqrt{\epsilon}\,\xi$.

26. The rectilinear motion of a body of constant mass m acted on by a force F is governed by Newton's law, $F = ma = m\, d^2s/dt^2$.

(i) If the applied force depends only on the velocity $v = ds/dt$ $(F = F(v))$, show that the equation is separable in the v and t variables or in the variables v and s.

(ii) If the force depends only on position $(F = F(s))$, show that one integration of Newton's law is possible by writing it as $mvdv = F(s)\, ds$. How is the second integral performed?

27. A body of mass m moves horizontally through a liquid with only a friction force opposing the motion acting on it. The motion is governed by $mdv/dt = F$ where $F = F(v)$ is the friction force. If the resisting force is proportional to the velocity of the body, determine $v(t)$ for $t > 0$ assuming $v(0)$ is given. How far will the object move and when will it stop?

28. If the friction force in the previous problem is proportional to the square of the velocity of the body, determine $v(t)$ for $t > 0$ and $s(t)$, the distance at time t. How far will the object move and when will it stop?

29. Oscillations of a body of mass m about an equilibrium position are described by Hooke's law: the restoring force is proportional to the displacement from equilibrium. If $s(t)$ is the displacement at time t from the equilibrium position, then $ms''(t) = -ks(t)$ where k is the Hooke's law constant. If $s(0) = a$ and $s'(0) = b$, determine $s(t)$.

30. An object in free fall near the earth's surface experiences two forces, gravity and air resistance. Define $s(t)$ to be the height of the object above the surface of the earth. If the force of air resistance is proportional to the speed of the object, then $s''(t) = -g - ks'(t)$ where $k > 0$ is a constant.

(i) If $s(0) = h$, $s'(0) = 0$, determine $s(t)$ and compare to the $k = 0$ special case.

(ii) Define $v(t) = s'(t)$ and $T(h,k) =$ time of fall from height h. Calculate $v(t)$ and $T(h,k)$.

(iii) Evaluate $\lim_{t\to\infty} v(t)$ and $\lim_{k\to 0} T(h,k)$. Interpret these limits.

5
Vectors

5.1 Basic concepts

Many physical quantities are entirely specified by a numerical value measured in appropriate units. For example, normal body temperature is 98.6°F, water weighs 1 gm/cm^3, a severe earthquake registers 8 on the logarithmic Richter scale. Quantities that are determined by a magnitude alone are called *scalars* or scalar quantities. Common examples are temperature, mass, volume, density, time, pressure, length, energy, and electric potential.

However, not all variables occurring in nature are scalars. An important class of measurements requires not only a magnitude but also a direction for a complete specification. For example in the determination of wind velocity, knowledge of the wind speed, say 15 km/h, is insufficient information, as the wind direction must also be given, say 15 km/h from the southwest. Wind speed is a magnitude, a scalar quantity, while velocity is speed plus direction. Variables of this type are called *vectors*. Other familiar examples are force, displacement, acceleration, angular velocity, and electric field.

The concept of a vector quantity has long been recognized, but the ability to deal directly with vectors in mathematical form was not developed until the turn of this century and was largely due to the American physicist Willard Gibbs (1839-1903).

Today a knowledge of vectors is fundamental in all scientific disciplines. In this chapter, the aim is to provide a motivated introduction to the subject, and for the most part, the discussion is restricted to vectors which arise in the physical world, namely, three-dimensional vectors. Initially, the emphasis will be largely geometrical.

As a preliminary, we sketch the conventional way of representing the position of a point in three-dimensional space. With a point O, the origin, chosen as a reference, three mutually perpendicular lines Ox, Oy, Oz (see Fig. 1.1), called *axes*, are constructed. The three mutually perpendicular planes, yOz, zOx, xOy, are called the *coordinate planes*. The position of a point $P = (x, y, z)$ is then specified, relative to O, by the three numbers x, y, z, its respective signed distances from these planes. This configuration is called a *cartesian frame* and can be imagined as arising from the

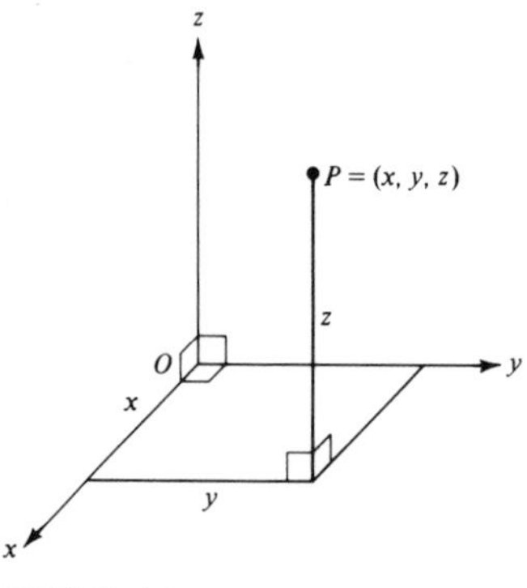

FIGURE 1.1

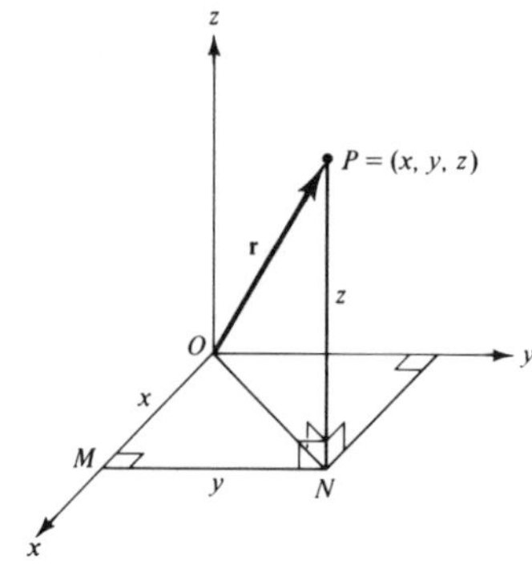

FIGURE 1.2

usual two-dimensional horizontal axes Ox, Oy by the addition of a third axis Oz vertically upward. Such a system of axes is called *right-handed*. (If Oz were drawn vertically downward, the system would be called left-handed, but it is conventional to use only right-handed axes.)

All points having coordinates $(x, y, 0)$ lie in the xOy plane, so that $z = 0$ is the equation of this plane. Likewise $x = 0$ and $y = 0$ are the equations of the planes yOz and zOx, respectively.

2

Armed with these basic ideas of three-dimensional analytic geometry, we return to the discussion of vectors, those quantities specified by both magnitude and direction. A vector can be represented in a three-dimensional coordinate frame Ox, Oy, Oz by drawing a line segment in the proper direction with length OP equal to the magnitude of the vector. The directed line segment OP, extending *from O to P*, represents the vector and is written $\overrightarrow{OP} = \mathbf{r}$. Boldface type or an arrow above is the standard notation for vectors. The length of the vector $\overrightarrow{OP}$ is denoted by $|\overrightarrow{OP}| = |\mathbf{r}|$, or more simply by $OP = r = |\mathbf{r}|$. If P is the point (x, y, z), then from Fig. 1.2 it follows that

$$OP = \sqrt{x^2 + y^2 + z^2}. \tag{1}$$

This is established by applying the Pythagorean theorem to each of the right-angled triangles OMN and ONP, to obtain

$$ON^2 = OM^2 + MN^2 = x^2 + y^2$$

and

$$OP^2 = ON^2 + NP^2 = x^2 + y^2 + z^2.$$

Two scalars are equal if they have the same magnitude. *For two vectors to be equal both their magnitudes and directions must be the same.* Certainly this definition fits with one's intuitive feel for the equality of vectors. (Compare with the earlier example of wind velocity.) Let $\overrightarrow{OP'} = \mathbf{r}'$, where $P' = (x', y', z')$; then $\overrightarrow{OP} = \overrightarrow{OP'}$ if and only if $x' = x$, $y' = y$, and $z' = z$. (Unless otherwise stated, only free vectors will be considered. This means that a vector is free to move under parallel displacements, and two vectors are equal if by translation they can be superposed exactly upon each other.)

Multiplication of a vector **a** by a positive scalar λ is defined to represent another vector having the same direction and a new magnitude of λ times the original length. Multiplication by a negative number reverses the direction of the vector and changes its magnitude accordingly (see Fig. 1.3). It is clear that for any two points, $\overrightarrow{OP} = -\overrightarrow{PO}$; that is, the vector *from O to P* is just the negative of the vector *from P to O*.

For completeness a *zero vector*, denoted by **0**, is any vector of zero length (direction is then quite irrelevant). The significance of **0** in vector equations is similar to that of the scalar zero in ordinary algebra.

Addition of vectors is defined to conform to the known experimental fact that the resultant of two forces obeys the *triangle law* (also known as the *parallelogram law*). This means that the sum of two vectors $\overrightarrow{OA}$ and $\overrightarrow{AB}$ is the vector $\overrightarrow{OB}$ (see Fig. 1.4), and we write

$$\overrightarrow{OB} = \overrightarrow{OA} + \overrightarrow{AB}. \tag{2}$$

(When the vectors to be added form the legs of a triangle, their sum is the third side.) If the point C is added to complete the parallelogram, then by the definition of the equality of vectors, $\overrightarrow{OA} = \overrightarrow{CB}$ and $\overrightarrow{AB} = \overrightarrow{OC}$, so that Eq. (2) can be rewritten as

$$\overrightarrow{OB} = \overrightarrow{CB} + \overrightarrow{OC}.$$

But from the triangle OCB

$$\overrightarrow{OB} = \overrightarrow{OC} + \overrightarrow{CB}.$$

Comparing the last two equations shows that vectors can be added in any order, or, more formally, vector addition is commutative.

In the form $\overrightarrow{OB} = \overrightarrow{OA} + \overrightarrow{OC}$ we see that addition of two vectors is equivalent to the physical problem of finding the resultant of two forces applied at a single point.

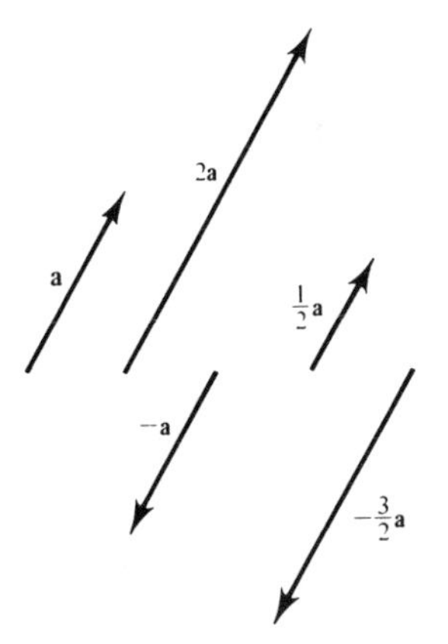

FIGURE 1.3

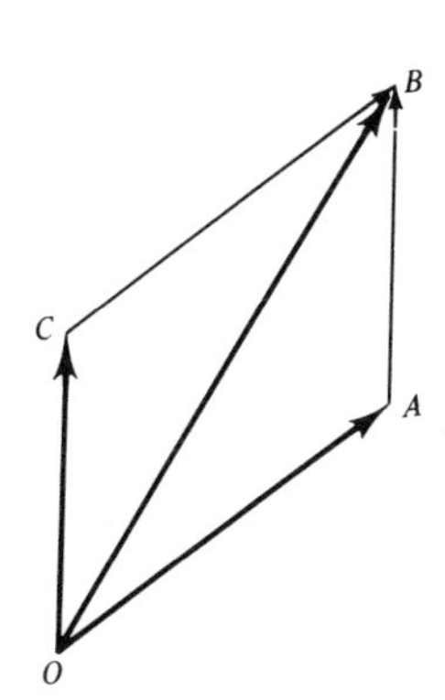

FIGURE 1.4

On this basis we would expect addition to be independent of order. The above geometrical argument merely verifies this fact.

The sum of two vectors $\overrightarrow{OC}$, $\overrightarrow{OA}$ is the diagonal $\overrightarrow{OB}$ of the parallelogram shown in Fig. 1.4; their difference is the other diagonal of this geometrical figure, i.e.,

$$\overrightarrow{OC} - \overrightarrow{OA} = \overrightarrow{AC}.$$

To establish this, we start with the addition formula

$$\overrightarrow{OC} = \overrightarrow{OA} + \overrightarrow{AC}.$$

By adding the vector $\overrightarrow{AO} = -\overrightarrow{OA}$ to both sides of this equation, i.e., transposing $\overrightarrow{OA}$ to the left-hand side, we obtain the desired result

$$\overrightarrow{OC} - \overrightarrow{OA} = \overrightarrow{AC}.$$

In terms of vectors **a**, **b**, the sum **a** + **b** and the difference **a** − **b** are shown in Fig. 1.5.

The addition of n vectors is obtained by repeated application of the triangle rule. As shown in Fig. 1.6,

$$\overrightarrow{OA_2} = \overrightarrow{OA_1} + \overrightarrow{A_1A_2}, \qquad \overrightarrow{OA_3} = \overrightarrow{OA_2} + \overrightarrow{A_2A_3},$$

and therefore

$$\overrightarrow{OA_3} = \overrightarrow{OA_1} + \overrightarrow{A_1A_2} + \overrightarrow{A_2A_3}.$$

Continuing in this way makes it evident that

$$\overrightarrow{OA_n} = \overrightarrow{OA_1} + \overrightarrow{A_1A_2} + \cdots + \overrightarrow{A_{n-1}A_n}. \tag{3}$$

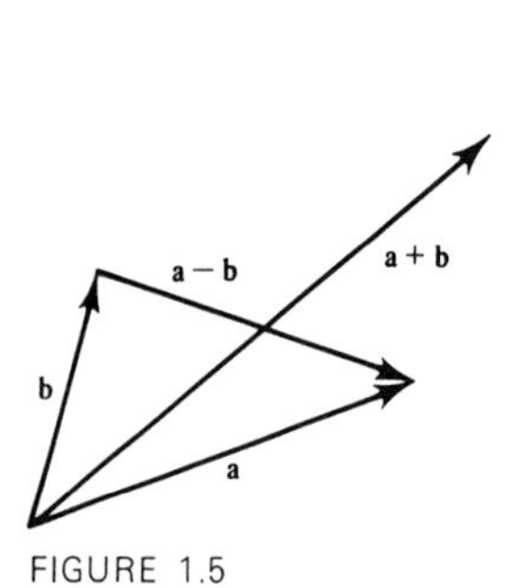

FIGURE 1.5

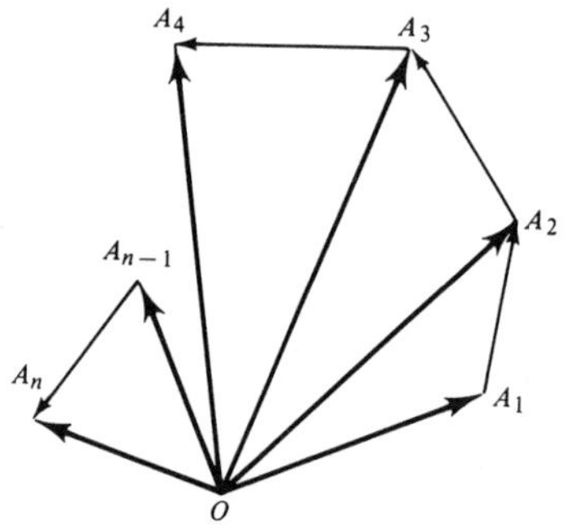

FIGURE 1.6

Example 1 Solve the vector equation $\mathbf{b} - \mathbf{x} = \mathbf{x} - \mathbf{a}$.

The ordinary rules of addition can be applied, so that

$$\mathbf{x} = \tfrac{1}{2}(\mathbf{a} + \mathbf{b}).$$

This solution has a simple geometrical interpretation, illustrated in Fig. 1.7. With

$$\overrightarrow{OA} = \mathbf{a}, \qquad \overrightarrow{OB} = \mathbf{b} \qquad \text{and} \qquad \overrightarrow{OC} = \mathbf{x}$$

the point C is located such that $\overrightarrow{AC} = \overrightarrow{CB}$. In other words C is the midpoint of AB.

A *unit vector* is any vector having magnitude 1 but arbitrary direction. Therefore, if $\mathbf{r}$ is a given vector,

$$\hat{\mathbf{r}} = \frac{\mathbf{r}}{|\mathbf{r}|} = \frac{\overrightarrow{OP}}{OP}$$

is a vector having unit length in the direction of $\overrightarrow{OP}$.

A *constant vector* has both fixed magnitude and direction. Constant unit vectors of particular importance in rectangular axes are those parallel to the positive x, y, and z directions. They will be denoted by $\mathbf{i}$, $\mathbf{j}$, and $\mathbf{k}$, respectively (Fig. 1.8). Let M and N be defined as the points $(x, 0, 0)$ and $(x, y, 0)$; then $\overrightarrow{OM} = x\mathbf{i}$, $\overrightarrow{MN} = y\mathbf{j}$, and $\overrightarrow{NP} = z\mathbf{k}$. Therefore

$$\overrightarrow{OP} = \overrightarrow{OM} + \overrightarrow{MN} + \overrightarrow{NP},$$

or

$$\mathbf{r} = \overrightarrow{OP} = x\mathbf{i} + y\mathbf{j} + z\mathbf{k}.$$

The scalars x, y, z are called the *components* of the vector $\mathbf{r}$ referred to the particular cartesian frame.

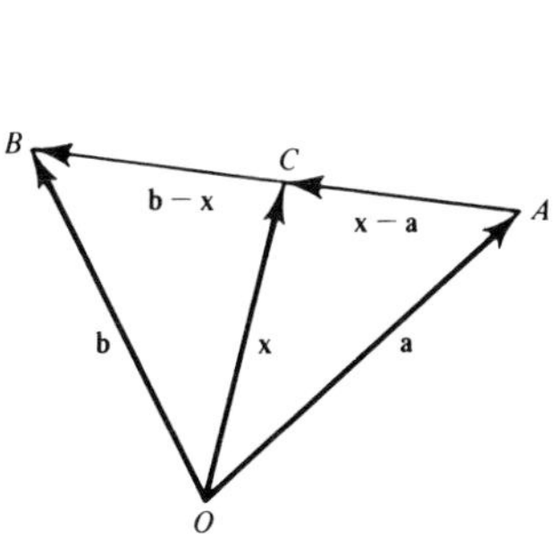

FIGURE 1.7

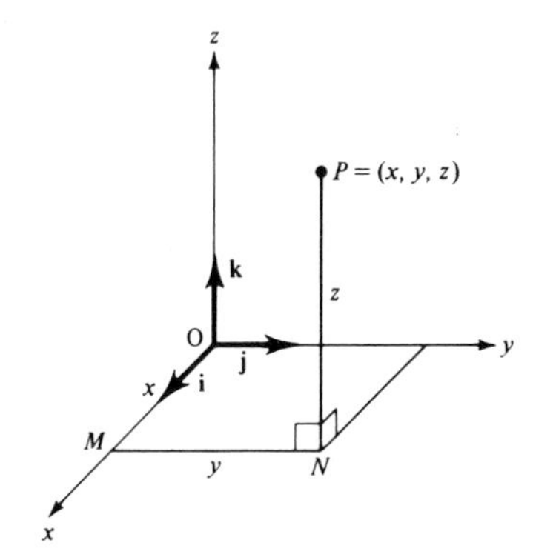

FIGURE 1.8

Any vector $\mathbf{a}$ can be written in component form

$$\mathbf{a} = a_1\mathbf{i} + a_2\mathbf{j} + a_3\mathbf{k}, \tag{4}$$

and a_1, a_2, and a_3 are called the x, y, and z components, respectively, of the vector $\mathbf{a}$.

By the law of vector addition, if

$$\mathbf{a} = a_1\mathbf{i} + a_2\mathbf{j} + a_3\mathbf{k} \qquad \text{and} \qquad \mathbf{b} = b_1\mathbf{i} + b_2\mathbf{j} + b_3\mathbf{k},$$

then

$$\mathbf{a} + \mathbf{b} = (a_1 + b_1)\mathbf{i} + (a_2 + b_2)\mathbf{j} + (a_3 + b_3)\mathbf{k}. \tag{5}$$

Similarly if λ and μ are scalars, then

$$\lambda\mathbf{a} + \mu\mathbf{b} = (\lambda a_1 + \mu b_1)\mathbf{i} + (\lambda a_2 + \mu b_2)\mathbf{j} + (\lambda a_3 + \mu b_3)\mathbf{k}. \tag{6}$$

In particular, note that

$$|\mathbf{a} - \mathbf{b}| = [(a_1 - b_1)^2 + (a_2 - b_2)^2 + (a_3 - b_3)^2]^{1/2},$$

is the distance between the two points (a_1, a_2, a_3) and (b_1, b_2, b_3). Two vectors $\mathbf{a}$ and $\mathbf{b}$ are equal to each other when $|\mathbf{a} - \mathbf{b}| = 0$, and this implies that their individual components are identical:

$$a_1 = b_1, \qquad a_2 = b_2, \qquad a_3 = b_3.$$

Example 2 If $\mathbf{a} = 2\mathbf{i} - \mathbf{j} + 2\mathbf{k}$ and $\mathbf{b} = 3\mathbf{i} + 4\mathbf{j} - 5\mathbf{k}$, find $\mathbf{a} + 2\mathbf{b}$, $|\mathbf{a} - \mathbf{b}|$, and a unit vector in the direction of $3\mathbf{a} - \mathbf{b}$.

These computations are straightforward applications of earlier results:

$$\begin{aligned} \mathbf{a} + 2\mathbf{b} &= 2\mathbf{i} - \mathbf{j} + 2\mathbf{k} + 2(3\mathbf{i} + 4\mathbf{j} - 5\mathbf{k}), \\ &= 8\mathbf{i} + 7\mathbf{j} - 8\mathbf{k}. \\ \mathbf{a} - \mathbf{b} &= 2\mathbf{i} - \mathbf{j} + 2\mathbf{k} - (3\mathbf{i} + 4\mathbf{j} - 5\mathbf{k}), \\ &= -\mathbf{i} - 5\mathbf{j} + 7\mathbf{k}, \end{aligned}$$

so that

$$|\mathbf{a} - \mathbf{b}| = [(-1)^2 + (-5)^2 + 7^2]^{1/2} = \sqrt{75} = 5\sqrt{3}.$$

A unit vector in the direction of $3\mathbf{a} - \mathbf{b}$ is $(3\mathbf{a} - \mathbf{b})/|(3\mathbf{a} - \mathbf{b})|$, and this is found to be

$$\frac{3\mathbf{i} - 7\mathbf{j} + 11\mathbf{k}}{\sqrt{179}}.$$

4

Apart from the fact that many physical variables are vectors, it is often convenient to think of a set of scalars (x, y, z) as the components of a vector. In such cases the number of scalars need not be limited to three, and a vector can have n components $(x_1, x_2, \ldots, x_n)$. Such vectors are called *n-dimensional vectors*, and while the geometrical representation is no longer useful, the algebra still follows the simple rules outlined earlier. In any problem involving sets of n scalars, it may be advantageous to think of them as the components of a vector.

Example 3 In a given year a manufacturer produces x_1 refrigerators, y_1 washers, z_1 driers, and w_1 ovens. In the same year a second manufacturer produces x_2 refrigerators, y_2 washers, z_2 driers, and w_2 ovens. If the production vectors of the two manufacturers $\mathbf{p}_i$, $(i = 1, 2)$ have components (x_i, y_i, z_i, w_i), interpret the vectors $\mathbf{p}_1 + \mathbf{p}_2$, $\mathbf{p}_1 - \mathbf{p}_2$.

The vector $\mathbf{p}_1 + \mathbf{p}_2$ has components $(x_1 + x_2,\ y_1 + y_2,\ z_1 + z_2,\ w_1 + w_2)$ and is the total-production vector. The components, in order, give the combined output of refrigerators, washers, driers, and ovens.

In a similar way $\mathbf{p}_1 - \mathbf{p}_2$ has components $(x_1 - x_2, y_1 - y_2, z_1 - z_2, w_1 - w_2)$, and each component represents the amount by which the output of the first manufacturer exceeds that of the second for each appliance.

Two simple examples will serve to illustrate the usefulness of vectors in kinematical and geometrical problems.

Example 4 A boat with a speed of U mph in still water must travel straight across a river flowing with uniform speed of V mph. In what direction should the boat head, and what is its actual speed? Is the trip always possible?

Assume that the direction the boat must head makes an angle α with the actual travel direction, and let x and y axes be chosen as indicated in the Fig. 1.9. The true velocity of the boat $\mathbf{w}$ is the vector sum of the velocity $\mathbf{u}$ in still water and the river velocity $\mathbf{v}$, that is,

$$\mathbf{w} = \mathbf{u} + \mathbf{v}.$$

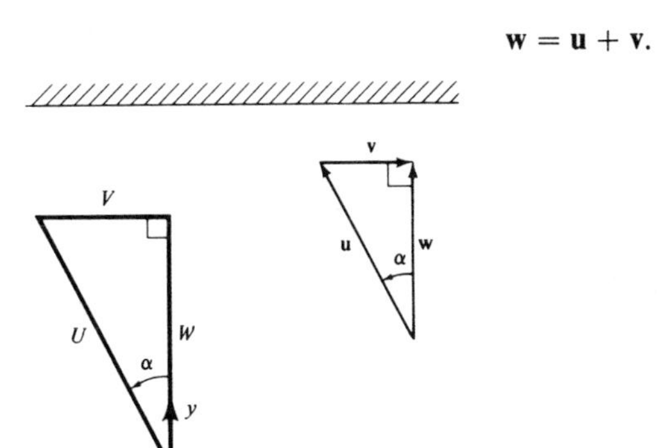

FIGURE 1.9

Let $|\mathbf{u}| = U$, $|\mathbf{v}| = V$, and $|\mathbf{w}| = W$, and let $\mathbf{i}$, $\mathbf{j}$ denote the usual unit vectors; then it follows that $\mathbf{u}$, $\mathbf{v}$, $\mathbf{w}$ have the component forms

$$\mathbf{u} = -U \sin \alpha\, \mathbf{i} + U \cos \alpha\, \mathbf{j}, \qquad \mathbf{v} = V\mathbf{i}, \qquad \mathbf{w} = W\mathbf{j}.$$

By equating the x and y components of $\mathbf{w} = \mathbf{u} + \mathbf{v}$ we obtain two scalar equations for the unknowns W and α;

$$W = U \cos \alpha, \qquad V = U \sin \alpha,$$

so that the required direction α is given by $\alpha = \sin^{-1}(V/U)$ and the actual speed is $W = \sqrt{U^2 - V^2}$. If $V > U$, W is no longer a real number and the trip straight across is not possible. Of course, even if $V > U$, it is possible for the boat to cross the river, but it would end up somewhere downstream.

Example 5 Prove that the lines joining the midpoints of the opposite edges of a tetrahedron intersect and bisect each other.

Take any point O as an origin, and let A, B, C, D be the vertices of a tetrahedron, with $\overrightarrow{OA} = \mathbf{a}$, $\overrightarrow{OB} = \mathbf{b}$, $\overrightarrow{OC} = \mathbf{c}$, $\overrightarrow{OD} = \mathbf{d}$, as shown in Fig. 1.10. If E and F are the midpoints of AB and CD (one pair of opposite edges of the tetrahedron), we first show that

$$\overrightarrow{OE} = \tfrac{1}{2}(\overrightarrow{OA} + \overrightarrow{OB}) = \tfrac{1}{2}(\mathbf{a} + \mathbf{b}).$$

This follows from the fact that

$$\overrightarrow{OE} = \overrightarrow{OB} + \overrightarrow{BE},$$

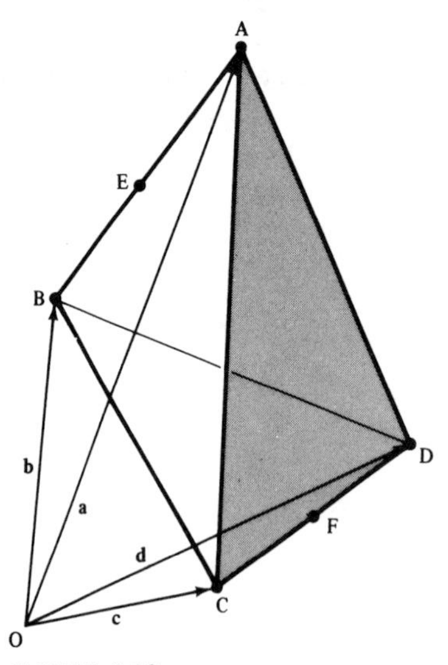

FIGURE 1.10

and, since E is the midpoint of BA (using Example 1),

$$\overrightarrow{BE} = \tfrac{1}{2}\overrightarrow{BA} = \tfrac{1}{2}(\overrightarrow{OA} - \overrightarrow{OB}).$$

Use of this expression for $\overrightarrow{BE}$ gives

$$\overrightarrow{OE} = \overrightarrow{OB} + \tfrac{1}{2}(\overrightarrow{OA} - \overrightarrow{OB}) = \tfrac{1}{2}(\overrightarrow{OA} + \overrightarrow{OB}) = \tfrac{1}{2}(\mathbf{a} + \mathbf{b}).$$

In the same manner, since F is the midpoint of CD,

$$\overrightarrow{OF} = \tfrac{1}{2}(\mathbf{c} + \mathbf{d}).$$

If P is the midpoint of EF, a similar argument yields

$$\begin{aligned}\overrightarrow{OP} &= \tfrac{1}{2}(\overrightarrow{OE} + \overrightarrow{OF}) = \tfrac{1}{2}[\tfrac{1}{2}(\mathbf{a} + \mathbf{b}) + \tfrac{1}{2}(\mathbf{c} + \mathbf{d})],\\ &= \tfrac{1}{4}(\mathbf{a} + \mathbf{b} + \mathbf{c} + \mathbf{d}).\end{aligned}$$

But the symmetry of this expression shows that each pair of opposite edges would lead to the same expression, and therefore P is the common point of intersection.

Exercises 5.1

1. Four points A, B, C, and D have coordinates given by $A = (1,1,0)$, $B = (-1,2,1)$, $C = (2,3,4)$ and $D = (0,-2,7)$. Find the vectors $\overrightarrow{AB}$, $\overrightarrow{BC}$, $\overrightarrow{CD}$, $\overrightarrow{DA}$, $\overrightarrow{AC}$ and $\overrightarrow{BD}$. Verify that $\overrightarrow{AB} + \overrightarrow{BC} = \overrightarrow{AC}$ and $\overrightarrow{AB} + \overrightarrow{BC} + \overrightarrow{CD} + \overrightarrow{DA} = \mathbf{0}$.

2. For the tetrahedron with vertices A, B, C and D of the previous problem, find the midpoints of the six edges and verify the result of Example 5.

3. If $A = (1,-2,5)$ and $B = (3,1,11)$, find $\overrightarrow{AB}$ and $\overrightarrow{BA}$. Determine the distances OA, OB and AB and verify that $AB < OA + OB$. Illustrate with a diagram.

4. If $\mathbf{a} = \mathbf{i} + 2\mathbf{j} - 2\mathbf{k}$ and $\mathbf{b} = 3\mathbf{i} - \mathbf{j} + \mathbf{k}$, find $\mathbf{a} - \mathbf{b}$, $\mathbf{a} + \mathbf{b}$, $\mathbf{a} + 2\mathbf{b}$, $2\mathbf{a} - \mathbf{b}$. Illustrate with a diagram and explain why all six vectors lie in the same plane. Find unit vectors in the directions of these six vectors.

5. Use a diagram to indicate the following vectors.

(i) $\mathbf{i} + \mathbf{j}$ (ii) $\mathbf{j} + \mathbf{k}$ (iii) $\mathbf{k} + \mathbf{i}$

(iv) $\mathbf{i} + \mathbf{j} + \mathbf{k}$ (v) $\mathbf{i} + \mathbf{j} - \mathbf{k}$ (vi) $-\mathbf{i} - 2\mathbf{j} + 3\mathbf{k}$.

6. If $\mathbf{a} = \mathbf{i} - 2\mathbf{j} + 3\mathbf{k}$, $\mathbf{b} = 7\mathbf{i} - \mathbf{j}$ and $\mathbf{c} = -\mathbf{i} + 3\mathbf{k}$, determine the following vectors. Which are unit vectors?

(i) $\mathbf{a} + \mathbf{b}$ (ii) $\mathbf{a} + \mathbf{b} + \mathbf{c}$ (iii) $\mathbf{b} - \mathbf{c}$

(iv) $\dfrac{\mathbf{a} + \mathbf{b}}{|\mathbf{c}|}$ (v) $\dfrac{\mathbf{a} + \mathbf{b} + \mathbf{c}}{|\mathbf{a} + \mathbf{b} + \mathbf{c}|}$ (vi) $\dfrac{\mathbf{a}}{|\mathbf{a}|} + \dfrac{\mathbf{b}}{|\mathbf{b}|} + \dfrac{\mathbf{c}}{|\mathbf{c}|}$

7. Use vector methods to show that the lines joining the midpoints of the sides of a quadrilateral (taken in order) form a parallelogram.

8. If $\mathbf{a}$ and $\mathbf{b}$ are two given vectors, verify the following inequalities and explain their geometrical significance.

(i) $|\mathbf{a} + \mathbf{b}| \leq |\mathbf{a}| + |\mathbf{b}|$ (ii) $|\mathbf{a} - \mathbf{b}| \geq \big|\,|\mathbf{a}| - |\mathbf{b}|\,\big|$.

9. Use vector methods to prove that the three medians of a triangle are concurrent. (The point of intersection is called the *centroid* of the triangle.)
10. Given two triangles ABC and $A'B'C'$ with G and G' as their respective centroids, show that $\overrightarrow{AA'} + \overrightarrow{BB'} + \overrightarrow{CC'} = 3\overrightarrow{GG'}$. Illustrate with a diagram.
11. If $ABCD$ is a parallelogram and E is the midpoint of AB, show that DE and AC trisect each other. Illustrate with a diagram.
12. Given $\overrightarrow{OA} = \mathbf{a}$, $\overrightarrow{OB} = \mathbf{b}$, suppose that P represents a point on the line AB. Show that, in general, $\overrightarrow{OP} = \mathbf{a} + \lambda(\mathbf{b} - \mathbf{a}) = \mathbf{b} - \mu(\mathbf{b} - \mathbf{a})$ where λ, μ are constants satisfying $\lambda = 1 - \mu$. Illustrate this result geometrically. Show that

$$\overrightarrow{OP} = \mu\mathbf{a} + \lambda\mathbf{b} = (1 - \lambda)\mathbf{a} + \lambda\mathbf{b} = \mu\mathbf{a} + (1 - \mu)\mathbf{b}.$$

Determine the location of P in the special cases $\lambda = 0$, $\mu = 0$ and $\lambda = \mu = 1/2$. Describe points P that satisfy $\lambda < 0$, $0 \leq \lambda \leq 1$, and $\lambda > 1$.
13. Let Ox_*, Oy_* be rectangular axes in the plane obtained from Ox, Oy by rotation of axes through an angle α (Fig. 1.11). Let $\mathbf{i}_*$, $\mathbf{j}_*$ be unit vectors in the directions Ox_*, Oy_* and $\mathbf{i}$, $\mathbf{j}$ the familiar unit vectors in the directions of Ox, Oy. Show that

$$\mathbf{i}_* = \mathbf{i}\cos\alpha + \mathbf{j}\sin\alpha, \qquad \mathbf{j}_* = -\mathbf{i}\sin\alpha + \mathbf{j}\cos\alpha.$$

Solve these equations for $\mathbf{i}$, $\mathbf{j}$ in terms of $\mathbf{i}_*$, $\mathbf{j}_*$.

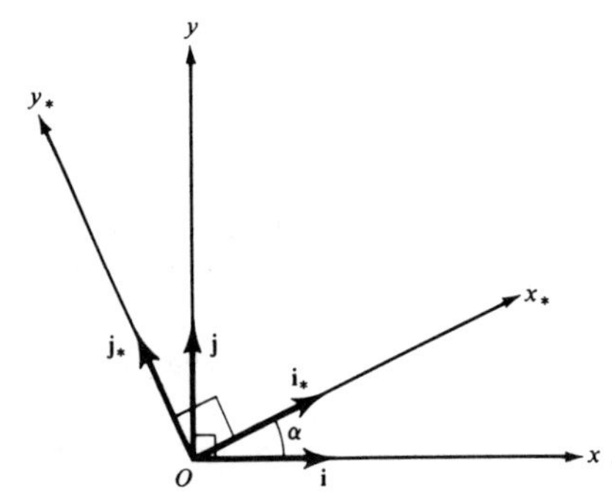

FIGURE 1.11

14. A given point $P = (x, y, z)$ is distant $r = \sqrt{x^2 + y^2 + z^2}$ from the origin. If α, β, γ are the angles the line OP makes with the axes Ox, Oy, Oz respectively, show that $x = r\cos\alpha$, $y = r\cos\beta$, and $z = r\cos\gamma$. Prove that $\cos^2\alpha + \cos^2\beta + \cos^2\gamma = 1$. (The three quantities $\cos\alpha$, $\cos\beta$ and $\cos\gamma$ are called the *direction cosines* of the line OP.)
15. Find the direction cosines for the lines OP determined by the following points:

(i) $P = (1, 1, 1)$ **(ii)** $P = (1, 2, 3)$ **(iii)** $P = (-1, 0, 2)$

16. Given a fixed point $P_0 = (x_0, y_0, z_0)$ and a general point $P = (x,y,z)$, define α, β and γ to be the angles the line P_0P makes with the Ox, Oy, Oz axes respectively. Show that $x - x_0 = r\cos\alpha$, $y - y_0 = r\cos\beta$, $z - z_0 = r\cos\gamma$ where $r = |\overrightarrow{P_0P}|$. Prove that $\cos^2\alpha + \cos^2\beta + \cos^2\gamma = 1$. (As before, $\cos\alpha$, $\cos\beta$ and $\cos\gamma$ are called the *direction cosines* of the line P_0P.)
17. Find the direction cosines for the lines P_0P determined by the following points:

(i) $P_0 = (1,1,1)$, $P = (2,3,4)$ **(ii)** $P_0 = (1,0,-1)$, $P = (2,1,0)$ **(iii)** $P_0 = (-1,0,2)$, $P = (1,1,1)$

18. A population vector $\mathbf{n} = (n_1, n_2, \ldots, n_m)$ is defined to be an m-dimensional vector whose ith component is the population of the ith species in an ecosystem $(i = 1, 2, \ldots, m)$. If $\mathbf{n}_1$ and $\mathbf{n}_2$ are two population vectors, what interpretation can be given to the vectors $\mathbf{n}_1 - \mathbf{n}_2$ and $(\mathbf{n}_1 + \mathbf{n}_2)/2$?
19. A "university" vector $\mathbf{u}$ is defined to be a three-dimensional vector having components (f, s, a), where f is the number of faculty, s the number of students and a the number of administrators. If $\mathbf{u}_1, \mathbf{u}_2, \mathbf{u}_3$ are three university vectors, what interpretation can be given to the vectors $\mathbf{u}_1 - \mathbf{u}_2$, $(\mathbf{u}_1 + \mathbf{u}_2 + \mathbf{u}_3)/3$?
20. A particle at the corner of a cube is acted on by forces of 1, 2, and 3 dynes, respectively, along the diagonals of the faces of the cube which meet at the particle. Find the resultant force on the particle.
21. Five forces act at one vertex A of a regular hexagon in the directions of the other vertices, their magnitudes being proportional to the distances of the vertices from A. Find the resultant force.
22. A man travelling east at speed U finds that the wind seems to blow directly from the north. On doubling his speed he finds it appears to come from the northeast. Find the velocity and direction of the wind.
23. A plane has a speed of U mph in calm air and has to fly to the northeast. If there is a west wind blowing at V mph, in what direction should the pilot head the plane, and what is his actual speed? Is this flight always possible? What about the return flight?

5.2 *Differentiation of vectors*

In many cases of interest vectors are not constant but depend on one or more scalar variables. In this section, and indeed in this entire chapter, attention will be limited to *vector functions of a single variable.* (Vector functions of more than one variable are the central topic of Chap. 8.)

A simple physical example not only motivates the discussion but is also easily generalized. Suppose at time t the position of a particle (or, if you prefer it, a satellite) has coordinates (x, y, z) with respect to a fixed rectangular frame. As the particle moves, its position will change with time. The coordinates x, y, and z will be functions of the single scalar variable t, so that

$$x = x(t), \qquad y = y(t), \qquad z = z(t). \tag{1}$$

These are *parametric equations* for the trajectory of the orbit in terms of the time parameter t. Given any value for t, we can locate the position (x, y, z) of the particle. A more concise description of the motion is obtained by introducing the *position vector*

$$\mathbf{r} = \mathbf{r}(t) = x(t)\mathbf{i} + y(t)\mathbf{j} + z(t)\mathbf{k}. \tag{2}$$

Suppose the position vector $\mathbf{r}(t)$ is known, how can the velocity of the particle be found? It was almost the identical problem which earlier (in Chap. 1) led to the concept of a derivative. The only essential difference here is that the function is now a vector rather than a scalar. Nevertheless the same ideas apply. If at time $t + \Delta t$ the position vector is $\mathbf{r} + \Delta\mathbf{r} = \mathbf{r}(t + \Delta t)$, then

$$\mathbf{r} + \Delta\mathbf{r} = \mathbf{r}(t + \Delta t) = x(t + \Delta t)\mathbf{i} + y(t + \Delta t)\mathbf{j} + z(t + \Delta t)\mathbf{k}. \tag{3}$$

In the time Δt the small displacement $\Delta\mathbf{r}$ is approximately tangent to the trajectory as shown in Fig. 2.1. Therefore the vector quantity

$$\frac{\Delta\mathbf{r}}{\Delta t} = \frac{\mathbf{r}(t + \Delta t) - \mathbf{r}(t)}{\Delta t}$$

is a measure of the average velocity over the time interval Δt. The instantaneous velocity $\mathbf{v}$ is merely the limit as $\Delta t \to 0$ of the average velocity, i.e., the *derivative* of $\mathbf{r}(t)$ with respect to t, or,

$$\mathbf{v} = \frac{d\mathbf{r}}{dt} = \lim_{\Delta t \to 0} \frac{\mathbf{r}(t + \Delta t) - \mathbf{r}(t)}{\Delta t}. \tag{4}$$

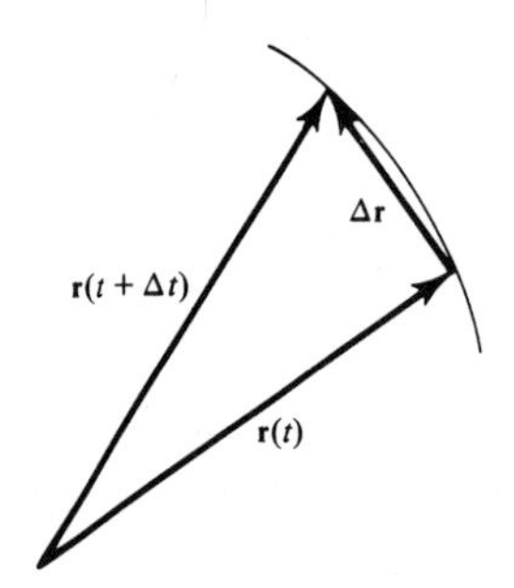

FIGURE 2.1

In practice, this limit is usually found by using the component form:

$$\mathbf{v} = \frac{d\mathbf{r}}{dt} = \mathbf{i}\lim_{\Delta t \to 0} \frac{x(t + \Delta t) - x(t)}{\Delta t} + \mathbf{j}\lim_{\Delta t \to 0} \frac{y(t + \Delta t) - y(t)}{\Delta t} + \mathbf{k}\lim_{\Delta t \to 0} \frac{z(t + \Delta t) - z(t)}{\Delta t},$$

$$= \frac{dx}{dt}\mathbf{i} + \frac{dy}{dt}\mathbf{j} + \frac{dz}{dt}\mathbf{k}, \tag{5}$$

the quantities dx/dt, dy/dt, dz/dt, being the *velocity components*.

Higher derivatives are found in the same way. For example, the acceleration vector, $\mathbf{a} = d\mathbf{v}/dt = d^2\mathbf{r}/dt^2$, is

$$\mathbf{a} = \frac{d\mathbf{v}}{dt} = \frac{d^2\mathbf{r}}{dt^2} = \frac{d^2x}{dt^2}\mathbf{i} + \frac{d^2y}{dt^2}\mathbf{j} + \frac{d^2z}{dt^2}\mathbf{k}. \tag{6}$$

The general procedure is now clear. *Differentiation* of vectors is performed by applying the rules of scalar differentiation to each cartesian component. *Integration* is just the inverse operation; an arbitrary constant vector arises in each integration. If $\mathbf{v}(t)$ is prescribed, then the solution of

$$\frac{d\mathbf{r}}{dt} = \mathbf{v} \quad \text{is} \quad \mathbf{r}(t) = \mathbf{r}(t_0) + \int_{t_0}^{t} \mathbf{v}(t)\, dt.$$

Example 1 The position of a particle at time t is $(x, y, z) = (t^2, \log t, e^{2t})$. Find its velocity and acceleration.

Here the position vector is

$$\mathbf{r} = t^2\mathbf{i} + \log t\,\mathbf{j} + e^{2t}\mathbf{k}.$$

The velocity and acceleration are found by differentiation, and therefore

$$\mathbf{v} = \frac{d\mathbf{r}}{dt} = 2t\mathbf{i} + \frac{1}{t}\mathbf{j} + 2e^{2t}\mathbf{k},$$

$$\mathbf{a} = \frac{d^2\mathbf{r}}{dt^2} = 2\mathbf{i} - \frac{1}{t^2}\mathbf{j} + 4e^{2t}\mathbf{k}.$$

Example 2 A particle moves around a circle of radius a with constant angular velocity ω. Find its velocity and acceleration.

This motion is called uniform circular motion. If at time $t = 0$ the motion starts at (a, O), then the angle θ (indicated in Fig. 2.2) is given by $\theta = \omega t$. Furthermore,

$$x = a\cos\theta = a\cos\omega t,$$
$$y = a\sin\theta = a\sin\omega t,$$

so that at time t the two-dimensional position vector for the particle is

$$\mathbf{r} = \mathbf{i}a\cos\omega t + \mathbf{j}a\sin\omega t.$$

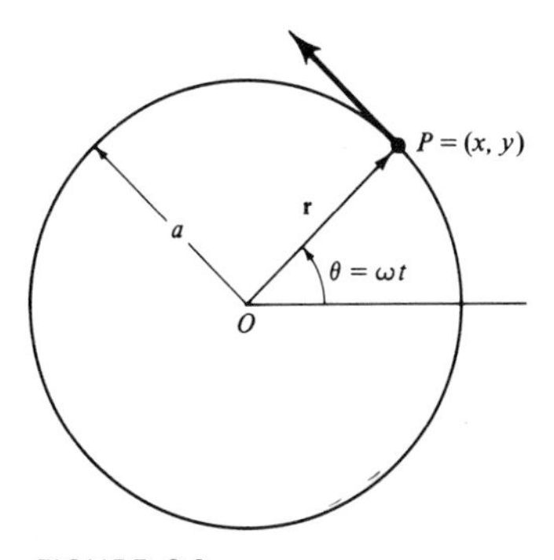

FIGURE 2.2

Note that $\mathbf{r}$ is constant in magnitude ($|\mathbf{r}| = a$) but its direction varies. Differentiation gives

$$\mathbf{v} = -\mathbf{i}a\omega \sin \omega t + \mathbf{j}a\omega \cos \omega t,$$
$$\mathbf{a} = -\mathbf{i}a\omega^2 \cos \omega t - \mathbf{j}a\omega^2 \sin \omega t = -\omega^2\mathbf{r}.$$

The velocity vector is tangent to the circle, and the acceleration is of constant magnitude $a\omega^2$, directed radially inward ($\mathbf{a} = -\omega^2\mathbf{r}$). This acceleration is the *centripetal acceleration*. The position vector specified by the general solution of the vector differential equation $d^2\mathbf{r}/dt^2 = -\omega^2\mathbf{r}$ is $\mathbf{r} = \mathbf{A}\cos \omega t + \mathbf{B}\sin \omega t$, where $\mathbf{A}$ and $\mathbf{B}$ are arbitrary vectors. (The reader should check that this solution will lead to circular motion only for special initial conditions.)

2

Newton's second law of motion, which states that the rate of change of *momentum* is equal to the externally applied force, can be formulated as the vector differential equation

$$\frac{d}{dt}(m\mathbf{v}) = \mathbf{F}. \tag{7}$$

Here m is the mass and $\mathbf{F}$ the external force. The individual components are determined by

$$\frac{d}{dt}\left(m\frac{dx}{dt}\right) = F_1, \qquad \frac{d}{dt}\left(m\frac{dy}{dt}\right) = F_2, \qquad \frac{d}{dt}\left(m\frac{dz}{dt}\right) = F_3, \tag{8}$$

F_1, F_2, and F_3 being the components of the external force $\mathbf{F}$.

Example 3 A shell is projected with velocity V at angle α above the horizontal. If air resistance is neglected and the only external force is gravity (assumed constant), discuss the motion.

This example is the classical projectile problem. Suppose the origin is chosen as the launch point, with the x axis as the horizontal and the y axis vertically upward (Fig. 2.3). Newton's second law is

$$\frac{d}{dt}(m\mathbf{v}) = \mathbf{F}.$$

Here the mass m is constant, and $\mathbf{F} = m\mathbf{g}$, where $\mathbf{g} = -g\mathbf{j}$, is the gravitational acceleration. (The gravitational force acts in the negative y direction.) In these circumstances, the preceding equation reduces to

$$\frac{d\mathbf{v}}{dt} = \mathbf{g},$$

and upon integration we obtain

$$\mathbf{v} = \mathbf{A} + \mathbf{g}t,$$

where **A** is a constant vector of integration. **A** is determined by the initial conditions at $t = 0$, namely,

$$\mathbf{v}(0) = \mathbf{A} = \mathbf{i}V\cos\alpha + \mathbf{j}V\sin\alpha.$$

Parametric equations for the trajectory can be found by a further integration:

$$\mathbf{r} = \mathbf{B} + \mathbf{A}t + \tfrac{1}{2}\mathbf{g}t^2.$$

In order that $\mathbf{r} = 0$ at $t = 0$, we must set $\mathbf{B} = 0$. The vector equation for the trajectory is then

$$\mathbf{r} = \mathbf{A}t + \tfrac{1}{2}\mathbf{g}t^2,$$

or in component form

$$x\mathbf{i} + y\mathbf{j} = \mathbf{i}Vt\cos\alpha + \mathbf{j}(Vt\sin\alpha - \tfrac{1}{2}gt^2).$$

Upon equating coefficients of the unit vectors, we obtain

$$x = Vt\cos\alpha, \qquad y = Vt\sin\alpha - \tfrac{1}{2}gt^2. \tag{9}$$

These are the parametric equations of the trajectory, a few of whose properties should be noted. The time of flight is the time until return to the level $y = 0$. With $y = 0$, (9) implies

$$t = 0 \qquad \text{or} \qquad t = \frac{2V\sin\alpha}{g}.$$

Since the first value corresponds to liftoff, the time of flight is given by $t = T = (2V\sin\alpha)/g$. At time T, the horizontal distance travelled,

$$R = \frac{V^2\sin 2\alpha}{g}, \tag{10}$$

is called the *range*. For a given launch velocity V, R is a function only of α, and the maximum range, V^2/g, corresponds to a projection angle of 45°.

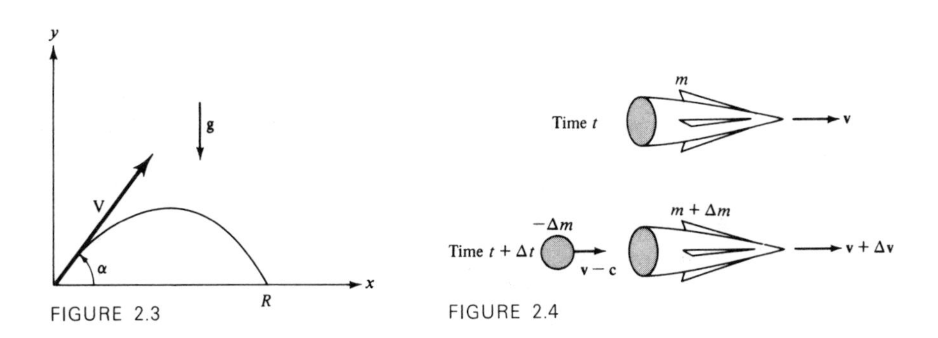

FIGURE 2.3

FIGURE 2.4

Elimination of the parameter t from Eqs. (9) shows that the shell trajectory is the parabola

$$y = x \tan \alpha - \frac{1}{2}\frac{gx^2}{V^2}\sec^2 \alpha.$$

Example 4 Find the equation governing a rocket of mass m, velocity $\mathbf{v}$, and exhaust velocity $\mathbf{c}$ moving under the action of no external forces.

Since there is no external force acting on the system, the total momentum must be conserved. Suppose that at time t the mass of the rocket plus fuel is m and it is moving with velocity $\mathbf{v}$, while at time $t + \Delta t$ the mass is $m + \Delta m$ and the velocity $\mathbf{v} + \Delta\mathbf{v}$. The mass of fuel used in this time $(-\Delta m)$ is moving relative to the rocket with the exhaust velocity $\mathbf{c}$, so that its actual velocity is $\mathbf{v} - \mathbf{c}$. (As shown in Fig. 2.4, Δm is clearly a negative increment.) The momentum balance required by Newton's law of motion is

$$(m + \Delta m)(\mathbf{v} + \Delta\mathbf{v}) + (-\Delta m)(\mathbf{v} - \mathbf{c}) = m\mathbf{v}.$$

Simplification, division by Δt, and the limit $\Delta t \to 0$ give

$$m\frac{d\mathbf{v}}{dt} = -\mathbf{c}\frac{dm}{dt}. \tag{11}$$

This equation plays a fundamental role in rocket propulsion. The acceleration of the rocket is proportional to the exhaust velocity and to the rate of fuel consumption. If $\mathbf{c}$ is constant, this equation can be rewritten

$$\frac{d\mathbf{v}}{dt} = -\mathbf{c}\frac{1}{m}\frac{dm}{dt}$$

and integrated to yield

$$\mathbf{v}(t) = \mathbf{v}(0) - \mathbf{c}\log\frac{m(t)}{m(0)}.$$

For example, suppose that the initial mass of a rocket is three-fourths fuel. If the exhaust velocity is $|\mathbf{c}| = 2$ km/sec and the rocket starts from rest, then at the end of the burn

$$|\mathbf{v}| = -2\log\tfrac{1}{4} = 2\log 4 \approx 2.8 \text{ km/sec}.$$

3

Differentiation or integration of a vector function of a single variable follows the usual rules, namely, scalar differentiation or integration of components. We can associate a space curve with any vector $\mathbf{f} = \mathbf{f}(t)$ by interpreting the function as the position vector of a particle at time t. However, this association, although often useful, is by no means necessary. If $\mathbf{f}$ is written in component form,

$$\mathbf{f} = f_1(t)\mathbf{i} + f_2(t)\mathbf{j} + f_3(t)\mathbf{k},$$

then

$$\frac{d\mathbf{f}}{dt} = \frac{df_1}{dt}\mathbf{i} + \frac{df_2}{dt}\mathbf{j} + \frac{df_3}{dt}\mathbf{k},$$

provided of course the derivative exists. In a similar way, higher derivatives, Taylor series, and numerical differentiation or integration of $\mathbf{f}(t)$ can be dealt with in a straightforward manner.

Exercises 5.2

1. Graph the circular helix $\mathbf{r}(t) = \cos t\,\mathbf{i} + \sin t\,\mathbf{j} + t\mathbf{k}$ for $t \geq 0$. Determine the velocity $d\mathbf{r}/dt$ and the acceleration $d^2\mathbf{r}/dt^2$ at time t. Show that the acceleration is a unit vector directed toward the vertical axis.

2. Graph the following trajectories in the plane or in three dimensions. Determine the velocity and acceleration at time t.

(i) $\mathbf{r}(t) = 2e^t\mathbf{i} + 4e^{2t}\mathbf{j} = (2e^t, 4e^{2t})$ **(ii)** $\mathbf{r}(t) = (\cosh t, \sinh t)$

(iii) $\mathbf{r}(t) = (t\cos t, t\sin t, t)$ **(iv)** $\mathbf{r}(t) = (t, t^2, t^4)$

3. For the following vector functions, determine $d\mathbf{f}/dt$ and $\int_0^t \mathbf{f}(t')\,dt'$.

(i) $\mathbf{f}(t) = t^2\mathbf{i} + t^3\mathbf{j} + t^4\mathbf{k}$ **(ii)** $\mathbf{f}(t) = te^t\mathbf{i} + te^{-t}\mathbf{j} + t^2\mathbf{k}$

(iii) $f(t) = \sin^2 t\,\mathbf{i} + \cos^2 t\,\mathbf{j} + \mathbf{k}$ **(iv)** $\mathbf{f}(t) = t\log t\,\mathbf{i} + t\mathbf{j} + t\mathbf{k}$

4. A particle moves in the plane with velocity $\mathbf{v}(t) = x\mathbf{i} - (2/x^2)\mathbf{j}$ at time t. Determine $\mathbf{r}(t)$ if the initial position is $\mathbf{r}(0) = \mathbf{i} + \mathbf{j}$. Plot the trajectory of the particle. If the particle has unit mass, determine the force $\mathbf{F}(t)$ acting on the particle at time t.

5. Find the general solution of the vector equation $d^2\mathbf{r}/dt^2 = \mathbf{A}t + \mathbf{B}$, where $\mathbf{A}$ and $\mathbf{B}$ are constant vectors.

6. A UFO is sighted and believed to have position vectors $\mathbf{r}_1$, $\mathbf{r}_2$, and $\mathbf{r}_3$ at the times $t - \Delta t$, t, and $t + \Delta t$, respectively. Use finite differences to estimate its velocity and acceleration vectors. How could you calculate $\int_{t-\Delta t}^{t+\Delta t} \mathbf{r}(t')\,dt'$ (approximately)?

7. A boy can throw a baseball 200 ft. How fast can he throw? How high can he throw?

8. A baseball is caught by an outfielder 6 seconds after it is hit. How high did it travel?

9. A plane is engaged in fighting a forest fire and wants to drop a carbon dioxide canister on a ground target. If air resistance is negligible and the plane flies over the target area at speed U and altitude h, where should the canister be released?

10. A shell is fired with an initial velocity of 600 m/sec in a direction making an angle of 25° with the horizontal. How far does it travel before it hits level ground? How high does it travel? If the firing velocity is subject to an error of $\pm 3\%$ and the firing angle has an error of $\pm 1\%$, estimate the possible deviation in the range from the calculated value.

11. Determine the range of a projectile fired at an angle α above the horizontal on an inclined plane which is at an angle β above the horizontal ($\alpha > \beta$). (*Hint:* Where does the trajectory intersect the line $y = (\tan\beta)x$?)

12. A shell is fired with initial velocity V at an angle α to the horizontal in a medium where the drag force (air resistance) is proportional to the velocity. Investigate the differential equations for the trajectory.

13. In the previous problem investigate the special case of vertical motion ($\alpha = \pi/2$). Calculate the time of ascent. Study also the problem of free fall with air resistance proportional to velocity.

14. A frisbee of radius a thrown horizontally with velocity U remains spinning with constant angular velocity Ω about a vertical axis. Find the velocity and acceleration of a point on the rim.

15. The wheels of a bicycle are 75 cm in diameter, the gear ratio between the crank and wheel axles is 2.5, and the length of a pedal arm is 20 cm. If the bicycle is travelling at 10 m/sec, find the velocity and acceleration of the pedal arm when it is at its highest point.

16. A small single-engine plane is flying horizontally with constant speed V, so that the center of its propeller Q moves along the x axis. Each blade of the propeller has length a and rotates with constant angular velocity Ω. Find the function $\mathbf{r}(t) = \overrightarrow{OP}$, where O is the origin and P is the tip of a propeller blade which at time $t = 0$ is at the point $(0, a, 0)$.

17. A rocket under the action of no forces moves in a straight line. The exhaust velocity is c, and the rocket loses mass at a constant rate. If the rocket starts with velocity v_0, show that when half its mass is used up, its velocity is $v_0 + c \log 2$.

18. A uniform chain of length l is coiled at the edge of a table. One end is gently pushed over the edge. Find the velocity of the chain when it leaves the table. How long does this take?

5.3 *Scalar Products*

There are two types of products in common use in vector analysis. The first of these is the *scalar product*. The scalar product, $\mathbf{a} \cdot \mathbf{b}$, of two vectors $\mathbf{a}$ and $\mathbf{b}$ is defined by

$$\mathbf{a} \cdot \mathbf{b} = |\mathbf{a}|\,|\mathbf{b}| \cos \theta, \tag{1}$$

where θ is the angle between $\mathbf{a}$ and $\mathbf{b}$ $(0 \leq \theta \leq \pi)$ (see Fig. 3.1). Because $\mathbf{a} \cdot \mathbf{b}$ is read "$\mathbf{a}$ dot $\mathbf{b}$," the name *dot product* is often used. Since $\cos(-\theta) = \cos \theta$, the roles of $\mathbf{a}$ and $\mathbf{b}$ can be interchanged to show that $\mathbf{a} \cdot \mathbf{b} = \mathbf{b} \cdot \mathbf{a}$. Note that the scalar product of two vectors is a scalar.

By definition $\mathbf{a} \cdot \mathbf{b}$ represents the magnitude of $\mathbf{a}$ ($|\mathbf{a}|$) times the magnitude of the *projection* of $\mathbf{b}$ on $\mathbf{a}$ ($|\mathbf{b}| \cos \theta$). This provides a simple geometrical interpretation of the scalar product.

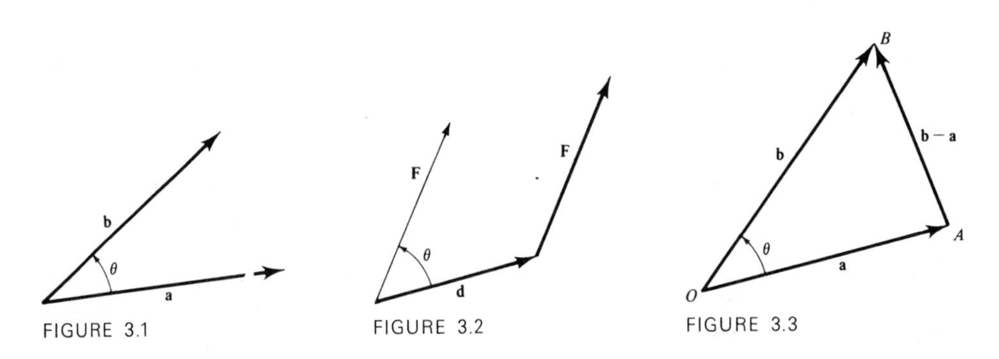

FIGURE 3.1 FIGURE 3.2 FIGURE 3.3

If $\mathbf{F}$ is a force and its point of application is displaced by a vector $\mathbf{d}$, then the scalar product $\mathbf{F} \cdot \mathbf{d} = |\mathbf{F}||\mathbf{d}|\cos\theta$ is the work done by the force (see Fig. 3.2).

If $\mathbf{a}$ and $\mathbf{b}$ are perpendicular vectors, $\theta = \pi/2$ and $\cos(\pi/2) = 0$, so that $\mathbf{a} \cdot \mathbf{b} = 0$.

Conversely, if $\mathbf{a} \cdot \mathbf{b} = 0$, and $|\mathbf{a}|$ and $|\mathbf{b}|$ are nonzero, then $\theta = \pi/2$ and the vectors are perpendicular.

2

It is desirable to have an expression for $\mathbf{a} \cdot \mathbf{b}$ in terms of the components of $\mathbf{a}$ and $\mathbf{b}$. This can be derived by applying the cosine law to the triangle OAB (Fig. 3.3), namely,

$$AB^2 = OA^2 + OB^2 - 2OA\ OB\cos\theta,$$

or in vector notation

$$\begin{aligned} |\mathbf{b} - \mathbf{a}|^2 &= |\mathbf{a}|^2 + |\mathbf{b}|^2 - 2|\mathbf{a}||\mathbf{b}|\cos\theta, \\ &= |\mathbf{a}|^2 + |\mathbf{b}|^2 - 2\mathbf{a} \cdot \mathbf{b}. \end{aligned}$$

The result of transposing terms is

$$\mathbf{a} \cdot \mathbf{b} = \tfrac{1}{2}(|\mathbf{a}|^2 + |\mathbf{b}|^2 - |\mathbf{b} - \mathbf{a}|^2). \tag{2}$$

If the component forms of the vectors $\mathbf{a}$ and $\mathbf{b}$ are

$$\mathbf{a} = a_1\mathbf{i} + a_2\mathbf{j} + a_3\mathbf{k} \qquad \text{and} \qquad \mathbf{b} = b_1\mathbf{i} + b_2\mathbf{j} + b_3\mathbf{k},$$

then

$$\mathbf{b} - \mathbf{a} = (b_1 - a_1)\mathbf{i} + (b_2 - a_2)\mathbf{j} + (b_3 - a_3)\mathbf{k}.$$

The use of these expressions in Eq. (2) gives

$$\begin{aligned} \mathbf{a} \cdot \mathbf{b} = \tfrac{1}{2}\{(a_1{}^2 + a_2{}^2 + a_3{}^3) + (b_1{}^2 + b_2{}^2 + b_3{}^2) - \\ [(b_1 - a_1)^2 + (b_2 - a_2)^2 + (b_3 - a_3)^2]\}, \end{aligned}$$

which, upon simplification, reduces to

$$\mathbf{a} \cdot \mathbf{b} = a_1b_1 + a_2b_2 + a_3b_3. \tag{3}$$

This may be taken as an alternative definition of a scalar product.

Example 1 Find the angle between the two vectors $\mathbf{i} + \mathbf{j} + \mathbf{k}$ and $2\mathbf{i} - 3\mathbf{j} + 6\mathbf{k}$.

This problem is readily solved by using the definition of the dot product, $\mathbf{a} \cdot \mathbf{b} = |\mathbf{a}||\mathbf{b}|\cos\theta$, and formula (3), which together imply

$$\cos\theta = \frac{\mathbf{a} \cdot \mathbf{b}}{|\mathbf{a}||\mathbf{b}|} = \frac{(1)(2) + (1)(-3) + (1)(6)}{(1^2 + 1^2 + 1^2)^{1/2}(2^2 + 3^2 + 6^2)^{1/2}} = \frac{5}{7\sqrt{3}}.$$

The result is $\theta \approx 66°$.

The unit vectors **i**, **j**, **k** have simple properties under scalar multiplication. For example, $\mathbf{i} \cdot \mathbf{i} = 1 \cdot 1 \cos 0 = 1$, and $\mathbf{i} \cdot \mathbf{j} = 1 \cdot 1 \cos(\pi/2) = 0$. The following relations are immediate consequences of the definition,

$$\mathbf{i} \cdot \mathbf{i} = \mathbf{j} \cdot \mathbf{j} = \mathbf{k} \cdot \mathbf{k} = 1, \tag{4}$$

$$\mathbf{i} \cdot \mathbf{j} = \mathbf{j} \cdot \mathbf{k} = \mathbf{k} \cdot \mathbf{i} = 0. \tag{5}$$

Note that $\mathbf{a} \cdot \mathbf{a} = |\mathbf{a}|^2 = a^2$, and if $\hat{\mathbf{n}}$ is any unit vector, $\hat{\mathbf{n}} \cdot \hat{\mathbf{n}} = 1$.

Scalar multiplication is distributive, i.e.,

$$\mathbf{a} \cdot (\mathbf{b} + \mathbf{c}) = \mathbf{a} \cdot \mathbf{b} + \mathbf{a} \cdot \mathbf{c}. \tag{6}$$

Geometrically this follows from the fact that the projection of the sum of two vectors is equal to the sum of the projections in a given direction. However, it can also be verified directly by using the component forms. Indeed, Eq. (6) is a consequence of the identity

$$\sum_{m=1}^{3} a_m(b_m + c_m) = \sum_{m=1}^{3} a_m b_m + \sum_{m=1}^{3} a_m c_m .$$

The dot product of two vectors **a** and **b** when expressed in component form may, using the distributive law, be written as a sum of nine terms:

$$\begin{aligned}\mathbf{a} \cdot \mathbf{b} &= (a_1\mathbf{i} + a_2\mathbf{j} + a_3\mathbf{k}) \cdot (b_1\mathbf{i} + b_2\mathbf{j} + b_3\mathbf{k}) \\ &= a_1b_1\mathbf{i} \cdot \mathbf{i} + a_1b_2\mathbf{i} \cdot \mathbf{j} + a_1b_3\mathbf{i} \cdot \mathbf{k} + \cdots.\end{aligned}$$

Six of these are zero according to (4) and (5), and those which remain constitute Eq. (3) once again.

Example 2 For what value of c are the vectors $2\mathbf{i} - 7\mathbf{j} + \mathbf{k}$ and $\mathbf{i} - \mathbf{j} + c\mathbf{k}$ perpendicular?

For the vectors to be perpendicular their scalar product must vanish, and therefore we require that

$$(2)(1) + (-7)(-1) + (1)(c) = 0$$

or

$$c = -9.$$

Example 3 Find the equation of the circle having the points $P_1 = (x_1, y_1)$, $P_2 = (x_2, y_2)$ as ends of a diameter.

With O as the origin in the xy plane and $P = (x, y)$ any point on the circle, the vectors concerned are

$$\begin{aligned}\overrightarrow{OP} &= \mathbf{r} = x\mathbf{i} + y\mathbf{j}, \\ \overrightarrow{OP_1} &= \mathbf{r}_1 = x_1\mathbf{i} + y_1\mathbf{j}, \\ \overrightarrow{OP_2} &= \mathbf{r}_2 = x_2\mathbf{i} + y_2\mathbf{j}.\end{aligned}$$

In order that P should be on the circle with P_1P_2 as diameter, P_1P and P_2P

must be perpendicular, as shown in Fig. 3.4. Therefore

$$\overrightarrow{P_1P} \cdot \overrightarrow{P_2P} = 0,$$

or

$$(\mathbf{r} - \mathbf{r}_1) \cdot (\mathbf{r} - \mathbf{r}_2) = 0,$$

which is the required equation. In cartesian form this is

$$(x - x_1)(x - x_2) + (y - y_1)(y - y_2) = 0.$$

Example 4 Prove that the three altitudes of a triangle have a common point of intersection. (This point is the orthocenter of the triangle, Fig. 3.5.)

Let the altitudes from A and B meet in O. It remains to prove that OC is perpendicular to AB. If $\overrightarrow{OA} = \mathbf{a}$, $\overrightarrow{OB} = \mathbf{b}$, and $\overrightarrow{OC} = \mathbf{c}$, then $\overrightarrow{BC} = \mathbf{c} - \mathbf{b}$,

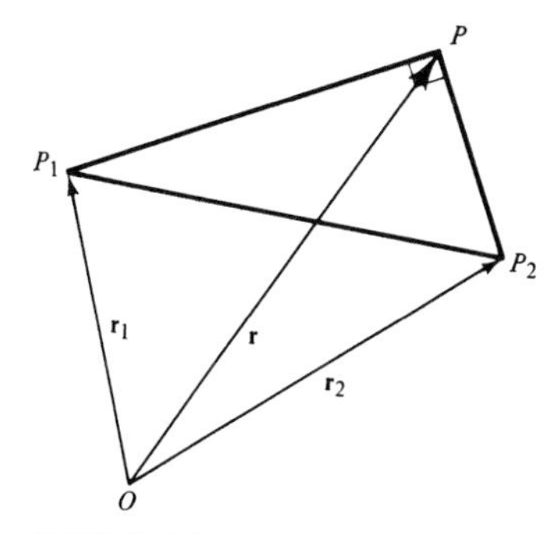

FIGURE 3.4

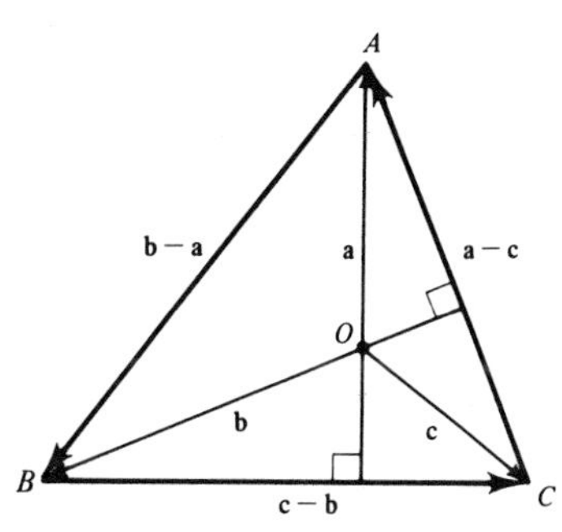

FIGURE 3.5

$\overrightarrow{CA} = \mathbf{a} - \mathbf{c}$, and $\overrightarrow{AB} = \mathbf{b} - \mathbf{a}$. By construction OA is perpendicular to BC, so that

$$\mathbf{a} \cdot (\mathbf{c} - \mathbf{b}) = 0$$

or

$$\mathbf{c} \cdot \mathbf{a} = \mathbf{a} \cdot \mathbf{b}. \tag{7}$$

Similarly OB is perpendicular to CA, and therefore

$$\mathbf{b} \cdot (\mathbf{a} - \mathbf{c}) = 0$$

or

$$\mathbf{b} \cdot \mathbf{c} = \mathbf{a} \cdot \mathbf{b}. \tag{8}$$

From Eqs. (7) and (8) it follows that

$$\mathbf{c} \cdot \mathbf{b} = \mathbf{c} \cdot \mathbf{a}$$

or

$$\mathbf{c} \cdot (\mathbf{b} - \mathbf{a}) = 0 = \overrightarrow{OC} \cdot \overrightarrow{AB}.$$

Therefore, $\overrightarrow{OC}$ is perpendicular to $\overrightarrow{AB}$, as required.

3

A formula for the *differentiation of a scalar product* follows most simply from the component form

$$\mathbf{a}(t) \cdot \mathbf{b}(t) = \mathbf{a} \cdot \mathbf{b} = \sum_{m=1}^{3} a_m(t) b_m(t),$$

in which case

$$\frac{d}{dt}(\mathbf{a} \cdot \mathbf{b}) = \sum_{m=1}^{3} a_m \frac{db_m}{dt} + \sum_{m=1}^{3} \frac{da_m}{dt} b_m,$$

$$= \mathbf{a} \cdot \frac{d\mathbf{b}}{dt} + \frac{d\mathbf{a}}{dt} \cdot \mathbf{b}. \tag{9}$$

This result can also be obtained from first principles by using the Δ process (the limit definition). Note that $d(\mathbf{a} \cdot \mathbf{a})/dt = 2\mathbf{a} \cdot d\mathbf{a}/dt$. In particular, if the magnitude of $\mathbf{a}$ is constant, this equation implies $\mathbf{a} \cdot d\mathbf{a}/dt = 0$, showing that in this circumstance $\mathbf{a}$ and $d\mathbf{a}/dt$ are perpendicular.

Exercises 5.3

1. Find the magnitudes of the vectors $\overrightarrow{OA}$, $\overrightarrow{OB}$ and the angle AOB.
(i) $A = (1,0,1)$, $B = (0,1,1)$ **(ii)** $A = (12,1,-12)$, $B = (8,4,1)$
(iii) $A = (-6,2,7)$, $B = (3,-5,4)$ **(iv)** $A = (2,3,6)$, $B = (-2,1,2)$
2. The points $A = (1,0,3)$, $B = (2,1,1)$ and $C = (1,1,1)$ are three vertices of a parallelogram which has AB and BC as adjacent edges. Find the coordinates of the fourth vertex D. Show that the diagonals AC and BD intersect at their midpoints.
3. Find a unit vector equally inclined to the three coordinate axes. Determine the angle between the vector and each axis.
4. What is the angle between two diagonals of a cube?
5. Draw in a diagram the vectors $\lambda\mathbf{i} + 3\mathbf{j} - \mathbf{k}$ and $\mathbf{i} + \lambda\mathbf{j} + 4\mathbf{k}$ for $-\infty < \lambda < \infty$. For what value of λ are the two vectors perpendicular?
6. Show that the line joining $(2,3,4)$ to $(1,2,3)$ is perpendicular to the line joining $(1,0,2)$ to $(2,3,-2)$. Do the lines intersect?
7. If $A = (a,0,0)$, $B = (0,b,0)$, and $C = (0,0,c)$, find the angles of the triangle ABC. What is the radius of the sphere through O, A, B, and C?
8. Show that any vector $\mathbf{V}$ can be written in the form $\mathbf{V} = (\mathbf{V} \cdot \mathbf{i})\mathbf{i} + (\mathbf{V} \cdot \mathbf{j})\mathbf{j} + (\mathbf{V} \cdot \mathbf{k})\mathbf{k}$.
9. Find the center and radius of the sphere $4(x^2 + y^2 + z^2) = 4(4x + y + 2z) + 15$. Write the equation of the sphere in vector notation. [*Hint:* Define $\mathbf{r} = x\mathbf{i} + y\mathbf{j} + z\mathbf{k} = (x, y, z)$ and $\mathbf{a} = (4, 1, 2)$.]
10. Find a unit vector which bisects the angle between two given vectors $\mathbf{a}$ and $\mathbf{b}$. Illustrate with the special case $\mathbf{a} = (1,0,3)$ and $\mathbf{b} = (5,2,1)$.
11. Find any two vectors which are equally inclined to $\mathbf{k}$, are perpendicular to each other, and perpendicular to $\mathbf{i} + \mathbf{j} + \mathbf{k}$.
12. Find the equation of a circular cone with vertex at the origin having a semivertical angle of 30° with its axis on the z axis. Solve using vector methods and then write the equation in terms of the cartesian coordinates.

13. If $\mathbf{a}$ is a given vector and $\mathbf{r} = \overrightarrow{OP}$ is the general radius vector, find the locus of P such that $|\mathbf{r} - \mathbf{a}| + |\mathbf{r} + \mathbf{a}| = k = \text{constant}$. Illustrate with the special cases $\mathbf{a} = (1,0,0)$, $k = 3$ and $\mathbf{a} = (1,1,1)$, $k = 5$.

14. A particle with position vector $\mathbf{r}(t)$ at time t moves in any manner on a fixed sphere centered at the origin. If $\mathbf{v}(t)$ and $\mathbf{a}(t)$ are the velocity and acceleration, show that $\mathbf{r} \cdot \mathbf{v} = 0$ and $\mathbf{r} \cdot \mathbf{a} + |\mathbf{v}|^2 = 0$. Verify these identities for the special case

$$\mathbf{r}(t) = (e^{-t}\cos t, e^{-t}\sin t, \sqrt{1 - e^{-2t}}).$$

15. Let $\mathbf{i}' = \frac{1}{3}(\mathbf{i} - 2\mathbf{j} + 2\mathbf{k})$ and $\mathbf{j}' = \frac{1}{3}(2\mathbf{i} + 2\mathbf{j} + \mathbf{k})$.

(i) Prove that $\mathbf{i}'$ and $\mathbf{j}'$ are perpendicular unit vectors.

(ii) Find a unit vector $\mathbf{k}'$ perpendicular to both $\mathbf{i}'$ and $\mathbf{j}'$.

(iii) Express $\mathbf{i} + \mathbf{j}$ in the $\mathbf{i}'$, $\mathbf{j}'$, $\mathbf{k}'$ system; i.e., find α, β, γ so that

$$\mathbf{i} + \mathbf{j} = \alpha\mathbf{i}' + \beta\mathbf{j}' + \gamma\mathbf{k}'.$$

16. If a particle moves in three dimensions with a velocity of constant magnitude, show that the acceleration vector is orthogonal to the velocity vector. (The acceleration is at right angles to the direction of motion.) Show the general result; for any vector function $\mathbf{f}(t)$,

$$\frac{d}{dt}|\mathbf{f}(t)|^2 = 2\mathbf{f} \cdot \frac{d\mathbf{f}}{dt}.$$

17. Plot the trajectory in the plane of a particle whose position at time t is given by $\mathbf{r}(t) = (t - \sin t)\mathbf{i} + (1 - \cos t)\mathbf{j}$. Show that there are times when the velocity is zero but that the acceleration is never zero. When is the speed greatest?

18. Newton's second law for a particle of constant mass m is $\mathbf{F} = m d\mathbf{v}/dt$ where $\mathbf{F}$ is the external force. Deduce that $\mathbf{F} \cdot \mathbf{v} = d(\frac{1}{2}mv^2)/dt$, the rate of working of the external forces is equal to the rate of change of the kinetic energy.

19. The earth is taken to be a sphere of radius R. If the z axis is the polar axis and the zOx plane has zero longitude, show that a point P having latitude λ (positive in the northern hemisphere) and longitude μ (measured to the east) has coordinates $P = (R\cos\lambda\cos\mu, R\cos\lambda\sin\mu, R\sin\lambda)$. If P' has latitude λ' and longitude μ', show that the straight-line distance PP' is

$$R\{2[1 - \cos\lambda\cos\lambda'\cos(\mu - \mu') - \sin\lambda\sin\lambda']\}^{1/2},$$

and the great-circle distance PP' is

$$R\cos^{-1}[\cos\lambda\cos\lambda'\cos(\mu - \mu') + \sin\lambda\sin\lambda'].$$

20. Determine the straight-line and great-circle distances between Montreal, Québec ($\lambda = 45.5°$, $\mu = 73.7°$) and Cambridge, Mass. ($\lambda' = 42.4°$, $\mu' = 71.2°$). Determine the greatest depth of the straight-line distance below ground assuming that the earth is a sphere with radius 6370 km.

5.4 Vector products

The second type of product useful in applications is the *vector product*. The vector product, $\mathbf{a} \times \mathbf{b}$, of two vectors $\mathbf{a}$ and $\mathbf{b}$ is defined as the vector

$$\mathbf{a} \times \mathbf{b} = |\mathbf{a}|\,|\mathbf{b}|\sin\theta\,\hat{\mathbf{n}}, \tag{1}$$

where θ is the angle between $\mathbf{a}$ and $\mathbf{b}$ ($0 \le \theta \le \pi$) and $\hat{\mathbf{n}}$ is a unit vector perpendicular to both of these in the direction a right-threaded screw would move turning from $\mathbf{a}$ to

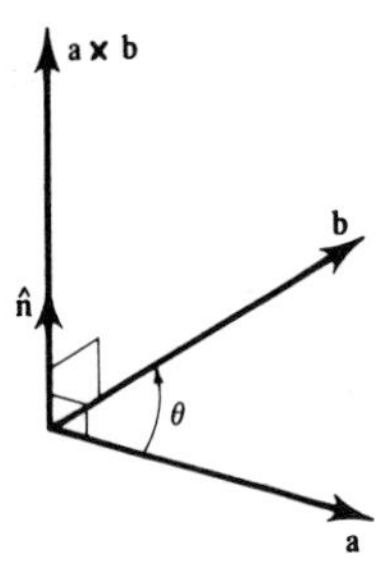

FIGURE 4.1

b (see Fig. 4.1). There are two directions perpendicular to both **a** and **b**, and the above convention determines which direction is to be used. The vector $\mathbf{b} \times \mathbf{a}$ has the same magnitude as $\mathbf{a} \times \mathbf{b}$, but the corresponding unit vector $\hat{\mathbf{n}}$ is in the opposite direction, so that

$$\mathbf{b} \times \mathbf{a} = -\mathbf{a} \times \mathbf{b}. \tag{2}$$

The order of multiplication is therefore important, as vector multiplication is not commutative.

We read $\mathbf{a} \times \mathbf{b}$ as "**a** cross **b**," and the name *cross product* is sometimes used instead of vector product. Note that the vector product of two vectors yields a vector.

Geometrically, the magnitude of $\mathbf{a} \times \mathbf{b}$, that is,

$$|\mathbf{a} \times \mathbf{b}| = |\mathbf{a}|\,|\mathbf{b}| \sin\theta,$$

is the area of parallelogram $OACB$ (see Fig. 4.2). The direction of $\mathbf{a} \times \mathbf{b}$ is normal to this area.

If **F** is a force and **r** is the vector from a fixed point O to any point on **F**, then

$$\mathbf{L} = \mathbf{r} \times \mathbf{F}$$

can be interpreted as the torque, or moment, of the force **F** about the point O. Since $|\mathbf{r} \times \mathbf{F}| = |\mathbf{r}|\,|\mathbf{F}| \sin\theta = l|\mathbf{F}|$, the magnitude of the torque is correctly given. Further, the direction of $\mathbf{r} \times \mathbf{F}$ is along a line perpendicular to both **r** and **F**, in precisely the direction a right-threaded screw would tend to move under the action of the force (see Fig. 4.3).

The vector product occurs naturally in many other physical phenomena, e.g., the Coriolis acceleration (to be discussed later), which plays a basic role in meteorology, and the Lorentz force in electrodynamics. This partially accounts for its common acceptance despite the rather peculiar distinction between right- and left-handed coordinates.

---

2 The unit vectors **i**, **j**, **k** have simple relationships under cross multiplication. For example, $\mathbf{j} \times \mathbf{k} = 1 \cdot 1 \sin(\pi/2)\mathbf{i} = \mathbf{i}$, $\mathbf{i} \times \mathbf{i} = \mathbf{0}$ as $\theta = 0$, and in general

$$\mathbf{j} \times \mathbf{k} = -\mathbf{k} \times \mathbf{j} = \mathbf{i}, \quad \mathbf{k} \times \mathbf{i} = -\mathbf{i} \times \mathbf{k} = \mathbf{j}, \quad \mathbf{i} \times \mathbf{j} = -\mathbf{j} \times \mathbf{i} = \mathbf{k},$$
$$\mathbf{i} \times \mathbf{i} = \mathbf{j} \times \mathbf{j} = \mathbf{k} \times \mathbf{k} = \mathbf{0}. \tag{3}$$

Note also (from the definition) that

$$\mathbf{a} \times \mathbf{a} = \mathbf{0},$$

for any vector $\mathbf{a}$. The vanishing of the cross product of two nonzero vectors implies that the vectors are parallel or antiparallel.

It is useful to determine the component form of $\mathbf{a} \times \mathbf{b}$ from the individual representations

$$\mathbf{a} = a_1\mathbf{i} + a_2\mathbf{j} + a_3\mathbf{k} \quad \text{and} \quad \mathbf{b} = b_1\mathbf{i} + b_2\mathbf{j} + b_3\mathbf{k}.$$

Let

$$\mathbf{a} \times \mathbf{b} = p_1\mathbf{i} + p_2\mathbf{j} + p_3\mathbf{k};$$

the problem is then to find p_1, p_2, and p_3 in terms of the components of $\mathbf{a}$ and $\mathbf{b}$.

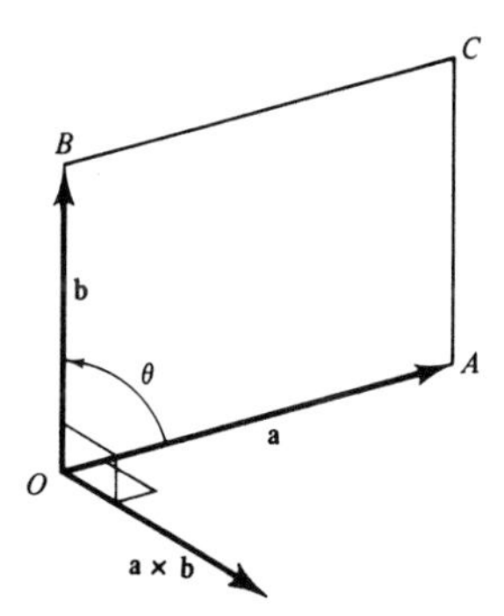

FIGURE 4.2

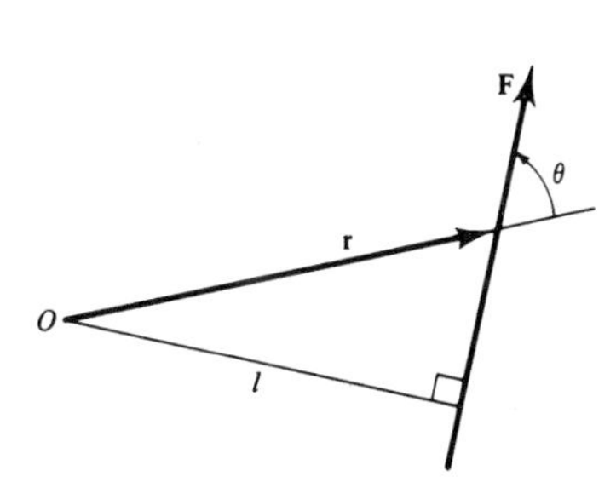

FIGURE 4.3

Since $\mathbf{a} \times \mathbf{b}$ is by definition perpendicular to both $\mathbf{a}$ and $\mathbf{b}$, the dot product of $\mathbf{a} \times \mathbf{b}$ with either $\mathbf{a}$ or $\mathbf{b}$ must vanish. This implies the two equations

$$p_1a_1 + p_2a_2 + p_3a_3 = 0$$

and

$$p_1b_1 + p_2b_2 + p_3b_3 = 0.$$

If these two equations are solved for p_1 and p_2, we find that

$$p_1 = \frac{(a_2b_3 - a_3b_2)p_3}{a_1b_2 - a_2b_1}, \qquad p_2 = \frac{(a_3b_1 - a_1b_3)p_3}{a_1b_2 - a_2b_1}.$$

A more symmetric way of writing this solution is

$$\frac{p_1}{a_2b_3 - a_3b_2} = \frac{p_2}{a_3b_1 - a_1b_3} = \frac{p_3}{a_1b_2 - a_2b_1} = \lambda, \tag{4}$$

where λ is some as yet unknown constant.

So far the only fact we have used is that $\mathbf{a} \times \mathbf{b}$ is perpendicular to both $\mathbf{a}$ and $\mathbf{b}$. The constant λ can be specified from information pertaining to the magnitude of the vector, and with the identity $\sin^2\theta = 1 - \cos^2\theta$, we obtain

$$|\mathbf{a}|^2|\mathbf{b}|^2\sin^2\theta = |\mathbf{a}|^2|\mathbf{b}|^2 - |\mathbf{a}|^2|\mathbf{b}|^2\cos^2\theta$$

or

$$|\mathbf{a} \times \mathbf{b}|^2 = |\mathbf{a}|^2|\mathbf{b}|^2 - (\mathbf{a} \cdot \mathbf{b})^2.$$

This equation is equivalent to

$$p_1{}^2 + p_2{}^2 + p_3{}^2 = (a_1{}^2 + a_2{}^2 + a_3{}^2)(b_1{}^2 + b_2{}^2 + b_3{}^2) - (a_1b_1 + a_2b_2 + a_3b_3)^2.$$

But by (4)

$$p_1{}^2 + p_2{}^2 + p_3{}^2 = \lambda^2[(a_2b_3 - a_3b_2)^2 + (a_3b_1 - a_1b_3)^2 + (a_1b_2 - a_2b_1)^2]. \quad (5)$$

With some effort, we can establish the identity

$$(a_1{}^2 + a_2{}^2 + a_3{}^2)(b_1{}^2 + b_2{}^2 + b_3{}^2) - (a_1b_1 + a_2b_2 + a_3b_3)^2 = (a_2b_3 - a_3b_2)^2 + (a_3b_1 - a_1b_3)^2 + (a_1b_2 - a_2b_1)^2,$$

substitution of which in Eq. (5) leads to the result $\lambda^2 = 1$. Therefore, $\lambda = \pm 1$, and the choice

$$\lambda = +1$$

corresponds to the right-handed convention. (For example, with $\mathbf{a} = \mathbf{j}$, $\mathbf{b} = \mathbf{k}$, in which case $\mathbf{a} \times \mathbf{b} = \mathbf{i}$, an easy calculation shows that $\lambda = +1$.)

The final result is

$$\mathbf{a} \times \mathbf{b} = (a_2b_3 - a_3b_2)\mathbf{i} + (a_3b_1 - a_1b_3)\mathbf{j} + (a_1b_2 - a_2b_1)\mathbf{k}. \quad (6)$$

A useful mnemonic is to express Eq. (6) as the 3×3 determinant

$$\mathbf{a} \times \mathbf{b} = \begin{vmatrix} \mathbf{i} & \mathbf{j} & \mathbf{k} \\ a_1 & a_2 & a_3 \\ b_1 & b_2 & b_3 \end{vmatrix}. \quad (7)$$

The determinant is expanded in the usual fashion, treating the vectors as ordinary entries. Note that the properties $\mathbf{a} \times \mathbf{b} = -\mathbf{b} \times \mathbf{a}$ and $\mathbf{a} \times \mathbf{a} = \mathbf{0}$ follow immediately from Eq. (6) or from (7) by using elementary properties of determinants.

From (6) it is also a simple matter to show that the vector product is distributive, i.e.,

$$\mathbf{a} \times (\mathbf{b} + \mathbf{c}) = \mathbf{a} \times \mathbf{b} + \mathbf{a} \times \mathbf{c}.$$

Example 1 If $\mathbf{a} = \mathbf{i} - 3\mathbf{j} + 5\mathbf{k}$, $\mathbf{b} = 2\mathbf{i} + \mathbf{j} - 7\mathbf{k}$, find $\mathbf{a} \times \mathbf{b}$ and the area of the triangle OAB, where $\overrightarrow{OA} = \mathbf{a}$, $\overrightarrow{OB} = \mathbf{b}$.

From Eq. (7) we find that

$$\mathbf{a} \times \mathbf{b} = \begin{vmatrix} \mathbf{i} & \mathbf{j} & \mathbf{k} \\ 1 & -3 & 5 \\ 2 & 1 & -7 \end{vmatrix} = \mathbf{i}\begin{vmatrix} -3 & 5 \\ 1 & -7 \end{vmatrix} - \mathbf{j}\begin{vmatrix} 1 & 5 \\ 2 & -7 \end{vmatrix} + \mathbf{k}\begin{vmatrix} 1 & -3 \\ 2 & 1 \end{vmatrix},$$

which reduces to

$$\mathbf{a} \times \mathbf{b} = 16\mathbf{i} + 17\mathbf{j} + 7\mathbf{k}.$$

The area of the triangle OAB is $\mathscr{A} = \tfrac{1}{2}|\mathbf{a} \times \mathbf{b}|$, so that

$$\mathscr{A} = \tfrac{1}{2}\sqrt{16^2 + 17^2 + 7^2} = \tfrac{1}{2}\sqrt{594}.$$

Example 2 If P_1, P_2, P_3 are three points in space with position vectors $\overrightarrow{OP_1} = \mathbf{r}_1$, $\overrightarrow{OP_2} = \mathbf{r}_2$, and $\overrightarrow{OP_3} = \mathbf{r}_3$, show that the area of the triangle $P_1P_2P_3$ is given by $\tfrac{1}{2}|\mathbf{r}_2 \times \mathbf{r}_3 + \mathbf{r}_3 \times \mathbf{r}_1 + \mathbf{r}_1 \times \mathbf{r}_2|$.

The vector product will be used to calculate the area of the triangle. From Fig. 4.4, it is seen that $\overrightarrow{P_1P_2} = \mathbf{r}_2 - \mathbf{r}_1$ and $\overrightarrow{P_1P_3} = \mathbf{r}_3 - \mathbf{r}_1$. If $\mathscr{A}$ is the area of the triangle $P_1P_2P_3$, then

$$\mathscr{A} = \tfrac{1}{2}|(\mathbf{r}_2 - \mathbf{r}_1) \times (\mathbf{r}_3 - \mathbf{r}_1)|.$$

Use of the distributive law yields

$$\mathscr{A} = \tfrac{1}{2}|\mathbf{r}_2 \times \mathbf{r}_3 - \mathbf{r}_2 \times \mathbf{r}_1 - \mathbf{r}_1 \times \mathbf{r}_3 + \mathbf{r}_1 \times \mathbf{r}_1|.$$

The last term is zero ($\mathbf{a} \times \mathbf{a} = \mathbf{0}$), and since $\mathbf{r}_2 \times \mathbf{r}_1 = -\mathbf{r}_1 \times \mathbf{r}_2$ and $\mathbf{r}_1 \times \mathbf{r}_3 = -\mathbf{r}_3 \times \mathbf{r}_1$, the second and third terms can be rewritten to give the symmetric expression

$$\mathscr{A} = \tfrac{1}{2}|\mathbf{r}_2 \times \mathbf{r}_3 + \mathbf{r}_3 \times \mathbf{r}_1 + \mathbf{r}_1 \times \mathbf{r}_2|.$$

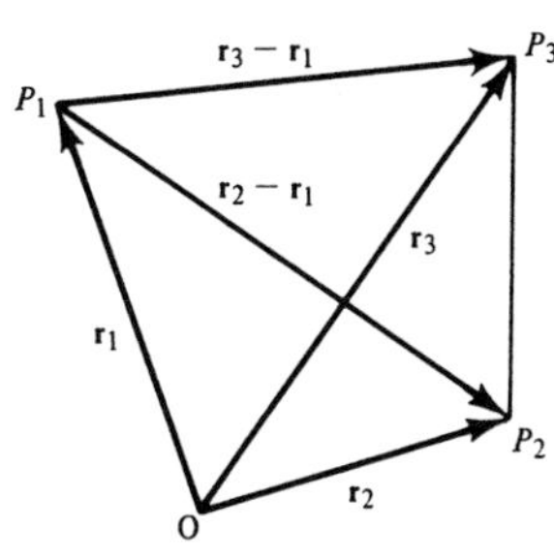

FIGURE 4.4

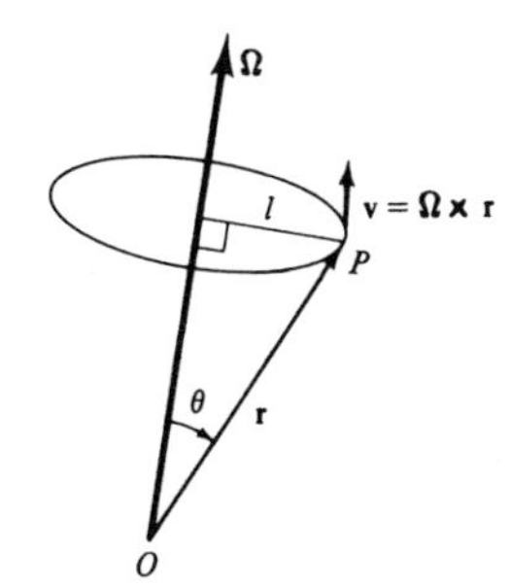

FIGURE 4.5

Example 3 A rigid body rotates about a fixed axis with constant angular velocity Ω. With O a point on this axis, show that the velocity of a point P in the body is

$$\mathbf{v} = \frac{d\mathbf{r}}{dt} = \boldsymbol{\Omega} \times \mathbf{r}, \tag{8}$$

where $\overrightarrow{OP} = \mathbf{r}$ and $\boldsymbol{\Omega}$ is the vector of magnitude Ω drawn along the axis of rotation in the direction in which a right-threaded screw would move (Fig. 4.5).

The speed of the point P is equal to Ωl, where l is the distance of P from the rotation axis. But $|\boldsymbol{\Omega} \times \mathbf{r}| = \Omega r \sin\theta = \Omega l$, and since $\boldsymbol{\Omega} \times \mathbf{r}$ is perpendicular to both $\boldsymbol{\Omega}$ and $\mathbf{r}$, it has precisely the correct magnitude and direction of the velocity of the point P.

---

3 The derivative of a cross product can be calculated directly from first principles. If $\mathbf{a} = \mathbf{a}(t)$, $\mathbf{b} = \mathbf{b}(t)$, and $\mathbf{c} = \mathbf{a} \times \mathbf{b}$, then

$$\begin{aligned}\Delta\mathbf{c} &= (\mathbf{a} + \Delta\mathbf{a}) \times (\mathbf{b} + \Delta\mathbf{b}) - \mathbf{a} \times \mathbf{b}, \\ &= \mathbf{a} \times \Delta\mathbf{b} + \Delta\mathbf{a} \times \mathbf{b} + \Delta\mathbf{a} \times \Delta\mathbf{b}.\end{aligned}$$

Therefore

$$\begin{aligned}\frac{d}{dt}(\mathbf{a} \times \mathbf{b}) &= \lim_{\Delta t \to 0} \frac{\mathbf{a} \times \Delta\mathbf{b} + \Delta\mathbf{a} \times \mathbf{b} + \Delta\mathbf{a} \times \Delta\mathbf{b}}{\Delta t} \\ &= \mathbf{a} \times \frac{d\mathbf{b}}{dt} + \frac{d\mathbf{a}}{dt} \times \mathbf{b}.\end{aligned} \tag{9}$$

Differentiation of vector products is similar to the usual product rule for finding derivatives, but care must be taken to preserve the order of the cross products. Equation (9) can also be obtained by first writing $\mathbf{a} \times \mathbf{b}$ in component form and then differentiating.

---

4 Newton's second law for the motion of a particle of constant mass is

$$m\frac{d\mathbf{v}}{dt} = \mathbf{F},$$

and an interesting form of this equation, involving vector products, is obtained by some simple manipulations. Vector premultiplication by $\mathbf{r}$ gives

$$m\mathbf{r} \times \frac{d\mathbf{v}}{dt} = \mathbf{r} \times \mathbf{F}$$

or, upon rearrangement,

$$m\left[\frac{d}{dt}(\mathbf{r} \times \mathbf{v}) - \frac{d\mathbf{r}}{dt} \times \mathbf{v}\right] = \mathbf{r} \times \mathbf{F}.$$

However $\mathbf{v} = d\mathbf{r}/dt$, and this implies that the second term in the above equation is zero ($\mathbf{v} \times \mathbf{v} = \mathbf{0}$), in which case

$$\frac{d}{dt}(\mathbf{r} \times m\mathbf{v}) = \mathbf{r} \times \mathbf{F}. \tag{10}$$

The quantity $\mathbf{r} \times m\mathbf{v}$ is called the *angular momentum* of the particle. Equation (10) states that the rate of change of angular momentum is equal to the *torque* produced by the external forces. In particular, if any component of $\mathbf{r} \times \mathbf{F}$ is zero, the corresponding component of angular momentum is conserved. We make use of this law of conservation of angular momentum in Sec. 5.8 when discussing planetary motion.

5

In many situations measurements are made in a coordinate frame which is rotating. The earth itself is rotating, so that all geophysical measurements fit into this category. It is necessary then to examine effects attributable to the rotation of a coordinate frame, and for this purpose we proceed to calculate the rate of change of a vector quantity in such a system. Although translation is a relative phenomenon, inasmuch as two observers moving apart at constant velocity cannot determine which one is actually in motion, rotation is an absolute property. Whether a body is revolving or not can be determined without relating it to the position of any other object because there are forces and accelerations associated with the process of rotation alone. One of the new effects due to rotation is the *Coriolis acceleration*, which plays a dominant role in determining large-scale atmospheric and oceanic circulations on the earth. At this stage, we will content ourselves with a derivation of the velocity and acceleration in a rotating frame.

Consider two rectangular frames $\mathscr{F}$ and $\mathscr{R}$ with a common origin and coincident at a certain time t (Fig. 4.6). Frame $\mathscr{F}$ is fixed in space, and frame $\mathscr{R}$ rotates with angular velocity $\boldsymbol{\Omega}$. The problem at issue is to determine the velocity of a particle whose position vector with respect to the moving coordinate system $\mathscr{R}$ is

$$\mathbf{r} = \mathbf{r}(t) = x\mathbf{i} + y\mathbf{j} + z\mathbf{k}, \tag{11}$$

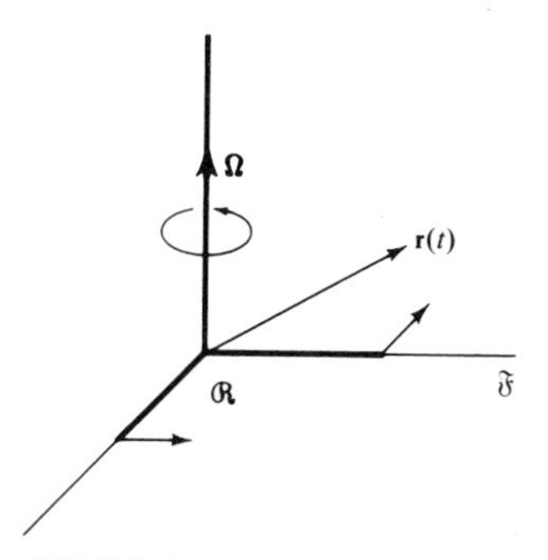

FIGURE 4.6

where x, y, z are functions of time. The position vector may also be described relative to frame $\mathscr{F}$ as

$$\mathbf{r} = x_* \mathbf{i}_* + y_* \mathbf{j}_* + z_* \mathbf{k}_* .$$

The difference between these descriptions is that unit vectors $\mathbf{i}$, $\mathbf{j}$, $\mathbf{k}$ in frame $\mathscr{R}$ are functions of time and change direction but not magnitude, whereas $\mathbf{i}_*$, $\mathbf{j}_*$, $\mathbf{k}_*$ are actually constant vectors. Accordingly, the velocity of the particle measured in the fixed frame is given either by

$$\mathbf{v}_{\mathscr{F}} = \frac{d\mathbf{r}}{dt} = \frac{dx_*}{dt}\mathbf{i}_* + \frac{dy_*}{dt}\mathbf{j}_* + \frac{dz_*}{dt}\mathbf{k}_*, \tag{12}$$

or by

$$\mathbf{v}_{\mathscr{F}} = \frac{d}{dt}(x\mathbf{i} + y\mathbf{j} + z\mathbf{k}),$$

so that

$$\mathbf{v}_{\mathscr{F}} = \frac{dx}{dt}\mathbf{i} + \frac{dy}{dt}\mathbf{j} + \frac{dz}{dt}\mathbf{k} + x\frac{d\mathbf{i}}{dt} + y\frac{d\mathbf{j}}{dt} + z\frac{d\mathbf{k}}{dt}. \tag{13}$$

In (13) we can identify

$$\mathbf{v}_{\mathscr{R}} = \frac{dx}{dt}\mathbf{i} + \frac{dy}{dt}\mathbf{j} + \frac{dz}{dt}\mathbf{k}, \tag{14}$$

as the velocity of point P relative to the moving reference frame $\mathscr{R}$. The remaining terms on the right-hand side express the motion of coordinate system $\mathscr{R}$ with respect to $\mathscr{F}$. They can be analyzed further, using the results of Example 3, where it was shown in (8) that

$$\frac{d\mathbf{i}}{dt} = \boldsymbol{\Omega} \times \mathbf{i}$$

is the velocity of point (1, 0, 0) in the rotating frame at any instant of time. By analogy,

$$\frac{d\mathbf{j}}{dt} = \boldsymbol{\Omega} \times \mathbf{j} \qquad \text{and} \qquad \frac{d\mathbf{k}}{dt} = \boldsymbol{\Omega} \times \mathbf{k},$$

so that

$$x\frac{d\mathbf{i}}{dt} + y\frac{d\mathbf{j}}{dt} + z\frac{d\mathbf{k}}{dt} = \boldsymbol{\Omega} \times (x\mathbf{i} + y\mathbf{j} + z\mathbf{k}) = \boldsymbol{\Omega} \times \mathbf{r}.$$

Equation (13) takes the form

$$\mathbf{v}_{\mathscr{F}} = \mathbf{v}_{\mathscr{R}} + \boldsymbol{\Omega} \times \mathbf{r}. \tag{15}$$

In words, this states that the velocity measured in frame $\mathscr{F}$ is the velocity measured in frame $\mathscr{R}$ *plus* the velocity of frame $\mathscr{R}$ relative to frame $\mathscr{F}$. Equation (15) can be written in a convenient notation as

$$\left(\frac{d}{dt}\right)_{\mathscr{F}} \mathbf{r} = \left(\frac{d}{dt} + \mathbf{\Omega} \times\right)_{\mathscr{R}} \mathbf{r}. \tag{16}$$

Indeed the analysis which leads to this formula is the same for any vector that replaces $\mathbf{r}$. Therefore, the rate of change of a vector in frame $\mathscr{F}$ is the rate of change of the vector in frame $\mathscr{R}$ plus $\mathbf{\Omega} \times$ the vector concerned. In particular, the acceleration measured in frame $\mathscr{F}$,

$$\mathbf{a}_{\mathscr{F}} = \left(\frac{d}{dt}\right)_{\mathscr{F}} \mathbf{v}_{\mathscr{F}},$$

may be related to the acceleration in frame $\mathscr{R}$ via the substitutions

$$\mathbf{a}_{\mathscr{F}} = \left(\frac{d}{dt} + \mathbf{\Omega} \times\right)_{\mathscr{R}} (\mathbf{v}_{\mathscr{R}} + \mathbf{\Omega} \times \mathbf{r}).$$

Therefore

$$\mathbf{a}_{\mathscr{F}} = \frac{d\mathbf{v}_{\mathscr{R}}}{dt} + 2\mathbf{\Omega} \times \mathbf{v}_{\mathscr{R}} + \frac{d\mathbf{\Omega}}{dt} \times \mathbf{r} + \mathbf{\Omega} \times (\mathbf{\Omega} \times \mathbf{r}),$$

or

$$\mathbf{a}_{\mathscr{F}} = \mathbf{a}_{\mathscr{R}} + 2\mathbf{\Omega} \times \mathbf{v}_{\mathscr{R}} + \frac{d\mathbf{\Omega}}{dt} \times \mathbf{r} + \mathbf{\Omega} \times (\mathbf{\Omega} \times \mathbf{r}). \tag{17}$$

Newton's second law for a particle of constant mass

$$m\mathbf{a}_{\mathscr{F}} = \mathbf{F}$$

becomes, *relative to the frame* $\mathscr{R}$,

$$\frac{d\mathbf{v}}{dt} + 2\mathbf{\Omega} \times \mathbf{v} + \frac{d\mathbf{\Omega}}{dt} \times \mathbf{r} + \mathbf{\Omega} \times (\mathbf{\Omega} \times \mathbf{r}) = \frac{\mathbf{F}}{m}. \tag{18}$$

The additional acceleration terms introduced by the rotation all vanish when $\mathbf{\Omega} = \mathbf{0}$. The first term $2\mathbf{\Omega} \times \mathbf{v}$, called the *Coriolis acceleration*, acts in a direction perpendicular to both $\mathbf{\Omega}$ and $\mathbf{v}$; the second is due to possible nonuniformities in the rotation, and the third is, in fact, the *centripetal acceleration* in a disguised form. Of these, the most interesting term is the Coriolis acceleration $2\mathbf{\Omega} \times \mathbf{v}$, which when transposed to the right side of (18) can be interpreted as the *Coriolis force* (per unit mass) $-2\mathbf{\Omega} \times \mathbf{v}$. Anyone who has tried to walk on a moving merry-go-round has felt the effects of the Coriolis force. This force acts to deflect motion in a direction perpendicular to both $\mathbf{\Omega}$ and $\mathbf{v}$ (see Fig. 4.7).

Similar forces act on all particles and objects moving on the rotating earth. For example, a long pendulum will gradually precess due to the Coriolis force. This movement of the so-called Foucault pendulum provides visible evidence of the earth's revolution that is independent of any astronomical observation.

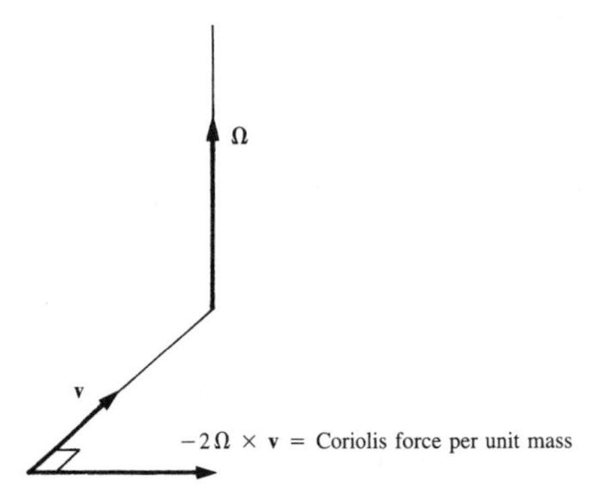

FIGURE 4.7

Exercises 5.4

1. Calculate $\mathbf{a} \cdot \mathbf{b}$ and $\mathbf{a} \times \mathbf{b}$ for the following vectors:

(i) $a = \mathbf{i} + \mathbf{j}$, $\mathbf{b} = \mathbf{j} + \mathbf{k}$
(ii) $\mathbf{a} = \mathbf{i} + \mathbf{j} + \mathbf{k}$, $\mathbf{b} = \mathbf{i} - \mathbf{j} + \mathbf{k}$
(iii) $\mathbf{a} = 2\mathbf{i} - \mathbf{j}$, $\mathbf{b} = \mathbf{i} + 2\mathbf{j}$
(iv) $\mathbf{a} = 2\mathbf{j}$, $\mathbf{b} = \mathbf{i} + \mathbf{k}$
(v) $\mathbf{a} = \mathbf{i} - 2\mathbf{j} - \mathbf{k}$, $\mathbf{b} = 3\mathbf{i} - 5\mathbf{j} + 7\mathbf{k}$
(vi) $\mathbf{a} = \mathbf{i} + 3\mathbf{j} + 5\mathbf{k}$, $\mathbf{b} = 5\mathbf{i} + 3\mathbf{j} - \mathbf{k}$

2. For the vectors in the previous problem, calculate the cosines and sines of the angles between the pairs of vectors.

3. If $\mathbf{a} = \mathbf{i} + \mathbf{j} + \mathbf{k}$, $\mathbf{b} = \mathbf{i} - \mathbf{j} - 9\mathbf{k}$ and $\mathbf{c} = 2\mathbf{i} + \mathbf{j} - 3\mathbf{k}$, calculate $\mathbf{a} \times \mathbf{b}$, $\mathbf{b} \times \mathbf{c}$ and $\mathbf{c} \times \mathbf{a}$. Determine the angles between the three pairs of vectors. Verify the identity

$$\mathbf{a} \times (\mathbf{b} + \mathbf{c}) = \mathbf{a} \times \mathbf{b} + \mathbf{a} \times \mathbf{c}.$$

4. Calculate the sine of the angle between the vectors $\mathbf{a} = 3\mathbf{i} + \mathbf{j} - 3\mathbf{k}$ and $\mathbf{b} = 2\mathbf{i} - \mathbf{j} + \mathbf{k}$. Find a unit vector $\mathbf{c}$ perpendicular to both $\mathbf{a}$ and $\mathbf{b}$ such that $\mathbf{a}$, $\mathbf{b}$ and $\mathbf{c}$ form a right-handed set. Illustrate with a diagram.

5. If $\mathbf{a} + \mathbf{b} + \mathbf{c} = \mathbf{0}$, show that $\mathbf{a} \times \mathbf{b} = \mathbf{b} \times \mathbf{c} = \mathbf{c} \times \mathbf{a}$. Give a geometrical interpretation of these results. (*Hint:* $\mathbf{a}$, $\mathbf{b}$ and $\mathbf{c}$ are coplanar.)

6. If $\mathbf{a}$ and $\mathbf{b}$ are two nonparallel vectors, show that the vectors $\mathbf{e}_1 = \mathbf{a}/|\mathbf{a}|$, $\mathbf{e}_2 = (\mathbf{a} \times \mathbf{b})/|\mathbf{a} \times \mathbf{b}|$ and $\mathbf{e}_3 = \mathbf{e}_1 \times \mathbf{e}_2$ form a right-handed set of mutually perpendicular unit vectors. Illustrate with the special case $\mathbf{a} = \mathbf{i} + \mathbf{j} + \mathbf{k}$, $\mathbf{b} = \mathbf{i} - \mathbf{j} + \mathbf{k}$.

7. Two unit vectors $\boldsymbol{\alpha} = \mathbf{i} \cos\theta_1 + \mathbf{j} \sin\theta_1$ and $\boldsymbol{\beta} = \mathbf{i} \cos\theta_2 + \mathbf{j} \sin\theta_2$ makes angles θ_1 and θ_2 with the x axis. By considering $\boldsymbol{\alpha} \cdot \boldsymbol{\beta}$ and $\boldsymbol{\alpha} \times \boldsymbol{\beta}$ deduce the trigonometric formulas for $\cos(\theta_1 - \theta_2)$ and $\sin(\theta_1 - \theta_2)$.

8. If $A = (3,2,1)$, $B = (2,1,3)$, and $C = (1,2,3)$, find the angles of the triangle ABC. What is the area of the triangle?

9. By vector methods, verify the law of sines for triangles $\sin A/a = \sin B/b = \sin C/c$, where A, B, C represent the angles at vertices A, B, C and a, b, c are the lengths of the opposite sides. Illustrate with the triangle in the previous example.

10. A tetrahedron has its vertices at the points $(1,1,1)$, $(1,2,3)$, $(1,1,2)$, and $(3,-1,2)$. Find the total surface area of the tetrahedron.

11. Given a tetrahedron with vertices A, B, C, D, define four vectors, $\mathbf{v}_1$, $\mathbf{v}_2$, $\mathbf{v}_3$ and $\mathbf{v}_4$ with magnitudes equal to the areas of the sides opposite the vertices A, B, C and D, respectively, and directions given by the outward normals of the sides. Verify that $\mathbf{v}_1 + \mathbf{v}_2 + \mathbf{v}_3 + \mathbf{v}_4 = \mathbf{0}$ and illustrate with the tetrahedron in the previous problem.

12. A body rotates at an angular velocity of 10 radians per second with the vector $2\mathbf{i} - 3\mathbf{j} + \mathbf{k}$ as axis of rotation. Determine the angular velocity vector $\boldsymbol{\Omega}$. Find the velocity of the body at the point $(1,2,3)$. What is the speed at that point?

13. A force of 50 dynes (gm cm sec^{-2}) acts at the point $(1,1,0)$ in the direction of the vector $\mathbf{i} + 2\mathbf{j} - \mathbf{k}$. Find the torque of this force about the origin.

14. For any vector $\mathbf{r}$, show that $2|\mathbf{r}|^2 = |\mathbf{r} \times \mathbf{i}|^2 + |\mathbf{r} \times \mathbf{j}|^2 + |\mathbf{r} \times \mathbf{k}|^2$. Write this identity in terms of the components of $\mathbf{r}$. How does this result relate to the direction cosines of the given vector, $\mathbf{r} = |\mathbf{r}|\ (\cos\alpha,\ \cos\beta,\ \cos\gamma)$?

15. If $\mathbf{b}(t) = \mathbf{a} \times d\mathbf{a}/dt$, show that $d\mathbf{b}/dt = \mathbf{a} \times d^2\mathbf{a}/dt^2$. Illustrate this calculation with the example $\mathbf{a}(t) = (\cos\omega t,\ \sin\omega t,\ t)$.

16. If $\mathbf{a}$ and $\mathbf{b}$ are given constant vectors and ω is a constant, describe the trajectory of a particle given by $\mathbf{r}(t) = \mathbf{a}\cos\omega t + \mathbf{b}\sin\omega t$. Verify the following:

(i) $\dfrac{d^2\mathbf{r}}{dt^2} + \omega^2\mathbf{r} = \mathbf{0}$ (ii) $\mathbf{r} \times \dfrac{d\mathbf{r}}{dt} = \omega\mathbf{a} \times \mathbf{b}$ (iii) $\left|\dfrac{d\mathbf{r}}{dt}\right|^2 + \omega^2|\mathbf{r}|^2 = \omega^2(|\mathbf{a}|^2 + |\mathbf{b}|^2)$

17. Plot the trajectory $\mathbf{r}(t) = \mathbf{a}\cos 2\pi t + \mathbf{b}\sin 2\pi t$ for $\mathbf{a} = (1,0,1)$, $\mathbf{b} = (1,1,-1)$. Examine the special cases of the three identities of the previous problem.

18. If $\mathbf{f}(t)$ and $\mathbf{g}(t)$ are vector functions, derive the results:

(i) $\displaystyle\int_{t_0}^{t_1} \mathbf{f}(t) \cdot \mathbf{g}'(t)\, dt = \mathbf{f}(t) \cdot \mathbf{g}(t)\Big]_{t_0}^{t_1} - \int_{t_0}^{t_1} \mathbf{g}(t) \cdot \mathbf{f}'(t)\, dt$

(ii) $\displaystyle\int_{t_0}^{t_1} \mathbf{f}(t) \times \mathbf{g}'(t)\, dt = \mathbf{f}(t) \times \mathbf{g}(t)\Big]_{t_0}^{t_1} + \int_{t_0}^{t_1} \mathbf{g}(t) \times \mathbf{f}'(t)\, dt.$

19. Forces P, $2P$, $3P$, and $4P$ act along the sides OA, AB, BC, and CO of a square whose vertices are O the origin, $A = (a,0)$, $B = (a,a)$, and $C = (0,a)$. Find the torque about O. Also find the locus of points about which the torque is zero.

20. A body is subject to a force P acting at the point $(1,1,1)$ in the direction of the vector $\mathbf{i} - \mathbf{j} + 2\mathbf{k}$ and a force $2P$ acting at the point $(-1,3,4)$ in the direction of the vector $\mathbf{i} + \mathbf{j} - \mathbf{k}$. Show that this system of forces is equivalent to a force $\mathbf{F}$ and a couple $\mathbf{G}$ at the origin and find $\mathbf{F}$ and $\mathbf{G}$.

21. If $\mathbf{f}(t)$ is any vector quantity, show that

$$\frac{d\mathbf{f}_{\mathcal{F}}}{dt} = \frac{d\mathbf{f}_{\mathcal{R}}}{dt} + \boldsymbol{\Omega} \times \mathbf{f}_{\mathcal{R}},$$

where the subscripts $\mathcal{F}$ and $\mathcal{R}$ refer to the fixed and rotating frames. Find $d\boldsymbol{\Omega}/dt$.

22. In a system rotating with constant angular velocity $\boldsymbol{\Omega}$ about the z axis, a particle of mass m at the point $(a,2a,3a)$ moves with speed V in the direction of the vector $\mathbf{i} + \mathbf{j} + \mathbf{k}$. Calculate the Coriolis and centrifugal forces acting on the particle. Illustrate with a diagram of the forces.

5.5 *Higher-order products*

Having investigated the properties of scalar and vector products, we are faced with the question of how to deal with products of more than two vectors. There are really only two situations which need to be considered, and these concern products of three vectors. For any three vectors **a**, **b**, and **c** these two products are $\mathbf{a} \cdot (\mathbf{b} \times \mathbf{c})$, the *triple scalar product*, and $\mathbf{a} \times (\mathbf{b} \times \mathbf{c})$, the *triple vector product*. [Note that a product such as $\mathbf{a} \cdot (\mathbf{b} \cdot \mathbf{c})$ does not arise, for the dot product of a vector **a** with a scalar $\mathbf{b} \cdot \mathbf{c}$ has no meaning.]

Triple scalar product The triple scalar product $\mathbf{a} \cdot (\mathbf{b} \times \mathbf{c})$ can be given a very simple geometrical interpretation. Let S be the area of the parallelogram for which **b** and **c** are adjacent sides, and let $\hat{\mathbf{n}}$ be the unit normal to this area in the right-hand sense (see Fig. 5.1). It has been shown that

$$\mathbf{b} \times \mathbf{c} = S\hat{\mathbf{n}}.$$

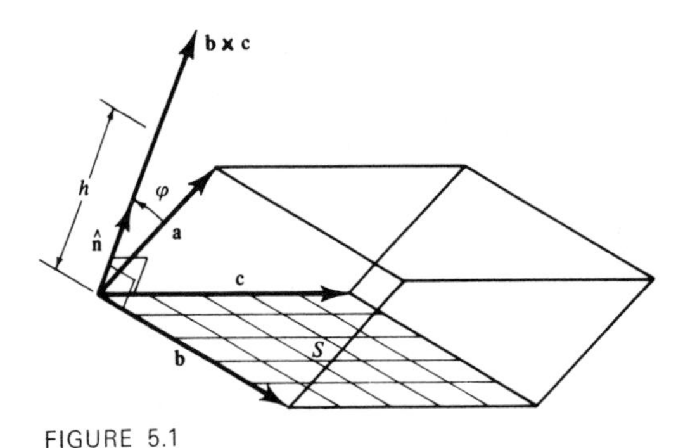

FIGURE 5.1

If φ is the angle between $\hat{\mathbf{n}}$ and **a**, then

$$\mathbf{a} \cdot \hat{\mathbf{n}} = |\mathbf{a}|\,|\hat{\mathbf{n}}|\cos\varphi = |\mathbf{a}|\cos\varphi = h,$$

where h is the length of the projection of **a** onto $\hat{\mathbf{n}}$. The result of combining the preceding equations is

$$\mathbf{a} \cdot (\mathbf{b} \times \mathbf{c}) = \mathbf{a} \cdot S\hat{\mathbf{n}} = Sh = V, \tag{1}$$

where V, the volume of the parallelepiped, is precisely the base area S times the height h.

Therefore, a triple scalar product can be thought of as representing the volume of a parallelepiped which has the three vectors as concurrent edges. (A negative value for V would simply mean that the ordering was left-handed.) In this interpretation, a triple scalar product with two vectors equal (or parallel) is zero, since the relevant volume is zero. Conversely, if $\mathbf{a} \cdot (\mathbf{b} \times \mathbf{c}) = 0$, the three vectors are coplanar. Furthermore, apart from sign, the order of the vectors is irrelevant (as the same volume V will result whatever the order).

These conclusions can also be deduced from the component form of the triple scalar product. If, in the usual notation,

$$\mathbf{a} = a_1\mathbf{i} + a_2\mathbf{j} + a_3\mathbf{k},$$
$$\mathbf{b} = b_1\mathbf{i} + b_2\mathbf{j} + b_3\mathbf{k},$$
$$\mathbf{c} = c_1\mathbf{i} + c_2\mathbf{j} + c_3\mathbf{k},$$

it follows that

$$\mathbf{b} \times \mathbf{c} = (b_2 c_3 - b_3 c_2)\mathbf{i} + (b_3 c_1 - b_1 c_3)\mathbf{j} + (b_1 c_2 - b_2 c_1)\mathbf{k}$$

and

$$\mathbf{a} \cdot (\mathbf{b} \times \mathbf{c}) = a_1(b_2 c_3 - b_3 c_2) + a_2(b_3 c_1 - b_1 c_3) + a_3(b_1 c_2 - b_2 c_1).$$

Alternatively, this expression can be written as a 3×3 determinant

$$\mathbf{a} \cdot (\mathbf{b} \times \mathbf{c}) = \begin{vmatrix} a_1 & a_2 & a_3 \\ b_1 & b_2 & b_3 \\ c_1 & c_2 & c_3 \end{vmatrix}. \tag{2}$$

The elementary properties of determinants yield some immediate results. Thus, if two rows are equal, the determinant is zero, which is equivalent to

$$\mathbf{a} \cdot (\mathbf{a} \times \mathbf{b}) = 0$$

and shows that the vectors $\mathbf{a}$ and $\mathbf{a} \times \mathbf{b}$ are perpendicular. This is otherwise obvious from the definition of $\mathbf{a} \times \mathbf{b}$. The interchange of two rows changes only the sign of a determinant, so that

$$\mathbf{a} \cdot (\mathbf{b} \times \mathbf{c}) = -\mathbf{b} \cdot (\mathbf{a} \times \mathbf{c}) = \mathbf{b} \cdot (\mathbf{c} \times \mathbf{a}) = \mathbf{c} \cdot (\mathbf{a} \times \mathbf{b}). \tag{3}$$

Since $\mathbf{c} \cdot (\mathbf{a} \times \mathbf{b}) = (\mathbf{a} \times \mathbf{b}) \cdot \mathbf{c}$ (the dot product of the two vectors $\mathbf{c}$ and $\mathbf{a} \times \mathbf{b}$ commute), Eq. (3) yields

$$\mathbf{a} \cdot (\mathbf{b} \times \mathbf{c}) = (\mathbf{a} \times \mathbf{b}) \cdot \mathbf{c}, \tag{4}$$

showing that dot and cross can be interchanged in a given triple scalar product without affecting the value. Since the meaning of $\mathbf{a} \cdot \mathbf{b} \times \mathbf{c}$ can only be that of $\mathbf{a} \cdot (\mathbf{b} \times \mathbf{c})$, the parentheses can be dropped without confusion.

Example 1 Find the volume of the parallelepiped which has the three vectors $\mathbf{a} = 2\mathbf{i} - \mathbf{j} - \mathbf{k}$, $\mathbf{b} = \mathbf{i} + \mathbf{j} + 3\mathbf{k}$, and $\mathbf{c} = -\mathbf{i} + \mathbf{j} + 5\mathbf{k}$ as concurrent edges.

The required volume is $V = \mathbf{a} \cdot \mathbf{b} \times \mathbf{c}$, that is,

$$V = \begin{vmatrix} 2 & -1 & -1 \\ 1 & 1 & 3 \\ -1 & 1 & 5 \end{vmatrix}.$$

Expansion by elements of the first row yields

$$V = 2\begin{vmatrix} 1 & 3 \\ 1 & 5 \end{vmatrix} + 1\begin{vmatrix} 1 & 3 \\ -1 & 5 \end{vmatrix} - 1\begin{vmatrix} 1 & 1 \\ -1 & 1 \end{vmatrix},$$

which upon simplification becomes

$$V = 4 + 8 - 2 = 10.$$

---

2 **Triple vector product**

The triple vector product $\mathbf{a} \times (\mathbf{b} \times \mathbf{c})$, frequently occurs in practice. The parentheses mean that $\mathbf{b} \times \mathbf{c}$ is evaluated first and then $\mathbf{a} \times (\mathbf{b} \times \mathbf{c})$ is calculated. The order is important, for in general $\mathbf{a} \times (\mathbf{b} \times \mathbf{c}) \neq (\mathbf{a} \times \mathbf{b}) \times \mathbf{c}$, so that the associative law of multiplication does not hold.

The basic result is that a triple vector product can be written

$$\mathbf{a} \times (\mathbf{b} \times \mathbf{c}) = (\mathbf{a} \cdot \mathbf{c})\mathbf{b} - (\mathbf{a} \cdot \mathbf{b})\mathbf{c}. \tag{5}$$

A formula like (5) is anticipated because the cross product of two vectors is perpendicular to both of them. In particular, $\mathbf{a} \times (\mathbf{b} \times \mathbf{c})$ will be a vector perpendicular to $\mathbf{b} \times \mathbf{c}$, which is, by definition, normal to the plane containing $\mathbf{b}$ and $\mathbf{c}$. This means that $\mathbf{a} \times (\mathbf{b} \times \mathbf{c})$ must lie in the plane of $\mathbf{b}$ and $\mathbf{c}$, as indicated by Eq. (5) (see Fig. 5.2).

The expansion formula (5) for a triple vector product can be established in several ways, and two are given below.

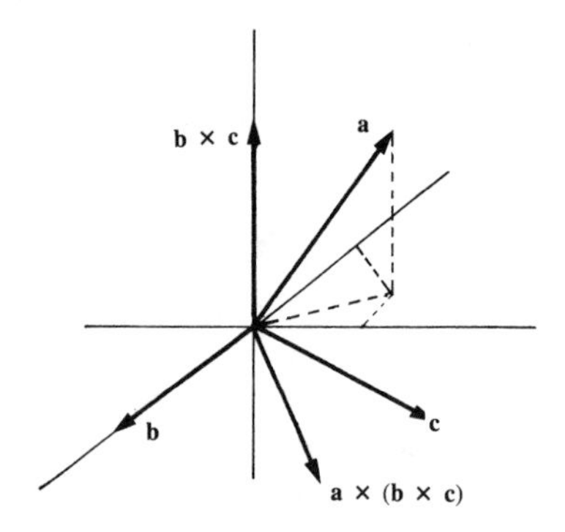

FIGURE 5.2

The first method is a direct verification by calculating components. Let us consider the x component of $\mathbf{a} \times (\mathbf{b} \times \mathbf{c})$ and denote it by $[\mathbf{a} \times (\mathbf{b} \times \mathbf{c})]_1$. This must equal $a_2(\mathbf{b} \times \mathbf{c})_3 - a_3(\mathbf{b} \times \mathbf{c})_2$, where the notation $(\mathbf{b} \times \mathbf{c})_2$, $(\mathbf{b} \times \mathbf{c})_3$ means the y and z components of $\mathbf{b} \times \mathbf{c}$, respectively. Since

$$(\mathbf{b} \times \mathbf{c})_3 = b_1 c_2 - b_2 c_1, \qquad (\mathbf{b} \times \mathbf{c})_2 = b_3 c_1 - b_1 c_3$$

and similarly

$$[\mathbf{a} \times (\mathbf{b} \times \mathbf{c})]_1 = a_2(\mathbf{b} \times \mathbf{c})_3 - a_3(\mathbf{b} \times \mathbf{c})_2,$$

it follows that

$$[\mathbf{a} \times (\mathbf{b} \times \mathbf{c})]_1 = a_2(b_1 c_2 - b_2 c_1) - a_3(b_3 c_1 - b_1 c_3).$$

A slight rearrangement yields

$$[\mathbf{a} \times (\mathbf{b} \times \mathbf{c})]_1 = (a_1 c_1 + a_2 c_2 + a_3 c_3) b_1 - (a_1 b_1 + a_2 b_2 + a_3 b_3) c_1,$$

where the term $a_1 b_1 c_1$ has been added to the first parentheses and subtracted in the second. This equation can be rewritten as

$$[\mathbf{a} \times (\mathbf{b} \times \mathbf{c})]_1 = (\mathbf{a} \cdot \mathbf{c}) b_1 - (\mathbf{a} \cdot \mathbf{b}) c_1, \tag{6}$$

which shows that the x components of the vector equation (5) are identical. In a similar way, the y and z components can be shown to be equal, and (5) is thereby established.

An alternative proof is based on a convenient choice of axes. We choose the x axis to be in the direction of **b** and the xy plane to be the plane of **b** and **c**. In general, **a** will then have nonzero x, y, and z components. In this special system of axes, the general forms of the three vectors are

$$\mathbf{a} = a_1 \mathbf{i} + a_2 \mathbf{j} + a_3 \mathbf{k}, \qquad \mathbf{b} = b_1 \mathbf{i}, \qquad \mathbf{c} = c_1 \mathbf{i} + c_2 \mathbf{j}.$$

It follows that

$$\mathbf{b} \times \mathbf{c} = b_1 \mathbf{i} \times (c_1 \mathbf{i} + c_2 \mathbf{j}) = b_1 c_2 \mathbf{k},$$

in which case

$$\begin{aligned} \mathbf{a} \times (\mathbf{b} \times \mathbf{c}) &= (a_1 \mathbf{i} + a_2 \mathbf{j} + a_3 \mathbf{k}) \times (b_1 c_2 \mathbf{k}), \\ &= a_2 b_1 c_2 \mathbf{i} - a_1 b_1 c_2 \mathbf{j}, \end{aligned} \tag{7}$$

since $\mathbf{j} \times \mathbf{k} = \mathbf{i}$, $\mathbf{i} \times \mathbf{k} = -\mathbf{j}$, and $\mathbf{k} \times \mathbf{k} = \mathbf{0}$. On calculating the right side of Eq. (5) we find that

$$\begin{aligned} (\mathbf{a} \cdot \mathbf{c})\mathbf{b} - (\mathbf{a} \cdot \mathbf{b})\mathbf{c} &= (a_1 c_1 + a_2 c_2) b_1 \mathbf{i} - (a_1 b_1)(c_1 \mathbf{i} + c_2 \mathbf{j}), \\ &= a_2 b_1 c_2 \mathbf{i} - a_1 b_1 c_2 \mathbf{j}. \end{aligned} \tag{8}$$

Equations (7) and (8) are identical, and (5) is established.

Example 2 If $\mathbf{a} = \mathbf{i} - \mathbf{j} + \mathbf{k}$, $\mathbf{b} = 2\mathbf{i} - \mathbf{j}$, and $\mathbf{c} = 3\mathbf{i} + 5\mathbf{j} - 7\mathbf{k}$, verify the identity $\mathbf{a} \times (\mathbf{b} \times \mathbf{c}) = (\mathbf{a} \cdot \mathbf{c})\mathbf{b} - (\mathbf{a} \cdot \mathbf{b})\mathbf{c}$.

First the interior cross product is evaluated:

$$\mathbf{b} \times \mathbf{c} = \begin{vmatrix} \mathbf{i} & \mathbf{j} & \mathbf{k} \\ 2 & -1 & 0 \\ 3 & 5 & -7 \end{vmatrix} = 7\mathbf{i} + 14\mathbf{j} + 13\mathbf{k}.$$

The triple product is then calculated in like manner:

$$\mathbf{a} \times (\mathbf{b} \times \mathbf{c}) = \begin{vmatrix} \mathbf{i} & \mathbf{j} & \mathbf{k} \\ 1 & -1 & 1 \\ 7 & 14 & 13 \end{vmatrix} = -27\mathbf{i} - 6\mathbf{j} + 21\mathbf{k}.$$

Alternatively, Eq. (5) can be used, and since

$$\mathbf{a} \cdot \mathbf{c} = 3 - 5 - 7 = -9$$

and

$$\mathbf{a} \cdot \mathbf{b} = 2 + 1 + 0 = 3,$$

it is found that

$$\begin{aligned}(\mathbf{a} \cdot \mathbf{c})\mathbf{b} - (\mathbf{a} \cdot \mathbf{b})\mathbf{c} &= -9(2\mathbf{i} - \mathbf{j}) - 3(3\mathbf{i} + 5\mathbf{j} - 7\mathbf{k}),\\ &= -27\mathbf{i} - 6\mathbf{j} + 21\mathbf{k},\end{aligned}$$

in agreement with the direct calculation.

3

In a vector product of two vectors, the order of multiplication is important. Of course, this is true for triple vector products, but then the position of the parentheses is also critical. For instance, $\mathbf{a} \times (\mathbf{b} \times \mathbf{c}) \neq (\mathbf{a} \times \mathbf{b}) \times \mathbf{c}$. In Example 2, the reader should verify that $(\mathbf{a} \times \mathbf{b}) \times \mathbf{c} = -19\mathbf{i} + 10\mathbf{j} - \mathbf{k}$, which is certainly not the same as $\mathbf{a} \times (\mathbf{b} \times \mathbf{c}) = -27\mathbf{i} - 6\mathbf{j} + 21\mathbf{k}$. Indeed, $(\mathbf{a} \times \mathbf{b}) \times \mathbf{c} = -\mathbf{c} \times (\mathbf{a} \times \mathbf{b}) = (\mathbf{a} \cdot \mathbf{c})\mathbf{b} - (\mathbf{b} \cdot \mathbf{c})\mathbf{a}$ is a vector in the plane of $\mathbf{a}$ and $\mathbf{b}$, while $\mathbf{a} \times (\mathbf{b} \times \mathbf{c}) = (\mathbf{a} \cdot \mathbf{c})\mathbf{b} - (\mathbf{a} \cdot \mathbf{b})\mathbf{c}$ is a vector in the plane of $\mathbf{b}$ and $\mathbf{c}$, so that these two vectors will be equal only in the exceptional circumstance that $\mathbf{a}$ is parallel to $\mathbf{c}$.

A few examples illustrate the usefulness of the results obtained in simplifying vector identities and solving vector equations.

Example 3 Prove that $\mathbf{a} \times (\mathbf{b} \times \mathbf{c}) + \mathbf{b} \times (\mathbf{c} \times \mathbf{a}) + \mathbf{c} \times (\mathbf{a} \times \mathbf{b}) = \mathbf{0}$.

The use of formula (5) yields

$$\begin{aligned}\mathbf{a} \times (\mathbf{b} \times \mathbf{c}) &= \qquad\qquad (\mathbf{a} \cdot \mathbf{c})\mathbf{b} - (\mathbf{a} \cdot \mathbf{b})\mathbf{c},\\ \mathbf{b} \times (\mathbf{c} \times \mathbf{a}) &= -(\mathbf{b} \cdot \mathbf{c})\mathbf{a} \qquad\qquad + (\mathbf{b} \cdot \mathbf{a})\mathbf{c},\\ \mathbf{c} \times (\mathbf{a} \times \mathbf{b}) &= (\mathbf{c} \cdot \mathbf{b})\mathbf{a} - (\mathbf{c} \cdot \mathbf{a})\mathbf{b}.\end{aligned}$$

When these three equations are added, terms on the right-hand side cancel in pairs and give the required result.

Example 4 Simplify $\mathbf{a} \times [\mathbf{a} \times (\mathbf{a} \times \mathbf{b})] \cdot \mathbf{c}$.

First decompose the triple vector product, treating $\mathbf{a} \times \mathbf{b}$ as a single entity, in which case

$$\begin{aligned}\mathbf{a} \times [\mathbf{a} \times (\mathbf{a} \times \mathbf{b})] &= (\mathbf{a} \cdot \mathbf{a} \times \mathbf{b})\mathbf{a} - (\mathbf{a} \cdot \mathbf{a})(\mathbf{a} \times \mathbf{b}),\\ &= -|\mathbf{a}|^2\mathbf{a} \times \mathbf{b},\end{aligned}$$

since $\mathbf{a} \cdot (\mathbf{a} \times \mathbf{b}) = 0$, because it is a triple scalar product having two vectors equal. With this simplification

$$\mathbf{a} \times [\mathbf{a} \times (\mathbf{a} \times \mathbf{b})] \cdot \mathbf{c} = -|\mathbf{a}|^2\mathbf{a} \times \mathbf{b} \cdot \mathbf{c} = -|\mathbf{a}|^2\mathbf{a} \cdot \mathbf{b} \times \mathbf{c}.$$

Example 5 Solve the vector equation $\mathbf{a} \times \mathbf{y} = \mathbf{b}$ when $\mathbf{a}$ and $\mathbf{b}$ are given.

First note that scalar multiplication of the equation by $\mathbf{a}$ yields

$$\mathbf{a} \cdot \mathbf{a} \times \mathbf{y} = 0 = \mathbf{a} \cdot \mathbf{b}.$$

It follows that $\mathbf{a} \cdot \mathbf{b} = 0$. A necessary condition for there to be *any* solution of the equation is that $\mathbf{a}$ and $\mathbf{b}$ are perpendicular.

The vectors $\mathbf{a}$, $\mathbf{b}$, and $\mathbf{a} \times \mathbf{b}$ form a set of three mutually perpendicular directions. As such, any three-dimensional vector can be represented as a linear combination of this triad. One natural way to proceed then is to look for a solution in the form

$$\mathbf{y} = \lambda \mathbf{a} + \mu \mathbf{b} + \nu \mathbf{a} \times \mathbf{b},$$

where scalars λ, μ, and ν are to be determined. Substitution of this expression in $\mathbf{a} \times \mathbf{y} = \mathbf{b}$ gives

$$\mathbf{a} \times (\lambda \mathbf{a} + \mu \mathbf{b} + \nu \mathbf{a} \times \mathbf{b}) = \mathbf{b}$$

or, by use of the identity (5),

$$\lambda \mathbf{a} \times \mathbf{a} + \mu \mathbf{a} \times \mathbf{b} + \nu[(\mathbf{a} \cdot \mathbf{b})\mathbf{a} - |\mathbf{a}|^2 \mathbf{b}] = \mathbf{b}.$$

The foregoing equation reduces to

$$\mu \mathbf{a} \times \mathbf{b} = (1 + \nu |\mathbf{a}|^2)\mathbf{b},$$

since $\mathbf{a} \times \mathbf{a} = \mathbf{0}$, and $\mathbf{a} \cdot \mathbf{b} = 0$. However, $\mathbf{b}$ and $\mathbf{a} \times \mathbf{b}$ are known to be perpendicular vectors, and the only way the preceding equation can be satisfied is to require that

$$\mu = 0, \qquad \nu = -\frac{1}{|\mathbf{a}|^2}.$$

Therefore the general solution is

$$\mathbf{y} = \lambda \mathbf{a} - \frac{\mathbf{a} \times \mathbf{b}}{|\mathbf{a}|^2},$$

and, since λ is an arbitrary scalar, the solution is not unique.

Exercises 5.5

1. If $\mathbf{a} = 4\mathbf{i} - \mathbf{j} + \mathbf{k}$, $\mathbf{b} = \mathbf{i} + 5\mathbf{j} - \mathbf{k}$ and $\mathbf{c} = -\mathbf{i} - \mathbf{j} + 6\mathbf{k}$, calculate the following:

(i) $\mathbf{a} \cdot (\mathbf{b} \times \mathbf{c})$ **(ii)** $\mathbf{a} \times (\mathbf{b} \times \mathbf{c})$ **(iii)** $(\mathbf{c} \times \mathbf{a}) \cdot \mathbf{b}$

(iv) $\mathbf{c} \times (\mathbf{a} \times \mathbf{b})$ **(v)** $\mathbf{c} \cdot (\mathbf{a} \times \mathbf{b})$ **(vi)** $\mathbf{a} \cdot (\mathbf{c} \times \mathbf{a})$

Why are the answers to **(i)**, **(iii)** and **(v)** the same?

2. Simplify the following expressions:

(i) $\mathbf{i} \times (\mathbf{j} \times \mathbf{k})$ **(ii)** $\mathbf{i} \cdot (\mathbf{j} \times \mathbf{k})$ **(iii)** $\mathbf{j} \times (\mathbf{j} \times \mathbf{k})$

(iv) $\mathbf{j} \cdot (\mathbf{j} \times \mathbf{k})$ **(v)** $(\mathbf{i} + \mathbf{j}) \times (\mathbf{j} + \mathbf{k})$ **(vi)** $(\mathbf{i} \times (\mathbf{j} + \mathbf{k})) \cdot \mathbf{j}$

3. Find the volume of the parallelepiped which has the vectors $\mathbf{a} = 2\mathbf{i} - \mathbf{j} + 3\mathbf{k}$, $\mathbf{b} = 3\mathbf{i} - 2\mathbf{j} + 3\mathbf{k}$, and $\mathbf{c} = \mathbf{i} + \mathbf{j} + \mathbf{k}$ as edges (one vertex is the origin). Determine the areas of the six faces and verify that the volume is equal to the area of a face multiplied by the distance to the opposite face. Illustrate with a diagram.

4. If $\mathbf{a} = 2\mathbf{i} - \mathbf{j} + 3\mathbf{k}$, $\mathbf{b} = 3\mathbf{i} - 2\mathbf{j} + 3\mathbf{k}$ and $\mathbf{c} = \mathbf{i} + \mathbf{j} + \mathbf{k}$, verify the following identities.

(i) $\mathbf{a} \cdot (\mathbf{b} \times \mathbf{c}) = \mathbf{b} \cdot (\mathbf{c} \times \mathbf{a}) = \mathbf{c} \cdot (\mathbf{a} \times \mathbf{b})$

(ii) $\mathbf{a} \times (\mathbf{b} \times \mathbf{c}) = (\mathbf{a} \cdot \mathbf{c})\mathbf{b} - (\mathbf{a} \cdot \mathbf{b})\mathbf{c}$

(iii) $\mathbf{a} \times (\mathbf{b} \times \mathbf{c}) + \mathbf{b} \times (\mathbf{c} \times \mathbf{a}) + \mathbf{c} \times (\mathbf{a} \times \mathbf{b}) = \mathbf{0}$

5. Verify the following identities:

(i) $\mathbf{a} \times \mathbf{b} = [(\mathbf{i} \times \mathbf{a}) \cdot \mathbf{b}]\mathbf{i} + [(\mathbf{j} \times \mathbf{a}) \cdot \mathbf{b}]\mathbf{j} + [(\mathbf{k} \times \mathbf{a}) \cdot \mathbf{b}]\mathbf{k}$

(ii) $\mathbf{a} \times [\mathbf{b} \times (\mathbf{c} \times \mathbf{d})] = (\mathbf{b} \cdot \mathbf{d})(\mathbf{a} \times \mathbf{c}) - (\mathbf{b} \cdot \mathbf{c})(\mathbf{a} \times \mathbf{d})$

(iii) $(\mathbf{a} \times \mathbf{b}) \cdot (\mathbf{c} \times \mathbf{d}) = (\mathbf{a} \cdot \mathbf{c})(\mathbf{b} \cdot \mathbf{d}) - (\mathbf{b} \cdot \mathbf{c})(\mathbf{a} \cdot \mathbf{d})$

(iv) $(\mathbf{a} \times \mathbf{b}) \cdot (\mathbf{c} \times \mathbf{d}) + (\mathbf{b} \times \mathbf{c}) \cdot (\mathbf{a} \times \mathbf{d}) + (\mathbf{c} \times \mathbf{a}) \cdot (\mathbf{b} \times \mathbf{d}) = 0$

In **(ii)**, **(iii)** and **(iv)**, examine the special cases $\mathbf{a} = \mathbf{c}$, $\mathbf{b} = \mathbf{d}$ and $\mathbf{a} = \mathbf{d}$, $\mathbf{b} = \mathbf{c}$.

6. The points A, B, C form a triangle with vertices given by the vectors $\mathbf{a}$, $\mathbf{b}$ and $\mathbf{c}$ relative to the origin. Verify that the vector $\mathbf{a} \times \mathbf{b} + \mathbf{b} \times \mathbf{c} + \mathbf{c} \times \mathbf{a}$ is perpendicular to the plane of the triangle. (*Hint:* The side AB corresponds to the vector $\mathbf{b} - \mathbf{a}$.)

7. Verify the identify $(\mathbf{a} \times \mathbf{b}) \cdot [(\mathbf{b} \times \mathbf{c}) \times (\mathbf{c} \times \mathbf{a})] = [\mathbf{a} \cdot (\mathbf{b} \times \mathbf{c})]^2$. (*Hint:* First prove that $(\mathbf{b} \times \mathbf{c}) \times (\mathbf{c} \times \mathbf{a}) = (\mathbf{a} \cdot \mathbf{b} \times \mathbf{c})\mathbf{c}$.)

8. If $\mathbf{a}$, $\mathbf{b}$ and $\mathbf{c}$ are noncoplanar, prove that any vector $\mathbf{d}$ can be written as a linear combination of $\mathbf{a}$, $\mathbf{b}$ and $\mathbf{c}$, that is, $x_1\mathbf{a} + x_2\mathbf{b} + x_3\mathbf{c} = \mathbf{d}$ where x_1, x_2 and x_3 are constants. Show that

$$x_1 = \frac{\mathbf{d} \cdot (\mathbf{b} \times \mathbf{c})}{\mathbf{a} \cdot (\mathbf{b} \times \mathbf{c})},\ x_2 = \frac{\mathbf{d} \cdot (\mathbf{c} \times \mathbf{a})}{\mathbf{b} \cdot (\mathbf{c} \times \mathbf{a})} \text{ and } x_3 = \frac{\mathbf{d} \cdot (\mathbf{a} \times \mathbf{b})}{\mathbf{c} \cdot (\mathbf{a} \times \mathbf{b})}.$$

9. If $\mathbf{a}$, $\mathbf{b}$ and $\mathbf{c}$ are noncoplanar and $\mathbf{d}$ is any vector, show that

$$(\mathbf{a} \cdot \mathbf{b} \times \mathbf{c})\mathbf{x} = (\mathbf{b} \times \mathbf{c})(\mathbf{a} \cdot \mathbf{x}) + (\mathbf{c} \times \mathbf{a})(\mathbf{b} \cdot \mathbf{x}) + (\mathbf{a} \times \mathbf{b})(\mathbf{c} \cdot \mathbf{x}).$$

Is this result valid if $\mathbf{a}$, $\mathbf{b}$ and $\mathbf{c}$ are coplanar? (*Hint:* Use result of previous problem. If $\mathbf{a}$, $\mathbf{b}$ and $\mathbf{c}$ are coplanar, write $\mathbf{c} = \lambda\mathbf{a} + \mu\mathbf{b}$.)

10. Given vectors $\mathbf{a}$, $\mathbf{b}$ and $\lambda \neq 0$, find a vector $\mathbf{x}$ such that $\lambda\mathbf{x} + \mathbf{x} \times \mathbf{a} = \mathbf{b}$. Illustrate this problem geometrically. (*Hint:* Express $\mathbf{x} = \alpha\mathbf{a} + \beta\mathbf{b} + \gamma(\mathbf{a} \times \mathbf{b})$.)

11. Given vectors $\mathbf{a}$, $\mathbf{b}$ and constants λ, μ, find vectors $\mathbf{x}$ and $\mathbf{y}$ such that $\lambda\mathbf{x} + \mu\mathbf{y} = \mathbf{a}$ and $\mathbf{x} \times \mathbf{y} = \mathbf{b}$. Interpret this problem geometrically. Show that $\mathbf{a} \cdot \mathbf{b} = 0$ is a necessary condition for a solution.

12. If $\mathbf{a}(t)$, $\mathbf{b}(t)$ and $\mathbf{c}(t)$ are functions of t, verify the following results.

(i) $\dfrac{d}{dt}[\mathbf{a} \cdot (\mathbf{b} \times \mathbf{c})] = \mathbf{a} \cdot \left(\mathbf{b} \times \dfrac{d\mathbf{c}}{dt}\right) + \mathbf{a} \cdot \left(\dfrac{d\mathbf{b}}{dt} \times \mathbf{c}\right) + \dfrac{d\mathbf{a}}{dt} \cdot (\mathbf{b} \times \mathbf{c})$

(ii) $\dfrac{d}{dt}[\mathbf{a} \times (\mathbf{b} \times \mathbf{c})] = \mathbf{a} \times \left(\mathbf{b} \times \dfrac{d\mathbf{c}}{dt}\right) + \mathbf{a} \times \left(\dfrac{d\mathbf{b}}{dt} \times \mathbf{c}\right) + \dfrac{d\mathbf{a}}{dt} \times (\mathbf{b} \times \mathbf{c})$

13. Use the result of part (i) of Exercise 12 to deduce a formula for the differentiation of 3×3 determinants, namely,

$$\frac{d}{dt}\begin{vmatrix} a_1 & a_2 & a_3 \\ b_1 & b_2 & b_3 \\ c_1 & c_2 & c_3 \end{vmatrix} = \begin{vmatrix} \dfrac{da_1}{dt} & \dfrac{da_2}{dt} & \dfrac{da_3}{dt} \\ b_1 & b_2 & b_3 \\ c_1 & c_2 & c_3 \end{vmatrix} + \begin{vmatrix} a_1 & a_2 & a_3 \\ \dfrac{db_1}{dt} & \dfrac{db_2}{dt} & \dfrac{db_3}{dt} \\ c_1 & c_2 & c_3 \end{vmatrix} + \begin{vmatrix} a_1 & a_2 & a_3 \\ b_1 & b_2 & b_3 \\ \dfrac{dc_1}{dt} & \dfrac{dc_2}{dt} & \dfrac{dc_3}{dt} \end{vmatrix}.$$

14. Verify the following results by two methods.

(i) $\dfrac{d}{dt}\begin{vmatrix} \cos t & \sin t \\ -\sin t & \cos t \end{vmatrix} = 0$ (ii) $\dfrac{d}{dt}\begin{vmatrix} t & 1 & 2 \\ 2 & t & 3 \\ 3 & 4 & t \end{vmatrix} = 3t^2 - 2$

15. Using the result of Exercise 12(i), simplify $\dfrac{d}{dt}\left[\mathbf{r}\cdot\left(\dfrac{d\mathbf{r}}{dt}\times\dfrac{d^2\mathbf{r}}{dt^2}\right)\right]$.

16. Two particles at $\mathbf{r}_1(t)$, $\mathbf{r}_2(t)$ have velocities $\mathbf{v}_1(t)$, $\mathbf{v}_2(t)$. Find the rate at which the area of the triangle formed by the origin and the two particles is changing. Examine the special case $\mathbf{v}_2(t) = \mathbf{0}$. Illustrate with a diagram showing the triangles at times t and $t + \Delta t$.

5.6 *Geometrical applications*

Vectors have been introduced and developed in association with familiar geometrical concepts, and many problems in three-dimensional geometry are efficiently solved by their use. In this section vectors are employed to develop the properties of lines and planes in three dimensions.

A *line* is specified by a point and a direction. Let O be the origin and let $P_1 = (x_1, y_1, z_1)$ be a definite point on the line, with

$$\overrightarrow{OP_1} = \mathbf{r}_1 = x_1\mathbf{i} + y_1\mathbf{j} + z_1\mathbf{k}.$$

The direction of a line through P_1 can be specified by a constant vector

$$\mathbf{v} = a\mathbf{i} + b\mathbf{j} + c\mathbf{k}.$$

Let $P = (x, y, z)$ be *any other point* on the line, with

$$\overrightarrow{OP} = \mathbf{r} = x\mathbf{i} + y\mathbf{j} + z\mathbf{k};$$

then

$$\overrightarrow{P_1P} = \overrightarrow{OP} - \overrightarrow{OP_1} = \mathbf{r} - \mathbf{r}_1$$

must be parallel to $\mathbf{v}$ (see Fig. 6.1). In other words, $\overrightarrow{P_1P}$ and $\mathbf{v}$ have the same direction, and one is just a multiple of the other, no matter what point P on the line is used.

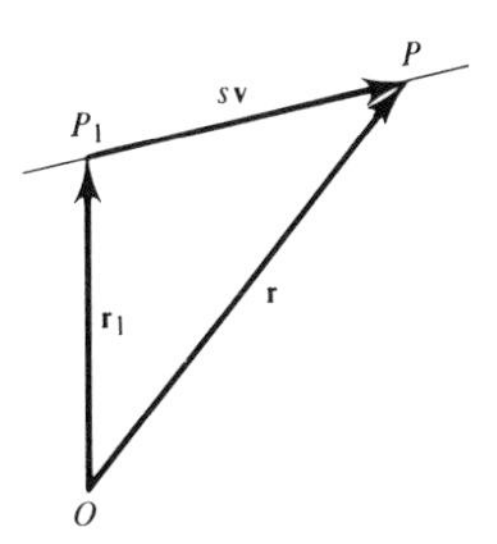

FIGURE 6.1

This implies

$$\mathbf{r} - \mathbf{r}_1 = s\mathbf{v}$$

or

$$\mathbf{r} = \mathbf{r}_1 + s\mathbf{v}, \tag{1}$$

where s is an arbitrary scalar. A definite value of s specifies a particular point on the line. If $\mathbf{v}$ is a unit vector, then $|\mathbf{r} - \mathbf{r}_1| = |s|$, so that s measures positive and negative distance along the line from P_1. (If $\mathbf{v}$ is not a unit vector, $s|\mathbf{v}|$ measures this distance.)

In component form the equation of the line, (1), is

$$x = x_1 + sa, \qquad y = y_1 + sb, \qquad z = z_1 + sc, \tag{2}$$

which can be rearranged to read

$$\frac{x - x_1}{a} = \frac{y - y_1}{b} = \frac{z - z_1}{c} = s. \tag{3}$$

A convenient alternative vector form of (1) is

$$(\mathbf{r} - \mathbf{r}_1) \times \mathbf{v} = \mathbf{0}. \tag{4}$$

A line can also be determined by two points, $P_1 = (x_1, y_1, z_1)$ and $P_2 = (x_2, y_2, z_2)$. If $\overrightarrow{OP_1} = \mathbf{r}_1$ and $\overrightarrow{OP_2} = \mathbf{r}_2$, the direction vector $\mathbf{v}$ is identified as

$$\mathbf{v} = \mathbf{r}_2 - \mathbf{r}_1,$$

and with this specification the equation of the line is

$$\mathbf{r} = \mathbf{r}_1 + s(\mathbf{r}_2 - \mathbf{r}_1). \tag{5}$$

The cartesian form of this is

$$\frac{x - x_1}{x_2 - x_1} = \frac{y - y_1}{y_2 - y_1} = \frac{z - z_1}{z_2 - z_1} = s. \tag{6}$$

The parameter values $s = 0$ and $s = 1$ correspond to the points P_1 and P_2.

Example 1 Find the equation of the line through the point $(1, 2, -3)$ in the direction $\mathbf{v} = \mathbf{i} - 3\mathbf{j} + 5\mathbf{k}$. Where does this line meet the xy plane?

The equation of the line is, by (3),

$$\frac{x - 1}{1} = \frac{y - 2}{-3} = \frac{z + 3}{5} = s.$$

This line meets the xy plane where $z = 0$, and the result of substituting this value in the foregoing is

$$x - 1 = 3/5, \qquad y - 2 = -9/5.$$

The point of intersection is therefore $(8/5, 1/5, 0)$ and corresponds to the value $s = 3/5$.

Example 2 Find the equation of the line joining the points (1, 1, 1) and (2, −1, 4).

Let $(x_1, y_1, z_1) = (1, 1, 1)$ and $(x_2, y_2, z_2) = (2, -1, 4)$; the required line is, by Eq. (6),

$$\frac{x-1}{2-1} = \frac{y-1}{-1-1} = \frac{z-1}{4-1} = s,$$

or

$$x = 1 + s, \qquad y = 1 - 2s, \qquad z = 1 + 3s.$$

A *plane* is determined by one point $P_1 = (x_1, y_1, z_1)$ and the direction of its normal vector

$$\mathbf{n} = a\mathbf{i} + b\mathbf{j} + c\mathbf{k}.$$

With respect to the origin, the position vector to P_1 (see Fig. 6.2) is

$$\overrightarrow{OP_1} = \mathbf{r}_1 = x_1\mathbf{i} + y_1\mathbf{j} + z_1\mathbf{k}.$$

The position vector to *any other point P on the plane* is

$$\overrightarrow{OP} = \mathbf{r} = x\mathbf{i} + y\mathbf{j} + z\mathbf{k}.$$

The vector $\overrightarrow{P_1P} = \mathbf{r} - \mathbf{r}_1$ lies completely on this surface, and *by definition* it must be perpendicular to $\mathbf{n}$; that is,

$$(\mathbf{r} - \mathbf{r}_1) \cdot \mathbf{n} = 0, \tag{7}$$

or

$$\mathbf{r} \cdot \mathbf{n} = d, \tag{8}$$

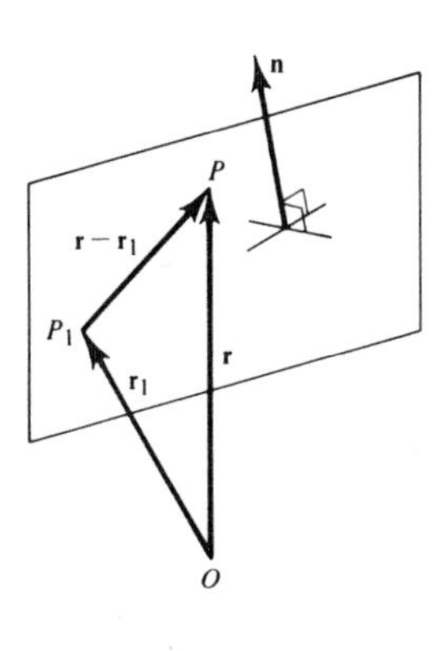

FIGURE 6.2

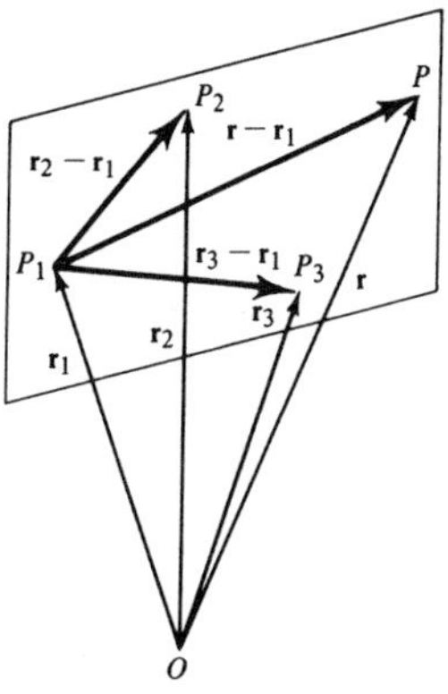

FIGURE 6.3

where $d = \mathbf{r}_1 \cdot \mathbf{n}$ is a constant. This is the equation of the plane, and in cartesian coordinates it can be written

$$a(x - x_1) + b(y - y_1) + c(z - z_1) = 0, \tag{9}$$

or

$$ax + by + cz = d, \tag{10}$$

where

$$ax_1 + by_1 + cz_1 = d.$$

Conversely, any linear equation represents a plane.

Example 3 Find the equation of the plane through three points P_1, P_2, P_3, where $\overrightarrow{OP_1} = \mathbf{r}_1$, $\overrightarrow{OP_2} = \mathbf{r}_2$, $\overrightarrow{OP_3} = \mathbf{r}_3$ (see Fig. 6.3).

The plane through P_1 which has the normal $\mathbf{n}$ is

$$(\mathbf{r} - \mathbf{r}_1) \cdot \mathbf{n} = 0.$$

The normal $\mathbf{n}$ may be found by taking the cross product of any two vectors which lie in the plane. In particular, $\mathbf{r}_2 - \mathbf{r}_1$ and $\mathbf{r}_3 - \mathbf{r}_1$ are such vectors, so that the normal is

$$\mathbf{n} = (\mathbf{r}_2 - \mathbf{r}_1) \times (\mathbf{r}_3 - \mathbf{r}_1) = \mathbf{r}_2 \times \mathbf{r}_3 - \mathbf{r}_2 \times \mathbf{r}_1 - \mathbf{r}_1 \times \mathbf{r}_3 + \mathbf{r}_1 \times \mathbf{r}_1 .$$

On making use of known results for the cross product, this expression can be written in the symmetric form

$$\mathbf{n} = \mathbf{r}_2 \times \mathbf{r}_3 + \mathbf{r}_3 \times \mathbf{r}_1 + \mathbf{r}_1 \times \mathbf{r}_2 .$$

The required plane is therefore

$$(\mathbf{r} - \mathbf{r}_1) \cdot \mathbf{n} = (\mathbf{r} - \mathbf{r}_1) \cdot (\mathbf{r}_2 \times \mathbf{r}_3 + \mathbf{r}_3 \times \mathbf{r}_1 + \mathbf{r}_1 \times \mathbf{r}_2) = 0.$$

Since a triple scalar product with two equal vectors is zero, the above equation can be further simplified and written in the symmetric form

$$\mathbf{r} \cdot \mathbf{r}_2 \times \mathbf{r}_3 + \mathbf{r} \cdot \mathbf{r}_3 \times \mathbf{r}_1 + \mathbf{r} \cdot \mathbf{r}_1 \times \mathbf{r}_2 = \mathbf{r}_1 \cdot \mathbf{r}_2 \times \mathbf{r}_3 .$$

---

3 Consider now the problem of finding the distance from an arbitrary point in space $P_2 = (x_2, y_2, z_2)$, with $\overrightarrow{OP_2} = \mathbf{r}_2 = x_2\mathbf{i} + y_2\mathbf{j} + z_2\mathbf{k}$, to the plane

$$\mathbf{r} \cdot \mathbf{n} = d,$$

as shown in Fig. 6.4. The solution is obtained by first finding the line through P_2 which is perpendicular to the plane. The required distance is then the length P_2M along this line from P_2 to the point of intersection M on the plane.

The equation of the line through P_2 and perpendicular to the plane is

$$\mathbf{r} = \mathbf{r}_2 + s\frac{\mathbf{n}}{|\mathbf{n}|}, \tag{11}$$

where s represents actual distance from the space point in this direction. The use of $\mathbf{n}/|\mathbf{n}|$ as the direction vector automatically assures that the line is a normal to the plane.

The distance P_2M is determined by calculating the specific value of s which corresponds to the point of intersection M. (Note that $s = 0$ gives point P_2.) Since the position vector to point M must satisfy both (11) and (8), the result of eliminating $\mathbf{r} \cdot \mathbf{n}$ between these is

$$\left(\mathbf{r}_2 + s\frac{\mathbf{n}}{|\mathbf{n}|}\right) \cdot \mathbf{n} = d,$$

so that

$$s = \frac{d - \mathbf{r}_2 \cdot \mathbf{n}}{|\mathbf{n}|} = P_2 M. \tag{12}$$

In cartesian form, the absolute distance is given as

$$|s| = \left| \frac{d - ax_2 - by_2 - cz_2}{(a^2 + b^2 + c^2)^{1/2}} \right|. \tag{13}$$

(The sign of s relates to the fact that P_2 can be on either side of the plane.)

This distance is also easily calculated by considering projections. For example, $\mathbf{r}_2 - \mathbf{r}$ is a vector from an arbitrary point in the plane to P_2. As shown in Fig. 6.5, the projection of this vector in the direction perpendicular to the plane given by the unit normal $\mathbf{n}/|\mathbf{n}|$ is the absolute distance. Therefore,

$$|s| = \left| \frac{(\mathbf{r}_2 - \mathbf{r}) \cdot \mathbf{n}}{|\mathbf{n}|} \right|,$$

or

$$|s| = \left| \frac{\mathbf{r}_2 \cdot \mathbf{n} - d}{|\mathbf{n}|} \right|,$$

which is the same as (12).

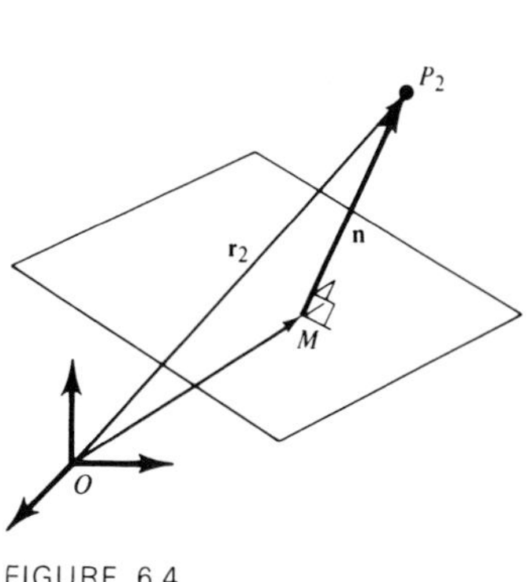

FIGURE 6.4

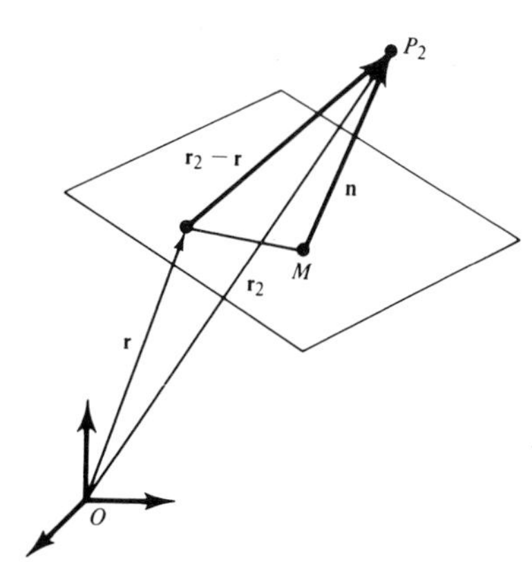

FIGURE 6.5

Example 4 Find the distance of the point $(1, -1, 1)$ from the plane $2x - 3y - 6z = 7$.
From formula (13) the desired distance is

$$|s| = \left| \frac{7 - 2(1) + 3(-1) + 6(1)}{(2^2 + 3^2 + 6^2)^{1/2}} \right| = \frac{8}{7}.$$

Example 5 Find the coordinates of the point where the line through the point $(1, 2, 3)$ in the direction $\mathbf{v} = \mathbf{i} - 4\mathbf{j} + \mathbf{k}$ cuts the plane $x + y + z = 1$.
The equations of the line are

$$\frac{x-1}{1} = \frac{y-2}{-4} = \frac{z-3}{1} = s$$

or

$$x = 1 + s, \qquad y = 2 - 4s, \qquad z = 3 + s. \tag{14}$$

The point of intersection with the plane is given by

$$(1 + s) + (2 - 4s) + (3 + s) = 1,$$

which yields

$$s = 5/2.$$

Upon substituting this particular value of s in (14), we obtain

$$x = 1 + 5/2, \qquad y = 2 - 4(5/2), \qquad z = 3 + 5/2,$$

so the coordinates of the intersection are $(7/2, -8, 11/2)$.

Example 6 Find the plane through the origin which contains the line of intersection of the planes $x + y + z = 1$ and $2x - y + 7z = 5$.
Consider the equation

$$x + y + z - 1 + \lambda(2x - y + 7z - 5) = 0. \tag{15}$$

With λ a constant, this equation is linear in x, y, and z and hence represents a plane. Further, for *any* value of λ the equation is identically satisfied by those values of x, y, and z for which both $x + y + z = 1$ and $2x - y + 7z = 5$. Accordingly, (15) represents the family of planes which contain the line of intersection of the two given planes. In the problem posed, λ is determined by the added condition that $(0, 0, 0)$ be a solution of Eq. (15). Therefore

$$-1 - 5\lambda = 0, \qquad \text{or} \qquad \lambda = -1/5.$$

The particular plane of the family is found by substituting this value of λ into (15), and the result is

$$x + y + z - 1 - 1/5(2x - y + 7z - 5) = 0$$

or

$$3x + 6y - 2z = 0.$$

Example 7 Find the shortest distance between the two skew lines $\mathbf{r} = \mathbf{r}_1 + s\mathbf{v}_1$, $\mathbf{r} = \mathbf{r}_2 + s\mathbf{v}_2$, shown in Fig. 6.6.

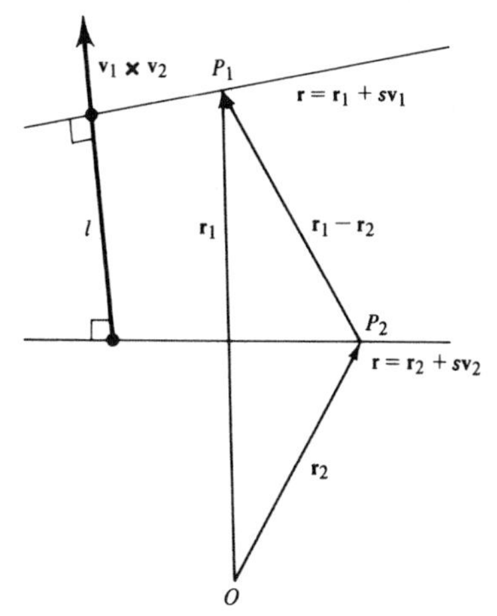

FIGURE 6.6

The shortest distance between two lines is the unique common perpendicular of the two lines. To see this, picture a circular cylinder that has one of the lines as its axis. If the radius of the cylinder is gradually increased, eventually it will just touch the second line. At this instant, the radius is the shortest distance between the lines and clearly perpendicular to both. With this result, an analytical expression for the shortest distance follows readily by vector methods. The direction of the shortest distance is $\mathbf{v}_1 \times \mathbf{v}_2$ since this vector is perpendicular to both $\mathbf{v}_1$ and $\mathbf{v}_2$. Therefore $(\mathbf{v}_1 \times \mathbf{v}_2)/|\mathbf{v}_1 \times \mathbf{v}_2|$ is a unit vector in this direction. The length of the shortest distance is obtained as the projection of $\overrightarrow{P_2P_1} = \mathbf{r}_1 - \mathbf{r}_2$ in this direction, i.e.,

$$l = \frac{(\mathbf{r}_1 - \mathbf{r}_2) \cdot \mathbf{v}_1 \times \mathbf{v}_2}{|\mathbf{v}_1 \times \mathbf{v}_2|}.$$

Exercises 5.6

1. Find the lines through the point $(2, 3, -1)$ parallel to the following vectors:
   (i) $3\mathbf{i} - 4\mathbf{j} + 5\mathbf{k}$ (ii) $2\mathbf{i} + 3\mathbf{j} - \mathbf{k}$ (ii) $\mathbf{i} + \mathbf{j} + \mathbf{k}$
2. Find the planes through the point $(2, 3, -1)$ perpendicular to the following vectors:
   (i) $3\mathbf{i} - 4\mathbf{j} + 5\mathbf{k}$ (ii) $2\mathbf{i} + 3\mathbf{j} - \mathbf{k}$ (iii) $\mathbf{i} + \mathbf{j} + \mathbf{k}$

3. Determine the lines joining the following pairs of points:
(i) $(1,2,3)$, $(3,2,1)$ (ii) $(1,0,-1)$, $(2,7,4)$ (iii) $(2,1,0)$, $(-1,-1,1)$
(iv) $(1,1,1)$, $(1,1,0)$ (v) $(1,0,0)$, $(1,1,1)$ (vi) $(2,5,8)$, $(3,6,9)$
4. Find the points where the lines in the previous problem meet the three coordinate planes.
5. Find the planes that pass through the origin and the pairs of points in Exercise 3.
6. Consider the plane $3x+4y+z=6$ and the point $(1,1,1)$.
(i) Find the line normal to the plane through the given point.
(ii) Find the parallel plane through the given point.
(iii) Find the orthogonal plane through the given point and the origin.
7. Consider the two planes $x-2y+4z=2$ and $3x+y-z=1$. Determine their point of intersection $(1,y_1,z_1)$ with the plane $x=1$. Write the equations of their line of intersection in the form

$$\frac{x-1}{1}=\frac{y-y_1}{b}=\frac{z-z_1}{c}$$

8. Where does the line $(x-3)/2=(y-5)/3=(z-1)/4$ intersect the plane $2x-y+3z=1$? Determine the angle between the line and the normal to the plane.
9. Calculate the perpendicular distance from the origin to the plane through the three points $(a,0,0)$, $(0,b,0)$ and $(0,0,c)$. Determine the equations of the line normal to the plane that passes through the origin. Where does this line intersect the plane? Illustrate with a diagram.
10. Find a unit vector parallel to the plane $3x-y+7z=5$ and perpendicular to the vector $\mathbf{i}-\mathbf{j}+2\mathbf{k}$. Is this vector unique? Find a third vector to complete a right-handed triad.
11. Find the equation of the plane through the point $(1,1,1)$ which is perpendicular to the line of intersection of the two planes $x-y-3z=-1$ and $x-3y+z=2$.
12. Find the distance of the point $(2,0,1)$ from the line $3(x+3)=2(y+2)=-6(z+1)$. Find the point on the line nearest to the given point. (Solve by algebraic methods and by minimizing $(x-2)^2+y^2+(z-1)^2$ for (x,y,z) on the line.)
13. Show that the points $(1,-1,3)$ and $(3,3,3)$ are at the same distance from the plane $5x+2y-7z+9=0$ but on opposite sides of it. Determine the line joining the two points and its point of intersection with the plane.
14. Under what conditions does the plane $ax+by+cz+d=0$ just touch the sphere $x^2+y^2+z^2=1$? (*Hint:* The normals to the plane and the sphere must be the same.)
15. Under what conditions does the line $(x-a)/\cos\alpha=(y-b)/\cos\beta=(z-c)/\cos\gamma$ just touch the sphere $x^2+y^2+z^2=1$? Draw the cone through (a,b,c) formed by these lines. (The parameters satisfy $a^2+b^2+c^2>1$ and $\cos^2\alpha+\cos^2\beta+\cos^2\gamma=1$.)
16. Find the points where the line $x-3y+z+3=3x-5y+4z-1=0$ cuts the sphere $x^2+y^2+z^2=9$.
17. Find the mirror image of the point (α, β, γ) in the plane $2x+y+z=6$ and hence find the image of the line $(x-1)/2=(y-2)/1=(z-3)/(-4)$ in this plane.
18. Show that the plane through the origin containing the vectors $\mathbf{a}$ and $\mathbf{b}$ is given by $\mathbf{r}\cdot(\mathbf{a}\times\mathbf{b})=0$. Interpret this equation geometrically.
19. Obtain the equation of the line through the point $\mathbf{r}_3$ which intersects both the lines $\mathbf{r}=\mathbf{r}_1+s\mathbf{v}_1$ and $\mathbf{r}=\mathbf{r}_2+s\mathbf{v}_2$, where $\mathbf{v}_1\neq\mathbf{v}_2$. Does this problem have a solution if $\mathbf{v}_1=\mathbf{v}_2$? Illustrate with a diagram.
20. Find the shortest distance between the lines

$$\frac{x-5}{-4}=\frac{y-1}{1}=\frac{z-2}{1} \quad \text{and} \quad \frac{x}{2}=\frac{y}{2}=\frac{z-8}{-3}.$$

21. The components of the unit vector in the direction of a given line are the *direction cosines* of the line, $\mathbf{v} = (\cos\alpha, \cos\beta, \cos\gamma)$ where $\cos^2\alpha + \cos^2\gamma + \cos^2\beta = 1$. If (x_1, y_1, z_1) is a point on the line, the equations of the line can be written in direction cosine form

$$\frac{x - x_1}{\cos\alpha} = \frac{y - y_1}{\cos\beta} = \frac{z - z_1}{\cos\gamma}.$$

Write the equations of the following lines in this form:
(i) $3(x-1) = 4(y-1) = 3(z-2)$
(ii) $x + y - z = 1,\ x - y + z = 1 \qquad (x_1, y_1, z_1) = (1, 1, 1)$

22. Determine the direction cosines of the line of intersection of the planes

$$a_1x + b_1y + c_1z = d_1 \qquad a_2x + b_2y + c_2z = d_2.$$

23. The equations of three planes are

$$a_1x + b_1y + c_1z = d_1, \qquad a_2x + b_2y + c_2z = d_2, \qquad a_3x + b_3y + c_3z = d_3.$$

With the notation

$$\mathbf{a} = a_1\mathbf{i} + a_2\mathbf{j} + a_3\mathbf{k}, \qquad \mathbf{b} = b_1\mathbf{i} + b_2\mathbf{j} + b_3\mathbf{k},$$
$$\mathbf{c} = c_1\mathbf{i} + c_2\mathbf{j} + c_3\mathbf{k}, \qquad \mathbf{d} = d_1\mathbf{i} + d_2\mathbf{j} + d_3\mathbf{k},$$

show that the above equations are equivalent to the single vector equation $x\mathbf{a} + y\mathbf{b} + z\mathbf{c} = \mathbf{d}$. Deduce that $(\mathbf{a} \cdot \mathbf{b} \times \mathbf{c})x = (\mathbf{d} \cdot \mathbf{b} \times \mathbf{c})$ and two similar results. Show that if $\mathbf{a} \cdot \mathbf{b} \times \mathbf{c} \neq 0$, the point of intersection of the three planes is

$$\frac{x}{\begin{vmatrix} d_1 & b_1 & c_1 \\ d_2 & b_2 & c_2 \\ d_3 & b_3 & c_3 \end{vmatrix}} = \frac{y}{\begin{vmatrix} a_1 & d_1 & c_1 \\ a_2 & d_2 & c_2 \\ a_3 & d_3 & c_3 \end{vmatrix}} = \frac{z}{\begin{vmatrix} a_1 & b_1 & d_1 \\ a_2 & b_2 & d_2 \\ a_3 & b_3 & d_3 \end{vmatrix}} = \frac{1}{\begin{vmatrix} a_1 & b_1 & c_1 \\ a_2 & b_2 & c_2 \\ a_3 & b_2 & c_3 \end{vmatrix}}.$$

(This result, known as *Cramer's rule*, generalizes to n linear equations in n unknowns.)

5.7 Properties of curves

Space curves are described most easily in terms of a parametric variable which may itself have a physical significance. In Sec. 5.2 the curve $\mathbf{r} = \mathbf{r}(t)$ represented the motion of a particle, $\mathbf{r}$ being the position vector at time t. Of course t need not be identified as a time variable, and any parameter can be used if it facilitates analysis as long as the appropriate functional changes are made where necessary. For example, under the transformation $t = f(u)$, the parametric description $\mathbf{r}(t)$ becomes $\mathbf{r}(f(u))$, and the derivatives with respect to the variables t and u are related by

$$d\mathbf{r}/du = f'(u)\, d\mathbf{r}/dt.$$

It is desirable and advantageous to use a parameter that relates to the curve in some fundamental manner, and the *arc length* s is a natural choice. Let $\mathbf{r} = x\mathbf{i} + y\mathbf{j} + z\mathbf{k}$ be the position vector to a point on a curve, as shown in Fig. 7.1; a formula for arc length s is readily obtained from the corresponding differential relationship

$$d\mathbf{r} = dx\,\mathbf{i} + dy\,\mathbf{j} + dz\,\mathbf{k}. \tag{1}$$

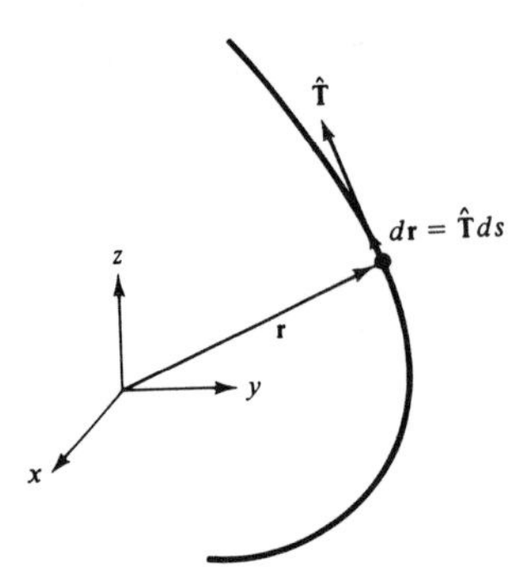

FIGURE 7.1

The increment $d\mathbf{r}$ is a vector along or tangent to the curve; its magnitude is, by definition, an infinitesimal element of arc length:

$$ds = |d\mathbf{r}|. \tag{2}$$

This, by (1), is the same as

$$ds^2 = |d\mathbf{r}|^2 = dx^2 + dy^2 + dz^2, \tag{3}$$

and the calculation is exactly equivalent to finding the distance between the two neighboring points on the curve, (x, y, z) and $(x + dx, y + dy, z + dz)$.

A minor modification of (2), which is now expressed as

$$\left|\frac{d\mathbf{r}}{ds}\right| = 1,$$

leads to an important but rather obvious conclusion, namely,

$$\hat{\mathbf{T}} = \frac{d\mathbf{r}}{ds} = \frac{dx}{ds}\mathbf{i} + \frac{dy}{ds}\mathbf{j} + \frac{dz}{ds}\mathbf{k}, \tag{4}$$

is the *unit tangent vector* to the curve.

The velocity of a particle whose trajectory is the curve under discussion can be determined when the position vector is given as a function of time,

$$\mathbf{r} = \mathbf{r}(t) = x(t)\mathbf{i} + y(t)\mathbf{j} + z(t)\mathbf{k}.$$

The particle moves a distance $d\mathbf{r}$, given in (1), in a time dt, and its velocity is therefore

$$\mathbf{v} = \frac{d\mathbf{r}}{dt} = \frac{dx}{dt}\mathbf{i} + \frac{dy}{dt}\mathbf{j} + \frac{dz}{dt}\mathbf{k}.$$

The speed $|\mathbf{v}| = v$ may be calculated from this or by dividing (3) by the time increment, and the result of either is

$$|\mathbf{v}|^2 = \left|\frac{ds}{dt}\right|^2 = \left(\frac{dx}{dt}\right)^2 + \left(\frac{dy}{dt}\right)^2 + \left(\frac{dz}{dt}\right)^2. \tag{5}$$

The velocity, which is directed along the curve, is easily expressed in terms of the tangent vector $\hat{\mathbf{T}}$ and the speed v by writing

$$\mathbf{v} = \frac{d\mathbf{r}}{dt} = \frac{d\mathbf{r}}{ds}\frac{ds}{dt}, \tag{6}$$

so that, from (4),

$$\mathbf{v} = \frac{ds}{dt}\hat{\mathbf{T}} = v\hat{\mathbf{T}}.$$

The actual distance moved by a particle along its trajectory is calculated by first solving (5) for ds/dt and then integrating to obtain

$$s - s_0 = \int_{t_0}^{t} \left[\left(\frac{dx}{dt}\right)^2 + \left(\frac{dy}{dt}\right)^2 + \left(\frac{dz}{dt}\right)^2\right]^{1/2} dt.$$

Here the position of the particle at time t_0 is s_0, and s is taken to increase as t increases. The two-dimensional formula corresponding to $dz/dt = 0$ was derived in Chap. 3.

Example 1 Find the arc length of the astroid $x^{2/3} + y^{2/3} = a^{2/3}$, shown in Fig. 7.2.

A convenient parametric description of this curve is $x = a\cos^3 t$, $y = a\sin^3 t$. As t ranges from 0 to 2π, the entire curve is described once. If t is indeed time, the parametric equations above give the position of a particle on its trajectory. The arc length of the astroid is then the distance that the particle travels in one complete orbit.

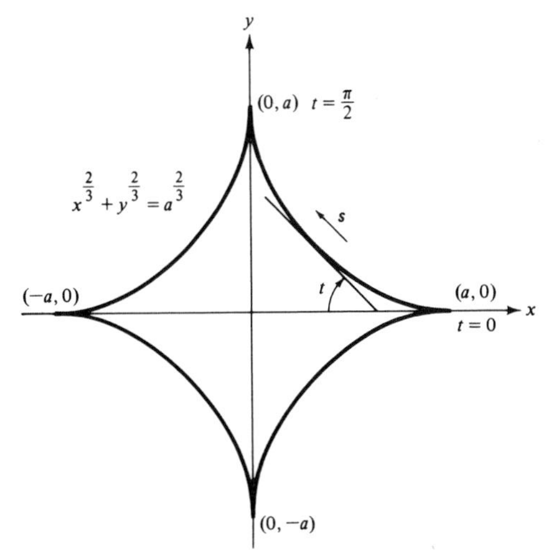

FIGURE 7.2

Differentiation yields

$$\frac{d\mathbf{r}}{dt} = \frac{dx}{dt}\mathbf{i} + \frac{dy}{dt}\mathbf{j} = -3a\cos^2 t \sin t\,\mathbf{i} + 3a\sin^2 t\cos t\,\mathbf{j},$$

from which it follows that

$$\left(\frac{ds}{dt}\right)^2 = 9a^2\sin^2 t\cos^2 t.$$

If we agree to measure s in the direction of increasing t, so that $s = 0$ at $t = 0$, we can write

$$\frac{ds}{dt} = 3a\sin t\cos t = \frac{3a}{2}\sin 2t.$$

From symmetry, the total arc length is 4 times the arc length in the first quadrant (where t ranges from 0 to $\pi/2$). Therefore, the perimeter of the curve is

$$s = 4\int_0^{\pi/2}\frac{3a}{2}\sin 2t\,dt = -3a\cos 2t\Big]_0^{\pi/2},$$

$$= 3a[1-(-1)] = 6a.$$

The unit tangent vector $\hat{\mathbf{T}} = \dfrac{d\mathbf{r}}{dt}\Big/\left|\dfrac{d\mathbf{r}}{dt}\right|$ is found to be

$$\hat{\mathbf{T}} = -\cos t\,\mathbf{i} + \sin t\,\mathbf{j}.$$

so that t is actually the angle between the tangent vector and the negative x direction.

Example 2 Find the arc length of the circular helix $x = a\cos\omega t$, $y = a\sin\omega t$, $z = b\omega t$, where a, b, and ω are constants.

This curve is obtained by winding a string at a given pitch around a circular cylinder $x^2 + y^2 = a^2$, as shown in Fig. 7.3. Once again s is measured from the point $t = 0$ in the direction of t increasing and since

$$\frac{d\mathbf{r}}{dt} = -a\omega\sin\omega t\,\mathbf{i} + a\omega\cos\omega t\,\mathbf{j} + b\omega\mathbf{k},$$

(3) becomes

$$s = \int_0^t\left|\frac{d\mathbf{r}}{dt}\right|dt,$$

$$= \int_0^t[(-a\omega\sin\omega t)^2 + (a\omega\cos\omega t)^2 + b^2\omega^2]^{1/2}\,dt,$$

$$= \omega(a^2+b^2)^{1/2}\int_0^t dt = (a^2+b^2)^{1/2}\omega t.$$

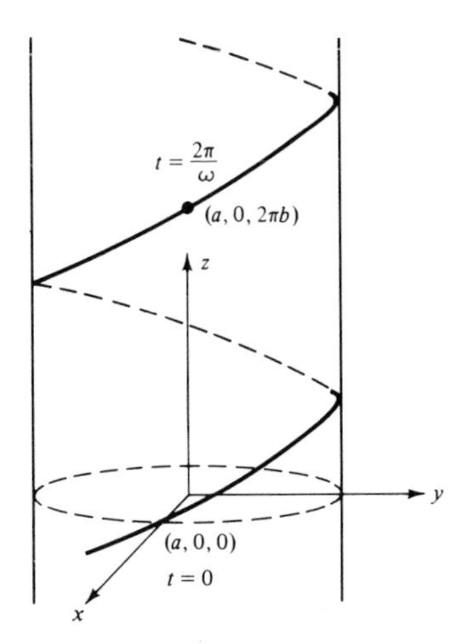

FIGURE 7.3

For example, the arc length between the points $(a, 0, 0)$ and $(a, 0, 2\pi b)$, which correspond to $t = 0$ and $t = 2\pi/\omega$, respectively, is $2\pi(a^2 + b^2)^{1/2}$. In particular, for $b = 0$ this reduces to $2\pi a$, as it must because the helix is then a circle of radius a, with the circumference $2\pi a$.

2

For the remainder of this section attention will be restricted to *plane curves*, i.e., curves which are in the xy plane. In order to facilitate the discussion certain geometrical quantities are introduced.

Let $\hat{\mathbf{T}}$ and $\hat{\mathbf{N}}$ be the *unit tangent and unit normal vectors* at a point on a plane curve (Fig. 7.4). If φ is the angle that $\hat{\mathbf{T}}$ makes with Ox, then $\varphi = \tan^{-1}(dy/dx)$ and

$$\hat{\mathbf{T}} = \cos\varphi\,\mathbf{i} + \sin\varphi\,\mathbf{j} \quad \text{and} \quad \hat{\mathbf{N}} = -\sin\varphi\,\mathbf{i} + \cos\varphi\,\mathbf{j}. \tag{7}$$

In particular,

$$\hat{\mathbf{T}} \times \hat{\mathbf{N}} = \mathbf{k}.$$

In Eq. (4) we showed that

$$\hat{\mathbf{T}} = \frac{d\mathbf{r}}{ds} = \frac{dx}{ds}\mathbf{i} + \frac{dy}{ds}\mathbf{j},$$

and upon comparing this expression with (7) the following relations are obtained:

$$\frac{dx}{ds} = \cos\varphi, \qquad \frac{dy}{ds} = \sin\varphi. \tag{8}$$

These formulas can also be deduced by examining the infinitesimals forming the triangle in Fig. 7.5.

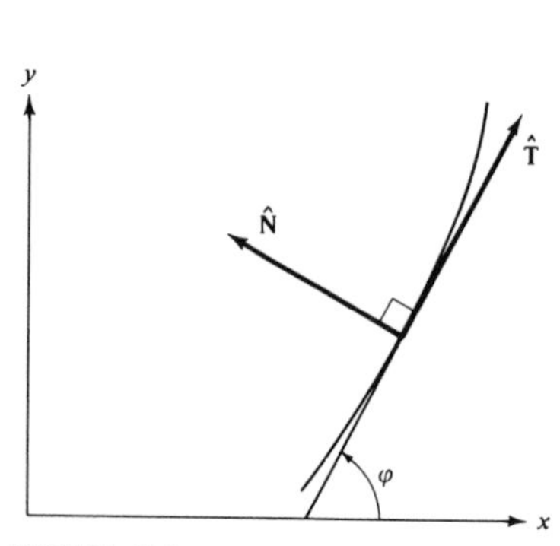

FIGURE 7.4

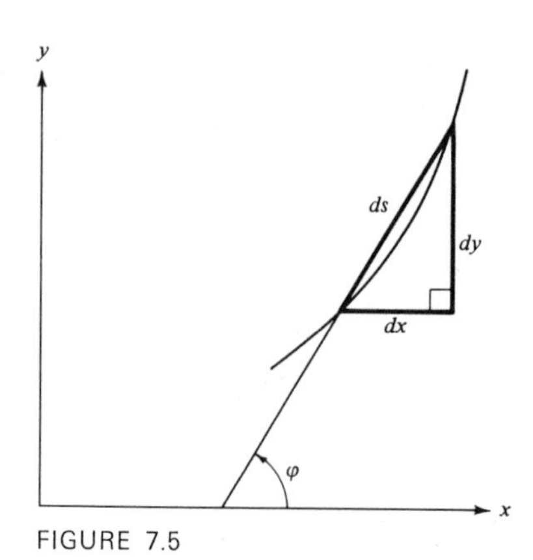

FIGURE 7.5

The quantities s and φ are of importance in determining properties of plane curves, and an equation $s = s(\varphi)$ is called the *intrinsic equation* of a curve. In general, the intrinsic equation of a curve is determined by integration, but in one simple case geometrical arguments suffice.

Example 3 Find the intrinsic equation of a circle of radius R.

If s is measured in a counterclockwise sense with $s = 0$ when $\varphi = \pi/2$, the arc length of the circle is $s = R\theta$, where θ is indicated in Fig. 7.6. But by elementary geometry $\varphi = \pi/2 + \theta$, and therefore

$$s = R\left(\varphi - \frac{\pi}{2}\right).$$

In particular $ds/d\varphi = R$ is the radius of the circle.

---

3 Suppose that by some means we know the intrinsic equation $s = s(\varphi)$ of a curve. At each point, $ds/d\varphi$ (which has the dimensions of a length) may be interpreted as the radius of some circle, as indicated in Example 3. This would be a meaningless

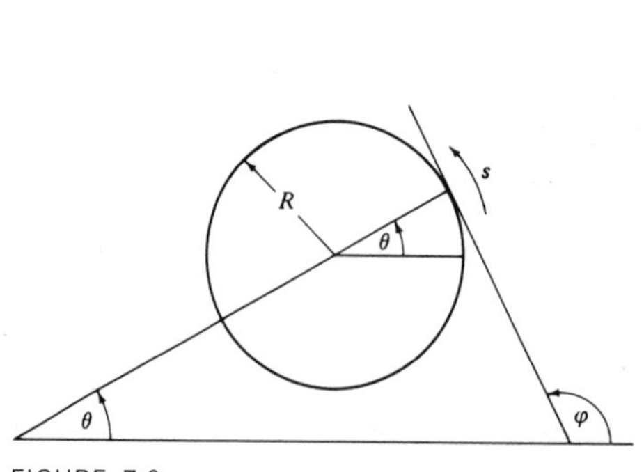

FIGURE 7.6

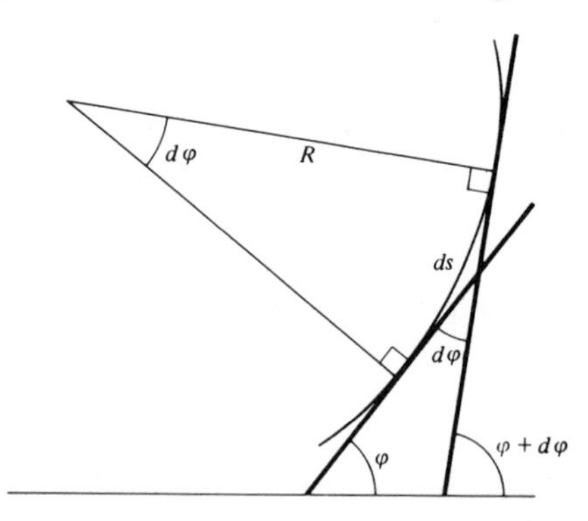

FIGURE 7.7

observation unless this circle related in some sense to the given curve at the particular point. Loosely speaking, it turns out that this circle is the one which locally fits the given curve most snugly at the point concerned.

To be more precise, consider two tangents to a curve at the points (s, φ) and $(s + ds, \varphi + d\varphi)$ (Fig. 7.7). In the limit as these points coincide, the circle which touches both tangents will have radius R because, as shown, $ds = R\, d\varphi$. Therefore, $R = |ds/d\varphi|$ is called the *radius of curvature* of the curve at the given point. (The absolute value is a convention used to avoid negative values of R.) It is sometimes more convenient to work with the *curvature* $\kappa = d\varphi/ds$, in which case the following relations hold:

$$\kappa = \frac{d\varphi}{ds}, \qquad R = \frac{1}{|\kappa|}. \tag{9}$$

The circle concerned is called the *circle of curvature*, and its center is called the *center of curvature* (see Fig. 7.8).

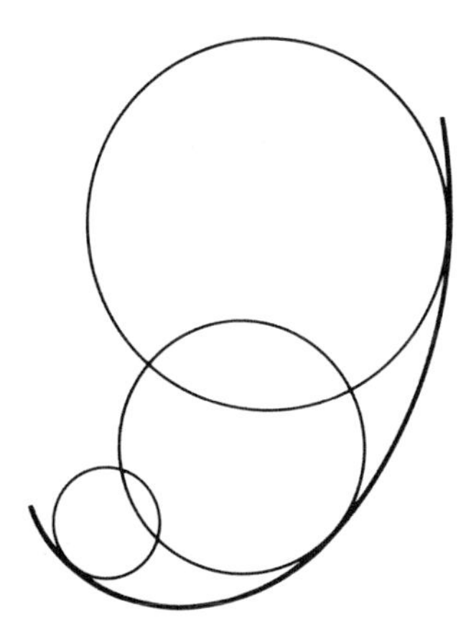

FIGURE 7.8

A circle can be drawn through any three points in a unique way. When these points are situated on a given curve and are allowed to coalesce, the limiting circle obtained is the circle of curvature. Therefore we expect the circle of curvature to have the same values of dy/dx and d^2y/dx^2 as the given curve (see Exercise 18). A large curvature, i.e., small radius of curvature, means the curve is turning rapidly, while a small curvature, i.e., large radius of curvature, means it is nearly straight. In the extreme limiting case of a straight line, the intrinsic equation is $\varphi = \text{constant}$, so that $\kappa = 0$ and R is infinite.

---

4 We now show how the curvature can be calculated from a knowledge of the cartesian equations for a curve. In the process of this derivation, formulas for the tangential and normal components of velocity and acceleration will be found.

Returning to Eq. (7), the unit tangent vector is

$$\hat{\mathbf{T}} = \cos\varphi\,\mathbf{i} + \sin\varphi\,\mathbf{j},$$

so that

$$\frac{d\hat{\mathbf{T}}}{ds} = (-\sin\varphi\,\mathbf{i} + \cos\varphi\,\mathbf{j})\,\frac{d\varphi}{ds}.$$

The definitions of $\hat{\mathbf{N}}$ and κ [Eqs. (7) and (9)] are used to rewrite this compactly as

$$\frac{d\hat{\mathbf{T}}}{ds} = \kappa\hat{\mathbf{N}}. \tag{10}$$

This equation enables us to resolve the acceleration vector of a moving particle in the directions tangential and normal to its path. If the trajectory is the curve $\mathbf{r} = \mathbf{r}(t)$, the velocity of the particle is

$$\mathbf{v} = \frac{d\mathbf{r}}{dt} = \frac{ds}{dt}\frac{d\mathbf{r}}{ds} = \frac{ds}{dt}\hat{\mathbf{T}} = v\hat{\mathbf{T}}. \tag{11}$$

The acceleration $\mathbf{a}$ is obtained by another differentiation with respect to time:

$$\mathbf{a} = \frac{d\mathbf{v}}{dt} = \frac{d}{dt}\left(\frac{ds}{dt}\hat{\mathbf{T}}\right),$$

or

$$\mathbf{a} = \frac{d^2 s}{dt^2}\hat{\mathbf{T}} + \frac{ds}{dt}\frac{d\hat{\mathbf{T}}}{ds}\frac{ds}{dt}.$$

This can be simplified by using (10), to read

$$\mathbf{a} = \frac{d^2 s}{dt^2}\hat{\mathbf{T}} + \kappa\left(\frac{ds}{dt}\right)^2\hat{\mathbf{N}}. \tag{12}$$

The two components of acceleration, shown in Fig. 7.9, are the linear acceleration in the tangential direction,

$$\frac{d^2 s}{dt^2} = \frac{dv}{dt} = v\,\frac{dv}{ds},$$

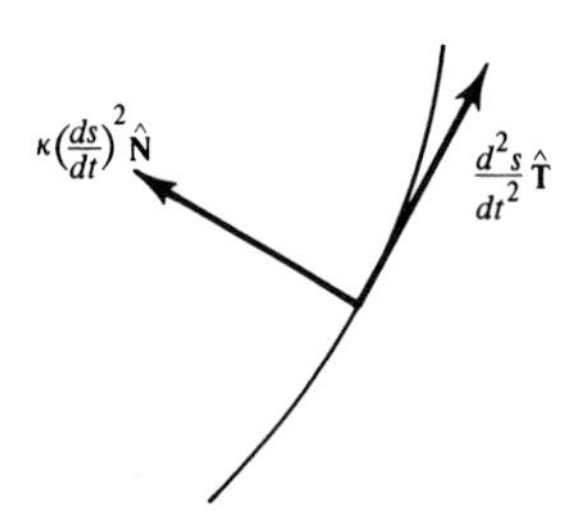

FIGURE 7.9

and the centripetal acceleration in the normal direction,

$$\kappa\left(\frac{ds}{dt}\right)^2 = \kappa v^2.$$

From these results we can obtain a fundamental formula for the calculation of curvature. Since

$$\mathbf{v} = v\hat{\mathbf{T}}$$

and, by (12),

$$\mathbf{a} = \frac{dv}{dt}\hat{\mathbf{T}} + \kappa v^2\hat{\mathbf{N}},$$

we find that

$$\mathbf{v} \times \mathbf{a} = v\hat{\mathbf{T}} \times \left(\frac{dv}{dt}\hat{\mathbf{T}} + \kappa v^2\hat{\mathbf{N}}\right) = \kappa v^3\hat{\mathbf{T}} \times \hat{\mathbf{N}}. \tag{13}$$

But

$$\begin{aligned}\hat{\mathbf{T}} \times \hat{\mathbf{N}} &= (\cos\varphi\,\mathbf{i} + \sin\varphi\,\mathbf{j}) \times (-\sin\varphi\,\mathbf{i} + \cos\varphi\,\mathbf{j}),\\ &= (\cos^2\varphi + \sin^2\varphi)\mathbf{k} = \mathbf{k}.\end{aligned}$$

The substitution of this in the preceding formula yields

$$\mathbf{v} \times \mathbf{a} = \kappa v^3\mathbf{k}$$

or,

$$\kappa = \frac{\mathbf{k}\cdot(\mathbf{v}\times\mathbf{a})}{|\mathbf{v}|^3}. \tag{14}$$

This fundamental formula for curvature is readily expressed in terms of vector components. Since

$$\mathbf{v} = \frac{dx}{dt}\mathbf{i} + \frac{dy}{dt}\mathbf{j} \qquad \text{and} \qquad \mathbf{a} = \frac{d^2x}{dt^2}\mathbf{i} + \frac{d^2y}{dt^2}\mathbf{j},$$

it follows that

$$\mathbf{k}\cdot(\mathbf{v}\times\mathbf{a}) = \frac{dx}{dt}\frac{d^2y}{dt^2} - \frac{dy}{dt}\frac{d^2x}{dt^2} \qquad \text{and} \qquad |\mathbf{v}|^3 = \left[\left(\frac{dx}{dt}\right)^2 + \left(\frac{dy}{dt}\right)^2\right]^{3/2}.$$

Therefore, Eq. (14) can be written in the form

$$\kappa = \frac{\dfrac{dx}{dt}\dfrac{d^2y}{dt^2} - \dfrac{dy}{dt}\dfrac{d^2x}{dt^2}}{\left[\left(\dfrac{dx}{dt}\right)^2 + \left(\dfrac{dy}{dt}\right)^2\right]^{3/2}}. \tag{15}$$

Since there is nothing in this analysis that really depends on a time coordinate, t can of course denote any parameter by which the curve is described. For example, when t is taken to be the variable x, the equation of the curve becomes explicit, $y = f(x)$. In this special case,

$$\frac{dx}{dt} = 1, \qquad \frac{d^2x}{dt^2} = 0, \qquad \frac{dy}{dt} = \frac{dy}{dx}, \qquad \frac{d^2y}{dt^2} = \frac{d^2y}{dx^2},$$

and Eq. (15) reduces to

$$\kappa = \frac{d^2y/dx^2}{[1 + (dy/dx)^2]^{3/2}}. \tag{16}$$

Formulas (15) and (16) are the ones most commonly used in calculations. Note that the dimensions of κ are, as expected, those of an inverse length.

Example 4 Find the curvature of the curve $y = \cosh x$.

Applying the formula,

$$\kappa = \frac{d^2y/dx^2}{[1 + (dy/dx)^2]^{3/2}} = \frac{\cosh x}{[1 + \sinh^2 x]^{3/2}} = \frac{1}{\cosh^2 x}.$$

The largest curvature occurs at the point $x = 0$, $y = 1$ where $\kappa = 1$.

Example 5 Find the curvature of the circle $x^2 + y^2 = R^2$.

Introducing the parametrization $x = R\cos t$, $y = R\sin t$,

$$\kappa = \frac{dx/dt\, d^2y/dt^2 - dy/dt\, d^2x/dt^2}{[(dx/dt)^2 + (dy/dt)^2]^{3/2}}$$

$$= \frac{(-R\sin t)(-R\sin t) - (R\cos t)(-R\cos t)}{[(-R\sin t)^2 + (R\cos t)^2]^{3/2}}$$

Simplifying, we find $\kappa = 1/R$, a very predictable result. The radius of curvature is the radius R of the circle.

Example 6 Find the radius of curvature of the parabola $y = x^2/4a$ at the origin (Fig. 7.10).

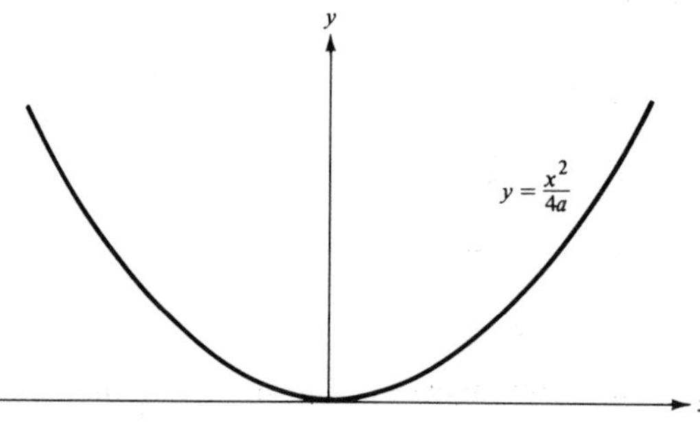

FIGURE 7.10

In this case, $dy/dx = x/2a = 0$ at $x = 0$, and $d^2y/dx^2 = 1/(2a)$. Use of formula (16) gives

$$\frac{1}{R} = \kappa = \frac{1/(2a)}{(1+0)^{3/2}} = \frac{1}{2a};$$

the radius of the curvature is $R = 2a$. At a general point (x, y) on the curve, the radius of curvature is $2a(1 + x^2/4a^2)^{3/2}$. As an exercise, the reader might check that a parametric representation, for example, $x = 2at$, $y = at^2$, and use of (15) lead to the same result.

Example 7 Investigate the motion of a particle of mass m on a smooth curve under the action of gravity.

Newton's second law states that the mass acceleration equals the applied force; therefore the components of these vectors in the tangential and normal directions can be equated (see Fig. 7.11). It follows that

$$m\frac{d^2s}{dt^2} = -mg \sin \varphi, \tag{17}$$

and

$$\frac{mv^2}{R} = N - mg \cos \varphi, \tag{18}$$

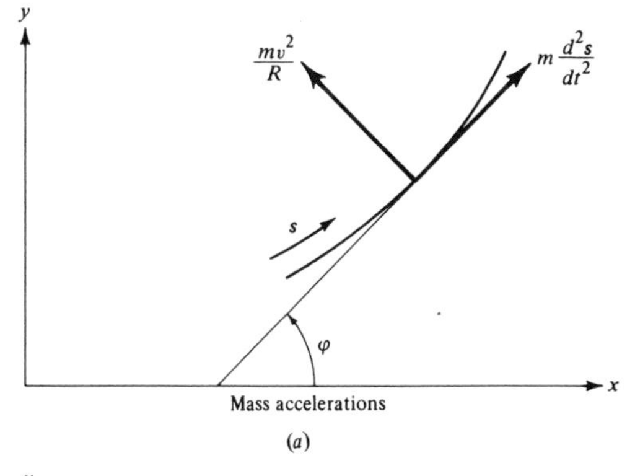

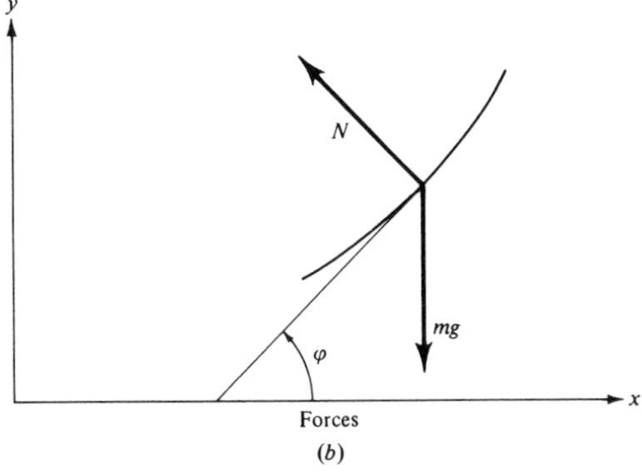

FIGURE 7.11

where N is the normal reaction of the curve on the particle. Equation (17) can be integrated; in order to do this, note first that

$$\frac{d^2s}{dt^2} = \frac{dv}{dt} = \frac{ds}{dt}\frac{dv}{ds} = v\frac{dv}{ds}$$

and by (8)

$$\sin\varphi = \frac{dy}{ds}.$$

Therefore (17) can be rewritten in the form

$$mv\frac{dv}{ds} = -mg\frac{dy}{ds}$$

or

$$\frac{d}{ds}(\tfrac{1}{2}mv^2 + mgy) = 0$$

and integrated to obtain

$$\tfrac{1}{2}mv^2 + mgy = \text{constant}. \tag{19}$$

This equation shows that the sum of the *kinetic energy* ($\frac{1}{2}mv^2$) and *potential energy* (mgy) is constant throughout the motion or, equivalently, *the total energy is conserved* in the absence of any dissipative action due to friction.

Equation (18) determines the normal reaction force N. If N vanishes or becomes negative, at any point of the motion, the particle will leave the curve (unless otherwise attached).

As an illustration, consider a circular track or loop of radius a and suppose that a marble is projected with velocity U at point O. If s is measured from $\varphi = 0$, where $s = 0$, as shown in Fig. 7.12, the intrinsic equation of the circle is $s = a\varphi$. Since $y = a(1 - \cos\varphi)$, the energy equation takes the form

$$\tfrac{1}{2}v^2 + ga(1 - \cos\varphi) = \tfrac{1}{2}U^2,$$

and it follows that

$$v^2 = U^2 - 2ga(1 - \cos\varphi).$$

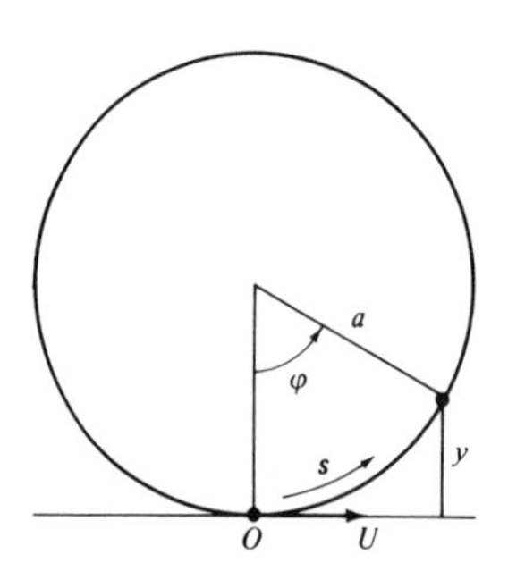

FIGURE 7.12

Three possible situations arise depending on the value of U, the initial velocity. With sufficient speed, the marble can maintain contact with the loop throughout a complete revolution. On the other hand, with insufficient initial energy, the marble will just oscillate back and forth about the lowest point. Lastly, the marble can ride the loop, lose contact, and fall off at some point (where $N = 0$). The mathematics contains all this information, but further elucidation is left to the exercises.

Exercises 5.7

1. For the curve $\mathbf{r}(t) = (1 + 2t)\mathbf{i} + (2 - 3t)\mathbf{j} + (1 + 6t)\mathbf{k}$, show that the unit tangent vector $\hat{\mathbf{T}}$ is constant. What does this imply about the curve? Find the arc length $s(t)$ from the point $\mathbf{r}(0) = (1,2,1)$ to the point $\mathbf{r}(t)$. Illustrate with a graph.
2. Verify that the curve $\mathbf{r}(t) = e^t\mathbf{i} + e^{-t}\mathbf{j} + \sqrt{2}\, t\mathbf{k}$ lies on the intersection of the cylinders $xy = 1$ and $x = e^{z/\sqrt{2}}$. Draw a graph of the curve for $t \geq 0$. Find the unit tangent vector $\hat{\mathbf{T}}(t)$ and calculate the arc length from $t = 0$ to $t = 1$.
3. Show that the circular helix $\mathbf{r}(t) = (a \cos \omega t,\ a \sin \omega t,\ bt)$ where a,b and ω are constants lies on the cylinder $x^2 + y^2 = a^2$. Draw a graph of the curve for $t \geq 0$. Calculate the arc length from $\mathbf{r}(0)$ to $\mathbf{r}(t)$.
4. A thin wire of total length 1000 cm is formed into a flexible coil that is a circular helix. If there are 10 turns to each centimeter of height and the radius of the helix is 3 cm, determine the equation of the coil. (Introduce coordinates x, y, z with the lowest point of the coil at $z = 0$ and with the axis of the coil along the z axis.) How tall is the coil?
5. Graph the following plane curves. Determine the unit tangent and unit normal vectors.
(i) $\mathbf{r}(t) = (at^2,\ 2at)$ **(ii)** $\mathbf{r}(t) = (a \cos t,\ b \sin t)$ **(iii)** $\mathbf{r}(t) = (a \cosh t,\ b \sinh t)$
6. If n is a positive integer, sketch the curves $y_n(x) = (1/n) \sin nx$ for $n = 1, 2, 3$. Show that each curve $y_n(x)$ has the same arc length between $x = 0$ and $x = \pi$.
7. Find the radius of curvature for each curve at the given point.
(i) $y = x^2$, $(1/2, 1/4)$ **(ii)** $y = \sin^2 x$, $(\pi/2, 1)$ **(iii)** $x = y^3$, $(1,1)$
(iv) $\dfrac{x^2}{a^2} + \dfrac{y^2}{b^2} = 1$, $(a,0)$ **(v)** $x^2 + xy + y^2 = 3$, $(1,1)$ **(vi)** $y = e^x$, $(0,1)$
8. For the curve $\mathbf{r}(t) = t\mathbf{i} + t^2\mathbf{j}$, find the unit tangent vector and the unit normal vector at $t = 1$. What are the x and y components of acceleration at $t = 1$? Illustrate with a graph.
9. What is the curvature at the origin for each branch of the curve $x^2 + xy + y^3 = 0$? (*Hint:* Solve for $x = x(y)$.)
10. Show that the curve $\mathbf{r}(t) = t \cos t\, \mathbf{i} + t \sin t\, \mathbf{j} + t\mathbf{k}$ lies on the surface of the cone $x^2 + y^2 = z^2$ and draw its graph for $t \geq 0$. Determine the unit tangent vector at a general point. What is the arc length of the curve between $t = 0$ and $t = \pi$? between $t = \pi$ and $t = 2\pi$?
11. Show that the curve $x = 2u$, $y = u^2$, $z = 1/3\, u^3$ lies on the intersection of the cylinder $x^2 = 4y$ and the surface $z = 1/6\, xy$. Determine the unit tangent vector at a general point. Find the length of the curve from $u = 0$ to $u = 1$.
12. Draw a graph of the curve $\mathbf{r}(t) = (t, t^2, t^3)$ for $t \geq 0$. Find the unit tangent vector $\hat{\mathbf{T}}$ at the points where the curve intersects the plane $2x + y - z = 0$.
13. A smooth roller-coaster track has the form of one turn of the circular helix $(a \cos \theta$, $a \sin \theta,\ b\theta)$. A car leaves the point where $\theta = 2\pi$ with zero velocity and moves under gravity

(acting in the negative z direction) to the point where $\theta = 0$. Using the fact that the sum of the kinetic and potential energies must remain constant, show that the velocity v at the point θ is $v^2 = 2gb(2\pi - \theta)$. Deduce that the time taken to reach $\theta = 0$ is $2[\pi(a^2+b^2)/gb]^{1/2}$.

14. Draw a graph of the catenary $y = c \cosh(x/c)$. Find the unit tangent and unit normal vectors at (x,y) and determine the curvature. Measuring the arc length s from the point $(0,c)$, show that $s = c \sinh(x/c)$ and $y^2 = s^2 + c^2$. Deduce the intrinsic equation of the catenary.

15. Draw a graph of the curve $y = c \log \sec(x/c)$ defined for $0 \le x < \pi c/2$. Find the unit tangent and unit normal vectors and the curvature at (x,y). Measuring arc length s from the point $(0,c)$, show that $s = c \log[\sec(x/c) + \tan(x/c)]$ and deduce the intrinsic equation.

16. At time t the position of a particle is given by

$$\mathbf{r}(t) = \mathbf{i}\int_0^t \cos u^2\, du + \mathbf{j}\int_0^t \sin u^2\, du.$$

If the arc length s is measured from the origin, show that $s = t$ and that the curvature $\kappa = 2t$. Deduce that the intrinsic equation of the path is $s^2 = \varphi$.

17. If the Taylor series for a function about $x = 0$ is of the form $y = a_2x^2 + a_3x^3 + \cdots$, show that the intrinsic equation has the form

$$s = \frac{\varphi}{2a_2} - \frac{3a_3}{8a_2^3}\varphi^2 + \cdots$$

for small φ.

18. The circle $(x-a)^2 + (y-b)^2 = R^2$ is chosen so that the values of y, dy/dx, and d^2y/dx^2 have the same values at the point (x_1, y_1) as for the curve $y = f(x)$. Deduce the results:

(i) $(x_1 - a)^2 + [f(x_1) - b]^2 = R^2$

(ii) $(x_1 - a) + f'(x_1)[f(x_1) - b] = 0$

(iii) $1 + [f'(x_1)]^2 + [f(x_1) - b]f''(x_1) = 0$

Solve these equations for a, b and R. In particular, show that the value of R is identical to the radius of curvature of $y = f(x)$ at the point (x_1, y_1).

19. If (x', y') are the coordinates of the center of curvature for a point (x,y) on a given curve, show that $x' = x + dy/d\varphi$ and $y' = y + dx/d\varphi$. Illustrate with a diagram.

20. Show that the center of curvature for the parabola $(at^2, 2at)$ is located at the point $(a(3t^2+2), -2at^3)$. Deduce that the cartesian equation for the locus of the center of curvature of the parabola (called the *evolute* of the curve) is the curve $27ay^2 = 4(x-2a)^3$. Sketch both curves for $a = 1$ and $a = 2$.

21. From the definition of curvature $\kappa = \dfrac{d\varphi}{ds}$ and $\varphi = \tan^{-1}\left(\dfrac{dy}{dx}\right)$, derive the formulas

$\kappa = \dfrac{d^2y}{dx^2}\bigg/\left(1 + \left(\dfrac{dy}{dx}\right)^2\right)^{3/2}$ and $\kappa = \dfrac{d^2x}{dy^2}\bigg/\left(1 + \left(\dfrac{dx}{dy}\right)^2\right)^{3/2}$ by differentiating.

$\left(\textit{Hint: } \dfrac{d\varphi}{ds} = \dfrac{d\varphi}{dx}\dfrac{dx}{ds} \text{ or } \dfrac{d\varphi}{ds} = \dfrac{d\varphi}{dy}\dfrac{dy}{ds}.\right)$

22. For a curve $x = x(t)$, $y = y(t)$ given parametrically, derive the formula for the curvature by writing $\kappa = \dfrac{d\varphi}{dt}\dfrac{dt}{ds}$, $\varphi = \tan^{-1}\left(\dfrac{dy}{dt}\dfrac{dt}{dx}\right) = \tan^{-1}\left(\dfrac{dy}{dt}\bigg/\dfrac{dx}{dt}\right)$ and differentiating.

23. If r and θ are polar coordinates, prove that the curvature of the curve $r = f(\theta)$ is given by

$$\kappa = \frac{f^2 + 2f'^2 - ff''}{(f^2 + f'^2)^{3/2}} = \frac{r^2 + 2(dr/d\theta)^2 - r(d^2r/d\theta^2)}{[r^2 + (dr/d\theta)^2]^{3/2}}$$

(*Hint:* $x = r\cos\theta = f(\theta)\cos\theta$, $y = r\sin\theta = f(\theta)\sin\theta$.)

24. Show that the curve $\mathbf{r}=\mathbf{r}(s)$ has curvature $\kappa = \pm|d^2\mathbf{r}/ds^2|$. (*Hint:* $\hat{\mathbf{T}} = d\mathbf{r}/ds$.)
25. Sketch the following curves given in polar coordinates. Calculate the curvature at $\theta = \pi/2$ and find the length of arc from $\theta = 0$ to $\theta = \pi$.

(i) $r = a\theta$ **(ii)** $r = a(1+\cos\theta)$ **(iii)** $r = a^\theta$

26. A thin string is wrapped around the circle $x^2+y^2=a^2$. One end of the string initially at the point $(a,0)$ is gradually unwound (Fig. 7.13). Show that when a length $a\theta$ of the string is unwound, the moving end is at the point $\mathbf{r} = a(\cos\theta + \theta\sin\theta)\mathbf{i} + a(\sin\theta - \theta\cos\theta)\mathbf{j}$. At this position show that P has described an arc of length equal to $(a/2)\theta^2$. By identifying θ with φ deduce that the intrinsic equation for the locus of P is $s = \frac{1}{2}a\varphi^2$.

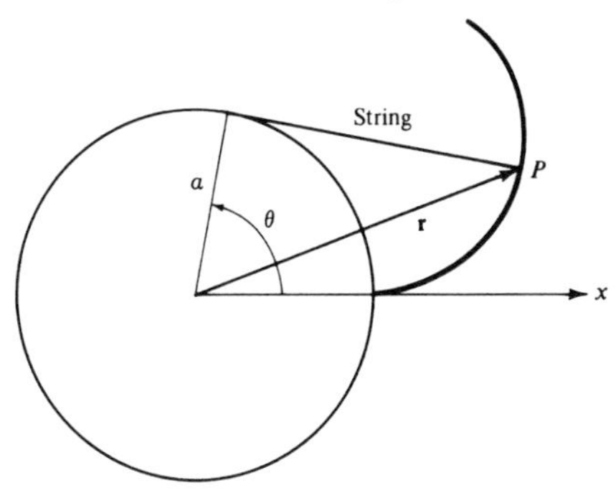

FIGURE 7.13

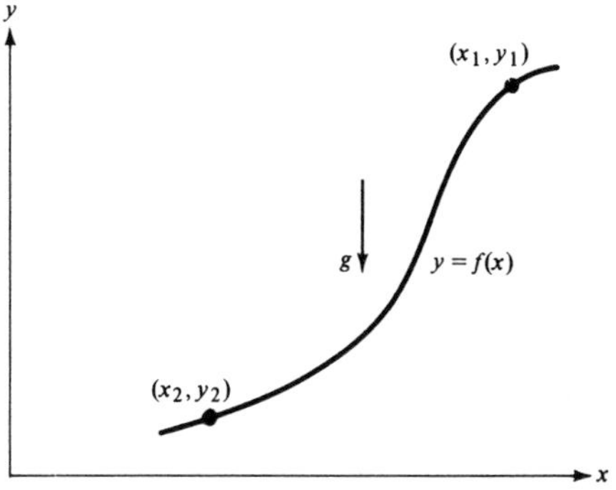

FIGURE 7.14

27. A bead slides down a smooth curve $y=f(x)$ under the action of gravity (Fig. 7.14). If it is released at the point (x_1,y_1) with zero velocity, show that the time t taken to reach the point (x_2,y_2) is given by

$$t = \int_{x_2}^{x_1} \left\{ \frac{1+[f'(x)]^2}{2g[f(x_1)-f(x)]} \right\}^{1/2} dx.$$

[To determine the *function* $f(x)$ which makes t a minimum is a classical problem which leads to a study of the calculus of variations. In this particular example t is minimized by choosing $f(x)$ to be an arc of a cycloid.]
28. On the basis of Exercise 27 (or otherwise), find the time of descent t if $(x_1,y_1)=(a,a)$ and $(x_2,y_2)=(0,0)$ when the path is a straight line and when the path is the quadrant of the circle $x^2+y^2-2ay=0$.
29. The curve defined by $x = \cos t + (\varepsilon/4)\cos 4t$, $y = \sin t + (\varepsilon/4)\sin 4t$, with $\varepsilon \ll 1$, is a slightly corrugated unit circle. Show that the length of the curve is
$L = \int_0^{2\pi}(1+\varepsilon^2+2\varepsilon\cos 3t)^{1/2}\,dt$ and therefore $L = 2\pi(1+\frac{1}{4}\varepsilon^2+\cdots)$. Find the curvature at the point $t=0$.
30. The family of curves $y(x,\epsilon) = (e^{\epsilon x}-1)/(e^{\epsilon}-1)$ all pass through the two points $(0,0)$ and $(1,1)$. Show that $\bar{y}(x) = \lim_{\epsilon\to 0} y(x,\epsilon) = x$. Find the radius of curvature $R(\epsilon)$ for $y=y(x,\epsilon)$ at $x=1$ and evaluate $\lim_{\epsilon\to 0} R(\epsilon)$.
31. A particle performs small amplitude oscillations (under the action of gravity) near a minimum of the curve $y=f(x)$ (Fig. 7.15). Show that the period of small oscillations is $2\pi(gf''(0))^{1/2}$. (*Hint:* From Example 7, $d^2s/dt^2 = -g\sin\varphi$, and for φ small, $\sin\varphi \approx \varphi$.)
32. A string of length a has one end fixed. A mass hangs from the other end and is projected horizontally with velocity U. Find the smallest value of U for which the mass can make a complete revolution without the string becoming slack.

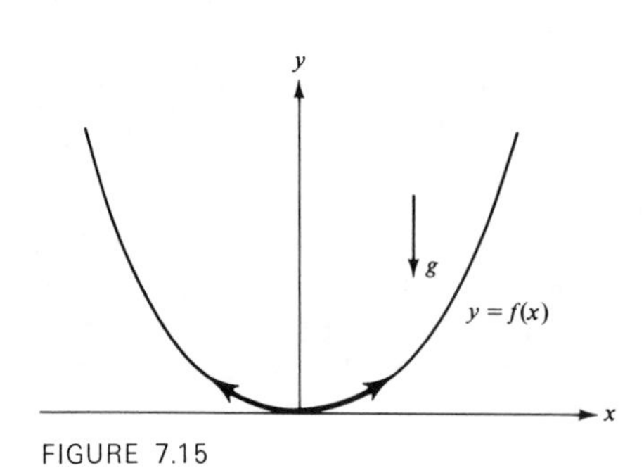

FIGURE 7.15

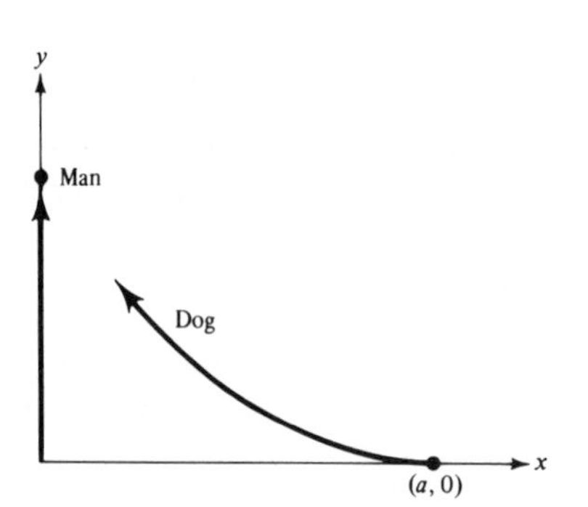

FIGURE 7.16

33. A man runs at constant speed U along the positive y axis and is pursued by a dog moving at constant speed λU. At each instant the dog heads straight for the man. At time $t = 0$ the man is located at the origin and the dog at the point $(a,0)$ (Fig. 7.16). If the dog is at the point (x,y) at time t, deduce the equations:

(i) $t = \dfrac{1}{U}\left(y - x\dfrac{dy}{dx}\right)$ **(ii)** $\lambda U \dfrac{dt}{dx} = -\left[1 + \left(\dfrac{dy}{dx}\right)^2\right]^{1/2}.$

From (i) and (ii) prove that

(iii) $\lambda x \dfrac{dp}{dx} = (1 + p^2)^{1/2},$ where $p = \dfrac{dy}{dx}.$

Integrate (iii) to show

(iv) $\dfrac{dy}{dx} = \dfrac{1}{2}\left[\left(\dfrac{x}{a}\right)^{1/\lambda} - \left(\dfrac{a}{x}\right)^{1/\lambda}\right].$

Integration of (iv) gives the path of the dog as

$$y = \frac{a}{2}\left\{\frac{\lambda}{\lambda + 1}\left[\left(\frac{x}{a}\right)^{1 + 1/\lambda} - 1\right] - \frac{\lambda}{\lambda - 1}\left[\left(\frac{x}{a}\right)^{1 - 1/\lambda} - 1\right]\right\}, \qquad \lambda \neq 1.$$

If $\lambda > 1$ (so that the dog's speed exceeds the man's speed), show that contact is made at $(0,Y)$, where $Y = a\lambda/(\lambda^2 - 1)$.

34. Examine Exercise 33 when the speed of the man and the dog are the same, that is, $\lambda = 1$. In particular, find the path of the dog.

35. As a result of especially high tides, a boat drifts from a harbor out to sea. In the next 4 days fishermen report its noon position to be 8 miles north, 10 miles east; 14 miles north, 18 miles east; 18 miles north, 26 miles east; 20 miles north, 35 miles east. Where should the coast guard go on the fifth day to retrieve the boat? (Approximate methods are needed, and some reasonable assumptions must be made.)

5.8 Planetary motion

In the early 1600s, Kepler (1571-1630) suggested the following three laws of planetary motion:

1. The planets describe ellipses about the sun as focus.

2. The radius vector drawn from the sun to a planet describes equal areas in equal times.
3. The squares of the periods of the planets are proportional to the cubes of the major axes of the ellipses.

Kepler proposed these laws after 20 years of studying the astronomical observations made by his teacher and colleague Tycho Brahe (1546-1601). One can only marvel at this feat of intellect and perserverance. The simplicity of these laws strongly motivated Newton, who, using his newly invented calculus, was led to propose his famous law of gravitation. Moreover, Newton was able to show that Kepler's laws are a consequence of the law of gravitation and thereby condensed lifetimes of effort into a succinct and elegant formalism.

Before these and related problems are analyzed in this section, two preliminary results are established.

The first order of business is to express the equations of conic sections (ellipse, parabola, or hyperbola) in polar coordinates. This task is most straightforward if we start from the focus-directrix definition of a conic, Fig. 8.1.

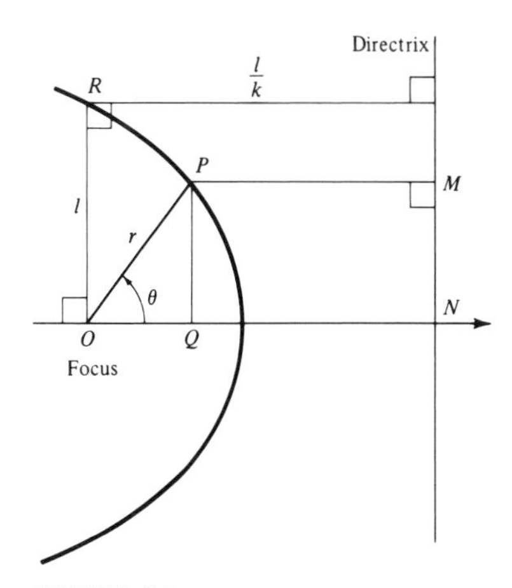

FIGURE 8.1

Let O be a fixed point (a focus) and MN a fixed line (a directrix). The locus of a point P is a conic if the ratio of the distance OP from the focus to the distance PM from the directrix is constant. In mathematical terms, this statement is

$$\frac{OP}{PM} = k = \text{constant}; \tag{1}$$

the positive constant k is the eccentricity of the conic.

Let (r, θ) be the polar coordinates of point P, $OP = r$, as shown in Fig. 8.1. It follows that

$$PM = ON - OQ = \frac{l}{k} - r\cos\theta,$$

where l is called the *semilatus rectum* (OR in Fig. 8.1). The polar equation of a conic, (1), is therefore

$$\frac{r}{l/k - r\cos\theta} = k,$$

which upon simplification can be written

$$\frac{l}{r} = 1 + k\cos\theta. \tag{2}$$

From this equation it is clear that r can be infinite only if there is a value of θ for which $1 + k\cos\theta = 0$. If $k < 1$, this is not possible and the curve will be confined; in this case the conic section is an ellipse. If $k > 1$, $r \to \infty$ as $\theta \to \pm\cos^{-1}(-1/k)$ and the conic is a hyperbola. For $k = 1$, the conic is a parabola. Equation (2) will prove useful in the subsequent discussion of Kepler's laws.

3

The second preliminary is to derive the acceleration vector in terms of radial and circumferential components in the plane, i.e., to resolve, the acceleration vector along the radius and perpendicular to it. If $\hat{\mathbf{r}}$ and $\hat{\boldsymbol{\theta}}$ are the unit vectors in the radial and circumferential directions (Fig. 8.2), then

$$\hat{\mathbf{r}} = \cos\theta\,\mathbf{i} + \sin\theta\,\mathbf{j}, \tag{3}$$

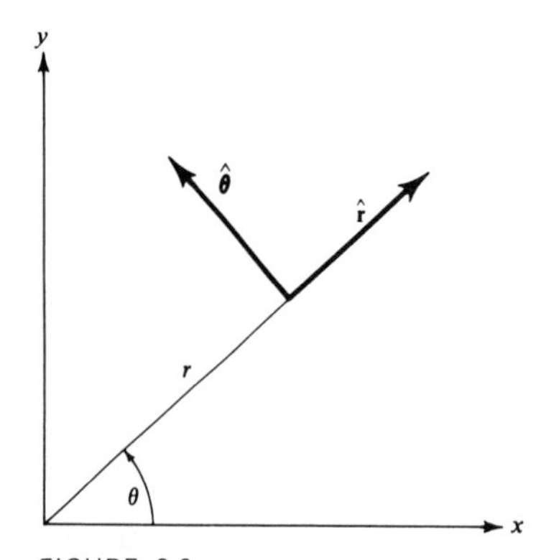

FIGURE 8.2

where $\mathbf{i}$ and $\mathbf{j}$ are the usual constant unit vectors parallel to the x and y axes. The unit vector $\hat{\boldsymbol{\theta}}$ is obtained from $\hat{\mathbf{r}}$ by replacing θ by $\theta + \pi/2$ in (3) so that

$$\hat{\boldsymbol{\theta}} = \cos\left(\theta + \frac{\pi}{2}\right)\mathbf{i} + \sin\left(\theta + \frac{\pi}{2}\right)\mathbf{j}$$

or

$$\hat{\boldsymbol{\theta}} = -\sin\theta\,\mathbf{i} + \cos\theta\,\mathbf{j}. \tag{4}$$

The unit vectors $\hat{\mathbf{r}}$ and $\hat{\boldsymbol{\theta}}$ vary in direction, and differentiation gives

$$\frac{d\hat{\mathbf{r}}}{d\theta} = -\sin\theta\,\mathbf{i} + \cos\theta\,\mathbf{j} = \hat{\boldsymbol{\theta}}, \tag{5}$$

and in a similar way

$$\frac{d\hat{\boldsymbol{\theta}}}{d\theta} = -\hat{\mathbf{r}}. \tag{6}$$

Imagine now a particle in motion in the plane with position vector $\mathbf{r} = \mathbf{r}(t)$. In polar coordinates, it is possible to consider this vector equation as two parametric equations: $r = r(t)$ and $\theta = \theta(t)$. The velocity and acceleration of the particle can be calculated from

$$\mathbf{r} = r\hat{\mathbf{r}}, \tag{7}$$

by differentiation with respect to t, which yields

$$\mathbf{v} = \frac{d\mathbf{r}}{dt} = \frac{dr}{dt}\hat{\mathbf{r}} + r\frac{d\hat{\mathbf{r}}}{dt}.$$

However, Eq. (5), which shows that

$$\frac{d\hat{\mathbf{r}}}{dt} = \frac{d\hat{\mathbf{r}}}{d\theta}\frac{d\theta}{dt} = \frac{d\theta}{dt}\hat{\boldsymbol{\theta}},$$

may be used in the expression for $\mathbf{v}$ to obtain

$$\mathbf{v} = \dot{r}\hat{\mathbf{r}} + r\dot{\theta}\hat{\boldsymbol{\theta}}, \tag{8}$$

where *a dot is used to denote a differentiation with respect to time.* [These results might have been anticipated by realizing that in time dt the particle moves through the distance elements dr and $r\,d\theta$ in the radial and the circumferential directions respectively. The ratio of distance to time increments provides (8) once again (see Fig. 8.3*a*).]

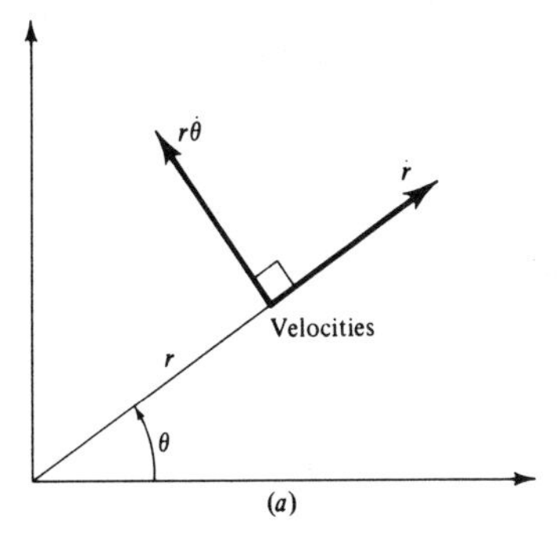

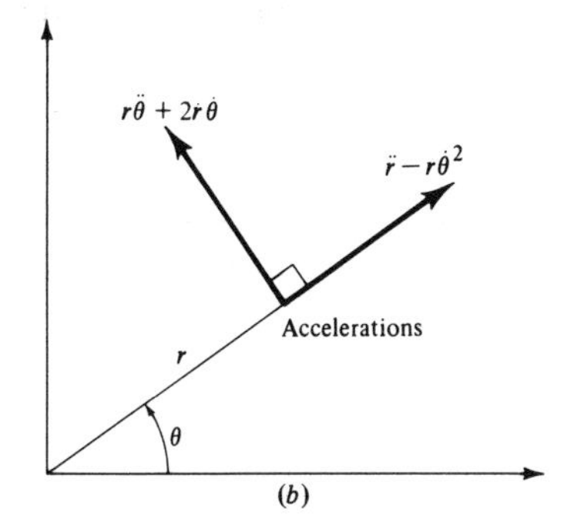

FIGURE 8.3

A second time differentiation gives the acceleration vector:

$$\mathbf{a} = \ddot{r}\hat{\mathbf{r}} + \dot{r}\dot{\theta}\frac{d}{d\theta}\hat{\mathbf{r}} + (r\ddot{\theta} + \dot{r}\dot{\theta})\hat{\boldsymbol{\theta}} + r\dot{\theta}^2\frac{d}{d\theta}\hat{\boldsymbol{\theta}}.$$

On making use of Eqs. (5) and (6) this expression reduces to

$$\mathbf{a} = (\ddot{r} - r\dot{\theta}^2)\hat{\mathbf{r}} + (r\ddot{\theta} + 2\dot{r}\dot{\theta})\hat{\boldsymbol{\theta}}, \tag{9}$$

so that $\ddot{r} - r\dot{\theta}^2$ and $r\ddot{\theta} + 2\dot{r}\dot{\theta}$ are the *radial and circumferential components of acceleration* (see Fig. 8.3*b*).

4

In order to discuss planetary motion, we set the origin O in the center of the sun and describe the position P of a planet at time t by polar coordinates (r, θ). The planet is acted on by the gravitational force of the sun whose magnitude varies with distance but whose direction is always radially inward. (This type of force is often called a *central force*.) We designate the attractive force per unit mass by $f(r)$. The equations of motion are obtained by equating mass accelerations to forces in both the radial and circumferential directions, and since there is no circumferential force, the results are

$$\ddot{r} - r\dot{\theta}^2 = -f(r), \tag{10}$$

$$r\ddot{\theta} + 2\dot{r}\dot{\theta} = 0. \tag{11}$$

These second-order differential equations describe the orbit. If at time $t = 0$, r, θ, $\dot{r}$, $\dot{\theta}$ are specified, the problem can be solved, at least numerically, to determine the subsequent motion of the planet. However, in the present situation the equations can be greatly simplified, as we shall now show.

Multiplication of Eq. (11) by r gives

$$r^2\ddot{\theta} + 2r\dot{r}\dot{\theta} = 0,$$

which can be written

$$\frac{d}{dt}(r^2\dot{\theta}) = 0.$$

Integration yields

$$r^2\dot{\theta} = h, \tag{12}$$

where the constant h represents the constant angular momentum (per unit mass) of the planet. But the element dA of polar area (see Chap. 3) is also given by

$$dA = \tfrac{1}{2}r^2\,d\theta,$$

so that (12) implies

$$\frac{dA}{dt} = \tfrac{1}{2}r^2\dot{\theta} = \frac{h}{2}. \tag{13}$$

This states that a planet sweeps out equal areas in equal times, which is *Kepler's second law*. [Note that (13) is not at all dependent on the form of $f(r)$.]

Equation (12) can now be used to simplify (10), and the replacement of $\dot{\theta}$ yields

$$\ddot{r} - \frac{h^2}{r^3} = -f(r). \tag{14}$$

The multiplication of this expression by $\dot{r}$ renders it an exact differential which can be integrated once:

$$\dot{r}\ddot{r} - \frac{h^2\dot{r}}{r^3} = -f(r)\dot{r},$$

or

$$\frac{d}{dt}\left(\frac{\dot{r}^2}{2} + \frac{h^2}{2r^2}\right) = -\frac{d}{dt}\int f(r)\,dr,$$

and finally

$$\frac{1}{2}\left(\dot{r}^2 + \frac{h^2}{r^2}\right) = -\int f(r)\,dr + K. \tag{15}$$

The left-hand side may be identified as the kinetic energy of the body (divided by the mass) since

$$\tfrac{1}{2}\left(\dot{r}^2 + \frac{h^2}{r^2}\right) = \tfrac{1}{2}(\dot{r}^2 + r^2\dot{\theta}^2) = \tfrac{1}{2}v^2.$$

Equation (15) is then a statement of the conservation of energy: the sum of the kinetic energy of a body, $\mathrm{KE} = \tfrac{1}{2}m(\dot{r}^2 + r^2\dot{\theta}^2)$, and its potential energy, $\mathrm{PE} = m\int f(r)\,dr$, is constant throughout the motion, i.e.,

$$\mathrm{KE} + \mathrm{PE} = \text{constant}.$$

The last integration to recover r as an explicit function of t is a straightforward calculation. Equation (15) is first solved for $\dot{r}$ as a function of r; variables are then separated, and the integrations, though nontrivial, are examples that have already been done. A simpler procedure, based on much hindsight and perhaps just a little foresight, is to change variables so that the inverse radius $u = 1/r$ in the course of the analysis becomes the dependent variable, whereas θ remains an independent variable. With this substitution, (12) may be written

$$\dot{\theta} = hu^2, \tag{16}$$

and this may be used at once to rewrite the radial velocity component as

$$\frac{dr}{dt} = \frac{dr}{du}\frac{du}{d\theta}\frac{d\theta}{dt} = -\frac{1}{u^2}\frac{du}{d\theta}hu^2 = -h\frac{du}{d\theta}.$$

Moreover

$$\ddot{r} = \frac{d}{dt}\left(\frac{dr}{dt}\right) = \frac{d}{dt}\left(-h\frac{du}{d\theta}\right),$$

$$= \frac{d}{d\theta}\left(-h\frac{du}{d\theta}\right)\dot{\theta} = -h^2u^2\frac{d^2u}{d\theta^2}.$$

Let

$$f(r) = f\left(\frac{1}{u}\right) = g(u)$$

then the result of substituting all these results in (14) is

$$\frac{d^2u}{d\theta^2} + u = \frac{1}{h^2u^2}\, g(u). \tag{17}$$

If the gravitational force is an inverse-square law,

$$f(r) = \frac{\mu}{r^2}, \tag{18}$$

where μ is a constant, it follows that $g(u) = f(1/u) = \mu u^2$ and (17) becomes

$$\frac{d^2u}{d\theta^2} + u = \frac{\mu}{h^2}.$$

The general solution of this differential equation is

$$u = \frac{1}{r} = C\cos(\theta - \alpha) + \frac{\mu}{h^2}, \tag{19}$$

where C and α are arbitrary constants. However, the polar equation for a conic is, by (2),

$$u = \frac{1}{r} = \frac{1}{l} + \frac{k}{l}\cos\theta$$

or, more generally,

$$\frac{1}{r} = \frac{1}{l} + \frac{k}{l}\cos(\theta - \alpha), \tag{20}$$

when the major axis is inclined at an angle α to the x axis (see Fig. 8.4). Therefore (19) represents a conic section with

$$C = \frac{k}{l}, \qquad h^2 = \mu l. \tag{21}$$

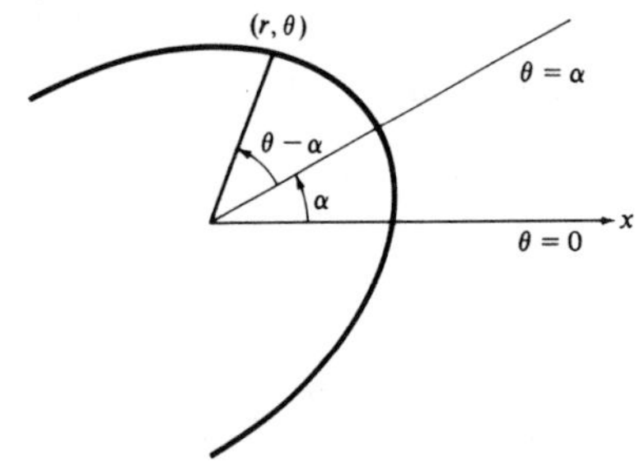

FIGURE 8.4

In other words, the orbits of heavenly bodies about the sun are all ellipses, parabolas, or hyperbolas. In particular, since the orbit of a planet is confined, the trajectory must be an ellipse with the sun as focus, and this is in agreement with *Kepler's first law*. It should be emphasized that the elliptical orbit is a consequence of the inverse-square law.

Conversely, if the orbit is an ellipse (or hyperbola or parabola) with the sun as focus, then the law of gravitational attraction must be an inverse-square law. Given

$$u = \frac{1}{l} + \frac{k}{l}\cos\theta,$$

it follows that

$$\frac{d^2u}{d\theta^2} + u = \frac{1}{l},$$

and by Eq. (17) we make the identification

$$f(r) = g(u) = \frac{h^2u^2}{l} = \frac{h^2}{lr^2}$$

or

$$f(r) = \frac{\mu}{r^2}.$$

In this sense, planetary motion suggests *Newton's law of gravitation*. This law, which is consistent with astronomical observations, states that any two particles in the universe attract each other with a force

$$\frac{Gm_1m_2}{r^2}, \tag{22}$$

m_1 and m_2 being the masses, r the distance between them, and G the gravitational constant. G has the dimensions of (Force)(Distance)^2(Mass)$^{-2}$ = $(MLT^{-2})L^2M^{-2} = M^{-1}L^3T^{-2}$ and has the value $G = 6.670 \times 10^{-8}$ cm^3/gm-sec^2. We can now identify the constant in (18) as $\mu = GM$, where M is the mass of the sun. While the validity of the inverse-square law is well documented by astronomical data, the motion of the elementary particles in nuclear physics involves other force laws.

Example 1 Estimate the mass of the earth given its radius $r = 6.44 \times 10^8$ cm, the gravitational acceleration at the surface $g = 981$ cm/sec^2, and the gravitational constant $G = 6.670 \times 10^{-8}$ cm^3/gm-sec^2.

A particle of unit mass ($m_1 = 1$) is subject to an acceleration g at the surface of the earth, and this by (22) must equal Gm/r^2, where $m_2 = m$ is the mass of the earth. Therefore

$$g = \frac{Gm}{r^2},$$

or

$$m = \frac{gr^2}{G} = \frac{(6.44)^2 \times 10^{16} \times 9.81 \times 10^2}{6.670 \times 10^{-8}},$$

$$= 61.0 \times 10^{26} \text{ gm}.$$

---

6 To deduce Kepler's third law, we return to Eq. (13), that is, $dA/dt = h/2$. This equation integrates to give

$$A = \frac{h}{2}T,$$

where A is the area described during the time T for one revolution of the planet. Since the orbit is an ellipse, $A = \pi ab$, where a and b are lengths of the semimajor and semiminor axes of the elliptical path, the planetary period is given by

$$T = \frac{2\pi ab}{h}. \tag{23}$$

The semilatus rectum l of an ellipse is given by $l = b^2/a$ (see Exercise 4), and since $h^2 = \mu l$, we deduce that

$$T^2 = \frac{4\pi^2 a^2 b^2}{\mu(b^2/a)} = \frac{4\pi^2 a^3}{\mu}.$$

Therefore

$$T^2 \propto a^3, \tag{24}$$

in agreement with *Kepler's third law*. Table 8.1 shows the observed values for the planets of the solar system.

TABLE 8.1

Planet	*Mean distance from sun, millions of km*	*Period*	$10^7T^2/a^3$
Mercury	57.9	87.967 days	2.988
Venus	108.3	224.701 days	2.979
Earth	149.7	365.256 days	2.983
Mars	228.0	1.881 years	2.983
Jupiter	778.7	11.862 years	2.979
Saturn	1427.6	29.458 years	2.982
Uranus	2872.6	84.015 years	2.978
Neptune	4501.2	164.788 years	2.978
Pluto	5906.1	247.697 years	2.978

Example 2 A particle is subject to a central force of attraction $f(r)$ per unit mass. Is a circular orbit about the center of force possible? Determine whether this orbit is stable.

The basic equations of motion are

$$\ddot{r} - r\dot{\theta}^2 = -f(r),$$
$$r\ddot{\theta} + 2\dot{r}\dot{\theta} = 0.$$

In order that $r = a$ be a solution of this system the first equation implies that, for consistency,

$$\dot{\theta} = \omega = \text{constant} \qquad \text{or} \qquad \theta = \omega t + \text{constant},$$

where

$$\omega^2 = \frac{f(a)}{a}.$$

A circular orbit *is* a possible trajectory when the angular velocity $\dot{\theta}$ is constant.

The concept of *stability* is of fundamental importance in many areas of applied mathematics, and planetary motion provides us with an opportunity to introduce some basic ideas.

We have just shown that one possible solution of the orbit problem is circular motion. Now we seek to determine whether it is also physically realizable. If the particle is slightly displaced, so that r and $\dot{\theta}$ have values close to a and ω, respectively, but not exactly equal to them, will the subsequent motion of the particle remain close to the circular orbit? If after a small displacement (usually called a *perturbation*) the ensuing motion does remain almost circular, the orbit is said to be *stable*. But when a small displacement causes the motion to deviate markedly from circular, the orbit is said to be *unstable*.

The mathematical formulation of this idea begins by writing

$$r(t) = a + \varepsilon s(t), \qquad \dot{\theta}(t) = \omega + \varepsilon \dot{\varphi}(t),$$

where the *small* parameter ε is introduced to exhibit the fact that the displacements are small. The parameter ε represents the amplitude of the perturbation. Under this transformation, Eqs. (10) and (11) become

$$\varepsilon\ddot{s} - (a + \varepsilon s)(\omega + \varepsilon\dot{\varphi})^2 = -f(a + \varepsilon s), \tag{25}$$

$$(a + \varepsilon s)\varepsilon\ddot{\varphi} + 2\varepsilon\dot{s}(\omega + \varepsilon\dot{\varphi}) = 0. \tag{26}$$

Of course, these are identically satisfied when $\varepsilon = 0$ [since $a\omega^2 = f(a)$]. Moreover, when ε is very small, terms which involve powers of ε higher than the first can be neglected (at least to a first approximation) and (25) and (26) can be written in much simpler forms

$$\ddot{s} - 2a\omega\dot{\varphi} - \omega^2 s = -f'(a)s, \tag{27}$$

and

$$a\ddot{\varphi} + 2\omega\dot{s} = 0. \tag{28}$$

Integration of the last equation gives

$$\dot{\varphi} = C - \frac{2\omega s}{a},$$

where C is an arbitrary constant. When this value of $\dot{\varphi}$ is substituted into Eq. (27), we obtain

$$\ddot{s} + [3\omega^2 + f'(a)]s = 2a\omega C. \tag{29}$$

This has the constant $2a\omega C[3\omega^2 + f'(a)]^{-1}$ as a particular solution. More importantly, the homogeneous equation has *sinusoidal* solutions if

$$3\omega^2 + f'(a) > 0,$$

that is,

$$\frac{3f(a)}{a} + f'(a) > 0. \tag{30}$$

In this case, s (the radial perturbation) remains bounded (oscillatory), and the orbit is stable. However, if $3f(a)/a + f'(a) < 0$, the particle deviates sharply from the original circular path and the general solution of (29) contains an exponentially growing term, which means that the orbit is unstable.

In particular let $f(r) = \mu/r^n$, in which case (30) becomes

$$\frac{3f(a)}{a} + f'(a) = \frac{\mu(3-n)}{a^{n+1}}.$$

For this force law, the circular orbit will be stable provided $n < 3$, and this analysis shows that the inverse-square law ($n = 2$) produces stable circular orbits. All of which is reassuring.

Exercises 5.8

1. The polar coordinates of a particle at time t are $r = t^2$, $\theta = t$. Calculate the velocity and acceleration at time t. Find the radial and circumferential components of acceleration when $t = 1$ and $t = 2$. Illustrate with a graph.

2. The position vector of a particle at time t is $\mathbf{r}(t) = (t^2 - 1)\mathbf{i} + 2t\mathbf{j}$. Draw a graph of the trajectory for $t \geq 0$. Find the radial and circumferential components of velocity and acceleration when $t = 0$, 1 and 2.

3. Verify that the unit vectors $\hat{\mathbf{r}}$ and $\hat{\boldsymbol{\theta}}$ both satisfy the differential equation $d^2\mathbf{v}/d\theta^2 + \mathbf{v} = \mathbf{0}$.

4. Show that the ellipse

$$\frac{x^2}{1} + \frac{y^2}{1-k^2} = a^2, \qquad k^2 < 1$$

has foci at $(\pm ak, 0)$ with $x = \pm a/k$ as the corresponding directrices. By moving the origin to a focus and converting to polar coordinates, show that the resulting equation is the same as that used in the text.

5. A particle of mass m moves under a central force along the spiral $r = e^{\theta}$. Show that the force is of magnitude $2mh^2/r^3$ and the speed of the particle is $h\sqrt{2}/r$ where h is the constant angular momentum per unit mass. Illustrate with a graph.

6. A particle of mass m moves under a central force along the following curves. Find the law of force and the speed of the particle in terms of h, the constant angular momentum per unit mass. Illustrate with graphs.

(i) $r = a\theta$ **(ii)** $r = e^{a\theta}$ **(iii)** $r = a \sin \theta$

7. If a particle describes an ellipse whose center is at the center of force, show that the force of attraction is linearly proportional to distance.

8. The period of Neptune is 164.8 years. Show that it is about 30 times as far from the sun as the earth is.

9. If the mean distance of Mars from the sun is 1.524 times that of the earth, find the length of the Martian year.

10. If a planet moves in a circular orbit given by $x(t) = r \cos \omega t$, $y(t) = r \sin \omega t$, the equations of motion are $d^2x/dt^2 = -kx/r^3$, $d^2y/dt^2 = -ky/r^3$ where k is a constant. Verify for this special case that the square of the period is proportional to the cube of the radius. (*Hint:* Substitute the formulas for $x(t)$ and $y(t)$ in the equations of motion.)

11. A satellite orbits the earth in 90 minutes. The moon at a distance of 370,000 km orbits in 27.3 days. Assuming that both orbits are circular, find the altitude of the satellite (the height above the surface of the earth). The radius of the earth is 6370 km.

12. The maximum distance of the moon from the earth (center to center) is 384,300 km and the minimum distance is 356,300 km. Determine the eccentricity of the moon's orbit.

13. Determine the altitude of the circular geostationary orbit, the earth orbit with period 24 hours.

14. The period of the earth's rotation about the sun is 365.256 days. It would be more convenient to have a period of exactly 365 days. How should the mean distance from the sun be changed to correct this anomaly?

15. Determine the minimum period of a satellite in earth orbit. (*Hint:* Take the altitude to be zero.)

16. If a planet were suddenly stopped in its orbit, supposed circular, show that it would fall into the sun in a time which is $\sqrt{2}/8$ times the period of the planet's revolution.

17. A particle describes an ellipse under a force per unit mass $w + \mu/r^2$ toward the focus. If it has velocity v_0 at a distance r_0 from the center of force, show that its period is

$$\frac{2\pi}{\sqrt{\mu}}\left(\frac{2}{r_0} - \frac{v_0^2}{\mu}\right)^{-3/2}.$$

18. At a certain instant a manned space capsule is travelling with speed v_0 at a distance r_0 and in a direction making an angle θ_0 with the radius vector. Show that in the subsequent motion

$$\left(\frac{dr}{dt}\right)^2 = v_0^2 - \frac{2\mu}{r_0} + \frac{2\mu}{r} - \frac{r_0^2 v_0^2 \sin^2\theta_0}{r^2}.$$

19. An alpha particle of charge $2q$ and mass m is projected with velocity v_∞ from infinity along a line which is at a perpendicular distance p from a nucleus of charge q'. [This means $f(r) = 2qq'/mr^2$ in our notation.] Show that if δ is the total angle of deflection of the alpha particle, then $\cot(\delta/2) = pmv_\infty^2/2qq'$.

6
Partial differentiation

6.1 Preliminaries

1

Until now we have worked almost exclusively with problems in which a given variable depends on just one other, i.e., a function of a single variable. In the real world such situations are the exceptions rather than the rule because almost all phenomena are complicated interactions of many competing processes. Any single quantity can then be viewed as depending on several other factors, and in this way we are led to the concept of *a function of several variables*. Of course, it often turns out that some intricate effect can be explained satisfactorily by a relationship between just two of the variables involved, all other considerations being rather minor and even negligible compared to this. Such a functional relationship provides a model of a complex phenomenon, one that sacrifices only a small measure of completeness for a tremendous simplification in the mathematical formalism. In essence, a function of several variables may be approximated by a function of fewer variables, a procedure that makes the solution of a problem a more probable and possible task. The art of constructing mathematical models involves judicious approximations based on an intimate knowledge of the phenomena under study. It is, perhaps, one of the most important and difficult but least teachable aspects of theoretical work, acquired mostly by the accumulation of experience.

In this chapter and subsequent ones, attention will be focused on how the one-variable calculus can be extended to handle problems involving many variables. To a lesser extent the reverse is also considered: how problems with many variables can be approximated or modelled by simpler functional relationships.

Examples of functions of many variables are plentiful, as illustrated by the following:

1. The temperature T in a given region depends on the position coordinates x, y, and z and the time t, so that $T = T(x, y, z, t)$ and the temperature is a function of four independent variables.
2. The height h of the earth's surface above mean sea level depends on latitude x and longitude y, so that $h = h(x, y)$ is a function of two independent variables.

3. The total sales S of a company depend on the x dollars spent on improvements and the y dollars spent on promotion, so that $S = S(x, y)$ is a function of two independent variables.

In general, a variable z which depends on n other variables $x_1, x_2, \ldots, x_n$ is written

$$z = f(x_1, x_2, \ldots, x_n). \tag{1}$$

The quantities $x_1, x_2, \ldots, x_n$ are called the *independent variables* and z the *dependent variable*. When $n = 1$ and there is only one independent variable, (1) reduces to the case studied in earlier chapters, $z = f(x_1)$, or in the more familiar notation $y = f(x)$.

2

The geometrical representation (or graph) of $y = f(x)$ proved to be a convenient and useful way of describing a function of a single variable. For a function of several variables $z = f(x_1, \ldots, x_n)$, an $(n + 1)$-dimensional space is required for a similar geometrical representation, and only for $n \leq 2$ is there any advantage to such an interpretation. However, functions of two independent variables ($n = 2$) illustrate almost all the basic features of the more general situation and for this reason we will concern ourselves mostly with this case.

For convenience, we write $x_1 = x$, $x_2 = y$, so that

$$z = f(x, y). \tag{2}$$

If x and y are now interpreted as rectangular coordinates in a horizontal plane, z as determined from (2) is the altitude above the plane. Accordingly, the totality of such points, each satisfying $z = f(x, y)$, represents a *surface* in xyz space (Fig. 1.1). The surface is a plane (see Sec. 5.6) if z is a linear function of x and y. For example,

$$z = 1 - x - y$$

is the plane having unit intercepts on the axes; i.e., it passes through the three points (1, 0, 0), (0, 1, 0), and (0, 0, 1) (Fig. 1.2).

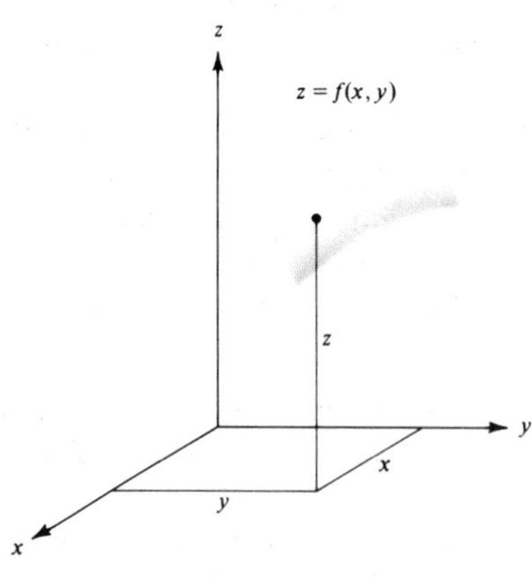

FIGURE 1.1

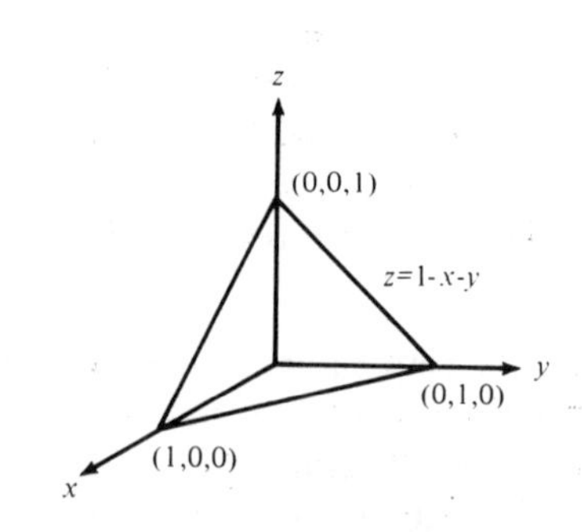

FIGURE 1.2

In Eq. (2), z is an *explicit* function of x and y. Sometimes the functional relationship is implicit, of the form

$$F(x, y, z) = 0.$$

In principle this can be solved for $z = z(x, y)$ to obtain an explicit form [or $x = x(y, z)$ or $y = y(z, x)$], but often it is simpler to work with the implicit form directly. For example, the equation $x^2 + y^2 + z^2 = a^2$ represents a sphere of radius a centered at the origin. The solutions of this equation are $z = \pm(a^2 - x^2 - y^2)^{1/2}$, where the $\pm$ signs correspond to the upper and lower hemispheres (Fig. 1.3). However, it is not often possible to solve implicit equations, and in the case of

$$xz + e^{z^2+y} + \sin(x + y + z) + 1 = 0$$

there is no simple explicit analytical form.

The choice of coordinate origin is arbitrary and made on the basis of overall convenience. A *translation of axes* is readily accomplished. If $\mathbf{r}$ is the position vector from an origin O and $\mathbf{r}'$ the position vector from another origin O', where $\overrightarrow{OO'} = \boldsymbol{a}$ (Fig. 1.4), then the coordinate systems are related by

$$\mathbf{r}' = \mathbf{r} - \boldsymbol{a}. \tag{3}$$

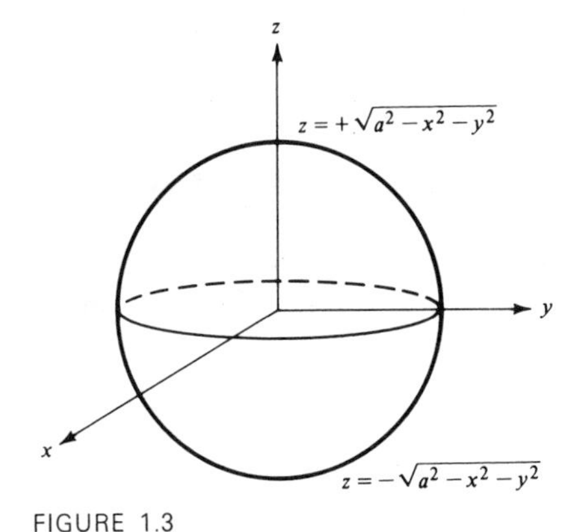

FIGURE 1.3

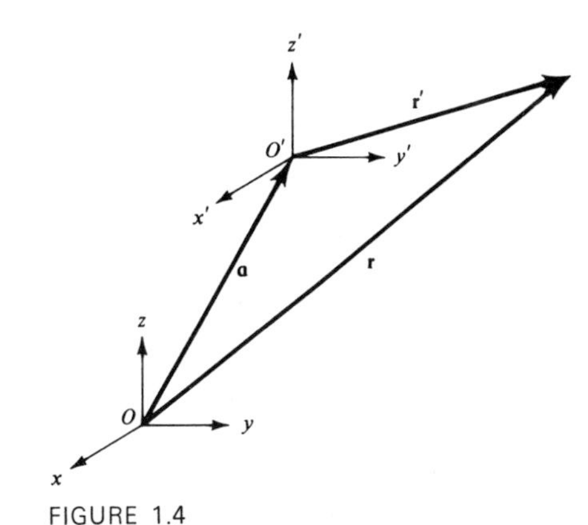

FIGURE 1.4

In component form, let

$$\mathbf{r} = x\mathbf{i} + y\mathbf{j} + z\mathbf{k},$$

$$\mathbf{r}' = x'\mathbf{i} + y'\mathbf{j} + z'\mathbf{k},$$

$$\boldsymbol{a} = a\mathbf{i} + b\mathbf{j} + c\mathbf{k};$$

then

$$x' = x - a, \qquad y' = y - b, \qquad z' = z - c. \tag{4}$$

The surface $z - c = f(x - a, y - b)$ may be interpreted as the surface $z = f(x, y)$ moved bodily (without rotation) the respective distances a, b, and c parallel to the x, y, and z axes.

Example 1 Find the center and radius of the sphere

$$x^2 + y^2 + z^2 - 4x + 2y - 6z + 13 = 0.$$

Upon completing the square in each variable, this equation takes the form

$$(x - 2)^2 - 4 + (y + 1)^2 - 1 + (z - 3)^2 - 9 + 13 = 0$$

or

$$(x - 2)^2 + (y + 1)^2 + (z - 3)^2 = 1.$$

This equation can be written

$$x'^2 + y'^2 + z'^2 = 1,$$

using the coordinates

$$x' = x - 2, \qquad y' = y + 1, \qquad z' = z - 3.$$

The original equation represents a sphere of radius 1 centered at $(2, -1, 3)$. This point is the origin in the x', y', z' coordinate system.

---

3 A *surface of revolution* is formed when a curve in the zx plane is rotated about the z axis. If the equation of the curve is $z = f(x)$, $y = 0$, then $z = f((x^2 + y^2)^{1/2})$ represents the surface of revolution. (The distance of a point on the curve to the z axis is constant as it revolves about this line, as shown in Fig. 1.5.) For example,

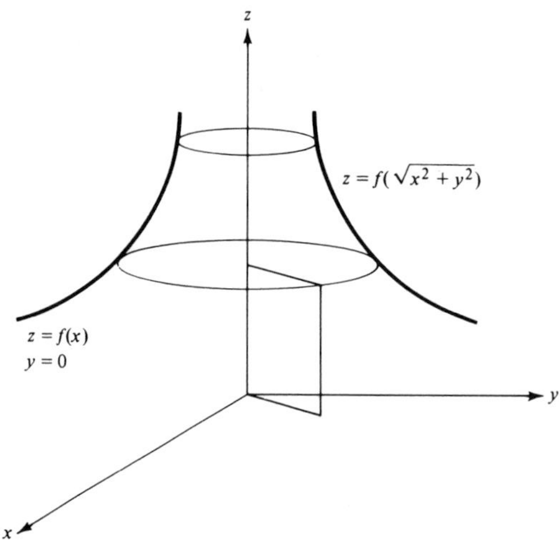

FIGURE 1.5

the rotation of the parabola $z = x^2$, $y = 0$, about the z axis produces the surface $z = x^2 + y^2$, a paraboloid of revolution (Fig. 1.6). Likewise, rotation of the circle $y = 0$, $(x - a)^2 + z^2 = b^2$, and $0 < b < a$, about Oz generates a doughnut-shaped surface called a *torus* (Fig. 1.7). The equation for the torus is then

$$[(x^2 + y^2)^{1/2} - a]^2 + z^2 = b^2$$

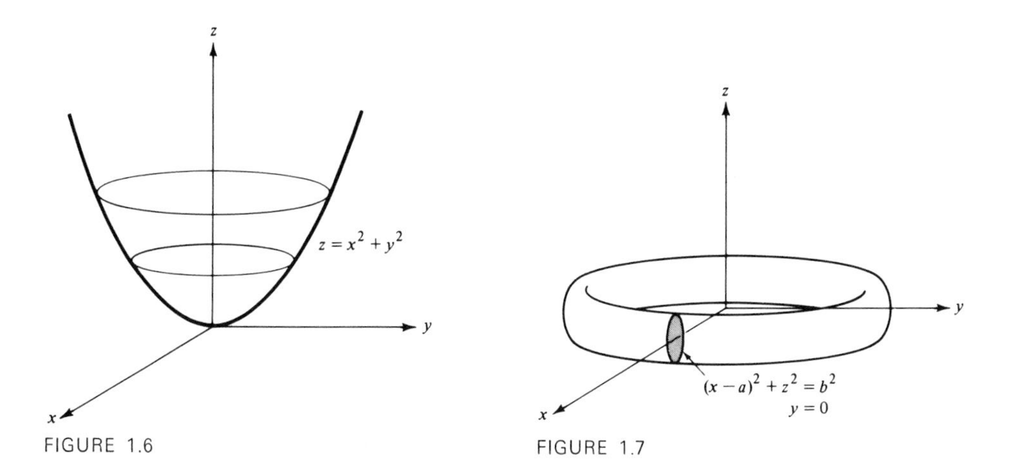

FIGURE 1.6

FIGURE 1.7

or

$$(x^2 + y^2 + z^2 + a^2 - b^2)^2 = 4a^2(x^2 + y^2).$$

Similar considerations apply to surfaces of revolution about the other coordinate axes.

Even a rough drawing is highly informative in describing a three-dimensional surface. A good procedure for sketching is to consider the curves formed by the intersection of the surface under consideration with planes perpendicular to the coordinate axes. In other words, the surface is "sliced" by horizontal and vertical planes, and the resulting curves of intersection are drawn. Since the surface is the sum total of such families of curves, a few of them are often sufficient to indicate the general shape. A few examples will illustrate the method.

Example 2 Sketch the surface

$$\frac{x^2}{a^2} + \frac{y^2}{b^2} - \frac{z^2}{c^2} = 1.$$

The horizontal plane $z = z_0$ cuts the surface in the curve

$$\frac{x^2}{a^2} + \frac{y^2}{b^2} = 1 + \frac{z_0^2}{c^2},$$

an ellipse (Fig. 1.8); its major and minor axes increase in length as z_0 increases. Similarly, the planes $x = x_0$ or $y = y_0$ cut the surface in hyperbolas. (This surface is called a *hyperboloid of one sheet*.)

Example 3 Sketch the surface $z = ye^{-|x|}$.

The plane $y = y_0$ cuts this surface in the curve $z = y_0 e^{-|x|}$. This family of curves has a jump in derivative at $x = 0$. Similarly, the plane $x = x_0$ cuts the surface in the straight line $z = ye^{-|x_0|}$. With this information and a minimum of imagination, the surface shape can now be sketched (Fig. 1.9). Note that since $f(x, y) = ye^{-|x|}$, $f(x, -y) = -f(x, y)$, and $f(-x, y) = f(x, y)$, a knowledge of the surface in the first quadrant enables us to determine it everywhere.

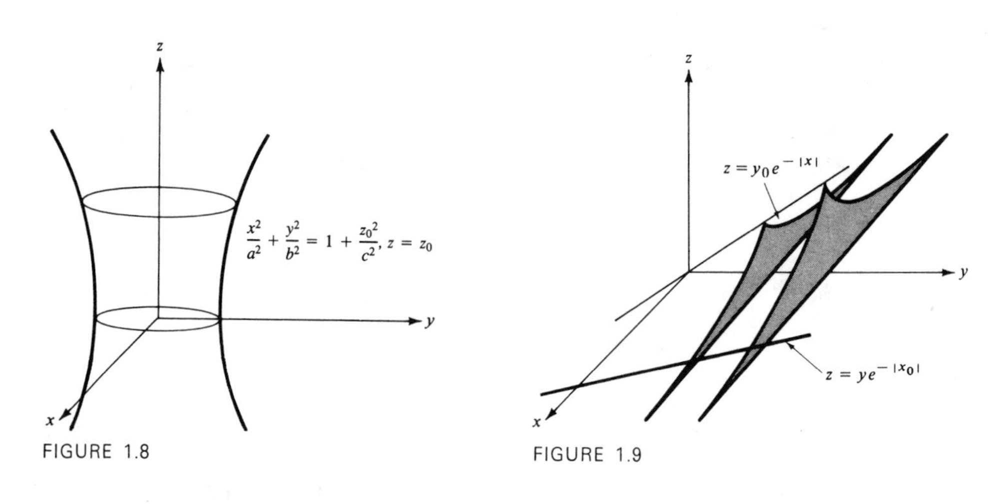

FIGURE 1.8

FIGURE 1.9

Although surfaces can be sketched for all elementary functions, the artistic requirements may seem formidable. For many purposes a much simpler procedure is entirely adequate. The basic idea is to describe the surface $z = f(x, y)$ by drawing the family of curves $f(x, y) = \text{constant} = z_0$. These are called *level curves* and are the curves where the surface $z = f(x, y)$ is cut by the series of planes $z = z_0$ projected onto the base plane $z = 0$.

For example, if z is the altitude above sea level and x is measured to the east and y to the north, the curves corresponding to z = 1,000, 2,000, 3,000, 4,000, 5,000, 6,000 ft can be drawn (Fig. 1.10) in the xy plane. Thus, instead of sketching the surface $z = f(x, y)$, we obtain a *topographic map* showing lines of constant elevation. Level curves are familiar in depicting atmospheric pressure, land elevations, precipitation amounts, population patterns, etc., and enable the layman to get a bird's-eye view of the surface under consideration. As a matter of fact, looking down on a mountain from a great height gives a view comparable to that obtained in Fig. 1.10.

Example 4 Sketch the level curves for the surface $z = ye^{-|x|}$.

This is the surface discussed in Example 3.

The level curves are $y = z_0 e^{|x|}$ and are sketched in Fig. 1.11 for $z_0 = -3$, -2, -1, 0, 1, 2, 3. The reader should compare this picture with the surface shown in Fig. 1.9.

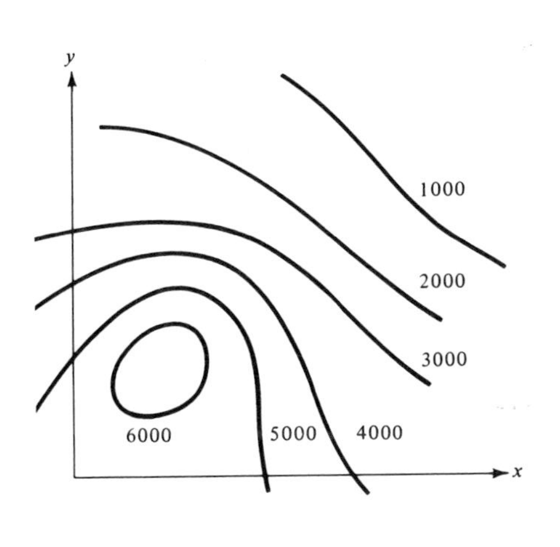

FIGURE 1.10

FIGURE 1.11

Some remarks on continuity are in order before discussing the differentiation of functions of several variables. Continuity is a simple concept which can easily be translated into a less simple but precise mathematical definition that generalizes the ideas presented in Chap. 1.

The function $z = f(x, y)$ is said to be *continuous* at a point $P = (x_0, y_0)$ if for any values of x and y "close" to (x_0, y_0), z remains "close" to $z_0 = f(x_0, y_0)$. This definition can be formalized as follows: a function $f(x,y)$ is continuous at (x_0,y_0) if, given any positive number ε, a number δ can be found such that

$$|f(x, y) - f(x_0, y_0)| < \varepsilon, \tag{5}$$

provided

$$0 < [(x - x_0)^2 + (y - y_0)^2]^{1/2} < \delta. \tag{6}$$

Neighboring points to (x_0, y_0) are taken to lie in a small circle of radius δ; in practice, δ depends on the point under consideration and the assigned value ε. This definition is a generalization of that given earlier for continuity of functions of one variable. From the geometrical viewpoint, continuity means that if we, or preferably a particle, were to move just a little from the given point, it would not hit a vertical wall or fall down a vertical cliff. A surface is continuous if a slight change in horizontal position implies an equally slight change in vertical elevation. To be even more simple-minded, continuity at a point means that a skier (on very short skis) can ski safely on a small neighborhood of the point. While Eqs. (5) and (6) give a formal definition of continuity, an alternative but equivalent definition provides a test which is frequently used in practice. A function $f(x, y)$ is continuous at (x_0, y_0) if both

$$f(x_0, y_0) \quad \text{is finite and} \quad \lim_{\substack{x \to x_0 \\ y \to y_0}} f(x, y) = f(x_0, y_0), \tag{7}$$

where x and y approach their limits x_0 and y_0 in any manner. A few examples illustrate the method.

Example 5 Is the function $f(x, y) = 1/(x^2 + y^2)$ continuous at the origin?
Clearly not, since $f(0, 0)$ is infinite.

Example 6 A function is defined by

$$f(x, y) = \begin{cases} x^2 + y^2 & \text{if } x^2 + y^2 \le 1; \\ \lambda - x^2 - y^2 & \text{if } x^2 + y^2 > 1. \end{cases}$$

Is $f(x, y)$ continuous at points on the unit circle $x^2 + y^2 = 1$?

Geometrically, the surface consists of two portions, each a paraboloid of revolution. These will be disjoint along $x^2 + y^2 = 1$ unless $\lambda = 2$. If $\lambda = 2$, the function is continuous at all points, but if $\lambda \ne 2$, it is discontinuous at all points on $x^2 + y^2 = 1$ (see Fig. 1.12).

Example 7 The function $z = f(x, y)$ is defined by

$$f = \begin{cases} \dfrac{x^2 - y^2}{x^2 + y^2} & \text{if } x \ne 0 \text{ and } y \ne 0; \\ 0 & \text{if } x = 0 \text{ and } y = 0. \end{cases}$$

Is this function continuous at (0, 0)?

In this case, $f(0, 0) = 0$ is finite by definition. To examine the behavior of the function close to the origin, it is helpful to express it in terms of polar coordinates. With $x = r \cos \theta$, $y = r \sin \theta$,

$$z = f(x, y) = \frac{r^2(\cos^2 \theta - \sin^2 \theta)}{r^2} = \cos 2\theta,$$

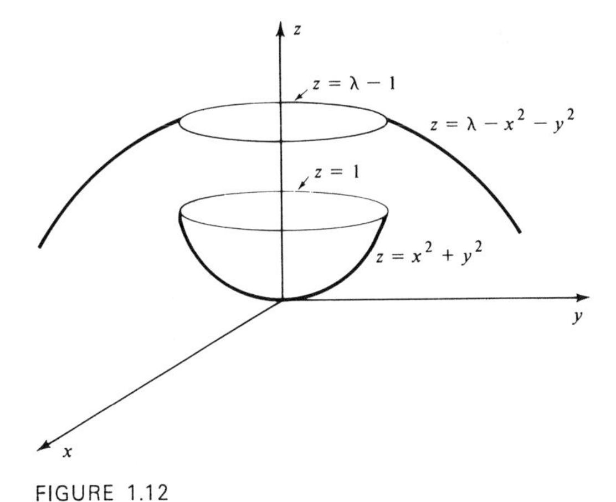

FIGURE 1.12

and the function actually depends on the angle coordinate alone. Therefore

$$\lim_{\substack{x\to 0\\ y\to 0}} f(x, y) = \lim_{r\to 0} f(x, y) = \cos 2\theta,$$

which is not zero (except for certain special directions of approach, $\theta = \pm\pi/4$, $\pm 3\pi/4$). The function does not have as a limit its assigned value at the origin and is therefore discontinuous at that point.

Exercises 6.1

1. Find the center and radius of each of the following spheres.

(i) $x^2 + y^2 + z^2 - 2x - 2z - 8 = 0$ **(ii)** $x^2 + y^2 + z^2 + x + y + z = 0$

(iii) $x^2 + y^2 + z^2 - 8x - 10y + 2z = 0$ **(iv)** $2x^2 + 2y^2 + 2z^2 + 6x + 4y = 12$

2. Determine the plane that contains the circle of intersection of the two spheres **(i)** and **(ii)** in the previous exercise. Illustrate with a diagram.

3. Simplify the following equations by translation of axes and describe the surfaces.

(i) $4x^2 + 9y^2 - z^2 + 8x - 36y + 2z = 1$ **(ii)** $x^2 + 4y^2 + 2x - 4y - 2z = 12$

(iii) $x^2 + 2y^2 + 4z^2 - 2x - 4y + 4z = 10$ **(iv)** $x^2 + y^2 - z^2 = 2(x + y - z)$

4. How does the ellipsoid $3x^2 + 4y^2 + 5z^2 = 60$ relate to the surface with equation $3(x-1)^2 + 4(y-2)^2 + 5(z-3)^2 = 60$? to the ellipsoid $4x^2 + 3y^2 + 5z^2 = 60$? Illustrate with diagrams.

5. Show that the surface with equation $|x| + |y| + |z| = 1$ represents an eight-faced polyhedron. Determine the six vertices and sketch the polyhedron.

6. Sketch the following cylinders. Determine the tangent planes at points where $y = 0$.

(i) $x^2 + y^2 = 1$ **(ii)** $y^2 + z^2 = 4$ **(iii)** $y^2 = 4x$

(iv) $(x-1)^2 + (y-1)^2 = 1$ **(v)** $2x^2 + (y-2)^2 = 4$ **(vi)** $4x^2 - y^2 = 4$

7. Sketch the following surfaces $z = z(x, y)$. Draw the level curves in the xy plane.

(i) $z = e^x$ **(ii)** $z = x^2 + y^2$ **(iii)** $z = 4x^2 + y^2$

(iv) $z^2 = x^2 + y^2$ **(v)** $z = e^{xy}$ **(vi)** $z = e^{-x^2-y^2}$

8. Sketch the surfaces of the one parameter family $x^2 + y^2 + z^2 = 2\alpha z - 1$. For what values of α is a surface defined? Does this family of spheres have an envelope?
9. Sketch the surface obtained by rotating the line $z = x$, $y = 0$ about the z axis and determine its equation.
10. Sketch the surfaces obtained by rotating the curve $y = e^x$, $z = 0$ about the x axis and about the y axis. Determine the equations of the two surfaces. Where do the two surfaces intersect?
11. The curve $y = \sin x$, $z = 0$, is rotated about the x axis. Find the resulting surface.
12. The line $x = 2y = 3z$ is rotated about the line $x = y = z$. Find the surface generated.
13. The curve $y = x^2$, $z = 0$ is rotated about the line $y = x$, $z = 0$. Find the equation of the surface of revolution formed.
14. What relationship does the curve $z = f(x)$, $y = 0$ have to the surface $z = f(x + y)$? Illustrate with diagrams of the surfaces $z = (x + y)^2$ and $z = e^{-(x+y)^2}$.
15. The pressure distribution on a plane surface is given by $p(x,y)$ measured in atmospheres at the point (x,y). Suppose that $p(x,y)$ satisfies $x^2 - 2px + y^2 + 1 = 0$. Sketch the level curves (isobars) in the xy plane.
16. The temperature $T(x,y)$ measured in °C at a point in the xy plane is given by $T(x,y) = 50(1 - \sin x \sin y)$. Sketch the level curves (isotherms). How does the temperature vary along the line $y = x$?
17. The height $h(x,y)$ in kilometers at a point on a volcanic island is approximated by $h(x,y) = 1 + (x^2 + 2y^2)e^{-(x^2+2y^2)}$. Sketch the level curves (contour lines). How does the height vary along the line $y = x$? Determine the maximum height on the island.
18. A function $f(x, y)$ is defined by $f(x,y) = 1 - |x| - |y|$ if $|x| + |y| \le 1$ and by $f(x,y) = |x| + |y| - 1$ if $|x| + |y| > 1$. Is this function continuous at all points? Sketch the level curves in the xy plane. Define $g(x) = f(x,0)$ and $h(y) = f(0,y)$. Show that g and h are not differentiable at $x = 1$ and $y = 1$, respectively. What happens at the origin?
19. Determine the values of the following functions $f(x,y)$ along the lines $y = mx$. Are the functions continuous at the origin?

(i) $f(x,y) = \dfrac{xy}{x^2 + y^2} \quad (x,y) \ne (0,0)$
$\qquad = 0 \quad (x,y) = (0,0)$

(ii) $f(x,y) = \dfrac{x^3 - y^3}{x^2 + y^2} \quad (x,y) \ne (0,0)$
$\qquad = 0 \quad (x,y) = (0,0)$

20. If $f(x, y) = (x^3 - y^3)/(x - y)$ for $y \ne x$ and $f(x, x) = 3x^2$, is $f(x, y)$ discontinuous at any point in the plane?
21. Give a definition and description of what is meant by saying that $f(x, y, z)$ is continuous at the point (x_0, y_0, z_0). Illustrate with $f(x, y, z) = x^2 + y^2 + z^2$.
22. If $f(x, y, z) = (x + y + z)^\alpha/(x^2 + y^2 + z^2)$ for $(x, y, z) \ne (0, 0, 0)$ and $f(0, 0, 0) = 0$, for what values of α is $f(x, y, z)$ continuous at the origin?

6.2 Partial derivatives

1 A derivative of $z = f(x, y)$ can be calculated in one of two rather natural ways: (i) by differentiating $f(x, y)$ with respect to x, keeping y constant; (ii) by differentiating $f(x, y)$ with respect to y, keeping x constant. Each of these is called a *partial derivative* of f, the former with respect to x and the latter with respect to y. Partial derivatives were briefly discussed at the end of Chap. 2.

Various notations are used for the first partial derivative of f with respect to x, and the two most common ones are

$$\frac{\partial f}{\partial x} \quad \text{and} \quad f_x,$$

read "partial dee f by dee x" and "f sub x," respectively. Sometimes the notation $D_x f$ is used for operational convenience, and in certain situations (especially in implicit functions where ambiguities are possible) the notation $\left(\frac{\partial f}{\partial x}\right)_y$ is used, the subscript y accentuating the fact that y is to be held constant.

In the same way, the notations $\frac{\partial f}{\partial y}$ or f_y, and to a lesser extent $D_y f$ or $\left(\frac{\partial f}{\partial y}\right)_x$, are used to denote the first partial derivative of f with respect to y.

The procedure for calculating partial derivatives is straightforward since in reality we treat $f(x, y)$ as a function of a single variable (and a parameter). The usual rules of differentiation are applicable, as illustrated by the following examples.

Example 1 If $f(x, y) = 2x^2 + y + 3xy^2 - x^3y^4$, calculate $\partial f/\partial x$ and $\partial f/\partial y$.

To calculate $\partial f/\partial x$, y is treated as a parameter *and held constant*, so that

$$\frac{\partial f}{\partial x} = 4x + 3y^2 - 3x^2y^4.$$

In a like manner

$$\frac{\partial f}{\partial y} = 1 + 6xy - 4x^3y^3.$$

Example 2 If $f(x, y) = x \sin y + ye^{xy}$, find $\partial f/\partial x$ and $\partial f/\partial y$.

In this case,

$$\frac{\partial f}{\partial x} = \sin y + y^2e^{xy}, \qquad \frac{\partial f}{\partial y} = x \cos y + (1 + xy)e^{xy}.$$

For a function of a single variable $y = f(x)$, the definition of the derivative is

$$f'(x) = \lim_{\Delta x \to 0} \frac{f(x + \Delta x) - f(x)}{\Delta x}. \tag{1}$$

For a function of two variables $z = f(x, y)$, y is kept constant in the calculation of $\partial f/\partial x$. Therefore, by analogy,

$$\frac{\partial}{\partial x} f(x, y) = \lim_{\Delta x \to 0} \frac{f(x + \Delta x, y) - f(x, y)}{\Delta x}. \tag{2}$$

Similarly $$\frac{\partial}{\partial y} f(x, y) = \lim_{\Delta y \to 0} \frac{f(x, y + \Delta y) - f(x, y)}{\Delta y}, \tag{3}$$

because x is a constant in this calculation.

---

2 First-order partial derivatives have a rather simple geometrical interpretation as slopes of tangent lines to the surface $z = f(x, y)$. To see this, let P and Q be two neighboring points on the surface the coordinates of which are $(x, y, f(x, y))$ and $(x + \Delta x, y, f(x + \Delta x, y))$. The point P is a height $z = f(x, y)$ above the xy plane, whereas the elevation of Q is $f(x + \Delta x, y)$. From Fig. 2.1 it is seen that MQ is the difference in the altitudes of P and Q, that is,

$$MQ = f(x + \Delta x, y) - f(x, y).$$

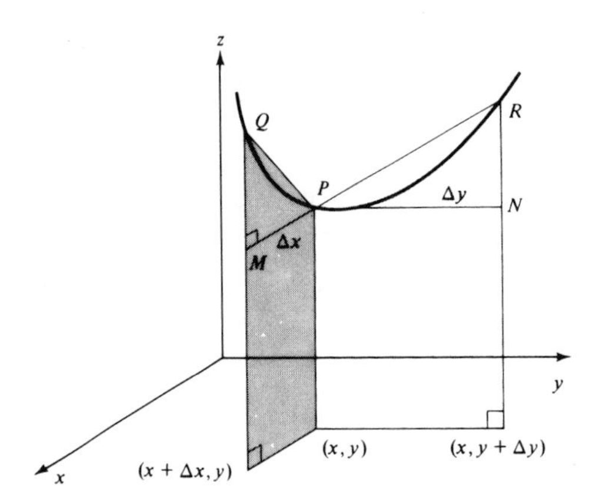

FIGURE 2.1

Since $PM = \Delta x$, it follows that

$$\frac{MQ}{PM} = \frac{f(x + \Delta x, y) - f(x, y)}{\Delta x}$$

is the slope of the secant to the curve formed by the intersection of the given surface and the plane $y =$ constant. The slope of the tangent to this curve is found by taking the limit $\Delta x \to 0$ and is represented by the partial derivative

$$\frac{\partial f}{\partial x} = \lim_{\Delta x \to 0} \frac{f(x + \Delta x, y) - f(x, y)}{\Delta x}.$$

In a similar way, the slope of the tangent at P to the curve PR (formed by intersection of the given surface and the plane $x =$ constant) is

$$\frac{\partial f}{\partial y} = \lim_{\Delta y \to 0} \frac{f(x, y + \Delta y) - f(x, y)}{\Delta y}.$$

3 The notion of a partial derivative can be generalized easily. For a function of n variables $z = f(x_1, x_2, \ldots, x_n)$, n partial derivatives (of the first order) can be found, namely, $\partial f/\partial x_1$, $\partial f/\partial x_2$, $\ldots$, $\partial f/\partial x_n$ or $f_{x_1}, f_{x_2}, \ldots, f_{x_n}$. The definition of $\partial f/\partial x_i$ is

$$\frac{\partial}{\partial x_i} f(x_1, x_2, \ldots, x_n)$$
$$= \lim_{\Delta x_i \to 0} \frac{f(x_1, \ldots, x_{i-1}, x_i + \Delta x_i, x_{i+1}, \ldots, x_n) - f(x_1, \ldots, x_{i-1}, x_i, x_{i+1}, \ldots, x_n)}{\Delta x_i}, \tag{4}$$

where all independent variables except x_i are held constant.

Example 3 If $f(x_1, x_2, x_3, x_4) = x_1 \sin(x_2 + 2x_3) + x_4{}^2 e^{3x_2} \log x_3$, calculate $\partial f/\partial x_1$, $\partial f/\partial x_2$, $\partial f/\partial x_3$, and $\partial f/\partial x_4$.

Each $\partial f/\partial x_i$ is found by varying only x_i and treating all other variables as constants. Routine calculations give

$$\frac{\partial f}{\partial x_1} = \sin(x_2 + 2x_3),$$

$$\frac{\partial f}{\partial x_2} = x_1 \cos(x_2 + 2x_3) + 3x_4{}^2 e^{3x_2} \log x_3,$$

$$\frac{\partial f}{\partial x_3} = 2x_1 \cos(x_2 + 2x_3) + \frac{x_4{}^2}{x_3} e^{3x_2},$$

$$\frac{\partial f}{\partial x_4} = 2x_4 e^{3x_2} \log x_3.$$

---

4 Derivatives of orders higher than 1 are defined in a similar way. For instance, the two first-order partial derivatives $\partial f/\partial x$ and $\partial f/\partial y$ of the function $f(x, y)$ (or f_x and f_y in the alternative notation) can themselves be differentiated with respect to either x or y. Differentiating $\partial f/\partial x$, we obtain the two second derivatives

$$\frac{\partial}{\partial x}\left(\frac{\partial f}{\partial x}\right) = \frac{\partial^2 f}{\partial x^2}, \quad \text{or} \quad (f_x)_x = f_{xx}, \tag{5}$$

and

$$\frac{\partial}{\partial y}\left(\frac{\partial f}{\partial x}\right) = \frac{\partial^2 f}{\partial y\, \partial x}, \quad \text{or} \quad (f_x)_y = f_{xy}. \tag{6}$$

In the same way, $\partial f/\partial y$ can be differentiated with respect to either x or y to yield

$$\frac{\partial}{\partial x}\left(\frac{\partial f}{\partial y}\right) = \frac{\partial^2 f}{\partial x\, \partial y}, \quad \text{or} \quad (f_y)_x = f_{yx}, \tag{7}$$

$$\frac{\partial}{\partial y}\left(\frac{\partial f}{\partial y}\right) = \frac{\partial^2 f}{\partial y^2}, \quad \text{or} \quad (f_y)_y = f_{yy}. \tag{8}$$

It would appear that for a function of two variables, there are four distinct second-order derivatives, namely, f_{xx}, f_{xy}, f_{yy}, and f_{yx}. However, we shall show in the next section that $f_{xy} = f_{yx}$ and, in general, the order in which the partial differentiations are performed is irrelevant provided f satisfies certain mild continuity conditions. The simple functions actually used to describe complex natural phenomena almost always fulfill these requirements.

Example 4 If $f(x, y) = e^{ny} \cos nx$ and n is a constant, show that $f_{xx} + f_{yy} = 0$.

Here

$$f_x = -ne^{ny} \sin nx, \qquad \text{and} \qquad f_y = ne^{ny} \cos nx,$$

$$f_{xx} = -n^2 e^{ny} \cos nx, \qquad f_{yy} = n^2 e^{ny} \cos nx.$$

Addition of the expressions for f_{xx} and f_{yy} leads to

$$f_{xx} + f_{yy} = 0.$$

[Note that $f_{xy} = \dfrac{\partial}{\partial y}(-ne^{ny} \sin nx) = -n^2 e^{ny} \sin nx$,

$f_{yx} = \dfrac{\partial}{\partial x}(ne^{ny} \cos nx) = -n^2 e^{ny} \sin nx$, so that $f_{xy} = f_{yx}$.]

Example 5 If $f(x, y) = g(x) + h(y)$, show that

$$\frac{\partial^2 f}{\partial x \, \partial y} = 0.$$

Differentiation with respect to y gives

$$\frac{\partial f}{\partial y} = h'(y), \qquad \text{since} \qquad \frac{\partial}{\partial y} g(x) = 0.$$

A second differentiation with respect to x then yields

$$\frac{\partial^2 f}{\partial x \, \partial y} = 0.$$

---

5 The definitions of higher-order derivatives and their calculation are routine. Thus

$$\frac{\partial^3 f}{\partial x^2 \, \partial y} = \frac{\partial}{\partial x}\left(\frac{\partial^2 f}{\partial x \, \partial y}\right)$$

As a concrete illustration, let $f(x, y) = e^{\alpha x + \beta y}$, in which case

$$\frac{\partial^{r+s} f}{\partial x^r \, \partial y^s} = \alpha^r \beta^s e^{\alpha x + \beta y},$$

r and s being positive integers.

Higher derivatives for functions of n variables are defined in an analogous manner. Thus if $z = f(x_1, x_2, \ldots, x_n)$,

$$\frac{\partial^2 f}{\partial x_i\, \partial x_j} = \frac{\partial}{\partial x_i}\left(\frac{\partial f}{\partial x_j}\right) = (f_{x_j})_{x_i} = f_{x_j x_i},$$

and in general

$$\frac{\partial^{p_1+p_2+\cdots+p_n} f}{\partial x_1{}^{p_1}\, \partial x_2{}^{p_2} \cdots \partial x_n{}^{p_n}}$$

means p_1 differentiations with respect to x_1, p_2 differentiations with respect to $x_2, \ldots,$ and p_n differentiations with respect to x_n.

Example 6 If $f(x_1, x_2, x_3, x_4) = x_1{}^4 x_2{}^3 x_3{}^2 x_4$, find

$$\frac{\partial^4 f}{\partial x_1\, \partial x_2\, \partial x_3\, \partial x_4}.$$

This fourth derivative is found by successive differentiation.

$$\frac{\partial f}{\partial x_4} = x_1{}^4 x_2{}^3 x_3{}^2,$$

$$\frac{\partial^2 f}{\partial x_3\, \partial x_4} = 2x_1{}^4 x_2{}^3 x_3,$$

$$\frac{\partial^3 f}{\partial x_2\, \partial x_3\, \partial x_4} = 6x_1{}^4 x_2{}^2 x_3,$$

$$\frac{\partial^4 f}{\partial x_1\, \partial x_2\, \partial x_3\, \partial x_4} = 24x_1{}^3 x_2{}^2 x_3.$$

(Calculation of the derivatives in a different order would lead to the same result!)

Exercises 6.2

1. For the following functions $f(x,y)$, evaluate $\partial f/\partial x$ and $\partial f/\partial y$.

(i) $x^3y - 2x^2y^4$ **(ii)** $xe^{-x^2-2y^2}$ **(iii)** $(x+2y)^{-1}$

(iv) $\log xy$ **(v)** x^y **(vi)** $(e^x - e^y)/(x-y)$

(vii) $(x+2y)(y-3x)$ **(viii)** $\tan(x^2+xy+y^2)$ **(ix)** $y|x| + x|y|$

2. Calculate the second partial derivatives of the following functions $f(x, y)$.

(i) x^2+y^2 **(ii)** $\sin(x-y)+\sin(x+y)$ **(iii)** $(1+x^2+y^2)^{1/3}$

(iv) $\text{erf}(x^2+y^2)$ **(v)** $\tan(x^3+y^3)$ **(vi)** $(x^3+x^2y+xy^2+y^3)^2$

(vii) e^{xy} **(viii)** $y \sin xy$ **(ix)** $\cosh(\log xy)$

3. Find all first and second derivatives of the following functions $f(x, y, z)$.

(i) $x^2+2y^2+3z^2$ **(ii)** $xy+yz+zx$

(iii) $(x+1)^{10}(y-2)^2(z+3)^3$ **(iv)** $xyz\sin xyz$

(v) $e^{x+2y}\log(3y-x)$ **(vi)** $yz^2+z^2x^3+x^3y^4$

Verify that $f_{xy}=f_{yx}$, $f_{xz}=f_{zx}$ and $f_{yz}=f_{zy}$.

4. For $x=r\cos\theta$, $y=r\sin\theta$ (polar coordinates), find all first and second derivatives of x and y with respect to r and θ.

5. For $r=(x^2+y^2)^{1/2}$, $\theta=\tan^{-1}(y/x)$, find all first and second derivatives of r and θ with respect to x and y.

6. For $x=r\cos\theta\sin\phi$, $y=r\sin\theta\sin\phi$, $z=r\cos\phi$ (spherical polar coordinates), find all first and second derivatives of x,y and z with respect to r, θ and ϕ.

7. For $r=(x^2+y^2+z^2)^{1/2}$, $\theta=\tan^{-1}(y/x)$, $\phi=\cos^{-1}[z/(x^2+y^2+z^2)^{1/2}]$, find all first and second derivatives of r, θ and ϕ with respect to x, y and z.

8. Prove that $f(r,\theta)=ar^n\cos n\theta+br^n\sin n\theta$ (a,b,n constants) is a solution of

$$\frac{\partial^2 f}{\partial r^2}+\frac{1}{r}\frac{\partial f}{\partial r}+\frac{1}{r^2}\frac{\partial^2 f}{\partial\theta^2}=0.$$

Deduce that $\sum_{n=1}^{\infty}[a_nr^n\cos n\theta+b_nr^n\sin n\theta]$ is a solution (a_n, b_n given constants) if the series and its derivatives converge.

9. Verify the following statements.

(i) $z(x,y)=\tan^{-1}(y/x)$ satisfies the equation $z_{xx}+z_{yy}=0$.

(ii) $u(x,t)=e^{-an^2t}\sin nx$ (a, n constants) satisfies $u_t=au_{xx}$.

(iii) $f(\theta,\phi)=F(\phi)\cos\theta+G(\phi)\sin\theta$ (F, G given functions) satisfies $f_{\theta\theta}+f=0$.

10. If $z(x,y)=[u(x,y)]^2$, prove the following identities.

(i) $z_x=2uu_x$, $z_y=2uu_y$ **(ii)** $z_{xx}=2(u_x{}^2+uu_{xx})$

(iii) $z_{xy}=2(u_xu_y+uu_{xy})$ **(iv)** $z_{yy}=2(u_y{}^2+uu_{yy})$

Check these identities for the special case $u(x,y)=x^2+y^2$.

11. Given $G(p,q,r)=(p^2+q^2+r^2)^{-1/2}$, verify the following results.

(i) $p\dfrac{\partial G}{\partial p}+q\dfrac{\partial G}{\partial q}+r\dfrac{\partial G}{\partial r}+G=0$ **(ii)** $\dfrac{\partial^2 G}{\partial p^2}+\dfrac{\partial^2 G}{\partial q^2}+\dfrac{\partial^2 G}{\partial r^2}=0.$

12. A function $f(x,y)$ satisfies the equation $\partial f/\partial x=0$ for all x,y. Show that this implies $f=F(y)$ where F is an arbitrary function of y.

13. If a function $u(x,y)$ satisfies the equation $\partial^2u/\partial x^2=0$ for all x,y, show that $u(x,y)$ must be of the form $u=xF(y)+G(y)$ where F and G are arbitrary functions of y. (*Hint:* $u_{xx}=(u_x)_x=0$.)

14. If $u(x,y)$ satisfies the equation $\partial^2u/\partial x\partial y=0$ for all x, y, show that $u(x,y)$ must be of the form $u=F(x)+G(y)$ where F and G are arbitrary functions. (*Hint:* $u_{xy}=(u_x)_y=0$.)

15. Use l'Hôpital's rule to prove the following results.

(i) $\displaystyle\lim_{h\to 0}\frac{f(x+h,y)-f(x-h,y)}{2h}=f_x(x,y)$

(ii) $\displaystyle\lim_{k\to 0}\frac{f(x,y+k)-f(x,y-k)}{2k}=f_y(x,y)$

(iii) $\displaystyle\lim_{h\to 0}\frac{f(x+h,y)-2f(x,y)+f(x-h,y)}{h^2}=f_{xx}(x,y)$

(iv) $\displaystyle\lim_{k\to 0}\frac{f(x,y+k)-2f(x,y)+f(x,y-k)}{k^2}=f_{yy}(x,y)$

(v) $\lim_{k\to 0}\left\{\frac{1}{k}\lim_{h\to 0}\frac{f(x+h, y+k)+f(x, y)-f(x+h, y)-f(x, y+k)}{h}\right\}=f_{xy}(x, y)$

16. Show that the function $u(x_1, x_2, \ldots, x_n)=\exp(\alpha_1 x_1+\alpha_2 x_2+\cdots+\alpha_n x_n)$ where $\alpha_1, \alpha_2, \ldots, \alpha_n$ are constants satisfies the following equations.

(i) $\frac{\partial u}{\partial x_1}+\frac{\partial u}{\partial x_2}+\cdots+\frac{\partial u}{\partial x_n}=(\alpha_1+\alpha_2+\cdots+\alpha_n)u$

(ii) $\frac{\partial^2 u}{\partial x_1^2}+\frac{\partial^2 u}{\partial x_2^2}+\cdots+\frac{\partial^2 u}{\partial x_n^2}=(\alpha_1^2+\alpha_2^2+\cdots+\alpha_n^2)u$

17. If $f(x_1, x_2, x_3, x_4)=x_1 x_2^2/(x_3+2x_4)$, evaluate the following derivatives.

(i) $f_{x_1x_2}$ **(ii)** $f_{x_3x_4}$ **(iii)** $f_{x_1x_2x_3x_4}$

18. If $f(x_1, x_2, \ldots, x_n)=x_1x_2^2 \ldots x_n^n$, evaluate the following derivatives.

(i) $f_{x_1x_2}$ **(ii)** $f_{x_3x_4}$ **(iii)** $f_{x_1x_2 \ldots x_n}$

6.3 Increments and differentials

The computation of partial derivatives outlined in Sec. 6.2 was little more than a routine exercise. Here we examine how the Δ process, fundamental to the differential calculus of one variable, applies to functions of many variables.

Let $z = f(x, y)$ once again and consider two points (x, y) and $(x+\Delta x, y+\Delta y)$ in the plane. The values of f at these points are $f(x, y)$ and $f(x+\Delta x, y+\Delta y)$, and the difference between these two values is $\Delta z = \Delta f$, where

$$\Delta z = \Delta f = f(x+\Delta x, y+\Delta y) - f(x, y). \tag{1}$$

The present aim is to find an alternative expression for Δf in terms of partial derivatives which is, in fact, a simple generalization of the mean-value theorem to functions of two variables. In deriving this result we lean heavily on the one-variable mean-value theorem, which is now restated for convenience.

If $f(x)$ has a derivative at every point in the interval $[x, x+\Delta x]$, then

$$f(x+\Delta x) - f(x) = f'(x+\theta\,\Delta x)\,\Delta x, \qquad 0<\theta<1. \tag{2}$$

If, in addition, $f'(x)$ is continuous in the interval, then

$$f'(x+\theta\,\Delta x) = f'(x)+\varepsilon,$$

where $\varepsilon \to 0$ as $\Delta x \to 0$. In this circumstance, Eq. (2) can be written

$$f(x+\Delta x) - f(x) = f'(x)\Delta x + \varepsilon\,\Delta x. \tag{3}$$

Now we return to the two-variable case, where it is assumed that f, f_x, and f_y are all continuous in some small region which contains the points (x, y) and $(x+\Delta x, y+\Delta y)$. By writing

$$\Delta f = [f(x+\Delta x, y+\Delta y) - f(x, y+\Delta y)] + [f(x, y+\Delta y) - f(x, y)], \tag{4}$$

we see that Δf can be expressed as the sum of two terms each of which can be handled by the one-variable mean-value theorem. In the first term, only the variable x changes, and it follows from (2) that

$$f(x + \Delta x, y + \Delta y) - f(x, y + \Delta y) = f_x(x + \theta_1 \Delta x, y + \Delta y) \Delta x, \qquad 0 < \theta_1 < 1.$$

Likewise, only the y variable changes in the difference $f(x, y + \Delta y) - f(x, y)$, and therefore

$$f(x, y + \Delta y) - f(x, y) = f_y(x, y + \theta_2 \Delta y) \Delta y, \qquad 0 < \theta_2 < 1.$$

Upon substituting for these two expressions in (4), we obtain

$$\Delta f = f_x(x + \theta_1 \Delta x, y + \Delta y) \Delta x + f_y(x, y + \theta_2 \Delta y) \Delta y, \qquad 0 < \theta_1, \theta_2 < 1, \tag{5}$$

which is one form of the *mean-value theorem* for a function of two variables.

Since f_x and f_y are assumed continuous at (x, y),

$$f_x(x + \theta_1 \Delta x, y + \Delta y) = f_x(x, y) + \varepsilon_1,$$

and

$$f_y(x, y + \theta_2 \Delta y) = f_y(x, y) + \varepsilon_2,$$

where ε_1 and ε_2 both tend to zero as Δx and Δy tend to zero. These formulas are used to establish a more desirable form of (5):

$$\Delta f = f_x(x, y) \Delta x + f_y(x, y) \Delta y + \varepsilon_1 \Delta x + \varepsilon_2 \Delta y. \tag{6}$$

A perfectly good approximation for the increment Δf, valid for small Δx and Δy, is obtained by neglecting the epsilon terms in the last expression, i.e.,

$$\Delta f \approx f_x \Delta x + f_y \Delta y. \tag{7}$$

The next example helps to illustrate these ideas.

Example 1 If $f(x, y) = 1 + x + 2y + 3x^2 - 4y^2$, find Δf from first principles.

Clearly

$$f(x + \Delta x, y + \Delta y) = 1 + (x + \Delta x) + 2(y + \Delta y) + 3(x + \Delta x)^2 - 4(y + \Delta y)^2,$$

so that

$$\begin{aligned} \Delta f &= f(x + \Delta x, y + \Delta y) - f(x, y) \\ &= (1 + 6x) \Delta x + (2 - 8y) \Delta y + 3(\Delta x)^2 - 4(\Delta y)^2. \end{aligned}$$

The comparison of formula (6) with this result shows that $f_x = 1 + 6x$, $f_y = 2 - 8y$, $\varepsilon_1 = 3 \Delta x$, and $\varepsilon_2 = -4 \Delta y$. If Δx and Δy are very small, their squares will be negligible and the approximation (7) holds,

$$\Delta f \approx (1 + 6x) \Delta x + (2 - 8y) \Delta y.$$

The previous analysis is squarely based on rewriting the quantity

$$\Delta f = f(x + \Delta x, y + \Delta y) - f(x, y)$$

as the sum of two differences,

$$\Delta f = [f(x + \Delta x, y + \Delta y) - f(x, y + \Delta y)] + [f(x, y + \Delta y) - f(x, y)].$$

A geometrical interpretation of this decomposition is that the total height increment Δz from P to S in Fig. 3.1 is achieved by the two successive steps indicated. Since P and V are at the same altitude (as are W and R), the height of S above P is

$$\Delta z = VS = VW + WS.$$

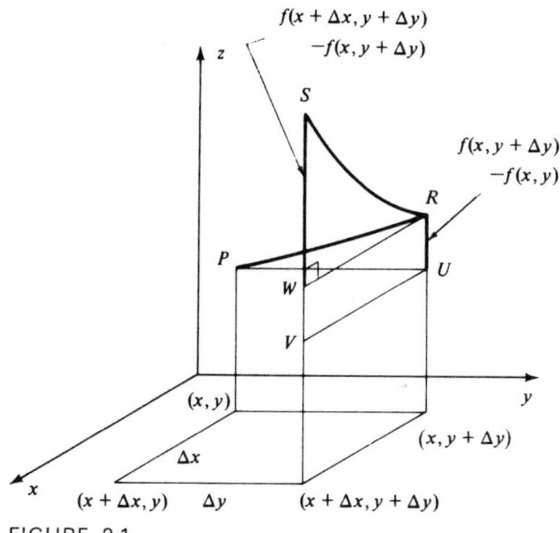

FIGURE 3.1

But since

$$VW = UR = f(x, y + \Delta y) - f(x, y)$$

and

$$WS = f(x + \Delta x, y + \Delta y) - f(x, y + \Delta y),$$

the replacement of these expressions in the preceding formula for Δz yields (4).

---

2 Equations (5) and (6) can be generalized to functions of n variables. Here the analysis is given only for the case $n = 3$. With $z = f(x_1, x_2, x_3)$

$$\Delta f = f(x_1 + \Delta x_1, x_2 + \Delta x_2, x_3 + \Delta x_3) - f(x_1, x_2, x_3), \tag{8}$$

and the breakdown is performed in three steps, namely,

$$f(x_1 + \Delta x_1, x_2 + \Delta x_2, x_3 + \Delta x_3) - f(x_1, x_2 + \Delta x_2, x_3 + \Delta x_3)$$
$$= f_{x_1}(x_1 + \theta_1 \Delta x_1, x_2 + \Delta x_2, x_3 + \Delta x_3)\, \Delta x_1, \qquad 0 < \theta_1 < 1;$$

$$f(x_1, x_2 + \Delta x_2, x_3 + \Delta x_3) - f(x_1, x_2, x_3 + \Delta x_3)$$
$$= f_{x_2}(x_1, x_2 + \theta_2 \Delta x_2, x_3 + \Delta x_3) \Delta x_2, \qquad 0 < \theta_2 < 1;$$
$$f(x_1, x_2, x_3 + \Delta x_3) - f(x_1, x_2, x_3)$$
$$= f_{x_3}(x_1, x_2, x_3 + \theta_3 \Delta x_3) \Delta x_3, \qquad 0 < \theta_3 < 1.$$

Upon adding, terms on the left cancel pairwise to give

$$\begin{aligned}\Delta f = &f_{x_1}(x_1 + \theta_1 \Delta x_1, x_2 + \Delta x_2, x_3 + \Delta x_3) \Delta x_1 \\ &+ f_{x_2}(x_1, x_2 + \theta_2 \Delta x_2, x_3 + \Delta x_3) \Delta x_2 \\ &+ f_{x_3}(x_1, x_2, x_3 + \theta_3 \Delta x_3) \Delta x_3, \qquad 0 < \theta_1, \theta_2, \theta_3 < 1,\end{aligned} \tag{9}$$

which is the three-variable counterpart of (5).

If the partial derivatives are continuous in a neighborhood of the point (x_1, x_2, x_3), (9) takes the form

$$\begin{aligned}\Delta f = &f_{x_1}(x_1, x_2, x_3) \Delta x_1 + f_{x_2}(x_1, x_2, x_3) \Delta x_2 + f_{x_3}(x_1, x_2, x_3) \Delta x_3 \\ &+ \varepsilon_1 \Delta x_1 + \varepsilon_2 \Delta x_2 + \varepsilon_3 \Delta x_3,\end{aligned} \tag{10}$$

where ε_1, ε_2, and $\varepsilon_3 \to 0$ as Δx_1, Δx_2, and $\Delta x_3 \to 0$.

3

As in the one-variable calculus, it is natural to make use of *differentials*. If $z = f(x, y)$, then

$$dz = df = f_x\, dx + f_y\, dy, \tag{11}$$

is the exact differential analog of the incremental approximation (7). The differentials dx, dy and dz represent relative increments on a tangent plane of the surface $z = f(x, y)$. In the more general case $z = f(x_1, \ldots, x_n)$

$$df = \sum_{i=1}^{n} f_{x_i}\, dx_i. \tag{12}$$

Example 2 Show that the differential expression $df = A\, dx + B\, dy$ implies that $A = f_x$ and $B = f_y$.

The given expression for the differential df, combined with (11), yields

$$(A - f_x)\, dx + (B - f_y)\, dy = 0.$$

Since x and y are independent variables, so are the elements dx and dy. In particular, if $dy = 0$ but $dx \neq 0$, this equation gives $A = f_x$. Likewise, with $dx = 0$ and $dy \neq 0$, we find that $B = f_y$.

Example 3 If $z = f(x, y)$ and $y = y(x)$, find dz/dx.

Here, z is really a function of the single variable x,

$$z = f(x, y(x)), \tag{13}$$

written in a more complicated form than usual. In any particular case we could convert this equation to, say,

$$z = G(x),$$

but this might be exceedingly complicated. For this reason, and because functions like that in (13) occur frequently, the derivative is computed directly. This is really quite an easy calculation, in effect a minor rearrangement of (11):

$$\frac{dz}{dx} = f_x + f_y \frac{dy}{dx}. \tag{14}$$

Here the partial derivatives f_x and f_y and the ordinary derivative dy/dx are known functions. In other words, the operations

$$\frac{dz}{dx} = \frac{d}{dx} f(x, y(x)) \qquad \text{and} \qquad \left(\frac{\partial}{\partial x} + \frac{dy}{dx}\frac{\partial}{\partial y}\right) f(x, y)$$

are equivalent:

$$\frac{d}{dx} f(x, y(x)) = \left(\frac{\partial}{\partial x} + \frac{dy}{dx}\frac{\partial}{\partial y}\right) f(x, y). \tag{15}$$

Example 4 If x and y are implicitly related by $f(x, y) = 0$, find dy/dx and d^2y/dx^2 in terms of the partial derivatives of f.

The relationship $f(x, y) = 0$ implicitly defines a curve $y = y(x)$ along which both f and df are zero. Therefore by (14)

$$\frac{df}{dx} = f_x + f_y \frac{dy}{dx} = 0, \qquad \text{or} \qquad \frac{dy}{dx} = -\frac{f_x}{f_y}. \tag{16}$$

The second derivative is found by repeating this procedure,

$$\frac{d}{dx}\left(\frac{df}{dx}\right) = \frac{d}{dx}\left(f_x + f_y \frac{dy}{dx}\right) = 0,$$

or

$$\frac{d}{dx}(f_x) + f_y \frac{d^2y}{dx^2} + \frac{dy}{dx}\frac{d}{dx}(f_y) = 0.$$

But since the operations $\dfrac{d}{dx}$ and $\dfrac{\partial}{\partial x} + \dfrac{dy}{dx}\dfrac{\partial}{\partial y}$ are equivalent, this becomes

$$\left(f_{xx} + \frac{dy}{dx} f_{xy}\right) + f_y \frac{d^2y}{dx^2} + \frac{dy}{dx}\left(f_{xy} + \frac{dy}{dx} f_{yy}\right) = 0.$$

If dy/dx is replaced by (16), the preceding equation can then be solved for d^2y/dx^2 to give

$$\frac{d^2y}{dx^2} = -\frac{1}{f_y^{\,3}}(f_{xx}f_y^{\,2} - 2f_xf_yf_{xy} + f_{yy}f_x^{\,2}).$$

The unproved identity $f_{xy} = f_{yx}$ has been used. (In *practice* these calculations are usually done by the methods outlined in Sec. 2.6. However, the general formulas are as given above.)

Example 5 Find, approximately, the change in $x^2 + xy + 2y^2$ when (x, y) changes from (1, 1) to (1.1, 0.9).

The approximation is based on the differential formula (11), wherein dx and dy are replaced by the small finite increments of change Δx and Δy [see (7)]. Here $f_x = 2x + y$, $f_y = x + 4y$, and at the point (1, 1) $f_x = 3$ and $f_y = 5$. Moreover, $\Delta x = 0.1$ and $\Delta y = -0.1$, so that

$$\Delta f \approx 3(0.1) + 5(-0.1) = -0.2.$$

The exact increment Δf is easily calculated in this particular case;

$$\begin{aligned}\Delta f &= f(1.1, 0.9) - f(1, 1)\\ &= 3.82 - 4\\ &= -0.18.\end{aligned}$$

The error incurred in the approximate calculation is 0.02.

4

The differential relation

$$dz = f_x\,dx + f_y\,dy, \tag{17}$$

for the surface $z = f(x, y)$ has important geometrical implications. If this equation is expressed as $-f_x\,dx - f_y\,dy + dz = 0$, it can be interpreted as the vanishing of the dot product of the vector

$$\mathbf{n} = -f_x\mathbf{i} - f_y\mathbf{j} + \mathbf{k},$$

with the vector element of arc length on the surface (Fig. 3.2)

$$d\mathbf{r} = dx\mathbf{i} + dy\mathbf{j} + dz\mathbf{k}.$$

In other words (17) can be expressed as

$$\mathbf{n} \cdot d\mathbf{r} = 0. \tag{18}$$

Since at the point $(x, y, f(x, y))$ $\mathbf{n}$ is perpendicular to every vector $d\mathbf{r}$ tangent to the surface $z = f(x, y)$, the vector $\mathbf{n}$ is *the normal to the surface*.

The normal line to the surface at the point $\mathbf{r}_0 = x_0\mathbf{i} + y_0\mathbf{j} + z_0\mathbf{k}$ is

$$\mathbf{r} = \mathbf{r}_0 + s\mathbf{n}, \tag{19}$$

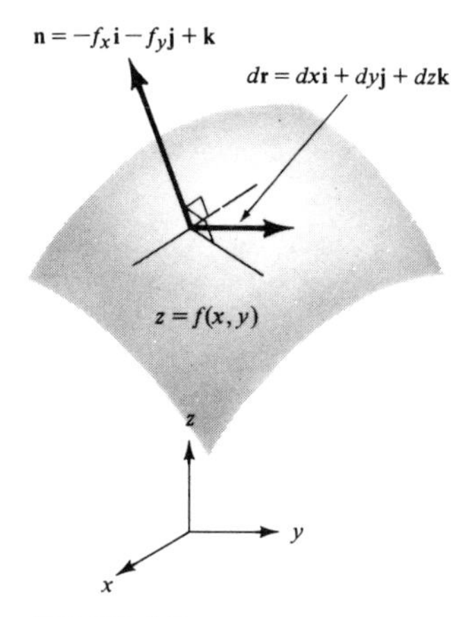

FIGURE 3.2

or in cartesian form

$$\frac{x - x_0}{-f_{x_0}} = \frac{y - y_0}{-f_{y_0}} = \frac{z - z_0}{1} = s, \tag{20}$$

(see Sec. 5.6). The notation f_{x_0} means f_x evaluated at (x_0, y_0), and f_{y_0} is f_y evaluated at (x_0, y_0).

The plane through $\mathbf{r}_0$ having $\mathbf{n}$ as normal is called *the tangent plane* to the surface at $\mathbf{r}_0$, and its equation is

$$(\mathbf{r} - \mathbf{r}_0) \cdot \mathbf{n} = 0, \tag{21}$$

or

$$(x - x_0)f_{x_0} + (y - y_0)f_{y_0} - (z - z_0) = 0. \tag{22}$$

In this form, that is, $f_{x_0}\Delta x + f_{y_0}\Delta y = \Delta z$, we see that the error implied by the approximation (7) represents geometrically the local deviation of the true surface from its tangent plane at a specific point. The deviation is zero for the differentials dx, dy, dz.

If the surface is given in the implicit form $F(x, y, z) = 0$, then

$$dF = F_x\, dx + F_y\, dy + F_z\, dz = 0.$$

Since $d\mathbf{r}$ is a vector element tangent to the surface, this equation can be written $\mathbf{n} \cdot d\mathbf{r} = 0$, and in this case the normal vector is

$$\mathbf{n} = F_x \mathbf{i} + F_y \mathbf{j} + F_z \mathbf{k}. \tag{23}$$

Example 6 Find the equation of the tangent plane to the surface $z = x^2 + 5xy - 2y^2$ at the point (1, 2, 3).

Here $z = f(x, y) = x^2 + 5xy - 2y^2$, and $f_x = 2x + 5y$, $f_y = 5x - 4y$. At the point (1, 2, 3) $f_x = 12$ and $f_y = -3$. Therefore, the equation of the required tangent plane is by (22)

$$12(x-1) - 3(y-2) - (z-3) = 0$$

or
$$12x - 3y - z = 3.$$

Example 7 Find the tangent plane and the normal line at the point $(2, -3, 3)$ for the hyperboloid of one sheet $x^2 + y^2 - z^2 = 4$.

With $F(x, y, z) = x^2 + y^2 - z^2 - 4$, we have $F_x = 2x$, $F_y = 2y$, and $F_z = -2z$; and so at the point $(2, -3, 3)$ the normal to the surface is by (23)

$$\mathbf{n} = 4\mathbf{i} - 6\mathbf{j} - 6\mathbf{k}.$$

Therefore, the normal line at this point is

$$\frac{x-2}{4} = \frac{y+3}{-6} = \frac{z-3}{-6} = s$$

or
$$\frac{x-2}{2} = \frac{y+3}{-3} = \frac{z-3}{-3} = 2s.$$

The equation of the tangent plane at $(2, -3, 3)$ is

$$4(x-2) - 6(y+3) - 6(z-3) = 0$$

or
$$2x - 3y - 3z = 4.$$

Note that the point $(2, -3, 3)$ must of course satisfy both the equations for the normal line and tangent plane.

Example 8 The two surfaces $F(x, y, z) = 0$, $G(x, y, z) = 0$ meet in a space curve. Find the tangent vector to this space curve.

Differentials are used to solve this problem. If $d\mathbf{r}$ is a vector element of the common curve, it must lie on both surfaces. As a consequence, it is perpendicular to the normal vectors of each surface, i.e.,

$$F_x\,dx + F_y\,dy + F_z\,dz = 0,$$
$$G_x\,dx + G_y\,dy + G_z\,dz = 0.$$

These equations are equivalent to

$$\frac{dx}{F_yG_z - F_zG_y} = \frac{dy}{F_zG_x - F_xG_z} = \frac{dz}{F_xG_y - F_yG_x}$$

which together determine the common curve. The vector

$$(F_yG_z - F_zG_y)\mathbf{i} + (F_zG_x - F_xG_z)\mathbf{j} + (F_xG_y - F_yG_x)\mathbf{k}$$

is then the required tangent to the space curve.

Finally, we return to the question of whether the cross derivatives f_{xy} and f_{yx} are equal and show that the order in which partial derivatives are taken is not important. A more deliberate statement is the following: If $f(x, y)$ is continuous with continuous second partial derivatives in the neighborhood of a point (x, y), then $f_{xy} = f_{yx}$ at the point (x, y). As might be anticipated, the proof entails showing that the order in which the double limit ($\Delta x \to 0$, $\Delta y \to 0$) is calculated is immaterial.

A slight change in notation is convenient, and instead of Δx and Δy we will use h and k, respectively.

We start with a formula that is a good finite-difference approximation to both second partial derivatives f_{xy} and f_{yx} [see part (v) of Exercise 15 in Sec. 6.2]:

$$\mathscr{F}(x, y) = \frac{1}{hk}[f(x+h, y+k) - f(x, y+k) - f(x+h, y) + f(x, y)]. \tag{24}$$

We treat the x variation first and for the sake of brevity let

$$\mathscr{G}(x) = f(x, y+k) - f(x, y), \tag{25}$$

in which case it is an easy matter to check that $\mathscr{F}$ can be written

$$\mathscr{F}(x, y) = \frac{1}{hk}[\mathscr{G}(x+h) - \mathscr{G}(x)]. \tag{26}$$

According to the mean-value theorem,

$$\mathscr{G}(x+h) - \mathscr{G}(x) = h\mathscr{G}'(x + \theta_1 h), \qquad 0 < \theta_1 < 1,$$

and upon differentiating (25) and substituting here for $\mathscr{G}'$ we obtain

$$\mathscr{G}(x+h) - \mathscr{G}(x) = h[f_x(x + \theta_1 h, y+k) - f_x(x + \theta_1 h, y)].$$

Once again the mean-value theorem is applied to the right-hand side of the last expression, and the result is

$$\mathscr{G}(x+h) - \mathscr{G}(x) = hkf_{xy}(x + \theta_1 h, y + \theta_2 k), \qquad 0 < \theta_2 < 1. \tag{27}$$

It follows that the original finite-difference formula (24) can be written concisely as

$$\mathscr{F}(x, y) = f_{xy}(x + \theta_1 h, y + \theta_2 k), \qquad 0 < \theta_1, \theta_2 < 1. \tag{28}$$

In an analogous manner, we can work on the y variation, and if

$$\mathscr{H}(y) = f(x+h, y) - f(x, y), \tag{29}$$

then

$$\mathscr{F}(x, y) = \frac{1}{hk}[\mathscr{H}(y+k) - \mathscr{H}(y)].$$

The mean-value theorem implies that

$$\mathscr{F}(x, y) = \frac{1}{h}\mathscr{H}'(y + \theta_3 k), \qquad 0 < \theta_3 < 1.$$

Upon replacing $\mathscr{H}$ here by (29) this becomes

$$\mathscr{F}(x, y) = \frac{1}{h}[f_y(x + h, y + \theta_3 k) - f_y(x, y + \theta_3 k)].$$

Once more the mean-value theorem is used to write this as

$$\mathscr{F}(x, y) = f_{yx}(x + \theta_4 h, y + \theta_3 k), \qquad 0 < \theta_3, \theta_4 < 1. \tag{30}$$

A comparison of (28) and (30) shows that

$$f_{xy}(x + \theta_1 h, y + \theta_2 k) = f_{yx}(x + \theta_4 h, y + \theta_3 k);$$

but f_{xy} and f_{yx} are continuous and the limit $h \to 0$ and $k \to 0$ yields

$$f_{xy}(x, y) = f_{yx}(x, y), \tag{31}$$

as desired. The cross derivatives are equal.

This result generalizes in a natural way. For instance, if the third-order derivatives are all continuous, then

$$\frac{\partial^3 f}{\partial x^2\, \partial y} = \frac{\partial^3 f}{\partial x\, \partial y\, \partial x} = \frac{\partial^3 f}{\partial y\, \partial x^2}.$$

The step by step arguments are as follows:

$$\frac{\partial^3 f}{\partial x^2\, \partial y} = \frac{\partial}{\partial x}\left(\frac{\partial^2 f}{\partial x\, \partial y}\right) = \frac{\partial}{\partial x}\left(\frac{\partial^2 f}{\partial y\, \partial x}\right) = \frac{\partial^3 f}{\partial x\, \partial y\, \partial x}$$

$$= \frac{\partial^2}{\partial x\, \partial y}\left(\frac{\partial f}{\partial x}\right) = \frac{\partial^2}{\partial y\, \partial x}\left(\frac{\partial f}{\partial x}\right) = \frac{\partial^3 f}{\partial y\, \partial x^2}.$$

Note that only the equivalence of the operations $\dfrac{\partial^2}{\partial x\, \partial y}$ and $\dfrac{\partial^2}{\partial y\, \partial x}$ has been used in this demonstration.

The point to be remembered is that partial differentiations can be performed in any order provided the derivatives concerned are continuous.

Exercises 6.3

1. Apply the Δ process to $z = f(x,y) = x^2 - 2y^2$ to find $f(1 + \Delta x,\ 1 + \Delta y) - f(1,1)$ in terms of Δx and Δy. Compute approximations for $f(1.02,\ 1.03)$ and $f(0.99, 1.01)$ and compare to the accurate values.

2. In the previous problem, determine the tangent plane to the surface $z = x^2 - 2y^2$ at the point $(1, 1, -1)$. Explain the relation of the tangent plane to the approximate calculations.

3. The mean value theorem applied to $z = f(x,y)$ states

$$\Delta f = f_x(x + \theta_1 \Delta x,\ y + \Delta y)\Delta x + f_y(x,\ y + \theta_2 \Delta y)\Delta y \qquad 0 < \theta_1,\ \theta_2 < 1$$

Find θ_1 and θ_2 for the following examples.

(i) $f(x, y) = x^2 + xy + y^2$ $\quad (x, y) = (0,0),\ \Delta x = 1,\ \Delta y = 2$

(ii) $f(x, y) = e^{x-y}$ $\quad (x, y) = (1,1),\ \Delta x = 1/2,\ \Delta y = 1/3$

(iii) $f(x, y) = \sin(x + y)$ $\quad (x, y) = (\pi/4,\ \pi/4),\ \Delta x = \pi/12,\ \Delta y = -\pi/12$

4. Evaluate the following limits and write them as partial derivatives.

(i) $\lim_{y\to 0} \dfrac{\sin(x+y)-\sin x}{y}$ (ii) $\lim_{x\to 0} \dfrac{e^{-(x+y)^2}-e^{-y^2}}{x}$ (iii) $\lim_{x\to 0}\lim_{y\to 0} \dfrac{e^{xy}-1}{xy}$

5. Find the differentials dz of the following functions $z = z(x, y)$.

(i) $x \sin y$ (ii) $(x^2-y^2)e^{x^2+y^2}$ (iii) $y \log(1+x) + x \log(1+y)$
(iv) $(x^3+y^3)^2$ (v) $\sec^2(x^2+2y^2)$ (vi) $e^{x+y}\cos(x-y)$

6. Determine the tangent planes and normal lines at the point $x=0$, $y=0$ on the surfaces $z = z(x, y)$ of the previous problem.

7. Calculate the differentials df for the following functions.

(i) $f(x, y) = y/x$ (ii) e^{xy} (iii) $x^2+y^2+z^2$
(iv) $(x+y+z)^3$ (v) $\cosh(x+y-z)$ (vi) $(x+y)(y+z)(z+x)$

8. If possible, find the functions $f(x,y)$ with the following first derivatives:

(i) $f_x = x, f_y = y$ (ii) $f_x = y, f_y = x^2$ (iii) $f_x = e^x \sin y, f_y = e^x \cos y$

9. A rectangular crate has sides of length 2m, 3m and 5m. If these sides are all increased by 1 cm, find the approximate increase of volume. Compare to the accurate value.

10. A triangle has an angle of 61° between sides of length 10.1 and 19.9 cm. Find the approximate area of the triangle. Compare to the exact value.

11. Find the approximate values of the following numbers using differentials. Compare to the accurate values.

(i) $[(3.02)^2+(3.97)^2]^{1/2}$ (ii) $(4.1)^{1/2}+(26.9)^{1/3}+(256.3)^{1/4}$ (iii) $(0.996)^3(1.006)^{-1}$

12. Three resistors having resistances r_1, r_2, and r_3 are connected in parallel, the resistance R of the circuit being determined by $R^{-1} = r_1^{-1} + r_2^{-1} + r_3^{-1}$. If the individual resistances are each subject to a small percentage error ε, find the approximate maximum percentage error in R.

13. A closed cylindrical oil-storage tank is to have an interior radius of 16 m and an interior height of 8 m. If it is to be constructed of metal 5 cm thick, find the approximate amount of metal needed. Would more metal be needed if the interior radius were 8 m and the tank had the same capacity?

14. Two curves are said to be *orthogonal* if their tangent lines at the points of intersection are perpendicular. Show that the two plane curves $f(x, y) = 0$ and $g(x, y) = 0$ are orthogonal if $f_x g_x + f_y g_y = 0$ at the point of intersection.

15. Show that all curves of the family $xy = a^2$ are orthogonal to the curves of the family $x^2 - y^2 = b^2$. Illustrate with a diagram.

16. Show that all curves of the family of circles $x^2 + y^2 = 2ax$ are orthogonal to the family $x^2 + y^2 = 2by$. Illustrate with a diagram.

17. Find dy/dx for the following implicit functions $y = y(x)$.

(i) $x^3+y^3 = 3xy$ (ii) $x^y = y^x$ (iii) $\sin x \cosh y = 2$
(iv) $(\cos x)^y = (\sin y)^x$ (v) $x^2+xy+y^2+x+y=1$ (vi) $y \tan^{-1} e^x = x e^{\tan^{-1} y}$
(vii) $x^2y^2 - x^2 - 4y^2 = 0$ (viii) $x^2+y^2 = e^{xy}$ (ix) $x + y^2 = \int_0^{y-x} \cos^2 t \, dt$

18. Find the tangent plane and normal line at the point $(1, -1, 1)$ on the surface $x^2 - 2y^2 + 3z^2 = 2$.

19. Prove that every tangent plane to the surface $z = xe^{x/y}$ passes through the origin. Interpret this result geometrically.

20. Find the tangent vector at the point (1,1,2) to the curve of intersection of the surfaces $z = x^2 + y^2$ and $2x^2 + 2y^2 - z^2 = 0$.

21. Find the point on the surface $z = x^2 + y^2$ which is closest to the point (1,1,0). Show that the line joining (1,1,0) to this closest point is normal to the surface.

22. Find the tangent plane and normal line to the surface $yz + zx + xy = 3$ at the point $(1, 1, 1)$.
23. Show that the tangent plane at any point on the surface $xyz = 1$ and the coordinate planes form a tetrahedron of constant volume.
24. If $f(x,y) = (x^3 + y^3)/(x - y)$ for $x \neq y$, and if $f(x,x) = 0$, show that f is discontinuous at the origin even though $f_x(0,0)$ and $f_y(0,0)$ are both finite. (This example shows that even if the two first partial derivatives exist at a point, the function need not be continuous.)
25. If $f(x, y) = xy(x^2 - y^2)/(x^2 + y^2)$ for $x^2 + y^2 \neq 0$ and if $f(0,0) = 0$, show that:
 (i) f is continuous at $(0,0)$ **(ii)** $f_x(0,y) = -y$ **(iii)** $f_{xy}(0,0) = -1$
 $f_y(x,0) = x$ $f_{yx}(0,0) = +1$
(f_{xy} and f_{yx} need not be the same at a point where they are discontinuous.)

6.4 *Directional derivatives and the gradient*

Imagine a climber at a point on a mountain $z = f(x, y)$, where z is altitude and x and y are horizontal coordinates. Obviously ascent or descent in his next step will depend on the direction in which the climber moves. In going from the point (x, y) to the point $(x + \Delta x, y + \Delta y)$ his change in altitude is $\Delta z = \Delta f = f(x + \Delta x, y + \Delta y) - f(x, y)$, while the horizontal distance moved is $\Delta s = [(\Delta x)^2 + (\Delta y)^2]^{1/2}$. An average rate of change of altitude with horizontal distance can be calculated from these increments:

$$\frac{\Delta z}{\Delta s} = \frac{\Delta f}{\Delta s} = \frac{f(x + \Delta x, y + \Delta y) - f(x, y)}{[(\Delta x)^2 + (\Delta y)^2]^{1/2}}, \tag{1}$$

and the instantaneous rate of change is obtained as the limit $\Delta x \to 0$ and $\Delta y \to 0$. This result will depend on the direction in which the climber moves on the surface, because this affects the ratio $\Delta y/\Delta x$.

Suppose the point $(x + \Delta x, y + \Delta y)$ is allowed to approach the point (x, y) from a *fixed* direction which makes an angle φ with the x axis, as shown in Fig. 4.1. In this case

$$\Delta x = \Delta s \cos \varphi, \qquad \Delta y = \Delta s \sin \varphi, \tag{2}$$

and Eq. (1) takes the particular form

$$\frac{\Delta f}{\Delta s} = \frac{f(x + \Delta s \cos \varphi, y + \Delta s \sin \varphi) - f(x, y)}{\Delta s}. \tag{3}$$

The limit of this ratio of increments as $\Delta s \to 0$ defines the *directional derivative* df/ds in the direction φ:

$$\frac{df}{ds} = \lim_{\Delta s \to 0} \frac{f(x + \Delta s \cos \varphi, y + \Delta s \sin \varphi) - f(x, y)}{\Delta s}. \tag{4}$$

The directional derivative measures the rate of change of the function along a particular ray emanating from a point. As such, the value of df/ds is clearly dependent on the

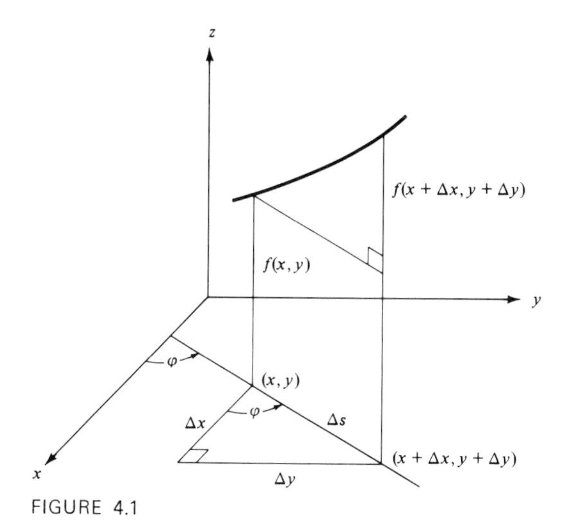

FIGURE 4.1

angle φ. The right side of (4) can be expressed in terms of partial derivatives by using the mean-value theorem [Eq. (6), Sec. 6.3], and the result is

$$\frac{df}{ds} = \lim_{\Delta s \to 0} \frac{f_x(x, y)\,\Delta s \cos\varphi + f_y(x, y)\Delta s \sin\varphi + \varepsilon_1\,\Delta s \cos\varphi + \varepsilon_2\,\Delta s \sin\varphi}{\Delta s},$$

where ε_1, $\varepsilon_2 \to 0$ as Δx, $\Delta y \to 0$ (or $\Delta s \to 0$). On evaluation of this limit we find that

$$\frac{df}{ds} = f_x \cos\varphi + f_y \sin\varphi. \tag{5}$$

In the special cases $\varphi = 0$ and $\varphi = \pi/2$, the directional derivative df/ds reduces to the usual partial derivatives f_x and f_y, respectively.

Equation (5) also follows directly from the differential formula

$$df = f_x\,dx + f_y\,dy$$

upon dividing by ds

$$\frac{df}{ds} = f_x \frac{dx}{ds} + f_y \frac{dy}{ds}.$$

Clearly

$$\frac{dx}{ds} = \cos\varphi, \qquad \frac{dy}{ds} = \sin\varphi.$$

Example 1 Calculate the directional derivative of the function e^{x+y^2} at the point (0, 1) in the direction of the vector $-\mathbf{i}+\mathbf{j}$.

The direction specified by the vector $-\mathbf{i}+\mathbf{j}$ corresponds to $\varphi = 3\pi/4$. With $f(x, y) = e^{x+y^2}$, $f_x = e^{x+y^2}$, and $f_y = 2ye^{x+y^2}$, then at the point (0, 1)

$$f_x = e, \qquad f_y = 2e.$$

The value of the desired directional derivative is obtained from (5):

$$\frac{df}{ds} = e\cos\frac{3\pi}{4} + 2e\sin\frac{3\pi}{4}$$

$$= -\frac{e}{\sqrt{2}} + \frac{2e}{\sqrt{2}} = \frac{e}{\sqrt{2}}.$$

Example 2 A climber is at the point $(-\frac{3}{2}, -1, \frac{3}{4})$ on a mountain whose height z is approximated by the formula $z = 5 - x^2 - 2y^2$, x and y being rectangular horizontal coordinates. He decides to follow the path of steepest ascent. In what direction should he start off?

Since $f(x, y) = 5 - x^2 - 2y^2$, $f_x = -2x$ and $f_y = -4y$; at the point $(-\frac{3}{2}, -1)$, $f_x = 3$ and $f_y = 4$. Therefore the value of the directional derivative df/ds at $(-\frac{3}{2}, -1)$ in the direction φ is

$$\frac{df}{ds} = 3\cos\varphi + 4\sin\varphi.$$

Since df/ds is really just a function of φ at the point in question, its extremal points are given by $\dfrac{d}{d\varphi}\left(\dfrac{df}{ds}\right) = 0$. But

$$\frac{d}{d\varphi}\left(\frac{df}{ds}\right) = -3\sin\varphi + 4\cos\varphi,$$

and this is clearly zero when $\varphi = \tan^{-1}\frac{4}{3}$ or $\varphi = \pi + \tan^{-1}\frac{4}{3}$. A check of the second derivative shows $\varphi = \tan^{-1}\frac{4}{3}$ to be a relative maximum of df/ds, while $\varphi = \pi + \tan^{-1}\frac{4}{3}$ is a relative minimum. In the direction of greatest increase $\cos\varphi = \frac{3}{5}$, $\sin\varphi = \frac{4}{5}$, and $(df/ds)_{\max} = 5$.

2

The preceding example is, in fact, a very special case of a far more fundamental concept, which we shall now discuss.

The directional derivative

$$\frac{df}{ds} = f_x\cos\varphi + f_y\sin\varphi = f_x\frac{dx}{ds} + f_y\frac{dy}{ds},$$

can be written as the scalar product of two vectors

$$\frac{df}{ds} = \hat{\mathbf{u}} \cdot \nabla f, \tag{6}$$

where

$$\hat{\mathbf{u}} = \cos\varphi\,\mathbf{i} + \sin\varphi\,\mathbf{j} = \frac{dx}{ds}\mathbf{i} + \frac{dy}{ds}\mathbf{j} \tag{7}$$

is a unit vector in the direction specified by φ and

$$\nabla f = f_x\mathbf{i} + f_y\mathbf{j}. \tag{8}$$

∇f is called the *gradient* of f (read "grad f"). If θ is the angle between the vectors $\hat{\mathbf{u}}$ and ∇f (Fig. 4.2), it follows that

$$\frac{df}{ds} = |\hat{\mathbf{u}}|\,|\nabla f|\cos\theta,$$

or

$$\frac{df}{ds} = |\nabla f|\cos\theta. \tag{9}$$

Therefore, the maximum value of the directional derivative at a given point occurs when $\cos\theta = 1$, or $\theta = 0$, and the *directional derivative* is a *maximum in the direction of the vector* ∇f. Furthermore, from Eq. (9),

$$\left(\frac{df}{ds}\right)_{\max} = |\nabla f|, \tag{10}$$

which means that the magnitude of ∇f is equal to the value of the maximum directional derivative.

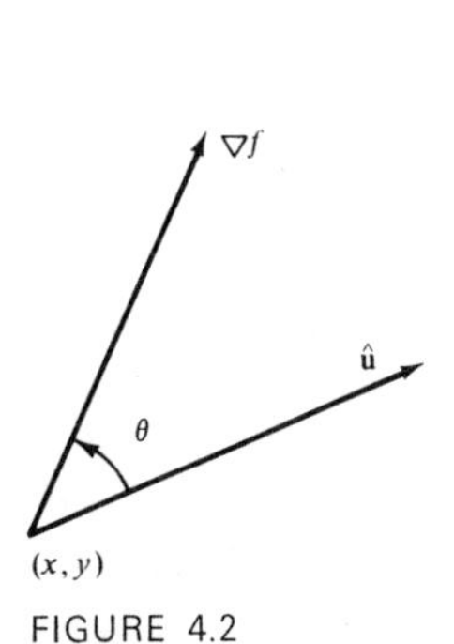

FIGURE 4.2

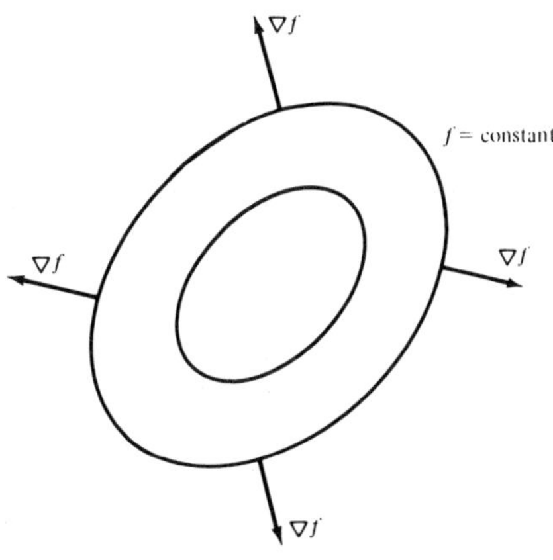

FIGURE 4.3

To summarize, the gradient of a scalar function $f(x, y)$, $\nabla f = f_x\mathbf{i} + f_y\mathbf{j}$, is a vector (a vector function) which at each point gives the *magnitude and direction of the maximum directional derivative* of the function. The study of vector functions will be the main topic of Chap. 8, where ∇f will play an important role.

The gradient of f can also be interpreted as the vector which at any point is perpendicular to the level curve of the surface $z = f(x, y)$ through that point,

$$f(x, y) = z_0, \tag{11}$$

where each level curve is specified by a different value of z_0 (Fig. 4.3). If z represents an altitude, then the points (x, y) on a particular level curve correspond to the same elevation z_0. Intuition suggests that the fastest way to gain altitude is to move perpendicular to the level curves (since moving along level curves produces no change in altitude). This can be established by noting that the differential df is zero along a level curve $f(x, y) = z_0$;

$$f_x\,dx + f_y\,dy = 0. \tag{12}$$

Thus, the vector ∇f is perpendicular to the vector $d\mathbf{r} = \mathbf{i}\,dx + \mathbf{j}\,dy$. But since $d\mathbf{r}$ is an element of arc length of the level curve, (12) shows that ∇f is normal to the level curve at each point. Analogous interpretations hold when f represents temperature, pressure, etc.

Example 3 A metal plate occupies the region $0 < x < 1$, $0 < y < 1$ of the xy plane. The temperature of the plate is known to be $T = xy(1 - x)(1 - y)$. In what direction should an insect at the point $(\frac{1}{4}, \frac{1}{3})$ move in order to cool off as quickly as possible?

The question involves the calculation of the temperature gradient ∇T at the point $(\frac{1}{4}, \frac{1}{3})$. Since

$$\nabla T = (1 - 2x)y(1 - y)\mathbf{i} + (1 - 2y)x(1 - x)\mathbf{j},$$

the evaluation at the point $(\frac{1}{4}, \frac{1}{3})$ gives

$$\nabla T = \tfrac{1}{9}\mathbf{i} + \tfrac{1}{16}\mathbf{j}.$$

Therefore, to *cool* off as rapidly as possible the insect should move in the direction of the vector $-\mathbf{i}/9 - \mathbf{j}/16$, that is, opposite to the gradient.

Example 4 A shark that detects the presence of blood will respond by moving continually in the direction of strongest scent. In a certain test conducted at the surface of the ocean, the concentration C of blood in parts per million of water is well approximated by $C(x, y) = e^{-(x^2+2y^2)/10^4}$ where x and y are horizontal coordinates measured in meters from the blood source. Find the shark's approach path from any point.

Since the shark follows the strongest scent, it will move at each instant in the direction of the vector ∇C. Differentiation shows that

$$\nabla C = 10^{-4}(-2x\mathbf{i} - 4y\mathbf{j})e^{-(x^2+2y^2)/10^4};$$

and if $d\mathbf{r} = dx\,\mathbf{i} + dy\,\mathbf{j}$ is the vector increment along the shark's approach path, it must be parallel to ∇C, that is, $d\mathbf{r} \times \nabla\mathbf{C} = \mathbf{0}$. It follows that

$$\frac{dx}{-2x} = \frac{dy}{-4y}$$

or

$$\frac{dy}{y} = \frac{2dx}{x},$$

and upon integration

$$y = Ax^2,$$

where A is a constant. All approach paths are parabolas (see Fig. 4.4), and a shark that begins his attack at the point (x_0, y_0) (which determines A) follows the path

$$y = y_0\left(\frac{x}{x_0}\right)^2.$$

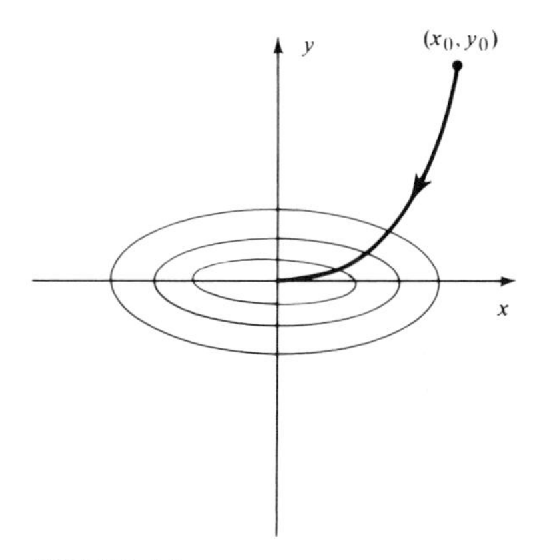

FIGURE 4.4

The concepts of directional derivative and gradient extend in a natural way to functions of n variables. Here we will be content to indicate the results for three independent variables.

Let w be a function of three independent variables x, y, and z, that is,

$$w = f(x, y, z). \tag{13}$$

The directional derivative is defined in terms of the values of w at two neighboring points (x, y, z) and $(x + \Delta x, y + \Delta y, z + \Delta z)$ as the limit

$$\frac{df}{ds} = \lim_{\Delta s \to 0} \frac{f(x + \Delta x, y + \Delta y, z + \Delta z) - f(x, y, z)}{\Delta s}, \tag{14}$$

where $\Delta s = [(\Delta x)^2 + (\Delta y)^2 + (\Delta z)^2]^{1/2}$ is the increment of length. In terms of differentials, this can be written

$$\frac{df}{ds} = \frac{f_x\, dx + f_y\, dy + f_z\, dz}{ds} = f_x \frac{dx}{ds} + f_y \frac{dy}{ds} + f_z \frac{dz}{ds}. \tag{15}$$

Furthermore, if

$$dx = l\, ds, \qquad dy = m\, ds, \qquad dz = n\, ds,$$

so that

$$l^2 + m^2 + n^2 = 1,$$

then

$$\hat{\mathbf{u}} = l\mathbf{i} + m\mathbf{j} + n\mathbf{k}$$

is the unit vector which specifies the direction under consideration. Equation (15) can now be expressed as

$$\frac{df}{ds} = lf_x + mf_y + nf_z \tag{16}$$

and even more succinctly as

$$\frac{df}{ds} = \hat{\mathbf{u}} \cdot \nabla f. \tag{17}$$

Here

$$\nabla f = f_x\mathbf{i} + f_y\mathbf{j} + f_z\mathbf{k} \tag{18}$$

is the three-dimensional gradient, and the directional derivative is taken in the direction of the unit vector $\hat{\mathbf{u}}$. The repetition of earlier arguments shows that $(df/ds)_{\max} = |\nabla f|$; that is, the direction for the maximum directional derivative is the direction of ∇f. Furthermore ∇f is normal to the level surfaces $f(x, y, z) = \text{constant}$ (see Sec. 6.3).

Example 5 Find the directional derivative of the function

$$f = (x - 1)^2 + 2(y + 1)^2 + 3(z - 2)^2 - 6$$

at the point (2, 0, 1) in the direction of the vector $\mathbf{i} - 2\mathbf{j} - 2\mathbf{k}$.

Here $\nabla f = 2(x - 1)\mathbf{i} + 4(y + 1)\mathbf{j} + 6(z - 2)\mathbf{k}$, and at (2, 0, 1), $\nabla f = 2\mathbf{i} + 4\mathbf{j} - 6\mathbf{k}$. A unit vector in the specified direction is

$$\hat{\mathbf{u}} = \frac{\mathbf{i} - 2\mathbf{j} - 2\mathbf{k}}{3},$$

and the required directional derivative is

$$\frac{df}{ds} = \hat{\mathbf{u}} \cdot \nabla f = \tfrac{1}{3}(2 - 8 + 12) = 2.$$

Note that $(df/ds)_{\max} = |\nabla f| = 2\sqrt{14}$.

Exercises 6.4

1. Find the gradient vectors ∇f for the following functions $f(x,y)$. Evaluate the gradients at $x = y = 1$ and verify that the gradients are orthogonal to the curves $f(x,y) = f(1,1)$.

(i) $x^2y + xy^2$ **(ii)** $\sin x \sin y$ **(iii)** $(x^2 + y^2 - 1)^{1/2}$
(iv) $x^{1/2} + y^{1/2}$ **(v)** $\log(x^3 + y^3)$ **(vi)** $e^{x^2 - y^2}$

2. Find the gradients ∇f for the following functions $f(x,y,z)$. Evaluate the gradients at $x = y = z = 1$ and verify that they are orthogonal to the surfaces $f(x,y,z) = f(1,1,1)$.

(i) $x^2 + y^2 + z^2$ **(ii)** $x^{-1} + y^{-1} + z^{-1}$ **(iii)** $yz + zx + xy$
(iv) $e^{xy^2z^3}$ **(v)** $\sin x \sin y \sin z$ **(vi)** $\cosh(x + y - z)$

3. Draw the level curves of the function $f(x, y) = e^{x^2+y^2}$ in the xy plane. Calculate the directional derivative at the point (2, 1) in the direction of $\mathbf{i} - \mathbf{j}$. What are the extreme values of the directional derivatives at this point?

4. What linear function $f(x, y)$ has at the point (1,0) the value 2 and the vector $2\mathbf{i} + \mathbf{j}$ as its gradient? The graph of $z = f(x,y)$ is a plane. Is it possible to draw on this plane a line of slope 3?

5. What is the geometrical significance for the surface $z = f(x,y)$ of the following statements?

(i) $\nabla f = \mathbf{0}$ at a given point (x_0, y_0)
(ii) $\nabla f = \mathbf{0}$ at all points (x,y)
(iii) $\nabla f = \mathbf{a}$ (constant vector) at all points (x,y)

6. Suppose that, at a certain point (x_0, y_0), the directional derivatives of $f(x,y)$ are known in the directions φ_1 and φ_2. Is this sufficient information to determine ∇f at this point? What is the corresponding result in three dimensions?

7. The directional derivative of a certain function $z = f(x,y)$ at the point $P_0 = (2,0)$ in the direction toward $P_1 = (2, -2)$ is 1 and it is -3 in the direction toward the origin. What are the values of dz/ds in the directions toward $P_2 = (2,1)$ and toward $P_3 = (3,2)$? Estimate $f(2.0, 0.1)$ and $f(2.1, 0.2)$ if $f(2,0) = 1$.

8. If $\nabla f = \mathbf{i} + \mathbf{j} + \mathbf{k}$, find the directional derivative of $f(x,y,z)$ in the following directions.

(i) $3\mathbf{i} + 2\mathbf{j} + \mathbf{k}$ **(ii)** $\mathbf{i} + \mathbf{j} - \mathbf{k}$ **(iii)** $\mathbf{i} + \mathbf{j} - 2\mathbf{k}$

9. If $w = xyz + \log xyz$, find the directional derivative dw/ds at the point (1,2,1) in the direction $\mathbf{i} + \mathbf{j}$. In what directions is the directional derivative at (1,2,1) equal to zero?

10. Given $f(x,y,z) = x^2 + 2y^2 + 3z^2 + yz - zx - 2xy$, find ∇f at the point $(1, -1, -2)$. Determine the extreme values of the directional derivatives at this point.

11. Find the rates of change of the function $f(x,y,z) = x^3 + y^3 + z^3$ at the point (1,2,3) in the radial direction and in the three axial directions. Determine the direction of the maximum rate of change.

12. Find the rate of change of the function $f(x,y,z) = x^2 + y^2 + z^2$ at the point (3,5,4) along the curve of intersection of the surfaces $2x^2 - y^2 + 2z^2 = 25$ and $x^2 - y^2 + z^2 = 0$.

13. At a common point of two surfaces $f(x, y, z) = 0$ and $g(x, y, z) = 0$ show that the angle ν between their normals is determined by

$$\cos \nu = \frac{\nabla f \cdot \nabla g}{|\nabla f|\,|\nabla g|}$$

and that their curve of intersection is in the direction $\nabla f \times \nabla g$.

14. Find the rate of change of the function $f(x, y, z) = 3x^2 + y^2 + z^2$ at the point $(-2, 1, 3)$ along the curve of intersection of the surfaces $x^2 + xz + y^2 = -1$ and $x^2y = 5z - 11$.

15. A hiker is climbing a mountain whose height is $z = 1000 - 2x^2 - 3y^2$. When he is at the point (1,1995) in what direction should he move in order to ascend as rapidly as possible? If he continues to move on a path of steepest ascent, show that the projection of this path on the xy plane is $y = x^{3/2}$.

16. Find the direction of steepest descent in the xy plane for the function $f(x, y) = 1000 - 2x^2 - y^2$. Show that the curve $x = t^2$, $y = t$ (as t increases) follows the direction of steepest descent.

17. The temperature distribution $T(x, y, z)$ at any point (x, y, z) is given by $T(x, y, z) = (x - 10)^2 + 2y^2 + 3(z - 2)^2$. Determine the rate of change of the temperature at the point $(9, 0, 3)$ in the direction of $(11, 2, 4)$. In which direction is the rate of change a maximum?

18. The temperature at a point in space is $T = 100 - x^2 - y^2 - 2z^2$. In what direction should one move from the point $(2, 1, 1)$ in order to cool off as rapidly as possible?

19. If $f = f(r, \theta)$, show that $\nabla f = \dfrac{\partial f}{\partial r}\hat{\mathbf{r}} + \dfrac{1}{r}\dfrac{\partial f}{\partial \theta}\hat{\boldsymbol{\theta}}$ where $\hat{\mathbf{r}}$ and $\hat{\boldsymbol{\theta}}$ are unit vectors in the radial and circumferential directions.

20. For the following functions $f(r, \theta)$, determine the gradients ∇f.

(i) $f(r, \theta) = re^{a\theta}$ **(ii)** $f(r, \theta) = r \sin 3\theta$ **(iii)** $f(r, \theta) = r/(1 + e \cos \theta)$

21. If $\mathbf{r} = x\mathbf{i} + y\mathbf{j}$ and $f(r) = \log r$, show that $\nabla f = \mathbf{r}/r^2$. If $\mathbf{r} = x\mathbf{i} + y\mathbf{j} + z\mathbf{k}$ and $f(r) = r^{-n}$, show that $\nabla f = -n\mathbf{r}/r^{n+2}$.

22. Verify that the second directional derivative of a function $f(x, y)$ is given by the formula $d^2f/ds^2 = f_{xx} \cos^2\varphi + 2f_{xy} \cos\varphi \sin\varphi + f_{yy} \sin^2\varphi$ where φ is the direction angle. Evaluate d^2f/ds^2 for the following functions.

(i) $x^2y + xy^2$ **(ii)** $\cosh(x + y)$ **(iii)** $e^x \cos y$

23. Large-scale atmospheric motions are well approximated by balancing the Coriolis force and the pressure gradient (this is called the geostropic balance), namely, $2\boldsymbol{\Omega} \times \mathbf{v} = -(\nabla p)/\rho$. Here $\boldsymbol{\Omega}$ is the angular velocity of the earth, $\mathbf{v}$ is the velocity of the air, ρ the density, and p the pressure. Deduce that $\mathbf{v} \cdot \nabla p = 0$ and that the wind direction is parallel to isobars. Show that in the northern hemisphere the geostrophic balance predicts a clockwise or counter-clockwise flow about a region of high or low pressure.

6.5 The chain rule

1 One of the basic rules of differentiation is the chain rule. It states that if $y = f(x)$ and $x = x(t)$, in which case $y = f(x(t))$, then

$$\frac{dy}{dt} = \frac{dy}{dx}\frac{dx}{dt}. \tag{1}$$

We will now derive a similar formula for functions of several variables.

If $z = f(x, y)$ and $x = x(r, s)$, $y = y(r, s)$, then $z = f(x(r, s), y(r, s))$ is actually a function of r and s. The problem is now to calculate the partial derivatives of f with respect to r and s; in doing this the partial derivatives f_x and f_y are basic. The resultant formulas, known as the *chain rule* for partial differentiation, are

$$\frac{\partial f}{\partial r} = \frac{\partial f}{\partial x}\frac{\partial x}{\partial r} + \frac{\partial f}{\partial y}\frac{\partial y}{\partial r}, \tag{2}$$

and

$$\frac{\partial f}{\partial s} = \frac{\partial f}{\partial x}\frac{\partial x}{\partial s} + \frac{\partial f}{\partial y}\frac{\partial y}{\partial s}, \tag{3}$$

or, in the alternative subscript notation,

$$f_r = f_x x_r + f_y y_r, \tag{4}$$

and

$$f_s = f_x x_s + f_y y_s. \tag{5}$$

To obtain these formulas, we could proceed via the mean-value theorem, but a simple and direct derivation based on differentials is more to the point.

From $z = f(x, y)$, $x = x(r, s)$, and $y = y(r, s)$, it follows that

$$dz = df = f_x\,dx + f_y\,dy, \tag{6}$$

and

$$dx = x_r\,dr + x_s\,ds, \tag{7}$$

$$dy = y_r\,dr + y_s\,ds. \tag{8}$$

The replacement of dx and dy in (6) yields

$$df = (f_x x_r + f_y y_r)\,dr + (f_x x_s + f_y y_s)\,ds. \tag{9}$$

However, f is a function of r and s, which means that

$$df = f_r\,dr + f_s\,ds. \tag{10}$$

Upon equating the coefficients of dr and ds in the last two expressions, we obtain the chain-rule formulas (4) and (5). The only essential difference between this demonstration and a proof of the chain rule (based on the mean-value theorem) is that a host of small ε's, which inevitably tend to zero anyway, is not included. Despite the breach of rigor, differentials will always be used because they provide an easy and correct method of finding relationships among partial derivatives.

Example 1 If $f(x, y) = \log x + \log y$ and $x = re^s$, $y = re^{-s}$, verify the chain-rule formulas.

Differentiation yields

$$f_x = \frac{1}{x}, \qquad f_y = \frac{1}{y},$$

$$x_r = e^s, \qquad x_s = re^s,$$

$$y_r = e^{-s}, \qquad y_s = -re^{-s},$$

and according to the chain rule,

$$\begin{aligned} f_r &= f_x x_r + f_y y_r, & f_s &= f_x x_s + f_y y_s, \\ &= \frac{1}{x} e^s + \frac{1}{y} e^{-s}, & &= \frac{1}{x} re^s + \frac{1}{y}(-re^{-s}), \\ &= \frac{1}{re^s} e^s + \frac{1}{re^{-s}} e^{-s}, & &= \frac{1}{re^s} re^s + \frac{1}{re^{-s}}(-re^{-s}), \\ &= \frac{2}{r}. & &= 0. \end{aligned}$$

In this particular case, f can easily be found in terms of r and s, namely,

$$\begin{aligned} f &= \log x + \log y, \\ &= \log xy, \\ &= \log r^2, \\ &= 2 \log r. \end{aligned}$$

Therefore $f_r = 2/r$ and $f_s = 0$, in agreement with the results found by using the chain rule.

A special case of the chain rule occurs when $z = f(x, y)$ and $x = x(t)$, $y = y(t)$, so that $z = f(x(t), y(t))$ is a function of a single variable t. To use the chain-rule formulas we identify $r = t$ and set $x_s = y_s = 0$ so that (4) becomes

$$f_t = f_x x_t + f_y y_t. \tag{11}$$

Since, x, y, and f are functions of t alone, we can write these derivatives as total (as opposed to partial) derivatives, for example, $x_t = dx/dt$. Equation (11) then takes the form

$$\frac{df}{dt} = \frac{\partial f}{\partial x}\frac{dx}{dt} + \frac{\partial f}{\partial y}\frac{dy}{dt}. \tag{12}$$

Of course this formula reduces to the chain rule for the one-variable case [Eq. (1)].

2

The chain rule can be extended to cover a more general situation. If z is a function of n independent variables $x_1, x_2, \ldots, x_n$, that is,

$$z = f(x_1, x_2, \ldots, x_n), \tag{13}$$

and each x_j, $j = 1, \ldots, n$, is a function of m other independent variables r_k, $k = 1, \ldots, m$, that is,

$$x_j = x_j(r_1, r_2, \ldots, r_m), \qquad j = 1, 2, \ldots, n, \tag{14}$$

then f can be considered as a function of the m independent variables $r_1, r_2, \ldots, r_m$. The generalized chain rule becomes

$$\frac{\partial f}{\partial r_k} = \sum_{j=1}^{n} \frac{\partial f}{\partial x_j}\frac{\partial x_j}{\partial r_k}, \qquad k = 1, 2, \ldots, m. \tag{15}$$

Once again differentials provide an immediate and simple proof. The differentials

$$dz = df = \sum_{j=1}^{n} \frac{\partial f}{\partial x_j}\, dx_j \qquad \text{and} \qquad dx_j = \sum_{k=1}^{m} \frac{\partial x_j}{\partial r_k}\, dr_k$$

can be combined to yield

$$df = \sum_{k=1}^{m} \left(\sum_{j=1}^{n} \frac{\partial f}{\partial x_j}\frac{\partial x_j}{\partial r_k} \right) dr_k .$$

But if f is regarded as a function of the variables r_k, then

$$df = \sum_{k=1}^{m} \frac{\partial f}{\partial r_k}\, dr_k .$$

The coefficients of dr_k, $k = 1, 2, \ldots, m$, in the last two formulas can be equated, and this leads to Eq. (15).

3

Example 2 Find an expression for $|\nabla f|^2$ in plane polar coordinates.

The problem is to transform the expression $f_x^2 + f_y^2$ into one which involves derivatives of f with respect to r and θ, where $x = r \cos \theta$, $y = r \sin \theta$. By the chain rule

$$f_x = f_r r_x + f_\theta \theta_x , \tag{16}$$

and

$$f_y = f_r r_y + f_\theta \theta_y . \tag{17}$$

Calculation of the partial derivatives of $r = (x^2 + y^2)^{1/2}$ and $\theta = \tan^{-1}(y/x)$, gives

$$r_x = \frac{x}{(x^2 + y^2)^{1/2}} = \cos\theta, \qquad r_y = \frac{y}{(x^2 + y^2)^{1/2}} = \sin\theta,$$

$$\theta_x = -\frac{y}{x^2 + y^2} = -\frac{\sin\theta}{r}, \qquad \theta_y = \frac{x}{x^2 + y^2} = \frac{\cos\theta}{r},$$

and substitution of these results into (16) and (17) yields

$$f_x = \cos\theta\, f_r - \frac{\sin\theta}{r} f_\theta,$$

$$f_y = \sin\theta\, f_r + \frac{\cos\theta}{r} f_\theta.$$

If the squares of these expressions are added to form $|\nabla f|^2 = f_x^2 + f_y^2$, we find that

$$|\nabla f|^2 = f_r^2 + \frac{1}{r^2} f_\theta^2.$$

Example 3 If $f(x, y) = x^2 - y^2$, $x = \cos t$, and $y = \sin t$, find df/dt.

Since $f_x = 2x$, $f_y = -2y$, it follows from (12) that

$$\begin{aligned}\frac{df}{dt} &= f_x \frac{dx}{dt} + f_y \frac{dy}{dt},\\ &= 2x(-\sin t) + (-2y)\cos t,\\ &= -4\sin t\cos t,\\ &= -2\sin 2t.\end{aligned}$$

Of course, this result can be calculated directly by finding f explicitly in terms of t, that is, $f(x(t), y(t)) = \cos^2 t - \sin^2 t = \cos 2t$.

---

4 The chain rule can be used to find second (and higher) derivatives. Consider the two-variable case with $z = f(x, y)$, $x = x(r, s)$, and $y = y(r, s)$ and let us find f_{rr} in terms of the derivatives of f with respect to x and y *and* the derivatives of x and y with respect to r and s. We begin by differentiating (4) with respect to r, using the usual product rule. This gives

$$f_{rr} = \frac{\partial}{\partial r}(f_x x_r + f_y y_r) = f_x x_{rr} + x_r \frac{\partial}{\partial r}(f_x) + f_y y_{rr} + y_r \frac{\partial}{\partial r}(f_y). \tag{18}$$

The second and fourth terms must be computed using the chain rule, and this is accomplished by applying Eqs. (4) and (5) to the functions f_x and f_y. The results are

$$\frac{\partial}{\partial r}(f_x) = (f_x)_x x_r + (f_x)_y y_r,$$

$$= f_{xx} x_r + f_{xy} y\,; \tag{19}$$

$$\frac{\partial}{\partial r}(f_y) = f_{xy} x_r + f_{yy} y_r. \tag{20}$$

The substitution of these expressions into (18) yields

$$f_{rr} = f_{xx} x_r^{\,2} + 2f_{xy} x_r y_r + f_{yy} y_r^{\,2} + f_x x_{rr} + f_y y_{rr}. \tag{21}$$

In the same way, it can be shown that

$$f_{ss} = f_{xx} x_s^{\,2} + 2f_{xy} x_s y_s + f_{yy} y_s^{\,2} + f_x x_{ss} + f_y y_{ss}, \tag{22}$$

and

$$f_{rs} = f_{xx} x_r x_s + f_{xy}(x_r y_s + x_s y_r) + f_{yy} y_r y_s + f_x x_{rs} + f_y y_{rs}. \tag{23}$$

Higher derivatives can be computed in a straightforward way although the algebra is messy. It is important to understand the basic procedure involved here: few people can memorize the results (or think it worthwhile).

Example 4 If $z = f(x, y)$ and $x = e^{r+s} + e^{r-s}$, $y = e^{r+s} - e^{r-s}$, calculate f_{rr} and f_{ss}.

This problem is a direct application of formulas (21) and (22). For this purpose it is necessary to determine the r and s derivatives of x and y, which are easily obtained:

$$x_r = e^{r+s} + e^{r-s} = x, \qquad y_r = e^{r+s} - e^{r-s} = y,$$
$$x_s = e^{r+s} - e^{r-s} = y, \qquad y_s = e^{r+s} + e^{r-s} = x.$$

It follows that

$$x_{rr} = x_r = x, \qquad y_{rr} = y_r = y,$$
$$x_{ss} = y_s = x, \qquad y_{ss} = x_s = y;$$

and on using these results, (21) and (22) become

$$f_{rr} = x^2 f_{xx} + 2xyf_{xy} + y^2 f_{yy} + xf_x + yf_y,$$
$$f_{ss} = y^2 f_{xx} + 2xyf_{xy} + x^2 f_{yy} + xf_x + yf_y.$$

---

5 Equations involving the partial derivatives of a function are called *partial differential equations.* Such equations govern a large variety of phenomena in the physical world, and a study of partial differential equations is an important topic of current applied mathematics research. At this stage, we can only indicate a few rather elementary ideas, but this topic will consume more of our attention as we proceed.

In studying the integration of an ordinary differential equation it was found that a first-order equation has an arbitrary constant in the solution. The solution of the comparable first-order partial differential equation, i.e., one containing only first-order partial derivatives, will have an arbitrary function. For example, in ordinary differential equations, the general solution of

$$\frac{du}{dx} = 0$$

is

$$u(x) = C,$$

where C is an arbitrary constant.

The corresponding solution $u(x, y)$ of the partial differential equation,

$$\frac{\partial u}{\partial x} = 0, \tag{24}$$

is

$$u = C(y). \tag{25}$$

Here C is an *arbitrary function* of the other variable y.

Example 5 Show that the function $u = f(x - ct)$ satisfies the partial differential equation $\dfrac{\partial u}{\partial t} + c\dfrac{\partial u}{\partial x} = 0$, where c is a constant. Conversely, show that the general solution of $\dfrac{\partial u}{\partial t} + c\dfrac{\partial u}{\partial x} = 0$ is $u = f(x - ct)$, where f is an arbitrary function.

Let $v = x - ct$, in which case $u = f(v)$. It follows that

$$\frac{\partial u}{\partial t} = \frac{du}{dv}\frac{\partial v}{\partial t} = f'(v)(-c) = -cf'(v)$$

and

$$\frac{\partial u}{\partial x} = \frac{du}{dv}\frac{\partial v}{\partial x} = f'(v).$$

The elimination of $f'(v)$ gives

$$\frac{\partial u}{\partial t} + c\frac{\partial u}{\partial x} = 0. \tag{26}$$

Conversely, we can show that $f(x - ct)$ is the solution of the preceding equation by changing from x and t to the new variables

$$\xi = x - ct, \qquad \eta = t.$$

In terms of this transformation, we find that

$$\frac{\partial u}{\partial t} = \frac{\partial u}{\partial \xi}\frac{\partial \xi}{\partial t} + \frac{\partial u}{\partial \eta}\frac{\partial \eta}{\partial t} = -c\frac{\partial u}{\partial \xi} + \frac{\partial u}{\partial \eta}$$

and

$$\frac{\partial u}{\partial x} = \frac{\partial u}{\partial \xi}\frac{\partial \xi}{\partial x} + \frac{\partial u}{\partial \eta}\frac{\partial \eta}{\partial x} = \frac{\partial u}{\partial \xi}.$$

Therefore, (26) transforms to

$$\frac{\partial u}{\partial \eta} = 0,$$

which has the solution

$$u = f(\xi)$$

or

$$u = f(x - ct),$$

where f is an arbitrary function.

This solution has an especially important physical interpretation. With x a distance and t a time, c has the dimensions of a velocity, and $f(x - ct)$ describes a *wave* moving with speed c. This may be seen by examining the function at different times, say at $t = 0$ when $u = f(x)$, and $t = t_0$ when $u = f(x - x_0)$, and $x_0 = ct_0$. Figure 5.1 shows that the function *translates* a distance $x_0 = ct_0$ in this time span. In other words, an observer moving with the wave speed will see no change with time.

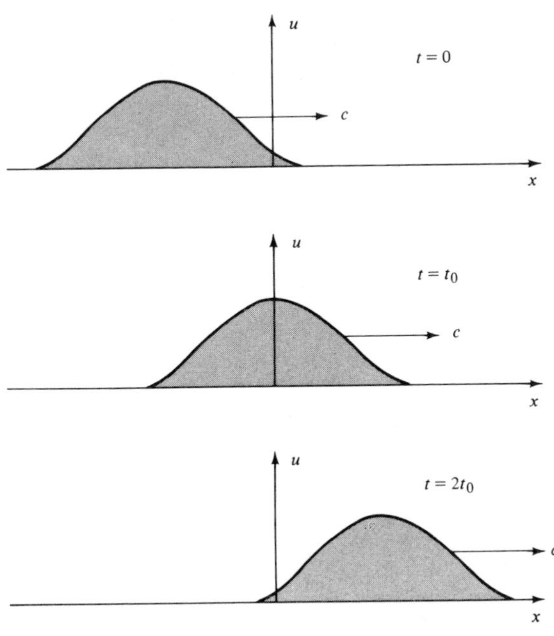

FIGURE 5.1

Example 6 If $u = f(x^2 - 3y^2)$, show that $3y\dfrac{\partial u}{\partial x} + x\dfrac{\partial u}{\partial y} = 0$.

With $v = x^2 - 3y^2$ and $u = f(v)$,

$$\frac{\partial u}{\partial x} = \frac{du}{dv}\frac{\partial v}{\partial x} = 2xf'(v),$$

and

$$\frac{\partial u}{\partial y} = \frac{du}{dv}\frac{\partial v}{\partial y} = -6yf'(v).$$

The elimination of $f'(v)$ between these two equations gives the desired result.

Exercises 6.5

1. In the following problems, calculate df/dt by the chain rule and by solving for f as an explicit function of t.

(i) $f(x, y) = x^2 + y^2$, $x = t - \dfrac{1}{t}$, $y = t + \dfrac{1}{t}$ **(ii)** $f = \dfrac{x^2y^2}{x^2 + y^2}$, $x = t$, $y = t^2$

(iii) $f = e^{x^2 - y^2}$, $x = \log t$, $y = t^2$ **(iv)** $f = e^{x^2 + y^2}$, $x = \cos t$, $y = \sin t$

(v) $f = \sin(2x + 3y)$, $x = e^t$, $y = e^{-t}$ **(vi)** $f = x^2 + y^2 + z^2$, $x = t\cos t$, $y = t\sin t$, $z = t$

2. For the function $f(x, y) = x^3y^4$, define a change of variables $x = r + s$, $y = r - s$. Calculate $\partial f/\partial r$ and $\partial f/\partial s$ by the chain rule and by solving for f as an explicit function of r and s.

3. For $z(x, y) = x^2 + 2y^2$, consider the change of variables $x = 3r + s$, $y = r - s$. Find $\partial z/\partial r$ and $\partial z/\partial s$ by the chain rule and by solving for z as an explicit function of r and s. Show that the second derivatives $\partial^2 z/\partial r^2$, $\partial^2 z/\partial r\partial s$ and $\partial^2 z/\partial s^2$ are constants.

4. An object moves so that at time t its position vector is $\mathbf{r} = (1 + t^2)\mathbf{i} + (1 - t)\mathbf{j} + (t + t^3)\mathbf{k}$. How fast is its distance from the origin changing when $t = 1$? Is this the speed of the object?

5. A particle of mass m moves on the surface $z = f(x, y)$. If $x(t)$ and $y(t)$ are the horizontal coordinates of the particle at time t, determine the velocity vector $\mathbf{v}(t)$. Show that the kinetic energy is

$$\frac{1}{2}m|\mathbf{v}|^2 = \frac{1}{2}m\left[(1 + f_x^2)\left(\frac{dx}{dt}\right)^2 + 2f_xf_y\frac{dx}{dt}\frac{dy}{dt} + (1 + f_y^2)\left(\frac{dy}{dt}\right)^2\right].$$

6. In the previous problem, determine the acceleration vector $\mathbf{a}(t)$. Illustrate both problems with the example $z = x^2 + y^2$, $x(t) = t\cos t$, $y(t) = t\sin t$.

7. Determine the inverses of the following changes of variables. (If $x = x(r, s)$, $y = y(r, s)$, solve for $r = r(x, y)$, $s = s(x, y)$.)

(i) $x = r + s$, $y = r - s$ **(ii)** $x = 2u + 3v$, $y = u + 2v$

(iii) $x = r\cos\theta$, $y = r\sin\theta$ **(iv)** $x = u\cosh v$, $y = u\sinh v$

(v) $x = \dfrac{2u}{u^2 + v^2}$, $y = \dfrac{-2v}{u^2 + v^2}$ **(vi)** $x = \dfrac{1}{2}(u^2 - v^2)$, $y = uv$

8. If $V(x, y, z) = f(x/z, y/z)$, show that $xV_x + yV_y + zV_z = 0$. Verify this identity in the special case $V(x, y, z) = \log[(x + y)/z]$.

9. If $V(x,y,z) = f(y - z,\ z - x,\ x - y)$, show that $V_x + V_y + V_z = 0$. Verify this identity in the special case $V(x,y,z) = (y - z)^2 + (z - x)^2 + (x - y)^2$.

10. Evaluate $f'(t)$ and $f''(t)$ by two methods in the following examples:

(i) $f(x,y) = e^x \cos y,\ x = t^2,\ y = 1 - t^3$ **(ii)** $f = e^{x^2+y^2+z^2},\ x = \cos t,\ y = \sin t,\ z = t$.

11. If $f = f(x,\ y)$ and $x = r - s,\ y = rs$, deduce the following results:

(i) $(r + s)\dfrac{\partial f}{\partial x} = r\dfrac{\partial f}{\partial r} - s\dfrac{\partial f}{\partial s}$ **(ii)** $(r + s)\dfrac{\partial f}{\partial y} = \dfrac{\partial f}{\partial r} + \dfrac{\partial f}{\partial s}$.

12. If $f = f(x,\ y)$ and $x = \frac{1}{2}\log(r^2 + s^2),\ y = \tan^{-1}(s/r)$, deduce the following results:

(i) $f_x{}^2 + f_y{}^2 = (r^2 + s^2)(f_r{}^2 + f_s{}^2)$ **(ii)** $f_{xx} + f_{yy} = (r^2 + s^2)(f_{rr} + f_{ss})$

13. If $f = f(x,\ y)$ and $x = p^2 - q^2 - 2pq,\ y = q$, show that $(p + q)\dfrac{\partial f}{\partial p} + (p - q)\dfrac{\partial f}{\partial q} = 0$ is equivalent to $\partial f/\partial y = 0$. Explain why $\partial f/\partial y$ is not the same as $\partial f/\partial q$.

14. If $f = f(x,y)$ and $x = e^r + e^{-s},\ y = e^s + e^{-r}$, deduce that

$$f_{rr} - 2f_{rs} + f_{ss} = x^2 f_{xx} - 2xyf_{xy} + y^2 f_{yy} + xf_x + yf_y.$$

15. If $u = u(x,\ t)$ and $\xi = x - ct,\ \eta = x + ct$ (c is a constant), show that $u_{xx} - (c^{-2})u_{tt} = 4u_{\xi\eta}$ and therefore the wave equation $u_{xx} - (c^{-2})u_{tt} = 0$ transforms to $u_{\xi\eta} = 0$. Deduce that the general solution can be written $u(x,\ t) = F(x - ct) + G(x + ct)$, F and G being arbitrary functions. Interpret this solution in terms of travelling waves.

16. If $z = xF(x + y) + yG(x + y)$ where F and G are arbitrary differentiable functions, deduce that $z_{xx} - 2z_{xy} + z_{yy} = 0$.

17. Transform the expression $u_{xx} + u_{yy}$ into one involving derivatives with respect to the polar coordinates r and θ where $x = r\cos\theta,\ y = r\sin\theta$.

18. Consider the polar coordinates ($x = r\cos\theta,\ y = r\sin\theta$) of a particle as functions of time t. Using Newton's dot notation for the time derivatives, express $\dot{x} = dx/dt$ and $\dot{y}$ in terms of $r,\ \theta,\ \dot{r}$ and $\dot{\theta}$. Interpret these derivatives on a graph of the trajectory of the particle.

19. For the moving particle of the previous problem, verify the following identities.

(i) $\ddot{x}\cos\theta + \ddot{y}\sin\theta = \ddot{r} - r\dot{\theta}^2$ **(ii)** $\ddot{y}\cos\theta - \ddot{x}\sin\theta = \dfrac{1}{r}\dfrac{d}{dt}(r^2\dot{\theta})$

Interpret these results as the components of acceleration in the radial direction and in the circumferential direction.

20. If $z = z(x,\ y)$ and $x = e^u \cos v,\ y = e^u \sin v$, verify the following identities:

(i) $\dfrac{\partial^2 z}{\partial u^2} = x^2\dfrac{\partial^2 z}{\partial x^2} + 2xy\dfrac{\partial^2 z}{\partial x\partial y} + y^2\dfrac{\partial^2 z}{\partial y^2} + x\dfrac{\partial z}{\partial x} + y\dfrac{\partial z}{\partial y}$

(ii) $\dfrac{\partial^2 z}{\partial v^2} = y^2\dfrac{\partial^2 z}{\partial x^2} - 2xy\dfrac{\partial^2 z}{\partial x\partial y} + x^2\dfrac{\partial^2 z}{\partial y^2} - x\dfrac{\partial z}{\partial x} - y\dfrac{\partial z}{\partial y}$

(iii) $\dfrac{\partial^2 z}{\partial u\partial v} = (x^2 - y^2)\dfrac{\partial^2 z}{\partial x\partial y} + xy\left(\dfrac{\partial^2 z}{\partial y^2} - \dfrac{\partial^2 z}{\partial x^2}\right) - y\dfrac{\partial z}{\partial x} + x\dfrac{\partial z}{\partial y}$

(iv) $\dfrac{\partial^2 z}{\partial u^2} + \dfrac{\partial^2 z}{\partial v^2} = (x^2 + y^2)\left(\dfrac{\partial^2 z}{\partial x^2} + \dfrac{\partial^2 z}{\partial y^2}\right)$

21. A function $f(x,y)$ is said to be *homogenous of degree n* if $f(\lambda x,\ \lambda y) = \lambda^n f(x,y)$ for *all* positive values of λ. Which of the following are homogeneous? Determine the degrees.

(i) $x^3 + 3x^2y + y^3$ **(ii)** $x + y + 1$ **(iii)** $(x^3 + y^3)/(x^2 + y^2)^{1/2}$

(iv) $x^{1/3} + y^{1/3}$ **(v)** $\log(x^2 + y^2) - 2\log x$ **(vi)** $x + y + \tan^{-1}(x + y)$

22. If $f(x,y)$ is homogeneous of degree n, show that f_x and f_y are homogeneous of degree $n - 1$ and verify that $xf_x + yf_y = nf$. Verify these results for the homogeneous functions in the previous problem.

23. If $f(x_1, x_2, \ldots, x_m)$ is homogeneous of degree n, show that

$$\left(\sum_{j=1}^{m} x_j \frac{\partial}{\partial x_j}\right)^r f = n(n-1)\cdots(n-r+1)f.$$

24. Defining a change of variables $\xi = x/y$ and $\eta = y$, show that the equation $x\,\partial u/\partial x + y\,\partial u/\partial y = nu$ transforms to the equation $\eta\,\partial u/\partial \eta = nu$ (n is a constant). Deduce that the solution u must be of the form $u(x,y) = y^n F(x/y)$ where F is an arbitrary function.
25. If a, b and c are constants, show that the function $u(x,y)$, which satisfies the partial differential equation $au_x + bu_y = cu$, must be of the form $u = e^{cx/a}F(bx - ay)$, F being an arbitrary function. (*Hint:* Use $\xi = bx - ay$ and $\eta = x$ as new independent variables.)
26. Use the result of Exercise 25 to show that the solution $u(x,t)$ of the equation $u_t + cu_x = -\beta^2 u$, with the initial condition $u(x, 0) = \sin x$, is $u = e^{-\beta^2 t}\sin(x - ct)$.
27. If $\xi = x - ct$ and $\eta = x + ct$, c being a constant, show that the wave equation

$$\frac{\partial^2 u}{\partial x^2} - \frac{1}{c^2}\frac{\partial^2 u}{\partial t^2} = 0$$

transforms to $\dfrac{\partial^2 u}{\partial \xi\,\partial \eta} = 0$. Deduce that the general solution of $\dfrac{\partial^2 u}{\partial x^2} - \dfrac{1}{c^2}\dfrac{\partial^2 u}{\partial t^2} = 0$ is $u(x,t) = F(x - ct) + G(x + ct)$, F and G being arbitrary functions. Interpret this solution in terms of travelling waves.
28. Use the result of Exercise 27 to show that the solution of $u_{xx} - (c^{-2})u_{tt} = 0$ which satisfies the initial conditions $u(x, 0) = \varphi(x)$ and $u_t(x, 0) = \psi(x)$ is

$$u = \frac{1}{2}[\varphi(x - ct) + \varphi(x + ct)] + \frac{1}{2c}\int_{x-ct}^{x+ct} \psi(\xi)\, d\xi.$$

Find the form of u in the following special cases:
(i) $\varphi(x) = e^{-x^2}$, $\psi(x) = 0$ **(ii)** $\varphi(x) = 0$, $\psi(x) = \cos x$.

6.6 *Functions with parameters*

Functions involving one or more parameters occur often enough to merit special attention. We previously considered the equation

$$y = f(x, \alpha), \tag{1}$$

where α is a parameter, and saw that this relationship provides one curve for every value of α. The totality of such curves is said to form a *family of curves.*

Example 1 Discuss the family of curves $x \cos \alpha + y \sin \alpha = 1$.

The parametric function describes a family of straight lines, each a unit distance from the origin. Figure 6.1 shows several of the lines corresponding to the values of α indicated. As more and more lines are added to the graph, an inscribed circle appears, which has the property of being tangent to every line of the family.

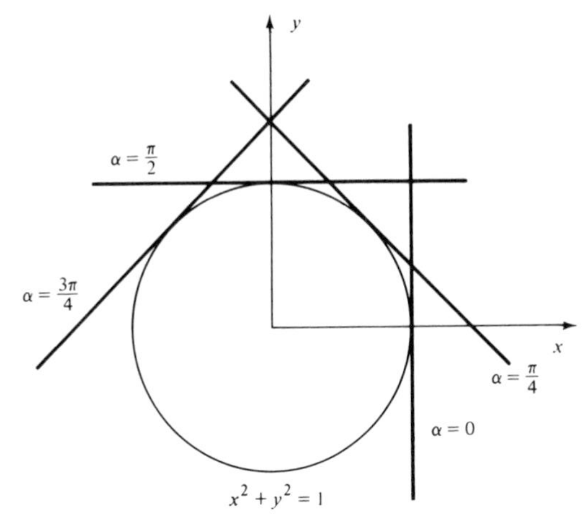

FIGURE 6.1

A curve is called an *envelope* if it simultaneously touches and is tangent to every member of a family of curves. The circle $x^2 + y^2 = 1$ is the envelope of the straight lines described by the above equation. In general, any curve is the envelope of its tangent lines.

Conditions for the existence of an *envelope* can be determined in a more general framework. To this end, the family of curves given by (1) is also described in the implicit form

$$F(x, y, \alpha) = 0. \tag{2}$$

The specification of the parameter α, say $\alpha = 1/2$, determines a single curve of the family. If (x, y) and $(x + dx, y + dy)$ are two points on a particular curve of the family, then

$$dF = 0 = F_x\, dx + F_y\, dy, \tag{3}$$

since (2) holds everywhere and α is constant. In other words, the formula

$$\frac{dy}{dx} = -\frac{F_x}{F_y}$$

gives the slope of the curve.

What about the envelope of this family? If one exists, there is a point on every curve of the family $F(x, y, \alpha) = 0$ at which the envelope will touch and be tangent to it. For the curves in Fig. 6.2 specified by α_0, α_1, α_2, that is, the curves given by $F(x, y, \alpha_0) = 0$, etc., this occurs at points P_0, P_1, and P_2, respectively. Clearly at *each* of these contact points the envelope must itself satisfy the same relationship that prescribes the individual curve, $F(x, y, \alpha_0) = 0$ at P_0; $F(x, y, \alpha_1) = 0$ at P_1; etc. In general, then, at the point of contact with an arbitrary curve of the family, the envelope also satisfies

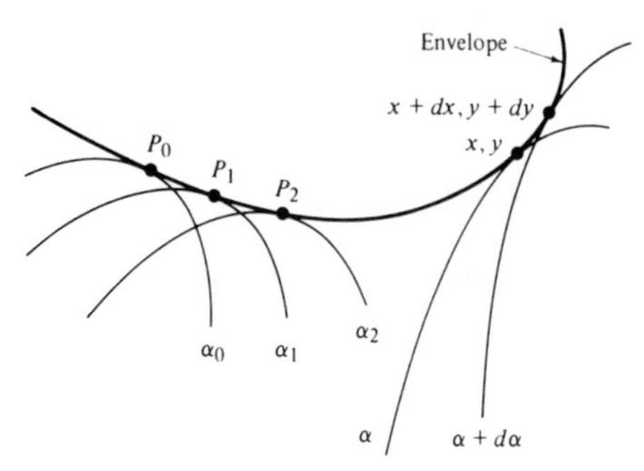

FIGURE 6.2

the equation of the curve $F(x, y, \alpha) = 0$. For another curve of the family there will be a different point of contact corresponding to a different value of α; this observation is important because it implies that the value of α changes from point to point *on the envelope*. As far as the envelope is concerned, α *must* be regarded as a function of position along this distinguished curve. Therefore, the equation of the envelope is

$$F(x, y, \alpha(x, y)) = 0. \tag{4}$$

(Note once again that if α is fixed, we obtain not the envelope but a member of the family which touches the envelope at one point. Only by adjusting α point by point do we determine the equation of the envelope.)

Now suppose that (x, y) and $(x + dx, y + dy)$ are two neighboring points on the envelope, the former corresponding to the value α and the latter associated with $\alpha + d\alpha$ (see Fig. 6.2). The increments of change along the envelope are related by (4), which implies

$$dF = 0 = F_x\,dx + F_y\,dy + F_\alpha\,d\alpha. \tag{5}$$

But the envelope has the same slope at (x, y) as the curve of the family through this point, and this means that (3) must also hold. This equation combined with (5) implies that

$$F_\alpha(x, y, \alpha) = 0, \tag{6}$$

at every point on the entire envelope. Therefore, the equation of the envelope is obtained by eliminating α between the two equations

$$F(x, y, \alpha) = 0 \qquad \text{and} \qquad F_\alpha(x, y, \alpha) = 0. \tag{7}$$

If these results are applied to the family of straight lines in Example 1, we find that

$$F(x, y, \alpha) = x \cos \alpha + y \sin \alpha - 1 = 0$$

and

$$F_\alpha(x, y, \alpha) = -x \sin \alpha + y \cos \alpha = 0.$$

The elimination of α shows that the equation of the envelope is

$$x^2 + y^2 = 1,$$

as indicated earlier.

Example 2 Find the envelope of the family of straight lines which are of unit length between their x and y intercepts (Fig. 6.3).

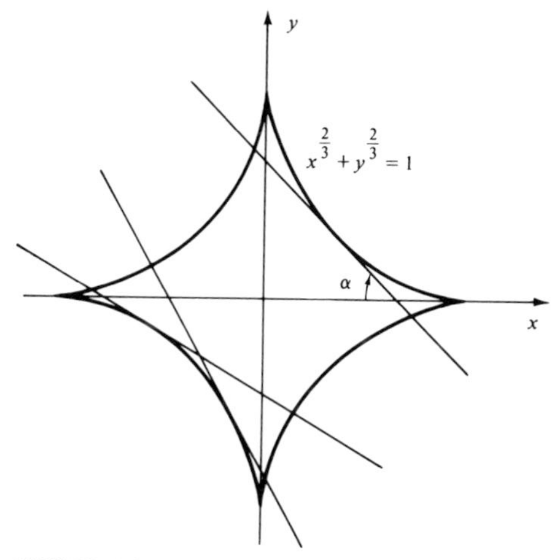

FIGURE 6.3

The line segment joining $(\cos\alpha, 0)$ and $(0, \sin\alpha)$ has unit length and the equation of the family is $x/(\cos\alpha) + y/(\sin\alpha) = 1$ or

$$F(x, y, \alpha) = x \sin\alpha + y \cos\alpha - \sin\alpha \cos\alpha = 0.$$

Moreover, $F_\alpha = 0$ implies that

$$x \cos\alpha - y \sin\alpha - (\cos^2\alpha - \sin^2\alpha) = 0.$$

Upon solving for x and y, we find that

$$x = \cos^3\alpha \qquad \text{and} \qquad y = \sin^3\alpha$$

are the parametric equations of the envelope. The result of eliminating α is a formula for the astroid

$$x^{2/3} + y^{2/3} = 1, \tag{8}$$

which is illustrated in Fig. 6.3.

---

2 Analogous situations occur for families of surfaces. The envelope of the one-parameter family of surfaces $F(x, y, z, \alpha) = 0$ is the surface which satisfies both

$$F(x, y, z, \alpha) = 0 \qquad \text{and} \qquad F_\alpha(x, y, z, \alpha) = 0. \tag{9}$$

The elimination of α will produce the equation of the surface envelope, if one exists.

Example 3 Find the envelope of the one-parameter family of surfaces

$$x^2 + y^2 + (z - \alpha)^2 - 1 = F(x, y, z, \alpha) = 0.$$

This is a family of spheres of unit radius centered about points on the z axis. Upon forming $F_\alpha(x, y, z, \alpha) = 0$, we obtain

$$(z - \alpha) = 0,$$

and the elimination of α between this and the defining relationship gives the equation of the envelope to be $x^2 + y^2 = 1$, a circular cylinder of unit radius whose axis is the z axis (Fig. 6.4). Each sphere of the family touches this surface; i.e., it just fits within the cylinder like a tennis ball in a can.

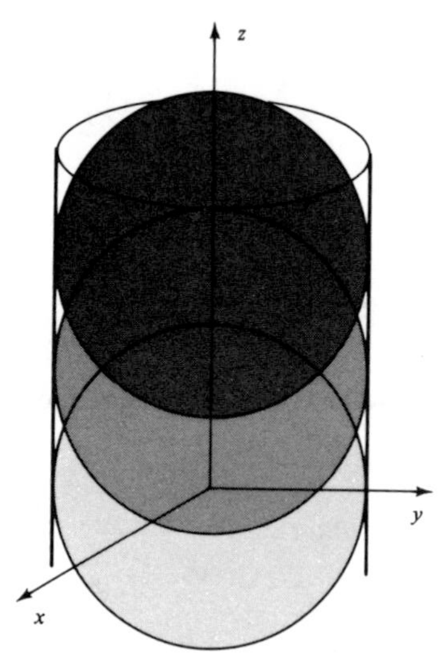

FIGURE 6.4

Example 4 Sound waves emanating from a point source propagate radially outward at speed c (assumed constant). If the source is held stationary in a wind of velocity equal to $-U\mathbf{i}$, find the family of sound waves and their envelope.

A concentrated sound pulse (a loud bang) emitted at time zero in a quiescent medium propagates through space as an ever expanding sphere whose radius at any time is ct. The location of the pulse for $t > 0$ is then

$$x^2 + y^2 + z^2 = c^2t^2.$$

If the pulse is created in a medium that flows with constant velocity U (in the negative x direction), it still expands spherically relative to the original source point which now drifts to the left in the flow. At time t, the initial source of the disturbance is located at $x = -Ut$, $y = 0$, $z = 0$, which means that the original sound pulse is the spherical surface about this point

$$(x + Ut)^2 + y^2 + z^2 = c^2t^2. \tag{10}$$

These convected sound waves constitute a family of spheres (Figs. 6.5 and 6.6). To discover whether the family has a real envelope (regarding t as the parameter), Eq. (10) must be differentiated with respect to t, which yields

$$U(x + Ut) = c^2t. \tag{11}$$

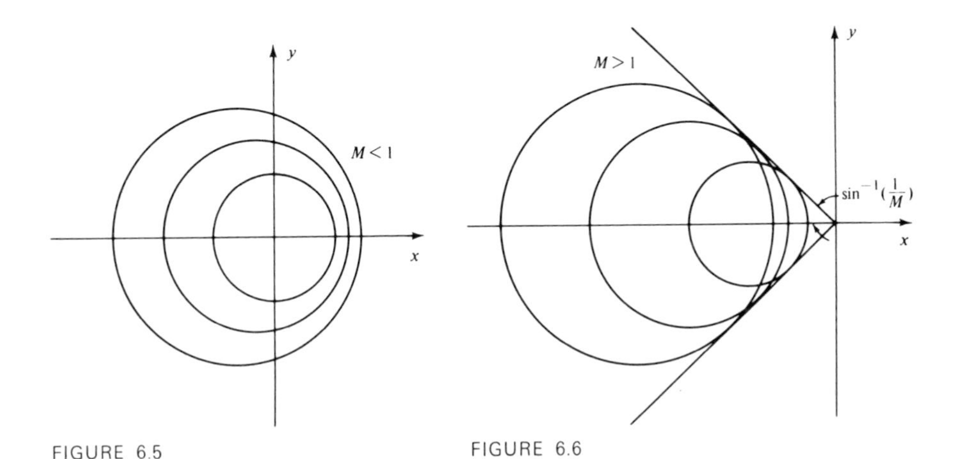

FIGURE 6.5

FIGURE 6.6

The elimination of t between (10) and (11) gives

$$y^2 + z^2 = \frac{1}{U^2/c^2 - 1}x^2 = \frac{x^2}{M^2 - 1}, \tag{12}$$

where $M = U/c$, the ratio of the speed of the stream to the speed of sound, is an important dimensionless parameter in aerodynamics called the *Mach number*. If $M < 1$, the flow is said to be *subsonic*, whereas for $M > 1$ it is *supersonic*.

Since the left-hand side of (12) is positive, so must the right be, and a real envelope can exist only for supersonic flow ($M > 1$), in which case the envelope is a cone (Fig. 6.6).

In the subsonic case, sound eventually reaches all regions in space, but in supersonic flow it is confined within a cone of semivertical angle $\sin^{-1}(1/M)$, called the *Mach cone*. Viewed in the moving coordinate frame of the noise source, this problem is a simple model for the sound field produced by a plane moving at sub- or supersonic speeds. Physically speaking, the envelope in the supersonic case corresponds to a weak shock wave.

Example 5 Find the normal velocity of a moving surface.

This geometrical problem arises in many practical situations, e.g., waves on the sea surface, weather fronts, expanding soap bubbles, and the melting of ice, to mention just a few.

Let the equation of the surface be $F(x, y, z, t) = 0$, where t is the time. Consider a point $P = (x, y, z)$ on the surface at time t, which an instant later at time $t + dt$ occupies a position $P' = (x + dx, y + dy, z + dz)$ on the displaced surface, as shown in Fig. 6.7.

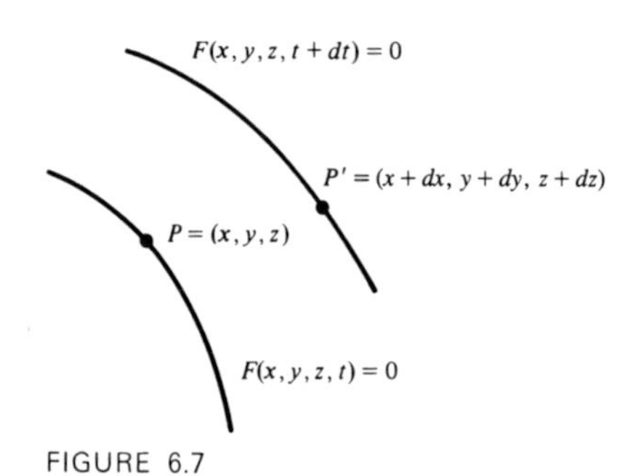

FIGURE 6.7

Since F remains identically zero,

$$dF = 0 = F_x\,dx + F_y\,dy + F_z\,dz + F_t\,dt,$$

which upon rearrangement becomes

$$F_x \frac{dx}{dt} + F_y \frac{dy}{dt} + F_z \frac{dz}{dt} + F_t = 0. \tag{13}$$

However, dx/dt, dy/dt, and dz/dt are the components of the actual velocity of the point P, that is,

$$\mathbf{v} = \frac{dx}{dt}\mathbf{i} + \frac{dy}{dt}\mathbf{j} + \frac{dz}{dt}\mathbf{k}. \tag{14}$$

Furthermore,

$$\nabla F = F_x\mathbf{i} + F_y\mathbf{j} + F_z\mathbf{k}$$

is the normal to the surface $F = 0$ (the unit normal is $\nabla F/|\nabla F|$), and therefore Eq. (13) can be written

$$F_t + \mathbf{v} \cdot \nabla F = 0, \tag{15}$$

or

$$\mathbf{v} \cdot \frac{\nabla F}{|\nabla F|} = -\frac{F_t}{|\nabla F|}.$$

In the latter form, the left-hand side is just the component in the direction of the normal of the actual velocity of a point on the surface. In other words, the normal velocity of the surface v_n is given by

$$v_n = -\frac{F_t}{|\nabla F|}.$$

Note that the point P in addition to moving with the surface is also free to move about on it, but the information given is not sufficient to determine the tangential velocity *on* the moving surface.

4

Definite integrals may contain one or more parameters, and, in fact, many of the functions of mathematical physics are defined in this manner. Consider the following prototype integral

$$F(\alpha) = \int_a^b f(x, \alpha)\, dx, \tag{16}$$

where a and b are the prescribed constant limits of integration and α a parameter. It is natural in the course of analysis that functions so defined must be differentiated, and we now show that

$$F'(\alpha) = \int_a^b \frac{\partial}{\partial \alpha} f(x, \alpha)\, dx. \tag{17}$$

The proof follows by forming the derivative in accordance with first principles:

$$\begin{aligned} \frac{F(\alpha + \Delta\alpha) - F(\alpha)}{\Delta\alpha} &= \int_a^b \frac{f(x, \alpha + \Delta\alpha) - f(x, \alpha)}{\Delta\alpha}\, dx, \\ &= \int_a^b \frac{\partial}{\partial\alpha} f(x, \alpha + \theta\, \Delta\alpha)\, dx, \qquad 0 < \theta < 1, \end{aligned}$$

where the mean-value theorem is employed in the last step. This implies that

$$\left| \frac{F(\alpha + \Delta\alpha) - F(\alpha)}{\Delta\alpha} - \int_a^b \frac{\partial}{\partial\alpha} f(x, \alpha)\, dx \right| = \left| \int_a^b \left[\frac{\partial}{\partial\alpha} f(x, \alpha + \theta\, \Delta\alpha) - \frac{\partial}{\partial\alpha} f(x, \alpha) \right] dx \right|,$$

and if $\partial f/\partial\alpha$ is assumed continuous, the stated result (17) follows upon taking the limit $\Delta\alpha \to 0$.

As an illustration of the practical value of this technique consider the known integral

$$\int_0^1 x^\alpha\, dx = \frac{1}{\alpha + 1}, \qquad \alpha > -1.$$

Differentiation with respect to α gives

$$\int_0^1 x^\alpha \log x\, dx = -\frac{1}{(\alpha + 1)^2},$$

and $n-1$ further α differentiations establish the more complicated formula

$$\int_0^1 x^\alpha (\log x)^n \, dx = \frac{(-1)^n n!}{(\alpha+1)^{n+1}}.$$

5

When improper integrals are involved, sufficient conditions that justify differentiation under the integral sign are much more rigid and more often than not also extremely difficult to check. In such situations, the applied mathematician takes an aggressive approach guided often by physical insight and mathematical intuition. In this spirit, two further examples are presented to illustrate the procedure.

Example 6 Evaluate $\int_0^\infty \frac{e^{-ax} - e^{-bx}}{x} \, dx, \qquad a, b > 0.$

Let $I(a, b) = \int_0^\infty \frac{e^{-ax} - e^{-bx}}{x} \, dx$, so that

$$\frac{\partial I}{\partial a} = -\int_0^\infty e^{-ax} \, dx = -\frac{1}{a}.$$

Upon integration, we obtain the formula

$$I(a, b) = -\log a + C(b), \tag{18}$$

where C is as yet an arbitrary function of b, the other parameter of the problem.
On the other hand,

$$\frac{\partial I}{\partial b} = \frac{1}{b}, \tag{19}$$

so that from (18) and (19)

$$\frac{\partial I}{\partial b} = C'(b) = \frac{1}{b}.$$

It follows that

$$C(b) = \log b + D, \tag{20}$$

where D is an arbitrary constant. Combining these results, we find that

$$I(a, b) = \log \frac{b}{a} + D.$$

The unknown constant is evaluated from the observation that $I(a, a) = 0$, a fact deduced by inspection. Therefore, $D = 0$, and we have succeeded in showing that the value of the original integral is

$$\int_0^\infty \frac{e^{-ax} - e^{-bx}}{x}\,dx = \log\frac{b}{a}, \qquad a, b > 0.$$

Example 7 Show that $\displaystyle\int_0^\infty \frac{\sin x}{x}\,dx = \frac{\pi}{2}$.

We begin with the function

$$I(\alpha) = \int_0^\infty e^{-\alpha x}\frac{\sin x}{x}\,dx, \qquad \alpha \geq 0.$$

It is intuitively clear, but more difficult to prove, that $I(\infty) = 0$. Obviously, $I(0)$ is the value of the integral in the problem statement. The derivative,

$$I'(\alpha) = -\int_0^\infty e^{-\alpha x}\sin x\,dx,$$

is a known integral (see integral 69 in Appendix Table 5.1) and has the value

$$I'(\alpha) = -\frac{1}{\alpha^2 + 1}.$$

Therefore

$$I(\alpha) = -\tan^{-1}\alpha + C,$$

where the constant C must be $\pi/2$ in order that $I(\infty) = 0$. It follows that

$$I(\alpha) = \int_0^\infty e^{-\alpha x}\frac{\sin x}{x}\,dx = \frac{\pi}{2} - \tan^{-1}\alpha,$$

and in particular

$$I(0) = \int_0^\infty \frac{\sin x}{x}\,dx = \frac{\pi}{2}.$$

---

6 Differentiation of an integral with respect to a parameter can be generalized by means of the chain rule. To illustrate this method we first find an expression for $\displaystyle\frac{d}{d\alpha}\int_{y(\alpha)}^{z(\alpha)} f(x, \alpha)\,dx$, where $y'(\alpha)$, $z'(\alpha)$ and $\displaystyle\frac{\partial}{\partial\alpha} f(x, \alpha)$ are assumed to be continuous functions of α.

Consider the function

$$I(z, y, \alpha) = \int_y^z f(x, \alpha)\,dx. \tag{21}$$

From the fundamental theorem of calculus

$$\frac{\partial I}{\partial z} = f(z, \alpha), \qquad \frac{\partial I}{\partial y} = -f(y, \alpha), \tag{22}$$

while differentiation with respect to the parameter α gives

$$\frac{\partial I}{\partial \alpha} = \int_y^z \frac{\partial}{\partial \alpha} f(x, \alpha)\, dx. \tag{23}$$

With $z = z(\alpha)$ and $y = y(\alpha)$ and $J(\alpha) = I(z(\alpha), y(\alpha), \alpha)$, application of the chain rule yields

$$\frac{dJ}{d\alpha} = \frac{\partial I}{\partial z}\frac{dz}{d\alpha} + \frac{\partial I}{\partial y}\frac{dy}{d\alpha} + \frac{\partial I}{\partial \alpha}.$$

If now (22) and (23) are used to replace the partial derivatives of I, we obtain

$$\frac{d}{d\alpha}\int_{y(\alpha)}^{z(\alpha)} f(x, \alpha)\, dx = f(z(\alpha), \alpha)\frac{dz}{d\alpha} - f(y(\alpha),\alpha)\frac{dy}{d\alpha} + \int_{y(\alpha)}^{z(\alpha)} \frac{\partial}{\partial \alpha} f(x, \alpha)\, dx. \tag{24}$$

In applying any technique of this type, a certain amount of imagination and experience is helpful.

Example 8 Show that $\dfrac{d}{d\alpha}\displaystyle\int_{3\alpha}^{\alpha^3} \frac{1}{x}\, dx = \frac{2}{\alpha}$.

In this case, the result can be checked by direct integration. Since

$$\int_{3\alpha}^{\alpha^3} \frac{1}{x}\, dx = \log x\Big]_{3\alpha}^{\alpha^3},$$

$$= \log \alpha^3 - \log 3\alpha,$$

$$= 2\log\alpha - \log 3,$$

we obtain

$$\frac{d}{d\alpha}\int_{3\alpha}^{\alpha^3} \frac{1}{x}\, dx = \frac{d}{d\alpha}(2\log\alpha - \log 3) = \frac{2}{\alpha}.$$

Alternatively, use of Eq. (24) provides another way of obtaining the same result. Here,

$$z(\alpha) = \alpha^3, \qquad y(\alpha) = 3\alpha, \qquad f(x, \alpha) = \frac{1}{x}, \qquad I = \int_y^z \frac{1}{x}\, dx,$$

so that

$$\frac{\partial I}{\partial z} = \frac{1}{z}, \qquad \frac{\partial I}{\partial y} = -\frac{1}{y}, \qquad \frac{\partial I}{\partial \alpha} = 0, \qquad \frac{dz}{d\alpha} = 3\alpha^2, \qquad \frac{dy}{d\alpha} = 3.$$

Therefore substitution of these data in (24) yields

$$\frac{d}{d\alpha}\int_{3\alpha}^{\alpha^3} \frac{1}{x}\,dx = \frac{1}{\alpha^3}3\alpha^2 - \frac{1}{3\alpha}3 + 0 = \frac{2}{\alpha}.$$

Exercises 6.6

1. Draw the straight lines of the one parameter family $x + \alpha^2 y = 2\alpha$ for $\alpha = 0, \pm 1/2, \pm 1$. Find their envelope.

2. Find the envelopes of the following one parameter families of straight lines. Illustrate with graphs.

(i) $y = \alpha x - \alpha^2$ **(ii)** $y = \alpha x + 2/\alpha$ **(iii)** $y = 3\alpha x - \alpha^3$

3. Determine first order differential equations which have the one parameter families of solutions given in the previous problem. (*Hint:* Eliminate α from the pair of equations $y = f(x, \alpha)$, $dy/dx = \partial f/\partial x$.) Verify that the envelopes also correspond to solutions of the differential equations.

4. Draw the circles of the one parameter family $(x - \alpha)^2 + y^2 = 1 - \alpha^2$ for $\alpha = 0, \pm 1/2, \pm 1$. Find the envelope of the circles.

5. Draw the parabolas of the one parameter family $y = \alpha(x - \alpha)^2$ for $\alpha = 0, \pm 1, \pm 2$ and find their envelope.

6. Given positive constants a, b, find the envelope of the one parameter family of straight lines $(x \cos\alpha)/a + (y \sin\alpha)/b = 1$.

7. Find the envelope of the family of ellipses $x^2/a^2 + y^2/b^2 = 1$ that all have area π. Draw a diagram to illustrate.

8. Draw the planes $x + \alpha y + \alpha^2 z = 0$ for $\alpha = 0, \pm 1, \pm 2$. Find the envelope of the family of planes.

9. Find the envelope of the family of spheres $(x - \alpha)^2 + (y - \alpha)^2 + (z - \alpha)^2 = 1$. Illustrate with a diagram.

10. Show that the envelope of the two parameter family of planes
$x \cos\alpha \sin\beta + y \sin\alpha \sin\beta + z \cos\beta = 1$ is the sphere $x^2 + y^2 + z^2 = 1$.

11. Determine the envelope of the three parameter family of planes
$ax + by + cz = (a^2 + b^2 + c^2)^{1/2}$. How is this problem related to the previous problem?

12. In Sec. 5.2, the trajectory of a shell fired with speed V at angle α to the horizontal is given by $y = x \tan\alpha - gx^2 \sec^2\alpha/2V^2$. Plot the trajectories for $\alpha = \pm \pi/3, \pm \pi/4, \pm \pi/6$. Determine the envelope of the trajectories. What is the significance of the envelope for the projectile problem?

13. An airplane is flying at Mach 2 at 12 000 m. What is the minimum distance of the plane from a ground observer at which it can be heard? Illustrate with a diagram.

14. A tidal wave has height $y = 20\,e^{-a(x-Vt)^2}$ at horizontal position x at time t. Find an expression for the normal velocity at the air-water interface. Considering t to be a parameter, find the envelope of the tidal wave. Interpret this envelope and illustrate with a diagram.

15. A vibrating violin string stretched between $x=0$ and $x=l$ has a lateral displacement $y(x,t) = a\sin(\pi x/l)\sin\omega t$ at time t at the point x. What is the maximum lateral velocity of the string? Considering t to be a parameter, find the envelope of the lateral displacement. Interpret and illustrate with a diagram.

16. By differentiating the integral $I_0(\alpha) = \int_0^\infty e^{-\alpha x}\,dx = 1/\alpha$ defined for $\alpha>0$, deduce that $I_n(\alpha) = \int_0^\infty x^n e^{-\alpha x}\,dx = n!/\alpha^{n+1}$ for any positive integer n. Obtain the same result by using integration by parts and mathematical induction.

17. Use the result of the previous problem to verify the following identities valid for $\alpha>\beta>0$.

(i) $\displaystyle\int_0^\infty e^{-\alpha x}\sin\beta x\,dx = \frac{\beta}{\alpha^2+\beta^2}$ (ii) $\displaystyle\int_0^\infty e^{-\alpha x}\cos\beta x\,dx = \frac{\alpha}{\alpha^2+\beta^2}$

(iii) $\displaystyle\int_0^\infty e^{-\alpha x}\sinh\beta x\,dx = \frac{\beta}{\alpha^2-\beta^2}$ (iv) $\displaystyle\int_0^\infty e^{-\alpha x}\cosh\beta x\,dx = \frac{\alpha}{\alpha^2-\beta^2}$

18. Evaluate the following derivatives by two methods.

(i) $\displaystyle\frac{\partial}{\partial\alpha}\int_\alpha^{\alpha^2}(x^2+\alpha^3)\,dx$ (ii) $\displaystyle\frac{\partial}{\partial\alpha}\int_{\alpha^2}^{\alpha}\frac{dx}{x+\alpha}$

19. From the known result $\int_0^1 x^\alpha\,dx = 1/(\alpha+1)$ valid for $\alpha>-1$, deduce the following:

(i) $\displaystyle\int_0^1 x^\alpha(\log x)^n\,dx = \frac{(-1)^n}{(\alpha+1)^{n+1}}$ (ii) $\displaystyle\int_0^1 \frac{x^\beta - x^\alpha}{\log x}\,dx = \log\frac{\beta+1}{\alpha+1}$ $(\beta>-1)$

(*Hint:* Differentiate and integrate with respect to α.)

20. Use the result $\displaystyle\int_0^\alpha \frac{dx}{\alpha^2+x^2} = \frac{\pi}{4\alpha}$ $(\alpha>0)$ to find $\displaystyle\int_0^\alpha \frac{dx}{(\alpha^2+x^2)^2}$ and $\displaystyle\int_0^\alpha \frac{dx}{(\alpha^2+x^2)^3}$.

21. Use the result $\displaystyle\int_0^\infty \frac{dx}{\alpha^2+x^2} = \frac{\pi}{2\alpha}$ $(\alpha>0)$ to evaluate $\displaystyle\int_0^\infty \frac{dx}{(\alpha^2+x^2)^n}$ for any positive integer n.

22. Evaluate $I(\alpha) = \int_0^1 \log(x^2+\alpha^2)\,dx$.

23. From the results $\int_0^\infty e^{-x^2}\,dx = \tfrac12\sqrt{\pi}$ and $\int_0^\infty xe^{-x^2}\,dx = \tfrac12$, make a change of variable to verify the following generalizations valid for $\alpha>0$.

(i) $\displaystyle\int_0^\infty e^{-\alpha x^2}\,dx = \frac12\sqrt{\frac{\pi}{\alpha}}$ (ii) $\displaystyle\int_0^\infty xe^{-\alpha x^2}\,dx = \frac{1}{2\alpha}$

24. Use the results of the previous problem to verify the following:

(i) $\displaystyle\int_0^\infty x^2e^{-\alpha x^2}\,dx = \frac{\pi^{1/2}}{4\alpha^{3/2}}$ (ii) $\displaystyle\int_0^\infty x^3e^{-\alpha x^2}\,dx = \frac{1}{2\alpha^2}$

25. Evaluate $\int_0^\infty x^n e^{-\alpha x^2}\,dx$ for all positive integers n.

26. Use Exercise 23 to verify the following results valid for $\beta>\alpha>0$.

(i) $\displaystyle\int_0^\infty \frac{e^{-\alpha x^2}-e^{-\beta x^2}}{x^2}\,dx = \sqrt{\pi}\,(\beta^{1/2}-\alpha^{1/2})$ (ii) $\displaystyle\int_0^\infty \frac{e^{-\alpha x^2}-e^{-\beta x^2}}{x}\,dx = \frac12\log\frac{\beta}{\alpha}$

27. If $I(\alpha,\beta) = \int_0^\infty e^{-\alpha x^2}\cos\beta x\,dx$, show that $\partial I/\partial\beta = -(\beta/2\alpha)I$. Deduce the result $I(\alpha,\beta) = \sqrt{\pi}\,e^{-\beta^2/4\alpha}/2\sqrt{\alpha}$.

28. Evaluate the following integrals:

(i) $\displaystyle\int_0^\infty xe^{-\alpha x^2}\cos\beta x\,dx$ (ii) $\displaystyle\int_0^\infty x^2e^{-\alpha x^2}\cos\beta x\,dx$.

29. Show that the function $I(x) = \int_0^\infty e^{-t^2 - x^2/t^2}\, dt$ satisfies $I'(x) = -2I(x)$ and use this result to evaluate $I(x)$. (*Hint:* Make a change of variables $u = 1/t$ in the integral $I'(x)$.)

30. Show that the function $J_0(x) = \dfrac{1}{\pi}\displaystyle\int_0^\pi \cos(x \sin\phi) d\phi$ satisfies $xJ_0'' + J_0' + xJ_0 = 0$. Evaluate $J_0(0)$ and $J_0'(0)$.

31. If $x(t) = \dfrac{1}{\omega}\displaystyle\int_0^t f(u) \sin \omega(t-u) du$, show that $x''(t) + \omega^2 x(t) = f(t)$. Evaluate $x(0)$ and $x'(0)$.

32. Find the function $f(x)$ that satisfies the *integral equation* $f(x) = 2 + \int_0^1 (x+t) f(t) dt$. (*Hint:* Write $f(x) = 2 + A_0 + A_1 x$ where $A_0 = \int_0^1 t f(t) dt$ and $A_1 = \int_0^1 f(t) dt$ and substitute in the integral equation.)

33. If $\phi(x)$ satisfies the integral equation $\phi(x) = f(x) + \lambda \int_0^{2\pi} \sin x \cos \alpha \phi(\alpha) d\alpha$, prove that $\phi(x) = f(x) + \lambda \int_0^{2\pi} \sin x \cos \alpha\, f(\alpha) d\alpha$.

34. Show that the general solution of the *integrodifferential equation* $f'(x) = 2 + \int_0^1 x f(t) dt$ is $f(x) = c + 2x + 3/5(c+1)x^2$ where c is an arbitrary constant. (*Hint:* Adapt the method used in Exercise **32**.)

6.7 *Implicit functions*

In principle an implicit function $F(x,y,z) = 0$ can be solved for one of the variables, say $z = f(x,y)$. However, in practice the explicit form of the solution cannot in general be found. If, for example, $F = x + y + z - 1 = 0$, then $z = 1 - x - y$; but in the more complicated case, $F = x^3 + z + \sin(y + z^2) + \log(y + z) = 0$, no simple closed-form analytical expression, $z = f(x, y)$, can be determined. Nevertheless, it is often required to calculate the derivatives f_x and f_y in terms of the derivatives of F; the essential procedure for doing this makes use of differentials.

Suppose the implicit function $F(x, y, z) = 0$ were solved for $z = f(x, y)$; then of course this would imply

$$dz = f_x\, dx + f_y\, dy = z_x\, dx + z_y\, dy. \tag{1}$$

But we can also write

$$dF = F_x\, dx + F_y\, dy + F_z\, dz = 0,$$

from which it follows that

$$dz = -\frac{F_x}{F_z} dx - \frac{F_y}{F_z} dy. \tag{2}$$

A comparison of these two forms for dz shows that

$$\frac{\partial z}{\partial x} = -\frac{F_x}{F_z}, \qquad \frac{\partial z}{\partial y} = -\frac{F_y}{F_z}. \tag{3}$$

Example 1 If the implicit function $F(x, y, z) = 0$ is solved consecutively for $x = x(y, z)$, $y = y(z, x)$, and $z = z(x, y)$, show that

$$\frac{\partial x}{\partial y}\frac{\partial y}{\partial z}\frac{\partial z}{\partial x} = -1.$$

The analogous forms of (2) for the differentials dx and dy are

$$dx = -\frac{F_y}{F_x}\,dy - \frac{F_z}{F_x}\,dz, \qquad dy = -\frac{F_z}{F_y}\,dz - \frac{F_x}{F_y}\,dx.$$

Therefore

$$\frac{\partial x}{\partial y} = -\frac{F_y}{F_x}, \qquad \frac{\partial y}{\partial z} = -\frac{F_z}{F_y}, \qquad \frac{\partial z}{\partial x} = -\frac{F_x}{F_z},$$

and on forming the triple product, we find that

$$\frac{\partial x}{\partial y}\frac{\partial y}{\partial z}\frac{\partial z}{\partial x} = \left(-\frac{F_y}{F_x}\right)\left(-\frac{F_z}{F_y}\right)\left(-\frac{F_x}{F_z}\right) = -1.$$

Sometimes there are ambiguities as to which variables are independent, which are dependent, and which are to be held constant in a particular differentiation. To avoid confusion, an additional subscript is often used to denote the role of the independent variables. For instance $\partial z/\partial x$ in Eq. (3) might be augmented to read $(\partial z/\partial x)_y$, to emphasize that this is the partial derivative of z with respect to x, y being held fixed.

Example 2 The equations $x = x(r, s)$ and $y = y(r, s)$ are solved for $r = r(x, y)$ and $s = s(x, y)$. Find the partial derivatives of r and s with respect to x and y in terms of the derivatives of x and y with respect to r and s.

In essence, r and s are the original independent variables, and the problem requires changing their roles and viewing them as dependent on the values of x and y. We wish then to find $\dfrac{\partial r}{\partial x}, \dfrac{\partial r}{\partial y}, \dfrac{\partial s}{\partial x}$, and $\dfrac{\partial s}{\partial y}$ in terms of $\dfrac{\partial x}{\partial r}, \dfrac{\partial x}{\partial s}, \dfrac{\partial y}{\partial r}$, and $\dfrac{\partial y}{\partial s}$. [Although there is really no ambiguity in this case, we could write $\partial r/\partial x$ and $\partial r/\partial y$ as $(\partial r/\partial x)_y$ and $(\partial r/\partial y)_x$.]

The equations $x = x(r, s)$ and $y = y(r, s)$ imply the differential relations

$$dx = \frac{\partial x}{\partial r}\,dr + \frac{\partial x}{\partial s}\,ds \qquad \text{and} \qquad dy = \frac{\partial y}{\partial r}\,dr + \frac{\partial y}{\partial s}\,ds.$$

These equations can be solved algebraically for dr and ds, in which case

$$dr = \frac{1}{\dfrac{\partial x}{\partial r}\dfrac{\partial y}{\partial s} - \dfrac{\partial x}{\partial s}\dfrac{\partial y}{\partial r}}\left(\frac{\partial y}{\partial s}\,dx - \frac{\partial x}{\partial s}\,dy\right),$$

$$ds = \frac{1}{\dfrac{\partial x}{\partial r}\dfrac{\partial y}{\partial s} - \dfrac{\partial x}{\partial s}\dfrac{\partial y}{\partial r}}\left(-\frac{\partial y}{\partial r}\,dx + \frac{\partial x}{\partial r}\,dy\right).$$

But the differentials dr, ds are also expressible as

$$dr = \frac{\partial r}{\partial x}\,dx + \frac{\partial r}{\partial y}\,dy \qquad \text{and} \qquad ds = \frac{\partial s}{\partial x}\,dx + \frac{\partial s}{\partial y}\,dy,$$

and by identifying the coefficients of dx, dy with the preceding expressions it follows that

$$\frac{\partial r}{\partial x} = \frac{\dfrac{\partial y}{\partial s}}{\dfrac{\partial x}{\partial r}\dfrac{\partial y}{\partial s} - \dfrac{\partial x}{\partial s}\dfrac{\partial y}{\partial r}}, \qquad \frac{\partial r}{\partial y} = -\frac{\dfrac{\partial x}{\partial s}}{\dfrac{\partial x}{\partial r}\dfrac{\partial y}{\partial s} - \dfrac{\partial x}{\partial s}\dfrac{\partial y}{\partial r}},$$

$$\frac{\partial s}{\partial x} = -\frac{\dfrac{\partial y}{\partial r}}{\dfrac{\partial x}{\partial r}\dfrac{\partial y}{\partial s} - \dfrac{\partial x}{\partial s}\dfrac{\partial y}{\partial r}}, \qquad \frac{\partial s}{\partial y} = \frac{\dfrac{\partial x}{\partial r}}{\dfrac{\partial x}{\partial r}\dfrac{\partial y}{\partial s} - \dfrac{\partial x}{\partial s}\dfrac{\partial y}{\partial r}}.$$

Note that in general $\dfrac{\partial r}{\partial x} \neq 1\Big/\dfrac{\partial x}{\partial r}$; i.e., $\left(\dfrac{\partial r}{\partial x}\right)_y \neq 1\Big/\left(\dfrac{\partial x}{\partial r}\right)_s$. To illustrate, let $x = re^s$ and $y = re^{-s}$, so that $r = (xy)^{1/2}$, $s = \tfrac{1}{2}\log(x/y)$. In this case $\partial x/\partial r = e^s$, $\partial r/\partial x = e^{-s}/2$, and obviously the two are not reciprocals.

Example 3 The pair of equations $z = f(x, y)$ and $g(x, y) = 0$ imply that $z = h(x)$. Find $dz/dx = h'(x)$ in terms of the partial derivatives of f and g.

The differential relations are

$$dz = f_x\,dx + f_y\,dy,$$
$$dg = 0 = g_x\,dx + g_y\,dy.$$

Elimination of dy gives

$$dz = \left(f_x - \frac{g_x f_y}{g_y}\right)dx,$$

and it follows that

$$\frac{dz}{dx} = h'(x) = \frac{f_x g_y - f_y g_x}{g_y}.$$

2

The combination of derivatives of two functions f and g, $f_x g_y - f_y g_x$, occurs frequently enough in practice to merit a special name, the *Jacobian* of f and g, and a special notation

$$\frac{\partial(f, g)}{\partial(x, y)} = \begin{vmatrix} f_x & f_y \\ g_x & g_y \end{vmatrix} = f_x g_y - f_y g_x. \tag{4}$$

Jacobians play an important role later in a change of variables in multiple integrals (see Sec. 7.2).

If the Jacobian of two functions f and g vanishes identically, the functions are strongly related to each other. To see this, first suppose that f and g are functionally related; i.e., there is a function φ which enables the value of f to be obtained from the value of g:

$$f = \varphi(g). \tag{5}$$

In this case,

$$f_x = \varphi'(g)g_x, \qquad f_y = \varphi'(g)g_y.$$

Therefore

$$\frac{\partial(f, g)}{\partial(x, y)} = \varphi'(g)g_x g_y - \varphi'(g)g_x g_y = 0, \tag{6}$$

and the Jacobian is zero. Conversely, if $\partial(f, g)/\partial(x, y) = 0$, it can be shown that $f = \varphi(g)$, (see Exercise 24).

As an illustration, consider the functions

$$f = x^2 - 2xy + y^2 - 3x + 3y + 1, \qquad g = e^{x-y}.$$

Since $f_x = 2x - 2y - 3$, $f_y = -2x + 2y + 3$, $g_x = e^{x-y}$, $g_y = -e^{x-y}$, it is relatively easy to show that the Jacobian is zero, $\partial(f, g)/\partial(x, y) = 0$, and, in fact, $f = (\log g)^2 - 3 \log g + 1$.

3

The calculation and manipulation of partial derivatives is important in thermodynamics, and an example illustrates the techniques involved.

Example 4 The reversible changes of a gas of pressure p, volume V, temperature T, internal energy E, and entropy S satisfy the differential relation

$$dS = \frac{1}{T}dE + \frac{p}{T}dV, \tag{7}$$

and the equation of state

$$pV = RT,$$

R being a constant. Show that $(\partial E/\partial V)_T = 0$ and interpret the result.

Note that there are no obviously "independent" variables in the problem; such a choice is really quite arbitrary and made for convenience. The possibility of confusion is apparent, and for this reason the more precise notation is adopted. Let us choose V and T as independent variables so that $S = S(V, T)$, $E = E(V, T)$. Since

$$dS = \left(\frac{\partial S}{\partial V}\right)_T dV + \left(\frac{\partial S}{\partial T}\right)_V dT$$

and

$$dE = \left(\frac{\partial E}{\partial V}\right)_T dV + \left(\frac{\partial E}{\partial T}\right)_V dT,$$

the increment dE can be substituted in Eq. (7) to obtain

$$dS = \left[\frac{1}{T}\left(\frac{\partial E}{\partial V}\right)_T + \frac{R}{V}\right] dV + \left[\frac{1}{T}\left(\frac{\partial E}{\partial T}\right)_V\right] dT$$

(where the equation of state has been used to eliminate p/T). Therefore

$$\left(\frac{\partial S}{\partial V}\right)_T = \frac{1}{T}\left(\frac{\partial E}{\partial V}\right)_T + \frac{R}{V},$$

$$\left(\frac{\partial S}{\partial T}\right)_V = \frac{1}{T}\left(\frac{\partial E}{\partial T}\right)_V,$$

and by cross differentiating to form

$$\frac{\partial^2 S}{\partial T\, \partial V} = \frac{\partial^2 S}{\partial V\, \partial T}$$

we obtain the formula

$$\frac{\partial}{\partial T}\left[\frac{1}{T}\left(\frac{\partial E}{\partial V}\right)_T + \frac{R}{V}\right] = \frac{\partial}{\partial V}\left[\frac{1}{T}\left(\frac{\partial E}{\partial T}\right)_V\right].$$

Since T and V are the designated independent variables, the preceding may be simplified to read

$$-\frac{1}{T^2}\left(\frac{\partial E}{\partial V}\right)_T = 0$$

or

$$\left(\frac{\partial E}{\partial V}\right)_T = 0. \tag{8}$$

This important result states that the internal energy E of a gas is a function of temperature only, $E = E(T)$.

Exercises 6.7

1. The function $z = z(x,y)$ is defined implicitly by the equation $x + y^2 + z^3 - xy = 2z$. Find $\partial z/\partial x$ and $\partial z/\partial y$ at the point $(1,1,1)$. Determine the tangent plane and normal line at that point and use the tangent plane to estimate $z(1.01, 1.03)$.

2. If $x = e^{yz} + z^2$, find $\partial z/\partial x$ and $\partial z/\partial y$ by implicit differentiation.

3. The equation $e^x + e^y + e^z = 3xyz$ defines y as a function of z and x. Find $\partial y/\partial z$ and $\partial y/\partial x$. Use these derivatives to determine the tangent plane and normal line at a general point (x_0, y_0, z_0). By a symmetry argument, evaluate $\partial x/\partial y$, $\partial x/\partial z$, $\partial z/\partial x$ and $\partial z/\partial y$ without differentiating again.

4. Evaluate $\partial z/\partial x$ and $\partial z/\partial y$ for the following implicit functions $z(x, y)$.

(i) $e^z + e^{-z} = z + 2xy$ **(ii)** $x^2 + y^2 + z^2 = 2\cosh z$ **(iii)** $\log(1 + xyz) = xyz$

5. For each implicit function in the previous problem, determine the tangent plane and normal line at the point (1, 1, 0) on each surface $z = z(x, y)$. Use the tangent planes to estimate $z(0.98, 1.01)$.

6. For the change of variables $x = r + s$, $y = 2r - s$, express dr and ds in terms of dx and dy. Find $\partial r/\partial x$, $\partial r/\partial y$, $\partial s/\partial x$ and $\partial s/\partial y$ in terms of r and s. For the function $f(r, s) = r/s$, find $\partial f/\partial x$ and $\partial f/\partial y$.

7. For the general linear change of variables $x = ar + bs$, $y = cr + ds$, express dr and ds in terms of dx and dy (a, b, c, d are given constants). Evaluate the Jacobians $\partial(x, y)/\partial(r, s)$ and $\partial(r, s)/\partial(x, y)$. What happens when either Jacobian is zero?

8. For the transformation to polar coordinates $x = r\cos\theta$ and $y = r\sin\theta$, prove that $\partial(x, y)/\partial(r, \theta) = r$ and $\partial(r, \theta)/\partial(x, y) = (x^2 + y^2)^{-1/2}$.

9. For the change of variables $x = r + s$, $y = r^2 + s^2$, express dr and ds in terms of dx and dy. Use these results to find $\partial r/\partial x$, $\partial r/\partial y$, $\partial s/\partial x$ and $\partial s/\partial y$ in terms of r and s. Evaluate the Jacobians $\partial(x, y)/\partial(r, s)$ and $\partial(r, s)/\partial(x, y)$.

10. The functions $x = x(r, s)$, $y = y(r, s)$ are defined implicitly by the equations $x^2 + y^2 + r^2 - 2s = 0$ and $x^3 - y^3 - r^3 + 3s = 1$. Evaluate $\partial x/\partial r$, $\partial y/\partial r$, $\partial x/\partial s$ and $\partial y/\partial s$ and the Jacobians $\partial(x, y)/\partial(r, s)$ and $\partial(r, s)/\partial(x, y)$.

11. The variables x and y are functions of r and s determined implicitly by the equations $x^2 - y^2 - 2r = 0$ and $xy - s = 0$. Find $\partial x/\partial r$, $\partial x/\partial s$, $\partial y/\partial r$ and $\partial y/\partial s$ and evaluate the Jacobians $\partial(x, y)/\partial(r, s)$ and $\partial(r, s)/\partial(x, y)$.

12. For the change of variables $r = \frac{1}{2}(x^2 - y^2)$, $s = xy$, plot the two families of curves $r = \alpha$, $s = \beta$ (α, β constants) in the xy plane. Show that the curves of the two families are orthogonal.

13. For the change of variables $r = \frac{1}{2}\log(x^2 + y^2)$, $s = \tan^{-1}(y/x)$, determine the Jacobians $\partial(x, y)/\partial(r, s)$ and $\partial(r, s)/\partial(x, y)$. Plot the families of curves $r = \alpha$, $s = \beta$ in the xy plane. Show that the curves of the two families are orthogonal.

14. The equations $x = uv$, $y = u + v$ and $z = u^2 - v^2$ define z as a function of x and y. Find $\partial z/\partial x$ and $\partial z/\partial y$. Estimate $z(0.98, 1.03)$ and compare to the accurate value by solving for $z = z(x, y)$ explicitly.

15. The equation $F(x, y, z) = 0$ defines z implicitly as a function of x and y. Find the second derivatives $\partial^2 z/\partial x^2$, $\partial^2 z/\partial x\partial y$ and $\partial^2 z/\partial y^2$ in terms of partial derivatives of F. Illustrate with the special case $x^2 + y^2 - z^2 = 0$.

16. If $z = z(x, y)$ is defined implicitly by $F(z/x, y/x) = 0$, verify that $x\,\partial z/\partial x + y\,\partial z/\partial y = z$. Illustrate with the special case $z = xe^{z/y}$.

17. If x and y are functions of r and s (and conversely) defined by the equations $F(x, y, r, s) = 0$ and $G(x, y, r, s) = 0$, verify the following identities.

(i) $\dfrac{\partial r}{\partial x}\dfrac{\partial x}{\partial r}+\dfrac{\partial r}{\partial y}\dfrac{\partial y}{\partial r}=1$ (ii) $\dfrac{\partial r}{\partial x}\dfrac{\partial x}{\partial s}+\dfrac{\partial r}{\partial y}\dfrac{\partial y}{\partial s}=0$

(iii) $\dfrac{\partial s}{\partial x}\dfrac{\partial x}{\partial r}+\dfrac{\partial s}{\partial y}\dfrac{\partial y}{\partial r}=0$ (iv) $\dfrac{\partial s}{\partial x}\dfrac{\partial x}{\partial s}+\dfrac{\partial s}{\partial y}\dfrac{\partial y}{\partial s}=1$

18. If the equations $x=x(r,s)$ and $y=y(r,s)$ are solved for r and s, so that $r=r(x,y)$ and $s=s(x,y)$, show that $\dfrac{\partial(x,y)}{\partial(r,s)}\dfrac{\partial(r,s)}{\partial(x,y)}=1$. Deduce that, if a change of variables has an inverse, then the Jacobian is not zero.

19. If $x=x(r,s)$, $y=y(r,s)$ and $r=r(u,v)$, $s=s(u,v)$ are two successive changes of variables, then x and y are defined as functions of u and v. Show that the Jacobians satisfy

$$\frac{\partial(x,y)}{\partial(r,s)}\frac{\partial(r,s)}{\partial(u,v)}=\frac{\partial(x,y)}{\partial(u,v)}.$$

If $u=x$ and $v=y$, deduce the result of the previous problem as a special case.

20. The equations $F(x,y,u,v)=0$ and $G(x,y,u,v)=0$ define u and v as functions of x and y. Prove that

$$\frac{\partial(F,G)}{\partial(x,y)}=\frac{\partial(F,G)}{\partial(u,v)}\frac{\partial(u,v)}{\partial(x,y)} \quad \text{or} \quad \frac{\partial(u,v)}{\partial(x,y)}=\frac{\partial(F,G)}{\partial(x,y)}\bigg/\frac{\partial(F,G)}{\partial(u,v)}.$$

21. Use the result of the previous problem to evaluate the Jacobian $\partial(u,v)/\partial(x,y)$ when $F(x,y,u,v)=x^2+y^2-u-v=0$ and $G(x,y,u,v)=x+y-u^2-v^2=0$.

22. If $f(x,y)=\sin(x^2+y^2)$ and $g(x,y)=\cos(x^2+y^2)$, show that $\partial(f,g)/\partial(x,y)=0$. Find a function ϕ such that $f(x,y)=\phi(g(x,y))$.

23. If $f(x,y)=\log y-\log x$ and $g(x,y)=xy/(x^2+y^2)$, show that $\partial(f,g)/\partial(x,y)=0$. Find a function ψ such that $g(x,y)=\psi(f(x,y))$.

24. If $f=f(x,y)$ and $g=g(x,y)$, show that $\dfrac{\partial g}{\partial y}\,df-\dfrac{\partial f}{\partial y}\,dg=\dfrac{\partial(f,g)}{\partial(x,y)}\,dx$. Illustrate this identity with the functions of the two previous problems. By considering f to be a function of the variables g and x, deduce that, if $\partial(f,g)/\partial(x,y)=0$, then there exists a function ϕ such that $f(x,y)=\phi(g(x,y))$.

25. The *equation of state* of a gas is $F(p,V,T)=0$ where p is the pressure (pascals), V the volume (m^3), and T the temperature (°K). This equation relates two independent thermodynamic variables to a third dependent variable. Prove the following results.

(i) $\left(\dfrac{\partial V}{\partial T}\right)_p=-\dfrac{F_T}{F_V}$ (ii) $\left(\dfrac{\partial^2 V}{\partial T^2}\right)_p=\dfrac{1}{F_V{}^3}(2F_TF_VF_{TV}-F_T{}^2F_{VV}-F_V{}^2F_{TT})$

26. For the equation of state $F(p,V,T)=0$, verify the following results.

(i) $\left(\dfrac{\partial V}{\partial T}\right)_p\left(\dfrac{\partial T}{\partial V}\right)_p=1$ (ii) $\left(\dfrac{\partial V}{\partial T}\right)_p\left(\dfrac{\partial T}{\partial p}\right)_V=-\left(\dfrac{\partial V}{\partial T}\right)_p$

(iii) $\left(\dfrac{\partial V}{\partial T}\right)_p\left(\dfrac{\partial T}{\partial p}\right)_V\left(\dfrac{\partial p}{\partial V}\right)_T=-1$

27. Find $(\partial V/\partial T)_p$ and $(\partial^2 V/\partial T^2)_p$ for the following equations of state. Verify the identities of the previous problem.

(i) $pV=RT$ (perfect gas) (ii) $p=\dfrac{RT}{V-\beta}-\dfrac{\alpha}{V^2}$ (van der Waals gas)

28. Third order Jacobians are defined in the following way. If $x=x(u,v,w)$, $y=y(u,v,w)$ and $z=z(u,v,w)$, then the Jacobian is the determinant

$$J = \frac{\partial(x, y, z)}{\partial(u, v, w)} = \begin{vmatrix} \frac{\partial x}{\partial u} & \frac{\partial x}{\partial v} & \frac{\partial x}{\partial w} \\ \frac{\partial y}{\partial u} & \frac{\partial y}{\partial v} & \frac{\partial y}{\partial w} \\ \frac{\partial z}{\partial u} & \frac{\partial z}{\partial v} & \frac{\partial z}{\partial w} \end{vmatrix}.$$

Find expressions for the differentials dx, dy and dz in terms of du, dv and dw. Show that these equations can be solved for du, dv and dw in a unique way if $J \neq 0$.

29. For the change of variables $x = u + v - w$, $y = u - v + w$, $z = u^2 + v^2 + w^2 - 2vw$, verify that the Jacobian $\partial(x, y, z)/\partial(u, v, w) = 0$. This implies that x, y and z are not independent functions. Determine $z = z(x, y)$ explicitly.

30. Calculate the Jacobians for the following changes of variables.

(i) $x = (u^2 - v^2)/2$, $y = uv$, $z = z$ (parabolic cylindrical coordinates)

(ii) $x = r \cos\theta$, $y = r \sin\theta$, $z = z$ (cylindrical polar coordinates)

(iii) $x = r \sin\phi \cos\theta$, $y = r \sin\phi \sin\theta$, $z = r \cos\phi$ (spherical polar coordinates)

(iv) $x = au \sin\phi \cos\theta$, $y = bu \sin\phi \sin\theta$, $z = cu \cos\phi$ (ellipsoidal polar coordinates)

31. The equation of state of a thermodynamic system $F(y, x_1, x_2, \ldots, x_n) = 0$ relates n independent properties $x_1, x_2, \ldots, x_n$ to the dependent property y. Verify the following relations between partial derivatives (where all variables except those directly involved are held constant). Assume i, j, k and l are distinct.

(i) $\dfrac{\partial y}{\partial x_i} = -F_{x_i}/F_y$ **(ii)** $\dfrac{\partial x_i}{\partial y} = -F_y/F_{x_i}$ **(iii)** $\dfrac{\partial x_i}{\partial x_j} = -F_{x_j}/F_{x_i}$

(iv) $\dfrac{\partial x_i}{\partial x_j}\dfrac{\partial x_j}{\partial x_i} = 1$ **(v)** $\dfrac{\partial x_i}{\partial x_j}\dfrac{\partial x_j}{\partial x_k}\dfrac{\partial x_k}{\partial x_i} = -1$ **(vi)** $\dfrac{\partial x_i}{\partial x_j}\dfrac{\partial x_j}{\partial x_k}\dfrac{\partial x_k}{\partial x_l}\dfrac{\partial x_l}{\partial x_i} = 1$

6.8 *Taylor's formula and approximate methods*

Suppose that at a point (a, b) the value of a function $f(x, y)$ and all its partial derivatives up to a certain order are known. How then can this information be used to determine the function, albeit approximately, at nearby points? For functions of one variable, this question was answered in Chap. 2 by deriving Taylor's formula:

$$f(a + h) = f(a) + \frac{h}{1!}f'(a) + \cdots + \frac{h^n}{n!}f^{(n)}(a) + \frac{h^{n+1}}{(n+1)!}f^{(n+1)}(a + \theta h), \qquad 0 < \theta < 1, \tag{1}$$

which for $n = 0$ reduces to a statement of the mean-value theorem:

$$f(a + h) = f(a) + hf'(a + \theta h), \qquad 0 < \theta < 1. \tag{2}$$

For functions of two (or more) independent variables there is a similar formula, which is most easily found by using (1) as a starting point.

The problem is to find $f(a + h, b + k)$ for small increments h, k, given $f(a, b)$ and all its partial derivatives up to a certain order at the point (a, b). With this aim in mind, we write

$$F(t) = f(a + ht, b + kt); \tag{3}$$

a, b, h, and k are all fixed, and t is the only variable. As t ranges from 0 to 1, the locus of the point $(a + ht, b + kt)$ is a straight line from the point (a, b) to $(a + h, b + k)$. Application of Taylor's formula (1) to $F(t)$, which is a function of one variable, gives

$$F(1) = F(0) + \frac{1}{1!}F'(0) + \cdots + \frac{1}{n!}F^{(n)}(0) + \frac{1}{(n+1)!}F^{(n+1)}(\theta), \qquad 0 < \theta < 1, \tag{4}$$

where

$$F(1) = f(a + h, b + k), \qquad F(0) = f(a, b), \tag{5}$$

and primes denote differentiations with respect to t.

Since $x = a + ht$, $y = b + kt$, $dx/dt = h$, and $dy/dt = k$, the chain rule gives the operational relationship

$$\frac{d}{dt} = h\frac{\partial}{\partial x} + k\frac{\partial}{\partial y}, \tag{6}$$

so that

$$F'(t) = hf_x(a + ht, b + kt) + kf_y(a + ht, b + kt).$$

In particular

$$F'(0) = hf_x(a, b) + kf_y(a, b),$$

which is the same as

$$F'(0) = hf_a(a, b) + kf_b(a, b) = \left(h\frac{\partial}{\partial a} + k\frac{\partial}{\partial b}\right)f(a, b). \tag{7}$$

Here the notation $f_a(a, b)$ means $f_x(x, y)$ evaluated at $x = a$, $y = b$, that is, $f_x(a, b)$; a similar meaning is attached to $f_b(a, b)$. Thus the operators $\dfrac{d}{dt}$ and $h\dfrac{\partial}{\partial a} + k\dfrac{\partial}{\partial b}$ are equivalent when evaluations are made at $t = 0$, and it follows that

$$F^{(r)}(0) = \left(h\frac{\partial}{\partial a} + k\frac{\partial}{\partial b}\right)^r f(a, b), \tag{8}$$

for any positive integer r. This formula enables us to rewrite (4) as

$$f(a + h, b + k) = f(a, b) + \frac{\left(h\dfrac{\partial}{\partial a} + k\dfrac{\partial}{\partial b}\right)}{1!}f(a, b) + \cdots + \frac{\left(h\dfrac{\partial}{\partial a} + k\dfrac{\partial}{\partial b}\right)^n}{n!}f(a, b)$$

$$+ \frac{\left(h\dfrac{\partial}{\partial a} + k\dfrac{\partial}{\partial b}\right)^{n+1}}{(n+1)!}f(a + \theta h, b + \theta k), \qquad 0 < \theta < 1, \tag{9}$$

and this is *Taylor's formula*, or the extended mean-value theorem for a function of two variables. The particular form for $n = 0$ is

$$f(a+h, b+k) = f(a, b) + hf_a(a+\theta h, b+\theta k) + kf_b(a+\theta h, b+\theta k), \qquad 0<\theta<1, \quad (10)$$

which is a symmetric version of the mean-value theorem derived in Sec. 6.3.

The geometrical interpretation of (10) is analogous to that for functions of one variable. The curve formed by cutting the surface $z = f(x, y)$ by a plane parallel to Oz containing the points $P = (a, b, f(a, b))$ and $P' = (a+h, b+k, f(a+h, b+k))$ is $z = f(a+ht, b+kt)$ (see Fig. 8.1). Equation (10) states that the tangent slope of this curve for at least one point between $t = 0$ and $t = 1$ is equal to the slope of the chord connecting the end points P and P'.

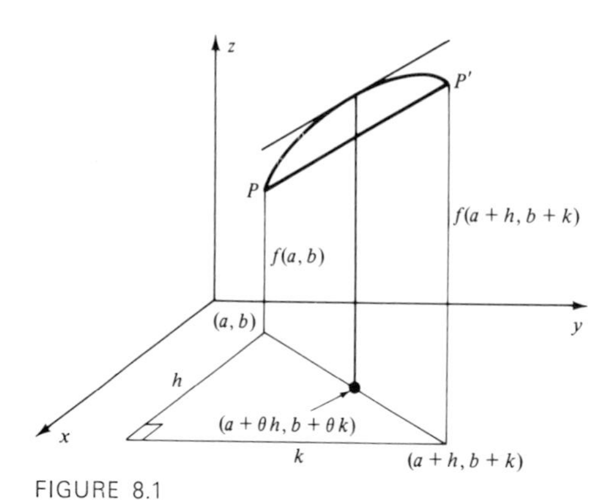

FIGURE 8.1

Taylor's formula without the remainder gives a sequence of local approximations to a function. Each successive term yields more and better information about the behavior of the function. In the limit as $n \to \infty$ (and the remainder presumably approaches zero), the infinite series implied by (9) is an exact representation of the function, its Taylor series expansion in two variables about the arbitrary point (a, b). Since $x - a = h$, $y - b = k$, (9) is really a power series expansion of the function about the point (a, b).

Example 1 Use Taylor's formula to find a quadratic approximation to the function $e^{(x-y)} + \cos(2x+3y)$ near the point $(0, 0)$.

The derivative calculations are straightforward and with $(a, b) = (0, 0)$; $h = x$, $k = y$:

$$f(x, y) = e^{x-y} + \cos(2x+3y), \qquad f(0, 0) = 2;$$
$$f_x = e^{x-y} - 2\sin(2x+3y), \qquad f_x(0, 0) = 1;$$

$$\begin{aligned}
f_y &= -e^{x-y} - 3\sin(2x+3y), & f_y(0,0) &= -1;\\
f_{xx} &= e^{x-y} - 4\cos(2x+3y), & f_{xx}(0,0) &= -3;\\
f_{xy} &= -e^{x-y} - 6\cos(2x+3y), & f_{xy}(0,0) &= -7;\\
f_{yy} &= e^{x-y} - 9\cos(2x+3y), & f_{yy}(0,0) &= -8.
\end{aligned}$$

Replacing them in (9) yields

$$f(x, y) = 2 + x - y + \tfrac{1}{2}(-3x^2 - 14xy - 8y^2) + \text{remainder}.$$

Note that in this particular example the identical result could have been found from the separate expansions of e^{x-y} and $\cos(2x + 3y)$, namely,

$$\begin{aligned}
e^{x-y} &= 1 + (x - y) + \tfrac{1}{2}(x - y)^2 + \cdots,\\
\cos(2x + 3y) &= 1 - \tfrac{1}{2}(2x + 3y)^2 + \cdots.
\end{aligned}$$

If these two series are added, we recover the formula quoted above, i.e., the first few terms of the Taylor series expansion of $f(x, y)$:

$$e^{x-y} + \cos(2x + 3y) = 2 + x - y + \tfrac{1}{2}(-3x^2 - 14xy - 8y^2) + \cdots.$$

Example 2 In the neighborhood of (0, 0, 1) find a quadratic approximation for z if $x + 2y + xy + 1 + z - 2e^{z-1} = 0$.

An approximate explicit form for the solution of this implicit equation can be found by means of Taylor's formula. A series approximation for $z = f(x, y)$ about the point (0, 0, 1) must be of the form

$$z = 1 + \alpha_1 x + \alpha_2 y + \beta_{11}x^2 + \beta_{12}xy + \beta_{22}y^2, \tag{11}$$

where only the terms to second order have been retained. The constants $\alpha_1, \alpha_2, \beta_{11}, \beta_{12}, \beta_{22}$, may be determined by implicit differentiation or alternatively by the direct substitution of (11) into the given implicit relation. Here we use the latter method.

If z is replaced everywhere in the implicit function, then

$$\begin{aligned}
x + 2y + xy + 1 + (1 + \alpha_1 x + \alpha_2 y + \beta_{11}x^2 + \beta_{12}xy + \beta_{22}y^2)&\\
- 2\exp(\alpha_1 x + \alpha_2 y + \beta_{11}x^2 + \beta_{12}xy + \beta_{22}y^2) &= 0.
\end{aligned}$$

The exponential function is expanded as a series ($e^u = 1 + u/1! + u^2/2! + \cdots$), and if only quadratic terms are retained, the above expression becomes

$$\begin{aligned}
x + 2y + xy + 1 + (1 + \alpha_1 x + \alpha_2 y + \beta_{11}x^2 + \beta_{12}xy + \beta_{22}y^2)&\\
- 2[1 + (\alpha_1 x + \alpha_2 y + \beta_{11}x^2 + \beta_{12}xy + \beta_{22}y^2) + \tfrac{1}{2}(\alpha_1 x + \alpha_2 y)^2] &= 0.
\end{aligned}$$

Upon collecting the coefficients, this reduces to

$$\begin{aligned}
(1 - \alpha_1)x + (2 - \alpha_2)y + (-\beta_{11} - \alpha_1{}^2)x^2&\\
+ (1 - \beta_{12} - 2\alpha_1\alpha_2)xy + (-\beta_{22} - \alpha_2{}^2)y^2 &= 0.
\end{aligned}$$

Every coefficient of a power term like $x^r y^s$ must be identically zero (recall the identity theorem for series), and it follows that

$$\alpha_1 = 1, \qquad \alpha_2 = 2, \qquad \beta_{11} = -1, \qquad \beta_{12} = -3, \qquad \beta_{22} = -4.$$

Therefore the desired explicit approximation valid in some small region about the point (0, 0) is

$$z = 1 + x + 2y - x^2 - 3xy - 4y^2.$$

Higher-order approximations can be obtained in a straightforward way, although the manipulations rapidly become more tedious. (In the above example the linear terms give the tangent-plane approximation to the surface, the quadratic terms give the first effects of curvature, etc.)

---

2 In applications of mathematics to scientific problems simplifications can often be made on the basis of the smallness or largeness of certain parameters. Perturbation methods are frequently used in such circumstances to generate approximate solutions. This is all the more important because the possibility of obtaining an exact solution is usually nil, and some (many!) approximations are mandatory.

Example 3 Find an approximate solution of the equation

$$y + \varepsilon \sin 2y = x,$$

where ε is a small number.

This problem was first posed in Sec. 1.11, where a solution was obtained by iteration. Here an approximation is developed using Taylor's formula.

If the parameter ε is regarded as just another independent variable, y can be interpreted as a function of two variables x and ε, that is,

$$y = y(x, \varepsilon).$$

Taylor's formula is used to express y as an infinite series expansion in powers of ε with x held fixed. In doing this, all derivatives with respect to ε that appear are written as partial derivatives, and the expansion about the point $\varepsilon = 0$ takes the form

$$y(x, \varepsilon) = y(x, 0) + \left.\frac{\partial y}{\partial \varepsilon}\right]_{\varepsilon=0} \varepsilon + \left.\frac{\partial^2 y}{\partial \varepsilon^2}\right]_{\varepsilon=0} \frac{\varepsilon^2}{2!} + \cdots + \left.\frac{\partial^n y}{\partial \varepsilon^n}\right]_{\varepsilon=0} \frac{\varepsilon^n}{n!} + \cdots.$$

It remains to evaluate the coefficients in this equation. The value of y at $\varepsilon = 0$, that is, $y(x, 0)$, is obtained by direct substitution in the original relation;

$$y(x, 0) + 0 \sin 2y(x, 0) = x,$$

or

$$y(x, 0) = x.$$

The value of the first partial derivative is a consequence of implicit differentiation:

$$\frac{\partial}{\partial \varepsilon}(y + \varepsilon \sin 2y) = \frac{\partial x}{\partial \varepsilon},$$

or since x is held fast,

$$\frac{\partial y}{\partial \varepsilon} + \sin 2y + 2\varepsilon \cos 2y \frac{\partial y}{\partial \varepsilon} = 0,$$

which in particular yields

$$\left.\frac{\partial y}{\partial \varepsilon}\right]_{\varepsilon=0} = -\sin 2y\Big]_{\varepsilon=0} = -\sin 2x.$$

Therefore to this order, we have

$$y(x, \varepsilon) \approx x - \varepsilon \sin 2x,$$

which agrees with the result derived earlier by iteration. The next coefficient is calculated by a further differentiation with respect to ε:

$$(1 + 2\varepsilon \cos 2y)\frac{\partial^2 y}{\partial \varepsilon^2} + 4 \cos 2y \frac{\partial y}{\partial \varepsilon} - 4\varepsilon \sin 2y\left(\frac{\partial y}{\partial \varepsilon}\right)^2 = 0.$$

Therefore, at $\varepsilon = 0$,

$$\left.\frac{\partial^2 y}{\partial \varepsilon^2}\right]_{\varepsilon=0} = -\left(4 \cos 2y \frac{\partial y}{\partial \varepsilon}\right)\Big]_{\varepsilon=0} = 2 \sin 4x,$$

and Taylor's formula to second order is

$$y(x, \varepsilon) \approx x - \varepsilon \sin 2x + \varepsilon^2 \sin 4x.$$

Example 4 Find the approximate solution of the equations $x - 1 = \varepsilon(x + y - 4)$, $y - 2 = -\varepsilon \log(2 - x)$ for ε small which reduces to $(1, 2)$ as $\varepsilon \to 0$.

Either the method of Example 3 can be used, or (as is more usual in practice) the expansions can be calculated ab initio. Here we illustrate the latter approach.

By inspection, when $\varepsilon = 0$, the solution is $x = 1$, $y = 2$, and therefore we anticipate (from Taylor's formula) that the perturbation expansions are of the form

$$x = 1 + \varepsilon x_1 + \varepsilon^2 x_2 + \varepsilon^3 x_3 + \cdots,$$
$$y = 2 + \varepsilon y_1 + \varepsilon^2 y_2 + \varepsilon^3 y_3 + \cdots.$$

Substitution of these expressions into the original equations gives

$$\begin{aligned}\varepsilon x_1 + \varepsilon^2 x_2 + \varepsilon^3 x_3 + \cdots &= \varepsilon[-1 + \varepsilon(x_1 + y_1) + \varepsilon^2(x_2 + y_2) + \cdots],\\ &= -\varepsilon + \varepsilon^2(x_1 + y_1) + \varepsilon^3(x_2 + y_2) + \cdots,\end{aligned}$$

and

$$\begin{aligned}\varepsilon y_1 + \varepsilon^2 y_2 + \varepsilon^3 y_3 + \cdots &= \varepsilon[(\varepsilon x_1 + \varepsilon^2 x_2 + \cdots) + \tfrac{1}{2}(\varepsilon x_1 + \varepsilon^2 x_2 + \cdots)^2 + \cdots],\\ &= \varepsilon^2 x_1 + \varepsilon^3(x_2 + \tfrac{1}{2}x_1{}^2) + \cdots,\end{aligned}$$

where we have made use of the known expansion

$$-\log(1 - t) = t + \frac{t^2}{2} + \cdots.$$

On sequentially equating the successive powers of ε to zero in both expressions we obtain

$$\begin{aligned}x_1 &= -1, & y_1 &= 0,\\ x_2 &= x_1 + y_1 = -1, & y_2 &= x_1 = -1,\\ x_3 &= x_2 + y_2 = -2, & y_3 &= x_2 + \tfrac{1}{2}x_1{}^2 = -\tfrac{1}{2}.\end{aligned}$$

Therefore, to this order of approximation [with an $O(\varepsilon^4)$ error] the solution is

$$\begin{aligned}x &\approx 1 - \varepsilon - \varepsilon^2 - 2\varepsilon^3,\\ y &\approx 2 - \varepsilon^2 - \tfrac{1}{2}\varepsilon^3.\end{aligned}$$

---

3 Finally, it should be mentioned that Taylor's formula can be extended, in a rather obvious way, to cover functions of more than two variables. With m independent variables, $f(x_1, x_2, \ldots, x_m)$ is expanded in the neighborhood of the point $(a_1, a_2, \ldots, a_m)$ as

$$\begin{aligned}f(a_1 + h_1, a_2 + h_2, \ldots, a_m + h_m)\\ = f(a_1, a_2, \ldots, a_m) + \left(h_1 \frac{\partial}{\partial a_1} + h_2 \frac{\partial}{\partial a_2} + \cdots + h_m \frac{\partial}{\partial a_m}\right) f(a_1, a_2, \ldots, a_m)\\ + \frac{1}{2!}\left(h_1 \frac{\partial}{\partial a_1} + h_2 \frac{\partial}{\partial a_2} + \cdots + h_m \frac{\partial}{\partial a_m}\right)^2 f(a_1, a_2, \ldots, a_m) + \cdots. \quad (12)\end{aligned}$$

This result is derived by reducing the problem to Taylor's formula for the function

$$F(t) = f(a_1 + h_1 t, a_2 + h_2 t, \ldots, a_n + h_m t). \quad (13)$$

The chain rule shows that

$$\frac{d}{dt} = h_1 \frac{\partial}{\partial a_1} + h_2 \frac{\partial}{\partial a_2} + \cdots + h_m \frac{\partial}{\partial a_m}, \quad (14)$$

and use of (4) leads to (12).

Exercises 6.8

1. Express the polynomial $f(x, y) = x^3 - y^2 + xy - x + 2y$ in powers of $x - 1$ and $y - 1$. Why does the Taylor series terminate? Is this a general result for polynomials?

2. For the function $f(x, y) = x^2 + xy + y^2$, define $F(t) = f(x + th, y + tk)$. Expand $F(t)$ in a Taylor series about $t = 0$ and use this expansion to find $f(x + h, y + k)$.

3. Use Taylor's formula to find linear approximations to the following functions valid in the neighborhoods of the points specified.

(i) $x^2 - y^2$, $(1, 1)$ **(ii)** e^{x+y}, $(0, 0)$ **(iii)** $\sin x + \sin y$, $(0, 0)$

(iv) $\log(1 + x + y)$, $(1, -1)$ **(v)** $xy(1 - x - y)$, $(1, 1)$ **(vi)** $\cos(x + y)$, $(1, -1)$

4. Use Taylor's formula to find quadratic approximations to the following functions valid in the neighborhoods of the points specified.

(i) $\dfrac{1 + x + y}{1 + x^2 + y^2}$, $(0, 0)$ **(ii)** $x^5 - 2y^3 + 3xy$, $(1, 2)$ **(iii)** $e^{xy} + e^{-xy}$, $(0, 0)$

5. Find the terms up to degree 2 in the Taylor expansions about $x = 0$, $y = 0$ of the following functions. Verify the results by simpler methods.

(i) $(1 + x + y)^{-1}$ **(ii)** $\cosh(x^2 + 2y^2)$ **(iii)** $e^{-(x+y)^2}$

(iv) $\cos x / \cos y$ **(v)** $\log(1 + x)\log(1 + y)$ **(vi)** $\tan^{-1}\dfrac{x + y}{1 - xy}$

6. Show that the Taylor formula approximation to $e^{-x^2 - 2y^2}$ near $(0,0)$ is of the form $a_0 + a_1x^2 + a_2y^2$, where higher-order terms have been neglected. What is the numerical error in using this approximation at the point $(\frac{1}{2}, \frac{1}{2})$?

7. Obtain quadratic approximations valid near $x = 0$, $y = 0$ for the following functions. Determine the errors in these approximations at $x = \frac{1}{2}$, $y = \frac{1}{2}$.

(i) $\dfrac{e^{x+y}}{e^x + e^y}$ **(ii)** $\dfrac{1}{(2 + x)(2 + y)}$ **(iii)** $\dfrac{1}{\cos x + \cos y}$

8. By multiplying or adding series together, verify the following identities.

(i) $e^{x+y} = e^x e^y$ **(ii)** $\log(1 + x)(1 + y) = \log(1 + x) + \log(1 + y)$

9. Given $z + e^z + x - 2y - 1 = 0$, find $z = f(x, y)$ in the approximate form

$$z = a_0 + a_1x + a_2y + a_{11}x^2 + a_{12}xy + a_{22}y^2,$$

where $f(0, 0) = a_0 = 0$.

10. Given $3x + ye^z + z = 1$, find $z = f(x, y)$ in the approximate form

$$z = a_0 + a_1x + a_2y + a_{11}x^2 + a_{12}xy + a_{22}y^2,$$

where $f(0, 0) = 1$.

11. The surface $z^3 - z(2y - x) + x + y - 2 = 0$ is cut by the line $x = y = 1$ in the three points $z = -1, 0, 1$. Obtain linear approximations to the surface of the form $z = a_0 + a_1(x - 1) + a_2(y - 1)$ at these three points. What is the geometrical interpretation of these approximations?

12. The error function is defined by $\operatorname{erf}(x) = 2\pi^{-1/2}\int_0^x e^{-t^2}\,dt$. Find an approximation to the function $f(x, y) = \operatorname{erf}(x + 2y)$ valid near $(0, 0)$ including fourth order terms. Find an approximation to $g(x, y, z) = \operatorname{erf}(x + 2y + 3z)$ valid near $(0, 0, 0)$ including fourth order terms.

13. Expand the following polynomials in powers of x, $y - \frac{1}{2}$ and $z - 1$.

(i) $x^2 + y^2 + z^2$ **(ii)** $(x + y + z)^2$ **(iii)** $xyz(1 - x - y - z)$

14. The function $y = y(x, \varepsilon)$ is the solution of the equation $y = e^{x - \varepsilon y}\cos x$. Obtain the first three terms of the perturbation expansion $y(x, \varepsilon) = y_0(x) + \varepsilon y_1(x) + \varepsilon^2 y_2(x) + \cdots$ valid for small ε $(\varepsilon \ll 1)$.

15. Obtain the exact solution of the equations $x - 1 = \varepsilon(y + x - 1)$ and $y - 2 = \varepsilon(x - y)$. Compare this solution with the perturbation solution correct to $O(\varepsilon^2)$.

16. Given $x - \alpha = \varepsilon f(x, y)$ and $y - \beta = \varepsilon g(x, y)$ show that for $\varepsilon \ll 1$,

$$x = \alpha + \varepsilon f(\alpha, \beta) + \varepsilon^2[f(\alpha, \beta)f_\alpha(\alpha, \beta) + g(\alpha, \beta)f_\beta(\alpha, \beta)] + \cdots,$$

$$y = \beta + \varepsilon g(\alpha, \beta) + \varepsilon^2[g(\alpha, \beta)g_\beta(\alpha, \beta) + f(\alpha, \beta)g_\alpha(\alpha, \beta)] + \cdots.$$

17. For the equation $y + \varepsilon \sin 2y = x$ (Example 3), assume an expansion of the solution of the form $y(x, \varepsilon) = y_0(x) + \varepsilon y_1(x) + \varepsilon^2 y_2(x) + \cdots$ to obtain the previous results.

18. If $z = e^{x+2\varepsilon y} + e^{y-3\varepsilon x} - e^{x+y}$ and ε is small, find a perturbation expansion for z which includes terms in ε^3. Use the perturbation expansion to estimate z at $x = y = 0$ and at $x = 1$, $y = -1$ with $\varepsilon = 1/10$. Determine the errors in the estimates.

19. Solve the pair of equations $x + y = \varepsilon x + 2$, $x - y = \varepsilon y$ by a perturbation expansion in powers of ε which includes terms in ε^3. Compare this with the exact solution.

20. The three surfaces $x - 1 = \varepsilon(y^2 + z^2)$, $y - 2 = \varepsilon(z^2 - x^2)$, $z - 3 = \varepsilon y \sin(\pi x/2)$ intersect at a point close to $(1, 2, 3)$ when ε is small. Find a better approximation for the intersection point. Determine the errors in the estimates when $\varepsilon = 1/10$.

21. Obtain correct to $O(\varepsilon)$ the solutions of the equations $x^2 + y^2 = 1 + \varepsilon(x^4 + y^4)$, $y = x(1 + \varepsilon)$ which lie close to the circle $x^2 + y^2 = 1$.

22. The Legendre polynomials $P_n(x)$ for n a non-negative integer are defined by the identity

$$F(x, t) = \frac{1}{\sqrt{1 - 2xt + t^2}} = \sum_{n=0}^{\infty} P_n(x)t^n.$$

(i) Show that $P_n(1) = 1$, $P_n(-1) = (-1)^n$ and evaluate $P_n(0)$.
(ii) Show that $P_0(x) = 1$, $P_1(x) = x$, $P_2(x) = 1/2(3x^2 - 1)$, $P_3(x) = 1/2(5x^3 - 3x)$.

23. By differentiating the identity in the previous problem with respect to x and t, deduce the following relations satisfied by the Legendre polynomials.

(i) $xP_n'(x) = nP_n(x) + P_{n-1}'(x)$ **(ii)** $(n + 1)P_{n+1}(x) = (2n + 1)xP_n(x) - nP_{n-1}(x)$

Verify by induction that $P_n(x)$ is a polynomial of degree n which is even when n is an even integer and odd when n is odd.

6.9 Maxima and minima problems

A smooth surface $z = f(x, y)$ may have *relative extrema*, points which are either local maxima or minima. At an extremal point (a, b) the tangent plane to the surface must be horizontal, i.e., parallel to the plane $z = 0$. This conclusion is evident geometrically, or it can be established from the analogous result for a function of one variable. After all, a relative extremal point of the surface must also be an extremal point for all curves on the surface which pass through it, including those made by vertical plane cuts. The tangent plane to the surface $z = f(x, y)$ at the point (a, b) is

$$z - f(a, b) = (x - a)f_a(a, b) + (y - b)f_b(a, b), \tag{1}$$

and since this plane is horizontal if (a, b) is a relative extremum, $\nabla f = \mathbf{0}$ there

or
$$f_a(a, b) = f_b(a, b) = 0. \tag{2}$$

These conditions are necessary for (a, b) to be a maximum or minimum point. [The notation $f_a(a, b)$, $f_b(a, b)$ is used for f_x, f_y evaluated at the point (a, b).] Points that satisfy both Eqs. (2) are called *stationary points* of the function $f(x, y)$. If in addition f increases with *any* small departure from the point (a, b), that is, $f(a+h, b+k) - f(a, b) > 0$ for small nonzero values of h and k, then f has a *minimum* at (a, b). If f decreases with *any* small departure from (a, b), that is, $f(a+h, b+k) - f(a, b) < 0$ for arbitrary small nonzero values of h and k, then f has a *maximum* at (a, b). It must be emphasized that for a point to be a maximum (minimum) f must decrease (increase) for *all* directions of departure from the point. Some elementary examples help to illustrate these ideas.

Example 1 Is the origin a relative maximum or minimum of the following functions?

(i) $1 + x^2 + 2y^2$, **(ii)** $1 - x^2 - 2y^2$, **(iii)** $x^2 - y^2$.

The necessary conditions for a relative extremal point at the origin are satisfied for each function, since $f_x(0, 0) = f_y(0, 0) = 0$. The nature of a point (a, b) [here $(0, 0)$] is decided by an examination of the sign of the difference $f(a+h, b+k) - f(a, b)$.

In case (i) $f(x, y) = 1 + x^2 + 2y^2$, and

$$f(h, k) - f(0, 0) = 1 + h^2 + 2k^2 - 1 = h^2 + 2k^2.$$

For all nonzero h and k this difference is positive and therefore $(0, 0)$ is a relative minimum. The minimum value of f is 1. Note that the level curves of the surface are ellipses (Fig. 9.1).

In case (ii) $f(x, y) = 1 - x^2 - 2y^2$, and

$$f(h, k) - f(0, 0) = 1 - h^2 - 2k^2 - 1 = -h^2 - 2k^2.$$

This difference is negative for all nonzero h and k, and therefore the point $(0, 0)$ is a maximum. The maximum value of f is 1, and the level curves are again ellipses (Fig. 9.2).

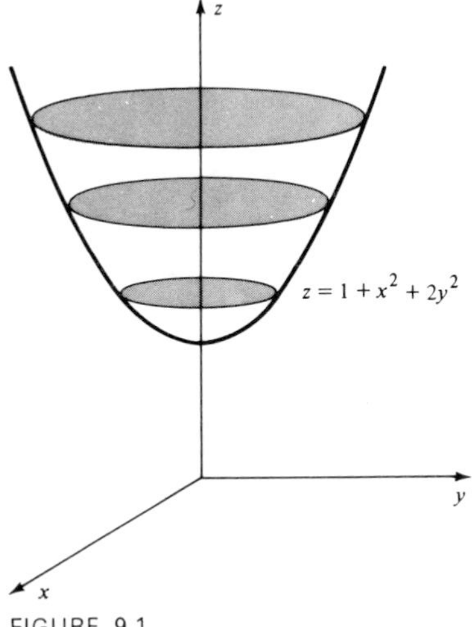

FIGURE 9.1

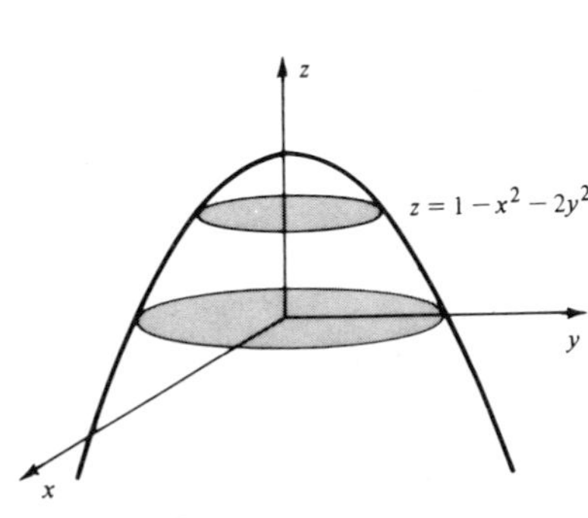

FIGURE 9.2

Case (iii) illustrates an interesting phenomenon absent in the one-variable case. With $f(x, y) = x^2 - y^2$,

$$f(h, k) - f(0, 0) = h^2 - k^2,$$

and this expression is positive for $h^2 > k^2$, zero if $h^2 = k^2$, and negative when $h^2 < k^2$. The point is therefore neither a "good" maximum nor minimum, and for geometrical reasons it is called a *saddle point* (or a minimax). This surface corresponds to the topography of a mountain pass (Fig. 9.3). The mountains are along the positive and negative x axis, and the valleys are along the positive and negative y axis, the y direction being the direction of the mountain pass. The level curves are hyperbolas.

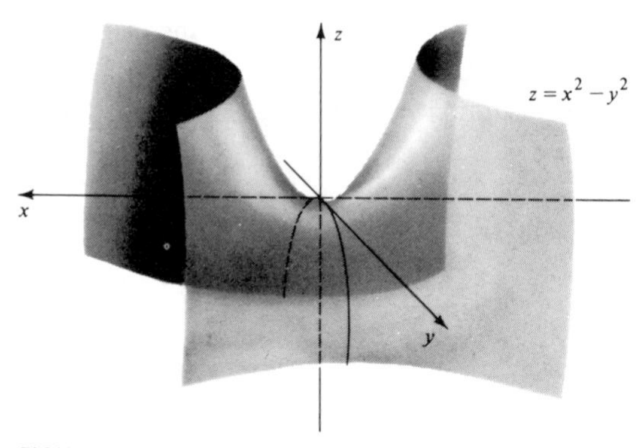

FIGURE 9.3

Example 2 Find stationary values of the function

$$f(x, y) = x^2 + 3xy + 3y^2 - 6x - 3y - 6.$$

Differentiation gives

$$f_x = 2x + 3y - 6,$$
$$f_y = 3x + 6y - 3.$$

The result of solving $f_x = 0, f_y = 0$ is that $(x, y) = (a, b) = (9, -4)$ is the only possible stationary point. On forming the difference $f(a + h, b + k) - f(a, b)$, we find

$$f(9 + h, -4 + k) - f(9, -4) = h^2 + 3hk + 3k^2$$

or

$$f(9 + h, -4 + k) - f(9, -4) = \left(h + \frac{3k}{2}\right)^2 + \frac{3}{4}k^2.$$

In the latter form the expression is the sum of two squares and therefore positive. Accordingly, the point $(9, -4)$ is a minimum, the minimum value of f being $f(9, -4) = -27$.

Each of the preceding examples is intentionally simple in that the sign of $f(a+h, b+k) - f(a, b)$ is very easy to determine. While this is the fundamental test in deciding whether the point is a maximum, minimum, or saddle point, it is desirable to have a more convenient analytical criterion, analogous to the sign of $f''(a)$ in one-variable problems. Such a criterion can be established by means of Taylor's formula.

Let (a, b) be a stationary point of the function $z = f(x, y)$ so that $f_a = f_b = 0$. Taylor's formula for this function is, by (9) of Sec. 6.8,

$$f(a+h, b+k) - f(a, b) = \tfrac{1}{2}(h^2 f_{aa} + 2hk f_{ab} + k^2 f_{bb}) + \cdots. \tag{3}$$

Suppose that at least one of the second derivatives f_{aa}, f_{ab}, f_{bb} is nonzero; then the sign of $f(a+h, b+k) - f(a, b)$, for h and k nonzero and small, is essentially determined by the sign of

$$\delta = h^2 f_{aa} + 2hk f_{ab} + k^2 f_{bb}. \tag{4}$$

Let $f_{aa} \neq 0$; then by completing the square Eq. (4) can be written

$$\delta = \frac{1}{f_{aa}}\,[(hf_{aa} + kf_{ab})^2 + (f_{aa}f_{bb} - f_{ab}{}^2)k^2]. \tag{5}$$

From this expression it is clear that $\delta > 0$ if $f_{aa} > 0$ *and* $f_{aa}f_{bb} - f_{ab}{}^2 > 0$ and that $\delta < 0$ if $f_{aa} < 0$ and $f_{aa}f_{bb} - f_{ab}{}^2 > 0$. On the other hand, if $f_{aa}f_{bb} - f_{ab}{}^2 < 0$, (5) can be written as the difference of two squares, which implies that the sign of δ will depend on the ratio of k/h. In this case, the point (a, b) is a saddle point. Of course these results apply even when $f_{aa} = 0$ as long as *one* of the other second partial derivatives is nonzero.

To summarize, the relative maxima and minima of the function $z = f(x, y)$ are located among the stationary points (a, b) for which

$$\nabla f = \mathbf{0} \qquad \text{or} \qquad f_a = f_b = 0. \tag{6}$$

If, in addition,

$$f_{aa}f_{bb} - f_{ab}{}^2 > 0, \qquad f_{aa} < 0 \qquad (\text{and of necessity } f_{bb} < 0), \tag{7}$$

the point is a relative maximum. On the other hand, if

$$f_{aa}f_{bb} - f_{ab}{}^2 > 0, \qquad f_{aa} > 0 \qquad (\text{and of necessity } f_{bb} > 0), \tag{8}$$

the point is a relative minimum. If, however,

$$f_{aa}f_{bb} - f_{ab}{}^2 < 0, \tag{9}$$

the point is a *saddle point*.

The test is inconclusive when $f_{aa}f_{bb} - f_{ab}{}^2 = 0$, and further considerations are necessary to decide the question. In such cases it is usually best to return to the basic criterion and calculate the sign of $f(a+h, b+k) - f(a, b)$.

The theory can be generalized to functions of more than two variables. If $z = f(x_1, \ldots, x_n)$, the stationary values of f occur where $\nabla f = \mathbf{0}$, but it is a more difficult matter to determine their nature. Fortunately, in practice, the decision can often be anticipated on physical grounds.

Example 3 In light of the above results reconsider Example 2.

Here

$$f(x, y) = x^2 + 3xy + 3y^2 - 6x - 3y - 6,$$

and the only possible stationary point is $(9, -4)$. Calculation of the second partial derivatives yields

$$f_{xx} = 2, \qquad f_{xy} = 3, \qquad f_{yy} = 6,$$

so that in general

$$f_{xx}f_{yy} - f_{xy}^2 = 12 - 9 = 3 > 0.$$

Therefore, according to (8), the point $(9, -4)$ is a relative minimum, as found before.

Example 4 Find the nature of the stationary points of $f(x, y) = x^3 - 12xy + 8y^3$.

The stationary points are determined from $\nabla f = \mathbf{0}$, that is,

$$f_x = 3x^2 - 12y = 0,$$
$$f_y = -12x + 24y^2 = 0.$$

The elimination of y gives $x(x^3 - 8) = 0$, and the solution of this simple equation implies that the only real stationary points are $(0, 0)$ and $(2, 1)$. Furthermore the second derivatives

$$f_{xx} = 6x, \qquad f_{xy} = -12, \qquad f_{yy} = 48y,$$

evaluated at the point $(0, 0)$ imply that

$$f_{xx} = 0, \qquad f_{xx}f_{yy} - {f_{xy}}^2 = -144 < 0;$$

whereas at $(2, 1)$

$$f_{xx} = 12, \qquad f_{xx}f_{yy} - {f_{xy}}^2 = 432 > 0.$$

Conditions (7) to (9) show that $(0, 0)$ is a saddle point and $(2, 1)$ is a minimum.

Example 5 A manufacturer wishes to make an *open* rectangular box of given volume V using the least possible material. Find the design specifications.

If the dimensions of the base of the box (Fig. 9.4) are x and y and z is the height, the specified volume is

$$V = xyz,$$

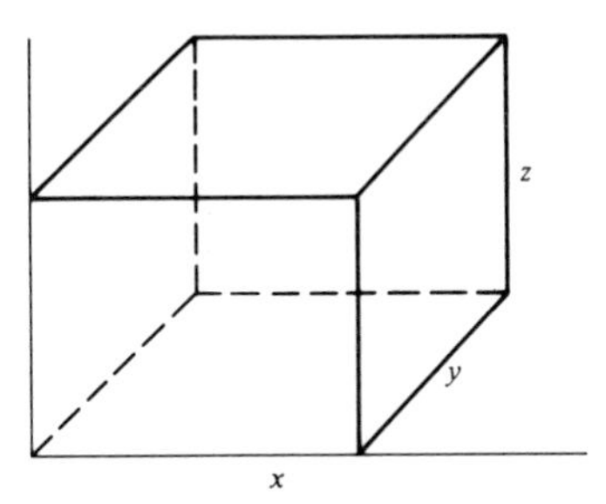

FIGURE 9.4

and the amount of material used is proportional to the surface area

$$S = xy + 2yz + 2zx.$$

The elimination of z between these equations gives

$$S = xy + 2V\left(\frac{1}{x} + \frac{1}{y}\right),$$

which is the function to be minimized. Differentiation yields

$$\frac{\partial S}{\partial x} = y - \frac{2V}{x^2}, \qquad \frac{\partial S}{\partial y} = x - \frac{2V}{y^2};$$

and upon equating these derivatives to zero we find that a stationary value of S occurs for

$$x = y = (2V)^{1/3}.$$

In this example consideration of very shallow and very tall boxes suggests that the stationary point is a relative minimum value of S. This can be formally verified by the second derivative test, and since

$$S_{xx} = \frac{4V}{x^3}, \qquad S_{xy} = 1, \qquad S_{yy} = \frac{4V}{y^3},$$

it follows that $S_{xx} = 2$ and $S_{xx}S_{yy} - S_{xy}{}^2 = 3$ at $x = y = (2V)^{1/3}$. These are the conditions for a minimum value of S. The optimal design specifications are $x = y = (2V)^{1/3} = 2z$, or a square box with a height equal to half the length of its base.

---

3 Extremal problems in more than one variable are usually quite complicated, and their solution may require simplifying approximations, a measure of ingenuity, numerical computation, and certainly hard work. These are typical characteristics of practical problems, which bear little resemblance to the textbook exercises like Example 2. A problem that brings us a little closer to reality is considered next from initial formulation to final resolution.

Example 6 A newly married couple decide to settle in that part of the country where they will see their respective parents as little as possible each year. Can mathematics help them?

The first step is to construct a simple but plausible model so that various ideas can be explored and tested. A more elaborate or sophisticated model can be developed if any success is achieved.

A reasonable assumption is that the number of visits per year and the length of each in days are correlated with the travel distance involved. If the travel distance is great, visits will be few in number but of extended duration. As the result of a little thought and trial and error the number of visits per year V by either set of parents is approximated by the simple relationship

$$V = 1 + \frac{1{,}000}{L},$$

where L is the distance between the parents and their child. For example, $L = 1{,}000$ km corresponds to $V = 2$, that is, two visits per year, whereas $L = 10$ km implies 101 visits per year. Of course, the formula is slightly inaccurate when $L = 0$, but the young couple has no intention of living in the same house with their parents. (No distinction is made between the parents as regards frequency or duration of their visits.)

A reasonable formula for the duration D of each visit in days is

$$D = \frac{1}{20}\left(1 + \frac{L}{10}\right),$$

so that $D = 5.05$ days for $L = 1{,}000$ km and $D = 1/10$ day $= 2.4$ hr for $L = 10$ km.

The number N of days per year that either set of parents will visit is the number of visits per year multiplied by the duration of each:

$$N = D \cdot V,$$

or, in particular,

$$N = \frac{1}{20}\left(1 + \frac{L}{10}\right)\left(1 + \frac{1{,}000}{L}\right) = \frac{1}{2}\left(10.1 + \frac{L}{100} + \frac{100}{L}\right).$$

However, the model is hardly accurate enough to justify using 10.1 instead of 10, and for algebraic and numerical convenience we adjust the preceding to read

$$N = \frac{1}{2}\left(10 + \frac{L}{100} + \frac{100}{L}\right).$$

Since this form makes it desirable to measure distance in units of 100 km, we introduce the variable $l = L/100$ and scale all distances accordingly (so that $l = 1.5$ means $L = 150$ km). Therefore

$$N = \frac{1}{2}\left(10 + l + \frac{1}{l}\right) = F(l). \tag{10}$$

Let the separation distance of the newlyweds to the homes of their respective parents be l_1 and l_2, as shown in Fig. 9.5. Here a coordinate system has been established with one axis passing through the homes of the parents and the origin situated at the midpoint of the line segment of length $2a$ that connects these positions. Point $P = (x, y)$ is the possible homesite of the young couple.

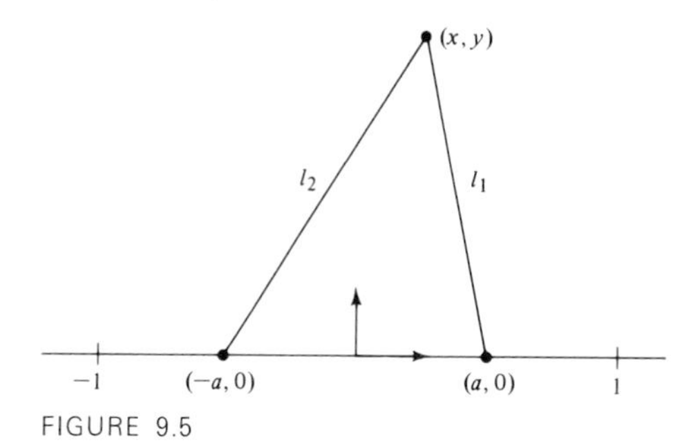

FIGURE 9.5

The total number of days per year that the newlyweds can expect to see the parents of both is the sum of the days each couple will visit:

$$T(x, y) = \frac{1}{2}\left(10 + l_1 + \frac{1}{l_1}\right) + \frac{1}{2}\left(10 + l_2 + \frac{1}{l_2}\right) = F(l_1) + F(l_2), \tag{11}$$

where
$$l_1^2 = (x - a)^2 + y^2; \qquad l_2^2 = (x + a)^2 + y^2. \tag{12}$$

The mathematical formulation is now complete, and it remains to find the relative minima of the function $T(x, y)$. (And then to see whether the solution, and by inference the model, has any merit whatsoever.)

It will be convenient to work almost exclusively in terms of l_1, l_2, and the function $F(l)$ defined in (10). The conditions for an extremum are then

$$\begin{aligned} T_x &= F'(l_1)\frac{\partial l_1}{\partial x} + F'(l_2)\frac{\partial l_2}{\partial x} = 0; \\ T_y &= F'(l_1)\frac{\partial l_1}{\partial y} + F'(l_2)\frac{\partial l_2}{\partial y} = 0. \end{aligned} \tag{13}$$

These equations are satisfied when either

$$F'(l_1) = 0 = F'(l_2), \tag{14}$$

or
$$\frac{\partial l_1}{\partial x}\frac{\partial l_2}{\partial y} - \frac{\partial l_1}{\partial y}\frac{\partial l_2}{\partial x} = 0, \tag{15}$$

and *all* possibilities must be examined.

Since $F'(l) = \frac{1}{2}(1 - 1/l^2)$, the stationary points given by (14) are located at

$$1 - \frac{1}{l_1^2} = 1 - \frac{1}{l_2^2} = 0,$$

or $l_1 = l_2 = 1$. (Only positive values are allowed.) The exact coordinates are obtained from (12):

$$x = 0, \qquad y = \pm(1 - a^2)^{1/2},$$

and this shows that $a \leq 1$ in order for the solution to exist. The interpretation of this result is made clear in Fig. 9.6. If the homes of both sets of parents are less than 200 km apart, the young couple should locate 100 km from each on the intersection of two circles with this radius, as indicated. Each of these points is a relative minimum corresponding to the value $T = 12$ days/year (which may be beyond endurance).

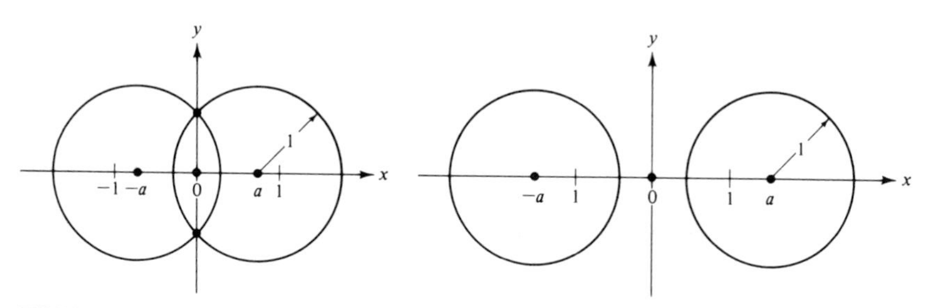

FIGURE 9.6

When $a \geq 1$, an extremum, if one exists, must be associated with the second criterion, (15). The following relationships, obtained from (12),

$$l_1 \frac{\partial l_1}{\partial x} = x - a, \qquad l_1 \frac{\partial l_1}{\partial y} = y, \qquad l_2 \frac{\partial l_2}{\partial x} = x + a, \qquad l_2 \frac{\partial l_2}{\partial y} = y,$$

may be used to rewrite Eq. (15) as

$$0 = l_1 l_2 \left(\frac{\partial l_1}{\partial x} \frac{\partial l_2}{\partial y} - \frac{\partial l_1}{\partial y} \frac{\partial l_2}{\partial x} \right) = (x - a)y - (x + a)y = -2ay.$$

In other words, $y = 0$, and the extremum lies on the x axis; its x coordinate can be determined from the first expression in (13). With $y = 0$, in which case $l_1 = |x - a|$, $l_2 = |x + a|$, that equation reduces to

$$\left(1 - \frac{1}{l_1^2}\right) \frac{x - a}{|x - a|} + \left(1 - \frac{1}{l_2^2}\right) \frac{x + a}{|x + a|} = 0. \tag{16}$$

Two distinct cases arise for positive x. For $x < a$, (16) becomes

$$\frac{1}{l_1^2} - \frac{1}{l_2^2} = 0,$$

which implies that $x = 0$. Therefore, the origin is a *possible* minimum. It is indeed a relative minimum for $a > 1$ but a saddle point for $a < 1$. This makes sense because as $a \to 1$, the two extrema found earlier coalesce to the origin, which remains a minimum as the unit circles in Fig. 9.6 separate for $a > 1$. At this point

$$T = \left(10 + a + \frac{1}{a}\right) \geq 12 \text{ days/year}.$$

The comparison with the first minimum obtained shows that the newlyweds benefit by having their respective parents live closer than 200 km to each other.

In the last case, $x > a$, Eq. (16) for the stationary points becomes

$$\frac{1}{l_1^2} + \frac{1}{l_2^2} = 2. \tag{17}$$

The solutions of this equation are of no consequence because they are saddle points.

The model problem is now solved, and reasonable results were obtained fairly easily. However, pain and labor are hidden in conclusions stated without proof. For example, how do we know that the solution of (17) is indeed a saddle point or for that matter that any of the identifications made are valid? The answer is that the derivative test has been applied in each instance and this involved extended calculations based on the formulas

$$T_{xx} = F''(l_1)\left(\frac{\partial l_1}{\partial x}\right)^2 + F'(l_1)\frac{\partial^2 l_1}{\partial x^2} + F''(l_2)\left(\frac{\partial l_2}{\partial x}\right)^2 + F'(l_2)\frac{\partial^2 l_2}{\partial x^2},$$

$$T_{yy} = F''(l_1)\left(\frac{\partial l_1}{\partial y}\right)^2 + F'(l_1)\frac{\partial^2 l_1}{\partial y^2} + F''(l_2)\left(\frac{\partial l_2}{\partial y}\right)^2 + F'(l_2)\frac{\partial^2 l_2}{\partial y^2},$$

$$T_{xy} = F''(l_1)\frac{\partial l_1}{\partial y}\frac{\partial l_1}{\partial x} + F'(l_1)\frac{\partial^2 l_1}{\partial x\,\partial y} + F''(l_2)\frac{\partial l_2}{\partial y}\frac{\partial l_2}{\partial x} + F'(l_2)\frac{\partial^2 l_2}{\partial x\,\partial y}.$$

The procedure is illustrated by classifying the point $(x, 0)$, which is the solution of (17). With some effort, it follows that

$$T_{xx} = \frac{2}{l_1^3} + \frac{2}{l_2^3} > 0, \qquad T_{yy} = \left(1 - \frac{1}{l_1^2}\right)\frac{1}{l_1} + \left(1 - \frac{1}{l_2^2}\right)\frac{1}{l_2}, \qquad T_{yx} = 0.$$

A little manipulation, using (17), enables us to write

$$T_{yy} = \left(\frac{1}{l_1 l_2} - 1\right)\left(\frac{1}{l_1} + \frac{1}{l_2}\right) = -\frac{1}{2}\left(\frac{1}{l_1} - \frac{1}{l_2}\right)^2\left(\frac{1}{l_1} + \frac{1}{l_2}\right) \leq 0.$$

Since $T_{xx} > 0$, $T_{yy} < 0$, $T_{xy} = 0$, the point in question is definitely a saddle point according to the derivative test. Statements made about the other stationary points are verified in a comparable manner.

4

Observations and experiments often lead to vast accumulations of data. A common situation is the measurement of just two variables x and y resulting in a set of n data points (x_i, y_i), $i = 1, \ldots, n$. On a suitable plot (perhaps using logarithmic scales) it is useful to assume that a linear relation exists between the variables x and y and to try to find a "best" straight line that fits the experimental data. Of course this assumption may be wishful thinking; on the other hand, it may provide an adequate approximation of a very complicated experiment beset with significant errors. A technique commonly used for this purpose is *the method of least squares.* The procedure is to choose the constants m and b of the straight line (see Fig. 9.7),

$$y = mx + b, \tag{18}$$

so that the sum of the squares $D = \sum_{i=1}^{n} d_i^2$ is a minimum. D represents a measure of the absolute deviations of the data points from the straight line. The distance d_i, parallel to Oy, between the line and measured ordinate y_i of the ith data point is

$$d_i = mx_i + b - y_i. \tag{19}$$

By the best straight line we mean that line for which

$$D = \sum_{i=1}^{n} (mx_i + b - y_i)^2, \tag{20}$$

is minimized as a function of m and b. (Other standards by which meaning is given to the word "best" are possible.) Since

$$\frac{\partial D}{\partial m} = 2\sum_{i=1}^{n} x_i(mx_i + b - y_i), \qquad \frac{\partial D}{\partial b} = 2\sum_{i=1}^{n} (mx_i + b - y_i),$$

the values of m and b which make D stationary satisfy the equations

$$m\sum_{i=1}^{n} x_i^2 + b\sum_{i=1}^{n} x_i = \sum_{i=1}^{n} x_i y_i, \qquad m\sum_{i=1}^{n} x_i + bn = \sum_{i=1}^{n} y_i.$$

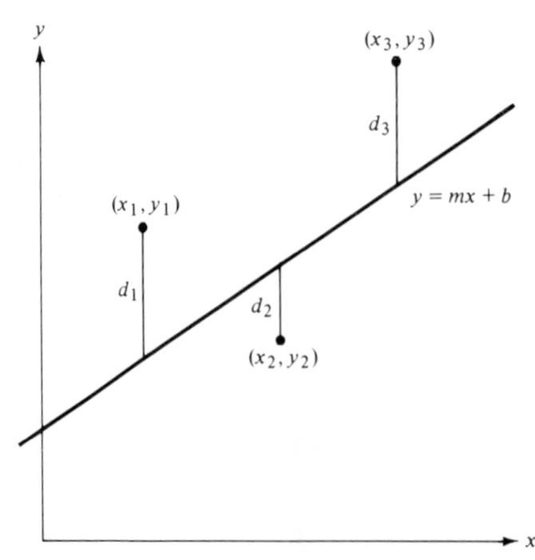

FIGURE 9.7

The solution of these linear equations is readily obtained:

$$m = \frac{n\sum_{i=1}^{n} x_i y_i - \left(\sum_{i=1}^{n} x_i\right)\left(\sum_{i=1}^{n} y_i\right)}{n\sum_{i=1}^{n} x_i^2 - \left(\sum_{i=1}^{n} x_i\right)^2}, \tag{21}$$

$$b = \frac{\left(\sum_{i=1}^{n} y_i\right)\left(\sum_{i=1}^{n} x_i^2\right) - \left(\sum_{i=1}^{n} x_i\right)\left(\sum_{i=1}^{n} x_i y_i\right)}{n\sum_{i=1}^{n} x_i^2 - \left(\sum_{i=1}^{n} x_i\right)^2}, \tag{22}$$

and with these values substituted in (18), we determine the best straight line.

Consideration of lines well away from the data points suggests that this is the minimum value for D, as can be checked analytically. In passing, it should be mentioned that least-square techniques are used in more general ways. For example, we might try to find the least-squares fit of the data to a quadratic polynomial; this and other generalizations are indicated in the exercises.

Example 7 Use the method of least squares to find the straight line which "best" fits the four points (0, 1), (2, 3), (3, 6), and (4, 8).

Here $n = 4$, $x_1 = 0$, $x_2 = 2$, $x_3 = 3$, $x_4 = 4$, $y_1 = 1$, $y_2 = 3$, $y_3 = 6$, and $y_4 = 8$; therefore

$$\sum_{i=1}^{4} x_i = 9, \qquad \sum_{i=1}^{4} y_i = 18, \qquad \sum_{i=1}^{4} x_i^2 = 29, \qquad \sum_{i=1}^{4} x_i y_i = 56$$

Use of formulas (21) and (22) gives $m = 62/35$ and $b = 18/35$, and the equation of the desired line is

$$y = 62/35\, x + 18/35.$$

5

A new class of extremal problems arises when the variables concerned are subject to one (or more) *constraints*. A geometrical problem of this type would be to find which point *on* the sphere $x^2 + y^2 + z^2 = 1$ is farthest from the point (1, 2, 3). This requires that the distance (squared for convenience)

$$f(x, y, z) = (x - 1)^2 + (y - 2)^2 + (z - 3)^2$$

be maximized subject to the constraint $g(x, y, z) = x^2 + y^2 + z^2 - 1 = 0$, which states that the solution is a point on the sphere. This particular problem can be solved by elementary geometry, but in general this is not the case.

There are two procedures for finding stationary values of a function $f(x, y, z)$ subject to a constraint $g(x\ y, z) = 0$.

1. We can (at least in principle) solve the constraint equation $g = 0$ for one of the variables, say $z = z(x, y)$, and then replace it in the function f. The extremal problem would then reduce to finding stationary values of $f(x, y, z(x, y))$, which is now a function of only two independent variables. The partial derivatives of f with respect to x and y are set equal to zero:

$$f_x + f_z z_x = 0, \qquad f_y + f_z z_y = 0.\dagger \tag{23}$$

Since

$$g(x, y, z) = 0, \tag{24}$$

and

$$dg = 0 = g_x\,dx + g_y\,dy + g_z(z_x\,dx + z_y\,dy)$$

or

$$0 = (g_x + z_x g_z)\,dx + (g_y + z_y g_z)\,dy,$$

it follows that

$$z_x = -\frac{g_x}{g_z}, \qquad z_y = -\frac{g_y}{g_z}. \tag{25}$$

These expressions are used to eliminate the partial derivatives z_x, z_y from (23), and the result is

$$f_x g_z - f_z g_x = 0, \qquad f_y g_z - f_z g_y = 0. \tag{26}$$

Together with the equation $g(x, y, z) = 0$, these three equations determine possible values of x, y, z, which make f stationary subject to the constraint $g = 0$.

2. An elegant alternative procedure is the brilliant and inspired method due to Lagrange, called the method of *Lagrange multipliers*.

Consider the function defined by

$$F(x, y, z, \lambda) = f(x, y, z) + \lambda g(x, y, z), \tag{27}$$

where λ, a variable, is called a Lagrange multiplier. The stationary values of F, considered now a function of four independent variables, are solutions of the system of equations

$$F_x = f_x + \lambda g_x = 0, \qquad F_y = f_y + \lambda g_y = 0, \qquad F_z = f_z + \lambda g_z = 0, \qquad F_\lambda = g = 0. \tag{28}$$

The result of eliminating λ from this set of equations is

$$f_x g_z - f_z g_x = 0, \qquad f_y g_z - f_z g_y = 0$$

and

$$g = 0.$$

† Here it pays to identify explicitly the variables held fixed in each partial differentiation. For example, in a more definitive notation, the first equation (23) reads

$$\left(\frac{\partial f}{\partial x}\right)_y = \left(\frac{\partial f}{\partial x}\right)_{y,z} + \left(\frac{\partial f}{\partial z}\right)_{x,y}\left(\frac{\partial z}{\partial x}\right)_y = 0.$$

In words, the partial derivative of $f(x, y, z(x, y))$ with respect to x, *holding y alone fixed*, equals the x partial derivative of $f(x, y, z)$ *holding y and z fixed* plus the additional term shown.

6.10 Numerical methods

Many pressing scientific and engineering problems are too complex to be handled by exact analytical methods. In such circumstances, it is necessary to use an approximate or numerical approach or some combination of both. Weather forecasting is an example where not only is the mathematical model of the atmosphere highly idealized but predictions are also based on a numerical integration of the equations of motion. The search for even more accurate and efficient numerical procedures is currently an active area of research, called *numerical analysis*. Here we shall discuss a few numerical problems involving functions of two variables.

First consider the *finite-difference simulation of partial derivatives*. (In Sec. 2.10 it was pointed out that numerical differentiation presupposes a certain smoothness of the function over the mesh size concerned.) If the value of a function $f(x, y)$ is known at the points (a, b) and $(a + h, b)$, then the basic definition of the partial derivative

$$f_a(a, b) = \lim_{h \to 0} \frac{f(a + h, b) - f(a, b)}{h}, \tag{1}$$

motivates a good numerical approximation when h is small, given by the *forward difference*

$$f_a(a, b) \approx \frac{f(a + h, b) - f(a, b)}{h}.$$

The accuracy of this approximation can be assessed from Taylor's formula

$$f(a + h, b) = f(a, b) + hf_a(a, b) + \frac{h^2}{2!} f_{aa}(a + \theta h, b), \qquad 0 < \theta < 1. \tag{2}$$

This can be written

$$\frac{f(a + h, b) - f(a, b)}{h} - f_a(a, b) = \frac{h}{2} f_{aa}(a + \theta h, b), \qquad 0 < \theta < 1;$$

therefore

$$f_a(a, b) = \frac{f(a + h, b) - f(a, b)}{h} + O(h), \tag{3}$$

where $O(h)$ means that the error is the size of the increment h.

Similarly, by using the *backward difference* we find that

$$f_a(a, b) = \frac{f(a, b) - f(a - h, b)}{h} + O(h). \tag{4}$$

A more accurate approximation can be obtained by using the arithmetic mean of the forward and backward differences called the *centered difference*, namely,

$$\begin{aligned} f_a(a, b) &\approx \frac{1}{2} \left[\frac{f(a + h, b) - f(a, b)}{h} + \frac{f(a, b) - f(a - h, b)}{h} \right], \\ &\approx \frac{f(a + h, b) - f(a - h, b)}{2h}. \end{aligned}$$

The error can be shown to be of order h^2, so that

$$f_a(a, b) = \frac{f(a + h, b) - f(a - h, b)}{2h} + O(h^2). \tag{5}$$

Finite-difference approximations to second partial derivatives are obtained in the same way. If $f(x, y)$ is known at the three points $(a - h, b)$, (a, b), and $(a + h, b)$, then since

$$f_{aa}(a, b) = \lim_{h \to 0} \frac{f_a(a + h, b) - f_a(a, b)}{h},$$

the use of finite-difference simulations for the first partial derivatives suggests that $f_{aa}(a, b)$ can be approximated by $(1/h^2)\,[f(a + h, b) - 2f(a, b) + f(a - h, b)]$. The use of Taylor's formula shows the error to be of order h^2; that is,

$$f_{aa}(a, b) = \frac{f(a + h, b) - 2f(a, b) + f(a - h, b)}{h^2} + O(h^2). \tag{6}$$

Similar ideas apply to finding other derivatives. For example, if $f(x, y)$ is known at the points $(a, b - k)$, (a, b) and $(a, b + k)$, the counterparts to (5) and (6) are

$$f_b(a, b) = \frac{f(a, b + k) - f(a, b - k)}{2k} + O(k^2), \tag{7}$$

and

$$f_{bb}(a, b) = \frac{f(a, b + k) - 2f(a, b) + f(a, b - k)}{k^2} + O(k^2). \tag{8}$$

If $f(x, y)$ is known at the four points (a, b), $(a + h, b)$, $(a, b + k)$, and $(a + h, b + k)$, then

$$f_{ab}(a, b) = \frac{1}{hk}[f(a + h, b + k) - f(a + h, b) - f(a, b + k) + f(a, b)] + O(h) + O(k), \tag{9}$$

and it is just as easy to simulate higher derivatives.

These ideas can be generalized to functions of more than two variables and developed via operational methods (see Exercise 8).

Example 1 The value of the function $f(x, y)$ is known at the three points (a, b), $(a + h, b)$, and $(a, b + k)$. Obtain a finite-difference approximation for the value of f at the point $(a + h, b + k)$ (Fig. 10.1).

Our procedure is to assume that

$$f(a + h, b + k) \approx A_0 f(a, b) + A_1 f(a + h, b) + A_2 f(a, b + k), \tag{10}$$

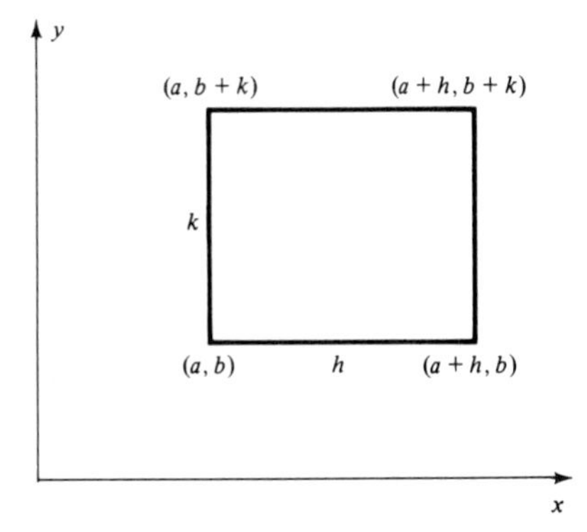

FIGURE 10.1

and to determine the constants A_0, A_1, and A_2 by the criterion that the Taylor expansions of both sides agree as closely as possible.

The Taylor series expansion for the left side of (10) for small h and k is

$$f(a+h, b+k) = f(a, b) + hf_a(a, b) + kf_b(a, b) + \cdots,$$

while the expansion of the right side yields

$$\begin{aligned} A_0 f(a, b) + A_1 f(a+h, b) + A_2 f(a, b+k) \\ = (A_0 + A_1 + A_2) f(a, b) + hA_1 f_a(a, b) + kA_2 f_b(a, b) + \cdots. \end{aligned}$$

The first three terms of these series can be made to agree by choosing

$$A_0 + A_1 + A_2 = 1, \qquad A_1 = 1, \qquad A_2 = 1.$$

(The series cannot be made identical beyond this stage, and this becomes the source of error in the numerical simulation.) The desired approximation is then

$$f(a+h, b+k) \approx f(a+h, b) + f(a, b+k) - f(a, b).$$

This approximation is equivalent to using the plane through the three given points $(a, b, f(a, b))$, $(a+h, b, f(a+h, b))$, and $(a, b+k, f(a, b+k))$ to estimate $f(a+h, b+k)$.

Example 2 It is desired to estimate the location of the eye of an offshore hurricane and its central pressure given simultaneous pressure measurements at the six locations indicated in Fig. 10.2.

Let x and y measure distances to the east and north in miles; the measured pressure p (in inches of mercury) at each location is tabulated below.

TABLE 10.1

Site	x	y	p
1	0	0	30.10
2	100	0	29.40
3	200	0	29.00
4	100	100	28.90
5	0	100	29.50
6	0	200	29.10

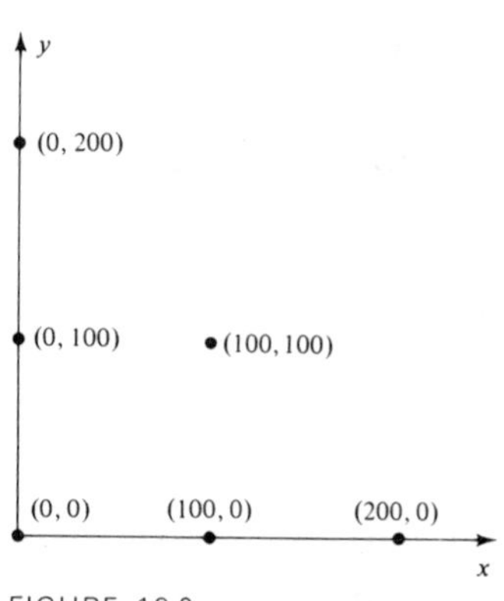

FIGURE 10.2

Six data points allow the determination of six constants, and for this reason we represent the pressure distribution as a second-degree polynomial,

$$p = a_0 + a_1 x + a_2 y + a_{11}x^2 + a_{12} xy + a_{22} y^2. \tag{11}$$

The constants $a_0, a_1, a_2, a_{11}, a_{12}, a_{22}$ are determined from the pressure measurements at the six sites. In order, these are

$$\begin{aligned}
30.10 &= a_0,\\
29.40 &= a_0 + 10^2 a_1 + 10^4 a_{11},\\
29.00 &= a_0 + 2\cdot 10^2 a_1 + 4\cdot 10^4 a_{11},\\
28.90 &= a_0 + 10^2(a_1 + a_2) + 10^4(a_{11} + a_{12} + a_{22}),\\
29.50 &= a_0 + 10^2 a_2 + 10^4 a_{22},\\
29.10 &= a_0 + 2\cdot 10^2 a_2 + 4\cdot 10^4 a_{22}.
\end{aligned}$$

The first three equations can be solved independently of the others, and the results are

$$a_0 = 30.10, \qquad a_1 = -0.85 \times 10^{-2}, \qquad a_{11} = 0.15 \times 10^{-4}.$$

Similarly the last two equations yield

$$a_2 = -0.7 \times 10^{-2}, \qquad a_{22} = 0.1 \times 10^{-4}.$$

The fourth equation can then be solved to give

$$a_{12} = 0.1 \times 10^{-4}.$$

With these specific values, Eq. (11) takes the form

$$p = 30.1 - 10^{-2}(0.85x + 0.70y) + 10^{-4}(0.15x^2 + 0.10xy + 0.10y^2),$$

and this provides a model of the hurricane at the time of the pressure measurements. The eye is the point of lowest pressure. This corresponds to the single minimum of the function $p(x, y)$, which is located where $\nabla p = \mathbf{0}$, that is,

$$3x + y = 850,$$
$$x + 2y = 700.$$

This implies $x = 200$, $y = 250$. Therefore an analysis using a quadratic approximation leads to the prediction that the eye of the storm is located at the point $(200, 250)$, the central pressure being $p(200, 250) = 28.375$.

The preceding example illustrates a method used in experimental situations where there are usually more than two variables and it is desired to find the maximum or minimum of some output, e.g., a particular chemical compound. A set of experiments are designed and the output is measured for each. From these data the optimal experimental conditions (if they exist) can be estimated by fitting the data to a quadratic function of the variables concerned. In practice this method is often used in combination with a least-squares analysis.

Another type of situation which can arise in practice is to find maxima and minima of a known but very complicated function $f(x, y)$. It is not a simple matter to form the derivatives f_x and f_y and to find where they vanish, let alone use the criterion for a maximum or minimum. An alternative approach is simply to compute the value of the function at a certain set of points and inspect its behavior. Refinement of the mesh size near suspected maxima and minima makes it possible to pin down the location and nature of any stationary points.

2

The task of locating the *simultaneous zeros of several functions in several variables* is at the heart of many extremal problems. Perhaps the most obvious technique for this purpose is to generalize the Newton-Raphson method (see Sec. 2.10).

In two variables, the problem is to solve the simultaneous equations

$$\begin{aligned} f(x, y) &= 0, \\ g(x, y) &= 0. \end{aligned} \tag{12}$$

Geometrically this corresponds to finding the points of intersection of the two curves $f = g = 0$. Let (x_0, y_0) be an initial guess for the common zero. Usually a better guess can then be obtained by some sort of iteration procedure. The Taylor series expansions of the functions about (x_0, y_0) are

$$f(x, y) = f(x_0, y_0) + (x - x_0)f_x(x_0, y_0) + (y - y_0)f_y(x_0, y_0) + \cdots,$$
$$g(x, y) = g(x_0, y_0) + (x - x_0)g_x(x_0, y_0) + (y - y_0)g_y(x_0, y_0) + \cdots.$$

If only linear terms are retained, we can estimate a new location for the root by setting the right-hand sides of the last equations equal to zero. This produces new values for the zeros; call them (x_1, y_1). The process is then repeated, using (x_1, y_1) as the initial guess, to obtain new approximations, etc. In other words the iteration is based on the formulas

$$\begin{aligned}(x_{n+1} - x_n)f_x(x_n, y_n) + (y_{n+1} - y_n)f_y(x_n, y_n) &= -f(x_n, y_n),\\ (x_{n+1} - x_n)g_x(x_n, y_n) + (y_{n+1} - y_n)g_y(x_n, y_n) &= -g(x_n, y_n).\end{aligned} \tag{13}$$

We omit any discussion of the convergence of this procedure and its rather obvious generalizations to larger systems of nonlinear equations.

Example 3 Use the Newton-Raphson method to find the zeros of

$$x^2 - y^2 - 3x + y + 4 = 0, \qquad 2xy - 3y - x + 3 = 0$$

in the first and fourth quadrant.

Let $f(x, y) = x^2 - y^2 - 3x + y + 4$ and $g(x, y) = 2xy - 3y - x + 3$ and use the iteration based on (13). The results of the iteration for the initial guesses $(x_0, y_0) = (2, 2)$ and $(x_0, y_0) = (2, 3)$ are tabulated. In each case the exact solution (1, 2) is attained after only four iterations.

TABLE 10.2

n	x_n	y_n	n	x_n	y_n
0	2.0000000	2.0000000	0	2.0000000	3.0000000
1	1.0999990	1.6999990	1	1.3846140	2.0769220
2	0.9875005	2.0374990	2	1.0153830	1.9538450
3	0.9998482	2.0004560	3	0.9997562	2.0007320
4	1.0000000	2.0000000	4	1.0000000	2.0000000
5	1.0000000	2.0000000	5	1.0000000	2.0000000

To find the solution in the fourth quadrant [the exact solution is $(2, -1)$] we list the results obtained by starting with the initial guesses $(x_0, y_0) = (2, 0)$ and $(x_0, y_0) = (2, -2)$. Again the convergence is rapid.

TABLE 10.3

n	x_n	y_n	n	x_n	y_n
0	2.0000000	0.0000000	0	2.0000000	−2.0000000
1	1.5000000	−1.5000000	1	1.9615380	−1.1923070
2	1.8750000	−1.0000000	2	1.9932680	−1.0086530
3	2.0012250	−0.9950981	3	1.9999610	−0.9999976
4	1.9999930	−1.0000040	4	2.0000000	−1.0000000
5	2.0000000	−1.0000000	5	2.0000000	−1.0000000

Many numerical techniques are available for solving *ordinary differential equations.* A few of the most commonly used methods are discussed now. (We will not attempt to enumerate all these methods, nor will we dwell on the merits and defects of particular techniques because such a discussion is more appropriate in a course on differential equations.)

Consider the first-order differential equation

$$\frac{dy}{dx} = f(x, y), \tag{14}$$

which is to be solved subject to the condition that $y = y_0$ at $x = x_0$. Only in certain special cases can the solution of (14) be found by exact analytical methods in terms of simple known functions. However, the form of the solution can always be pictured by drawing the direction field (see Sec. 2.11) and approximated by power-series expansions (see Secs. 4.4 and 4.8).

Let us try to find an approximate solution $y = y(x)$ of (14) by numerical means denoting by y_r the value of y at $x = x_r = x_0 + rh$, where h is the mesh size. The simple-minded approach is to replace dy/dx at the point x_r by the forward difference $(y_{r+1} - y_r)/h$ so that the differential equation is replaced by the difference equation

$$y_{r+1} = y_r + hf(x_r, y_r). \tag{15}$$

Since x_0 and y_0 are given, (15) enables us to calculate sequentially the value of y at the discrete points $x_r = x_0 + rh$, with $r = 0, 1, 2, \ldots$. Moreover, we anticipate that the accuracy of this approximate solution improves as h decreases. Indeed, use of Taylor's formula shows that

$$y_{r+1} - y_r - hf(x_r, y_r) = O(h^2), \tag{16}$$

so that halving the step size h will actually decrease the error by a factor of 4. (However, the increased number of subdivisions detracts somewhat from the ideal gain in accuracy.)

A minor refinement can greatly improve the accuracy, and the possible variations are endless. For example, we can write

$$\frac{y_{r+1} - y_r}{h} = \tfrac{1}{2}[f(x_r, y_r) + f(x_{r+1}, y_{r+1})]$$

and replace the unknown y_{r+1} on the right-hand side [in $f(x_{r+1}, y_{r+1})$] by

$$\begin{aligned} y_{r+1} &= y_r + h\left(\frac{dy}{dx}\right)\Bigg]_{x=x_r}, \\ &= y_r + hf(x_r, y_r). \end{aligned}$$

The finite-difference simulation then becomes

$$y_{r+1} = y_r + \frac{h}{2}[f(x_r, y_r) + f(x_{r+1}, y_r + hf(x_r, y_r))]. \tag{17}$$

Given x_0 and y_0, this equation determines y_1, y_2, etc., in succession. It is not at all clear that this represents any improvement over (15). However, an analysis of the error by use of Taylor's formula shows it to be $O(h^3)$. Equation (17) is a more accurate approximation, but the price paid for this increased precision is a more complicated recurrence relation. It should be noted that when f is independent of y, (17) reduces to the usual trapezoidal rule for numerical integration (see Sec. 3.10).

Finally, we state a method extensively used for its accuracy and relative simplicity, the four-point *Runge-Kutta formula*:

$$y_{r+1} = y_r + \frac{h}{6}(f_{r1} + 2f_{r2} + 2f_{r3} + f_{r4}), \tag{18}$$

where

$$\begin{aligned} f_{r1} &= f(x_r, y_r), \\ f_{r2} &= f\left(x_r + \frac{h}{2}, y_r + \frac{h}{2}f_{r1}\right), \\ f_{r3} &= f\left(x_r + \frac{h}{2}, y_r + \frac{h}{2}f_{r2}\right), \\ f_{r4} &= f(x_r + h, y_r + hf_{r3}). \end{aligned} \tag{19}$$

The error can be shown to be $O(h^5)$, which implies that halving the step size reduces the error by a factor of 32. Formula (18) is really a generalization of Simpson's rule (see Sec. 3.10), to which it reduces if f is independent of y.

We give one example to illustrate these methods.

Example 4 Apply the above three numerical schemes to find the solution of the differential equation

$$\frac{dy}{dx} = x^2 - y + x + 1, \qquad \text{with } y(0) = 1.$$

In particular calculate $y(1)$.

TABLE 10.4

x	y			
	Exact solution	*Method (i)*	*Method (ii)*	*Method (iii)*
		Mesh size $h = 0.1$		
0.0000000	1.0000000	1.0000000	1.0000000	1.0000000
0.1000000	1.0051626	1.0000000	1.0055000	1.0051627
0.2000000	1.0212692	1.0110000	1.0219275	1.0212695
0.3000000	1.0491818	1.0339000	1.0501444	1.0491821
0.4000000	1.0896799	1.0695100	1.0909307	1.0896804
0.5000000	1.1434693	1.1185590	1.1449922	1.1434699
0.6000000	1.2111884	1.1817031	1.2129680	1.2111890
0.7000000	1.2934147	1.2595328	1.2954360	1.2934155
0.8000000	1.3906710	1.3525795	1.3929196	1.3906719
0.9000000	1.5034303	1.4613215	1.5058922	1.5034313
1.0000000	1.6321205	1.5861894	1.6347825	1.6321216
		Mesh size $h = 0.05$		
0.0000000	1.0000000	1.0000000	1.0000000	1.0000000
0.0500000	1.0012706	1.0000000	1.0013125	1.0012706
0.1000000	1.0051626	1.0026250	1.0052454	1.0051626
0.1500000	1.0117920	1.0079937	1.0119147	1.0117920
0.2000000	1.0212692	1.0162190	1.0214307	1.0212693
0.2500000	1.0336992	1.0274081	1.0338985	1.0336992
0.3000000	1.0491818	1.0416627	1.0494178	1.0491818
0.3500000	1.0678119	1.0590796	1.0680837	1.0678119
0.4000000	1.0896799	1.0797506	1.0899865	1.0896800
0.4500000	1.1148718	1.1037631	1.1152121	1.1148719
0.5000000	1.1434693	1.1311999	1.1438424	1.1434694
0.5500000	1.1755502	1.1621399	1.1759551	1.1755502
0.6000000	1.2111884	1.1966579	1.2116241	1.2111884
0.6500000	1.2504542	1.2348250	1.2509200	1.2504543
0.7000000	1.2934147	1.2767088	1.2939095	1.2934147
0.7500000	1.3401334	1.3223733	1.3406564	1.3401335
0.8000000	1.3906710	1.3718797	1.3912213	1.3906711
0.8500000	1.4450851	1.4252857	1.4456618	1.4450851
0.9000000	1.5034303	1.4826464	1.5040326	1.5034304
0.9500000	1.5657590	1.5440141	1.5663860	1.5657590
1.0000000	1.6321205	1.6094384	1.6327716	1.6321206

The three schemes are given by (i) Eq. (15); (ii) Eq. (17); and (iii) Eq. (18). The exact solution of this differential equation is

$$y = x^2 - x + 2 - e^{-x},$$

and the numerical approximations can be checked against this function. The accurate value of $y(1)$ is 1.632 120 56. In Table 10.4 the exact solution and the solutions found by using methods (i) to (iii) are given; two mesh sizes, $h = 0.1$ and $h = 0.05$, were used. The results show the Runge-Kutta method, corresponding to method (iii), to be in excellent agreement with the exact solution for all values of x.

Exercises 6.10

1. If $f(0,0) = 1$, $f(0.1,0) = 1.1$, $f(0,0.1) = 0.95$, estimate $f(0.05,0.05)$, $f_x(0,0)$ and $f_y(0,0)$.
2. If the value of a function $f(x,y)$ is known at the three points (a,b), $(a+h,b)$, and $(a,b+k)$, find an approximation for ∇f at the point (a, b). If the value of f is known at the two additional points $(a-h,b)$ and $(a,b-k)$, find an improved estimate for ∇f.
3. If the value of a function $f(x,y)$ is known at six points (rh, sk), where r and s are nonnegative integers such that $r + s \leqslant 2$, find the best estimates for all the first and second partial derivatives at the point $(0,0)$. What can you say about the errors?
4. The value of a function $f(x,y,z)$ is known at the four points (a,b,c), $(a+h,b,c)$, $(a,b+k,c)$, and $(a,b,c+l)$. Estimate ∇f at the point (a,b,c).
5. Show that the difference equation

$$u(x+h,\ t) - u(x,\ t) - \frac{h}{k}\,[u(x,\ t+k) - u(x,\ t)] = 0$$

simulates the partial differential equation $\partial u/\partial t - \partial u/\partial x = 0$. In particular when $h = k$, show that this simulation is exact.
6. Obtain a finite-difference simulation of the equation $\partial u/\partial t = \partial^2 u/\partial x^2$.
7. The function $f(x,y)$ satisfies the equation $\partial^2 f/\partial x^2 + \partial^2 f/\partial y^2 = 0$ and is known at the four points $(\pm h, 0)$ and $(0, \pm h)$; show that

$$f(0,0) \approx 1/4[f(h,0) + f(-h,0) + f(0,h) + f(0-h)].$$

8. For a function of two variables $f(x,y)$ the operators Δ_x, Δ_y, E_x, E_y, D_x, and D_y are defined by:

$$\Delta_x f(x,y) = f(x+h,y) - f(x,y), \qquad \Delta_y f(x,y) = f(x,y+k) - f(x,y),$$
$$E_x f(x,y) = f(x+h,y), \qquad E_y f(x,y) = f(x,y+k),$$
$$D_x f(x,y) = \frac{\partial}{\partial x} f(x,y), \qquad D_y f(x,y) = \frac{\partial}{\partial y} f(x,y).$$

Show that:
(i) $E_x = 1 + \Delta_x$, $E_y = 1 + \Delta_y$ **(ii)** $E_x = e^{hD_x}$, $E_y = e^{kD_y}$
9. The snowfall measured at six locations A to F is indicated in the table and Fig. 10.3. Estimate the magnitude and location of maximum snowfall and find the extent of the region which received snow.

Location	Snowfall, in.
$A = (0, 0)$	5.2
$B = (50, 0)$	8.7
$C = (100, 0)$	7.2
$D = (0, 50)$	8.2
$E = (0, 100)$	1.2
$F = (50, 50)$	11.2

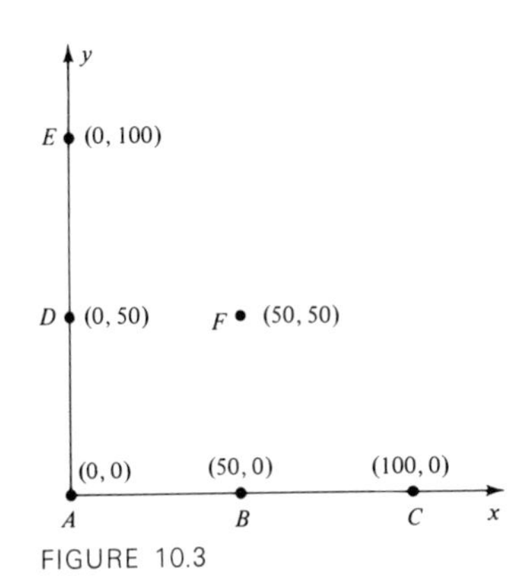

FIGURE 10.3

10. Find the smallest positive value of x (correct to three decimal places) which satisfies $y = \sin x$, $xy = 1$. Illustrate this problem geometrically.

11. Find the points of intersection (correct to three decimal places) of the curves $y = e^x$, $x^2 + y^2 = 1$.

12. Find the solution (correct to three decimal places) of the simultaneous equations

$$10x^3 - 2y^3 - 10y^2 + 1 = 0, \qquad x^3 - 3y^2 + 5x - 5 = 0$$

which is close to $x = 6$, $y = 8$.

13. For each of the following differential equations assume $y(0) = 1$ and find $y(1)$ by a numerical method.

(i) $\dfrac{dy}{dx} = x - y^2$; **(ii)** $\dfrac{dy}{dx} = x + \sin y$; **(iii)** $\dfrac{dy}{dx} = e^{-xy}$.

14. If $dy/dx = 1 + 2xy$ and $y(0) = 1$, find $y(1)$ by the three numerical methods described in the text and compare your results with the exact solution $y = e^{x^2}[1 + (\sqrt{\pi}/2)\ \mathrm{erf}(x)]$.

15. If $dy/dx = x - y$ with $y(0) = 1$, find $y(1)$ by the three numerical methods described in the text and compare your results with the exact solution $y = x - 1 + 2e^{-x}$.

16. In Exercises 14 and 15 use a power-series approximation $y = \sum_{r=0}^{\infty} a_r x^r$ (retaining terms in x^6) to solve the differential equations concerned and compare the results with the previous numerical and exact calculations.

17. Derive a simple finite-difference simulation for the second-order differential equation $y'' = f(x, y)$ given y and y' at one point $x = x_0$. Apply this method to $y'' = x - y$ with $y(0) = 1$, $y'(0) = 1$, to find $y(\pi/2)$. Compare your numerical result with the exact solution $y = x + \cos x$.

18. Use Taylor's formula to show that the error in the four-point Runge-Kutta method is of order h^5. Verify that this method reduces to Simpson's rule if $\partial f/\partial y = 0$.

7
Multiple integrals

7.1 Double integrals; iterated integrals

The definite integral arose from the calculation of an area as the limit of a sequence of approximations based on incremental rectangular slices, a sum of infinitesimal elements. For example, the area between the lines $x = a$, $x = b$, $y = 0$ and the curve $y = f(x)$, shown in Fig. 1.1, where $f(x)$ is assumed positive for convenience, is represented by the integral

$$I = \int_a^b f(x)\,dx. \tag{1}$$

The generic form which reveals the basic method of calculation is

$$I = \lim_{\substack{n\to\infty \\ \max|\Delta x_i|\to 0}} \sum_{i=1}^{n} f(\xi_i)\,\Delta x_i, \tag{2}$$

where $x_{i-1} \le \xi_i \le x_i$, $\Delta x_i = x_i - x_{i-1}$, $x_0 = a$, and $x_n = b$ (Fig. 1.2). Although

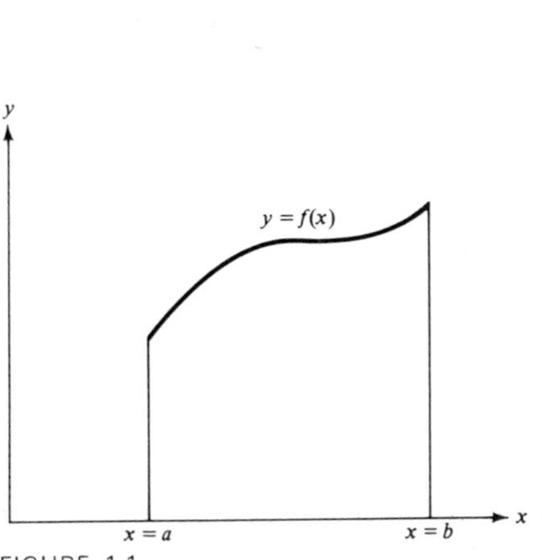

FIGURE 1.1

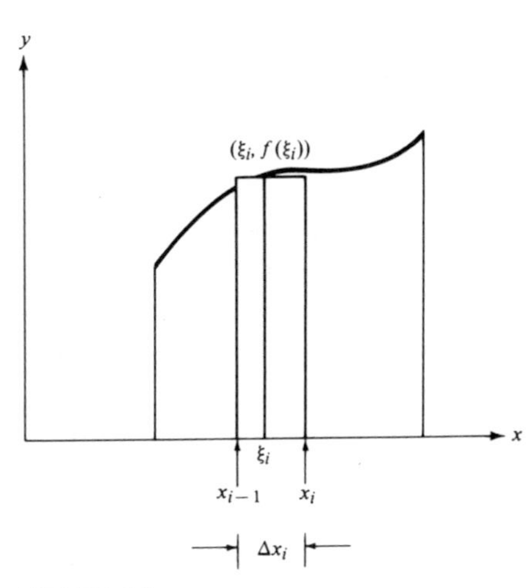

FIGURE 1.2

a great number of problems can be formulated as definite integrals and then solved explicitly, many still fall beyond the scope of the analytical methods developed so far. The calculation of an arbitrary volume, such as that of a potato, is representative of these; in fact, the techniques available at present are adequate only for bodies of revolution or other highly specialized volumes. However, it must be apparent that the basic idea of constructing approximations from small increments of volume is sound and in the limit will lead once again to the exact answer. This generalization of the single definite integral is the primary objective of the present chapter.

Let us begin the discussion by trying to find a volume such as that shown in Fig. 1.3. This volume is bounded below by a specified region R in the xy plane, above by the surface $z = f(x, y)$, and laterally by the vertical "wall" which passes through the perimeter of R. [To avoid "negative" volumes it will be assumed that $f(x, y) > 0$, but this condition will soon be relaxed.]

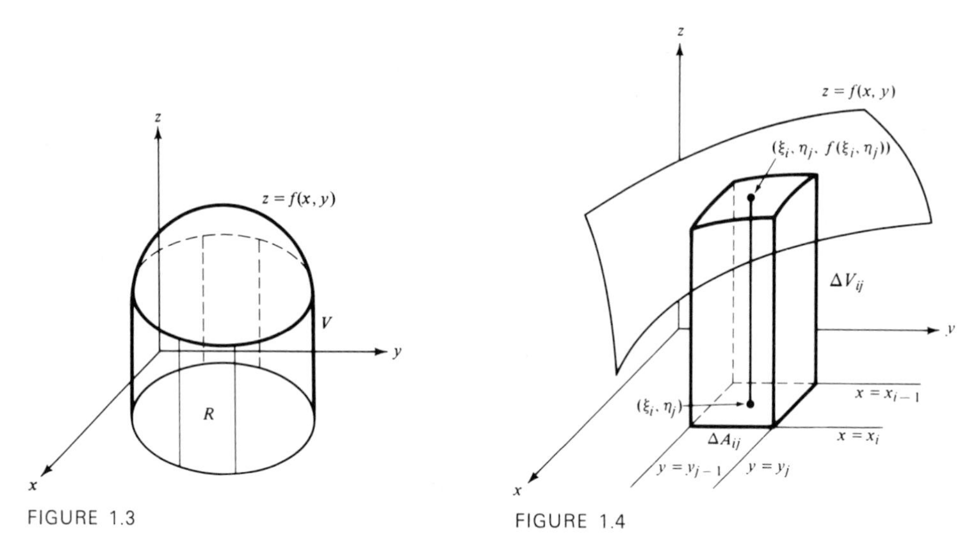

FIGURE 1.3

FIGURE 1.4

As always when confronted with a new problem, a return to a fundamental principle is indicated. This was enunciated at the outset of Chap. 1: *devise an approximate simple method of calculation which in a limiting sense will provide the exact answer.* It is *the* basic approach, and it works once again.

If we subdivide the region R into small areas by drawing the sets of lines $x = x_i$, $i = 1, 2, \ldots, n$, and $y = y_j$, $j = 1, 2, \ldots, p$, we can approximate the exact element of volume above the incremental area $\Delta A_{ij} = \Delta x_i \, \Delta y_j$ by the volume of the solid rectangular parallelepiped having the constant height $f(\xi_i, \eta_j)$ and the same base area ΔA_{ij}. In other words,

$$\Delta V_{ij} = f(\xi_i, \eta_j) \, \Delta x_i \, \Delta y_j, \tag{3}$$

where the point (ξ_i, η_j) is *any point* such that $x_{i-1} \le \xi_i \le x_i$ and $y_{j-1} \le \eta_j \le y_j$ (see Fig. 1.4). The summation of all such elements of volume is an approximation to the desired volume, an approximation that is systematically improved by using finer and finer rulings. In order to exclude from this summation all the incremental rectangles which lie completely outside R, we simply set $f(x, y) = 0$ when x, y is not a point of R. Therefore, in the limit as the size of the subdivisions become zero,

$$\lim_{\substack{n, p \to \infty \\ \max(|\Delta x_i|, |\Delta y_j|) \to 0}} \sum_{i=1}^{n} \sum_{j=1}^{p} f(\xi_i, \eta_j) \Delta x_i \Delta y_j \tag{4}$$

gives the precise value of the desired volume. The expression (4) is called the double integral of the function $f(x, y)$ over the region R and is written

$$\iint_R f(x, y)\, dA \qquad \text{or} \qquad \iint_R f(x, y)\, dx\, dy. \tag{5}$$

Functions $f(x, y)$ for which the limit (4) exists are said to be integrable over the region R. From a geometrical viewpoint a function $f(x, y)$ is integrable over a region R if the associated volume is finite. The particular case $f(x, y) = 1$ is especially noteworthy because the volume in question is the base area R multiplied by a unit height:

$$V = \iint_R f(x, y)\, dA = 1 \cdot \iint_R dA = A.$$

In other words, the *area* A of a finite region R is also represented as a double integral which is nothing more than the limiting sum of all the infinitesimal increments $dx\, dy$. This is the method of area calculation applied in Sec. 1.1. The relationship between a double integral and a single integral which represents the same area will be explained shortly.

These concepts are obvious generalizations of the one-variable case. However, the study of double integrals is no idle exercise in generalizations. Such integrals arise naturally in the real world, and apart from the initial motivation of area and volume calculations, many other physical problems necessitate the use of double integrals. Some of these applications will be discussed in Sec. 7.5.

2

Having defined what is meant by a double integral, the question arises how its numerical value can be calculated. The basic method, which we will now describe, is to reduce a double integral to the successive evaluation of two single integrals. This means that all the techniques developed in our earlier work on integration are directly applicable to the task at hand.

In expression (4), let us suppose that the summation over i is performed first and that this is followed by the summation over j. Since y is kept fixed in the first summation (over i), it is really a count of elemental rectangles over a strip *parallel* to the x axis (Fig. 1.5*a*). The corresponding elemental volumes together form a slice, a thin sliver of the body which is parallel to the x axis and of thickness Δy_j. The second summation (over j) is then a count in the y direction of adjacent strips. (In this way, the volume slices are reconstituted into the complete body.) In other words one way of evaluating the limits is to rewrite (4) as

$$\lim_{\substack{p\to\infty \\ \max|\Delta y_j|\to 0}} \sum_{j=1}^{p} \Delta y_j \left[\lim_{\substack{n\to\infty \\ \max|\Delta x_i|\to 0}} \sum_{i=1}^{n} f(\xi_i, \eta_j)\, \Delta x_i \right]. \tag{6}$$

Each of the limiting summations in this expression corresponds to an ordinary integration. The first or inner i summation is equivalent to an integration in the variable x, and likewise the j summation gives rise to a y integration. The limits of integration will depend on the boundary of the region R. For a fixed value of y, the x integration range extends from the extreme left to the extreme right. If $x = \psi_L(y)$ and $x = \psi_R(y)$ designate the left- and right-hand boundary curves of the region R, as shown in Fig. 1.5a, then $\psi_L(y) \le x \le \psi_R(y)$ is the range of integration and $\psi_L(y)$ and $\psi_R(y)$ are, re-

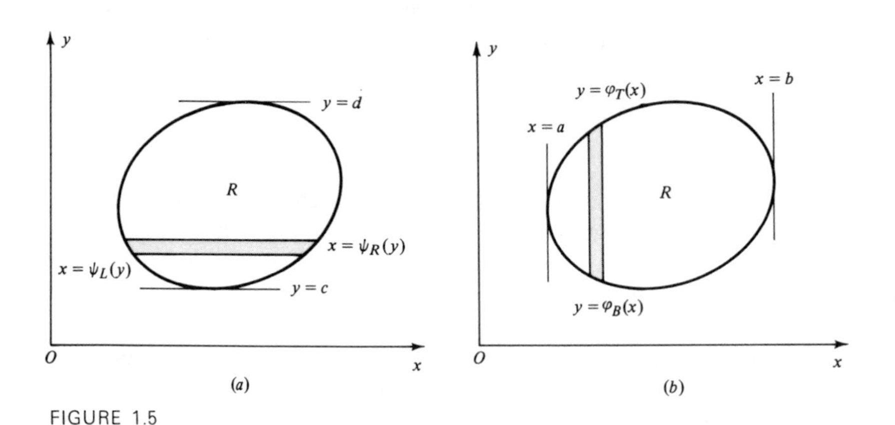

FIGURE 1.5

spectively, the lower and upper limits of integration for the inner sum in (6). The limits of the y integration, the outer sum in the equation, are the lowest and highest values of y on the perimeter of R, that is, $c \le y \le d$. In a manner of speaking, x sweeps from left to right for y constant, and the line segment so formed is then swept over all of R from bottom to top. By this means (6) can be written

$$\int_c^d dy \left[\int_{x=\psi_L(y)}^{x=\psi_R(y)} f(x, y)\, dx \right]. \tag{7}$$

Alternatively, we could first sum over j, that is, integrate with respect to y, and then complete the summation in the index i (integration with respect to x). Formula (4) for the double integral then becomes

$$\int_a^b dx \left[\int_{y=\varphi_B(x)}^{y=\varphi_T(x)} f(x, y)\, dy \right], \tag{8}$$

7. Evaluate the following integrals. Sketch the regions of integration.

(i) $\int_0^2 dy \int_0^1 (xy+3)\,dx$ **(ii)** $\int_{-1}^1 du \int_0^u (u-v)\,dv$ **(iii)** $\int_0^1 dx \int_{1-x}^{1+x} (2y+3x)\,dy$

(iv) $\int_0^1 dy \int_1^y (x^2+2y^2)\,dx$ **(v)** $\int_0^{\pi/2} dr \int_r^{\sin r} (r+s)\,ds$ **(vi)** $\int_0^1 d\theta \int_{-\theta}^{\theta} \theta\phi\,d\phi$

8. Evaluate the following integrals where R is the region bounded by $|x|+|y|=1$.

(i) $\iint_R (2x+3y)\,dx\,dy$ **(ii)** $\iint_R (x^2+y^2)\,dx\,dy$ **(iii)** $\iint_R e^{x+y}\,dx\,dy$

9. Sketch the regions of integration and reverse the order of the following integrals.

(i) $\int_0^1 dx \int_{x^2}^x f(x, y)\,dy$ **(ii)** $\int_0^{\pi/4} dx \int_{\sin x}^{\cos x} f(x, y)\,dy$ **(iii)** $\int_0^1 dy \int_{1-y}^{2-y} f(x, y)\,dx$

(iv) $\int_0^1 dr \int_{1-r}^1 \phi(r, s)\,ds$ **(v)** $\int_0^1 dy \int_y^{y^{1/3}} f(x, y)\,dx$ **(vi)** $\int_a^{2a} dx \int_x^{2x} f(x, y)\,dy$

10. Verify the identity $\int_0^x dt \int_0^t f(u)\,du = \int_0^x (x-u)f(u)\,du$.

11. Evaluate the following integrals by changing the order of integration. Sketch the regions of integration.

(i) $\int_0^1 dy \int_y^1 \sin x^2\,dx$ **(ii)** $\int_0^1 dx \int_x^{2-x} \frac{x}{y}\,dy$ **(iii)** $\int_0^1 dy \int_y^1 \frac{x^3}{x^2+y^2}\,dx$

12. By sketching the region of integration, show that

$$\int_0^1 dy \int_{1-y}^{1+y} f(x, y)\,dx = \int_0^1 dx \int_{1-x}^1 f(x, y)\,dy + \int_1^2 dx \int_{x-1}^1 f(x, y)\,dy.$$

13. Sketch the region of integration for $I(a) = \int_0^a dx \int_0^x e^{-y^2}\,dy = (\sqrt{\pi}/2)\int_0^a \text{erf}(x)\,dx$. By changing the order of integration, show that
$I(a) = a\int_0^a e^{-y^2}\,dy - ½(1-e^{-a^2}) = (\sqrt{\pi}a/2)\text{erf}(a) - ½(1-e^{-a^2})$.
Show that $\int_0^\infty [1-\text{erf}(x)]\,dx = \pi^{-1/2}$.

14. A plane conductor shaped like the quadrant of a circle $x^2+y^2 \le a^2$, $x \ge 0$, $y \ge 0$, has charge density $\sigma(x, y) = xy$. Find the total charge on the conductor.

15. Find the volume bounded by the cylinder $x^2+y^2=1$ and the planes $z=x+2y+3$ and $z=0$.

16. Find the volumê under the surface $z=1-2x^2-3y^2$ and above the plane $z=0$.

17. Find the volume above the region contained by the parabolas $y^2=4ax$, $x^2=(a/2)y$ and below the surface $z=(x^2+y^2)/a$. Verify the dimensions of the answer.

18. Find the volume of the wedge of material cut out from the semicircular cylinder $y=+\sqrt{a^2-x^2}$, $y=0$, by the planes $z=\pm y\tan\alpha$.

19. Calculate $\iint_R \frac{dx\,dy}{[xy(1-x-y)]^{1/2}}$, where R is the triangle bounded by $x=0$, $y=0$, $x+y=1$.

20. If R is the triangular region $x \ge 0$, $y \ge 0$, $x+y \le 1$, for what values of α does $\iint_R \frac{dx\,dy}{(1-x-y)^\alpha}$ converge?

21. Calculate the values of $\iint_R e^{-|x|-|y|}\,dx\,dy$ and $\iint_R e^{-|x+y|}\,dx\,dy$ when R is the interior of the square $x=\pm a$, $y=\pm a$. How do these integrals behave as $a \to \infty$?

22. Prove that

$$\int_0^1 dx \int_0^1 \frac{x-y}{(x+y)^3}\, dy = 1/2, \qquad \int_0^1 dy \int_0^1 \frac{x-y}{(x+y)^3}\, dx = -1/2.$$

[Note that in this case the integrand is infinite at the point (0, 0) and the double integral does not exist. Each of the iterated integrals exists, but they are unequal. This is a somewhat pathological situation.]

7.2 Polar coordinates and change of variables

It is natural to introduce double integrals in association with rectangular coordinates x and y, in which case the fundamental element of plane area is $dA = dx\, dy$. However, there are situations where other coordinate systems are far more convenient, and it is then desirable to transform double integrals or to formulate them directly in terms of new variables. Probably the most common example is polar coordinates, which will be given special attention before more general transformations are discussed.

In polar coordinates it is a relatively simple matter to derive an expression for the basic infinitesimal element of area dA. In Fig. 2.1 the finite area element ΔA, between the circles of radii r and $r + \Delta r$ and the radial lines of angles θ and $\theta + \Delta\theta$, is the difference between the arcs of two circular sectors;

$$\begin{aligned}\Delta A &= \tfrac{1}{2}(r + \Delta r)^2\, \Delta\theta - \tfrac{1}{2} r^2\, \Delta\theta,\\ &= r\, \Delta r\, \Delta\theta + \tfrac{1}{2}(\Delta r)^2\, \Delta\theta.\end{aligned}$$

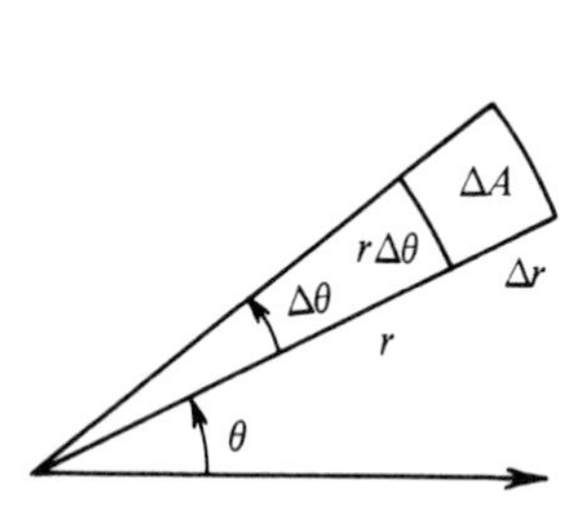

FIGURE 2.1

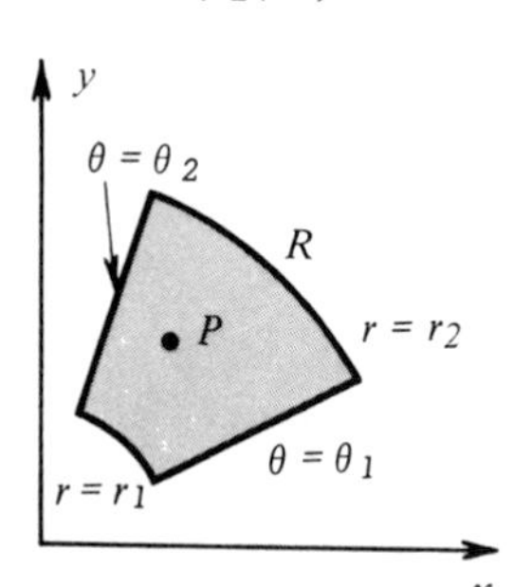

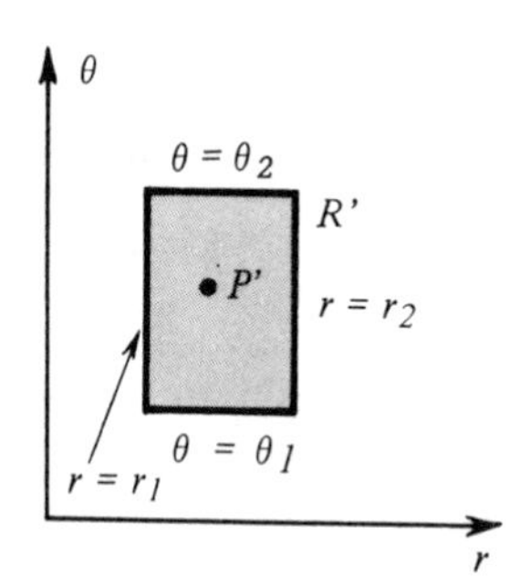

FIGURE 2.2

For small increments, the second term on the right in this equation is negligible compared to the first, so that in the limit the basic infinitesimal element of area is

$$dA = r\, dr\, d\theta. \tag{1}$$

This result is not unexpected since the infinitesimal element of area dA is essentially a rectangle with sides of length dr and $r\, d\theta$. Equation (1) implies that in the polar coordinates the general double integral

$$\iint_R f(x, y)\, dx\, dy$$

becomes

$$\iint_R F(r, \theta) r \, dr \, d\theta, \tag{2}$$

where $F(r, \theta) = f(r \cos \theta, r \sin \theta)$. (Both expressions represent the same volume in space.) Of course, when such an integral is evaluated as an iterated integral, the limits of integration appropriate to the r and θ ranges for R must be found. They are easily determined by considering the allowed variation of r for a fixed θ (or vice versa) and then the total variation of θ (or r) between its minimum and maximum values.

Alternatively, the two families of curves $r = \text{constant}$ (circles) and $\theta = \text{constant}$ (rays) become two sets of perpendicular lines when r and θ are viewed as *new rectangular* coordinates (Fig. 2.2). The change of variable is then interpreted as a *transformation* of an area in the xy plane into an area in the $r\theta$ plane. In other words, the transformation $x = r \cos \theta$, $y = r \sin \theta$ takes a point $P = (x, y)$ within a region R into a point $P' = (r, \theta)$ within a new region R'. The limits of integration in the r, θ integral are then determined from the boundary of R' in the usual way.

If the region R is confined between the curves $r = r_1(\theta)$, $r = r_2(\theta)$ and the rays $\theta = \theta_1$, $\theta = \theta_2$ (see Fig. 2.3), it is convenient to integrate with respect to r first, whereupon Eq. (2) takes the form

$$\int_{\theta_1}^{\theta_2} d\theta \int_{r_1(\theta)}^{r_2(\theta)} F(r, \theta) r \, dr. \tag{3}$$

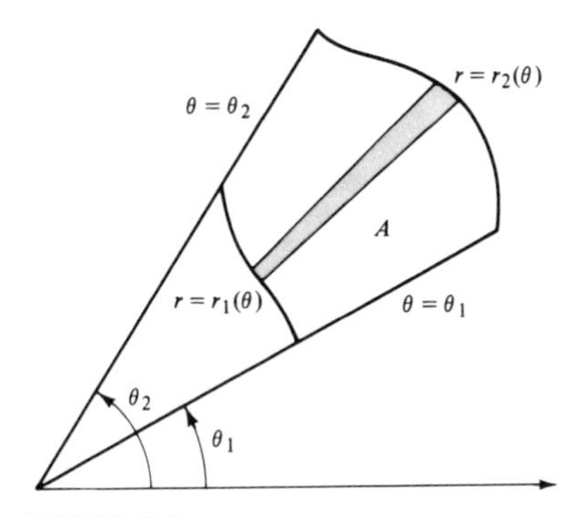

FIGURE 2.3

In particular, when $F = 1$, this expression represents the area A and the integration can be performed explicitly, to give

$$A = \int_{\theta_1}^{\theta_2} \left[\frac{1}{2} r^2\right]_{r_1(\theta)}^{r_2(\theta)} d\theta = \frac{1}{2} \int_{\theta_1}^{\theta_2} \{[r_2(\theta)]^2 - [r_1(\theta)]^2\} \, d\theta, \tag{4}$$

a result established earlier in Chap. 3.

The area A of a general region R can be viewed as the sum of all differential elements, whatever coordinate system is used, so that, for example,

$$A = \iint\limits_R dA = \iint\limits_R dx\,dy = \iint\limits_R r\,dr\,d\theta. \tag{5}$$

Example 1 Find the area of one petal of the rose $r = a \sin 3\theta$.

As θ varies from 0 to $\pi/3$, r starts at zero, reaches its maximum value, a, when $\theta = \pi/6$, and returns to zero when $\theta = \pi/3$. This implies that the range $0 \leq \theta \leq \pi/3$ includes one petal of the rose (Fig. 2.4); its area is

$$A = \int_0^{\pi/3} d\theta \int_0^{a \sin 3\theta} r\,dr.$$

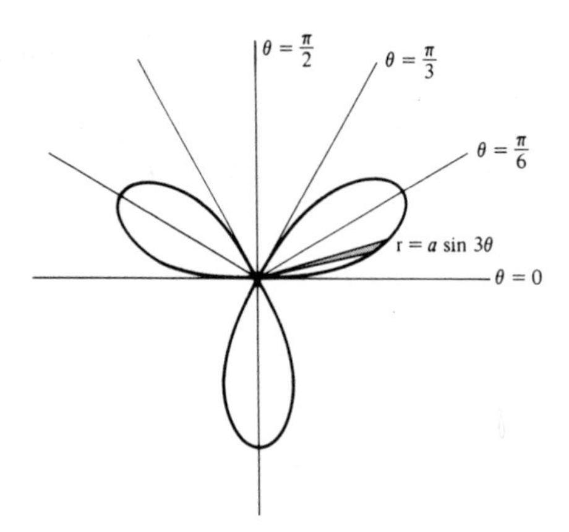

FIGURE 2.4

Here r varies along a ray from zero to its value on the rose, $a \sin 3\theta$, and the ray sweeps out one petal as it moves from $\theta = 0$ to $\theta = \pi/3$. The r integration yields

$$A = \frac{a^2}{2} \int_0^{\pi/3} \sin^2 3\theta\,d\theta,$$

and it follows that

$$A = \frac{a^2}{4} \int_0^{\pi/3} (1 - \cos 6\theta)\,d\theta = \frac{\pi a^2}{12}.$$

Example 2 Evaluate $I = \iint e^{(y-x)/(y+x)}\,dx\,dy$ over the interior of the triangle which has vertices at the points (0, 0), (1, 0), and (0, 1).

Some transformation is clearly needed to make a start on this problem. The advantage of polar coordinates is that the exponent becomes independent of r, and, accordingly, the r integration can certainly be done exactly. In polar coordinates the integral becomes

$$I = \int_0^{\pi/2} d\theta \int_0^{1/(\sin\theta + \cos\theta)} e^{(\sin\theta - \cos\theta)/(\sin\theta + \cos\theta)} r\, dr,$$

where the limits of integration can be read off from Fig. 2.5. (Find the typical r variation and then let the ray sweep out the area.) Integration with respect to r yields

$$\int_0^{\pi/2} e^{(\sin\theta - \cos\theta)/(\sin\theta + \cos\theta)} \left[\tfrac{1}{2} r^2\right]_0^{1/(\sin\theta + \cos\theta)} d\theta$$

$$= \int_0^{\pi/2} \frac{1}{2(\sin\theta + \cos\theta)^2} e^{(\sin\theta - \cos\theta)/(\sin\theta + \cos\theta)}\, d\theta.$$

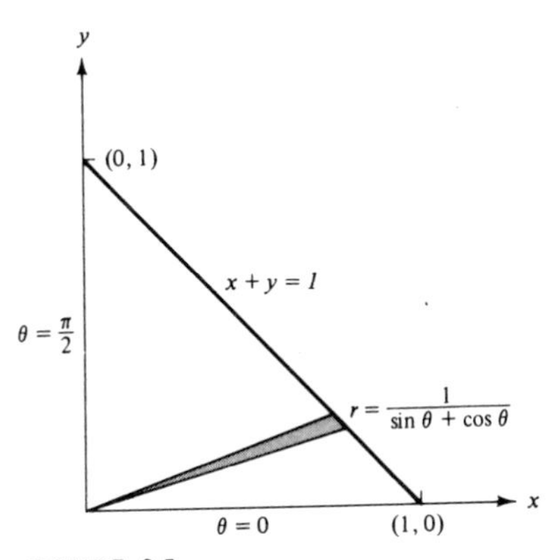

FIGURE 2.5

This latter integral can be evaluated by means of the transformation

$$u = \frac{\sin\theta - \cos\theta}{\sin\theta + \cos\theta},$$

in which case

$$du = \frac{2d\theta}{(\sin\theta + \cos\theta)^2}.$$

When $\theta = 0$ and $\pi/2$, $u = -1$ and $+1$, respectively, and the final integral takes the simple form

$$I = \frac{1}{4}\int_{-1}^{+1} e^u\, du = \frac{1}{4}\left(e - \frac{1}{e}\right) = 0.5876.$$

Example 3 A cylindrical hole of radius b is drilled symmetrically through a metal sphere of radius $a \geq b$. Find the volume of metal removed.

The equation of the sphere is $x^2 + y^2 + z^2 = a^2$, and the cylindrical hole is bounded by $x^2 + y^2 = b^2$. The volume V of metal removed is (see Fig. 2.6)

$$V = 2 \iint_R \sqrt{a^2 - x^2 - y^2}\, dx\, dy,$$

where R is the region $x^2 + y^2 \leq b^2$. (The factor 2 accounts for the volume below the plane $z = 0$.) While it is possible, but not easy, to evaluate this integral in its present form, a simpler procedure is to exploit the circular symmetry and to use polar coordinates r and θ. In this example the change of variable has the added advantage that the limits of integration become constants, since r ranges from 0 to b and θ varies from 0 to 2π, (Fig. 2.7). The expression for V becomes

$$V = \int_0^{2\pi} d\theta \int_0^b 2\sqrt{a^2 - r^2}\, r\, dr = \left(\int_0^{2\pi} d\theta\right) \cdot \left(\int_0^b 2\sqrt{a^2 - r^2}\, r\, dr\right),$$

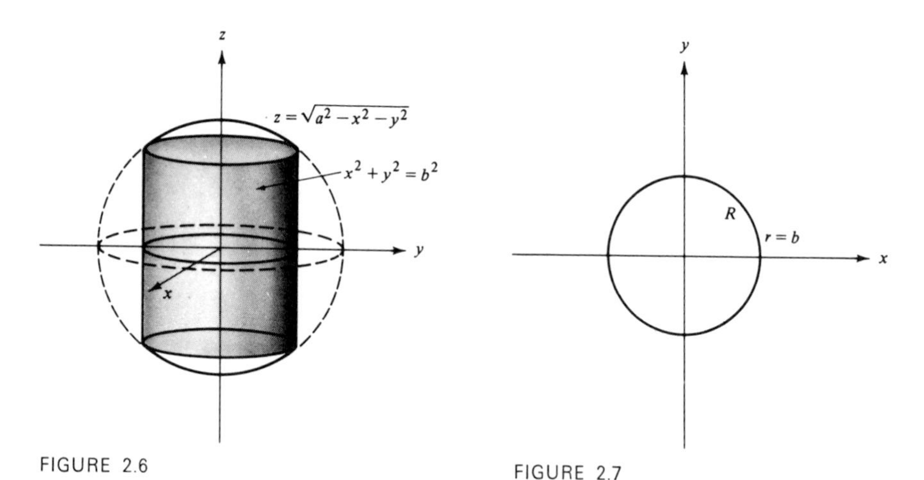

FIGURE 2.6

FIGURE 2.7

where the double integral is actually the product of two single definite integrals. Integration yields

$$V = 2\pi[-\tfrac{2}{3}(a^2 - r^2)^{3/2}]_0^b,$$

$$= \frac{4\pi}{3}[a^3 - (a^2 - b^2)^{3/2}].$$

This answer can be checked by noting certain limiting cases: (i) if $b = a$, all the metal is removed and $V = 4\pi a^3/3$, as it should; (ii) if $b = 0$, there is obviously no hole and $V = 0$. Furthermore, since a and b have dimensions of length, the formula for V is dimensionally correct.

The integral $\int_0^\infty e^{-x^2}\,dx$ occurred frequently in Chaps. 3 and 4. At that stage we were unable to find its value save by numerical methods. Curiously enough the simplest method of evaluating this integral exactly makes use of double integrals.

If

$$I = \int_0^\infty e^{-x^2}\,dx,$$

then equally well $I = \int_0^\infty e^{-y^2}\,dy$, since the name of the integration variable is of no consequence. Therefore,

$$I^2 = \left(\int_0^\infty e^{-x^2}\,dx\right)\cdot\left(\int_0^\infty e^{-y^2}\,dy\right),$$

$$= \int_0^\infty \int_0^\infty e^{-(x^2+y^2)}\,dx\,dy,$$

and this ingenious use of notation enables us to interpret the result as a double integral over the area of the entire first quadrant. Moreover, it is now possible to evaluate this double integral by means of a transformation to polar coordinates. Since the region of integration is the entire first quadrant and $x^2 + y^2 = r^2$, it follows that

$$I^2 = \int_0^{\pi/2} d\theta \int_0^\infty e^{-r^2} r\,dr.$$

But $e^{-r^2} r\,dr = -\tfrac{1}{2}\,d(e^{-r^2})$, and it follows that

$$I^2 = [\theta]_0^{\pi/2}[-\tfrac{1}{2}\,e^{-r^2}]_0^\infty = \frac{\pi}{4}.$$

Therefore

$$I = \int_0^\infty e^{-x^2}\,dx = \frac{\sqrt{\pi}}{2}, \tag{6}$$

or in terms of the error function

$$\operatorname{erf}(\infty) = \frac{2}{\pi^{1/2}}\,I = 1.$$

Example 4 In a crude mathematical model of the Gulf Stream, the northward velocity in miles per hour is given by the formula $v = 3e^{-(x/10)^2-(4z)^2}$, where x is measured in miles to the east and z is measured in miles vertically upward. (The ocean occupies the region $-\infty < x < \infty$, $-\infty < z < 0$.) Estimate the south-north volume of water transported in 1 hr.

The velocity profile v is sketched in Fig. 2.8. In 1 hr the volume of fluid which crosses a small cross-sectional area $dx\,dz$ perpendicular to the flow is $v\,dx\,dz$. The integration of all such contributions yields the volume transported in 1 hr:

$$\iint v\,dx\,dz = \int_{-\infty}^0 dz \int_{-\infty}^\infty 3e^{-(x/10)^2-(4z)^2}\,dx.$$

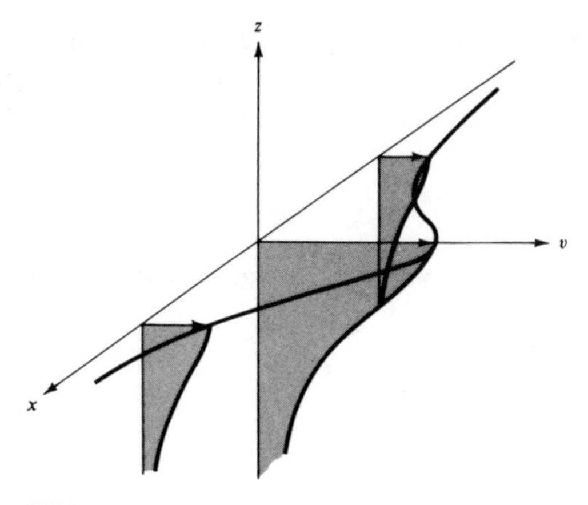

FIGURE 2.8

But for $\alpha > 0$,

$$\int_0^\infty e^{-\alpha^2 s^2}\, ds = \frac{1}{\alpha}\int_0^\infty e^{-\alpha^2 s^2}\, d(\alpha s) = \frac{1}{\alpha}\int_0^\infty e^{-t^2}\, dt$$

so that according to (6),

$$\int_0^\infty e^{-\alpha^2 s^2}\, ds = \frac{\sqrt{\pi}}{2\alpha} = \frac{1}{2}\int_{-\infty}^\infty e^{-\alpha^2 s^2}\, ds.$$

It follows that

$$\iint v\, dx\, dz = 3\int_{-\infty}^0 e^{-(4z)^2}\, dz \int_{-\infty}^\infty e^{-(x/10)^2}\, dx,$$

$$= 3\left(\frac{1}{2}\frac{\sqrt{\pi}}{4}\right)\cdot(\sqrt{\pi}\,10),$$

$$= \frac{15\pi}{4}\ \text{miles}^3/\text{hr}.$$

In other words, the Gulf Stream transports approximately 12 miles^3 of water across the plane $y = 0$ in 1 hr.

3

Vector analysis can be used to investigate the effect of a general change of variables on a double integral. Let

$$x = x(u, v), \qquad y = y(u, v), \tag{7}$$

be the transformation to the new integration variables u and v. If either u or v is held fixed, the preceding equations describe a curve in the xy plane. The curves for all constant values of u and v form the grid lines of a new coordinate system akin to the circles and rays of polar coordinates. Consider then the four curves corresponding to fixed values of u, v, $u + du$, and $v + dv$. These curves intersect in four points P, Q, R, S (Fig. 2.9) whose position vectors with respect to the origin are $\mathbf{r}(u, v)$, $\mathbf{r}(u + du, v)$, $\mathbf{r}(u, v + dv)$, and $\mathbf{r}(u + du, v + dv)$, respectively. If dA is the area element of the infinitesimal parallelogram $PQSR$, then (see Sec. 5.4)

$$dA = |\overrightarrow{PQ} \times \overrightarrow{PR}|. \tag{8}$$

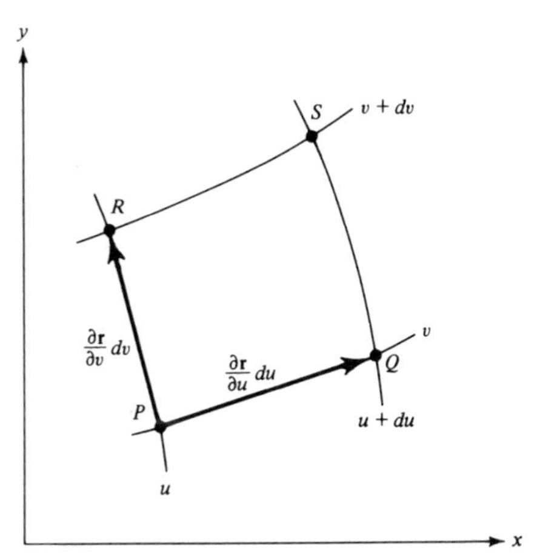

FIGURE 2.9

However,

$$\overrightarrow{PQ} = \mathbf{r}(u + du, v) - \mathbf{r}(u, v) = \frac{\partial \mathbf{r}}{\partial u}\, du;$$

likewise

$$\overrightarrow{PR} = \frac{\partial \mathbf{r}}{\partial v}\, dv.$$

Therefore, Eq. (8) can be rewritten in the form

$$dA = \left| \frac{\partial \mathbf{r}}{\partial u} \times \frac{\partial \mathbf{r}}{\partial v} \right| du\, dv,$$

and this gives the element of area in the new coordinate grid. If $|\partial \mathbf{r}/\partial u \times \partial \mathbf{r}/\partial v|$ is expressed in terms of the x and y components, we find that

$$\begin{aligned} \left| \frac{\partial \mathbf{r}}{\partial u} \times \frac{\partial \mathbf{r}}{\partial v} \right| &= \left| \left(\frac{\partial x}{\partial u}\mathbf{i} + \frac{\partial y}{\partial u}\mathbf{j} \right) \times \left(\frac{\partial x}{\partial v}\mathbf{i} + \frac{\partial y}{\partial v}\mathbf{j} \right) \right|, \\ &= \left| \frac{\partial x}{\partial u}\frac{\partial y}{\partial v} - \frac{\partial x}{\partial v}\frac{\partial y}{\partial u} \right|, \end{aligned}$$

and this is precisely the Jacobian of the transformation (see Sec. 6.7). The final expression for dA becomes

$$dA = |x_u y_v - x_v y_u| \, du \, dv = \left| \frac{\partial(x, y)}{\partial(u, v)} \right| du \, dv. \tag{9}$$

A check on this result is obtained by returning to the case of polar coordinates, where $x = r \cos \theta$, $y = r \sin \theta$. Upon identifying r and θ with u and v, the Jacobian of the transformation is

$$\left| \frac{\partial(x, y)}{\partial(r, \theta)} \right| = |(\cos \theta)(r \cos \theta) - (-r \sin \theta)(\sin \theta)| = r.$$

This calculation implies that the element of area $dx \, dy$ must be replaced (in polar coordinates) by the expression $r \, dr \, d\theta$, a result found earlier by geometrical methods.

In practice the desirability of making a change of variables in a double integral is governed by two considerations: does it facilitate the evaluation of one or more of the integrals involved and how simple do the limits of integration become?

Example 5 Evaluate $I = \iint_R (1 - x^2/a^2 - y^2/b^2)^{3/2} \, dx \, dy$, where R is the region enclosed by the ellipse $x^2/a^2 + y^2/b^2 = 1$.

One method of attack on this problem is first to transform R into a circle by means of the change of variables

$$x = au, \qquad y = bv,$$

in which case the associated Jacobian is

$$\frac{\partial(x, y)}{\partial(u, v)} = ab.$$

The region $x^2/a^2 + y^2/b^2 \leq 1$ transforms into the region $u^2 + v^2 \leq 1$, and the double integral converts to

$$I = ab \iint_{u^2 + v^2 \leq 1} (1 - u^2 - v^2)^{3/2} \, du \, dv.$$

In this form a further transformation to polar coordinates in the uv plane is irresistible. With $u = \rho \cos \varphi$ and $v = \rho \sin \varphi$, so that $\partial(u, v)/\partial(\rho, \varphi) = \rho$, $(1 - u^2 - v^2)^{3/2} = (1 - \rho^2)^{3/2}$, we find that

$$I = ab \int_0^{2\pi} d\varphi \int_0^1 (1 - \rho^2)^{3/2} \rho \, d\rho.$$

Each of these integrations is straightforward, and the calculation yields

$$I = 2\pi ab[-\tfrac{1}{5}(1 - \rho^2)^{5/2}]_0^1 = \frac{2\pi ab}{5}.$$

Example 6 Show that the area in the first quadrant enclosed by the four curves $\alpha y = x^3$, $\beta y = x^3$, $\gamma x = y^3$, and $\delta x = y^3$ is $\frac{1}{2}(\delta^{1/2} - \gamma^{1/2})(\beta^{1/2} - \alpha^{1/2})$, where $\beta > \alpha > 0$, $\delta > \gamma > 0$.

The problem is to evaluate $A = \iint_R dx\, dy$, where R is the region sketched in Fig. 2.10. The integration is difficult because R is not readily described in x, y coordinates. In this case the limits of integration are complicated while the integrand is rather trivial. By means of the change of variables

$$u = \frac{x^3}{y}, \qquad v = \frac{y^3}{x},$$

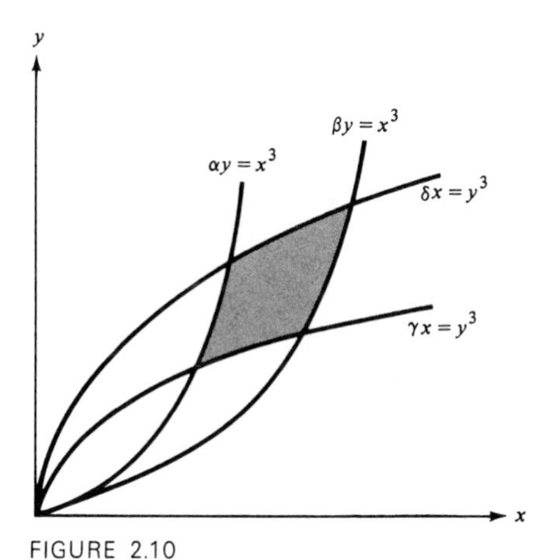

FIGURE 2.10

the limits of integration can be made constants and *at the same time* the integrand remains manageable. On solving for x and y in terms of u and v it is found that

$$x = u^{3/8} v^{1/8}, \qquad y = u^{1/8} v^{3/8};$$

the Jacobian of the transformation is

$$\frac{\partial(x, y)}{\partial(u, v)} = \frac{\partial x}{\partial u}\frac{\partial y}{\partial v} - \frac{\partial x}{\partial v}\frac{\partial y}{\partial u} = \tfrac{3}{8}u^{-5/8} v^{1/8} \cdot \tfrac{3}{8}u^{1/8} v^{-5/8} - \tfrac{1}{8}u^{3/8} v^{-7/8} \cdot \tfrac{1}{8}u^{-7/8} v^{3/8}$$

or

$$\frac{\partial(x, y)}{\partial(u, v)} = \tfrac{1}{8} u^{-1/2} v^{-1/2}.$$

In terms of the new variables, u ranges from α to β and v varies from γ to δ, and the desired area is

$$A = \int_\gamma^\delta dv \int_\alpha^\beta \tfrac{1}{8} u^{-1/2} v^{-1/2}\, du.$$

This double integral presents no difficulties, and upon evaluation we obtain

$$A = \tfrac{1}{2}(\delta^{1/2} - \gamma^{1/2})(\beta^{1/2} - \alpha^{1/2}).$$

As a check on this answer, it is obviously correct if $\alpha = \beta$ or $\gamma = \delta$. Furthermore, since the parameters have dimensions (length)^2, the formula for A is dimensionally correct.

Exercises 7.2

1. What is the area within the cardioid $r = a(1 + \cos\theta)$?
2. Find the area of one petal of the rose $r = a\sin n\theta$ where n is a positive integer.
3. Find the area common to the circles $r = 2$ and $r = 4\cos\theta$.
4. For the region R defined by $x^2 + y^2 \le a^2$, $x \ge 0$, $y \ge 0$, evaluate the following integrals.

(i) $\displaystyle\iint_R xy\, dA$ **(ii)** $\displaystyle\iint_R (x^2 + y^2)^{-1/2}\, dA$ **(iii)** $\displaystyle\iint_R (x + y)\, dA$

5. Use polar coordinates to evaluate $\displaystyle\int_0^1 dy \int_0^{(1-y^2)^{1/2}} e^{-(x^2+y^2)}\, dx.$
6. Graph the folium of Descartes $x^3 + y^3 = 3axy$ $(a > 0)$ and show that the curve is asymptotic to the line $x + y + a = 0$. Find the area of the loop in the first quadrant.
7. Find the area of one loop of the lemniscate $r^2 = a^2\cos 2\theta$.
8. Show that the area below the curve $r = e^\theta$, $0 \le \theta \le \pi$, and above the x axis can be expressed as either

$$\int_0^\pi d\theta \int_0^{e^\theta} r\, dr \qquad \text{or} \qquad \int_0^1 r\, dr \int_0^\pi d\theta + \int_1^{e^\pi} r\, dr \int_{\log r}^\pi d\theta.$$

Calculate the area from each expression.
9. Show that $\iint_R x^5 y^7\, dx\, dy = 0$, where R is the interior of the ellipse $x^2 + 4y^2 = 1$. (Don't integrate unless you really have to!)
10. Calculate the double integral $\iint_R (x + y)^3 \cos^2(x - y)\, dx\, dy$, R being the parallelogram with vertices at $(\pi, 0)$, $(3\pi, 2\pi)$, $(2\pi, 3\pi)$, and $(0, \pi)$.
11. Make the change of variables $u = x + y$, $v = y - x$ to show $\iint_R f(x + y)\, dx\, dy = \int_{-1}^1 f(u)\, du$ where R is the region $|x| + |y| \le 1$. Evaluate the following integrals.

(i) $\displaystyle\iint_R (x + y)^2\, dx\, dy$ **(ii)** $\displaystyle\iint_R e^{x+y}\, dx\, dy$ **(iii)** $\displaystyle\iint_R \log(1 + x + y)\, dx\, dy$

12. If $q > p > 0$ and $n > m > 0$, find the area in the first quadrant contained by the curves $xy = p$, $xy = q$ and the lines $y = mx$, $y = nx$. Verify the dimensions of the answer.
13. If $b > a > 0$ and $d > c > 0$, find the area in the first quadrant contained by the curves $y^3 = ax^2$, $y^3 = bx^2$, $x^4 = cy^3$ and $x^4 = dy^3$. Verify the dimensions of the answer.
14. Evaluate the following integrals over the annular region $a^2 \le x^2 + y^2 \le b^2$.

(i) $\displaystyle\iint_R x\, dxdy$ **(ii)** $\displaystyle\iint_R (x^2 + y^2)^{1/2}\, dxdy$ **(iii)** $\displaystyle\iint_R (x^2 + y^2)\, dxdy.$

15. Use the transformation $u = x - y$, $v = x + y$ to evaluate the integral $\iint_R e^{(y-x)/(y+x)}\, dxdy$ where R is the triangle with vertices at $(0,0)$, $(1,0)$ and $(0,1)$.

16. Sketch the region of integration in the integral $\int_1^2 dx \int_{-x}^{x} e^{y/x}\, dy$. Use the transformation $u = 1/x$, $v = y/x$ to evaluate the integral. Sketch the region of integration in the uv plane.

17. Evaluate $\iint\limits_R \dfrac{dx\, dy}{(1 + x^2 + y^2)^2}$ where R is a triangle with vertices at the points $(0, 0)$, $(2, 0)$, and $(1, \sqrt{3})$.

18. Find the volume under the surface $z = 1/(x^2 + y^2)$, above the plane $z = 0$, and between the two cylinders $x^2 + y^2 = 1$ and $x^2 + y^2 = 4$.

19. Calculate the volume common to the sphere $x^2 + y^2 + z^2 = a^2$ and the cylinder $x^2 + y^2 = ax$.

20. Use double integrals to find the volume of a sphere.

21. Show that the volume common to the cylinders $x^2 + y^2 = 2ax$ and $z^2 = 2ax$ is $(128/15)a^3$.

22. What is the volume contained by the three surfaces $z = 0$, $z = x^2/p^2 + y^2/q^2$, and $x^2/a^2 + y^2/b^2 = 1$? Verify the dimensions of the answer.

23. Find the values of α for which the integral $I(\alpha) = \iint\limits_R (x^2 + y^2)^\alpha\, dxdy$ converges where R is the interior of the circle $x^2 + y^2 = 1$. Evaluate $I(\alpha)$ for these values of α.

24. Calculate $\iint\limits_R e^{-(x^2 - y^2)}\, dxdy$, where R is the infinite sector $-\beta \le \theta \le \beta$, $|\beta| < \pi/4$. What happens as $\beta \to \pi/4$?

25. A thin metal sheet has the shape of the cardioid $r = a(1 + \cos\theta)$. If the density at any point is proportional to its distance from the origin, find the total mass and the average density.

26. A hurricane has the pressure distribution $p(x, y)$ (atmospheres) given by $p(x, y) = p_\infty + (p_0 - p_\infty)e^{-(x^2 + y^2)}$ where $p_0 < p_\infty$ are constants. Find the total suction force of the hurricane.

27. A glacier which occupies the region $0 > z > -\sqrt{10^{-2} - x^2}$ moves parallel to Oy with velocity $v(x, z) = 10^{-3}[1 - 10^2(x^2 + z^2)]$, where all lengths are measured in kilometres and the velocity is measured in kilometres per year. Find the volume of ice moved by the glacier in a year.

28. Evaluate the following improper integrals or show they diverge.

(i) $\displaystyle\int_0^\infty \int_0^\infty \frac{dxdy}{1 + x^2 + y^2}$ (ii) $\displaystyle\int_0^\infty \int_0^\infty xye^{-x^2 - y^2}\, dxdy$ (iii) $\displaystyle\int_0^\infty \int_0^\infty e^{-(x+y)^2}\, dxdy$

29. The *gamma function* $\Gamma(n)$ is defined by $\Gamma(n) = \int_0^\infty x^{n-1}e^{-x}\, dx$ for $n > 0$. Use integration by parts to show that $\Gamma(n + 1) = n\Gamma(n)$. Since $\Gamma(1) = 1$, conclude that $\Gamma(n + 1) = n!$ if n is a non-negative integer.

30. By a change of variables, show that $\Gamma(n) = 2\int_0^\infty x^{2n-1}e^{-x^2}\, dx$ and conclude that $\Gamma(1/2) = \sqrt{\pi}$. Evaluate $\Gamma(3/2)$ and $\Gamma(5/2)$.

31. Form the double integral $\Gamma(m)\Gamma(n) = 4\int_0^\infty \int_0^\infty x^{2m-1}y^{2n-1}e^{-x^2-y^2}\, dxdy$. By transforming to polar coordinates, show that the *beta function*

$$B(m, n) = 2\int_0^{\pi/2} \cos^{2m-1}\theta\, \sin^{2n-1}\theta\, d\theta = \frac{\Gamma(m)\Gamma(n)}{\Gamma(m + n)}.$$

32. Use the beta function to evaluate $\iint\limits_R x^m y^n\, dxdy$ where $m, n > -1$ and R is the region $x \ge 0$, $y \ge 0$, $x + y \le a$.

33. Use the beta function to show that the area bounded by the curve $x^{1/p} + y^{1/q} = 1$ for $p, q > 0$ which lies in the first quadrant is given by $\Gamma(p + 1)\Gamma(q + 1)/\Gamma(p + q + 1)$.

7.3 Triple integrals

Now that the theoretical foundation of double integrals has been discussed, it is relatively straightforward to generalize the procedure to *triple integrals* (and even integrals of higher order).

To be specific, let us find the total mass of a certain body of volume V when the material density $\rho(x, y, z)$ is known as a function of position. As before, the strategy is to divide and conquer. The volume V is subdivided into small volume elements ΔV_{ijk} (Fig. 3.1) by slicing the body with the sets of planes $x = x_i$, $i = 1, 2, \ldots, n$; $y = y_j$, $j = 1, 2, \ldots, p$; $z = z_k$, $k = 1, 2, \ldots, q$. The incremental mass Δm_{ijk} of the volume element ΔV_{ijk} is approximated by

$$\Delta m_{ijk} = \rho(\xi_i, \eta_j, \zeta_k)\, \Delta x_i\, \Delta y_j\, \Delta z_k, \tag{1}$$

where $\Delta x_i = x_i - x_{i-1}$, $\Delta y_j = y_j - y_{j-1}$, and $\Delta z_k = z_k - z_{k-1}$ and (ξ_i, η_j, ζ_k) is *any* point in ΔV_{ijk}. The approximate mass within the volume V is obtained by adding all the increments Δm_{ijk} and to facilitate this count we set $\rho(x, y, z) = 0$ at any point not in V. The exact mass results from the limit:

$$\lim_{\substack{n,\, p,\, q \to \infty \\ \max(|\Delta x_i|,\, |\Delta y_j|,\, |\Delta z_k|) \to 0}} \sum_{i=1}^{n} \sum_{j=1}^{p} \sum_{k=1}^{q} \rho(\xi_i, \eta_j, \zeta_k) \Delta x_i\, \Delta y_j\, \Delta z_k. \tag{2}$$

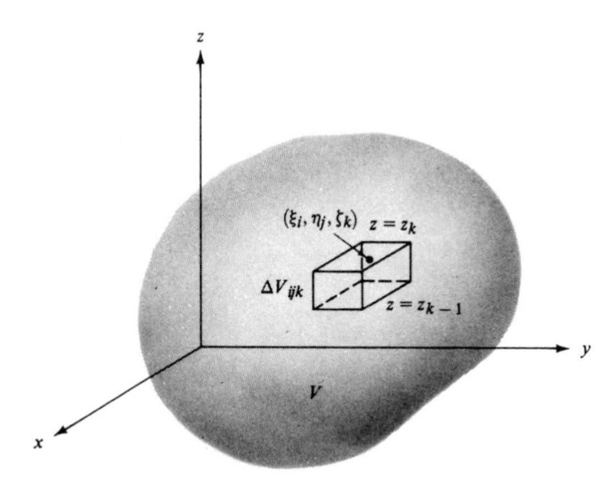

FIGURE 3.1

This expression is called the *triple integral* of the function $\rho(x, y, z)$ throughout the volume V, and it is written

$$\iiint_V \rho(x, y, z)\, dV \qquad \text{or} \qquad \iiint_V \rho(x, y, z)\, dx\, dy\, dz. \tag{3}$$

If the density is constant, say $\rho = 1$, the mass calculation is equivalent to finding the volume of the body:

$$\iiint_V \rho\, dV = 1 \cdot \iiint_V dV = V.$$

While (2) provides the basis for the numerical calculation of a triple integral, analytical evaluations are made by expressing this equation as three iterated integrals. Once again this corresponds to summing the increments in a special order, a procedure which also determines the limits on each integral. For example, if the sum is taken first with respect to the index k, then over j, and completed with the i summation, the integrations are performed in the order z first, y second, and x third. Since the limits of integration depend *only* on the geometrical configuration and not on the nature of the integrand, they are readily found by visualizing the process in which the infinitesimal volumes are assembled to form the volume V. In the order stated above, the first z integration is an accumulation of elemental volumes within V, for fixed x and y, into a vertical column. The two positions where this column pierces the outer surface of V give the appropriate limits for this integration. In the next integration with respect to y, all such vertical pillars, for a fixed x, are formed into a volume slice of thickness dx. The positions of the first and last column yield the limits of integration for the y integral. Finally all these slices are assembled into the volume via the x integral, whose limits of integration are the two tangent planes perpendicular to the x axis which touch V at its opposite extremities. In simple terms, points are added into a vertical line segment, the line segments are accumulated to form a plane slice, and the slices stacked to become the volume.

Figure 3.2 shows a simple volume which has the property that every line parallel to the x, y, or z axis cuts it at most twice. If $z = \chi_B(x, y)$ and $z = \chi_T(x, y)$ are the bottom and top parts of this surface, if $y = \varphi_B(x)$ and $y = \varphi_T(x)$ are the bottom and top boundary curves for the projection of V onto the xy plane, and if $x = a$ and $x = b$ are the extreme values for x, then the triple integral for the mass is equivalent to the iterated integral

$$\iiint_V \rho(x, y, z)\, dV = \int_a^b dx \int_{\varphi_B(x)}^{\varphi_T(x)} dy \int_{\chi_B(x,y)}^{\chi_T(x,y)} \rho(x, y, z)\, dz. \tag{4}$$

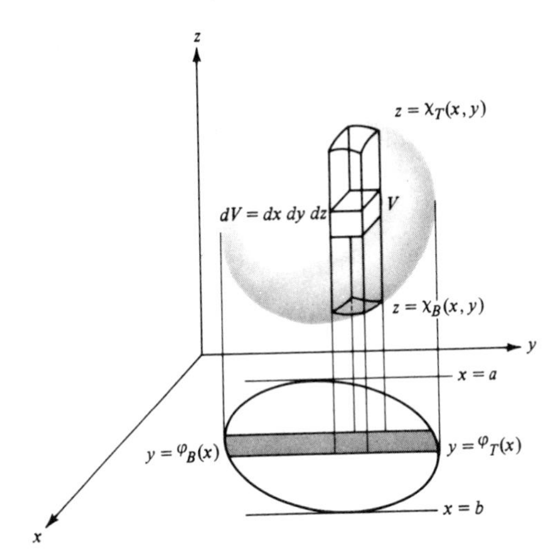

FIGURE 3.2

In particular for $\rho = 1$, (4) reduces to

$$V = \iiint dV = \int_a^b dx \int_{\varphi_B(x)}^{\varphi_T(x)} f(x, y)\, dy$$

with $f(x, y) = \chi_T(x, y) - \chi_B(x, y)$. This double integral for the volume of a solid of height f, wherein the volume element is $dV = f(x, y)\, dx\, dy$, was derived directly in the last section.

Example 1 Evaluate $I = \int_1^2 dx \int_0^1 dy \int_0^{\pi/2} x^2 y^3 \sin z\, dz$.

Since all the limits of integration are constant, the volume concerned is a rectangular block. The integral can be written as the product

$$I = \left(\int_1^2 x^2\, dx\right) \cdot \left(\int_0^1 y^3\, dy\right) \cdot \left(\int_0^{\pi/2} \sin z\, dz\right),$$

$$= \tfrac{7}{3} \cdot \tfrac{1}{4} \cdot 1 = \tfrac{7}{12}.$$

Example 2 Evaluate $I = \iiint_V x\, dV$, where V is the volume contained by the four planes $x = 0$, $y = 0$, $z = 0$, $x/a + y/b + z/c = 1$.

If the integrations are performed in the order z, y, x, the desired integral can be written as an iterated integral of the form

$$I = \int_0^a dx \int_0^{b(1-x/a)} dy \int_0^{c(1-x/a-y/b)} x\, dz.$$

Here the accumulation of infinitesimal cubes begins in the vertical direction, and the pillar formed extends from the base plane $z = 0$ to the slanted plane $z = c(1 - x/a - y/b)$ (see Fig. 3.3). In the y integration these columns are then stacked side by side between the base lines $y = 0$ and the line $y = b(1 - x/a)$, as shown in Fig. 3.4. Finally, the volume V is formed as the aggregate of the vertical slices from $x = 0$ to $x = a$. On performing the z integration we obtain

$$I = \int_0^a dx \int_0^{b(1-x/a)} cx\left(1 - \frac{x}{a} - \frac{y}{b}\right) dy,$$

and the y integration yields

$$I = \int_0^a \frac{bc}{2}\, x\left(1 - \frac{x}{a}\right)^2 dx.$$

The final result is

$$I = \frac{a^2bc}{24}.$$

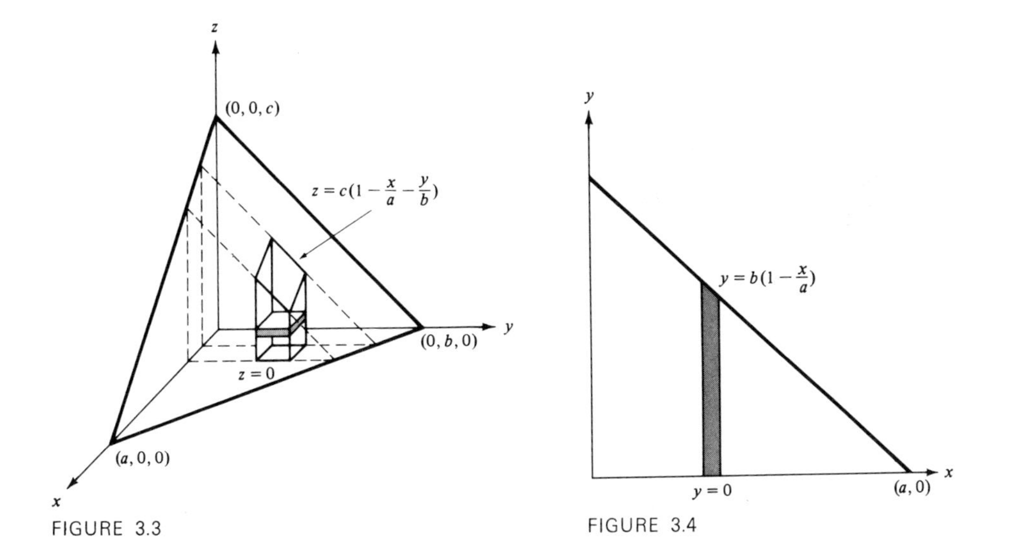

FIGURE 3.3

FIGURE 3.4

The integrations may be taken in a different order, say with respect to x first, y second, and z third. The limits of integration are easily determined, and the appropriate iterated integral for this choice is

$$I = \int_0^c dz \int_0^{b(1-z/c)} dy \int_0^{a(1-y/b-z/c)} x\, dx.$$

Of course, the answer is the same.

Example 3 Calculate the volume between the surfaces

$$x^2 + y^2 + z^2 = 2a^2 \qquad \text{and} \qquad z = (x^2 + y^2)/a,$$

for $a > 0$, which lies in the region $z > 0$.

The two surfaces intersect where $x^2 + y^2 = az = 2a^2 - z^2$, that is, for

$$z = a \qquad \text{or} \qquad z = -2a.$$

Only the value $z = a$ corresponds to a real curve, i.e., the circle $x^2 + y^2 = a^2$. As shown in Fig. 3.5, the z integration extends from the paraboloid to the surface of the sphere, $(x^2 + y^2)/a \le z \le (2a^2 - x^2 - y^2)^{1/2}$, so that the desired volume is

$$V = \iiint\limits_V dx\, dy\, dz = \iint\limits_R dx\, dy \int_{(x^2+y^2)/a}^{(2a^2-x^2-y^2)^{1/2}} dz.$$

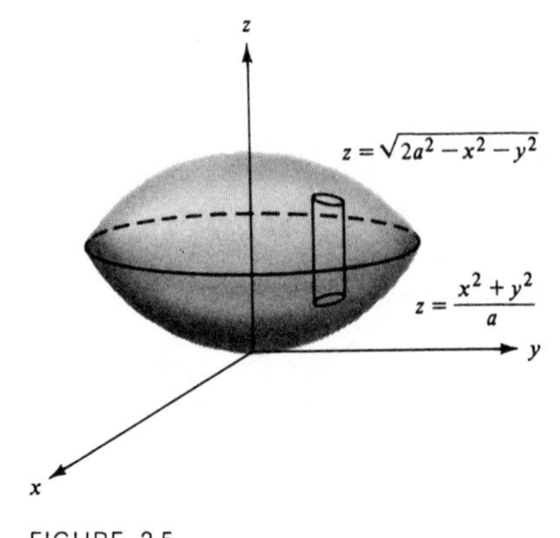

FIGURE 3.5

The first integration (with respect to z) is simple, and the result is

$$V = \iint_R \left[(2a^2 - x^2 - y^2)^{1/2} - \frac{x^2 + y^2}{a}\right] dx\, dy,$$

where R is the circular region $x^2 + y^2 \leq a^2$. This double integral is best treated by going to polar coordinates, in which case

$$V = \int_0^{2\pi} d\theta \int_0^a \left[(2a^2 - r^2)^{1/2} - \frac{r^2}{a}\right] r\, dr.$$

These integrations are also straightforward, and we find that

$$V = 2\pi\left[-\frac{1}{3}(2a^2 - r^2)^{3/2} - \frac{r^4}{4a}\right]_0^a,$$

$$= 2\pi a^3(2^{3/2}/3 - 7/12).$$

Example 4 Show that $\iiint_V x^4y^2z^3\, dV = 0$, where V is the interior of the sphere $x^2 + y^2 + z^2 = a^2$.

This integral can be evaluated directly, e.g., from the equivalent iterated integral

$$\int_{-a}^{a} dx \int_{-(a^2-x^2)^{1/2}}^{(a^2-x^2)^{1/2}} dy \int_{-(a^2-x^2-y^2)^{1/2}}^{(a^2-x^2-y^2)^{1/2}} x^4y^2z^3\, dz$$

or by using spherical coordinates (to be described later in this section). However, we can avoid *all* integrations by noting that the integrand is an odd function of z, in which case the z integration between the limits $\pm(a^2 - x^2 - y^2)^{1/2}$ yields zero automatically.

Problems which have symmetry about an axis (axisymmetric problems) are most efficiently described in *cylindrical coordinates*. In this system, the position of a point is specified by (r, θ, z), r and θ being the ordinary polar coordinates in the xy plane and z the third rectangular coordinate (Fig. 3.6). This means that the transformations

$$x = r \cos \theta, \qquad y = r \sin \theta, \qquad z = z, \tag{5}$$

relate cylindrical to cartesian coordinates. The surface $r =$ constant is a circular cylinder whose axis is the z axis; $\theta =$ constant is a plane through the z axis; and $z =$ constant is a plane perpendicular to the z axis. The intersections of these families of surfaces suitably spaced form the basic grid for the cylindrical coordinate system.

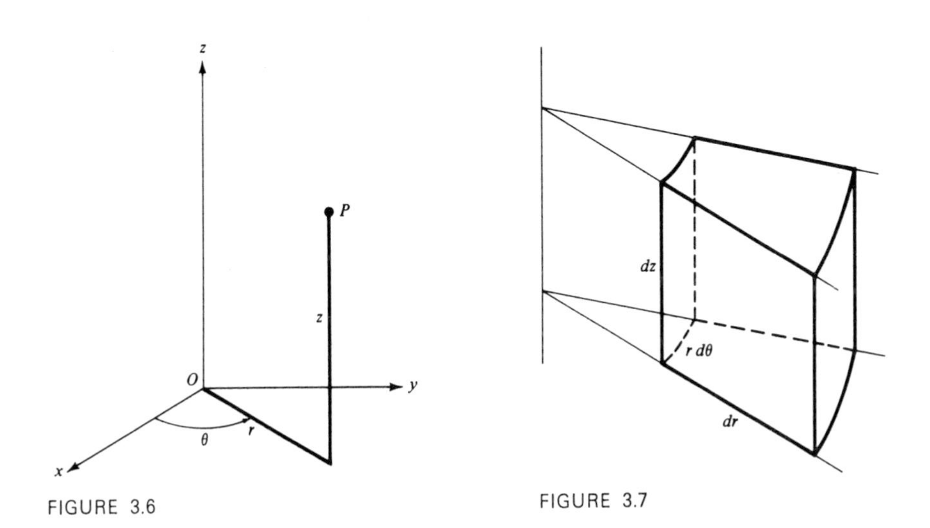

FIGURE 3.6

FIGURE 3.7

The element of volume dV in cylindrical coordinates is the infinitesimal volume contained by the surfaces of constant r, $r + dr$, θ, $\theta + d\theta$, z, and $z + dz$, as shown in Fig. 3.7. This rectangular increment is an infinitesimal box of sides dr, $r\,d\theta$, and dz, so that the element of volume is given by

$$dV = r\,dr\,d\theta\,dz. \tag{6}$$

If all such volume elements are added to form V, we write

$$V = \iiint dV = \iiint_V r\,dr\,d\theta\,dz,$$

which implies that

$$\iiint_V dx\,dy\,dz = \iiint_V r\,dr\,d\theta\,dz.$$

Moreover, the conversion of any triple integral from cartesian to cylindrical coordinates is then

$$\iiint_V f(x, y, z)\, dx\, dy\, dz = \iiint_V F(r, \theta, z) r\, dr\, d\theta\, dz, \tag{7}$$

where $F(r, \theta, z) = f(r \cos \theta, r \sin \theta, z)$, and the limits of integration are correctly transformed.

3

Example 5 A centrifuge is a cylindrical tank of radius a and height h that rotates about a vertical axis with constant angular velocity Ω. The tank is filled with a gas for which the equation of state is $\rho = Kp$, where ρ is the density, p the pressure, and K a constant (corresponding to a perfect gas under isothermal conditions). Find the mass of the contained gas and the pressure p at any point.

Using cylindrical coordinates, the mass of gas m in the container (Fig. 3.8), is given by

$$m = \iiint_V \rho\, dV = \iiint_V \rho r\, dr\, d\theta\, dz,$$

$$= \int_0^h dz \int_0^{2\pi} d\theta \int_0^a \rho r\, dr. \tag{8}$$

In order to evaluate (8), ρ must be found as an explicit function of position. This is accomplished by considering the force equilibrium on a small volume

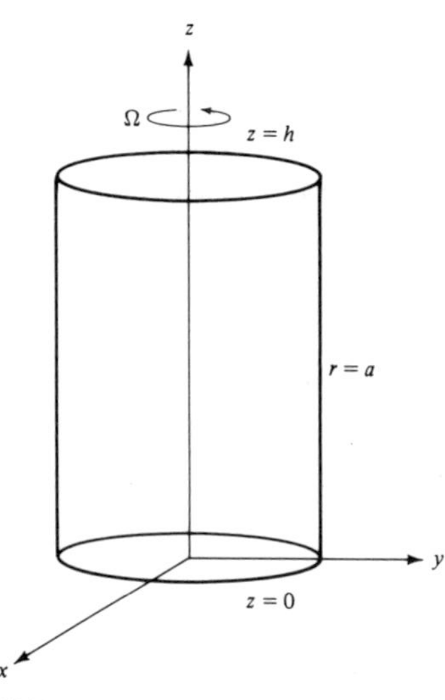

FIGURE 3.8

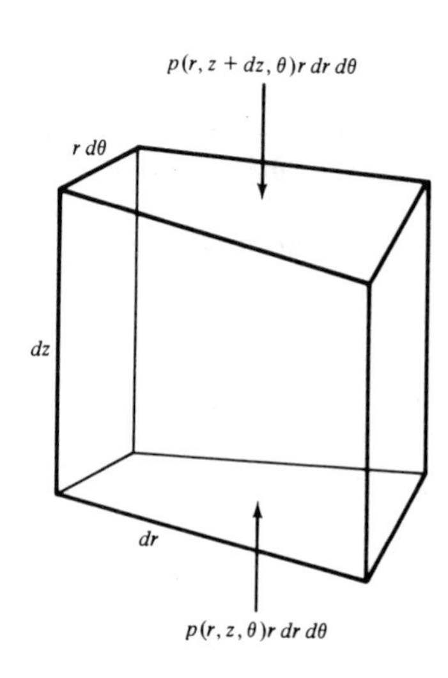

FIGURE 3.9

element of sides dr, $r\,d\theta$, and dz (see Fig. 3.9). The pressures $p(r, z, \theta)$, $p(r, z + dz, \theta)$ on the faces z, $z + dz$ give rise to an upward vertical force

$$[p(r, z, \theta) - p(r, z + dz, \theta)]\,dr\,r\,d\theta = -\frac{\partial p}{\partial z}\,r\,dr\,d\theta\,dz = -\frac{\partial p}{\partial z}\,dV,$$

where dV is the volume element concerned. This force balances the downward gravitational force $\rho g\,dV$ and therefore

$$\frac{\partial p}{\partial z} = -\rho g. \tag{9}$$

In a similar manner, the pressure differences on the surfaces of constant r and $r + dr$ give rise to an outward radial force equal to $-\partial p/\partial r\,dV$. In equilibrium, this force must cancel the centrifugal force $\rho\Omega^2 r\,dV$, and it follows that

$$\frac{\partial p}{\partial r} = \rho\Omega^2 r. \tag{10}$$

When similar arguments are applied to the circumferential direction, we draw the conclusion

$$\frac{\partial p}{\partial \theta} = 0. \tag{11}$$

The equation of state written in differential form is

$$d\rho = K\,dp = K\left(\frac{\partial p}{\partial r}\,dr + \frac{\partial p}{\partial \theta}\,d\theta + \frac{\partial p}{\partial z}\,dz\right),$$

and the substitution for the partial derivatives of the pressure using (9) to (11) leads to

$$d\rho = K(\rho\Omega^2 r\,dr - \rho g\,dz)$$

or
$$\frac{d\rho}{\rho} = K(\Omega^2 r\,dr - g\,dz) = Kd(\tfrac{1}{2}\Omega^2 r^2 - gz).$$

This exact differential can be integrated to yield

$$\log \rho = K(\tfrac{1}{2}\Omega^2 r^2 - gz) + \log \rho_0,$$

where $\rho = \rho_0$ at the point $r = z = 0$; on taking exponentials we obtain an explicit formula for the density distribution

$$\rho = \rho_0\,e^{K(\frac{1}{2}\Omega^2 r^2 - gz)}. \tag{12}$$

The mass calculation formulated in (8) can now be completed. Since

$$m = \int_0^h dz \int_0^{2\pi} d\theta \int_0^a \rho_0\,e^{K(\frac{1}{2}\Omega^2 r^2 - gz)} r\,dr$$

and the limits of integration are all constants, it follows that

$$m = \rho_0\left(\int_0^h e^{-Kgz}\,dz\right)\left(\int_0^{2\pi} d\theta\right)\left(\int_0^a e^{\frac{1}{2}K\Omega^2r^2}r\,dr\right),$$

$$= \frac{2\pi\rho_0}{gK^2\Omega^2}(e^{\frac{1}{2}K\Omega^2a^2} - 1)(1 - e^{-Kgh}). \tag{13}$$

The pressure p is determined from the equation of state

$$p = \frac{\rho}{K},$$

so that

$$p = \frac{\rho_0}{K}\,e^{K(\frac{1}{2}\Omega^2r^2 - gz)}.$$

The constant ρ_0 is related to the total mass m by (13) and may be replaced by it, in which case the final result for the pressure at any point of the centrifuge is given by

$$p = \frac{mgK\Omega^2}{2\pi}\,\frac{e^{K(\frac{1}{2}\Omega^2r^2 - gz)}}{(e^{\frac{1}{2}K\Omega^2a^2} - 1)(1 - e^{-Kgh})}.$$

---

4 In problems with spherical symmetry, the geometry can be exploited by introducing another special system of coordinates called *spherical coordinates*. In this description (see Fig. 3.10) the position of a point $P = (R, \varphi, \theta)$ is specified by R, the length of the radius vector $\overrightarrow{OP}$, φ, the angle OP makes with the z axis, and θ, the usual polar angle in the xy plane. The radial coordinate R is always positive, and the surface R = constant is a sphere centered at the origin. The angle φ, which is measured from the

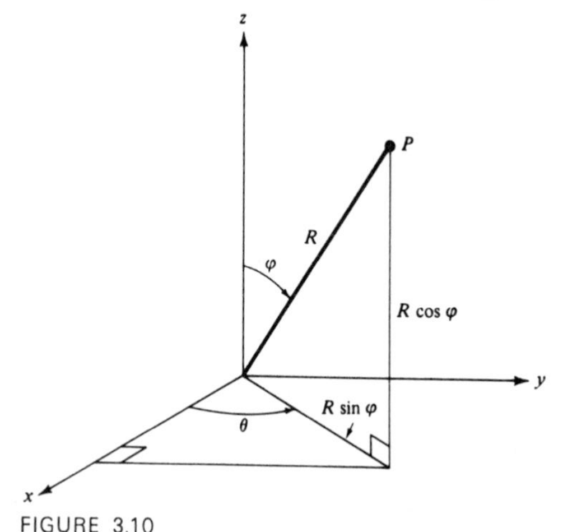

FIGURE 3.10

positive z axis, ranges from 0 at the north pole to π at the south pole. The surface $\varphi =$ constant is a circular cone which has the z axis as its centerline. The angle θ, which is often referred to as the *azimuthal angle*, ranges from 0 to 2π, and the surface $\theta =$ constant is a plane through the z axis. By elementary geometry the relationships between spherical and cartesian coordinates are

$$x = R \sin \varphi \cos \theta, \qquad y = R \sin \varphi \sin \theta, \qquad z = R \cos \varphi. \tag{14}$$

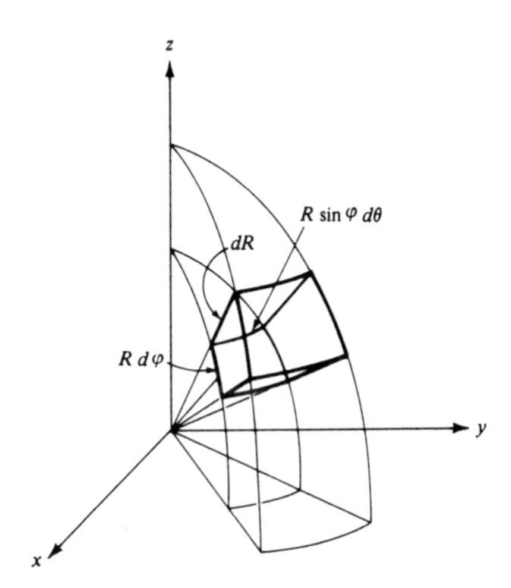

FIGURE 3.11

The increment dV in spherical coordinates is the infinitesimal rectangular element of volume between the surfaces of constant R, $R + dR$, φ, $\varphi + d\varphi$, θ, and $\theta + d\theta$ (Fig. 3.11). This rectangular element has sides of lengths dR, $R\,d\varphi$, and $R \sin \varphi \, d\theta$, and hence the element of volume is

$$dV = R^2 \sin \varphi \, dR \, d\varphi \, d\theta. \tag{15}$$

Equivalent formulas for a volume V are

$$V = \iiint_V dx\,dy\,dz = \iiint_V R^2 \sin \varphi \, dR \, d\varphi \, d\theta,$$

where, of course, the relevant limits of integration must be correctly given. The conversion of any triple integral from cartesian to spherical coordinates is then

$$\iiint_V f(x, y, z)\,dx\,dy\,dz = \iiint_V F(R, \varphi, \theta) R^2 \sin \varphi \, dR \, d\varphi \, d\theta, \tag{16}$$

where $F(R, \varphi, \theta) = f(R \sin \varphi \cos \theta, R \sin \varphi \sin \theta, R \cos \varphi)$.

Example 6 Find the volume cut out from a sphere of radius a by a circular cone of semivertical angle α whose vertex is at the center of the sphere.

The problem is to calculate $\iiint dV$ throughout the volume shown in Fig. 3.12. While it is not impossible to find this volume by integration in rectangular coordinates, it is far easier to use spherical coordinates (try it both ways!). One immediate advantage of formulating the problem in this manner is that all the limits of integration are constants, and, in fact, the required volume is

$$V = \int_0^{2\pi} d\theta \int_0^{\alpha} d\varphi \int_0^a R^2 \sin\varphi \, dR.$$

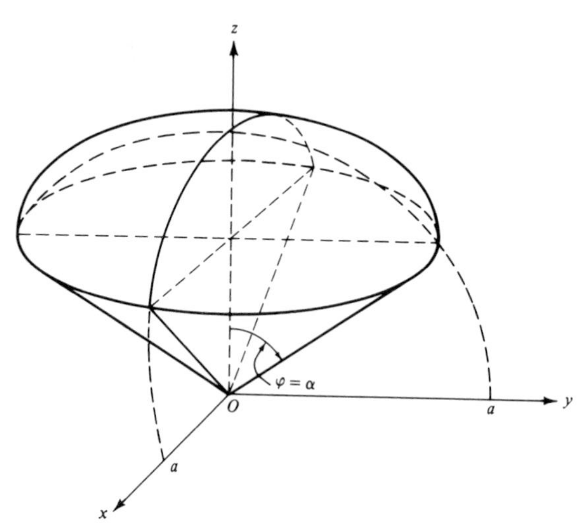

FIGURE 3.12

The evaluation is straightforward:

$$V = \left(\int_0^{2\pi} d\theta\right)\left(\int_0^{\alpha} \sin\varphi \, d\varphi\right)\left(\int_0^a R^2 \, dR\right),$$

$$= \frac{2\pi a^3}{3}(1 - \cos\alpha).$$

In particular if the cone has semivertical angle $\alpha = \pi$, this volume is the entire sphere, namely, $4\pi a^3/3$.

---

5 Consider next the general change of variables in a triple integral

$$x = x(u, v, w), \qquad y = y(u, v, w), \qquad z = z(u, v, w),$$

or equivalently

$$\mathbf{r} = \mathbf{r}(u, v, w), \tag{17}$$

and the conversion of an arbitrary triple integral $\iiint_V f(x, y, z)\, dV$ to the new coordinates.

The main point is to find the correct form for the infinitesimal volume element dV in terms of the differentials du, dv, dw. The volume increment enclosed by the surfaces of constant u, $u + du$, v, $v + dv$, w, and $w + dw$ is a small parallelepiped of sides

$$\overrightarrow{PQ} = \frac{\partial \mathbf{r}}{\partial u}\, du, \qquad \overrightarrow{PR} = \frac{\partial \mathbf{r}}{\partial v}\, dv, \qquad \overrightarrow{PS} = \frac{\partial \mathbf{r}}{\partial w}\, dw,$$

shown in Fig. 3.13. [Note, for example, $\overrightarrow{PQ} = \mathbf{r}(u + du, v, w) - \mathbf{r}(u, v, w) = (\partial \mathbf{r}/\partial u)du$.] Since the triple scalar product $\mathbf{a} \cdot \mathbf{b} \times \mathbf{c}$ represents the volume of a parallelepiped with sides $\mathbf{a}$, $\mathbf{b}$, $\mathbf{c}$ (see Sec. 5.5), it follows that

$$dV = \left| \frac{\partial \mathbf{r}}{\partial u} \cdot \frac{\partial \mathbf{r}}{\partial v} \times \frac{\partial \mathbf{r}}{\partial w} \right| du\, dv\, dw,$$

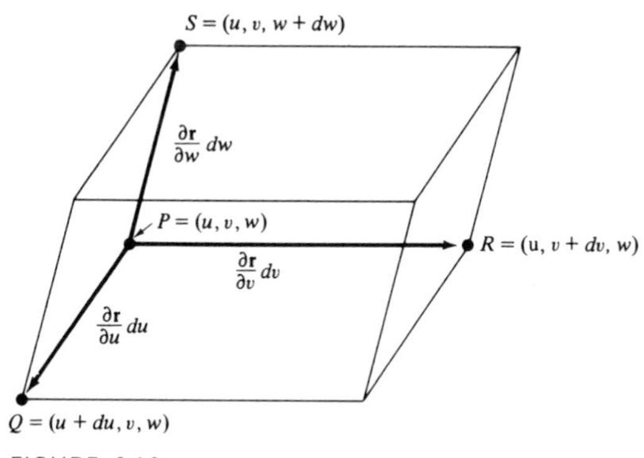

FIGURE 3.13

where the triple scalar product is in fact a third-order Jacobian. This is seen by writing the triple product in determinant form:

$$\frac{\partial \mathbf{r}}{\partial u} \cdot \frac{\partial \mathbf{r}}{\partial v} \times \frac{\partial \mathbf{r}}{\partial w} = \begin{vmatrix} \dfrac{\partial x}{\partial u} & \dfrac{\partial x}{\partial v} & \dfrac{\partial x}{\partial w} \\ \dfrac{\partial y}{\partial u} & \dfrac{\partial y}{\partial v} & \dfrac{\partial y}{\partial w} \\ \dfrac{\partial z}{\partial u} & \dfrac{\partial z}{\partial v} & \dfrac{\partial z}{\partial w} \end{vmatrix} = J,$$

so that

$$dV = |J|\, du\, dv\, dw. \tag{18}$$

The transformation of a general triple integral is then

$$\iiint_V f(x, y, z)\, dx\, dy\, dz = \iiint_V F(u, v, w)|J|\, du\, dv\, dw, \tag{19}$$

where $F(u, v, w) = f(x(u, v, w), y(u, v, w), z(u, v, w))$, and the limits of integration are converted appropriately.

Formula (18) can be used to check the expression for dV in cylindrical coordinates (or vice versa). With $x = r\cos\theta$, $y = r\sin\theta$, $z = z$, $(u = r, v = \theta, w = z)$ we find that

$$J = \begin{vmatrix} \cos\theta & -r\sin\theta & 0 \\ \sin\theta & r\cos\theta & 0 \\ 0 & 0 & 1 \end{vmatrix} = r$$

and

$$dV = r\, dr\, d\theta\, dz,$$

in agreement with the result in (6).

The transformation to spherical coordinates with $x = R\sin\varphi\cos\theta$, $y = R\sin\varphi\sin\theta$, $z = R\cos\varphi$, involves the determinant

$$J = \begin{vmatrix} \sin\varphi\cos\theta & R\cos\varphi\cos\theta & -R\sin\varphi\sin\theta \\ \sin\varphi\sin\theta & R\cos\varphi\sin\theta & R\sin\varphi\cos\theta \\ \cos\varphi & -R\sin\varphi & 0 \end{vmatrix},$$

and upon evaluation we find that

$$J = R^2 \sin\varphi$$

or

$$dV = R^2 \sin\varphi\, dR\, d\varphi\, d\theta,$$

in agreement with (15).

Example 7 Find the volume of the ellipsoid $x^2/a^2 + y^2/b^2 + z^2/c^2 = 1$.

The volume is $\iiint dx\, dy\, dz$, where the integration is throughout the region $x^2/a^2 + y^2/b^2 + z^2/c^2 \leq 1$. If we introduce the transformation $x = au$, $y = bv$, $z = cw$, the integration domain becomes $u^2 + v^2 + w^2 \leq 1$. The Jacobian of the transformation is $J = abc$, so that, by (19),

$$\iiint dx\, dy\, dz = abc \iiint du\, dv\, dw.$$

Since $\iiint du\, dv\, dw = 4\pi/3$ because the volume in uvw space is that of a unit sphere, it follows that the volume of the ellipsoid is $(4\pi/3)\, abc$.

Of course, it is possible to generalize the whole affair and consider a general multiple integral $\iint \cdots \int f(x_1, x_2, \ldots, x_n)\, dx_1, dx_2, \ldots, dx_n$; such integrals do arise in applications, but we will not discuss them.

Exercises 7.3

1. Evaluate the following integrals for the rectangular volume V, $0 \le x \le 1$, $0 \le y \le 2$, $0 \le z \le 3$.

(i) $\iiint\limits_V x^3 y^2 z \, dV$ **(ii)** $\iiint\limits_V e^{x+y-z} \, dV$ **(iii)** $\iiint\limits_V \sin x \sin y \sin z \, dV$

2. A crystal having six plane faces has the form of a truncated prism of square cross section, as shown in Fig. 3.14. If $OA = 1$ cm, $AD = 3$ cm, $CF = 2$ cm, and $OG = 4$ cm, find the volume of the crystal.

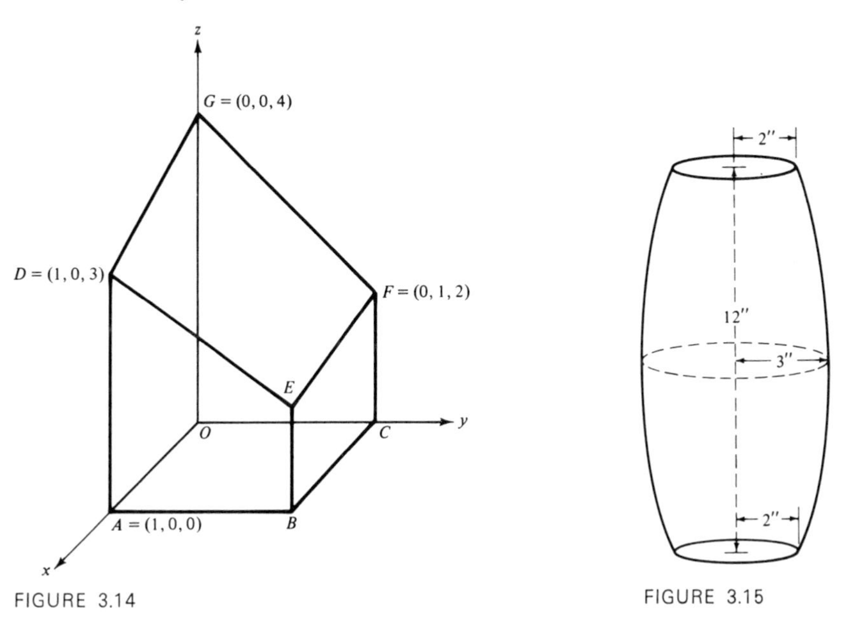

FIGURE 3.14

FIGURE 3.15

3. Find the volume in the first octant between the two planes $x + y + z = 1$ and $2x + 3y + z = 6$.

4. Evaluate the following triple integrals.

(i) $\int_{-1}^{1} dx \int_{0}^{1} dy \int_{-1}^{x} dz$ **(ii)** $\int_{0}^{1} dx \int_{0}^{x} dy \int_{0}^{x-y} dz$ **(iii)** $\int_{0}^{1} dy \int_{1}^{y} dz \int_{0}^{z} \log x \, dx$

(iv) $\int_{0}^{1} dz \int_{0}^{z} dy \int_{0}^{y} xe^{z} \, dx$ **(v)** $\int_{0}^{1} dx \int_{0}^{(1-x^2)^{1/2}} dy \int_{0}^{y} dz$ **(vi)** $\int_{0}^{1} dx \int_{0}^{x} dy \int_{y^2}^{y} z \, dz$

5. Rewrite each integral in Exercise 4 so that the first integration is with respect to y, the second with respect to z, and the third with respect to x.

6. Find the volume of the tetrahedron formed by the coordinate planes and the plane

$$3x + 4y + 6z = 12.$$

7. Find the volume of the ellipsoid of revolution $x^2/a^2 + (y^2 + z^2)/b^2 = 1$. Use this result and a simple geometrical argument to derive the formula for the volume of the ellipsoid $x^2/a^2 + y^2/b^2 + z^2/c^2 = 1$. (*Hint:* Compare the heights in the z direction of the two surfaces.)

8. Show that the volume bounded by the cylinders $y^2 = x$, $y = x^2$ and the planes $z = 0$, $x + y + z = 2$ is $\int_0^1 dx \int_{x^2}^{\sqrt{x}} dy \int_0^{2-x-y} dz$, and calculate this value.

9. Evaluate $\int_{-1}^{1} dz \int_{0}^{z} dy \int_{y-z}^{y+z} (x+y+z)\, dz$. Describe the volume of integration.

10. Calculate $\iiint_V \cos x \cos y \cos z\, dx\, dy\, dz$, where V is the volume $|x|+|y|+|z| \leq \pi$. (*Hint*: Use the symmetry properties of the integrand.)

11. Find the volume enclosed by the hyperboloid $x^2/a^2 + y^2/b^2 - z^2/c^2 = 1$ and the planes $z = \pm\, d$.

12. To test technique, use cylindrical coordinates to find the volume of a sphere and spherical coordinates to find the volume of a cylinder.

13. Evaluate the following integrals where V is the interior of the sphere $x^2 + y^2 + z^2 = a^2$

(i) $\iiint_V x^2\, dxdydz$ **(ii)** $\iiint_V (x^2 + y^2 + z^2)\, dxdydz$ **(iii)** $\iiint_V \frac{1}{x^2+y^2+z^2}\, dxdydz$

14. For what values of α does the integral $I(\alpha) = \iiint_V (x^2 + y^2 + z^2)^\alpha\, dxdydz$ converge if V is the interior of the sphere $x^2 + y^2 + z^2 = a^2$?

15. Evaluate the following integrals where V is the region of the positive octant within the ellipsoid $x^2/a^2 + y^2/b^2 + z^2/c^2 = 1$.

(i) $\iiint_V x\, dxdydz$ **(ii)** $\iiint_V xy\, dxdydz$ **(iii)** $\iiint_V xyz\, dxdydz$

16. Indicate the regions of integration and evaluate the following integrals.

(i) $\int_0^a dz \int_0^{(a^2-z^2)^{1/2}} dy \int_0^{(a^2-y^2-z^2)^{1/2}} x^2\, dx$

(ii) $\int_0^2 dz \int_0^{(2z-z^2)^{1/2}} dy \int_0^{(2z-z^2-y^2)^{1/2}} (x^2+y^2+z^2)^{-1/2}\, dx$

17. Evaluate $\iiint_V (x+y-z)(-x+y+z)(x-y+z)\, dxdydz$, where V is the region $0 \leq x+y-z \leq 1, 0 \leq -x+y+z \leq 1, 0 \leq x-y+z \leq 1$.

18. Find the mass of a spherical shell whose internal radius is a and external radius b if the density is proportional to distance from the center. Determine the average density.

19. Find the mass inside the sphere $x^2 + y^2 + z^2 = 1$ if the density is proportional to:

(i) Distance from the z axis **(ii)** Distance from the xy plane.

20. A volcano is represented by the surface shape $z = he^{-(1/4h)(x^2+y^2)^{1/2}}$, $z>0$. After an eruption in which a volume V of lava adheres to the mountain it has a similar shape. Find the percentage change in the height of the volcano.

21. A symmetrical coffee percolator holds 24 cups when full. The interior design is a circular cross section which tapers from a radius of 3 in. at the center position to 2 in. at the base and top (these being 12 in. apart). The bounding surface is parabolic (see Fig. 3.15). Where should the marking be placed to indicate the 6-cup level?

22. A hollow sphere of radius a is half filled with liquid of constant density. If it rotates with angular velocity Ω about a vertical diameter, how fast must the rotation rate be to just expose the lowest point of the sphere? (*Hint:* The free surface is of the form $z = \Omega^2(x^2 + y^2)/2g + \text{constant}$.)

23. A spherical planet has an atmosphere in which the density decreases exponentially with height above the surface. Find the total mass of the atmosphere. For the planet earth, half of the mass of the atmosphere lies below 5500 meters. Determine the average density of this portion of the atmosphere in terms of the surface density ρ_0.

24. Show that

$$\iiint_V (\mathbf{a}\cdot\mathbf{r})(\mathbf{b}\cdot\mathbf{r})(\mathbf{c}\cdot\mathbf{r})\,dxdydz = \frac{(\alpha\beta\gamma)^2}{8(\mathbf{a}\cdot\mathbf{b}\times\mathbf{c})},$$

where $\mathbf{a}$, $\mathbf{b}$, $\mathbf{c}$ are constant vectors, $\mathbf{r}$ is the position vector $x\mathbf{i}+y\mathbf{j}+z\mathbf{k}$, and V is defined by $0\le\mathbf{a}\cdot\mathbf{r}\le\alpha$, $0\le\mathbf{b}\cdot\mathbf{r}\le\beta$, $0\le\mathbf{c}\cdot\mathbf{r}\le\gamma$.

25. Find a formula for the distance between two points whose cylindrical coordinates are (r, θ, z) and (r', θ', z'). By considering these two points to be close together deduce that the element of arc length ds is determined by $ds^2 = dr^2 + r^2\,d\theta^2 + dz^2$. Could this result have been anticipated on geometrical grounds?

26. Repeat Exercise 25 for spherical coordinates and show that

$$ds^2 = dR^2 + R^2\,d\varphi^2 + R^2\sin^2\varphi\,d\theta^2.$$

27. The u, v, w coordinate system defined by $x = x(u, v, w)$, $y = y(u, v, w)$, $z = z(u, v, w)$ is said to be an *orthogonal coordinate system* if the vectors $\partial\mathbf{r}/\partial u$, $\partial\mathbf{r}/\partial v$ and $\partial\mathbf{r}/\partial w$ are orthogonal. Verify that cylindrical polar and spherical polar coordinates are orthogonal.

28. In an orthogonal coordinate system u, v, w, show that

$$(d\mathbf{r})^2 = d\mathbf{r}\cdot d\mathbf{r} = \left|\frac{\partial\mathbf{r}}{\partial u}\right|^2 (du)^2 + \left|\frac{\partial\mathbf{r}}{\partial v}\right|^2 (dv)^2 + \left|\frac{\partial\mathbf{r}}{\partial w}\right|^2 (dw)^2.$$

Illustrate this identity with cylindrical and spherical coordinates.

29. Verify the following results where V is the entire three dimensional space.

(i) $\iiint_V e^{-(x^2+y^2+z^2)}\,dxdydz = \pi^{3/2}$ **(ii)** $\iiint_V e^{-(\alpha^2x^2+\beta^2y^2+\gamma^2z^2)}\,dxdydz = \pi^{3/2}/|\alpha\beta\gamma|$

30. Evaluate the improper integral. $\displaystyle\int_{-\infty}^{\infty}\int_{-\infty}^{\infty}\int_{-\infty}^{\infty}\frac{e^{-(x^2+y^2+z^2)}-1}{(x^2+y^2+z^2)^2}\,dxdydz.$

7.4 Surface integrals

It is often necessary to integrate a function over a curved surface, and such integrals are called, appropriately enough, *surface integrals*. Simple applications relate to the surface area of a body and other quantities which depend on it, like heat loss from a satellite, lift on an airfoil, osmosis through a membrane, drag on a swimming fish, etc. Although the full significance of surface integrals will become more apparent in Chap. 8, the theory and techniques are developed here.

Let $z = \varphi(x, y)$ describe the particular surface S under consideration and take $f(x, y, z)$ to be a prescribed function, so that its value on S is $f(x, y, \varphi(x, y))$. The surface integral of f over S is defined in the usual manner, as the limiting sum of infinitesimal elements. To this end, we divide S completely into small surface elements ΔS_{ij} (Fig. 4.1). A systematic method of doing this is to slice S by the planes $x = x_i$, for $i = 1, 2, \ldots, n$, and $y = y_j$, for $j = 1, 2, \ldots, p$. This grid is made finer by using a greater number of more closely packed planes. If $(\xi_i, \eta_j, \zeta_{ij})$, where $\zeta_{ij} = \varphi(\xi_i, \eta_j)$, is any point in the surface element ΔS_{ij} (where for counting purposes, f is set equal to zero at any point not in S), then the surface integral is defined by

$$\iint_S f(x, y, z)\,dS = \lim_{\substack{n,\,p\to\infty\\ \max(d_{ij})\to 0}} \sum_{i=1}^{n}\sum_{j=1}^{p} f(\xi_i, \eta_j, \zeta_{ij})\,\Delta S_{ij}, \tag{1}$$

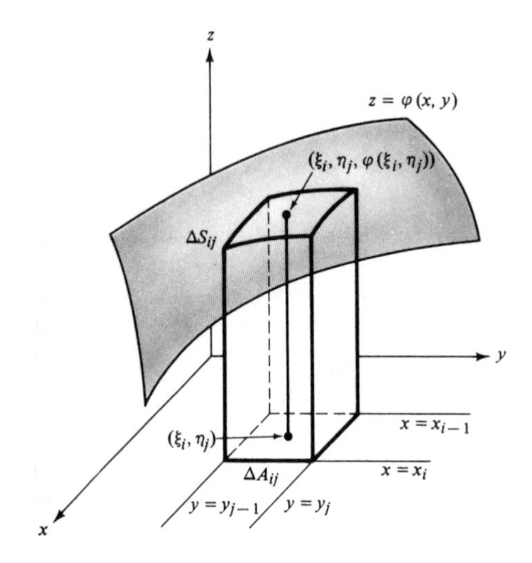

FIGURE 4.1

where d_{ij} is the maximum length of any transversal of ΔS_{ij}. For the special case $f = 1$, Eq. (1) determines the area of the curved surface S. Note also that when S is a region R of the xy plane, that is, $\varphi = 0$, (1) reduces to an ordinary double integral over a plane area as described earlier in Sec. 7.1. In fact, one way to calculate a surface integral is to reduce it to an equivalent integral over a plane surface, and this is the method we shall discuss first.

2

Suppose the surface S is simple in that it can be uniquely projected onto the xy plane. This will be the case if any line parallel to the z axis cuts S exactly once. (If this is not the case, S can be divided into portions which have this property and each portion treated by the method to be described.) In the limit, each of the infinitesimal surface elements is essentially flat, or planar, but tilted with respect to the horizontal.

Let A, B, C, D be points on the surface $z = \varphi(x, y)$ whose vertical projections are the points A_0, B_0, C_0, D_0 in the xy plane, where in particular $A_0 = (x, y)$, $B_0 = (x + dx, y)$, $C_0 = (x, y + dy)$ and $D_0 = (x + dx, y + dy)$, as shown in Fig. 4.2. The four points A_0, B_0, C_0, and D_0 form an infinitesimal rectangle of area $dx\, dy$; the points A, B, C, D form an infinitesimal parallelogram of area dS on the surface S. These two incremental areas are clearly related to each other, and this connection is easily established by using some elementary vector analysis.

Since dS is a parallelogram with sides $\overrightarrow{AB}$ and $\overrightarrow{AC}$, it follows (see Sec. 5.4) that

$$dS = |\overrightarrow{AB} \times \overrightarrow{AC}|. \tag{2}$$

However, each of the vectors $\overrightarrow{AB}$ and $\overrightarrow{AC}$ is easily calculated in terms of the function $z = \varphi(x, y)$. For instance,

$$\overrightarrow{AB} = \overrightarrow{AM} + \overrightarrow{MB},$$

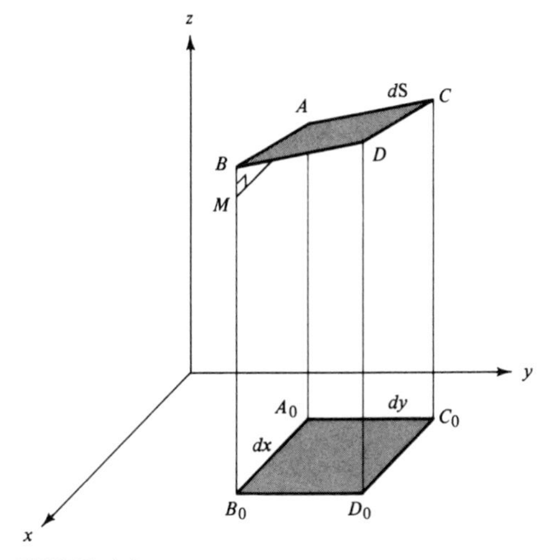

FIGURE 4.2

where M is illustrated in Fig. 4.2 and $\overrightarrow{AM}$ is a vector in the x direction of magnitude dx,

$$\overrightarrow{AM} = dx\,\mathbf{i}.$$

Also $\overrightarrow{MB}$ is in the z direction and of magnitude $\varphi(x + dx,\ y) - \varphi(x,\ y) = \varphi_x\,dx$, so that

$$\overrightarrow{MB} = \varphi_x\,dx\,\mathbf{k},$$

and therefore

$$\overrightarrow{AB} = dx\,\mathbf{i} + \varphi_x\,dx\,\mathbf{k}.$$

In the same way, it can be shown that

$$\overrightarrow{AC} = dy\,\mathbf{j} + \varphi_y\,dy\,\mathbf{k},$$

whereupon

$$\overrightarrow{AB} \times \overrightarrow{AC} = (dx\,\mathbf{i} + \varphi_x\,dx\,\mathbf{k}) \times (dy\,\mathbf{j} + \varphi_y\,dy\,\mathbf{k}) = (-\varphi_x\,\mathbf{i} - \varphi_y\mathbf{j} + \mathbf{k})\,dx\,dy.$$

Substitution into (2) yields the desired result:

$$dS = (1 + \varphi_x{}^2 + \varphi_y{}^2)^{1/2}\,dx\,dy. \tag{3}$$

An alternative derivation begins with the formula for the normal vector to the surface (see Sec. 6.3),

$$\mathbf{n} = -\varphi_x\mathbf{i} - \varphi_y\mathbf{j} + \mathbf{k}.$$

The angle γ between this normal and the vertical (Fig. 4.3) is given by

$$\cos\gamma = \frac{\mathbf{n}\cdot\mathbf{k}}{|\mathbf{n}|\,|\mathbf{k}|} = (1 + \varphi_x{}^2 + \varphi_y{}^2)^{-1/2},$$

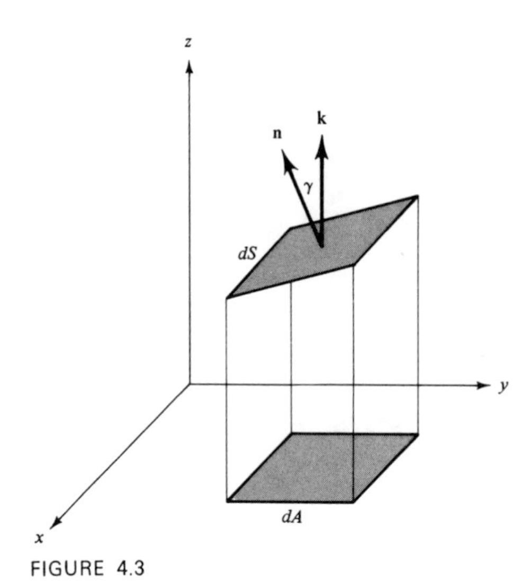

FIGURE 4.3

and this is also the angle of inclination of the element dS with its plane or shadow projection dA. This implies that

$$dA = \cos \gamma \, dS$$

or

$$dS = (1 + \varphi_x{}^2 + \varphi_y{}^2)^{1/2} \, dA,$$

which is the same as (3) when $dA = dx\,dy$.

Now that the element of surface area has been related to its projection onto a plane, it is straightforward to reduce a general surface integral to an equivalent double integral over the projected plane area:

$$\iint_S f(x, y, z)\, dS = \iint_R f(x, y, \varphi(x, y))(1 + \varphi_x{}^2 + \varphi_y{}^2)^{1/2}\, dx\, dy. \tag{4}$$

Here R is the projection of S onto the xy plane (Fig. 4.4).

However, in some problems it may be more convenient to project the surface S onto a plane other than the xy plane. It is relatively simple now to derive formulas for this purpose which correspond to (4). For instance, if the surface given by $y = \psi(z, x)$ is projected onto the zx plane, the surface integral can be written

$$\iint_S f(x, y, z)\, dS = \iint_{R_{zx}} f(x, \psi(z, x), z)\,(1 + \psi_z{}^2 + \psi_x{}^2)^{1/2}\, dz\, dx, \tag{5}$$

where R_{zx} is the projection of S onto the zx plane (Fig. 4.5). A projection onto the yz plane necessitates another slight modification of notation.

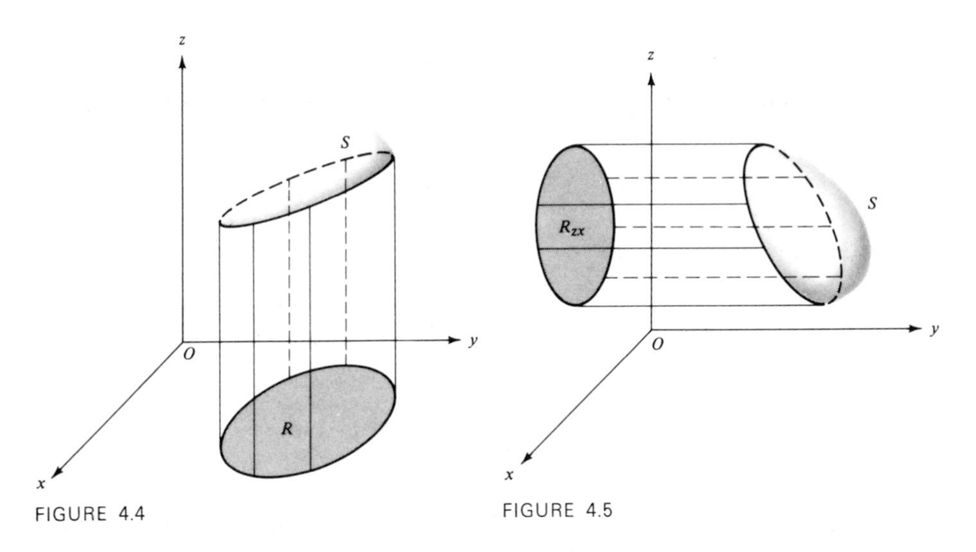

FIGURE 4.4

FIGURE 4.5

Example 1 Find the area of the plane $lx + my + nz + p = 0$ that lies within the cylinder $x^2 + y^2 = a^2$, as shown in Fig. 4.6.

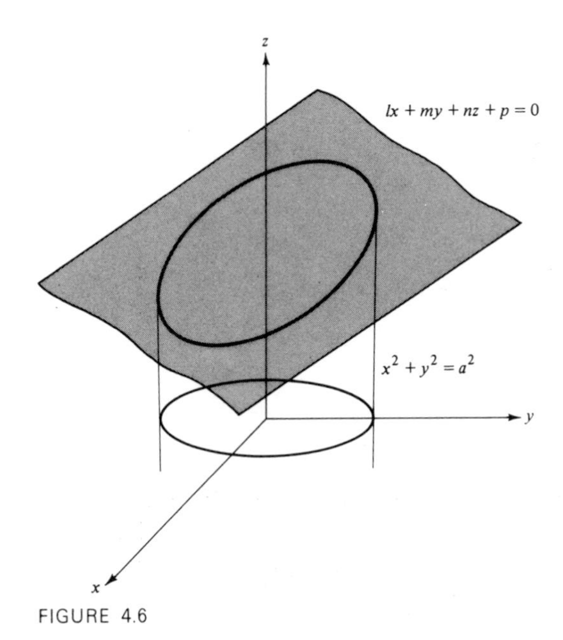

FIGURE 4.6

We must calculate $\iint_S dS$, which is accomplished by projecting onto the xy plane. Since $z = \varphi(x, y) = -(lx + my + p)/n$, partial differentiation yields

$$z_x = \varphi_x = \frac{-l}{n}, \qquad z_y = \varphi_y = \frac{-m}{n},$$

and from (4) it follows that

$$S = \iint_S dS = \iint_R \left[1 + \left(\frac{-l}{n}\right)^2 + \left(\frac{-m}{n}\right)^2\right]^{1/2} dx\,dy,$$

$$= \left(1 + \frac{l^2}{n^2} + \frac{m^2}{n^2}\right)^{1/2} \iint_R dx\,dy,$$

where R is the area $x^2 + y^2 \leq a^2$. Since the area of a circle of radius a is πa^2, the substitution of this result into the preceding equation shows that

$$S = \frac{\pi a^2}{|n|}(l^2 + m^2 + n^2)^{1/2}.$$

Example 2 Evaluate $I = \iint_S x^2y^2z^2\,dS$ over the curved surface of the cone $x^2 + y^2 = z^2$ which lies between $z = 0$ and $z = 1$, as shown in Fig. 4.7.

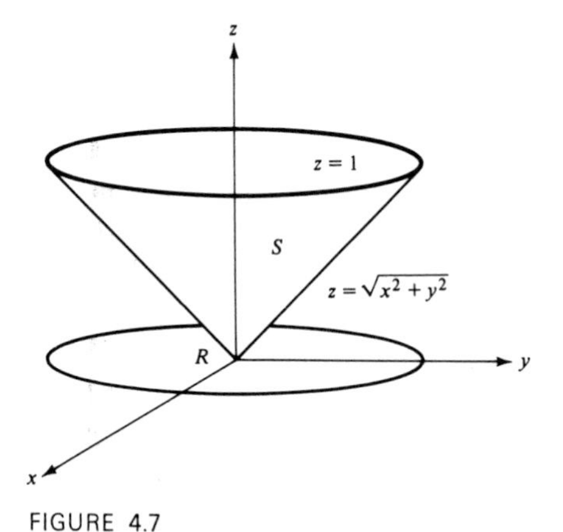

FIGURE 4.7

Again we project the area onto the xy plane. With $z^2 = x^2 + y^2$, $z_x = x/z$, and $z_y = y/z$, it follows that

$$(1 + z_x{}^2 + z_y{}^2)^{1/2} = \left(1 + \frac{x^2 + y^2}{z^2}\right)^{1/2} = \sqrt{2}.$$

The required surface integral is then

$$\iint_S x^2y^2z^2\, dS = \iint_R x^2y^2(x^2 + y^2)\sqrt{2}\, dx\, dy,$$

where R is the interior of the circle $x^2 + y^2 = 1$. This double integral is evaluated most simply by transforming to polar coordinates, $x = r\cos\theta$, $y = r\sin\theta$, in which case

$$I = \iint_S x^2y^2z^2\, dS = \int_0^{2\pi} d\theta \int_0^1 \sqrt{2}\, r^7 \cos^2\theta \sin^2\theta\, dr.$$

Integration with respect to r gives

$$I = \frac{\sqrt{2}}{8}\int_0^{2\pi} \cos^2\theta \sin^2\theta\, d\theta.$$

But since $\cos^2\theta\sin^2\theta = \frac{1}{8}(1 - \cos 4\theta)$, we can write

$$I = \frac{\sqrt{2}}{64}\int_0^{2\pi} (1 - \cos 4\theta)\, d\theta,$$

which is readily evaluated. The final result is

$$I = \frac{\pi\sqrt{2}}{32}.$$

Example 3 Find $I_n = \iint_S y^n\, dS$ over the hemisphere $y = \sqrt{1 - z^2 - x^2}$ for $n > -1$.

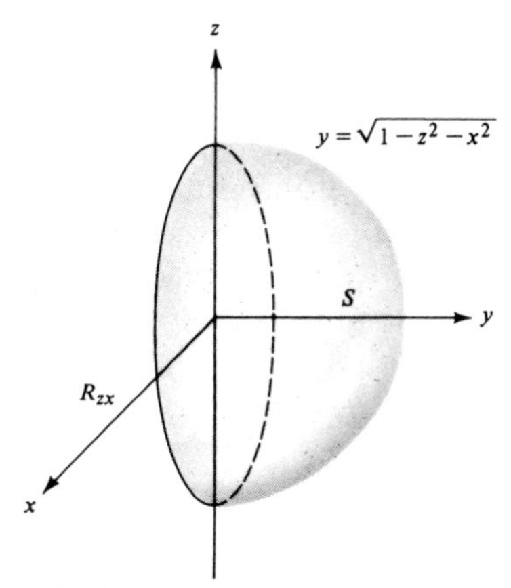

FIGURE 4.8

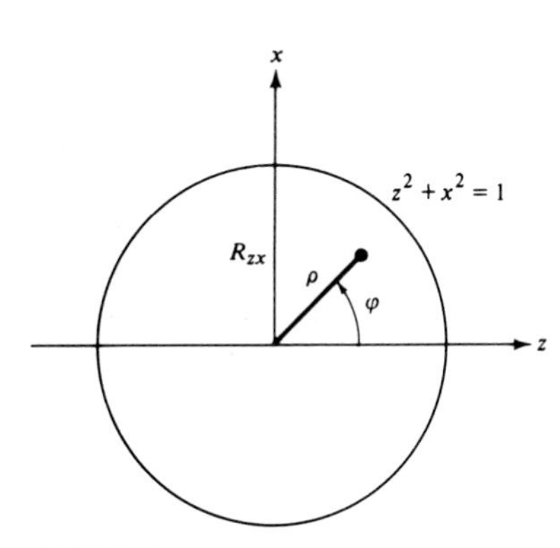

FIGURE 4.9

A projection onto the xy plane would necessitate splitting S into two regions. To avoid this we project instead onto the zx plane (Fig. 4.8). Since $y_z = -z/y$ and $y_x = -x/y$, the element of surface area is

$$dS = (1 + y_z^2 + y_x^2)^{1/2}\, dz\, dx = (1 - z^2 - x^2)^{-1/2}\, dz\, dx.$$

The surface integral is therefore

$$I_n = \iint_S y^n\, dS = \iint_{R_{zx}} (1 - z^2 - x^2)^{n/2} (1 - z^2 - x^2)^{-1/2}\, dz\, dx,$$

where R_{zx} is the interior of the circle $z^2 + x^2 = 1$, $y = 0$. A transformation to polar coordinates in the zx plane (see Fig. 4.9) with $z = \rho \cos \varphi$, $x = \rho \sin \varphi$, reduces the double integral over R_{zx} to a form that can be calculated explicitly:

$$\begin{aligned} I_n = \iint_S y^n\, dS &= \iint_{R_{zx}} (1 - z^2 - x^2)^{(n-1)/2}\, dz\, dx, \\ &= \int_0^{2\pi} d\varphi \int_0^1 (1 - \rho^2)^{(n-1)/2}\, \rho\, d\rho, \\ &= 2\pi \left[-\frac{1}{n+1} (1 - \rho^2)^{(n+1)/2} \right]_0^1. \end{aligned}$$

Therefore, for $n > -1$,

$$I_n = \iint_S y^n\, dS = \frac{2\pi}{n+1}.$$

---

3 We have seen that surface integrals can be evaluated by transforming them into equivalent double integrals over plane areas, i.e., by projection. However, if the surface S is particularly simple, it is usually easier to evaluate the surface integral directly by introducing coordinates appropriate for the particular geometry. A noteworthy case is when S is a sphere or a portion of a sphere, in which case spherical coordinates (R, φ, θ) give a natural representation for the element of surface area. If $R = a$ is the fixed radius of the sphere (Fig. 4.10), the element of area between the curves of constant angles φ, $\varphi + d\varphi$, θ, $\theta + d\theta$, that is, lines of constant latitude and longitude, is

$$dS = (a \sin \varphi\, d\theta)(a\, d\varphi)$$

or

$$dS = a^2 \sin \varphi\, d\varphi\, d\theta. \tag{6}$$

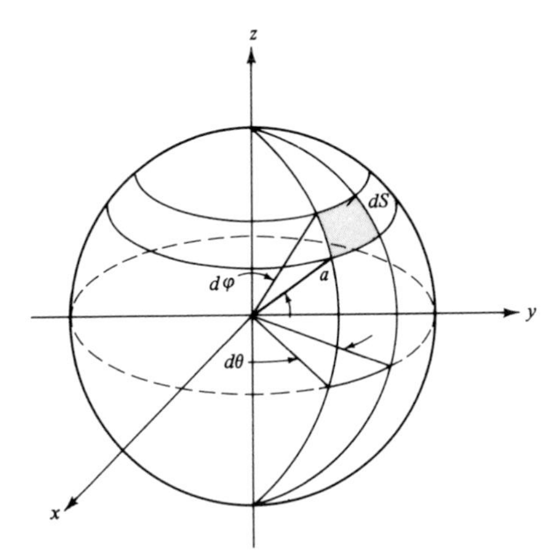

FIGURE 4.10

The surface integral over the entire sphere can then be expressed as

$$\iint_S f(x, y, z)\, dS = a^2 \int_0^{2\pi} d\theta \int_0^{\pi} F(a, \varphi, \theta) \sin \varphi \, d\varphi, \tag{7}$$

where $F(a, \varphi, \theta) = f(a \sin \varphi \cos \theta, a \sin \varphi \sin \theta, a \cos \varphi)$.

Example 4 Find the surface area of a sphere of radius a.

Although the result, $4\pi a^2$, is well known, this example demonstrates the convenience of spherical coordinates. The area of the whole sphere is

$$S = \iint_S dS = \int_0^{2\pi} d\theta \int_0^{\pi} a^2 \sin \varphi \, d\varphi,$$

where φ ranges from 0 to π and θ from 0 to 2π. Integration yields

$$S = [\theta]_0^{2\pi} \, [-a^2 \cos \varphi]_0^{\pi} = 4\pi a^2.$$

Example 5 A hollow spherical conductor of radius a has a constant surface charge density σ. Find the force exerted on a unit charge placed at a distance b from the center of the sphere (see Fig. 4.11).

It is assumed that the inverse-square law holds: two positive charges q_1 and q_2 a distance D apart repel each other with a force of magnitude $q_1 q_2 / D^2$.

The magnitude of the force increment on the unit charge $q_1 = 1$ at B due to the charge element $q_2 = \sigma \, dS$ on an area dS of the sphere is $\sigma \, dS/(PB)^2$. This

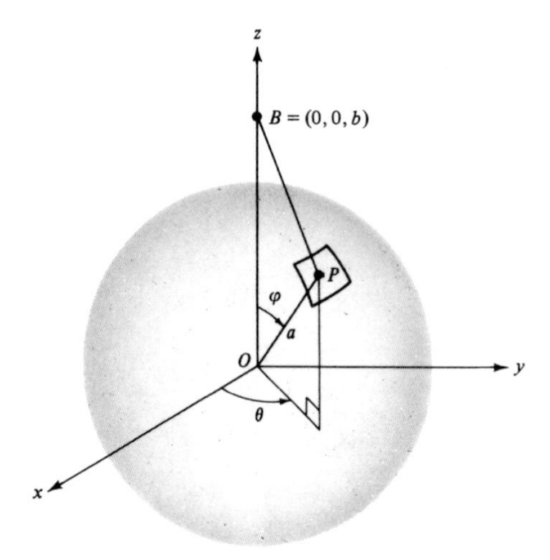

FIGURE 4.11

element of force $d\mathbf{F}$ is in the direction of the vector $\overrightarrow{PB}$, so that

$$d\mathbf{F} = \frac{\sigma\, dS}{(PB)^2} \frac{\overrightarrow{PB}}{|\overrightarrow{PB}|} = \frac{\sigma\, dS\, \overrightarrow{PB}}{|\overrightarrow{PB}|^3}.$$

Since the orientation of the cartesian coordinate frame is ours to choose, it is of great benefit to select the z axis to be in the direction OB, where O is the center of the sphere. (We will not preclude the possibility that $b < a$, that is, B lies within the sphere.)

The total force $\mathbf{F}$ on the charge B is found by adding the force increments from each element dS, and this implies integration over the entire surface of the sphere:

$$\mathbf{F} = \iint d\mathbf{F} = \iint_S \frac{\sigma \overrightarrow{PB}\, dS}{|\overrightarrow{PB}|^3}. \tag{8}$$

Symmetry arguments indicate that the x and y components of $\mathbf{F}$ will be zero and only the z component may be nonzero, and this fact will be verified by the subsequent analysis.

If the spherical coordinates of point P are (a, φ, θ), the vectors $\overrightarrow{OP}$, $\overrightarrow{OB}$, and $\overrightarrow{PB} = \overrightarrow{OB} - \overrightarrow{OP}$ are given by the expressions

$$\overrightarrow{OP} = a \sin\varphi \cos\theta\, \mathbf{i} + a \sin\varphi \sin\theta\, \mathbf{j} + a\cos\varphi\, \mathbf{k},$$

$$\overrightarrow{OB} = b\mathbf{k},$$

$$\overrightarrow{PB} = -a \sin\varphi \cos\theta\, \mathbf{i} - a \sin\varphi \sin\theta\, \mathbf{j} + (b - a\cos\varphi)\mathbf{k}.$$

Moreover $|\overrightarrow{PB}| = (a^2 - 2ab\cos\varphi + b^2)^{1/2}$, $dS = a^2 \sin\varphi\, d\varphi\, d\theta$; substitution of these expressions into the integral for $\mathbf{F}$ yields

$$\mathbf{F} = \int_0^{2\pi} d\theta \int_0^{\pi} \frac{\sigma(-a\sin\varphi\cos\theta\,\mathbf{i} - a\sin\varphi\sin\theta\,\mathbf{j} + (b - a\cos\varphi)\mathbf{k})a^2\sin\varphi\, d\varphi}{(a^2 - 2ab\cos\varphi + b^2)^{3/2}}.$$

The θ integration can be performed immediately, and because

$$\int_0^{2\pi} \cos\theta\, d\theta = \int_0^{2\pi} \sin\theta\, d\theta = 0,$$

the x and y components of $\mathbf{F}$ vanish, as anticipated. With these simplifications the expression for $\mathbf{F}$ reduces to

$$\mathbf{F} = 2\pi\sigma a^2\mathbf{k} \int_0^{\pi} \frac{(b - a\cos\varphi)\sin\varphi\, d\varphi}{(a^2 - 2ab\cos\varphi + b^2)^{3/2}}.$$

The change of variable $t = \cos\varphi$ brings this equation to the form

$$\mathbf{F} = 2\pi\sigma a^2\mathbf{k} \int_{-1}^{+1} \frac{(b - at)\, dt}{(a^2 - 2abt + b^2)^{3/2}}, \tag{9}$$

which is readily evaluated from the simpler result

$$\int_{-1}^{+1} \frac{dt}{(a^2 - 2abt + b^2)^{1/2}} = \left[-\frac{1}{ab}(a^2 - 2abt + b^2)^{1/2}\right]_{-1}^{+1},$$

$$= \frac{1}{ab}(|b + a| - |b - a|).$$

If this formula is written

$$\int_{-1}^{+1} \frac{dt}{(a^2 - 2abt + b^2)^{1/2}} = \begin{cases} \dfrac{2}{b} & \text{for} \quad b > a; \\[2ex] \dfrac{2}{a} & \text{for} \quad b < a, \end{cases}$$

the integral can be differentiated with respect to b, to give

$$\int_{-1}^{+1} \frac{(b - at)\, dt}{(a^2 - 2abt + b^2)^{3/2}} = \begin{cases} \dfrac{2}{b^2} & \text{for} \quad b > a; \\[2ex] 0 & \text{for} \quad b < a, \end{cases}$$

which is exactly (9). The final result for $\mathbf{F}$ is then

$$\mathbf{F} = \begin{cases} \dfrac{4\pi\sigma a^2}{b^2}\mathbf{k} & \text{for} \quad b > a; \\[2ex] \mathbf{0} & \text{for} \quad b < a. \end{cases}$$

This proves that there is *zero force* on a particle when it is *inside* the sphere ($b < a$), whereas for a particle *outside* the sphere ($b > a$) the effect of the uniform charge distribution is the same as if the total charge, $4\pi\sigma a^2$, were placed at the *center of the sphere*. (This calculation also applies to gravitational forces.)

5

So far, surfaces have usually been represented in terms of cartesian coordinates, for example, $z = \varphi(x, y)$. It is also possible and frequently convenient to describe a surface by a parametric representation. Two parameters are needed for this purpose since a surface has two degrees of freedom. For example, the equations

$$x = a \sin u \cos v, \qquad y = a \sin u \sin v, \qquad z = a \cos u, \tag{10}$$

with u and v as independent parameters, represent a sphere whose equation,

$$x^2 + y^2 + z^2 = a^2,$$

is obtained by elimination of u and v. Indeed, u and v are identical to the spherical coordinates φ and θ, respectively. If u is kept constant and only v is allowed to vary, the point (x, y, z) moves on a curve of constant latitude. If v is kept constant and only u varies, the locus of a point (x, y, z) lies on a curve of constant longitude (Fig. 4.12).

The set of equations

$$x = x(u, v), \qquad y = y(u, v), \qquad z = z(u, v), \tag{11}$$

with u and v as parameters, represents an arbitrary surface. This can be seen by eliminating u and v from the three equations, a procedure that leads to an equation of the form $F(x, y, z) = 0$. In terms of the radius vector $\mathbf{r} = x\mathbf{i} + y\mathbf{j} + z\mathbf{k}$, Eqs. (11) for the surface are written compactly as

$$\mathbf{r} = \mathbf{r}(u, v). \tag{12}$$

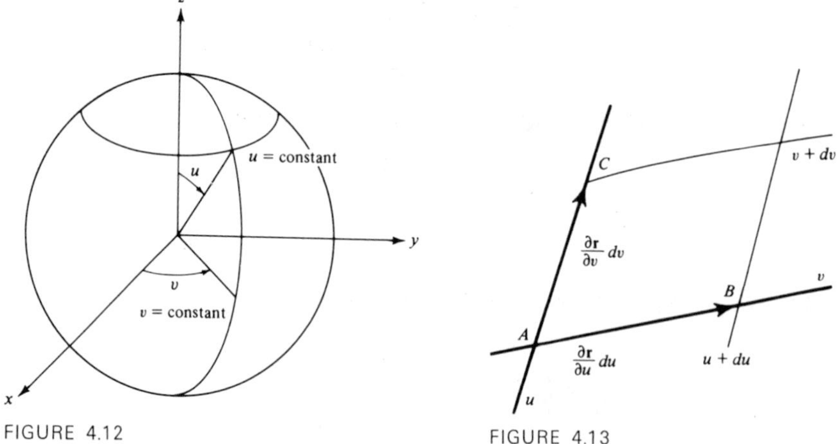

FIGURE 4.12

FIGURE 4.13

From the parametric equations for a surface it is possible to establish a formula for dS the element of surface area. (In the case of a sphere this element was found earlier.) In general, dS is obtained by calculating the area between the curves corresponding to constant u, $u + du$, v, and $v + dv$. For infinitesimal areas this element will be essentially planar and have an area

$$dS = |\overrightarrow{AB} \times \overrightarrow{AC}|, \tag{13}$$

where the vectors are the sides of the differential parallelogram shown in Fig. 4.13. The points A, B, and C correspond to the vector positions $\mathbf{r}$, $\mathbf{r} + \frac{\partial \mathbf{r}}{\partial u} du$ and $\mathbf{r} + \frac{\partial \mathbf{r}}{\partial v} dv$, respectively. Therefore $\overrightarrow{AB} = \frac{\partial \mathbf{r}}{\partial u} du$, $\overrightarrow{AC} = \frac{\partial \mathbf{r}}{\partial v} dv$, and the replacement by these expressions in (13) yields

$$dS = \left| \frac{\partial \mathbf{r}}{\partial u} \times \frac{\partial \mathbf{r}}{\partial v} \right| du\, dv. \tag{14}$$

In particular for the unit sphere

$$\mathbf{r} = a \sin u \cos v\, \mathbf{i} + a \sin u \sin v\, \mathbf{j} + a \cos u\, \mathbf{k},$$

so that

$$\frac{\partial \mathbf{r}}{\partial u} \times \frac{\partial \mathbf{r}}{\partial v} = \begin{vmatrix} \mathbf{i} & \mathbf{j} & \mathbf{k} \\ a \cos u \cos v & a \cos u \sin v & -a \sin u \\ -a \sin u \sin v & a \sin u \cos v & 0 \end{vmatrix},$$

$$= a^2(\sin^2 u \cos v\, \mathbf{i} + \sin^2 u \sin v\, \mathbf{j} + \sin u \cos u\, \mathbf{k}).$$

When magnitudes are taken, we obtain

$$\left| \frac{\partial \mathbf{r}}{\partial u} \times \frac{\partial \mathbf{r}}{\partial v} \right| = a^2(\sin^4 u \cos^2 v + \sin^4 u \sin^2 v + \sin^2 u \cos^2 u)^{1/2},$$

which upon simplification becomes

$$\left| \frac{\partial \mathbf{r}}{\partial u} \times \frac{\partial \mathbf{r}}{\partial v} \right| = a^2 \sin u.$$

Therefore Eq. (14) yields the relation

$$dS = a^2 \sin u\, du\, dv,$$

a result that was established earlier in this section by geometrical arguments.

Exercises 7.4

1. Evaluate the following integrals where S is the portion of the plane $3x + 2y - z + 1 = 0$ above the square $0 \leq x \leq 1$, $0 \leq y \leq 1$.

(i) $\iint_S dS$ **(ii)** $\iint_S (y + 2z)\, dS$ **(iii)** $\iint_S xyz\, dS$

2. Evaluate the following integrals where S is the portion of the plane $x+y+z=\sqrt{2}$ above the circle $x^2+y^2 \leq 1$.

(i) $\iint_S dS$ **(ii)** $\iint_S (x+y)\, dS$ **(iii)** $\iint_S z\, dS$

3. Evaluate $\iint_S x^2y^2\, dS$ over the total surface (including the ends) of the cylinder $x^2+y^2=a^2$, $z=0$, $z=h$. Verify the dimensions of the answer.

4. Find the surface area of the cylinder $x^2+y^2=a^2$ which lies within the cylinder $x^2+z^2=a^2$.

5. Calculate the surface area of the sphere $x^2+y^2+z^2=a^2$ which lies within the cylinder $x^2+y^2=ax$. What area of the cylinder lies within the sphere?

6. What is the surface area of the circular torus obtained by rotating the circle $x=0$, $(y-b)^2+z^2=a^2$, $a<b$, around the z axis?

7. Find the surface area on a sphere contained between two lines of longitude and two lines of latitude. Calculate the surface area of the earth between longitudes 30° and 60° and latitudes 30° and 45° ($r=6370$ km).

8. Evaluate the following integrals over the hemispherical surface $z=(a^2-x^2-y^2)^{1/2}$.

(i) $\iint_S x^2\, dS$ **(ii)** $\iint_S z\, dS$ **(iii)** $\iint_S z^2\, dS$

9. Evaluate $\iint_S (x^2+y^2+z^2)\, dS$ and $\iint_S (y^2z^2+z^2x^2+x^2y^2)\, dS$, where S is the sphere $x^2+y^2+z^2=a^2$.

10. Find the area of that portion of the sphere $x^2+y^2+z^2=a^2$ which is within the paraboloid $2cz=x^2+y^2$.

11. A solid is bounded by the cone $z^2=2xy$ and the cylinder $x^{1/2}+y^{1/2}=1$. Prove that the volume of the solid is $2\sqrt{2}/45$ and find the surface area of the conical boundary.

12. Find the surface area of the spheroid $(x^2+y^2)/a^2+z^2/c^2=1$, $c\leq a$. Verify the special case $c=a$.

13. Find the area of the sphere $x^2+y^2+z^2=a^2$ that is outside the cylinders $x^2+y^2-ax=0$ and $x^2+y^2+ax=0$. (*Hint:* Project onto the xy plane and use polar coordinates.)

14. Find the area of that portion of the sphere $x^2+y^2+z^2=a^2$ which is within the cylinder $x^2+y^2=b^2$ where $b<a$. Verify the special case $b=a$.

15. Find the area of the surface $z=xy$ in the positive octant that is within the cylinder $x^2+y^2=a^2$.

16. Find the cartesian equations of the following surfaces described parametrically.

(i) $x=u,\ y=v,\ z=1-u-v$ **(ii)** $x=u,\ y=v,\ z=\cos u \cos v$

(iii) $x=u\cos v,\ y=u\sin v,\ z=u^2$ **(iv)** $x=e^u,\ y=e^v,\ z=e^{u+v}$

(v) $x=u\cosh v,\ y=u\sinh v,\ z=u^2$ **(vi)** $x=a\sin u\cos v,\ y=b\sin u\sin v,\ z=c\cos u$

17. For the surfaces in the previous problem, calculate the element of surface area

$$|\partial\mathbf{r}/\partial u \times \partial\mathbf{r}/\partial v|\, du\, dv.$$

18. Show that $\mathbf{r}=u\cos v\,\mathbf{i}+u\sin v\,\mathbf{j}+f(u)\mathbf{k}$, $u_1\leq u\leq u_2$, $0\leq v\leq 2\pi$, is a surface of revolution and that the area of this surface is $2\pi\int_{u_1}^{u_2} u\{1+[f'(u)]^2\}^{1/2}\, du$. Use this formula to find the surface area of a circular cone.

19. A snow pile of volume V and exposed surface area S melts at a rate given by $dV/dt=-\alpha S$, where α is a constant. Assume that during melting the pile maintains the shape $z=h-(x^2+y^2)/h$, $z>0$, where $h=h(t)$. How long does it take for a snow pile of height h_0 to disappear? (*Hint:* Find $V=V(h)$ and $S=S(h)$ and a differential equation for $h(t)$.)

20. In a bay where the difference between high and low tides is 6 meters, a small island has the land elevation $z=10^2[1-(x^2+y^2)/10^6]$. All lengths are in meters, and the level $z=0$ corresponds to low tide. Find the ratio of the dry surface areas at high and low tides.

21. A spherical shell of radius a centered at the origin has a surface density at any point P given by $\rho = \rho_0(1 + \varepsilon \sin\varphi)$, where φ is the angle between OP and a fixed line. Find the total mass of the shell.

7.5 *Applications*

For the most part this chapter has dealt with analytical techniques for the evaluation of multiple integrals. Now we turn our attention to some applications, and in a sense this section is the follow-up of the earlier discussion (Sec. 3.8). However, we are now equipped to handle a far greater variety of problems.

As a preliminary, consider first the torque produced by a force $\mathbf{F}$ that acts at a distance $\mathbf{r}$ from a given point. If this point is the origin, the torque vector, as described in Sec. 5.4, is $\mathbf{r} \times \mathbf{F}$. Moreover, the torque about another point $\bar{\mathbf{r}}$ is the cross product of the new distance, $\mathbf{r} - \bar{\mathbf{r}}$, to the point of application of the force $\mathbf{F}$, that is, $(\mathbf{r} - \bar{\mathbf{r}}) \times \mathbf{F}$.

We now apply this result to a body acted upon by a uniform gravitational field. Since the force on an element of mass dm is $\mathbf{g}\,dm$, where $\mathbf{g}$ is the constant gravitational acceleration vector, the torque $d\mathbf{L}$ about an arbitrary point $\bar{\mathbf{r}}$ produced by this increment of force is $(\mathbf{r} - \bar{\mathbf{r}}) \times \mathbf{g}\,dm$. The total torque $\mathbf{L}$ on a rigid body is obtained by adding all such differential elements of torque, and the sum is the triple integral

$$\mathbf{L} = \iiint d\mathbf{L} = \iiint (\mathbf{r} - \bar{\mathbf{r}}) \times \mathbf{g}\,dm.$$

Since $\mathbf{g}$ is assumed to be a constant vector, this can be rewritten as

$$\mathbf{L} = \mathbf{g} \times \iiint (\bar{\mathbf{r}} - \mathbf{r})\,dm.$$

The torque about a point is, in general, nonzero, and this reflects common experience: a body suspended by a string will twist under the action of gravity and eventually find an equilibrium position corresponding to zero torque. It is well known that there is a position of perfect balance and, in effect, a body suspended from this point will not move. The value of $\bar{\mathbf{r}}$ which makes the total torque vector zero can be calculated quite easily. Upon setting $\mathbf{L} = \mathbf{0}$ in the last equation we find that

$$\bar{\mathbf{r}} \iiint dm = \iiint \mathbf{r}\,dm.$$

Therefore

$$\bar{\mathbf{r}} = \frac{\iiint \mathbf{r}\,dm}{\iiint dm} = \frac{1}{M} \iiint \mathbf{r}\,dm, \tag{1}$$

and this position vector $\bar{\mathbf{r}}$ is called *the center of mass* of the body whose total mass is M. A body fixed at its center of mass has no tendency to turn under the action of gravity. The object can then be balanced perfectly from a string attached anywhere along the vertical line through $\bar{\mathbf{r}} = \bar{x}\mathbf{i} + \bar{y}\mathbf{j} + \bar{z}\mathbf{k}$ (the most convenient spot being the point in the top surface).

In terms of the mass density ρ, $dm = \rho\, dV$, and Eq. (1) becomes

$$\bar{\mathbf{r}} = \frac{\iiint\limits_V \rho \mathbf{r}\, dV}{\iiint\limits_V \rho\, dV}, \tag{2}$$

where V is the volume of the body concerned. In component form, this is

$$\bar{x} = \frac{\iiint\limits_V \rho x\, dV}{\iiint\limits_V \rho\, dV}, \qquad \bar{y} = \frac{\iiint\limits_V \rho y\, dV}{\iiint\limits_V \rho\, dV}, \qquad \bar{z} = \frac{\iiint\limits_V \rho z\, dV}{\iiint\limits_V \rho\, dV}. \tag{3}$$

For a surface S which has a surface mass density σ so that $dm = \sigma\, dS$, Eq. (1) takes the form

$$\bar{\mathbf{r}} = \frac{\iint\limits_S \sigma \mathbf{r}\, dS}{\iint\limits_S \sigma\, dS}. \tag{4}$$

The center of mass also plays a fundamental role in dynamical systems. For example, the momentum of an element of mass dm is $dm\,(d\mathbf{r}/dt)$, which makes the total momentum $\mathbf{p}$ of a body

$$\mathbf{p} = \iiint \frac{d\mathbf{r}}{dt}\, dm.$$

For a fixed or rigid mass distribution, this can be written

$$\mathbf{p} = \frac{d}{dt} \iiint \mathbf{r}\, dm = \frac{d}{dt}\left(\bar{\mathbf{r}} \iiint dm\right) = \frac{d}{dt}(\bar{\mathbf{r}}M)$$

or

$$\mathbf{p} = M\frac{d\bar{\mathbf{r}}}{dt},$$

where $M = \iiint dm$ is the total mass. In other words, the momentum of a body is equal to its total mass multiplied by the velocity of the center of mass. From a statical and dynamical viewpoint the body can be imagined as concentrated at its mass center, which is located by (2) or (4).

Example 1 Find the center of mass of a uniform solid hemisphere (Fig. 5.1).

The region $x^2 + y^2 + z^2 \leq a^2$, $z \geq 0$, defines a hemisphere whose radius is a. If ρ is the constant density, Eq. (3) becomes

$$\bar{z} = \frac{\iiint\limits_V z\, dV}{\iiint\limits_V dV}.$$

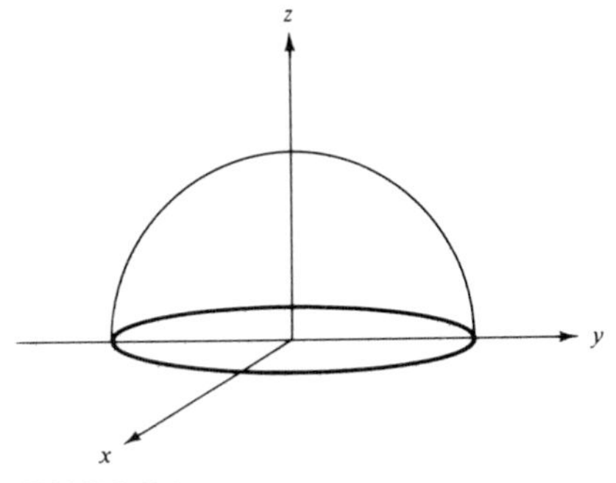

FIGURE 5.1

There is no need to calculate $\bar{x}$ and $\bar{y}$, since the symmetry of the problem ensures that $\bar{x} = \bar{y} = 0$. The geometry suggests that spherical coordinates (R, φ, θ) are convenient for calculations, and with $z = R\cos\varphi$ it follows that

$$\bar{z} = \frac{\int_0^{2\pi} d\theta \int_0^{\pi/2} d\varphi \int_0^a R\cos\varphi\, R^2 \sin\varphi\, dR}{\int_0^{2\pi} d\theta \int_0^{\pi/2} d\varphi \int_0^a R^2 \sin\varphi\, dR}.$$

Each of the integrations can be performed separately:

$$\bar{z} = \frac{[R^4/4]_0^a [\tfrac{1}{2}\sin^2\varphi]_0^{\pi/2} [\theta]_0^{2\pi}}{[R^3/3]_0^a [-\cos\varphi]_0^{\pi/2} [\theta]_0^{2\pi}},$$

and the final result is

$$\bar{z} = \frac{3a}{8}.$$

Example 2 Find the center of mass of the uniform solid formed by the planes $x = 0$, $y = 0$, $z = 0$, and $x/a + y/b + z/c = 1$, shown in Fig. 5.2.

Let ρ be the constant density. We begin by calculating

$$\bar{z} = \frac{\iiint\limits_V \rho z\, dV}{\iiint\limits_V \rho\, dV} = \frac{\iiint\limits_V z\, dV}{\iiint\limits_V dV}.$$

If the order of integration is taken as z first, y second, and x third, the equivalent iterated integrals are

$$\bar{z} = \frac{\int_0^a dx \int_0^{b(1-x/a)} dy \int_0^{c(1-x/a-y/b)} z\, dz}{\int_0^a dx \int_0^{b(1-x/a)} dy \int_0^{c(1-x/a-y/b)} dz}.$$

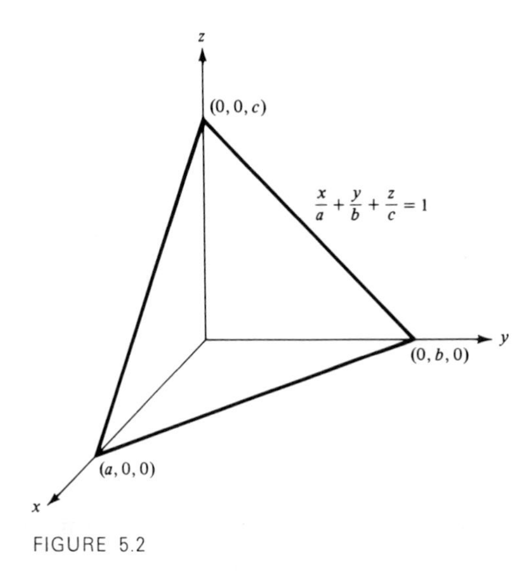

FIGURE 5.2

The result of the z integration is

$$\bar{z}=\frac{\int_0^a dx \int_0^{b(1-x/a)} \frac{1}{2}c^2(1-x/a-y/b)^2\, dy}{\int_0^a dx \int_0^{b(1-x/a)} c(1-x/a-y/b)\, dy},$$

and the integrations with respect to y give

$$\bar{z}=\frac{c}{3}\frac{\int_0^a(1-x/a)^3\, dx}{\int_0^a(1-x/a)^2\, dx}.$$

Finally, the remaining x integrations yield

$$\bar{z}=\frac{c}{4}.$$

Since x, y and z are associated symmetrically with the constants a, b and c, similar calculations of $\bar{x}$ and $\bar{y}$ must produce $\bar{x}=a/4$, $\bar{y}=b/4$. Therefore, the center of mass is located at the point $(a/4,\ b/4,\ c/4)$.

2 **Example 3** A thin uniform shell having the shape $z=1-x^2-y^2$, $z>0$, forms the roof of a building (Fig. 5.3). Locate the position of its center of mass.

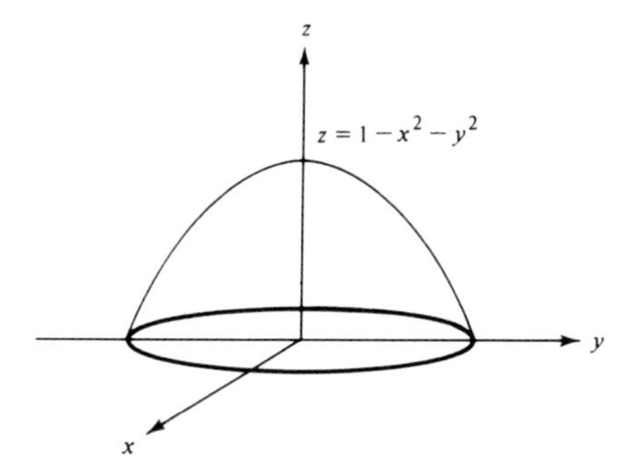

FIGURE 5.3

Let σ be the constant mass per unit area of the shell. The symmetry of the problem implies that the center of mass lies on Oz (that is, $\bar{x} = \bar{y} = 0$). Application of Eq. (4) shows that

$$\bar{z} = \frac{\iint_S \sigma z \, dS}{\iint_S \sigma \, dS} = \frac{\iint_S z \, dS}{\iint_S dS},$$

since σ is constant.

The surface integrals in this expression are calculated by projecting onto the xy plane. With $z = 1 - x^2 - y^2$, $z_x = -2x$, $z_y = -2y$, it follows that

$$\begin{aligned} dS &= (1 + z_x{}^2 + z_y{}^2)^{1/2} \, dx \, dy, \\ &= [1 + 4(x^2 + y^2)]^{1/2} \, dx \, dy. \end{aligned}$$

Substitution of this result in the surface integrals gives

$$\bar{z} = \frac{\iint_R (1 - x^2 - y^2)[1 + 4(x^2 + y^2)]^{1/2} \, dx \, dy}{\iint_R [1 + 4(x^2 + y^2)]^{1/2} \, dx \, dy},$$

where R is the interior of the circle $x^2 + y^2 = 1$. These look complicated, but the forms of the integrands and the shape of the region R suggest that a transformation to polar coordinates will be helpful. If this is done, we obtain

$$\bar{z} = \frac{\int_0^{2\pi} d\theta \int_0^1 (1 - r^2)(1 + 4r^2)^{1/2} r \, dr}{\int_0^{2\pi} d\theta \int_0^1 (1 + 4r^2)^{1/2} r \, dr}.$$

The 2π factors from the θ integrations cancel, and the further substitution $u = r^2$ gives

$$\bar{z} = \frac{\int_0^1 (1 - u)(1 + 4u)^{1/2} \, du}{\int_0^1 (1 + 4u)^{1/2} \, du}.$$

This, however, succumbs to routine integration, and the result is

$$\bar{z} = \frac{[\frac{5}{24}(1 + 4u)^{3/2} - \frac{1}{40}(1 + 4u)^{5/2}]_0^1}{[\frac{1}{6}(1 + 4u)^{3/2}]_0^1},$$

which upon simplification becomes

$$\bar{z} = \frac{307 - 15\sqrt{5}}{620} \approx 0.44.$$

---

3 The kinetic energy E is a quantity of great importance in dynamical systems. In the simplest case, a particle of mass m and velocity $\mathbf{v}$ has a kinetic energy given by

$$E = \tfrac{1}{2}m|\mathbf{v}|^2, \tag{5}$$

and this formula is used as the basis for more difficult calculations. For example, consider a rigid body that rotates with constant angular velocity Ω about a fixed axis (Fig. 5.4). An element of mass dm will contribute $\frac{1}{2}\,dm|\mathbf{v}|^2$ to the total kinetic

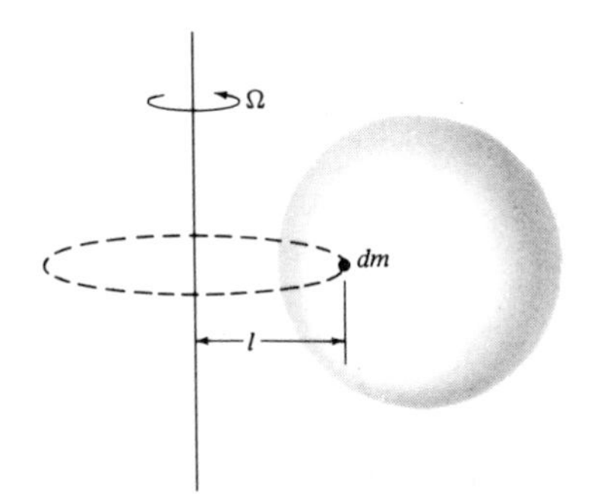

FIGURE 5.4

energy of the body. Furthermore, if l is the distance of dm from the axis of rotation, then $|\mathbf{v}| = \Omega l$, because each element of mass is in circular motion. The energy differential is then $dE = \frac{1}{2}\Omega^2 l^2\,dm$ and the total kinetic energy is obtained by summing over all volume elements, i.e., by an integration throughout the body:

$$E = \iiint dE = \iiint \tfrac{1}{2}\Omega^2 l^2\,dm.$$

Since Ω is constant, this can be expressed as

$$E = \tfrac{1}{2}I\Omega^2, \tag{6}$$

where

$$I = \iiint l^2\,dm \tag{7}$$

is called the *moment of inertia* of the body about the given axis. It can be seen from (7) that I has the dimensions ML^2. Accordingly the relationship

$$k^2 = \frac{\iiint l^2 \, dm}{\iiint dm} = \frac{I}{M}, \tag{8}$$

defines a length k, called the *radius of gyration* of the body about the axis concerned.

Special cases of (7) for which the rotation axis is identical with one of the three coordinate axes are

$$I_1 = \iiint (y^2 + z^2) \, dm, \qquad I_2 = \iiint (z^2 + x^2) \, dm, \qquad I_3 = \iiint (x^2 + y^2) \, dm. \tag{9}$$

Here I_1, I_2, I_3 are the moments of inertia about the x, y, z axes, respectively.

Example 4 A solid uniform rectangular box has sides of length a, b, and c. Find the moment of inertia about any one of its edges.

If ρ is the constant material density, the element of mass is

$$dm = \rho \, dx \, dy \, dz.$$

We may suppose that the box occupies the region $0 \le x \le a$, $0 \le y \le b$, $0 \le z \le c$, as illustrated in Fig. 5.5. Equation (9) is used to calculate the moment of inertia about Ox, in which case $l^2 = y^2 + z^2$ and

$$I_1 = \int_0^a dx \int_0^b dy \int_0^c \rho(y^2 + z^2) \, dz.$$

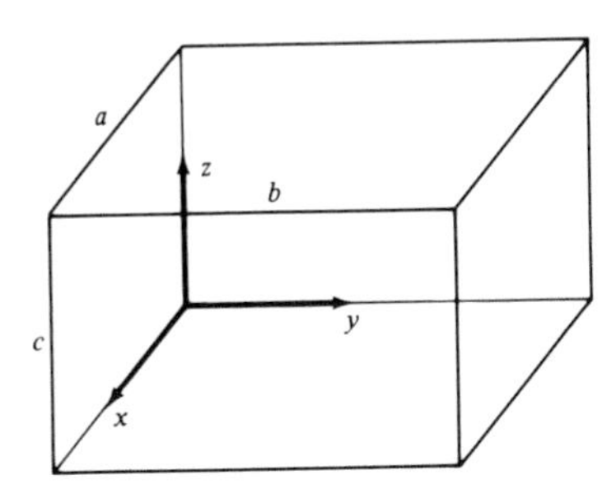

FIGURE 5.5

The limits of integration are constant, and a straightforward calculation shows that

$$I_1 = \rho a \left(\frac{b^3 c}{3} + \frac{bc^3}{3} \right) = \rho abc \, \frac{b^2 + c^2}{3}.$$

The total mass M of the box is the product of the density ρ and the volume abc:

$$M = \rho abc.$$

If k_1 denotes the radius of gyration about the x axis so that $I_1 = Mk_1{}^2$, then upon substitution for I_1 and M from the preceding equations we find that

$$k_1{}^2 = \tfrac{1}{3}(b^2 + c^2).$$

In the same way it can be shown that

$$k_2{}^2 = \tfrac{1}{3}(c^2 + a^2), \qquad k_3{}^2 = \tfrac{1}{3}(a^2 + b^2),$$

where k_2 and k_3 are the radii of gyration about the y and z axes, respectively.

Example 5 Find the moment of inertia of a uniform spherical shell about a diameter.

By symmetry arguments the moment of inertia about any diameter will be the same. Therefore we need only calculate the moment of inertia about the z axis, I_3. If σ is the surface density, then $dm = \sigma\, dS$ and (9) becomes

$$I_3 = \iint_S (x^2 + y^2)\sigma\, dS,$$

where S is the entire surface of the sphere (Fig. 5.6). The integration is readily accomplished in spherical coordinates. Since $dS = a^2 \sin\varphi\, d\varphi\, d\theta$ and $x^2 + y^2 = a^2 \sin^2\varphi$, the transformation yields

$$I_3 = \int_0^{2\pi} d\theta \int_0^{\pi} (a^2 \sin^2\varphi)\sigma a^2 \sin\varphi\, d\varphi.$$

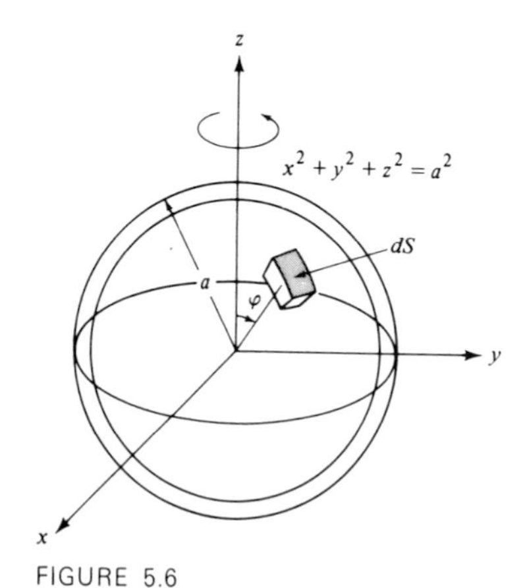

FIGURE 5.6

The θ integration can be performed immediately, and the result is

$$I_3 = 2\pi\sigma a^4 \int_0^{\pi} \sin^3\varphi\, d\varphi.$$

The remaining integral can be evaluated by writing

$$\sin^3\varphi\, d\varphi = \sin^2\varphi \sin\varphi\, d\varphi = -(1 - \cos^2\varphi)d(\cos\varphi),$$

whereupon

$$\int_0^\pi \sin^3 \varphi \, d\varphi = [-\cos \varphi + \tfrac{1}{3} \cos^3 \varphi]_0^\pi = \tfrac{4}{3}.$$

Therefore the moment of inertia is

$$I_3 = \tfrac{8}{3}\pi\sigma a^4.$$

The total mass of the spherical shell is $M = 4\pi a^2 \sigma$ (surface area times the surface density) and since, by definition, $I_3 = Mk_3{}^2$, it follows that

$$k_3{}^2 = \tfrac{2}{3}a^2.$$

An alternative and crafty technique is to exploit the full symmetry of the problem, which in effect means that $I_1 = I_2 = I_3 = I$. These quantities are given by

$$I_1 = \iint_S (y^2 + z^2)\sigma \, dS, \qquad I_2 = \iint_S (z^2 + x^2)\sigma \, dS, \qquad I_3 = \iint_S (x^2 + y^2)\sigma \, dS,$$

which when added yield

$$3I = 2 \iint_S (x^2 + y^2 + z^2) \, dS.$$

But $x^2 + y^2 + z^2 = a^2$ on the surface of the sphere, so that

$$3I = 2a^2\sigma \iint_S dS.$$

Moreover, since $\iint dS = S = 4\pi a^2$, the final result is again

$$I = \tfrac{8}{3}\pi\sigma a^4.$$

Example 6 Discuss the motion of a rigid body that is free to swing about a horizontal axis under the action of gravity.

Let M be the mass of the body, l the distance from the fixed axis O to the center of mass G, I the moment of inertia about the horizontal axis through O, g the gravitational acceleration, and θ the angle between OG and the vertical, as sketched in Fig. 5.7.

In the absence of any dissipative action, i.e., friction of any kind, the total energy of the body—kinetic plus potential—must remain constant during the motion.

The kinetic energy is the energy of rotation about the horizontal axis and is given by (6), where Ω is replaced by the instantaneous angular velocity $d\theta/dt$. The potential energy is the weight of the body multiplied by the vertical displacement of the mass center, that is, $-Mgl \cos \theta$. Therefore the energy equation is

$$\tfrac{1}{2}I \left(\frac{d\theta}{dt}\right)^2 - Mgl \cos \theta = \text{constant}.$$

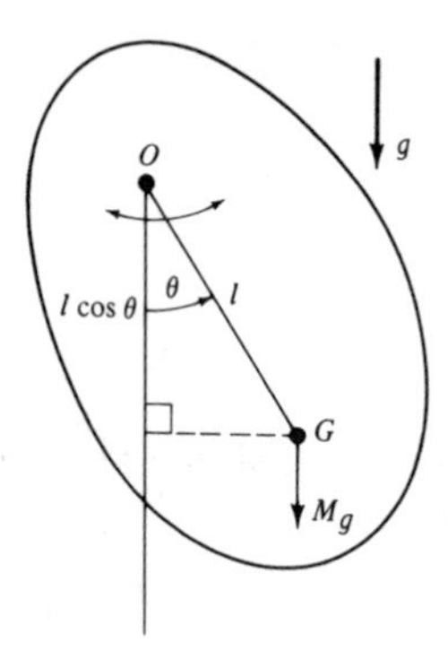

FIGURE 5.7

Moreover if the maximum deflection angle is taken to be $\theta = \alpha$ (where $d\theta/dt = 0$), this can be written

$$\left(\frac{d\theta}{dt}\right)^2 = \frac{2Mgl}{I}(\cos\theta - \cos\alpha), \tag{10}$$

a differential equation for the determination of θ as a function of t. The integration is nontrivial because it gives rise to an elliptic integral. However, for small oscillations about equilibrium, when OG remains nearly vertical, the solution is easily found. This particular analysis begins with the differentiation of (10), to obtain

$$\frac{d^2\theta}{dt^2} + \frac{Mgl}{I}\sin\theta = 0,$$

a generally valid result. When θ is small, $\sin\theta \approx \theta$ and the preceding equation is well approximated by

$$\frac{d^2\theta}{dt^2} + \frac{Mgl}{I}\theta = 0. \tag{11}$$

This differential equation has already been discussed many times, and its general solution is

$$\theta = A \sin\left(\frac{Mgl}{I}\right)^{1/2} t + B\cos\left(\frac{Mgl}{I}\right)^{1/2} t.$$

For small amplitudes, the motion is simple harmonic with period $2\pi(I/Mgl)^{1/2}$.

Example 7 A homogeneous solid cylinder of total mass M occupies the region $x^2 + y^2 \le a^2$, $0 \le z \le h$. Find the gravitational force on a particle of mass M' located at the point $(0, 0, b)$, where $b > h$.

Since the density ρ of the cylinder is constant, its total mass M is given by

$$M = \rho\pi a^2 h. \tag{12}$$

Newton's law of gravitation (see page 473) states that an element of mass dm located at a point $P = (x, y, z)$ (Fig. 5.8) attracts a mass M' at the point $B = (0, 0, b)$ with a force of magnitude $GM'dm/(PB)^2$, where G is the gravitational constant. Therefore, the vector element of force is given by

$$d\mathbf{F} = \frac{GM'\overrightarrow{BP}\, dm}{|\overrightarrow{PB}|^3}.$$

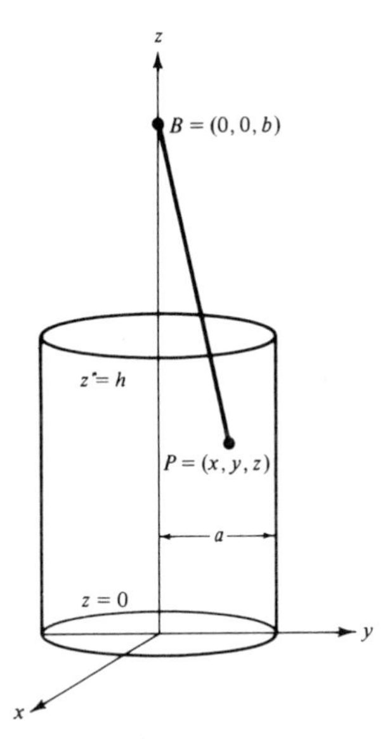

FIGURE 5.8

Since $\overrightarrow{OP} = x\mathbf{i} + y\mathbf{j} + z\mathbf{k}$ and $\overrightarrow{OB} = b\mathbf{k}$, it follows that $\overrightarrow{BP} = \overrightarrow{OP} - \overrightarrow{OB} = x\mathbf{i} + y\mathbf{j} + (z - b)\mathbf{k}$ and the preceding equation can be rewritten as

$$d\mathbf{F} = \frac{GM'[x\mathbf{i} + y\mathbf{j} + (z - b)\mathbf{k}]\, dm}{[x^2 + y^2 + (z - b)^2]^{3/2}}.$$

The differential element of mass is $dm = \rho\, dx\, dy\, dz$, and the total force is found by integrating over the volume of the cylinder:

$$\mathbf{F} = \iiint d\mathbf{F} = G\rho M' \iiint_V \frac{x\mathbf{i} + y\mathbf{j} + (z - b)\mathbf{k}}{[x^2 + y^2 + (z - b)^2]^{3/2}}\, dx\, dy\, dz,$$

where V is the region $x^2 + y^2 \le a^2$, $0 \le z \le h$.

On transforming to cylindrical coordinates, this integral becomes

$$\mathbf{F} = G\rho M \int_0^h dz \int_0^{2\pi} d\theta \int_0^a \frac{r\cos\theta\, \mathbf{i} + r\sin\theta\, \mathbf{j} + (z - b)\mathbf{k}}{[r^2 + (z - b)^2]^{3/2}}\, r\, dr.$$

The θ integration annihilates the x and y components of $\mathbf{F}$ (this could have been anticipated by symmetry arguments), and the integral reduces to

$$\mathbf{F} = G\rho M' 2\pi\mathbf{k} \int_0^h dz \int_0^a \frac{(z-b)r\,dr}{[r^2+(z-b)^2]^{3/2}}.$$

The r integration gives

$$\mathbf{F} = G\rho M' 2\pi\mathbf{k} \int_0^h \left\{\frac{b-z}{[a^2+(z-b)^2]^{1/2}} - 1\right\} dz,$$

and the final result is

$$\mathbf{F} = G\rho M' 2\pi\mathbf{k}[-(a^2+(z-b)^2)^{1/2} - z]_0^h.$$

If ρ is eliminated from this formula, using (12), we obtain

$$\mathbf{F} = -\frac{2GMM'}{a^2}\mathbf{k}\left(1 + \frac{1}{h}\{[a^2+(b-h)^2]^{1/2} - (a^2+b^2)^{1/2}\}\right).$$

In the special case when b is very much larger than either a or h, that is, $b \gg a$ and $b \gg h$, use of the binomial expansion shows that to a first approximation,

$$\mathbf{F} \approx -\frac{GMM'\mathbf{k}}{b^2}.$$

We conclude that at large distances the attractive force is the same as if *both* objects were point masses separated by a distance b.

Exercises 7.5

1. Find the position of the center of mass of a uniform triangular lamina.

2. Locate the center of mass of a solid homogeneous pyramid of height h which has a square cross section.

3. Find the center of mass of a solid homogeneous tetrahedron of height h which has an equilateral cross section.

4. Find the position of the center of mass of a uniform hemispherical shell of interior radius a and exterior radius b. Find the center of mass of a thin hemispherical shell.

5. Find the centroid of the portion of the uniform spherical shell $a^2 \le x^2+y^2+z^2 \le b^2$ in the positive octant. Examine the limit $b \to a$.

6. The plane $z=b$ cuts the sphere $x^2+y^2+z^2 \le a^2$ into two volumes $(0<b<a)$. Find the mass of the portion with $b \le z \le a$ for the following mass densities $\rho(x, y, z)$.

(i) $\rho(x, y, z) = \rho_0$ **(ii)** $\rho(x, y, z) = (x^2+y^2+z^2)^{1/2}$ **(iii)** $\rho(x, y, z) = (x^2+y^2+z^2)^{-1/2}$

7. A circular plate has a density proportional to distance from its center. Find the total mass of the plate.

8. At a point on a square plate, the density is proportional to the square of the distance of the point from the center. Find the total mass using rectangular coordinates and polar coordinates. Which method is easier?

9. Find the position of the center of mass of a uniform lamina whose boundary is the cardioid $r = a(1 + \cos\theta)$.

10. Find the moment of inertia of a homogeneous solid sphere about a diameter.

11. Find the moment of inertia of a homogeneous solid spherical shell with interior radius a and exterior radius b about a diameter.

12. A uniform solid circular cone has height h and base radius a. Find the position of the center of mass and the moment of inertia about the axis of symmetry.

13. Find the moments of inertia about the coordinate axes of the homogeneous solid formed by the planes $x = 0$, $y = 0$, $z = 0$, $x/a + y/b + z/c = 1$.

14. Calculate the radius of gyration about a symmetry axis of the solid contained in the region $a^2 \leq x^2 + y^2 + z^2 \leq b^2$ if the density is proportional to distance from the origin.

15. If the moment of inertia of a body of mass M about an axis through the center of mass is I, prove that the moment of inertia about an axis parallel to the first one is $I + Md^2$, where d is the distance between the axes.

16. A uniform rod of length l, pinned at one end, performs small-amplitude oscillations under the action of gravity. Find the period of the motion. (*Hint:* Use the method of Example 6.)

17. A homogeneous solid sphere *rolls* straight down a plane inclined at an angle α to the horizontal. Show that the acceleration of the sphere is $(5/7)g \sin\alpha$. (*Hint:* Use conservation of energy including rotational kinetic energy.)

18. In a simplified model of a hurricane the velocity is taken to be purely in the circumferential direction and of magnitude $v(r, z) = \Omega r e^{-z/h - r/a}$, where r and z are cylindrical coordinates measured from the eye of the hurricane at sea level. If the density ρ of the atmosphere is $\rho(z) = \rho_0 e^{-z/h}$, find the total kinetic energy of the motion and locate where the velocity has a maximum value.

19. Fluid of constant density ρ is contained between two cylinders, radii a_1 and a_2, which rotate at angular velocities Ω_1 and Ω_2. Every fluid particle moves in the circumferential direction with velocity $v(r) = A/r + Br$. The constants A and B are chosen to make $v(a_1) = \Omega_1 a_1$ and $v(a_2) = \Omega_2 a_2$. If the fluid is confined between the planes $z = 0$ and $z = h$ (see Fig. 5.9), find the total kinetic energy of the fluid. (*Hint:* $E = \iiint_V \frac{1}{2}\rho|\mathbf{v}|^2\, dV$.)

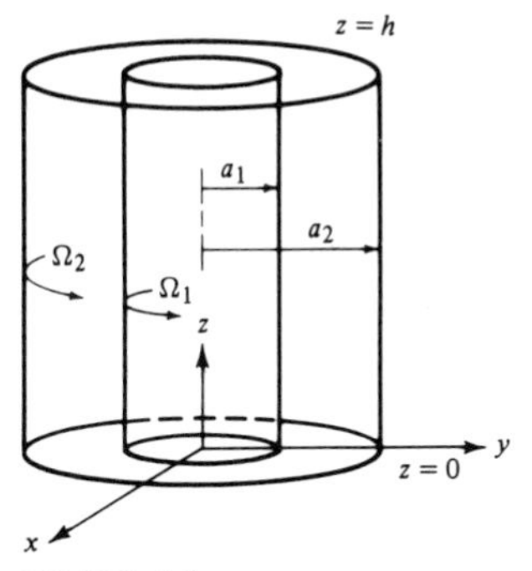

FIGURE 5.9

20. A thin uniform disk of radius a has a uniform charge density σ. Find the force on a unit charge a distance b from the disk on the axis of symmetry.

21. A uniform surface charge lies in the region $z = 0$ for $x^2 + y^2 > a^2$ and $z = +(a^2 - x^2 - y^2)^{1/2}$ for $x^2 + y^2 \leq a^2$. Find the force on a unit charge placed at the point $(0, 0, b)$.

22. Show that the moment of inertia I about a line through the origin in the direction of the unit vector $\hat{\mathbf{u}} = l\mathbf{i} + m\mathbf{j} + n\mathbf{k}$ is

$$I = Al^2 + Bm^2 + Cn^2 - 2Dmn - 2Enl - 2Flm,$$

where

$$A = \iiint (y^2 + z^2)\, dm, \qquad B = \iiint (z^2 + x^2)\, dm, \qquad C = \iiint (x^2 + y^2)\, dm,$$

$$D = \iiint yz\, dm, \qquad E = \iiint zx\, dm, \qquad F = \iiint xy\, dm.$$

(A, B and C are the moments of inertia about the x, y, and z axes. The quantities D, E, and F are called *products of inertia*.)

7.6 *Approximate methods*

Most multiple integrals cannot be evaluated exactly by analytical means, and it is imperative that approximate procedures be developed to extract the information contained in these complicated formulas and make it comprehensible. A similar situation arose in the case of one-dimensional integrals (see Sec. 3.11), and the techniques employed there can be extended and applied to multiple integrals. In the interest of simplicity, attention will be restricted to double integrals, i.e., integrals of the form $\iint_R f(x, y)\, dx\, dy$.

Two essentially different approaches are possible. We can proceed immediately with a direct numerical attack and approximate the integral by a finite sum of terms in accordance with any one of many recipes. Alternatively, analytical simplifications can be utilized—end runs, so to speak, about the complexities—and this approach will be discussed first.

It should be realized that the creation of approximate methods is to some extent an "art form." There are no hard and fast rules; the aim is always to attain maximum accuracy with a minimum of effort.

An examination of the integrand usually motivates the choice of an appropriate method. For example, we might decide that a Taylor series will provide a good approximation to an integrand over the entire region of integration R or that an integrand which contains a small parameter is susceptible to a perturbation expansion. Of course, the success or failure of any analytical approach is judged by the relative ease with which the approximating functions can be integrated and by the accuracy attained. These ideas are best illustrated by examples.

Example 1 Calculate the approximate value of $\int_0^1 \int_0^1 e^{-x^2 y^3}\, dx\, dy$.

A most natural approach to this problem is to develop the integrand as a Taylor series expansion about the origin. From the known convergent series

$$e^{-u} = 1 - \frac{u}{1!} + \frac{u^2}{2!} - \frac{u^3}{3!} + \cdots$$

it follows that

$$e^{-x^2y^3} = 1 - \frac{x^2y^3}{1!} + \frac{x^4y^6}{2!} - \frac{x^6y^9}{3!} + \cdots,$$

$$= \sum_{r=0}^{\infty} \frac{(-1)^r x^{2r} y^{3r}}{r!}.$$

The substitution of this series for the integrand, followed by term-by-term integration, gives

$$\int_0^1 \int_0^1 e^{-x^2y^3}\,dx\,dy = \sum_{r=0}^{\infty} \int_0^1 \int_0^1 \frac{(-1)^r x^{2r} y^{3r}}{r!}\,dx\,dy,$$

$$= \sum_{r=0}^{\infty} \frac{(-1)^r}{r!(2r+1)(3r+1)}.$$

This series is rapidly convergent, and the retention of only a few terms provides an accurate estimate of the exact value. For instance, if only the first six terms are retained, we get

$$\int_0^1 \int_0^1 e^{-x^2y^3}\,dx\,dy \approx 1 - \frac{1}{1!3\cdot 4} + \frac{1}{2!5\cdot 7} - \frac{1}{3!7\cdot 10} + \frac{1}{4!9\cdot 13} - \frac{1}{5!11\cdot 16},$$

$$\approx 1 - \frac{1}{12} + \frac{1}{70} - \frac{1}{420} + \frac{1}{2{,}808} - \frac{1}{21{,}120},$$

$$\approx 0.9289,$$

a value that is correct to four decimal places.

Example 2 Find the approximate value of $\int_0^1 \int_0^1 \frac{dx\,dy}{5 + x^4 + y^4}$.

Crude bounds for the value of the integral can be obtained from the following inequality, which is valid for every point in the region of integration:

$$\frac{1}{7} \le \frac{1}{5 + x^4 + y^4} \le \frac{1}{5}.$$

It follows that

$$\int_0^1 \int_0^1 \frac{dx\,dy}{7} \le \int_0^1 \int_0^1 \frac{dx\,dy}{5 + x^4 + y^4} \le \int_0^1 \int_0^1 \frac{dx\,dy}{5}$$

or

$$\frac{1}{7} \le \int_0^1 \int_0^1 \frac{dx\,dy}{5 + x^4 + y^4} \le \frac{1}{5}.$$

For some purposes, these bounds may be sufficient information about the integral. To obtain a more accurate estimate, we again employ a Taylor series expansion:

$$\frac{1}{5 + x^4 + y^4} = \frac{1}{5} - \frac{x^4 + y^4}{5^2} + \frac{(x^4 + y^4)^2}{5^3} - \frac{(x^4 + y^4)^3}{5^4} + \frac{(x^4 + y^4)^4}{5^5} - \cdots.$$

Since $x^4 + y^4 \le 2$ in the region of integration, this series is convergent [as may be proved by comparison with the geometric series $\sum(2/5)^n$]. Term-by-term integration yields

$$\int_0^1 \int_0^1 \frac{dx\,dy}{5 + x^4 + y^4} = \frac{1}{5} - \frac{1}{5^2}\left(\frac{1}{5} + \frac{1}{5}\right) + \frac{1}{5^3}\left(\frac{1}{9} + \frac{2}{5 \cdot 5} + \frac{1}{9}\right)$$
$$- \frac{1}{5^4}\left(\frac{1}{13} + \frac{3}{9 \cdot 5} + \frac{3}{5 \cdot 9} + \frac{1}{13}\right)$$
$$+ \frac{1}{5^5}\left(\frac{1}{17} + \frac{4}{13 \cdot 5} + \frac{6}{9 \cdot 9} + \frac{4}{5 \cdot 13} + \frac{1}{17}\right) - \cdots.$$

These few terms of the series give the approximate value of the integral to be 0.1861, a result correct to four decimal places.

2 **Example 3** Find the area confined by the lines $\theta = 0$, $\theta = \pi/4$, and the curve $r = 2\cos(\theta + \varepsilon r)$, where ε is a small number.

Since the element of area in polar coordinates is $r\,dr\,d\theta$, the desired area is

$$A = \iint_R r\,dr\,d\theta,$$

where R is the region shown in Fig. 6.1.

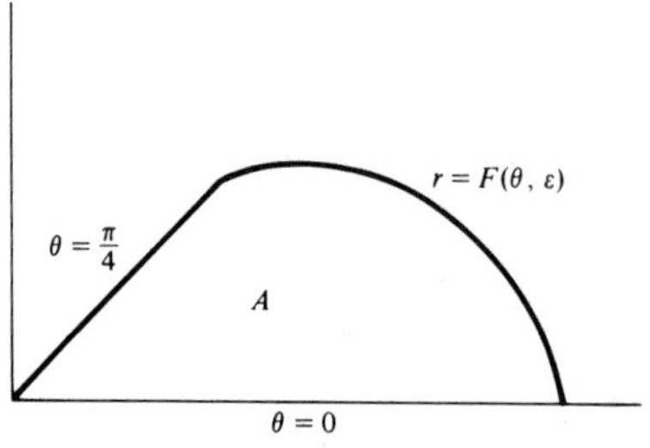

FIGURE 6.1

The difficulty in this problem is that part of the boundary is described implicitly. However, by exploiting the fact that ε is a small number, it is possible to develop an explicit equation for r as a perturbation series of the form

$$r = F(\theta, \varepsilon) = r_0(\theta) + \varepsilon r_1(\theta) + \varepsilon^2 r_2(\theta) + \cdots.$$

This is done by substituting the expansion into the defining equation, $r = 2\cos(\theta + \varepsilon r)$, which yields

$$\begin{aligned} r_0 + \varepsilon r_1 + \varepsilon^2 r_2 + \cdots &= 2\cos(\theta + \varepsilon r_0 + \varepsilon^2 r_1 + \cdots), \\ &= 2\cos\theta\cos(\varepsilon r_0 + \varepsilon^2 r_1 + \cdots) \\ &\quad -2\sin\theta\sin(\varepsilon r_0 + \varepsilon^2 r_1 + \cdots). \end{aligned}$$

The right-hand side is then expressed as a series in powers of ε by using the expansions

$$\cos x = 1 - \frac{x^2}{2!} + \cdots \qquad \text{and} \qquad \sin x = x - \frac{x^3}{3!} + \cdots.$$

It follows that

$$r_0 + \varepsilon r_1 + \varepsilon^2 r_2 + \cdots = 2\cos\theta\,(1 - \tfrac{1}{2}\varepsilon^2 r_0^2 + \cdots) - 2\sin\theta\,(\varepsilon r_0 + \varepsilon^2 r_1 + \cdots),$$

where only the first few terms of each expansion have been retained. Successive powers of ε are now equated, to obtain

$$r_0 = 2\cos\theta, \qquad r_1 = -2r_0\sin\theta, \qquad r_2 = -r_0^2\cos\theta - 2r_1\sin\theta,$$

which imply

$$r_0 = 2\cos\theta, \qquad r_1 = -4\sin\theta\cos\theta, \qquad r_2 = 4\cos\theta\,(3\sin^2\theta - 1).$$

Accordingly, the explicit functional form for the boundary curve is

$$r = F(\theta, \varepsilon) = 2\cos\theta + \varepsilon(-4\sin\theta\cos\theta) + \varepsilon^2[4\cos\theta\,(3\sin^2\theta - 1)] + \cdots.$$

Truncation at any power of ε provides an approximate but explicit formula for the almost circular perimeter. We can make use of this result in the calculation of the area by substituting this expansion for the function $F(\theta, \varepsilon)$ in the integral

$$A = \int_0^{\pi/4} d\theta \int_0^{F(\theta,\varepsilon)} r\,dr = \frac{1}{2}\int_0^{\pi/4} F^2\,d\theta.$$

Therefore

$$A = \frac{1}{2}\int_0^{\pi/4} [2\cos\theta - 4\varepsilon\sin\theta\cos\theta + 4\varepsilon^2\cos\theta\,(3\sin^2\theta - 1) + \cdots]^2\,d\theta,$$

or

$$A = \int_0^{\pi/4} [2\cos^2\theta - 8\varepsilon\sin\theta\cos^2\theta + 8\varepsilon^2\cos^2\theta\,(4\sin^2\theta - 1) + \cdots]\,d\theta.$$

The result of term-by-term integration is

$$A = \tfrac{1}{4}(\pi + 2) - \tfrac{2}{3}\varepsilon(4 - \sqrt{2}) - 2\varepsilon^2 + \cdots.$$

Example 4 The State House dome is to be resurfaced with gold leaf. The legislature believes that the dome (Fig. 6.2) is a perfect hemisphere of the form $x^2 + y^2 + z^2 = 1$, $z > 0$, and authorizes funds for an amount of gold leaf just sufficient to resurface the area 2π. One of the prospective contractors makes some careful measurements and discovers that the dome is slightly imperfect and actually has the shape $x^2/(1 + \alpha) + y^2/(1 + \beta) + z^2 = 1$, $z > 0$, where $|\alpha|$ and $|\beta|$ are very small numbers. After measuring the values of α and β he suspects that there might be some gold leaf left over after completion of the job which can be sold for additional profit. Is he right?

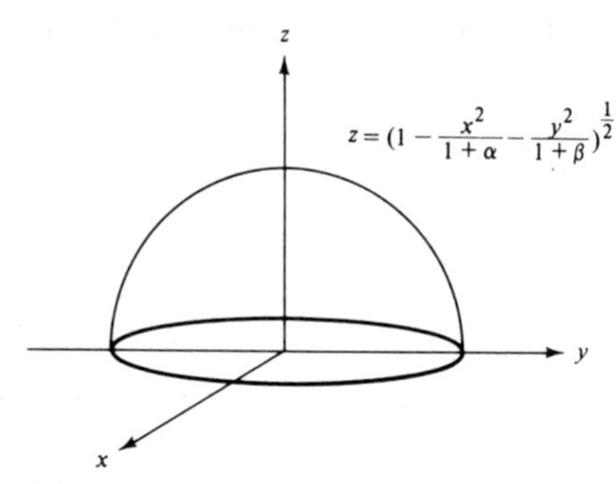

FIGURE 6.2

The question at issue is whether the surface area of

$$z = [1 - x^2/(1 + \alpha) - y^2/(1 + \beta)]^{1/2}$$

is smaller or larger than 2π. The area must be calculated, and this will be done via a projection onto the xy plane, in which case the element of surface area is

$$dS = (1 + z_x{}^2 + z_y{}^2)^{1/2}\, dx\, dy.$$

However $z^2 = 1 - x^2/(1 + \alpha) - y^2/(1 + \beta)$, and partial differentiation gives

$$zz_x = \frac{-x}{1 + \alpha}, \qquad zz_y = \frac{-y}{1 + \beta}.$$

Therefore

$$(1 + z_x{}^2 + z_y{}^2)^{1/2} = \left\{1 + \frac{1}{z^2}\left[\frac{x^2}{(1 + \alpha)^2} + \frac{y^2}{(1 + \beta)^2}\right]\right\}^{1/2},$$

or, by making use of the equation for the surface,

$$(1 + z_x{}^2 + z_y{}^2)^{1/2} = \left[\frac{1 - \alpha x^2/(1 + \alpha)^2 - \beta y^2/(1 + \beta)^2}{1 - x^2/(1 + \alpha) - y^2/(1 + \beta)}\right]^{1/2}.$$

It follows that the desired surface area S, which is a function of α and β, is

$$S(\alpha, \beta) = \iint\limits_{R} \left[\frac{1 - \alpha x^2/(1+\alpha)^2 - \beta y^2/(1+\beta)^2}{1 - x^2/(1+\alpha) - y^2/(1+\beta)}\right]^{1/2} dx\, dy,$$

where R is the region $x^2/(1+\alpha) + y^2/(1+\beta) \leq 1$ (see Fig. 6.3). This is a *very* difficult integration. Fortunately, we need only determine $S(\alpha, \beta)$ for α and β small, and a perturbation expansion is ideally suited for this purpose. This is found by determining the two-variable Taylor series of the function

$$S(\alpha, \beta) = S(0, 0) + \alpha S_\alpha(0, 0) + \beta S_\beta(0, 0) + \text{higher powers of } \alpha \text{ and } \beta,$$

where $S_\alpha(0, 0)$ and $S_\beta(0, 0)$ are the first partial derivatives of S with respect to α and β evaluated at the point $(\alpha, \beta) = (0, 0)$. However, in order to find the

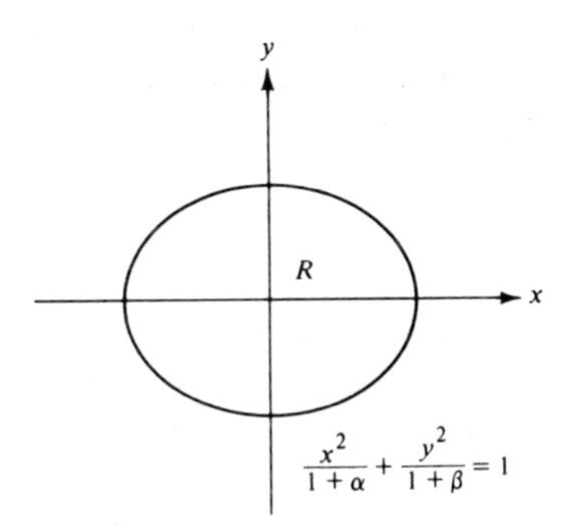

FIGURE 6.3

various partial derivatives of $S(\alpha, \beta)$ it is helpful—almost obligatory—to express the integral in a form that minimizes the algebra involved. For example, the calculation is greatly facilitated by eliminating the parameters from the limits of integration by a suitable change of variables. This is accomplished easily with variables [illegible] by

$$x = (1+\alpha)^{1/2} X, \qquad y = (1+\beta)^{1/2} Y.$$

Since the Jacobian of the transformation is

$$\frac{\partial(x, y)}{\partial(X, Y)} = (1+\alpha)^{1/2}(1+\beta)^{1/2},$$

$S(\alpha, \beta)$ can be rewritten as

$$S(\alpha, \beta) = \iint\limits_{A} \left[\frac{1 + \alpha(1 - X^2) + \beta(1 - Y^2) + \alpha\beta(1 - X^2 - Y^2)}{1 - X^2 - Y^2}\right]^{1/2} dX\, dY,$$

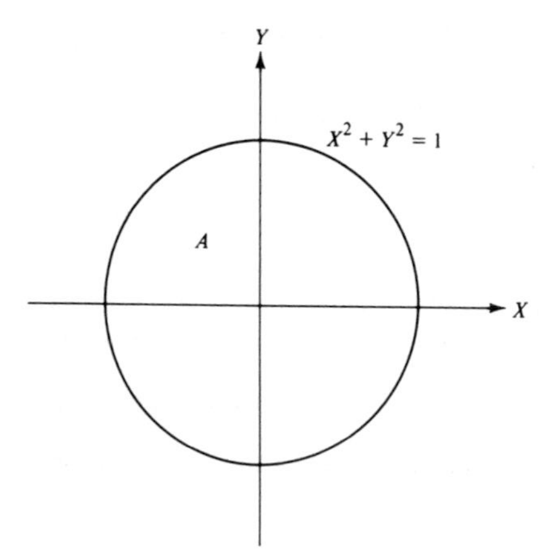

FIGURE 6.4

where A is now the interior of the circle $X^2 + Y^2 = 1$ and is no longer dependent on the parameters (Fig. 6.4). The circular symmetry suggests a final reduction to polar coordinates, $X = r\cos\theta$, $Y = r\sin\theta$, and this implies

$$S(\alpha, \beta) = \int_0^{2\pi} d\theta \int_0^1 \left[\frac{1 + \alpha(1 - r^2\cos^2\theta) + \beta(1 - r^2\sin^2\theta) + \alpha\beta(1 - r^2)}{1 - r^2}\right]^{1/2} r\,dr.$$

The function and its first partial derivatives can now be determined at the origin. The value of $S(0, 0)$ is simple, and $S_\alpha(0, 0)$ and $S_\beta(0, 0)$ can be found by differentiating under the integral signs (see Sec. 6.6). The relevant double integrals are as follows:

$$S(0, 0) = \int_0^{2\pi} d\theta \int_0^1 \frac{r\,dr}{(1 - r^2)^{1/2}} = 2\pi \int_0^1 \frac{r\,dr}{(1 - r^2)^{1/2}} = 2\pi,$$

$$S_\alpha(0, 0) = \int_0^{2\pi} d\theta \int_0^1 \frac{1 - r^2\cos^2\theta}{2(1 - r^2)^{1/2}}\, r\,dr = \frac{\pi}{2}\int_0^1 \frac{2 - r^2}{(1 - r^2)^{1/2}}\, r\,dr = \frac{2\pi}{3},$$

$$S_\beta(0, 0) = \int_0^{2\pi} d\theta \int_0^1 \frac{1 - r^2\sin^2\theta}{2(1 - r^2)^{1/2}}\, r\,dr = \frac{\pi}{2}\int_0^1 \frac{2 - r^2}{(1 - r^2)^{1/2}}\, r\,dr = \frac{2\pi}{3}.$$

The first few terms of the Taylor series are then

$$S(\alpha, \beta) = 2\pi\left(1 + \frac{\alpha + \beta}{3} + \text{higher powers of } \alpha \text{ and } \beta\right),$$

and these are sufficient to answer the problem posed. Clearly, if $\alpha + \beta < 0$, the surface of the dome is slightly less than the value 2π for the perfect hemisphere, and in this situation the contractor will have a surplus of gold leaf after the job has been completed. On the other hand, if $\alpha + \beta > 0$, the gold leaf supplied by the legislature is not sufficient to complete the job.

In this example, it is possible to perform the r integration for arbitrary α and β, but the remaining θ integration is a complicated elliptic integral which has not been characterized as a "simple known" function. However, in the special case $\alpha = \beta$, the exact value of S can be found because the dome is then a surface of revolution about the z axis.

---

3 We turn now to a discussion of numerical techniques. The basic idea is to return to the definition of the double integral (see Sec. 7.1) and to employ the finite sum

$$\sum_{i=1}^{n} \sum_{j=1}^{p} f(\xi_i, \eta_j)\, \Delta x_i\, \Delta y_j, \tag{1}$$

as an approximation (Fig. 6.5), which in the limiting sense becomes exact. It is possible with the aid of a computer to use a very large number of points and subdivisions and thereby attain good accuracy. In practice, somewhat more orderly and efficient procedures are used, which are for the most part generalizations of the trapezoidal and parabolic rules for one-dimensional integrals.

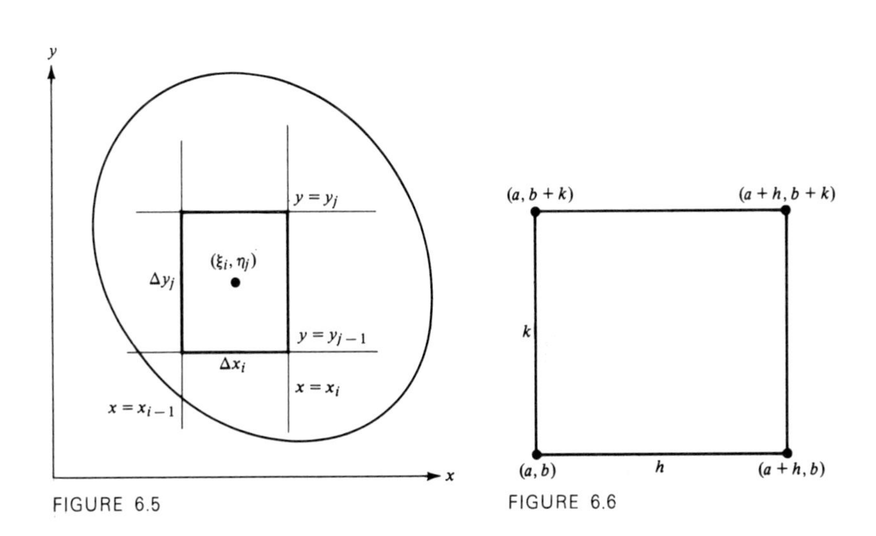

FIGURE 6.5

FIGURE 6.6

Consider the application of the trapezoidal rule to the integral

$$I = \iint_R f(x, y)\, dx\, dy,$$

where R is the rectangular region $a \le x \le a + h,\ b \le y \le b + k$ (Fig. 6.6). (The rectangular region will be interpreted shortly as one element of a grid that covers an arbitrary region.) The integral I can be expressed as the iterated integral

$$I = \int_b^{b+k} dy \int_a^{a+h} f(x, y)\, dx, \tag{2}$$

and the calculation requires two separate integrations. First we evaluate

$$\varphi(y) = \int_a^{a+h} f(x, y)\, dx$$

and then

$$I = \int_b^{b+k} \varphi(y)\, dy.$$

If each of these single integrals is approximated by the ordinary trapezoidal rule, page 300, then

$$\varphi(y) \approx \frac{h}{2}\,[f(a, y) + f(a + h, y)]$$

and

$$I \approx \frac{k}{2}\,[\varphi(b) + \varphi(b + k)].$$

The result of eliminating φ in these equations is

$$I \approx \frac{hk}{4}\,[f(a, b) + f(a + h, b) + f(a, b + k) + f(a + h, b + k)], \tag{3}$$

and this provides the desired generalization of the trapezoidal rule to double integrals.

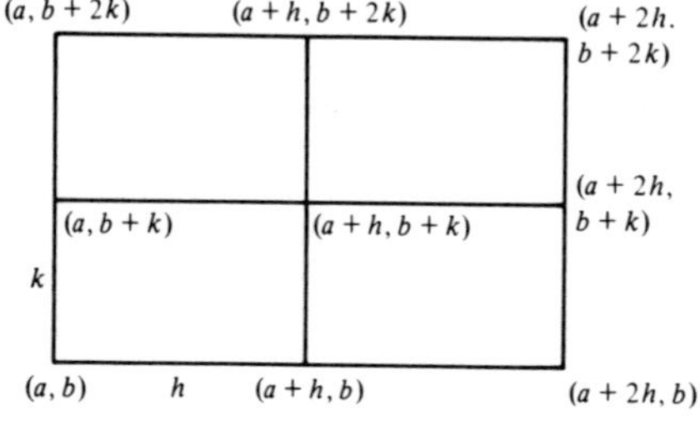

FIGURE 6.7

In a similar way, it is possible to generalize Simpson's rule. For this purpose we consider the double integral over a grid rectangle of sides $2h$ and $2k$ (see Fig. 6.7):

$$J = \int_b^{b+2k} dy \int_a^{a+2h} f(x, y)\, dx. \tag{4}$$

This expression can also be rewritten in terms of the two separate integrals:

$$\psi(y) = \int_a^{a+2h} f(x, y)\, dx \qquad \text{and} \qquad J = \int_b^{b+2k} \psi(y)\, dy.$$

Application of the one-dimensional form of Simpson's rule to each of these integrals gives

$$\psi(y) \approx \frac{h}{3}\,[f(a, y) + 4f(a + h, y) + f(a + 2h, y)],$$

and
$$J \approx \frac{k}{3}\,[\psi(b) + 4\psi(b+k) + \psi(b+2k)].$$

The elimination of ψ from these equations yields

$$J \approx \frac{hk}{9}\{f(a,b) + f(a+2h,b) + f(a,b+2k) + f(a+2h,b+2k)$$
$$+ 4[f(a+h,b) + f(a+h,b+2k) + f(a,b+k) + f(a+2h,b+k)]$$
$$+ 16f(a+h,b+k)\}. \quad (5)$$

Perhaps the easiest way of visualizing the generalization of the trapezoidal rule and Simpson's rule is by means of the weights associated with each vertex, as indicated in Fig. 6.8.

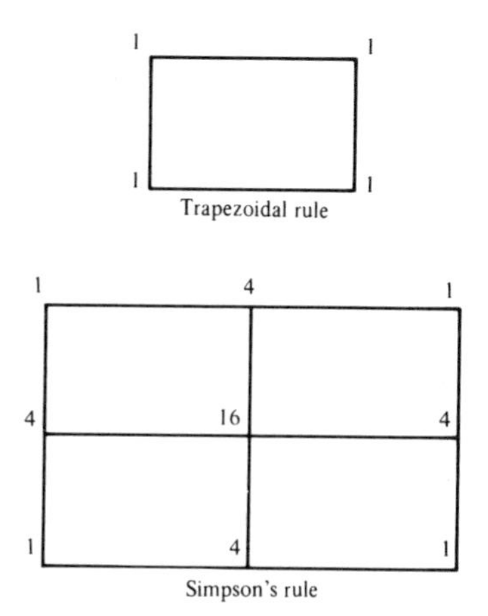

FIGURE 6.8

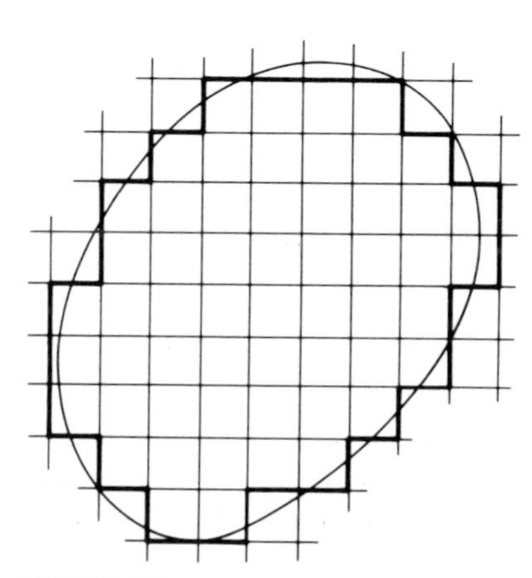

FIGURE 6.9

In practice, the region R is covered by a mesh or grid, with neighboring points set a distance h apart in the x direction and k apart in the y direction. Formula (3) or (5) can then be applied to each incremental rectangle. If the boundary of the region R is curved, it can be approximated by the sides of the rectangles which form the grid (Fig. 6.9), though this geometrical approximation is a source of additional numerical errors. Of course, as h and k become smaller, the accuracy of the approximations improves.

Example 5 Integrate numerically to find the value of

$$I = \int_0^1 \int_0^1 (x+y)^2 \, dx \, dy.$$

In order to check procedure, this example is chosen because the integral can be evaluated exactly:

$$\int_0^1 \int_0^1 (x + y)^2 \, dx \, dy = \int_0^1 \int_0^1 (x^2 + 2xy + y^2) \, dx \, dy,$$

$$= 7/6 = 1.166 \ldots .$$

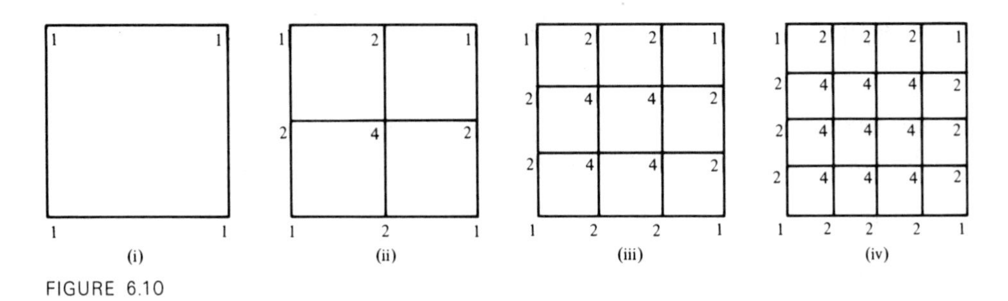

FIGURE 6.10

The trapezoidal rule (3) is applied for grids that are made progressively smaller: (i) $h = k = 1$, (ii) $h = k = 1/2$, (iii) $h = k = 1/3$, and (iv) $h = k = 1/4$. These correspond to the subdivisions illustrated in Fig. 6.10. The results of numerical integration for I are

(i) 1.500, **(ii)** 1.250, **(iii)** 1.204, **(iv)** 1.188.

The weights 2 or 4 indicate that some vertices are members of more than one rectangle. Even a coarse mesh gives reasonable results. but accuracy is increased significantly as h and k diminish.

The same calculation is made using Simpson's rule in the cases (i) $h = k = 1/2$, (ii) $h = k = 1/4$, and (iii) $h = k = 1/6$. The different weights, shown in Fig. 6.11, indicate that some vertices are again common to more than one rectangle. The results for I give in each case the precise value $7/6$ since the integrand is parabolic in x and y, and in such situations Simpson's rule is exact.

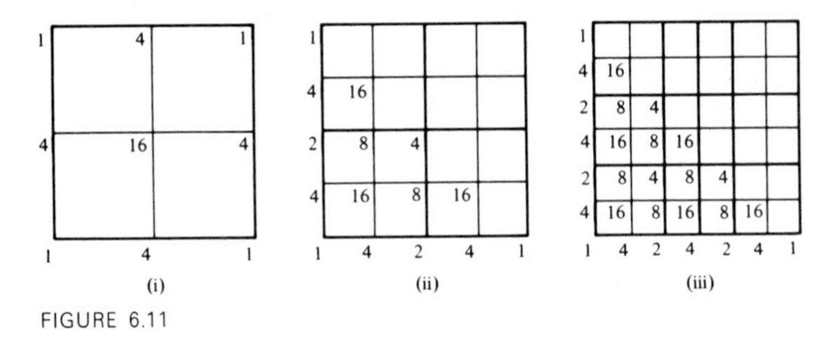

FIGURE 6.11

Exercises 7.6

1. Use Taylor's series to find the approximate value of $\iint_R \sin xy\, dx\, dy$, where R is the region within the lines $x = 0$, $x = 1$, $y = 0$, $y = 1$.
2. Find, correct to three decimal places, the value of $\iint_R e^{-xy}\, dx\, dy$, where R is the region defined by $y = \pm x$ and $x = 1$.
3. Calculate the approximate value of $\iint_R (10 + x + y^2)^{1/3}\, dx\, dy$, where R is the interior of a square which has vertices at (0, 0), (1, 0), (0, 1), (1, 1).
4. Evaluate $\iiint_V e^{-xy^2z^3}\, dx\, dy\, dz$, where V is the cube with sides $x = \pm 1$, $y = \pm 1$, $z = \pm 1$.
5. Find the approximate value of $\iint_R \cos(xy^2)\, dx\, dy$, where R is the interior of the region formed by the lines $x = \pm 1$, $y = \pm 1$.
6. If ε is small, find, correct to order ε^2, the area under the curve $y = \sin(x + \varepsilon y)$ and above the x axis between $x = 0$ and $x = \pi$.
7. If ε is small, calculate, correct to order ε^2, $\iint_R \log(x + \varepsilon y^2)\, dx\, dy$, where R is the region enclosed by the lines $x = 1$, $x = 2$, $y = 0$, $y = 1$.
8. A slightly corrugated sphere has the equation $R = a(1 + \varepsilon \sin \varphi \sin \theta)$, where ε is a small parameter and (R, φ, θ) are spherical coordinates. Find the volume enclosed.
9. A vibrating metal panel has its edges fixed in the plane $z = 0$ along the lines $x = 0$, $x = \pi$, $y = 0$, $y = \pi$; and the surface of the panel at time t has the form $z = \varepsilon \sin x \sin y \cos t$, where ε is small. Find the surface area of the panel at any time correct to $0(\varepsilon^2)$.

10. Show that $$\iint_{x^2+y^2\le\varepsilon^2} \frac{dx\, dy}{x^2 + y^2 + \varepsilon^2} = \pi \log 2.$$

11. Show that
$$\int_{-\infty}^{\infty}\int_{-\infty}^{\infty} f(x, y)e^{-\lambda(x^2+y^2)}\, dx\, dy \approx \frac{\pi f(0, 0)}{\lambda},$$
for large positive values of λ.
12. A nearly circular cylinder has longitudinal corrugations of the form $r = 1 + \varepsilon \sin n\theta$, n being an integer and ε a small parameter. Find, correct to order ε^2, the area of the curved surface between the planes $z = 0$ and $z = 1$. [Here (r, θ, z) are cylindrical coordinates.]
13. Obtain an expression for the gravitational force at any point due to a solid uniform body whose shape is $x^2 + y^2 + z^2/(1 - \varepsilon) = 1$ when ε is small. (The earth is such a body.)
14. If a, b, c, and d are functions of a parameter α and $K(\alpha) = \int_a^b dx \int_c^d f(x, y, \alpha)\, dy$, show that $K'(\alpha) = b' \int_c^d f(b, y, \alpha)\, dy - a' \int_c^d f(a, y, \alpha)\, dy + d' \int_a^b f(x, d, \alpha)\, dx - c' \int_a^b f(x, c, \alpha)\, dx + \int_a^b dx \int_c^d \frac{\partial}{\partial\alpha} f(x, y, \alpha)\, dy$.
15. The area A in Example 3 is $A(\varepsilon) = \int_0^{\pi/4} d\theta \int_0^{F(\theta,\, \varepsilon)} r\, dr$, where $r = F(\theta, \varepsilon)$ is determined from $r = 2\cos(\theta + \varepsilon r)$. Calculate $A(0)$, $A'(0)$, and $A''(0)$ directly. In this way reproduce the result found in the text.
16. Two nearly circular closed curves are represented by the equations $r = a_1 + \varepsilon f_1(\theta)$ and $r = a_2 + \varepsilon f_2(\theta)$, where ε is very small. Use a perturbation method to find the approximate area between the two curves.
17. Two nearly spherical closed surfaces are given by $R = a_1 + \varepsilon f_1(\theta, \varphi)$ and $R = a_2 + \varepsilon f_2(\theta, \varphi)$, where ε is very small. Calculate the approximate volume between the two surfaces.

18. Use the generalized trapezoidal rule and Simpson's rule to calculate $\iint_R (x-y)^2\, dx\, dy$ over the region contained by the lines $x=0$, $x=1$, $y=0$, $y=1$.

19. Evaluate $\iint_R xy\, dx\, dy$ and $\iint_R (x^2+y^2)\, dx\, dy$, where R is the region contained by the lines $x=0$, $x=1$, $y=0$, $y=1$. Show that Simpson's rule gives the exact answer in both cases.

20. Use $h=k=\pi/8$ in the trapezoidal rule and Simpson's rule to find the approximate value of $\iint_R \sin x \sin y\, dx\, dy$, where R is a square with vertices at $(0, 0)$, $(\pi/2, 0)$, $(0, \pi/2)$, $(\pi/2, \pi/2)$. Compare these estimates with the exact result.

21. Use a numerical method to calculate the approximate value of $\iint_R (x^2+2x^3y^3+3y^2)\, dx\, dy$, where R is the region enclosed by $|x|+|y|=1$.

22. How would you devise a numerical method to evaluate $\int_\alpha^\beta d\theta \int_a^b f(r, \theta)\, dr$? In particular use your method to find the area of a circle.

23. R is a square region with vertices at $(1, 1)$, $(-1, 1)$, $(-1, -1)$, and $(1, -1)$. Use a numerical method to find the approximate value of the following integrals:

(i) $\iint_R e^{-x^2-y^2}\, dx\, dy$;

(ii) $\iint_R \sin(x^2+xy+y^2)\, dx\, dy$;

(iii) $\iint_R \cos^2\left(\dfrac{x+y}{1+x^2+y^2}\right) dx\, dy$;

(iv) $\iint_R \operatorname{erf}(|x|^{1/2}+|y|^{1/2})\, dx\, dy$.

24. Obtain the three-dimensional counterpart to the trapezoidal rule, i.e., estimate $\iiint_V f(x, y, z)\, dx\, dy\, dz$, where $a \le x \le a+h$, $b \le y \le b+k$, $c \le z \le c+l$.

25. Obtain the three-dimensional generalization of Simpson's rule for

$$\iiint_V f(x, y, z)\, dx\, dy\, dz, \qquad \text{where } V \text{ is the volume} \qquad \begin{array}{l} a \le x \le a+2h, \\ b \le y \le b+2k, \\ c \le z \le c+2l. \end{array}$$

8
Vector calculus

8.1 Vector fields

1

In Chap. 5 a distinction was made between scalar and vector quantities. At that stage, the vectors concerned were either constants or functions of a single scalar variable. While such vectors are useful in many geometrical and dynamical problems, the scope of their applications is limited. Many of the vectors in the physical world are functions of more than one variable, and a study of such vectors and their relationship with the calculus is the principal aim of this chapter.

An example of a vector function of more than one variable is provided by the measurement of wind velocity $\mathbf{v}$. The magnitude and direction of $\mathbf{v}$ will depend on the position (x, y, z) and the time t, which means that $\mathbf{v} = \mathbf{v}(x, y, z, t)$. Such a function is said to constitute a *vector field*, because a vector is associated with each point in space. For the sake of simplicity we will suppose that $\mathbf{v}$ is independent of t, in which case the vector field is said to be *steady* (as opposed to *unsteady*). In this example the vector represents the magnitude and direction of the wind velocity at the point concerned. A hypothetical velocity field is sketched for steady airflow over a mountain range in Fig. 1.1.

The notion of a direction field is helpful in visualizing the flow. If at a suitable number of arbitrary points in the field elements $d\mathbf{r}$ are drawn in the local directions of $\mathbf{v}$, a family of curves can be constructed, or inferred, whose tangents are always parallel to the flow. These are called *streamlines* of the steady flow. The condition

FIGURE 1.1

that $d\mathbf{r}$ and $\mathbf{v}$ be parallel, that is, $d\mathbf{r} \times \mathbf{v} = \mathbf{0}$, implies that the equations of any streamline shown in Fig. 1.2 are

$$\frac{dx}{v_1} = \frac{dy}{v_2} = \frac{dz}{v_3}. \tag{1}$$

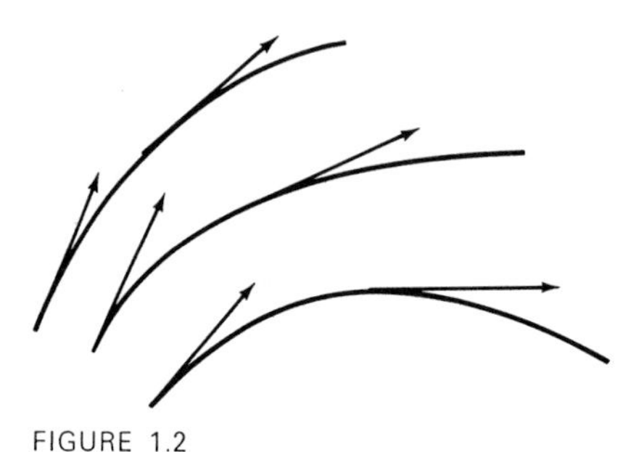

FIGURE 1.2

In other words, an element of fluid remains on the same streamline during steady fluid motion. Note that two streamlines can never cross; if they did, the velocity at the point of intersection would not be single-valued. Clearly, this is a physical absurdity because a particle cannot move in two directions at once.

Example 1 Sketch the flow for the two-dimensional velocity field

$$\mathbf{v} = \Omega(-y\mathbf{i} + x\mathbf{j}).$$

With $\mathbf{r} = x\mathbf{i} + y\mathbf{j}$, it follows that

$$\mathbf{v} \cdot \mathbf{r} = 0.$$

Therefore the streamlines are always perpendicular to the two-dimensional radius vector, which shows that they are concentric circles (Fig. 1.3). Since $|\mathbf{v}| = \Omega\sqrt{x^2 + y^2} = \Omega r$, the velocity field corresponds to a rigid-body rotation about the z axis, that is, $\mathbf{v} = \mathbf{\Omega} \times \mathbf{r}$, where $\mathbf{\Omega} = \Omega\mathbf{k}$ (see Sec. 5.4).

Alternatively, the family of streamlines can be calculated directly from (1):

$$\frac{dx}{-\Omega y} = \frac{dy}{\Omega x}.$$

This is equivalent to

$$x\,dx + y\,dy = 0,$$

which can be rewritten as the exact differential

$$d(x^2 + y^2) = 0.$$

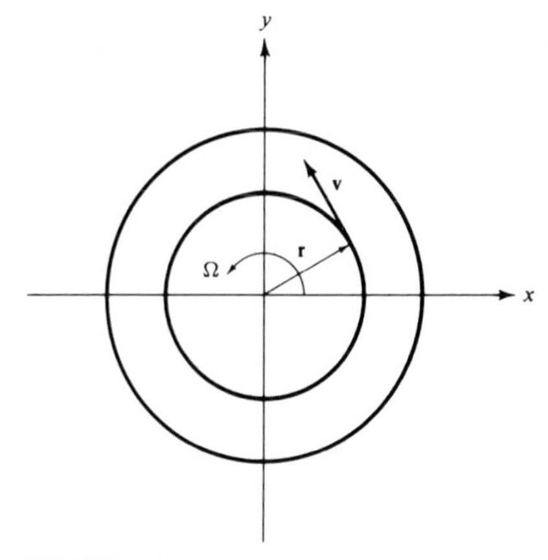

FIGURE 1.3

Integration yields

$$x^2 + y^2 = \text{constant},$$

which again shows that the streamlines are circles.

2

Another illustration of a vector field is provided by Newton's law of gravitation (see Sec. 5.8). In this case, a particle of mass M situated at the origin attracts a mass m located at the vector position $\mathbf{r}$ with a force given by

$$\mathbf{F} = -\frac{GMm}{|\mathbf{r}|^3}\mathbf{r}.$$

The field is shown in Fig. 1.4. The curves tangent to $\mathbf{F}$ at each point are called the *lines of force*, and here they are rays through the origin.

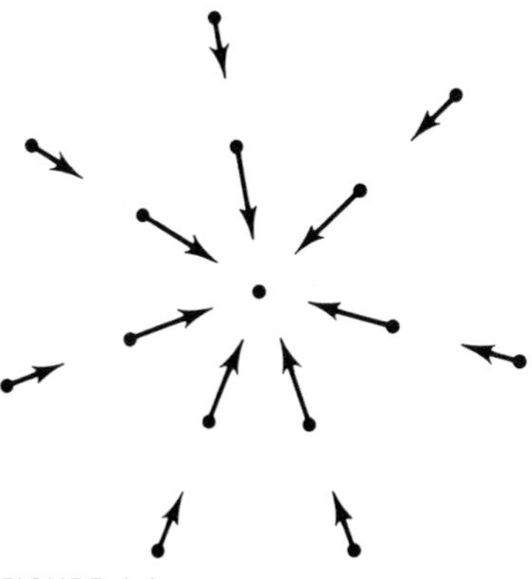

FIGURE 1.4

In general, curves which are everywhere tangent to the vector field $\mathbf{F}(x, y, z)$ are called *field lines*. If $\mathbf{F}$ is identified as a velocity $\mathbf{v}$, the special terms velocity field and streamlines are used in place of vector field and field lines, respectively. If $\mathbf{F}$ is a force, the corresponding words used are force field and lines of force.

A scalar function of many variables $\varphi = \varphi(x, y, z)$ is said to form a *scalar field*, and the surfaces $\varphi =$ constant are called *level surfaces*. If φ represents temperature, a level surface is an isothermal surface, one of constant temperature. If pressure is the scalar field under consideration, the level surfaces are termed isobaric surfaces.

A vector field can be derived from any scalar field φ by the gradient operation

$$\mathbf{F} = \nabla\varphi. \tag{2}$$

At each point, the vector $\mathbf{F}$ is the maximum directional derivative of φ. Field lines are everywhere perpendicular to the level surfaces of constant potential, $\varphi(x, y, z) =$ constant. This conclusion follows from the differential formula relating incremental changes on the constant-potential surface:

$$d\varphi = \varphi_x\, dx + \varphi_y\, dy + \varphi_z\, dz = 0.$$

In other words,

$$\nabla\varphi \cdot d\mathbf{r} = 0,$$

which by (2) can be written

$$\mathbf{F} \cdot d\mathbf{r} = 0.$$

Therefore $\mathbf{F}$, which is the tangent vector of a field line, and $d\mathbf{r}$, the element of length on the surface, are perpendicular vectors, and this is equivalent to the first statement.

Sometimes a scalar field can be determined from a prescribed vector field. For example, the gravitational force, $\mathbf{F} = (-GMm/|\mathbf{r}|^3)\mathbf{r}$, can be expressed as

$$\mathbf{F} = \nabla\varphi = \nabla\frac{GMm}{r}. \tag{3}$$

In this case, the function φ is called the gravitational potential. A vector field $\mathbf{F}$ that can be written in this way as the gradient of a scalar φ, that is $\mathbf{F} = \nabla\varphi$, is called a *conservative field*, and φ is the *potential* for the field.

However, a vector field $\mathbf{F}$ cannot necessarily be derived from a potential φ, and to illustrate this point we return to Example 1, where

$$\mathbf{F} = \Omega(-y\mathbf{i} + x\mathbf{j}).$$

If there were a potential φ associated with this vector $\mathbf{F}$, then, by definition,

$$\mathbf{F} = \frac{\partial\varphi}{\partial x}\mathbf{i} + \frac{\partial\varphi}{\partial y}\mathbf{j}$$

and it would follow that

$$\frac{\partial\varphi}{\partial x} = -\Omega y, \qquad \frac{\partial\varphi}{\partial y} = \Omega x.$$

But differentiation of the first equation with respect to y and the second with respect to x leads to an inconsistency,

$$\frac{\partial^2 \varphi}{\partial y\, \partial x} = -\Omega, \qquad \frac{\partial^2 \varphi}{\partial x\, \partial y} = \Omega.$$

Accordingly, in this case, we conclude that it is not possible to express **F** as a gradient, $\nabla\varphi$.

For the moment we leave aside any further discussion of the conditions under which a given vector field can be written as the gradient of a scalar, but we return to this question in the next section.

Example 2 Discuss the vector field $\mathbf{F} = x\mathbf{i} - y\mathbf{j}$.

A quick picture of the vector field is obtained by drawing the vector **F** at a few selected points, as sketched in Fig. 1.5. The field lines are obtained from Eq. (1):

$$\frac{dx}{x} = \frac{dy}{-y}.$$

This can be rewritten as

$$d(xy) = 0$$

and integrated immediately to show that the field lines are the family of hyperbolas, $xy =$ constant.

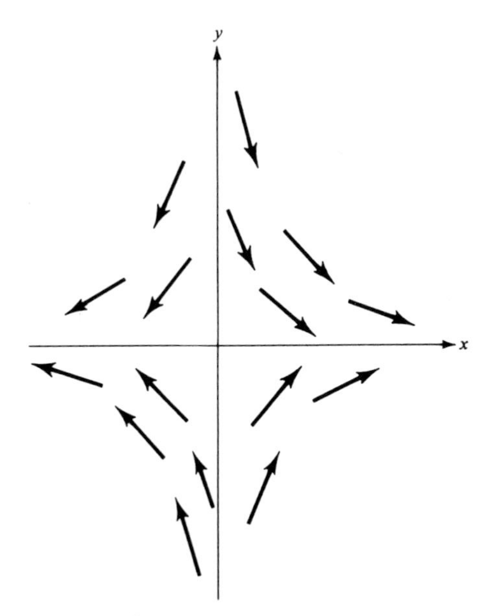

FIGURE 1.5

Note that $\mathbf{F} = \mathbf{0}$ at the origin. If $\mathbf{F}$ is a force field, a particle at the origin remains there because it experiences no force. The origin is then an equilibrium point of the force field. If $\mathbf{F}$ represents a velocity field, the origin is called a stagnation point in the flow, i.e., a point of no motion.

In this example, $\mathbf{F}$ is indeed a conservative field. To demonstrate this, we construct a function $\varphi(x, y)$ such that $\mathbf{F} = \nabla\varphi$, or

$$\frac{\partial\varphi}{\partial x} = x, \qquad \frac{\partial\varphi}{\partial y} = -y.$$

Integration of the first equation shows that

$$\varphi = \frac{x^2}{2} + f(y),$$

where f is an arbitrary function. Substitution in the second equation yields

$$f'(y) = -y$$

or, upon integration,

$$f(y) = \frac{-y^2}{2} + C,$$

where C is an arbitrary constant. It follows that the potential is

$$\varphi = \tfrac{1}{2}(x^2 - y^2) + C.$$

It is a simple matter to check that the field lines, $xy = \text{constant}$, and the level curves (or equipotential curves), $\varphi = \text{constant}$, are indeed perpendicular (see Fig. 1.6).

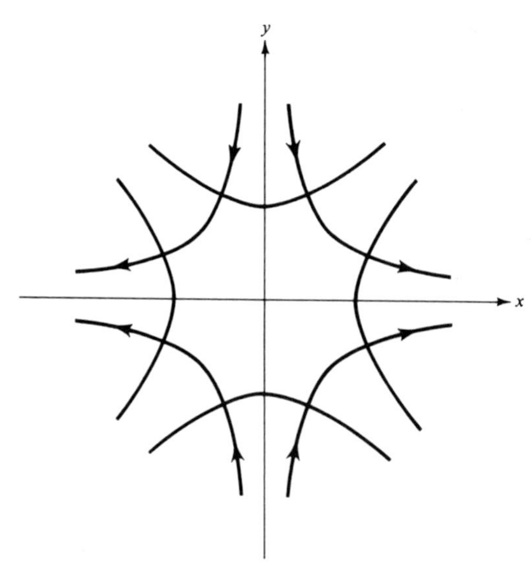

FIGURE 1.6

Example 3 Derive an expression for the velocity field produced by a point source placed in an infinite fluid of constant density as shown in Fig. 1.7.

By symmetry, the fluid velocity is a function of the spherical radius only, of the form

$$\mathbf{v} = v(r)\hat{\mathbf{r}},$$

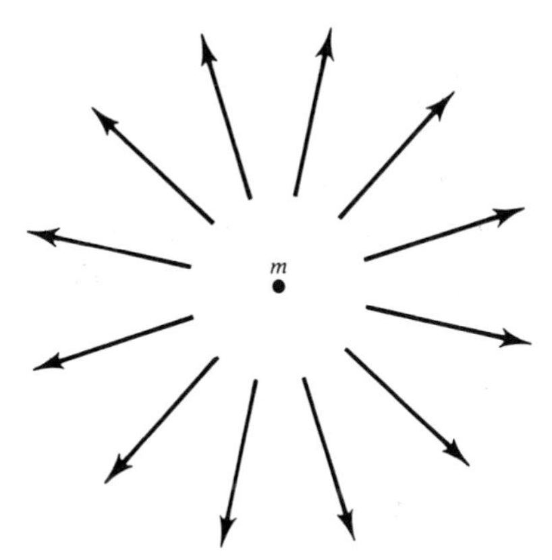

FIGURE 1.7

where $\hat{\mathbf{r}} = \mathbf{r}/r$ is a unit vector in the radial direction. The functional dependence is a consequence of the conservation of mass in the flow, which, simply put, means that there can be no local accumulation of a fluid of constant density (otherwise the density would change). Therefore, the rate at which fluid is introduced at the source must equal the rate at which fluid crosses any concentric sphere of radius r which surrounds it. The mathematical formulation of this statement is the essential step in the solution process. Let $4\pi m$ be the volume of fluid per unit time injected at the source, which is then said to be of strength m. The rate at which this fluid crosses a sphere at radius r is given by the area of the sphere, $4\pi r^2$, multiplied by the radial velocity $v(r)$ at this position, that is, $4\pi r^2 v(r)$. Conservation of mass asserts that these rates are equal:

$$4\pi r^2 v(r) = 4\pi m,$$

so that

$$v(r) = \frac{m}{r^2}$$

and

$$\mathbf{v} = \frac{m}{|\mathbf{r}|^3}\,\mathbf{r},$$

where $\mathbf{r} = x\mathbf{i} + y\mathbf{j} + z\mathbf{k}$. The streamlines are rays from the source point, and the magnitude of the velocity is given by the inverse-square law. If m is negative, the flow is inward; a negative source is called a *sink*. Despite the fact that sources and sinks are rather idealized concepts, they prove to be extremely useful in the simulation of quite complicated fluid flows.

The velocity field can be expressed in terms of a velocity potential given by

$$\varphi = -\frac{m}{r},$$

so that

$$\mathbf{v} = \nabla\left(\frac{-m}{r}\right).$$

This is in complete analogy with the derivation of a gravitational potential in Eq. (3). Evidently, certain ideal fluid flows have mathematical formulations identical with those of gravitational problems. As a matter of fact, this formalism also applies to an electric charge of strength q which produces an electric field $\mathbf{E}$ given by

$$\mathbf{E} = \frac{q\mathbf{r}}{|\mathbf{r}|^3}.$$

In this case, the electrostatic potential, defined by $\mathbf{E} = -\nabla\varphi$, is

$$\varphi = \frac{q}{r}.$$

The equipotential surfaces ($\varphi = \text{constant}$) are spheres with center the origin.

Example 4 Find the flow field induced by a *dipole* of strength μ placed in an infinite fluid.

A source of strength m and a sink of strength m placed a distance l apart are said to form a dipole when the distance l between them tends to zero in such a way that $\mu = ml$ remains constant. The constant μ is called the strength of the dipole.

For convenience, the z axis is chosen to be the axis of the dipole in that the source is located at the point $(0, 0, l/2)$ and the corresponding sink is placed at $(0, 0, -l/2)$, as shown in Fig. 1.8. The induced velocity or preferably the com-

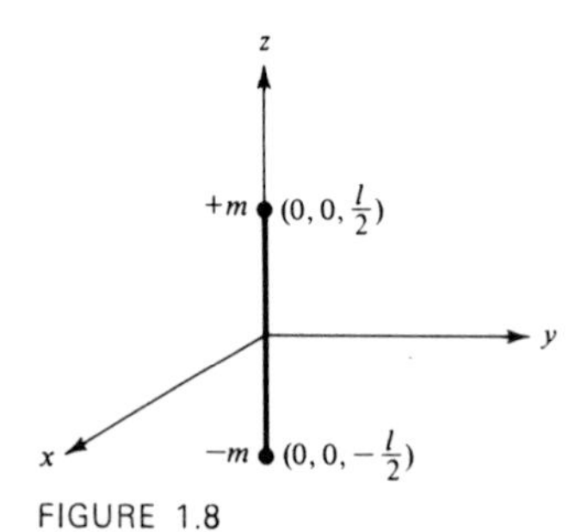

FIGURE 1.8

bined velocity potential due to the presence of both the source and sink is the sum of the individual contributions, so that

$$\varphi = \frac{-m}{[x^2 + y^2 + (z - l/2)^2]^{1/2}} + \frac{m}{[x^2 + y^2 + (z + l/2)^2]^{1/2}}.$$

This expression can be rewritten in the form

$$\varphi = \frac{ml}{l}\left\{\frac{1}{[x^2 + y^2 + (z + l/2)^2]^{1/2}} - \frac{1}{[x^2 + y^2 + (z - l/2)^2]^{1/2}}\right\},$$

which facilitates the limit $l \to 0$, $ml = \mu$ a constant. Since

$$\lim_{l \to 0} \varphi = \mu \lim_{l \to 0} \frac{1}{l}\left\{\left[x^2 + y^2 + \left(z + \frac{l}{2}\right)^2\right]^{-1/2} - \left[x^2 + y^2 + \left(z - \frac{l}{2}\right)^2\right]^{-1/2}\right\},$$

$$= \mu \frac{\partial}{\partial z}(x^2 + y^2 + z^2)^{-1/2},$$

the potential for a dipole field is

$$\varphi = -\frac{\mu z}{(x^2 + y^2 + z^2)^{3/2}}.$$

The corresponding velocity $\mathbf{v}$, obtained by calculating $\nabla\varphi$, is

$$\mathbf{v} = \mu \frac{3zx\mathbf{i} + 3zy\mathbf{j} - (x^2 + y^2 - 2z^2)\mathbf{k}}{(x^2 + y^2 + z^2)^{5/2}}.$$

The streamlines are sketched in Fig. 1.9.

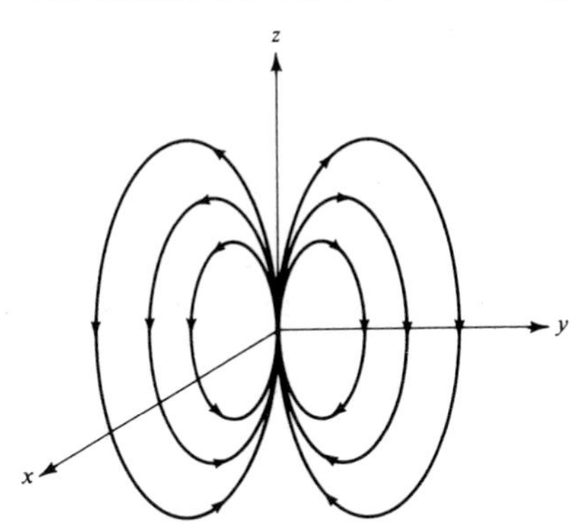

FIGURE 1.9

The same mathematical analysis applies to dipole fields in many branches of science. The magnetic field of the earth, for example, is largely dipole in character, although its origin is still a matter of study.

Exercises 8.1

1. For the following velocity vectors $\mathbf{v} = \mathbf{v}(x, y)$, sketch the velocity fields and find the equations of the streamlines through the point (1, 1).

(i) $\mathbf{v} = -x\mathbf{i} + y\mathbf{j}$ (ii) $\mathbf{v} = (1 + x)\mathbf{i} + y\mathbf{j}$ (iii) $\mathbf{v} = e^{-y}\mathbf{i} + e^{-x}\mathbf{j}$

2. Sketch the field lines for the following plane vector fields $\mathbf{v} = \mathbf{v}(x, y)$. Find the equations of the streamlines through the origin.

(i) $\mathbf{i} + 2\mathbf{j}$ (ii) $x\mathbf{i} + 2y\mathbf{j}$ (iii) $2y\mathbf{i} - x\mathbf{j}$
(iv) $y\mathbf{i} + \mathbf{j}$ (v) $y\mathbf{i} + \sin x\mathbf{j}$ (vi) $y^2\mathbf{i} + x^2\mathbf{j}$

3. What is the general form of a vector field $\mathbf{v}(x, y, z)$ if its direction is always parallel to $\mathbf{i} + \mathbf{j} + \mathbf{k}$ but its magnitude varies from point to point? if its direction is always away from the origin with variable magnitude?

4. The velocity of a fluid through a channel, $-h<y<h$, is $\mathbf{v} = K(h^2 - y^2)\mathbf{i}$, where K is a constant. Sketch the streamlines and find the maximum velocity of the fluid.

5. Find the field lines for $\mathbf{F} = y\mathbf{i} + x\mathbf{j}$. Show that $\mathbf{F}$ is a conservative field and obtain its potential φ, where $\mathbf{F} = \nabla\varphi$. Verify that the field lines are perpendicular to the level curves of φ.

6. What restrictions must be placed on the constants a, b, c, d in order that the vector field $\mathbf{F} = (ax + by)\mathbf{i} + (cx + dy)\mathbf{j}$ can be expressed as the gradient of a scalar function $\varphi(x, y)$? What is the corresponding result in three dimensions?

7. Describe by diagrams and simple mathematical models the streamline patterns you would expect to find for the following.

(i) The wake of a jet engine (ii) A steady wind around a tall building
(iii) The wind in a hurricane (iv) The solar wind

8. Calculate the vector fields $\mathbf{F}(x, y, z)$ corresponding to the following potential functions.

(i) $\phi(x, y, z) = xy^2z^3$ (ii) $\sin(x - y + 2z)$ (iii) $2x^2 + y^2 + 3z^2$
(iv) $x + yz + z^2x^2$ (v) ye^{xz} (vi) $(x^2 - y^2)\log(z^2 + x^2)$

9. Show that the vector field $\mathbf{F}(x, y) = x^2y\mathbf{i} + xy^2\mathbf{j}$ is not conservative. Determine the streamlines and find a conservative vector field with the same streamlines.

10. Find the potential functions for the following conservative vector fields.

(i) $3x^2y\mathbf{i} + (x^3 + y)\mathbf{j}$ (ii) $x\mathbf{i} + y^2\mathbf{j} + z^3\mathbf{k}$ (iii) $2xy\mathbf{i} + (x^2 - z)\mathbf{j} - y\mathbf{k}$

11. Obtain the electrostatic potential due to point charges $2q$ at (0, 0, 0), $-q$ at $(0, 0, a)$, and $-q$ at $(0, 0, -a)$. Calculate the potential at the point $(a, 2a, 3a)$ and find the direction of the field line through this point.

12. What is the velocity potential due to a source $4m$ and a source $-m$ a distance d apart? Deduce that one of the equipotential surfaces is a sphere. Are there any stagnation points in the flow field?

13. Show that the potential φ due to a dipole μ can be expressed in the form $\varphi = -(\mu\cos\Phi)/R^2$, where (R, Φ, θ) are spherical coordinates, the notation Φ being introduced to avoid confusion with the potential φ.

14. Use Exercise 13 to verify that the potential $\varphi = -UR\cos\Phi - (\mu\cos\Phi)/R^2$ represents the flow produced by a dipole placed in a stream which moves with speed U in the negative z direction. When $\mu = \frac{1}{2}Ua^3$, show that there is no flow through the surface of the sphere $R = a$.

15. When fluid is ejected perpendicularly from the z axis in a uniform manner, the resulting flow remains parallel to the xy plane. If $2\pi m$ is the volume ejected per unit time per unit length of the z axis, show that the induced velocity $\mathbf{v}$ is $\mathbf{v} = m\mathbf{r}/|r|^2$, where $\mathbf{r} = x\mathbf{i} + y\mathbf{j}$. (This configuration is called a line source of strength m.)

16. Use Exercise 15 to obtain the potential $\varphi = m \log r$ for a line source.
17. Two line sources parallel to Oz have strengths m and $-m$ and cut the xy plane at $(l/2, 0)$ and $(-l/2, 0)$, respectively. Find the velocity potential for this configuration. In the limit $l \to 0$, with $ml = \mu$ finite, this system becomes a two-dimensional dipole. Show that the potential of the dipole is $\varphi = (\mu \cos \theta)/r$, where r and θ are polar coordinates in the xy plane.
18. Verify that the streamlines of a two-dimensional dipole form a family of circles. How does this family relate to the family of equipotential curves?
19. If $\varphi = \varphi(x, y, z)$ is a potential function for the vector field $\mathbf{F}$, show that the streamlines of the vector field are orthogonal to the equipotential surfaces $\varphi =$ constant.

8.2 *Line integrals*

Differentiation and integration of a vector function can always be performed by dealing separately with its components. Since each component is a scalar function, all the techniques developed in Chaps. 6 and 7 are directly applicable. In fact, this procedure has already been adopted in the earlier applications of Sec. 7.5. For example, if $\mathbf{F}(x, y, z)$ is a vector whose component form is

$$\mathbf{F} = F_1(x, y, z)\mathbf{i} + F_2(x, y, z)\mathbf{j} + F_3(x, y, z)\mathbf{k},$$

then, for the reasons given,

$$\frac{\partial \mathbf{F}}{\partial x} = \frac{\partial F_1}{\partial x}\mathbf{i} + \frac{\partial F_2}{\partial x}\mathbf{j} + \frac{\partial F_3}{\partial x}\mathbf{k},$$

and other partial derivatives can be calculated in an equally straightforward manner.

In a similar vein, the evaluation of the vector volume integral $\mathbf{G} = \iiint_V \mathbf{F}(x, y, z)\, dV$, where V is a prescribed volume, can be found by writing

$$\mathbf{G} = \mathbf{i} \iiint_V F_1(x, y, z)\, dV + \mathbf{j} \iiint_V F_2(x, y, z)\, dV + \mathbf{k} \iiint_V F_3(x, y, z)\, dV.$$

Each of the scalar integrals is then calculated separately.

One type of integral, called a *line integral*, merits special attention, and the remainder of this section is devoted to its study.

Let $\mathbf{F}$ be a vector field and $P_0 = (x_0, y_0, z_0)$ and $P_1 = (x_1, y_1, z_1)$ be two fixed points joined by a prescribed curve C (Fig. 2.1). The line integral of $\mathbf{F}$ from P_0 to P_1 along the curve C is written

$$\int_{P_0}^{P_1} \mathbf{F} \cdot d\mathbf{r} = \int_{P_0}^{P_1} \mathbf{F} \cdot \hat{\mathbf{T}}\, ds, \tag{1}$$

where $\hat{\mathbf{T}}$ is the unit tangent vector to the curve and ds is the element of arc length. The interpretation of a line integral, its basic limit definition, is that of an ordinary integral of the scalar function $\mathbf{F} \cdot \hat{\mathbf{T}}$ over the arc length from P_0 to P_1. Sometimes the following alternative notations are used:

$$\int_C \mathbf{F} \cdot d\mathbf{r} \qquad \text{or} \qquad \int_C (F_1\, dx + F_2\, dy + F_3\, dz).$$

The representation of the work done by a force as a line integral provides an immediate illustration of physical importance. The work done by the force **F** moving through a vector displacement $d\mathbf{r}$ is $\mathbf{F} \cdot d\mathbf{r}$ (see Sec. 5.3). The total work done by the force **F** in moving from P_0 to P_1 *along* C is then the addition of all infinitesimal contributions, that is, $\int_{P_0}^{P_1} \mathbf{F} \cdot d\mathbf{r}$.

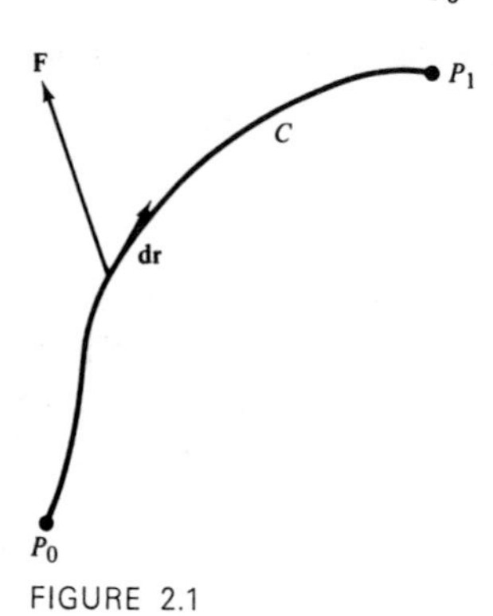

FIGURE 2.1

The simplest way to evaluate a line integral is to describe the curve C in parametric form

$$\mathbf{r} = \mathbf{r}(t),$$

where the parametric values $t = t_0$ and $t = t_1$ correspond to points P_0 and P_1. Equation (1) can then be expressed as

$$\int_{P_0}^{P_1} \mathbf{F} \cdot d\mathbf{r} = \int_{t_0}^{t_1} \left(\mathbf{F} \cdot \frac{d\mathbf{r}}{dt} \right) dt, \tag{2}$$

and the right-hand integral here is an ordinary scalar integral of the type discussed in Chap. 3.

A common parametric form of a line integral is obtained by writing

$$\mathbf{F} = F_1(x, y, z)\mathbf{i} + F_2(x, y, z)\mathbf{j} + F_3(x, y, z)\mathbf{k}$$

and

$$\mathbf{r}(t) = x(t)\mathbf{i} + y(t)\mathbf{j} + z(t)\mathbf{k},$$

so that

$$\mathbf{F} \cdot \frac{d\mathbf{r}}{dt} = F_1 \frac{dx}{dt} + F_2 \frac{dy}{dt} + F_3 \frac{dz}{dt}.$$

The substitution of this expression in (2) yields

$$\int_{P_0}^{P_1} \mathbf{F} \cdot d\mathbf{r} = \int_{t_0}^{t_1} \left(F_1 \frac{dx}{dt} + F_2 \frac{dy}{dt} + F_3 \frac{dz}{dt} \right) dt. \tag{3}$$

In general, the line integral of a vector function **F** will depend on the end points P_0 and P_1 *and* the path C connecting these points. Later in this section we will see that there is an important special class of vector fields for which the line integral does not depend on the curve C, but this is the exception rather than the rule.

Example 1 If $\mathbf{F} = xy\mathbf{i} + (y^2 + 1)\mathbf{j}$, calculate $\int_{P_0}^{P_1} \mathbf{F} \cdot d\mathbf{r}$, where $P_0 = (0, 0, 0)$ and $P_1 = (1, 1, 0)$ when C is (i) the straight line from P_0 to P_1; (ii) the x axis to the point $(1, 0, 0)$ then the line $x = 1$; (iii) the curve $y^2 = x$.

The calculations refer entirely to the plane $z = 0$. In case (i), the curve C is the line $y = x$, shown in Fig. 2.2 (i), for which a convenient parametric description is

$$x = t, \qquad y = t.$$

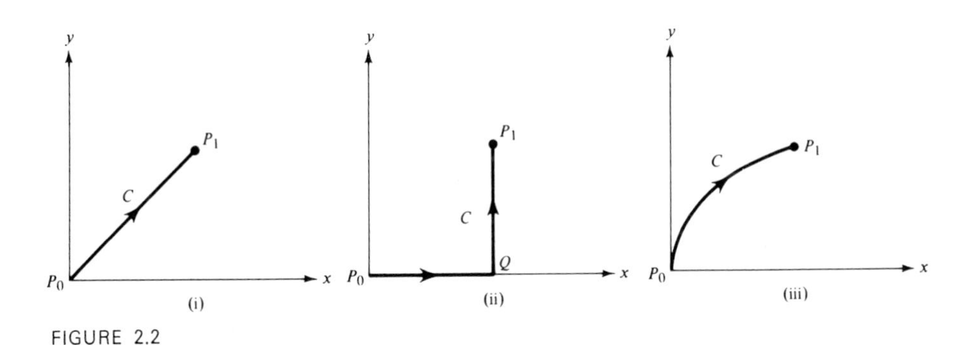

FIGURE 2.2

Here $t = 0$ and $t = 1$ correspond to the points (0, 0) and (1, 1) respectively. On this curve

$$\mathbf{F} = t^2\mathbf{i} + (t^2 + 1)\mathbf{j},$$

and

$$\frac{d\mathbf{r}}{dt} = \mathbf{i} + \mathbf{j}.$$

It follows from (2) that

$$\int_{P_0}^{P_1} \mathbf{F} \cdot d\mathbf{r} = \int_0^1 [t^2\mathbf{i} + (t^2 + 1)\mathbf{j}] \cdot (\mathbf{i} + \mathbf{j})\, dt,$$

$$= \int_0^1 (2t^2 + 1)\, dt = 5/3.$$

In case (ii), the curve C consists of two straight-line segments [Fig. 2.2(ii)], which can be treated separately by decomposing the line integral as

$$\int_{P_0}^{P_1} \mathbf{F} \cdot d\mathbf{r} = \int_{P_0}^{Q} \mathbf{F} \cdot d\mathbf{r} + \int_{Q}^{P_1} \mathbf{F} \cdot d\mathbf{r},$$

where $Q = (1, 0)$. Along P_0Q, $y = 0$, so that $\mathbf{F} = \mathbf{j}$ and $d\mathbf{r} = \mathbf{i}\,dx$, where x ranges from 0 to 1. On this segment x itself plays the role of a parameter. These relations imply that

$$\int_{P_0}^{Q} \mathbf{F} \cdot d\mathbf{r} = \int_0^1 \mathbf{j} \cdot \mathbf{i}\,dx = 0.$$

On the other hand, along QP_1, $x = 1$, $d\mathbf{r} = dy\mathbf{j}$, $\mathbf{F} = y\mathbf{i} + (y^2 + 1)\mathbf{j}$, and y ranges from 0 to 1, so that

$$\int_{Q}^{P_1} \mathbf{F} \cdot d\mathbf{r} = \int_0^1 [y\mathbf{i} + (y^2 + 1)\mathbf{j}] \cdot \mathbf{j}\,dy,$$

$$= \int_0^1 (y^2 + 1)\,dy = 4/3.$$

On adding the two contributions, we obtain

$$\int_{P_0}^{P_1} \mathbf{F} \cdot d\mathbf{r} = 4/3.$$

In case (iii), $y^2 = x$, and a convenient parameterization is $x = t^2$, $y = t$, where $t = 0$ and $t = 1$ correspond to P_0 and P_1 respectively [see Fig. 2.2(iii)]. On the integration path, $\mathbf{r} = t^2\mathbf{i} + t\mathbf{j}$, $d\mathbf{r} = (2t\mathbf{i} + \mathbf{j})\,dt$ and $\mathbf{F} = t^3\mathbf{i} + (t^2 + 1)\mathbf{j}$, and it follows that

$$\int_{P_0}^{P_1} \mathbf{F} \cdot d\mathbf{r} = \int_0^1 [t^3\mathbf{i} + (t^2 + 1)\mathbf{j}] \cdot (2t\mathbf{i} + \mathbf{j})\,dt,$$

$$= \int_0^1 (2t^4 + t^2 + 1)\,dt,$$

$$= 2/5 + 1/3 + 1 = 26/15.$$

Note that each of the three different paths of integration gives a different answer, confirming our earlier speculation that the expression $\int_{P_0}^{P_1} \mathbf{F} \cdot d\mathbf{r}$ depends on the path connecting P_0 and P_1.

Example 2 Calculate $\int_C \mathbf{F} \cdot d\mathbf{r}$ when $\mathbf{F} = (x - z)\mathbf{i} + (1 - xy)\mathbf{j} + y\mathbf{k}$, and C is the path (i) the straight line from (0, 0, 0) to (1, 1, 1) and (ii) the space curve $x = t$, $y = t^2$, $z = t^3$ from (0, 0, 0) to (1, 1, 1).

On the first straight-line path [Fig. 2.3(i)] $x = t$, $y = t$, $z = t$, so that

$$\mathbf{r} = t\mathbf{i} + t\mathbf{j} + t\mathbf{k} \qquad \text{and} \qquad \mathbf{F} = (1 - t^2)\mathbf{j} + t\mathbf{k}.$$

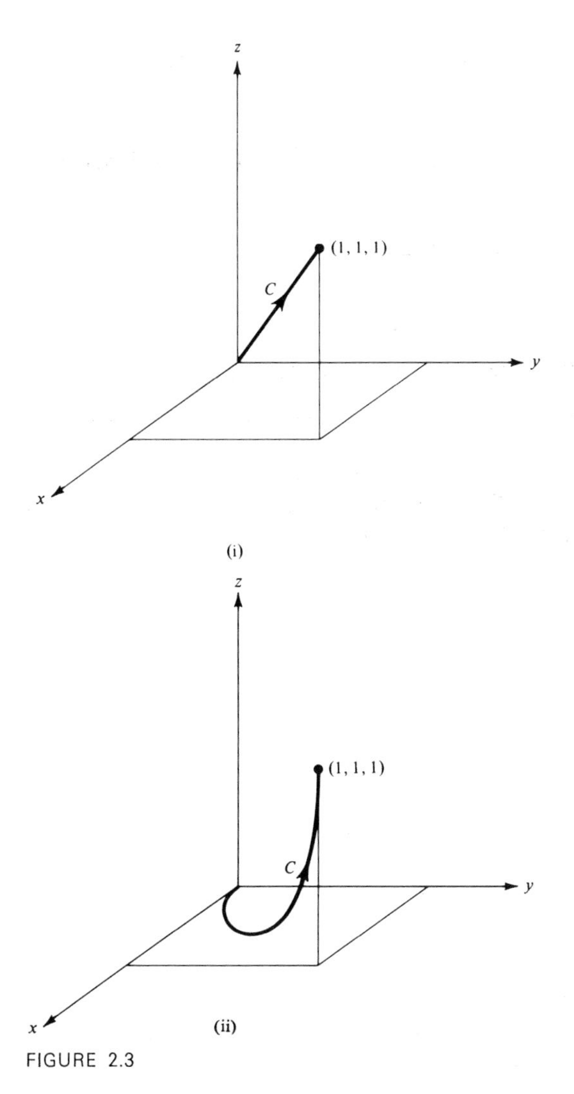

FIGURE 2.3

Since t ranges from 0 to 1 from beginning to end, the required line integral is

$$\int_C \mathbf{F} \cdot d\mathbf{r} = \int_0^1 [(1 - t^2)\mathbf{j} + t\mathbf{k}] \cdot (\mathbf{i} + \mathbf{j} + \mathbf{k})\, dt,$$

$$= \int_0^1 (1 + t - t^2)\, dt,$$

$$= 1 + \tfrac{1}{2} - \tfrac{1}{3} = \tfrac{7}{6}.$$

Along the second path specified [see Fig. 2.3(ii)],

$$\mathbf{r} = t\mathbf{i} + t^2\mathbf{j} + t^3\mathbf{k}, \qquad \text{and} \qquad \mathbf{F} = (t - t^3)\mathbf{i} + (1 - t^3)\mathbf{j} + t^2\mathbf{k}.$$

As before, $t = 0$ corresponds to (0, 0, 0) and $t = 1$ to the point (1, 1, 1), and we obtain

$$\int_C \mathbf{F} \cdot d\mathbf{r} = \int_0^1 [(t - t^3)\mathbf{i} + (1 - t^3)\mathbf{j} + t^2\mathbf{k}] \cdot (\mathbf{i} + 2t\mathbf{j} + 3t^2\mathbf{k})\, dt,$$

$$= \int_0^1 (t^4 - t^3 + 3t)\, dt,$$

$$= 1/5 - 1/4 + 3/2 = 29/20.$$

Again we see that the line integral depends on the path.

3

It is often necessary to consider situations in which the integration path C is a closed curve, so that the end point P_1 is identical to the initial point P_0 (see Fig. 2.4). To emphasize this fact, the closed line integral is written

$$\oint_C \mathbf{F} \cdot d\mathbf{r},$$

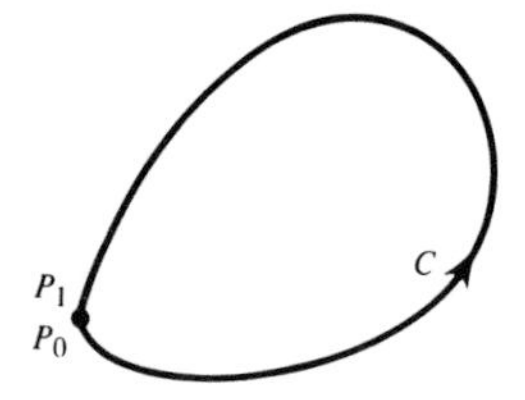

FIGURE 2.4

but the evaluation of such integrals presents no new difficulties. For plane curves it is conventional to take the positive direction of C, so that the interior of the closed curve is always to the left, as shown.

Example 3 Calculate $\oint_C \mathbf{F} \cdot d\mathbf{r}$ where $\mathbf{F} = 2y\mathbf{i} + 3x\mathbf{j}$ and C is the circle $x^2 + y^2 = 1$ taken in the counterclockwise sense (see Fig. 2.5).

A simple parametric form for C is $x = \cos t$, $y = \sin t$, or

$$\mathbf{r} = \mathbf{i} \cos t + \mathbf{j} \sin t,$$

so that

$$d\mathbf{r} = (-\mathbf{i} \sin t + \mathbf{j} \cos t)\, dt.$$

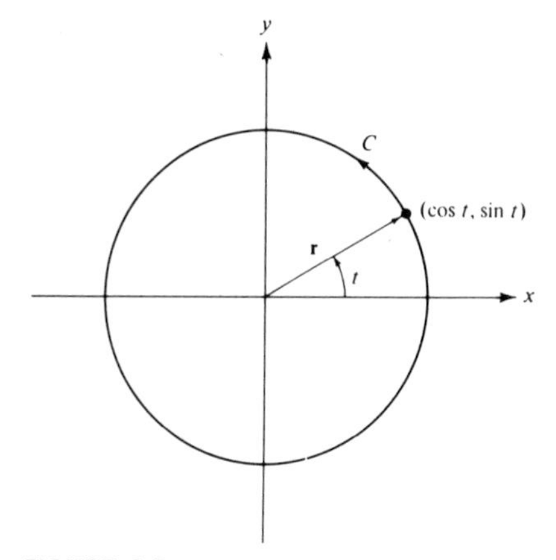

FIGURE 2.5

On the curve C, $\mathbf{F} = 2 \sin t\, \mathbf{i} + 3 \cos t\, \mathbf{j}$, and it follows that

$$\begin{aligned}\mathbf{F} \cdot d\mathbf{r} &= (2 \sin t\, \mathbf{i} + 3 \cos t\, \mathbf{j}) \cdot (-\mathbf{i} \sin t + \mathbf{j} \cos t)\, dt,\\ &= (3 \cos^2 t - 2 \sin^2 t)\, dt = (\tfrac{1}{2} + \tfrac{5}{2} \cos 2t)\, dt.\end{aligned}$$

The circle C is traversed once as t ranges from 0 to 2π, and therefore

$$\begin{aligned}\oint_C \mathbf{F} \cdot d\mathbf{r} &= \int_0^{2\pi} (\tfrac{1}{2} + \tfrac{5}{2} \cos 2t)\, dt,\\ &= \left[\frac{t}{2} + \frac{5}{4} \sin 2t\right]_0^{2\pi} = \pi.\end{aligned}$$

---

4 We turn now to a discussion of vector fields $\mathbf{F}$, for which the line integral $\int_C \mathbf{F} \cdot d\mathbf{r}$ does *not* depend on the path. This is important because it is a characteristic property of conservative fields which describe many natural phenomena. We will show that if $\mathbf{F} = \nabla\varphi$, then $\int_C \mathbf{F} \cdot d\mathbf{r}$ is path-independent, and conversely.

Assume first that there is a single-valued potential φ such that $\mathbf{F} = \nabla\varphi$. It follows that

$$\int_C \mathbf{F} \cdot d\mathbf{r} = \int_C \nabla\varphi \cdot d\mathbf{r} = \int_C \left(\frac{\partial\varphi}{\partial x}\, dx + \frac{\partial\varphi}{\partial y}\, dy + \frac{\partial\varphi}{\partial z}\, dz\right),$$

which can be expressed simply as

$$\int_C \mathbf{F} \cdot d\mathbf{r} = \int_C d\varphi = \int_{P_0}^{P} d\varphi = \varphi(P) - \varphi(P_0). \tag{4}$$

Equation (4) asserts that the line integral of a conservative field is merely the difference in the values of the potential φ at the two end points. One interpretation from the physical viewpoint is that work done by the conservative force $\mathbf{F} = \nabla\varphi$, in moving from P_0 to P is independent of the path and is equal to the potential difference between P and P_0. A similar result applies to any field derived from a potential, e.g., electrostatic, magnetostatic, or gravitational attraction.

In order to establish the converse of this result, we must show that if a line integral $\int_{P_0}^{P} \mathbf{F} \cdot d\mathbf{r}$ is independent of the path C that joins any two arbitrary points P_0 and P in a region, then $\mathbf{F}$ is a conservative field. In other words, there exists a potential φ such that $\mathbf{F} = \nabla\varphi$.

To prove this we begin by taking $P_0 = (x_0, y_0, z_0)$ as a fixed point and $P = (x, y, z)$ as a variable or movable point in the region. Since $\int_{P_0}^{P} \mathbf{F} \cdot d\mathbf{r}$ is, by assumption, path-independent (and P_0 is assumed fixed), this integral defines a function of x, y, and z, that is,

$$\varphi(x, y, z) = \int_{P_0}^{P} \mathbf{F} \cdot d\mathbf{r}. \tag{5}$$

In this expression, *any* integration path between P_0 and P is allowed; an especially convenient choice consists of line segments parallel to the coordinate axes. We choose, in particular, the path made up of three lines P_0Q, QR, and RP, which are parallel to Ox, Oy, and Oz, respectively (see Fig. 2.6). In other words, $Q = (x, y_0, z_0)$

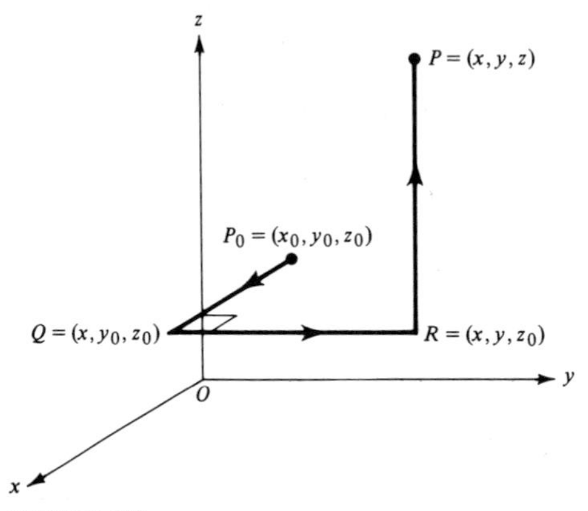

FIGURE 2.6

and $R = (x, y, z_0)$. Since $\mathbf{F} \cdot d\mathbf{r} = F_1\,dx$ on P_0Q, $\mathbf{F} \cdot d\mathbf{r} = F_2\,dy$ on QR, and $\mathbf{F} \cdot d\mathbf{r} = F_3\,dz$ on RP, Eq. (5) can be rewritten as the sum of three ordinary integrals, namely,

$$\varphi(x, y, z) = \int_{x_0}^{x} F_1(x', y_0, z_0)\,dx' + \int_{y_0}^{y} F_2(x, y', z_0)\,dy' + \int_{z_0}^{z} F_3(x, y, z')\,dz'. \tag{6}$$

Since the first two integrals here are independent of z, the result of differentiating with respect to this variable is

$$\begin{aligned} \frac{\partial\varphi(x, y, z)}{\partial z} &= \frac{\partial}{\partial z}\int_{z_0}^{z} F_3(x, y, z')\,dz', \\ &= F_3(x, y, z). \end{aligned} \tag{7}$$

(This is just an application of the formula $\dfrac{\partial}{\partial x}\displaystyle\int_a^x f(t)\,dt = f(x)$ derived on page 219.) If the straight-line paths are chosen in different orders, then by direct analogy we find that

$$\frac{\partial \varphi(x, y, z)}{\partial y} = F_2(x, y, z), \tag{8}$$

and

$$\frac{\partial \varphi(x, y, z)}{\partial x} = F_1(x, y, z). \tag{9}$$

But since $P = (x, y, z)$ is an arbitrary point in the region, Eqs. (7) to (9) are equivalent to

$$\mathbf{F} = \nabla\varphi, \tag{10}$$

which is the desired conclusion.

The fact that a line integral is path-independent is entirely equivalent to the statement that the line integral around any closed path is zero. To see this, let C_1 and C_2 be any two paths joining P_0 to P_1 shown in Fig. 2.7*a*. If the line integral is path-independent, then it follows by subtraction that

$$\int_{C_1} \mathbf{F}\cdot d\mathbf{r} - \int_{C_2} \mathbf{F}\cdot d\mathbf{r} = 0.$$

However, in this form the sign of the second integral can be changed by reversing the direction of integration over C_2 (which is equivalent to changing the sign of $d\mathbf{r}$). The effect is to make the integration over C_1 and C_2 an integration over a closed path C, shown in Fig. 2.7*b*. The equation then takes the form

$$\int_{C_1} \mathbf{F}\cdot d\mathbf{r} - \int_{C_2} \mathbf{F}\cdot d\mathbf{r} = \int_{C_1} \mathbf{F}\cdot d\mathbf{r} + \int_{-C_2} \mathbf{F}\cdot d\mathbf{r} = \oint_C \mathbf{F}\cdot d\mathbf{r} = 0, \tag{11}$$

which shows that the integral around every closed path vanishes. Moreover, the arguments that led to (10) imply that $\mathbf{F}$ is a conservative field, that is, $\mathbf{F} = \nabla\varphi$. Therefore, we conclude that (10) and (11) are equivalent. If $\oint_C \mathbf{F}\cdot d\mathbf{r} = 0$ for every closed curve C in a region, then $\mathbf{F} = \nabla\varphi$ and conversely.

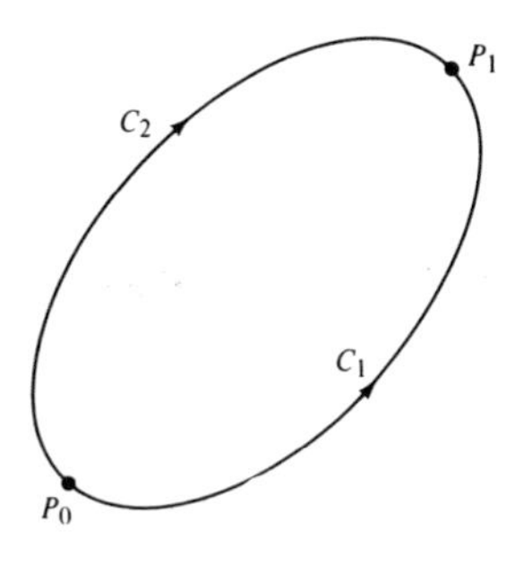

(*a*)

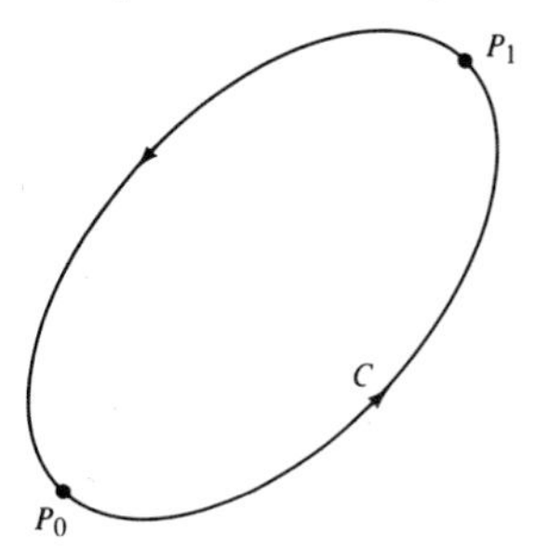

(*b*)

FIGURE 2.7

6

A related question of practical importance concerns the determination of the potential φ when a vector field $\mathbf{F}$ is known to be conservative. The equivalent mathematical problem is to find the solution, if one exists, of the scalar equations

$$\frac{\partial \varphi}{\partial x} = F_1, \qquad \frac{\partial \varphi}{\partial y} = F_2, \qquad \frac{\partial \varphi}{\partial z} = F_3, \tag{12}$$

where the functions F_1, F_2, and F_3 are prescribed. A complete answer to this question will be given in Sec. 8.4; at present any particular problem is best handled by direct integration. (It should be noted that it is a practical impossibility to verify $\oint_C \mathbf{F} \cdot d\mathbf{r} = 0$ for *every* closed path C.) Two examples illustrate the procedure.

Example 4 If $\mathbf{F} = y\mathbf{i} - z\mathbf{j} + x\mathbf{k}$, is there a potential φ such that $\mathbf{F} = \nabla\varphi$?

Let us assume that the function φ exists, in which case

$$\frac{\partial \varphi}{\partial x} = y, \qquad \frac{\partial \varphi}{\partial y} = -z, \qquad \frac{\partial \varphi}{\partial z} = x. \tag{13}$$

Integration of the first equation of this set, with y and z held constant, gives

$$\varphi = xy + g(y, z),$$

where g is an arbitrary function of y and z. Partial differentiation of this expression with respect to y yields

$$\frac{\partial \varphi}{\partial y} = x + \frac{\partial g}{\partial y},$$

and since, from (13),

$$\frac{\partial \varphi}{\partial y} = -z,$$

it follows that

$$x = -z - \frac{\partial}{\partial y} g(y, z).$$

But this equation implies that x is a function of y and z and contradicts the fact that x, y, and z are independent variables. We conclude that there is *no* potential φ for this particular vector field $\mathbf{F}$. [A similar contradiction would be reached by starting with any one of the equations in (13).]

Example 5 Find the potential φ so that

$$\nabla\varphi = \mathbf{F} = (y^2 + 2xz^2 - 1)\mathbf{i} + 2xy\mathbf{j} + (2x^2z + z^3)\mathbf{k}.$$

As in Example 4, we assume there is a function φ and proceed to integrate the equations. The gradient relation implies the three scalar equations

$$\frac{\partial \varphi}{\partial x} = y^2 + 2xz^2 - 1, \tag{14}$$

$$\frac{\partial \varphi}{\partial y} = 2xy, \tag{15}$$

$$\frac{\partial \varphi}{\partial z} = 2x^2 z + z^3. \tag{16}$$

We obtain from the integration of (14)

$$\varphi = y^2 x + x^2 z^2 - x + g(y, z), \tag{17}$$

where g is an arbitrary function of y and z. Using this to calculate $\partial\varphi/\partial y$, we can compare the result with (15), so that

$$2xy + \frac{\partial g}{\partial y} = 2xy$$

or

$$\frac{\partial g}{\partial y} = 0.$$

Integration gives

$$g = G(z),$$

which shows that g is really an arbitrary function of z only. With this information (17) can now be written

$$\varphi = y^2 x + x^2 z^2 - x + G(z). \tag{18}$$

Finally, $\partial\varphi/\partial z$ is calculated from this and compared with formula (16), to show that

$$2x^2 z + G'(z) = 2x^2 z + z^3$$

or

$$G'(z) = z^3.$$

Integration yields

$$G(z) = \tfrac{1}{4}z^4 + C,$$

where C is an arbitrary constant, and substitution for G in (18) gives an explicit formula for the potential,

$$\varphi = y^2 x + x^2 z^2 - x + \tfrac{1}{4}z^4 + C.$$

The existence of a potential function shows that the force field in this problem is conservative.

It should be emphasized that conservative vector fields are the exception rather than the rule because the existence of a potential is a rather severe restriction. However, in many physical problems the vector fields are indeed conservative, and this fact greatly facilitates their analysis, mainly because a knowledgė of the potential avoids any necessity of evaluating line integrals by virtue of the formula

$$\int_{P_0}^{P_1} \mathbf{F} \cdot d\mathbf{r} = \varphi(P_1) - \varphi(P_0).$$

Exercises 8.2

1. For the vector field $\mathbf{F} = (2x + y)\mathbf{i} + (3x - 2y)\mathbf{j}$, evaluate the line integral $\int_c \mathbf{F} \cdot d\mathbf{r}$ along the following curves C from (0, 0) to (1, 1).

(i) $y = x$ **(ii)** $y = x^2$ **(iii)** $y = \sin(\pi x/2)$

2. When $\mathbf{F} = xy\mathbf{i} - y\mathbf{j} + z\mathbf{k}$, calculate $\int_{P_0}^{P_1} \mathbf{F} \cdot d\mathbf{r}$ along the straight-line path from P_0 to P_1 in the following cases:

(i) $P_0 = (0, 0, 0)$, $P_1 = (1, 0, 0)$ **(ii)** $P_0 = (1, 0, 0)$, $P_1 = (1, 2, 0)$ **(iii)** $P_0 = (1, 2, 0)$, $P_1 = (1, 2, 3)$

(iv) $P_0 = (0, 0, 0)$, $P_1 = (1, 2, 3)$ **(v)** $P_0 = (1, 0, 0)$, $P_1 = (1, 2, 3)$ **(vi)** $P_0 = (0, 0, 0)$, $P_1 = (1, 2, 0)$

Can you deduce answers (iv) to (vi) knowing answers (i) to (iii)? Explain.

3. Evaluate $\int_c (x + 2y)\, dx + xy\, dy$ along the following curves from $(-1, 0)$ to $(2, 1)$.

(i) $y = (x + 1)/3$ **(ii)** $y = (x + 1)^2/9$ **(iii)** $y = (x + 1)^3/27$

4. Evaluate the following closed line integrals where C is the circle $x^2 + y^2 = a^2$ taken in the positive (counter-clockwise) sense.

(i) $\oint_c y\, dx + x\, dy$ **(ii)** $\oint_c y\, dx - x\, dy$ **(iii)** $\oint_c \dfrac{x\, dy - y\, dx}{x^2 + y^2}$

5. A particle is moved counterclockwise around the square $0 \leq x \leq 1$, $0 \leq y \leq 1$, $z = 0$ under the action of the force field $\mathbf{F} = (x - y^2)\mathbf{i} + (2y + x^2)\mathbf{j} + x\mathbf{k}$. Calculate the work done.

6. Evaluate $\int_c \mathbf{F} \cdot d\mathbf{r}$ for the following vector fields $\mathbf{F}(x, y, z)$ where C is the curve $y = x^2$, $z = y^2$ from (0, 0, 0) to (2, 4, 16).

(i) $\mathbf{F} = x\mathbf{i} + y\mathbf{j} + z\mathbf{k}$ **(ii)** $x^2\mathbf{i} + y^2\mathbf{j} + z^2\mathbf{k}$ **(iii)** $yz\mathbf{i} + zx\mathbf{j} + xy\mathbf{k}$

7. Explain why the following integrals along any curve C from P_0 to P_1 are independent of the path of integration.

(i) $\int_c f(x)dx + g(y)\, dy$ **(ii)** $\int_c f(xy)(y\, dx + x\, dy)$ **(iii)** $\int_c f(x^2 + y^2)(x\, dx + y\, dy)$.

8. Show that $\mathbf{F} = y\mathbf{i} + x\mathbf{j} + y\mathbf{k}$ is not a conservative field. Nevertheless, there are certain paths C for which $\oint_c \mathbf{F} \cdot d\mathbf{r} = 0$. Can you find one?

9. When possible, find potential functions $\varphi(x, y, z)$ corresponding to the following vector fields $\mathbf{F}(x, y, z)$. In each case, evaluate $\oint_c \mathbf{F} \cdot d\mathbf{r}$ where C is the intersection of the cylinder $y = x^2$ and the plane $y = z$ from (0, 0, 0) to (1, 1, 1).

(i) $x\mathbf{i} + y\mathbf{j} + z^2\mathbf{k}$ (ii) $y\mathbf{i} + x\mathbf{j} + xy\mathbf{k}$ (iii) $(y + z^2)\mathbf{i} + (yz + zx)\mathbf{k}$
(iv) $y\mathbf{i} + 2z\mathbf{j} + xy\mathbf{k}$ (v) $e^x\mathbf{i} + \tan y\mathbf{j} + \log z\mathbf{k}$ (vi) $e^x[(y + z)\mathbf{i} + \mathbf{j} + \mathbf{k}]$

10. A particle is moved from the origin to the point (a, b, c) in a force field $\mathbf{F} = (x + y)\mathbf{i} + (x - z)\mathbf{j} + (z - y)\mathbf{k}$. Show that the work done depends only on a, b and c, and find this value.

11. A particle moves from the point (0, 0) to the point (1, 0) along the curve $y = \alpha x(1 - x)$ in the force field $(y^2 + 1)\mathbf{i} + (x + y)\mathbf{j}$. Find α so that the work done is a minimum.

12. A frictional force is constant in magnitude and always acts in a direction which opposes any motion. Verify that the work done against friction in moving a body from one location to another is proportional to the length of the path. (*Hint:* $\mathbf{F} \cdot d\mathbf{r} = \mathbf{F} \cdot (d\mathbf{r}/ds)\, ds$.)

13. The gravitational force acting on a mass m at height z above the surface of the earth is $\mathbf{F}(x, y, z) = -mgr_0^2/(r_0 + z)^2\mathbf{k}$ where r_0 is the radius of the earth. Verify that this is a conservative force and determine the potential function.

14. If φ and ψ are scalar functions and C is any closed path, deduce that

$$\oint_C \varphi\nabla\psi \cdot d\mathbf{r} + \oint_C \psi\nabla\varphi \cdot d\mathbf{r} = 0.$$

15. Evaluate $\int_C \mathbf{F} \times d\mathbf{r}$ for $\mathbf{F} = (x + y)\mathbf{i} + (y + z)\mathbf{j} + (z + x)\mathbf{k}$ along the following curves joining (0, 0, 0) and (1, 1, 1).
(i) $x = y = z$ (ii) $x = y$, $z = x^2$ (iii) $y = x^2$, $z = x^3$

16. Find a geometrical interpretation for $\int_{P_0}^{P_1} \mathbf{r} \times d\mathbf{r}$ where the path lies in a plane containing the origin. (*Hint:* Interpret $|\mathbf{r} \times d\mathbf{r}|$ as an area.)

17. Calculate $\oint_C \mathbf{F} \cdot d\mathbf{r}$, $\oint_C \mathbf{F} \times d\mathbf{r}$ and $\oint_C \mathbf{F} \times \mathbf{r} \cdot d\mathbf{r}$ when $\mathbf{F} = (y - z)\mathbf{i} + (z - x)\mathbf{j} + (x - y)\mathbf{k}$ and C is the circle $x^2 + y^2 = 1$, $z = 2$.

18. Define the *vector line integral*

$$\int_C f(x, y, z)\, d\mathbf{r} = \mathbf{i} \int_C f(x, y, z)\, dx + \mathbf{j} \int_C f(x, y, z)\, dy + \mathbf{k} \int_C f(x, y, z)\, dz.$$

Evaluate $\int_C f\, d\mathbf{r}$ for the following scalar functions $f(x, y, z)$ along the straight line path from $P_0 = (1, -1, 1)$ to $P_1 = (2, 3, 4)$.
(i) xy^2z^2 (ii) $x^2 + y^2 + z^2$ (iii) e^{x+y}

19. Define the *vector volume integral* $\iiint_V \mathbf{F}\, dV = \mathbf{i} \iiint_V F_1\, dV + \mathbf{j} \iiint_V F_2\, dV + \mathbf{k} \iiint_V F_3\, dV$. Evaluate $\iiint_V \mathbf{F}\, dV$ for the following vector functions $\mathbf{F}(x, y, z)$ where V is the rectangular volume $0 \le x \le a$, $0 \le y \le b$, $0 \le z \le c$.
(i) $x^2\mathbf{i} + y^2\mathbf{j} + z^2\mathbf{k}$ (ii) $(x + y)\mathbf{i} + (x + z)\mathbf{k}$ (iii) $yz\mathbf{i} + zx\mathbf{j} + xy\mathbf{k}$

8.3 Vector operators

1 There are two basic types of vector integrals. One is the closed line integral, $\oint_C \mathbf{F} \cdot d\mathbf{r}$, studied in the last section. The other is the surface integral

$$\iint_S \mathbf{F} \cdot \hat{\mathbf{n}}\, dS, \tag{1}$$

where $\hat{\mathbf{n}}$ is the outward unit normal to the completely closed surface S which contains a finite volume of space, as shown in Fig. 3.1. Sometimes the vector element of surface area, $d\mathbf{S} = \hat{\mathbf{n}}\, dS$, is used to contract the notation in (1) to

$$\iint_S \mathbf{F} \cdot d\mathbf{S}.$$

In order to emphasize that S is a closed surface, the integral will be written

$$\oiint_S \mathbf{F} \cdot \hat{\mathbf{n}}\, dS = \oiint_S \mathbf{F} \cdot d\mathbf{S}.$$

The surface integral (1) is called the *flux* of the vector field $\mathbf{F}$ through surface S. The terminology derives from its original application and interpretation regarding the flow of a fluid. Since fluid phenomena are part of everyday life, it is quite natural, and very productive, to describe, visualize, and analogize other physical processes in such familiar terms. For this reason, the words flow, flux, streamline, source, sink, dipole, etc., are used almost universally in the study of electric, magnetic, and gravitational fields, current and charge distributions, and all diffusive processes governing heat, temperature, and salinity, to name only a few.

Consider, then, the flux or flow of a fluid through an arbitrary closed surface S, which is imagined fixed in space. Let $\mathbf{v}$ denote the fluid velocity and ρ be the fluid density, both of which may vary in space and time, i.e.,

$$\mathbf{v} = \mathbf{v}(x, y, z, t), \qquad \rho = \rho(x, y, z, t). \tag{2}$$

The flux through the surface is determined by examining first the outward flow through an element dS, as shown in Fig. 3.2. In a time increment, dt, the fluid that passes through dS with normal velocity $\mathbf{v} \cdot \hat{\mathbf{n}}$ fills a small cylinder of length

$$dl = \mathbf{v} \cdot \hat{\mathbf{n}}\, dt.$$

The volume dV of fluid in this cylindrical element is the product of area dS and length dl, or

$$dV = \mathbf{v} \cdot \hat{\mathbf{n}}\, dS\, dt.$$

Therefore, the mass increment dm which flows *out* through dS *in a time dt* is given by

$$dm = \rho\, dV = \rho(\mathbf{v} \cdot \hat{\mathbf{n}})\, dS\, dt.$$

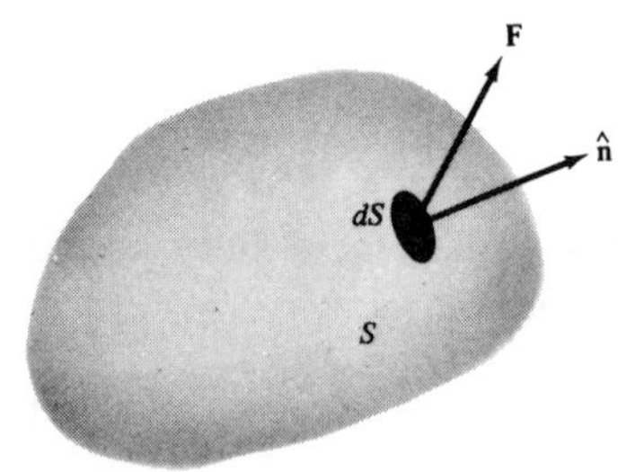

FIGURE 3.1

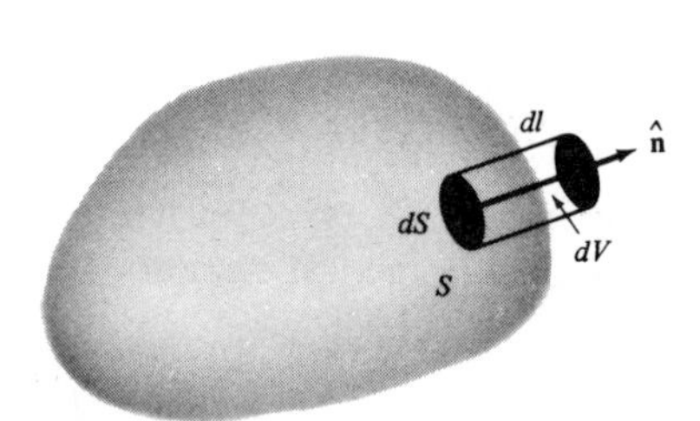

FIGURE 3.2

The sum of all the mass increments dm, one for each surface element dS, yields the total mass increment dM that flows out of the entire closed surface *in the infinitesimal time span dt*. (Note that the increments dm are in a sense doubly small because even when added together, they still constitute only a mass increment dM.) The summation or surface integral then yields

$$dM = dt \oiint_S \rho(\mathbf{v} \cdot \hat{\mathbf{n}})\, dS, \tag{3}$$

so that the *rate* at which mass passes out through S is

$$\frac{dM}{dt} = \oiint_S \rho(\mathbf{v} \cdot \hat{\mathbf{n}})\, dS. \tag{4}$$

This surface integral describes the flux of the mass transport vector $\rho\mathbf{v}$ through S.

In general, the flux of a vector $\mathbf{F}$ given by (1) describes the "flow" of the field. This flux is positive when the contribution to the surface integral from the regions where $\mathbf{F} \cdot \hat{\mathbf{n}}$ is positive exceeds the contribution from regions of negative $\mathbf{F} \cdot \hat{\mathbf{n}}$. Simply put, the flux is computed by totalling the field lines coming out of S and subtracting from this the number of lines which enter the volume. A quantitative measure of the number of lines threading through dS is $\mathbf{F} \cdot \hat{\mathbf{n}}\, dS$.

The calculation of the flux is an exercise in surface integration (see Sec. 7.4); only one example will be considered.

Example 1 If $\mathbf{F} = x^3\mathbf{i} + y^3\mathbf{j} + z^3\mathbf{k}$, calculate $I = \oiint_S \mathbf{F} \cdot \hat{\mathbf{n}}\, dS$, where S is the sphere $x^2 + y^2 + z^2 = a^2$, shown in Fig. 3.3.

The unit normal for the sphere is

$$\hat{\mathbf{n}} = \frac{1}{a}(x\mathbf{i} + y\mathbf{j} + z\mathbf{k}),$$

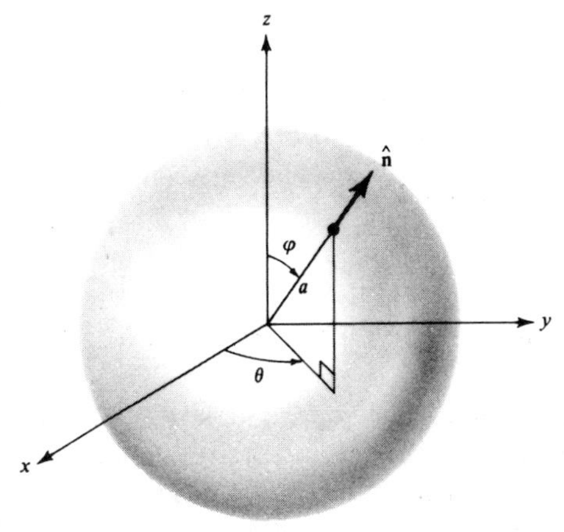

FIGURE 3.3

and it is straightforward to show that

$$\mathbf{F} \cdot \hat{\mathbf{n}} = \frac{x^4 + y^4 + z^4}{a}.$$

The flux is obtained by the evaluation of the surface integral

$$I = \oiint_S \frac{x^4 + y^4 + z^4}{a} \, dS,$$

over $x^2 + y^2 + z^2 = a^2$.

The complete symmetry of the configuration implies that

$$\oiint_S x^4 \, dS = \oiint_S y^4 \, dS = \oiint_S z^4 \, dS,$$

and it follows that

$$I = \frac{3}{a} \oiint_S z^4 \, dS.$$

Spherical coordinates provide a simple means of calculating this integral. Since, $dS = a^2 \sin \varphi \, d\varphi \, d\theta$ and $z = a \cos \varphi$, the expression for I becomes

$$I = \frac{3}{a} \int_0^{2\pi} d\theta \int_0^{\pi} (a^4 \cos^4 \varphi) a^2 \sin \varphi \, d\varphi.$$

Integration yields

$$I = 3a^5 [\theta]_0^{2\pi} [-\tfrac{1}{5} \cos^5 \varphi]_0^{\pi} = \frac{12\pi}{5} a^5.$$

2

The examination of the local behavior of a vector field leads to formulas that are of fundamental physical importance. Moreover, such a study anticipates the principal integral theorems of vector calculus. The concepts of line integration and flux are the main instruments in this analysis.

Any smooth vector field $\mathbf{F}(x, y, z)$ can be represented by its Taylor series in the neighborhood of a point (which is taken to be the origin for convenience):

$$\mathbf{F}(x, y, z) = \mathbf{A}_0 + \mathbf{A}_1 x + \mathbf{A}_2 y + \mathbf{A}_3 z + \cdots, \tag{5}$$

where

$$\mathbf{A}_0 = \mathbf{F}(0, 0, 0), \quad \mathbf{A}_1 = \frac{\partial}{\partial x} \mathbf{F}(0, 0, 0), \quad \mathbf{A}_2 = \frac{\partial}{\partial y} \mathbf{F}(0, 0, 0), \quad \mathbf{A}_3 = \frac{\partial}{\partial z} \mathbf{F}(0, 0, 0). \tag{6}$$

The higher-order terms of the Taylor series indicated by the dots *will be neglected* because our concern is with local behavior, and this means that the analysis can be restricted to a very (vanishingly) small region which encloses the origin.

First the flux of $\mathbf{F}$ over a small surface enclosing the origin is calculated. For simplicity, this surface is taken to be a sphere of radius a centered at the origin (Fig. 3.4). The unit normal to the sphere is

$$\hat{\mathbf{n}} = \frac{x\mathbf{i} + y\mathbf{j} + z\mathbf{k}}{a},$$

and this formula and approximation (5) together imply that

$$\oiint_S \mathbf{F} \cdot \hat{\mathbf{n}}\, dS = \oiint_S (\mathbf{A}_0 + \mathbf{A}_1 x + \mathbf{A}_2 y + \mathbf{A}_3 z) \cdot \frac{x\mathbf{i} + y\mathbf{j} + z\mathbf{k}}{a}\, dS, \tag{7}$$

where the integration is to be taken over the closed spherical surface.

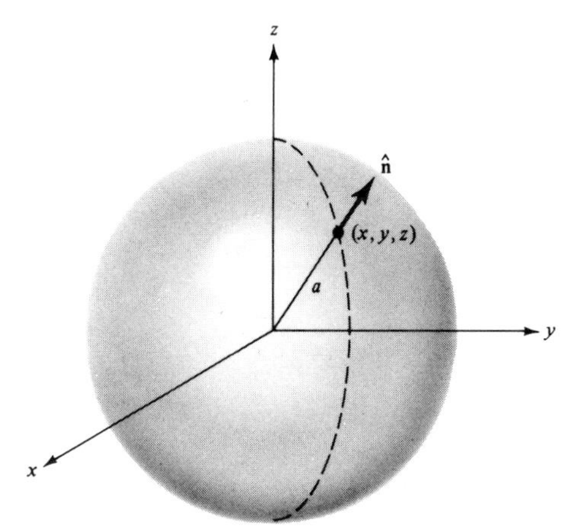

FIGURE 3.4

Inspection of (7) shows that the problem reduces to the evaluation of a series of scalar surface integrals, all but three of which are zero. Expressions of the form $\oiint_S x\, dS$ and $\oiint_S yz\, dS$ vanish because the symmetry of the spherical surface means an exact cancellation of positive and negative contributions. Symmetry considerations also imply that the nonzero integrals, $\iint_S x^2\, dS$, $\iint_S y^2\, dS$, and $\iint_S z^2\, dS$, are all equal. The result of incorporating these simplifications in Eq. (7) is

$$\oiint_S \mathbf{F} \cdot \hat{\mathbf{n}}\, dS = (\mathbf{A}_1 \cdot \mathbf{i} + \mathbf{A}_2 \cdot \mathbf{j} + \mathbf{A}_3 \cdot \mathbf{k}) \oiint_S \frac{z^2}{a}\, dS. \tag{8}$$

The remaining integral, $\oiint_S z^2/a\, dS$, is readily calculated using spherical coordinates, in which case:

$$\begin{aligned}
\oiint_S \frac{z^2}{a}\, dS &= \int_0^{2\pi} d\theta \int_0^{\pi} (a \cos^2 \varphi) a^2 \sin \varphi\, d\varphi, \\
&= [\theta]_0^{2\pi} \left[-\frac{a^3}{3} \cos^3 \varphi \right]_0^{\pi}, \\
&= \tfrac{4}{3}\pi a^3.
\end{aligned}$$

Substitution of this result into (8) yields

$$\oiint_S \mathbf{F} \cdot \hat{\mathbf{n}}\, dS = \tfrac{4}{3}\pi a^3(\mathbf{A}_1 \cdot \mathbf{i} + \mathbf{A}_2 \cdot \mathbf{j} + \mathbf{A}_3 \cdot \mathbf{k}) = V(\mathbf{A}_1 \cdot \mathbf{i} + \mathbf{A}_2 \cdot \mathbf{j} + \mathbf{A}_3 \cdot \mathbf{k}),$$

since $V = \tfrac{4}{3}\pi a^3$ is the volume of the sphere enclosed by S. Therefore, the *flux per unit volume* through the small sphere about the origin is

$$\frac{1}{V}\oiint_S \mathbf{F} \cdot \hat{\mathbf{n}}\, dS = \mathbf{A}_1 \cdot \mathbf{i} + \mathbf{A}_2 \cdot \mathbf{j} + \mathbf{A}_3 \cdot \mathbf{k}. \tag{9}$$

Although this equation is exact only for vector fields that are linear in x, y, and z, it can be shown that the higher-order terms in the Taylor series do not alter this result in the limit as the radius of the sphere tends to zero. In fact, the surface S need not even be spherical, and the most general form of the flux equation (9) is

$$\lim_{V\to 0}\left(\frac{1}{V}\oiint_S \mathbf{F} \cdot \hat{\mathbf{n}}\, dS\right) = \mathbf{A}_1 \cdot \mathbf{i} + \mathbf{A}_2 \cdot \mathbf{j} + \mathbf{A}_3 \cdot \mathbf{k}. \tag{10}$$

This limiting flux per unit volume is called the *divergence* of the vector field at the point concerned:

$$\operatorname{div} \mathbf{F} = \lim_{V\to 0}\left(\frac{1}{V}\oiint_S \mathbf{F} \cdot \hat{\mathbf{n}}\, dS\right). \tag{11}$$

Accordingly,

$$\operatorname{div} \mathbf{F} = \mathbf{A}_1 \cdot \mathbf{i} + \mathbf{A}_2 \cdot \mathbf{j} + \mathbf{A}_3 \cdot \mathbf{k},$$

and since by Eq. (6), $\mathbf{A}_1 = \partial\mathbf{F}/\partial x$, $\mathbf{A}_2 = \partial\mathbf{F}/\partial y$, and $\mathbf{A}_3 = \partial\mathbf{F}/\partial z$, each evaluated at the origin, it follows that

$$\operatorname{div} \mathbf{F} = \mathbf{i} \cdot \frac{\partial \mathbf{F}}{\partial x} + \mathbf{j} \cdot \frac{\partial \mathbf{F}}{\partial y} + \mathbf{k} \cdot \frac{\partial \mathbf{F}}{\partial z}. \tag{12}$$

However, $\mathbf{i} \cdot \mathbf{F} = F_1$, $\mathbf{j} \cdot \mathbf{F} = F_2$, and $\mathbf{k} \cdot \mathbf{F} = F_3$, so that (12) can be written

$$\operatorname{div} \mathbf{F} = \frac{\partial F_1}{\partial x} + \frac{\partial F_2}{\partial y} + \frac{\partial F_3}{\partial z}.$$

There is absolutely nothing special about the origin in this derivation: that point was chosen solely for convenience of calculation. Therefore, Eq. (10) really defines the divergence at any point (which is selected by the limit $V \to 0$), and its component form is generally valid:

$$\operatorname{div} \mathbf{F} = \frac{\partial F_1}{\partial x} + \frac{\partial F_2}{\partial y} + \frac{\partial F_3}{\partial z}. \tag{13}$$

A useful alternative notation for div **F** follows by expressing it as the scalar product of the vector *operator*

$$\nabla = \mathbf{i}\frac{\partial}{\partial x} + \mathbf{j}\frac{\partial}{\partial y} + \mathbf{k}\frac{\partial}{\partial z}$$

and the vector **F**. Since

$$\nabla \cdot \mathbf{F} = \left(\mathbf{i}\frac{\partial}{\partial x} + \mathbf{j}\frac{\partial}{\partial y} + \mathbf{k}\frac{\partial}{\partial z}\right) \cdot \mathbf{F} = \mathbf{i} \cdot \frac{\partial \mathbf{F}}{\partial x} + \mathbf{j} \cdot \frac{\partial F}{\partial y} + \mathbf{k} \cdot \frac{\partial \mathbf{F}}{\partial z},$$

$$= \frac{\partial}{\partial x}(\mathbf{i} \cdot \mathbf{F}) + \frac{\partial}{\partial y}(\mathbf{j} \cdot \mathbf{F}) + \frac{\partial}{\partial z}(\mathbf{k} \cdot \mathbf{F}),$$

it follows that

$$\text{div } \mathbf{F} = \nabla \cdot \mathbf{F} = \frac{\partial F_1}{\partial x} + \frac{\partial F_2}{\partial y} + \frac{\partial F_3}{\partial z}.$$

(The term $\nabla \cdot \mathbf{F}$ is read "del dot **F**" or simply as "div **F**.")

If x_1, x_2, x_3 are the three independent cartesian coordinates corresponding to unit vectors $\mathbf{i}_1, \mathbf{i}_2$, and $\mathbf{i}_3$, Eq. (12) can be written

$$\text{div } \mathbf{F} = \sum_{n=1}^{3} \mathbf{i}_n \cdot \frac{\partial}{\partial x_n} \mathbf{F}. \tag{14}$$

Here x_1, x_2, x_3 are identified as x, y, z, and $\mathbf{i}_1, \mathbf{i}_2, \mathbf{i}_3$ replace **i**, **j**, **k**.

The divergence of a vector is interpreted as a scalar measure of the local flux of the field emanating from any small region of space. Wherever div **F** is positive, the field lines tend to "diverge" from each other; where they converge, div **F** is negative, as shown in Fig. 3.5. The divergence is calculated by routine differentiation, as illustrated in the next example.

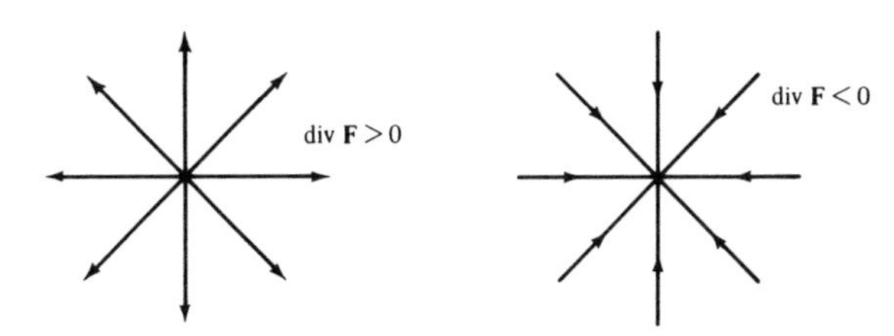

FIGURE 3.5

Example 2 Calculate div **F** when (i) $\mathbf{F} = x\mathbf{i} + y\mathbf{j} + z\mathbf{k}$, (ii) $\mathbf{F} = x^2y\mathbf{i} + y^2z\mathbf{j} + z^2x\mathbf{k}$, and (iii) $\mathbf{F} = yz\mathbf{i} + 2zx\mathbf{j} + 3xy\mathbf{k}$.

The use of formula (13) gives the following results:

(i) $\operatorname{div} \mathbf{F} = \frac{\partial x}{\partial x} + \frac{\partial y}{\partial y} + \frac{\partial z}{\partial z} = 3;$

(ii) $\operatorname{div} \mathbf{F} = \frac{\partial}{\partial x}(x^2 y) + \frac{\partial}{\partial y}(y^2 z) + \frac{\partial}{\partial z}(z^2 x) = 2(xy + yz + zx);$

(iii) $\operatorname{div} \mathbf{F} = \frac{\partial}{\partial x}(yz) + \frac{\partial}{\partial y}(2zx) + \frac{\partial}{\partial z}(3xy) = 0.$

3

Just as the local-flux calculation leads to the concept of the divergence of a vector, line integration gives rise to another quantity which describes an important characteristic of a vector field. It is called the *curl* of a vector, written curl **F**, and is itself a vector which relates to the local curling, twisting, or rotation of a vector field—very much as the angular velocity $\mathbf{\Omega}$ describes the rotation of a solid body.

We begin by calculating the value of $\oint_C \mathbf{F} \cdot d\mathbf{r}$, where C is a small circle centered at the origin whose radius will eventually approach zero. Vectors are completely described by their three cartesian components, and with this in mind we choose C consecutively to be the circles C_1: $y^2 + z^2 = a^2$, $x = 0$; C_2: $z^2 + x^2 = a^2$, $y = 0$; and C_3: $x^2 + y^2 = a^2$, $z = 0$, as shown in Fig. 3.6. In each case, the path direction is chosen in accordance with the convention governing right-handed coordinate systems. Of course, only one calculation need be made, say $\oint_{C_3} \mathbf{F} \cdot d\mathbf{r}$, because the other two can be obtained by a simple notational exchange.

Since the limit process, $a \to 0$, will be invoked at some stage, it is sufficient to describe the vector field $\mathbf{F}(x, y, z)$ by the linear terms of its Taylor series expansion about the origin,

$$\mathbf{F}(x, y, z) = \mathbf{A}_0 + \mathbf{A}_1 x + \mathbf{A}_2 y + \mathbf{A}_3 z, \tag{15}$$

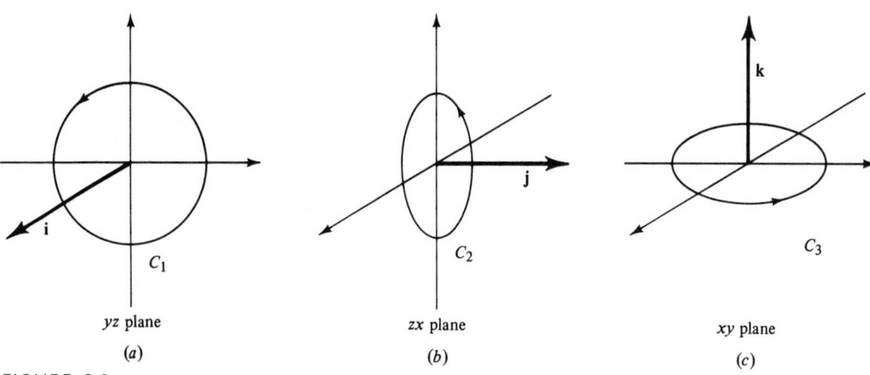

FIGURE 3.6

where the coefficients are defined in (6). A suitable parametric description of the circle C_3, for which $z = 0$, is

$$\mathbf{r} = a \cos \theta \, \mathbf{i} + a \sin \theta \, \mathbf{j},$$

in which case

$$d\mathbf{r} = (-a \sin \theta \, \mathbf{i} + a \cos \theta \, \mathbf{j}) \, d\theta.$$

Therefore

$$\mathbf{F} \cdot d\mathbf{r} = (\mathbf{A}_0 + \mathbf{A}_1 \, a \cos \theta + \mathbf{A}_2 \, a \sin \theta) \cdot (-a \sin \theta \, \mathbf{i} + a \cos \theta \, \mathbf{j}) \, d\theta,$$

and since θ ranges from 0 to 2π in one circuit about C_3, it follows that

$$\oint_{C_3} \mathbf{F} \cdot d\mathbf{r} = \int_0^{2\pi} (\mathbf{A}_0 + \mathbf{A}_1 a \cos \theta + \mathbf{A}_2 \, a \sin \theta) \cdot (-a \sin \theta \, \mathbf{i} + a \cos \theta \, \mathbf{j}) \, d\theta. \quad (16)$$

The integrals involved are all elementary:

$$\int_0^{2\pi} \cos \theta \, d\theta = \int_0^{2\pi} \sin \theta \, d\theta = \int_0^{2\pi} \sin \theta \cos \theta \, d\theta = 0;$$

$$\int_0^{2\pi} \cos^2 \theta \, d\theta = \int_0^{2\pi} \sin^2 \theta \, d\theta = \pi;$$

and these results allow the reduction of (16) to

$$\oint_{C_3} \mathbf{F} \cdot d\mathbf{r} = \pi a^2 (\mathbf{A}_1 \cdot \mathbf{j} - \mathbf{A}_2 \cdot \mathbf{i}) = \pi a^2 \left(\frac{\partial F_2}{\partial x} - \frac{\partial F_1}{\partial y} \right).$$

We choose to write this as

$$\frac{1}{S} \oint_{C_3} \mathbf{F} \cdot d\mathbf{r} = \mathbf{k} \cdot \mathbf{G}, \quad (17)$$

where $S = \pi a^2$ is the area of circle C_3 and by definition the z component of the vector $\mathbf{G}$ is

$$\mathbf{k} \cdot \mathbf{G} = \frac{\partial F_2}{\partial x} - \frac{\partial F_1}{\partial y}.$$

In the limit $a \to 0$ or $S \to 0$, formula (17) is an exact relationship for an arbitrary vector field *at any point*, so that a vector $\mathbf{G}$ at (x, y, z), or at least its third component, is uniquely given by

$$\mathbf{k} \cdot \mathbf{G} = \frac{\partial F_2}{\partial x} - \frac{\partial F_1}{\partial y} = \lim_{S \to 0} \frac{1}{S} \oint_{C_3} \mathbf{F} \cdot d\mathbf{r}. \quad (18)$$

Clearly, line integrals about C_1 and C_2 must lead to the analogous formulas which can be obtained from (18) by a cyclical permutation of subscripts and variables:

$$\begin{aligned} \mathbf{i} \cdot \mathbf{G} &= \left(\frac{\partial F_3}{\partial y} - \frac{\partial F_2}{\partial z} \right) = \lim_{S \to 0} \frac{1}{S} \oint_{C_1} \mathbf{F} \cdot d\mathbf{r}, \\ \mathbf{j} \cdot \mathbf{G} &= \left(\frac{\partial F_1}{\partial z} - \frac{\partial F_3}{\partial x} \right) = \lim_{S \to 0} \frac{1}{S} \oint_{C_2} \mathbf{F} \cdot d\mathbf{r}. \end{aligned} \quad (19)$$

We now have sufficient information to identify the vector **G** explicitly:

$$\mathbf{G} = \left(\frac{\partial F_3}{\partial y} - \frac{\partial F_2}{\partial z}\right)\mathbf{i} + \left(\frac{\partial F_1}{\partial z} - \frac{\partial F_3}{\partial x}\right)\mathbf{j} + \left(\frac{\partial F_2}{\partial x} - \frac{\partial F_1}{\partial y}\right)\mathbf{k}.$$

This is the vector we seek, and it defines curl **F**; that is,

$$\text{curl } \mathbf{F} = \left(\frac{\partial F_3}{\partial y} - \frac{\partial F_2}{\partial z}\right)\mathbf{i} + \left(\frac{\partial F_1}{\partial z} - \frac{\partial F_3}{\partial x}\right)\mathbf{j} + \left(\frac{\partial F_2}{\partial x} - \frac{\partial F_1}{\partial y}\right)\mathbf{k}. \tag{20}$$

Moreover, the limiting integrals in (18) and (19) are special cases of the more general formula

$$\lim_{S \to 0} \frac{1}{S} \oint_C \mathbf{F} \cdot d\mathbf{r} = \hat{\mathbf{n}} \cdot \text{curl } \mathbf{F}, \tag{21}$$

where C is a plane circle or any plane closed curve of area S whose unit normal in the positive direction is $\hat{\mathbf{n}}$.

4

A physical interpretation for curl **F** can be obtained from Eq. (18), where **F** is taken to represent a velocity field. In fact, to emphasize this we set $\mathbf{F} = \mathbf{v}$. Consider then the expression

$$\oint_{C_3} \mathbf{v} \cdot d\mathbf{r} = \int_0^{2\pi a} \mathbf{v} \cdot \hat{\mathbf{T}} \, ds = \int_0^{2\pi a} v_T \, ds, \tag{22}$$

where $\hat{\mathbf{T}}$ is the unit tangent vector, s the arc length of the circle, and v_T the tangential component of the velocity (see Fig. 3.7). Application of the integral mean-value theorem, $\int_a^b f(x)\,dx = (b-a)f(\xi)$, enables us to write

$$\int_0^{2\pi a} v_T \, ds = 2\pi a \bar{v}_T,$$

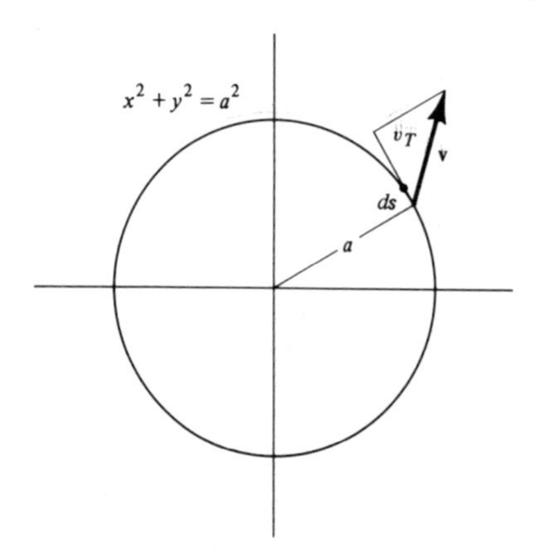

FIGURE 3.7

where $\bar{v}_T$ is some average tangential velocity on the perimeter. We associate an angular velocity about axis Oz with this circular motion by the relationship

$$\Omega_3 = \frac{\bar{v}_T}{a},$$

so that $\int_0^{2\pi a} v_T\, ds = 2\pi a^2 \Omega_3$ or, by (22),

$$\oint_{C_3} \mathbf{v} \cdot d\mathbf{r} = 2\pi a^2 \Omega_3.$$

In the light of this result and the definition for $\mathbf{k} \cdot \text{curl}\, \mathbf{v}$ given in (17) we obtain

$$\frac{1}{\pi a^2} 2\pi a^2 \Omega_3 = \mathbf{k} \cdot \text{curl}\, \mathbf{v}$$

or

$$\Omega_3 = \tfrac{1}{2}\mathbf{k} \cdot \text{curl}\, \mathbf{v}.$$

In other words, $\mathbf{k} \cdot \text{curl}\, \mathbf{v}$ (and hence the z component of curl $\mathbf{F}$) is actually equal to twice the local angular velocity produced by the velocity field $\mathbf{v}$ about the z axis. Since any axis can be chosen instead of the z axis, it follows that

$$\boldsymbol{\Omega} = \tfrac{1}{2}\, \text{curl}\, \mathbf{v}, \tag{23}$$

where

$$\boldsymbol{\Omega} = \Omega_1 \mathbf{i} + \Omega_2 \mathbf{j} + \Omega_3 \mathbf{k}$$

is the local angular velocity of the fluid in the neighborhood of any point. The physical significance of the term curl is therefore one of rotation, and in the older texts the nomenclature rot $\mathbf{v}$ = curl $\mathbf{v}$ is employed. Some possible local configurations are sketched in Fig. 3.8.

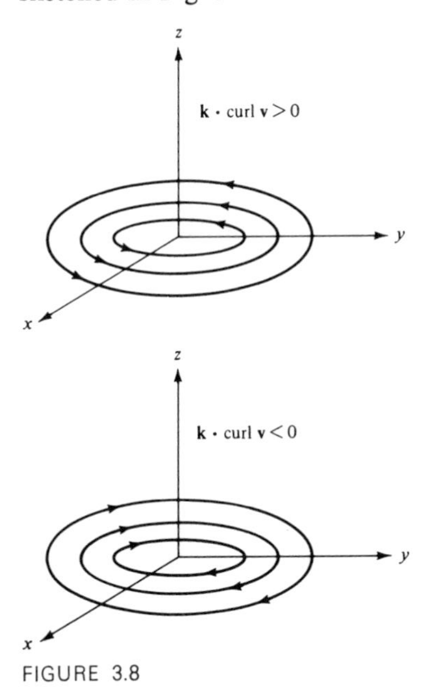

FIGURE 3.8

5

In operational form, the quantity curl **F** can also be written $\nabla \times \mathbf{F}$ (read "del cross **F**"). To see this, we write

$$\nabla \times \mathbf{F} = \left(\mathbf{i}\frac{\partial}{\partial x} + \mathbf{j}\frac{\partial}{\partial y} + \mathbf{k}\frac{\partial}{\partial z}\right) \times (F_1\mathbf{i} + F_2\mathbf{j} + F_3\mathbf{k}).$$

Upon formal simplification this becomes

$$\nabla \times \mathbf{F} = \left(\frac{\partial F_3}{\partial y} - \frac{\partial F_2}{\partial z}\right)\mathbf{i} + \left(\frac{\partial F_1}{\partial z} - \frac{\partial F_3}{\partial x}\right)\mathbf{j} + \left(\frac{\partial F_2}{\partial x} - \frac{\partial F_1}{\partial y}\right)\mathbf{k} = \text{curl}\,\mathbf{F}, \tag{24}$$

which is identical to (20).

The mnemonic for vector products is also quite useful in this context:

$$\nabla \times \mathbf{F} = \begin{vmatrix} \mathbf{i} & \mathbf{j} & \mathbf{k} \\ \dfrac{\partial}{\partial x} & \dfrac{\partial}{\partial y} & \dfrac{\partial}{\partial z} \\ F_1 & F_2 & F_3 \end{vmatrix}. \tag{25}$$

Another convenient representation is

$$\nabla \times \mathbf{F} = \mathbf{i} \times \frac{\partial \mathbf{F}}{\partial x} + \mathbf{j} \times \frac{\partial \mathbf{F}}{\partial y} + \mathbf{k} \times \frac{\partial \mathbf{F}}{\partial z},$$

which in the general notation of (14) becomes

$$\nabla \times \mathbf{F} = \sum_{n=1}^{3} \mathbf{i}_n \times \frac{\partial}{\partial x_n}\mathbf{F}. \tag{26}$$

Example 3 Find curl **F** when (i) $\mathbf{F} = x\mathbf{i} + y\mathbf{j} + z\mathbf{k}$, (ii) $\mathbf{F} = x^2y\mathbf{i} + y^2z\mathbf{j} + z^2x\mathbf{k}$, and (iii) $\mathbf{F} = yz\mathbf{i} + 2zx\mathbf{j} + 3xy\mathbf{k}$.

Use of formula (24) yields the following results:

(i) $\text{curl}\,\mathbf{F} = \left(\frac{\partial}{\partial y}z - \frac{\partial}{\partial z}y\right)\mathbf{i} + \left(\frac{\partial}{\partial z}x - \frac{\partial}{\partial x}z\right)\mathbf{j} + \left(\frac{\partial}{\partial x}y - \frac{\partial}{\partial y}x\right)\mathbf{k} = \mathbf{0};$

(ii) $\text{curl}\,\mathbf{F} = -y^2\mathbf{i} - z^2\mathbf{j} - x^2\mathbf{k};$

(iii) $\text{curl}\,\mathbf{F} = x\mathbf{i} - 2y\mathbf{j} + z\mathbf{k}.$

The preceding discussion shows how the vector operators div and curl arise naturally in a physical context. To summarize, the formulas obtained are

$$\text{grad}\,\varphi = \nabla\varphi = \frac{\partial \varphi}{\partial x}\mathbf{i} + \frac{\partial \varphi}{\partial y}\mathbf{j} + \frac{\partial \varphi}{\partial z}\mathbf{k},$$

$$\operatorname{div} \mathbf{F} = \nabla \cdot \mathbf{F} = \frac{\partial F_1}{\partial x} + \frac{\partial F_2}{\partial y} + \frac{\partial F_3}{\partial z},$$

$$\operatorname{curl} \mathbf{F} = \nabla \times \mathbf{F} = \left(\frac{\partial F_3}{\partial y} - \frac{\partial F_2}{\partial z}\right)\mathbf{i} + \left(\frac{\partial F_1}{\partial z} - \frac{\partial F_3}{\partial x}\right)\mathbf{j} + \left(\frac{\partial F_2}{\partial x} - \frac{\partial F_1}{\partial y}\right)\mathbf{k}.$$

Some unity of representation is apparent in the more general notation:

$$\nabla\varphi = \sum_{n=1}^{3} \mathbf{i}_n \frac{\partial}{\partial x_n} \varphi, \qquad \nabla \cdot \mathbf{F} = \sum_{n=1}^{3} \mathbf{i}_n \cdot \frac{\partial}{\partial x_n} \mathbf{F}, \qquad \nabla \times \mathbf{F} = \sum_{n=1}^{3} \mathbf{i}_n \times \frac{\partial}{\partial x_n} \mathbf{F}.$$

Exercises 8.3

1. Calculate the divergence and curl of each of the following vector functions $\mathbf{F}(x, y, z)$.

(i) $x\mathbf{i} + (y + z)\mathbf{j} + (x + y + z)\mathbf{k}$ **(ii)** $f(x)\mathbf{i} + g(y)\mathbf{j} + h(z)\mathbf{k}$

(iii) $f(y, z)\mathbf{i} + g(z, x)\mathbf{j} + h(x, y)\mathbf{k}$ **(iv)** $\sin(y + z)\mathbf{i} + e^z\mathbf{j} + xy\mathbf{k}$

(v) $(x^2 - yz)\mathbf{i} + (y^2 - zx)\mathbf{j} + (z^2 - xy)\mathbf{k}$ **(vi)** $(x + y + z)(x\mathbf{i} + y\mathbf{j} + z\mathbf{k})$

2. For the examples in the previous problem, verify that div curl $\mathbf{F} = 0$. Is this a general result?

3. Calculate the flux $\iint_S \mathbf{F} \cdot \hat{\mathbf{n}}\, dS$ and the line integral $\oint_C \mathbf{F} \cdot d\mathbf{r}$ for the following vector fields $\mathbf{F}(x, y, z)$ where S is the triangular area of the plane $x + 2y + 3z = 6$ with vertices on the coordinate axes and C is the boundary of this triangle.

(i) $x\mathbf{i} + y\mathbf{j} + z\mathbf{k}$ **(ii)** $y^2\mathbf{i} + \mathbf{j} + x^2\mathbf{k}$

4. Use spherical coordinates to evaluate the flux $\oiint_S \mathbf{F} \cdot \hat{\mathbf{n}}\, dS$ of the following vector fields $\mathbf{F}(x, y, z)$ where S is the spherical surface $x^2 + y^2 + z^2 = a^2$.

(i) $x\mathbf{i} + y\mathbf{j} + z\mathbf{k}$ **(ii)** $(x^2 + y^2 + z^2)^n(x\mathbf{i} + y\mathbf{j} + z\mathbf{k})$

5. For the vector fields in the previous problem, evaluate the flux $\oiint_S \mathbf{F} \cdot \hat{\mathbf{n}}\, dS$ where S is the spherical shell with boundary $x^2 + y^2 + z^2 = a^2$, $x^2 + y^2 + z^2 = b^2$ $(a<b)$.

6. Calculate the flux $\oiint_S \mathbf{F} \cdot \hat{\mathbf{n}}\, dS$ of the following vector fields $\mathbf{F}(x, y, z)$ where S is the closed surface bounded by the cylinder $x^2 + y^2 = a^2$ and the planes $z = 0$, $z = h$.

(i) $x\mathbf{i} + y\mathbf{j} + z\mathbf{k}$ **(ii)** $xy\mathbf{i} + yz\mathbf{j}$ **(iii)** $z(x + y)\mathbf{k}$

7. If $\mathbf{G} = (x + y)\mathbf{i} + (y + z)\mathbf{j} + (z + x)\mathbf{k}$, calculate $\iiint_V \operatorname{div} \mathbf{G}\, dV$ and $\iiint_V \operatorname{curl} \mathbf{G}\, dV$ when V is any closed volume.

8. Calculate the flux $\oiint_S \mathbf{F} \cdot \hat{\mathbf{n}}\, dS$ when $\mathbf{F} = m\mathbf{r}/|\mathbf{r}|^3$ and S is a sphere centered at the origin. Perform the same calculation when S is a cube centered at the origin. Why are the two results identical?

9. If $\mathbf{v} = \nabla\varphi$, show that $\oiint_S \mathbf{v} \cdot \hat{\mathbf{n}}\, dS = \oiint_S \frac{\partial\varphi}{\partial n}\, dS$. Use this result to calculate $\oiint_S \mathbf{v} \cdot \hat{\mathbf{n}}\, dS$ when $\mathbf{v}$ is the velocity field due to a dipole and S is a sphere centered at the dipole. Could this result have been anticipated without any calculation? Is the answer dependent on S?

10. If $\mathbf{F} = \mathbf{A}_0 + \mathbf{A}_1 x + \mathbf{A}_2 y + \mathbf{A}_3 z$, where $\mathbf{A}_0$, $\mathbf{A}_1$, $\mathbf{A}_2$, and $\mathbf{A}_3$ are constant vectors, calculate $\oiint_S \mathbf{F} \cdot \hat{\mathbf{n}}\, dS$ when S is the surface of the rectangular block $x = \pm a$, $y = \pm b$, $z = \pm c$. Deduce that $\operatorname{div} \mathbf{F} = \dfrac{1}{8abc} \oiint_S \mathbf{F} \cdot \hat{\mathbf{n}}\, dS$.

11. With $\mathbf{F}$ the vector of the previous problem, calculate $\oint_C \mathbf{F} \cdot d\mathbf{r}$ when C is any rectangle of area S centered at the origin in the xy plane. Verify that $\mathbf{k} \cdot \operatorname{curl} \mathbf{F} = \dfrac{1}{S} \oint_C \mathbf{F} \cdot d\mathbf{r}$.

12. If $\mathbf{F} = 3x^2y^2 \sin z\, \mathbf{i} + 2x^3y \sin z\, \mathbf{j} + (x^3y^2 \cos z + e^z)\mathbf{k}$, show that $\oint_C \mathbf{F} \cdot d\mathbf{r} = 0$ for any closed path C.

13. Calculate $\int_A^B \mathbf{F} \cdot d\mathbf{r}$ and $\int_A^B \mathbf{F} \times d\mathbf{r}$ when $\mathbf{F} = x\mathbf{i} + y\mathbf{j} + z\mathbf{k}$ and the path of integration is $\mathbf{r}(t) = \cos t\, \mathbf{i} + \sin t\, \mathbf{j} + t\mathbf{k}$ and the points A and B correspond to $t = 0$ and $t = 2\pi$. Do these results change if the straight-line path is chosen?

14. If C is the ellipse $x^2 + y^2 = 1$, $x + y + z = 0$ and $\mathbf{F} = yz^2\mathbf{i} + xz^2\mathbf{j} + \alpha xyz\mathbf{k}$, calculate $\oint_C \mathbf{F} \cdot d\mathbf{r}$. What is special about the case $\alpha = 2$?

15. Calculate the flux $\oiint \mathbf{F} \cdot \hat{\mathbf{n}}\, dS$ for the following vector fields where S is the closed surface formed by the paraboloid $z = 1 - x^2 - 2y^2$ and the plane $z = 0$.

(i) $\mathbf{F} = x\mathbf{i} + y\mathbf{j} + z\mathbf{k}$ **(ii)** $x\mathbf{i} + z\mathbf{j} + y\mathbf{k}$

16. A two-dimensional velocity field is of the form $\mathbf{v} = f(r)\hat{\boldsymbol{\theta}}$ where $r = (x^2 + y^2)^{1/2}$ and $\hat{\boldsymbol{\theta}} = -\sin\theta\, \mathbf{i} + \cos\theta\, \mathbf{j}$ is the unit vector in the circumferential direction in the xy plane. Find expressions for div $\mathbf{v}$ and curl $\mathbf{v}$.

17. When a fluid of constant density flows through a circular pipe, its velocity is $\mathbf{v} = U(1 - r^2/a^2)\mathbf{k}$, where r and z are cylindrical coordinates, a is the radius of the pipe, and U is the center velocity (see Fig. 3.9). Find the rate at which mass is transported down the pipe. If the pipe remains circular at each cross section but its radius changes slowly along the pipe, show that Ua^2 will remain constant.

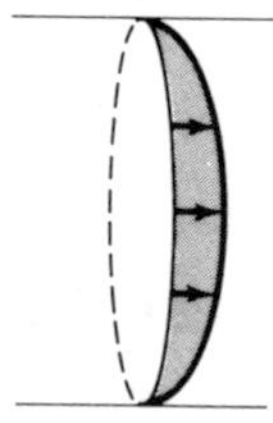

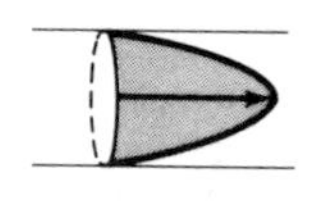

FIGURE 3.9

8.4 Vector identities

Many formulas relate the operators grad, div, and curl; a few of the most important are established in this section. These identities and the manipulative skill acquired in their derivation will greatly facilitate further development and application of vector calculus.

First let us restate the basic definitions given in Sec. 8.3:

$$\operatorname{div} \mathbf{F} = \nabla \cdot \mathbf{F} = \mathbf{i} \cdot \frac{\partial \mathbf{F}}{\partial x} + \mathbf{j} \cdot \frac{\partial \mathbf{F}}{\partial y} + \mathbf{k} \cdot \frac{\partial \mathbf{F}}{\partial z},$$

$$= \frac{\partial F_1}{\partial x} + \frac{\partial F_2}{\partial y} + \frac{\partial F_3}{\partial z}; \tag{1}$$

$$\operatorname{curl} \mathbf{F} = \nabla \times \mathbf{F} = \mathbf{i} \times \frac{\partial \mathbf{F}}{\partial x} + \mathbf{j} \times \frac{\partial \mathbf{F}}{\partial y} + \mathbf{k} \times \frac{\partial \mathbf{F}}{\partial z},$$

$$= \left(\frac{\partial F_3}{\partial y} - \frac{\partial F_2}{\partial z}\right)\mathbf{i} + \left(\frac{\partial F_1}{\partial z} - \frac{\partial F_3}{\partial x}\right)\mathbf{j} + \left(\frac{\partial F_2}{\partial x} - \frac{\partial F_1}{\partial y}\right)\mathbf{k}; \tag{2}$$

$$\operatorname{grad} \varphi = \nabla\varphi = \mathbf{i}\frac{\partial \varphi}{\partial x} + \mathbf{j}\frac{\partial \varphi}{\partial y} + \mathbf{k}\frac{\partial \varphi}{\partial z}. \tag{3}$$

Each of the operators can be represented in a general unifying notation as a particular case of

$$\nabla * = \sum_{n=1}^{3} \mathbf{i}_n * \frac{\partial}{\partial x_n} = \mathbf{i} * \frac{\partial}{\partial x} + \mathbf{j} * \frac{\partial}{\partial y} + \mathbf{k} * \frac{\partial}{\partial z} \tag{4}$$

where $*$ is replaced by a blank space, $\cdot$, or $\times$ to define grad, div, and curl respectively. Note that the operators div and curl apply only to vector functions and yield a scalar and a vector, respectively. However, grad operates on a scalar only, to give a vector function.

The combination div grad $\varphi = \nabla \cdot (\nabla\varphi)$ occurs often in practice and warrants the special notation $\nabla^2\varphi$ (read "del squared φ"). Let $\mathbf{F} = \nabla\varphi$; then the utilization of (1) and (3) shows that

$$\nabla^2\varphi = \operatorname{div}\operatorname{grad}\varphi = \nabla \cdot \mathbf{F} = \frac{\partial}{\partial x}\left(\frac{\partial \varphi}{\partial x}\right) + \frac{\partial}{\partial y}\left(\frac{\partial \varphi}{\partial y}\right) + \frac{\partial}{\partial z}\left(\frac{\partial \varphi}{\partial z}\right)$$

or

$$\nabla^2\varphi = \frac{\partial^2 \varphi}{\partial x^2} + \frac{\partial^2 \varphi}{\partial y^2} + \frac{\partial^2 \varphi}{\partial z^2}. \tag{5}$$

The notation $\nabla^2\mathbf{F}$, which is also used, is defined by

$$\nabla^2\mathbf{F} = \mathbf{i}\,\nabla^2 F_1 + \mathbf{j}\,\nabla^2 F_2 + \mathbf{k}\,\nabla^2 F_3.$$

Laplace's equation,

$$\nabla^2\varphi = 0,$$

arises remarkably often—in the theory of electricity, magnetism, optics, gravitation, diffusion, fluid dynamics, acoustics, and elasticity, to name a few.

The following formulas for the decomposition of div $\varphi\mathbf{F} = \nabla \cdot (\varphi\mathbf{F})$ and curl $\varphi\mathbf{F} = \nabla \times (\varphi\mathbf{F})$, are very useful:

$$\nabla \cdot \varphi\mathbf{F} = \varphi\nabla \cdot \mathbf{F} + \nabla\varphi \cdot \mathbf{F}, \tag{6}$$

$$\nabla \times \varphi\mathbf{F} = \varphi\nabla \times \mathbf{F} + \nabla\varphi \times \mathbf{F}. \tag{7}$$

Verification in each case is a direct consequence of the definition. For example, by (4)

$$\nabla \cdot \varphi\mathbf{F} = \sum_{n=1}^{3} \mathbf{i}_n \cdot \frac{\partial}{\partial x_n}(\varphi\mathbf{F});$$

upon differentiation and proper arrangement, this becomes

$$\nabla \cdot \varphi\mathbf{F} = \varphi \sum_{n=1}^{3} \mathbf{i}_n \cdot \frac{\partial \mathbf{F}}{\partial x_n} + \sum_{n=1}^{3} \frac{\partial \varphi}{\partial x_n} \mathbf{i}_n \cdot \mathbf{F}$$

The first summation on the right is merely div $\mathbf{F}$; the second is the scalar product of grad φ and $\mathbf{F}$, and (6) follows immediately. Equation (7) is deduced in the same way by simply replacing dots by crosses as in (4). (Note that the order of the vectors must be maintained.)

Two very important identities valid for any vector $\mathbf{F}$ and scalar φ are

$$\text{div curl } \mathbf{F} = \nabla \cdot \nabla \times \mathbf{F} = 0, \tag{8}$$

$$\text{curl grad } \varphi = \nabla \times \nabla\varphi = \mathbf{0}. \tag{9}$$

These results also follow from the basic definitions. For example,

$$\text{div curl } \mathbf{F} = \frac{\partial}{\partial x}\left(\frac{\partial F_3}{\partial y} - \frac{\partial F_2}{\partial z}\right) + \frac{\partial}{\partial y}\left(\frac{\partial F_1}{\partial z} - \frac{\partial F_3}{\partial x}\right) + \frac{\partial}{\partial z}\left(\frac{\partial F_2}{\partial x} - \frac{\partial F_1}{\partial y}\right)$$

and since the cross derivatives cancel in pairs, the right-hand side is zero and (8) is established. In the same way,

$$\text{curl grad } \varphi = \mathbf{i}\left[\frac{\partial}{\partial y}\left(\frac{\partial \varphi}{\partial z}\right) - \frac{\partial}{\partial z}\left(\frac{\partial \varphi}{\partial y}\right)\right] + \mathbf{j}\left[\frac{\partial}{\partial z}\left(\frac{\partial \varphi}{\partial x}\right) - \frac{\partial}{\partial x}\left(\frac{\partial \varphi}{\partial z}\right)\right]$$

$$+ \mathbf{k}\left[\frac{\partial}{\partial x}\left(\frac{\partial \varphi}{\partial y}\right) - \frac{\partial}{\partial y}\left(\frac{\partial \varphi}{\partial x}\right)\right].$$

Since each component is identically zero, (9) is verified.

2

Other important formulas are

$$\nabla \times (\nabla \times \mathbf{F}) = \nabla(\nabla \cdot \mathbf{F}) - \nabla^2\mathbf{F}, \tag{10}$$

$$\nabla \cdot (\mathbf{F} \times \mathbf{G}) = \mathbf{G} \cdot (\nabla \times \mathbf{F}) - \mathbf{F} \cdot (\nabla \times \mathbf{G}), \tag{11}$$

$$\nabla \times (\mathbf{F} \times \mathbf{G}) = (\nabla \cdot \mathbf{G})\mathbf{F} - (\nabla \cdot \mathbf{F})\mathbf{G} + (\mathbf{G} \cdot \nabla)\mathbf{F} - (\mathbf{F} \cdot \nabla)\mathbf{G}, \tag{12}$$

$$\nabla(\mathbf{F} \cdot \mathbf{G}) = \mathbf{F} \times (\nabla \times \mathbf{G}) + \mathbf{G} \times (\nabla \times \mathbf{F}) + (\mathbf{F} \cdot \nabla)\mathbf{G} + (\mathbf{G} \cdot \nabla)\mathbf{F}. \tag{13}$$

In Eqs. (12) and (13)

$$(\nabla \cdot \mathbf{G})\mathbf{F} = \left(\frac{\partial G_1}{\partial x} + \frac{\partial G_2}{\partial y} + \frac{\partial G_3}{\partial z}\right)\mathbf{F} = \sum_{n=1}^{3} \mathbf{F}\left(\mathbf{i}_n \cdot \frac{\partial \mathbf{G}}{\partial x_n}\right),$$

and

$$(\mathbf{G} \cdot \nabla)\mathbf{F} = G_1 \frac{\partial \mathbf{F}}{\partial x} + G_2 \frac{\partial \mathbf{F}}{\partial y} + G_3 \frac{\partial \mathbf{F}}{\partial z} = \sum_{n=1}^{3} (\mathbf{G} \cdot \mathbf{i}_n) \frac{\partial \mathbf{F}}{\partial x_n}.$$

Clearly these expressions are quite different, which points up the fact that the scalar products $\nabla \cdot \mathbf{G}$ and $\mathbf{G} \cdot \nabla$ *are not* equivalent operations. Care must be exercised in the manipulation of identities, and the operator ∇ is definitely not to be treated as an ordinary vector.

Example 1 Verify Eq. (12).

From the definition of the curl,

$$\nabla \times (\mathbf{F} \times \mathbf{G}) = \sum_{n=1}^{3} \mathbf{i}_n \times \frac{\partial}{\partial x_n}(\mathbf{F} \times \mathbf{G});$$

after completing the vector differentiation this becomes

$$\nabla \times (\mathbf{F} \times \mathbf{G}) = \sum_{n=1}^{3} \mathbf{i}_n \times \left(\mathbf{F} \times \frac{\partial \mathbf{G}}{\partial x_n}\right) + \sum_{n=1}^{3} \mathbf{i}_n \times \left(\frac{\partial \mathbf{F}}{\partial x_n} \times \mathbf{G}\right).$$

Each of these triple vector products can be expanded by means of the vector identity $\mathbf{a} \times (\mathbf{b} \times \mathbf{c}) = (\mathbf{a} \cdot \mathbf{c})\mathbf{b} - (\mathbf{a} \cdot \mathbf{b})\mathbf{c}$, and we obtain

$$\nabla \times (\mathbf{F} \times \mathbf{G}) = \sum_{n=1}^{3} \left(\mathbf{i}_n \cdot \frac{\partial \mathbf{G}}{\partial x_n}\right)\mathbf{F} - \sum_{n=1}^{3} (\mathbf{i}_n \cdot \mathbf{F}) \frac{\partial \mathbf{G}}{\partial x_n}$$
$$+ \sum_{n=1}^{3} (\mathbf{i}_n \cdot \mathbf{G}) \frac{\partial \mathbf{F}}{\partial x_n} - \sum_{n=1}^{3} \left(\mathbf{i}_n \cdot \frac{\partial \mathbf{F}}{\partial x_n}\right)\mathbf{G}.$$

In this form the individual sums can be rewritten in terms of the operator ∇, and the result is

$$\nabla \times (\mathbf{F} \times \mathbf{G}) = (\nabla \cdot \mathbf{G})\mathbf{F} - (\mathbf{F} \cdot \nabla)\mathbf{G} + (\mathbf{G} \cdot \nabla)\mathbf{F} - (\nabla \cdot \mathbf{F})\mathbf{G}.$$

The same method is used to establish (10), (11), and (13).

Example 2 If $\mathbf{\Omega}$ is a constant vector and $\mathbf{r} = x\mathbf{i} + y\mathbf{j} + z\mathbf{k}$, show that curl $\mathbf{\Omega} \times \mathbf{r} = 2\mathbf{\Omega}$.

Since the vector field $\mathbf{\Omega} \times \mathbf{r}$ corresponds to the velocity of a body in rigid rotation with angular velocity $\mathbf{\Omega}$, the result quoted should be anticipated in view of the interpretation given the curl operation in Sec. 8.3.

The formal calculation is based on (12), from which it follows that

$$\nabla \times (\mathbf{\Omega} \times \mathbf{r}) = (\nabla \cdot \mathbf{r})\mathbf{\Omega} - (\nabla \cdot \mathbf{\Omega})\mathbf{r} + (\mathbf{r} \cdot \nabla)\mathbf{\Omega} - (\mathbf{\Omega} \cdot \nabla)\mathbf{r}.$$

However, $\mathbf{\Omega}$ is a constant vector, so that the second and third terms are zero. Moreover, $\nabla \cdot \mathbf{r} = 3$ and $(\mathbf{\Omega} \cdot \nabla)\mathbf{r} = \mathbf{\Omega}$, and substitution of these results in the preceding equation yields

$$\text{curl}(\mathbf{\Omega} \times \mathbf{r}) = 3\mathbf{\Omega} - \mathbf{\Omega} = 2\mathbf{\Omega},$$

as desired.

3 **Example 3** Let $f(x, y, z)$ be a scalar function and $\mathbf{F}(x, y, z)$ a vector function. Show that the normal and tangential components of $\mathbf{F}$ to the level surface, $f =$ constant, are, respectively,

$$\frac{(\mathbf{F} \cdot \nabla f)\, \nabla f}{|\nabla f|^2} \quad \text{and} \quad \frac{\nabla f \times (\mathbf{F} \times \nabla f)}{|\nabla f|^2}.$$

Since the vector ∇f is perpendicular to the level surface $f(x, y, z) =$ constant, the unit normal is

$$\hat{\mathbf{n}} = \frac{\nabla f}{|\nabla f|}. \tag{14}$$

Let $\mathbf{F} = \mathbf{F}_1 + \mathbf{F}_2$, where $\mathbf{F}_1$ and $\mathbf{F}_2$ are vectors that are tangent and normal to the level surface at the point under consideration (see Fig. 4.1). It follows that

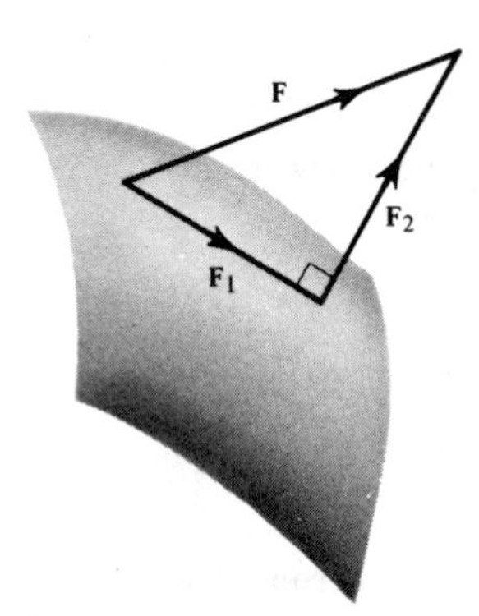

FIGURE 4.1

$\mathbf{F}_2$ is the projection of $\mathbf{F}$ in the direction of $\hat{\mathbf{n}}$, and in mathematical terms $\mathbf{F}_2 = (\mathbf{F} \cdot \hat{\mathbf{n}})\hat{\mathbf{n}}$. This equation can be rewritten, using (14), as

$$\mathbf{F}_2 = \frac{(\mathbf{F} \cdot \nabla f)\, \nabla f}{|\nabla f|^2},$$

and this identifies the normal component of $\mathbf{F}$ explicitly. The calculation of $\mathbf{F}_1$ follows from

$$\mathbf{F}_1 = \mathbf{F} - \mathbf{F}_2.$$

Substitution of the formula for $\mathbf{F}_2$ yields

$$\mathbf{F}_1 = \mathbf{F} - \frac{(\mathbf{F} \cdot \nabla f)\, \nabla f}{|\nabla f|^2}$$

or

$$\mathbf{F}_1 = \frac{1}{|\nabla f|^2} [(\nabla f \cdot \nabla f)\mathbf{F} - (\nabla f \cdot \mathbf{F})\, \nabla f].$$

The last expression can be identified as a triple vector product, that is, $\mathbf{a} \times (\mathbf{b} \times \mathbf{c}) = (\mathbf{a} \cdot \mathbf{c})\mathbf{b} - (\mathbf{a} \cdot \mathbf{b})\mathbf{c}$, and upon contraction we obtain

$$\mathbf{F}_1 = \frac{\nabla f \times (\mathbf{F} \times \nabla f)}{|\nabla f|^2}.$$

This completes the solution.

---

4 It was shown earlier that div curl $\mathbf{F} = 0$; the converse of this result is now established. A vector whose divergence is zero can always be expressed as the curl of some other vector: if

$$\operatorname{div} \mathbf{F} = 0,$$

then there is a vector $\mathbf{A}$ such that

$$\mathbf{F} = \operatorname{curl} \mathbf{A}. \tag{15}$$

The vector $\mathbf{A}$, called a *vector potential* for $\mathbf{F}$, is not uniquely determined. To see this, note that if $\mathbf{A}$ is a solution of (15), then so is $\mathbf{A} + \nabla\varphi$ (where φ is *any* scalar function) because curl $\nabla\varphi = \mathbf{0}$.

In order to establish (15), scalar functions A_1, A_2, A_3, which are the components of $\mathbf{A}$, are constructed such that

$$\frac{\partial A_3}{\partial y} - \frac{\partial A_2}{\partial z} = F_1, \qquad \frac{\partial A_1}{\partial z} - \frac{\partial A_3}{\partial x} = F_2, \qquad \frac{\partial A_2}{\partial x} - \frac{\partial A_1}{\partial y} = F_3, \tag{16}$$

where

$$\frac{\partial F_1}{\partial x} + \frac{\partial F_2}{\partial y} + \frac{\partial F_3}{\partial z} = 0. \tag{17}$$

To make matters as simple as possible a shortcut is tried, and it is assumed that $A_3 = 0$. (This hypothesis exploits the fact that the solution is not unique anyway.) It follows that the first and second equations of (16) can then be integrated, to yield

$$A_1 = \int_{z_0}^{z} F_2(x, y, z')\, dz' + M(x, y),$$

$$A_2 = -\int_{z_0}^{z} F_1(x, y, z')\, dz' + N(x, y),$$

where M and N are arbitrary scalar functions. Together, these equations imply that

$$\frac{\partial A_2}{\partial x} - \frac{\partial A_1}{\partial y} = -\int_{z_0}^{z} \left[\frac{\partial}{\partial x} F_1(x, y, z') + \frac{\partial}{\partial y} F_2(x, y, z')\right] dz' + \frac{\partial N}{\partial x} - \frac{\partial M}{\partial y}.$$

Equation (17) is now used to simplify the integrand involved, and the result is an exact differential

$$\frac{\partial A_2}{\partial x} - \frac{\partial A_1}{\partial y} = \int_{z_0}^{z} \frac{\partial}{\partial z'} F_3(x, y, z')\, dz' + \frac{\partial N}{\partial x} - \frac{\partial M}{\partial y},$$

the integration of which yields

$$\frac{\partial A_2}{\partial x} - \frac{\partial A_1}{\partial y} = F_3(x, y, z) - F_3(x, y, z_0) + \frac{\partial N}{\partial x} - \frac{\partial M}{\partial y}. \tag{18}$$

The functions M and N can be chosen in many ways to make the right-hand side of (18) exactly equal to $F_3(x, y, z)$. For the choice $M = 0$ and $N(x, y) = \int_{x_0}^{x} F_3(x', y, z_0)\, dx'$, the components of the vector potential $\mathbf{A}$ are

$$\begin{aligned} A_1 &= \int_{z_0}^{z} F_2(x, y, z')\, dz', \\ A_2 &= -\int_{z_0}^{z} F_1(x, y, z')\, dz' + \int_{x_0}^{x} F_3(x', y, z_0)\, dx', \\ A_3 &= 0. \end{aligned} \tag{19}$$

Example 4 If $\mathbf{F} = (x - y)\mathbf{i} + (y + xz)\mathbf{j} + (y - 2z)\mathbf{k}$, show that $\operatorname{div} \mathbf{F} = 0$ and find a vector potential $\mathbf{A}$.

For this particular function, $\partial F_1/\partial x = 1$, $\partial F_2/\partial y = 1$, $\partial F_3/\partial z = -2$, so that

$$\operatorname{div} \mathbf{F} = \frac{\partial F_1}{\partial x} + \frac{\partial F_2}{\partial y} + \frac{\partial F_3}{\partial z} = 1 + 1 - 2 = 0.$$

The calculation of a vector potential $\mathbf{A}$ follows from Eq. (19):

$$\begin{aligned} A_1 &= \int_0^z (y + xz')\, dz' = yz + \frac{xz^2}{2}, \\ A_2 &= -\int_0^z (x - y)\, dz' + \int_0^x y\, dx' = xy + yz - zx, \\ A_3 &= 0. \end{aligned}$$

Therefore, one solution is

$$\mathbf{A} = \left(yz + \frac{xz^2}{2}\right)\mathbf{i} + (xy + yz - zx)\mathbf{j}.$$

As a check, the calculation of curl $\mathbf{A}$ does give precisely the vector $\mathbf{F}$. (Of course $\mathbf{B} = \mathbf{A} + \nabla\varphi$, where φ is arbitrary, is also a solution.)

5

It is of more immediate interest at this stage to establish the converse to the identity curl $\nabla\varphi = \mathbf{0}$: if curl $\mathbf{F} = \mathbf{0}$, then there is a scalar potential φ such that $\mathbf{F} = \nabla\varphi$. Note first, that the vector equation curl $\mathbf{F} = \mathbf{0}$ implies the vanishing of its three components, i.e.,

$$\frac{\partial F_3}{\partial y} - \frac{\partial F_2}{\partial z} = 0, \qquad \frac{\partial F_1}{\partial z} - \frac{\partial F_3}{\partial x} = 0, \qquad \frac{\partial F_2}{\partial x} - \frac{\partial F_1}{\partial y} = 0. \tag{20}$$

We now proceed to demonstrate that there is a scalar function φ such that $\nabla\varphi = \mathbf{F}$ by actually constructing it. (See Sec. 8.6 for an alternative nonconstructive proof.)

Let

$$\frac{\partial \varphi}{\partial z} = F_3(x, y, z), \tag{21}$$

so that integration gives

$$\varphi(x, y, z) = \int_{z_0}^{z} F_3(x, y, z')\, dz' + P(x, y), \tag{22}$$

where $P(x, y)$ is an arbitrary function and (x_0, y_0, z_0) denotes a fixed point. The aim now is to show that the function $P(x, y)$ can be chosen so that Eqs. (20) are satisfied *and* $\varphi_x = F_1$, $\varphi_y = F_2$. [These together with (21) would then be equivalent to the statement $\mathbf{F} = \nabla\varphi$.]

Differentiation of (22) with respect to either x or y yields

$$\frac{\partial \varphi}{\partial x} = \int_{z_0}^{z} \frac{\partial}{\partial x} F_3(x, y, z')\, dz' + \frac{\partial P}{\partial x}$$

and

$$\frac{\partial \varphi}{\partial y} = \int_{z_0}^{z} \frac{\partial}{\partial y} F_3(x, y, z')\, dz' + \frac{\partial P}{\partial y}.$$

However, the first two equations in (20) allow $\partial F_3/\partial x$ and $\partial F_3/\partial y$ to be replaced by $\partial F_1/\partial z$ and $\partial F_2/\partial z$, respectively, in which case the preceding expressions become

$$\frac{\partial \varphi}{\partial x} = \int_{z_0}^{z} \frac{\partial}{\partial z'} F_1(x, y, z')\, dz' + \frac{\partial P}{\partial x},$$

$$\frac{\partial \varphi}{\partial y} = \int_{z_0}^{z} \frac{\partial}{\partial z'} F_2(x, y, z')\, dz' + \frac{\partial P}{\partial y}.$$

These substitutions allow the integrals to be calculated exactly, the results being

$$\begin{aligned} \frac{\partial \varphi}{\partial x} &= F_1(x, y, z) - F_1(x, y, z_0) + \frac{\partial P}{\partial x}, \\ \frac{\partial \varphi}{\partial y} &= F_2(x, y, z) - F_2(x, y, z_0) + \frac{\partial P}{\partial y}. \end{aligned} \tag{23}$$

Finally, the function P *may be chosen* so that

$$\frac{\partial P}{\partial x} = F_1(x, y, z_0), \qquad \frac{\partial P}{\partial y} = F_2(x, y, z_0), \tag{24}$$

in which case (23) reduces to $\partial\varphi/\partial x = F_1$ and $\partial\varphi/\partial y = F_2$. Such a function P is readily found by integrating the second equation in (24) to obtain

$$P = \int_{y_0}^{y} F_2(x, y', z_0)\, dy' + Q(x). \tag{25}$$

The function $Q(x)$ is then determined by equating the value of $\partial P/\partial x$ from (25) with the first of Eq. (24). This leads to the identity

$$\frac{d}{dx} Q(x) = F_1(x, y, z_0) - \int_{y_0}^{y} \frac{\partial}{\partial x} F_2(x, y', z_0)\, dy',$$

which since $\partial F_2/\partial x = \partial F_1/\partial y$, can be simplified to yield

$$\frac{d}{dx} Q(x) = F_1(x, y_0, z_0),$$

so that

$$Q(x) = \int_{x_0}^{x} F_1(x', y_0, z_0)\, dx'. \tag{26}$$

Therefore, a symmetrical formula for the scalar potential φ, obtained by combining (22), (25), and (26), is

$$\varphi(x, y, z) = \int_{x_0}^{x} F_1(x', y_0, z_0)\, dx' + \int_{y_0}^{y} F_2(x, y', z_0)\, dy' + \int_{z_0}^{z} F_3(x, y, z')\, dz'. \tag{27}$$

It is worth noting that this result is identical to Eq. (6) of Sec. 8.2, where it was shown that when the line integral $\int_{P_0}^{P} \mathbf{F} \cdot d\mathbf{r}$ is path-independent, $\mathbf{F} = \nabla\varphi$. In the light of the above result, we conclude that the following three statements are entirely equivalent to each other:

1. $\mathbf{F} = \nabla\varphi$.
2. curl $\mathbf{F} = \mathbf{0}$.
3. $\int_{P_0}^{P} \mathbf{F} \cdot d\mathbf{r}$ is independent of the path.

Until now, there was no a priori way of knowing whether a vector field had an associated scalar potential; only by a direct attack, a difficult and extended construction of the function, could it be determined whether this was possible. However, the relatively easy calculation of curl **F** will tell, once and for all, whether **F** is a conservative field or not, i.e., whether $\mathbf{F} = \nabla\varphi$.

Example 5 If $\mathbf{F} = y^2\mathbf{i} - z^3\mathbf{j} + x^4\mathbf{k}$, is there a function φ such that $\mathbf{F} = \nabla\varphi$?

A simple calculation gives curl $\mathbf{F} = 3z^2\mathbf{i} - 4x^3\mathbf{j} - 2y\mathbf{k}$. Since this is not identically zero, **F** is not a conservative field and it is *not* possible to find a scalar potential φ.

Example 6 If possible, find a potential φ for

$$\mathbf{F} = (y^2 + 2xz^2 - 1)\mathbf{i} + 2xy\mathbf{j} + (2x^2z + z^3)\mathbf{k}.$$

This is the same problem considered in Sec. 8.2., Example 5. In this case, the procedure now is to calculate the curl, and since

$$\operatorname{curl} \mathbf{F} = \mathbf{0},$$

a potential φ exists. It is determined by line integration, which is equivalent to Eq. (27):

$$\varphi(x, y, z) = \int_0^x F_1(x', 0, 0)\,dx' + \int_0^y F_2(x, y', 0)\,dy' + \int_0^z F_3(x, y, z')\,dz',$$

where for convenience the choice $x_0 = y_0 = z_0 = 0$ is made. Substitution of the particular vector $\mathbf{F}$ in this formula gives

$$\varphi(x, y, z) = \int_0^x (-1)\,dx' + \int_0^y 2xy'\,dy' + \int_0^z (2x^2z' + z'^3)\,dz',$$

and finally

$$\varphi(x, y, z) = -x + xy^2 + x^2z^2 + \frac{z^4}{4}.$$

Apart from an arbitrary constant, which can always be added to φ, this is the same function found earlier.

Exercises 8.4

1. Verify the following identities for the position vector $\mathbf{r}$ in three dimensions.

(i) $\operatorname{div} \mathbf{r} = 3$ **(ii)** $\operatorname{curl} \mathbf{r} = \mathbf{0}$ **(iii)** $\nabla r = \mathbf{r}/r$
(iv) $\operatorname{div} \nabla r = 2/r$ **(v)** $\nabla \times (\mathbf{r}/r) = \mathbf{0}$ **(vi)** $\nabla^2(\log r) = 1/r^2$.

2. If $\mathbf{r} = x\mathbf{i} + y\mathbf{j} + z\mathbf{k}$, find div $\mathbf{F}$ and curl $\mathbf{F}$ for the following vector fields $\mathbf{F}(x, y, z)$.

(i) $\mathbf{F} = r^n\mathbf{r}$ **(ii)** $\mathbf{F} = \mathbf{k} \times \mathbf{r}$ **(iii)** $\mathbf{F} = \mathbf{r} \times (\mathbf{i} \times \mathbf{r})$

3. Calculate div $\mathbf{F}$, curl $\mathbf{F}$, $\nabla^2\mathbf{F}$, curl curl $\mathbf{F}$ and grad div $\mathbf{F}$ for the following vector fields $\mathbf{F}(x, y, z)$.

(i) $yz\mathbf{i} + zx\mathbf{j} + xy\mathbf{k}$ **(ii)** $(x + y)\mathbf{i} + (y + z)\mathbf{j} + (z + x)\mathbf{k}$ **(iii)** $x^3yz\mathbf{i} + y^3zx\mathbf{j} + z^3xy\mathbf{k}$

4. When div $\mathbf{F} = 0$, the vector $\mathbf{F}$ is said to be divergence-free or solenoidal. Show that the following vectors $\mathbf{F}$ are solenoidal and find a vector potential $\mathbf{A}$, where $\mathbf{F} = \nabla \times \mathbf{A}$:

(i) $\mathbf{i} - \mathbf{j} + \mathbf{k}$ **(ii)** $y\mathbf{i} + (y + x)\mathbf{j} - z\mathbf{k}$
(iii) $yz\mathbf{i} + zx\mathbf{j} + xy\mathbf{k}$ **(iv)** $(x^2 + yz)\mathbf{i} + (y^2 + zx)\mathbf{j} - 2z(x + y)\mathbf{k}$

5. A vector $\mathbf{F}$ for which curl $\mathbf{F} = \mathbf{0}$ is said to be conservative, curl-free, or irrotational. Show that the following vectors $\mathbf{F}$ are irrotational and find the corresponding scalar potential φ, where $\mathbf{F} = \nabla\varphi$:

(i) $\mathbf{i} - 2\mathbf{j} + 3\mathbf{k}$ **(ii)** $x^2\mathbf{i} - y\mathbf{j} + z^3\mathbf{k}$
(iii) $(1 + yz)\mathbf{i} + (2y + zx)\mathbf{j} + (3z^2 + xy)\mathbf{k}$ **(iv)** $xy(2yz\mathbf{i} + 2zx\mathbf{j} + xy\mathbf{k})$

6. Suppose that a vector field $\mathbf{F}(x, y, z)$ is solenoidal and irrotational (zero divergence and zero curl). Show that its scalar potential function $\phi(x, y, z)$ satisfies Laplace's equation $\nabla^2\phi = 0$.

7. If $\mathbf{r}$ is the radius vector in three dimensions and $f(r)$ is any scalar function of the magnitude $r = |\mathbf{r}|$, calculate $\boldsymbol{\nabla} f(r)$, $\operatorname{div} f(r)\mathbf{r}$, and $\operatorname{curl} f(r)\mathbf{r}$.

8. If $\mathbf{r}$ is the radius vector and $\mathbf{a}$ is any constant vector, verify the following identities.

(i) $\boldsymbol{\nabla} \cdot (\mathbf{a} \times \mathbf{r}) = 0$ (ii) $\boldsymbol{\nabla} \cdot (r^2\mathbf{a}) = 2\mathbf{r} \cdot \mathbf{a}$ (iii) $\boldsymbol{\nabla} \cdot [(\mathbf{a} \cdot \mathbf{r})\mathbf{r}] = 4\mathbf{a} \cdot \mathbf{r}$

(iv) $\boldsymbol{\nabla} \times (\mathbf{a} \times \mathbf{r}) = 2\mathbf{a}$ (v) $\boldsymbol{\nabla} \times (r^2\mathbf{a}) = 2\mathbf{r} \times \mathbf{a}$ (vi) $(\mathbf{a} \times \boldsymbol{\nabla}) \times \mathbf{r} = -2\mathbf{a}$

9. If $\mathbf{r}$ is the radius vector and $\mathbf{a}$, $\mathbf{b}$ are constant vectors, prove the following identities.

(i) $\nabla \times \left(\dfrac{\mathbf{a} \times \mathbf{r}}{r^3}\right) = 3\dfrac{(\mathbf{a} \cdot \mathbf{r})}{r^5}\mathbf{r} - \dfrac{\mathbf{a}}{r^3}$ (ii) $\mathbf{a} \cdot \boldsymbol{\nabla}\left(\mathbf{b} \cdot \boldsymbol{\nabla}\left(\dfrac{1}{r}\right)\right) = \dfrac{3(\mathbf{a} \cdot \mathbf{r})(\mathbf{b} \cdot \mathbf{r})}{r^5} - \dfrac{\mathbf{a} \cdot \mathbf{b}}{r^3}$

(iii) $\boldsymbol{\nabla}[(\mathbf{a} \times \mathbf{r}) \cdot (\mathbf{b} \times \mathbf{r})] = \mathbf{a} \times (\mathbf{r} \times \mathbf{b}) + \mathbf{b} \times (\mathbf{r} \times \mathbf{a})$

10. For vector fields $\mathbf{F}(x, y, z)$ and $\mathbf{G}(x, y, z)$, verify the following identities.

(i) $\boldsymbol{\nabla} \times (\boldsymbol{\nabla} \times \mathbf{F}) = \boldsymbol{\nabla}(\boldsymbol{\nabla} \cdot \mathbf{F}) - \boldsymbol{\nabla}^2\mathbf{F}$

(ii) $\boldsymbol{\nabla} \cdot (\mathbf{F} \times \mathbf{G}) = \mathbf{G} \cdot (\boldsymbol{\nabla} \times \mathbf{F}) - \mathbf{F} \cdot (\boldsymbol{\nabla} \times \mathbf{G})$

(iii) $\boldsymbol{\nabla}(\mathbf{F} \cdot \mathbf{G}) = \mathbf{F} \times (\boldsymbol{\nabla} \times \mathbf{G}) + \mathbf{G} \times (\boldsymbol{\nabla} \times \mathbf{F}) + (\mathbf{F} \cdot \boldsymbol{\nabla})\mathbf{G} + (\mathbf{G} \cdot \boldsymbol{\nabla})\mathbf{F}$.

11. If $\mathbf{r} = x\mathbf{i} + y\mathbf{j} + z\mathbf{k}$ and $\phi(x, y, z)$ is any scalar function, verify the following identities.

(i) $\boldsymbol{\nabla} \cdot (\mathbf{r} \times \boldsymbol{\nabla}\phi) = 0$ (ii) $\nabla^2(\mathbf{r} \times \boldsymbol{\nabla}\phi) = \mathbf{r} \times \nabla^2(\boldsymbol{\nabla}\phi)$

12. If $\phi(x, y, z)$ and $\psi(x, y, z)$ are scalar functions, verify the following identities.

(i) $\boldsymbol{\nabla} \cdot (\phi\boldsymbol{\nabla}\psi) = \phi\boldsymbol{\nabla}^2\psi + \boldsymbol{\nabla}\phi \cdot \boldsymbol{\nabla}\psi$ (ii) $\boldsymbol{\nabla} \cdot (\phi\boldsymbol{\nabla}\psi + \psi\boldsymbol{\nabla}\phi) = \boldsymbol{\nabla}^2(\phi\psi)$

(iii) $\boldsymbol{\nabla} \cdot (\phi\boldsymbol{\nabla}\psi - \psi\boldsymbol{\nabla}\phi) = \phi\boldsymbol{\nabla}^2\psi - \psi\boldsymbol{\nabla}^2\phi$

13. Verify the following identities for the scalar functions $\phi(x, y, z)$ and $\psi(x, y, z)$.

(i) $\boldsymbol{\nabla} \times (\phi\boldsymbol{\nabla}\psi) = \boldsymbol{\nabla}\phi \times \boldsymbol{\nabla}\psi$ (ii) $\boldsymbol{\nabla} \times (\phi\boldsymbol{\nabla}\psi + \psi\boldsymbol{\nabla}\phi) = \mathbf{0}$

(iii) $\boldsymbol{\nabla} \times (\phi\boldsymbol{\nabla}\psi - \psi\boldsymbol{\nabla}\phi) = 2\boldsymbol{\nabla}\phi \times \boldsymbol{\nabla}\psi$

14. If $\phi(x, y, z)$ and $\psi(x, y, z)$ are scalar functions, verify that

$$\nabla^2(\phi\psi) = \phi\nabla^2\psi + \psi\nabla^2\phi + 2\boldsymbol{\nabla}\phi \cdot \boldsymbol{\nabla}\psi.$$

15. If $\phi(x, y, z)$ is a solution of Laplace's equation $\nabla^2\phi = 0$ and $\mathbf{a}$ is any constant vector, show that $\boldsymbol{\nabla}(\mathbf{a} \cdot \boldsymbol{\nabla}\phi) + \boldsymbol{\nabla} \times (\mathbf{a} \times \boldsymbol{\nabla}\phi) = \mathbf{0}$.

16. A vector function $\mathbf{F}(x, y, z)$ and a scalar function $\phi(x, y, z)$ are related by $\boldsymbol{\nabla}\phi = \boldsymbol{\nabla} \times \mathbf{F}$. Deduce that $\nabla^2\phi = 0$ and $\nabla^2\mathbf{F} = -\boldsymbol{\nabla}(\boldsymbol{\nabla} \cdot \mathbf{F})$.

17. A vector function $\mathbf{F}(x, y, z)$ is defined in terms of three scalar functions $\phi_1(x, y, z)$, $\phi_2(x, y, z)$ and $\phi_3(x, y, z)$ by $\mathbf{F} = \phi_3(\boldsymbol{\nabla}\phi_1 \times \boldsymbol{\nabla}\phi_2)$. If ϕ_1, ϕ_2 and ϕ_3 are functionally dependent, show that $\boldsymbol{\nabla} \cdot \mathbf{F} = 0$. (*Hint:* If $\phi_3 = \phi_3(\phi_1, \phi_2)$, then $\partial(\phi_1, \phi_2, \phi_3)/\partial(x, y, z) = 0$.)

18. Newton's law of motion applied to a perfect fluid is $\rho\mathbf{a} = \boldsymbol{\nabla}p$. This equation relates the density $\rho(x, y, z)$ and pressure $p(x, y, z)$ to the acceleration $\mathbf{a}(x, y, z)$ at a point (x, y, z) in the fluid. Verify the following identities.

(i) $\mathbf{a} \cdot (\boldsymbol{\nabla} \times \mathbf{a}) = 0$ (ii) $\boldsymbol{\nabla} \times (\rho\mathbf{a}) = \mathbf{0}$ (iii) $\rho(\boldsymbol{\nabla} \times \mathbf{a}) = \mathbf{a} \times \boldsymbol{\nabla}\rho$

19. Maxwell's equations relating the electric field $\mathbf{E}$ and the magnetic field $\mathbf{H}$ in a region of space containing no charges and no currents are the following:

$$\operatorname{div} \mathbf{E} = 0, \ \operatorname{div} \mathbf{H} = 0, \ \operatorname{curl} \mathbf{E} = -\frac{1}{c}\frac{\partial \mathbf{H}}{\partial t}, \ \operatorname{curl} \mathbf{H} = \frac{1}{c}\frac{\partial \mathbf{E}}{\partial t}.$$

The constant c is the speed of light. Verify the following identities.

(i) $\boldsymbol{\nabla} \times (\boldsymbol{\nabla} \times \mathbf{E}) = -\dfrac{1}{c^2}\dfrac{\partial^2\mathbf{E}}{\partial t^2}$ (ii) $\boldsymbol{\nabla} \times (\boldsymbol{\nabla} \times \mathbf{H}) = -\dfrac{1}{c^2}\dfrac{\partial^2\mathbf{H}}{\partial t^2}$

(iii) $\boldsymbol{\nabla}^2\mathbf{E} = \dfrac{1}{c^2}\dfrac{\partial^2\mathbf{E}}{\partial t^2}$ (iv) $\boldsymbol{\nabla}^2\mathbf{H} = \dfrac{1}{c^2}\dfrac{\partial^2\mathbf{H}}{\partial t^2}$

8.5 *The divergence theorem*

Without question the most significant and far-reaching result in the entire vector calculus is the divergence theorem, also known as Gauss' theorem. The mathematical statement of the theorem is

$$\oiint_S \mathbf{F} \cdot \hat{\mathbf{n}}\, dS = \iiint_V \operatorname{div} \mathbf{F}\, dV = \iiint_V \nabla \cdot \mathbf{F}\, dV, \tag{1}$$

where $\mathbf{F}$ is an arbitrary vector, S a closed surface whose interior volume is V, and $\hat{\mathbf{n}}$ the outward unit normal, as shown in Fig. 5.1. In other words, the flux of $\mathbf{F}$ through a closed surface S is identical to the volume integral of div $\mathbf{F}$ taken throughout V.

The divergence theorem appears more formidable than it should at this stage. A very plausible and convincing argument can be based on the discussion of Sec. 8.3, where it was shown that

$$\operatorname{div} \mathbf{F} = \lim_{\Delta V \to 0} \frac{1}{\Delta V} \oiint_{\Delta S} \mathbf{F} \cdot \hat{\mathbf{n}}\, dS, \tag{2}$$

ΔS being the surface area of an incremental volume ΔV. The crucial step is to relate this strictly local property of the field to a finite volume of space. However, *any* volume V can be conceptually subdivided or sliced into small increments, say as the rectangular blocks in Fig. 5.2, and (2) can be applied to each of these in its equivalent form

$$\oiint_{\Delta S} \mathbf{F} \cdot \hat{\mathbf{n}}\, dS = (\operatorname{div} \mathbf{F} + \varepsilon)\, \Delta V, \tag{3}$$

(where $\varepsilon \to 0$ as $\Delta V \to 0$). If all the contributions are added together, the right-hand side of this equation, in the limit $\Delta V \to 0$, becomes

$$\iiint_V \operatorname{div} \mathbf{F}\, dV.$$

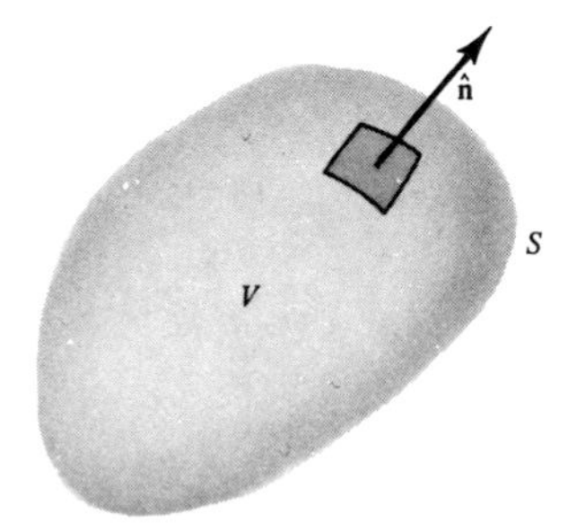

FIGURE 5.1

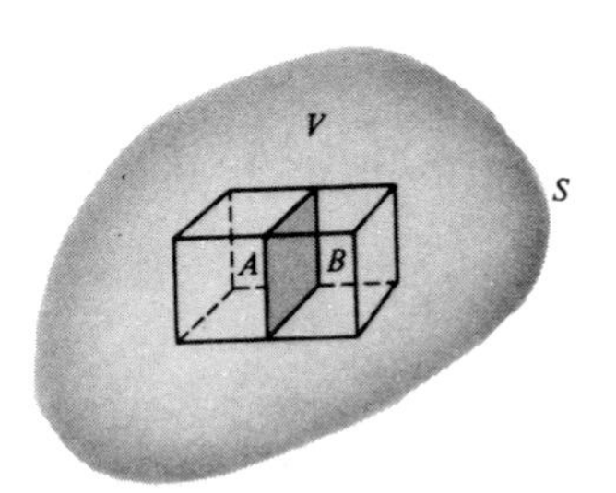

FIGURE 5.2

Moreover, the net flux through every surface element *within* the exterior surface S is zero because what *emerges* from one volume increment enters another. If dS is the common surface element of two interior blocks labelled A and B in Fig. 5.2, then the outward normal to the first is in fact the inward normal to the second. The outward flux $\mathbf{F} \cdot \hat{\mathbf{n}}\, dS$ from A is therefore the *inward* flux to B, and these terms cancel each other in the grand summation. Only the flux through the surface elements of the outer closed surface S survives this cancellation, and the summation yields the equivalent of $\oiint_S \mathbf{F} \cdot \hat{\mathbf{n}}\, dS$, which completes the derivation of (1). This argument can be made the basis of a proof, but the conclusion is so important that another more formal derivation is given next.

2

For simplicity, assume that the volume V bounded by S is cut at most twice by any line parallel to the coordinate axes, as shown in Fig. 5.3. (This requirement will be relaxed later.) In particular, let $z = \chi_T(x, y)$ and $z = \chi_B(x, y)$ be the "top" and "bottom" portions of S.

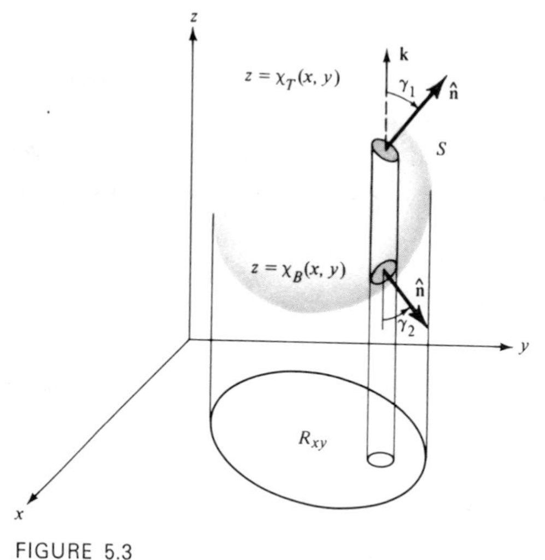

FIGURE 5.3

Consider the quantity $\iiint_V \dfrac{\partial F_3}{\partial z}\, dV$, which is one of the terms on the right-hand side of (1). This can be integrated with respect to z to obtain

$$\iiint_V \frac{\partial F_3}{\partial z}\, dV = \iint_{R_{xy}} dx\, dy \int_{\chi_B(x,y)}^{\chi_T(x,y)} \frac{\partial F_3}{\partial z}\, dz,$$

$$= \iint_{R_{xy}} [F_3(x, y, \chi_T(x, y)) - F_3(x, y, \chi_B(x, y))]\, dx\, dy, \tag{4}$$

where R_{xy} is the projection of S onto the xy plane. The preceding formula is now related to the only term on the left-hand side of (1) that involves F_3, that is,

$\oiint_S F_3\mathbf{k}\cdot\hat{\mathbf{n}}\,dS$. This surface integral can be evaluated by finding the contributions from the top and bottom parts of S. On the top, $z = \chi_T(x, y)$ and $\hat{\mathbf{n}}\cdot\mathbf{k}\,dS = \cos\gamma_1\,dS = dA$ while at the bottom $z = \chi_B(x, y)$ and $\hat{\mathbf{n}}\cdot\mathbf{k}\,dS = -\cos\gamma_2\,dS = -\,dA$, where the angles γ_1 and γ_2 are shown in Fig. 5.3. With $dA = dx\,dy$, it follows that

$$\oiint_S F_3\mathbf{k}\cdot\hat{\mathbf{n}}\,dS = \iint_{R_{xy}} [F_3(x, y, \chi_T(x, y)) - F_3(x, y, \chi_B(x, y))]\,dx\,dy, \tag{5}$$

and on comparing (4) and (5) we obtain

$$\oiint_S F_3\mathbf{k}\cdot\hat{\mathbf{n}}\,dS = \iiint_V \frac{\partial F_3}{\partial z}\,dV.$$

In a similar way it can be shown that

$$\oiint_S F_1\mathbf{i}\cdot\hat{\mathbf{n}}\,dS = \iiint_V \frac{\partial F_1}{\partial x}\,dV$$

and

$$\oiint_S F_2\mathbf{j}\cdot\hat{\mathbf{n}}\,dS = \iiint_V \frac{\partial F_2}{\partial y}\,dV.$$

Addition of the last three equations gives the divergence theorem:

$$\oiint_S (F_1\mathbf{i} + F_2\mathbf{j} + F_3\mathbf{k})\cdot\hat{\mathbf{n}}\,dS = \iiint_V \left(\frac{\partial F_1}{\partial x} + \frac{\partial F_2}{\partial y} + \frac{\partial F_3}{\partial z}\right) dV,$$

or

$$\oiint_S \mathbf{F}\cdot\hat{\mathbf{n}}\,dS = \iiint_V \operatorname{div}\mathbf{F}\,dV.$$

The restriction on the configuration can be removed by subdividing the arbitrary volume V into smaller volumes, each of which has the property stated. For example, if the volume is that shown in Fig. 5.4, we can introduce a common surface S_c that divides V into two regions V_1 and V_2. Each of these has the property that it is cut by a line parallel to the axes in at most two points. The previous proof of the divergence theorem applies to each volume V_1 and V_2:

$$\oiint_{S_1+S_c} \mathbf{F}\cdot\hat{\mathbf{n}}\,dS = \iiint_{V_1} \operatorname{div}\mathbf{F}\,dV, \qquad \oiint_{S_2+S_c} \mathbf{F}\cdot\hat{\mathbf{n}}\,dS = \iiint_{V_2} \operatorname{div}\mathbf{F}\,dV,$$

where S_1 and S_2 are indicated in Fig. 5.4. On adding these two equations, the two contributions from interior surface S_c cancel since the normal vectors are in opposite directions and the result is

$$\oiint_{S_1+S_2} \mathbf{F}\cdot\hat{\mathbf{n}}\,dS = \iiint_{V_1+V_2} \operatorname{div}\mathbf{F}\,dV.$$

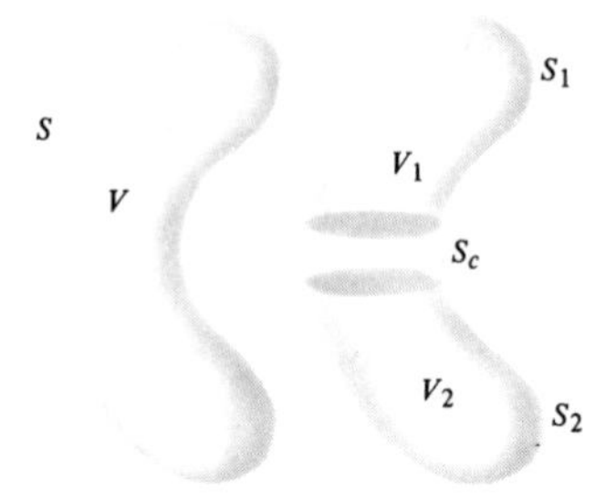

FIGURE 5.4

But $S_1 + S_2$ is the entire exterior surface S, and $V_1 + V_2$ is the complete volume V, bounded by S. Therefore, the preceding formula can be rewritten as

$$\oiint_S \mathbf{F} \cdot \hat{\mathbf{n}}\, dS = \iiint_V \operatorname{div} \mathbf{F}\, dV,$$

which is (1) once again. In this way, using as many subdivisions as necessary, the divergence theorem can be established for any closed region.

Example 1 By direct calculation check the validity of the divergence theorem in each of the following cases: (i) $\mathbf{F} = x^3\mathbf{i} + y^3\mathbf{j} + z^3\mathbf{k}$ when V is the volume of the sphere $x^2 + y^2 + z^2 = a^2$ and (ii) $\mathbf{F} = x\mathbf{i} + y\mathbf{j} + z\mathbf{k}$ when V is the volume of the cube $0 \le x, y, z \le l$.

In the case (i)

$$\operatorname{div} \mathbf{F} = \frac{\partial}{\partial x}(x^3) + \frac{\partial}{\partial y}(y^3) + \frac{\partial}{\partial z}(z^3) = 3(x^2 + y^2 + z^2).$$

To calculate $\iiint_V \operatorname{div} \mathbf{F}\, dV$, where V is the prescribed sphere, it is convenient to use spherical coordinates (R, φ, θ), so that $\operatorname{div} \mathbf{F} = 3R^2$ and $dV = R^2 \sin \varphi dR\, d\varphi\, d\theta$. Therefore

$$\iiint_V \operatorname{div} \mathbf{F}\, dV = \int_0^{2\pi} d\theta \int_0^{\pi} d\varphi \int_0^a 3R^4 \sin \varphi\, dR,$$

and upon integration

$$\iiint_V \operatorname{div} \mathbf{F}\, dV = [\theta]_0^{2\pi}[-\cos \varphi]_0^{\pi}[3/5 R^5]_0^a = 12/5\, \pi a^5.$$

The surface integral $\oiint \mathbf{F} \cdot \hat{\mathbf{n}}\, dS$ has been calculated previously (Example 1 of Sec. 8.3). The value obtained there, $12\pi a^5/5$, confirms the divergence theorem.

In case (ii) $\mathbf{F} = x\mathbf{i} + y\mathbf{j} + z\mathbf{k}$ and $\operatorname{div}\mathbf{F} = 3$, so that the calculation is very simple:

$$\iiint_V \operatorname{div}\mathbf{F}\,dV = 3\iiint_V dV = 3V.$$

But the volume of a cube of side l is $V = l^3$, so that

$$\iiint_V \operatorname{div}\mathbf{F}\,dV = 3l^3.$$

To verify the divergence theorem in this case it is necessary to show that $\oiint \mathbf{F}\cdot\hat{\mathbf{n}}\,dS = 3l^3$. We begin by calculating the contribution from the two faces $x = 0$, $x = l$. Since $\hat{\mathbf{n}} = -\mathbf{i}$ on the surface $x = 0$ and $\hat{\mathbf{n}} = \mathbf{i}$ on $x = l$ (see Fig. 5.5), it follows that

$$\mathbf{F}\cdot\hat{\mathbf{n}} = \begin{cases} -\mathbf{F}\cdot\mathbf{i} = 0 & \text{at} \quad x = 0; \\ \;\;\,\mathbf{F}\cdot\mathbf{i} = l & \text{at} \quad x = l. \end{cases}$$

There is no flux across the face $x = 0$, but at $x = l$ we find that

$$\iint_S \mathbf{F}\cdot\hat{\mathbf{n}}\,dS = l\iint_S dS = l^3,$$

since the area of any one face of the cube is l^2.

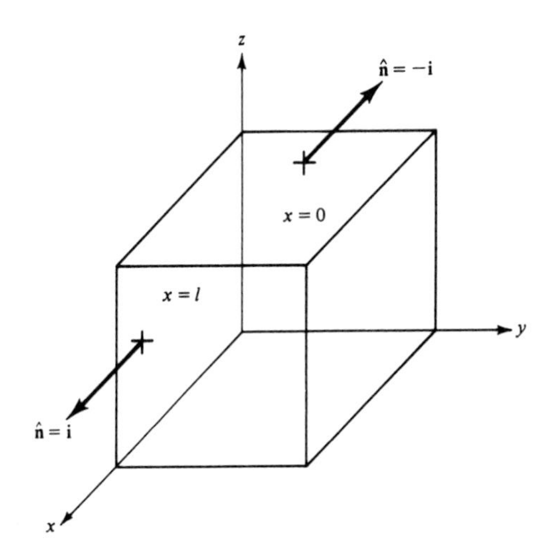

FIGURE 5.5

Similar considerations apply to the planes perpendicular to Oy and Oz. Each contributes l^3 to the flux, and the final result is

$$\oiint_S \mathbf{F} \cdot \hat{\mathbf{n}}\, dS = 3l^3,$$

as desired.

3

Example 2 Derive an equation that governs the temperature distribution $T(x, y, z, t)$ in a solid.

A great deal of knowledge was accumulated by experimentation before the nature of heat as thermal energy and the gross mechanisms of its transfer were understood. The basic facts are these:

1. The amount of heat required to raise the temperature of a small sample of material T degrees is proportional to T and the mass of the sample. If the volume element is dV and ρ is the material density, the heat energy that must be added to this increment to raise it T degrees is $\rho cT\, dV$. Here c, the proportionality factor, is called the *specific heat* and is a property of the material which we assume to be a constant.

2. Heat is conducted from hot to cold (Fig. 5.6). More specifically, the rate at which heat is conducted across an element of surface area dS whose normal direction is $\hat{\mathbf{n}}$ is given by $-k\hat{\mathbf{n}} \cdot \nabla T\, dS$. The negative sign accounts for the flow

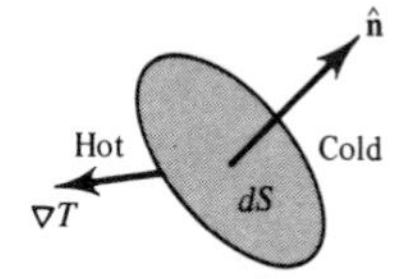

FIGURE 5.6

in the direction antiparallel to the gradient; the proportionality factor k, known as the *coefficient of thermal conductivity*, is a property of the material and is also assumed to be constant.

In the simplest circumstances, the law governing the disposition of thermal energy states that the total energy in this form is conserved. Therefore, the net thermal energy that enters an arbitrary volume V of material through its surface S must equal the amount of heat absorbed by the material as its temperature increases. (There is no loss or gain of heat in the absence of a source or sink of such energy, i.e., the conversion of other energy to this form.) The equation that governs the temperature is obtained by formulating the principle of energy conservation into mathematical symbolism. Suppose that the temperature of a material element $\rho\, dV$ increases by ΔT degrees in a time span Δt. The increase in thermal energy in this element is $c\, \Delta T\, \rho\, dV$, and this, of course, depends on the location of the infinitesimal volume within V. The total increase in heat energy of the entire volume in this time is

$$\iiint_V c\rho \, \Delta T \, dV.$$

Where does this energy come from? According to the conservation principle, it enters through the surface S from the surrounding material, and we assume this is by conduction only. The amount of heat which *enters* V through a surface element dS in time Δt is

$$k\hat{\mathbf{n}} \cdot \nabla T \, dS \, \Delta t.$$

Therefore, the total amount of heat that *enters* through S in this small time is

$$\oiint_S k\hat{\mathbf{n}} \cdot \nabla T \, dS \, \Delta t.$$

The conservation principle asserts that

$$\left(\oiint_S k\hat{\mathbf{n}} \cdot \nabla T \, dS\right) \Delta t = \iiint_V c\rho \, \Delta T \, dV,$$

or in the limit $\Delta t \to 0$ and $\Delta T/\Delta t \to \partial T/\partial t$,

$$\oiint_S k\hat{\mathbf{n}} \cdot \nabla T \, dS = \iiint_V c\rho \frac{\partial T}{\partial t} \, dV. \tag{6}$$

The divergence theorem can now be applied to eliminate any reference to the arbitrary volume V. Since by (1),

$$\oiint_S k\hat{\mathbf{n}} \cdot \nabla T \, dS = \iiint_V \nabla \cdot (k \, \nabla T) \, dV,$$

Eq. (6) can be written entirely as a volume integral:

$$\iiint_V \nabla \cdot (k \, \nabla T) \, dV = \iiint_V \frac{\partial}{\partial t}(c\rho T) \, dV,$$

or

$$\iiint_V \left[\nabla \cdot (k \, \nabla T) - \frac{\partial}{\partial t}(c\rho T)\right] dV = 0. \tag{7}$$

The fact that the volume of integration in this expression is arbitrary implies that the integrand must itself be zero. To see this, assume the contrary, in which case there must be some region where the integrand as a continuous function is of one sign, say positive. However, if V is chosen to be entirely within the region where the integrand is positive, the triple integral in (7) would then be strictly positive, contradicting the assertion that it is indeed identically zero for all V. We conclude that the integrand *must* be zero everywhere:

$$\frac{\partial}{\partial t}(\rho c T) = \frac{\partial}{\partial x}\left(k\frac{\partial T}{\partial x}\right) + \frac{\partial}{\partial y}\left(k\frac{\partial T}{\partial y}\right) + \frac{\partial}{\partial z}\left(k\frac{\partial T}{\partial z}\right).$$

Since ρ, c, and k are assumed to be constant, this can be written

$$\frac{\partial T}{\partial t} = \kappa\left(\frac{\partial^2 T}{\partial x^2} + \frac{\partial^2 T}{\partial y^2} + \frac{\partial^2 T}{\partial z^2}\right), \tag{8}$$

where $\kappa = k/\rho c$ is called the *coefficient of thermometric conductivity*. Equation (8), called the *heat equation*, describes an entire range of diffusive phenomena.

4

A number of useful results follow from the divergence theorem; two such formulas are derived here.

First the divergence theorem is applied to the vector $\mathbf{F} = \psi\mathbf{a}$, where $\mathbf{a}$ is an arbitrary constant vector and ψ is any scalar function. Equation (1) becomes

$$\oint\!\!\!\!\oint_S \psi\mathbf{a}\cdot\hat{\mathbf{n}}\,dS = \iiint_V \operatorname{div}\psi\mathbf{a}\,dV. \tag{9}$$

The identity

$$\boldsymbol{\nabla}\cdot\psi\mathbf{a} = \psi\boldsymbol{\nabla}\cdot\mathbf{a} + \boldsymbol{\nabla}\psi\cdot\mathbf{a},$$

which when $\mathbf{a}$ is a constant reduces to

$$\boldsymbol{\nabla}\cdot\psi\mathbf{a} = \boldsymbol{\nabla}\psi\cdot\mathbf{a},$$

allows (9) to be expressed as

$$\mathbf{a}\cdot\oint\!\!\!\!\oint_S \hat{\mathbf{n}}\psi\,dS = \mathbf{a}\cdot\iiint_V \boldsymbol{\nabla}\psi\,dV$$

or

$$\mathbf{a}\cdot\left(\oint\!\!\!\!\oint_S \hat{\mathbf{n}}\psi\,dS - \iiint_V \boldsymbol{\nabla}\psi\,dV\right) = 0.$$

This equation states that the scalar product of two vectors is zero; and since $\mathbf{a}$ is an *arbitrary* constant vector, it follows that the second vector concerned *must* be zero. This establishes the fundamental formula

$$\oint\!\!\!\!\oint_S \hat{\mathbf{n}}\psi\,dS = \iiint_V \boldsymbol{\nabla}\psi\,dV. \tag{10}$$

Next the divergence theorem is applied to the vector $\mathbf{F}\times\mathbf{a}$ so that

$$\oint\!\!\!\!\oint_S \mathbf{F}\times\mathbf{a}\cdot\hat{\mathbf{n}}\,dS = \iiint_V \operatorname{div}(\mathbf{F}\times\mathbf{a})\,dV. \tag{11}$$

The vector identity for the divergence of a cross product (see equation (11), Sec. 8.4)

$$\operatorname{div} \mathbf{F} \times \mathbf{a} = \mathbf{a} \cdot \operatorname{curl} \mathbf{F} - \mathbf{F} \cdot \operatorname{curl} \mathbf{a},$$

which for $\mathbf{a}$ constant becomes

$$\operatorname{div} \mathbf{F} \times \mathbf{a} = \mathbf{a} \cdot \operatorname{curl} \mathbf{F},$$

can be used to simplify (11). It follows that

$$\mathbf{a} \cdot \left(\oiint_S \hat{\mathbf{n}} \times \mathbf{F}\, dS - \iiint_V \operatorname{curl} \mathbf{F}\, dV \right) = 0,$$

or, since $\mathbf{a}$ is an arbitrary vector

$$\oiint_S \hat{\mathbf{n}} \times \mathbf{F}\, dS = \iiint_V \operatorname{curl} \mathbf{F}\, dV. \tag{12}$$

The three equations (1), (10), and (12) can be written in a unified symbolic manner:

$$\oiint_S \hat{\mathbf{n}} * (\quad)\, dS = \iiint_V \nabla * (\quad)\, dV, \tag{13}$$

where $*$ may be a cross, a dot, or a blank space when the parentheses are filled with a vector $\mathbf{F}$ or a scalar, ψ, as the case may be.

One important application of formulas (10) and (12) is in a theoretical derivation of the principle of Archimedes: "When a body that is partially or totally immersed in a liquid is in static equilibrium, the buoyancy force is equal to the weight of the displaced liquid and acts at the center of mass of the displaced liquid." An immediate conclusion is that in equilibrium the buoyancy force $\mathbf{B}$ and the weight of the body $\mathbf{W}$ (the force of gravity) are of equal magnitudes but oppositely directed (see Fig. 5.7).

To establish this principle we determine the buoyancy force $\mathbf{B}$ exerted on the body by the surrounding fluid. This necessitates calculation of the total force on the surface of the body due to the fluid pressure. The pressure p acts perpendicular to each element of surface dS along the inward normal (Fig. 5.8), so that the force on this area incre-

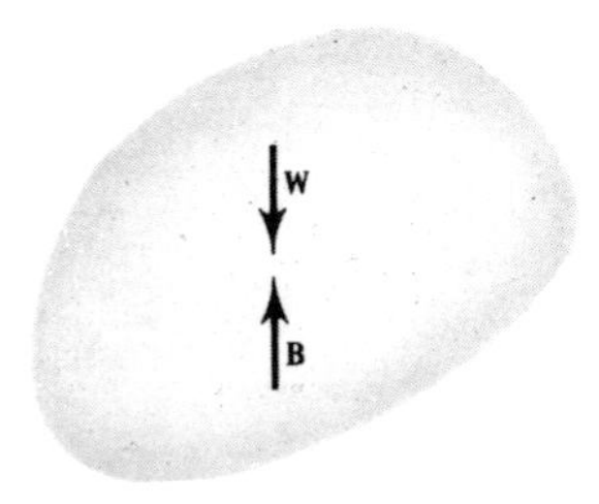

FIGURE 5.7

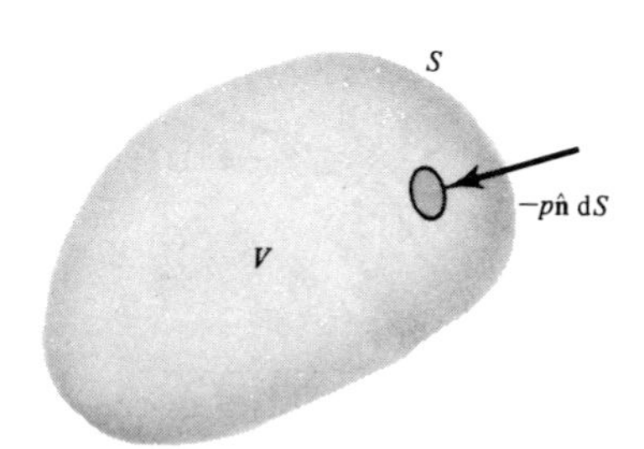

FIGURE 5.8

ment is $-p\hat{\mathbf{n}}\,dS$. The buoyancy force is the sum total of these force increments and is given by the integral

$$\mathbf{B} = -\oiint_S p\hat{\mathbf{n}}\,dS. \tag{14}$$

This, according to (10), can be transformed into a volume integral:

$$\mathbf{B} = -\iiint_V \nabla p\,dV. \tag{15}$$

The pressure in a fluid at rest is related to the fluid density ρ by the hydrostatic law

$$\nabla p = \rho\mathbf{g}, \tag{16}$$

where $\mathbf{g}$ is the gravitational acceleration. The substitution of (16) into (15) yields

$$\mathbf{B} = -\iiint_V \rho\mathbf{g}\,dV.$$

However, $\rho\,dV = dm$ is in fact the mass of the displaced fluid that filled this volume increment. Consequently, the preceding integral can be written

$$\mathbf{B} = -\iiint_V \mathbf{g}\,dm = -\mathbf{g}\iiint_V dm = -\mathbf{W}, \tag{17}$$

where $\mathbf{W}$ is the weight of fluid displaced. Therefore, in equilibrium,

$$\mathbf{W} + \mathbf{B} = \mathbf{0},$$

which is Archimedes' principle.

We proceed to calculate the torque on the body produced by the pressure force on its surface S (see Fig. 5.9). The torque $\mathbf{L}$ about a vector position $\mathbf{r}_0$, as defined in Sec. 5.4, is

$$\mathbf{L} = \oiint_S (\mathbf{r} - \mathbf{r}_0) \times (-p\hat{\mathbf{n}})\,dS = \oiint_S \hat{\mathbf{n}} \times p(\mathbf{r} - \mathbf{r}_0)\,dS.$$

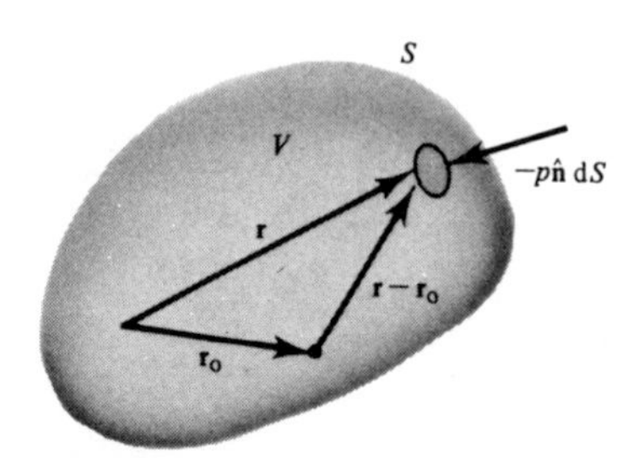

FIGURE 5.9

By using (12), this equation can be transformed to a volume integral:

$$\mathbf{L} = \iiint_V \operatorname{curl}[p(\mathbf{r} - \mathbf{r}_0)]\, dV.$$

But

$$\operatorname{curl} p(\mathbf{r} - \mathbf{r}_0) = p \operatorname{curl}(\mathbf{r} - \mathbf{r}_0) + \nabla p \times (\mathbf{r} - \mathbf{r}_0),$$

and since $\operatorname{curl}(\mathbf{r} - \mathbf{r}_0) = \mathbf{0}$, it follows that

$$\mathbf{L} = \iiint_V \nabla p \times (\mathbf{r} - \mathbf{r}_0)\, dV.$$

However, the pressure is hydrostatic, and (16) lets us express this as

$$\mathbf{L} = -\iiint_V (\mathbf{r} - \mathbf{r}_0) \times \mathbf{g}\rho\, dV.$$

Since $\mathbf{g}$ is constant and $\rho dV = dm$, this implies that the torque $\mathbf{L}$ about a point $\mathbf{r}_0$ vanishes when

$$\mathbf{r}_0 \iiint_V dm = \iiint_V \mathbf{r}\, dm,$$

in which case the buoyancy force passes through $\mathbf{r}_0 = \bar{\mathbf{r}}$, the center of mass of the displaced fluid.

Archimedes' principle establishes the conditions for a body to be in equilibrium. It can also be used to decide whether an equilibrium position is a stable configuration, a problem of much practical importance.

Consider a symmetrical body that is tilted very slightly from its equilibrium position of flotation. In general, the location of center of buoyancy will move, say from H to H', as shown in Fig. 5.10. The buoyancy and gravitational forces remain of equal magnitude but no longer act along the same vertical line. As a consequence, these forces then produce a torque on the body, which tends either to restore the body to its equilibrium position or to accentuate the displacement. The equilibrium is said to be stable in the former case and unstable in the latter.

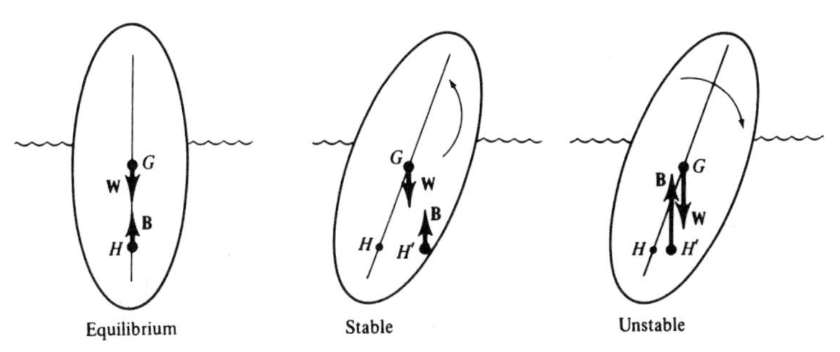

FIGURE 5.10

Example 3 A space capsule of mass M has a hemispherical base of radius a and a conical top of height h (Fig. 5.11). On splashdown the capsule should float in an "upright position," so that only the spherical surface is immersed in the water. What design specifications are necessary?

Consider the extreme case when the capsule floats so that its entire hemispherical bottom is submerged. From Archimedes' principle we know that the weight and buoyancy forces must balance, and this implies

$$Mg = \tfrac{2}{3}\rho\pi a^3 g$$

or

$$M = \tfrac{2}{3}\rho\pi a^3,$$

where ρ is the density of seawater. In order to ensure that only the spherical part of the capsule or a fraction thereof is under water, the mass of the capsule must be less than the extreme value just calculated. For example, with $a = 1$ m and $\rho = 10^3$ kg/m^3, M must be less than about 2100 kg.

In this particular case, the question of stability can be answered rather simply by purely geometrical arguments because the submerged surface is completely spherical, so that the center of buoyancy H is *always* the center of the sphere.

As shown in Fig. 5.12 the forces tend to restore the capsule to its equilibrium position when the center of mass of the body lies *below* the center of the hemisphere.

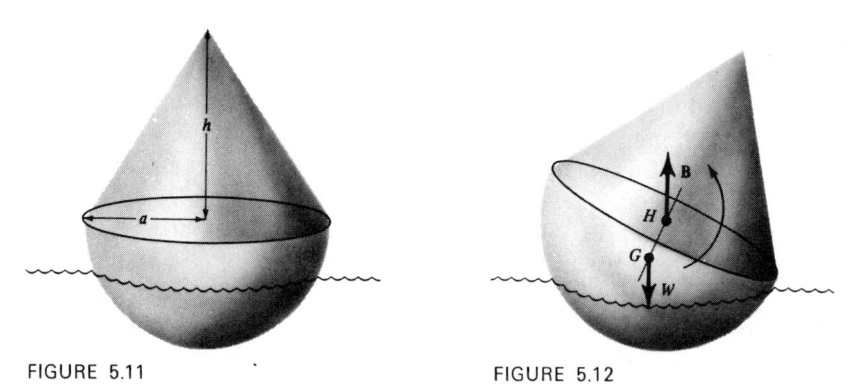

FIGURE 5.11 FIGURE 5.12

Exercises 8.5

1. Verify the divergence theorem for the following vector functions $\mathbf{F}(x, y, z)$ where V is the unit sphere $x^2 + y^2 + z^2 = 1$. (Evaluate the surface and volume integrals independently.)

(i) $\mathbf{F} = x\mathbf{i} + y\mathbf{j} + z\mathbf{k}$ **(ii)** $x^2(\mathbf{i} + \mathbf{j} + \mathbf{k})$ **(iii)** $(x^2 + y^2 + z^2)(x\mathbf{i} + y\mathbf{j} + z\mathbf{k})$

2. Verify the divergence theorem for the following vector functions $\mathbf{F}(x, y, z)$ where V is the rectangular cube bounded by the coordinate planes and the planes $x = a$, $y = b$ and $z = c$.

(i) $\mathbf{F} = x\mathbf{i} + y\mathbf{j} + z\mathbf{k}$ **(ii)** $x^2yz\mathbf{i} + xy^2z\mathbf{j} + xyz^2\mathbf{k}$ **(iii)** $xy\mathbf{i} + yz\mathbf{j} + zx\mathbf{k}$

3. Verify the divergence theorem for the following functions $\mathbf{F}(x, y, z)$ where S is the closed surface formed by the cylinder $x^2 + y^2 = a^2$ and the planes $z = 0$, $z = h$. Check the dimensions of the answers.

(i) $\mathbf{F} = x\mathbf{i} + y\mathbf{j} + z\mathbf{k}$ (ii) $x^2\mathbf{i} + y^2\mathbf{j} + z^2\mathbf{k}$ (iii) $x\mathbf{i} - y\mathbf{j}$

4. Show that the volume V of any region bounded by the closed surface S can be expressed as $V = \frac{1}{3} \oint\!\!\oint_S \mathbf{r} \cdot \mathbf{n}\, dS$. Verify that this agrees with the calculations in exercises **1(i)**, **2(i)** and **3(i)**.

5. Evaluate the flux $\oint\!\!\oint_S \mathbf{F} \cdot \mathbf{n}\, dS$ for the following vector fields $\mathbf{F}(x, y, z)$ where S is the closed surface formed by the hemisphere $z = (a^2 - x^2 - y^2)^{1/2}$ and the plane $z = 0$.

(i) $\mathbf{F} = x\mathbf{i} + y\mathbf{j} + z\mathbf{k}$ (ii) $x^2y\mathbf{i} + xy^2\mathbf{j}$ (iii) $xyz(\mathbf{i} + \mathbf{j} + \mathbf{k})$

6. Find div $\mathbf{G}$ when $\mathbf{G} = (x\mathbf{i} + y\mathbf{j} + z\mathbf{k})/(x^2 + y^2 + z^2)$ and calculate $\iiint_V \text{div}\, \mathbf{G}\, dV$, where V is the volume between the two spheres $x^2 + y^2 + z^2 = a^2$ and $x^2 + y^2 + z^2 = b^2$, $b > a$.

7. If $\mathbf{F}$ and ψ are arbitrary vector and scalar functions and S is any closed surface, prove that $\oint\!\!\oint_S \hat{\mathbf{n}} \cdot \text{curl}\, \mathbf{F}\, dS$ and $\oint\!\!\oint_S \hat{\mathbf{n}} \times \nabla\psi\, dS$ are both zero.

8. Use the identity div $f\mathbf{F} = f$ div $\mathbf{F} + \nabla f \cdot \mathbf{F}$ to deduce the formula

$$\oint\!\!\oint_S f\mathbf{F} \cdot \hat{\mathbf{n}}\, dS = \iiint_V (f \,\text{div}\, \mathbf{F} + \nabla f \cdot \mathbf{F})\, dV.$$

9. Prove that $\iiint_V \nabla^2 f\, dV = \oint\!\!\oint_S \frac{\partial f}{\partial n}\, dS$ where $\frac{\partial f}{\partial n}$ is the directional derivative of the scalar function $f(x, y, z)$ in the outward normal direction on the closed surface S.

10. Establish the identity

$$\nabla \cdot (f\nabla g - g\nabla f) = f\nabla^2 g - g\nabla^2 f$$

and deduce that

$$\oint\!\!\oint_S \left(f\frac{\partial g}{\partial n} - g\frac{\partial f}{\partial n}\right) dS = \iiint_V (f\nabla^2 g - g\nabla^2 f) dV.$$

11. If $r = (x^2 + y^2 + z^2)^{1/2}$, use the result of the previous problem to verify that

$$\oint\!\!\oint_S \left[\frac{1}{r}\nabla g - g\nabla\left(\frac{1}{r}\right)\right] \cdot \hat{\mathbf{n}}\, dS = \iiint_V \frac{\nabla^2 g}{r}\, dV + k$$

where $k = 4\pi g(0, 0, 0)$ if V encloses the origin and $k = 0$ otherwise.

12. Show that the position of the center of mass of a uniform volume V enclosed by a surface S can be expressed as

$$\bar{\mathbf{r}} = \frac{1}{2V} \oint\!\!\oint_S r^2 \hat{\mathbf{n}}\, dS.$$

13. Use the divergence theorem to establish the identity

$$\oint\!\!\oint_S [\mathbf{G} \times (\nabla \times \mathbf{F}) - \mathbf{F} \times (\nabla \times \mathbf{G})] \cdot \hat{\mathbf{n}}\, dS = \iiint_V [\mathbf{F} \cdot \nabla \times (\nabla \times \mathbf{G}) - \mathbf{G} \cdot \nabla \times (\nabla \times \mathbf{F})]\, dV.$$

14. The temperature T in a body is zero over a closed surface S surrounding a volume V. Show that

$$\iiint_V \frac{\partial}{\partial t}\left(\frac{T^2}{2}\right) dV = -\iiint_V \kappa |\nabla T|^2\, dV.$$

(*Hint:* $\partial T/\partial t = k\nabla^2 T$ implies $\partial(\frac{1}{2}T^2)/\partial t = kT\nabla^2 T = k[\nabla \cdot (T\nabla T) - |\nabla T|^2]$.)

15. In an ocean of depth h the density of seawater at a depth z is well approximated by the formula $\rho = \rho_s + (\rho_b - \rho_s)\, z/h$, where ρ_s and ρ_b refer to the surface and bottom densities. In a series of oceanographic measurements it is planned to position three spherical floats of radius a at depths $h/4$, $h/2$, and $3h/4$. How should the floats be weighted so that they will remain at these positions? Assume $a \leq h/4$.

16. The electric potential function $\phi(x, y, z)$ at the point $P(x, y, z)$ due to a point charge q at $P_1(x_1, y_1, z_1)$ is given by $\phi(x, y, z) = q/r$ where $r = [(x - x_1)^2 + (y - y_1)^2 + (z - z_1)^2]^{1/2}$. Calculate the electric field $\mathbf{E} = -\nabla\phi$ due to the point charge and draw the field lines. Verify that $\oiint_S \mathbf{E} \cdot \hat{\mathbf{n}}\, dS = 4\pi q$ where S is any surface enclosing the point charge and the integral is zero otherwise.

17. The potential $\phi(x, y, z)$ at the point $P(x, y, z)$ due to a continuous distribution of charge over a volume V with density ρ is given by $\phi = \iiint_V \rho/r\, dV$. Verify that

$\oiint_S \mathbf{E} \cdot \hat{\mathbf{n}}\, dS = 4\pi \iiint_V \rho\, dV$ (Gauss' law) and $\nabla^2\phi = -4\pi\rho$ (Poisson's equation).

8.6 The theorems of Green and Stokes

According to the divergence theorem, the flux of a vector over a closed surface S equals the volume integral of its divergence:

$$\oiint_S \mathbf{F} \cdot \hat{\mathbf{n}}\, dS = \iiint_V \operatorname{div} \mathbf{F}\, dV. \tag{1}$$

An important special case of this result, called *Green's theorem*, arises when $\mathbf{F}$ is taken as a two-dimensional vector field and the volume V is chosen to be a cylinder. Specifically, let

$$\mathbf{F} = F_1(x, y)\mathbf{i} + F_2(x, y)\mathbf{j}, \tag{2}$$

and V be the volume of the cylinder of unit height, bounded by $z = 0$ and $z = 1$, whose cross section is a region R of the xy plane, as shown in Fig. 6.1.

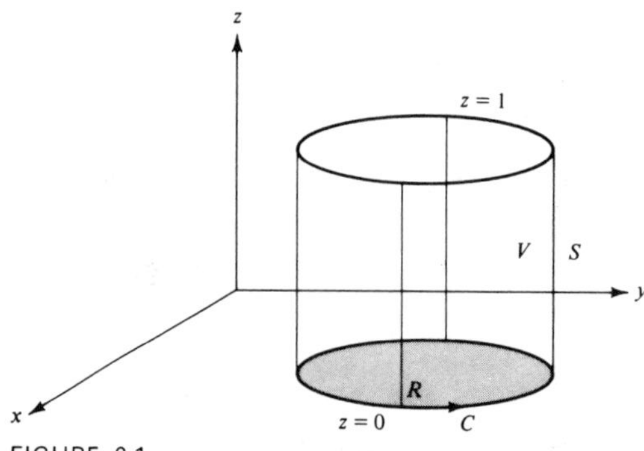

FIGURE 6.1

For this configuration, $dV = dx\,dy\,dz$, and because the integrand does not depend on z, the volume integral in (1) can be rewritten as an area integral over R:

$$\iiint_V \operatorname{div} \mathbf{F}\, dV = \iint_R \left(\frac{\partial F_1}{\partial x} + \frac{\partial F_2}{\partial y}\right) dx\,dy \int_0^1 dz,$$

or

$$\iiint_V \operatorname{div} \mathbf{F}\, dV = \iint_R \left(\frac{\partial F_1}{\partial x} + \frac{\partial F_2}{\partial y}\right) dx\,dy. \tag{3}$$

Next the particular value of the surface integral in (1) is calculated. On the respective planes $z = 0$, $z = 1$, the unit normal is $\hat{\mathbf{n}} = -\mathbf{k}$, $\hat{\mathbf{n}} = \mathbf{k}$, and since $\mathbf{F} \cdot \hat{\mathbf{n}} = 0$ in each case, it follows that there is no contribution to the flux from the two end plates. On the lateral surface of the cylinder, the element of surface area dS can be identified as a rectangular strip of unit height and width ds. As shown in Fig. 6.2, ds is an element of arc length of the curve C which bounds the plane region R. Therefore

$$dS = 1\,ds,$$

and with this substitution, the lateral surface integral can be represented as a line integral:

$$\oiint_S \mathbf{F} \cdot \hat{\mathbf{n}}\, dS = \oint_C \mathbf{F} \cdot \hat{\mathbf{n}}\, ds. \tag{4}$$

A more convenient form is obtainable. Since $\mathbf{k}$, $\hat{\mathbf{n}}$, and the unit tangent vector $\hat{\mathbf{T}}$ to the curve C form a right-handed triad (see Fig. 6.3), it follows that

$$\hat{\mathbf{n}} = \hat{\mathbf{T}} \times \mathbf{k}. \tag{5}$$

The replacement of $\hat{\mathbf{n}}$ in the line integral in (4) yields

$$\oiint_S \mathbf{F} \cdot \hat{\mathbf{n}}\, dS = \oint_C (F_1\mathbf{i} + F_2\mathbf{j}) \cdot \hat{\mathbf{T}} \times \mathbf{k}\, ds$$

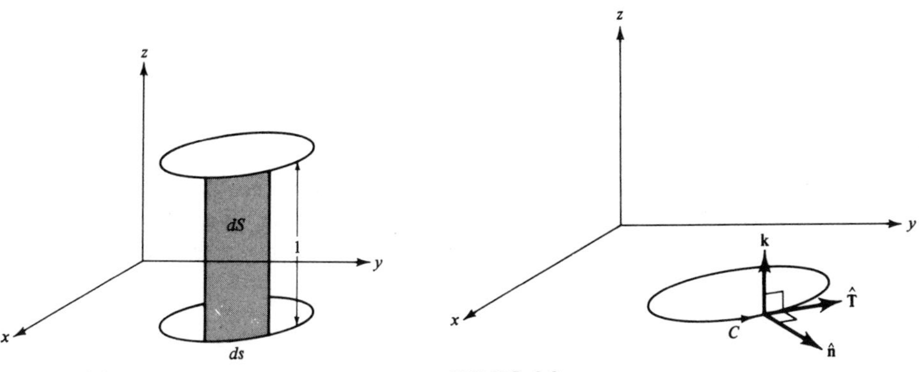

FIGURE 6.2

FIGURE 6.3

or, by a slight rearrangement,

$$\oiint_S \mathbf{F} \cdot \hat{\mathbf{n}}\, dS = \oint_C \mathbf{k} \times (F_1\mathbf{i} + F_2\mathbf{j}) \cdot \hat{\mathbf{T}}\, ds.$$

However, the *vector* element of arc length of the bounding curve C is $d\mathbf{r} = \hat{\mathbf{T}}\, ds$, so that the preceding equation is equivalent to

$$\oiint_S \mathbf{F} \cdot \hat{\mathbf{n}}\, dS = \oint_C (F_1\mathbf{j} - F_2\mathbf{i}) \cdot d\mathbf{r}. \tag{6}$$

Note that Eq. (5), in specifying $\hat{\mathbf{T}}$, implies that the positive direction of integration about C is counterclockwise.

We may now equate (3) and (6) to obtain one form of Green's theorem:

$$\iint_R \left(\frac{\partial F_1}{\partial x} + \frac{\partial F_2}{\partial y}\right) dx\, dy = \oint_C (F_1\, dy - F_2\, dx), \tag{7}$$

where the relations $\mathbf{i} \cdot d\mathbf{r} = dx$ and $\mathbf{j} \cdot d\mathbf{r} = dy$ have been used.

Equation (7) is usually written

$$\iint_R \left(\frac{\partial N}{\partial x} - \frac{\partial M}{\partial y}\right) dx\, dy = \oint_C (M\, dx + N\, dy), \tag{8}$$

which is obtained by setting $F_1 = N$, $F_2 = -M$.

In either form Green's theorem shows that the line integral around a closed plane curve C can be expressed as a double integral over the enclosed area R. (An alternative method of deriving (8) is indicated in the exercises.)

In the special case $M = -y$, $N = 0$, Eq. (8) reads

$$\iint_R dx\, dy = -\oint_C y\, dx,$$

and for $M = 0$, $N = x$ we find that

$$\iint_R dx\, dy = \oint_C x\, dy.$$

Each of these expressions relates the area A of the region R to a closed line integral about its perimeter (see Fig. 6.4). This is often a particularly easy way to calculate plane areas; a more symmetric form for this purpose is obtained by adding the two equations, whereupon

$$A = \frac{1}{2}\oint_C (x\, dy - y\, dx). \tag{9}$$

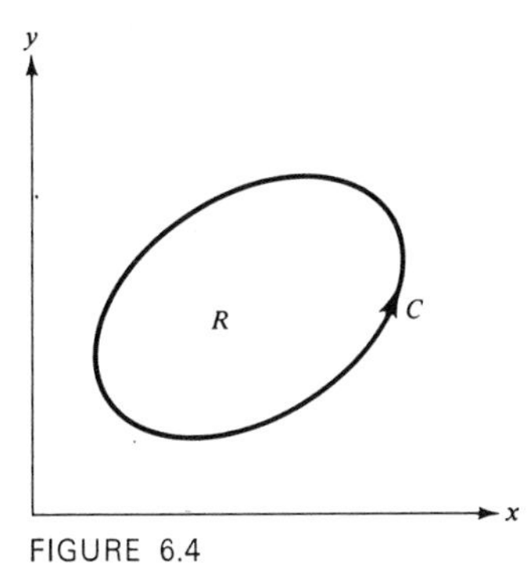

FIGURE 6.4

Example 1 Show that the value $\oint_C(2y\,dx + 3x\,dy)$ is the area enclosed by the curve C. Set $M = 2y$ and $N = 3x$ in (8) to obtain

$$\oint_C (2y\,dx + 3x\,dy) = \iint_R \left[\frac{\partial}{\partial x}(3x) - \frac{\partial}{\partial y}(2y)\right] dx\,dy,$$

$$= \iint_R dx\,dy = A,$$

where A is the area enclosed by C. [Compare this calculation with the direct computation (Example 3, Sec. 8.2) in the special case when C is the circle $x^2 + y^2 = 1$.]

Example 2 Use Green's theorem to evaluate $\oint_C[(x - xy)\,dx + (y^3 + 1)\,dy]$, where C is the boundary of a square with vertices at the points (1, 0), (2, 0), (2, 1), and (1, 1), as shown in Fig. 6.5.

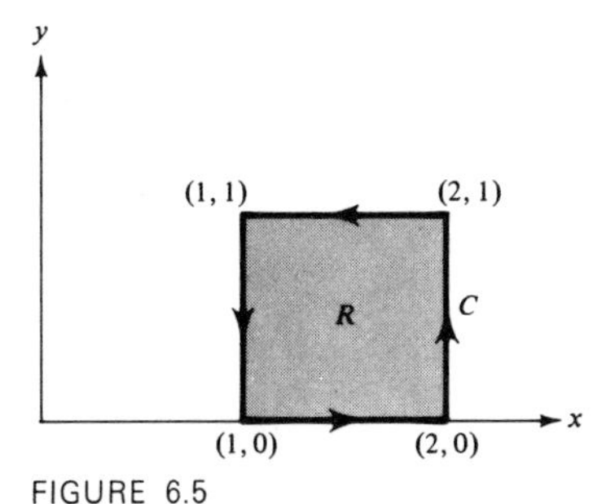

FIGURE 6.5

Green's theorem, Eq. (8), is used with the identifications $M = x - xy$ and $N = y^3 + 1$, so that

$$\oint_C [(x - xy)\,dx + (y^3 + 1)\,dy] = \iint\limits_R \left[\frac{\partial}{\partial x}(y^3 + 1) - \frac{\partial}{\partial y}(x - xy)\right] dx\,dy,$$

$$= \iint\limits_R x\,dx\,dy.$$

The value of the double integral is easily obtained:

$$\iint\limits_R x\,dx\,dy = \int_0^1 dy \int_1^2 x\,dx = [y]_0^1[\tfrac{1}{2}x^2]_1^2 = \tfrac{3}{2}.$$

Therefore,

$$\oint_C [(x - xy)\,dx + (y^3 + 1)\,dy] = \tfrac{3}{2}.$$

Example 3 Find the area enclosed by the astroid $x^{2/3} + y^{2/3} = a^{2/3}$.

The area is calculated by line integration, using Eq. (9):

$$A = \frac{1}{2}\oint_C (x\,dy - y\,dx),$$

where C is indicated in Fig. 6.6.

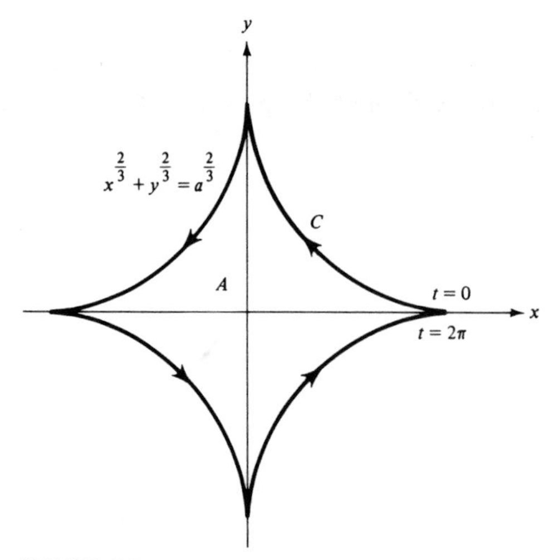

FIGURE 6.6

The line integral is evaluated by utilizing the parametric representation of the curve

$$x = a\cos^3 t, \qquad y = a\sin^3 t.$$

The curve C is traversed once as t ranges from 0 to 2π and the line integral can then be rewritten as

$$A = \frac{1}{2}\int_0^{2\pi} [a\cos^3 t\,(3a\sin^2 t\cos t) - a\sin^3 t\,(-3a\cos^2 t\sin t)]\,dt,$$

$$= \frac{3a^2}{2}\int_0^{2\pi} \sin^2 t\cos^2 t\,dt.$$

But

$$\sin^2 t\cos^2 t = \tfrac{1}{8}(1 - \cos 4t),$$

and the substitution of this in the foregoing allows its integration:

$$A = \frac{3a^2}{16}\left[t - \tfrac{1}{4}\sin 4t\right]_0^{2\pi} = \frac{3\pi a^2}{8}.$$

2

The derivation of Green's theorem showed that it is really just a special case, a two-dimensional form, of the divergence theorem. However, Green's theorem can itself be generalized in a quite different manner to obtain a fundamental relationship that is not at all obvious. Known as *Stokes' theorem*, it relates the line integral of a vector **F** around a closed curve C, which is not necessarily plane, to the integral of the *normal* component of curl **F** over *any open surface* S whose perimeter is C (see Fig. 6.7). In mathematical terms, Stokes' theorem is

$$\oint_C \mathbf{F}\cdot d\mathbf{r} = \iint_S \hat{\mathbf{n}}\cdot \operatorname{curl}\mathbf{F}\,dS, \tag{10}$$

where $\hat{\mathbf{n}}$ is the unit normal to S drawn in the right-handed sense with respect to the positive direction for C.

Before giving a formal derivation of (10), it is instructive to understand how and why this formula might be anticipated. For this purpose, we return to Eq. (8) and now replace M and N by F_1 and F_2:

$$\oint_C (F_1 dx + F_2 dy) = \iint_R \left(\frac{\partial F_2}{\partial x} - \frac{\partial F_1}{\partial y}\right) dx\,dy,$$

which in vector notation is

$$\oint_C \mathbf{F}\cdot d\mathbf{r} = \iint_R \mathbf{k}\cdot \operatorname{curl}\mathbf{F}\,dx\,dy, \tag{11}$$

where R is a surface in the xy plane, with perimeter C and unit normal $\mathbf{k}$ (Fig. 6.8). Equation (11), which is a form of Green's theorem, is closely related to the original integral definition of curl $\mathbf{F}$, and for this reason it should come as no surprise.

A plausible derivation of Stokes' theorem can be based on the special result in (11). For this purpose the curved surface S with perimeter C is completely divided into many smaller regions, as shown in Fig. 6.9. By this means, every small element of surface area dS is made essentially planar, so that (11) is applicable to each if $\mathbf{k}$ is replaced by the local unit normal $\hat{\mathbf{n}}$. However, when all the equations for all increments are added together, the line integrals along the interior boundaries cancel identically (because each is traversed twice in opposite directions). Therefore, the only line integral

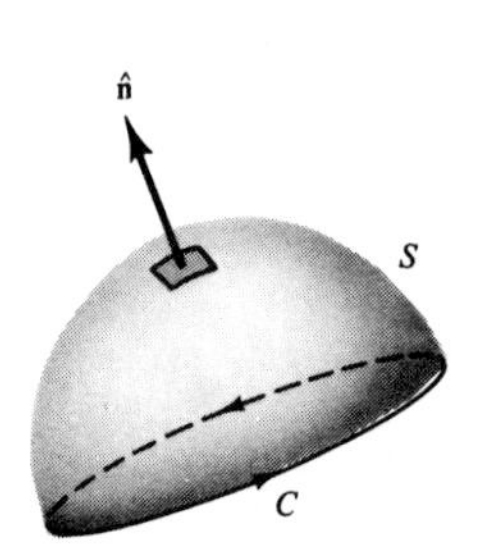

FIGURE 6.7

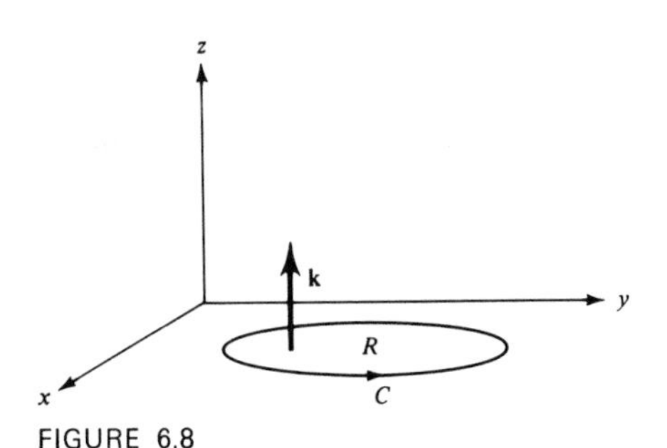

FIGURE 6.8

segments that survive this process constitute a closed integral about the outer perimeter C; that is, $\oint_C \mathbf{F} \cdot d\mathbf{r}$. But the addition of the surface integrals gives $\iint_S \hat{\mathbf{n}} \cdot \text{curl}\, \mathbf{F}\, dS$, and by equating the two results Stokes' theorem is obtained. This argument can be made quite precise, but rather than proceed along these lines an alternative approach is adopted.

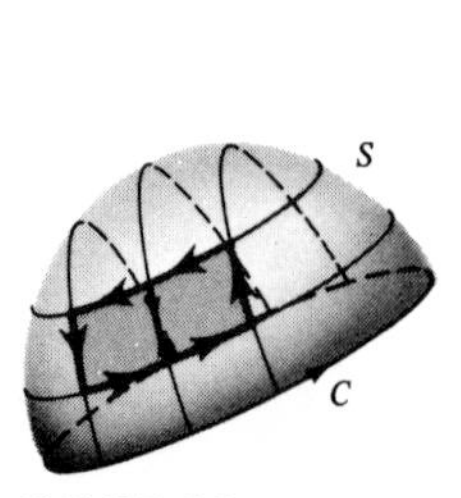

FIGURE 6.9

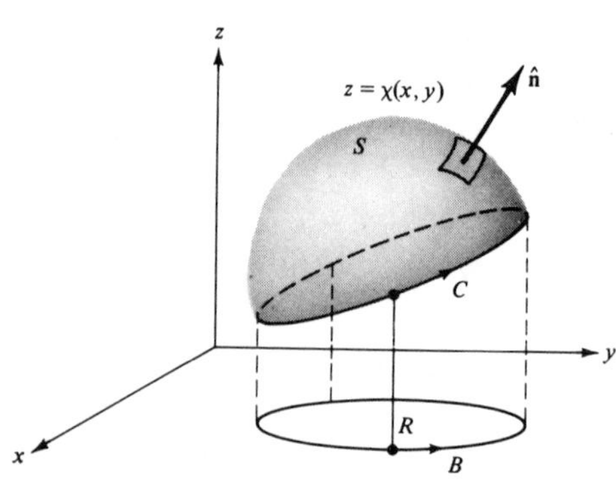

FIGURE 6.10

A formal derivation of Stokes' theorem will be given for the configuration shown in Fig. 6.10. The basic idea is to make use of Green's theorem for the plane region R bounded by the closed curve B, where R and B are the projections of S and C onto the xy plane. Let S be the surface $z = \chi(x, y)$ and C the curve $x = x(t)$, $y = y(t)$, $z = z(t)$, where t ranges from t_0 to t_1. Since C is a part of S it follows that $z(t) = \chi(x(t), y(t))$.

Consider the first term $\oint_C F_1\, dx$ of the line integral $\oint_C \mathbf{F} \cdot d\mathbf{r}$, which in terms of the parameter t becomes

$$\oint_C F_1(x, y, z)\, dx = \int_{t_0}^{t_1} F_1(x(t), y(t), z(t)) \frac{dx}{dt}\, dt,$$

or since C lies on the surface $z = \chi(x, y)$,

$$\oint_C F_1(x, y, z)\, dx = \int_{t_0}^{t_1} F_1(x(t), y(t), \chi(x(t), y(t))) \frac{dx}{dt}\, dt.$$

However, the right-hand side of the last equation is equivalent to a line integral taken around B (which is the projection of C onto the xy plane), so that

$$\oint_C F_1(x, y, z)\, dx = \oint_B F_1(x, y, \chi(x, y))\, dx. \tag{12}$$

Green's theorem applies to contour B because it is a *plane* closed curve. Therefore, by (8), with $M = F_1$, $N = 0$ and $C = B$ we obtain

$$\oint_B F_1(x, y, \chi(x, y))\, dx = \iint_R \left(- \frac{\partial F_1}{\partial y} - \frac{\partial F_1}{\partial z} \frac{\partial \chi}{\partial y} \right) dx\, dy, \tag{13}$$

where R is the projection of S onto the xy plane. Now the process is reversed and (13) is rewritten as a surface integral over S. For this purpose we recall that

$$(\hat{\mathbf{n}} \cdot \mathbf{k})\, dS = dA = dx\, dy, \tag{14}$$

where
$$\hat{\mathbf{n}} = \frac{- \partial\chi/\partial x\, \mathbf{i} - \partial\chi/\partial y\, \mathbf{j} + \mathbf{k}}{\sqrt{1 + (\partial\chi/\partial x)^2 + (\partial\chi/\partial y)^2}},$$

is the unit normal to S. Since

$$\frac{\hat{\mathbf{n}} \cdot \mathbf{j}}{\hat{\mathbf{n}} \cdot \mathbf{k}} = - \frac{\partial \chi}{\partial y}, \tag{15}$$

Eqs. (14) and (15) can be used to rewrite (13) as

$$\oint_B F_1(x, y, \chi(x, y))\, dx = \iint_S \left(- \frac{\partial F_1}{\partial y} + \frac{\partial F_1}{\partial z} (\hat{\mathbf{n}} \cdot \mathbf{j})/(\hat{\mathbf{n}} \cdot \mathbf{k}) \right) (\hat{\mathbf{n}} \cdot \mathbf{k})\, dS,$$

$$= \iint_S \hat{\mathbf{n}} \cdot \left(\mathbf{j} \frac{\partial F_1}{\partial z} - \mathbf{k} \frac{\partial F_1}{\partial y} \right) dS.$$

According to (12), this is equivalent to

$$\oint_C F_1(x, y, z)\, dx = \iint_S \hat{\mathbf{n}} \cdot \left(\mathbf{j}\frac{\partial F_1}{\partial z} - \mathbf{k}\frac{\partial F_1}{\partial y}\right) dS. \tag{16}$$

Analogous results can be derived for each of the other components of $\mathbf{F}$:

$$\oint_C F_2(x, y, z)\, dy = \iint_S \hat{\mathbf{n}} \cdot \left(\mathbf{k}\frac{\partial F_2}{\partial x} - \mathbf{i}\frac{\partial F_2}{\partial z}\right) dS, \tag{17}$$

$$\oint_C F_3(x, y, z)\, dz = \iint_S \hat{\mathbf{n}} \cdot \left(\mathbf{i}\frac{\partial F_3}{\partial y} - \mathbf{j}\frac{\partial F_3}{\partial x}\right) dS. \tag{18}$$

The addition of (16), (17), and (18) yields

$$\oint_C (F_1\, dx + F_2\, dy + F_3\, dz) = \iint_S \hat{\mathbf{n}} \cdot \left[\mathbf{i}\left(\frac{\partial F_3}{\partial y} - \frac{\partial F_2}{\partial z}\right) + \mathbf{j}\left(\frac{\partial F_1}{\partial z} - \frac{\partial F_3}{\partial x}\right) + \mathbf{k}\left(\frac{\partial F_2}{\partial x} - \frac{\partial F_1}{\partial y}\right)\right] dS,$$

which in vector notation is

$$\oint_C \mathbf{F} \cdot d\mathbf{r} = \iint_S \hat{\mathbf{n}} \cdot \operatorname{curl} \mathbf{F}\, dS, \tag{19}$$

and the derivation is complete. Stokes' theorem relates the line integral of a vector $\mathbf{F}$ around a closed curve C to the flux of curl $\mathbf{F}$ over *any* open surface S whose perimeter is C. The surface S can always be subdivided into regions which project uniquely onto the coordinate planes, and in this way the analysis just given can be applied to a general surface. Of course, the result is again Eq. (19).

If S_1 and S_2 are two surfaces with the perimeter C, then together they form a closed surface of volume V. Since div curl $\mathbf{F} = 0$, the divergence theorem may be used to convert the volume integral, $\iiint \operatorname{div} \operatorname{curl} \mathbf{F}\, dV = 0$, into surface integrals over S_1 and S_2; this provides another simple explanation of why the surface S in (19) is indeed *any* open surface with perimeter C. (For details, see Exercise 25.)

Example 4 Verify Stokes' theorem when $\mathbf{F} = y\mathbf{i} + 2z\mathbf{j} + 3x\mathbf{k}$ and S is the hemispherical surface $z = +\sqrt{1 - x^2 - y^2}$, as shown in Fig. 6.11.

First let us calculate $\oint_C \mathbf{F} \cdot d\mathbf{r}$, where C is defined by $x = \cos t$, $y = \sin t$, $z = 0$ and t ranges from 0 to 2π. On this curve (the equatorial circle)

$$\mathbf{F} \cdot d\mathbf{r} = (\mathbf{i} \sin t + 3\mathbf{k} \cos t) \cdot (-\mathbf{i} \sin t + \mathbf{j} \cos t)\, dt = -\sin^2 t\, dt,$$

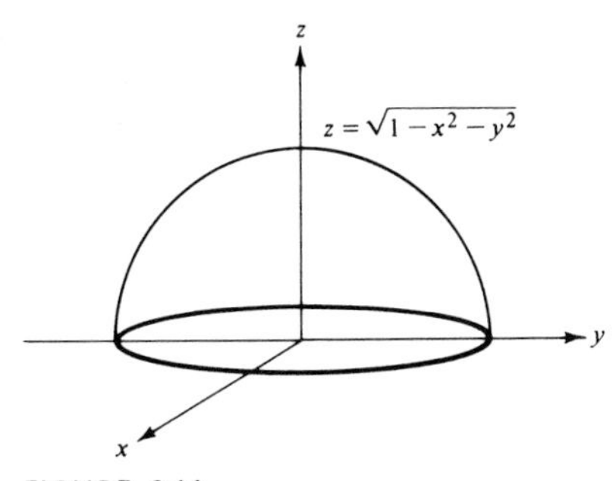

FIGURE 6.11

so that

$$\oint_C \mathbf{F} \cdot d\mathbf{r} = -\int_0^{2\pi} \sin^2 t \, dt = -\pi.$$

In order to evaluate $\iint_S \hat{\mathbf{n}} \cdot \text{curl}\, \mathbf{F}\, dS$ over the hemisphere S we first calculate

$$\text{curl}\, \mathbf{F} = \begin{vmatrix} \mathbf{i} & \mathbf{j} & \mathbf{k} \\ \dfrac{\partial}{\partial x} & \dfrac{\partial}{\partial y} & \dfrac{\partial}{\partial z} \\ y & 2z & 3x \end{vmatrix} = -2\mathbf{i} - 3\mathbf{j} - \mathbf{k}.$$

Since the unit normal to S is $\hat{\mathbf{n}} = x\mathbf{i} + y\mathbf{j} + z\mathbf{k}$,

$$\hat{\mathbf{n}} \cdot \text{curl}\, \mathbf{F} = -2x - 3y - z,$$

and therefore

$$\iint_S \hat{\mathbf{n}} \cdot \text{curl}\, \mathbf{F}\, dS = -\iint_S (2x + 3y + z)\, dS.$$

The first two terms of this integral are zero (by symmetry arguments), whereupon

$$\iint_S \hat{\mathbf{n}} \cdot \text{curl}\, \mathbf{F}\, dS = -\iint_S z\, dS.$$

This integration is most easily performed by transforming to spherical coordinates, and the result,

$$\iint_S \hat{\mathbf{n}} \cdot \text{curl}\, \mathbf{F}\, dS = -\int_0^{2\pi} d\theta \int_0^{\pi/2} \sin\varphi \cos\varphi \, d\varphi = -\pi,$$

is identical to that obtained earlier for the line integral.

As the last important point, we note that Stokes' theorem provides an alternative and simple proof of the fact that if curl $\mathbf{F} = \mathbf{0}$ everywhere, then $\oint_C \mathbf{F} \cdot d\mathbf{r} = 0$ for all closed paths C, and conversely. These conclusions, which were established earlier in Sec. 8.4, follow immediately from Eq. (19).

Exercises 8.6

1. Verify Green's theorem $\oint_C M\,dx + N\,dy = \iint_R (\partial N/\partial x - \partial M/\partial y)\,dxdy$ for the following functions $M(x, y)$, $N(x, y)$ where C is the circle $x^2 + y^2 = 1$. (Evaluate the two integrals independently.)

(i) $M = y^2, N = x$ **(ii)** $M = y, N = -x$ **(iii)** $M = x^2 + y^2, N = x^2 - y^2$

2. Verify Green's theorem for the following functions $M(x, y)$, $N(x, y)$ where R is the rectangular region $1 \leq x \leq 2$, $1 \leq y \leq 3$.

(i) $M = e^y, N = e^x$ **(ii)** $M = y^2, N = x^2$ **(iii)** $M = (x + y)^2, N = (x - y)^2$

3. If C is the closed curve in the first quadrant defined by $y = x^2$ and $y^2 = x$, verify Green's theorem for the following integrals.

(i) $\oint_C y^2\,dx + x^2\,dy$ **(ii)** $\oint_C 3xy\,dx + x^2\,dy$ **(iii)** $\oint_C xe^y\,dx + ye^x\,dy$

4. Give a direct proof of Green's theorem by integrating the expressions $\iint_R \frac{\partial N}{\partial x}\,dxdy$ and $\iint_R \frac{\partial M}{\partial y}\,dxdy$.

5. If F_T and F_N are the tangential and normal components of $\mathbf{F}$ on a plane closed curve C, find equivalent double integrals for $\oint_C F_T\,ds$ and $\oint_C F_N\,ds$, where s denotes the arc length of C.

6. If C' is a curve interior to C and R is the region between C and C', show that

$$\iint_R \left(\frac{\partial N}{\partial x} - \frac{\partial M}{\partial y}\right) dxdy = \oint_C (M\,dx + N\,dy) - \oint_{C'} (M\,dx + N\,dy),$$

where the directions of C and C' are indicated in Fig. 6.12.

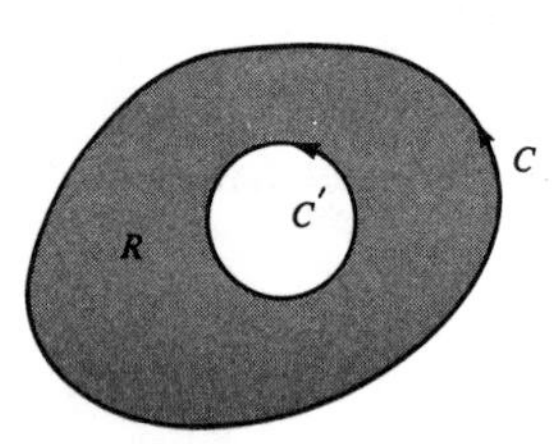

FIGURE 6.12

7. Generalize the result of the previous problem to cases where there is more than one interior curve.

8. Verify Green's theorem for the following functions $M(x, y)$, $N(x, y)$ where R is the annular region $a^2 \leq x^2 + y^2 \leq b^2$.

(i) $M = xy, N = x^2 - y^2$ **(ii)** $M = x^2 + y^2, N = 0$ **(iii)** $M = x^2 + y^2, N = x^2 - y^2$

9. By the formulas $A = \oint_C x\,dy = -\oint_C y\,dx = \frac{1}{2}\oint_C (x\,dy - y\,dx)$, find the areas bounded by the following curves.

(i) $\dfrac{x^2}{a^2} + \dfrac{y^2}{b^2} = 1$ **(ii)** $x^{1/2} + y^{1/2} = a^{1/2}$ **(iii)** $x^{2/3} + y^{2/3} = a^{2/3}$

10. In polar coordinates (r, θ), show that $x\,dy - y\,dx = r^2\,d\theta$. By means of graphs, show that $\frac{1}{2}\oint_C r^2\,d\theta$ represents the area bounded by the simple closed contour C.

11. Use the result of the previous problem to find the areas bounded by the following curves.

(i) $r = a(1 + \cos\theta)$ **(ii)** $r = a\sin 2\theta$ **(iii)** $r^2 = a^2\cos 2\theta$

12. By evaluating either $\oint_C x\,dy$ or $-\oint_C y\,dx$, find the areas enclosed by the following curves.

(i) $(x-2)^2 + (y-2)^2 = 4$ **(ii)** $y = x^2,\ x = y^2$ **(iii)** $y = x^3,\ x = y^3$

13. Verify that the formula for area after a change of variables $x = x(u, v)$, $y = y(u, v)$ is given by $\iint_R dx\,dy = \iint_{R'} \partial(x, y)/\partial(u, v)\,du\,dv$ where the region R in the xy plane is mapped to the region R' in the uv plane. (*Hint:* In the line integral for the area, write $dx = (\partial x/\partial u)\,du + (\partial x/\partial v)\,dv$, $dy = (\partial y/\partial u)\,du + (\partial y/\partial v)\,dv$ and apply Green's theorem in the uv coordinates.)

14. Verify Stokes' theorem $\oint_C \mathbf{F}\cdot d\mathbf{r} = \iint_S \hat{\mathbf{n}}\cdot \text{curl}\,\mathbf{F}\,dS$ for the following vector functions $\mathbf{F}(x, y, z)$ where S is the hemispherical surface $z = (a^2 - x^2 - y^2)^{1/2}$.

(i) $\mathbf{F} = z\mathbf{i} + x\mathbf{j} + y\mathbf{k}$ **(ii)** $yz\mathbf{i} + zx\mathbf{j} + xy\mathbf{k}$ **(iii)** $(x^2 + y^2 + z^2)(\mathbf{i} + \mathbf{j} + \mathbf{k})$

15. For the vector functions in the previous problem, calculate $\iint_S \hat{\mathbf{n}}\cdot \text{curl}\,\mathbf{F}\,dS$ where S is the circular region $x^2 + y^2 \le a^2$ in the $z = 0$ plane. Why are the answers equal?

16. Verify Stokes' theorem for the vector function $\mathbf{F}(x, y, z) = (x + y)\mathbf{i} + 2(x + y)\mathbf{j} + z^2\mathbf{k}$ where S is the portion of the sphere $x^2 + y^2 + z^2 = 25$ above the plane $z = 4$.

17. Evaluate $\oint_C x\,dx + (x + y)\,dy + (x + y + z)\,dz$ where C is the curve of intersection of the cylinder $x^2 + y^2 = 1$ and the plane $z = x$.

18. Evaluate $\oint_C [(y + z)\,dx + (z + x)\,dy + (x + y)\,dz]$, where C is the curve of intersection of the surfaces $x^2 + y^2/2 + z^2/3 = 1$, $z = x^2 + 2y^2$.

19. Evaluate $\oint_C y^2\,dx + xy\,dy + xz\,dz$ where C is the curve of intersection of the cylinder $x^2 + y^2 = 2ax$ and the plane $z = x$.

20. By direct calculation show that $\oint_C (z\,dx + x\,dy + y\,dz) = \pi\sqrt{3}$, where C is the circle $x + y + z = 0$, $x^2 + y^2 + z^2 = 1$. Obtain the same result by using Stokes' theorem.

21. If $f(x, y, z)$ and $g(x, y, z)$ are scalar functions, verify the following identities.

(i) $\oint_C f\nabla g\cdot d\mathbf{r} = -\oint_C g\nabla f\cdot d\mathbf{r}$ **(ii)** $\oint_C f\nabla g\cdot d\mathbf{r} = \iint_S \hat{\mathbf{n}}\cdot \nabla f\times\nabla g\,dS$

22. Use Stokes' theorem to deduce that $\oint_C \psi\,d\mathbf{r} = \iint_S \hat{\mathbf{n}}\times\nabla\psi\,dS$. (*Hint:* Let $\mathbf{F} = \psi\mathbf{a}$ in Stokes' theorem and suppose $\mathbf{a}$ is a constant vector.)

23. If $\oint_C \mathbf{F}\times d\mathbf{r} = \mathbf{0}$ for all closed curves C, show that $\mathbf{F}$ must be a constant vector. (*Hint:* The x component of $\oint_C \mathbf{F}\times d\mathbf{r}$ is $\mathbf{i}\cdot\oint_C \mathbf{F}\times d\mathbf{r} = \oint_C (\mathbf{i}\times\mathbf{F})\cdot d\mathbf{r}$.)

24. If φ and ψ are arbitrary scalar functions, show that there is a vector $\mathbf{A}$ such that curl $\mathbf{A} = \nabla\varphi\times\nabla\psi$. Calculate $\iint_S \nabla\varphi\times\nabla\psi\cdot\hat{\mathbf{n}}\,dS$ when $\varphi = (x + y + z)^2$, $\psi = x^2 - y^2 + z^2$, and S is the curved surface of the hemisphere $x^2 + y^2 + z^2 = 1$, $z \ge 0$.

25. Let S_1, S_2 be two surfaces with the same perimeter C. Since $S = S_1 + S_2$ is a closed surface of volume V, use the divergence theorem

$$\iiint_V \operatorname{div} \operatorname{curl} \mathbf{F} \, dV = \oiint_{S=S_1+S_2} \hat{\mathbf{n}} \cdot \operatorname{curl} \mathbf{F} \, dS$$

to show that

$$\iint_{S_1} \hat{\mathbf{n}} \cdot \operatorname{curl} \mathbf{F} \, dS = \iint_{S_2} \hat{\mathbf{n}} \cdot \operatorname{curl} \mathbf{F} \, dS.$$

(In the last equation, $\hat{\mathbf{n}}$ is the unit normal to S_1 or S_2 defined by traversing the contour C in the positive sense. In this context the outward normal to V on S_2 is $-\hat{\mathbf{n}}$.)

8.7 Fluid motion

Thus far, the discussion has for the most part emphasized the mathematical rather than the physical significance of the vector calculus. To some extent this is unavoidable until a certain minimal capability is acquired, but the real justification for any mathematical analysis lies in showing that the results are truly useful. We therefore proceed to examine one very common physical situation, the motion of a fluid, i.e., a liquid or a gas. The aim is to show how the vector calculus enables basic physical laws to be translated into mathematical statements (and how relatively simple models of complex phenomena are devised).

In essence a fluid motion is specified by a knowledge of the velocity vector $\mathbf{v}$ and the density ρ as functions of position x, y, z and time t, that is,

$$\mathbf{v} = \mathbf{v}(x, y, z, t), \qquad \text{and} \qquad \rho = \rho(x, y, z, t). \tag{1}$$

(A myriad of rather minor approximations is already involved here.) Relative to fixed cartesian axes the velocity $\mathbf{v}$ will be written

$$\mathbf{v} = v_1 \mathbf{i} + v_2 \mathbf{j} + v_3 \mathbf{k}, \tag{2}$$

where v_1, v_2, and v_3 are the velocity components in the x, y, and z directions.

An eminently sensible law governing the macroscopic fluid phenomena of everyday life is the *conservation of mass*, which states that mass is neither created nor destroyed. The mathematical formulation of this rule is developed from its implications regarding the flow through an arbitrary closed surface S lying entirely within the fluid medium (Fig. 7.1). The fixed reference or control surface S, which contains a volume V, is

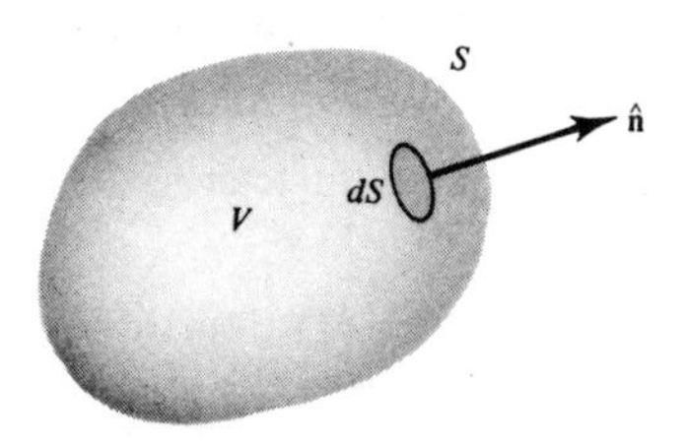

FIGURE 7.1

an imaginary theoretical device, a purely geometrical entity, that in no way affects the fluid motion. In other words the fluid passes through S unhindered. The conservation of mass asserts, then, that the rate at which fluid *enters* V through S must equal the rate at which mass *accumulates* within V. It remains to translate this word equation into the appropriate symbolism. The total mass within the volume V at any time t is

$$M = \iiint_V \rho \, dV, \tag{3}$$

and its time rate of change is

$$\frac{dM}{dt} = \frac{d}{dt} \iiint_V \rho \, dV = \iiint_V \frac{\partial \rho}{\partial t} \, dV, \tag{4}$$

since the volume V is fixed. One half of the formulation of the word equation is now complete, and we turn to the other. It was shown in Sec. 8.3 (see page 689) that the element of mass dM which flows *out* through S in a time dt is

$$dt \oiint_S \rho \mathbf{v} \cdot \hat{\mathbf{n}} \, dS,$$

and as a consequence, the *rate* at which mass *enters* through S is

$$\frac{dM}{dt} = - \oiint_S \rho \mathbf{v} \cdot \hat{\mathbf{n}} \, dS. \tag{5}$$

This provides the last half of the mathematical translation. The rates described in (4) and (5) must be equal, and consequently the mathematical statement of the conservation of mass is

$$\iiint_V \frac{\partial \rho}{\partial t} \, dV + \oiint_S \rho \mathbf{v} \cdot \hat{\mathbf{n}} \, dS = 0. \tag{6}$$

Equation (6) is true for an arbitrary closed surface S which contains a volume V. A more useful relationship is found by using the divergence theorem to convert the second surface integral into a volume integral; in this way we find that

$$\iiint_V \left(\frac{\partial \rho}{\partial t} + \operatorname{div} \rho \mathbf{v} \right) dV = 0. \tag{7}$$

However, since V is an arbitrary volume and the functions concerned are *assumed* to be continuous, it follows that the integrand must be zero:

$$\frac{\partial \rho}{\partial t} + \operatorname{div} \rho \mathbf{v} = 0. \tag{8}$$

Equation (8), a partial differential equation, is equivalent to the principle of conservation of mass; it is a much more convenient and accessible mathematical statement because it does not involve any particular volume.

---

2 As a second application, we consider the far less obvious natural law discovered by Newton. This governs the change of momentum, and its word statement with respect to a control surface S is as follows:

> The rate of change of momentum within $V =$ the rate of change of momentum flowing into V through $S +$ the forces exerted on the mass within V by the rest of the universe. (9)

Once again, the individual terms of this equation must be described symbolically to translate it into a formal mathematical statement. These are considered sequentially.

The fluid of density ρ and velocity $\mathbf{v}$ within the volume element dV at any instant has a momentum equal to $\rho\mathbf{v}\,dV$ (momentum equals mass times velocity). Therefore, the total momentum $\mathbf{h}$ within volume V is

$$\mathbf{h} = \iiint_V \rho\mathbf{v}\,dV,$$

and its time rate of change is given by

$$\frac{d\mathbf{h}}{dt} = \iiint_V \frac{\partial}{\partial t}(\rho\mathbf{v})\,dV. \tag{10}$$

In a time dt, a mass of fluid $\rho\mathbf{v}\cdot\hat{\mathbf{n}}\,dS\,dt$ flows out of S through dS. The momentum of this fluid is $\mathbf{v}(\rho\mathbf{v}\cdot\hat{\mathbf{n}}\,dS\,dt)$. Accordingly, $\oint\!\!\oint_S \mathbf{v}(\rho\mathbf{v}\cdot\hat{\mathbf{n}})\,dS$ is the *rate* at which momentum flows *out* through S; the rate at which momentum flows *in* through S is

$$-\oint\!\!\oint_S \mathbf{v}\,(\rho\mathbf{v}\cdot\hat{\mathbf{n}})\,dS. \tag{11}$$

In the absence of frictional effects (an approximation) the presence of the exterior fluid is felt through the pressure p, which acts along the inward normal to S. Therefore, the surface force on S due to the exterior fluid is

$$-\oint\!\!\oint_S p\hat{\mathbf{n}}\,dS. \tag{12}$$

If the external force per unit mass is $\mathbf{F}$, for example, the gravitation acceleration vector $\mathbf{g}$, the body force on a small mass increment $\rho\,dV$ is $\mathbf{F}\rho\,dV$. Therefore the total body force acting on V is

$$\iiint_V \rho\mathbf{F}\,dV. \tag{13}$$

On substituting expressions (10) to (13) into Eq. (9) we obtain an integral statement of the momentum balance:

$$\iiint_V \frac{\partial}{\partial t}(\rho \mathbf{v})\, dV = -\oint\!\!\oint_S \mathbf{v}(\rho\mathbf{v}\cdot\hat{\mathbf{n}})\, dS - \oint\!\!\oint_S p\hat{\mathbf{n}}\, dS + \iiint_V \rho\mathbf{F}\, dV. \tag{14}$$

This vector relationship is equivalent to Newton's second law applied to a fixed but arbitrary volume V.

In order to reduce (14) to a more useful form that is independent of V, the surface integrals are replaced by volume integrals according to Eq. (10) of Sec. 8.5. For example, $\oint\!\!\oint_S p\hat{\mathbf{n}}\, dS$ is replaced by $\iiint_V \nabla p\, dV$, in which case (14) becomes

$$\iiint_V \frac{\partial}{\partial t}(\rho \mathbf{v})\, dV = -\oint\!\!\oint_S \mathbf{v}(\rho\mathbf{v}\cdot\hat{\mathbf{n}})\, dS - \iiint_V \nabla p\, dV + \iiint_V \rho\mathbf{F}\, dV. \tag{15}$$

At this stage it is simpler to proceed by examining the component equations separately. The x component is

$$\iiint_V \frac{\partial}{\partial t}(\rho v_1)\, dV = -\oint\!\!\oint_S v_1(\rho\mathbf{v}\cdot\hat{\mathbf{n}})\, dS - \iiint_V \frac{\partial p}{\partial x}\, dV + \iiint_V \rho F_1\, dV. \tag{16}$$

In this form, the divergence theorem can be applied to the remaining surface integral to yield

$$\iiint_V \left\{\frac{\partial}{\partial t}(\rho v_1) + \operatorname{div}[v_1(\rho\mathbf{v})] + \frac{\partial p}{\partial x} - \rho F_1\right\} dV = 0. \tag{17}$$

Since V is arbitrary and all functions are again assumed continuous, the integrand itself must be zero. Another simplification results by making use of the identity

$$\operatorname{div} v_1\rho\mathbf{v} = v_1 \operatorname{div} \rho\mathbf{v} + \nabla v_1 \cdot \rho\mathbf{v},$$

so that the vanishing of the integrand in (17) implies that

$$\rho\frac{\partial v_1}{\partial t} + v_1\left(\frac{\partial \rho}{\partial t} + \operatorname{div} \rho\mathbf{v}\right) + \nabla v_1 \cdot \rho\mathbf{v} = -\frac{\partial p}{\partial x} + \rho F_1. \tag{18}$$

However, Eq. (8), the conservation of mass, shows that the second term on the left in (18) is zero, in which case it reduces to

$$\rho\left(\frac{\partial v_1}{\partial t} + v_1\frac{\partial v_1}{\partial x} + v_2\frac{\partial v_1}{\partial y} + v_3\frac{\partial v_1}{\partial z}\right) = -\frac{\partial p}{\partial x} + \rho F_1. \tag{19}$$

A similar analysis for the y and z components gives

$$\rho\left(\frac{\partial v_2}{\partial t} + v_1\frac{\partial v_2}{\partial x} + v_2\frac{\partial v_2}{\partial y} + v_3\frac{\partial v_2}{\partial z}\right) = -\frac{\partial p}{\partial y} + \rho F_2, \tag{20}$$

$$\rho\left(\frac{\partial v_3}{\partial t} + v_1 \frac{\partial v_3}{\partial x} + v_2 \frac{\partial v_3}{\partial y} + v_3 \frac{\partial v_3}{\partial z}\right) = -\frac{\partial p}{\partial z} + \rho F_3. \tag{21}$$

Alternatively, the three scalar equations (19) to (21) can be written compactly as a vector equation,

$$\rho\left[\frac{\partial \mathbf{v}}{\partial t} + (\mathbf{v} \cdot \nabla)\mathbf{v}\right] = -\nabla p + \rho \mathbf{F}. \tag{22}$$

Equations (8) and (22) are fundamental to fluid phenomena, and with appropriate modifications, viscous, thermal, rotational, and electromagnetic effects can be incorporated. In this way they form the starting point for theoretical investigations in a wide range of subjects: meteorology, oceanography, acoustics, plasmas, aerodynamics, and even certain aspects of traffic flow and astrophysics. The pursuit of these topics takes us beyond "Calculus: An Introduction to Applied Mathematics," a venture for which we have now completed our preparation.

Exercises 8.7

1. Show that Eq. (8) can be rewritten in the form

$$\frac{D\rho}{Dt} + \rho \operatorname{div} \mathbf{v} = 0,$$

where the operator

$$\frac{D}{Dt} = \frac{\partial}{\partial t} + \mathbf{v} \cdot \nabla$$

is called the substantial derivative. (D/Dt corresponds to a time rate of change following the same fluid particle.)

2. Show that

$$\frac{D}{Dt} = \frac{\partial}{\partial t} + v_1 \frac{\partial}{\partial x} + v_2 \frac{\partial}{\partial y} + v_3 \frac{\partial}{\partial z}.$$

3. Verify that Eq. (22) can be expressed as

$$\frac{D\mathbf{v}}{Dt} = -\frac{1}{\rho} \nabla p + \mathbf{F}.$$

4. The vorticity vector $\boldsymbol{\omega}$ in a fluid is defined by $\boldsymbol{\omega} = \operatorname{curl} \mathbf{v}$. Show that $\operatorname{div} \boldsymbol{\omega} = 0$ and deduce that $\oiint_S \boldsymbol{\omega} \cdot \hat{\mathbf{n}}\, dS = 0$.

5. For a fluid of constant density show that Eq. (8) reduces to $\operatorname{div} \mathbf{v} = 0$. In particular, if $\boldsymbol{\omega} = 0$, show that the velocity potential φ (where $\mathbf{v} = \nabla\varphi$) must be a solution of Laplace's equation, $\nabla^2\varphi = 0$.

6. Derive the vector identity $(\mathbf{v} \cdot \nabla)\mathbf{v} = \nabla(\frac{1}{2}\mathbf{v} \cdot \mathbf{v}) - \mathbf{v} \times \boldsymbol{\omega}$, where $\boldsymbol{\omega} = \nabla \times \mathbf{v}$.

7. If a fluid has constant density ρ_0 and zero vorticity ($\nabla \times \mathbf{v} = \mathbf{0}$), show that the equations for conservation of mass and momentum become

$$\nabla^2 \varphi = 0 \qquad \text{and} \qquad \nabla\left(\frac{\partial \varphi}{\partial t} + \tfrac{1}{2}|\nabla \varphi|^2 + \frac{p}{\rho_0}\right) = \mathbf{F}.$$

In the special case when the external forces are conservative ($\mathbf{F} = -\nabla \psi$), deduce Bernoulli's equation:

$$\frac{\partial \varphi}{\partial t} + \tfrac{1}{2}|\nabla \varphi|^2 + \frac{p}{\rho_0} + \psi = \text{constant}.$$

8. For a fluid of constant density, obtain the vorticity equation

$$\partial \boldsymbol{\omega}/\partial t + (\mathbf{v} \cdot \nabla)\boldsymbol{\omega} = (\boldsymbol{\omega} \cdot \nabla)\mathbf{v} + \text{curl } \mathbf{F}, \text{ where } \boldsymbol{\omega} = \nabla \times \mathbf{v}.$$

9. The circulation Γ of a fluid around a closed curve C is defined by $\Gamma = \oint_C \mathbf{v} \cdot d\mathbf{r}$. Deduce that $\Gamma = \iint_S \boldsymbol{\omega} \cdot \hat{\mathbf{n}}\, dS$ and show that, if $\boldsymbol{\omega} = \mathbf{0}$, the circulation is everywhere zero.

Appendix

A.1 Some algebraic formulas

$$n! = n(n-1)(n-2)\cdots 1$$

$$(a+b)^n = a^n + na^{n-1}b + \frac{n(n-1)}{2!}a^{n-2}b^2 + \frac{n(n-1)(n-2)}{3!}a^{n-3}b^3 + \cdots + nab^{n-1} + b^n$$

$$(a \pm b)^2 = a^2 \pm 2ab + b^2$$

$$a^2 - b^2 = (a-b)(a+b)$$

$$a^3 - b^3 = (a-b)(a^2 + ab + b^2)$$

$$a^3 + b^3 = (a+b)(a^2 - ab + b^2)$$

$$a^n - b^n = (a-b)(a^{n-1} + a^{n-2}b + \cdots + b^{n-1})$$

$$(a+b+c)^2 = a^2 + b^2 + c^2 + 2ab + 2bc + 2ca$$

$$a^4 + a^2b^2 + b^4 = (a^2 + ab + b^2)(a^2 - ab + b^2)$$

$$a^4 + b^4 = (a^2 + 2^{1/2}ab + b^2)(a^2 - 2^{1/2}ab + b^2)$$

Solutions of

$$ax^2 + bx + c = 0$$

are

$$x = \frac{1}{2a}[-b \pm (b^2 - 4ac)^{1/2}].$$

The roots are real when $b^2 - 4ac \geq 0$ and imaginary when $b^2 - 4ac < 0$.

$$a^0 = 1; \qquad a^x a^y = a^{x+y}; \qquad a^x/a^y = a^{x-y}; \qquad a^{-x} = 1/a^x; \qquad (a^x)^y = a^{xy}.$$

If

$$\frac{a}{b} = \frac{c}{d}$$

then

$$\frac{a+b}{b} = \frac{c+d}{d}; \qquad \frac{a-b}{a+b} = \frac{c-d}{c+d}; \qquad \frac{a-b}{b} = \frac{c-d}{d}.$$

A.2 Determinants

A determinant of order 2 is a number assigned to a square array of four numbers as follows:

$$\begin{vmatrix} a_1 & b_1 \\ a_2 & b_2 \end{vmatrix} = a_1 b_2 - a_2 b_1.$$

The solution of the linear equations

$$a_1 x + b_1 y = c_1, \qquad a_2 x + b_2 y = c_2$$

is then

$$x = \frac{\begin{vmatrix} c_1 & b_1 \\ c_2 & b_2 \end{vmatrix}}{\begin{vmatrix} a_1 & b_1 \\ a_2 & b_2 \end{vmatrix}}, \qquad y = \frac{\begin{vmatrix} a_1 & c_1 \\ a_2 & c_2 \end{vmatrix}}{\begin{vmatrix} a_1 & b_1 \\ a_2 & b_2 \end{vmatrix}}.$$

A determinant of third order is a number assigned to a square array of nine numbers:

$$\begin{vmatrix} a_1 & b_1 & c_1 \\ a_2 & b_2 & c_2 \\ a_3 & b_3 & c_3 \end{vmatrix} = a_1 b_2 c_3 + a_2 b_3 c_1 + a_3 b_1 c_2 - a_3 b_2 c_1 - a_2 b_1 c_3 - a_1 b_3 c_2.$$

It follows that

$$\begin{vmatrix} a_1 & b_1 & c_1 \\ a_2 & b_2 & c_2 \\ a_3 & b_3 & c_3 \end{vmatrix} = a_1 \begin{vmatrix} b_2 & c_2 \\ b_3 & c_3 \end{vmatrix} - b_1 \begin{vmatrix} a_2 & c_2 \\ a_3 & c_3 \end{vmatrix} + c_1 \begin{vmatrix} a_2 & b_2 \\ a_3 & b_3 \end{vmatrix}.$$

The interchange of any two rows or any two columns changes the sign of the determinant so that

$$\begin{vmatrix} a_1 & b_1 & c_1 \\ a_2 & b_2 & c_2 \\ a_3 & b_3 & c_3 \end{vmatrix} = - \begin{vmatrix} a_2 & b_2 & c_2 \\ a_1 & b_1 & c_1 \\ a_3 & b_3 & c_3 \end{vmatrix} = \begin{vmatrix} a_2 & b_2 & c_2 \\ a_3 & b_3 & c_3 \\ a_1 & b_1 & c_1 \end{vmatrix}$$

$$= - \begin{vmatrix} b_1 & a_1 & c_1 \\ b_2 & a_2 & c_2 \\ b_3 & a_3 & c_3 \end{vmatrix} = \begin{vmatrix} b_1 & c_1 & a_1 \\ b_2 & c_2 & a_2 \\ b_3 & c_3 & a_3 \end{vmatrix}.$$

The solution of the system of three linear equations

$$a_1 x + b_1 y + c_1 z = d_1,$$
$$a_2 x + b_2 y + c_2 z = d_2,$$
$$a_3 x + b_3 y + c_3 z = d_3$$

is

$$x = \frac{\begin{vmatrix} d_1 & b_1 & c_1 \\ d_2 & b_2 & c_2 \\ d_3 & b_3 & c_3 \end{vmatrix}}{\begin{vmatrix} a_1 & b_1 & c_1 \\ a_2 & b_2 & c_2 \\ a_3 & b_3 & c_3 \end{vmatrix}}, \qquad y = \frac{\begin{vmatrix} a_1 & d_1 & c_1 \\ a_2 & d_2 & c_2 \\ a_3 & d_3 & c_3 \end{vmatrix}}{\begin{vmatrix} a_1 & b_1 & c_1 \\ a_2 & b_2 & c_2 \\ a_3 & b_3 & c_3 \end{vmatrix}}, \qquad z = \frac{\begin{vmatrix} a_1 & b_1 & d_1 \\ a_2 & b_2 & d_2 \\ a_3 & b_3 & d_3 \end{vmatrix}}{\begin{vmatrix} a_1 & b_1 & c_1 \\ a_2 & b_2 & c_2 \\ a_3 & b_3 & c_3 \end{vmatrix}}.$$

A.3 Some trigonometric formulas

$$\sin\theta\csc\theta = \cos\theta\sec\theta = \tan\theta\cot\theta = 1$$

$$\tan\theta = \frac{\sin\theta}{\cos\theta}, \qquad \cot\theta = \frac{\cos\theta}{\sin\theta}$$

$$\sec^2\theta = 1 + \tan^2\theta, \qquad \csc^2\theta = 1 + \cot^2\theta$$

$$\sin^2\theta + \cos^2\theta = 1$$

$$\sin(\theta \pm \varphi) = \sin\theta\cos\varphi \pm \cos\theta\sin\varphi$$

$$\cos(\theta \pm \varphi) = \cos\theta\cos\varphi \mp \sin\theta\sin\varphi$$

$$\tan(\theta \pm \varphi) = \frac{\tan\theta \pm \tan\varphi}{1 \mp \tan\theta\tan\varphi}$$

$$\cot(\theta \pm \varphi) = \frac{\cot\theta\cot\varphi \mp 1}{\cot\varphi \pm \cot\theta}$$

$$\sin 2\theta = 2\sin\theta\cos\theta$$

$$\sin 3\theta = 3\sin\theta - 4\sin^3\theta$$

$$\sin 4\theta = 8\cos^3\theta\sin\theta - 4\cos\theta\sin\theta$$

$$\cos 2\theta = 2\cos^2\theta - 1 = \cos^2\theta - \sin^2\theta$$

$$\cos 3\theta = 4\cos^3\theta - 3\cos\theta$$

$$\cos 4\theta = 8\cos^4\theta - 8\cos^2\theta + 1$$

$$\sin\theta \pm \sin\varphi = 2\sin\tfrac{1}{2}(\theta \pm \varphi)\cos\tfrac{1}{2}(\theta \mp \varphi)$$

$$\cos\theta + \cos\varphi = 2\cos\tfrac{1}{2}(\theta + \varphi)\cos\tfrac{1}{2}(\theta - \varphi)$$

$$\cos\theta - \cos\varphi = -2\sin\tfrac{1}{2}(\theta + \varphi)\sin\tfrac{1}{2}(\theta - \varphi)$$

$$\tan\theta \pm \tan\varphi = \frac{\sin(\theta \pm \varphi)}{\cos\theta\cos\varphi}, \qquad \cot\theta \pm \cot\varphi = \pm\frac{\sin(\theta \pm \varphi)}{\sin\theta\sin\varphi}$$

$$\sin\frac{\theta}{2} = \pm\left(\frac{1 - \cos\theta}{2}\right)^{1/2}, \qquad \cos\frac{\theta}{2} = \pm\left(\frac{1 + \cos\theta}{2}\right)^{1/2}$$

$$\tan\frac{\theta}{2} = \frac{1 - \cos\theta}{\sin\theta} = \frac{\sin\theta}{1 + \cos\theta} = \pm\left(\frac{1 - \cos\theta}{1 + \cos\theta}\right)^{1/2}$$

$$\frac{\sin\theta \pm \sin\varphi}{\cos\theta + \cos\varphi} = \tan\tfrac{1}{2}(\theta \pm \varphi)$$

Sine law:

$$\frac{a}{\sin\theta} = \frac{b}{\sin\varphi} = \frac{c}{\sin\psi}$$

Cosine law:

$$a^2 = b^2 + c^2 - 2bc\cos\theta$$

$$b^2 = c^2 + a^2 - 2ca\cos\varphi$$

$$c^2 = a^2 + b^2 - 2ab\cos\psi$$

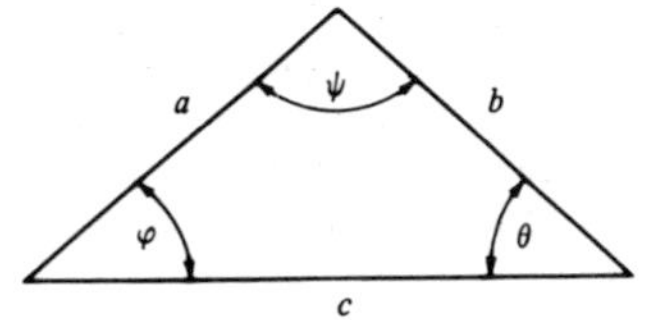

TABLE 3.1

Trigonometric functions

deg	*rad*	*sin*	*cos*	*tan*	*cot*		
0.0	0	0	1	0	∞	1.57080	90.0
1.0	0.01745	0.01745	0.99985	0.01746	57.28996	1.55334	89.0
2.0	0.03491	0.03490	0.99939	0.03492	28.63625	1.53589	88.0
3.0	0.05236	0.05234	0.99863	0.05241	19.08114	1.51844	87.0
4.0	0.06981	0.06976	0.99756	0.06993	14.30067	1.50098	86.0
5.0	0.08727	0.08716	0.99619	0.08749	11.43005	1.48353	85.0
6.0	0.10472	0.10453	0.99452	0.10510	9.51436	1.46608	84.0
7.0	0.12217	0.12187	0.99255	0.12278	8.14435	1.44862	83.0
8.0	0.13963	0.13917	0.99027	0.14054	7.11537	1.43117	82.0
9.0	0.15708	0.15643	0.98769	0.15838	6.31375	1.41372	81.0
10.0	0.17453	0.17365	0.98481	0.17633	5.67128	1.39626	80.0
11.0	0.19199	0.19081	0.98163	0.19438	5.14455	1.37881	79.0
12.0	0.20944	0.20791	0.97815	0.21256	4.70463	1.36136	78.0
13.0	0.22689	0.22495	0.97437	0.23087	4.33148	1.34390	77.0
14.0	0.24435	0.24192	0.97030	0.24933	4.01078	1.32645	76.0
15.0	0.26180	0.25882	0.96593	0.26795	3.73205	1.30900	75.0
16.0	0.27925	0.27564	0.96126	0.28675	3.48741	1.29154	74.0
17.0	0.29671	0.29237	0.95630	0.30573	2.27085	1.27409	73.0
18.0	0.31416	0.30902	0.95106	0.32492	3.07768	1.25664	72.0
19.0	0.33161	0.32557	0.94552	0.34433	2.90421	1.23918	71.0
20.0	0.34907	0.34202	0.93969	0.36397	2.74748	1.22173	70.0
21.0	0.36652	0.35837	0.93358	0.38386	2.60509	1.20428	69.0
22.0	0.38397	0.37461	0.92718	0.40403	2.47509	1.18682	68.0
23.0	0.40143	0.39073	0.92050	0.42447	2.35585	1.16937	67.0
24.0	0.41888	0.40674	0.91355	0.44523	2.24604	1.15192	66.0
25.0	0.43633	0.42262	0.90631	0.46631	2.14451	1.13446	65.0
26.0	0.45379	0.43837	0.89879	0.48773	2.05030	1.11701	64.0
27.0	0.47124	0.45399	0.89101	0.50953	1.96261	1.09956	63.0
28.0	0.48869	0.46947	0.88295	0.53171	1.88073	1.08210	62.0
29.0	0.50615	0.48481	0.87462	0.55431	1.80405	1.06465	61.0
30.0	0.52360	0.50000	0.86603	0.57735	1.73205	1.04720	60.0
31.0	0.54105	0.51504	0.85717	0.60086	1.66428	1.02974	59.0
32.0	0.55851	0.52992	0.84805	0.62487	1.60033	1.01229	58.0
33.0	0.57596	0.54464	0.83867	0.64941	1.53986	0.99484	57.0
34.0	0.59341	0.55919	0.82904	0.67451	1.48256	0.97738	56.0
35.0	0.61087	0.57358	0.81915	0.70021	1.42815	0.95993	55.0
36.0	0.62832	0.58779	0.80902	0.72654	1.37638	0.94248	54.0
37.0	0.64577	0.60182	0.79864	0.75355	1.32704	0.92502	53.0
38.0	0.66323	0.61566	0.78801	0.78129	1.27994	0.90757	52.0
39.0	0.68068	0.62932	0.77715	0.80978	1.23490	0.89012	51.0
40.0	0.69813	0.64279	0.76604	0.83910	1.19175	0.87266	50.0
41.0	0.71558	0.65606	0.75471	0.86929	1.15037	0.85521	49.0
42.0	0.73304	0.66913	0.74314	0.90040	1.11061	0.83776	48.0
43.0	0.75049	0.68200	0.73135	0.93252	1.07237	0.82030	47.0
44.0	0.76794	0.69466	0.71934	0.96569	1.03553	0.80285	46.0
45.0	0.78540	0.70711	0.70711	1.00000	1.00000	0.78540	45.0
		cos	*sin*	*cot*	*tan*	*rad*	*deg*

A.4 Analytic geometry

Equations of a straight line:

1. $y = mx + b$; m is the slope; b is the y intercept.
2. $y - y_1 = m(x - x_1)$; m is the slope; the line passes through point $P_1 = (x_1, y_1)$.
3. $\dfrac{y - y_1}{y_2 - y_1} = \dfrac{x - x_1}{x_2 - x_1}$;

$m = (y_2 - y_1)/(x_2 - x_1)$ is the slope; the line passes through points $P_1 = (x_1, y_1)$ and $P_2 = (x_2, y_2)$. $[(x_2 - x_1)^2 + (y_2 - y_1)^2]^{1/2}$ is the distance between P_2 and P_1.

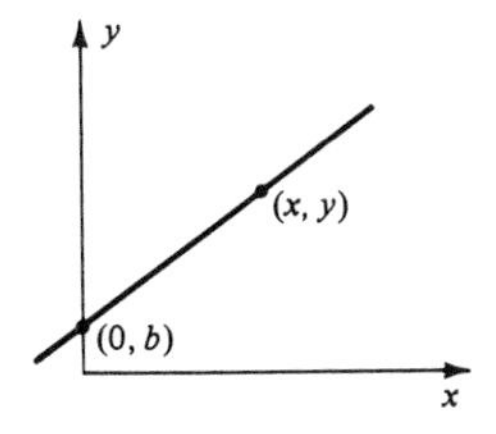

4. $Ax + By + C = 0$; $m = -A/B$ is the slope; $-C/B$ is the y intercept.

The equation of a circle of radius a centered at (x_o, y_o) is

$$(x - x_o)^2 + (y - y_o)^2 = a^2.$$

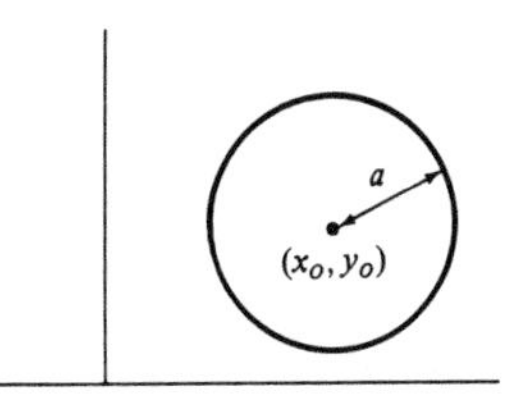

The equation of an ellipse centered at (x_o, y_o) with major axis $2a$ and minor axis $2b$ is

$$\frac{(x - x_o)^2}{a^2} + \frac{(y - y_o)^2}{b^2} = 1.$$

The distance from the center to either focus is $(a^2 - b^2)^{1/2}$; the eccentricity is $k = (a^2 - b^2)^{1/2}/a$; the latus rectum AB is $2b^2/a$; the sum of distances from the foci to point P is a constant; $PF + PF_* = 2a$.

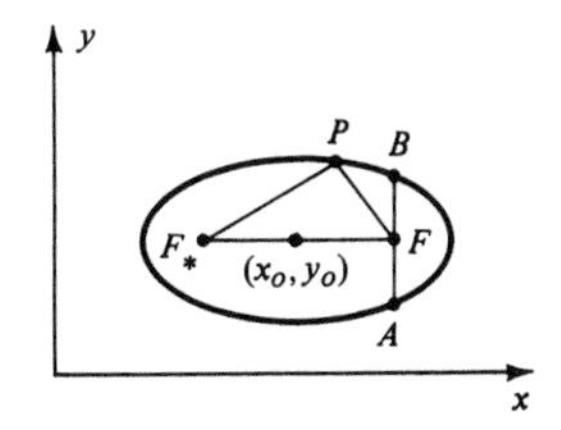

CONIC SECTIONS

If the locus of a point P is such that its distance from a fixed point F (the focus) is a constant multiple k (the eccentricity) of the distance to a fixed straight line (the directrix), then the curve is a conic. The definition

$$\frac{PF}{PM} = k$$

implies

$$x^2 + y^2 = k^2(d - x)^2$$

or

$$r = \frac{kd}{1 + k\cos\theta}$$

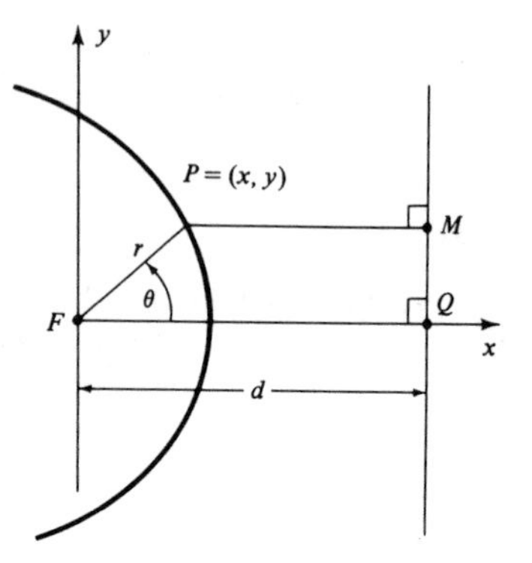

The conic is:

1. An ellipse when $k < 1$.
2. A hyperbola when $k > 1$.
3. A parabola when $k = 1$.

A.5 *Integrals*

TABLE 5.1

Table of integrals

1. $\int x^n\,dx = \frac{x^{n+1}}{n+1},\ n \neq -1$

2. $\int \frac{dx}{x} = \log|x|$

3. $\int a^x\,dx = \frac{a^x}{\log a}$

4. $\int \tan x\,dx = -\log|\cos x|$

5. $\int \cot x\,dx = \log|\sin x|$

6. $\int \sec x\,dx = \log|\sec x + \tan x|$

7. $\int \frac{dx}{x-a} = \log|x - a|$

8. $\int \frac{dx}{x^2+a^2} = \frac{1}{a}\tan^{-1}\frac{x}{a}$

9. $\int \frac{x\,dx}{x^2+a^2} = \tfrac{1}{2}\log(x^2 + a^2)$

10. $\int \frac{dx}{x^2-a^2} = \frac{1}{2a}\log\left|\frac{x-a}{x+a}\right|$

11. $\int \frac{dx}{(a^2-x^2)^{1/2}} = \sin^{-1}\frac{x}{a}$

12. $\int \frac{dx}{(x^2 \pm a^2)^{1/2}} = \log|x + (x^2 \pm a^2)^{1/2}|$

13. $\int (a^2 - x^2)^{1/2}\,dx = \tfrac{1}{2}\left[x(a^2 - x^2)^{1/2} + a^2\sin^{-1}\frac{x}{a}\right]$

14. $\int x(a^2 - x^2)^{1/2}\,dx = -\tfrac{1}{3}(a^2 - x^2)^{3/2}$

15. $\int (x^2 \pm a^2)^{1/2}\,dx = \tfrac{1}{2}x(x^2 \pm a^2)^{1/2} \pm \frac{a^2}{2}\log|x + (x^2 \pm a^2)^{1/2}| \qquad a > 0$

16. $\int \frac{dx}{(a^2 - x^2)^{3/2}} = \frac{x}{a^2(a^2 - x^2)^{1/2}}$

17. $\int \frac{x\,dx}{(a^2 \pm x^2)^{1/2}} = \pm(a^2 \pm x^2)^{1/2}$

18. $\int x(x^2 \pm a^2)^{1/2}\,dx = \frac{1}{3}(x^2 \pm a^2)^{3/2}$

19. $\int \frac{dx}{(x^2 \pm a^2)^{3/2}} = \frac{\pm x}{a^2(x^2 \pm a^2)^{1/2}}$

20. $\int \sin ax\,dx = -\frac{1}{a}\cos ax$

21. $\int \sin^2 ax\,dx = \frac{1}{2a}(ax - \frac{1}{2}\sin 2ax)$

22. $\int \cos ax\,dx = \frac{1}{a}\sin ax$

23. $\int \cos^2 ax\,dx = \frac{1}{2a}(ax + \frac{1}{2}\sin 2ax)$

24. $\int \sec^2 x\,dx = \tan x$

25. $\int \sec x \tan x\,dx = \sec x$

26. $\int \sin ax \sin bx\,dx = \frac{\sin(a-b)x}{2(a-b)} - \frac{\sin(a+b)x}{2(a+b)} \qquad a \neq b$

27. $\int \cos ax \cos bx\,dx = \frac{\sin(a-b)x}{2(a-b)} + \frac{\sin(a+b)x}{2(a+b)} \qquad a \neq b$

28. $\int e^{ax}\,dx = \frac{e^{ax}}{a}$

29. $\int xe^{ax}\,dx = \frac{xe^{ax}}{a} - \frac{e^{ax}}{a^2}$

30. $\int \log x\,dx = x(\log x - 1)$

31. $\int x \log x\,dx = x^2\left(\frac{\log x}{2} - \frac{1}{4}\right)$

32. $\int \sinh ax\,dx = \frac{1}{a}\cosh ax$

33. $\int \cosh ax\,dx = \frac{1}{a}\sinh ax$

34. $\int \tan^3 x\,dx = \frac{1}{2}\tan^2 x + \log\cos x$

35. $\int \cot^3 x\,dx = -\frac{1}{2}\cot^2 x - \log\sin x$

36. $\int \tan^2 x\,dx = \tan x - x$

37. $\int \cot^2 x\,dx = -\cot x - x$

38. $\int x \sin x\,dx = \sin x - x\cos x$

39. $\int x^2 \sin x\,dx = 2x\sin x - (x^2 - 2)\cos x$

40. $\int x^3 \sin x\,dx = (3x^2 - 6)\sin x - (x^3 - 6x)\cos x$

41. $\int x \cos x\,dx = \cos x + x\sin x$

42. $\int x^2 \cos x\,dx = 2x\cos x + (x^2 - 2)\sin x$

43. $\int x^3 \cos x\,dx = (3x^2 - 6)\cos x + (x^3 - 6x)\sin x$

44. $\int \sin^{-1} x\,dx = x\sin^{-1} x + (1 - x^2)^{1/2}$

45. $\int \cos^{-1} x\,dx = x\cos^{-1} x - (1 - x^2)^{1/2}$

46. $\int \tan^{-1} x\,dx = x\tan^{-1} x - \frac{1}{2}\log(1 + x^2)$

47. $\int \cot^{-1} x\,dx = x\cot^{-1} x + \frac{1}{2}\log(1 + x^2)$

48. $\int \sec^{-1} x\,dx = x\sec^{-1} x - \log[x + (x^2 - 1)^{1/2}]$

TABLE 5.1 (Continued)

49. $\int \csc^{-1} x\, dx = x \csc^{-1} x + \log[x + (x^2 - 1)^{1/2}]$

50. $\int x \sin^{-1} x\, dx = \frac{1}{4}[(2x^2 - 1)\sin^{-1} x + x(1 - x^2)^{1/2}]$

51. $\int x \cos^{-1} x\, dx = \frac{1}{4}[(2x^2 - 1)\cos^{-1} x - x(1 - x^2)^{1/2}]$

52. $\int x \tan^{-1} x\, dx = \frac{1}{2}(x^2 + 1)\tan^{-1} x - \frac{x}{2}$

53. $\int (\sin^{-1} x)^2\, dx = x(\sin^{-1} x)^2 - 2x + 2(1 - x^2)^{1/2} \sin^{-1} x$

54. $\int (\cos^{-1} x)^2\, dx = x(\cos^{-1} x)^2 - 2x - 2(1 - x^2)^{1/2} \cos^{-1} x$

55. $\int \sin(\log x)\, dx = \frac{1}{2} x \sin(\log x) - \frac{1}{2} x \cos(\log x)$

56. $\int \cos(\log x)\, dx = \frac{1}{2} x \sin(\log x) + \frac{1}{2} x \cos(\log x)$

57. $\int \tanh x\, dx = \log \cosh x$

58. $\int \coth x\, dx = \log \sinh x$

59. $\int \operatorname{sech} x\, dx = \tan^{-1}(\sinh x)$

60. $\int \operatorname{csch} x\, dx = \log\left(\tanh \frac{x}{2}\right)$

61. $\int \frac{dx}{x(a + bx)^{1/2}} = \frac{1}{a^{1/2}} \log \frac{(a + bx)^{1/2} - a^{1/2}}{(a + bx)^{1/2} + a^{1/2}}$

62. $\int \frac{x\, dx}{a + bx} = \frac{x}{b} - \frac{a}{b^2} \log(a + bx)$

63. $\int \frac{x\, dx}{(a + bx)^2} = \frac{1}{b^2}\left[\log(a + bx) + \frac{a}{a + bx}\right]$

64. $\int \frac{x^2\, dx}{a + bx} = \frac{1}{b^3}\left[\frac{1}{2}(a + bx)^2 - 2a(a + bx) + a^2 \log(a + bx)\right]$

65. $\int \frac{dx}{x(a + bx)} = -\frac{1}{a} \log \frac{a + bx}{x}$

66. $\int \frac{dx}{x(a + bx)^2} = \frac{1}{a(a + bx)} - \frac{1}{a^2} \log \frac{a + bx}{x}$

67. $\int \frac{dx}{a + be^{rx}} = \frac{x}{a} - \frac{1}{ar} \log(a + be^{rx})$

68. $\int e^{ax} \log x\, dx = \frac{e^{ax} \log x}{a} - \frac{1}{a} \int \frac{e^{ax}}{x}\, dx$

69. $\int e^{ax} \sin bx\, dx = e^{ax} \frac{a \sin bx - b \cos bx}{a^2 + b^2}$

70. $\int e^{ax} \cos bx\, dx = e^{ax} \frac{a \cos bx + b \sin bx}{a^2 + b^2}$

71. $$\int \frac{dx}{a + bx^2} = \begin{cases} \dfrac{1}{(ab)^{1/2}} \tan^{-1} \dfrac{x(ab)^{1/2}}{a} & ab > 0; \\ \dfrac{1}{(-ab)^{1/2}} \tanh^{-1} \dfrac{x(-ab)^{1/2}}{a} & ab < 0 \end{cases}$$

72. $\int \frac{dx}{a^2 + b^2 x^2} = \frac{1}{ab} \tan^{-1} \frac{bx}{a}$

73. $\int \frac{x\, dx}{a + bx^2} = \frac{1}{2b} \log(a + bx^2)$

If $X = a + bx + cx^2$ and $q = 4ac - b^2$:

74. $$\int \frac{dx}{X} = \begin{cases} \dfrac{2}{\sqrt{q}} \tan^{-1} \dfrac{2cx+b}{\sqrt{q}} & q > 0 \\[2ex] \dfrac{1}{\sqrt{-q}} \log \dfrac{2cx+b-\sqrt{-q}}{2cx+b+\sqrt{-q}}, & q < 0 \end{cases}$$

75. $$\int \frac{dx}{X^2} = \frac{2cx+b}{qX} + \frac{2c}{q} \int \frac{dx}{X}$$

76. $$\int \frac{x\,dx}{X} = \frac{1}{2c} \log X - \frac{b}{2c} \int \frac{dx}{X}$$

77. $$\int \frac{x\,dx}{X^2} = -\frac{bx+2a}{qX} - \frac{b}{q} \int \frac{dx}{X}$$

78. $$\int \frac{dx}{X^{1/2}} = \begin{cases} \dfrac{1}{c^{1/2}} \log \left(X^{1/2} + xc^{1/2} + \dfrac{b}{2c^{1/2}} \right) & c > 0 \\[2ex] \dfrac{1}{(-c)^{1/2}} \sin^{-1} \left(\dfrac{-2cx-b}{(b^2-4ac)^{1/2}} \right) & c < 0 \end{cases}$$

79. $$\int X^{1/2}\,dx = \frac{(2cx+b)X^{1/2}}{4c} + \frac{q}{8c} \int \frac{dx}{X^{1/2}}$$

80. $$\int \frac{x\,dx}{X^{1/2}} = \frac{X^{1/2}}{c} - \frac{b}{2c} \int \frac{dx}{X^{1/2}}$$

TABLE 5.2

Definite integrals

1. $$\int_0^\infty x^n e^{-ax}\,dx = \frac{n!}{a^{n+1}}$$

2. $$\int_1^\infty \frac{dx}{x^a} = \frac{1}{a-1} \qquad a > 1$$

3. $$\int_0^\infty \frac{dx}{(1+x)x^a} = \pi \csc a\pi \qquad 0 < a < 1$$

4. $$\int_0^\infty \frac{x^{a-1}}{1+x}\,dx = \frac{\pi}{\sin a\pi} \qquad 0 < a < 1$$

5. $$\int_0^\infty \frac{a\,dx}{a^2+x^2} = \begin{cases} \dfrac{\pi}{2} & a > 0 \\[1ex] 0 & a = 0 \\[1ex] -\dfrac{\pi}{2} & a < 0 \end{cases}$$

6. $$\int_0^{\pi/2} \cos^n x\,dx = \int_0^{\pi/2} \sin^n x\,dx = \begin{cases} \dfrac{1 \cdot 3 \cdot 5 \cdots (n-1)}{2 \cdot 4 \cdot 6 \cdots n} \dfrac{\pi}{2} & \text{for } n \text{ even} \\[2ex] \dfrac{2 \cdot 4 \cdot 6 \cdots (n-1)}{1 \cdot 3 \cdot 5 \cdots n} & \text{for } n \text{ odd} \end{cases}$$

7. $$\int_0^\infty \frac{\sin x}{x}\,dx = \frac{\pi}{2}$$

8. $$\int_0^\infty \frac{\cos ax}{1+x^2}\,dx = \frac{\pi}{2} e^{-|a|}$$

TABLE 5.2 (Continued)

9. $\int_0^{\pi} \sin nx \sin mx \, dx = \int_0^{\pi} \cos nx \cos mx \, dx = 0 \qquad m, n \text{ integers} \qquad m \neq n$

10. $\int_0^{\pi} \sin^2 nx \, dx = \int_0^{\pi} \cos^2 nx \, dx = \dfrac{\pi}{2}$

11. $\int_0^{\infty} \dfrac{\sin^2 x}{x^2} dx = \dfrac{\pi}{2}$

12. $\int_0^{\infty} \sin x^2 \, dx = \int_0^{\infty} \cos x^2 \, dx = \dfrac{1}{2}\left(\dfrac{\pi}{2}\right)^{1/2}$

13. $\int_0^{\infty} \dfrac{\sin x}{x^{1/2}} dx = \int_0^{\infty} \dfrac{\cos x}{x^{1/2}} dx = \left(\dfrac{\pi}{2}\right)^{1/2}$

14. $\int_0^{\infty} e^{-ax} \, dx = \dfrac{1}{a} \qquad a > 0$

15. $\int_0^{\infty} \dfrac{e^{-ax} - e^{-bx}}{x} dx = \log \dfrac{b}{a} \qquad a, b > 0$

16. $\int_0^{\infty} e^{-a^2x^2} \, dx = \dfrac{1}{2a} \pi^{1/2}$

17. $\int_0^{\infty} xe^{-x^2} \, dx = 1/2$

18. $\int_0^{\infty} x^2e^{-x^2} \, dx = \dfrac{\pi^{1/2}}{4}$

19. $\int_0^{\infty} x^2e^{-ax^2} \, dx = \dfrac{\pi^{1/2}}{4} a^{-3/2}$

20. $\int_0^{\infty} e^{-a^2x^2} \cos x \, dx = \dfrac{\pi}{2|a|} e^{-1/4a^2} \qquad a \neq 0$

21. $\int_0^{\infty} x[\mathrm{erf}(ax)]e^{-b^2x^2} \, dx = \dfrac{a}{2b^2}(a^2 + b^2)^{-1/2}$

22. $\int_0^{\infty} x^2[\mathrm{erf}(ax)]e^{-b^2x^2} \, dx = \dfrac{1}{2\pi^{1/2}}\left[\dfrac{a}{b^2(a^2 + b^2)} + \dfrac{1}{b^3} \tan^{-1} \dfrac{a}{b}\right]$

23. $\int_0^{\infty} [\mathrm{erf}(ax)]e^{-b^2x^2} \, dx = \dfrac{1}{(\pi b^2)^{1/2}} \tan^{-1} \dfrac{a}{b}$

24. $\int_0^{\infty} [1 - \mathrm{erf}(ax)] \, dx = \dfrac{1}{a\pi^{1/2}}$

25. $\int_0^{1} (\log x)^n \, dx = (-1)^n n!$

26. $\int_0^{1} \dfrac{\log x}{1 + x} dx = -\dfrac{\pi^2}{12}$

27. $\int_0^{1} \dfrac{\log x}{(1 - x^2)^{1/2}} dx = -\dfrac{\pi}{2} \log 2$

28. $\int_0^{\pi/2} \log(\tan x) \, dx = 0$

29. $\int_0^{\pi} x \log(\sin x) \, dx = -\dfrac{\pi^2}{2} \log 2$

A.6 Vector formulas

$$\mathbf{a} \cdot \mathbf{b} \times \mathbf{c} = \mathbf{b} \cdot \mathbf{c} \times \mathbf{a} = \mathbf{c} \cdot \mathbf{a} \times \mathbf{b}$$
$$\mathbf{a} \times (\mathbf{b} \times \mathbf{c}) = (\mathbf{a} \cdot \mathbf{c})\mathbf{b} - (\mathbf{a} \cdot \mathbf{b})\mathbf{c}$$
$$(\mathbf{a} \times \mathbf{b}) \times (\mathbf{c} \times \mathbf{d}) = (\mathbf{a} \times \mathbf{b} \cdot \mathbf{d})\mathbf{c} - (\mathbf{a} \times \mathbf{b} \cdot \mathbf{c})\mathbf{d}$$
$$(\mathbf{a} \times \mathbf{b}) \cdot (\mathbf{c} \times \mathbf{d}) = (\mathbf{a} \cdot \mathbf{c})(\mathbf{b} \cdot \mathbf{d}) - (\mathbf{a} \cdot \mathbf{d})(\mathbf{b} \cdot \mathbf{c})$$

$$\nabla(\varphi + \psi) = \nabla\varphi + \nabla\psi$$
$$\nabla \cdot (\mathbf{a} + \mathbf{b}) = \nabla \cdot \mathbf{a} + \nabla \cdot \mathbf{b}$$
$$\nabla \times (\mathbf{a} + \mathbf{b}) = \nabla \times \mathbf{a} + \nabla \times \mathbf{b}$$
$$\nabla \cdot \varphi\mathbf{a} = \nabla\varphi \cdot \mathbf{a} + \varphi \nabla \cdot \mathbf{a}$$
$$\nabla \times \varphi\mathbf{a} = \nabla\varphi \times \mathbf{a} + \varphi \nabla \times \mathbf{a}$$

$$\nabla(\mathbf{a} \cdot \mathbf{b}) = \mathbf{a} \cdot \nabla\mathbf{b} + \mathbf{b} \cdot \nabla\mathbf{a} + \mathbf{a} \times (\nabla \times \mathbf{b}) + \mathbf{b} \times (\nabla \times \mathbf{a})$$

$$\nabla \times (\mathbf{a} \times \mathbf{b}) = \mathbf{a}\nabla \cdot \mathbf{b} - \mathbf{b}\nabla \cdot \mathbf{a} + \mathbf{b} \cdot \nabla\mathbf{a} - \mathbf{a} \cdot \nabla\mathbf{b}$$

$$\nabla \cdot (\mathbf{a} \times \mathbf{b}) = \mathbf{b} \cdot \nabla \times \mathbf{a} - \mathbf{a} \cdot \nabla \times \mathbf{b}$$

$$\nabla \times (\nabla \times \mathbf{a}) = \nabla\nabla \cdot \mathbf{a} - \nabla^2\mathbf{a} \qquad \nabla \cdot \mathbf{r} = 3$$

$$\nabla \times \nabla\varphi = \mathbf{0} \qquad \nabla \cdot \nabla \times \mathbf{a} = 0 \qquad \nabla \times \mathbf{r} = \mathbf{0}$$

$$\iiint \nabla\varphi \, dV = \oiint \hat{\mathbf{n}} \, \varphi \, dS$$

$$\iiint \nabla \cdot \mathbf{F} \, dV = \oiint \hat{\mathbf{n}} \cdot \mathbf{F} \, dS$$

$$\iiint \nabla \times \mathbf{F} \, dV = \oiint \hat{\mathbf{n}} \times \mathbf{F} \, dS$$

$$\iint \hat{\mathbf{n}} \times \nabla\varphi \, dS = \oint \varphi \, d\mathbf{r}$$

$$\iint \hat{\mathbf{n}} \cdot \nabla \times \mathbf{F} \, dS = \oint \mathbf{F} \cdot d\mathbf{r}$$

Answers to Odd-numbered Exercises

Chapter 1 Limits

Exercises 1.1

1. The averages are $(26 + 52)/2 = 39$ ft.^{2}, 39.75 ft.^{2}, 40.25 ft.^{2}; the area is 39.58 ft.^{2}. The errors in the bounds tend to cancel in this particular case but the accuracy of the average does not necessarily improve at every stage of the calculation.
3. Divide the maximum width of the figure into n equal parts and use the largest height in each interval to construct a rectangle. The sum of the areas of these rectangles is an upper bound to the area of the figure. As the integer n is made larger, the approximation becomes more accurate.
5. Use a larger number of subdivisions of the interior angle, each of magnitude $360°/n = 2\pi/n$ where n is an integer. Use exterior polygons to construct an upper bound.
7. Form similar triangles from rays and tangent lines to the circle as shown.

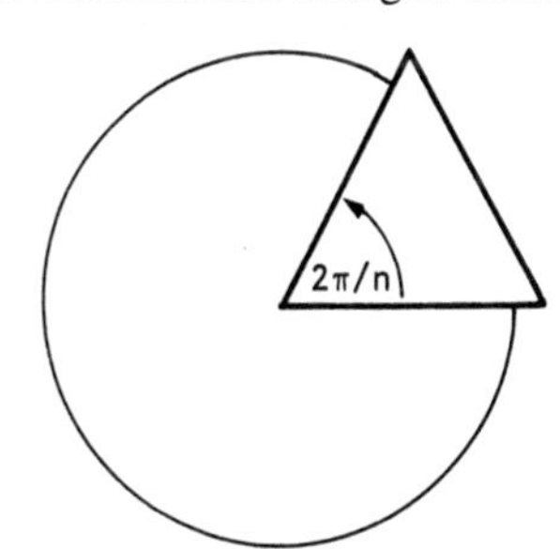

The perimeter of a regular exterior polygon of n sides constructed this way is $2nr\tan \pi/n$ which is an upper bound for the perimeter of a circle. (For example, circumscribing triangle and square have perimeters, $6\sqrt{3}r$, $8r$.)

Exercises 1.2

1. (i) $C = 5/9(F - 32)$ (ii) $z = \sqrt{x^2 + y^2}$ (iii) $V = xyz$ (iv) $S = 2(xy + yz + zx)$
3. (i) $-\infty < x < \infty, -\infty < y < \infty$ (ii) $-\infty < x < \infty, -\infty < y \leq 1$
   (iii) $-\infty < x \leq 1, 0 \leq y < \infty$
5. $f(1) = m + b$, $f(2) = 2m + b$; $m + b = 2$, $2m + b = 3$ imply $m = 1$, $b = 1$.
7. $y = \sqrt{1 - x^2}$; $A = 4xy = 4x\sqrt{1 - x^2}$
   $= 4\sqrt{x^2 - x^4} = 4\sqrt{1/4 - (1/2 - x^2)^2}$.
   Maximum value occurs when
   $x^2 = 1/2$ or $x = \pm 1/\sqrt{2}$.
   Maximum area is 2.

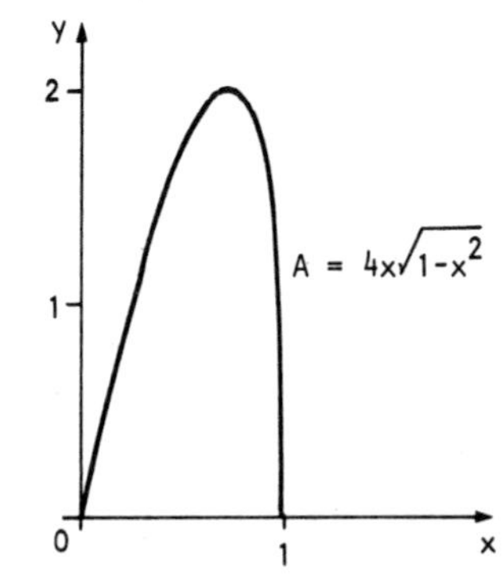

9. $y = x + 1$.

11. If $y = f(x) = a_0 + a_1x + a_2x^2 + \cdots + a_nx^n$, then
$f(cx) = a_0 + ca_1x + c^2a_2x^2 + \cdots + c^na_nx^n$. If $c = 0$, $y = f(0) = a_0$.

13. The quadratic equation $x^2 = mx + b$ has a unique solution if $m^2 - 4b = 0$.

15. $y = \pm 2x$. (For what values of m does $mx = x^2 + 1$ have a unique solution?)

17. To graph $y = -f(x)$, reflect graph of $y = f(x)$ about x axis. For $y = f(-x)$, reflect graph about y axis.

19. With $a = 1/2$, plot $y = x^2$, $y = (2x)^2$, $y = \frac{1}{2}x^2$, $y = 2x^2$.

21. $y - 0.81 = \dfrac{1.21 - 0.81}{1.1 - 0.9}(x - 0.9)$, $y = 0.81 + 2(x - 0.9) = 2x - 0.99$; at $x = 1$, $y = 1.01$ and error is 0.01. With closer points $y = 2x - 0.9975$; error at $x = 1$ is 0.0025.

23. **(i)** $(-1, 1)$, 1 **(ii)** $(1/2, -3/2)$, $\sqrt{5/2}$ **(iii)** $(-2, -2)$, 2 **(iv)** $(1, 2)$, 0

25. Define $c = (a^2 - b^2)^{1/2}$. If $((x-c)^2 + y^2)^{1/2} + ((x+c)^2 + y^2)^{1/2} = 2a$, then
$[(x-c)^2 + y^2]^{1/2} = 2a - [(x+c)^2 + y^2]^{1/2}$ and
$x^2 - 2cx + c^2 + y^2 =$
$4a^2 - 4a[(x-c)^2 + y^2]^{1/2} +$
$x^2 + 2cx + c^2 + y^2$ and
$a[(x+c)^2 + y^2]^{1/2} = a^2 + cx$. Finally,
squaring $(a^2 - c^2)x^2 + a^2y^2 = a^2(a^2 - c^2)$
or $\dfrac{x^2}{a^2} + \dfrac{y^2}{b^2} = 1$.

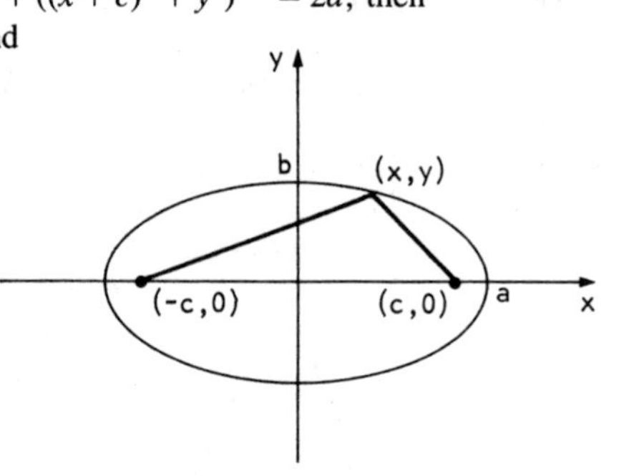

27. $H(x-5) = \begin{cases} 0 & x < 5 \\ 1 & x \geq 5 \end{cases}$, $H(x^2 - x) = \begin{cases} 0 & 0 < x < 1 \\ 1 & x \leq 0, x \geq 1 \end{cases}$. Show that $H(x) - H(-x) = 1$ if $x > 0$, 0 if $x = 0$, -1 if $x < 0$.

29. **(i)** $|ab| = ab\,\mathrm{sgn}(ab) = a\,\mathrm{sgn}(a)\, b\,\mathrm{sgn}(b) = |a|\,|b|$.

31. $|x_1 + x_2 + x_3 + \cdots + x_n| \leq |x_1| + |x_2 + x_3 + \cdots + x_n| \leq |x_1| + |x_2| + |x_3 + \cdots + x_n| \leq \cdots$
Equality holds if all numbers are non-negative or all are non-positive.

33. **(i)** $\sin 2x = \sin(x + x) = \sin x \cos x + \cos x \sin x = 2 \sin x \cos x$.
(ii) $\sin 3x = \sin x \cos 2x + \cos x \sin 2x = \sin x(1 - 2\sin^2 x) + 2\sin x(1 - \sin^2 x)$
(iii) $\sin x_1 + \sin x_2 = \sin\left(\dfrac{x_1 + x_2}{2} + \dfrac{x_1 - x_2}{2}\right) + \sin\left(\dfrac{x_1 + x_2}{2} - \dfrac{x_1 - x_2}{2}\right)$

35. **(i)**

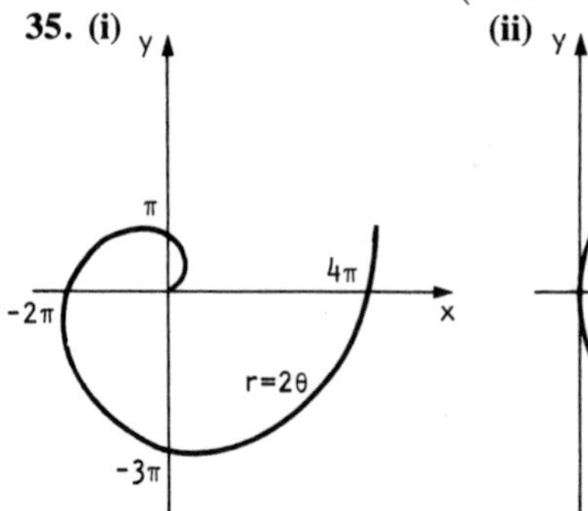

(ii)

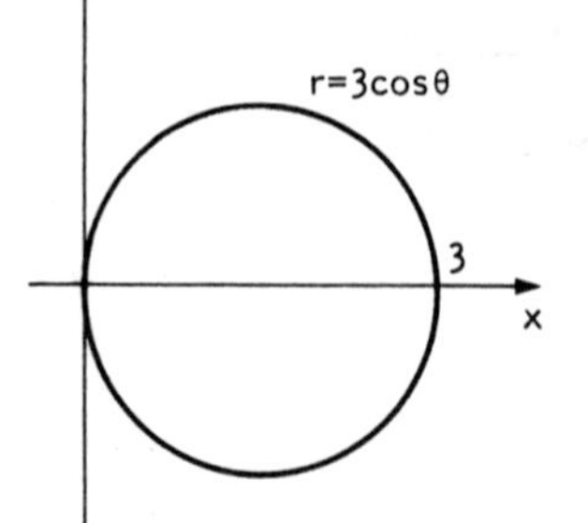

(iii)

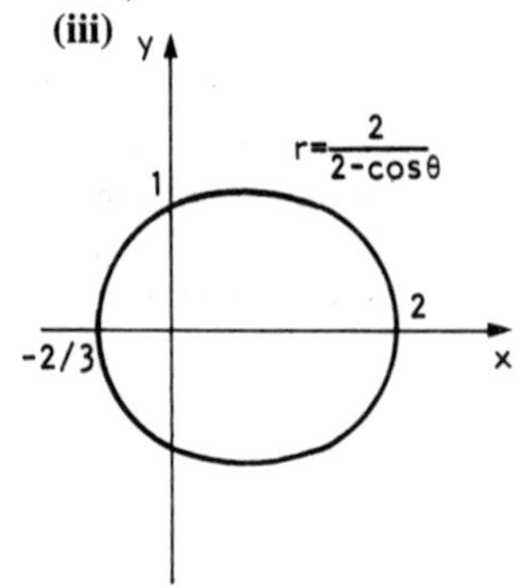

37. **(i)** $\sec^2 x = \dfrac{1}{\cos^2 x} = \dfrac{\cos^2 x + \sin^2 x}{\cos^2 x} = 1 + \tan^2 x$

(iii) $\tan(x_1 + x_2) = \dfrac{\sin(x_1 + x_2)}{\cos(x_1 + x_2)} = \dfrac{\sin x_1 \cos x_2 + \cos x_1 \sin x_2}{\cos x_1 \cos x_2 - \sin x_1 \sin x_2}$

$= \dfrac{(\sin x_1/\cos x_1) + (\sin x_2/\cos x_2)}{1 - (\sin x_1/\cos x_1)(\sin x_2/\cos x_2)}$

(iv) $\dfrac{\sin x}{1 + \cos x} = \dfrac{2 \sin x/2 \cos x/2}{1 + (2 \cos^2 x/2 - 1)}$

(v) $\sin 2x = 2 \sin x \cos x = \dfrac{2 \tan x}{\sec^2 x} = \dfrac{2 \tan x}{1 + \tan^2 x}$

39. **(i)** $u = x^2 + 2x,\ y = u^3$ or $u = (x^2 + 2x)^{1/3},\ y = u^9$, etc.
(ii) $u = x^3,\ y = u^2 + u + 1$ or $u = x^6,\ y = u + u^{1/2} + 1$
(iii) $u = 1 + x^2,\ y = u^{-2}$ **(iv)** $u = 1 - x^2,\ y = \sin u$ **(v)** $u = \cos(1 + x),\ y = u^2$
(vi) $u = \sec x,\ y = u^{-2}$

Exercises 1.3

1. **(i)** If $f(x) = \dfrac{1 - x^2}{1 - x}$, $f(0.9) = 1.9$, $f(0.99) = 1.99$, $f(0.999) = 1.999$, and $\lim_{x \to 1} f(x) = 2$.
(ii) 3 **(iii)** 1

3. Use $\dfrac{\sin x}{x} = \dfrac{-\sin(-x)}{-(-x)} = \dfrac{\sin(-x)}{-x}$ and define $x = -z$; $\lim_{x \to 0-} \dfrac{\sin x}{x} = \lim_{z \to 0+} \dfrac{\sin z}{z} = 1$.

5. Limit = 2. Use $\dfrac{\sin(2 \sin x)}{x} = \dfrac{\sin(2 \sin x)}{2 \sin x} \dfrac{2 \sin x}{x}$.

7. **(i)** 1 **(ii)** -1 **(iii)** ∞ **(iv)** 0 **(v)** 3 **(vi)** 0

9. **(iv)** is unbounded $\left(\tan\left(-\dfrac{\pi}{4}\right) = -1\right)$;
(v) is not defined at $x = \dfrac{\pi}{2}, \dfrac{3\pi}{2}, \cdots$; all others are defined and bounded for $-\infty < x < \infty$.

11. $R(x) = \dfrac{(x - 1)(x + 1)}{x(2x - 1)(x + 2)}$ **(i)** $1/2$ **(ii)** $-3/10$ **(iii)** $3/10$

13. $\lim_{x \to 0} x\, R(x) = \lim_{x \to 0} \dfrac{x^2 + 2x + 5}{(x - 1)(x + 2)} = -\dfrac{5}{2}$; $\lim_{x \to 1} (x - 1)\, R(x) = \lim_{x \to 1} \dfrac{x^2 + 2x + 5}{x(x + 2)} = \dfrac{8}{3}$
$\lim_{x \to -2} \dfrac{x^2 + 2x + 5}{x(x - 1)} = \dfrac{5}{6}$.

15. $$R(x) = \frac{a_n x^n + a_{n-1} x^{n-1} + \cdots + a_0}{b_m x^m + b_{m-1} x^{m-1} + \cdots + b_0} = \frac{a_n}{b_m} x^{n-m} \frac{\left(1 + \dfrac{a_{n-1}}{a_n}\left(\dfrac{1}{x}\right) + \cdots + \dfrac{a_0}{a_n}\left(\dfrac{1}{x}\right)^n\right)}{\left(1 + \dfrac{b_{m-1}}{b_m}\left(\dfrac{1}{x}\right) + \cdots + \dfrac{b_0}{b_m}\left(\dfrac{1}{x}\right)^m\right)}$$

If $|x|$ is large, $R(x) \approx \dfrac{a_n}{b_m} x^{n-m}$ and $\dfrac{R(x) - \dfrac{a_n}{b_m} x^{n-m}}{\dfrac{a_n}{b_m} x^{n-m}} \approx \left(\dfrac{a_{n-1}}{a_n} - \dfrac{b_{m-1}}{b_m}\right)\dfrac{1}{x}$.

17. $R(1/z) = \dfrac{(1/z)^3 - 1}{1/z - 2} = \dfrac{z^3 - 1}{z^2(2z - 1)}$ is singular at $z = 0$.

19. $\sin\frac{1}{x} = 0$ where $x = \frac{1}{n\pi}$ and $\sin\frac{1}{x} = \pm 1$ where $x = \frac{1}{\left(n+\frac{1}{2}\right)\pi}$ (n any integer). The function $\sin(1/x)$ oscillates very rapidly between $+1$ and -1 as x approaches 0. The function $x\sin(1/x)$ oscillates with decreasing amplitude as $x \to 0$ and $\lim_{x\to 0} x\sin(1/x) = 0$.

Exercises 1.4

1. (i) $\frac{s(t+\Delta t)-s(t)}{\Delta t} = 2,\ v(t) = \lim_{\Delta t\to 0}\frac{s(t+\Delta t)-s(t)}{\Delta t} = 2$

(ii) $\frac{s(t+\Delta t)-s(t)}{\Delta t} = 3-2t,\ v(t) = 3-2t$

(iii) $\frac{s(t+\Delta t)-s(t)}{\Delta t} = -1+2t+\Delta t,\ v(t) = -1+2t.$

(i)

S
S=3+2t
3
2
1
1
2
t

(ii)

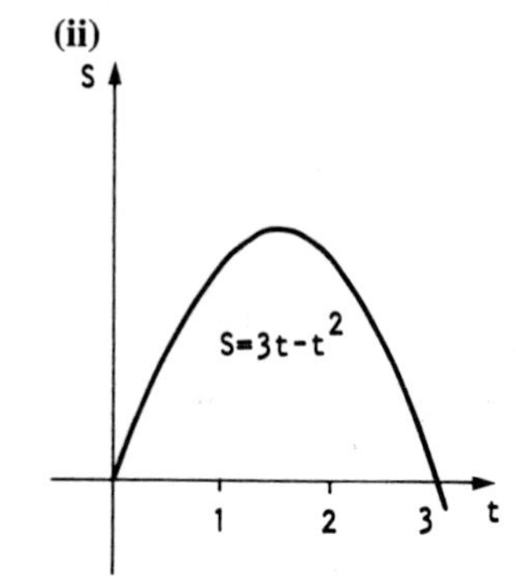

(iii)

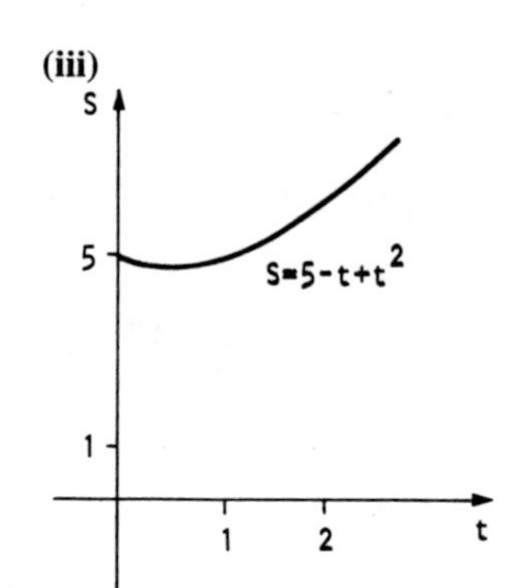

3. (i) $\frac{n(t+\Delta t)-n(t)}{\Delta t} = 10^3,\ \frac{dn}{dt} = 10^3$

(ii) $\frac{n(t+\Delta t)-n(t)}{\Delta t} = 10^4(5-2t-\Delta t),\ \frac{dn}{dt} = 10^4(5-2t)$

(iii) $\frac{n(t+\Delta t)-n(t)}{\Delta t} = -10^5+10^4(2t+\Delta t),\ \frac{dn}{dt} = -10^5+10^4(2t).$

5. $s(t) = 0$ when $t = \sqrt{\frac{2h}{g}}$.

7. $t(x) = \sqrt{\frac{2x}{g}},\ x(t) = \frac{1}{2}gt^2.$

9. $mgs + \frac{1}{2}mv^2 = mg\left[v(0)t - \frac{1}{2}gt^2\right] + \frac{1}{2}m[v(0)-gt]^2 = \frac{1}{2}m(v(0))^2.$

11. $s(t) = \begin{cases} \frac{1}{2}at^2 & 0 \leqslant t \leqslant 3 \\ \frac{9}{2}a + 3a(t-3) & 3 < t \leqslant 10 \end{cases}\quad v(t) = \begin{cases} at & 0 \leqslant t \leqslant 3 \\ 3a & 3 < t \leqslant 10 \end{cases}\quad a(t) = \begin{cases} a & 0 \leqslant t \leqslant 3 \\ 0 & 3 < t \leqslant 10 \end{cases}$

Since $s(10) = 100$, $a = \dfrac{200}{51}$ m sec^{-2}; $s(3) = \dfrac{300}{17}$ m, $s(5) = \dfrac{700}{17}$ m; $v(t) = 10$ m sec^{-1} when $t = \dfrac{51}{20}$ sec.

13.

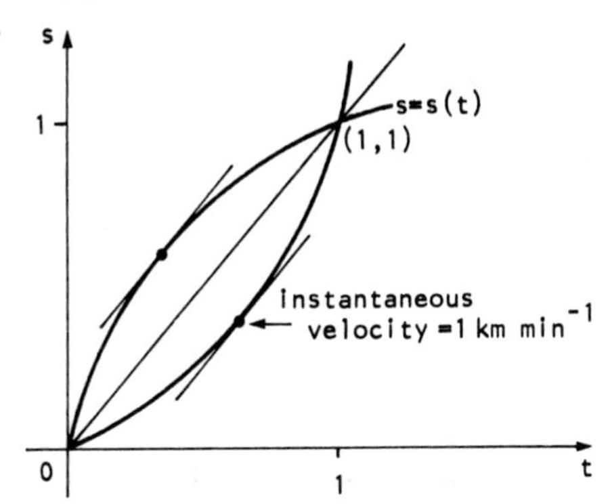

15. $s(t) = 50t - \dfrac{1}{2}at^2$, $v(t) = 50 - at$; $v(t) = 0$ when $t = \dfrac{50}{a}$;

$s\left(\dfrac{50}{a}\right) = 1000 - 50\left(\dfrac{50}{a}\right) - \dfrac{1}{2}a\left(\dfrac{50}{a}\right)^2 = \dfrac{2500}{2a}$; $a = 1.25$, $t = 40$ sec.

17. With $a_i = (v_{i+1} - v_i)/(t_{i+1} - t_i)$, the accelerations $a_1, a_2, \ldots, a_{10}$ are -3.41, -5.92, -6.81, -5.91, -3.42, 0, 3.42, 6.81, 5.92, 3.41.
With $a_i = 4(s_{i+1} - 2s_i + s_{i-1})/(t_{i+1} - t_{i-1})^2$, $a_1, a_2, \ldots, a_{10}$ are -3.41, -5.92, -6.81, -5.91, -3.42, 0, 3.42, 6.81, 5.92, 3.41.

Exercises 1.5

1. $|(3x+1) - 7| = |3x - 6| = 3|x - 2| < \varepsilon$ when $|x - 2| < \varepsilon/3$. Choose $\delta = \varepsilon/3$.
If $\delta = \varepsilon/3$, $|x - 2| < \delta$ implies
$|(3x+1) - 7| < \varepsilon$.

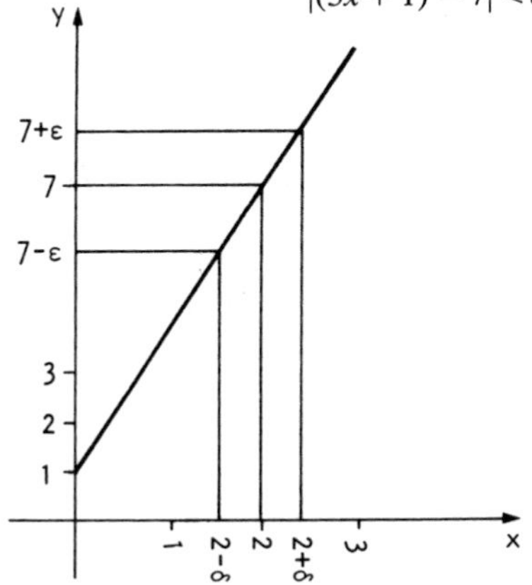

3. **(i)** $|(f(x)+g(x))-(A+B)| \leqslant |f(x)-A|+|g(x)-B|$ (triangle inequality).
(ii) $|(f(x)-g(x))-(A-B)| \leqslant |f(x)-A|+|g(x)-B|$
(iii) $\left|\dfrac{f(x)}{g(x)}-\dfrac{A}{B}\right| = \left|\dfrac{(f(x)-A)B+A(B-g(x))}{g(x)B}\right| \leqslant \dfrac{1}{|g(x)|}|f(x)-A| + \dfrac{|A|}{|B|}\dfrac{|B-g(x)|}{|g(x)|}$

5. $ax_0^2+bx_0+c$.

7. **(i)** 3 **(ii)** 1 **(iii)** $-\sin x$ **(iv)** 0 **(v)** $-1/a^2$ **(vi)** 0 **(vii)** $-1/a^3$
(viii) 0 **(ix)** 0

9. **(i)** asymptotes $x=1$, $x=-1$, $y=1$ **(ii)** $x=1$, $y=2$ **(iii)** $x=0$, $x=-2$, $y=x$

(i) **(ii)**

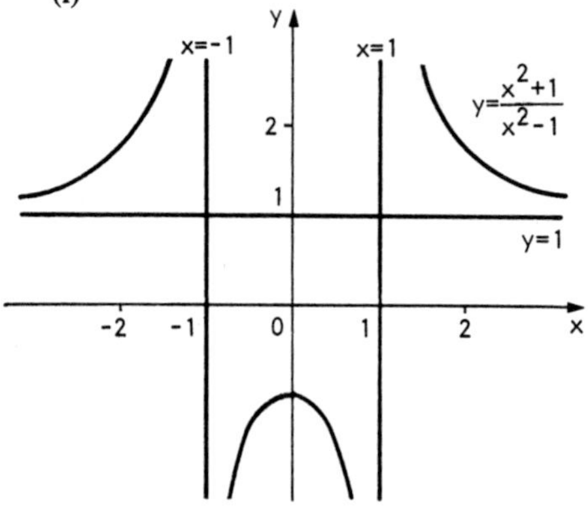

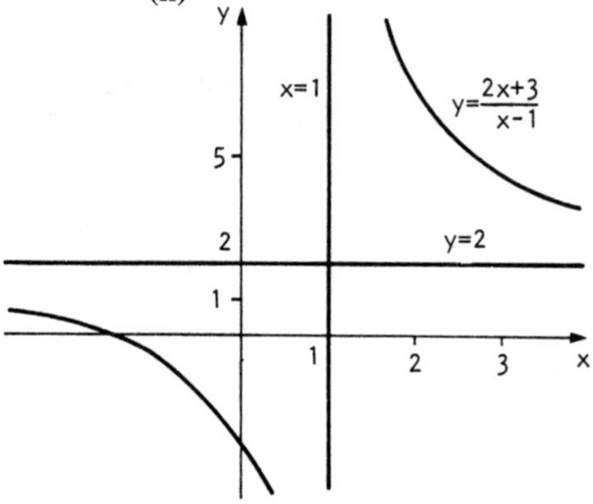

11. Since $\lim_{x\to a}(f(x)-g(x)) = \lim_{x\to a} f(x) - \lim_{x\to a} g(x) = \lim_{x\to a} f(x) - M \geqslant 0$, results follow; $f(x)=1$, $g(x)=\dfrac{1+x^2}{2+x^2}$, $a=\infty$.

13. **(i)** $\left|\dfrac{1}{x^2}-\dfrac{1}{x_0^2}\right| = \dfrac{|x+x_0|}{|x^2x_0^2|}|x-x_0|$. **(ii)** $|x^{1/2}-x_0^{1/2}| = \left|\dfrac{x-x_0}{x^{1/2}+x_0^{1/2}}\right|$
(iii) $||x|-|x_0|| = |x-x_0|$.

15. Since, for any $\varepsilon>0$, $f(x_0)-\varepsilon<f(x)<f(x_0)+\varepsilon$ when $|x-x_0|<\delta$, choose $\varepsilon<f(x_0)$.

17. $x=1$; $\lim_{x\to 1} f(x)=1$.

Exercises 1.6

1. **(i)** 2 **(ii)** $6x$ **(iii)** $2x-1$ **(iv)** $5x^4$ **(v)** $4x+3$ **(vi)** $x/3$ **(vii)** a
(viii) $2ax+b$ **(ix)** $3ax^2+2bx+c$

3. $\dfrac{dy}{dx} = \lim_{h\to 0}\dfrac{\sqrt{x+h}-\sqrt{x}}{h} = \lim_{h\to 0}\dfrac{\sqrt{x+h}-\sqrt{x}}{h}\dfrac{\sqrt{x+h}+\sqrt{x}}{\sqrt{x+h}+\sqrt{x}} = \lim_{h\to 0}\dfrac{1}{\sqrt{x+h}+\sqrt{x}} = \dfrac{1}{2\sqrt{x}}$.
Derivative is not defined at $x=0$.

5.

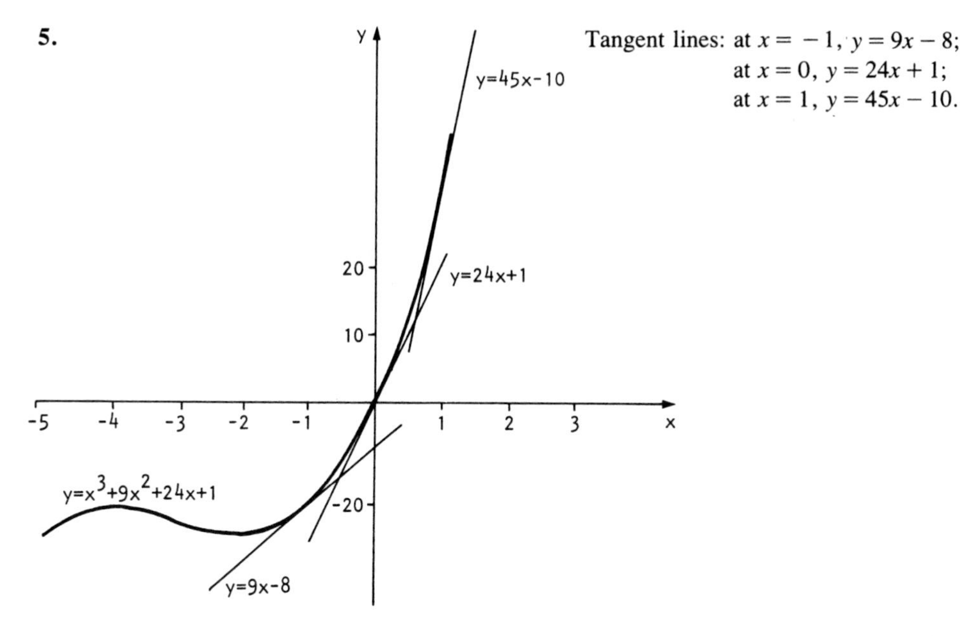

Tangent lines: at $x = -1$, $y = 9x - 8$;
at $x = 0$, $y = 24x + 1$;
at $x = 1$, $y = 45x - 10$.

7. **(i)** Use $\cos k(x + \Delta x) = \cos kx \cos k\Delta x - \sin kx \sin k\Delta x$.
(ii) Use $\sin(x + a) = \sin x \cos a + \cos x \sin a$
(iii) Use $\cos(x + a) = \cos x \cos a - \sin x \sin a$.

9. **(i)** $\dfrac{d}{d\theta}\cos^2\theta = \dfrac{d}{d\theta}\dfrac{1}{2}(1 + \cos 2\theta) = \dfrac{d}{d\theta}\dfrac{\cos 2\theta}{2} = -\sin 2\theta = -2\sin\theta\cos\theta$

(ii) $\dfrac{d}{d\theta}\sin^2\theta = \dfrac{d}{d\theta}(1 - \cos^2\theta) = -\dfrac{d}{d\theta}\cos^2\theta = 2\sin\theta\cos\theta$.

11. $y = x^2 - 3x + 2$.

13. $x = \pm(\sqrt{33} - 5)^{1/2}/2$. (Solve $6x^2 + 15 = 3/x^2$).

15. **(i)** $2(x + a)$ **(ii)** $(x + a)^{-1/2}/2$ **(iii)** $-(x + a)^{-2}$

(iv) $-2(x + a)^{-3}$ **(v)** $\cos\left(x + \dfrac{\pi}{2}\right) = -\sin x$ **(vi)** $-\sin\left(x + \dfrac{\pi}{2}\right) = -\cos x$.

17. **(i)** $3a^3x^2$ **(ii)** $-a^{-1}x^{-2}$ **(iii)** $\frac{1}{2}\,a(ax)^{-1/2} = a^{1/2}/2x^{1/2}$ **(iv)** $3\cos 3x$
(v) $-5\sin 5x$ **(vi)** $a\sec^2 ax$

Exercises 1.7

1. **(i)** Given $\varepsilon > 0$, $\left|3 - \dfrac{1}{n} - 3\right| < \varepsilon$ when $n > N = \dfrac{1}{\varepsilon}$.

(ii) Given $\varepsilon > 0$, $\left|\dfrac{n}{n+2} - 1\right| = \left|\dfrac{2}{n+2}\right| < \varepsilon$ when $n > N = \dfrac{1}{\varepsilon} - 1$.

(iii) Given $\varepsilon > 0$, $\left|\dfrac{1}{\sqrt{n^2+1}}\right| < \varepsilon$ when $n > N = \dfrac{1}{\varepsilon}$.

3. **(i)** $\left|\frac{1}{5n}\right| < \varepsilon$ for $n > \frac{1}{5\varepsilon}$ **(ii)** $\left|\frac{n}{2n+1} - \frac{1}{2}\right| < \varepsilon$ for $n > \frac{1}{4\varepsilon}$

(iii) $\frac{1}{2^n} < \varepsilon$ for $2^n > \frac{1}{\varepsilon}$ or $n > \log_2\left(\frac{1}{\varepsilon}\right)$.

5. **(i)** $1/2$ **(ii)** $1/3$ **(iii)** 1 **(iv)** 0 **(v)** 0 **(vi)** divergent **(vii)** 0 **(viii)** $-1/3$ **(ix)** 0

7. Since $\lim_{n\to\infty} (a - a_n) = 0$, $\lim_{n\to\infty} (a^2 - a_n{}^2) = \lim_{n\to\infty} (a - a_n) \lim_{n\to\infty} (a + a_n) = 0$.

9. $\lim_{n\to\infty} \left(\frac{a_{n+1}}{a_n} - 1\right) = \lim_{n\to\infty} \frac{a_{n+1} - a_n}{a_n} = \lim_{n\to\infty} (a_{n+1} - a_n) \lim_{n\to\infty} \frac{1}{a_n} = 0.$

11. $a_1 = 1$, $a_2 = 1.4142136, \ldots, a_{10} = 1.6180165$; $a = \lim_{n\to\infty} a_n$ satisfies $a = \sqrt{1+a}$ or

$a^2 - a - 1 = 0$, solution is $\frac{1+\sqrt{5}}{2} = 1.618034$.

13. $a_1 = 1$, $a_2 = 1.7320508, \ldots, a_{10} = 2.9935697$; $a = \lim_{n\to\infty} a_n$ satisfies $a = (3a)^{1/2}$; solution is $a = 3$.

Exercises 1.8

1. $a_1 = a$, $a_2 = a + d$, $a_3 = a + 2d, \ldots$; $s_n = a + (a + d) + (a + 2d) + \cdots + (a + (n-1)d)$; sum s_n by reversing terms or verify formula by induction.

3. $a_1 = 1$, $a_2 = 2r$, $a_3 = 3r^2, \ldots$; $s_n = 1 + 2r + 3r^2 + \cdots + nr^{n-1}$; verify that

$s_n - rs_n = 1 + r + r^2 + \cdots + r^{n-1} - nr^n = \frac{1 - r^n}{1 - r} - nr^n = \frac{1 - (n+1)r^n + nr^{n+1}}{1 - r}$.

5. If $n = 1$, $(1 + x)^1 = 1 + x$ is a polynomial of degree 1. Suppose $(1 + x)^n$ is a polynomial degree n for some positive integer n, i.e., $(1 + x)^n = a_0 + a_1x + \cdots + a_nx^n$ where $a_n \neq 0$. Then $(1 + x)^{n+1} = (1 + x)[a_0 + a_1x + \cdots + a_nx^n] = a_0 + (a_1 + a_0)x + \cdots + (a_n + a_{n-1})x^n + a_nx^n$ is a polynomial of degree $n + 1$. Induction is complete.

7. $s_n = \frac{1}{1\cdot 2} + \frac{1}{2\cdot 3} + \cdots + \frac{1}{n(n+1)} = \left(\frac{1}{1} - \frac{1}{2}\right) + \left(\frac{1}{2} - \frac{1}{3}\right) + \cdots + \left(\frac{1}{n} - \frac{1}{n+1}\right) = 1 - \frac{1}{n+1}$

If $n = 1$, $s_1 = \frac{1}{1\cdot 2} = 1 - \frac{1}{1+1}$. If $s_n = 1 - \frac{1}{n+1}$ for some positive integer n, then

$s_{n+1} = 1 - \frac{1}{n+1} + \frac{1}{(n+1)(n+2)} = 1 - \frac{1}{n+2} = 1 - \frac{1}{(n+1)+1}$.

9. Try $s_n = a_1n + a_2n^2 + a_3n^3 + \frac{1}{4}n^4$; using known values of s_1, s_2 and s_3, solve for the constants; then verify $s_n = n^2(n+1)^2/4$ by induction.

11. Use Exercise 10.

13. $\cos x$ is a polynomial of degree 1 and $\cos 2x = 2\cos^2 x - 1$ is a polynomial of degree 2 in the variable $\cos x$. If $\cos(n-1)x$ and $\cos nx$ are polynomials of degree $(n-1)$ and n, then $\cos(n+1)x = 2\cos x \cos nx + \cos(n-1)x$ is a polynomial of degree $n+1$.

15. For $n = 1$, $(1 + x)^1 = 1 + x^1$. Assume expansion holds for positive integer n; then

$(1 + x)^{n+1} = (1 + x)(1 + x)^n = (1 + x)[1 + nx + \frac{n(n-1)}{2!}x^2 + \cdots + nx^{n-1} + x^n]$

$$= 1 + (n+1)x + \left(\frac{n(n-1)}{2!} + n\right) x^2 + \cdots + (n+1)x^n + x^{n+1}$$

$$= 1 + (n+1)x + \frac{(n+1)n}{2!} x^2 + \cdots + (n+1)x^n + x^{n+1}.$$

17. $\left(1 + \frac{1}{n}\right)^n = 1 + n\left(\frac{1}{n}\right) + \frac{n(n-1)}{2!}\left(\frac{1}{n}\right)^2 + \frac{n(n-1)(n-2)}{3!}\left(\frac{1}{n}\right)^3 + \cdots + \left(\frac{1}{n}\right)^n$

$$= 1 + 1 + \frac{1}{2!}\left(1 - \frac{1}{n}\right) + \frac{1}{3!}\left(1 - \frac{1}{n}\right)\left(1 - \frac{2}{n}\right) + \cdots + \frac{1}{n!}\left(1 - \frac{1}{n}\right)\left(1 - \frac{2}{n}\right) \cdots \left(1 - \frac{n-1}{n}\right).$$

Conclude that $2 < \left(1 + \frac{1}{n}\right)^n < 1 + 1 + \frac{1}{2!} + \frac{1}{3!} + \cdots + \frac{1}{n!}$

$$< 1 + 1 + \frac{1}{2} + \frac{1}{2^2} + \cdots + \frac{1}{2^{n+1}} < 3.$$

19. $a_1 = h + rh$, $a_2 = rh + r^2h = r(1+r)h, \ldots, a_n = r^{n-1}(1+r)h$;
$s_n = (1+r)h + r(1+r)h + \cdots r^{n-1}(1+r)h = (1+r)h(1-r^n)/(1-r)$.
Total distance travelled is $s = \lim_{n\to\infty} s_n = (1+r)h/(1-r)$.

Exercises 1.9

1.

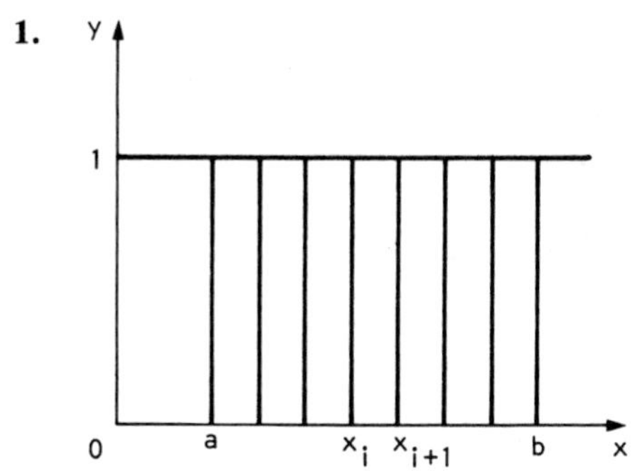

The approximation with interior rectangles is exact.

$$A = \lim_{n\to\infty} \underline{A}_n = \underline{A}_n = \sum_{i=0}^{n-1} (x_{i+1} - x_i) = x_n - x_0 = b - a.$$

3. (i) $\frac{1}{4}\left[1 + \frac{3}{2} + 2 + \frac{5}{2}\right] < \int_0^1 (2x+1)\,dx < \frac{1}{4}\left[\frac{3}{2} + 2 + \frac{5}{2} + 3\right]$ or $\frac{7}{4} < A < \frac{9}{4}$

(ii) $\frac{\pi}{4}\left[0 + \frac{1}{\sqrt{2}} + \frac{1}{\sqrt{2}} + 0\right] < \int_0^{\pi} \sin x\,dx < \frac{\pi}{4}\left[\frac{1}{\sqrt{2}} + 1 + 1 + \frac{1}{\sqrt{2}}\right]$ or
$\frac{\pi\sqrt{2}}{4} < A < \frac{\pi}{4}(2 + \sqrt{2})$

(iii) $\frac{1}{4}\left[\frac{4}{5} + \frac{2}{3} + \frac{4}{7} + \frac{1}{2}\right] < \int_1^2 \frac{1}{x}\,dx < \frac{1}{4}\left[1 + \frac{4}{5} + \frac{2}{3} + \frac{4}{7}\right]$, $\frac{533}{840} < A < \frac{638}{840}$

(iv) $\frac{4}{4}\left[0 + 1 + \sqrt{2} + \sqrt{3}\right] < \int_0^4 \sqrt{x}\,dx < \frac{4}{4}\left[1 + \sqrt{2} + \sqrt{3} + 2\right]$,
$1 + \sqrt{2} + \sqrt{3} < A < 3 + \sqrt{2} + \sqrt{3}$.

5. Explain why $\int_0^X (ax^2 + bx + c)\,dx = a\int_0^X x^2\,dx + b\int_0^X x\,dx + c\int_0^X 1\,dx$ and quote previous results.

7. In the diagram, $x_i = i\dfrac{X}{n}$ and $x_{i+1} - x_i = \dfrac{X}{n}$.

$$\int_0^X x^3\,dx = \lim_{n\to\infty}\sum_{i=0}^{n-1} x_i^3 \frac{X}{n}$$
$$= \lim_{n\to\infty}\frac{X^4}{n^4}\sum_{i=0}^{n-1} i^3$$
$$= \lim_{n\to\infty}\frac{X^4}{n^4}\frac{n^2(n-1)^2}{4} = \frac{X^4}{4}.$$

9. **(i)** $3/2$ **(ii)** $56/3$ **(iii)** $268/9$

11. $\int_0^2 (x - x^2/2)\,dx = 2/3$

13. $V = \int_0^H \dfrac{\pi R^2}{H^2} z^2\,dz = \dfrac{1}{3}\pi R^2 H.$

Exercises 1.10

1. **(i)** $s(t) = 2t + \dfrac{3}{2}t^2$ **(ii)** $s(t) = t + \dfrac{1}{2}t^2 + \dfrac{1}{3}t^3$ **(iii)** $s(t) = 4t + 2t^2 + \dfrac{1}{3}t^3$

3. $s(2) = \dfrac{17}{3}$, $s(5) - s(2) = \dfrac{635}{12}$; $v(t) = 1 + \dfrac{t}{2} + t^2$, $v(2) = 6$, $v(5) = \dfrac{57}{2}$.

5. $s = \dfrac{4}{3}t^3$ for $0 \leqslant t \leqslant 5$, $s = \dfrac{500}{3} + 100(t-5) - \dfrac{15}{2}(t-5)^2$ for $5 < t$.

Exercises 1.11

1. $y = a(x+1)$; when $x = -1$, $y = 0$.

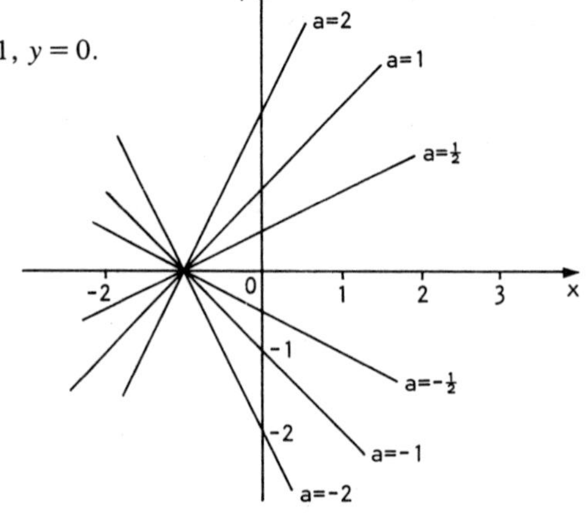

3.

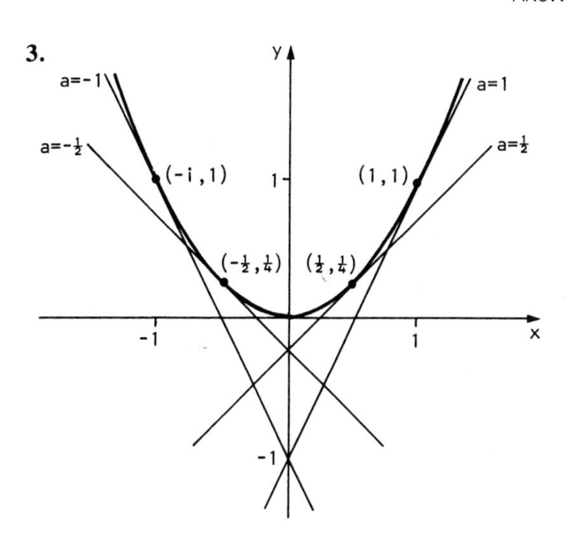

The line $y = 2ax - a^2$ is tangent to the parabola $y = x^2$ at the point (a, a^2).

5. The line $y = mx + b$ intersects the hyperbola $y = 1/x$ when $mx^2 + bx - 1 = 0$, i.e., $x = \dfrac{-b \pm \sqrt{b^2 + 4m}}{2m}$. The line is tangent when $b^2 + 4m = 0$. The family of tangent lines is given by $y = -\dfrac{b^2}{4}x + b$.

7. $y(z) \approx 1$, valid when z/ε is small; $y(z) \approx -1$, valid when εz is small.

9. Use the results for fractional exponents.

11. Position and atmospheric pressure.

13. $1 - \varepsilon/5 + 2\varepsilon^2/25 + \cdots$, 0.998 (follow Exercise 12).

15. $x_1 = 2$, $x_2 = 2 - \varepsilon/2$, $x_3 = 2 - \varepsilon(4 - \varepsilon)/(8 - \varepsilon)$.

Exercises 1.12

1. **(i)** ML^{-3} **(ii)** MLT^{-1} **(iii)** MLT^{-2} **(iv)** $ML^{-1}T^{-2}$ **(v)** ML^2T^{-2}
(vi) T^{-1} **(vii)** ML^2T^{-1} **(viii)** ML^2 **(ix)** ML^2T^{-1} **(x)** ML^2T^{-2}

3. $L^3M^{-1}T^{-2}$

5. $V_0 = 1$ liter, $t_0 = 1$ sec; $\dfrac{V(t)}{V_0} = 40 - 3\dfrac{t}{t_0} - \left(\dfrac{t}{t_0}\right)^2$.

7. $g = 9.80621$ m sec^{-2}

9. $[a] = L^{-1}T^{-1}$, $[b] = [T^{-1}]$, $[c] = LT^{-1}$; $v = 0$ when $s = \left(-b + \sqrt{b^2 - 4ac}\right)/2a$; $[s] = L$.

11. If $y = f(x - t)$, then y is constant when $x - t$ is constant.

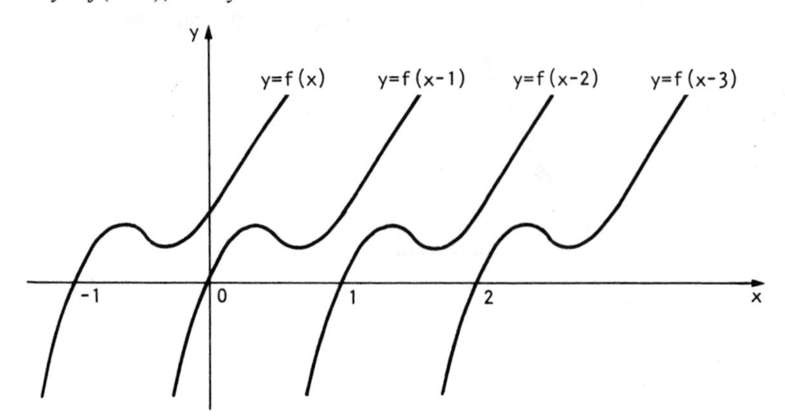

13. $c = \omega/k$.

15. (i) If $[x] = [a] = [b] = L$, then $\left[\int_a^b x\,dx\right] = L^2 = \left[\frac{1}{2}(b^2 - a^2)\right]$.

Chapter 2 Differentiation

Exercises 2.1

1. $\lim_{h\to 0} \frac{(x+h)^3 - x^3}{h} = \lim_{h\to 0} \frac{3hx^2 + 3h^2x + h^3}{h} = 3x^2,$

$\lim_{h\to 0} \frac{(x+h)^4 - x^4}{h} = \lim_{h\to 0} \frac{4hx^3 + 6h^2x^2 + 4h^3x + h^4}{h} = 4x^3,$

$\lim_{h\to 0} \frac{(x+h)^5 - x^5}{h} = \lim_{h\to 0} \frac{5hx^4 + 10h^2x^3 + 10h^3x^2 + 5h^4x + h^5}{h} = 5x^4.$

3. $\frac{3}{2}x^{1/2}$, $\frac{5}{2}x^{3/2}$, $\frac{7}{2}x^{5/2}$

5. To obtain **(i)**, replace Δx by $-\Delta x$; **(ii)** is the average of the first two equations. Graphically, compare the slopes of three chords.

7. $\frac{d}{dx}\frac{1}{x^n} = \lim_{h\to 0} \frac{1}{h}\left[\frac{1}{(x+h)^n} - \frac{1}{x^n}\right] = \lim_{h\to 0} \frac{1}{h}\left(\frac{x^n - (x+h)^n}{x^n(x+h)^n}\right) = -\frac{n}{x^{n+1}}.$

For any integer n (positive, negative or zero), $\frac{d}{dx}x^n = nx^{n-1}$.

9. (ii) $\frac{d}{dx}xf(x) = \lim_{h\to 0} \frac{(x+h)f(x+h) - xf(x)}{h} = \lim_{h\to 0} x\frac{(f(x+h) - f(x))}{h} + \lim_{h\to 0} f(x+h) = xf'(x) + f(x)$

(iii) use $(f(x+\Delta x))^2 - (f(x))^2 = [f(x+\Delta x) + f(x)][f(x+\Delta x) - f(x)]$.

11. $x = \pm 1/\sqrt{3}$; equate derivatives and solve for x.

13. The points of intersection satisfy $mx + b = x^2 + 1$ or $x^2 - mx + (1 - b) = 0$. The points coincide if $m^2 + 4(b - 1) = 0$, $b = 1 - m^2/4$. The family of tangent lines is $y = mx + (1 - m^2/4)$. By calculus methods, equating slopes to find $m = 2x$ and $y = x^2 + 1 = \left(\frac{m}{2}\right)^2 + 1 = m\left(\frac{m}{2}\right) + b$. As before, $b = 1 - \frac{m^2}{4}\cdot$ If $b = 0$, $m = \pm 2$, tangent lines are $y = \pm 2x$.

15. **(i)** $-2\sin 2\theta$ **(ii)** $2\sin\theta\cos\theta$ **(iii)** $-\sin\theta - \cos\theta$ **(iv)** $1 - 1/x^2$
(v) $4x(a^2 + x^2)$ **(vi)** $-2\sin 2\theta$

17. $\sin(x + dx) = \sin x + (\cos x)\,dx$; $\sin 31° \approx \frac{1}{2} + \frac{\sqrt{3}}{2}\frac{\pi}{180} = 0.515115$, accurate value 0.515038; absolute error 0.000077, relative error 0.000149.

19. $y = 2x - 1$, error $E(x) = x^2 - (2x - 1) = (x - 1)^2$; $E(1.05) = 0.0025$, $E(1.1) = 0.01$.

21. **(i)** 3.92, 3.9204; 4.12, 4.1209 **(ii)** 7.76, 7.762392; 8.36, 8.365427
(iii) 0.99, 0.98995; 1.015, 1.01489 **(iv)** 0.68989, 0.6897259; 0.73535, 0.73496
(v) 0.31011, 0.31027; 0.26465, 0.265038 **(vi)** 2.00, 2.000408; 2.00, 2.000874.

23. $dV = 2.8045 \times 10^9\ (\text{km})^3$.

Exercises 2.2

1. **(i)** $2 + 2x$ **(ii)** $3 + 6x + 3x^2$ **(iii)** $-8x + 16x^3$ **(iv)** $7x^6 - 5x^4 + 3x^2 - x$
(v) $8x(x^2 + 1)^3$ **(vi)** $5(3x + 3x^2)^4(3 + 6x)$ **(vii)** $8x - \frac{1}{2}x^3$ **(viii)** $4x/(x^2 + 1)^3$
(ix) $-8x(x^2 + 1)^{-5}$

3. $a\sec ax\tan ax$, $-a\csc ax\cot ax$, $-a\csc^2 ax$

5. $\frac{d}{dx}(x^{1/m})^m = m(x^{1/m})^{m-1}\frac{d}{dx}x^{1/m} = 1$, $\frac{d}{dx}x^{1/m} = \frac{1}{m}x^{1/m-1}$.

7. **(i)** $\frac{1}{3}x^{-2/3}$ **(ii)** $\frac{3}{4}x^{-3/4}$ **(iii)** $\frac{1}{2}x^{-1/2} - \frac{1}{2}x^{-3/2}$ **(iv)** $\frac{3}{5}x^{-2/5} + 1 + \frac{7}{5}x^{2/5}$
(v) $x^{-1/2} + x^{-3/4} + x^{-5/6}$ **(vi)** $1 - x^{-2}$

9. **(i)** $4x^3$ **(ii)** $(1 + x)(2x + x^2)^{-1/2}$ **(iii)** $3x^2(1 + x^3)^{-2}$ **(iv)** $-(3 - 2x)^{-1/2}$
(v) $5x(1 + 2x^2)^{3/2}$ **(vi)** $-20x(1 + x^2)^{-11}$ **(vii)** $3x^2(1 + x^2)^{-3} - 6x^4(1 + x^2)^{-4}$
(viii) 1 **(ix)** -1

11. $\frac{\pi}{180}\cos x$, $-\frac{\pi}{180}\sin x$, $\frac{\pi}{180}\sec^2 x$, $-\frac{\pi}{180}\csc x\cot x$, $\frac{\pi}{180}\sec x\tan x$, $-\frac{\pi}{180}\csc^2 x$.

13. **(i)** $-\frac{(1 + 2x)}{(x + x^2)^2}$ **(ii)** $\frac{s}{(1 - s^2)^{3/2}}$ **(iii)** $\frac{1}{(1 - t)^2}\cos\frac{t}{1 - t}$

(iv) $\frac{1}{4}t^{-1/2}\cos t^{1/2}(\sin t^{1/2} + 1)^{-1/2}$ **(v)** $3\cos 3x - 4\cos 4x$

(vi) $(1 - 2x^2)(1 - x^2)^{-3/2}$ **(vii)** $\frac{1}{2}(1 + \sin x)^{-3/2}\cos x$ **(viii)** $\frac{2\sec^2 x}{(1 - \tan^2 x)^2}$

(ix) $\left(\frac{1 + x}{1 - x}\right)^{-1/2}\frac{1}{(1 - x)^2}$

15. $\frac{d}{dx}\left(1 + nx + \frac{n(n - 1)}{2!}x^2 + \cdots + nx^{n-1} + x^n\right) = n + n(n - 1)x + \cdots + n(n - 1)x^{n-2} + nx^{n-1}$

$= n\left[1 + (n - 1)x + \frac{(n - 1)(n - 2)}{2!}x^2 + \cdots + (n - 1)x^{n-2} + x^{n-1}\right] = n(1 + x)^{n-1}$

$= \frac{d}{dx}(1 + x)^n$.

17. $\dfrac{d}{dx}\dfrac{f(x)}{f(x)} = \dfrac{f(x)f'(x) - f(x)f'(x)}{(f(x))^2} = 0,\ \dfrac{d}{dx}f(x)\dfrac{1}{f(x)} = f'(x)\dfrac{1}{f(x)} - f(x)\dfrac{f'(x)}{(f(x))^2} = 0.$

Exercises 2.3

1. **(i)** $\cos x + \sin x,\ -\sin x + \cos x$ **(ii)** $2x\cos x^2,\ 2\cos x^2 - 4x^2\sin x^2$
(iii) $-5\sin 10x,\ -50\cos 10x$ **(iv)** $2x\sin(4-x^2),\ 2\sin(4-x^2) - 4x^2\cos(4-x^2)$
(v) $\sec^2 x,\ 2\sec^2 x\tan x$ **(vi)** $-\csc^2 x,\ 2\csc^2 x\cot x$ **(vii)** $2\sin 4x,\ 8\cos 4x$
(viii) $\sec^2 x - \csc^2 x,\ 2\sec^2 x\tan x + 2\csc^2 x\cot x$
(ix) $x^2\cos x + 2x\sin x,\ (2-x^2)\sin x + 4x\cos x$

3. **(i)** $x = (2n+1)\pi/6$, maximum for n even, minimum for n odd; $x = n\pi/3$
(ii) $x = n\pi/2$, maximum for n even, minimum for n odd; $x = (2n+1)\pi/2$
(iii) maximum at $x = -2/\sqrt{3}$, minimum at $x = 2/\sqrt{3}$; $x = 0$
(iv) maximum at $x = 0$; $x = \pm 1$
(v) maximum at $x = 1$, minimum at $x = -1$; $x = 0,\ \pm\sqrt{6}$
(vi) minimum at $x = 0$, $x = \pm 1$

5. An even function satisfies $f(x) = f(-x)$. Differentiating, $f'(x) = \dfrac{d}{dx}f(-x) = -f'(-x)$ by the chain rule. Conclude that $f'(x)$ is an odd function. Similarly for an odd function, $f(x) = -f(-x)$ and $f'(x) = f'(-x)$.

7. **(i)** Correct unless relative maximum is zero **(ii)** correct
(iii) not correct, consider $y = (x^2 - 1)^2$.

9.

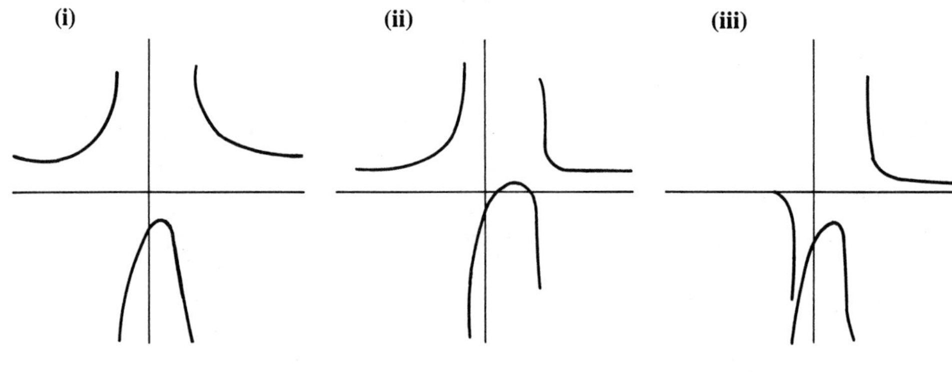

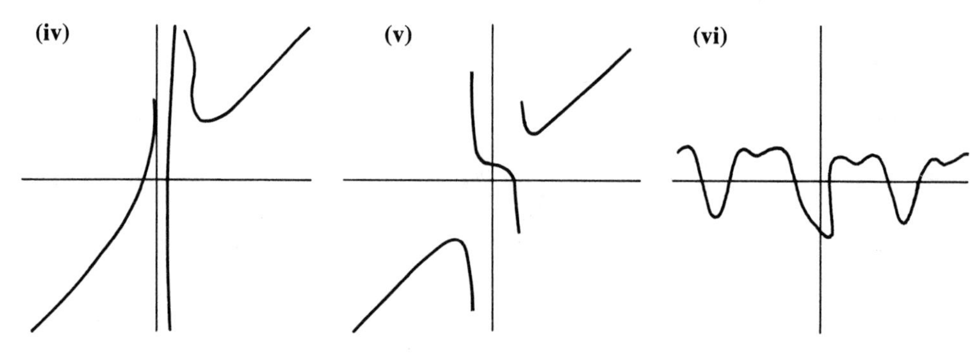

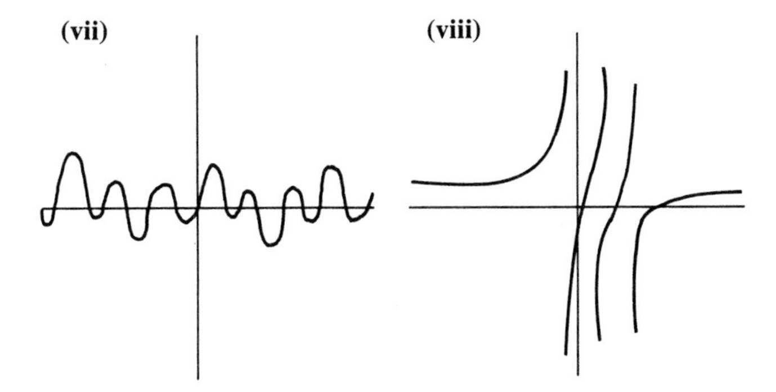

11. Use product rule.
13. Use induction. The function $1/x^2$ is more singular at $x = 0$ than the function $1/x$, etc.
17. Solve $y'' = 3ax + b = 0$, $y' = 3ax^2 + 2bx + c = 0$.
19. **(i)** $y = x$, $y = -x$ **(ii)** $x = 1$, $x = -1$, $y = 0$ **(iii)** $x = 0$, $y = 0$, $x + y = 0$
(iv) $y = x$, $y = -x$ **(v)** $y = x$ **(vi)** $y = 1$ (write $x^2 = y^3/(1 - y)$)
21. If $E(x)$ has degree n, $E'(x)$ has degree $n - 1$ and $E(x) \neq E'(x)$.

Exercises 2.4

1. $20/3$ sec.
3. $v(0) = 21$ m/sec, $v(1) = 11.2$, $v(2) = 1.4$; $a(0) = a(1) = a(2) = -9.8$ m/sec^2. Maximum height 22.5 m, when $v = 0$.
5. **(i)** Chain rule **(ii)** $\dfrac{d^2y}{dt^2} = \dfrac{d}{dt}\left(\dfrac{dy}{dx}\dfrac{dx}{dt}\right) = \dfrac{d}{dt}\left(\dfrac{dy}{dx}\right)\dfrac{dx}{dt} + \dfrac{dy}{dx}\dfrac{d^2x}{dt^2} = \dfrac{d^2y}{dx^2}\left(\dfrac{dx}{dt}\right)^2 + \dfrac{dy}{dt}\dfrac{d^2x}{dt^2}$.
7. $\dfrac{dy}{dx} = \dfrac{1}{2\sqrt{x}}\dfrac{dx}{dt}$; $x = \dfrac{1}{4}$, $y = \dfrac{1}{2}$, tangent line $y = x + \dfrac{1}{4}$.
9. 2.7183
11. $P_n(x) = 1 - x + \dfrac{x^2}{2!} - \dfrac{x^3}{3!} + \cdots + (-1)^n \dfrac{x^n}{n!}$, $F(1) = 0.3679$
13. **(i)** $-x^3 + x$ **(ii)** $x^3 - 3x^2 - x + 3$ **(iii)** $x^3 - 2x^2$ **(iv)** $x^3 - 4x$
15. $\dfrac{dr}{dt} = \dfrac{dr}{dV}\dfrac{dV}{dt} = \dfrac{1}{4\pi r^2}\left(-\dfrac{1}{1 + t^2}\right)$
17. $r(t) = [10^2 + (10\sqrt{3} - 350t)^2]^{1/2}$; minimum separation (10 km) occurs at $t = \sqrt{3}/35$ hrs.
19. $v(t) = \omega A \cos \omega t$, $a(t) = -\omega^2 A \sin \omega t$
21. $c_0 = a_0 + a_1 a + \cdots + a_n a^n$, $c_1 = a_1 + 2a_2 a + \cdots + n a_n a^{n-1}$,
$c_2 = \dfrac{1}{2!}(2a_2 + 3 \cdot 2a_3 a + \cdots + n(n-1)a_n a^{n-2})$, $\ldots$, $c_n = a_n$.
23. $\sin x = \dfrac{1}{2} + \dfrac{\sqrt{3}}{2}\left(x - \dfrac{\pi}{6}\right) - \dfrac{1}{4}\left(x - \dfrac{\pi}{6}\right)^2 - \dfrac{\sqrt{3}}{12}\left(x - \dfrac{\pi}{6}\right)^3$; using four terms,
$\sin 40° = \sin\left(\dfrac{\pi}{6} + \dfrac{\pi}{18}\right) = 0.642767$; accurate value is 0.642788.

25. **(i)** $2udu + 2vdv$ **(ii)** $2uv^2du + 2u^2v\,dv$ **(iii)** $\dfrac{1}{\sqrt{u^2+v^2}}(2u\,du + 2v\,dv)$

(iv) $\dfrac{vdu - udv}{(u+v)^2}$ **(v)** $\cos(uv)[udv + vdu]$ **(vi)** $\sec^2\dfrac{u}{v}\left[\dfrac{vdu - udv}{v^2}\right]$

27. **(i)** 2.12, 2.1216 **(ii)** 1, 0.9998 **(iii)** -1.003, -1.003003
(iv) 0.2651, 0.2653 **(v)** 1.0698, 1.0723 **(vi)** 1.4142, 1.4140

29. $\Delta A = (4\pi r + 2\pi h)\Delta r + 2\pi r\Delta h$, $\dfrac{\Delta A}{A} = \dfrac{(2r+h)\Delta r + r\Delta h}{r^2 + rh}$

31. $\dfrac{\Delta T}{T} = \dfrac{-15}{3600} = \dfrac{1}{2}\dfrac{\Delta l}{l}$. Shorten pendulum by one part in 120 or 0.83%.

33. $-r\sin\theta[1 + r(l^2 - r\sin^2\theta)^{-1/2}\cos\theta]\Delta\theta$.

Exercises 2.5

1. $y = -\dfrac{1}{3}x^2 - \dfrac{2}{3}x + 1$

3. $y = 10\left(x^2 + \dfrac{14}{10}x\right) + 5 = 10\left(x + \dfrac{7}{10}\right)^2 + \dfrac{1}{10} \geq \dfrac{1}{10}$.

5. -1, 3; 1; approximate roots -2.28, 0.66, 4.62

7. **(i)** $2 - x - x^2 = \dfrac{9}{4} - \left(x + \dfrac{1}{2}\right)^2$; maximum value $\dfrac{9}{4}$ at $x = -\dfrac{1}{2}$.

(ii) $x^2 + 5x + 1 = \left(x + \dfrac{5}{2}\right)^2 - \dfrac{21}{4}$; minimum $-\dfrac{21}{4}$ at $x = -\dfrac{5}{2}$.

(iii) $2x^2 - 3x + 1 = 2\left(x - \dfrac{3}{4}\right)^2 - \dfrac{1}{8}$; minimum $-\dfrac{1}{8}$ at $x = \dfrac{3}{4}$.

9. If $p>0$, $f(x)$ is monotone increasing and has exactly one real root. If $p<0$, condition for three real roots is $f(x)<0$ at stationary point $x = (-p/3)^{1/2}$. At this point, $f(x) = \dfrac{2}{3}p\left(-\dfrac{p}{3}\right)^{1/2} + q<0$ if $q< -\dfrac{2}{3}p\left(-\dfrac{p}{3}\right)^{1/2}$ or $q^2< -\dfrac{4}{27}p^3$.

11. Maximum at $x = 2 - 1/\sqrt{3}$, minima at $x = 2 + 1/\sqrt{3}$ and $x = 1/2$; yes, $f'(1) = 2$.

13. Area $= 4xy = 4x/(x^2 + 1)$, maximum area is 2 at $x = 1$. Inscribed circle tangent to curves at $x = \pm(2^{1/3} - 1)^{1/2}$, $y = 2^{-1/3}$; equation of circle $x^2 + y^2 = 2^{1/3} - 1 + 2^{-2/3}$.

15. $x = 4l/(\pi + 4)$, where a length x is used to form the square and l is the total length of wire. Hint: combined area $A = \dfrac{x^2}{16} + \dfrac{(l-x)^2}{4\pi}$.

17. $V = \dfrac{2\sqrt{3}}{27}\pi l^3$, $h = \dfrac{l}{\sqrt{3}}$, $r = \sqrt{\dfrac{2}{3}}\,l$

19. To maximize net profits, 25¢ is absorbed by the firm and 25¢ passed on the purchaser. The book would then cost \$17.75 but the firm's profit would be reduced \$5012.50.

21. If $a \leq 1/3$, $\theta = 0$; if $a > 1/3$, $\theta = \cos^{-1}[(-1 - 8(9a^2 - 1)/9a]$ where θ is the angular position (measured from the positive x-axis) at which the rescuer enters the water. Hint: The time until rescue is $t = R[\theta + 3(1 + a^2 + 2a\cos\theta)^{1/2}]/V$ where V is the running speed.

23. $3600\,v/(6 + 3v + v^2/3)$ cars, $v = \sqrt{18}$ m/sec.

25. $\dfrac{\sin\theta_1}{\sin\theta_2} = \dfrac{v_1}{v_2}$ in the notation of Fig. 5.2. This problem is the same as that of Example 5 when c_1 and c_2 are replaced by $1/v_1$ and $1/v_2$.

27. $\$21.167(10)^5$

29. All the Kuwait oil goes to London while the oil in Galveston and Caracas goes to New York. The minimum cost is $101,000. Hint: If x thousand barrels go from Kuwait to New York and y thousand go from Galveston to New York, the total cost is $C(x, y) = 10[9x - 5y + 10{,}850]$. This must be minimized subject to the constraints $0 \le x \le 200$, $0 \le y \le 150$, $150 \le x + y \le 250$. This gives $x = 0$, $y = 150$.

Exercises 2.6

1. **(i)** $x = y^{1/2}$ or $x = -y^{1/2}$ **(ii)** $x = \dfrac{-1 \pm (1 + 4y)^{1/2}}{2}$ **(iii)** $x = y^3$

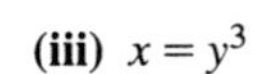

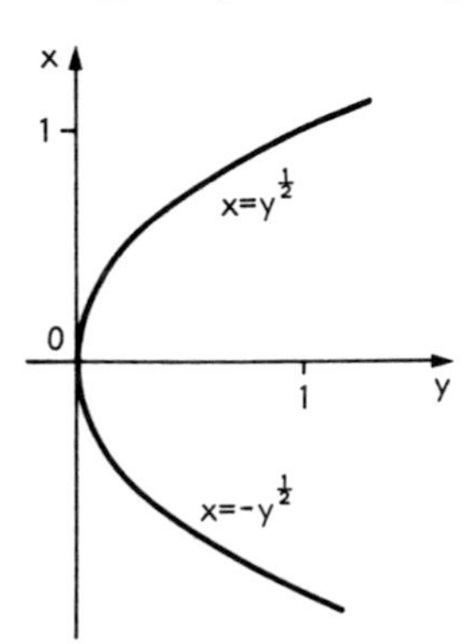

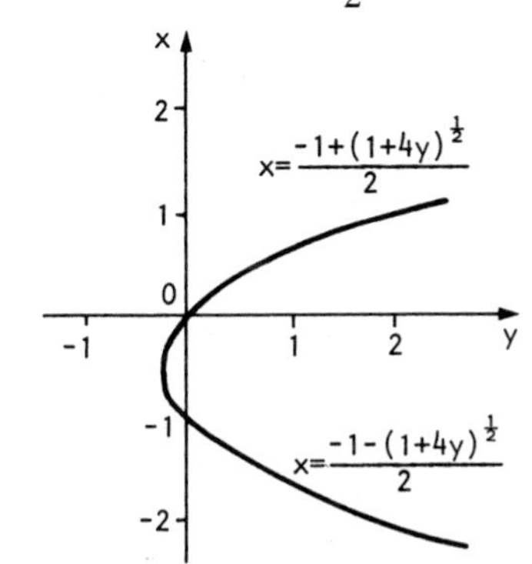

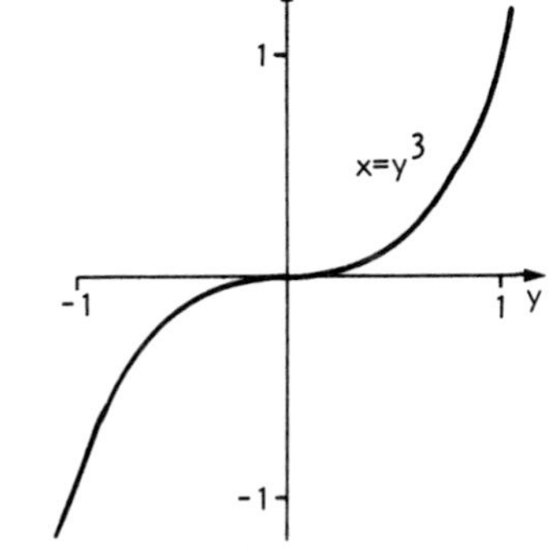

3. **(i)** $\dfrac{dy}{dx} = -\dfrac{y + 4x^{3/2}y^{1/2}}{x + 2x^{1/2}y^{1/2}} = \left(\dfrac{dx}{dy}\right)^{-1}$ **(ii)** $-\dfrac{b^2x}{a^2y}$ **(iii)** $\dfrac{16x}{y(3y - 16)}$

(iv) $-\dfrac{2xy + y^2}{x^2 + 2xy}$ **(v)** $-\left(\dfrac{y}{x}\right)^{1/2}$ **(vi)** $\dfrac{2x(1 + y^2)}{1 - 2y - 2y^3}$ **(vii)** $\dfrac{3x^2 + y^2}{1 - 2xy}$

(viii) $-\dfrac{y}{x}$ **(ix)** $\dfrac{\sin(x + y) - y}{x - \sin(x + y)}$

(For example, in **(i)**, $y' + \dfrac{1}{2}(xy)^{-1/2}[1 + xy'] + 2x = 0$; solve for y'.)

5. Slope 0 at $(\pm a, 0)$; 1 at $\left(-\dfrac{a\sqrt{2}}{4}, \dfrac{a\sqrt{2}}{4}\right)$, $\left(\dfrac{a\sqrt{2}}{4}; -\dfrac{a\sqrt{2}}{4}\right)$; -1 at $\left(\dfrac{a\sqrt{2}}{4}, \dfrac{a\sqrt{2}}{4}\right)$, $\left(-\dfrac{a\sqrt{2}}{4}, -\dfrac{a\sqrt{2}}{4}\right)$.

7. $\dfrac{dr}{dh} = \dfrac{-r}{2r + h}$, $\dfrac{dh}{dr} = -\dfrac{(2r + h)}{r}$. If the surface area remains constant, an increase in h must be accompanied by a decrease in r.

9. If $y = x + \sqrt{x^2 + 1}$, $(y - x)^2 = x^2 + 1$ or $y^2 - 2yx = 1$. Solving $x = (y^2 - 1)/2y$. If $y = x - \sqrt{x^2 + 1}$, similarly $x = (y^2 - 1)/2y$.

11. $\frac{dy}{dx} = \frac{dy}{dt}\frac{dt}{dx} = \frac{g'(t)}{f'(t)}, \frac{dx}{dy} = \frac{f'(t)}{g'(t)},$
$\frac{d^2y}{dx^2} = \frac{d}{dt}\frac{g'(t)}{f'(t)}\frac{dt}{dx} = \frac{f'(t)g''(t) - g'(t)f''(t)}{(f'(t))^3},$
$\frac{d^2x}{dy^2} = \frac{g'(t)f''(t) - f'(t)g''(t)}{(g'(t))^3}$

13. $\frac{dy}{dx} = \frac{x(2-x)}{(1-x)^2}, \frac{d^2y}{dx^2} = \frac{2}{(1-x)^3}$

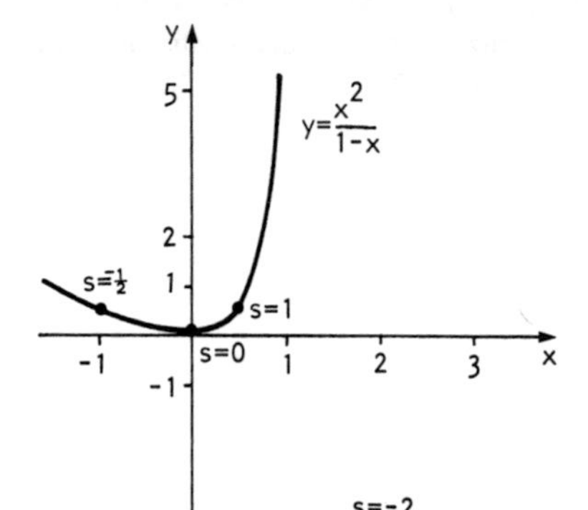

15. Note that $\sin(\sin^{-1}x + \cos^{-1}x) = \sin(\sin^{-1}x)\cos(\cos^{-1}x) + \cos(\sin^{-1}x)\sin(\cos^{-1}x) = x^2 + (1-x^2)^{1/2}(1-x^2)^{1/2} = 1.$

17. $\cot^{-1}x = \frac{\pi}{2} - \tan^{-1}x;\ \frac{d}{dx}\cot^{-1}x = -\frac{d}{dx}\tan^{-1}x = \frac{-1}{1+x^2}.$

19. **(i)** $\tan(\tan^{-1}x + \tan^{-1}a) = \frac{\tan(\tan^{-1}x) + \tan(\tan^{-1}a)}{1 - \tan(\tan^{-1}x)\tan(\tan^{-1}a)} = \frac{x+a}{1-xa}.$
In **(ii)**, **(iii)**, **(iv)** consider triangles.

21. **(i)** π **(ii)** $\pi/3$ **(iii)** $-\pi/2$ **(iv)** $\pi/4$ **(v)** $\pi/2$ **(vi)** $-\pi/2$
(vii) $\pi/4$ **(viii)** 0.46365^r **(ix)** 0

23. **(i)** $x^2 + y^2 = 2ax$
(ii) $(x^2+y^2)^2 = 3ay(x^2+y^2) - 4ay^3$ (use $\sin 3\theta = 3\sin\theta - 4\sin^3\theta$)
(iii) $x^2 + y^2 = a(x^2+y^2)^{1/2} - ay$ **(iv)** $(x^2+y^2)^{1/2} = a\tan^{-1}\frac{y}{x}$
(v) $(x^2+y^2)^{1/2}\tan^{-1}\frac{y}{x} = a$ **(vi)** $3\left(x - \frac{a}{3}\right)^2 + 4y^2 = \frac{4}{3}a^2$

25. $-4.42°$/sec, 0.48 m/sec. (Apply the cosine law to the triangle in Fig. 6.3).

27. **(i)** ½ **(ii)** ½ **(iii)** ½

29. $5\pi(4+x^2)$ km/min where x is measured along the coast from the point nearest to the lighthouse; 80π km/min.

Exercises 2.7

1. $f'(\xi) = 4\xi^3 - 2\xi = 2\xi(2\xi^2 - 1) = 0$ when $\xi = 0, \ \pm 1/\sqrt{2}$.

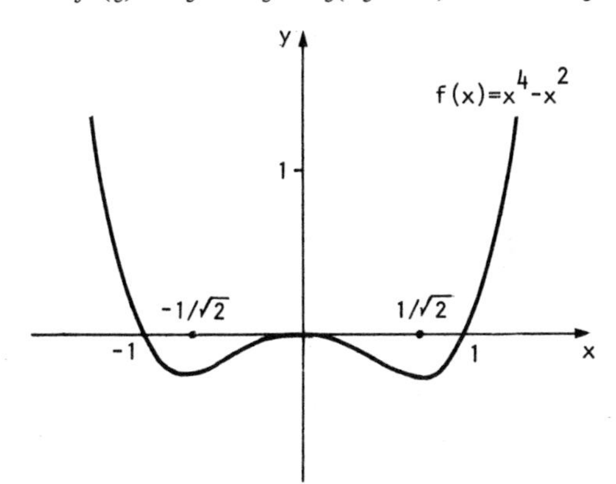

3. **(i)** $\sqrt{7}$ **(ii)** $\pi/4$ **(iii)** $(1 - 4/\pi)^{1/2}$ **(iv)** ± 0.4817 **(v)** 0 **(vi)** $1/\sqrt{3}$

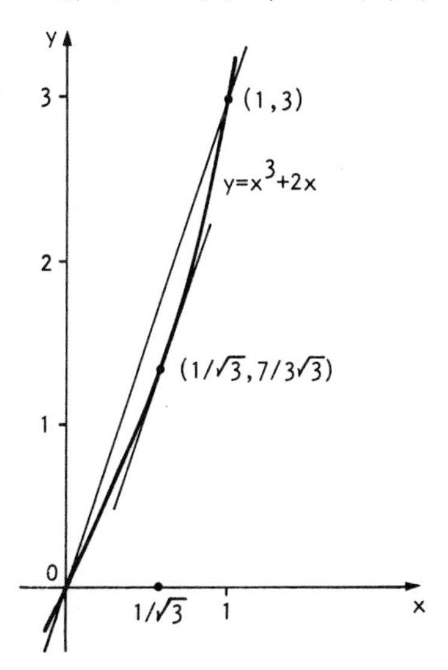

5. Since $f(x) - f(a) = (x - a)f'(\xi)$; if $f'(\xi) > 0$, then $f(x) > f(a)$ for $x > a$.

7. **(i)** 0 **(ii)** -1 **(iii)** -1 **(iv)** -2 **(v)** $-\pi$ **(vi)** 80 **(vii)** -2 **(viii)** $-1/2$ **(ix)** a **(x)** -2 **(xi)** a/b **(xii)** 1

9. There is a root of $f'(x)$ between any two consecutive roots of $f(x)$. If $f(x)$ has n distinct real roots, then $f'(x)$ has $n - 1$ distinct real roots, $f''(x)$ has $n - 2$ distinct real roots, etc. If $f(x)$ has a root of multiplicity p, then $f(x) = (x - a)^p g(x)$ and $f'(x)$ has the same root with multiplicity $p - 1$.

11. $f(x) = 1 - 2x + 3x^2 - 2x^3 + x^4$

13. **(i)** 1 **(ii)** 2 **(iii)** 0 **(iv)** 0 **(v)** 1 **(vi)** 1 **(vii)** 2 **(viii)** -2 **(ix)** $1/3$ **(x)** $+\infty$ **(xi)** -1 **(xii)** 0

15. **(i)** $\sin x = x - \dfrac{x^3}{3!} + \cdots + (-1)^n \dfrac{x^{2n+1}}{(2n+1)!} + (-1)^{n+1} \dfrac{x^{2n+2}}{(2n+2)!} \sin\xi,\ 0<\xi<x$

(ii) $\cos x = 1 - \dfrac{x^2}{2!} + \cdots + (-1)^n \dfrac{x^{2n}}{(2n)!} + (-1)^{n+1} \dfrac{x^{2n+1}}{(2n+1)!} \cos\xi,\ 0<\xi<x$

(iii) $\dfrac{1}{1+x} = 1 - x + x^2 - \cdots + (-1)^n x^n + (-1)^{n+1} \dfrac{x^{n+1}}{1+x},\ x \neq -1$

Bounds are $\dfrac{|x|^{2n+2}}{(2n+2)!}, \dfrac{|x|^{2n+1}}{(2n+1)!}, \dfrac{|x|^{n+1}}{1-|x|}$ (if $|x|<1$) and $\dfrac{|x|^{n+1}}{|x|-1}$ (if $|x|>1$).

17. $(a^2+x)^{1/n} = a^{2/n}\left(1+\dfrac{x}{a^2}\right)^{1/n} = a^{2/n}\left[1+\dfrac{1}{n}\dfrac{x}{a^2}+\dfrac{1-n}{2n^2}\left(\dfrac{x}{a^2}\right)^2+\cdots\right]$

(i) $(0.98)^{1/3} = (1-0.02)^{1/3} = 1 + \dfrac{1}{3}(-0.02) + \dfrac{(-2)}{2.3^2}(-0.02)^2 + \cdots = 0.9933$

(ii) $(1.44+0.01)^{1/2} \approx 1.204$ **(iii)** $(16+0.1)^{1/4} \approx 2.003$ **(iv)** $(1.22)^{1/3} \approx 1.0685$

19. **(i)** 0 **(ii)** 0 **(iii)** 1/3 **(iv)** 1/2 **(v)** ∞ **(vi)** 1

For example, in **(i)**, $\lim\limits_{x\to\infty} x\left[\left(1+\dfrac{1}{x^2}\right)^{1/2} - \left(1-\dfrac{1}{x^2}\right)^{1/2}\right]$

$$= \lim_{x\to\infty} x\left[\left(1+\frac{1}{2x^2}+\cdots\right) - \left(1-\frac{1}{2x^2}+\cdots\right)\right] = 0;$$

in **(iv)**, $\lim\limits_{x\to\infty} x^2\left[\left(1+\dfrac{1}{x^2}\right)^{-1} - \left(1+\dfrac{1}{x}\right)^{-1}\right]$

$$= \lim_{x\to\infty} x^2\left[\left(1-\frac{1}{x^2}+\cdots\right) - \left(1-\frac{1}{x}+\frac{1}{x^2}-\cdots\right)\right] = \infty.$$

Exercises 2.8

1. $e^x \approx 1 + \dfrac{x}{1!} + \dfrac{x^2}{2!} + \dfrac{x^3}{3!}$; 1.105167, 1.22133, 1.645833, 2.666667; accurate values 1.105171, 1.221403, 1.648721, 2.718282.

3. **(i)** $-3e^{-3x}$ **(ii)** $(\sin x + x\cos x)e^{x\sin x}$ **(iii)** $-2(x+1)e^{-(x+1)^2}$

(iv) $2f(x)f'(x)e^{(f(x))^2}$ **(v)** $-f'(x)\sin f(x)e^{\cos f(x)}$ **(vi)** $f'(x)\sec^2 f(x)e^{\tan f(x)}$

5. Use $e^x = e^{x/2}e^{x/2} = (e^{x/2})^2 \geq 0$ and $e^x e^{-x} = 1$ to conclude $e^x > 0$ for all x. By the mean value theorem $e^{x_1} - e^{x_2} = (x_1 - x_2)e^{\xi}$ for some ξ in $x_2<\xi<x_1$. Since $e^{\xi}>0$, $e^{x_1}-e^{x_2}>0$ if $x_1>x_2$.

7. **(i)** maximum at $x=0$, inflection points at $x = \pm 1/\sqrt{2}$

(ii) maximum at $x = 1/\sqrt{2}$, minimum at $x = -1/\sqrt{2}$, inflection points at $x = 0, \pm\sqrt{3/2}$

(iii) maxima at $x = \pm 1$, minimum at $x = 0$, inflection points at $x = \pm(5 \pm \sqrt{17})^{1/2}/2$

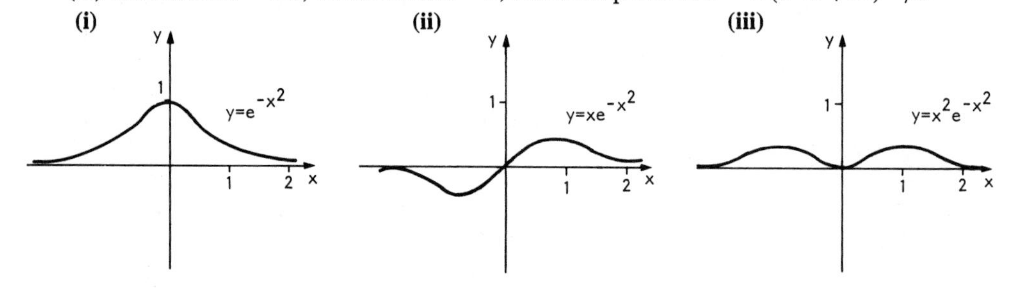

9.

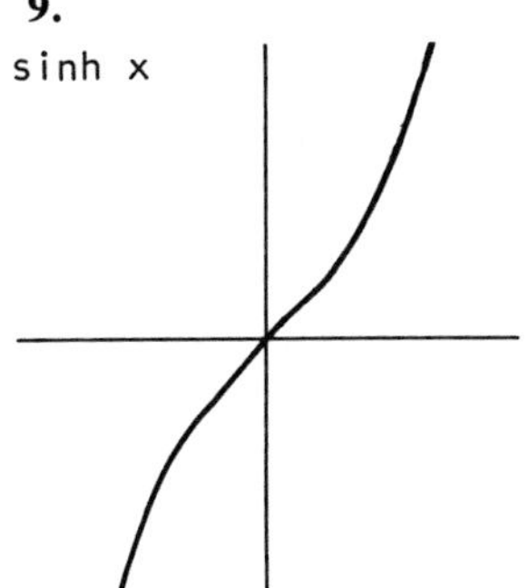

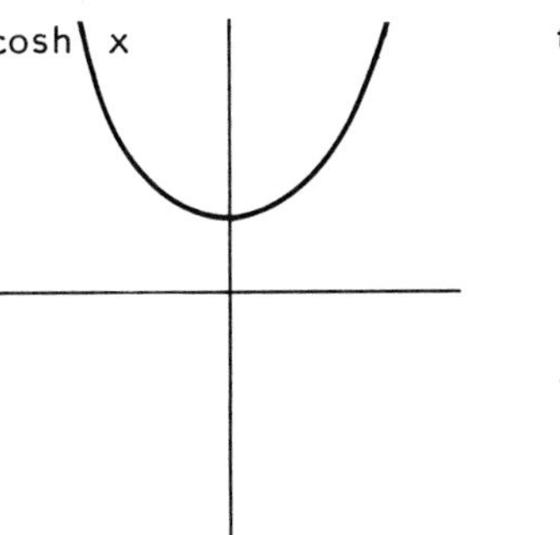

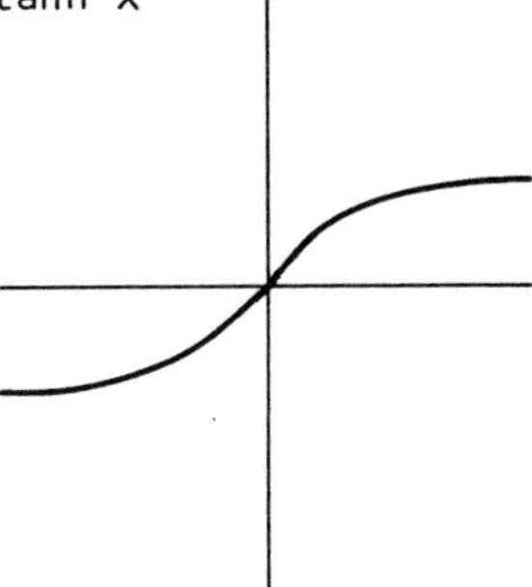

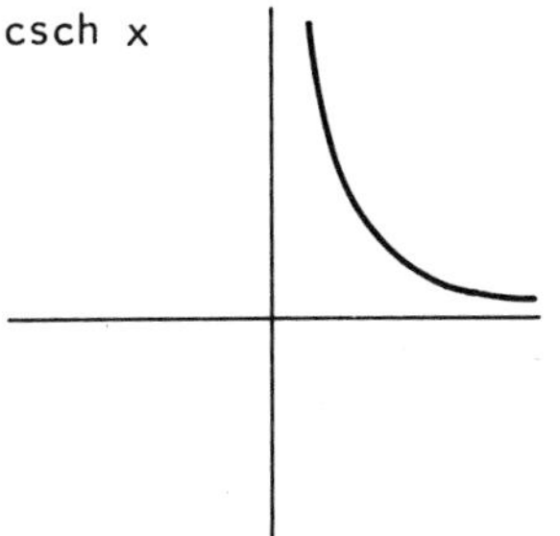

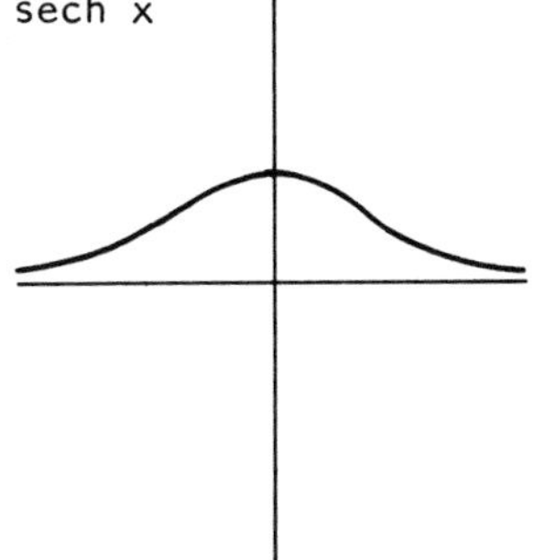

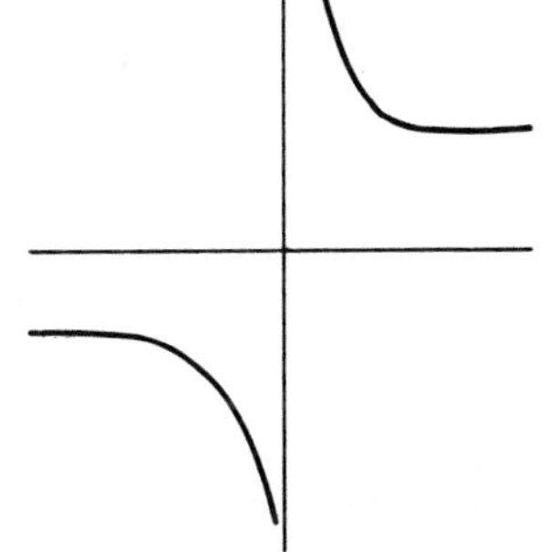

11. $\cosh x = 1 + \dfrac{x^2}{2!} + \dfrac{x^4}{4!} + \cdots + \dfrac{x^{2n}}{2n!} + \cdots,$

$\sinh x = x + \dfrac{x^3}{3!} + \dfrac{x^5}{5!} + \cdots + \dfrac{x^{2n+1}}{(2n+1)!} + \cdots.$

13. **(i)** $\cosh x + \sinh x = \dfrac{e^x + e^{-x}}{2} + \dfrac{e^x - e^{-x}}{2} = e^x$

(iii) $\cosh^2 x - \sinh^2 x = (\cosh x + \sinh x)(\cosh x - \sinh x) = e^x e^{-x} = 1$

(iv) $\operatorname{sech}^2 x = \dfrac{1}{\cosh^2 x}(\cosh^2 x - \sinh^2 x) = 1 - \tanh^2 x$

(vi) $\tanh^2 x = \dfrac{\sinh^2 x}{\cosh^2 x} = \dfrac{\sinh^2 x}{1 + \sinh^2 x} < 1.$

15. **(i)** $3e^{3x}$ **(ii)** $3x^2 e^{\cos x} - x^3 \sin x\, e^{\cos x}$ **(iii)** $2 \sinh x \cosh x$
(iv) $2x \sinh(1 + x^2)$ **(v)** $\frac{3}{2} e^x (e^x + 1)^{1/2}$ **(vi)** $2 \tanh x \operatorname{sech}^2 x$
(vii) $-e^{-x} \cos(e^{-x})$ **(viii)** $\frac{1}{2} \cosh x\, (\sinh x)^{-1/2}$ **(ix)** $\sinh x\, e^{\cosh x}$

17. $y'' = \dfrac{1}{a}(1 + (y')^2)^{1/2}$. Hint: $\cosh x = (1 + \sinh^2 x)^{1/2}$, $\dfrac{1}{a}\cosh\dfrac{x}{a} = \dfrac{1}{a}\left(1 + \sinh^2\dfrac{x}{a}\right)$.

19. **(i)** $\dfrac{e^x - \sinh x}{\cosh y}$ **(ii)** $\dfrac{-e^{-x}}{\operatorname{sech}^2(x+y)} - 1$ **(iii)** $\dfrac{1}{2y \sinh y^2}$ **(iv)** $\dfrac{\cos(x+2y)}{2e^{2y} - 2\cos(x+2y)}$

(v) $\dfrac{e^{x+y}}{(1-y^2)^{-1/2} - e^{x+y}}$ **(vi)** $\dfrac{2x\, e^{x^2-y^2} - \sin(x+y)}{\sin(x+y) + 2y\, e^{x^2-y^2}}$

(For example, in **(ii)**, $\operatorname{sech}^2(x+y)[1 + y'] = -e^{-x}$.)

21. (i) 0 (ii) 0 (iii) ∞ (iv) 0 (v) 1 (vi) 1 (vii) 0 (viii) 0 (ix) 0

23. $e^{-x} = 1 - \frac{x}{1!} + \frac{x^2}{2!} - \cdots + (-1)^n \frac{x^n}{n!} + \cdots;$

$e^{-x^2} = 1 - \frac{x^2}{1!} + \frac{x^4}{2!} - \cdots + (-1)^n \frac{x^{2n}}{n!} + \cdots;$

$e^{1/x} = 1 + \frac{1}{1!x} + \frac{1}{2!x^2} + \cdots + \frac{1}{n!x^n} + \cdots$

25. 64%, 87.04%; $t_{1/2} = 6.78$ hours.

27. 4053.3 m.

29. If $\lambda_1 \ll \lambda_2$, then $\lambda = \frac{\lambda_1\lambda_2}{\lambda_1+\lambda_2} = \lambda_1 \frac{\lambda_2}{\lambda_2(1+\lambda_1/\lambda_2)} = \lambda_1\left(1 - \frac{\lambda_1}{\lambda_2} + \left(\frac{\lambda_1}{\lambda_2}\right)^2 - \cdots\right)$

$$= \lambda_1 - \frac{\lambda_1^2}{\lambda_2} + \cdots$$

Therefore, $\lambda \approx \lambda_1$ and the first order correction is $-\lambda_1{}^2/\lambda_2$.

31. The height $h(t)$ of water at time t satisfies $dh/dt = \alpha Q$ where α is a constant and $Q = \beta(h_0 - h)$ where β is a constant and h_0 is the shutoff height. Solving,

$\frac{dh}{dt} = \frac{d}{dt}(h - h_0) = -\alpha\beta(h - h_0)$ and $h - h_0 = ke^{-\alpha\beta t}$ or $h(t) = h_0 + (h(0) - h_0)e^{-\alpha\beta t}$

where $h(0)$ is the height at $t = 0$.

Exercises 2.9

1. $\log(x+h) = \log x\left(1 + \frac{h}{x}\right) = \log x + \log\left(1 + \frac{h}{x}\right) = \log x + \frac{h}{x} - \frac{1}{2}\left(\frac{h}{x}\right)^2 + \frac{1}{3}\left(\frac{h}{x}\right)^3 - \cdots;$

$\log 6.97 \approx \log 7 - \frac{0.03}{7} = 1.94162$, $\log 7.05 \approx \log 7 + \frac{0.05}{7} = 1.95305$; accurate values are 1.9416152, 1.9530276.

3. (i) $2/x$ (ii) $1/x - \tan x$ (iii) $e^x/(1+e^x)$ (iv) $3f'(x)/f(x)$
(v) $-f'(x)\tan f(x)$ (vi) $-f'(\cos x)\sin x/f(\cos x)$

5.

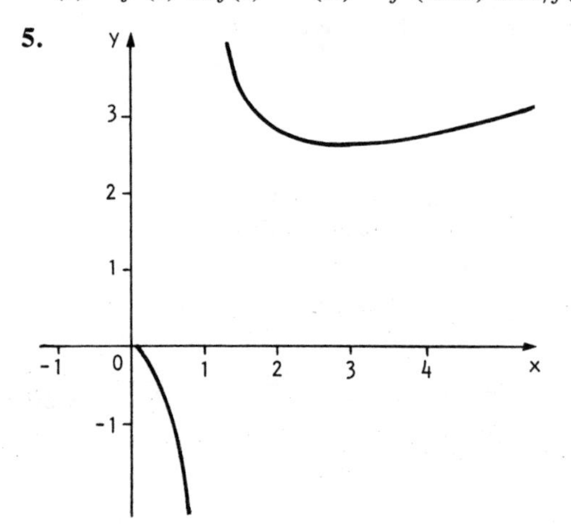

Slope is $y' = \frac{1}{\log x} - \frac{1}{(\log x)^2}$; slope at $x = 0$ is 0.

7. **(i)** 0.433 **(ii)** $(\log 2)^{-1}$ **(iii)** 1.2416 **(iv)** 1.2447

9. **(i)** Multiply first inequality by $1+x$; **(ii)** integrate first inequality.

11. **(i)** $x^2(3\log x+1)$ **(ii)** $-\dfrac{1}{x}\sin(\log x)$ **(iii)** $(\log 10)10^x$ **(iv)** $\dfrac{n}{x}(\log x)^{n-1}$

(v) $-\dfrac{2x}{1-x^2}$ **(vi)** $(x^2+1)^{-1/2}$ **(vii)** $\dfrac{1}{x}\sinh(\log x)$ **(viii)** $\dfrac{1}{x}\operatorname{sech}^2(\log x)$

(ix) $(\log 2)(\log 3)\, 3^x\, 2^{3^x}$

13. Use $\lim_{n\to\infty}\left(1+\dfrac{x}{n}\right)^n = e^x$ and $x = \lim_{n\to\infty} n\log\left(1+\dfrac{x}{n}\right)$

(i) $(0.99)^{100} = (1-0.01)^{100} \approx e^{-1} \approx 0.36788$, accurate value 0.36603

(ii) $e^2 = 7.389,\ 7.3160$ **(iii)** $e^{1/2} = 1.6487,\ 1.6483$ **(iv)** $e^6 = 403.43,\ 401.02$

(v) 2, 1.998 **(vi)** $-1,\ -1.0005$

15. $\dfrac{1}{x} = \lim_{h\to 0}\dfrac{\log(x+h)-\log x}{h} = \lim_{h\to 0}\dfrac{\log(1+h/x)}{h}$.

Let $t = \dfrac{1}{x}$, then $t = \lim_{h\to 0}\dfrac{\log(1+ht)}{h} = \lim_{h\to 0}\log(1+ht)^{1/h}$ and $e^t = \lim_{h\to 0}(1+ht)^{1/h}$.

17. **(i)** 1 **(ii)** $+\infty$ **(iii)** 0 **(iv)** 1 **(v)** 0 **(vi)** 0 **(vii)** 1 **(viii)** $\log(a/b)$

(ix) $\log(a/b)/\log(c/d)$

Hints: **(i)** By l'Hôpital, $\lim_{x\to 1/3}\dfrac{\log 3x}{3x-1} = \lim_{x\to 1/3}\dfrac{1/x}{3} = 1$

(ii) If $z = 1/2 + t$, $\lim_{z\to 1/2+}\dfrac{\tan z\pi}{\log(2z-1)} = \lim_{t\to 0+}\dfrac{\tan(\pi/2+t\pi)}{\log(2t)}$

$$= \lim_{t\to 0+}\frac{-\cot t\pi}{\log(2t)} = \lim_{t\to 0+}\frac{\cos t\pi}{\sin(t\pi)\log(2t)} = +\infty$$

(iii) If $x = \dfrac{\pi}{2}+t$, $\lim_{x\to\pi/2+}|\tan x|^{\tan x} = \lim_{t\to 0+}|\cot t|^{-\cot t} = 0$ ($\lim_{t\to 0+}\cot t = +\infty$)

(iv) Set $x = 1+t$ **(v)** $\lim_{x\to\infty}\dfrac{\log|\log x|}{x^{1/2}} = \lim_{x\to\infty}\dfrac{2}{x^{5/2}\log x} = 0$

(vi) $\lim_{x\to\infty}\dfrac{x}{e^{\beta x}} = \lim_{x\to\infty}\dfrac{1}{\beta e^{\beta x}} = 0$ if $\beta > 0$

(vii) $\lim_{x\to\infty}\dfrac{\log(1+1/x)}{\sin 1/x} = \lim_{x\to\infty}\dfrac{1/x - 1/2(1/x)^2 + \cdots}{1/x - 1/3!(1/x)^3 + \cdots} = 1$

(viii) $\lim_{x\to 0}\dfrac{a^x-b^x}{x} = \lim_{x\to 0}\dfrac{(\log a)a^x - (\log b)b^x}{1} = \log a - \log b = \log(a/b)$

(ix) $\lim_{x\to 0}\dfrac{a^x-b^x}{c^x-d^x} = \lim_{x\to 0}\dfrac{(\log a)a^x-(\log b)b^x}{(\log c)c^x-(\log d)d^x} = \dfrac{\log(a/b)}{\log(c/d)}$

19. $\lim_{n\to\infty} P\left(1+\dfrac{r}{100n}\right)^{Nn} = Pe^{rN/100}$

21. 5.25% simple interest; $(1.0525)^1 > (1.0125)^4 > (1.004)^{12}$

23. To calculate a and b, two pieces of information are needed; for example, the two values of x corresponding to two values of y.

Exercises 2.10

1. $\dfrac{f(x)-f(x-h)}{h}=f'(x)-\dfrac{h}{2}f''(\xi),\ x-h<\xi<x$ (compare to equation (6)).

3. Expanding each $f(a+rh)$ with $r=-2,\ -1,\ 0,\ 1,\ 2$ about $x=a$ by means of Taylor's formula, $c_0f(a+2h)+\cdots+c_4f(a-2h)=(c_0+c_1+c_2+c_3+c_4)f(a)+$
$(2c_0+c_1-c_3-2c_4)f'(a)+(4c_0+c_1+c_3+4c_4)\dfrac{f''(a)}{2}+(8c_0+c_1-c_3-8c_4)\dfrac{f'''(a)}{3!}+\cdots$
Equate to $f'''(a)$ and solve for the coefficients;
$c_0=\dfrac{1}{2h^3},\ c_1=-\dfrac{1}{h^3},\ c_2=0,\ c_3=\dfrac{1}{h^3},\ c_4=-\dfrac{1}{2h^3}.$

5. $\dfrac{y(0.55)-y(0.45)}{0.1}=-0.438$ or equivalently using three point Lagrange interpolation polynomial; $\dfrac{y(0.55)-2y(0.50)+y(0.45)}{(0.05)^2}=-1.12.$

7. $x=\mathscr{L}_n(y)=\dfrac{(y-y_1)(y-y_2)\cdots(y-y_n)}{(y_0-y_1)(y_0-y_2)\cdots(y_0-y_n)}x_0+\cdots+\dfrac{(y-y_0)\cdots(y-y_{n-1})}{(y_n-y_0)\cdots(y_n-y_{n-1})}x_n;$
$y=2x^2-\dfrac{3}{4}x;\ x=\dfrac{1}{117}(484y-128y^2);\ y(1)=\dfrac{5}{4},\ x(1)=\dfrac{356}{117};$ the value for $x(1)$ is very poor since $x=y^{1/3}$ has an infinite slope at the origin.

9. **(i)** 1.89549 **(ii)** 0.24866 **(iii)** 0.598767

11. $z^3+z=1$; to show uniqueness, plot $y=z^3$ and $y=1-z$; $y=0.136466$.

13. 1.32472

15. If $g'(x)<1$, then $x_{n+1}-x_n<x_n-x_{n-1}$ and iteration may converge.

17. $x=\pi-\epsilon\pi+\epsilon^2\pi^2-\epsilon^3\left(\dfrac{\pi^3}{6}+\pi\right)+\cdots$

19. $y_{n+1}-y_n=hF(x_n,y_n);\ y_{n+1}-y_n=h(x_n+y_n),\ y_0=0,\ x_0=0$

21. **(i)** $y=2x+1$ **(ii)** $y=\dfrac{x^2}{2}+\dfrac{x}{2}+1$ **(iii)** $y=-\dfrac{1}{90}x^3+\dfrac{49}{90}x^2+\dfrac{42}{90}x+1$
(iv) $y=-\dfrac{2}{90}x^4+\dfrac{19}{90}x^3-\dfrac{5}{90}x^2+\dfrac{78}{90}x+1.$

Exercises 2.11

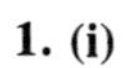

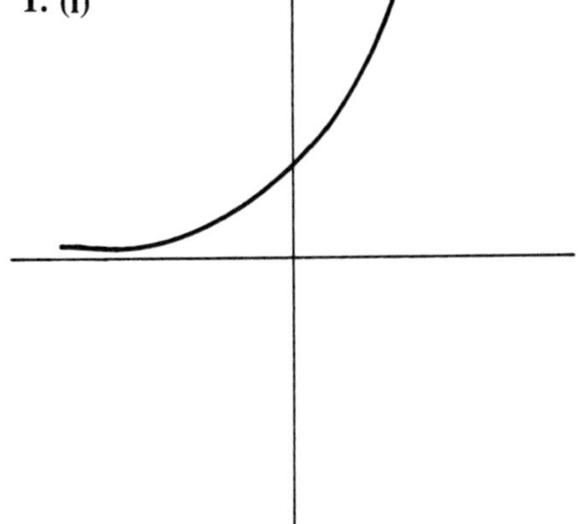

(iii)

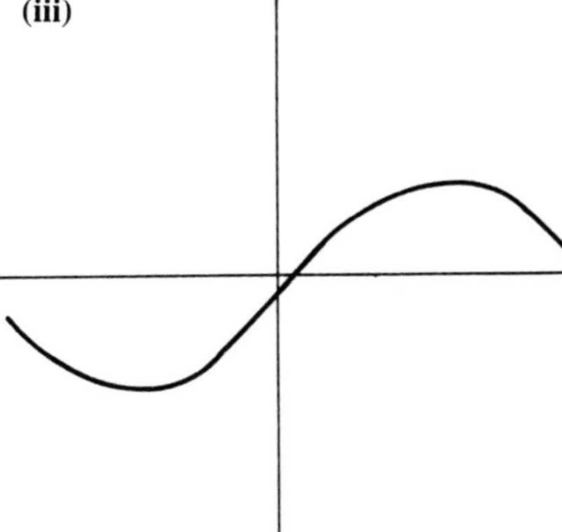

(iv)

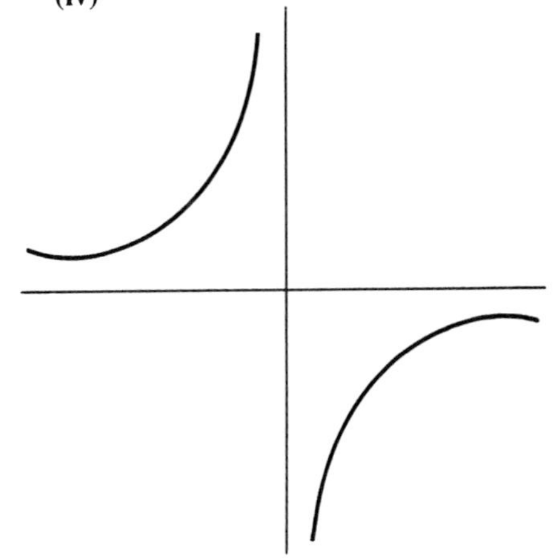

3. (i)

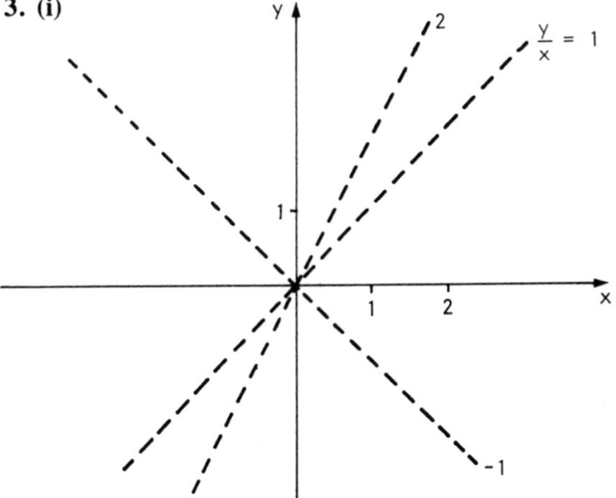

Solutions through (1, 0) are
(i) $y = 0$,
(ii) $x^2 + y^2 = 1$,
(iii) $y = 0$

5. From the graph, show $\lim_{x\to-\infty}(x+y)=-1$.

7.

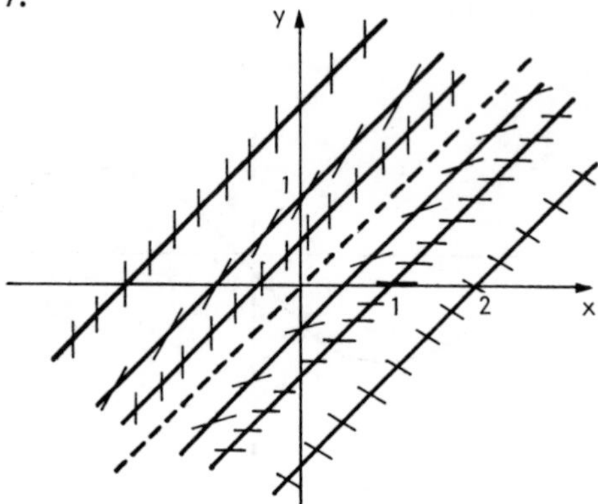

9. If $y=0$ at some point, then $y'=0$ at that point and solution is $y(x)=0$.

11. **(i)** 2, 2, 0, 0 **(ii)** 0, 0, 1, 1 **(iii)** $12x^2+4y^2$, $4x^2+12y^2$, $8xy$, $8xy$
(iv) $-\sin x\cos y$, $-\sin x\cos y$, $-\cos x\sin y$, $-\cos x\sin y$ **(v)** ye^x, 0, e^x, e^x
(vi) $-1/x^2$, $-1/y^2$, 0, 0 **(vii)** $\sinh(x+y)$, $\sinh(x+y)$, $\sinh(x+y)$, $\sinh(x+y)$
(viii) $e^x\cosh y$, $e^x\cosh y$, $e^x\sinh y$, $e^x\sinh y$
(ix) $2\,\text{sech}^2(x^2-y^2)[1-4x^2\tanh(x^2-y^2)]$, $-2\,\text{sech}^2(x^2-y^2)[1+4y^2\tanh(x^2-y^2)]$, $8\,xy\,\text{sech}^2(x^2-y^2)\tanh(x^2-y^2)$.

13. **(i)** $6x^5-6x^2y+y$, $2y-2x^3+x$ **(ii)** x/ω, y/ω **(iii)** $2x/e^\omega$, $2y/e^\omega$
(iv) $\omega y\cosh xy$, $\omega x\cosh xy$ **(v)** $-\omega/x$, $-\omega/y$
(vi) $-2x\omega/(e^\omega+x^2+y^2)$, $-2y\omega/(e^\omega+x^2+y^2)$

15. $\partial u/\partial t=-c\cos(x-ct)$, $\partial u/\partial x=\cos(x-ct)$.

17. Solve the quadratic equation for $z=z(x,y)$.

19. **(i)** $2e^{2t}-2e^{-2t}$ **(ii)** $(t^2-1)/t(t^2+1)$ **(iii)** $\sinh t$ **(iv)** 0

Chapter 3 Integration

Exercises 3.1

1. $\int_a^x [pf(x)+qg(x)]\,dx = \int_a^x pf(x)\,dx+\int_a^x qg(x)\,dx = p\int_a^x f(x)\,dx+q\int_a^x g(x)\,dx$

3. **(i)** $\int_a^b G(x)\,dx$ **(ii)** $\int_1^2 \frac{dx}{x}$ **(iii)** $\int_0^1 \frac{dx}{1+x^2}$

Hint: In **(iii)**, write $\lim_{n\to\infty}\frac{1}{n}\left[\frac{1}{1+(^1\!/_n)^2}+\frac{1}{1+(^2\!/_n)^2}+\cdots+\frac{1}{1+(^n\!/_n)^2}\right]$

5. **(i)** false; for example, $\int_0^2 (1-x)\,dx = 0$.

(ii) true since, if $f(x) \neq 0$ at some point, then $\int_a^b (f(x))^2\,dx > 0$ assuming $f(x)$ is continuous.

(iii) true since $xe^x > x$ on $0 < x \leq 2$ **(iv)** true (draw a diagram).

7.

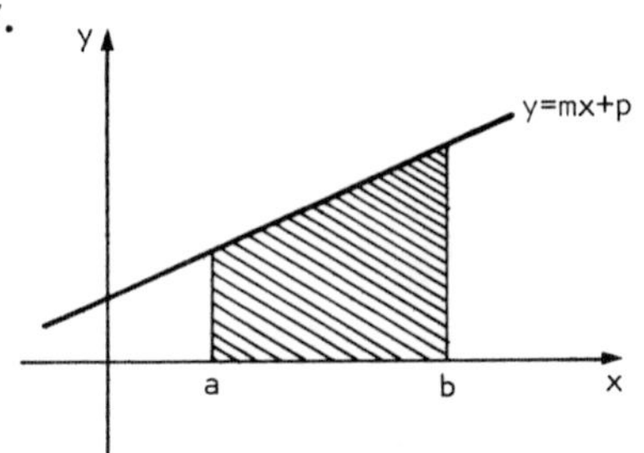

Area of trapezoid

$= \frac{(b-a)}{2}[(ma+p)+(mb+p)]$

$= \left(\frac{1}{2}mb^2 + pb\right) - \left(\frac{1}{2}ma^2 + pa\right)$

If $ma + p < 0$, formula is correct, but it represents a *difference* of areas.

9. If $t = x - \Omega/2$, then $f(t) = -f(-t)$, i.e., f is an odd function. Since $f(x)$ is periodic with period Ω,

$\int_0^{\Omega} f(x)\,dx = \int_{-\Omega/2}^{\Omega/2} f(x)\,dx = 0.$

If $f(x) = \sin x$, then $\Omega = 2\pi$

and $\int_0^{2\pi} \sin x\,dx = 0.$

(Verify $\sin(x-\pi) = -\sin(\pi - x)$.)

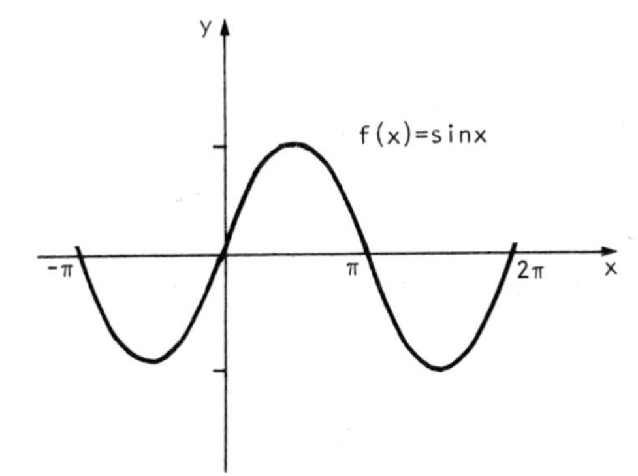

11. **(i)** 10 **(ii)** 5/6 **(iii)** 3

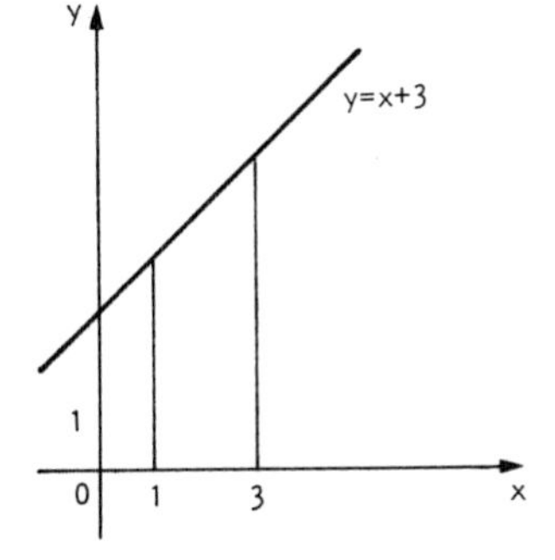

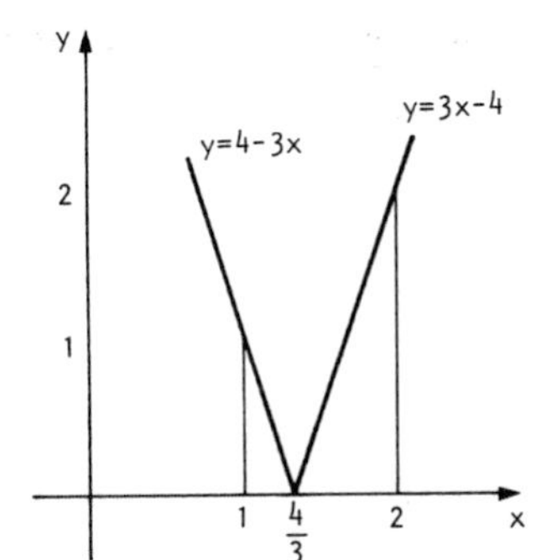

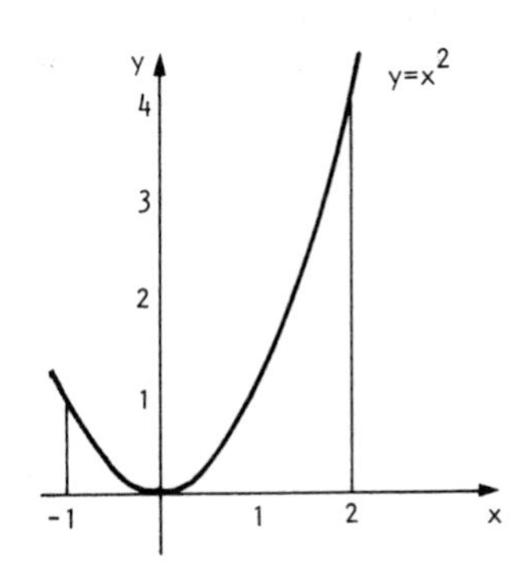

13. $\int_{-3}^{2} (x+3)(x-2)(x-5)\,dx = 1375/12$, $\int_{2}^{5} (x+3)(x-2)(x-5)\,dx = -117/4$;
the areas are $1375/12$ and $117/4$.

15. Graph the integrands. For **(i)** and **(ii)**, see Exercise **9**. In **(iii)**, explain why
$\int_0^{\pi/2} \sin x\,dx = \int_0^{\pi/2} \cos x\,dx.$

17. **(i)** 3 $\quad \left(\text{use } \dfrac{x^2}{1+x^2} \le 1 \text{ or } \dfrac{x^2}{1+x^2} \le x^2\right)$

(ii) $\dfrac{1}{2}$ $\quad \left(\dfrac{1}{1+x^4} \le \dfrac{1}{2} \text{ for } x \ge 1\right)$

(iii) $\dfrac{\pi}{4}$ $\quad (\cos^2 x \le 1)$

(iv) e^{-4} $\quad \left(\dfrac{e^{-x^2}}{x^2-1} \le \dfrac{e^{-4}}{3} \text{ for } 2 \le x \le 5\right)$

(v) $1 - e^{-2\pi}$ $\quad (e^{-x} \sin x \le e^{-\pi})$ $\qquad$ **(vi)** $\dfrac{\pi}{2}$ $\quad \left(\dfrac{\cos x}{1+x^2} \le 1\right)$.

19.

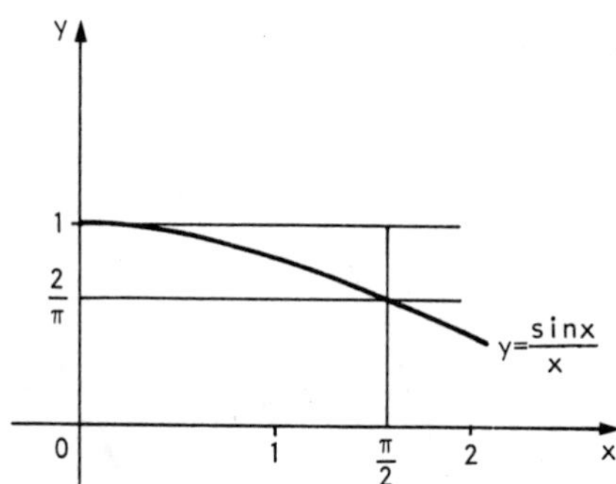

On $0 < x < \dfrac{\pi}{2}$, $\dfrac{2}{\pi} < \dfrac{\sin x}{x} < 1$.
Areas of bounding rectangles are 1 and $\dfrac{\pi}{2}$.

21. For example, if $F(x) = x^2$, $F(x_i) - F(x_{i-1}) = x_i^2 - x_{i-1}^2 = (x_1 + x_{i-1})(x_i - x_{i-1})$.
Since $F'(\xi_i) = 2\xi_i$, we have $\xi_i = (x_i + x_{i-1})/2$ and
$\sum_{i=1}^{n} 2\xi_i(x_i - x_{i-1}) = \sum_{i=1}^{n} (x_i^2 - x_{i-1}^2) = b^2 - a^2$ or $\int_a^b 2x\,dx = b^2 - a^2$.
The finite sum is exactly equal to the integral if ξ_i is chosen this way in each subinterval.

23. Use upper and lower sums.

25. Use $g(x) = 1$ in previous result, equality if $f(x) =$ constant.
If $f(x) = x$, $\dfrac{\pi^4}{4} = \left(\int_0^{\pi} x\,dx\right)^2 \le \pi \int_0^{\pi} x^2\,dx = \dfrac{\pi^4}{3}$.
If $f(x) = \sin x$, $4 = \left(\int_0^{\pi} \sin x\,dx\right)^2 \le \pi \int_0^{\pi} \sin^2 x\,dx = \dfrac{\pi^2}{2}$.

27. **(i)** $\int_a^b \cosh x\, dx = \int_a^b \frac{e^x + e^{-x}}{2}\, dx = \frac{1}{2}(e^b - e^a) - \frac{1}{2}(e^{-b} - e^{-a})$

$= \frac{1}{2}(e^b - e^{-b}) - \frac{1}{2}(e^a - e^{-a}) = \sinh b - \sinh a.$

(ii) Use $\sinh x = \frac{e^x - e^{-x}}{2}$ and integrate as in **(i)**.

Exercises 3.2

1. $\lim_{n\to\infty} \frac{1}{n}\sum_{i=1}^{n} y_i = \lim_{n\to\infty} \frac{\Delta x}{b-a}\sum_{i=1}^{n} y_i = \frac{1}{b-a}\lim_{n\to\infty}\sum_{i=1}^{n} y_i\Delta x = \frac{1}{b-a}\int_a^b f(x)\, dx.$

3. **(i)** $\lim_{n\to\infty} \frac{1}{n}\left[\frac{1}{1+1/n} + \frac{1}{1+2/n} + \cdots + \frac{1}{1+n/n}\right] = \int_0^1 \frac{dx}{1+x} = \log 2$

(ii) $\int_0^1 2^x\, dx = \int_0^1 e^{x \log 2}\, dx = (\log 2)^{-1}$

5. If $f(x)>0$ at some point $x=c$, then for some $\varepsilon>0$ we have $f(x)>0$ in the interval $c-\varepsilon<x<c+\varepsilon$ and $\int_{c-\varepsilon}^{c+\varepsilon} f(x)\, dx>0$. This is a contradiction. Similarly, $f(x)$ cannot be negative at any point. Conclude $f(x)=0$ for all x.

7. **(i)** $\frac{d}{dx}(xe^x - e^x + C) = xe^x$ **(ii)** $\frac{d}{dx}(\sin^2 x + C) = 2\sin x\cos x = \sin 2x$

(iii) $\frac{d}{dx}\left(\frac{x^2}{2}\log x - \frac{x^2}{4} + C\right) = x\log x.$

9. **(i)** $\frac{d}{dx}\log(\log x) = \frac{1}{\log x}\frac{d}{dx}\log x = \frac{1}{x\log x}$

(ii) $\frac{d}{dx}\left(x\log(a^2+x^2) - 2x + 2a\tan^{-1}\frac{x}{a}\right)$

$= \log(a^2+x^2) + \frac{2x^2}{a^2+x^2} - 2 + \frac{2}{1+(x/a)^2} = \log(a^2+x^2)$

(iii) $\frac{d}{dx}\left(\cos^{-1}\left(\frac{1}{\sqrt{2}}\cos x\right)\right) = \frac{-1}{(1-\cos^2 x/2)^{1/2}}\left(-\frac{1}{\sqrt{2}}\sin x\right)$

$= \tan x\left(1 + 2\tan^2 x\right)^{-1/2}$

11. **(i)** $\int^x x(ax^2+b)^{n-1}\, dx = \frac{1}{2an}(ax^2+b)^n + C$

(ii) $\int^x x\sin ax^2\, dx = -\frac{1}{2a}\cos ax^2 + C$

(iii) $\int^x \frac{dx}{(ax+b)^{3/2}} = \frac{-2}{a(ax+b)^{1/2}} + C$

(iv) $\int^x \frac{a}{a^2+x^2}\, dx = \tan^{-1}\frac{x}{a} + C.$

13. **(i)** $y = 2\log x + \dfrac{x^3}{3} - \cos x + C$ **(ii)** $y = \sin^{-1}x + C$

(iii) $y = \dfrac{x^2}{2} + \dfrac{x^3}{3} + \dfrac{x^4}{2} + C$ **(iv)** $y = xe^x + C$

15. **(i)** e^{-t^2} **(ii)** $\dfrac{2t}{1+t^2} - \dfrac{1}{1+t^2}$ **(iii)** $2f(2t) - f(t)$

Hint: In **(ii)**, if $F(t) = \int_a^t \dfrac{dx}{1+x}$, then $\int_t^{t^2} \dfrac{dx}{1+x} = F(t^2) - F(t)$.

17. $s(t) = a\sin\omega t$; period is $2\pi/\omega$. Dimensions of a and ω are L and T^{-1}.

19. $s = \int_{1.5}^{6.2} \dfrac{30\,e^{2t}}{1+e^{2t}}\,dt = 15\log(1+e^{2t})\Big|_{1.5}^{6.2} = 140.3$ (m).,
average velocity $= 140.3/4.7 = 29.85$ m/sec.

21. $v(t) = 20 - 9.8t$, $s(t) = 20t - 4.9t^2$, maximum height is 20.4 (m) reached when $t = 20/9.8 = 2.04$ sec. Ball returns to ground when $t = 20/4.9 = 4.08$ sec.

Exercises 3.3

1. **(i)** $\dfrac{x^3}{3} + x^2 + C$ **(ii)** $\dfrac{x^5}{5} - \dfrac{2}{3}x^3 + x + C$ **(iii)** $\dfrac{2}{3}x^{3/2} - \dfrac{x^2}{2} + C$

3. **(i)** $-\dfrac{1}{3}\cos 3x + C$ **(ii)** $\dfrac{1}{8}\sin^8 x + C$

(iii) $-\dfrac{3}{4}\cos x + \dfrac{1}{12}\cos 3x + C$ **(iv)** $\dfrac{3x}{8} - \dfrac{1}{4}\sin 2x + \dfrac{1}{32}\sin 4x + C$

(v) $\dfrac{3}{4}\sin x + \dfrac{1}{12}\sin 3x + C$ **(vi)** $\dfrac{3x}{8} + \dfrac{1}{4}\sin 2x + \dfrac{1}{32}\sin 4x + C$

Hints: **(iii)** $\sin^3 x = \dfrac{3}{4}\sin x - \dfrac{1}{4}\sin 3x$

(iv) $\sin^4 x = \dfrac{3}{8} - \dfrac{1}{2}\cos 2x + \dfrac{1}{8}\cos 4x$

(v) $\cos^3 x = \dfrac{3}{4}\cos x + \dfrac{1}{4}\cos 3x$

(vi) $\cos^4 x = \dfrac{3}{8} + \dfrac{1}{2}\cos 2x + \dfrac{1}{8}\cos 4x$.

Alternatively, $\sin^3 x = \sin x(1 - \cos^2 x)$, $\cos^3 x = \cos x(1 - \sin^2 x)$.

5. **(i)** $\log|x+3| + C$ **(ii)** $\dfrac{1}{3}\log|3x+1| + C$ **(iii)** $\dfrac{1}{a}\log|ax+b| + C$

(iv) $\log\left|\dfrac{x-2}{x-1}\right| + C$ **(v)** $\log|x| - \dfrac{1}{2}\log|1-x^2| + C$ **(vi)** $-\dfrac{1}{4}(\log|1-x^2|)^2 + C$

Hints: **(v)** $\dfrac{1}{x(1-x)^2} = \dfrac{1}{x} + \dfrac{x}{1-x^2}$

(vi) define $u = \log(1-x^2)$.

7. **(i)** $\frac{1}{3}(2x+3)^{3/2}+C$ **(ii)** $\frac{2}{3a}(ax+b)^{3/2}+C$ **(iii)** $\frac{(ax+b)^{n+1}}{a(n+1)}+C$

(iv) $-\frac{1}{2}\log|\cos(2x+1)|+C$ **(v)** $\frac{1}{9}e^{3x^3}+C$ **(vi)** $e^{\sin x}+C$ **(vii)** $\frac{1}{4}\sin^4 x+C$

(viii) $\frac{1}{3}\tan^3 x+C$ **(ix)** $\tan x - x + C$ (use $\tan^2 x = \sec^2 x - 1$).

9. **(i)** $-\frac{1}{2}e^{-s^2}+C$ $(u=s^2)$ **(ii)** $\frac{(\tan^{-1}x)^2}{2}+C$ $(u=\tan^{-1}x)$

(iii) $\frac{1}{4}\tan^4 x+C$ $(u=\tan x)$ **(iv)** $\frac{1}{2}(\sin^{-1}x)^2+C$ $(u=\sin^{-1}x)$

(v) $e^{\sin x}+C$ $(u=\sin x)$ **(vi)** $\sin(x^2+4)^{1/2}+C$ $(u=(x^2+4)^{1/2})$.

11. **(i)** $\frac{1}{4}\sin 2x+\frac{x}{2}\cos 2a+C$ **(ii)** $-\frac{1}{4}\sin 2x+\frac{x}{2}\cos 2a+C$

13. **(i)** $\frac{-1}{2(a-b)}\cos(a-b)x-\frac{1}{2(a+b)}\cos(a+b)x+C$

(ii) $\frac{1}{2(a-b)}\sin(a-b)x-\frac{1}{2(a+b)}\sin(a+b)x+C$

(iii) $\frac{1}{2(a-b)}\sin(a-b)x+\frac{1}{2(a+b)}\sin(a+b)x+C$

15. **(i)** $\frac{1}{2}\log|x^2+4x+1|-\frac{1}{\sqrt{3}}\log\left|\frac{x+2-\sqrt{3}}{x+2+\sqrt{3}}\right|+C$

(ii) $\frac{1}{5}\log\left|\frac{x-2}{x-3}\right|+C$ **(iii)** $\frac{1}{4}\log\left|\frac{e^x-1}{e^x+3}\right|+C$

(iv) $\frac{1}{4}\log\left|\frac{x^2+1}{x^2+3}\right|+C$ **(v)** $\frac{1}{\sqrt{3}}\tan^{-1}\left(\frac{z-1}{\sqrt{3}}\right)+C$ **(vi)** $\frac{-1}{1+\sin\theta}+C$

Hints: **(i)** $\frac{x}{x^2+4x+1}=\frac{x+2}{x^2+4x+1}-\frac{2}{x^2+4x+1}$ **(iii)** define $u=e^x$

(iv) $u=x^2+2$ **(v)** $(z-2)^2+2z=(z-1)^2+3$ **(vi)** $\cos^2\theta=1-\sin^2\theta$.

Exercises 3.4

1. **(i)** $\frac{1}{25}\int_0^{21}(u^{1/2}-u^{-1/2})\,du=\frac{12}{25}\sqrt{21}$

(ii) $\frac{1}{8}\int_1^5(u^{5/3}+2u^{2/3}+u^{-1/3})\,du=\frac{135}{64}(5)^{2/3}-\frac{123}{320}$

(iii) $\frac{1}{3}\int_1^9 u^{1/2}\,du=\frac{52}{9}$ **(iv)** $4\int_1^2\frac{(u-1)^2}{u}\,du=\frac{47}{15}-4\log 2$

(v) $4\int_0^1\frac{du}{2+9u}=\frac{4}{9}\log{}^{11}\!/_2$ **(vi)** $\int_{\log 2}^{\log 3}\frac{du}{u}=\log(\log 3)-\log(\log 2)$

3. **(i)** $\frac{39}{14}2^{1/3}-\frac{135}{56}\left(\frac{3}{2}\right)^{1/3}$ (define $u=1+x$)

(ii) $x=1+u^2$, $\int_1^{\sqrt{2}}2(1+u^2)\,du=(10\sqrt{2}-8)/3$

(iii) $u=x^2+1$, $\frac{3}{2}(3^{1/3}-1)$ **(iv)** $u=1+x$, $\frac{45(3^{1/3})+9}{28}$

(v) $u=1+x$, $2^{3/2}/3$ **(vi)** $u=x^4$, 0 (original integrand is an odd function)

5. Make a change of variables $u = a - x$ in the second integral. The functions $f(x)$ and $f(a-x)$ take on the same values in reverse order on the interval $(0, a)$.

7. $\displaystyle\int \frac{t^2-1}{t^4+1}\,dt = \int \left(1 - \frac{1}{t^2}\right)\frac{dt}{t^2 + 1/t^2} = \int \left(1 - \frac{1}{t^2}\right)\frac{dt}{(t+1/t)^2 - 2} = \frac{1}{2\sqrt{2}}\log\left|\frac{t + 1/t - \sqrt{2}}{t + 1/t + \sqrt{2}}\right| + C$

9. **(i)** $\displaystyle\frac{x^2+1}{2}\tan^{-1}x - \frac{x}{2} + C$ (define $u = \tan^{-1}x$, $dv = x\,dx$)

(ii) $\displaystyle x\tan^{-1}x - \frac{1}{2}\log(x^2+1) + C$ **(iii)** $x\sin^{-1}x + (1-x)^{1/2} + C$

(iv) $x\tan x + \log|\cos x| + C$ **(v)** $\displaystyle\left(\frac{x^2}{a} - \frac{2x}{a^2} + \frac{2}{a^3}\right)e^{ax} + C$

(vi) $x(\log x^2) - 2x\log x + 2x + C$ (define $u = (\log x)^2$, $dv = dx$)

(vii) $\displaystyle\frac{2}{3}x^{3/2}\log x - \frac{4}{9}x^{3/2} + C$ (define $u = \log x$, $dv = x^{1/2}\,dx$)

(viii) $\displaystyle\frac{x}{2}[\sin(\log x) - \cos(\log x)] + C$ (define $u = \sin(\log x)$, $dv = dx$)

(ix) $\displaystyle\frac{3}{5}x^{3/2}\cos x^{1/2} + \frac{1}{5}x^{3/2}\sin x^{1/2} + C$ (define $u = \cos x^{1/2}$, $dv = x^{1/2}\,dx$)

11. **(i)** $\displaystyle\frac{e^{2x}}{13}(2\cos 3x + 3\sin 3x) + C$ **(ii)** $\displaystyle\frac{-e^{-x}}{2}(\sin x + \cos x) + C$

(iii) $\displaystyle\frac{e^{3x}}{34}(3\sin 5x - 5\cos 5x) + C$

13. $\displaystyle\frac{\pi}{32 + 4\pi} < I < \frac{\pi}{32}$ or $0.07049 < I < 0.09817$.

15. If $\displaystyle I_n = \int \frac{dx}{(x^2+a^2)^n}$, then $\displaystyle I_n = \frac{1}{a^2}I_{n-1} - \frac{1}{a^2}\int \frac{x^2}{(x^2+a^2)^n}\,dx$;

$$\int \frac{x^2\,dx}{(x^2+a^2)^n} = \frac{1}{2}\int x\frac{d(x^2+a^2)}{(x^2+a^2)^n} = -\frac{1}{(2n-1)}\int x\,d(x^2+a^2)^{-(n-1)}$$

Result follows by integration by parts.

17. In **(ii)**, write $\sin^n x = \sin^{n-2}x - \sin^{n-2}x\cos^2 x$ and

$$\int \sin^{n-2}x\cos^2 x\,dx = \int \cos x\,d\left(\frac{\sin^{n-1}x}{n-1}\right)$$

19. $\displaystyle\frac{2n(2n-2)\cdots 4\cdot 2}{(2n+1)(2n-1)\cdots 5\cdot 3}$ (use Exercise **17(ii)** and **(iii)**).

21. $\displaystyle\int x^n\cos ax\,dx = \frac{x^n}{a}\sin ax - \frac{n}{a}\int x^{n-1}\sin ax\,dx,$

$$\int x^n\sin ax\,dx = -\frac{x^n}{a}\cos ax + \frac{n}{a}\int x^{n-1}\cos ax\,dx$$

$$\int x^2\cos 3x\,dx = \frac{x^2}{3}\sin^3 x + \frac{2x}{9}\cos 3x - \frac{2}{27}\sin 3x + C,$$

$$\int x^2\sin 3x\,dx = -\frac{x^2}{3}\cos 3x + \frac{2x}{9}\sin 3x + \frac{2}{27}\cos 3x + C$$

Exercises 3.5

1. **(i)** $\frac{1}{2}\tan^{-1}\frac{x}{2}+C$ **(ii)** $\frac{-1}{x-2}+C$

(iii) $\frac{1}{3}\tan^{-1}\frac{2x+1}{3}+C$ **(iv)** $\frac{1}{2}\log\left|\frac{3x-2}{3x}\right|+C$

(v) $\frac{2}{\sqrt{3}}\tan^{-1}\left[\frac{\sqrt{3}}{2}\left(x-\frac{1}{2}\right)\right]+C$ **(vi)** $\frac{1}{11}\log\left|\frac{5x-1}{x+2}\right|+C$

3. **(i)** $5\log|x-1|-3\log|x-2|+C$ **(ii)** $\log\left|\frac{x+1}{x}\right|-\frac{1}{x}+C$

(iii) $\frac{1}{5}\log|x-2|-\frac{1}{10}\log(x^2+1)+\frac{2}{5}\tan^{-1}x+C$ **(iv)** $\frac{1}{20}\log\left|\frac{2x-5}{2x+5}\right|+C$

(v) $2\log\left|\frac{x-1}{x+2}\right|-\frac{3}{x-1}+C$

(vi) $\sqrt{3}\tan^{-1}\frac{2}{\sqrt{3}}\left(x+\frac{1}{2}\right)+\log|x+1|-\frac{1}{2}\log|x^2+x+1|+C$

(vii) $-\frac{1}{x}-\tan^{-1}x+C$ **(viii)** $\frac{1}{6}\log\left|\frac{3+x}{3-x}\right|+C$ **(ix)** $\frac{-1}{e^x-1}+C$

5. **(i)** $\frac{1}{3}\log|x+1|-\frac{1}{6}\log|x^2-x+1|+\frac{1}{\sqrt{3}}\tan^{-1}\frac{2x-1}{\sqrt{3}}+C$

(ii) $\frac{1}{4\sqrt{2}}\log\left|\frac{x^2+\sqrt{2}x+1}{x^2-\sqrt{2}x+1}\right|+\frac{\sqrt{2}}{4}[\tan^{-1}(\sqrt{2}x-1)+\tan^{-1}(\sqrt{2}x+1)]+C$

(iii) $\int\frac{x^4}{1+x^4}dx=\int dx-\int\frac{dx}{1+x^4}=x-$ **(ii)**

Hints: $\frac{1}{1+x^3}=\frac{1}{3}\left[\frac{1}{x+1}+\frac{2-x}{x^2-x+1}\right],$

$\frac{1}{1+x^4}=\frac{1}{2\sqrt{2}}\left[\frac{\sqrt{2}-x}{x^2-\sqrt{2}x+1}+\frac{\sqrt{2}+x}{x^2+\sqrt{2}x+1}\right]$

7. **(i)** $\sin^{-1}\frac{x}{2}+C$ **(ii)** $\frac{1}{3}\tan^{-1}\frac{x}{3}+C$

(iii) $\frac{x}{2}(1-x)^{1/2}+\frac{1}{2}\sin^{-1}x+C$ **(iv)** $\frac{x}{2}(25-x^2)^{1/2}+\frac{25}{2}\sin^{-1}\frac{x}{5}+C$

(v) $(1-x^2)^{-1/2}$ **(vi)** $\frac{x}{a^2}(x^2+a^2)^{-1/2}+C$

(vii) $\frac{1}{128}\tan^{-1}\frac{x}{4}+\frac{x}{32(16+x^2)}+C$ **(viii)** $-\frac{2}{\sqrt{5}}\log\left|\frac{\sqrt{5+x}+\sqrt{5}}{x}\right|+C$

(ix) $\frac{3}{4}(9+x^2)^{2/3}+C$ **(x)** $8\sin^{-1}\frac{x}{4}-\frac{x}{2}(16-x^2)^{1/2}+C$

(xi) $\frac{1}{2}\sin^{-1}\frac{2}{3}x$ **(xii)** $-\frac{1}{3}\log\left|\frac{3+(9-4x^2)^{1/2}}{2x}\right|+C$

9. $z=\tan x$ is discontinuous at $x=\pi/2$. If the integral were evaluated from $-\pi/2$ to $\pi/2$, the correct value π would be obtained.

11. **(i)** $-\frac{1}{2}\log|4-x^2|+C$ **(ii)** $\frac{1}{8}(x^2+1)^4+C$ (define $u=x^2+1$)

(iii) $-\frac{2}{3}(1-x)^{3/2}+\frac{4}{5}(1-x)^{5/2}-\frac{2}{7}(1-x)^{7/2}+C$ (define $u=1-x$)

(iv) $\frac{1}{5}(x^2+1)^{5/2}-\frac{1}{3}(x^2+1)^{3/2}+C$ (define $u=x^2+1$)

(v) $\frac{5}{9}(1+x)^{9/5}-\frac{5}{4}(1+x)^{4/5}+C$ **(vi)** $\frac{1}{2}\sin^{-1}\left(x-\frac{1}{2}\right)+C$

(vii) $-3e^{-x^{1/3}}[x^{5/3}+5x^{4/3}+20x+60x^{2/3}+120x^{1/3}+120]+C$ (define $u=x^{1/3}$)

(viii) $x\log(1+x^2)-2x+2\tan^{-1}x+C$ (define $u=\log(1+x^2)$, $dv=dx$)

(ix) $\frac{1}{2}(1+x^2)[\log(1+x^2)-1]$ (define $u=1+x^2$)

(x) $\frac{1}{2}\log\left|\frac{x(x+2)}{(x+1)^2}\right|+C$

(xi) $\frac{x^2}{2}+x-\log|x^2+x+1|+C$ $\left(\frac{x^3+2x^2}{x^2+x+1}=x+1-\frac{2x+1}{x^2+x+1}\right)$

(xii) $x\log[x+(x^2-1)^{1/2}]-(x^2-1)^{1/2}+C$

13. **(i)** 9π $\left(\text{define } u=\frac{x-5}{6}\right)$ **(ii)** 8π $\left(u=\frac{x-3}{4}\right)$ **(iii)** π $\left(u=\frac{x-2}{2}\right)$

15. Use $\sinh x=2\sinh\frac{x}{2}\cosh\frac{x}{2}$, $\cosh x=\cosh^2\frac{x}{2}+\sinh^2\frac{x}{2}$.

Exercises 3.6

1. $4\sqrt{3}/9$. Using strips parallel to Ox, area $=2\int_0^{1/\sqrt{3}}(1-3y^2)\,dy$. Using strips parallel to Oy, area $=2\int_0^1\left(\frac{x}{5}\right)^{1/2}dx+2\int_1^{5/3}\left[\left(\frac{x}{5}\right)^{1/2}-\left(\frac{x-1}{2}\right)^{1/2}\right]dx$.

3. $\int_0^2 x^2e^{-x}\,dx=2(1-5e^{-2})$; $\int_0^\infty x^2e^{-x}\,dx=2$

5. Area $=2\int_0^\pi[(2\pi^2-y^2)^{1/2}-y^2/\pi]\,dy=\pi^2\left(\frac{1}{3}+\frac{\pi}{2}\right)$

7. Use equation (3), $A=\int\frac{1}{2}r^2\,d\theta$ or areas of circle and ellipse.

(i) πa^2 **(ii)** $\pi a^2/4$ **(iii)** $3\pi a^2/2$ **(iv)** $\pi(a^2+b^2/2)$
(v) $4\pi a^2/3$ (determine axes of ellipse) **(vi)** $\pi^3a^2/48$

9. Area $=2\int_0^{3\pi/4}\frac{1}{2}(1-\cos\theta)^2\,d\theta+2\int_{7\pi/4}^{2\pi}\frac{1}{2}(1-\cos\theta)^2\,d\theta=\frac{3\pi}{2}-2\sqrt{2}$

11. Use equation (8); **(i)** $\frac{8}{27}\left[\left(\frac{11}{2}\right)^{3/2}-1\right]$ **(ii)** $1+\frac{1}{2}\log\frac{3}{2}$

(iii) $\frac{1}{2}\left[\sqrt{5}+\frac{1}{2}\log(2+\sqrt{5})\right]$ **(iv)** $2\sqrt{5}$

(v) $\log\left(1+\frac{2}{\sqrt{3}}\right)$ **(vi)** $\frac{8}{27}\left(10^{3/2}-1\right)$

13. Use equation (12); **(i)** $8(2\sqrt{2}-1)$ **(ii)** $2\sqrt{5}+8\log\left(\frac{1+\sqrt{2}}{2}\right)$

(iii) $\sqrt{2}(e-1)$ **(iv)** $\sqrt{2}$

Exercises 3.7

1. **(i)** $\pi\left(\frac{m^2}{3}+mb+b^2\right)$ **(ii)** $\frac{\pi c^2}{2k}(e^{4k}-e^{2k})$ **(iii)** $(\pi c)^2/8k$

3. $\frac{4}{3}\pi a^3 = 4\pi\int_0^a x(a^2-x^2)^{1/2}\,dx$

5. $2\pi,\ \pi a^4\left(\frac{1}{a_1}-\frac{1}{a_2}\right)$

7. **(i)** $\pi\int_{2-\sqrt{2}}^{2+\sqrt{2}}\left[\left(\frac{4-x}{2}\right)^2-\frac{1}{x^2}\right]dx=\frac{4\pi}{3}\sqrt{2}$

(ii) $\pi\int_{1-1\sqrt{2}}^{1+1\sqrt{2}}\left[(4-2y)^2-\frac{1}{y^2}\right]dy=\frac{8\pi}{3}\sqrt{2}$

(iii) $8\pi\int_0^{2\pi}(1-\cos\theta)^3\,d\theta=40\pi^2$

9. $\frac{\pi}{2}(e^2+1)$.

11. $2\pi R^2(1-\cos\Phi)$, $2.54\times10^8\ (\text{km})^2$.

13. **(i)** $2\pi(\sqrt{2}+\log(1+\sqrt{2}))$ **(ii)** $\frac{\pi}{4}[2\sqrt{5}+\log(2+\sqrt{5})]$

(iii) $\frac{2\pi}{9}[37\sqrt{37}-10\sqrt{10}]$

(iv) $\pi\left[e(1+e^2)^{1/2}-\sqrt{2}+\log\left(\frac{e+(1+e^2)^{1/2}}{1+\sqrt{2}}\right)\right]$

15. **(i)** $e=0$ (circle) **(ii)** $e=0$

(iii) $e=0$ (circle of infinite radius) or undefined

(iv) $e=\sqrt{3}/2$ **(v)** $e=1/2$ **(vi)** $e=1/2$

17. (See Exercise **8**) $10^6/3\ \text{cm}^3$, $\int_{z_1}^{z_2}(h-z)^2\,dz=\frac{1}{3}[(h-z_1)^3-(h-z_2)^3]$

19. Volume $=4\int_{-1}^{1}(1-x^2)\,dx=\frac{16}{3}$

Exercises 3.8

1. Mass $=\pi R^2\rho_0\int_0^H\frac{dz}{1+z/H}=\rho_0\pi R^2H\log 2$, average density $\bar{\rho}=\text{mass}/\pi R^2H=\rho_0\log 2$,

center of mass $\bar{z}=\frac{1}{M}\iiint z\rho\,dV=\frac{\pi R^2\rho_0}{M}\int_0^H\frac{z\,dz}{1+z/H}=\frac{(1-\log 2)H}{\log 2}\approx(0.443)H$.

3. $(4-10e^{-1})\pi\beta R^3 = 2\pi\beta \int_0^R r(R^2-r^2)^{1/2}\, e^{-(R^2-r^2)^{1/2}/R}\, dr$

5. $5\rho_0\, l/4,\ 14l/25.$

7. $\bar{x} = \bar{y} = -8a/3(2+3\pi)$

9. $2000\ \mathrm{m}^3 = \int_0^{50} 20\left(1+\frac{2x}{50}\right) dx.$ (Answer is obvious since average depth is 2 m.)

11. Force $= 2\int_{a-h}^{a} \rho g(z-a+h)(a^2-z^2)^{1/2}\, dz$

$$= \rho g\left[\frac{1}{3}(3a^2-2ah+h^2)[(2a-h)h]^{1/2} - a^2(a-h)\cos^{-1}\left(1-\frac{h}{a}\right)\right]$$

13. $\bar{\rho} = \frac{1}{V}\iiint \rho\, dV = \frac{1}{\pi R^2 H}\int_0^R 2\pi r H\rho_0\left(1+\frac{r}{R}\right) dr = \frac{5}{3}\rho_0.$

Kinetic energy $= \frac{1}{2}\int v^2\, dm = \pi\rho_0\Omega^2 H\int_0^R \frac{r^3\, dr}{1+r/R} = \pi\rho_0 R^4 H\Omega^2[5/6 - \log 2]$

15. Kinetic energy $= \pi\rho H\Omega^2\int_0^R x^3\left(1-\frac{x}{R}\right) dx = \frac{1}{20}\pi\rho H\Omega^2 R^4$

Exercises 3.9

1. (i) $\lim_{\varepsilon\to 0+}\int_\varepsilon^2 \frac{1}{(2-x)^{1/2}}\, dx = \lim_{\varepsilon\to 0+} -2(2-x)\Big|_\varepsilon^2 = \lim_{\varepsilon\to 0+} 2(2-\varepsilon)^{1/2} = 2\sqrt{2}$, convergent

(ii) divergent $\left(\text{examine} \int_{-1}^0 \frac{1}{x^2}\, dx \text{ and } \int_0^1 \frac{1}{x^2}\, dx\right)$

(iii) $\lim_{s\to\infty}\int_1^s \frac{dx}{x^{2/3}} = \lim_{s\to\infty} 3(s^{1/3}-1) = \infty$, divergent

(iv) 1, convergent **(v)** -1, convergent **(vi)** divergent
(vii) $\log 2$, convergent **(viii)** divergent **(ix)** divergent **(x)** π, convergent
(xi) 1/2, convergent **(xii)** π, convergent

3. (i) $\int_1^2 \frac{dx}{x(\log x)^{1/2}} = 2(\log x)^{1/2}\Big|_1^2 = 2(\log 2)^{1/2}$

(ii) $\int_0^\infty \frac{dx}{(1+x)^4} = -\frac{1}{3(1+x)^3}\Big|_0^\infty = \frac{1}{3}$

(iii) divergent $\left(\frac{2x}{\pi} < \sin x < x \text{ for } 0 < x < \frac{\pi}{2}\right)$

(iv) $\pi - 2\tan^{-1}\sqrt{2}$ (let $u = x^2-1$)

(v) $1 < \int_0^1 \frac{x}{\sin x}\, dx < \csc 1$ $\left(\frac{\sin x}{x} > \sin 1 \text{ for } 0 < x < 1\right)$

(vi) 1 **(vii)** -1 **(viii)** π (define $x = \sin\theta$)

(ix) $\int_0^1 \frac{dx}{(2x+1)^{1/3}} + \int_1^\infty \frac{dx}{(3x^6)^{1/3}} < \int_0^\infty \frac{dx}{(2x+x^6)^{1/3}} < \int_0^1 \frac{dx}{(2x)^{1/3}} + \int_1^\infty \frac{dx}{x^2};\ \frac{5}{12}3^{2/3} - \frac{3}{4} < I < \frac{3}{2^{4/3}} + 1.$

(x) divergent (examine integrand near $x = 0$)

(xi) $-\log 2 < \int_0^\infty \frac{\sin x}{(x+1)(x+2)}\, dx < \log 2$ $(-1 \le \sin x \le 1)$

(xii) 8/5 $(\cos^3 x = (1-\sin^2 x)\cos x)$

5. Use partial fractions; change of variable is $x = 1/t$.

7. $I = \int_0^{\pi/2} \log(\sin x)\,dx = \int_0^{\pi/2} \log(\cos x)\,dx$ since integrands take on same values on interval.

$$I = \frac{1}{2}\int_0^{\pi/2} [\log(\sin x) + \log(\cos x)]\,dx = \frac{1}{2}\int_0^{\pi/2} [\log(2\,\sin x\,\cos x) - \log 2]\,dx$$

$$= \frac{1}{2}\int_0^{\pi/2} \log(\sin 2x)\,dx - \frac{\pi}{4}\log 2.$$ Define $u = 2x$.

$$I = \frac{1}{4}\int_0^{\pi} \log(\sin u)\,du - \frac{\pi}{4}\log 2 = \frac{1}{2}\int_0^{\pi/2} \log(\sin u)\,du - \frac{\pi}{4}\log 2$$

$= \frac{1}{2}I - \frac{\pi}{4}\log 2$. Result follows. $\int_0^{\pi/2} \log(a\,\sin x)\,dx = \frac{\pi}{2}\log\left(\frac{a}{2}\right)$,

$\int_0^{\pi/2} \log(a\,\tan x)\,dx = \frac{\pi}{2}\log a$.

9. $\int_{-\infty}^{\infty} f(t)g(a-t)\,dt = \lim_{R,S\to\infty} \int_{-S}^{R} f(t)g(a-t)\,dt$. Define $u = a - t$.

$\int_{-\infty}^{\infty} (a-t)e^{-t^2}\,dt = \int_{-\infty}^{\infty} te^{-(a-t)^2}\,dt$.

11. Define $u = ax$. If $a>0$ $\int_0^{\infty} \frac{\sin ax}{x}\,dx = \int_0^{\infty} \frac{\sin u}{u}\,du$. If $a<0$, $\int_0^{-\infty} \frac{\sin u}{u}\,du$.

13. 0; yes, but limit is taken in a very particular way. Only finite if b is proportional to a, i.e., if $\lim_{b,a\to 0+} \frac{b}{a}$ is finite.

15. $V = \pi \int_0^{\infty} e^{-2x}\,dx = \pi/2$, $S = 2\pi \int_0^{\infty} e^{-x}(1 + e^{-2x})^{1/2}\,dx = \pi[\sqrt{2} + \log(1 + \sqrt{2})]$;

$V = \pi a^2 \int_0^{\infty} e^{-2bx}\,dx = \pi a^2/2b$, $S = 2\pi a \int_0^{\infty} e^{-bx}(1 + ab)^2 e^{-2bx})^{1/2}\,dx$

$$= \pi\frac{a}{b}(1 + a^2b^2)^{1/2} + \frac{\pi}{b^2}\log[ab + (1 + a^2b^2)^{1/2}]$$

17. $\frac{1}{2}mv_0{}^2 + \int_{s_0}^{\infty} F(s)\,ds$

19. mgr_0; 11.24 km/sec

Chapter 4 Series

Exercises 4.1

1. **(i)** $(1^2 + 2) + (2^2 + 4) + (3^2 + 6) + (4^2 + 8) + (5^2 + 10) + (6^2 + 12) = 133$

(ii) $\frac{1}{1} + \frac{1}{2} + \frac{1}{3} + \frac{1}{4} + \frac{1}{5} = \frac{137}{60}$

(iii) $1^1 + 2^2 + 3^3 = 32$ **(iv)** $\frac{1}{2} + \frac{1}{2^2} = \frac{3}{4}$

(v) $\dfrac{1}{2^2+3}+\dfrac{1}{3^2+3}+\dfrac{1}{4^2+3}+\dfrac{1}{5^2+3}=\dfrac{502}{1596}$

(vi) $\dfrac{1}{2^0}+\dfrac{1}{2^2}+\dfrac{1}{2^4}+\cdots+\dfrac{1}{2^{2n}}+\cdots=\dfrac{4}{3}$

3. **(i)** $\sum_{n=0}^{\infty}\dfrac{(-1)^n}{(2n)!}$, 0.54 **(ii)** $\sum_{n=0}^{\infty}\dfrac{(-1)^n}{(n+2)^4}$, 0.053

(iii) $\sum_{n=0}^{\infty}\dfrac{(-1)^n}{(n+2)^{2n+4}}$, 0.061 **(iv)** $\sum_{n=1}^{\infty}\dfrac{n!}{(2n)!}$, 0.59

(v) $\sum_{n=0}^{\infty}\dfrac{(0.1)^n}{n!}$, 1.11 **(vi)** $\sum_{n=0}^{\infty}\dfrac{(-1)^n}{2^{2n}}$, 0.80

5. Divergent, $1+(1/2+1/2)+(1/3+1/3+1/3)+\cdots=1+1+1+\cdots=\infty$. Sum of first $n(n+1)/2$ terms is n. Since $44(45/2)=990$, sum of first 1000 terms is $44+10/45$. Since $1413\,(1414)/2=998{,}991$, sum of first million terms is $1413+1009/1414$.

7. **(i)** 1, $\sum_{n=1}^{\infty}\dfrac{1}{n(n+1)}=\sum_{n=1}^{\infty}\left(\dfrac{1}{n}-\dfrac{1}{n+1}\right)$

(ii) $\dfrac{1}{5}$, $\dfrac{1}{(n+4)(n+5)}=\dfrac{1}{n+4}-\dfrac{1}{n+5}$

(iii) $\dfrac{1}{2}$, $\dfrac{1}{(2n-1)(2n+1)}=\dfrac{1/2}{2n-1}-\dfrac{1/2}{2n+1}$

(iv) 1, $(1-r)r^{n-1}=r^{n-1}-r^n$

(v) 1, $\dfrac{2n+1}{n^2(n+1)^2}=\dfrac{1}{n^2}-\dfrac{1}{(n+1)^2}$

(vi) $\dfrac{1}{2}$, $\dfrac{1}{n(n+1)(n+2)}=\dfrac{1}{n(n+1)}-\dfrac{1}{(n+1)(n+2)}$

9. **(i)** Use $\sum_{n=1}^{N}(a_n+a_{n+1})=2\left(\sum_{n=1}^{N}a_n\right)+a_{N+1}-a_1$ and $\lim_{N\to\infty}a_{N+1}=0$

(ii) $\sum_{n=1}^{N}(a_n+b_n)=\sum_{n=1}^{N}a_n+\sum_{n=1}^{N}b_n$, $\left|\sum_{n=1}^{N}(a_n+b_n)-(s+t)\right|\le\left|\sum_{n=1}^{N}a_n-s\right|+\left|\sum_{n=1}^{N}b_n-t\right|$.

11. **(v)** and **(ix)** divergent since $\lim_{n\to\infty}\dfrac{n^n}{(n+1)^n}=e^{-1}$ and $\lim_{n\to\infty}e^{-n}\cosh n=\dfrac{1}{2}$. All others convergent since $\lim_{n\to\infty}c_n=0$ and $\{c_n\}$ is decreasing sequence.

13. Use equations (13) and (14) in each case.

15. **(i)** $\sum_{n=1}^{\infty}2a_n$ **(ii)** $\sum_{k=1}^{\infty}(0)=0$ **(iii)** $\sum_{k=1}^{\infty}[(2r)^{k+1}+(3r)^{k+2}]$

(iv) $\sum_{n=1}^{\infty}\left[\dfrac{1}{(n+1)^2}-\dfrac{1}{(n+2)(n+1)}\right]=\sum_{n=1}^{\infty}\dfrac{1}{(n+1)^2(n+2)}$

Exercises 4.2

1. **(i)** convergent, $\dfrac{1}{n2^n}<\dfrac{1}{2^n}$ for $n>1$ **(ii)** cgt, $\dfrac{1}{n^2+1}<\dfrac{1}{n^2}$

(iii) dgt, $\dfrac{1}{(n+1)^{1/2}}>\dfrac{1}{2n^{1/2}}$ **(iv)** dgt, $\dfrac{n+1}{n^2+2}>\dfrac{1}{n}$ for $n>2$

(v) dgt, $\dfrac{n}{(2n+1)^{3/2}} > \dfrac{1}{4n^{1/2}}$ for $n>1$ **(vi)** cgt, $\dfrac{1}{(n+4)^{4/3}} < \dfrac{1}{n^{4/3}}$

(vii) cgt, $\dfrac{n^2+4}{n^4-6} < \dfrac{16}{10n^2}$ for $n>2$ **(viii)** cgt, $\dfrac{e^n}{(n+2)^n} < \left(\dfrac{e}{3}\right)^n$ for $n>1$

(ix) cgt, $\dfrac{1}{(n+2)(n+3)} < \dfrac{1}{n^2}$.

3. **(i)** $\displaystyle\sum_{n=0}^{\infty}\left(\frac{e}{\pi}\right)^n = \frac{\pi}{\pi - e}$ **(ii)** $\displaystyle\sum_{n=1}^{\infty}\frac{(-1)^{n+1}}{\sqrt{n}}$, cgt alternating series

(iii) $\displaystyle\sum_{n=1}^{\infty}\frac{\log(n+1)}{n+1}$, dgt by comparison test, $\dfrac{\log(n+1)}{n+1} > \dfrac{1}{n+1}$ for $n>2$

(iv) $\displaystyle\sum_{n=1}^{\infty}\frac{n}{n+1}$, dgt since $\displaystyle\lim_{n\to\infty}\frac{n}{n+1} = 1$ **(v)** $\displaystyle\sum_{n=0}^{\infty}\frac{(-1)^n}{2n+1}$, cgt alternating series

(vi) $\displaystyle\sum_{n=1}^{\infty}(-1)^{n+1}\sin\frac{1}{n}$, cgt alternating series

5. Differentiate $\displaystyle\sum_{n=0}^{\infty} r^n = \frac{1}{1-r}$ to obtain $\displaystyle\sum_{n=0}^{\infty} n\, r^{n-1} = \frac{1}{(1-r)^2}$. Multiply by r to obtain $\displaystyle\sum_{n=1}^{\infty} nr^n = \frac{r}{(1-r)^2}$. Differentiate this identity to obtain the second result.

7. **(i)** conditionally convergent by alternating series test; not absolutely convergent by comparison test $\left(\dfrac{\log n}{n} > \dfrac{1}{n} \text{ for } n>2\right)$

(ii) dgt, $\displaystyle\lim_{n\to\infty}\frac{n}{n+1} = 1$ **(iii)** abs. cgt by ratio test

(iv) abs. cgt, $\dfrac{1}{n^2}\left(1+\dfrac{1}{n}\right)^n < \dfrac{e}{n^2}$ **(v)** cgt, conditionally since $\displaystyle\lim_{n\to\infty} c_n = 0$

(vi) divergent since $\displaystyle\lim_{n\to\infty} a_n \neq 0$ **(vii)** conditionally cgt, $\cos n\pi = (-1)^n$

(viii) absolutely cgt by comparison test, $\dfrac{2^n}{1+2^{2n}} < \dfrac{1}{2^n}$

(ix) conditionally cgt since $\displaystyle\lim_{n\to\infty}\tanh n = 1$.

9. **(i)** convergent, $$\lim_{n\to\infty}\frac{a_{n+1}}{a_n} = \lim_{n\to\infty}\frac{2^{n+1}(n+1)!}{(2n+2)!}\frac{(2n)!}{2^n n!} = \lim_{n\to\infty}\frac{2(n+1)}{(2n+2)(2n+1)} = 0$$

(ii) dgt, $\displaystyle\lim_{n\to\infty}\left|\frac{a_{n+1}}{a_n}\right| = \lim_{n\to\infty}\frac{n+1}{10} = \infty$ **(iii)** cgt **(iv)** cgt

(v) cgt **(vi)** cgt **(vii)** cgt **(viii)** cgt **(ix)** cgt

11. Given $\varepsilon>0$, then $(r-\varepsilon)^n < |a_n| < (r+\varepsilon)^n$ for $n>N(\varepsilon)$. By comparison with a geometric series, $\displaystyle\sum_{n=1}^{\infty}|a_n|$ converges for $r<1$ and diverges if $r>1$; **(i)**, **(ii)**, **(iii)** all convergent with $r = 1/2,\ 0,\ 2/3$.

13. Given $\varepsilon>0$, then $A-\varepsilon < na_n < A+\varepsilon$ and $(A-\varepsilon)/n < a_n < (A+\varepsilon)/n$.

(i) $\displaystyle\lim_{n\to\infty} n\sin\frac{\pi}{n} = \pi$ **(ii)** $\displaystyle\lim_{n\to\infty} n\log\left(1+\frac{2}{n}\right) = 2$ **(iii)** $\displaystyle\lim_{n\to\infty}\frac{n^2\log n}{1+n\log n^2} = \frac{1}{2}$

15. Since $\sum_{n=1}^{N}\frac{1}{n}\approx\int_1^N\frac{1}{x}\,dx=\log N$ and $\sum_{n=2}^{N}\frac{1}{n\log n}\approx\int_2^N\frac{dx}{x\log x}\approx\log(\log N)$, approximately $e^{100}\approx 2.7\times 10^{43}$ and $e^{e^{100}}$ terms are required.

Exercises 4.3

1. **(i)** $\sum_{n=0}^{\infty}\left(\frac{x}{4}\right)^n=\left(1-\frac{x}{4}\right)^{-1}$, $|x|<4$ **(ii)** $\sum_{n=0}^{\infty}(-5x)^n=(1+5x)^{-1}$, $|x|<1/5$

(iii) $\sum_{n=0}^{\infty}e^{nx}=(1-e^x)^{-1}$, $x<0$ **(iv)** $\sum_{n=0}^{\infty}\frac{(-1)^n}{(2n)!}\left(\frac{x}{2}\right)^{2n}=\cos\frac{x}{2}$, $-\infty<x<\infty$

(v) $\sum_{n=0}^{\infty}(n+1)x^n=(1-x)^{-2}$, $|x|<1$ **(vi)** $\sum_{n=0}^{\infty}(n+1)^2x^n=\frac{1-x}{(1-x)^3}$, $|x|<1$

(vii) $1+\sum_{n=1}^{\infty}\frac{x^n}{n}=1-\log(1-x)$, $|x|<1$

(viii) $\sum_{n=0}^{\infty}(-1)^n\frac{(\tan x)^{2n+1}}{(2n+1)!}=\sin(\tan x)$, $|x|<\frac{\pi}{2}$.

3. **(i)** Since $\frac{d}{dx}\sin^{-1}x=(1-x^2)^{-1/2}=\sum_{n=0}^{\infty}\frac{(2n)!}{2^{2n}(n!)^2}x^{2n}$,

$$\sin^{-1}x=\sum_{n=0}^{\infty}\frac{(2n)!}{2^{2n}(n!)^2}\frac{x^{2n+1}}{(2n+1)}$$

(ii) $\cos^{-1}x=\pi/2-\sin^{-1}x$ **(iii)** $\sum_{n=0}^{\infty}\frac{x^{2n}}{(2n)!}$ **(iv)** $\sum_{n=0}^{\infty}(-1)^n\frac{x^{2n+1}}{2n+1}$.

(v) $\tan x=x+\frac{x^3}{3}+\frac{2x^5}{15}+\frac{17x^7}{315}+\frac{62x^9}{2835}+\cdots$ **(vi)** $\tanh x=x-\frac{x^3}{3}+\frac{2x^5}{15}-\frac{17x^7}{315}+\cdots$

5. $f^{(4)}(0)=4!a_4$, $f^{(5)}(0)=5!a_5$. **(i)** 0, 2^5 **(ii)** 3^4, 0 **(iii)** 24, -120
(iv) -6, 24 **(v)** $-15/16$, $-105/32$ **(vi)** 120, 720

7. $1/(1-x)^2=\frac{d}{dx}1/(1-x)=\frac{d}{dx}(1+x+x^2+\cdots+x^n+\cdots)=1+2x+\cdots+nx^{n-1}+\cdots$,

$$\log(1-x)=-\int_0^x\frac{dt}{1-t}=-\int_0^x(1+t+t^2+\cdots+t^n+\cdots)\,dt$$
$$=-x-\frac{x^2}{2}-\cdots-\frac{x^{n+1}}{n+1}-\cdots.$$

9. $-x^2-\frac{5}{12}x^4-\frac{47}{180}x^6+\cdots$ (Multiply the two series and collect equal powers of x.)

11. **(i)** $\frac{2}{\pi^{1/2}}\left(x-\frac{x^3}{3\cdot 1!}+\frac{x^5}{5\cdot 2!}-\frac{x^7}{7\cdot 3!}+\cdots\right)$, cgt for all x

(ii) see **3(i)**

(iii) $\sum_{n=0}^{\infty}(-1)^n\frac{1\cdot 3\cdots(2n-1)}{2\cdot 4\cdots 2n}\frac{x^{2n+1}}{2n+1}=\sum_{n=0}^{\infty}(-1)^n\frac{(2n)!}{2^{2n}(n!)^2}\frac{x^{2n+1}}{2n+1}$

(iv) $\sum_{n=0}^{\infty}\frac{x^{2n+1}}{2n+1}$

13. **(i)** $\frac{1}{1-x}-\frac{1}{1-x/2}=(1+x+x^2+\cdots+x^n+\cdots)-\left(1+\frac{x}{2}+\left(\frac{x}{2}\right)^2+\cdots+\left(\frac{x}{n}\right)^n+\cdots\right)$

$$=1+\frac{1}{2}x+\frac{3}{4}x^2+\frac{7}{8}x^3+\cdots+\left(1-\frac{1}{2^n}\right)x^n+\cdots,\ \text{convergent for } |x|<1$$

(ii) $\dfrac{2x+1}{x^2+4x+3}=\dfrac{5}{6}\dfrac{1}{1+x/3}-\dfrac{1}{2}\dfrac{1}{1+x}=\sum_{n=0}^{\infty}\left[\dfrac{5}{6}\left(-\dfrac{1}{3}\right)^n-\dfrac{1}{2}(-1)^n\right]x^n$, *cgt* for $|x|<1$

(iii) $\dfrac{1}{3}\left[\dfrac{1}{x-1}-\dfrac{1}{x+2}\right]=\sum_{n=0}^{\infty}\left[\left(-\dfrac{1}{3}\right)(-1)^n-\left(\dfrac{1}{6}\right)\left(-\dfrac{1}{2}\right)^n\right]x^n$, *cgt* for $|x|<1$.

15. $(1+x^2)^{1/2} = 1+\dfrac{(1/2)}{1!}x^2+\dfrac{(1/2)(-1/2)}{2!}x^4+\dfrac{(1/2)(-1/2)(-3/2)}{3!}x^6+\cdots$

$= 1+\dfrac{1}{2}x^2+\sum_{n=2}^{\infty}(-1)^{n+1}\dfrac{1.3\cdots(2n+3)}{2^n n!}x^{2n}$

$(1+x^2)^{-1/2} = 1+\dfrac{(-1/2)}{1!}x^2+\dfrac{(-1/2)(-3/2)}{2!}x^4+\dfrac{(-1/2)(-3/2)(-5/2)}{3!}x^6+\cdots$

$= 1+\sum_{n=1}^{\infty}(-1)^n\dfrac{1\cdot 3\cdots(2n-1)x^{2n}}{2^n n!}$

Verify that $x(1+x^2)^{-1/2}=\dfrac{d}{dx}(1+x^2)^{1/2}$ by comparing series.

17. (i) $x=\sin^{-1}y=y+\dfrac{y^3}{6}+\cdots$ **(ii)** $x=\tan y=y+\dfrac{y^3}{3}+\cdots$

(iii) $x=e^y-1=\dfrac{y}{1!}+\dfrac{y^2}{2!}+\cdots$

19. Differentiate $\log y=\dfrac{\log(1+x)}{x}=1-\dfrac{x}{2}+\dfrac{x^2}{3}-\dfrac{x^3}{4}+\cdots$ to obtain

$\dfrac{1}{y}\dfrac{dy}{dx}=-\dfrac{1}{2}+\dfrac{2}{3}x-\dfrac{3}{4}x^2+\cdots$ and $\dfrac{1}{y}\dfrac{d^2y}{dx^2}-\dfrac{1}{y^2}\left(\dfrac{dy}{dx}\right)^2=\dfrac{2}{3}-\dfrac{3}{2}x+\cdots$.

Set $x=0$ and $y(0)=e$ to find $y'(0)=-e/2$, $y''(0)=11e/12$;

then $y(x)=e-\dfrac{e}{2}x+\dfrac{11e}{24}x^2-\cdots$.

21. (i) $\log\cos x=\log\left[1-\left(\dfrac{x^2}{2!}-\dfrac{x^4}{4!}+\dfrac{x^6}{6!}-\cdots\right)\right]$

$=-\left(\dfrac{x^2}{2!}-\dfrac{x^4}{4!}+\cdots\right)-\dfrac{1}{2}\left(\dfrac{x^2}{2!}-\dfrac{x^4}{4!}+\cdots\right)^2-\cdots;-\dfrac{x^2}{2}-\dfrac{x^4}{12}-\dfrac{x^6}{45}-\cdots$

(ii) $-\dfrac{x^2}{6}-\dfrac{x^4}{180}+\left[\dfrac{1}{5!}-\dfrac{1}{3(6)^3}\right]x^6+\cdots$

(iii) $\log\dfrac{\tan x}{x}=\log\dfrac{\sin x}{x}-\log\cos x$.

23. Integrate $1-\dfrac{x^2}{2}<\cos x\leq 1$ to find $x-\dfrac{x^3}{3!}<\sin x<x$, etc.

25. Since $1+(\alpha-1)x<(1+x)^{\alpha-1}<1+(\alpha-1)x+\dfrac{(\alpha-1)(\alpha-2)}{2}x^2$,

integrate to find $x+\dfrac{(\alpha-1)}{2}x^2<\dfrac{(1+x)^\alpha-1}{\alpha}<x+\dfrac{(\alpha-1)}{2}x^2+\dfrac{(\alpha-1)(\alpha-2)}{3!}x^3$

or $1+\alpha x+\dfrac{\alpha(\alpha-1)}{2!}x^2+\dfrac{\alpha(\alpha-1)(\alpha-2)}{3!}x^3<(1+x)^\alpha<1+\alpha x+\dfrac{\alpha(\alpha-1)}{2!}x^2$.

Integrating repeatedly, if $|x|<1$, $(1+x)^\alpha=\lim_{n\to\infty}\sum_{k=0}^{n}\dfrac{\alpha(\alpha-1)\cdots(\alpha-k+1)}{k!}x^k$

Exercises 4.4

1. Differentiate $y'' = y + xy'$, $e^{x^2/2} = 1 + \frac{x^2}{2} + \frac{x^4}{8} + \cdots$

3. $|x| < \infty$, $Y'(x) = \sum_{n=0}^{\infty} \frac{2^n n!}{(2n)!} x^{2n} = 1 + \sum_{n=1}^{\infty} \frac{2^n n!}{2n!} x^{2n} = 1 + x\,Y(x)$.

5. $y = a_0\left[1 + \frac{x^2}{1\cdot 2} + \left(1 + \frac{1}{1\cdot 2}\right)\frac{x^4}{3\cdot 4} + \cdots\right] + a_1\left[x + \frac{x^3}{2\cdot 3} + \left(1 + \frac{1}{2\cdot 3}\right)\frac{x^5}{4\cdot 5} + \cdots\right]$,
convergent for $|x| < 1$ by the ratio test.

7. $J(x) = \int_0^X \left(1 - \frac{t^4}{2!} + \frac{t^8}{4!} - \frac{t^{12}}{6!} + \cdots\right) dt = x - \frac{x^5}{5\cdot 2!} + \frac{x^9}{9\cdot 4!} - \frac{x^{13}}{13\cdot 6!} + \cdots; J\left(\frac{\pi}{4}\right) = 0.756.$

9. $s(p) = 2\int_0^{1/\sqrt{p}} (1 + 4p^2x^2)\,dx = \frac{1}{p}\int_0^{2/\sqrt{p}} (1 + u^2)^{1/2}\,du$

$$= \frac{1}{p}\left[\frac{u}{2}(1 + u^2)^{1/2} + \frac{1}{2}\log(u + \sqrt{u^2 + 1})\right]_0^{2\sqrt{p}}$$

$$= \frac{1}{p}\left[\sqrt{p(1 + 4p)} + \frac{1}{2}\log(2\sqrt{p} + \sqrt{4p + 1})\right]$$

For p small, $s(p) = \frac{1}{2}\int_0^{2/\sqrt{p}} \left[1 + \frac{1}{2}u^2 + \frac{(1/2)(-1/2)}{2!}u^4 + \cdots\right] du$

$$= \frac{2}{\sqrt{p}}\left[1 + \frac{2}{3}p - \frac{2}{5}p^2 + \cdots\right].$$

For p large, $s(p) \approx \frac{1}{p}\sqrt{p + 4p^2} + \frac{1}{2}p\log(4\sqrt{p}) = 2\left(1 + \frac{1}{4p}\right)^{1/2} + \frac{1}{2}p\log(4\sqrt{p})$

$= 2 + \frac{1}{2}p\log(4\sqrt{p}) + \frac{1}{4p} + \cdots$

11. $1 - \frac{1}{5}\varepsilon + \frac{2}{25}\varepsilon^2 + \cdots$ (Write $x = 1 + \varepsilon x_1 + \varepsilon^2 x_2 + \cdots$, substitute and equate powers of ε.)

This is the Taylor series of $x = x(\varepsilon)$; $\frac{dx}{d\varepsilon}(0) = -\frac{1}{5}$, etc.

13. $\frac{1}{2}\log 3; f'(x) = 1 + x^2 + \cdots + x^{2n} + \cdots = (1 - x^2)^{-1}$

and $f(x) = \int_0^x \frac{1}{1 - t^2}\,dt = \frac{1}{2}\log\left(\frac{1 + x}{1 - x}\right)$, $s = f(1/2) = \frac{1}{2}\log 3$.

15. 0.69315.
The second series is much more rapidly convergent due to increasing powers of $1/2$.

17. $1/24$; all terms except first cancel in pairs.

19. $\sum_{n=1}^{\infty} \frac{1}{n^2} = \sum_{n=1}^{\infty}\left[\frac{1}{n^2} - \frac{1}{n(n+1)}\right] + 1 = 1 + \sum_{n=1}^{\infty} \frac{1}{n^2(n+1)}$

$$= \frac{5}{4} + \sum_{n=1}^{\infty}\left[\frac{1}{n^2(n+1)} - \frac{1}{n(n+1)(n+2)}\right] = \frac{5}{4} + 2\sum_{n=1}^{\infty} \frac{1}{n^2(n+1)(n+2)}$$

The last series converges very rapidly; with five terms, the sum is 1.64127 compared to the exact value $\pi^2/6 = 1.644934$.

21. **(i)** $-\log 2+\sum_{n=1}^{\infty}\frac{(-1)^n(n-2)}{n(n^2+2)}$ **(ii)** $1-\sum_{n=1}^{\infty}\frac{1}{n^3+n^2}$ **(iii)** $-\frac{\pi^2}{12}+2\sum_{n=1}^{\infty}\frac{(-1)^{n+1}}{n^2(n+2)}$

Hints: **(i)** use $\log 2=\sum_{n=1}^{\infty}\frac{(-1)^{n+1}}{n}$ **(ii)** $1=\sum_{n=1}^{\infty}\frac{1}{n(n+1)}$

$$\textbf{(iii)}\ \sum_{n=1}^{\infty}\frac{(-1)^n}{n^2}=-\frac{1}{1^2}+\frac{1}{2^2}-\frac{1}{3^2}+\frac{1}{4^2}-\cdots$$
$$=-\left(\frac{1}{1^2}+\frac{1}{2^2}+\frac{1}{3^2}+\cdots\right)+\frac{2}{2^2}\left(\frac{1}{1^2}+\frac{1}{2^2}+\frac{1}{3^2}+\cdots\right)$$
$$=-\frac{1}{2}\left(\frac{1}{1^2}+\frac{1}{2^2}+\frac{1}{3^2}+\cdots\right)=-\frac{\pi^2}{12}.$$

Exercises 4.5

1. For x small, $I(x)=x-\frac{x^3}{3}+\frac{x^5}{5}-\frac{x^7}{7}+\cdots$; $I(0.1)=0.0997$.

For x large, $I(x)=\int_0^{\infty}\frac{dt}{1+t^2}-\int_x^{\infty}\frac{dt}{1+t^2}=\frac{\pi}{2}-\int_x^{\infty}\frac{1}{t^2}\left[1-\frac{1}{t^2}+\frac{1}{t^4}-\cdots\right]dt$

$=\frac{\pi}{2}-\frac{1}{x}+\frac{1}{3x^3}-\frac{1}{5x^5}+\cdots$; $I(100)=1.561$.

Accurate values are $\tan^{-1}(0.1)=0.0996687$ and $\tan^{-1}100=1.5607967$.

3. $Ei(0)=\int_0^{\infty}\frac{e^{-z}}{z}\,dz=\infty$ (singularity at origin).

$Ei(x)=\int_0^{\infty}\frac{e^{-z}}{z}\,dz=-\frac{e^{-z}}{z}\Big|_x^{\infty}-\int_x^{\infty}\frac{e^{-z}}{z^2}\,dz=\frac{e^{-x}}{x}-\int_x^{\infty}\frac{e^{-z}}{z^2}\,dz$. Repeating,

$Ei(x)=e^{-x}\left[\frac{1}{x}-\frac{1}{x^2}+\frac{2}{x^3}-\frac{3!}{x^4}+\cdots+(-1)^n\frac{n!}{x^{n+1}}\right]-(-1)^n(n+1)!\int_x^{\infty}\frac{e^{-z}}{z^{n+2}}\,dz.$

By ratio test, series does not converge but it is asymptotic, $\lim_{x\to\infty}x^nR_n(x)=0$.

5. Follow method of Example 3, Section 3.10 with $Ei(x)=Ei(3)-\int_3^x\frac{e^{-z}}{z}\,dz$.

7. Use repeated integration by parts. If $u=1/z$ and $dv=ze^{-z^2}\,dz$

$\operatorname{erf} x=1-\frac{2}{\pi^{1/2}}\int_x^{\infty}\frac{z}{z}e^{-z^2}\,dz=1-\frac{2}{\pi^{1/2}}\left[-\frac{e^{-z^2}}{2z}\Big|_x^{\infty}-\frac{1}{2}\int_x^{\infty}\frac{e^{-z^2}}{z^2}\,dz\right]$

$=1-\frac{e^{-x^2}}{x\pi^{1/2}}+\frac{1}{\pi^{1/2}}\int_x^{\infty}\frac{e^{-z^2}}{z^2}\,dz.$

9. $I(x)=e^{-x^3}\left[\frac{1}{3x^2}-\frac{2}{3^2x^5}+\frac{2\cdot 5}{3^3x^8}-\frac{2\cdot 5\cdot 8}{3^4x^{11}}+\cdots+\frac{(-1)^n2\cdot 5\cdots(3n-1)}{3^{n+1}x^{3n+2}}\right]$

$-(-1)^n\frac{2\cdot 5\cdots(3n-2)}{3^{n+1}}\int_x^{\infty}\frac{e^{-t^3}}{t^{3n+3}}\,dt.$

Hint: write $I(x)=\int_x^{\infty}\frac{t^2}{t^2}e^{-t^3}\,dt$ and integrate by parts. $I(10)\approx e^{-1000}/300$.

11. **(i)** $\sum_{n=0}^{\infty}(-1)^n\frac{(4n)!}{\lambda^{4n}}$ **(ii)** $e^{-\lambda}\left[\frac{1}{\lambda}-\frac{2!}{\lambda^2}+\frac{3!}{\lambda^3}-\frac{4!}{\lambda^4}+\cdots\right]$ **(iii)** $\sum_{n=0}^{\infty}(-1)^n\frac{(4n+2)!}{\lambda^{4n+2}}$

Exercises 4.6

1. **(i)** $Eu(x) = u\left(x+\frac{1}{2}\right) = \left(x+\frac{1}{2}\right)^2 + 2\left(x+\frac{1}{2}\right) + 5 = x^2 + 3x + \frac{25}{4}$

(ii) $\Delta u(x) = u\left(x+\frac{1}{2}\right) - u(x) = x + \frac{5}{4}$

(iii) $\Delta^2 u(x) = u(x+1) - 2u\left(x+\frac{1}{2}\right) + u(x) = \frac{1}{2}$

(iv) $D\Delta u(x) = u'\left(x+\frac{1}{2}\right) - u'(x) = 1$

(v) $E^{5/3}u(x) = u\left(x+\frac{5}{6}\right) = x^2 + \frac{11}{3}x + \frac{265}{36}$

3. **(i)** $f(a+2h) + f(a+h) - 2f(a)$

(ii) $f(a+2h) - f(a+h) - f\left(a+\frac{h}{2}\right) + f'(a) - f'\left(a+\frac{3h}{2}\right) + f'\left(a+\frac{h}{2}\right)$

5.

x	x^4	Δx^4	$\Delta^2 x^4$	$\Delta^3 x^4$	$\Delta^4 x^4$	$\Delta^5 x^4$
1	1	15	50	60	24	0
2	16	65	110	84	24	0
3	81	175	194	108	24	0
4	256	369	302	132	24	0
5	625	671	434	156	24	0
6	1296	1105	590	180	24	
7	2401	1695	770	204		
8	4096	2465	974			
9	6561	3439				
10	10000					

7.

x	e^x	Δe^x	$\Delta^2 e^x$	$\Delta^3 e^x$	$\Delta^4 e^x$
0.0	1.0000	0.6487	0.4209	0.2729	0.1773
0.5	1.6487	1.0696	0.6938	0.4502	
1.0	2.7183	1.7634	1.1440		
1.5	4.4817	2.9074			
2.0	7.3891				

9. 1.4760; construct difference table and evaluate $f(0.48) = f(0.45 + (0.6)(0.05))$ using equation (15).

11. $\frac{n}{3}(48n^3 + 80n^2 - 6n - 47)$; use equation (17) to calculate $S_n = \sum_{k=0}^{n-1} (4k+1)(4k+5)^2$

13. 0.6744; construct difference table and use equation (15); accurate value is 0.67451.

15. $\nabla = 1 - E^{-1} = \frac{\Delta}{1+\Delta} = \sum_{n=0}^{\infty} (-1)^n \Delta^{n+1}$

Exercises 4.7

3.

θ	$K(\theta)$	$\Delta K(\theta)$	$\Delta^2 K(\theta)$	$\Delta^3 K(\theta)$	$\Delta^4 K(\theta)$
10°	1.5828	0.0372	0.0286	0.0066	0.0060
20°	1.6200	0.0658	0.0352	0.0126	
30°	1.6858	0.1010	0.0478		
40°	1.7868	0.1488			
50°	1.9356				

5. 5.234; use equation (15) Sec. 4.6 for each value.

7. Use equation (19) with $f(x) = \dfrac{1}{x^4}$

9. Use $E = 1 + \Delta$.

11. Use Exercise 10.

13. **(i)** $-1 \le x < 1$ **(ii)** $-1 < x < 1$; use Stirling's formula, equation (23).

Exercises 4.8

1. **(i)** second order linear **(ii)** third order **(iii)** first **(iv)** first linear **(v)** second **(vi)** third linear **(vii)** fifth linear **(viii)** sixth linear **(ix)** first

3.

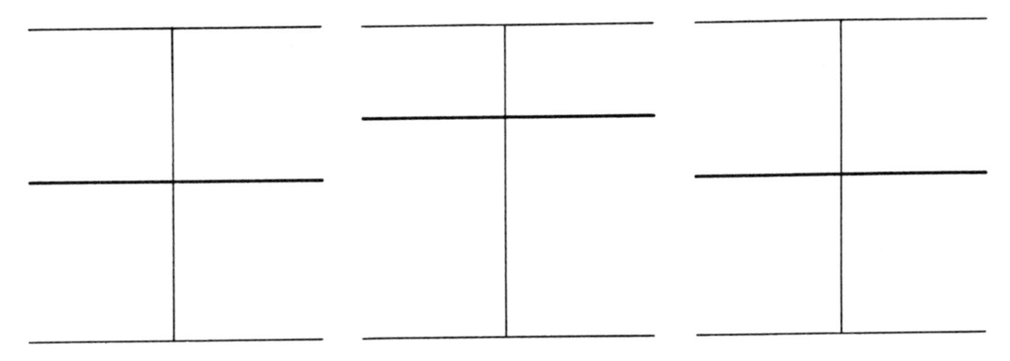

5. $y = \dfrac{3}{2}x^2 + x^2 \log x - \dfrac{1}{2}\cdot$ Hint: $\left(\dfrac{y}{x^2}\right)' = \dfrac{1}{x^3} + \dfrac{1}{x}$

7. **(i)** $x^2 + y^2 = C$ **(ii)** $y = Cx$ **(iii)** $y = Ce^{e^x}$

(iv) $y = Ce^{\sin x}$ **(v)** $(y-1)e^y - \dfrac{1}{2}e^{-x^2} = C$ **(vi)** $\dfrac{1}{2}\theta + \dfrac{1}{4}\sin 2\theta + \cos t = C$

(vii) $y = Ce^{x - x^2/2}$ **(viii)** $y = \tan\left(\dfrac{x^3}{3} + C\right)$ **(ix)** $r = Ce^{-\frac{\alpha}{\omega}\cos \omega t}$

9. **(i)** $x = y\log y - Cy$ **(ii)** $\dfrac{2x}{x-y} + \log\left|\dfrac{x-y}{x^2}\right| = C$

(iii) $\log\left|\sec\left(\dfrac{y}{x}\right) + \tan\left(\dfrac{y}{x}\right)\right| = \log|x| + C$

(iv) $y^2 = 2x^2 \log|x| + Cx^2$ **(v)** $y = xe^{1+Cx}$

(vi) $\dfrac{y-3x}{y-2x} = C x^5$ (If $y' = f\left(\dfrac{y}{x}\right)$ and $y = vx$, then $v + xv' = f(v)$ or $\dfrac{v'}{f(v)-v} = \dfrac{1}{x}\cdot$ This equation for $v = v(x)$ is separated.)

11. $y_{n+1} - y_n = h[g(x_n) - p(x_n)y_n]$, y_0 given; $y_{n+1} - y_n = h[e^{-x_n} - y_n]$, $y_0 = 1$. Construct a table of values with $h = 1/2$, $h = 1/4$, etc.

13. Let $x = \alpha$ to find $y(\alpha, \varepsilon) = y_0(\alpha) + \varepsilon\, y_1(\alpha) + \varepsilon^2 y_2(\alpha) + \cdots$. Since $y(\alpha, \varepsilon) = \beta$, we must have $y_0(\alpha) = \beta$, $y_1(\alpha) = 0$, $y_2(\alpha) = 0$, ... by equating coefficients of powers of ε. In general, $y_n(x) = \dfrac{1}{n!}\dfrac{\partial^n}{\partial\varepsilon^n} y(x, \varepsilon)$.

15. The exact solution is $y = 1 - e^{-x/\varepsilon}$; the perturbation method gives $y = 1$. The difficulty is that $e^{-x/\varepsilon}$ does not have a Taylor expansion in powers of ε.

17. **(i)** $m = 1, -3$ **(ii)** $2, -2$ **(iii)** $6, 2$ **(iv)** $\pm 3\sqrt{-1}$ **(v)** $2, 1, 0$ **(vi)** 1

19. $V(t) = r + (V(O) - r)e^{-\frac{k}{a}t} = 1.2 + (1.8)e^{-2t}$; $V(t) = 1.5$ when $t = 0.896$ s.

21. $x(t) = a(1 - e^{-kt})$, $A(t) = ae^{-kt}$

23. $x(t) = \dfrac{ab(1 - e^{(b-a)kt})}{a - b\, e^{(b-a)kt}}\cdot$ If $a < b$, $\lim_{t\to\infty} x(t) = a$; if $b < a$, $\lim_{t\to\infty} x(t) = b$.

25. $y = A\, e^{x/\sqrt{\varepsilon}} + B\, e^{-x/\sqrt{\varepsilon}}$ (define $z = x/\sqrt{\varepsilon}$).

27. $m\dfrac{dv}{dt} = F(v) = -mkv$, $v(t) = v(0)\, e^{-kt}$, $x(t) = x(0) + \dfrac{v(0)}{k}(1 - e^{-kt})$. The object moves distance $\dfrac{v(0)}{k}$; it never comes to a stop.

29. $s(t) = a\cos\left(\sqrt{\dfrac{k}{m}}\,t\right) + b\sqrt{\dfrac{m}{k}}\sin\left(\sqrt{\dfrac{k}{m}}\,t\right)$

Chapter 5 Vectors

Exercises 5.1

1. $\overrightarrow{AB} = \overrightarrow{OB} - \overrightarrow{OA} = -2\mathbf{i} + \mathbf{j} + \mathbf{k}$, $\overrightarrow{BC} = \overrightarrow{OC} - \overrightarrow{OB} = 3\mathbf{i} + \mathbf{j} + 3\mathbf{k}$,
$\overrightarrow{CD} = \overrightarrow{OD} - \overrightarrow{OC} = -2\mathbf{i} - 5\mathbf{j} + 3\mathbf{k}$, $\overrightarrow{DA} = \overrightarrow{OA} - \overrightarrow{OD} = \mathbf{i} + 3\mathbf{j} - 7\mathbf{k}$,
$\overrightarrow{AC} = \overrightarrow{OC} - \overrightarrow{OA} = \mathbf{i} + 2\mathbf{j} + 4\mathbf{k}$, $\overrightarrow{BD} = \overrightarrow{OD} - \overrightarrow{OB} = \mathbf{i} - 4\mathbf{j} + 6\mathbf{k}$.

3. $\overrightarrow{AB} = (2, 3, 6)$, $\overrightarrow{BA} = (-2, -3, -6)$; $\overrightarrow{OA} = \sqrt{30}$, $OB = \sqrt{131}$, $AB = 7$.

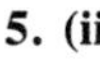

5. (ii)

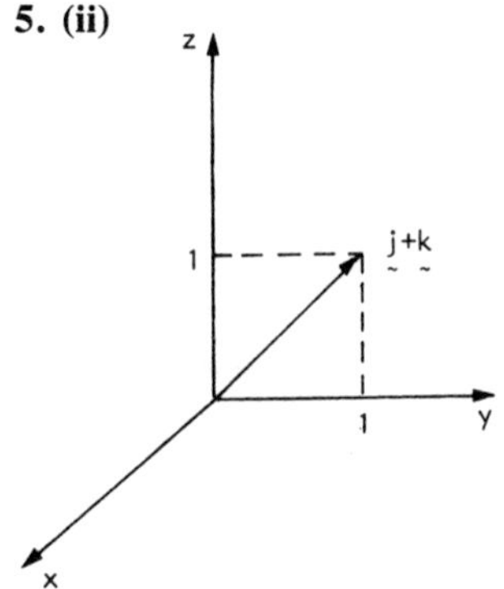

(iv)

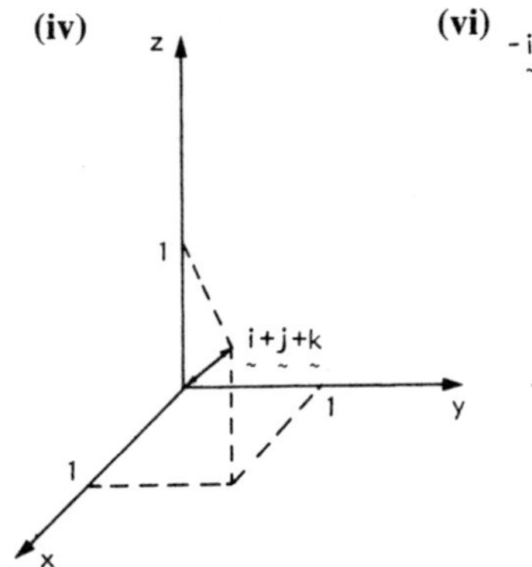

(vi)

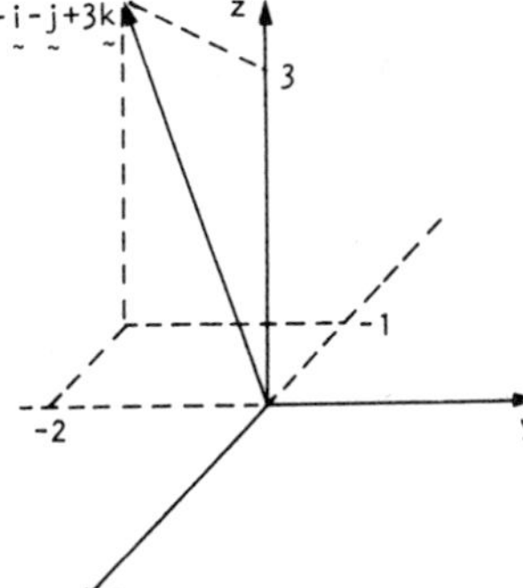

7.

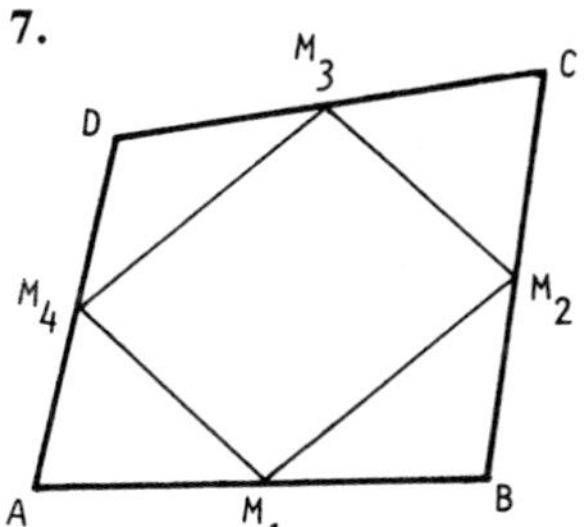

If $A = (a_1, a_2)$, $B = (b_1, b_2)$, $C = (c_1, c_2)$, $D = (d_2, d_2)$,

then $\overrightarrow{M_1M_2} = \left(\dfrac{b_1 + c_1}{2}, \dfrac{b_2 + c_2}{2}\right) - \left(\dfrac{a_1 + b_1}{2}, \dfrac{a_2 + b_2}{2}\right)$

$= \left(\dfrac{c_1 - a_1}{2}, \dfrac{c_2 - a_2}{2}\right)$.

Show that $\overrightarrow{M_1M_2} = \overrightarrow{M_4M_3}$ and $\overrightarrow{M_2M_3} = \overrightarrow{M_1M_4}$.

9. If $A = (a_1, a_2)$, $B = (b_1, b_2)$ and $C = (c_1, c_2)$, show that the medians all pass through the point $\left(\dfrac{a_1 + b_1 + c_1}{3}, \dfrac{a_2 + b_2 + c_2}{3}\right)$.

11.

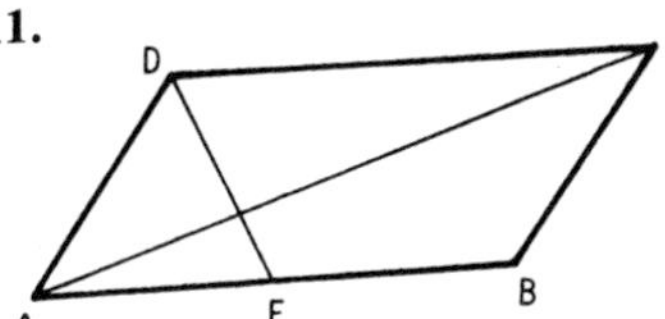

Using $b_1 - a_1 = c_1 - d_1$ and $d_1 - a_1 = c_1 - b_1$, show that

$\dfrac{2}{3} d_1 + \dfrac{1}{3}\left(\dfrac{a_1 + b_1}{2}\right) = \dfrac{1}{3} a_1 + \dfrac{2}{3} c_1$;

similarly for other components.

13. $\mathbf{i} = \mathbf{i}_* \cos\alpha - \mathbf{j}_* \sin\alpha$, $\mathbf{j} = \mathbf{i}_* \sin\alpha + \mathbf{j}_* \cos\alpha$

15. (i) $\dfrac{1}{\sqrt{3}}, \dfrac{1}{\sqrt{3}}, \dfrac{1}{\sqrt{3}}$ **(ii)** $\dfrac{1}{\sqrt{14}}, \dfrac{2}{\sqrt{14}}, \dfrac{3}{\sqrt{14}}$ **(iii)** $-\dfrac{1}{\sqrt{5}}, 0, \dfrac{2}{\sqrt{5}}$

17. (i) $\dfrac{1}{\sqrt{14}}, \dfrac{2}{\sqrt{14}}, \dfrac{3}{\sqrt{14}}$ **(ii)** $\dfrac{1}{\sqrt{3}}, \dfrac{1}{\sqrt{3}}, \dfrac{1}{\sqrt{3}}$ **(iii)** $\dfrac{2}{\sqrt{6}}, \dfrac{1}{\sqrt{6}}, -\dfrac{1}{\sqrt{6}}$

19. The components of $\mathbf{u}_1 - \mathbf{u}_2$ give the numbers of faculty, students and administrators at the first university minus those at the second. The components of $(\mathbf{u}_1 + \mathbf{u}_2 + \mathbf{u}_3)/3$ give the average numbers of faculty, students and administrators.

21. Let k be the proportionality factor. Then the five forces are:

$\mathbf{F}_1 = k\,\overrightarrow{OA_1} = ka\left(\frac{1}{2}\mathbf{i} - \frac{\sqrt{3}}{2}\mathbf{j}\right)$,

$\mathbf{F}_2 = k\,\overrightarrow{OA_2} = ka\left(\frac{3}{2}\mathbf{i} - \frac{\sqrt{3}}{2}\mathbf{j}\right)$,

$\mathbf{F}_3 = k\,\overrightarrow{OA_3} = ka\,2\mathbf{i}$,

$\mathbf{F}_4 = k\,\overrightarrow{OA_4} = ka\left(\frac{3}{2}\mathbf{i} + \frac{\sqrt{3}}{2}\mathbf{j}\right)$

$\mathbf{F}_5 = k\,\overrightarrow{OA_5} = ka\left(\frac{1}{2}\mathbf{i} + \frac{\sqrt{3}}{2}\mathbf{j}\right)$.

The resultant force is

$\mathbf{R} = \sum_{n=1}^{5} \mathbf{F}_n = ka\,6\mathbf{i}$.

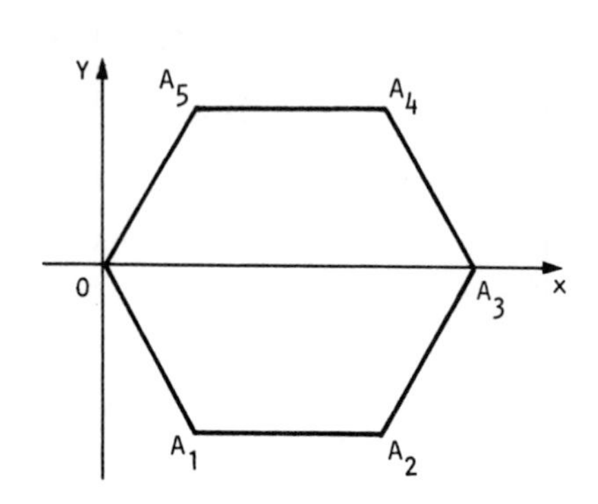

23. At an angle $\alpha = \sin^{-1}(P/U\sqrt{2})$ to the north of east;
$V/\sqrt{2} + (U^2 - V^2/2)^{1/2}$ if $U > V$; $V/\sqrt{2} \pm (U^2 - V^2/2)^{1/2}$
if $V > U \geq V/\sqrt{2}$; only if $U > V/\sqrt{2}$;
return flight is possible if $U > V$.
Hint: Let P be the actual speed and α the direction, then $U^2 = P^2 + V^2 - \sqrt{2}PV$ and $U \sin\alpha = P/\sqrt{2}$.

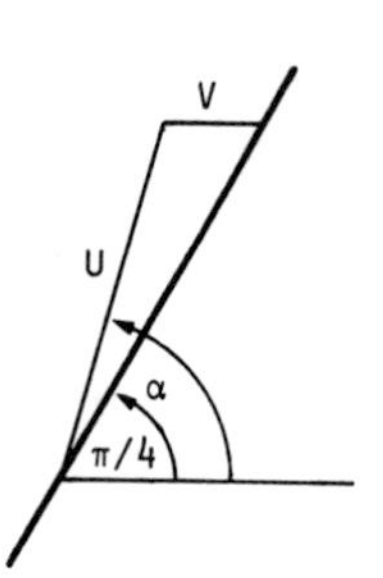

Exercises 5.2

1. $\frac{dr}{dt} = -\sin t\,\mathbf{i} + \cos t\,\mathbf{j} + \mathbf{k},\ \frac{d^2\mathbf{r}}{dt^2} = -\cos t\,\mathbf{i} - \sin t\,\mathbf{j},\ \left|\frac{d^2\mathbf{r}}{dt^2}\right| = 1.$

3. **(i)** $2t\,\mathbf{i} + 3t^2\mathbf{j} + 4t^3\mathbf{k},\ \frac{t^3}{3}\mathbf{i} + \frac{t^4}{4}\mathbf{j} + \frac{t^5}{5}\mathbf{k}$

(ii) $e^t(1+t)\mathbf{i} + e^{-t}(1-t)\mathbf{j} + 2t\,\mathbf{k},\ (te^t - e^t + 1)\mathbf{i} + (1 - te^{-t} - e^{-t})\mathbf{j} + \frac{t^3}{3}\mathbf{k}$

(iii) $\sin 2t\,\mathbf{i} - \sin 2t\,\mathbf{j},\ \left(\frac{t}{2} - \frac{\sin 2t}{4}\right)\mathbf{i} + \left(\frac{\sin 2t}{4} + \frac{t}{2}\right)\mathbf{j} + t\,\mathbf{k}$

(iv) $(1 + \log t)\mathbf{i} + \mathbf{j} + \mathbf{k},\ \left(\frac{t^2}{2}\log t - \frac{t^2}{4}\right)\mathbf{i} + \frac{t^2}{2}\mathbf{j} + \frac{t^2}{2}\mathbf{k}$

5. $\mathbf{r}(t) = \mathbf{A}\frac{t^3}{6} + \mathbf{B}\frac{t^2}{2} + \mathbf{C}t + \mathbf{D}$; $\mathbf{C}$ and $\mathbf{D}$ are arbitrary constant vectors.

7. 80 ft/sec, 100 ft (using $g = -32$ ft/sec^2).

9. A distance $U\sqrt{2h/g}$ before the target.

11. Solve $y = (\tan\beta)x = x\tan\alpha - \frac{gx^2}{V^2}\sec^2\alpha$; $x = \frac{2V^2}{g}\frac{\tan\alpha - \tan\beta}{\sec^2\alpha}$,
$y = x\tan\beta$, the range is $\sqrt{x^2 + y^2} = x\sec\beta$; maximum range is $V^2/g(1 + \sin\beta)$ when $\tan\alpha = \tan\beta + \sec\beta$.

13. $y'' = -g - ky'$; solving $y'(t) = y'(0)e^{-kt} - \frac{g}{k}(1 - e^{-kt})$, $y(t) = \left(y'(0) + \frac{g}{k}\right)\left(\frac{1 - e^{-kt}}{k}\right) - \frac{gt}{k}$.

Time of ascent is $\frac{1}{k}\log\left(\frac{g + ky'(0)}{g}\right)$ found by solving $y'(t) = 0$.

For free fall, solve $y'' = -g + ky'$, $y(0) = h$, $y'(0) = 0$.

15. $|\mathbf{v}| = 12.1$ m/sec, $|\mathbf{a}| = 22.5$ m/sec^2, use Example 2.

17. (Use Example 4) $\quad m\frac{dv}{dt} = -c\frac{dm}{dt}$, $v = v_0 - c\log\frac{m(t)}{m(0)}$.

Exercises 5.3

1. **(i)** $\sqrt{2}, \sqrt{2}, \pi/3$ **(ii)** $17, 9, \cos^{-1}\frac{88}{153}$ **(iii)** $\sqrt{89}, 5\sqrt{2}, \pi/2$ **(iv)** $7, 3\cos^{-1}\frac{11}{21}$

3. $\frac{1}{\sqrt{3}}(\mathbf{i} + \mathbf{j} + \mathbf{k})$, $\cos^{-1}\left(\frac{1}{\sqrt{3}}\right)$

5. $\lambda = 1$

7. $\cos^{-1}\frac{a^2}{[(a^2 + b^2)(a^2 + c^2)]^{1/2}}$, $\cos^{-1}\frac{b^2}{[(b^2 + a^2)(b^2 + c^2)]^{1/2}}$, $\cos^{-1}\frac{c^2}{[(c^2 + a^2)(c^2 + b^2)]^{1/2}}$;

$\frac{1}{2}(a^2 + b^2 + c^2)^{1/2}$

9. $(2, \frac{1}{2}, 1)$, 3; $4\mathbf{r}\cdot\mathbf{r} = 4(\mathbf{r}\cdot\mathbf{a}) + 15$.

11. Any multiples of $\frac{1 \pm \sqrt{3}}{2}\mathbf{i} + \left(\frac{1 \pm \sqrt{3}}{2}\right)\mathbf{j} - \mathbf{k}$. Hint: The magnitudes of the vectors are arbitrary so that $(\lambda + 1)\mathbf{i} - \lambda\mathbf{j} - \mathbf{k}$ and $(\lambda' + 1)\mathbf{i} - \lambda'\mathbf{j} - \mathbf{k}$ are two vectors perpendicular to $\mathbf{i} + \mathbf{j} + \mathbf{k}$. The remaining conditions yield $2 + \lambda + \lambda' + 2\lambda\lambda' = 0$ (orthogonality) and $1/[(\lambda + 1)^2 + \lambda^2 + 1]^{1/2} = 1/[(\lambda' + 1)^2 + \lambda'^2 + 1]^{1/2}$ or $(\lambda - \lambda')(\lambda + \lambda' + 1) = 0$ (equal inclination to $\mathbf{k}$). Since $\lambda \neq \lambda'$, it follows that $\lambda^2 + \lambda - \frac{1}{2} = 0$.

13. An ellipsoid of revolution (or ellipse in two dimensions) with $\mathbf{a}$ and $-\mathbf{a}$ as foci.

15. **(ii)** $\mathbf{k}' = \pm\frac{1}{3}(2\mathbf{i} - \mathbf{j} - 2\mathbf{k})$. (If $\mathbf{k}' = x\mathbf{i} + y\mathbf{j} + z\mathbf{k}$, solve $x - 2y + 2z = 0$, $2x + 2y + z = 0$, $x^2 + y^2 + z^2 = 1$.)

17. $\mathbf{v}(t) = (1 - \cos t)\mathbf{i} + \sin t\,\mathbf{j}$, $\mathbf{a}(t) = \sin t\,\mathbf{i} + \cos t\,\mathbf{j}$; $\mathbf{v}(t) = 0$ when $t = 2n\pi$, maxima of $|\mathbf{v}| = (2 - 2\cos t)^{1/2}$ occur at $t = (2n + 1)\pi$; $|\mathbf{a}(t)| = 1$.

19. Use $|\overrightarrow{PP}'| = |\overrightarrow{OP}' - \overrightarrow{OP}|$ for the straight line distance and $R\theta$ where

$\cos\theta = \frac{\overrightarrow{OP}\cdot\overrightarrow{OP}'}{R^2}$ for the great circle distance.

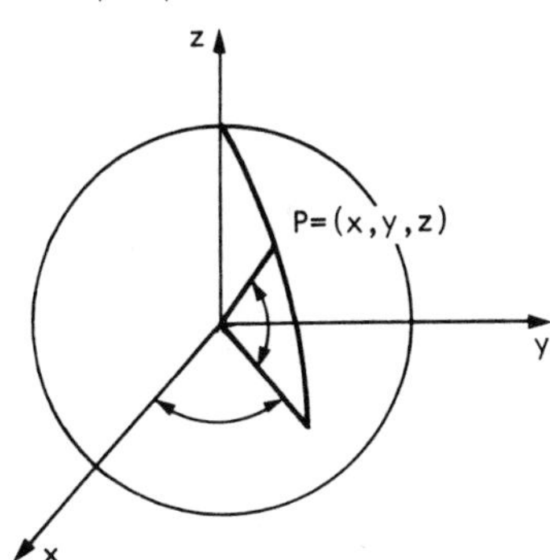

Exercises 5.4

1. **(i)** $1, \mathbf{i} - \mathbf{j} + \mathbf{k}$ **(ii)** $1, 2\mathbf{i} - 2\mathbf{k}$ **(iii)** $0, 5\mathbf{k}$ **(iv)** $0, 2\mathbf{i} - 2\mathbf{k}$
(v) $6, -19\mathbf{i} - 10\mathbf{j} + \mathbf{k}$ **(vi)** $9, -18\mathbf{i} + 26\mathbf{j} - 12\mathbf{k}$

3. $-8\mathbf{i} + 10\mathbf{j} - 2\mathbf{k}$, $12\mathbf{i} - 15\mathbf{j} + 3\mathbf{k}$, $4\mathbf{i} - 5\mathbf{j} + \mathbf{k}$; $\cos^{-1}\frac{5}{6}$, $\cos^{-1}\frac{1}{3}$, $\cos^{-1}\frac{1}{2}$.

5. Each vector is directed perpendicular to the plane containing **a**, **b** and **c**. Their magnitudes represent twice the area of the triangle whose sides are **a**, **b** and **c**.

7. $\boldsymbol{\alpha} \cdot \boldsymbol{\beta} = |\boldsymbol{\alpha}|\,|\boldsymbol{\beta}| \cos(\theta_1 - \theta_2) = \cos(\theta_1 - \theta_2) = \cos\theta_1 \cos\theta_2 + \sin\theta_1 \sin\theta_2$,
$|\boldsymbol{\alpha} \times \boldsymbol{\beta}| = |\boldsymbol{\alpha}|\,|\boldsymbol{\beta}| \sin(\theta_1 - \theta_2) = \sin(\theta_1 - \theta_2) = \sin\theta_1 \cos\theta_2 - \sin\theta_2 \cos\theta_1$.

9. The sides **a**, **b**, **c** satisfy $\mathbf{a} + \mathbf{b} + \mathbf{c} = 0$. Use Exercise 5 and $\sin A = |\mathbf{b} \times \mathbf{c}|/2$ (Area), etc.

13. $\mathbf{F} = \frac{50}{\sqrt{6}}(\mathbf{i} + 2\mathbf{j} - \mathbf{k})$; torque $= \mathbf{r} \times \mathbf{F} = \frac{50}{\sqrt{6}}(-\mathbf{i} + \mathbf{j} + \mathbf{k})$.

15. Use $\frac{d\mathbf{a}}{dt} \times \frac{d\mathbf{a}}{dt} = \mathbf{0}$

17.

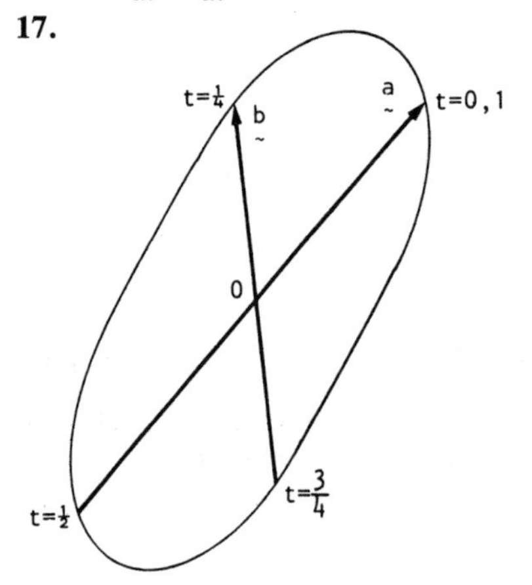

19. Torque $= \mathbf{r} \times \mathbf{F} = (a, 0) \times (0, 2P) + (0, a) \times (-3P, 0) = 5aP\mathbf{k}$; $2x - 2y + 5a = 0$.

21. $\frac{d\mathbf{\Omega}}{dt} = \mathbf{0}$. (Use the method in the text.)

Exercises 5.5

1. **(i)** 125 **(ii)** $\mathbf{i} + 13\mathbf{j} + 9\mathbf{k}$ **(iii)** 125 **(iv)** $-51\mathbf{i} - 3\mathbf{j} - 9\mathbf{k}$ **(v)** 125 **(vi)** 0
(i), (iii) and (v) represent the same volume.

3. $\mathbf{a} \cdot (\mathbf{b} \times \mathbf{c}) = 5$; area of **a**, **b** face is $|\mathbf{a} \times \mathbf{b}| = \sqrt{19}$; distance from (1, 1, 0) to the plane $3x + 3y - z = 0$ is $5/\sqrt{19}$.

5. **(i)** Check components; $(\mathbf{i} \times \mathbf{a}) \cdot \mathbf{b} = a_2b_3 - a_3b_2$, etc.
(ii) First decompose $\mathbf{b} \times (\mathbf{c} \times \mathbf{d}) = (\mathbf{b} \cdot \mathbf{d})\mathbf{c} - (\mathbf{b} \cdot \mathbf{c})\mathbf{d}$
(iii) Consider $(\mathbf{a} \times \mathbf{b}) \cdot (\mathbf{c} \times \mathbf{d})$ as the triple scalar product $\mathbf{a} \cdot (\mathbf{b} \times (\mathbf{c} \times \mathbf{d}))$ and decompose $\mathbf{b} \times (\mathbf{c} \times \mathbf{d})$ as in **(ii)**
(iv) Use result in **(ii)**

7. First decompose the triple vector product $(\mathbf{b} \times \mathbf{c}) \times (\mathbf{c} \times \mathbf{a})$ as in **5(iii)**.

9. Write $\mathbf{x} = \alpha(\mathbf{b} \times \mathbf{c}) + \beta(\mathbf{c} \times \mathbf{a}) + \gamma(\mathbf{a} \times \mathbf{b})$. Determine α, β, γ by scalar and vector multiplication by $\mathbf{a}$, etc. Result is valid $(0 = 0)$ if $\mathbf{a}$, $\mathbf{b}$, $\mathbf{c}$ coplanar.

11. Write $\mathbf{x} = \alpha\mathbf{a} + \beta(\mathbf{a} \times \mathbf{b})$ to determine $\mathbf{x} = k\mathbf{a} + \mu \dfrac{\mathbf{a} \times \mathbf{b}}{|\mathbf{a}|^2}$, $\mathbf{y} = \dfrac{1}{\mu}\left[(1 - \lambda k)\mathbf{a} - \mu\lambda \dfrac{\mathbf{a} \times \mathbf{b}}{|\mathbf{a}|^2}\right]$

where k is arbitrary. Since $\mathbf{b}$ is orthogonal to $\mathbf{x}$ and $\mathbf{y}$ and $\mathbf{a}$ is in the plane defined by $\mathbf{x}$ and $\mathbf{y}$, we must have $\mathbf{a} \cdot \mathbf{b} = 0$.

13. Compare to **12(i)**.

15. $\mathbf{r} \cdot \left(\dfrac{d\mathbf{r}}{dt} \times \dfrac{d^2\mathbf{r}}{dt^2}\right)$.

Exercises 5.6

1. **(i)** $\dfrac{x-2}{3} = \dfrac{y-3}{-4} = \dfrac{z+1}{5}$ **(ii)** $\dfrac{x-2}{2} = \dfrac{y-3}{3} = \dfrac{z+1}{-1}$ **(iii)** $\dfrac{x-2}{1} = \dfrac{y-3}{1} = \dfrac{z+1}{1}$

3. **(i)** $\dfrac{x-1}{2} = \dfrac{z-3}{-2}$, $y = 2$ **(ii)** $\dfrac{x-1}{1} = \dfrac{y}{7} = \dfrac{z+1}{5}$ **(iii)** $\dfrac{x-2}{3} = \dfrac{y-1}{2} = \dfrac{z}{-1}$

5. **(i)** $x - 2y + z = 0$ **(ii)** $7x - 6y + 7z = 0$ **(iii)** $x - 2y + z = 0$ **(iv)** $x - y = 0$
(v) $y - z = 0$ **(vi)** $x - 2y + z = 0$

7. $y_1 = -7/2$, $z_1 = -3/2$; $\dfrac{x-1}{1} = \dfrac{y + 7/2}{-13/2} = \dfrac{z + 3/2}{-7/2}$

9. $\dfrac{x}{a} + \dfrac{y}{b} + \dfrac{z}{c} = 1$, $d = (a^{-2} + b^{-2} + c^{-2})^{-1/2}$; $ax = by = cz$ (normal line).

11. $5x + 2y + z = 8$. (Use the cross product $(1, -1, -3) \times (1, -3, 1) = (-10, -4, -2)$.)

13. At $(1, -1, 3)$, $5x + 2y - 7z + 9 = -9$; at $(3, 3, 3)$, $5x + 2y - 7z + 9 = 9$; $\dfrac{x-1}{2} = \dfrac{y+1}{4}$,

$z = 3$; $(2, 1, 3)$.

15. Set $\dfrac{x-a}{\cos\alpha} = \dfrac{y-b}{\cos\beta} = \dfrac{z-c}{\cos\gamma} = t$;
intersection with sphere implies $(a + t\cos\alpha)^2 + (b + t\cos\beta)^2 + (c + t\cos\gamma)^2 = 1$.
Condition for unique solution is discriminant $= 0$ or
$(a\cos\alpha + b\cos\beta + c\cos\gamma)^2 = a^2 + b^2 + c^2 - 1$.

17. $(4 - (\alpha + 2\beta + 2\gamma)/3,\ 2 - (2\alpha - 2\beta + \gamma)/3,\ 2 - (2\alpha + \beta - 2\gamma)/3)$;
$\dfrac{x - 1/3}{4} = \dfrac{y - 5/3}{2} = \dfrac{z - 8/3}{-13}$. The line through (α, β, γ) perpendicular to the plane is $x = \alpha + 2s$, $y = \beta + s$, $z = \gamma + s$. For $s = 1 - (2\alpha + \beta + \gamma)/6$, the line intersects the plane and twice this value of s gives the mirror image. For the second part, find either the images of any two points on the line or the image of the general point to determine the image line.

19. $(\mathbf{r} - \mathbf{r}_3) \cdot (\mathbf{r}_3 - \mathbf{r}_1) \times \mathbf{v}_1 = (\mathbf{r} - \mathbf{r}_3) \cdot (\mathbf{r}_3 - \mathbf{r}_2) \times \mathbf{v}_2 = 0$ or
$\mathbf{r} = \mathbf{r}_3 + s\,\{(\mathbf{r}_3 - \mathbf{r}_1) \times \mathbf{v}_1]\} \times \{(\mathbf{r}_3 - \mathbf{r}_2) \times \mathbf{v}_2\}$. The plane through $\mathbf{r}_3$ that contains the first line has $(\mathbf{r}_3 - \mathbf{r}_1) \times \mathbf{v}_1$ as its normal and the plane through $\mathbf{r}_3$ that contains the second line has $(\mathbf{r}_3 - \mathbf{r}_2) \times \mathbf{v}_2$ as its normal. These two planes determine the required line. Alternatively, since we know that the vectors $(\mathbf{r}_3 - \mathbf{r}_1) \times \mathbf{v}_1$ and $(\mathbf{r}_3 - \mathbf{r}_2) \times \mathbf{v}_2$ are both perpendicular to the desired line, its direction is found by taking the cross product. If $\mathbf{v}_1 = \mathbf{v}_2$, the problem has a solution if $\mathbf{r}_3$ is in the plane formed by the two lines.

21. **(i)** $\cos\alpha = \frac{4}{\sqrt{41}}, \cos\beta = \frac{3}{\sqrt{41}}, \cos\gamma = \frac{4}{\sqrt{41}}$

(ii) $\frac{x-1}{0} = \frac{y-1}{1} = \frac{z-1}{1}, \alpha = \frac{\pi}{2}, \beta = \gamma = 0.$

23. Use scalar multiplication by $\mathbf{b} \times \mathbf{c}$, etc.

Exercises 5.7

1. $\hat{\mathbf{T}} = \left(\frac{2}{7}, -\frac{3}{7}, \frac{6}{7}\right)$; curve is a straight line, $s(t) = 7t$.

3. $s(t) = \int_0^t (a^2\,\omega^2 \sin^2\omega t + a^2\,\omega^2 \cos^2\omega t + b^2)^{1/2}\,dt = t(a^2\,\omega^2 + b^2)^{1/2}$

5. **(i)** $y^2 = 4ax$, $\hat{\mathbf{T}} = \left(\frac{t}{\sqrt{t^2+1}}, \frac{1}{\sqrt{t^2+1}}\right)$, $\hat{\mathbf{N}} = \left(-\frac{1}{\sqrt{t^2+1}}, \frac{t}{\sqrt{t^2+1}}\right)$

(ii) $\frac{x^2}{a^2} + \frac{y^2}{b^2} = 1$, $(a^2 \sin^2 t + b^2 \cos^2 t)^{-1}\,(-a \sin t,\, b \cos t)$,
$(a^2 \sin^2 t + b^2 \cos^2 t)^{-1}\,(-b \cos t,\, -a \sin t)$

(iii) $\frac{x^2}{a^2} - \frac{y^2}{b^2} = 1$, $(a^2 \sinh^2 t + b^2 \cosh^2 t)^{-1}\,(a \sinh t,\, b \cosh t)$,
$(a^2 \sinh^2 t + b^2 \cosh^2 t)^{-1}\,(-b \cosh t,\, a \sinh t)$.

7. **(i)** $\sqrt{2}$ **(ii)** ½ **(iii)** $\frac{5}{3}\sqrt{10}$ **(iv)** b^2/a **(v)** $3\sqrt{2}$ **(vi)** $2\sqrt{2}$

9. $\frac{1}{2}, \sqrt{2}$. (Solve for $x = -\frac{y}{2} \pm \frac{y}{2}(1-4y)^{1/2}$ so that near the origin $x \approx -y^2$ or $x \approx -y + y^2$.)

11. $\hat{\mathbf{T}} = \frac{1}{2+u^2}(2, 2u, u^2)$, $s = \int_0^1 (2+u^2)\,du = \frac{7}{3}$.

13. Kinetic energy + potential energy $= \frac{1}{2} m\,(a^2 + b^2)\left(\frac{d\theta}{dt}\right)^2 + mgb\theta = 2\pi mgb$. Solve for $\frac{d\theta}{dt}$; integration with $\theta = 2\pi$ at $t = 0$ gives desired result.

15. $\hat{\mathbf{T}} = \left(\sec\frac{x}{c}\right)^{-1}\left(1, \tan\frac{x}{c}\right)$, $\hat{\mathbf{N}} = \left(\sec\frac{x}{c}\right)^{-1}\left(-\tan\frac{x}{c}, 1\right)$; $\kappa = (c \sec\frac{x}{c})^{-1}$;

$s = c \log\left[\sec\frac{x}{c} + \tan\frac{x}{c}\right] = c \log[\sec\phi + \tan\phi]$ since $\tan\phi = \frac{dy}{dx} = \tan\frac{x}{c}$.

17. Eliminate x from series expansions of $s = \int_0^x \left(1 + \left(\frac{dy}{dx}\right)^2\right)^{1/2} dx$, $\phi = \tan^{-1}\frac{dy}{dx}$.

19. From the figure, $x' = x - R \sin\phi$, $y' = y + R \cos\phi$. But $R = ds/d\phi$ and $\cos\phi = dx/ds$, $\sin\phi = dy/ds$ which leads to the result.

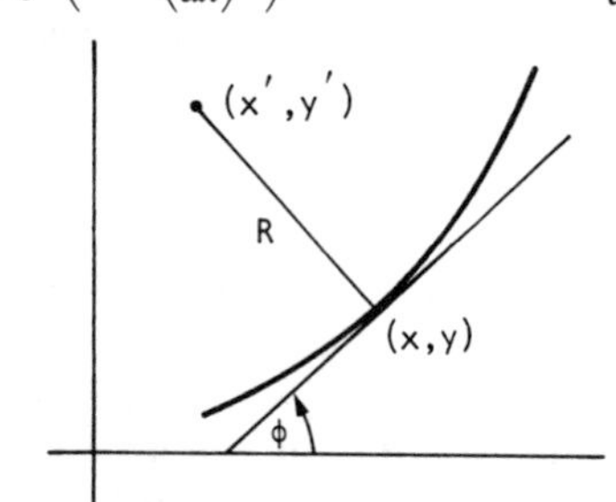

23. Use θ as the parameter in $x = r\cos\theta = f(\theta)\cos\theta$, $y = f(\theta)\sin\theta$.

25. **(i)** $s = \dfrac{a}{2}\left[\pi(\pi^2+1)^{1/2} + \log(\pi + (\pi^2+1)^{1/2})\right]$, $\kappa = \dfrac{1}{a(\theta^2+2)^{1/2}}$

$\left(\text{use } \left(\dfrac{ds}{d\theta}\right)^2 = \left(\dfrac{dr}{d\theta}\right)^2 + r^2\right)$

(ii) $8a$, $\dfrac{3}{4a}\sec\left(\dfrac{\theta}{2}\right)$ **(iii)** $\dfrac{\sqrt{1+(\log a)^2}}{\log a}(a^{\pi}-1)$, $a^{-\theta}$

27. The sum of the kinetic and potential energies must remain constant during the motion $mgf(x_1) = mgf(x) + \dfrac{1}{2}mv^2$.

29. $(1+4\varepsilon)/(1+\varepsilon)^2$

31. Use Exercise 17. Near $x = 0$, $y = \dfrac{f''(0)}{2}x^2 + \cdots$ and $s = f''(0)\phi + \cdots$.

33. Equations (i) and (ii) follow from direction and speed considerations respectively; (iii) is obtained by eliminating t between (i) and (ii). For the last part of the problem, find y for $x = 0$.

35. 20 miles north, 45 miles south. This result is based on use of the Lagrange interpolation formula for $\mathbf{r}(t)$. Note $\mathbf{r}(0)$, $\mathbf{r}(1)$, $\mathbf{r}(2)$, $\mathbf{r}(3)$ and $\mathbf{r}(4)$ are known and it is required to approximate $\mathbf{r}(5)$.

Exercises 5.8

1. $\mathbf{v}(t) = \dot{r}\hat{\mathbf{r}} + r\dot{\theta}\hat{\boldsymbol{\theta}} = 2t\hat{\mathbf{r}} + t^2\hat{\boldsymbol{\theta}}$; $\mathbf{v}(1) = 2\hat{\mathbf{r}} + \hat{\boldsymbol{\theta}}$, $\mathbf{v}(2) = 4\hat{\mathbf{r}} + 4\hat{\boldsymbol{\theta}}$;
$\mathbf{a}(t) = (\ddot{r} - r\dot{\theta}^2)\hat{\mathbf{r}} + (r\ddot{\theta} + 2\dot{r}\dot{\theta})\hat{\boldsymbol{\theta}} = (2-t^2)\hat{\mathbf{r}} + 4t\hat{\boldsymbol{\theta}}$; $\mathbf{a}(1) = \hat{\mathbf{r}} + 4\hat{\boldsymbol{\theta}}$, $\mathbf{a}(2) = -2\hat{\mathbf{r}} + 8\hat{\boldsymbol{\theta}}$.

3. Combine equations (5) and (6).

5. Substitute $u = e^{-\theta}$ in equations (17) and (15).

7. The equation of an ellipse $\dfrac{x^2}{a^2} + \dfrac{y^2}{b^2} = 1$ becomes $u^2 = \dfrac{\cos^2\theta}{a^2} + \dfrac{\sin^2\theta}{b^2}\cdot$ Using (17), the result follows.

9. 1.881 years (Use (24).)

11. 57 km. Use (24) to calculate a and then subtract the radius of the earth to obtain the altitude.

13. 35,840 km. (Use (24).)

15. 84.4 minutes.

17. Use equation (15) to find $\dfrac{1}{2}v^2 = \dfrac{\mu}{r} + \dfrac{1}{2}v_0^2 - \dfrac{\mu}{r_0}\cdot$ Since $v = \dfrac{dr}{dt}$, find $\dfrac{dt}{dr}$ and integrate over one period.

19. Identify $p = b$ in the hyperbola $\dfrac{x^2}{a^2} - \dfrac{y^2}{b^2} = 1$ and note that $b^2 = la = a^2(k-1)$. Use the facts $\cot\dfrac{\delta}{2} = \dfrac{b}{a} = \dfrac{l}{p}$,

$h^2 = V_\infty^2 p^2 = \mu l$, $\mu = \dfrac{2qq'}{m}\cdot$

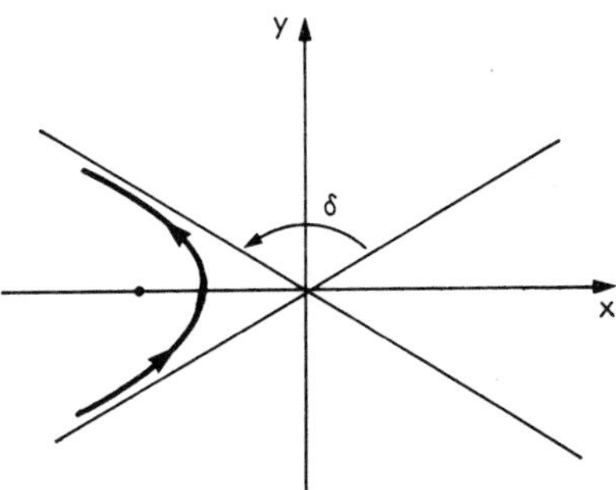

Chapter 6 Partial Differentiation

Exercises 6.1

1. **(i)** $(1, 0, 1)$, $\sqrt{10}$ **(ii)** $(-\tfrac{1}{2}, -\tfrac{1}{2}, -\tfrac{1}{2})$, $\sqrt{3}/2$ **(iii)** $(4, 5, -1)$, $\sqrt{42}$
(iv) $(-\tfrac{3}{2}, -1, 0)$, $\sqrt{37}/2$

3. **(i)** $4x'^2 + 9y'^2 - z'^2 = 40$ where $x' = x + 1$, $y' = y - 2$, $z' = z - 1$, hyperboloid of one sheet
(ii) $x'^2 + 4y'^2 - 2z' = 0$ where $x' = x + 1$, $y' = y - \tfrac{1}{2}$, $z' = z + 7$, cone
(iii) $x'^2 + 2y'^2 + 4z'^2 = 14$ where $x' = x - 1$, $y' = y - 1$, $z' = z + \tfrac{1}{2}$, hyperboloid of one sheet
(iv) $x'^2 + y'^2 - z'^2 = 1$ where $x' = x - 1$, $y' = y - 1$, $z' = z - 1$, hyperboloid of one sheet.

5. The eight planes $\pm x \pm y \pm z = 1$ form the faces of the octahedron. The six vertices are $(\pm 1,0,0)$, $(0, \pm 1,0)$, $(0,0, \pm 1)$.

7. 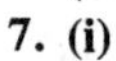
(i)

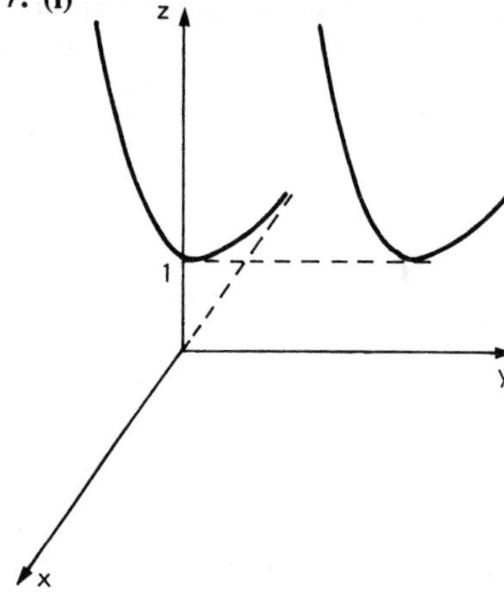

(ii)

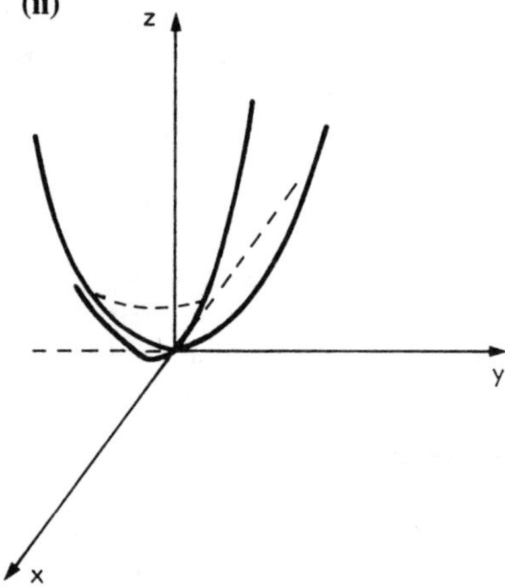

(iii)

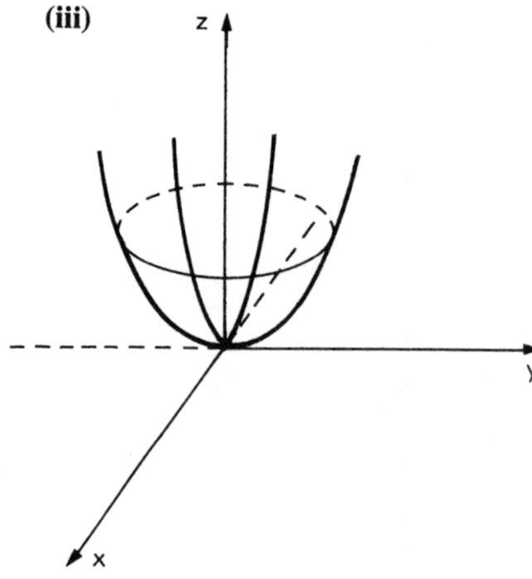

(iv)

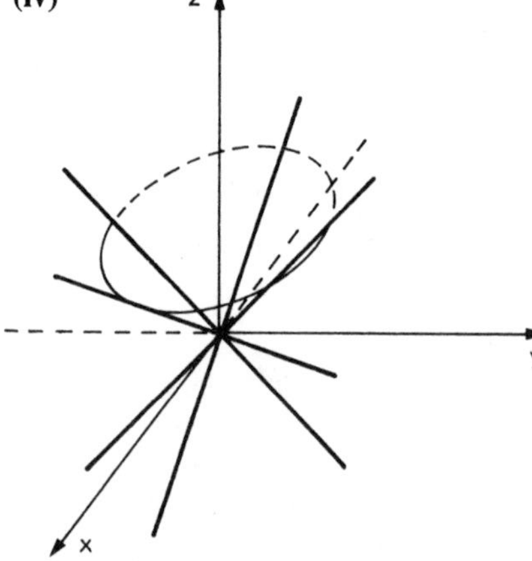

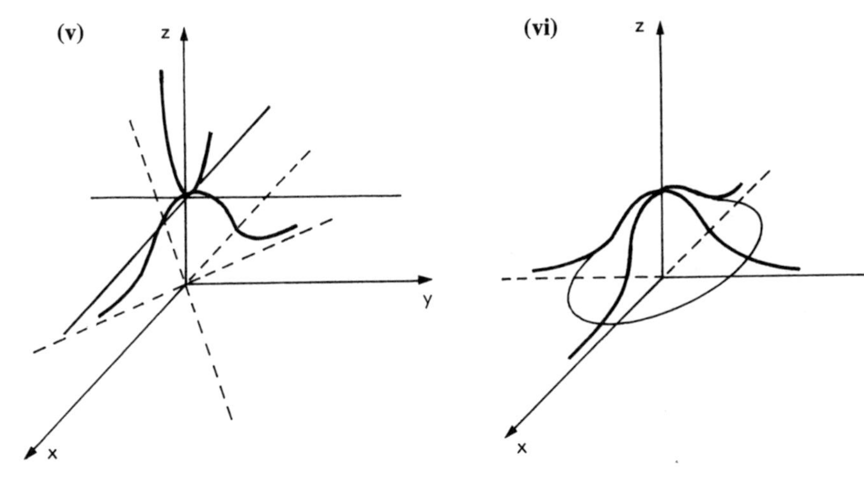

9. $z^2 = x^2 + y^2$
11. $y^2 + z^2 = \sin^2 x$.
13. $x + y + ((y-x)^2 + 2z^2)^{1/2} = \frac{1}{2}[x + y - ((y-x)^2 + 2z^2)^{1/2}]^2$. Hint: First rotate axes in the xy plane so that $y = x$, $z = 0$ becomes the new x-axis.
15. A family of circles centered at $(p,0)$ of radius $\sqrt{p^2 - 1}$.
17. Level curves are ellipses $x^2 + 2y^2 =$ constant. Along $y = x$, $h(x, x) = 1 + 3x^2\,e^{-3x^2}$. Maximum height is $1 + e^{-1} = 1.368$ km.
19. **(i)** $f(x, mx) = \dfrac{m}{1+m^2}$, $\lim\limits_{x\to 0} f(x, mx) = \dfrac{m}{1+m^2}$, no.
(ii) $f(x, mx) = x\,\dfrac{1-m^2}{1+m^2}$, $\lim\limits_{x\to 0} f(x, mx) = 0$, yes.
21. $f(x, y, z)$ is continuous at (x_0, y_0, z_0) if $f(x_0, y_0, z_0)$ is finite and $\lim\limits_{x,y,z\to x_0,y_0,z_0} f(x, y, z) = f(x_0, y_0, z_0)$ where x, y and z approach their limits independently.

Exercises 6.2

1. **(i)** $3x^2y - 4xy^4,\ x^3 - 8x^2y^3$ **(ii)** $(1-2x^2)e^{-x^2-2y^2},\ -4xye^{-x^2-2y^2}$
(iii) $-(x+2y)^{-2},\ -2(x+2y)^{-2}$ **(iv)** $1/x,\ 1/y$ **(v)** $yx^{y-1},\ x^y \log x$
(vi) $\dfrac{e^x(x-y-1)+e^y}{(x-y)^2},\ \dfrac{e^y(y-x-1)+e^x}{(x-y)^2}$ **(vii)** $-6x - 5y,\ 4y - 5x$
(viii) $(2x+y)\sec^2(x^2+xy+y^2),\ (2y+x)\sec^2(x^2+xy+y^2)$
(ix) $y\,\mathrm{sgn}\,x + |y|,\ |x| + x\,\mathrm{sgn}\,y$.
3. **(i)** $f_x = 2x, f_{xx} = 2, f_y = 4y, f_{yy} = 4, f_{xy} = 0$, etc.
(ii) $f_x = y + z, f_y = x + z, f_z = y + x, f_{xx} = f_{yy} = f_{zz} = 0, f_{xy} = f_{yx} = f_{yz} = \cdots = 1$
(iii) $f_x = (10)(x+1)^9(y-2)^2(z+3)^3, f_{xx} = 90(x+1)^8(y-2)^2(z+3)^3$,
$f_{xy} = 20(x+1)^9(y-2)(z+3)^3$

(iv) $f_x = yz \sin xyz + xy^2z^2 \cos xyz$, $f_{xx} = 2y^2z^2 \cos xyz - xy^3z^3 \sin xyz$

(v) $f_x = e^{x+2y} \log(3y - x) - \dfrac{e^{x+2y}}{3y - x}$, $f_{xx} = e^{x+2y} \log(3y - x) - 2\dfrac{e^{x+2y}}{3y - x} + \dfrac{e^{x+2y}}{(3y - x)^2}$

(vi) $f_x = 3z^2x^2 + 3x^2y^4$, $f_{xx} = 6z^2x + 6xy^4$, $f_{xy} = 12x^2y^3 = f_{yx}$

5. $r_x = x(x^2 + y^2)^{-1/2}$, $r_y = y(x^2 + y^2)^{-1/2}$, $\theta_x = -y(x^2 + y^2)^{-1}$, $\theta_y = x(x^2 + y^2)^{-1}$,
$r_{xx} = y^2(x^2 + y^2)^{-3/2}$, $r_{xy} = -xy(x^2 + y^2)^{-3/2}$, $r_{yy} = x^2(x^2 + y^2)^{-3/2}$,
$\theta_{xx} = -2xy(x^2 + y^2)^{-2}$, $\theta_{xy} = (y^2 - x^2)(x^2 + y^2)^{-2}$, $\theta_{yy} = -2xy(x^2 + y^2)^{-2}$

7. $r_x = x(x^2 + y^2 + z^2)^{-1/2}$, $r_{xx} = (y^2 + z^2)(x^2 + y^2 + z^2)^{-3/2}$, $r_{xy} = -xy(x^2 + y^2 + z^2)^{-3/2}$;
$\theta_x = -y(x^2 + y^2)^{-1}$, $\theta_y = x(x^2 + y^2)^{-1}$, $\theta_z = 0$, $\theta_{xx} = 2xy(x^2 + y^2)^{-2}$;
$\phi_x = -xz(x^2 + y^2)^{-1/2}(x^2 + y^2 + z^2)^{-1}$, $\phi_y = -yz(x^2 + y^2)^{-1/2}(x^2 + y^2 + z^2)^{-1}$.

9. (i) $z_x = -y(x^2 + y^2)^{-1}$, $z_y = x(x^2 + y^2)^{-1}$; $z_{xx} = 2xy(x^2 + y^2)^{-2}$, $z_{yy} = -2xy(x^2 + y^2)^{-2}$.
Conclude that $z_{xx} + z_{yy} = 0$.

11. $G_p = p(p^2 + q^2 + r^2)^{-1/2}$, $G_{pp} = (q^2 + r^2)(p^2 + q^2 + r^2)^{-3/2}$, etc.

13. From Exercise 12, u_x is a function of y only, $u_x = F(y)$.
Therefore, $\dfrac{\partial}{\partial x}(u - xF(y)) = u_x - F(y) = 0$ and $u - xF(y)$ is a function of y only;
$u(x, y) = xF(y) + G(y)$.

15. Differentiate numerator and denominator with respect to h (or k) and set $h = 0$.
(i) $\lim_{h\to 0} \dfrac{f(x + h, y) - f(x - h, y)}{2h} = \dfrac{f_x(x, y) - (-f_x(x, y))}{2} = f_x(x, y)$.

17. (i) $\dfrac{2x_2}{x_3 + 2x_4}$ (ii) $\dfrac{4x_1x_2^2}{(x_3 + 2x_4)^2}$ (iii) $\dfrac{8x_2}{(x_3 + 2x_4)^3}$

Exercises 6.3

1. $f(1 + \Delta x, 1 + \Delta y) - f(1, 1) \approx f_x\Delta x + f_y\Delta y = 2\Delta x - 4\Delta y$; -1.08, -1.06; accurate values -1.0814, -1.0601.

3. (i) $\theta_1 = \theta_2 = 1/2$
(ii) Solve $e^{1/6} - 1 = \frac{1}{2}e^{\theta_1/2} - \frac{1}{3}e^{-\theta_2/3}$ (θ_1, θ_2 not unique)
(iii) $\theta_1 = \theta_2$ arbitrary

5. (i) $\sin y\, dx + x \cos y\, dy$
(ii) $2e^{x^2+y^2}[x(1 + x^2 - y^2)dx + y(x^2 - y^2 - 1)dy]$
(iii) $\left[\dfrac{y}{1+x} + \log(1 + y)\right]dx + \left[\log(1 + x) + \dfrac{x}{1+y}\right]dy$
(iv) $6(x^3 + y^3)(x^2dx + y^2dy)$
(v) $4 \sec^2(x^2 + 2y^2) \tan(x^2 + 2y^2) \{x\, dx + 2y\, dy\}$
(vi) $e^{x+y} \{[\cos(x - y) - \sin(x - y)]dx + [\cos(x - y) + \sin(x - y)]dy\}$

7. (i) $df = -\dfrac{y}{x^2}dx + \dfrac{1}{x}dy$ (ii) $ye^{xy}dx + xe^{xy}dy$ (iii) $2x\, dx + 2y\, dy + 2z\, dz$
(iv) $3(x + y + z)^2(dx + dy + dz)$ (v) $\sinh(x + y - z)(dx + dy - dz)$
(vi) $(y + z)(z + 2x + y)dx + (z + x)(x + 2y + z)dy + (x + y)(y + 2z + x)dz$

9. $\Delta V \approx \dfrac{\partial V}{\partial x}\Delta x + \dfrac{\partial V}{\partial y}\Delta y + \dfrac{\partial V}{\partial z}\Delta z = yz\Delta x + xz\Delta y + xy\Delta z$. When $x = 2$, $y = 3$, $z = 5$ and $\Delta x = \Delta y = \Delta z = 0.1$, $\Delta V \approx 0.31\ \text{m}^3$. Accurate value is $0.311001\ \text{m}^3$.

11. **(i)** 4.988 **(ii)** 9.0225 **(iii)** 0.982
(Consider the functions (i) $\sqrt{x^2 + y^2}$, (ii) $x^{1/2} + y^{1/3} + z^{1/4}$ and (iii) x^3/y.)

13. $V = \pi r^2 h$, $\Delta V = 2\pi r h\ \Delta r + \pi r^2 \Delta h$, volume of metal required is $\Delta V + \pi r^2 \Delta h = 192\pi/5\ \text{m}^3$. (Extra term $\pi r^2 \Delta h$ is volume of metal in base.) No.

15. Use Exercise 14 with $f(x, y) = xy - a^2 = 0$ and $g(x, y) = x^2 - y^2 - b^2 = 0$.

17. **(i)** $\dfrac{y - x^2}{y^2 - x}$ **(ii)** $\dfrac{y(x \log y - y)}{x(y \log x - x)}$ **(iii)** $-\cot x \coth y$ **(iv)** $\dfrac{y \tan x + \log \sin y}{\log \cos x - x \cot y}$

(v) $-\dfrac{(2x + y + 1)}{x + 2y + 1}$ **(vi)** $\left[e^{\tan^{-1} y} - \dfrac{ye^x}{1 + e^{2x}}\right] \div \left[\tan^{-1} x - \dfrac{xe^{\tan^{-1} y}}{1 + y^2}\right]$

(vii) $\dfrac{-4x}{y(x^2 - 4)^2}$ **(viii)** $\dfrac{2x - ye^{xy}}{xe^{xy} - 2y}$ **(ix)** $\dfrac{1 + \cos^2(y - x)}{\cos^2(y - x) - 2y}$

19. Tangent plane through $(x_0, y_0, x_0 e^{x_0/y_0})$ is

$$z - x_0 e^{x_0/y_0} = e^{x_0/y_0}\left[\left(1 + \frac{x_0}{y_0}\right)(x - x_0) - \frac{x_0^2}{y_0^2}(y - y_0)\right] \text{ or } z = e^{x_0/y_0}\left[\left(1 + \frac{x_0}{y_0}\right)x - \frac{x_0^2}{y_0^2}y\right],$$

a plane through origin. The surface is a cone with vertex at origin.

21. $(1/2, 1/2, 1/2)$. Minimize $(x - 1)^2 + (y - 1)^2 + z^2 = (x - 1)^2 + (y - 1)^2 + (x^2 + y^2)^2$.

23. Tangent plane at $\left(x_0, y_0, \dfrac{1}{x_0 y_0}\right)$ is $z - \dfrac{1}{x_0 y_0} = -\dfrac{1}{x_0^2 y_0}(x - x_0) - \dfrac{1}{x_0 y_0^2}(y - y_0)$ or

$z = \dfrac{3x_0 y_0 - y_0 x - x_0 y}{x_0^2 y_0^2}$. Intersections with axes are $\left(0, 0, \dfrac{3}{x_0 y_0}\right)$, $(3x_0, 0, 0)$ and $(0, 3y_0, 0)$.

Volume of tetrahedron $= \dfrac{1}{6}\, 27 \dfrac{x_0 y_0}{x_0 y_0} = \dfrac{27}{6}$.

25. **(i)** Use polar coordinates **(ii)** Calculate directly

(iii) $f_{xy}(0, 0) = \lim\limits_{y \to 0} \dfrac{f_x(0, y) - f_x(0, 0)}{y} = -1$, $f_{yx}(0, 0) = \lim\limits_{x \to 0} \dfrac{f_y(x, 0) - f_y(0, 0)}{x} = 1$

Exercises 6.4

1. On curve $f(x, y) = f(1, 1)$, the tangent direction is given by $(f_y, -f_x)$ which is orthogonal to (f_x, f_y).

(i) $(2xy + y^2)\mathbf{i} + (x^2 + 2xy)\mathbf{j}$ **(ii)** $\cos x \sin y\ \mathbf{i} + \sin x \cos y\ \mathbf{j}$

(iii) $\dfrac{x}{(x^2 + y^2 - 1)^{1/2}}\mathbf{i} + \dfrac{y}{(x^2 + y^2 - 1)^{1/2}}\mathbf{j}$ **(iv)** $\dfrac{1}{2x^{1/2}}\mathbf{i} + \dfrac{1}{2y^{1/2}}\mathbf{j}$

(v) $\dfrac{3x^2}{x^3 + y^3}\mathbf{i} + \dfrac{3y^2}{x^3 + y^3}\mathbf{j}$ **(vi)** $2xe^{x^2 - y^2}\mathbf{i} - 2ye^{x^2 - y^2}\mathbf{j}$

3. Level curves $x^2 + y^2 = \text{constant}$; $\sqrt{2}\, e^5$ (Use equation (6)); $\pm\sqrt{20}\, e^5$

5. **(i)** The tangent plane at $(x_0, y_0, f(x_0, y_0))$ is horizontal.
(ii) The function $f(x, y)$ is a constant.
(iii) The surface $z = f(x, y)$ is a plane with normal vector $\mathbf{a}$.

7. $-1, \frac{1}{\sqrt{5}}$; $f(2, 0.1) \approx f(2, 0) + (0.1)(-1) = 0.9$, $f(2.1, 0.2) \approx f(2, 0) + \frac{1}{\sqrt{5}}\frac{\sqrt{5}}{10} = 1.1$.
(From given information, $z_x = 3$ and $z_y = -1$.)
9. $9\sqrt{2}/4$. Any direction perpendicular to $2\mathbf{i} + \mathbf{j} + 2\mathbf{k}$. (Use eqn. (17))
11. $108/\sqrt{14}$; 3, 12, 27; $(3/\sqrt{882}, 12/\sqrt{882}, 27/\sqrt{882})$
13. ∇f and ∇g are normal to the surfaces $f = 0$ and $g = 0$ respectively.
15. $-4\mathbf{i} - 6\mathbf{j}$. (Solve $\frac{dy}{dx} = \frac{f_y}{f_x} = \frac{3y}{2x}$ to obtain $y = x^{3/2}$.)
17. $\frac{dT}{ds} = \frac{2}{3}$, in direction of $(-2, 0, 6)$
19. Find the directional derivatives in the radial and circumferential directions (Section 5.8);
$f_x\mathbf{i} + f_y\mathbf{j} = f_r(r_x\mathbf{i} + r_y\mathbf{j}) + f_\theta(\theta_x\mathbf{i} + \theta_y\mathbf{j})$
$= f_r\hat{\mathbf{r}} + \frac{1}{r}f_\theta\hat{\boldsymbol{\theta}}$ since $r_x\mathbf{i} + r_y\mathbf{j} = \frac{x}{(x^2+y^2)^{1/2}}\mathbf{i} + \frac{y}{(x^2+y^2)^{1/2}}\mathbf{j} = \hat{\mathbf{r}}$

and $\theta_x\mathbf{i} + \theta_y\mathbf{j} = -\frac{x}{x^2+y^2}\mathbf{i} + \frac{y}{x^2+y^2}\mathbf{j} = \hat{\boldsymbol{\theta}}$.

21. $f_x\mathbf{i} + f_y\mathbf{j} = \frac{x}{x^2+y^2}\mathbf{i} + \frac{y}{x^2+y^2}\mathbf{j} = \frac{\mathbf{r}}{r^2}$.
23. $\mathbf{v}\cdot\nabla p = -\rho\mathbf{v}\cdot(2\boldsymbol{\Omega} \times \mathbf{v}) = 0$. Since ∇p is perpendicular to isobars, $\mathbf{v}$ is parallel.

Exercises 6.5

1. **(i)** $4t - \frac{4}{t^3}$ **(ii)** $\frac{2t^3(2+t^2)}{(1+t^2)^2}$ **(iii)** $\left(\frac{2x}{t} - 4yt\right)e^{x^2-y^2}$
(iv) 0 **(v)** $(2e^t - 3e^{-t})\sin(2x + 3y)$ **(vi)** $4t$
3. $\frac{\partial z}{\partial r} = 22r + 2s$, $\frac{\partial z}{\partial s} = 2r + 6s$, $\frac{\partial^2 z}{\partial r^2} = 22$, $\frac{\partial^2 z}{\partial r \partial s} = 2$, $\frac{\partial^2 z}{\partial s^2} = 6$.
5. Use $\mathbf{v} = \left(\frac{dx}{dt}, \frac{dy}{dt}, \frac{dz}{dt}\right)$ and $\frac{dz}{dt} = f_x\frac{dx}{dt} + f_y\frac{dy}{dt}$.
7. **(i)** $r = (x+y)/2$, $s = (x-y)/2$ **(ii)** $u = 2x - 3y$, $v = 2y - x$
(iii) $r = (x^2+y^2)^{1/2}$, $\theta = \tan^{-1}(y/x)$ **(iv)** $u = (x^2-y^2)^{1/2}$, $v = \tanh^{-1}(y/x)$
(v) $u = \frac{2x}{x^2+y^2}$, $v = -\frac{2y}{x^2+y^2}$ **(vi)** $u = [x + (x^2+y^2)^{1/2}]^{1/2}$, $v = \frac{y}{[x + (x^2+y^2)^{1/2}]^{1/2}}$.
9. Let $r = y - z$, $s = z - x$, $t = x - y$ and use the chain rule.
11. Use the chain rule $f_r = f_x x_r + f_y y_r$, etc.
13. To evaluate $\partial f/\partial y$, x is held constant, to evaluate $\partial f/\partial q$, p is held constant.
15. The differential operators $\frac{\partial}{\partial x}$ and $\frac{\partial}{\partial t}$ are equivalent to $\frac{\partial}{\partial \xi} + \frac{\partial}{\partial \eta}$ and $c\left(\frac{\partial}{\partial \xi} - \frac{\partial}{\partial \eta}\right)$ respectively by the chain rule. Applying these operators twice,
$\frac{\partial^2 u}{\partial x^2} = \left(\frac{\partial}{\partial \xi} + \frac{\partial}{\partial \eta}\right)^2 u = \frac{\partial^2 u}{\partial \xi^2} + 2\frac{\partial^2 u}{\partial \xi \partial \eta} + \frac{\partial^2 u}{\partial \eta^2}$,
$\frac{1}{c^2}\frac{\partial^2 u}{\partial t^2} = \left(\frac{\partial}{\partial \xi} - \frac{\partial}{\partial \eta}\right)^2 u = \frac{\partial^2 u}{\partial \xi^2} - 2\frac{\partial^2 u}{\partial \xi \partial \eta} + \frac{\partial^2 u}{\partial \eta^2}$.
The general solution is the sum of two waves; $F(x - ct)$ and $G(x + ct)$ represent waves moving to the right and left respectively, both with speed c.

17. $u_{xx} + u_{yy} = u_{rr} + \dfrac{1}{r} u_r + \dfrac{1}{r^2} u_{\theta\theta}$.

19. Differentiate $\dot{x} = \dot{r} \cos\theta - r\dot{\theta} \sin\theta$, $\dot{y} = \dot{r} \sin\theta + r\dot{\theta} \cos\theta$ to obtain **(i)** and **(ii)**.
The acceleration vector $\mathbf{a}(t) = (\ddot{x}, \ddot{y})$ has radial component
$\mathbf{a}\cdot\hat{\mathbf{r}} = (\ddot{x}, \ddot{y})\cdot(\cos\theta, \sin\theta) = \ddot{x} \cos\theta + \ddot{y} \sin\theta$ and transverse component
$\mathbf{a}\cdot\hat{\boldsymbol{\theta}} = (\ddot{x}, \ddot{y})\cdot(-\sin\theta, \cos\theta) = -\ddot{x} \sin\theta + \ddot{y} \cos\theta$.

21. **(i)** $n = 3$ **(ii)** $n = 2$ **(iii)** $n = 1/3$ **(iv)** $n = 0$.

23. Repeat the procedure in Exercise 22.

25. Follow method of Exercise 24.

27. **(i)** $e^{-(x^2+c^2t^2)} \cosh(2cxt)$ **(ii)** $\dfrac{1}{c} \cos x \sin ct$.

Exercises 6.6

1. $xy = 1$ (Eliminate α between $F(x, y, \alpha) = 0$ and $F_\alpha(x, y, \alpha) = 0$)

3. **(i)** $y = xy' - (y')^2$ **(ii)** $y = xy' + 2/y'$ **(iii)** $y = xy' - (y')^3/27$
In **(i)**, envelope $y = x^2/4$ satisfies the differential equation since $y' = x/2$ and $xy' - (y')^2 = x(x/2) - (x/2)^2 = x^2/4 = y$. Similarly for **(ii)** and **(iii)**.

5. $y = 0$ and $y = 4x^3/27$.

7. Area $= \pi ab = \pi$. Family of ellipses $\dfrac{x^2}{a^2} + a^2y^2 = 1$ has envelope the straight lines $\pm x \pm y = 1$.

9. $x^2 + y^2 + z^2 - yz - zx - xy = 3/2$, a circular cylinder of radius 1 whose axis is the line $x = y = z$.

11. $x^2 + y^2 + z^2 = 1$; α, β can be determined in terms of a, b, c;
$\cos\alpha \sin\beta = a(a^2 + b^2 + c^2)^{-1/2}$, $\sin\alpha \sin\beta = b(a^2 + b^2 + c^2)^{-1/2}$,
$\cos\beta = c(a^2 + b^2 + c^2)^{-1/2}$.

13. 24,000 m.

15. $a\omega$, $y = a \sin(\pi x/l)$

17. **(i)**
$$\int_0^\infty e^{-\alpha x} \sin\beta x \, dx = \sum_{n=0}^\infty \int_0^\infty e^{-\alpha x}(-1)^n \frac{(\beta x)^{2n+1}}{(2n+1)!} dx$$
$$= \sum_{n=0}^\infty (-1)^n \frac{\beta^{2n+1}}{\alpha^{2n+2}} = \frac{\beta/\alpha^2}{1 + (\beta/\alpha)^2} = \frac{\beta}{\alpha^2 + \beta^2}$$

19. **(i)** $\dfrac{\partial}{\partial\alpha} \displaystyle\int_0^1 x^\alpha \, dx = \int_0^1 x^\alpha \log x \, dx$

(ii) $\displaystyle\int_{t=\alpha}^\beta \left[\int_0^1 x^t dx\right] dt = \int_0^1 \frac{x^\beta - x^\alpha}{\log x} dx.$

21. $\pi(n-1)!/2^n \alpha^{2n-1}$ (Differentiate $n - 1$ times.)

23. Define $u = \sqrt{\alpha}\, x$.

25. $\displaystyle\int_0^\infty x^{2n+1}e^{-\alpha x^2} dx = \frac{n!}{2\alpha^{n+1}}, \int_0^\infty x^{2n} e^{-\alpha x^2} dx = \frac{1\cdot 3\cdot 5 \cdots (2n-1)}{2^{n+1}\alpha^n} \sqrt{\frac{\pi}{\alpha}}.$

27. Use integration by parts. Solve the differential equation and use $I(\alpha, 0) = \sqrt{\pi}/2\sqrt{\alpha}$.

29. $I(x) = \dfrac{\sqrt{\pi}}{2} e^{-2x}$

31. $x(0) = x'(0) = 0.$

33. $\phi(x) = f(x) + \lambda K \sin x$ where $K = \int_0^{2\pi} \phi(\alpha) \cos\alpha \, d\alpha$. Since $\int_0^{2\pi} \sin x \cos x \, dx = 0$, it follows that $K = \int_0^{2\pi} f(\alpha) \cos \alpha \, d\alpha$.

Exercises 6.7

1. $\dfrac{\partial z}{\partial x} = 0, \dfrac{\partial z}{\partial y} = 1$; tangent plane $y - z = 0$, normal line $x = 1, y + z = 2$; $z(1.01, 1.03) \approx 1.03$.

3. $\dfrac{\partial y}{\partial z} = \dfrac{3xy - e^z}{e^y - 3zx}, \dfrac{\partial y}{\partial x} = \dfrac{3yz - e^x}{e^y - 3zx}, \dfrac{\partial x}{\partial y} = \dfrac{3xz - e^y}{e^x - 3zy}$, etc.

Tangent plane $y - y_0 = \dfrac{\partial y}{\partial x}(x - x_0) + \dfrac{\partial y}{\partial z}(z - z_0)$; normal line $\dfrac{x - x_0}{-y_x} = \dfrac{y - y_0}{1} = \dfrac{z - z_0}{-y_z}$.

5. **(i)** $2x + 2y + z = 4$; $x - 1 = y - 1 = 2z$; $z(0.98, 1.01) \approx -0.02$
(ii) $x + y = 2$; $x = y$, $z = 0$; $z(0.98, 1.01)$ cannot be estimated by differentials since $z_x = z_y = \infty$ at $(1, 1)$
(iii) $z = 0$; $x = y = 1$; $z(0.98, 1.01) \approx 0$.

7. $dr = \dfrac{d\,dx - b\,dy}{ad - bc}, ds = -\dfrac{c\,dx + a\,dy}{ad - bc}, \dfrac{\partial(x, y)}{\partial(r, s)} = ad - bc, \dfrac{\partial(r, s)}{\partial(x, y)} = \dfrac{1}{ad - bc}$.

If $ad - bc = 0$, y is a multiple of x; $\dfrac{\partial(r, s)}{\partial(x, y)}$ cannot be zero.

9. $dr = \dfrac{1}{r - s}\left(\dfrac{1}{2} dy - s\,dx\right), ds = \dfrac{1}{r - s}(r\,dx - \dfrac{1}{2} dy)$;

$\dfrac{\partial r}{\partial x} = -\dfrac{s}{r - s}, \dfrac{\partial r}{\partial y} = \dfrac{1}{2(r - s)}, \dfrac{\partial s}{\partial x} = \dfrac{r}{r - s}, \dfrac{\partial s}{\partial y} = -\dfrac{1}{2(r - s)}; \dfrac{\partial(r, s)}{\partial(x, y)} = -\dfrac{1}{2}, \dfrac{\partial(x, y)}{\partial(r, s)} = -2.$

11. $\dfrac{\partial x}{\partial r} = \dfrac{x}{x^2 + y^2}, \dfrac{\partial x}{\partial s} = \dfrac{y}{x^2 + y^2}, \dfrac{\partial y}{\partial r} = \dfrac{-y}{x^2 + y^2}, \dfrac{\partial y}{\partial s} = \dfrac{x}{x^2 + y^2}; \dfrac{\partial(x, y)}{\partial(r, s)} = \dfrac{1}{x^2 + y^2}, \dfrac{\partial(r, s)}{\partial(x, y)} = x^2 + y^2$

13. $\dfrac{\partial(x, y)}{\partial(r, s)} = x^2 + y^2, \dfrac{\partial(r, s)}{\partial(x, y)} = (x^2 + y^2)^{-1}$; circles $x^2 + y^2 = e^{2\alpha}$ are orthogonal to the lines $y = x \tan \beta$.

15. Since $\dfrac{\partial z}{\partial x} = \dfrac{F_x}{F_z}$ and $\dfrac{\partial}{\partial x} g(x, y, z(x, y)) = \dfrac{\partial g}{\partial x} + \dfrac{\partial g}{\partial z}\dfrac{\partial z}{\partial x}$, it follows that

$\dfrac{\partial}{\partial x}\left(\dfrac{\partial z}{\partial x}\right) = \dfrac{\partial}{\partial x}\left(-\dfrac{F_x}{F_z}\right) - \dfrac{F_x}{F_z}\dfrac{\partial}{\partial z}\left(-\dfrac{F_x}{F_z}\right) = \{2F_xF_zF_{xz} - F_x^2F_{zz} - F_z^2F_{xx}\}/F_z^3,$

$z_{xy} = \{F_yF_zF_{xz} + F_xF_zF_{yz} - F_xF_yF_{zz} - F_z^2F_{xy}\}/F_z^3$, $z_{yy} = \{2F_yF_zF_{yz} - F_y^2F_{zz} - F_z^2F_{yy}\}/F_z^3$

17. Use differentials; $dF = F_x dx + F_y dy + F_r dr + F_s ds = 0$, $dx = x_r dr + x_s ds$, etc.

19. Use the chain rule on terms in the Jacobian $\partial(x, y)/\partial(u, v)$; $\dfrac{\partial x}{\partial u} = \dfrac{\partial x}{\partial r}\dfrac{\partial r}{\partial u} + \dfrac{\partial x}{\partial s}\dfrac{\partial s}{\partial u}$, etc.

21. $(x - y)/(v - u)$

23. $\psi(f) = e^f/(1 + e^{2f})$

25. Use the results of Exercise 15.

27. **(i)** $\dfrac{R}{P}$, 0 **(ii)** $\dfrac{RV^2}{pV(3V-2\beta)-\alpha}$, $\dfrac{2R^2V^3}{[pV(3V-2\beta)-\alpha]^3}[3pV(V-\beta)-2\alpha]$

29. Since $x = u + (v - w)$ and $y = u - (v - w)$,

$$z = u^2 + (v-w)^2 = \left(\frac{x+y}{2}\right)^2 + \left(\frac{x-y}{2}\right)^2 = \frac{x^2+y^2}{2}$$

31. **(iv)**, **(v)** and **(vi)** follow from **(iii)**.

Exercises 6.8

1. $2 + 3(x-1) + (y-1) + 3(x-1)^2 + (x-1)(y-1) - (y-1)^2 + (x-1)^3$.
All partial derivatives of order greater than three are identically zero. Yes.

3. **(i)** $2(x-1) - 2(y-1)$ **(ii)** $1 + x + y$ **(iii)** $x + y$ **(iv)** $x + y$
(v) $-1 - 2(x-1) - 2(y-1)$ **(vi)** 1

5. **(i)** $1 - x - y + x^2 + 2xy + y^2$ **(ii)** 1 **(iii)** $1 - x^2 - 2xy - y^2$ **(iv)** $1 - \dfrac{x^2}{2} + \dfrac{y^2}{2}$
(v) xy **(vi)** $\tan^{-1}x + \tan^{-1}y = x + y + \cdots$

7. **(i)** $\dfrac{1}{2} + \dfrac{1}{4}x + \dfrac{1}{4}y + xy$, error $= 0.176$

(ii) $\dfrac{1}{4} - \dfrac{x}{8} - \dfrac{y}{8} + \dfrac{x^2}{16} - \dfrac{xy}{16} + \dfrac{y^2}{16}$, error $= 0.0194$

(iii) $1 + \dfrac{x^2}{2} + \dfrac{y^2}{2}$, error $= 0.680$.

9. $-\dfrac{1}{2}x + y - \dfrac{1}{16}x^2 + \dfrac{1}{4}xy - \dfrac{1}{4}y^2$

11. $z = -1 - \dfrac{3}{2}(y-1)$, $z = (x-1) + (y-1)$, $z = 1 - (x-1) + \dfrac{1}{2}(y-1)$. These are the tangent planes to the surface at the points $(1, 1, -1)$, $(1, 1, 0)$ and $(1, 1, 1)$ respectively.

13. **(i)** $\dfrac{5}{4} + \left(y - \dfrac{1}{2}\right) + (z-1) + x^2 + \left(y - \dfrac{1}{2}\right)^2 + (z-1)^2$

(ii) $\dfrac{9}{4} + 3x + 3\left(y - \dfrac{1}{2}\right) + 3(z-1) + x^2 + \left(y - \dfrac{1}{2}\right)^2 + (z-1)^2 + 2x\left(y - \dfrac{1}{2}\right) + 2x(z-1) + 2\left(y - \dfrac{1}{2}\right)(z-1)$

(iii) $-\dfrac{1}{4}x - \dfrac{1}{2}x^2 - x\left(y - \dfrac{1}{2}\right) - \dfrac{3}{4}x(z-1) - x^2\left(y - \dfrac{1}{2}\right) - 2x\left(y - \dfrac{1}{2}\right)(z-1) - \dfrac{1}{2}x^2(z-1) - x\left(y - \dfrac{1}{2}\right)^2 - \dfrac{x}{2}(z-1)^2 - x^2\left(y - \dfrac{1}{2}\right)(z-1) - x\left(y - \dfrac{1}{2}\right)^2(z-1) - x\left(y - \dfrac{1}{2}\right)(z-1)^2$.

15. Exact solution $x = \dfrac{1 + 2\varepsilon - \varepsilon^2}{1 - 2\varepsilon^2}$, $y = \dfrac{2 - \varepsilon - \varepsilon^2}{1 - 2\varepsilon^2}$; perturbation solution $x = 1 + 2\varepsilon + \varepsilon^2 + \cdots$, $y = 2 - \varepsilon + 3\varepsilon^2 + \cdots$

17. $y_0(x) = x$, $y_1(x) = -\sin 2x$, $y_2(x) = \sin 4x$.

19. Exact solution $x = \dfrac{2(1+\varepsilon)}{2 - \varepsilon^2}$, $y = \dfrac{2}{2 - \varepsilon^2}$;

perturbation solution is $x = 1 + \varepsilon + \dfrac{\varepsilon^2}{2} + \dfrac{\varepsilon^3}{2} + \ldots$, $y = 1 + \dfrac{\varepsilon^2}{2} + \ldots$

21. $x = \pm\frac{1}{\sqrt{2}}\left(1 - \frac{\varepsilon}{4}\right), y = \pm\frac{1}{\sqrt{2}}\left(1 + \frac{3\varepsilon}{4}\right)$

23. The second equation follows by differentiating with respect to t and equating powers of t^n. To obtain the first equation differentiate with respect to x and make use of the second equation.

Exercises 6.9

1. **(i)** $(1/2, -1/10)$ minimum **(ii)** $(-1, 1)$ min **(iii)** $(1/2, 5/4)$ max
(iv) $(-1, 0)$ saddle point **(v)** $(0, 0)$ max **(vi)** $(11/7, -15/7)$ min.
(For example, in **(i)**, $x^2 + 5y^2 - x + y = \left(x - \frac{1}{2}\right)^2 + 5\left(y + \frac{1}{10}\right)^2 - \frac{3}{10}$
(iv) $(x + y + 1)(x - y + 1) = (x + 1)^2 - y^2$.)

3. **(i)** $(1/2, 1/2, -1/2)$ minimum; $h^2 + k^2 + \ell^2 + 1/4$
(rewrite as $\left(x - \frac{1}{2}\right)^2 + \left(y - \frac{1}{2}\right)^2 + \left(z + \frac{1}{2}\right)^2 + 1/4$)
(ii) $(-2, 1/2, -1/4)$ minimum; $h^2 + k^2 + 2\ell^2 + 5/8$
(iii) $x = n_1\pi$, $y = n_2\pi$, $z = n_3\pi$ (n_1, n_2, n_3 integers). If n_1, n_2, n_3 all even or zero, point is a maximum; if all odd, minimum; otherwise saddle point;
$\cos x + \cos y + \cos z = \cos n_1\pi + \cos n_2\pi + \cos n_3\pi - 1/2\{(-1)^{n_1}(x - n_1\pi)^2 + (-1)^{n_2}(y - n_2\pi)^2 + (-1)^{n_3}(z - n\pi)^2\} + \ldots$
(iv) $(-1/13, -2/13, -7/13)$ saddle point; $81/169 + h^2 + 2k^2 + 2\ell^2 - hk$.

5. $(2^{3/2}, 0, -2^{-1/2})$ maximum, $(-2^{3/2}, 0, 2^{-1/2})$ minimum. (Calculate second derivatives.)

7. $\frac{\partial(F, G)}{\partial(x, y)} = 0$. (Either use differentials or consider the function $z + \lambda_1 F + \lambda_2 G$.)

9. Equilateral triangle and square.

11. If curve $g(x, y) = 0$ is given parametrically by $x = x(t)$, $y = y(t)$,
consider $\frac{dg}{dt} = g_x\frac{dy}{dt} + g_y\frac{dy}{dt} = 0$ and $\frac{df}{dt}$. Condition for stationary point is $\frac{df}{dt} = 0$.

13. $x = 2$, $y = 1$, $z = 2/3$; maximum volume

15. $d^4/(a^{4/3} + b^{4/3} + c^{4/3})^3$

17. $8\,abc/3\sqrt{3}$

19. $\sqrt{35}/7$ (distance to $(5/7, -1/7, 3/7)$).

21. Generalize Exercise 18.

23. $x = \ell/3$, $\theta = \pi/3$. The semi-circular design is preferable since it contains an area ℓ^2/π, while the straight side design has a maximum area equal to $\sqrt{3}\,\ell^2/12$. (Find the maximum value of the cross-sectional area $A = x(\ell - 2x)\sin\theta + \frac{1}{2}x^2\sin 2\theta$.)

25. At the centroid, the vector position $\bar{\mathbf{r}} = \frac{1}{n}\sum_{i=1}^{n}\mathbf{r}_i$ where the vectors $\mathbf{r}_i$ are the positions of the n patients. (Minimize $\sum_{i=1}^{n}|\mathbf{r} - \mathbf{r}_i|^2$.)

27. Examine second derivatives. If $a = 1$, expand $T(x, y)$ in powers of x, y;
$T(x, y) = \frac{1}{2}\left[10 + (x-1)^2 + y^2 + \frac{1}{(x-1)^2 + y^2}\right] + \frac{1}{2}\left[10 + (x+1)^2 + y^2 + \frac{1}{(x+1)^2 + y^2}\right]$
$= 12 + 2x^2 + 2y^2 + \cdots$ Therefore, (0, 0) is a relative minimum.

29. **(i)** $y = \frac{4}{3}x + \frac{3}{2}$ **(ii)** $y = -\frac{27}{28}x + \frac{10}{7}$ **(iii)** $y = \frac{9}{5}x - \frac{13}{20}$ **(iv)** $y = \frac{7}{10}x + \frac{6}{5}$

31. $a_0 = 3\int_0^1 (10x^2 - 12x + 3)\, f(x)\, dx$, $a_1 = 12\int_0^1 (-15x^2 + 16x - 3)\, f(x)\, dx$,
$a_2 = 30\int_0^1 (6x^2 - 6x + 1)\, f(x)\, dx$. For $f(x) = x^3$, $a_0 = \frac{1}{20}$, $a_1 = -\frac{3}{5}$, $a_2 = \frac{3}{2}$.

33. 30 standard, 20 luxury models. (Maximize profit $p(x, y) = 150x + 250y$ from x standard and y luxury models where $20x + 30y \leq 1200$, $10x + 20y \leq 700$, $x \geq 0$, $y \geq 0$.)

35. No. Sketch the regions defined by the inequalities.

37. Raise no sheep, 40 hogs, 40 cattle with maximum profit \$560.

Chapter 7 Multiple Integrals

Exercises 7.1

1. It represents the volume contained between the surfaces $z = f_1(x, y)$ and $z = f_2(x, y)$ above and below the region R in the xy plane.

3. $\int_0^{f(x)} dy = f(x)$

5. **(i)** $\frac{1}{3}(b^3 - a^3)(\tan^{-1} d - \tan^{-1} c)$
(ii) $(be^b - e^b - ae^a + e^a)(de^d - e^d - ce^c + e^c)$
(iii) $(d - c)(\cos a - \cos b) + (b - a)(\sin d - \sin a)$

7. **(i)** 7 **(ii)** $1/3$ **(iii)** 4 **(iv)** $-5/12$ **(v)** $1 + \frac{\pi}{8} - \frac{\pi^3}{24}$ **(vi)** 0

9. **(i)** $\int_0^1 dy \int_y^{y^{1/3}} f(x, y)\, dx$

(ii) $\int_{1/\sqrt{2}}^1 dy \int_0^{\cos^{-1} y} f(x, y)\, dx + \int_0^{1/\sqrt{2}} dy \int_0^{\sin^{-1} y} f(x, y)\, dx$

(iii) $\int_0^1 dx \int_{1-x}^1 f(x, y)\, dy + \int_1^2 dx \int_0^{2-x} f(x, y)\, dy$

(iv) $\int_0^1 ds \int_{1-s}^1 \phi(r, s)\, dr$ **(v)** $\int_0^1 dx \int_x^{x^3} f(x, y)\, dy$

(vi) $\int_a^{2a} dy \int_a^y f(x, y)\, dx + \int_{2a}^{4a} dy \int_{y/2}^{2a} f(x, y)\, dx$

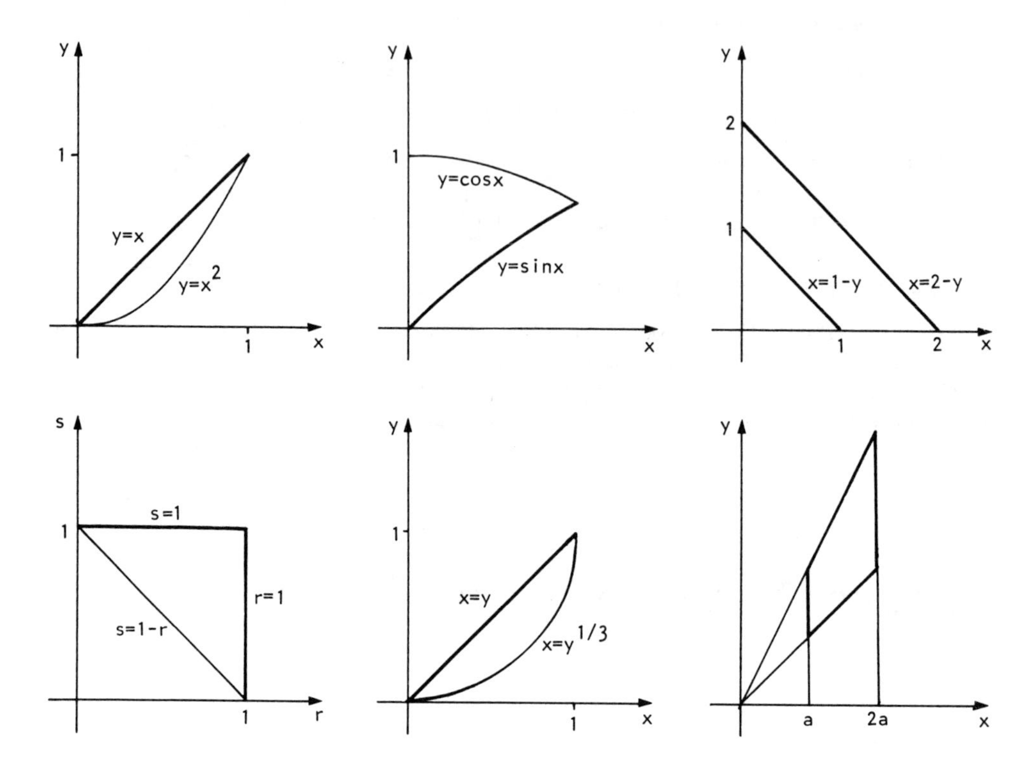

11. **(i)** $\frac{1}{2}(1 - \cos 1)$. $\left(\text{Sketch the region of integration and show that}\right.$

$$\left.\int_0^1 dy \int_y^1 \sin x^2\, dx = \int_0^1 dx \int_0^x \sin x^2\, dy.\right)$$

(ii) $2 \log 2 - 1$. $\left(\text{Show} \int_0^1 dx \int_x^{2-x} \frac{x}{y}\, dy = \int_0^1 \frac{dy}{y} \int_0^y x\, dx + \int_1^2 \frac{dy}{y} \int_0^{2-y} x\, dx.\right)$

(iii) $\frac{\pi}{12} \left(\text{Use} \int_0^1 dy \int_y^1 \frac{x^3}{x^2 + y^2}\, dx = \int_0^1 x^3 dx \int_0^x \frac{dy}{x^2 + y^2}\right)$

13. $I(a) = \int_0^a e^{-y^2} dy \int_y^a dx = \int_0^a (a - y)\, e^{-y^2} dy = a \int_0^a e^{-y^2} dy - \tfrac{1}{2}(1 - e^{-a^2})$;

$$\int_0^\infty \left[1 - erf(x)\right] dx = \lim_{a\to\infty} \int_0^a \left[1 - erf(x)\right] dx = \lim_{a\to\infty} \left[a - \frac{2}{\sqrt{\pi}} I(a)\right] = \pi^{-1/2}$$

15. $3\pi = \int_{-1}^1 dx \int_{-(1-x^2)^{1/2}}^{(1-x^2)^{1/2}} (x + 2y + 3)\, dy$

17. $\frac{6a^3}{7} = \int_0^a dx \int_{2x^2/a}^{\sqrt{4ax}} \frac{x^2 + y^2}{a}\, dy$

19. 2π; use $\iint_R \frac{dx\, dy}{(xy(1 - x - y))^{1/2}} = \int_0^1 \frac{dx}{x^{1/2}} \int_0^{1-x} \frac{dy}{(y(1 - x - y))^{1/2}}$ and let $y = (1 - x) \sin^2\theta$ in the first integration.

21. $4(1-e^{-a})^2$, $2(2a-1-e^{-2a})$. As $a\to\infty$, the first integral converges to the value 4 and the second integral diverges. (The first integral is $4\int_0^a dx\int_0^a e^{-(x+y)}\,dy$ and the second integral is $2\int_{-a}^a dx\int_{-x}^a e^{-(x+y)}\,dy$.)

Exercises 7.2

1. $\dfrac{3\pi a^2}{2}=\displaystyle\iint_R r\,dr\,d\theta=2a^2\int_0^\pi\frac{(1+\cos\theta)^2}{2}\,d\theta$

3. Circles intersect when $\cos\theta=1/2$, $\theta=\pm\pi/3$; $\dfrac{8}{3}\pi-2\sqrt{3}=\dfrac{4\pi}{3}+2\displaystyle\int_{\pi/3}^{\pi/2}\frac{16\cos^2\theta}{2}\,d\theta.$

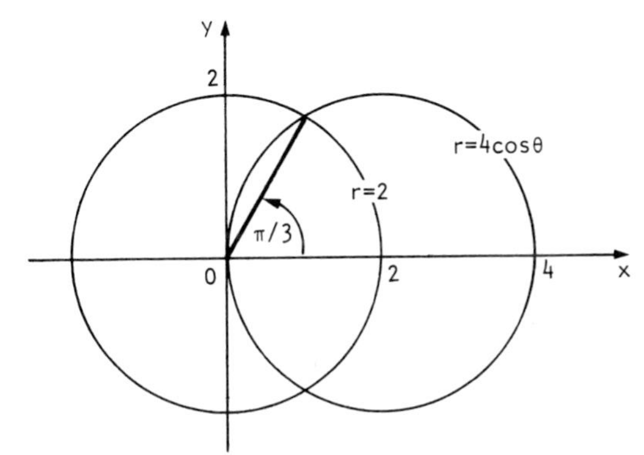

5. $\dfrac{\pi}{4}\left(1-\dfrac{1}{e}\right)=\displaystyle\int_0^{\pi/2}d\theta\int_0^1 re^{-r^2}\,dr$

7. $\dfrac{a^2}{2}=2\displaystyle\int_0^{\pi/4}\frac{a^2}{2}\cos 2\theta\,d\theta$

9. If $f(x)$ is odd, then $\displaystyle\int_{-a}^a f(x)\,dx=0$.

11. **(i)** $2/3$ **(ii)** $e-1/e$ **(iii)** $2\log 2-2$

13. $\dfrac{6}{35}(b^{7/6}-a^{7/6})(d^{5/6}-c^{5/6})$; use the method of Example 6. If x and y have dimensions of length, then constants a, b, c, d have dimensions of length and answer has dimensions $(\text{length})^{7/6}(\text{length})^{5/6}=(\text{length})^2$.

15. $\dfrac{1}{4}\left(e-\dfrac{1}{e}\right)$; compare this method with Example 2.

17. $\dfrac{\sqrt{3}}{2}\tan^{-1}\left(\dfrac{1}{2}\right)$. In polar coordinates, the integral is $\displaystyle\int_0^{\pi/3}d\theta\int_0^{r(\theta)}\frac{r\,dr}{(1+r^2)^2}$ where $r(\theta)=2\sqrt{3}/(\sin\theta+\sqrt{3}\cos\theta)$.

19. $\left(\dfrac{2\pi}{3}-\dfrac{8}{9}\right)a^3=2\displaystyle\iint_R(a^2-x^2-y^2)^{1/2}\,dx\,dy=2\int_{-\pi/2}^{\pi/2}d\theta\int_0^{a\cos\theta}r(a^2-r^2)^{1/2}\,dr.$

21. Evaluate $2\iint_R (2ax)^{1/2}\, dx\, dy$ where R is the interior of the circle $x^2+y^2 = 2ax$ or $r = 2a\cos\theta$.

23. $\alpha > -1$; $\iint_R (x^2+y^2)^{\alpha} dx\, dy = \int_0^{2\pi} d\theta \int_0^1 r^{2\alpha+1}\, dr = 2\pi/(2\alpha+2)$

25. $\frac{5}{3}\pi k a^3$ where the density $\rho = kr$; average density $\bar{\rho} = \frac{10}{9} ka$ (use Exercise 1).

27. $\frac{\pi}{4} 10^{-5}$ km^3; use the method of Example 4.

29. $\Gamma(n+1) = \int_0^{\infty} x^n e^{-x}\, dx = \int_0^{\infty} x\, x^{n-1} e^{-x}\, dx$; let $u = x$, $dv = x^{n-1} e^{-x}\, dx$.

31. $\Gamma(m)\,\Gamma(n) = 4\int_{r=0}^{\infty}\int_{\theta=0}^{\pi} r^{2m+2n-1} e^{-r^2} \cos^{2m-1}\theta \sin^{2n-1}\theta\, dr\, d\theta$; use Exercise 30.

33. Let $x = u^p$, $y = v^q$; then $\iint dx\, dy = pq \iint u^{p-1} v^{q-1}\, du\, dv$ and use Exercise 32.

Exercises 7.3

1. **(i)** $3 = \int_0^1 x^3\, dx \int_0^2 y^2\, dy \int_0^3 z\, dz$ **(ii)** $(e-1)(e^2-1)(1-e^{-3})$

(iii) $(1-\cos 1)(1-\cos 2)(1-\cos 3)$

3. 5; show that the volume under the plane $\frac{x}{a}+\frac{y}{b}+\frac{z}{c} = 1$ is $\frac{abc}{6}$

5. **(i)** $\int_{-1}^1 dx \int_{-1}^x dz \int_0^1 dy$ **(ii)** $\int_0^1 dx \int_0^1 dz \int_0^{x-z} dy$

(iii) $\int_0^1 \log x\, dx \int_1^x dz \int_0^z dy$ **(iv)** $\int_0^1 x\, dx \int_x^1 e^z dz \int_x^z dy$

(v) $\int_0^1 dx \int_0^{(1-x^2)^{1/2}} dz \int_z^{(1-x^2)^{1/2}} dy$

7. Volume $= \int_{-a}^a \pi(y^2+z^2) dx = \pi b^2 \int_{-a}^a \left(1-\frac{x^2}{a^2}\right) dx = \frac{4}{3}\pi a b^2$; corresponding heights at (x, y) are $z = b\left(1-\frac{x^2}{a^2}-\frac{y^2}{b^2}\right)^{1/2}$ on ellipsoid of revolution and $z = c\left(1-\frac{x^2}{a^2}-\frac{y^2}{b^2}\right)^{1/2}$ for the general ellipsoid; corresponding volume is $\frac{4}{3}\pi a b^2\left(\frac{c}{b}\right) = \frac{4}{3}\pi abc$.

9. 0

11. $\pi abd\left(2+\frac{2d^2}{3c^2}\right)$. The volume above and below the ellipse $\frac{x^2}{a^2}+\frac{y^2}{b^2} = 1$ in the $z = 0$ plane is $2\pi abd$. The rest of the volume is $2\iint_R \left[d - c\left(\frac{x^2}{a^2}+\frac{y^2}{b^2}-1\right)^{1/2}\right] dx\, dy$ where R is the annular region $1 \le \frac{x^2}{a^2}+\frac{y^2}{b^2} \le 1+\frac{d^2}{c^2}$; let $x' = \frac{x}{a}$, $y' = \frac{y}{b}$ and use polar coordinates.

13. **(i)** $\frac{4\pi}{15}a^5$ **(ii)** $\frac{4\pi}{5}a^5$ **(iii)** $4\pi a$; use spherical coordinates

15. **(i)** $\pi\frac{a^2bc}{16}$ **(ii)** $\frac{a^2b^2c}{30}$ **(iii)** $\frac{a^2b^2c^2}{48}$;
let $x = au,\ y = bv,\ z = cw$ and then use spherical coordinates in (u, v, w) space.

17. $\frac{1}{32}$; let $u = x + y - z,\ v = -x + y + z,\ w = x - y + z$.

19. **(i)** $\frac{K\pi^2}{4}$ **(ii)** $\frac{K\pi}{2}$ where K is the proportionality factor (use spherical coordinates)

21. 3,506 in. from the base; solve for the appropriate root of the equation
$2\int_0^{6-h}\left(3-\frac{y^2}{36}\right)^2 dy = \int_0^6\left(3-\frac{y^2}{36}\right)^2 dy$. (Equation of intersection of surface with xy plane is $x = 3 - \frac{y^2}{36}$; element of area is $\pi x^2\,dy$.)

23. $\frac{4\pi\rho_0}{K^3}(2 + 2aK + a^2K^2)$ where $\rho = \rho_0 e^{-K(R-a)}$ and a is the radius of the planet.
$\left(\text{The mass is } 4\pi\rho_0\int_a^\infty R^2\,e^{-K(R-a)}dR.\right)$ For earth, integrate from a to $a + 5.5$(km) to determine K.

25. $(PP')^2 = (r\cos\theta - r'\cos\theta')^2 + (r\sin\theta - r'\sin\theta')^2 + (z' - z)^2$
$= r^2 + r'^2 - 2rr'\cos(\theta - \theta') + (z - z')^2$. Yes; consider the mutually perpendicular differential elements of cylindrical coordinates.

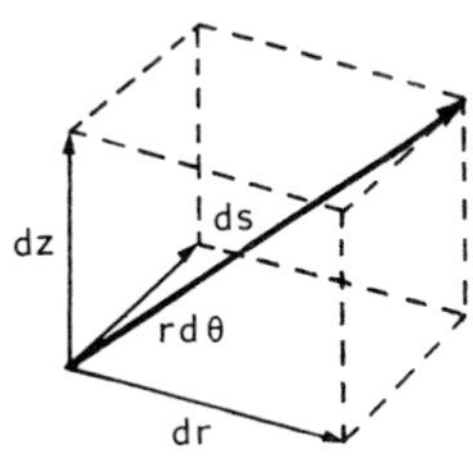

27. For cylindrical polars, $\frac{\partial \mathbf{r}}{\partial r} = (\cos\theta,\ \sin\theta,\ 0)$, $\frac{\partial \mathbf{r}}{\partial \theta} = (-r\sin\theta,\ r\cos\theta,\ 0)$, $\frac{\partial \mathbf{r}}{\partial z} = (0,0,1)$. For spherical polars, $\frac{\partial \mathbf{r}}{\partial R} = (\sin\phi\cos\theta,\ \sin\phi\sin\theta,\ \cos\phi)$,
$\frac{\partial \mathbf{r}}{\partial \theta} = (-R\sin\phi\sin\theta,\ R\sin\phi\cos\theta,\ 0)$, $\frac{\partial \mathbf{r}}{\partial \phi} = (R\cos\phi\cos\theta,\ R\cos\phi\sin\theta,\ -R\sin\theta)$.

29. **(i)** Use spherical polars **(ii)** Define $u = \alpha x,\ v = \beta y,\ w = \gamma z$ and use (i).

Exercises 7.4

1. **(i)** $\sqrt{14} = \int_{x=0}^{1}\int_{y=0}^{1} \sqrt{14}\, dx\, dy$

(ii) $\dfrac{15\sqrt{14}}{2} = \int_{x=0}^{1}\int_{y=0}^{1} (6x + 5y + 2)\sqrt{14}\, dx\, dy$

(iii) $13\sqrt{14}/12 = \int_{x=0}^{1}\int_{y=0}^{1} (3x^2y + 2xy^2 + xy)\, dx\, dy$

(The surface is $z = \phi(x,y) = 3x + 2y + 1$ and $(1 + \phi_x^2 + \phi_y^2)^{1/2} = \sqrt{14}$.)

3. $\dfrac{\pi a^5 h}{4} + \dfrac{\pi a^6}{12}$; on the curved surface, use $dS = a d\theta dz$, $x = a\cos\theta$, $y = a\sin\theta$; on the ends, use $x = r\cos\theta$, $y = r\sin\theta$, $dS = r dr d\theta$.

5. $(2\pi - 4)a^2$, $4a^2$; use spherical and cylindrical coordinates respectively.

7. $a^2(\theta_2 - \theta_1)(\cos\phi_1 - \cos\phi_2) = \int_{\theta_1}^{\theta_2}\int_{\theta_1}^{\theta_2} a^2 \sin\phi\, d\phi\, d\phi$ where θ is longitude measured to the east and ϕ is co-latitude measured from the north pole; $4.40 \times 10^6 (\text{km})^2$

9. $4\pi a^4$, $\dfrac{2}{5}\pi a^6$; use spherical coordinates.

11. Volume $= \dfrac{2\sqrt{2}}{45} = 2\sqrt{2}\iint_R \sqrt{xy}\, dx\, dy$ where R is the region $x^{1/2} + y^{1/2} \le 1$, $x \ge y \ge 0$;

surface area $= \sqrt{2}\iint_R \left[\left(\dfrac{y}{x}\right)^{1/2} + \left(\dfrac{x}{y}\right)^{1/2}\right] dx\, dy = \dfrac{2\sqrt{2}}{3}$

13. $8a^2$; compare to Exercise 5.

15. $\dfrac{\pi}{6}[(1 + a^2)^{3/2} - 1] = \iint_R \sqrt{1 + x^2 + y^2}\, dx\, dy$; use polar coordinates.

17. **(i)** $\sqrt{3}\, du\, dv$ **(ii)** $(1 + \sin^2 u \cos^2 v + \cos^2 u \sin^2 v)^{1/2}\, du\, dv$

(iii) $u(1 + 4u^2)^{1/2}\, du\, dv$ **(iv)** $e^{u+v}(1 + e^{2u} + e^{2v})^{1/2}\, du\, dv$

(v) $u[1 + 4u^2(\cosh^2 v + \sinh^2 v)]^{1/2}\, du\, dv$

(vi) $\sin u[(b^2c^2\cos^2 v + c^2a^2\sin^2 v)\sin^2 u + a^2b^2\cos^2 u]^{1/2}\, du\, dv$

19. $\dfrac{9h_0}{124a}(5\sqrt{5} + 1)$. Show $V = \dfrac{\pi}{2}h^3$, $S = \dfrac{\pi}{6}h^2(5\sqrt{5} - 1)$ and from $\dfrac{dV}{dt} = -\alpha S$ find a differential equation for $h(t)$.

21. $4\pi\rho_0 a^2\left(1 + \dfrac{\pi\varepsilon}{4}\right) = \iint_S \rho\, dS$; use spherical coordinates.

Exercises 7.5

1. At the centroid of the triangle use elementary strips parallel to each base to show that the center of mass must be on each median.

3. At height $h/3$ above the base. This result can be deduced from Exercise 1.

5. $\bar{x} = \bar{y} = \bar{z} = \dfrac{3}{8}\dfrac{b^4 - a^4}{b^3 - a^3}$; $\lim_{b\to a}\bar{x} = \dfrac{a}{2}$, etc. This result follows from Exercise 4.

7. $\dfrac{2\pi}{3}ka^3 = \int_0^{2\pi}\int_0^{a} kr^2\, dr\, d\theta$ where density $\rho = kr$

9. $\bar{x} = \frac{5a}{6}$, $\bar{y} = 0$; use $\bar{x} = \frac{\iint x\,dx\,dy}{\iint dx\,dy} = \frac{\iint r^2 \cos\theta\,dr\,d\theta}{\iint r\,dr\,d\theta}$; use Exercise 7.2.1.

11. $\frac{2}{5} M \frac{b^5 - a^5}{b^3 - a^3}$; follow Exercise 10.

13. $I_1 = \frac{M}{10}(b^2 + c^2)$, $I_2 = \frac{M}{10}(c^2 + a^2)$, $I_3 = \frac{M}{10}(a^2 + b^2)$ where $M = \rho_0 \frac{abc}{6}$ is the mass of the tetrahedron; use limits in Example 7.3.2.

15. Following Example 6, period $= 2\pi(I/Mg\ell)^{1/2} = 2\pi(\ell/3g)^{1/2}$ where $I = M\ell^2/3$. (This moment of inertia can be derived as a limit of Example 4 when $b = \ell$ and $a \to 0$, $c \to 0$ with $M = \rho abc$).

17. $\frac{\pi}{8} \rho_0 \Omega^2 ha^4 = \pi \rho_0 \Omega^2 \int_0^\infty e^{-3z/h}\,dz \int_0^\infty r^3 e^{-2r/a}\,dr$; velocity is largest on circle $r = a$, $z = 0$ (solve $v_r = 0$).

19. By symmetry, force is in $\mathbf{k}$ direction where $\mathbf{k}$ is a unit vector perpendicular to the disc. Following Example 7 and using polar coordinates,

$$\mathbf{F} = \rho\mathbf{k} \int_0^{2\pi} \int_0^a \frac{b}{(b^2 + r^2)^{1/2}}\,r\,dr\,d\theta = 2\pi\sigma\mathbf{k}\left[1 - \frac{b}{(a^2 + b^2)^{1/2}}\right].$$

21. $I = \iiint \ell^2\,dm$ where $\ell = |\mathbf{r} \times \hat{\mathbf{u}}|$ is the perpendicular distance of the point $\mathbf{r} = x\mathbf{i} + y\mathbf{j} + z\mathbf{k}$ from the line concerned; $\mathbf{r} \times \hat{\mathbf{u}} = (ny - mz)\,\mathbf{i} + (\ell z - nx)\,\mathbf{j} + (mx - \ell y)\,\mathbf{k}$.

Chapter 8 Vector Calculus

Exercises 8.1

1. **(i)** $xy = 1$ **(ii)** $1 + x = 2y$ **(iii)** $e^{-x} + e^{-y} = 2e^{-1}$

In each case, solve $\frac{dy}{dx} = \frac{v_2}{v_1}$, $y(1) = 1$.

3. $f(x,y,z)\,(\mathbf{i} + \mathbf{j} + \mathbf{k})$; $f(x,y,z)\,(x\mathbf{i} + y\mathbf{j} + z\mathbf{k})$

5. $x^2 - y^2 = C$, $\phi = xy + k$; the streamlines satisfy $\frac{dy}{dx} = \frac{x}{y}$ and, on the level curves of ϕ, $\frac{d}{dx} = -\frac{y}{x}$. Since the product of the slopes is -1, the two curves intersect at right angles.

9. Show that $\frac{\partial\phi}{\partial x} = x^2y$, $= \frac{\partial\phi}{\partial y} = xy^2$ is impossible by evaluating $\frac{\partial^2\phi}{\partial x\partial y}$; streamlines are $y = cx$; field $x\mathbf{i} + y\mathbf{j}$ has same streamlines.

11. $\phi(x,y,z) = \frac{2q}{(x^2 + y^2 + z^2)^{1/2}} - \frac{q}{[x^2 + y^2 + (z + a)^2]^{1/2}} - \frac{q}{[x^2 + y^2 + (z + a)]^{1/2}}$ using $\phi = \frac{q}{r}$ for a point charge at distance r; $\phi(a, 2a, 3a) = \frac{q}{a}\left[\frac{2}{\sqrt{14}} - \frac{1}{3} - \frac{1}{\sqrt{21}}\right]$;

$$\nabla\phi\,(a, 2a, 3a) = \left(\frac{1}{7\sqrt{14}} - \frac{1}{27} - \frac{1}{21\sqrt{21}}\right)(\mathbf{i} + 2\mathbf{j}) + \left(\frac{3}{7\sqrt{14}} - \frac{2}{27} - \frac{4}{21\sqrt{21}}\right)\mathbf{k}$$

13. From Example 4, $\phi = \dfrac{-\mu z}{(x^2+y^2+z^2)^{3/2}} = -\dfrac{\mu \cos\phi}{R^2}$

15. Use the methods of Example 3 to show that $2\pi r v(r) = 2\pi m$ so that $\mathbf{v}(r) = \dfrac{m\mathbf{r}}{|\mathbf{r}|^2}$.

17. Use the limiting process discussed in Example 4.

19. Any curve $x = x(t)$, $y = y(t)$, $z = z(t)$ on equipotential surface $\phi(x,y,z) = \text{constant}$ satisfies $\nabla\phi \cdot \left(\dfrac{dx}{dt}, \dfrac{dy}{dt}, \dfrac{dz}{dt}\right) = 0$. Conclude streamlines in direction $\nabla\phi$ are orthogonal to $\phi = \text{constant}$.

Exercises 8.2

1. **(i)** 2 **(ii)** $^7/_3$ **(iii)** $3 - {}^4/_\pi$

3. **(i)** 5 **(ii)** $^{21}/_5$ **(iii)** $^{117}/_{42}$

5. $\oint \boldsymbol{F}\cdot d\boldsymbol{r} = 2$

7. **(i)** $\int_C f(x)dx + \int_C g(y)dy$ **(ii)** $\int_C f(u)du$ **(iii)** $\int_C f(u)\dfrac{du}{2}$

9. **(i)** $\dfrac{x^2}{2} + \dfrac{y^2}{2} + \dfrac{z^3}{3}, \dfrac{4}{3}$ **(ii)** $\dfrac{7}{5}$ **(iii)** $\dfrac{19}{15}$ **(iv)** $\dfrac{26}{15}$

(v) $e^x - \log|\cos y| + z(\log|z| - 1),\ e - 2 - \log\cos 1$ **(vi)** $(y+z)e^x$, $2e$

11. $\int \mathbf{F}\cdot d\mathbf{r} = \dfrac{1}{30}\alpha^2 - \dfrac{1}{6}\alpha + 1 = \dfrac{1}{30}\left(\alpha - \dfrac{5}{2}\right)^2 + \dfrac{19}{24}$; minimum is $\dfrac{19}{24}$ at $\alpha = \dfrac{5}{2}$

13. $\phi(x,y,z) = \dfrac{mgr_0^2}{r_0 + z}$; verify $\mathbf{F} = \nabla\phi$

15. **(i)** 0 **(ii)** 0 **(iii)** $-{}^1/_4$

17. -2π, $\pi(\mathbf{j} - \mathbf{i})$, 0

19. **(i)** $\dfrac{abc}{3}(a^2\mathbf{i} + b^2\mathbf{j} + c^2\mathbf{k})$ **(ii)** $\dfrac{abc}{2}[(a+b)\,\mathbf{i} + (a+c)\,\mathbf{k}]$

(iii) $\dfrac{abc}{4}[(bc\mathbf{i} + ac\mathbf{j} + ab\mathbf{k}]$

Exercises 8.3

1. **(i)** 3, $-\mathbf{j}$ **(ii)** $f'(x) + g'(y) + h'(z)$, $\mathbf{0}$

(iii) 0, $\left(\dfrac{\partial h}{\partial y} - \dfrac{\partial g}{\partial z}\right)\mathbf{i} + \left(\dfrac{\partial f}{\partial z} - \dfrac{\partial h}{\partial x}\right)\mathbf{j} + \left(\dfrac{\partial g}{\partial x} - \dfrac{\partial f}{\partial y}\right)\mathbf{k}$

(iv) 0, $(x - e^z)\mathbf{i} + [\cos(y+z) - y]\,\mathbf{j} - \cos(y+z)\mathbf{k}$ **(v)** $2(x+y+z)$, $\mathbf{0}$

(vi) $4(x+y+z)$, $(z-y)\mathbf{i} + (x-z)\mathbf{j} + (y-x)\mathbf{k}$

3. **(i)** $\iint_S \mathbf{F}\cdot\hat{\mathbf{n}}\,dS = \iint_S \dfrac{x+2y+3z}{\sqrt{14}}\,dS = \dfrac{6}{\sqrt{14}}\iint_S dS = \dfrac{6}{\sqrt{14}}\iint_S \dfrac{\sqrt{14}}{3}\,dx\,dy = 18,$

$\oint_C \mathbf{F}\cdot d\mathbf{r} = \oint_C x\,dx + y\,dy + z\,dz = \left.\dfrac{x^2+y^2+z^2}{2}\right|_{P_0}^{P_0} = 0$

(ii) $\dfrac{61}{288}, -\dfrac{7}{36}$

5. **(i)** $4\pi(b^3 - a^3)$ **(ii)** $4\pi(b^{2n+3} - a^{2n+3})$
7. $3V$, $-(\mathbf{i}+\mathbf{j}+\mathbf{k})V$. Use div $\mathbf{G} = 3$, curl $\mathbf{G} = -(\mathbf{i}+\mathbf{j}+\mathbf{k})$
9. 0. Yes, since there is no net flow through S. This holds for any closed surface S. (Use $\hat{\mathbf{n}} \cdot \nabla\phi = \dfrac{\partial\phi}{\partial n}$ and the result of Example 4, Section 8.1.)
11. $4\,ab(\mathbf{A}_1 \cdot \mathbf{j} - \mathbf{A}_2 \cdot \mathbf{i})$ where $x = \pm\, a$, $y = \pm\, b$, $z = 0$ defines the rectangle. (Show that $\oint_C \mathbf{F} \cdot d\mathbf{r} = (\mathbf{A}_1 \cdot \mathbf{j}) \oint_C x\,dy + (\mathbf{A}_2 \cdot \mathbf{i}) \oint_C y\,dx = 4ab\,[\mathbf{A}_1 \cdot \mathbf{j} - \mathbf{A}_2 \cdot \mathbf{i}] = 4\,ab\,\mathbf{k} \cdot \text{curl}\,\mathbf{F}$.)
13. $2\pi^2$, $2\pi(\mathbf{j}+\mathbf{k})$. For the straight line path, $\int_A^B \boldsymbol{F}\cdot d\boldsymbol{r}$ does not change since $\mathbf{F} = \nabla\left(\dfrac{x^2+y^2+z^2}{2}\right)$ but $\int_A^B \mathbf{F} \times d\mathbf{r} = 2\pi\,\mathbf{j}$
15. **(i)** $\dfrac{2\pi}{3\sqrt{2}}$ **(ii)** $\dfrac{\pi\sqrt{2}}{4}$
17. $\rho\dfrac{\pi U a^2}{2}$ where ρ is the constant density. Since the mass flow is constant, Ua^2 is constant.

Exercises 8.4

1. (i) - (vi) Use the definitions
3. **(i)** div $\mathbf{F} = 0$, curl $\mathbf{F} = \mathbf{0}$, $\nabla^2\mathbf{F} = \mathbf{0}$, curl curl $\mathbf{F} = \mathbf{0}$, grad div $\mathbf{F} = \mathbf{0}$
(ii) div $\mathbf{F} = 3$, curl $\mathbf{F} = -\,\mathbf{i} - \mathbf{j} - \mathbf{k}$, $\nabla^2\mathbf{F} = \mathbf{0}$, curl curl $\mathbf{F} = \mathbf{0}$, grad div $\mathbf{F} = \mathbf{0}$
(iii) div $\mathbf{F} = 3\,xyz\,(x+y+z)$, curl $\mathbf{F} = (z^3 - y^3)\,x\mathbf{i} + (x^3 - z^3)\,y\mathbf{j} + (y^3 - x^3)z\mathbf{k}$, $\nabla^2\mathbf{F} = 6\,xyz(\mathbf{i}+\mathbf{j}+\mathbf{k})$, curl curl $\mathbf{F} = 3\,yz(y+z)\mathbf{i} + 3\,zx(z+x)\mathbf{j} + 3\,xy(x+y)\mathbf{k}$, grad (div $\mathbf{F}$) $= 3\,yz(2x+y+z)\mathbf{i} + 3\,zx(x+2y+z)\mathbf{j} + 3xy(x+y+2z)\mathbf{k}$.
5. **(i)** $x - 2y + 3z$ **(ii)** $\dfrac{1}{3}x^3 - \dfrac{1}{2}y^2 + \dfrac{1}{4}z^4$ **(iii)** $x + xyz + y^2 + z^3$ **(iv)** x^2y^2z

Solve $\dfrac{\partial\phi}{\partial x} = F_1$, $\dfrac{\partial\phi}{\partial y} = F_2$, $\dfrac{\partial\phi}{\partial z} = F_3$
7. $\dfrac{f'(r)}{r}\mathbf{r}$, $rf'(r) + 3f(r)$, 0
9. Use the definitions
11. Verify directly from components
13. Use Equation (7), $\nabla \times \phi\mathbf{F} = \phi\nabla \times \mathbf{F} + \nabla\phi \times \mathbf{F}$.
15. First check for one component.
17. Show $\nabla \cdot \mathbf{F} = \dfrac{\partial(\phi_1, \phi_2, \phi_3)}{\partial(x, y, z)} = 0$
19. **(i)** $\nabla \times (\nabla \times \mathbf{E}) = \nabla \times \left(-\dfrac{1}{c}\dfrac{\partial \mathbf{H}}{\partial t}\right) = -\dfrac{1}{c}\dfrac{\partial}{\partial t}(\nabla \times \mathbf{H}) = -\dfrac{1}{c^2}\dfrac{\partial^2 \mathbf{E}}{\partial t^2}$

(iii) $\nabla^2\mathbf{E} = \nabla(\nabla \cdot \mathbf{E}) - \nabla \times (\nabla \times \mathbf{E}) = \dfrac{1}{c^2}\dfrac{\partial^2 \mathbf{E}}{\partial t^2}$; use Exercise 10(i).

Exercises 8.5

1. **(i)** 4π **(ii)** 0 **(iii)** 4π
In (i), $\mathbf{F} = \mathbf{r}$, $\hat{\mathbf{n}} = \mathbf{r}$, div $\mathbf{F} = 3$ and $\oint\!\!\oint_S \mathbf{F} \cdot \mathbf{r}\, dS = \oint\!\!\oint_S dS = \iiint_V 3dV = 4\pi$

3. **(i)** $3\pi a^2 h$ **(ii)** $\pi a^2 h^2$ **(iii)** 0

5. **(i)** $2\pi a^2$ **(ii)** 0 **(iii)** 0

7. Use div(curl $\mathbf{F}$) = 0, curl (grad ψ) = $\mathbf{0}$

9. Since $\nabla^2 f = \nabla \cdot \nabla f$, write $\nabla f = \mathbf{F}$ in the divergence theorem and recall $\dfrac{\partial f}{\partial n} = \nabla f \cdot \hat{\mathbf{n}}$.

11. Note that, if $r \neq 0$, $\nabla^2\left(\dfrac{1}{r}\right) = 0$; but for a small sphere centered at the origin,

$$\iiint_V g\ \nabla^2\left(\frac{1}{r}\right) dV \approx g(0,0,0) \iiint_V \nabla \cdot \nabla \left(\frac{1}{r}\right) dV = g(0,0,0) \oiint_S \nabla \left(\frac{1}{r}\right) \cdot \hat{n}\ dS = g(0,0,0) \oiint_S \left(-\frac{1}{r^2}\right) dS$$

$= -\ 4\pi\ g(0,0,0)$.

13. Use Equation (11) of Section 8.4

15. $\dfrac{4}{3}\ \pi a^3 \rho_s g \left[1 + \left(\dfrac{\rho_b}{\rho_s} - 1\right)\dfrac{n}{4}\right]$ where $n = 1,2,3$; the weight of the float at depth z_0 must be $\displaystyle\iiint_V \left[\rho_s + (\rho_b - \rho_s)\frac{z}{h}\right] g\ dV$ where V is the interior of the sphere $x^2 + y^2 + z^2 = a^2$.

17. Use Exercise 16 and add up results for small subvolumes of V;
$\displaystyle\oiint_S \mathbf{E} \cdot \hat{\mathbf{n}}\ dS = -\iiint_V \nabla^2 \phi\ dV = 4\pi \iiint_V \rho\ dV$; since this holds for arbitrarily small volumes V, conclude $\nabla^2 \phi = -4\pi\rho$.

Exercises 8.6

1. **(i)** π **(ii)** -2π **(iii)** 0

3. **(i)** 0 **(ii)** $^{-3}/_{20}$ **(iii)** 0

5. If $\mathbf{F} = M\mathbf{i} + N\mathbf{j}$, $\hat{\mathbf{T}} = \dfrac{dx}{ds}\mathbf{i} + \dfrac{dy}{ds}\mathbf{j}$, $\hat{\mathbf{N}} = -\dfrac{dy}{ds}\mathbf{i}, + \dfrac{dx}{ds}\mathbf{j}$, then $\oint_C F_T\, ds = \oint_C \mathbf{F} \cdot \hat{\mathbf{T}}\ ds$

$$= \oint_C M\ dx + N\ dy = \iint_R \left(\frac{\partial N}{\partial x} - \frac{\partial M}{\partial y}\right) dx\ dy,\ \oint_C F_N ds = \oint_C N\ dx - M\ dy = -\iint_R \left(\frac{\partial M}{\partial x} + \frac{\partial N}{\partial y}\right) dx\ dy.$$

7. Insert enough barriers to form a closed region.

9. **(i)** πab **(ii)** $\dfrac{2}{3}a^2$ **(iii)** $\dfrac{3}{8}\ \pi a^2$

In (ii), define $x = a\cos^4\theta$, $y = a\sin^4\theta$; in (iii), $x = a\cos^3\theta$, $y = a\sin^3\theta$.

11. **(i)** $\dfrac{3}{2}\ \pi a^2$ **(ii)** $\dfrac{\pi}{4}a^2$ **(iii)** a^2

13.
$$\iint_R dx\ dy = \frac{1}{2}\oint_C (x\ dy - y\ dx) = \frac{1}{2}\oint_C \left(x\frac{\partial y}{\partial u} - y\frac{\partial x}{\partial u}\right) du + \left(x\frac{\partial y}{\partial v} - y\frac{\partial x}{\partial v}\right) dv$$

$$= \frac{1}{2}\iint_{R'}\left[\frac{\partial}{\partial u}\left(x\frac{\partial y}{\partial v} - y\frac{\partial x}{\partial v}\right) - \frac{\partial}{\partial v}\left(x\frac{\partial y}{\partial u} - y\frac{\partial x}{\partial u}\right)\right] du\ dv = \frac{1}{2}\iint_{R'} 2\left[\frac{\partial x}{\partial u}\frac{\partial y}{\partial v} - \frac{\partial x}{\partial v}\frac{\partial y}{\partial u}\right] du\ dv$$

$$= \iint_{R'} \frac{\partial(x,y)}{\partial(u,v)}\ du\ dv$$

15. **(i)** πa^2 **(ii)** 0 **(iii)** 0;
in each case, both surface intergrals are equal to $\oint_C \mathbf{F} \cdot d\mathbf{r}$ by Stokes' theorem.

17. $\iint_S \hat{\mathbf{n}} \cdot \nabla \times \mathbf{F}\, dS = \iint_S \frac{1}{\sqrt{2}}\, dS = \iint_R \frac{1}{\sqrt{2}} \sqrt{2}\, dx\, dy = \pi$; to evaluate $\oint_C \mathbf{F} \cdot d\mathbf{r}$ use $x = \cos\theta,\ y = \sin\theta,\ z = \cos\theta$.

19. $\oint_C \mathbf{F} \cdot d\mathbf{r} = \iint_S \frac{y}{\sqrt{2}}\, dS = 0$ by symmetry.

21. In **(i)**, use $\nabla(fg) = f\,\nabla g + g\,\nabla f$ and $\oint_C \nabla(fg) \cdot d\mathbf{r} = 0$; in **(ii)** apply Stokes' theorem to $f\,\nabla g$.

23. Following Example 22, $\mathbf{a} \cdot \oint_C \mathbf{F} \times d\mathbf{r} = \oint_C (\mathbf{a} \times \mathbf{F}) \cdot d\mathbf{r} = 0$ for all C implies curl $(\mathbf{a} \times \mathbf{F}) = \mathbf{0}$ for any constant vector $\mathbf{a}$. Choose $\mathbf{a} = \mathbf{i}, \mathbf{j}, \mathbf{k}$ to find $\frac{\partial F_1}{\partial x} = \frac{\partial F_1}{\partial y} = \frac{\partial F_1}{\partial z} = 0$ and similarly for F_2, F_3.

25. Illustrate with a diagram.

Exercises 8.7

1. Use the identity div $\rho\boldsymbol{\nu} = \rho$ div $\boldsymbol{\nu} + \nabla\rho \cdot \boldsymbol{\nu}$

3. $\frac{D}{Dt}\boldsymbol{\nu} = \frac{\partial \boldsymbol{\nu}}{\partial t} + \boldsymbol{\nu} \cdot \nabla\boldsymbol{\nu} = -\frac{1}{\rho}\nabla p + \mathbf{F}.$

5. Use $\frac{D\rho}{Dt} = 0$ and Exercise 1. If $\mathbf{w} = \text{curl}\ \boldsymbol{\nu} = \mathbf{0}$, then $\boldsymbol{\nu} = \nabla\phi$ and $\nabla^2\phi = \text{div}\ \boldsymbol{\nu} = 0$.

7. Use Exercise 6 to rewrite the momentum equation (22).

9. Use Stokes' theorem.

Index

Some Physical Constants

Radius of the earth at the equator	=	6,378.39 km = 3,963.34 miles
Radius of the earth at the pole	=	6,356.91 km = 3,949.99 miles
One degree of latitude (at 40°C)	=	69 miles
Mean density of the earth	=	5.522 gm/cm^3 = 344.7 lb/ft^3
Constant of gravitation	=	6.673×10^{-8} $cm^3/gm\text{-}sec^2$
Gravitational acceleration (sea level, 45° latitude)	=	980.621 cm/sec^2 = 32.1725 ft/sec^2
Standard atmospheric pressure	=	1.013×10^6 $dynes/cm^2$ = 14.69 $lb/in.^2$
Speed of light in vacuum	=	2.997925×10^{10} cm/sec
Speed of sound in dry air at 0°C	=	331.36 m/sec = 1,087.1 ft/sec
Gas constant	=	8.3143×10^7 erg/deg-mole
Density of water at 3.98°C	=	1.0000 gm/cm^3
Density of mercury at 0°C	=	13.596 gm/cm^3
Density of dry air at 0°C	=	1.2929×10^{-3} gm/cm^3
Maximum weight of 1 ft^3 of water	=	62.425 lb

Conversion Factors

Acre	=	0.0016 $mile^2$ = 4.3560×10^4 ft^2
Centimeter	=	0.0328 ft = 0.3937 in.
Cubic foot	=	7.4805 gal = 0.0283 m^3
Degree	=	0.0175 rad
Foot	=	0.3048 m
Gallon	=	0.1337 ft^3 = 3.7853 liters
Inch	=	2.5400 cm
Kilogram	=	2.2046 lb (avdp)
Kilometer	=	0.6214 mile
Liter	=	0.2642 gal = 0.03532 ft^3
Meter	=	1.9036 yd = 3.2808 ft = 39.3701 in.
Mile	=	1.6093 km
Ounce	=	1.8047 $in.^3$ = 29.5737 cm^3
Pound (avdp)	=	0.4536 kg = 453.5924 gm
Pound force	=	4.4482 newtons
Quart	=	0.9463 liter
Radian	=	57.2958 degrees
Yard	=	0.9144 m